Translational Medicine Cancer

Edited by
Robert A. Meyers

Related Titles

Tsimberidou, A., Kurzrock, R., Anderson, K.C. (eds.)

Targeted Therapy in Translational Cancer Research

2015
ISBN: 978-1-118-46857-9,
also available in digital formats

Tobias, J.S., Hochhauser, D.

Cancer and its Management
7th Edition

2015
ISBN: 978-1-118-46873-9,
also available in digital formats

Brenner, M.K., Hung, M. (eds.)

Cancer Gene Therapy by Viral and Non-viral Vectors

2014
ISBN: 978-1-118-50162-7,
also available in digital formats

Tiwari, A., Choi, J., Patra, H.K. (eds.)

Advanced Theranostics Materials

2015
ISBN: 978-1-118-99829-8,
also available in digital formats

Translational Medicine Cancer

Advances in Molecular Biology and Medicine

Edited by
Robert A. Meyers

Volume 1

WILEY-VCH
Verlag GmbH & Co. KGaA

The Editor

Dr. Robert A. Meyers
Editor-in-Chief
Ramtech Limited
122, Escalle Lane
Larkspur, CA 94939
USA

Cover

Figure with kind permission from Brigitte L. Thériault, Helen Dimaras, Brenda L Gallie and Timothy W Corson (2014): The genomic landscape of retinoblastoma: a review. Clinical & Experimental Ophthalmology, Volume 42, Issue 1: 33–52.

Library of Congress Card No.: applied for

British Library Cataloguing-in-Publication Data
A catalogue record for this book is available from the British Library.

Bibliographic information published by the Deutsche Nationalbibliothek
The Deutsche Nationalbibliothek lists this publication in the Deutsche Nationalbibliografie; detailed bibliographic data are available on the Internet at <http://dnb.d-nb.de>.

Print ISBN: 978-3-527-33569-5
ePDF ISBN: 978-3-527-68186-0
ePub ISBN: 978-3-527-68187-7
Mobi ISBN: 978-3-527-68185-3

Cover Design Adam Design, Weinheim, Germany

Typesetting SPi Global, Chennai, India

Printing and Binding Markono Print Media Pte Ltd, Singapore

Printed on acid-free paper.

Contents

Preface

In 2015, there will be more than 14,000,000 cases of cancer diagnosed worldwide and an estimated 1,658,370 new cancer cases diagnosed and 589,430 cancer deaths in the United States alone. But, cancer prevention and the latest research has resulted in tremendous gains in both longer survival and quality of life such that cancer death rates in the United States have declined 20% from their peak in 1991. Today, there are a record 14.5 million cancer survivors alive in the United States. There is now a race to both prevent and cure cancer once and for all. For the next 10 years, it is forecast that cancer research efforts will focus on areas such as (i) tumor biology and especially cancer stem cells, exploring the theory that cancer stem cells are biologically different from regular cancer cells, have the ability to self-renew, and are more resistant to chemotherapy leading to therapy that attacks these cells, (ii) faster and less expensive next-generation sequencing, liquid biopsies to more accurately detect circulating tumor DNA and microRNA and metabolomics in blood and determine effective treatment, and cancer prevention vaccines, (iii) immunotherapies, some of which reprogram the body's own immune cells to attack cancer, (iv) development of new drugs and methods to carry these drugs to cancerous tumors using very new approaches including nanotechnology, and also (v) cancer etiology, which informs prevention. Given the explosion in understanding how genetics, lifestyle, individual behaviors, social environment, policy, and health-care interventions can reduce the burden of cancer, new insights and integration of knowledge can help shape prevention strategies and lead to reductions in cancer incidence. All of these approaches are leading toward precision personalized medicine and eventual reduction or even elimination of cancer as a fatal disease. Our cancer translational medicine compendium covers the latest developments in all of these areas in detail.

Our compendium is written for university undergraduates, graduate students, faculty, and investigators at research institutes. Our Board of 11 Nobel Laureates approved our overall approach and content and our selection of articles was validated and enhanced by reviewers from major research institutions and each manuscript was then reviewed by scientists from other universities or research institutes. There is a glossary of keywords with definitions at the beginning of each article for the benefit of students. The result is that there are 28 peer reviewed articles at a length of over 750 pages and as such is the largest in depth, up-to-date treatment presently available.

Our compendium is organized into three sections: **Cancer Biology**; **Molecular Diagnostics and Imaging**; and **Cancer Therapy**. The section on **Cancer Biology** begins with an article on cancer etiology and an article on cancer metastasis theory and articles defining the biology of cancer stem cells and their niches and then intracellular signaling and also tumor immunology. An article on chromatin remodeling brings in the concept of epigenetics where the traditional dogma was that the underlying genetic deregulation emerged through an accumulation of individual mutations that inactivate tumor suppressor genes and activate oncogenes. During recent years, it has become evident that epigenetic changes are potently contributing to tumor formation. And there are articles on specific cancers to illustrate the above-mentioned concepts. The section on **Molecular diagnostics and Imaging** covers biomarkers that can be used to detect cancer in different stages and also the relatively new field of metabolomic imaging for cancer detection, which involves the global variations of metabolites, with which malignancy conditions can be evaluated by profiling the entire measurable metabolome and the application of metabolomics for *in vivo* cancer imaging. Also, this section covers the emerging field of microfluidics in nanomedicine, which is capable of serving as contrast agents for different imaging modalities as well as targeting specific sites for delivery of drugs/genes. The section on **Cancer Therapy** includes strategies for targeting cancer stem cells and therapeutics against cancer stem cells; immunologic therapies such as immune checkpoint inhibitors and also immunotherapy with autologous cells; very new and experimental approaches such as carbon nanotubes for enhanced biopharmaceutical delivery and gold and iron oxide nanoparticles with antibody guides to find and destroy cancer cells; RNAi in cancer therapy; role of the microbiota in cancer therapy; radioactive microbes; and surveys of broadly applicable therapeutic approaches such as targeted therapy: genomic approaches; cancer chemotherapy; and cancer pharmacogenomics.

Our participating authors are from some of the most active cancer research institutions including the National Cancer Institute, MD Anderson Cancer Center, Harvard University, Massachusetts General Hospital, the Massachusetts Institute of Technology, Stanford University, University of Oxford, Washington University School of Medicine, Baylor College of Medicine, Pennsylvania State University, Albert Einstein College of Medicine, Cornell University, University of Michigan, Inserm, Fox Chase Cancer Center, and the University of California.

Our team hopes that you, the reader, will benefit from our hard work – finding the content useful in your research and educational. We thank our Managing Editor, Sarah Mellor as well as our Executive Editors, Gregor Ciccetti and Frank Otmar Weinreich for both their advice and hard work in the course of this project.

Larkspur, CA, USA
January 2016

Robert A. Meyers
RAMTECH Limited

Part I
Cancer Biology

Translational Medicine Cancer: Advances in Molecular Biology and Medicine,
First Edition. Edited by Robert A. Meyers.

1
Cancer Etiology

Graham A. Colditz and Chetan Vakkalagadda
Washington University School of Medicine and Barnes-Jewish Hospital, The Division of Public Health Sciences, Department of Surgery, and the Siteman Cancer Center, Saint Louis, Missouri 63110, USA

Translational Medicine Cancer: Advances in Molecular Biology and Medicine,
First Edition. Edited by Robert A. Meyers.

Keywords

Causes
Substances and exposures that lead to cancer are called carcinogens. They may impact on DNA directly or indirectly by increasing the rate at which cells divide

Lifestyle
Lifestyle factors, including nutrition, physical activity and tobacco use are components of people live their lives in relation to the risk of chronic diseases

Tobacco
The nicotine-rich leaves of an American plant which, after curing, is smoked or chewed

Obesity
Obesity is defined by the World Health Organization as a body-mass index (BMI; a measure of adiposity) of 30 $kg\,m^{-2}$, or greater. The BMI is a measure of a person's body weight relative to their height

Causal
A causal relationship conveys the inference that changing a given factor will lead to a change in the population burden of disease. This may be either by reducing the number of cases or by making disease occur later than it would have

Prevention
The elimination of causes of disease from the population, or reducing exposure to them, so that the risk of disease is either reduced. Prevention thus aims to lower the risk of disease

Prospective cohort study
Epidemiological studies of people who either have or have not been exposed to a suspected risk factor, and who are subsequently observed over time for the development of the disease of interest. A link between the suspected risk factor and the disease is indicated when the exposed and unexposed participants have a significantly different frequency of future development of the disease

Risk (of disease)
The probability that a disease will develop in an individual during a specified time period

Understanding cancer etiology informs prevention. Given the recent explosion in understanding how genetics, lifestyle, individual behaviors, social environment, policy, and healthcare interventions can reduce the burden of cancer, new insights and an integration of knowledge can help to shape prevention strategies and lead to reductions in cancer incidence and improved population health. Inherited genetic predisposition may account for 5–10% of cancers. Currently, over 50% of cancers can be prevented on the basis of presently available knowledge. Age and gender show relations to most cancers, and race/ethnicity may be a marker for either genetic or lifestyle factors. Lifestyle (including smoking, diet, obesity, lack of exercise, alcohol, and excess sun exposure) contributes to the excess burden of cancer, while reproductive factors influence the risk of female cancer and occupational and environmental exposure also contribute to risk. Viruses and other infections also cause a large percentage of cancers in low- and middle-income countries. Further refinements in understanding the "molecular pathway" from lifestyle to cancer incidence will help to refine prevention priorities and guide policy and practice to reduce the burden of cancer.

1 Introduction

1.1 The Evolving Landscape of Cancer

At the start of 2012, approximately 13.7 million Americans had been diagnosed with cancer at some point in their lives. This number continues to increase, and an additional 1.65 million people are expected to be newly diagnosed with cancer in 2015 [1]. This already-large yet increasing burden of cancer is not just limited to the United States. Across the world in 2012, there were 14 million new cases of cancer with costs of diagnosis and treatment amounting to US$ 290 billion [2]. These global numbers are projected to rise to 22 million new cases in 2030, generating a monetary burden of US$ 458 billion. Already, more cases are diagnosed annually in low- and middle-income countries than in high-income countries [3]. The burden of cancer presents a major cause for concern among healthcare workers, public health researchers, and governments everywhere.

It has been for half a century that lifestyle, occupational, environmental and other external factors contribute significantly to the cancer burden. Occupational exposures have been largely regulated to reduce exposure and the disease burden experienced by workers. A 1964 report from a World Health Organization (WHO) panel of cancer prevention experts affirmed that "extrinsic factors" are a key contributor to the majority of cancers [4]. This report was the first official statement from experts to make the wide-ranging claim that cancer could be caused by external causes. But, with what is now known the panel's decision appears conservative in its estimates. Epidemiologists now agree that, given ideal conditions, at least 50% of cancers could be prevented [5, 6].

In addition to the WHO report, 1964 is well-known in public health for a landmark determination regarding smoking and lung cancer in the United States. Cigarette smoking became more widespread during the early 1900s with the inclusion of cigarettes in rations to soldiers in World War I and World War II. With much advertising increasing social acceptability and pressure on society to smoke, cigarette smoking became highly prevalent in the US by the mid-twentieth century, when more than half of all adult men smoked. This rise in smoking was accompanied by a rise in scientific inquiry into the health effects of smoking. Studies performed during the 1950s had begun to link cigarette smoking and lung cancer, and in 1962 the British Royal College of Physicians released a report, "*Smoking and Health,*" which stated that cigarette smoking is a cause of lung cancer [7]. Luther Terry, the Surgeon General of the United States from 1961 to 1965, created an American committee upon release of the British report to further explore the findings. In 1964, the Surgeon General's office released the *US Surgeon General's Report on Smoking and Health,* which concluded, based on seven prospective studies starting as early as 1951, that there is a causal relationship between smoking and lung cancer [8, 9]. This was not the first report to be produced by the US government highlighting this association [9], but it was by far the most famous. The release of the Surgeon's General report precipitated a nationwide movement reducing the use of cigarettes and tobacco and, as a consequence, sales of cigarettes were increasingly regulated by the government; indeed, starting in 1966 all cigarette packets were legally required to show a warning label. Cigarette smoking rates peaked in 1964, but have decreased steadily since that time (Fig. 1). While lung cancer remains a major public health issue (see Sect. 2.4), the tobacco-prevention efforts that started after publication of the 1964 Surgeon General's Report showed that to prevent the external causes of cancers could also be effective.

The risk of lung cancer increases in line with the number of cigarettes smoked per day. Among men and women who smoked 40 cigarettes per day, the relative risk of

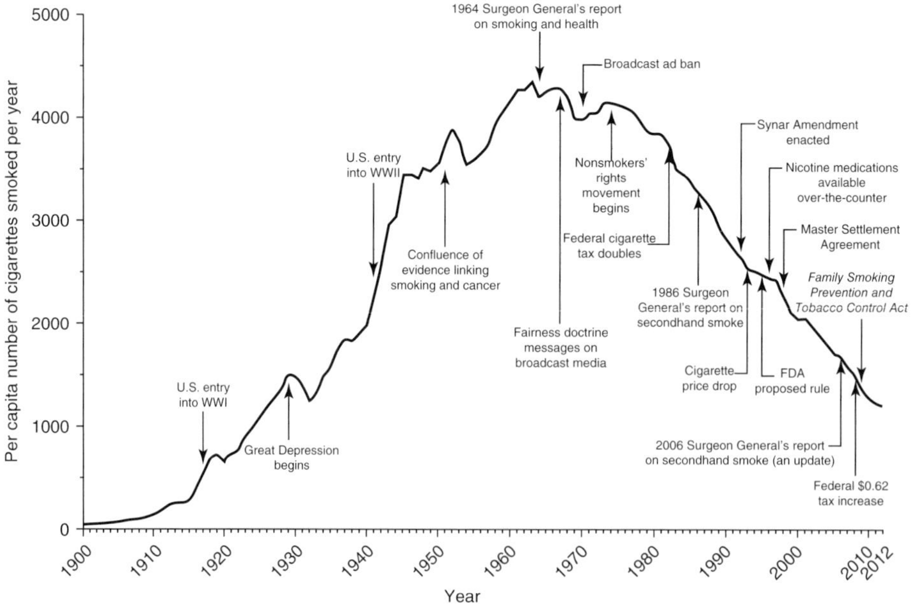

Fig. 1 Smoking rates in the United States during the twentieth century. Reproduced from Ref. [11].

lung cancer compared to "never-smokers" was 40 or more [10]. Whilst smoking causes 80–90% of all lung cancers, mortality due to lung cancer in the US has decreased by over 30% since peaking during the early 1990s; hence, prevention programs are clearly effective. Trends in lung cancer mortality have also shown that the time course between changes in a causal factor (e.g., smoking) and population changes in the cancer burden may be substantial, in part reflecting the timing of exposure in the pathway to carcinogenesis.

Updates to reports from the US Surgeon General continue to document the health effects of smoking and the benefits of quitting. More cancers have been added to the list of those caused by smoking, with reports summarizing the epidemiologic evidence, biologic mechanisms, carcinogenic products contained in tobacco smoke, and the population burden of cancer caused by smoking [11]. A review of the concepts of causal inference from epidemiologic data is presented in Chapter 1 of the 2004 US Surgeon General's report [12], while a history of the causal inference methods evolving from 1964 to 2014 is presented in Chapter 3 of the 2014 report [11]. These reviews, and the approach used by the International Agency for Research on Cancer (and the US Institute of Medicine), clearly support the use of terms such as "cause" for factors such as smoking, alcohol, lack of physical activity, estrogen plus progestin therapy, and other lifestyle factors that have been studied predominantly through observational epidemiologic investigations, as described in the following sections.

Programs and public health initiatives have substantially reduced the prevalence of smoking in the US, and a framework for ending the "tobacco epidemic" has been set forth. However, sustained public health programs are needed to achieve the potential reductions in cancer through tobacco control; unfortunately, these are beyond the scope of the present chapter but they have been clearly summarized elsewhere [13].

Beyond tobacco being the leading preventable cause of cancer, other occupational, environmental, lifestyle and social forces drive cancer risk. Evidence on the causes of cancer, and the potential mechanisms that link these causes to cancer and may inform prevention, are reviewed in the following sections.

1.2
The Cancer Burden

The incidence rates of several cancers in the United States from 2007 to 2011, and the projected number of new cases in 2015, are listed in Table 1 By a disproportionate amount, those cancers with the greatest burden on the United States population are (and continue to be) cancers of the breast, prostate, lung/bronchus and colon/rectum, which account for 50% or more of all cancer cases in men and women. Moreover, all of these cancers have established causes, many of which are modifiable and hence may provide pathways for their prevention.

2
The Leading Cancers

2.1
Breast Cancer

Breast cancer is one of the leading cancers diagnosed in women in the US and in women worldwide; in fact, it accounts for 25% of all cancers diagnosed among women worldwide. It is estimated that 12% of all women in the US will develop breast cancer at some point in their lifetimes, and that

Tab. 1 Recent incidence rates[a)] and projected new cases[b)] of leading cancers, USA.

Cancer	*Incidence rate per 100 000 (2007–2011)*		*Total new cases (projected) (2015)*
	Males	*Females*	*Both sexes*
Prostate	142.1	n/a	220 800
Breast (female)	n/a	122.8	231 840
Breast (male)			2 350
Lung/bronchus	78.6	54.6	221 200
Colorectal	50.0	37.8	132 700
Bladder	36.7	9.1	74 000
Melanoma	25.1	15.8	73 870
Non-Hodgkin lymphoma	23.2	16.1	71 850
Leukemia	16.6	10.2	54 270
Pancreas	13.8	10.8	46 420
Uterine corpus	n/a	25.0	48 960
Total	535.8	419.1	1 658 370

a) Standardized cancer incidence from Cancer in North America 2014 [14].
b) Case number projection from Ref. [15].

this percentage stands to increase. The incidence of breast cancer increases with age, rising rapidly through the premenopausal years, with dramatically slower increases after menopause. The major risk factors for breast cancer include reproductive factors (age at menarche, age at first birth, parity, and age at menopause), alcohol intake, ionizing radiation treatments, obesity, oral contraceptive use, and postmenopausal hormones [16, 17].

Breast cancer incidence rates among Asian, Hispanic and American Indian women in the US are considerably lower than those of (non-Hispanic) white women [18]. Migrant status often drives variation in risk; indeed, a large body of literature shows increases in breast cancer incidence following migration from a low-risk country to the US. For example, Ziegler *et al.* [19] noted a sixfold gradient in the risk of breast cancer among Asian women, depending on their time since migration. Asian American women with three or four grandparents born in the West were at highest risk, whereas women who had been born in rural areas of Asia and whose length of residence in the US was a decade or less were at lowest risk. These findings strongly suggest that factors associated with the lifestyle or environment of the destination country influence breast cancer risk, and are consistent with a positive relationship between the length of time in the destination country and adoption of that country's lifestyle. For example, among immigrants the fertility rate and average number of children born tend to converge to the rates of the destination country [20, 21].

Within the US, African American women have a higher breast cancer mortality than Caucasian women and a lower overall lifetime risk. Through the premenopausal years, African American women have a higher incidence of breast cancer and are more likely to be diagnosed with triple-negative or aggressive subtypes of breast cancer. The incidence is higher in African American women than Caucasian women [22].

Although a family history of breast cancer is an accepted risk factor, the proportion of breast cancer estimated as being due to rare, highly penetrant genes such as *BRCA1* and *BRCA2* is less than 10% [23], with currently documented mutations in *BRCA1* and *BRCA2* explaining 15% of all breast cancer familial risk [24]. Carriers of *BRCA1* mutations are estimated to have a 57% chance of developing breast cancer by the age of 70 years.

Percent mammographic density (PMD) is one of the strongest risk factors for breast cancer, and is predictive of breast cancer risk for at least 10 years in the future. Women with the highest mammographic density are at a four- to sixfold greater risk of breast cancer than those with the lowest density [25, 26].

Benign breast disease is related to the subsequent risk for breast cancer. Proliferative lesions carry approximately double the risk of normal breast tissue, while atypical hyperplasia on benign biopsy carries a four- to fivefold increase in risk that is bilateral [27]. In other words, these benign lesion serve as a marker of subsequent risk of breast cancer [28].

2.1.1 Obesity

Reflecting the global epidemic of obesity, an increasing proportion of breast cancers are due to adult weight gain and obesity after menopause. This effect of obesity on female breast cancer incidence varies according to the time point in a woman's life that she is obese. Obesity in childhood, adolescence, and young adult years is inversely related to the premenopausal incidence of breast cancer [29]. A higher body mass index (BMI) in childhood is related to lower peak height growth velocity and a lower lifetime risk for breast cancer [30, 31]. On the other hand, obesity after menopause leads to higher circulating estrogen levels and causes postmenopausal breast cancer [32], and also contributes to a poor prognosis for women with postmenopausal breast cancer [33].

Proposed mechanisms for the association of obesity with breast cancer center on hormones released by adipose tissue. The inverse association between obesity and breast cancer before menopause may be due to the effects of childhood adiposity, slower growth in height, lower breast density [34], and decreased risk through the premenopausal years. After menopause, adipose tissue is a major production source for estrogen, a driver of postmenopausal breast cancer. Adipose tissue, a major producer of sex steroids such as estrogen [35], contains high levels of the enzyme aromatase that produces estrogen; indeed, aromatase levels are increased in adipose tissue as a function of increasing age. Combined data acquired from prospective studies of estrogen levels have shown that high levels of estrogen are directly related to incidence, and that this is strongest for receptor-positive breast cancer [36]. Furthermore, in that analysis a strong relationship with hormone levels persisted, even after statistically controlling for BMI. Such an analysis points to estrogen being the mediator in the pathway from adiposity to breast cancer incidence. Further support for this pathway has been obtained from prevention trials with selective estrogen receptor modulators (e.g., tamoxifen and raloxifene) which block estrogen receptors and significantly reduce the risk of breast cancer [37].

More recently, other biologic pathways have emerged linking adipose tissue to breast cancer progression. One such theory centers on adiponectin, the hormone produced and secreted by adipocytes. The proposed mechanism states that adiponectin induces the expression of growth factors such as vascular endothelial growth factor (VEGF) and fibroblast growth factor (FGF), and enhances the migration of breast cancer cells [38]. Both, growth factor expression and cell migration contribute to the progression of breast cancer. Another proposed pathway derives from the chronic inflammation induced by obesity and the recruitment of inflammatory cells, which in turn promote to breast cancer [39]. These multiple pathways support the proposal that hormones secreted by fat cells are directly related to the progression of breast cancer. Local effects of adipose tissue in the breast producing estrone and estradiol also represent possible pathways for adiposity to impact on breast cancer etiology and progression.

A logical next question would be that, if the hormones produced by adipose tissue drive postmenopausal breast cancer, would a reduction in the volume of adipose tissue translate to a lower occurrence of breast cancer? Mechanistically, studies performed in mice have shown that calorie restriction decreases the inflammation seen with obesity [39]. Consequently, as inflammation is associated with several different cancers, it can be surmised that attempts to lose weight can indeed help in decreasing cancer progression. A prospective analysis based on the Nurses' Health Study cohort shows that sustained weight loss after menopause is associated with a reduced incidence of breast cancer [40]. In this case, women who lost 10 kg or more and maintained such loss halved their risk of breast cancer. Meta-analyses have also shown that bariatric surgery, which leads to substantial and sustained weight loss, is associated with lowered risks of breast cancer [41]. This suggests that the weight loss in and of itself – and not necessarily the method of achieving it – is protective against breast cancer.

2.1.2 Maternal Age and Menarche

The results of several epidemiological studies and statistical analyses have shown that breast cancer risk accumulation is modified by reproductive events, and their timing. These include: (i) having children at a young age; (ii) later-onset menarche; and (iii) earlier-onset menopause. A particular subtype of breast cancer, namely progesterone receptor (PR)-negative (PR−) cancer, shows a direct relationship between later age at first pregnancy and breast cancer risk [42]; PR+ cancers do not show this association, however. Age at menarche marks the onset of monthly cycling of female hormones; breast cancer incidence (or breast tissue aging in Pike's model) [43] is then increased by 8.5% per additional year between menarche and menopause [44]. This annual increase in risk slows to 2.5% per year after menopause, and is also slowed by pregnancy. In addition, the closer the pregnancies are together, the lower the lifetime risk [45, 46].

The mechanistic basis for these findings centers on breast tissue aging [43]. Rodent models have shown that pregnancy confers a protective effect against breast cancer by causing a differentiation of the breast tissue, making it less susceptible to carcinogenesis [47]. Subsequent studies have demonstrated molecular changes induced by pregnancy. Specifically, pregnancy induces decreases in the number of hormone-sensitive luminal cells and a downregulation of the Wnt signaling pathway in basal stem cells and/or progenitor cells in the FVB/NHanHsd mouse model, making breast tissue less susceptible to carcinogens [48]. Following a protocol-timed mating at 42 days with genetically homogeneous FVB/NHanHsd mice, mammary epithelial cell subpopulations were isolated from parous and age-matched virgin mice. A reduced expression of Wnt4 in the mammary cells from parous mice corresponded to a decrease in the proportion of Wnt4-secreting estrogen/progesterone receptor-positive cells. Recombinant Wnt4 rescued the proliferation defect *in vitro*, supporting a causal link to parity-induced alterations of basal stem/progenitor cell properties and a long-term protection from first pregnancy [49]. Other studies have shown that human chorionic gonadotropin (hCG), a key hormone during pregnancy, mimics this protective effect, a finding also shown in women [50].

Chromosomal studies have added an additional layer of support for the differentiation of breast cells with first pregnancy. Nulliparous breast epithelial cells contain large, euchromatic nuclei, while parous breast cells contain smaller, heterochromatic (i.e., epigenetically silenced) nuclei [51]. Histone methylation, a well-known mechanism of gene silencing, is observed more in the breasts of parous women than in those of nulliparous women. Taken together, these molecular studies provide laboratory-based support for pathways between the interval from menarche to first pregnancy, parity, and the risk of breast cancer.

2.1.3 Alcohol Intake

Various epidemiological studies have shown conclusively that alcohol use is associated with an increased risk of breast cancer and, indeed, alcohol is classified by the International Agency for Research on Cancer as a breast carcinogen [52]. The risk of breast cancer increases directly with the amount of alcohol consumed, with even moderate alcohol consumption causing a clear increase [53, 54]. Mechanistically, randomized trials of alcohol feeding have shown increases in female hormones when taking alcohol compared to placebo (both in premenopausal and postmenopausal women) [55, 56]. An increased production of reactive oxygen species (ROS) such as hydroxyl radicals has also been proposed as an underlying factor [57]. Alcohol intake before the first pregnancy has also been shown to increase the risk of premalignant lesions, and the longer the interval from menarche to first birth the greater the adverse effect of alcohol intake [58]. While some evidence suggests that higher blood folate levels may counter the adverse effects of alcohol [59], this remains to be confirmed.

2.1.4 Postmenopausal Hormone Use

Exogenous hormones (i.e., hormone therapy), which is used for the relief of menopausal symptoms and to prevent osteoporosis, have also been purported to provide other health benefits among postmenopausal women. As noted above when discussing endogenous hormone levels in the context of obesity, estrogen is directly related to breast cancer risk, with the vast majority of prospective epidemiologic studies having shown that a longer duration of estrogen use alone increases the risk of both breast and endometrial cancer [60]. The collaborative reanalysis included individual patient data acquired from 51 epidemiological studies (over 50 000 breast cancer patients and 100 000 non-cancer women). Current use of hormone therapy, or its use within the previous four years, was associated with a significant increase in risk of breast cancer compared to women who had never used hormone therapy [60]. For those women who had stopped hormone therapy for more than five years, however, there was no significant excess risk of breast cancer. Importantly, the addition of progestins to counter the adverse effects of estrogen on the endometrium also increased the risk of breast cancer, as demonstrated in prospective epidemiologic studies [61] and the Women's Health Initiative randomized trial [62]. Growing evidence has indicated that the timing of therapy initiation modifies the risk, with use closer to menopause (typical of traditional evidence from epidemiologic studies) and among lean women showing greater adverse effects [63]. Data from the Women's Health Initiative randomized trial, in which older women aged up to 79 years were randomized to hormone therapy, suggested that initiating use at an older age may not convey the same level of increased risk.

Many investigations have been conducted to elucidate the mechanistic basis for the link between progesterone and breast cancer. The main hypothesis is that progesterone exerts its carcinogenic effects through the receptor activator of NF-κB ligand (RANKL) [64]. The progesterone–RANKL axis is the main mediator of breast tissue proliferation in both mouse and human, and an inhibition of RANKL signaling has been shown to reduce the incidence of breast cancer metastasis [65, 66]. Therefore, RANKL represents a good short-term target for cancers caused by the use of progesterone as hormone replacement therapy (HRT), with a long-term goal being to eliminate progesterone as HRT entirely. Based not only on evidence acquired from studies conducted in women but also from biologic pathways, the International Agency for Research on Cancer classified combination estrogen plus progestin as a human carcinogen in 2007 [67].

2.1.5 Physical Activity

Many studies have now shown that women who are physically active throughout their lifetime have a lower risk of breast cancer. Indeed, women who routinely perform 3 h or more exercise each week have a 20% lower risk of breast cancer than those who are minimally active or inactive [68]. In the large Nurses' Health Study II, regular physical activity between the ages of 12 and 22 years was shown to provide the greatest protection against premenopausal breast cancer when compared to activity in older groups, with the most active women having a 25% lower risk of breast cancer through their premenopausal years. The IARC classifies a lack of physical activity as a cause of breast cancer [32].

2.1.6 Other Factors

Various other casual factors for breast cancer have been identified, the most common being ionizing radiation. Evidence acquired from the follow-up of atomic bomb survivors has pointed to the timing of exposure, and also age at exposure, as important when defining the magnitude of the adverse effect of exposure to this carcinogen [69]. Exposure to ionizing radiation in childhood and adolescence carries the greatest risk of breast cancer.

A current use of oral contraceptives is associated with a transient increase in the risk of breast cancer [70]. Lactation reduces the risk of breast cancer [71], and adolescent diets high in fruits, vegetables and fiber may also lower risk [72, 73]. Plasma markers of plant-based diets such as carotenoids are inversely related to the risk of breast cancer [74]. Cigarette smoking also increases the risk of breast cancer and death from the condition [11].

2.2 Colorectal Cancer

Colorectal cancer is the third most common cancer in the Western world [2], and the fourth most common in the US (Table 1). Only 10% of colon cancers are caused by genetic predispositions such as hereditary non-polyposis colorectal cancer (HNPCC) and familial adenomatous polyposis (FAP); the remaining 90% are caused by environmental and lifestyle factors, some of which are modifiable and include diet, cigarette smoking, obesity and physical activity, aspirin, and hormone therapy. Colon polyps, precursor lesions for colon cancer are often investigated to appreciate the time course of risk factors and thus the prevention of colon cancer.

2.2.1 Diet

Whilst eating a well-balanced diet is the optimal way of reducing the risk of several different types of cancer, specific factors have become associated with colon cancer.

Red Meat Intake Various studies have consistently demonstrated a positive association between a higher consumption of red or processed meat and an increased risk of colon cancer [75, 76]. Both, recent and long-term consumption of red meat confer an increased risk of developing colon cancer, especially when compared to poultry or fish intake [77]. In fact, there is an inverse association between the intake of fish and the incidence of colorectal cancers [78]. Proposed mechanisms for the association between red meat consumption and increased colon cancer include heme iron, nitrates from processed meats, and heterocyclic amine intake [79], all of which are known carcinogens. Heme iron in particular has been associated with DNA damage and a hyperproliferation of epithelial cells in the colon via a hydrogen peroxide mechanism [80]. Studies performed in rats have shown that fish intake is associated with less DNA damage and fewer inflammatory markers when compared to a controlled diet [81]. Together, these data indicate that red meat intake is a modifiable risk factor for colon cancer.

Calcium Several studies have demonstrated the benefits of sufficient dietary calcium with regards to the risk of colorectal cancer. Data obtained from the Nurses' Health Study and Health Professionals Follow-Up Study have indicated that calcium intakes of 700 mg per day or higher conferred protection against developing distal colorectal cancer but not necessarily from proximal colorectal cancer. Notably, the benefit was linear up to levels of about 700 mg per day, but leveled off at higher calcium intakes [82]. Subsequent studies corroborated this protective effect of calcium, particularly in the form of calcium supplements [83, 84]. Although the amount of calcium designated as "sufficient" remains uncertain, the pooled analysis of prospective cohort studies suggests that benefit decreases above 1200 mg per day [85]. In a randomized trial of patients who had undergone polyp removal, a daily supplement of 1200 mg calcium led to a significant reduction in the risk of subsequent polyps [86], and the benefit persisted when the supplementation was stopped [87].

The mechanism underlying calcium's chemopreventive effects most likely centers on the high sensitivity of colonic epithelial cells to calcium. Whereas, normal colonic epithelial cells will proliferate in the presence of trace amounts of calcium, higher levels of calcium (as are normal for most other tissues) cause the cells' proliferative machinery to shut down.

Vitamin D Whilst a dietary intake of vitamin D is known to be a protective factor against many cancers, studies from the IARC have indicated that vitamin D is most protective against the development of colon cancer [88]. In 2011, a meta-analysis conducted by the US Preventive Services Task Force confirmed this finding, where each 10 $nmol\,l^{-1}$ increase in the blood level of vitamin D was associated with a 6% decrease in colorectal cancer risk [89].

Mechanisms connecting vitamin D and decreased colon cancer risk are centered on the vitamin D receptor (VDR) which, when bound to vitamin D, interacts with DNA and signaling pathways [90]. In this case, the major pathway is the Wnt-Beta catenin pathway, which promotes colorectal oncogenesis. The suppression of beta-catenin signaling by VDR results in decreased oncogenesis [91]. Another pathway affected by VDR is the inflammatory pathway, and bowel inflammation (especially in conditions such as Crohn's disease) are a major risk factor for colon cancer. VDR suppresses the production of inflammatory markers and downregulates T_H1 types of T cells in favor of T_{reg} and T_H2 cells, which are less inflammatory [92]. While the epidemiology on the specifics of vitamin D levels and the magnitude of protection against colon cancer remains unclear, molecular biology studies have shown that claims for such a relationship have strong biologic basis.

Alcohol A long history of studies relates higher alcohol intake to an increased risk of colon cancer and to an increased risk of colorectal adenomas [93]. A dose–response relationship has been observed across several studies [94], with some evidence suggesting that even moderate drinkers (one drink per day) are at a higher risk of colon cancer than nondrinkers. These data have been extensively reviewed and summarized by the IARC [52], which has concluded that the association does not vary by alcoholic beverage type, gender, or smoking status. The effects of alcohol appear to be exacerbated by low levels of folate and methionine. In the Health Professionals Follow-Up Study, an increased risk of colon cancer was observed among both current and past drinkers, but only among those who also had low intakes of methionine or folate [95]. Because alcohol is an antagonist of methyl-group metabolism, it may imbalance DNA methylation, which may in turn contribute to the process of carcinogenesis.

Folate Folate is a vitamin found in green leafy vegetables, and low levels of folate intake have long been associated with colon and rectal cancers [93, 96]. Beneficial effects of folate supplementation were demonstrated through a reduction of dysplasia in patients with chronic ulcerative colitis, which is a risk factor for colon cancer [97]. Findings for a reduction in the risk of polyps, a precursor lesion for colon cancer, suggest that the timing of exposure may again be important for the prevention of colon cancer. Prospective data from the Nurses' Health Study and Health Professionals Follow-Up Study have shown that folate intake may be beneficial over decades before the diagnosis of colon cancer, but does not necessarily protect against colorectal cancer in the short term, which is consistent with protecting against early molecular changes [98]. Other studies, though typically much shorter in duration, have demonstrated a possible increased association between folate intake and colorectal adenomas [99]. Taken together, these data suggest that folate may play both a protective role against colorectal cancer and also a potential carcinogenic role. Further research is required to disentangle the mechanisms and timing of exposure.

2.2.2 Smoking

Smoking causes many cancers, and colorectal cancer is no exception. Prospective epidemiologic studies have shown that smokers have a higher incidence of colon and rectal cancers, with a higher risk of colon cancer than rectal cancer compared to never-smokers [11, 100, 101]. Additionally, smokers of 40 cigarettes per day or more, have a 50% higher incidence of rectal cancer than those who have never smoked. Current smokers have a higher incidence of proximal colon cancer compared to distal colon cancer or rectal cancers. Importantly, timing matters again in the carcinogenesis of colorectal cancer. Exposure to smoking for decades before a diagnosis drives the risk, which is consistent with smoking acting as an initiator for colorectal cancer [102].

Mechanistically, the silencing of tumor suppressor genes through cigarette smoke has been proposed to link smoking to colon cancer incidence. Many colorectal cancers show a silencing of KRAS and BRAF, two major tumor suppressors, and this is also associated with cigarette smoking [103, 104]. As summarized in the Surgeon General's Report, other carcinogens have also been found in cigarette smoke [11]. In particular, polycyclic aromatic hydrocarbons and heterocyclic amines in cigarette smoke are associated with increased rates of tumorigenesis [105].

2.2.3 Obesity

Obesity appears to influence the development of adenomatous polyps, as well as the progression of polyps to malignancy. Both, men and women with excess body weight are at increased risk of polyps and colon cancer, though the magnitude of increased risk may be slightly higher among men than women [94, 106]. Men with high levels of abdominal fat appear to be at a particularly high risk [107].

The underlying mechanism for the relationship between obesity and colon cancer is unknown. However, it has been hypothesized that elevated insulin or other growth-related factors may be mediating the influence of obesity, and C-peptide (a marker of insulin secretion) has been shown to be directly related to risk in prospective blood studies [108]. Abdominal obesity, in particular, is a strong determinant of hyperinsulinemia, a condition which may promote colonic tumor growth.

2.2.4 Physical Activity

Regular physical activity is associated with decreased risks of colon cancer, as summarized in recent meta-analyses for both colon polyps and colon cancer [109, 110]. The level of physical activity is important as it is protective against colon cancer in a dose-dependent manner, with a higher activity leading to a lower risk [109]. The protective benefit of physical activity seems strong even in individuals who are obese, which suggests it is an independent factor for colon cancer unrelated to obesity.

Many different mechanisms have all been shown to support the idea that physical activity protects against colon cancer. One mechanism relates to insulin resistance, in that physical inactivity leads to insulin resistance where higher levels of insulin are needed to normalize blood glucose levels. Insulin and insulin-like growth factors (IGFs) in turn promote colorectal tumor growth by inducing mRNA expression of VEGF, which is responsible for the initiation and growth of several cancers, in the colon [111]. Consistent with this mechanism, prospective studies of glucose and insulin have demonstrated direct relationships to colon cancer incidence [108]. Exercise also limits the expression of inflammatory factors such as nitric oxide synthase [112], thereby reducing risk. Many more mechanisms have been proposed, related to prostaglandins, immune responses, and other biological factors [113]. Currently, research is ongoing to identify new possible pathways underlying the protective effects of physical activity against colon cancer, potentially opening new pathways for chemoprevention.

2.2.5 Aspirin

Combined data from observational studies have shown that the use of aspirin is protective against colorectal cancer [114], and this finding has been supported by the follow-up of randomized trials [115]. A randomized trial of aspirin in carriers of hereditary risk (Lynch syndrome) showed a significant reduction in colorectal cancer risk of approximately 40% [115, 116].

2.2.6 Hormone Therapy

In women, the apparent smaller adverse effect of obesity is hypothesized to be due to higher estrogen levels. Data acquired from the randomized trial of hormone therapy (Women's Health Initiative) showed consistency with observational data [117] and a lower risk of colon cancer among those receiving hormone therapy [118].

2.3 Prostate Cancer

Prostate cancer is a leading cancer diagnosed among men, with the introduction of widespread screening leading to the detection of potentially nonlethal tumors. Today, investigations continue to focus and refine the present understanding of the factors that drive fatal prostate cancer. African American men are at a significantly higher risk than Caucasian men, and this is evident from the age of 35 years onwards, based on national estimates of age-specific incidence [119]. Several non-modifiable factors predispose men to higher risks of prostate cancer. One is genetics: men with *BRCA1* and *BRCA2* mutations, which are commonly associated with breast cancer, also have an increased risk of developing prostate cancer [120], and single nucleotide polymorphisms (SNPs) convey risk [121], including those in developmentally important genes such as *HOXB13* also increase risk for prostate cancer [122]. Yet, some risk factors are within the control of men that might modulate the risk of their developing prostate cancer, the most notable being diet (e.g., meat and tomato intake) and a high body weight. Despite extensive investigations, however, the evidence for modifiable risk factors in prostate cancer is less well established than for other cancers.

2.3.1 Meat Intake

Over the years, the intake of animal fat and meat has been shown to be associated with prostate cancer in several different prospective studies. In 2001, the risk for metastatic prostate cancer was shown to be increased with the intake of red meat and processed meats [123]. In subsequent studies, the possible contributory factors that would increase susceptibility to prostate cancer were narrowed down, with heme iron, nitrite/nitrate from meat, grilled/barbecued meat, and benzo(a)pyrene each being associated with elevated risks of advanced prostate cancer, in addition to red and processed meats. The same factors (except for nitrite/nitrate exposure) were also found to be associated with total prostate cancer [124]. Several mechanisms have been proposed as to how these compounds may contribute to prostate cancer. Heme iron has been suggested to cause free-radical damage in the prostate as well in other organs, which could contribute to the progression of prostate cancer [125]. *N*-nitroso compounds are also a possible contributor to the increased carcinogenesis that occurs with red meat consumption; the exogenous formation of these compounds results from the ingestion of nitrite-preserved meats, while their endogenous formation results from the heme iron content of red meat

[124]. Thus, limiting the consumption of meats and animal fats could curb the effects of chemicals causing prostate cancer.

2.3.2 Lycopene

Lycopene is found in tomatoes, and has potent anti-oxidant effects. Currently, it is widely believed that eating tomato products helps to protect against prostate cancer. Whereas, older studies yielded contradictory conclusions as to the protective effects of lycopene in prostate cancer, more recent data from prospective studies have indicated that larger amounts of lycopene are inversely associated with the incidence of both total and lethal prostate cancer [126]. The results of this prospective study showed that higher levels of lycopene intake were associated with lower levels of angiogenic markers. This, in turn, suggested that lycopene's protective effect against prostate cancer could be centered on preventing tumor angiogenesis, which is a separate mechanism from the antioxidant hypothesis.

2.3.3 Obesity

Obesity is a well-known risk factor for prostate cancer mortality, as demonstrated by several meta-analyses [127]. Consistent with excess mortality, obesity is associated with more aggressive prostate cancers [128, 129].

Endocrine factors are responsible for the association between obesity and prostate cancer. A comparison of the serum of lean and obese men showed the latter to contain higher levels of leptin and other factors, that led to an increased proliferation, invasion, and migration of tumors, as well as an increased activity of proteins needed for epithelial–mesenchymal transition [130]. These characteristics of metastatic tumors support the relationship between obesity and increased tumor aggression.

2.4 Lung Cancer

Smoking is the primary cause of lung cancer, accounting for up to 90% of all cases, though other factors such as nutritional intake also play a role. As noted in Sect.1, lung cancer incidence is decreasing across the US due to reductions in smoking rates.

2.4.1 Smoking

The link between cigarette smoking and lung cancer is undisputed. Numerous studies have demonstrated the strong association between cigarette smoking or secondhand smoke ("passive" smoking) and lung cancer [11, 131]. Cigar smoking and other forms of smoking have also been shown to lead to lung cancer [132].

There are myriad mechanisms by which cigarette smoking leads to lung cancer. Carcinogens found in tars in cigarettes or in cigarette smoke, including compounds such as benzo[a]pyrene and nitrosamines [133], lead to DNA damage. Cigarette carcinogens are also associated with mutations or the silencing of important tumor suppressor genes, such as *TP53*, *KRAS*, and *P16* [134]. There is also a genetic component to lung cancer, with some smokers or even nonsmokers being more likely to develop lung cancer if they have mutations in genes such as *CYP1A1* and *GSTM1* [135]. Others genes that play a role in lung cancer have been reviewed in reports of the US Surgeon General.

As discussed in Sect. 1, since the mid-1960s cigarette smoking rates have declined in men, and lung cancer deaths have declined in the US by one-third since the early 1990s. The delay between these two effects reflects in part the dual impact of smoking on lung cancer, namely that: (i) the age when starting to smoke modifies the risk [136]; and (ii) the risk declines steadily over many years after stopping [137]. These characteristics of the relationships between smoking and lung cancer highlight the importance of considering where in the pathway a lifestyle factor might operate to increase cancer risk, and where preventive efforts should be focused [138].

2.4.2 Fruit/Vegetable Intake

An increased consumption of fruits and vegetables is associated with a decreased risk of lung cancer [139]. Mechanisms proposed for this relationship include beta-carotene, a vitamin found in carrots, pumpkins, and other fruits and vegetables. While dietary studies have shown this relationship with fruits and vegetables consistently, the use of a supplement in a chemoprevention trial such as CARET (a randomized controlled trial comparing beta-carotene supplementation with a placebo) failed to show any benefit of the supplementation. A combined analysis of four chemoprevention trials demonstrated an increase in lung cancer risk among current smokers who were randomized to beta-carotene supplementation [140]. Baseline fruit and vegetable intake before randomization was related to a lower risk of lung cancer during the Alpha-Tocopherol, Beta-Carotene Cancer Prevention (ATBC) trial. This again highlights the challenges of timing a preventive intervention to reduce the risk [141]. Other insights on this issue have arisen from a nutritional prevention trial in China, where the benefits were most evident among men who were younger at randomization. This suggested that intervening in the pathway to cancer at an early stage of carcinogenesis could reduce risk but that, among older men, where a greater risk had already accumulated, the nutritional intervention did not produce any benefit [142].

3 Medical Interventions Interrupting Pathways to Cancer

Medical interventions based on insights from etiology typically use drugs to interrupt a pathway to carcinogenesis. Alternatively, for infectious causes of cancer, vaccines, or drugs to treat infections can reduce risk. These are discussed in turn.

As noted in Sect. 2.1, selective estrogen receptor modulators reduce the incidence of breast cancer by half in women who are at high risk of the disease, or who are eligible for the prevention of osteoporosis. Likewise, aspirin has been shown to significantly reduce the incidence of colon cancer in patients at high genetic risk [116]. Observational data support the prevention benefits of aspirin across a number of cancer sites beyond colon. The benefits for colon and other cancer risk reduction over 20 or more years of aspirin use are substantial [143]. Although oral contraceptives increase the risk of breast cancer among current users, they provide a long-term reduction in the risks of ovarian cancer and endometrial cancer.

Infections cause approximately 7% of cancers in high-income countries, and approximately 25% of cancers in low and middle-income countries [144]. The IARC has classified numerous infectious agents as carcinogens for humans [145], including:

- Stomach: *Helicobacter pylori*
- Liver cancer: Hepatitis B virus, hepatitis C virus (HCV), *Opisthorchis viverrini*, *Clonorchis sinensis*
- Cervix uteri: Human papillomavirus (HPV), with or without HIV
- Anogenital: (penile, vulva, vagina, anus): HPV with or without HIV
- Nasopharynx: Epstein–Barr virus (EBV)
- Oropharynx: HPV with or without tobacco or alcohol consumption

- Kaposi's sarcoma: Human herpesvirus type 8 with or without HIV
- Non-Hodgkin lymphoma: *H. pylori*, EBV with or without HIV, HCV, human T-cell lymphotropic virus type 1
- Hodgkin's lymphoma: EBV with or without HIV
- Bladder: *Schistosoma haematobium*.

Hepatitis B vaccination reduces the incidence of liver cancer when compared to vaccinated and unvaccinated birth cohorts [146]. Currently, HPV vaccination is recommended for the prevention of cervical cancer and anal cancer, and conveys long-term protection against these cancers caused by viruses [147].

3.1 Balancing Risks and Benefits with Chemoprevention

Breast cancer chemoprevention demonstrates the ability to counter the estrogen stimulation of breast cancers through selective estrogen receptor modulators and aromatase inhibitors that reduce estrogen production. The balance of risks and benefits for chemoprevention depends on several factors, including a woman's age, race, and risk of breast cancer; whether she has a uterus; and the type of medication (tamoxifen or raloxifene) [148]. Tamoxifen increases the risk of endometrial cancer, stroke, pulmonary embolism, deep-vein thrombosis and cataracts, but decreases the risk of bone fractures [149]. Compared with tamoxifen, raloxifene (which is only approved for postmenopausal women) has a lower risk of endometrial cancer, cataracts and thromboembolic events [37]. In white women under the age of 50 years, the benefits of tamoxifen are likely to outweigh the risks when the five-year risk of breast cancer is at least 1.5% [150]. Younger black women at high risk of breast cancer also benefit from tamoxifen, although black women aged 40–50 years may need a higher level of breast cancer risk to derive a net benefit. In women aged 50 years or more, the benefit/risk ratio tends to be better for raloxifene than for tamoxifen among women with a uterus, and similar for the two drugs among women without a uterus [148]. The level of breast cancer risk that is required to derive a net benefit increases with age, and also tends to be higher for black women than for white women [148].

Aspirin (see Sect. 2.2.5) also demonstrates these issues, with increased bleeding complications occurring in a subset of users. Again, this risk must be balanced against the drug's long-term use to achieve protection against colon cancer [151].

4 Conclusions

Multiple modifiable risk factors have been identified that together account for the majority of cancers as experienced in high-income countries [152]. Tobacco and obesity account for approximately 30% each. While the US Surgeon General Reports on smoking and health set out clear criteria for causal relations between smoking and cancer, the number of cancers caused by smoking often surprises. The 2014 report [11] lists the following cancers as caused by smoking;

- Lung
- Bladder
- Cervix
- Colorectal
- Esophagus
- Renal cell and renal pelvis
- Liver

- Acute myeloid leukemia
- Larynx
- Oral cavity and pharynx
- Pancreas
- Gastric

Notably, smoking reduces the risk of endometrial cancer.

Obesity accounts for up to 20% of all cancer deaths, but for a higher proportion of incident cancers. Cancers of the breast, endometrium, esophagus, kidney, and colon [32], of the thyroid gland, liver, gallbladder and pancreas, as well as hematologic malignancies, are directly related to overweight and obesity [153].

Diet, physical activity, occupational factors, alcohol, reproductive factors and ionizing radiation each also contribute as preventable causes of cancer. Ionizing radiation drives lung cancer and melanoma risk, which is a rapidly increasing cancer among white populations worldwide. Importantly, infection has a greater impact on cancer incidence in low- and middle-income countries, but as these economies evolve to higher standards of living the incidences of many of the cancers associated with obesity, lack of physical activity and smoking will increase. Clearly, an understanding of the pathways and mechanisms involved is necessary to inform prevention methods, to guide the timing of interventions across a patient's lifetime, and to stratify risk so that interventions which carry both risks and benefits can be targeted to high-risk subgroups, so that the benefits may outweigh the risks.

The location of the causes of cancer during the time course of disease development, from its early initiation to the late promotion of tumor growth, has major implications for prevention and the time over which reductions in incidence will be observed after preventive programs have been implemented, and these issues have been addressed in greater detail elsewhere [138, 152]. A further refinement of the present understanding of molecular pathways from lifestyle to cancer incidence will help to identify new pathways for prevention, to refine prevention priorities, and to guide policy and practice in order to reduce the worldwide burden of cancer.

References

1. Siegel, R.L., Miller, K.D., and Jemal, A. (2015) Cancer statistics, 2015. *CA Cancer J. Clin.*, **65** (1), 5–29.
2. International Agency for Research on Cancer (2012) GLOBOCAN 2012: Estimated Cancer Incidence, Mortality and Prevalence Worldwide in 2012. International Agency for Research on Cancer, World Health Organization. Available at: *http://globocan.iarc.fr/Pages/fact_sheets_cancer.aspx* (accessed 26 January 2014).
3. Bray, F. and Moller, B. (2006) Predicting the future burden of cancer. *Nat. Rev. Cancer*, **6** (1), 63–74.
4. World Health Organization (1964) *Prevention of Cancer: Report of a WHO Expert Committee*, World Health Organization, Geneva.
5. Doll, R. and Peto, R. (1981) The causes of cancer: quantitative estimates of avoidable risks of cancer in the United States today. *J. Natl Cancer Inst.*, **66** (6), 1191–1308.
6. Colditz, G.A., Sellers, T.A., and Trapido, E. (2006) Epidemiology – identifying the causes and preventability of cancer? *Nat. Rev. Cancer*, **6** (1), 75–83.
7. Anonymous (1962) *Smoking and Health: A Report of the Royal College of Physicians on Smoking in Relation to Cancer of the Lung and other Diseases*, Pitman Medical Publishing Co. Ltd, London.
8. United States (1964) *Surgeon General's Advisory Committee on Smoking and Health, United States*, United States Public Health Service, Office of the Surgeon General.
9. Burney, L.E. (1957) Excessive cigarette smoking. *Public Health Rep.*, **72** (9), 786.

10. Thun, M.J., Carter, B.D., Feskanich, D., Freedman, N.D. *et al.* (2013) 50-year trends in smoking-related mortality in the United States. *N. Engl. J. Med.*, **368** (4), 351–364.
11. U.S. Department of Health and Human Services (2014) *The Health Consequences of Smoking—50 Years of Progress: A Report of the Surgeon General*, U.S. Department of Health and Human Services, Centers for Disease Control and Prevention, National Center for Chronic Disease Prevention and Health Promotion, Office on Smoking and Health, Atlanta.
12. U.S. Department of Health and Human Services (2004) *The Health Consequences of Smoking: A Report of the Surgeon General*, Centers for Disease Control and Prevention, Washington.
13. Koh, H.K. and Sebelius, K.G. (2012) Ending the tobacco epidemic. *J. Am. Med. Assoc.*, **308** (8), 767–768.
14. Copeland, G., Lake, A., Firth, R., Wohler, B., Wu, X.C., Russell, C., Zakaria, D., Schymura, M., Hofferkamp, J., Sherman, R., Kohler, B. (Eds.) (2014) *Cancer in North America: 2007–2011. Combined Cancer Incidence for the United States, Canada and North America*, Vol. I, North American Association of Central Cancer Registries, Inc., Springfield, IL.
15. American Cancer Society (2015) Cancer Facts & Figures, 2015. American Cancer Society, Atlanta.
16. Colditz, G.A. and Bohlke, K. (2014) Priorities for the primary prevention of breast cancer. *CA Cancer J. Clin.*, **64** (3), 186–194.
17. Colditz, G.A., Bohlke, K., and Berkey, C.S. (2014) Breast cancer risk accumulation starts early: prevention must also. *Breast Cancer Res. Treat.*, **145** (3), 567–579.
18. Sporn, M.B. and Roberts, A.B. (1983) Role of retinoids in differentiation and carcinogenesis. *Cancer Res.*, **43**, 3034–3040.
19. Ziegler, R.G., Hoover, R.N., Pike, M.C., Hildesheim, A. *et al.* (1993) Migration patterns and breast cancer risk in Asian-American women. *J. Natl Cancer Inst.*, **85** (22), 1819–1827.
20. Bouchardy, C. (1993) Cancer in Italian migrant populations: France. *IARC Sci. Publ.*, **123**, 149–159.
21. Young, C. (1991) Changes in demographic behaviour of migrants in Australia and the transition between generations. *Popul. Stud.*, **45**, 67–89.
22. Newman, L.A. (2014) Breast cancer disparities: high-risk breast cancer and African ancestry. *Surg. Oncol. Clin. N. Am.*, **23** (3), 579–592.
23. Whittemore, A.S. (1999) Genetically tailored preventive strategies: an effective plan for the twenty-first century? *Cancer Epidemiol. Biomarkers Prev.*, **8** (8), 649–658.
24. Couch, F.J., Nathanson, K.L., and Offit, K. (2014) Two decades after BRCA: setting paradigms in personalized cancer care and prevention. *Science*, **343** (6178), 1466–1470.
25. Yaghjyan, L., Colditz, G.A., Collins, L.C., Schnitt, S.J. *et al.* (2011) Mammographic breast density and subsequent risk of breast cancer in postmenopausal women according to tumor characteristics. *J. Natl Cancer Inst.*, **103** (15), 1179–1189.
26. Tamimi, R.M., Byrne, C., Colditz, G.A., and Hankinson, S.E. (2007) Endogenous hormone levels, mammographic density, and subsequent risk of breast cancer in postmenopausal women. *J. Natl Cancer Inst.*, **99** (15), 1178–1187.
27. London, S.J., Connolly, J.L., Schnitt, S.J., and Colditz, G.A. (1992) A prospective study of benign breast disease and the risk of breast cancer. *J. Am. Med. Assoc.*, **267** (7), 941–944.
28. Dyrstad, S.W., Yan, Y., Fowler, A.M., and Colditz, G.A. (2015) Breast cancer risk associated with benign breast disease: systematic review and meta-analysis. *Breast Cancer Res. Treat.*, **149** (3), 569–575.
29. Baer, H.J., Tworoger, S.S., Hankinson, S.E., and Willett, W.C. (2010) Body fatness at young ages and risk of breast cancer throughout life. *Am. J. Epidemiol.*, **171** (11), 1183–1194.
30. Berkey, C.S., Gardner, J.D., Frazier, A.L., and Colditz, G.A. (2000) Relation of childhood diet and body size to menarche and adolescent growth in girls. *Am. J. Epidemiol.*, **152** (5), 446–452.
31. Berkey, C.S., Frazier, A.L., Gardner, J.D., and Colditz, G.A. (1999) Adolescence and breast carcinoma risk. *Cancer*, **85** (11), 2400–2409.
32. International Agency for Research on Cancer (2002) *Weight Control and Physical*

Activity, IARC Handbook on Cancer Prevention, vol. 6, International Agency for Research on Cancer, Lyon.

33. Chan, D.S., Vieira, A.R., Aune, D., Bandera, E.V. *et al.* (2014) Body mass index and survival in women with breast cancer – systematic literature review and meta-analysis of 82 follow-up studies. *Ann. Oncol.*, **25** (10), 1901–1914.
34. Andersen, Z.J., Baker, J.L., Bihrmann, K., Vejborg, I. *et al.* (2014) Birth weight, childhood body mass index, and height in relation to mammographic density and breast cancer: a register-based cohort study. *Breast Cancer Res.*, **16** (1), R4.
35. Kershaw, E.E. and Flier, J.S. (2004) Adipose tissue as an endocrine organ. *J. Clin. Endocrinol. Metab.*, **89** (6), 2548–2556.
36. Key, T., Appleby, P., Barnes, I., and Reeves, G. (2002) Endogenous sex hormones and breast cancer in postmenopausal women: reanalysis of nine prospective studies. *J. Natl Cancer Inst.*, **94** (8), 606–616.
37. Vogel, V.G., Costantino, J.P., Wickerham, D.L., Cronin, W.M. *et al.* (2010) Update of the National Surgical Adjuvant Breast and Bowel Project Study of Tamoxifen and Raloxifene (STAR) P-2 Trial: preventing breast cancer. *Cancer Prev. Res. (Phila.)*, **3** (6), 696–706.
38. Jia, Z., Liu, Y., and Cui, S. (2014) Adiponectin induces breast cancer cell migration and growth factor expression. *Cell Biochem. Biophys.*, **70** (2), 1239–1245.
39. Bhardwaj, P., Du, B., Zhou, X.K., Sue, E. *et al.* (2013) Caloric restriction reverses obesity-induced mammary gland inflammation in mice. *Cancer Prev. Res. (Phila.)*, **6** (4), 282–289.
40. Eliassen, A.H., Colditz, G.A., Rosner, B., Willett, W.C. *et al.* (2006) Adult weight change and risk of postmenopausal breast cancer. *J. Am. Med. Assoc.*, **296** (2), 193–201.
41. Tee, M.C., Cao, Y., Warnock, G.L., Hu, F.B. *et al.* (2013) Effect of bariatric surgery on oncologic outcomes: a systematic review and meta-analysis. *Surg. Endosc.*, **27** (12), 4449–4456.
42. Colditz, G.A. and Rosner, B.A. (2006) What can be learnt from models of incidence rates? *Breast Cancer Res.*, **8** (3), 208.
43. Pike, M.C., Krailo, M.D., Henderson, B.E., Casagrande, J.T. *et al.* (1983) "Hormonal" risk factors, "breast tissue age" and the age-incidence of breast cancer. *Nature*, **303**, 767–770.
44. Colditz, G.A. and Rosner, B. (2000) Cumulative risk of breast cancer to age 70 years according to risk factor status: data from the Nurses' Health Study. *Am. J. Epidemiol.*, **152** (10), 950–964.
45. Rosner, B. and Colditz, G.A. (1996) Nurses' Health Study: log-incidence mathematical model of breast cancer incidence. *J. Natl Cancer Inst.*, **88** (6), 359–364.
46. Rosner, B., Colditz, G.A., and Willett, W.C. (1994) Reproductive risk factors in a prospective study of breast cancer: the Nurses' Health Study. *Am. J. Epidemiol.*, **139** (8), 819–835.
47. Russo, J., Tay, L.K., and Russo, I.K. (1982) Differentiation of the mammary gland and susceptibility to carcinogenesis. *Breast Cancer Res. Treat.*, **2**, 5–73.
48. Medina, D. (2013) Pregnancy protection of breast cancer: new insights reveal unanswered questions. *Breast Cancer Res.*, **15** (3), 103.
49. Meier-Abt, F., Milani, E., Roloff, T., Brinkhaus, H. *et al.* (2013) Parity induces differentiation and reduces Wnt/Notch signaling ratio and proliferation potential of basal stem/progenitor cells isolated from mouse mammary epithelium. *Breast Cancer Res.*, **15** (2), R36.
50. Bernstein, L., Hanisch, R., Sullivan-Halley, J., and Ross, R.K. (1995) Treatment with human chorionic gonadotropin and risk of breast cancer. *Cancer Epidemiol. Biomarkers Prev.*, **4** (5), 437–440.
51. Russo, J., Santucci-Pereira,. J, de Cicco, R.L., Sheriff, F. et al. (2012) Pregnancy-induced chromatin remodeling in the breast of postmenopausal women. *Int. J. Cancer*, **131** (5), 1059–1070.
52. IARC Working Group on the Evaluation of Carcinogenic Risks to Humans (2007: Lyon, France) (2010) *Alcohol Consumption and Ethyl Carbamate*, IARC Monographs on the Evaluation of Carcinogenic Risks to Humans, Vol **96**. International Agency for Research on Cancer, Distributed by WHO Press, Lyon.
53. Willett, W.C., Stampfer, M.J., Colditz, G.A., Rosner, B.A. *et al.* (1987) Moderate alcohol consumption and the risk of breast cancer. *N. Engl. J. Med.*, **316** (19), 1174–1180.

54. Smith-Warner, S.A., Spiegelman, D., Yaun, S.-S., Adami, H.O. *et al.* (1998) Alcohol and breast cancer in women: a pooled analysis of cohort studies. *J. Am. Med. Assoc.*, **279**, 535–540.
55. Ginsburg, E.S., Mello, N.K., Mendelson, J.H., Barbieri, R.L. *et al.* (1996) Effects of alcohol ingestion on estrogens in postmenopausal women. *J. Am. Med. Assoc.*, **276**, 1747–1751.
56. Reichman, M.E., Judd, J.T., Longcope, C., Schatzkin, A. *et al.* (1993) Effects of alcohol consumption on plasma and urinary hormone concentrations in premenopausal women. *J. Natl Cancer Inst.*, **85**, 722–727.
57. Wright, R.M., McManaman, J.L., and Repine, J.E. (1999) Alcohol-induced breast cancer: a proposed mechanism. *Free Radic. Biol. Med.*, **26** (3-4), 348–354.
58. Liu, Y., Colditz, G.A., Rosner, B., Berkey, C.S. *et al.* (2013) Alcohol intake between menarche and first pregnancy: a prospective study of breast cancer risk. *J. Natl Cancer Inst.*, **105** (20), 1571–1578.
59. Zhang, S.M., Willett, W.C., Selhub, J., Hunter, D.J. *et al.* (2003) Plasma folate, vitamin B6, vitamin B12, homocysteine, and risk of breast cancer. *J. Natl Cancer Inst.*, **95** (5), 373–380.
60. Collaborative Group on Hormonal Factors in Breast Cancer (1997) Breast cancer and hormone replacement therapy: collaborative reanalysis of data from 51 epidemiologic studies of 52 705 women with breast cancer and 108 411 women without breast cancer. *Lancet*, **350**, 1047–1059.
61. Colditz, G.A., Hankinson, S.E., Hunter, D.J., Willett, W.C. *et al.* (1995) The use of estrogens and progestins and the risk of breast cancer in postmenopausal women. *N. Engl. J. Med.*, **332** (24), 1589–1593.
62. Chlebowski, R.T., Anderson, G.L., Gass, M., Lane, D.S. *et al.* (2010) Estrogen plus progestin and breast cancer incidence and mortality in postmenopausal women. *J. Am. Med. Assoc.*, **304** (15), 1684–1692.
63. Beral, V., Reeves, G., Bull, D., and Green, J. (2011) Breast cancer risk in relation to the interval between menopause and starting hormone therapy. *J. Natl Cancer Inst.*, **103** (4), 296–305.
64. Fernandez-Valdivia, R. and Lydon, J.P. (2012) From the ranks of mammary progesterone mediators, RANKL takes the spotlight. *Mol. Cell. Endocrinol.*, **357** (1-2), 91–100.
65. Gonzalez-Suarez, E., Jacob, A.P., Jones, J., Miller, R. *et al.* (2010) RANK ligand mediates progestin-induced mammary epithelial proliferation and carcinogenesis. *Nature*, **468** (7320), 103–107.
66. Tan, W., Zhang, W., Strasner, A., Grivennikov, S. *et al.* (2011) Tumour-infiltrating regulatory T cells stimulate mammary cancer metastasis through RANKL-RANK signalling. *Nature*, **470** (7335), 548–553.
67. IARC Working Group on the Evaluation of Carcinogenic Risks to Humans, World Health Organization (2007) *Combined Estrogen-Progestogen Contraceptives and Combined Estrogen-Progestogen Menopausal Therapy*, IARC Monographs on the Evaluation of Carcinogenic Risks to Humans, vol. 91, International Agency for Research on Cancer; Distributed by WHO Press, Lyon.
68. Bernstein, L., Patel, A.V., Ursin, G., Sullivan-Halley, J. *et al.* (2005) Lifetime recreational exercise activity and breast cancer risk among black women and white women. *J. Natl Cancer Inst.*, **97** (22), 1671–1679.
69. Land, C.E., Tokunaga, M., Koyama, K., Soda, M. *et al.* (2003) Incidence of female breast cancer among atomic bomb survivors, Hiroshima and Nagasaki, 1950–1990. *Radiat. Res.*, **160** (6), 707–717.
70. Collaborative Group on Hormonal Factors in Breast Cancer (1996) Breast cancer and hormonal contraceptives: further results. *Contraception*, **54** (Suppl.), 1S–106S.
71. Collaborative Group on Hormonal Factors in Breast C (2002) Breast cancer and breastfeeding: collaborative reanalysis of individual data from 47 epidemiological studies in 30 countries, including 50 302 women with breast cancer and 96 973 women without the disease. *Lancet*, **360** (9328), 187–195.
72. Berkey, C.S., Willett, W.C., Tamimi, R.M., Rosner, B. *et al.* (2013) Vegetable protein and vegetable fat intakes in pre-adolescent and adolescent girls, and risk for benign breast disease in young women. *Breast Cancer Res. Treat.*, **141** (2), 299–306.
73. Liu, Y., Colditz, G.A., Cotterchio, M., Boucher, B.A. *et al.* (2014) Adolescent

dietary fiber, vegetable fat, vegetable protein, and nut intakes and breast cancer risk. *Breast Cancer Res. Treat.*, **145** (2), 461–470.
74. Eliassen, A.H., Hendrickson, S.J., Brinton, L.A., Buring, J.E. *et al.* (2012) Circulating carotenoids and risk of breast cancer: pooled analysis of eight prospective studies. *J. Natl Cancer Inst.*, **104** (24), 1905–1916.
75. Willett, W.C., Stampfer, M.J., Colditz, G.A., Rosner, B.A. *et al.* (1990) Relation of meat, fat, and fiber intake to the risk of colon cancer in a prospective study among women. *N. Engl. J. Med.*, **323** (24), 1664–1672.
76. Chan, D.S., Lau, R., Aune, D., Vieira, R. *et al.* (2011) Red and processed meat and colorectal cancer incidence: meta-analysis of prospective studies. *PLoS One*, **6** (6), e20456.
77. Chao, A., Thun, M.J., Connell, C.J., McCullough, M.L. *et al.* (2005) Meat consumption and risk of colorectal cancer. *J. Am. Med. Assoc.*, **293** (2), 172–182.
78. Norat, T., Bingham, S., Ferrari, P., Slimani, N. *et al.* (2005) Meat, fish, and colorectal cancer risk: the European Prospective Investigation into cancer and nutrition. *J. Natl Cancer Inst.*, **97** (12), 906–916.
79. Cross, A.J., Ferrucci, L.M., Risch, A., Graubard, B.I. *et al.* (2010) A large prospective study of meat consumption and colorectal cancer risk: an investigation of potential mechanisms underlying this association. *Cancer Res.*, **70** (6), 2406–2414.
80. Ishikawa, S., Tamaki, S., Ohata, M., Arihara, K. *et al.* (2010) Heme induces DNA damage and hyperproliferation of colonic epithelial cells via hydrogen peroxide produced by heme oxygenase: a possible mechanism of heme-induced colon cancer. *Mol. Nutr. Food Res.*, **54** (8), 1182–1191.
81. Barros, K.V., Xavier, R.A., Abreu, G.G., Martinez, C.A. *et al.* (2010) Soybean and fish oil mixture increases IL-10, protects against DNA damage and decreases colonic inflammation in rats with dextran sulfate sodium (DSS) colitis. *Lipids Health Dis.*, **9**, 68.
82. Wu, K., Willett, W.C., Fuchs, C.S., Colditz, G.A. *et al.* (2002) Calcium intake and risk of colon cancer in women and men. *J. Natl Cancer Inst.*, **94** (6), 437–446.
83. McCullough, M.L., Robertson, A.S., Rodriguez, C., Jacobs, E.J. *et al.* (2003) Calcium, vitamin D, dairy products, and risk of colorectal cancer in the Cancer Prevention Study II Nutrition Cohort (United States). *Cancer Causes Control*, **14** (1), 1–12.
84. Park, Y., Leitzmann, M.F., Subar, A.F., Hollenbeck, A. *et al.* (2009) Dairy food, calcium, and risk of cancer in the NIH-AARP Diet and Health Study. *Arch. Intern. Med.*, **169** (4), 391–401.
85. Cho, E., Smith-Warner, S.A., Spiegelman, D., Beeson, W.L. *et al.* (2004) Dairy foods, calcium, and colorectal cancer: a pooled analysis of 10 cohort studies. *J. Natl Cancer Inst.*, **96** (13), 1015–1022.
86. Baron, J., Beach, M., van Mandel, J., Stolk, R. *et al.*, and Calcium Polyp Prevention Study Group (1999) Calcium supplements for the prevention of colorectal adenomas. *N. Engl. J. Med.*, **340**, 101–107.
87. Grau, M.V., Baron, J.A., Sandler, R.S., Wallace, K. *et al.* (2007) Prolonged effect of calcium supplementation on risk of colorectal adenomas in a randomized trial. *J. Natl Cancer Inst.*, **99** (2), 129–136.
88. IARC Working Group on Vitamin D (2008) *Vitamin D and Cancer: A Report of the IARC Working Group on Vitamin D*, IARC Working Group Reports, vol. **5**, International Agency for Research on Cancer.
89. Chung, M., Lee, J., Terasawa, T., Lau, J. *et al.* (2011) Vitamin D with or without calcium supplementation for prevention of cancer and fractures: an updated meta-analysis for the U.S. Preventive Services Task Force. *Ann. Intern. Med.*, **155** (12), 827–838.
90. Byers, S.W., Rowlands, T., Beildeck, M., and Bong, Y.S. (2012) Mechanism of action of vitamin D and the vitamin D receptor in colorectal cancer prevention and treatment. *Rev. Endocr. Metab. Disord.*, **13** (1), 31–38.
91. Shah, S., Islam, M.N., Dakshanamurthy, S., Rizvi, I. *et al.* (2006) The molecular basis of vitamin D receptor and beta-catenin crossregulation. *Mol. Cell*, **21** (6), 799–809.
92. Mullin, G.E. and Dobs, A. (2007) Vitamin D and its role in cancer and immunity: a prescription for sunlight. *Nutr. Clin. Pract.*, **22** (3), 305–322.
93. Giovannucci, E., Stampfer, M.J., Colditz, G.A., Rimm, E.B. *et al.* (1993) Folate, methionine, and alcohol intake and risk

of colorectal adenoma. *J. Natl Cancer Inst.*, **85** (11), 875–884.
94. World Cancer Research Fund, American Institute for Cancer Research (1997) *Food, Nutrition and the Prevention of Cancer: A Global Perspective*, American Institute for Cancer Research, Washington.
95. Giovannucci, E., Rimm, E.B., Ascherio, A., Stampfer, M.J. *et al.* (1995) Alcohol, low-methionine–low-folate diets, and risk of colon cancer in men. *J. Natl Cancer Inst.*, **87** (4), 265–273.
96. Freudenheim, J.L., Graham, S., Marshall, J.R., Haughey, B.P. *et al.* (1991) Folate intake and carcinogenesis of the colon and rectum. *Int. J. Epidemiol.*, **20**, 368–374.
97. Lashner, B.A., Heidenreich, P.A., Su, G.L., Kane, S.V. *et al.* (1989) Effect of folate supplementation on the incidence of dysplasia and cancer in chronic ulcerative colitis. A case-control study. *Gastroenterology*, **97** (2), 255–259.
98. Lee, J.E., Willett, W.C., Fuchs, C.S., Smith-Warner, S.A. *et al.* (2011) Folate intake and risk of colorectal cancer and adenoma: modification by time. *Am. J. Clin. Nutr.*, **93** (4), 817–825.
99. Cole, B.F., Baron, J.A., Sandler, R.S., Haile, R.W. *et al.*, and Polyp Prevention Study Group (2007) Folic acid for the prevention of colorectal adenomas: a randomized clinical trial. *J. Am. Med. Assoc.*, **297** (21), 2351–2359.
100. Cheng, J., Chen, Y., Wang, X., Wang, J. *et al.* (2015) Meta-analysis of prospective cohort studies of cigarette smoking and the incidence of colon and rectal cancers. *Eur. J. Cancer Prev.*, **24** (1), 6–15.
101. Wei, E.K., Giovannucci, E., Wu, K., Rosner, B. *et al.* (2004) Comparison of risk factors for colon and rectal cancer. *Int. J. Cancer*, **108** (3), 433–442.
102. Giovannucci, E., Colditz, G.A., Stampfer, M.J., Hunter, D. *et al.* (1994) A prospective study of cigarette smoking and risk of colorectal adenoma and colorectal cancer in U.S. women. *J. Natl Cancer Inst.*, **86** (3), 192–199.
103. Samowitz, W.S., Albertsen, H., Sweeney, C., Herrick, J. *et al.* (2006) Association of smoking, CpG island methylator phenotype, and V600E BRAF mutations in colon cancer. *J. Natl Cancer Inst.*, **98** (23), 1731–1738.
104. Rosenberg, D.W., Yang, S., Pleau, D.C., Greenspan, E.J. *et al.* (2007) Mutations in BRAF and KRAS differentially distinguish serrated versus non-serrated hyperplastic aberrant crypt foci in humans. *Cancer Res.*, **67** (8), 3551–3554.
105. Voutsinas, J., Wilkens, L.R., Franke, A., Vogt, T.M. *et al.* (2013) Heterocyclic amine intake, smoking, cytochrome P450 1A2 and N-acetylation phenotypes, and risk of colorectal adenoma in a multiethnic population. *Gut*, **62** (3), 416–422.
106. Renehan, A.G., Tyson, M., Egger, M., Heller, R.F. *et al.* (2008) Body-mass index and incidence of cancer: a systematic review and meta-analysis of prospective observational studies. *Lancet*, **371** (9612), 569–578.
107. Giovannucci, E., Ascherio, A., Rimm, E.B., Colditz, G.A. *et al.* (1995) Physical activity, obesity, and risk for colon cancer and adenoma in men. *Ann. Intern. Med.*, **122** (5), 327–334.
108. Wei, E.K., Ma, J., Pollak, M.N., Rifai, N. *et al.* (2006) C-peptide, insulin-like growth factor binding protein-1, glycosylated hemoglobin, and the risk of distal colorectal adenoma in women. *Cancer Epidemiol. Biomarkers Prev.*, **15** (4), 750–755.
109. Wolin, K.Y., Yan, Y., Colditz, G.A., and Lee, I.M. (2009) Physical activity and colon cancer prevention: a meta-analysis. *Br. J. Cancer*, **100** (4), 611–616.
110. Wolin, K.Y., Yan, Y., and Colditz, G.A. (2011) Physical activity and risk of colon adenoma: a meta-analysis. *Br. J. Cancer*, **104** (5), 882–885.
111. Warren, R.S., Yuan, H., Matli, M.R., Ferrara, N. *et al.* (1996) Induction of vascular endothelial growth factor by insulin-like growth factor 1 in colorectal carcinoma. *J. Biol. Chem.*, **271** (46), 29483–29488.
112. Aoi, W., Naito, Y., Takagi, T., Kokura, S. *et al.* (2010) Regular exercise reduces colon tumorigenesis associated with suppression of iNOS. *Biochem. Biophys. Res. Commun.*, **399** (1), 14–19.
113. Quadrilatero, J. and Hoffman-Goetz, L. (2003) Physical activity and colon cancer. A systematic review of potential mechanisms. *J. Sports Med. Phys. Fitness*, **43** (2), 121–138.
114. Bosetti, C., Rosato, V., Gallus, S., Cuzick, J. *et al.* (2012) Aspirin and cancer risk: a

quantitative review to 2011. *Ann. Oncol.*, **23** (6), 1403–1415.
115. Flossmann, E. and Rothwell, P.M. (2007) Effect of aspirin on long-term risk of colorectal cancer: consistent evidence from randomised and observational studies. *Lancet*, **369** (9573), 1603–1613.
116. Burn, J., Gerdes, A.M., Macrae, F., Mecklin, J.P. *et al.* (2011) Long-term effect of aspirin on cancer risk in carriers of hereditary colorectal cancer: an analysis from the CAPP2 randomised controlled trial. *Lancet*, **378** (9809), 2081–2087.
117. Grodstein, F., Martinez, M., Platz, E., Giovannucci, E. *et al.* (1998) Postmenopausal hormone use and risk of colorectal cancer and adenoma. *Ann. Intern. Med.*, **128**, 705–712.
118. Rossouw, J.E., Anderson, G.L., Prentice, R.L., LaCroix, A.Z. *et al.* (2002) Risks and benefits of estrogen plus progestin in healthy postmenopausal women: principal results From the Women's Health Initiative randomized controlled trial. *J. Am. Med. Assoc.*, **288** (3), 321–333.
119. Platz, E.A., Rimm, E.B., Willett, W.C., Kantoff, P.W. *et al.* (2000) Racial variation in prostate cancer incidence and in hormonal system markers among male health professionals. *J. Natl Cancer Inst.*, **92**, 2009–2017.
120. The Breast Cancer Linkage Consortium (1999) Cancer risks in BRCA2 mutation carriers. *J. Natl Cancer Inst.*, **91** (15), 1310–1316.
121. Eeles, R.A., Olama, A.A., Benlloch, S., Saunders, E.J. et al. (2013) Identification of 23 new prostate cancer susceptibility loci using the iCOGS custom genotyping array. *Nat. Genet.*, **45** (4), 385–391, e381–382.
122. Ewing, C.M., Ray, A.M., Lange, E.M., Zuhlke, K.A. *et al.* (2012) Germline mutations in HOXB13 and prostate-cancer risk. *N. Engl. J. Med.*, **366** (2), 141–149.
123. Michaud, D.S., Augustsson, K., Rimm, E.B., Stampfer, M.J. *et al.* (2001) A prospective study on intake of animal products and risk of prostate cancer. *Cancer Causes Control*, **12**, 557–567.
124. Sinha, R., Park, Y., Graubard, B.I., Leitzmann, M.F. *et al.* (2009) Meat and meat-related compounds and risk of prostate cancer in a large prospective cohort study in the United States. *Am. J. Epidemiol.*, **170** (9), 1165–1177.
125. Tappel, A. (2007) Heme of consumed red meat can act as a catalyst of oxidative damage and could initiate colon, breast and prostate cancers, heart disease and other diseases. *Med. Hypotheses*, **68** (3), 562–564.
126. Zu, K., Mucci, L., Rosner, B.A., Clinton, S.K. *et al.* (2014) Dietary lycopene, angiogenesis, and prostate cancer: a prospective study in the prostate-specific antigen era. *J. Natl Cancer Inst.*, **106** (2), djt430.
127. MacInnis, R.J. and English, D.R. (2006) Body size and composition and prostate cancer risk: systematic review and meta-regression analysis. *Cancer Causes Control*, **17** (8), 989–1003.
128. Cao, Y. and Ma, J. (2011) Body mass index, prostate cancer-specific mortality, and biochemical recurrence: a systematic review and meta-analysis. *Cancer Prev. Res. (Phila.)*, **4** (4), 486–501.
129. Parker, A.S., Thiel, D.D., Bergstralh, E., Carlson, R.E. *et al.* (2013) Obese men have more advanced and more aggressive prostate cancer at time of surgery than non-obese men after adjusting for screening PSA level and age: results from two independent nested case-control studies. *Prostate Cancer Prostatic Dis.*, **16** (4), 352–356.
130. Price, R.S., De Cavazos, D.A., Angel, R.E., Hursting, S.D. *et al.* (2012) Obesity-related systemic factors promote an invasive phenotype in prostate cancer cells. *Prostate Cancer Prostatic Dis.*, **15** (2), 135–143.
131. Peto, R., Darby, S., Deo, H., Silcocks, P. *et al.* (2000) Smoking, smoking cessation, and lung cancer in the UK since 1950: combination of national statistics with two case-control studies. *Br. Med. J.*, **321** (7257), 323–329.
132. Shapiro, J.A., Jacobs, E.J., and Thun, M.J. (2000) Cigar smoking in men and risk of death from tobacco-related cancers. *J. Natl Cancer Inst.*, **92** (4), 333–337.
133. Hecht, S. (1999) Tobacco smoke carcinogens and lung cancer. *J. Natl Cancer Inst.*, **91** (14), 1194–1210.
134. Cooper, W.A., Lam, D.C., O'Toole, S.A., and Minna, J.D. (2013) Molecular biology of lung cancer. *J. Thorac Dis.*, **5** (Suppl. 5), S479–S490.

135. Ihsan, R., Chauhan, P.S., Mishra, A.K., Yadav, D.S. *et al.* (2011) Multiple analytical approaches reveal distinct gene-environment interactions in smokers and non-smokers in lung cancer. *PLoS One*, **6** (12), e29431.
136. Kenfield, S.A., Wei, E.K., Stampfer, M.J., Rosner, B.A. *et al.* (2008) Comparison of aspects of smoking among the four histological types of lung cancer. *Tob. Control*, **17** (3), 198–204.
137. Kenfield, S.A., Stampfer, M.J., Rosner, B.A., and Colditz, G.A. (2008) Smoking and smoking cessation in relation to mortality in women. *J. Am. Med. Assoc.*, **299** (17), 2037–2047.
138. Wei, E.K., Wolin, K.Y., and Colditz, G.A. (2010) Time course of risk factors in cancer etiology and progression. *J. Clin. Oncol.*, **28** (26), 4052–4057.
139. World Cancer Research Fund (2007) *Food, Nutrition, Physical Activity, and the Prevention of Cancer: A Global Perspective*, AICR, Washington.
140. Tanvetyanon, T. and Bepler, G. (2008) Beta-carotene in multivitamins and the possible risk of lung cancer among smokers versus former smokers: a meta-analysis and evaluation of national brands. *Cancer*, **113** (1), 150–157.
141. Colditz, G.A. and Taylor, P.R. (2010) Prevention trials: their place in how we understand the value of prevention strategies. *Annu. Rev. Public Health*, **31**, 105–120.
142. Qiao, Y.L., Dawsey, S.M., Kamangar, F., Fan, J.H. *et al.* (2009) Total and cancer mortality after supplementation with vitamins and minerals: follow-up of the Linxian General Population Nutrition Intervention Trial. *J. Natl Cancer Inst.*, **101** (7), 507–518.
143. Rothwell, P.M., Fowkes, F.G., Belch, J.F., Ogawa, H. *et al.* (2011) Effect of daily aspirin on long-term risk of death due to cancer: analysis of individual patient data from randomised trials. *Lancet*, **377** (9759), 31–41.
144. de Martel, C., Ferlay, J., Franceschi, S., Vignat, J. *et al.* (2012) Global burden of cancers attributable to infections in 2008: a review and synthetic analysis. *Lancet Oncol.*, **13** (6), 607–615.
145. IARC (2011) *A Review of Carcinogen – Part B: Biological Agents*, Monographs on the Evaluation of Carcinogenic Risks to Humans, vol. 100, International Agency for Research on Cancer, Lyon.
146. Chang, M.H., You, S.L., Chen, C.J., Liu, C.J. *et al.* (2009) Decreased incidence of hepatocellular carcinoma in hepatitis B vaccinees: a 20-year follow-up study. *J. Natl Cancer Inst.*, **101** (19), 1348–1355.
147. Bogaards, J.A., Coupe, V.M., Xiridou, M., Meijer, C.J. *et al.* (2011) Long-term impact of human papillomavirus vaccination on infection rates, cervical abnormalities, and cancer incidence. *Epidemiology*, **22** (4), 505–515.
148. Freedman, A.N., Yu, B., Gail, M.H., Costantino, J.P. *et al.* (2011) Benefit/risk assessment for breast cancer chemoprevention with raloxifene or tamoxifen for women age 50 years or older. *J. Clin. Oncol.*, **29** (17), 2327–2333.
149. Fisher, B., Costantino, J.P., Wickerham, D.L., Redmond, C.K. *et al.* (1998) Tamoxifen for prevention of breast cancer – report of the National Surgical Adjuvant Breast and Bowel Project P-1. *J. Natl Cancer Inst.*, **90**, 1371–1388.
150. Gail, M.H., Costantino, J.P., Bryant, J., Croyle, R. *et al.* (1999) Weighing the risks and benefits of tamoxifen treatment for preventing breast cancer. *J. Natl Cancer Inst.*, **91** (21), 1829–1846.
151. Cuzick, J., Thorat, M.A., Bosetti, C., Brown, P.H. *et al.* (2015) Estimates of benefits and harms of prophylactic use of aspirin in the general population. *Ann. Oncol.*, **26** (1), 47–57.
152. Colditz, G.A., Wolin, K.Y., and Gehlert, S. (2012) Applying what we know to accelerate cancer prevention. *Sci. Transl. Med.*, **4** (127), 127rv4.
153. Wolin, K.Y., Carson, K., and Colditz, G.A. (2010) Obesity and cancer. *Oncologist*, **15** (6), 556–565.

2
Cancer Metastasis Redux

Gary A. Clawson
Pennsylvania State University, Department of Pathology and Gittlen Cancer Research Laboratories, Hershey Medical Center, Hershey, PA 17033, USA

Keywords

Circulating tumor cell CTC
A cancer cell that has dislodged from the primary tumor into the vascular system, which can be found in blood.

Translational Medicine Cancer: Advances in Molecular Biology and Medicine,
First Edition. Edited by Robert A. Meyers.

Cancer stem cell CSC
A cancer cell found within cancer cell populations that possess characteristics normally associated with normal stem cells, primarily retaining the ability to differentiate into all cell types found in the cancer.

Epithelial-to-mesenchymal transition EMT
This represent a process in which epithelial cells lose cell–cell adhesion (polarity, etc.), and become mesenchymal cells, thereby gaining migratory/invasive properties and other characteristics.

Macrophage–tumor cell fusion MTF
Cells which results from the fusion of macrophages with cancer cells.

Tumor microenvironment TME
The tumor microenvironment refers to the local environment in which a solid cancer exists, including blood vessels and lymphatics, fibroblasts, extracellular matrix, insatiate fluids, and immune cells.

Most of the morbidity and mortality associated with cancer derives from its metastatic dissemination. The combination of steps involved in the establishment of a metastatic lesion at a distant site is traditionally referred to as the "metastatic cascade," which is the subject of this chapter. This cascade is generally thought of as consisting of two phases: the physical movement of cancer cells, followed by their colonization and growth at the distant sites.

1 Introduction and Historical Perspective

1.1 Introduction

Most of the morbidity and mortality associated with cancer derives from its metastatic dissemination, but many aspects of the process are not well understood. The combination of steps involved in the establishment of a metastatic lesion (a metastasis, which means "displacement" in Greek) is generally referred to as the "metastatic cascade" [1]. This metastatic cascade can be thought of as a trip with a typical itinerary, including departure, travel, and arrival (Phase I), followed by accommodations and activities (Phase II). As model concepts for metastasis have developed, Phase I encompasses the physical journey of a circulating tumor cell (CTC); intravasation, physical translocation, and extravasation of CTCs into distant tissues (Fig. 1). Phase II encompasses the adaptation of translocated CTCs to their new accommodations (either as solitary cells or as small foci of cells termed "micrometastases"), followed by subsequent proliferation into metastatic cancers. Much is now known with respect to the mechanisms that are active in Phase I, whereas many aspects of Phase II remain macaronic at best (a juxtaposition of phenomena). Taking a cue from the Master

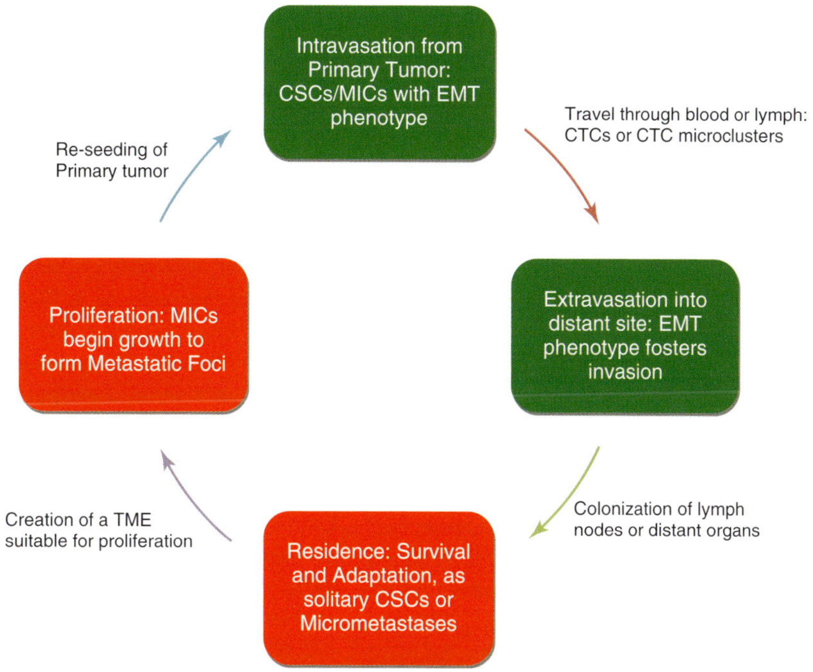

Fig. 1 The canonical metastatic cascade. Green blocks denote efficient steps (Phase I) in the metastatic cascade, whereas red blocks denote inefficient steps (Phase II). The process begins with MICs locally entering the blood or lymphatic vasculature (intravasation), thus becoming CTCs. CTCs then travel through blood or lymph to "distant" sites, which include sentinel lymph nodes, and so on. The CTCs then leave the blood or vascular system (extravasation) and invade the adjacent tissue. MICs must then somehow adapt to their new environment, and produce a tumor microenvironment, subsequently supporting proliferation of MICs to produce metastatic lesions.

of Disguise, this Perspective will explore some of the Who, Why, Where, and How of cancer metastasis.

1.2 Historical Perspective: Metastatic Inefficiency

It has long been recognized that once cancer cells leave the primary tumor, a very small proportion of them successfully colonize distant sites and produce metastases [2, 3], and historical views on this have evolved over the decades [4–6]. Paget was probably the first to recognize that the distribution of metastases in breast cancer was nonrandom, and proposed the "Seed and Soil" hypothesis, recognizing the need for a conducive tumor microenvironment (TME) for the successful development of metastatic lesions [7]. Others argued that blood flow considerations could explain the ultimate distribution of metastases [8], and it was subsequently recognized that a synthesis of these two factors was really needed to explain results with different cancers [9]. The Seed and Soil hypothesis is still evolving [10], and developing additional complexities [11] including tumor "self-seeding" from distant metastases [12–14].

Fidler's early work showed that, after the introduction (non-physiologic) of tumor cells into the vascular system, less than 0.1% of them survived and produced metastases [15]. Pioneering studies performed by Luzzi *et al.* showed that the early stages (Phase I) of metastases were completed efficiently, whereas the growth phases of metastases (Phase II) proceeded very inefficiently (Fig. 1), focusing attention on the regulation of growth at a distant site as the critical factor [3, 16]. Inherently similar results showing the efficient extravasation of cancer cells in a chick embryo model were also described [17].

The efficiency of Phase I actually raises a concern, since there are now a number of reports which describe the physical translocation of CTCs into the circulation (blood or lymph) during surgical procedures [18–22]. In many instances, their forced release appears to have a prognostic relevance for patients; for example, the hematogenous dissemination of cancer cells during surgery for lung cancer has been reported [20, 21] and was related to clinical prognosis [21], while similar observations have been reported for esophageal, stomach and colorectal cancers [23–25]. In a related vein, there are also reports of direct physical seeding of cancer cells during biopsies [26, 27].

Additional features of the metastatic process also became clear. There is a selectivity inherent in the metastatic process [28] and it was recognized that different subpopulations of tumor cells had varying malignant potentials [29, 30], in essence presaging the concepts of CTC heterogeneity and dormancy [30].

2 The Who, Why, and How of Metastasis

2.1 CTCs

The first report documenting the existence of CTCs appears to date back to 1869 ([31]; taken from Wikipedia, where there is an interesting account of it. The journal is variously listed as *Med. J. Aust.*, *Aust. Med.*, or *Aust. Med. J.*). There have since been hundreds of reports describing the clinical utility of CTC counts, and this is now being confirmed in large cohort analyses. In breast cancer, for example, CTC counts have prognostic value for progression-free survival and overall survival [32], and their measurement has significant value in monitoring patient response to therapy [32, 33]. However, it has also become clear that CTCs comprise a very heterogeneous population of cells.

CellSearch, the clinical assay developed for measuring CTCs, relies on the epithelial cell adhesion molecule (Gene ID#4072; EpCAM)-mediated enrichment of CTCs, followed by staining for cytokeratins and the common leukocyte antigen CD45. A CTC thus must be EpCAM-positive (to be captured), and stain immunofluorescently for cytokeratins. The CTC must also be negative for CD45 (and have an intact nucleus). The CD45-negative staining was mandated by the fact that CTCs are outnumbered by peripheral blood mononuclear cells by a factor of about 10^6. CTCs are generally found as solitary cells, but they can also travel as aggregates or microclusters [34–38], which seem to have significance in *in vivo* models.

At this juncture, it is already clear that CellSearch is seriously flawed: CTC detection is compromised by the epithelial-to-mesenchymal transition (EMT) [39, 40], as well as the fact that some CTCs appear to be CD45-positive. In spite of these important shortcomings, CellSearch still has clinical utility as noted, probably because of a stochastic relationship between CTCs and those cells capable of ultimately forming a metastatic lesion. Perhaps the most important point to make regarding the heterogeneity of CTCs is that it is not known which subpopulation(s) of CTCs actually represent the cells capable of initiating metastases.

2.2
Potential Importance of the EMT in Metastasis

The constellation of mesenchymal phenotypic changes which characterize the EMT have now been recognized as integral to metastatic dissemination of many (probably all) cancers [41–43]. As was noted, induction of the EMT in non-cancer stem cell (CSC) populations produces cells with stem cell properties, potentially providing a route by which CSCs arise [44, 45], and the EMT provides a link between CTCs and CSCs [46], with additional ramifications for the development of drug resistance [47]. A major proportion of CTCs demonstrate an EMT phenotype. In one study on breast cancer, for example, more than 50% of CTCs dual-stained for epithelial/EMT markers in early-stage patients, and this increased to more than 95% in patients with metastatic disease [48]. Many other reports have described EMT markers for CTCs for many different cancers [38, 49–52], and in fact mesenchymal markers are enriched in CTCs from breast cancer patients [35] and mesenchymal CTC levels are elevated in patients who are refractory to therapy [42].

It is important to note that the EMT conversion really needs to be coupled with a subsequent "reconversion" to an epithelial phenotype, a process referred to as the mesenchymal-to-epithelial transition [53]. However, it is not clear how seemingly "random" sets of heterogeneous genetic mutations/lesions culminate in a prototypical EMT phenotype; likewise, how the reconversion back to an epithelial phenotype occurs is equally unclear [54].

While the EMT is often considered to be an integral contributor to metastasis [42, 53], this is not universally accepted. It has been argued, for example, that miscommunications between cancer cells and the TME, which could arise from genetic lesions, are the main perpetrators of progression [55]. Others question whether the EMT is a "necessary" conversion of epithelial cells during metastasis as opposed to a reflection of myeloid/macrophage mesenchymal traits of metastasis-initiating cells (MICs) [56, 57]. This topic will be discussed in more detail in the following sections of the chapter.

2.3
Genetic Clues from Metastases

Some basic genetic features of metastatic lesions also provide clues to mechanisms operative during metastasis. Duesberg *et al.* have suggested that metastases should be considered as autonomous "species" that differ from each other (and the primary tumors) in intrinsically generated karyotypic variations [58]. Others have shown that distant metastasis occurs late during the genetic evolution within the primary cancer [59], and that the selection of clonally expanding cell subpopulations responsible for metastases seems to occur after dissemination has taken place [60]. This suggests that the analysis of circulating "metastatic precursor cells" may be a more fruitful approach to understanding metastasis and/or dormancy than examining metastatic lesions *per se* [61].

2.4 Heterogeneity of CTC Populations

In addition, although EMT markers – as well as CSC markers [62, 63] – are prominent features of CTCs, it is also clear that CTCs represent a very heterogeneous collection of cells [64–66]. In fact, a new wave of single-cell ("omic") analyses of CTC populations is underway, and is clearly demonstrating the inherent heterogeneity of CTCs, while confirming the importance of the mesenchymal/EMT phenotype in metastatic progression [50, 67].

Given the heterogeneity of CTC populations, it is still not clear which subpopulations comprise the MICs. Two recent reports have addressed this critical issue by using CTCs from breast cancer patients in xenograft models. Zhang *et al.* [68] identified a subpopulation of EpCAM-negative cells, which expressed a characteristic marker panel, and showed an ability to metastasize to the brain and lung [68], with enhanced brain metastatic capability when the marker panel was overexpressed. In contrast, Baccelli *et al.* identified an EpCAM-positive subpopulation, which expressed CSC markers, and showed the ability to metastasize to bone, lung and liver [69]. These authors also found that the number of CTCs in this subpopulation – but not the overall number of EpCAM-positive CTCs – was related to overall survival and increased numbers of metastases. Clearly, there is a critical need for further studies to define which CTCs are of importance, and which are not.

3 Models for the Metastatic Cascade

There are several basic hypotheses to explain the origin/nature of MICs [70].

3.1 Cancer Stem Cell (CSC) Model

In the CSC model, CSCs remain permanently seated at the top of the hierarchical pyramid of cancer cells. The basic idea is that CSCs undergo symmetric division to maintain the CSC subpopulation, which represents the MICs, as well as irreversible asymmetric division to generate non-CSCs, which constitute the majority of the cancer but have reduced (or absent) tumorigenic potential. CSCs were first functionally noted in lung cancer cell populations [30], although their molecular and operational characteristics were not really defined until the mid-1990s or early 2000s for a number of solid tumors [71, 72]. CSCs possess many traits such as self-renewal, motility, invasiveness, resistance to apoptosis, and they often remain quiescent; these traits make them ideal candidates for MICs.

Perhaps the most vexing question in this CSC model of metastasis is how does a suitable TME become established at a distant site? There is evidence that exosomes may help to establish a suitable TME in sentinel lymph nodes [73], but it is much more difficult to envision how sufficient concentrations of exosomes could be delivered systemically to "all" locations to prepare unknown distant sites for metastasis.

3.2 Clonal Evolution Model

The Clonal Evolution hypothesis proposes that successive mutations accumulate in cancer cells, which generate clonal outgrowths that variably survive/adapt within the TME, forming a hierarchy of subpopulations of genetically diverse cancer cells [74]. This hierarchy is probably somewhat substratose, but the diversity is certainly compatible with the heterogeneity of CTC populations. It is also likely that mechanisms other than mutations may generate important subpopulations (see below). These clonal outgrowths/subpopulations are generally thought of as competitors, but in some genetic mouse models there is evidence that interclonal cooperation is necessary to maintain the tumor phenotype [75].

3.3 CSC-Plasticity Model

The ideas inherent in the CSC and Clonal Evolution hypotheses have converged into a model of tumorigenicity proposed by Weinberg and colleagues [70], which could be termed the CSC-Plasticity Model. This model recognizes a plasticity between CSC and non-CSC subpopulations, such that non-CSCs can re-acquire a CSC phenotype. Given the importance of the EMT phenotype in many aspects of the metastatic cascade, it is noteworthy that the EMT has been shown to induce the conversion of non-CSCs into CSCs [44].

3.4 Macrophage-Tumor Cell Fusion (MTF) Model

In addition to what might be termed the "canonical" models for metastasis [70, 76], there is also an alternative theory to explain metastasis, which merits discussion. Pawelek and coworkers have long espoused the idea that macrophage-tumor cell fusions (MTFs) occur in primary tumors, and that these MTFs are responsible for metastatic dissemination [77–81], a concept also raised by others [56, 57]. The MTF hypothesis is based on extensive experimental data from animal models, and is also tied to the EMT; it has even been suggested that the EMT might better be referred to as an epithelial-myeloid transition [82]. The MTF hypothesis has received some recent support with human samples. Lazova *et al.* [83] identified a melanoma in the brain of a patient, and were able to confirm its MTF nature. The one caveat here was its designation as "metastatic": a primary tumor was not identified, and primary melanomas can arise in the central nervous system (CNS) [84, 85].

In support of the MTF concept, MTFs have been identified in the blood of melanoma, colorectal and pancreatic cancer patients [86], and have been cultured and extensively characterized in melanoma patients [87] and pancreatic ductal adenocarcinoma patients (G.A. Clawson *et al.*, manuscript in preparation). Melanoma MTFs express epithelial, EMT, melanocytic and macrophage (pan- and M2-polarized) markers, as well as the procarcinogenic cytokines, particularly macrophage migration inhibitory factor (MIF), and show various features that are far different from those expected for CTCs. Single-cell RNASeq analyses have revealed that virtually all cultured pancreatic cancer MTFs express EMT markers (e.g., vimentin) as well as pancreatic stem cell markers [e.g., aldehyde dehydrogenase 1 family member A1 (Gene ID#216; ALDH1A1, CD44)]. MTFs have also been identified in primary tumors, often enriched at the periphery [87] (G.A. Clawson *et al.*, manuscript in preparation). Thus, MTFs exist in primary tumors and can be cultured from the blood of cancer patients.

While these recent developments provide support for the MTF hypothesis of metastasis, it seems unlikely that most metastatic lesions are composed solely of MTFs (or cells derived therefrom). For example, while DNA ploidy is related to cancer progression [88–92], the abnormal ploidy distribution of cultured melanoma and pancreatic MTFs does not mirror that of primary or metastatic lesions [87]. Furthermore, the immunophenotyping of primary or metastatic lesions shows that MTFs, while consistently present, constitute a variable subpopulation of the cancer cells [87].

3.5 MTF/CSC-Plasticity Model

Considerations such as those above have led to a potential synthesis (Fig. 2) of the disparate hypotheses for metastasis [93]. The novel twist here is that MTFs play a critical role in the *earliest* stages of metastasis. The idea is that MTFs form at the periphery of primary cancers, and in this regard tumor-associated macrophages are known to play critical roles in tumor development, and to display an M2-polarized phenotype, with the secretion of procarcinogenic cytokines, particularly MIF. These MTFs would acquire "professional-grade" invasive capabilities, and could readily enter the bloodstream or lymphatic system, disseminate widely to distant sites, and extravasate and colonize them. These resident MTFs could then secrete a variety of cytokines

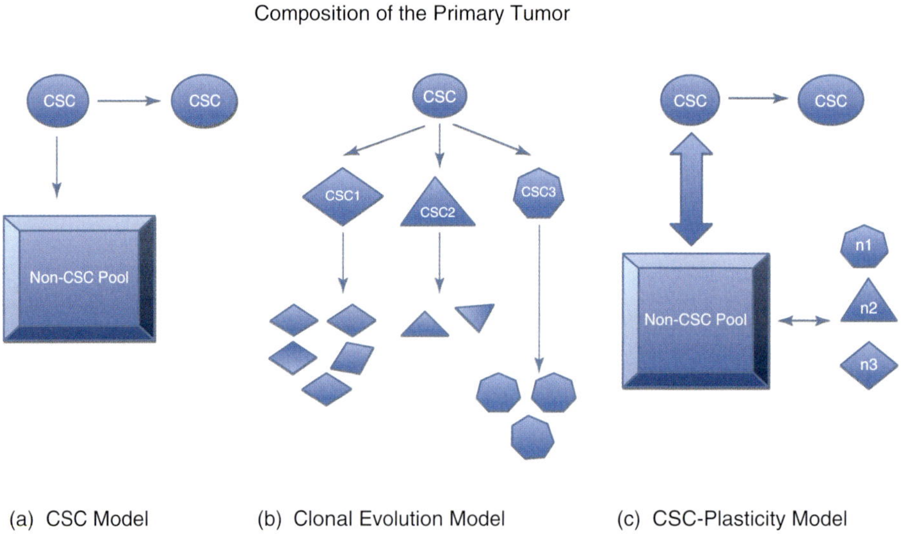

Fig. 2 Composition of the primary tumor. The CSC Model (A) has a population of CSCs, which either divide symmetrically to maintain a population of CSC, or divide asymmetrically to produce a population of non-CSCs, which constitute the vast majority of the tumor. The Clonal Evolution Model (B) starts with a population of CSCs, which undergo successive mutations (irreversible) to develop clonal outgrowths or subpopulations of CSCs (CSC1, CSC2, etc.), which then proliferate at different rates to constitute the bulk and phenotype of the tumor. These subpopulations are generally viewed as competing, but there is some recent evidence to suggest they may need to cooperate to maintain the tumor in some models. The CSC-Plasticity Model (C) incorporates elements of both A and B, but the formation of non-CSCs or various subclones (n1, n2, etc.) is not considered irreversible, so that emergence of events such as the EMT can reproduce CSCs from non-CSCs. The MTF model could fit into this basic scheme, with the fusion representing a clonal evolution event (replacing a mutation event).

(particularly MIF), with the effect of producing a TME that would be conducive to the subsequent establishment of metastatic lesions. Thus, the "job" of the MTFs is to prepare a "niche" for potential colonization by CSCs. The events proposed within the CSC Plasticity model could then proceed as described. In this scenario, metastatic lesions would only be produced when a disseminating CSC arrived at a distant site suitably prepared by the MTFs (Fig. 3). An attractive feature of this hypothesis is that it provides a ready explanation for how a TME niche becomes prepared at a distant site. In fact, MTFs which form in primary cancers may represent a specialized CSC subpopulation. Macrophages are known to proliferate locally in response to various circumstances [94], and differentiated macrophages are capable of self-renewal [95], raising the interesting possibility that MTFs might represent a subpopulation of CSCs which develop/evolve over the course of cancer progression. The inherent inefficiency of the latter steps (Phase II) in the metastatic cascade – particularly the failure of micrometastases to produce macroscopic metastases – are compatible in straightforward manner with initial colonization of distant sites by MTFs. Ultimately, the rate-limiting step would be the growth of micrometastases into

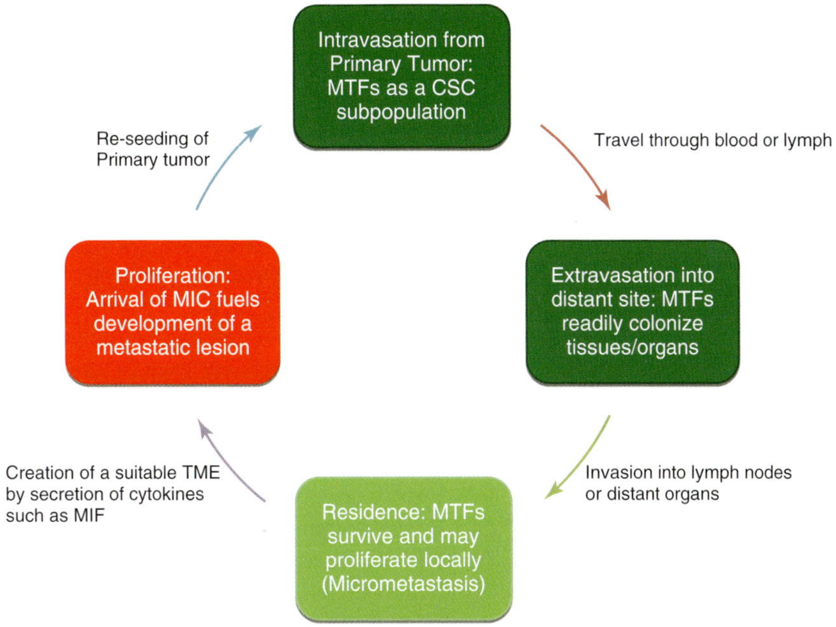

Fig. 3 A speculative role for macrophage-tumor cell fusions (MTFs) in the initial stages of the metastatic cascade. Green blocks denote efficient steps in the metastatic cascade, whereas red blocks denote inefficient steps. In this model, MTFs form in the primary lesions, as a subpopulation of CSCs. They acquire motility and invasive properties, enter the blood or lymphatic vasculature, travel to distant sites, and subsequently exit into adjacent tissues. They then survive and may proliferate locally, producing a TME conducive to growth of MICs. The final step, proliferation, remains inefficient, since it is dependent on an MIC landing in the pre-prepared "niche." It should be noted that the production of a metastatic "niche" by MTFs may actually influence the nature of the CSCs, which can become MICs.

metastatic lesions: this step would depend on the unlikely event of a conventional CSC arriving at the pre-prepared site.

The existence of "dormancy" is a perplexing problem that is readily compatible with the MTF/CSC-Plasticity explanation for metastasis. Dormancy is generally described as occurring at the solitary cell stage [96–98]. While dormancy also seems to be compatible with the CSC-Plasticity theory [99, 100] and to reflect a control point for metastasis development [101], it seems fundamentally unclear why the CSC went to the site in the first place, and/or why the CSC survives and prospers in the second place. However, if the solitary cells were actually MTFs they might reasonably be expected to not proliferate continually, perhaps dividing a few times, and to slowly remodel the tissue into a TME suitable for proliferation of CSCs.

4
The Where: Lymphatics and Metastasis

The central tendency is to describe metastasis as occurring through the bloodstream, a topic which is covered in many reviews [74, 102, 103]. However, lymphatic dissemination should not be short-changed; it is becoming clear that lymphatic vessels are an important route of dissemination, perhaps even more important than blood vessels, especially initially. Straightforward clinical observations often seem to indicate that carcinomas use the lymphatic system preferentially to metastasize, although the extensive interconnections with the blood vasculature provide ready access to hematogenous routes [5].

Lymphatic vessels form a system of channels that return interstitial fluids, as well as cells, to the bloodstream. An understanding of molecular mechanisms regulating lymphatic vasculature has lagged far behind that of the blood system, but specific properties of lymphatic endothelial cells are now being detailed (see Ref. [104]). With regard to lymphangiogenesis, several critical mediators have been identified: Specific mediators such as vascular endothelial growth factor C (Gene ID#7424; VEGF-C) and fas-related tyrosine kinase 4 (FLT4; aka VEGFR3; Gene ID#2324) are essential for lymphangiogenesis [105]. Nonetheless, many features of lymphatic spread of cancers remain poorly understood.

The development of new lymphatic vessels, as well as significant remodeling of existing lymphatics, appears to play a significant role in metastasis [106]. Many studies have been conducted in a variety of cancers which clearly indicate the importance of lymphatic density and lymphangiogenesis in patient outcomes (see Ref. [106]). In fact, cancer-induced lymphangiogenesis facilitates metastasis to sentinel lymph nodes, while subsequent lymphangiogenesis in draining lymph nodes seems to contribute to subsequent distant spread. Sentinel lymph nodes actually demonstrate an enhanced lymphangiogenesis even before metastasis has occurred [105]. The concept of a "lymphovascular niche" was proposed by Hirakawa [107] to depict the development of the expanded lymphatic network: enlargement of the sinusoidal lymphatic endothelium in essence might produce a TME which supports the migration, residence/survival, or proliferation of cancer cells. There is also evidence to suggest that lymphatics do not only represent passive channels for cancer cell movement, but rather that they also actively recruit cancer cells to lymph nodes, and promote CSC survival [108]. Indeed, endothelial cells have been shown to undergo a phenotypic switch analogous to the EMT, which has been termed the endothelial-to-mesenchymal transition. This latter process plays an important role in cancer progression [109]; indeed, it seems also to play a significant role in the production of mesenchymal stem cells [110, 111].

5 An Interesting Journey

Much progress has been made in understanding the many features involved in the "metastatic cascade," whereby cancer cells disseminate from primary tumors to distant "displaced" sites. Some features of CTCs seem clear, such as the need for mesenchymal traits and the heterogeneity of CTC populations. However, these features raised additional questions, notably: (i) which subpopulation(s) of CTCs are actually important for establishment of metastatic lesion; (ii) what is/are their origins; and (iii) how does a suitable TME become established at distant sites which facilitates growth of the metastasis? It is an interesting journey for all concerned, for CTCs and scientists alike.

References

1. Klein, C.A. (2008) Cancer. The metastasis cascade. *Science*, **321**, 1785–1787.
2. Weiss, L. (1990) Metastatic inefficiency. *Adv. Cancer Res.*, **54**, 159–211.
3. Cameron, M.D., Schmidt, E.E., Kerkvliet, N., Nadkarni, K.V. *et al.* (2000) Temporal progression of metastasis in lung: cell survival, dormancy, and location dependence of metastatic inefficiency. *Cancer Res.*, **60**, 2541–2546.
4. Poste, G. and Fidler, I. (1980) The pathogenesis of cancer metastasis. *Nature*, **283**, 139–146.
5. Fidler, I.J. (1990) Critical factors in the biology of human cancer metastasis: twenty-eighth GHA Clowes memorial award lecture. *Cancer Res.*, **50**, 6130.
6. Fidler, I.J. (1999) Critical determinants of cancer metastasis: rationale for therapy. *Cancer Chemother. Pharmacol.*, 43 Suppl., S3–S10.
7. Paget, S. (1889) The distribution of secondary growths in cancer of the breast. *Lancet*, **1**, 571–573.
8. Ewing, J. (1928) *Neoplastic Diseases. A Treatise on Tumors*. W.B. Saunders Co., Philadelphia, London.
9. Weiss, L. (1992) Comments on hematogenous metastatic patterns in humans as revealed by autopsy. *Clin. Exp. Metastasis*, **10**, 191–199.
10. Ribelles, N., Santonja, A., Pajares, B., Llacer, C., and Alba, E. (2014) The seed and soil hypothesis revisited: current state of knowledge of inherited genes on prognosis in breast cancer. *Cancer Treat. Rev.*, **40**, 293–299.
11. Comen, E.A. (2012) Tracking the seed and tending the soil: evolving concepts in metastatic breast cancer. *Discovery Med.*, **14**, 97–104.
12. Kim, M.Y., Oskarsson, T., Acharyya, S., Nguyen, D.X. *et al.* (2009) Tumor self-seeding by circulating cancer cells. *Cell*, **139**, 1315–1326.
13. Comen, E.A. and Norton, L. (2012) Self-seeding in cancer. *Recent Results Cancer Res.*, **195**, 13–23.
14. Comen, E.A., Norton, L., and Massague, J. (2011) Clinical implications of cancer self-seeding. *Nat. Rev. Clin. Oncol.*, **8**, 369–377.
15. Fidler, I. (1970) Metastasis: quantitative analysis of distribution and fate of tumor emboli labeled with ^{125}I-5-iodo-2′-deoxyuridine. *J. Natl Cancer Inst.*, **45**, 773–782.
16. Luzzi, K., MacDonald, I.C., Schmidt, E.E., Kerkvliet, N. *et al.* (1998) Multistep nature of metastatic inefficiency: dormancy of solitary cells after successful extravasation and limited survival of early micrometastases. *Am. J. Pathol.*, **153**, 865–873.
17. Koop, S., MacDonald, I.C., Luzzi, K., Schmidt, E.E. *et al.* (1995) Fate of melanoma cells entering the microcirculation: over 80% survive and extravasate. *Cancer Res.*, **55**, 2520–2523.
18. Camara, O., Kavallaris, A., Noschel, H., Rengsberger, M. *et al.* (2006) Seeding of epithelial cells into circulation during

surgery for breast cancer: the fate of malignant and benign mobilized cells. *World J. Surg. Oncol.*, **4**, 67–75.

19. Sawabata, N., Okumura, M., Utsumi, T., Inoue, M. *et al.* (2007) Circulating tumor cells in peripheral blood caused by surgical manipulation of non-small-cell lung cancer: pilot study using an immunocytology method. *Gen. Thorac. Cardiovasc. Surg.*, **55**, 189–192.
20. Ge, M.J., Shi, D., Wu, O.C., Wang, M., and Li, L.B. (2006) Observation of circulating tumour cells in patients with non-small cell lung cancer by real-time fluorescent quantitative reverse transcriptase-polymerase chain reaction in perioperative period. *J. Cancer Res. Clin. Oncol.*, **132**, 248–256.
21. Dong, Q., Huang, J., Zhou, Y., Li, L. *et al.* (2002) Hematogenous dissemination of lung cancer cells during surgery: quantitative detection by flow cytometry and prognostic significance. *Lung Cancer*, **37**, 293–301.
22. Engilbertsson, H., Aaltonen, K.E., Bjornsson, S., Kristmundsson, T. *et al.* (2014) TURBT can cause seeding of cancer cells into the bloodstream. *J. Urol.*, **193**, 53–57.
23. Broll, R., Lembcke, K., Stock, C., Zingler, M. *et al.* (1996) Tumor cell dissemination in bone marrow and peritoneal cavity. An immunocytochemical study of patients with stomach or colorectal carcinoma (in German). *Langenbecks Arch. Chir.*, **381**, 51–58.
24. Guller, U., Zajac, P., Schnider, A., Bosch, B. *et al.* (2002) Disseminated single tumor cells as detected by real-time quantitative polymerase chain reaction represent a prognostic factor in patients undergoing surgery for colorectal cancer. *Ann. Surg.*, **236**, 768–775.
25. Liu, Z., Jiang, M., Zhao, J., and Ju, H. (2007) Circulating tumor cells in perioperative esophageal cancer patients: quantitative assay system and potential clinical utility. *Clin. Cancer Res.*, **13**, 2992–2997.
26. Ishizuna, K., Ota, D., Okamoto, J., Fukuuchi, A. *et al.* (2011) A case of mucinous carcinoma of the breast in which needle tract seeding was diagnosed by preoperative diagnostic imaging. *Breast Cancer*, **18**, 324–327.
27. Shyamala, K., Girish, H.C., and Murgod, S. (2014) Risk of tumor cell seeding through biopsy and aspiration cytology. *J. Int. Soc. Prev. Community Dent.*, **4**, 5–11.
28. Fidler, I.J. (1973) The selection of successive tumor lines for metastasis. *Nat. New Biol.*, **242**, 148–149.
29. Nicholson, G.L., Brunson, K.W., and Fidler, I.J. (1978) Specificity of arrest, survival, and growth of selected metastatic variant cell lines. *Cancer Res.*, **38**, 4105–4111.
30. Stackpole, C.W. (1981) Distinct lung-colonizing and lung-metastasizing cell populations in B16 mouse melanoma. *Nature*, **289**, 798–800.
31. Ashworth, T.R. (1869) A case of cancer in which cells similar to those in the tumours were seen in the blood after death. *Med. J. Aust.*, **14**, 146–147.
32. Bidard, F.C., Peeters, D.J., Fehm, T., Nole, F. *et al.* (2014) Clinical validity of circulating tumor cells in patients with metastatic breast cancer: a pooled analysis of individual patient data. *Lancet Oncol.*, **15**, 406–414.
33. Pierga, J.Y., Hajage, D., Bachelot, T., Delaloge, S. *et al.* (2012) High independent prognostic and predictive value of circulating tumor cells compared with serum tumor markers in a large prospective trial in first-line chemotherapy for metastatic breast cancer. *Ann. Oncol.*, **23**, 618–624.
34. Liotta, L., Saidel, M.G., and Kleinerman, J. (1976) The significance of hematogenous tumor cell clumps in the metastatic process. *Cancer Res.*, **36**, 889–894.
35. Yu, M., Bardia, A., Wittner, B.S., Stott, S.L. *et al.* (2013) Circulating breast tumor cells exhibit dynamic changes in epithelial and mesenchymal composition. *Science*, **339**, 580–584.
36. Stott, S.L., Hsu, C.H., Tsukrov, D.I., Yu, M. *et al.* (2010) Isolation of circulating tumor cells using a microvortex-generating herringbone-chip. *Proc. Natl Acad. Sci. USA*, **107**, 18392–18397.
37. Cho, E.H., Wendel, M., Luttgen, M., Yoshioka, C. *et al.* (2012) Characterization of circulating tumor cell aggregates identified in patients with epithelial tumors. *Phys. Biol.*, **9**, 016001.
38. Friedlander, T.W., Ngo, V.T., Dong, H., Premasekharan, G. *et al.* (2014) Detection and characterization of invasive circulating tumor cells derived from men with

metastatic castration-resistant prostate cancer. *Int. J. Cancer*, **134**, 2284–2293.

39. Grover, P.K., Cummins, A.G., Price, T.J., Roberts-Thomson, I.C. Hardingham, J.E. (2014) Circulating tumour cells: the evolving concept and the inadequacy of their enrichment by EpCAM-based methodology for basic and clinical cancer research. *Ann. Oncol.*, **25**, 1506–1516.
40. Gorges, T.M., Tinhofer, I., Drosch, M., Rose, L. *et al.* (2012) Circulating tumour cells escape from EpCAM-based detection due to epithelial-to-mesenchymal transition. *BMC Cancer*, **12**, 178.
41. Rhim, A., Mirek, E., Aiello, N., Maitra, A. *et al.* (2012) EMT and dissemination precede pancreatic tumor formation. *Cell*, **148**, 349–361.
42. Thiery, J.P. and Lim, C.T. (2013) Tumor dissemination: an EMT affair. *Cancer Cell*, **23**, 273.
43. Gibbons, D.L., Lin, W., Creighton, C.J., Rizvi, Z.H. *et al.* (2009) Contextual extracellular cues promote tumor cell EMT and metastasis by regulating miR-200 family expression. *Genes Dev.*, **23**, 2140–2151.
44. Mani, S.A., Guo, W., Liao, M.-J., Eaton, E.N. *et al.* (2008) The epithelial-mesenchymal transition generates cells with properties of stem cells. *Cell*, **133**, 704–715.
45. Pattabiraman, D.R. and Weinberg, R.A. (2014) Tackling the cancer stem cells – what challenges do they pose? *Nat. Rev. Drug Discovery*, **13**, 497–512.
46. Ksiazkiewicz, M., Markiewicz, A., and Zaczek, A.J. (2012) Epithelial-mesenchymal transition: a hallmark in metastasis formation linking circulating tumor cells and cancer stem cells. *Pathobiology*, **79**, 195–208.
47. Singh, A. and Settleman, J. (2010) EMT, cancer stem cells and drug resistance: an emerging axis of evil in the war on cancer. *Oncogene*, **29**, 4741–4751.
48. Kallergi, G., Papadaki, M.A., Politaki, E., Mavroudis, D. *et al.* (2011) Epithelial to mesenchymal transition markers expressed in circulating tumour cells of early and metastatic breast cancer patients. *Breast Cancer Res.*, **13**, R59.
49. Kasimir-Bauer, S., Hoffmann, O., Wallwiener, D., Kimmig, R., and Fehm, T. (2012) Expression of stem cell and epithelial-mesenchymal transition markers in primary breast cancer patients with circulating tumor cells. *Breast Cancer Res.*, **14**, R15.
50. Chen, C.L., Mahalingam, D., Osmulski, P., Jadhav, R.R. *et al.* (2013) Single-cell analysis of circulating tumor cells identifies cumulative expression patterns of EMT-related genes in metastatic prostate cancer. *Prostate*, **73**, 813–826.
51. Armstrong, A.J., Marengo, M.S., Oltean, S., Kemeny, G. *et al.* (2011) Circulating tumor cells from patients with advanced prostate and breast cancer display both epithelial and mesenchymal markers. *Mol. Cancer Res.*, **9**, 997–1007.
52. Balasubramanian, P., Lang, J.C., Jatana, K.R., Miller, B. *et al.* (2012) Multiparameter analysis, including EMT markers, on negatively enriched blood samples from patients with squamous cell carcinoma of the head and neck. *PLoS One*, **7**, e42028.
53. Thiery, J.P. (2002) Epithelial–mesenchymal transitions in tumour progression. *Nat. Rev. Cancer*, **2**, 442–454.
54. Weinberg, R.A. (2014) *The Biology of Cancer*, 2nd edn, Garland Science, New York.
55. Tarin, D. (2011) Cell and tissue interactions in carcinogenesis and metastasis and their clinical significance. *Semin. Cancer Biol.*, **21**, 72–82.
56. Seyfried, T.N. and Huysentruyt, L.C. (2013) On the origin of cancer metastasis. *Crit. Rev. Oncogen.*, **18**, 43–73.
57. Huysentruyt, L.C. and Seyfried, T.N. (2010) Perspectives on the mesenchymal origin of metastatic cancer. *Cancer Metastasis Rev.*, **29**, 695–707.
58. Duesberg, P., Iacobuzio-Donahue, C., Brosnan, J.A., McCormack, A. *et al.* (2012) Origin of metastases: subspecies of cancers generated by intrinsic karyotypic variations. *Cell Cycle*, **11**, 1151–1166.
59. Yachida, S., Jones, S., Bozic, I., Antal, T. *et al.* (2010) Distant metastasis occurs late during the genetic evolution of pancreatic cancer. *Nature*, **467**, 1114–1117.
60. Klein, C.A., Blankenstein, T.J., Schmidt-Kittler, O., Petronio, M. *et al.* (2002) Genetic heterogeneity of single disseminated tumour cells in minimal residual cancer. *Lancet*, **360**, 683–689.
61. Klein, C.A. (2008) The direct molecular analysis of metastatic precursor cells

in breast cancer: a chance for a better understanding of metastasis and for personalised medicine. *Eur. J. Cancer*, **44**, 2721–2725.

62. Raimondi, C., Gradilone, A., Naso, G., Vincenzi, B. *et al.* (2011) Epithelial–mesenchymal transition and stemness features in circulating tumor cells from breast cancer patients. *Breast Cancer Res. Treat.*, **130**, 449–455.
63. Scatena, R., Bottoni, P., and Giardina, B. (2013) Circulating tumor cells and cancer stem cells: a role for proteomics in defining the interrelationships between function, phenotype and differentiation with potential clinical applications. *Biochim. Biophys. Acta*, **1835**, 129–143.
64. Lustberg, M.B., Balasubramanian, P., Miller, B., Garcia-Villa, A. *et al.* (2014) Heterogeneous atypical cell populations are present in blood of metastatic breast cancer patients. *Breast Cancer Res.*, **16**, R23.
65. Khoja, L., Shenjere, P., Hodgson, C., Hodgetts, J. *et al.* (2014) Prevalence and heterogeneity of circulating tumor cells in metastatic cutaneous melanoma. *Melanoma Res.*, **24**, 40–46.
66. Markiewicz, A., Ksiazkiewicz, M., Seroczynska, B., Skokowski, J. *et al.* (2013) Heterogeneity of mesenchymal markers expression-molecular profiles of cancer cells disseminated by lymphatic and hematogenous routes in breast cancer. *Cancers (Basel)*, **5**, 1485–1503.
67. Powell, A.A., Talasaz, A.H., Zhang, H., Coram, M.A. *et al.* (2012) Single cell profiling of circulating tumor cells: transcriptional heterogeneity and diversity from breast cancer cell lines. *PLoS One*, **7**, e33788.
68. Zhang, L., Ridgway, L.D., Wetzel, M.D., Ngo, J. *et al.* (2013) The identification and characterization of breast cancer CTCs competent for brain metastasis. *Sci. Transl. Med.*, **5**, ra48.
69. Baccelli, I., Schneeweiss, A., Riethdorf, S., Stenzinger, A. *et al.* (2013) Identification of a population of blood circulating tumor cells from breast cancer patients that initiates metastasis in a xenograft assay. *Nat. Biotechnol.*, **31**, 539–544.
70. Marjanovic, N.D., Weinberg, R.A., and Chaffer, C.L. (2013) Cell plasticity and heterogeneity in cancer. *Clin. Chem.*, **59**, 168–179.
71. Al-Hajj, M., Wicha, M.S., Benito-Hernandez, A., Morrison, S.J., and Clarke, M.F. (2003) Prospective identification of tumorigenic breast cancer cells. *Proc. Natl Acad. Sci. USA*, **100**, 3983–3988.
72. Singh, S.K., Hawkins, C., Clarke, I.D., Squire, J.A. *et al.* (2004) Identification of human brain tumor initiating cells. *Nature*, **432**, 396–401.
73. Hood, J.L., San Roman, S., and Wickline, S.A. (2011) Exosomes released by melanoma cells prepare sentinel lymph nodes for tumor metastasis. *Cancer Res.*, **71**, 3792–3801.
74. Baccelli, I. and Trumpp, A. (2012) The evolving concept of cancer and metastasis stem cells. *J. Cell Biol.*, **198**, 281–293.
75. Cleary, A.S., Leonard, T.L., Gestl, S.A., and Gunther, E.J. (2014) Tumour cell heterogeneity maintained by cooperating subclones in Wnt-driven mammary cancers. *Nature*, **508**, 113–117.
76. Scheel, C. and Weinberg, R.A. (2012) Cancer stem cells and epithelial–mesenchymal transition: concepts and molecular links. *Semin. Cancer Biol.*, **22**, 396–403.
77. Pawelek, J.M. (2000) Tumour cell hybridization and metastasis revisited. *Melanoma Res.*, **10**, 507–514.
78. Pawelek, J.M. (2005) Tumour-cell fusion as a source of myeloid traits in cancer. *Lancet Oncol.*, **6**, 988–993.
79. Pawelek, J.M. and Chakraborty, A.K. (2008) The cancer cell–leukocyte fusion theory of metastasis. *Adv. Cancer Res.*, **101**, 397–444.
80. Pawelek, J.M. and Chakraborty, A.K. (2008) Fusion of tumour cells with bone marrow-derived cells: a unifying explanation for metastasis. *Nat. Rev. Cancer*, **8**, 377–386.
81. Lazova, R., Chakraborty, A.K., and Pawelek, J.M. (2011) Leukocyte-cancer cell fusion: initiator of the Warburg effect in malignancy? *Adv. Exp. Med. Biol.*, **714**, 151–172.
82. Schramm, H.M. (2014) Should EMT of cancer cells be understood as epithelial–myeloid transition. *J. Cancer*, **5**, 125–132.
83. Lazova, R., LeBerge, G.S., Duvall, E., Spoelstra, N. *et al.* (2013) A melanoma brain metastasis with a donor–patient hybrid genome following bone marrow

transplantation: first evidence for fusion in human cancer. *PLoS One*, **8**, e66731.
84. Liubinas, S.V., Maartens, N., and Drummond, K.J. (2010) Primary melanocytic neoplasms of the central nervous system. *J. Clin. Neurosci.*, **17**, 1227–1232.
85. Brat, D.J., Giannini, C., Scheithauer, B.W., and Burger, P.C. (1999) Primary melanocytic neoplasms of the central nervous system. *Am. J. Surg. Pathol.*, **23**, 745–754.
86. Clawson, G.A., Kimchi, E., Patrick, S.D., Xin, P. *et al.* (2012) Circulating tumor cells in melanoma patients. *PLoS One*, **7**, e41052.
87. Clawson, G.A., Matters, G.L., Xin, P., Imamura-Kawasawa, Y. *et al.* (2014) Macrophage–tumor cell fusions from peripheral blood of melanoma patients. *PLoS One*, **10**, e0134320.
88. Kamphues, C., Al-Abadi, H., Durr, A., Al-Abadi, N. *et al.* (2013) DNA index as a strong prognostic factor in patients with adenocarcinoma of the pancreatic head: results of a 5-year prospective study. *Pancreas*, **42**, 807–812.
89. Eskelinen, M., Lipponen, P., Marin, S., Haapasalo, H. *et al.* (1992) DNA ploidy, S-phase fraction, and G2 fraction as prognostic determinants in human pancreatic cancer. *Scand. J. Gastroenterol.*, **27**, 39–43.
90. Tsavaris, N., Kavantzas, N., Tsigritis, K., Xynos, I.D. *et al.* (2009) Evaluation of DNA ploidy in relation with established prognostic factors in patients with locally advanced (unresectable) or metastatic pancreatic adenocarcinoma: a retrospective analysis. *BMC Cancer*, **9**, 264.
91. Xynos, I.D., Kavantzas, N., Tsaousi, S., Zacharakis, M. *et al.* (2013) Factors influencing survival in Stage IV colorectal cancer: the influence of DNA ploidy. *ISRN Gastroenterol.*, **2013**, 490578.
92. Hedley, D.W., Leary, J.A., and Kirsten, F. (1985) Metastatic adenocarcinoma of unknown primary site: abnormalities of cellular DNA content and survival. *Eur. J. Cancer Clin. Oncol.*, **21**, 185–189.
93. Clawson, G.A. (2013) Cancer. Fusion for moving. *Science*, **342**, 699–700.
94. Haase, J., Weyer, U., Immig, K., Kloting, N. *et al.* (2014) Local proliferation of macrophages in adipose tissue during obesity-induced inflammation. *Diabetologia*, **57**, 562–571.
95. Sieweke, M.H. and Allen, J.E. (2013) Beyond stem cells: self-renewal of differentiated macrophages. *Science*, **342**, 1242974.
96. Naumov, G.N., MacDonald, I.C., Chambers, A.F., and Groom, A.C. (2001) Solitary cancer cells as a possible source of tumour dormancy? *Semin. Cancer Biol.*, **11**, 271–276.
97. Stein, W.D. and Litman, T. (2006) Data on the recurrence of breast tumors fit a model in which dormant cells are subject to slow attrition but can randomly awaken to become malignant. *Cell Cycle*, **5**, 2348–2353.
98. Townson, J.L. and Chambers, A.F. (2006) Dormancy of solitary metastatic cells. *Cell Cycle*, **5**, 1744–1750.
99. Bleau, A.M., Agliano, A., Larzabal, L., Lopez, A., and Calvo, A. (2014) Metastatic dormancy: a complex network between cancer stem cells and their microenvironment. *Histol. Histopathol.*, **29**, 1499–1510.
100. Kleffel, S. and Schatton, T. (2013) Tumor dormancy and cancer stem cells: two sides of the same coin? *Adv. Exp. Med. Biol.*, **734**, 145–179.
101. Hahnfeldt, P. (2013) The host support niche as a control point for tumor dormancy: implications for tumor development and beyond. *Adv. Exp. Med. Biol.*, **734**, 19–35.
102. Chaffer, C.L. and Weinberg, D. (2011) A perspective on cancer cell metastasis. *Science*, **331**, 1559–1564.
103. Chambers, A.F., Groom, A.C., and MacDonald, I.C. (2002) Dissemination and growth of cancer cells at metastatic sites. *Nat. Rev. Cancer*, **2**, 563–572.
104. Kerjaschki, D. (2014) The lymphatic vasculature revisited. *J. Clin. Invest.*, **124**, 874–877.
105. Alitalo, A. and Detmar, M. (2012) Interaction of tumor cells and lymphatic vessels in cancer progression. *Oncogene*, **31**, 4499–4508.
106. Stacker, S.A., Williams, S.P., Karnezis, T., Shayan, R. *et al.* (2014) Lymphangiogenesis and lymphatic vessel remodelling in cancer. *Nat. Rev. Cancer*, **14**, 159–172.
107. Hirakawa, S. (2009) From tumor lymphangiogenesis to lymphovascular niche. *Cancer Sci.*, **100**, 983–989.

108. Karaman, S. and Detmar, M. (2014) Mechanisms of lymphatic metastasis. *J. Clin. Invest.*, **124**, 922–928.

109. Potenta, S., Zeisberg, E.M., and Kalluri, R. (2008) The role of endothelial-to-mesenchymal transition in cancer progression. *Br. J. Cancer*, **99**, 1375–1379.

110. Medici, D., Shore, E.M., Lounev, V.Y., Kaplan, F.S. *et al.* (2010) Conversion of vascular endothelial cells into multipotent stem like cells. *Nat. Med.*, **16**, 1400–1406.

111. Medici, D. and Kalluri, R. (2012) Endothelial–mesenchymal transition and its contribution to the emergence of stem cell phenotype. *Semin. Cancer Biol.*, **22**, 379–384.

3
Cancer of the Prostate Gland: Molecular Genetics

Sharanjot Saini and Rajvir Dahiya
University of California, Department of Urology, Veterans Affairs Medical Center, Urology Research Center (112F), 4150 Clement Street, San Francisco, CA 94121, USA

Translational Medicine Cancer: Advances in Molecular Biology and Medicine,
First Edition. Edited by Robert A. Meyers.

Keywords

Prostate cancer
A form of cancer that develops in the prostate gland in the male reproductive system. Most prostate cancers are slow-growing (indolent) and can be managed by active surveillance. However, there are cases of aggressive prostate cancers, and cancer cells may metastasize (spread) from the prostate to other parts of the body, particularly the bones and lymph nodes.

Androgen receptor
An androgen receptor (AR) is a nuclear hormone receptor that binds to androgens and mediates AR signalling, with key roles in both normal prostate development and prostate cancer. Since prostate cancer growth is largely dependent on androgens, therapies focused on blocking the AR signaling pathway are used for prostate cancer treatment.

Castration-resistant prostate cancer
Castration-resistant prostate cancer (CRPC) is defined as prostate cancer progression despite androgen-deprivation therapy. Previously also called "hormone-refractory prostate cancer" or "androgen-independent prostate cancer," the term "castration-resistant" has replaced "hormone refractory" because, while they are no longer responsive to castration treatment (reduction of available androgens by chemical or surgical means), these cancers still rely on androgens for their growth.

Genomic rearrangements
Gross alterations in the genome such as duplications, deletions, insertions, inversions, and translocations.

Copy number alterations (CNAs)
These are genetic gains or losses that occur during carcinogenesis. The prostate cancer genome typically displays more CNAs than most other cancer types.

Mutation
A mutation is a change in the DNA sequence of a gene.

A key step towards understanding the clinical heterogeneity of prostate cancer is deciphering the molecular genetics underlying the disease. Recently, with the advent of advanced genome sequencing technologies, significant progress has been made on deciphering the complete genomic landscape of prostate cancer. During the past few years, a series of studies has employed genome-wide analyses to large patient datasets to delineate molecular alterations characteristic of prostate cancer. These studies have deciphered large-scale complex genomic rearrangements occurring in the prostate cancer genome, and also defined important signaling pathways underlying prostate cancer. Importantly, these studies suggest that there are distinct molecular subtypes of prostate cancer that have the potential for translating into better patient stratification and clinical management. In this review, the current state of understanding of prostate cancer molecular genetics will be discussed, highlighting recent advances in the field that have shed critical new insights into the prostate cancer genome.

1 Introduction

Prostate cancer is the most commonly diagnosed male malignancy and the second leading cause of cancer-related death among men [1]. In the United States, an estimated 242 000 new cases and 28 000 deaths were reported in 2012 [1]. The disease course of prostate cancer is remarkably heterogeneous; some tumors are indolent while others are very aggressive and often metastasize to bone and other organs, causing significant morbidity and mortality [2]. The enormous disease heterogeneity poses significant clinical challenges for the effective management of prostate cancer patients. A keys step towards unraveling this heterogeneity is to decipher the molecular genetics underlying the disease. Although prostate cancer is associated with a longer natural history than most other cancers, a complete understanding of the prostate cancer genome has been lacking to date. Currently, the genomes of several other cancers, including breast [3, 4], lung [5], and skin [6], have been completely sequenced and have provided new leads for both diagnosis and therapy [7].

Prostate cancer is characterized by an enormous genomic complexity, including structural rearrangements, copy number alterations (CNAs) and mutations. Although various studies have identified key genetic abnormalities driving the initiation, progression and metastasis of prostate cancer, the complete genome has not yet been characterized. Recently, with the advent of advanced genome sequencing technology, however, there has been a surge in studies focused on deciphering the complete genomic landscape of prostate cancer. The results of these studies have shed new light on the key initiating and driving events for prostate carcinogenesis, and also point to distinct molecular subtypes of prostate cancer with differing vulnerabilities to current therapeutics. These studies have vastly improved the present understanding of the molecular underpinnings of the disease, and offer potential to vastly increase the efficacy of current targeted strategies for prostate cancer diagnosis, prognosis and treatment [8]. In this review, the current state of our understanding of prostate cancer molecular genetics will be discussed, highlighting recent advances in the field that have shed critical new insights into the prostate cancer genome.

2 Cancer of the Prostate: Biology and Subtypes

2.1 Prostate Gland Biology

The prostate gland is an endodermal tissue that arises during late embryogenesis through ductal budding from the anterior urogenital sinus epithelium. The human prostate has a poorly demarcated zonal architecture, corresponding to central, periurethral transitional and peripheral zones, that contain acini located within a fibromuscular stroma [9, 10]. The acini are formed primarily from three differentiated epithelial cell types: luminal, basal, and neuroendocrine (NE) cells. These cells express distinct cellular markers:

- The luminal epithelial cells are polarized, columnar cells that line the lumen and produce prostatic secretions. They express low-molecular-weight cytokeratins (CKs) CK8/CK18, high levels of the androgen receptor (AR), and the prostate-specific homeobox gene *NKX3.1* [11, 12].
- Basal cells are elongated cells attached to the basement membrane that separates the lumen from the stroma. They express high-molecular-weight cytokeratins CK5/CK14, low levels of AR, and a p53 superfamily member p63 [13, 14]. These cells form a protective barrier for luminal cells and are believed to contain prostate stem cells [14, 15].
- The NE cells are rare, AR-negative, and secrete neuropeptides and growth factors for luminal cell growth. These cells express endocrine markers such as chromogranin A and synaptophysin [9, 16].

In addition to these three major cell types, a small population of "transit-amplifying" or "intermediate prostate epithelial cells" has also been reported [14, 17, 18]. Similar to human prostate, the mouse prostate has similar histological subtypes of epithelial cells but possesses a different glandular structure, being composed of ventral, lateral, dorsal and anterior lobes. The dorsal and lateral lobes are often combined as the "dorsolateral lobe," which is similar to the peripheral zone of the human prostate [9, 19]. The different lobes have distinct patterns of ductal branching, histological appearance, gene expression, and secretory protein expression [9, 20, 21].

Prostate embryonic and postnatal development is dependent on multiple critical regulators of differentiation, such as AR, transcription factor NKX3.1, and Forkhead box A1 (FOXA1). Androgens and AR signaling is critical for prostate architecture and cellular function. Androgen/AR ablation in the early stages of murine embryonic development inhibits prostatic development. NKX3.1 is a prostate-specific, homeodomain transcription factor that plays an important role in prostate organogenesis. The genetic deletion of *NKX3.1* leads to decreased ductal branching and altered secretory protein production [12, 16]. *FOXA1* regulates the expression of AR-regulated genes, and its deletion leads to abnormal pubertal ducts composed of immature luminal epithelial cells with abundant stroma [16, 22].

2.2 Benign Prostatic Hyperplasia (BPH), Prostatic Intraepithelial Neoplasia (PIN), and Prostate Cancer

In humans, the incidence of hyperplastic lesions increases with age, leading to benign prostatic hyperplasia (BPH), a common nonmalignant condition found in older men. BPH often arises from the transition zone of the prostate and is not a precursor of prostate cancer. In contrast, the majority of malignant prostate carcinomas arise from the outermost peripheral zone of the human prostate.

Prostatic intraepeithelial neoplasia (PIN) lesions, often characterized by luminal epithelial hyperplasia and a reduction in basal cells, are widely accepted as a precursor for prostate cancer [23–27]. PIN lesions – like prostate cancer – are often multifocal, predominantly located in the peripheral zone, and are associated with progressive phenotypic and genotypic alterations that are intermediate between benign epithelium and cancer [28]. These lesions are graded as either low-grade PINs

(LGPINs) or high-grade PINs (HGPINs) according to their degree of atypia. LGPINs represent areas of proliferative glandular epithelial cells with variable size, enlarged nuclei, non-prominent nucleoli, and an intact layer of basal cells [23]. In contrast, in HGPINs the epithelial cells display prominent nucleoli, nuclear hyperchromasia, and a fragmented/absent layer of basal cells.

Epidemiological, morphological and molecular evidence suggests that HGPINs represent a precursor of prostate carcinoma, though this has been not been demonstrated conclusively [9, 23–25]. The incidence and extent of HGPIN lesions increases with advancing age, much like prostate cancer [23, 24, 29]. The genetic and molecular alterations are similar in HGPIN and prostate cancer; frequent genomic alterations in both conditions include losses of chromosome 8p, 10q, 16q, and 18q and gains of 8q, 7, 10, 12, and Y [30–32]. About 16% of HGPIN lesions harbor a TMPRSS2-ERG fusion gene, which is a common molecular alteration observed in a majority of prostate cancers [33, 34]. Both, HGPIN and prostate cancer exhibit similar expression patterns of specific genes, including a decreased expression of prostate-specific NKX3.1 and p27, and an overexpression of p16, Bcl-2, MYC and alpha-methylacyl-CoA racemase (AMACR) [33]. PIN lesions show a progressive loss of some markers, such as prostate-specific antigen (PSA), prostatic acid phosphatase, cytoskeletal proteins and annexin I protein [29, 35], and a progressive increase of markers including c-erbB-2 (Her-2/neu) and c-erbB-3 oncoproteins, c-met proto-oncogene, and also hypermethylation of the promoter region of glutathione-*S*-transferase pi (*GSTP1*) [29]. Owing to its strong association with prostatic carcinoma, PIN has a high predictive value as a marker for adenocarcinoma when either multifocal or extensive [29].

Prostate cancer displays significant phenotypic heterogeneity. A majority of human prostate cancers (>95%) are adenocarcinomas with a predominantly luminal phenotype. In fact, negative staining with basal cell markers CK5/CK14, p63 and elevated staining with the luminal marker AMACR have been used as diagnostic criteria for prostate cancer [9, 36]. Apart from the most common acinar adenocarcinomas, a spectrum of prostate cancer subtypes exists [9, 37, 38]. Other subtypes of prostate cancer, such as ductal carcinoma, mucinous carcinoma and signet cell carcinoma, are extremely rare. A variant-NE prostate cancer, derived from NE cells, is a significant histological variant that is represented in <2% of cases. Prostate cancer is highly heterogeneous and usually presents multifocal lesions [39]. At the histological level, prostate cancer represent a range of differentiation phenotypes, often graded with the Gleason system [9, 40]. Within a given section of prostate cancer tissue, individual neoplastic lesions may be genetically distinct [9, 41, 42].

Recent evidence has suggested that, although prostate adenocarcinomas are predominantly luminal in nature, prostate cancer can originate in luminal cells, basal cells or NE cells [14]. Goldstein *et al.* showed that cells within the basal fraction can regenerate benign prostate tissue in immunodeficient mice [43]. Further, the introduction of oncogenic alterations (an increased expression of AKT, ERG, and AR) in the basal cells induced a disease that recapitulated the histological and molecular features of human prostate cancer, with a loss of basal cells and expansion of luminal cells expressing PSA and AMACR [43]. The results of the latter study support a basal cell origin of prostate cancer. Since basal cells are androgen-independent, it has been suggested that the "basal cell origin" of prostate cancer underlies frequent recurrence and androgen independence [43]. The results of a recent study also suggested that alternative histological phenotypes of prostate cancer can arise from a clonal cell of origin. In this case it was shown very elegantly that cancer initiated in basal cells could evolve to advanced prostate adenocarcinoma that was maintained by luminal-like tumor-propagating cells [44].

As focal regions of NE differentiation are also commonly observed following recurrence after prostatectomy and androgen deprivation therapy (ADT), it has been suggested that prostate cancer may originate in the NE cells [45, 46]. A few studies have also suggested that "intermediate prostate epithelial cells" are observed in prostate cancer and form the basis of prostate cancer origin [14, 47, 48]. Despite a series of investigations, however, conclusions regarding the cellular origin of prostate cancer have not been definitive, highlighting the extreme heterogeneity associated with the disease.

3 Prostate Cancer: A Challenge for Clinicians

One major clinical challenge of prostate cancer is posed by an inability of current diagnostic methods (including PSA screening and histopathological grading) to distinguish between aggressive and indolent tumors that require different treatments. Levels of PSA, a kallikrein-related serine protease present in normal prostatic secretions, are often elevated in prostate cancer [49, 50], and PSA testing has enabled clinicians to detect prostate tumors while they are still small, low-grade, and localized. In the past, patients with elevated PSA levels have generally undergone biopsy to assess the presence of prostate cancer. However, recent recommendations have suggested later and less frequent PSA screenings since PSA lacks specificity and its levels cannot accurately discern the degree of prostate cancer aggressiveness [51, 52]. Histopathological grading employing the Gleason scoring system has also been used as an important prognostic indicator of prostate cancer [9, 40, 53]. Based on the architectural pattern and tumor histology, the Gleason scheme classifies tumors from grade 1 to 5 (from most to least differentiated), according to their relative degree of tumor differentiation. The Gleason grade is usually heterogeneous within or between different tumor foci, so that a combined score which is the sum of the two most prevalent patterns within the gland can be assigned [9, 40, 53]. Although pathological grading is a powerful prognostic indicator, it cannot accurately predict the aggressiveness of the disease, as tumors with similar histological patterns may have different clinical outcomes [9, 54]. Hence, there is a critical need to identify alternate disease biomarkers with better diagnostic and prognostic potential.

The second major clinical challenge with prostate cancer is the high rate of recurrence associated with the disease [9]. Conventional treatment regimens for localized prostate cancer include surgery (radical prostatectomy) or radiation therapy, but in advanced prostate cancer these regimens are followed/substituted by ADT to reduce the tumor burden. Usually, this treatment is effective initially and reduces circulating PSA to low levels; however, the disease will subsequently recur and progresses to a hormone-refractory/castration-resistant state [55, 56] that is essentially untreatable [56, 57]. About 40% of men with localized prostate cancer suffer from relapse after the initial therapy [58], often due to aberrant reactivation. Despite an improved understanding of prostate cancer biology, the molecular mechanisms underlying the recurrence of the disease have not been fully delineated, and serve as a deterrent for developing effective therapies.

A third major clinical challenge is posed by the significant morbidity and mortality associated with metastatic prostate cancer. Advanced prostate cancer is associated with metastatic dissemination, typically to bones, which is primarily responsible for significant patient morbidity as well as mortality associated with the disease. Almost 80% of cases of advanced metastatic disease have bone metastasis [59], with characteristic osteoblastic rather than osteolytic lesions [9, 60]. Despite the huge prevalence of bone tropism, the current understanding of the molecular mechanisms that govern metastatic dissemination of prostate cancer remains fragmentary and there is no effective therapy for metastatic prostate cancer. Although, during the past few years, novel therapies with proven survival benefits have been developed [61, 62], the increase in survival rates have been marginal. Moreover, individual patients respond differently to these therapies, highlighting the genetic heterogeneity underlying the disease [63]. Clearly, an understanding of the molecular and genetic mechanisms that underlie the progression and metastasis of prostate cancer is key to developing better therapeutic strategies for the disease.

4 Genetic Landscape of Prostate Cancer

It has been realized that the prostate cancer genetic landscape is enormously complex and underlies the extreme clinical and phenotypic heterogeneity of the disease [9, 16, 54]. This complexity is being increasingly recognized with large-scale prostate cancer genomic studies utilizing advanced sequencing technologies [7, 16, 54, 63–71]. Such studies have detailed the complete genomic landscape of about 75 prostate cancers and hundreds of prostate cancer exomes [66]. Integrative genomic techniques, including array-comparative genomic hybridization, exome sequencing and methylation profiling, have also yielded significant information [72] on the complex genetic landscape of both primary and metastatic prostate cancer [7, 9, 16, 54, 67–69, 72]. Collectively, the findings of these studies broadly suggest the following prevailing themes in the prostate cancer genome:

1) The prostate cancer genome undergoes frequent large-scale genomic rearrangements that are nonrandom and deregulate important signaling pathways. Most strikingly, the prostate cancer genome frequently undergoes complex chromosomal rearrangements, called "chromoplexy," primarily involving known cancer-associated genes [64, 66, 67, 71]. Also, recurrent gene fusions involving E26 transformation-specific (ETS) transcription factors (*ERG*) and an androgen-regulated gene (*TMPRSS2*) are common in prostate cancer.
2) Somatic CNAs or genetic gains and losses are frequent in prostate cancer [72, 73]. Typically, genomic deletions are more numerous than gains, and frequently involve key regulatory genes such as phosphatase and tensin homolog (*PTEN*) at 10q23, *NKX3.1* at 8p21, and *MYC* at 8q24. These CNAs are focal ($\leq$1–5 Mb), and often affect a small part of the primary prostate

cancer genome and a larger region of metastatic prostate tumors [54, 72, 73].

3) The frequency of recurrent gene mutations is low compared with that of other cancers. In localized cancers, point mutations are most common in *SPOP* (speckle-type POZ protein), which encodes the substrate-recognition component of a ubiquitin ligase, *TP53* and *PTEN*, albeit with relatively low frequency (only about 5–10% for each gene). Following treatment with therapies targeting the androgen axis, metastatic lesions show amplifications and mutations of the *AR* gene.

Genomic studies have also led to an increased understanding of the heterogeneity and clonality of primary and metastatic prostate cancers. The results of these studies also suggest the presence of distinct prostate cancer genetic subgroups. This molecular subclassification has the potential to stratify patients into defined subtypes that will ultimately aid in improved therapeutic strategies against the disease [54].

5 Genome-Wide Analysis

In recent years, several attempts have been made at the genome-wide analysis of prostate cancer, and have enormously expanded the present understanding of the prostate cancer genome. These analyses suggest extensive genome rearrangements, CNAs and mutations underlying prostate cancer, as summarized below.

5.1 Genome Rearrangements

Genomic analyses have identified extensive structural genomic rearrangements in prostate cancer, suggesting that they may comprise a major mechanism driving prostate carcinogenesis. Of considerable importance has been the discovery of chromosomal rearrangements that activate the ETS family member of transcription factors (*ERG*, *ETV1*, *ETV4*) in prostate cancer [9, 63, 74]. In 2005, Tomlins *et al.* first reported gene rearrangements involving *ETS* genes (usually *ERG*, v-ets erythroblastosis virus *E26* oncogene homolog) with an androgen-regulated 5′ gene (*TMPRSS2*, transmembrane serine protease isoform 2) in about 50% of prostate tumors [63], allowing ETS transcription factors to be controlled by the androgen-responsive promoter. Tumors harboring *ERG*-rearrangements form a distinct subtype of prostate cancer with a distinct expression signature (Fig. 1) [74, 75]. *ERG*-rearranged tumors are associated with deletions of chromosomal regions 10q, 17p and 3p14, while those without the rearrangement are significantly enriched for 6q deletion, 7q gain and 16q deletion [54, 72, 76]. Since the initial discovery of *ETS* fusions in prostate cancer, recent studies have identified additional fusion events involving additional *ETS* family members and novel 5′ fusion partners. Also, non-*ETS*-based fusions such as those involving *RAF* kinase (for a review, see Ref. [77]) have been identified. Rearrangements often result in new fusion proteins that are constitutively active or have an altered function or cellular localization [54, 77]. Recent whole-genome analyses have provided new insights into the complexity of genome rearrangements prevalent in prostate cancer, and these are reviewed here.

Recently, Berger *et al.* [67] performed paired-end, whole-genome sequencing of seven primary human prostate cancers and corresponding normal tissues [67]. Strikingly, the study results suggested that

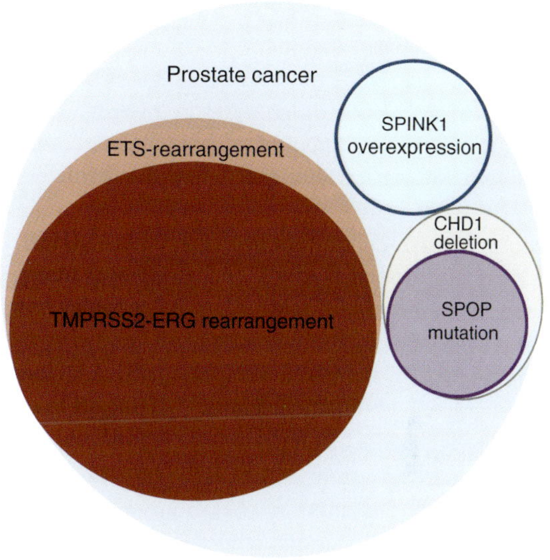

Fig. 1 Major molecular subtypes of prostate cancer. Recent genomic studies suggest distinct molecular subtypes of prostate cancer based on the key molecular alterations. The major molecular alteration is ETS rearrangements, most commonly *TMPRSS2-ERG* fusion gene (ETS^+) rearrangements. *TMPRSS2-ERG* fusions are present in about 50% of primary and advanced prostate cancer. ETS^- tumors are characterized based on the mutational status of *CHD1*, *SPOP* mutation, and *SPINK1* overexpression. SPOP mutations occur in approximately 10% of prostate cancer mutually exclusive with ETS rearrangements and are associated with deletions of *CHD1*.

the prostate cancer genome undergoes enormous genomic rearrangements that are complex and much more common than known previously. It was found that each cancer genome harbors an average of 90 inter- and intra-chromosomal arrangements, most of which were complex chains of balanced rearrangements (i.e., "copy-neutral" or "closed-chain") in which multiple chromosomal regions exchange breakpoint arms with no loss of genetic material. These rearrangements occurred within or adjacent to known cancer genes. Further, three tumors contained *TMPRSS2:ERG* fusion [7, 67], a rearrangement between *TMPRSS2* and the *ETS*-related gene *ERG* that has been reported previously in about 50% of primary prostate cancers [7, 78]. In *TMPRSS2:ERG*-positive tumors, the observed widespread chromosomal rearrangement breakpoints were precise and enriched near regions of open chromatin, AR and *ERG* DNA binding. In contrast, in tumors lacking *ETS* fusions the rearrangement breakpoints were inversely correlated with these regions. This suggests a link between chromatin or transcriptional regulation and the genesis of genomic aberrations in prostate cancer [67].

Baca *et al.* [64] sequenced the genomes of 57 prostate tumors and matched normal tissues and identified large chains of rearrangements (the phenomenon of chromoplexy) that coordinately affect multiple chromosomes in prostate cancer [64, 79]. These analyses suggested that chromoplexy disrupts multiple cancer genes coordinately, and may contribute to the aggressive clinical behavior of high-grade prostate tumors. This pattern of structural rearrangements is similar to the phenomenon of random breakage and fusion of genomic segments (termed chromothripsis) that has been observed in other cancers [79–81]. Chromoplexy observed in prostate cancer involves unclustered, chained rearrangements in multiple chromosomes that are relatively less frequent, whereas chromothripsis typically involves more frequent, locally clustered (within one or two chromosomes) rearrangements involving one or two chromosomes [79, 80]. The study results also suggested a molecular subclassification of prostate tumors based on the status of *TMPRSS2-ERG* fusion gene (*ETS*$^+$) and the mutational status of *CHD1*, a chromodomain helicase involved in chromatin remodeling (Fig. 1). *TMPRSS2-ERG*-positive tumors with wild-type *CHD1* have predominantly inter-chromosomal rearrangements, while the fusion gene-negative tumors harboring *CHD1* deletion displayed a higher frequency of intra-chromosomal rearrangements (resembling chromothripsis rather than chromoplexy) [79, 80]. This pattern of rearrangement also emphasizes the role of open chromatin and active transcription in determining genomic rearrangements. Further, the clonal hierarchy of genomic lesions in prostate tumors was characterized, which suggests a pathway of oncogenic events along which chromoplexy may drive prostate carcinogenesis [64]. These analyses suggested that a consensus path of tumor evolution begins with events such as the loss of *NKX3.1* or the fusion of *TMPRSS2* and *ERG*. The path proceeds with the loss of cyclin-dependent kinase inhibitor 1B (*CDKN1B*), *TP53* and *PTEN*, and other progression-associated lesions [64].

Weischenfeldt *et al.* [71] compared the genomic landscape of 11 early-onset cases with that of seven elderly-onset prostate cancer cases, and reported that early-onset cases harbored a prevalence of balanced structural rearrangements, with a specific abundance of androgen-regulated *ETS* gene fusions, including *TMPRSS2:ERG*. In contrast, elderly-onset cases displayed primarily non-androgen-associated structural rearrangements [71].

Both intra- and inter-chromosome genomic rearrangements result from the incorrect repair of double-stranded DNA (dsDNA) breaks that occur during DNA replication or transcription [54, 82]. These rearrangements are non-random, locus-specific, and often driven by transcription factors facilitating the juxtaposition of chromosomal loci and recruiting DNA repair enzymes to these loci [83]. For example, androgen signaling promotes the corecruitment of AR and topoisomerase II beta (TOP2B) to sites of *TMPRSS2-ERG* genomic breakpoints, triggering recombinogenic TOP2B-mediated dsDNA breaks. Furthermore, androgen stimulation resulted in the *de-novo* production of *TMPRSS2-ERG* fusion transcripts in a process that required TOP2B and components of the dsDNA break repair machinery [84]. Moreover, it has been demonstrated that in *TMPRSS2:ERG*-rearranged tumors, the rearrangement breakpoints are enriched near regions of open chromatin, AR and tERG binding sites, whereas in the *TMPRSS2:ERG* fusion-negative tumors the breakpoints are inversely correlated with these regions [67, 83].

5.2 Copy Number Alterations

Copy number alterations (CNAs) are genetic gains or losses that occur during carcinogenesis. The prostate cancer genome typically displays more CNAs than most other cancer types [54, 73], with more genomic deletions than amplifications. CNAs are often focal (≤1–5 Mb) [54] and affect a small fraction of the primary prostate cancer genome. However, these alterations are markedly increased in castration-resistant tumors compared to primary prostate tumors, while metastatic prostate tumor genomes typically display more CNAs affecting a larger portion of the genome [65, 67, 68, 72, 85].

Various groups conducting annotation of the prostate cancer genome have consistently reported deletions at chr 8p, 5q, 16q, 6q, 13q, 10q, 17p, 12p and 2q, whereas frequent copy number gains are observed at 8q, 7 and 3q [72, 86–92]. It has been suggested, and supported, that CNAs may frequently target multiple genes in common deleted/amplified regions that may cooperate to produce a more aggressive disease [93]. Most frequent genomic deletions are seen on chromosomes 8p, 10q and 13q that include tumor suppressor genes such as *NKX3.1*, *PTEN* and *RB1*. The frequent amplification of chromosomes X, 7, 8q and 9q regions that include the *AR* and *MYC* oncogenes are prevalent in castration-resistant prostate cancer (CRPC).

In a landmark study, Taylor *et al.* [72] performed a large-scale integrative analysis of the transcriptomes and CNAs of prostate cancer that has provided new insights into the complete prostate cancer genome (Fig. 2). In this case, a global analysis of CNA data from 194 primary and metastatic prostate cancers, as well as 12 cell lines/xenografts, revealed a broad diversity in CNA levels. Metastatic tumors and cell/xenograft lines harbored the greatest number of whole chromosome, chromosome arm, and focal amplifications and deletions. Primary tumors displayed a wide range of CNA levels, from tumors appearing metastatic-like in profile to those with fundamentally diploid genomes. Regions of recurrent CNAs were identified that showed 30 focal amplifications and 36 focal deletions, as well as recurrent gains and losses of seven chromosome arms (Fig. 2). The most frequent alteration in the prostate oncogenome was the loss of chr 8p [72], a common abnormality in many epithelial tumors that harbors the *NKX3.1* homeobox gene [94]. Interestingly, *NKX3.1* mRNA expression did not correlate with copy-number loss, suggesting the possibility of alternative tumor suppressors in this region. Consistent with prior studies, peaks of deletion were found targeting *PTEN* on 10q23.31, *RB1* on 13q14.2, *TP53* on 17p31.1, and the interstitial 21q22.2-3 deletion spanning *ERG* and *TMPRSS2*. Other broader deletions included 12p13.31-p12.3, which spans *ETV6* and *DUSP16* in addition to *CDKN1B* [72, 88]. The most common amplified loci included *MYC* on 8q24.21 and a novel oncogene *NCOA2* (nuclear receptor coactivator) on 8q13.3. Focal amplifications of *AR* (Xq12) were also common in metastatic tumors. Other gains spanned discontinuous regions of 7q, including genes such as *BRAF* and *EZH2*. In addition, less frequent gains of region 5p13.3-p13.1 spanning *AMACR*, *RICTOR*, and *SKP2*, as well as 47 other genes and two microRNAs, were observed [72]. Additionally, the androgen-driven *TMPRSS2-ERG* fusion was associated with a previously unrecognized, prostate-specific deletion at chromosome 3p14 that implicates *FOXP1*, *RYBP* and *SHQ1* as potential cooperative tumor suppressors. Most importantly, copy number data could robustly define clusters of low- and high-risk disease beyond that achieved by the Gleason score [72]. A comparison of CNAs in primary and metastatic tumors suggested that late-stage tumors resemble a subset of highly altered primary

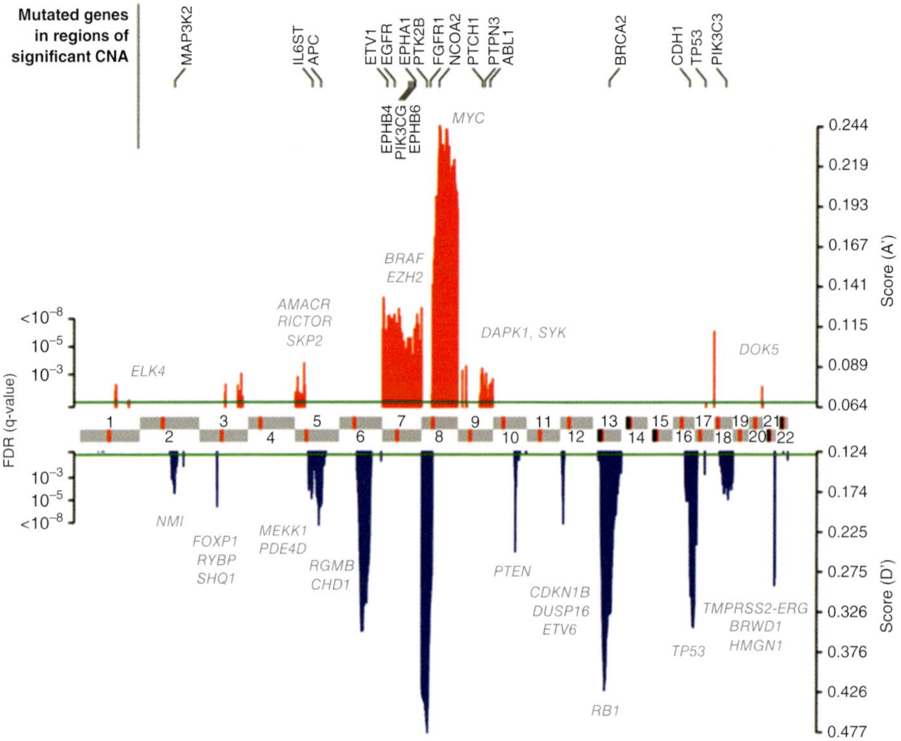

Fig. 2 A global view of the prostate cancer genome. Significant genomic aberrations in the prostate oncogenome. Regions of amplification (red) or deletion (blue) with false discovery rate (FDR) ≤10% are plotted, with chromosomes indicated at the center and centromeres in red. Genes in which somatic non-synonymous mutations were detected are listed on top (black). Additional genes of interest targeted by copy-number alterations alone are also indicated (gray). Reprinted with permission from Ref. [72]; © 2013, Elsevier.

tumors that was associated with a greater risk of relapse. Further, two additional studies compared CNAs in primary versus castration-resistant disease and observed increased CNAs in the latter category [65, 68, 72].

In another study, Robbins *et al.* [95] performed a high-resolution copy number analysis of focal metastatic prostate tumors. Homozygous deletions were detected at chromosomal regions 8p22, 10q23.31, 13q13.1, 13q14.11, 13q14.12, and key genes mapping within these regions included *PTEN*, *BRCA2*, *C13ORF15* and *SIAH3*. In addition, focal high-level amplifications were detected at regions 5p13.2-p12, 14q21.1, 7q22.1 and Xq12, including genes *SKP2*, *FOXA1* and *AR* [95]. Berger *et al.* reported that four out of seven primary prostate tumors harbored events disrupting either *PTEN* (unbalanced events) or *MAGI2* (balanced events), a PTEN interacting protein implicated in prostate tumorigenesis. [67]. A copy number loss on chromosome (chr) 5q21 harboring ATP-dependent chromatin-remodeling enzyme *CHD1*, was also recently reported [65, 68, 72] in prostate cancer, which was significantly associated with *ETS* gene family fusion-negative tumors. A marked overexpression of *SPINK1* (serine peptidase inhibitor, Kazal type 1), which encodes a secreted serine protease inhibitor, defines another aggressive molecular subtype of *ETS* fusion-negative prostate cancers (see Fig. 1) that constitutes about 10% of all prostate cancers [96]. Recent studies have suggested that *SPINK1* may be a therapeutic target in a subset of patients with $SPINK1^{+}/ETS^{-}$ prostate cancer [97].

5.3 Point Mutations

Prostate cancer has a low somatic mutation rate [69, 72, 98]. Although several mutations may occur in each prostate tumor genome, only a fraction of these (about 20 per genome) affect protein stability or function [54]. The mutation rates in CRPC is only modestly higher (two mutations per megabase) than those of primary tumors (0.9–1.5 mutations per megabase) [65, 67, 68, 85]. However, a subtype of prostate cancers exhibit "hypermutated" genomes [69], most likely due to a mutation of the DNA mismatch repair enzyme *MSH6* [7, 54, 65, 69]. Prostate cancer mutations cause aberrant activation of oncogenes (including *KRAS* and *BRAF*) and tumor suppressor genes (including *PTEN*, *RB*, *TP53* and *PI3KCA*) [54, 65, 68, 72, 99, 100]. Wnt pathway mutations potentially contribute to the development of castration resistance in prostate cancer [69]. In addition to these well-studied mutations, recent investigations have delineated novel recurrent mutations underlying prostate cancer, as summarized below.

Two recent studies, reporting the first whole-exome sequencing of large prostate cancer cohorts [65, 68, 85], have shed new light on the mutational landscape of the prostate cancer genome. Barbieri *et al.* [65, 85] sequenced the exomes of 112 primary tumors, providing a snapshot of prostate cancer genomes at diagnosis before treatment. In a second study by Grasso *et al.* [68, 85], the authors sequenced the exomes of 50 lethal, heavily pretreated metastatic CRPCs obtained at rapid autopsy (including three different foci from the same patient) and 11 treatment-naive, high-grade localized prostate cancers. Both studies reported recurrent mutations in genes regulating AR function, including *FOXA1* and *CDKN1B*. In addition, mutations were reported in multiple chromatin- and histone-modifying genes that are found in AR complexes (including *MLL2* and *ASXL2*). Genes for other transcriptional regulators, such as the transcriptional initiation component *MED12*, were also found to be mutated in both primary and CRPC tumors. AR plays a role in mediating castration resistance, and *AR* alterations were found only in CRPC [68]. Other proteins that physically interact with the AR, such as the *ERG* gene fusion product *UTX* (also known as *KDM6A*) and *ASXL1*, were found to be mutated in CRPC [68]. Mutations in *ETS2*, which is deleted in approximately one-third of CRPCs (commonly through *TMPRSS2:ERG* fusions), were also identified in CRPC [68].

Most importantly, these studies identified novel subtypes of prostate cancer that are *ETS* gene family fusion-negative [65, 68, 85]. Grasso *et al.* identified alterations in the *CHD1* gene (mutations or rearrangements) encoding a chromatin-remodeling factor as another *ETS* fusion-negative class of prostate cancer [68]. Barbieri *et al.* [65, 85] identified a new molecular subtype of prostate cancer characterized by *SPOP* mutations in *ETS* fusion-negative tumors (Fig. 1). *SPOP* encodes the substrate binding subunit of a Cullin-based E3 ubiquitin ligase [101–103] that is mutated in 6–15% of primary prostate tumors, with the mutations involving the SPOP substrate-binding cleft. Tumors with *SPOP* mutations were enriched in recurrent somatic deletions at chromosomal regions 5q21 and 6q21, which encode genes including the chromatin-modifying enzyme *CHD1*, the transcription factor *FOXO3*, and the tumor suppressor *PRDM1* [65, 85]. However, SPOP mutant tumors lacked *ETS* rearrangement as well as the well-characterized mutations in *PTEN*, *TP53* and *PIK3CA*. Further, it was demonstrated recently that *SPOP* mutations alter AR signaling in prostate cancer cells by influencing the stability of the transcriptional coactivator steroid receptor coactivator 3/nuclear receptor coactivator 3 (SRC3/NCOA3) protein [104]. In conclusion, recent studies have defined distinct subtypes of prostate cancer with characteristic molecular profiles, and offer the potential for a better therapeutic stratification of prostate cancer [54].

6
Prominent Molecular Alterations in Prostate Cancer Development and Progression

Genomic lesions in prostate cancer are very diverse, with considerable molecular heterogeneity between tumors. An appreciation of the heterogeneity comes from the observation that even different sites within the same tumor often exhibit unique genomic profiles [70]. However, recurrent alterations in specific signaling pathways underlie prostate cancer development and progression [7, 9, 16, 54, 62, 72, 76, 77, 85]. An integrative analysis of mutations, CNAs and expression changes has revealed alterations in the phosphoinositide 3-kinase (PI3K), RAS/RAF and AR pathways in nearly all metastatic prostate cancers, and a high frequency of primary prostate cases (Fig. 3) [72]. This large-scale study suggested the common alteration of three key pathways in prostate cancer – PI3K, RAS/RAF and retinoblastoma (RB) – with frequencies ranging from 34% to 43% in primary tumors versus 74% to 100% in metastases (Fig. 3b). The key signaling pathways that are important therapeutic targets for prostate cancer are discussed below.

6.1
Phosphoinositide 3-Kinase/PTEN/AKT Pathway

The PI3K/PTEN/AKT pathway is a critical regulator of various processes that are altered in cancer, including cell growth, proliferation, survival and motility [105]. PI3K transduces signals generated by receptor and nonreceptor tyrosine kinases by phosphorylating membrane-bound phospholipids in the plasma membrane. This leads to a recruitment of the protein Serine/Threonine kinase, AKT, to the cell membrane where it becomes activated. In contrast, the tumor-suppressor PTEN (phosphatase and tensin homolog deleted on chromosome 10) opposes PI3K signaling. PI3K/AKT signaling is activated in cancers by several different mechanisms [106, 107]. In prostate cancer, the PI3K pathway is frequently altered in both primary tumors (ca. 42%) and metastases (100%) [63, 72], often by a variety of mechanisms.

A loss of *PTEN* function, through deletion, mutation or reduced expression, is the major route for PI3K pathway activation in prostate cancer [67, 72, 95, 108]. Almost half of primary prostate cancers and all metastatic tumors have a focal *PTEN* deletion or chromosome 10 aneuploidy

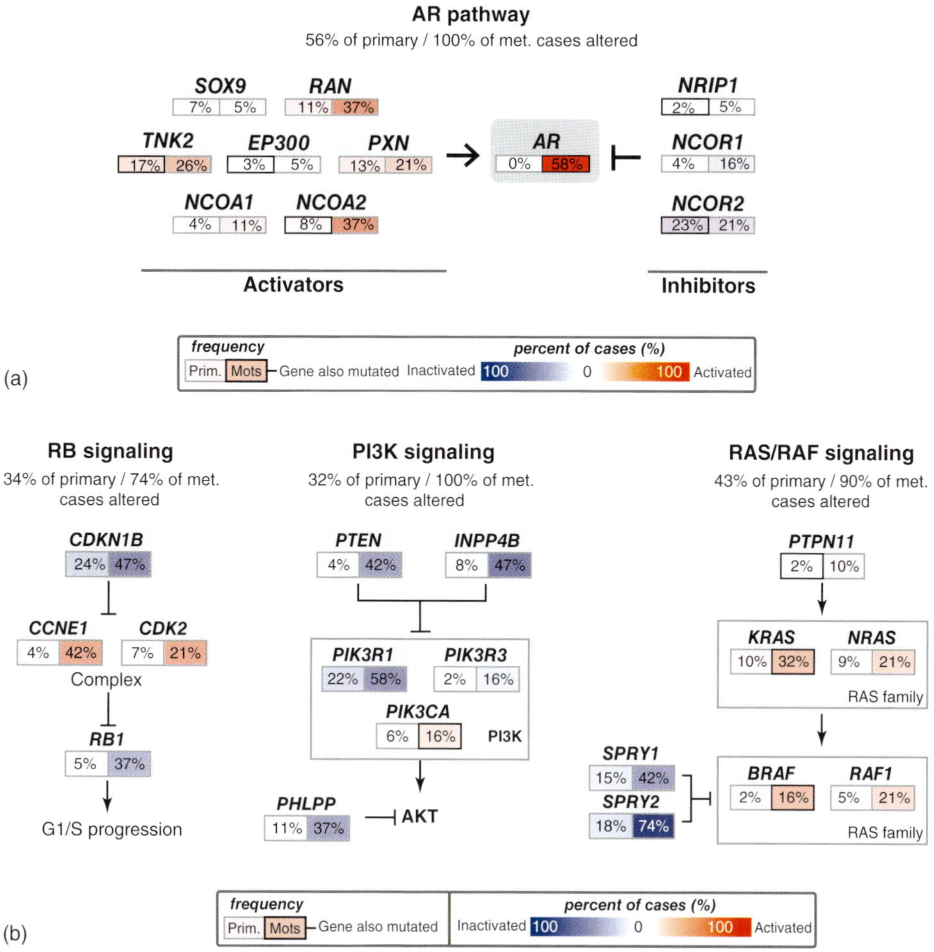

Fig. 3 Commonly altered pathways in prostate cancer. Four of the most commonly altered pathways in both primary and metastatic prostate cancers: AR, RB, PI3K, and RAS/RAF signaling. Additional important pathways are described in the text (see Sect. 6). (a) Alterations in androgen signaling components where frequencies are shown for *AR* and selected activators in primary and metastatic tumors; (b) RB, PI3K, and RAS/RAF signaling pathway alterations in prostate cancer. Alteration frequencies are shown for individual genes and for the entire pathway in primary and metastatic tumors. Alterations are defined as those having outlier expression (significant up- or downregulation) compared to the distribution of expression in normal prostate samples, or by somatic mutations, and are interpreted as activation (red) or inactivation (blue) of protein function. Reprinted with permission from Ref. [72]; © 2013, Elsevier.

[72], with 5–10% of primary cancers possessing inactivating mutations of *PTEN* [54, 65]. *PTEN* deletion is associated with a poor prognosis in prostate cancer, and is associated with a higher Gleason grade, disease progression, recurrence and death [99, 109–111]. *PTEN* loss is involved in prostate cancer initiation, and can promote progression to invasive prostate cancer and CRPC when combined with other genetic

lesions, such as *ERG* rearrangements and *NKX3.1* haploinsufficiency [78, 112, 113]. Germline loss of *PTEN* in heterozygous mutants or conditional *PTEN* deletion in the prostate epithelium leads to PIN lesions [9, 114, 115] that develop into invasive lesions when combined with other prominently altered pathways, such as *ERG* rearrangement [78, 112]. In addition to initiation, *PTEN* loss also alters AR signaling [9]. A frequent loss of *PTEN*, loss-of-function mutations, and rearrangements in a PTEN-interacting protein, MAGI2 (membrane-associated guanylate kinase WW and PDZ domain-containing 2) have also been recently reported in prostate cancer [54, 67]. MAGI2 is known to stabilize and interact with PTEN, enhancing its ability to suppress AKT activation [164, 165].

Large-scale integrative analyses of the prostate cancer oncogenome showed that, in addition to *PTEN* alterations, changes in phosphatases *INPP4B* and *PHLPP* (that were recently implicated in PI3K regulation), the *PIK3CA* gene (encoding a catalytic subunit of PI3K), and the PIK3CA regulatory subunits *PIK3R1* and *PIK3R3* [116–119] underlie the observed frequent alteration of this axis in prostate cancer [72] (Fig. 3b). Phosphatases *INPP4B* and *PHLPP* also regulate AKT, and their loss of expression is correlated with a higher Gleason score and a biochemical recurrence of prostate cancer [120, 121]. *PIK3CA* on chr 3q26.32 encodes the catalytic subunit of PI3K. Amplification and point mutations in *PIK3CA* also result in overactivation of the pathway, and have been detected in both primary and metastatic tumors [65, 72, 122]. The PI3K pathway also interacts with the AR signaling pathway in a reciprocal feedback loop [123]. AR inhibition activates AKT signaling by reducing levels of the AKT phosphatase PHLPP. A combined pharmacologic inhibition of PI3K and AR signaling caused prostate cancer regression in a Pten-deficient mouse model and in human prostate cancer xenografts [123]. In conclusion, the PI3K pathway is of central importance in prostate cancer pathogenesis, and targeted inhibitors of the pathway – either singly or in combination with AR inhibitors – are currently undergoing clinical investigation for the treatment of patients with advanced prostate cancer [63].

6.2 Androgen Signaling Pathway

The androgen receptor (AR) is a nuclear hormone receptor that binds to androgens and mediates AR signaling, and has key roles in both normal prostate development and prostate cancer [9, 124]. As prostate cancer growth is largely dependent on androgens, therapies focused on blocking the AR signaling pathway have been employed. Although these therapies are effective initially, the disease subsequently progresses to a hormone-refractory/castration-resistant state [55, 56] due to reactivation of the AR pathway. Several AR reactivation mechanisms have been identified, including *AR* gene amplifications, gain-of-function mutations, development of splice variants, or overexpression of its coactivators. Alternatively, tumor tissues endogenously express androgen-synthetic enzymes, leading to *de-novo* androgen synthesis (for reviews, see Refs [9, 125]).

Alterations in AR signaling through mutation, gene amplification, and/or overexpression are common but occur exclusively in metastatic samples (ca. 60%) [54, 72, 125]. Taylor *et al.* reported that an AR pathway analysis, including several known AR coactivators and corepressors, showed that alterations occur in 56% of primary tumors and in 100% of metastases (Fig. 3a) [72]. A further study using next-generation sequencing reported AR alterations in 60% of CRPC cases [63, 126]. In contrast, AR gene alterations were not observed in any of the clinically localized prostate tumors analyzed by exome sequencing [65]. *AR* gene amplification can occur either through focal amplification [127] or by gain of the entire X-chromosome on which *AR* resides [54, 72, 128]. Activating *AR* gene mutations also occur frequently in advanced prostate cancer [68, 72, 95].

Apart from AR, other components of the AR signaling pathway are also altered in both primary prostate cancer and metastatic cases, indicating the critical role of this pathway throughout the course of prostate cancer (Fig. 3a) [72]. Taylor *et al.* reported a frequent gain of the oncogenic nuclear receptor coactivator gene *NCOA2* (also known as *SRC2/TIF2/GRIP1*) in 8% of primary tumors and 37% of metastases by CNAs or mutations. *NCOA2* overexpression primes the AR response to reduced androgen levels [72]. Mutations in AR interacting proteins, including FOXA1, MLL (Mixed-Lineage Leukemia), ASXL1-3 and UTX, have also been reported in CRPC and may underlie treatment resistance [68]. Frequently observed *SPOP* mutations alter AR signaling in prostate cancer by influencing the stability of the AR transcriptional coactivator SRC3/NCOA3 [104]. The AR signaling pathway also interacts with oncogenic signaling pathways (e.g., PI3K/AKT) [123] that may underlie its functional redundancy and resistance to AR-targeted therapies [66].

AR signaling promotes genomic rearrangements that are observed widely in prostate cancer, including the prevalent *TMPRSS2:ERG* rearrangement [66]. Rearrangement breakpoints are common close to AR-binding sites [67], which suggests that AR-mediated transcriptional programs are predisposed to genomic rearrangements. AR signaling promotes *TMPRSS2:ERG* fusion by the corecruitment of AR and TOP2B to DNA, resulting in TOP2B-mediated, double-strand breaks (DSBs) [54, 84, 129]. Also, in response to genotoxic stress, AR promotes genomic rearrangements by recruiting two types of enzyme that include an activation-induced cytidine deaminase and the LINE-1 repeat-encoded ORF2 endonuclease. These enzymes synergistically generate site-selective DSBs at juxtaposed translocation loci that are ligated by the non-homologous end-joining pathway for specific translocations [129]. These data suggest that AR signaling could drive the common genomic alterations observed in prostate cancer, and is a critical pathway for therapeutic intervention in prostate cancer.

6.3 RAS/RAF/MAPK Pathway

The RAS/RAF/mitogen-activated protein kinase (MAPK) pathway plays a critical role in various human cancers. In prostate cancer, MAPK signaling is frequently activated, particularly in advanced disease [9]. Although *RAS/RAF* mutations are rare in prostate cancer, this pathway is frequently altered in advanced prostate cancer (Fig. 3b) [9, 65, 68, 72]. A stable expression of *RAS* effector-loop mutants that activate the RAS/MAP kinase pathway was sufficient to reduce the androgen requirements of prostate cancer cells for growth and tumorigenicity [130]. A combination of oncogenic mutations of *K-RAS* and beta-catenin drove the rapid progression of prostate tumorigenesis to invasive carcinoma, characterized by a synergistic elevation of AR, cyclooxygenase-2 and c-Myc [131]. A small percentage of advanced prostate tumors also contain recurrent rearrangements in the RAF pathway [132]. Further, tumors involving *RAS*/*RAF* are *ETS* fusion-negative, similar to tumors expressing *SPINK1* and mutant *SPOP* [66]. Although uncommon, alterations in the RAS/RAF/MAPK pathway are considered to be important targets for existing kinase inhibitors [66].

6.4 SRC Signaling Pathway

The SRC family of kinases (SFKs) are nonreceptor tyrosine kinases that are responsible for signal transduction during key cellular processes, including proliferation, differentiation, apoptosis, migration and adhesion [133, 134]. The activity of these kinases is often augmented in prostate cancer, correlating with disease severity/metastatic potential [133–136]. An increased SFK activity has been reported in hormone-independent prostate cancer, leading to poor prognosis, hormone relapse and reduced overall survival [137]. SRC, the prototypical member of SRC family kinases [138–140], is aberrantly activated in prostate cancer [133]. SRC signaling is implicated in androgen-induced proliferation of prostate cancer cells [133, 141], progression to an androgen-independent state, and metastasis [142–144]. *SRC* inhibition in prostate cancer cell lines leads to a significantly decreased proliferation, invasion and migration *in vitro* [133, 145–148]. The *in-vivo* inhibition of *SRC* led to decreased prostate cancer growth and metastasis in xenografts [133, 149] and in orthotopic [149] prostate mouse models. This kinase also plays a role in positively regulating osteoclast physiology, and thus is implicated in prostate bone metastasis [150, 151]. Another SFK, LYN (Lck/Yes novel tyrosine kinase), which originally was identified as a hematopoietic specific kinase [152], has been specifically implicated in prostate tumor growth and progression [149]. LYN is involved in numerous signaling pathways, including the PI3K pathway [134, 152, 153], and regulates apoptosis [154, 155], proliferation [156] and migration [157]. LYN is overexpressed in prostate cancer cell lines and primary prostatic tumors [134, 153, 157]. *LYN*$^{-/-}$ mice manifest prostate gland morphogenesis defects, which suggests an important role for LYN in normal prostate development and implications in prostate cancer [134, 153]. SRC kinases represent attractive therapeutic targets, and several SFK inhibitors are currently being tested clinically for prostate cancer treatment.

6.5 JAK/STAT Signaling Pathway

The Janus Kinase/signal transducers and activators of transcription (JAK/STAT) signaling pathway is required for various cellular processes such as proliferation, differentiation, survival and migration [158]. This pathway is constitutively activated in prostate cancer, leading to tumor cell proliferation and an inhibition of apoptosis mediated by STAT3, a member of STAT family of transcription factors [158, 159]. Mutations in the breast cancer susceptibility cancer 1 (*BRCA1*) gene have also been shown to increase predisposition to prostate cancer [160]. In prostate cancer, BRCA1 binds to STAT3 to promote JAK/STAT pathway activation, leading to an induction of cell proliferation and inhibition of apoptotic cell death [158, 161]. AR can also bind to STAT3, leading to an activation of the JAK/STAT pathway

[158, 159]. Stat3 is also activated by protein kinase Cε (PKCε), which is involved in prostate cancer development, progression and metastasis [162, 163]. An inhibition of the JAK/STAT3 signaling pathway suppressed prostate cancer growth and induced apoptosis [158], implicating the inhibition of this pathway as a therapeutic modality for prostate cancer.

7 Conclusions and Perspectives

During recent years, genomic studies have heralded a new understanding of the genetic landscape of prostate cancer. Genome-wide studies of both primary and metastatic prostate cancer, employing recently developed sequencing technologies, have helped to define the molecular alterations characteristic of prostate cancer. These studies have not only reinforced prevailing concepts but have also identified novel players in prostate cancer pathogenesis. Of note, large-scale complex genomic rearrangements have been deciphered that occur in the prostate cancer genome, while important signaling pathways that underlie prostate cancer initiation, progression and metastasis have been identified. Most importantly, major strides have been taken towards the molecular subclassification of prostate cancer, providing the potential to transition prostate cancer from a poorly understood heterogeneous disease into distinct molecular subtypes, each associated with distinct risk profiles [54]. As genomic-based classification offers hope for a more informed clinical decision-making, such understanding will translate into improved prostate cancer clinical management. Although there is an improved understanding of the distinct genomic alterations that underlie prostate cancer, with the potential of molecular subclassification, additional studies are clearly required. It has been speculated that, with an increased molecular understanding of prostate cancer, it will become possible to consider prostate cancer as a collection of cancers, much like breast cancer or acute myelogenous leukemia, as being definable by characteristic molecular changes [63], and this will in turn translate into effective tailored therapies for each genetic subtype. However, in order to achieve this goal many challenges must first be addressed, including a better understanding of the disease heterogeneity and treatment resistance [63]. It has been suggested that meta-analyses of existing genetic information may help to identify aberrations promoting prostate tumorigenesis [63]. In addition, it will be essential to perform more detailed integrative prostate cancer genomic analyses which will have the potential to yield more complete information on the genetic underpinnings of the disease. Finally, therapeutic regimens targeting key perturbed pathways will need to be tested clinically.

References

1. Siegel, R., Naishadham, M.A., and Jemal, A. (2012) Cancer statistics, 2012. *CA Cancer J. Clin.*, **62** (1), 10–29.
2. Loberg, R.D., Logothetis, C.J., Keller, E.T., and Pienta, K.J. (2005) Pathogenesis and treatment of prostate cancer bone metastases: targeting the lethal phenotype. *J. Clin. Oncol.*, **23** (32), 8232–8241.
3. Ding, L., Ellis, M.J., Li, S., Larson, D.E. *et al.* (2010) Genome remodelling in a basal-like breast cancer metastasis and xenograft. *Nature*, **464** (7291), 999–1005.
4. Shah, S.P., Morin, R.D., Khattra, J., Prentice, L. *et al.* (2009) Mutational evolution in a lobular breast tumour profiled at single nucleotide resolution. *Nature*, **461** (7265), 809-813.
5. Lee, W., Jiang, Z., Liu, J., Haverty, P.M. *et al.* (2010) The mutation spectrum

revealed by paired genome sequences from a lung cancer patient. *Nature*, **465** (7297), 473–477.
6. Pleasance, E.D., Cheetham, R.K., Stephens, P.J., McBride, D.J. *et al.* (2010) A comprehensive catalogue of somatic mutations from a human cancer genome. *Nature*, **463** (7278), 191–196.
7. Kumar, A., Shendure, J., and Nelson, P.S. (2011) Genome interrupted: sequencing of prostate cancer reveals the importance of chromosomal rearrangements. *Genome Med.*, **3** (4), 23.
8. MacVicar, G.R. and Hussain, M.H. (2013) Emerging therapies in metastatic castration-sensitive and castration-resistant prostate cancer. *Curr. Opin. Oncol.*, **25** (3), 252–260.
9. Shen, M.M., Abate-Shen. C. (2010) Molecular genetics of prostate cancer: new prospects for old challenges. *Genes Dev.*, **24** (18), 1967–2000.
10. Timms, B.G. (2008) Prostate development: a historical perspective. *Differentiation*, **76** (6), 565–577.
11. Bethel, C.R., Faith, D., Li, X., Guan, B. *et al.* (2006) Decreased NKX3.1 protein expression in focal prostatic atrophy, prostatic intraepithelial neoplasia, and adenocarcinoma: association with Gleason score and chromosome 8p deletion. *Cancer Res.*, **66** (22), 10683–10690.
12. Bhatia-Gaur, R., Donjacour, A.A., Sciavolino, P.J., Kim, M. *et al.* (1999) Roles for Nkx3.1 in prostate development and cancer. *Genes Dev.*, **13** (8), 966–977.
13. Signoretti, S., Waltregny, D., Dilks, J., Isaac, B. *et al.* (2000) p63 is a prostate basal cell marker and is required for prostate development. *Am. J. Pathol.*, **157** (6), 1769–1775.
14. Xin, L. (2013) Cells of origin for cancer: an updated view from prostate cancer. *Oncogene*, **32** (32), 3655–3663.
15. El-Alfy, M., Pelletier, G., Hermo, L.S., and Labrie, F. (2000) Unique features of the basal cells of human prostate epithelium. *Microsc. Res. Tech.*, **51** (5), 436–446.
16. Schrecengost, R. and Knudsen, K.E. (2013) Molecular pathogenesis and progression of prostate cancer. *Semin. Oncol.*, **40** (3), 244–258.
17. De Marzo, A.M., Meeker, A.K., Epstein, J.I., and Coffey, D.S. (1998) Prostate stem cell compartments: expression of the cell cycle inhibitor p27Kip1 in normal, hyperplastic, and neoplastic cells. *Am. J. Pathol.*, **153** (3), 911–919.
18. Litvinov, I.V., Vander Griend, D.J., Xu, Y., Antony, L. *et al.* (2006) Low-calcium serum-free defined medium selects for growth of normal prostatic epithelial stem cells. *Cancer Res.*, **66** (17), 8598–8607.
19. Berquin, I.M., Min, Y., Wu, R., Wu, H. *et al.* (2005) Expression signature of the mouse prostate. *J. Biol. Chem.*, **280** (43), 36442–36451.
20. Cunha, G.R., Donjacour, A.A., Cooke, P.S., Mee, S. *et al.* (1987) The endocrinology and developmental biology of the prostate. *Endocr. Rev.*, **8** (3), 338–362.
21. Donjacour, A.A., Cunha, G.R., and Sugimura, Y. (1987) Heterogeneity of structure and function in the mouse prostate. *Prog. Clin. Biol. Res.*, **239** 583–600.
22. Gao, N., Ishii, K., Mirosevich, J., Kuwajima, S. *et al.* (2005) Forkhead box A1 regulates prostate ductal morphogenesis and promotes epithelial cell maturation. *Development*, **132** (15), 3431–3443.
23. Bostwick, D.G. (1989) The pathology of early prostate cancer. *CA Cancer J. Clin.*, **39** (6), 376–393.
24. Chrisofos, M., Papatsoris, A.G., Lazaris, A., and Deliveliotis, C. (2007) Precursor lesions of prostate cancer. *Crit. Rev. Clin. Lab. Sci.*, **44** (3), 243–270.
25. DeMarzo, A.M., Nelson, W.G., Isaacs, W.B., and Epstein, J.I. (2003) Pathological and molecular aspects of prostate cancer. *Lancet*, **361** (9361), 955–964.
26. McNeal, J.E. and Bostwisk, D.G. (1986) Intraductal dysplasia: a premalignant lesion of the prostate. *Hum. Pathol.*, **17** (1), 64–71.
27. McNeal, J.E., Bostwisk, D.G., Kindrachuk, R.A., Redwine, E.A. *et al.* (1986) Patterns of progression in prostate cancer. *Lancet*, **1** (8472), 60–63.
28. Henrique, R., Jerónimo, C., Teixeira, M.R., Hoque, M.O. *et al.* (2006) Epigenetic heterogeneity of high-grade prostatic intraepithelial neoplasia: clues for clonal progression in prostate carcinogenesis. *Mol. Cancer Res.*, **4** (1), 1–8.
29. Bostwick, D.G. and Cheng, L. (2012) Precursors of prostate cancer. *Histopathology*, **60** (1), 4–27.
30. Emmert-Buck, M.R., Vocke, C.D., Pozzatti, R.O., Duray, P.H. *et al.* (1995) Allelic loss

on chromosome 8p12-21 in microdissected prostatic intraepithelial neoplasia. *Cancer Res.*, **55** (14), 2959–2962.
31. Jenkins, R.B., Qian, J., Lee, H.K., Huang, H. *et al.* (1998) A molecular cytogenetic analysis of 7q31 in prostate cancer. *Cancer Res.*, **58** (4), 759–766.
32. Qian, J., Bostwick, D.G., Takahashi, S., Borell, T.J. *et al.* (1995) Chromosomal anomalies in prostatic intraepithelial neoplasia and carcinoma detected by fluorescence in situ hybridization. *Cancer Res.*, **55** (22), 5408–5414.
33. Montironi, R., Mazzucchelli, R., Lopez-Beltran, A., Scarpelli, M. *et al.* (2011) Prostatic intraepithelial neoplasia: its morphological and molecular diagnosis and clinical significance. *BJU Int.*, **108** (9), 1394–1401.
34. Mosquera, J.M., Perner, S., Genega, E.M., Sanda, M. *et al.* (2008) Characterization of TMPRSS2-ERG fusion high-grade prostatic intraepithelial neoplasia and potential clinical implications. *Clin. Cancer Res.*, **14** (11), 3380–3385.
35. Kang, J.S., Calvo, B.F., Maygarden, S.J., Caskey, L.S. *et al.* (2002) Dysregulation of annexin I protein expression in high-grade prostatic intraepithelial neoplasia and prostate cancer. *Clin. Cancer Res.*, **8** (1), 117–123.
36. Humphrey, P.A. (2007) Diagnosis of adenocarcinoma in prostate needle biopsy tissue. *J. Clin. Pathol.*, **60** (1), 35–42.
37. Fine, S.W. (2012) Variants and unusual patterns of prostate cancer: clinicopathologic and differential diagnostic considerations. *Adv. Anat. Pathol.*, **19** (4), 204–216.
38. Grignon, D.J. (2004) Unusual subtypes of prostate cancer. *Mod. Pathol.*, **17** (3), 316–327.
39. Squire, J.A., Park, P.C., Yoshimoto, M., Alami, J. *et al.* (2011) Prostate cancer as a model system for genetic diversity in tumors. *Adv. Cancer Res.*, **112** 183–216.
40. Mellinger, G.T., Gleason, D., Bailar, J. III, (1967) The histology and prognosis of prostatic cancer. *J. Urol.*, **97** (2), 331–337.
41. Bostwick, D.G., Shan, A., Qian, J., Darson, M. *et al.* (1998) Independent origin of multiple foci of prostatic intraepithelial neoplasia: comparison with matched foci of prostate carcinoma. *Cancer*, **83** (9), 1995–2002.
42. Macintosh, C.A., Stower, M., Reid, N., and Maitland, N.J. (1998) Precise microdissection of human prostate cancers reveals genotypic heterogeneity. *Cancer Res.*, **58** (1), 23–28.
43. Goldstein, A.S., Huang, J., Guo, C., Garraway, I.P. *et al.* (2010) Identification of a cell of origin for human prostate cancer. *Science*, **329** (5991), 568–571.
44. Stoyanova, T., Cooper, A.R., Drake, J.M., Liu, X. *et al.* (2013) Prostate cancer originating in basal cells progresses to adenocarcinoma propagated by luminal-like cells. *Proc. Natl Acad. Sci. USA*, **110** (50), 20111–20116.
45. Komiya, A., Suzuki, H., Imamoto, T., Kamiya, N. *et al.* (2009) Neuroendocrine differentiation in the progression of prostate cancer. *Int. J. Urol.*, **16** (1), 37–44.
46. Yuan, T.C., Veeramani, S., and Lin, M.F. (2007) Neuroendocrine-like prostate cancer cells: neuroendocrine transdifferentiation of prostate adenocarcinoma cells. *Endocr. Relat. Cancer*, **14** (3), 531–547.
47. van Leenders, G.J., Gage, W.R., Hicks, J.L., van Balken, B. *et al.* (2003) Intermediate cells in human prostate epithelium are enriched in proliferative inflammatory atrophy. *Am. J. Pathol.*, **162** (5), 1529–1537.
48. Verhagen, A.P., Ramaekers, F.C., Aalders, T.W., Schaafsma, H.E. *et al.* (1992) Colocalization of basal and luminal cell-type cytokeratins in human prostate cancer. *Cancer Res.*, **52** (22), 6182–6187.
49. Lilja, H. (2008) Testing new PSA subforms to enhance the accuracy of predicting cancer risk and disease outcome in prostate cancer. *Clin. Chem.*, **54** (7), 1248–1249.
50. Lilja, H., Ulmert, D., and Vickers, A.J. (2008) Prostate-specific antigen and prostate cancer: prediction, detection and monitoring. *Nat. Rev. Cancer*, **8** (4), 268–278.
51. Brooks, D.D., Wolf, A., Smith, R.A., Dash, C. *et al.* (2010) Prostate cancer screening 2010: updated recommendations from the American Cancer Society. *J. Natl Med. Assoc.*, **102** (5), 423–429.
52. Wolf, A.M., Wender, R.C., Etzioni, R.B., Thompson, I.M. *et al.* (2010) American Cancer Society guideline for the early detection of prostate cancer: update 2010. *CA Cancer J. Clin.*, **60** (2), 70–98.

53. Epstein, J.I. (2010) An update of the Gleason grading system. *J. Urol.*, **183** (2), 433–440.
54. Schoenborn, J.R., Nelson, P., and Fang, M. (2013) Genomic profiling defines subtypes of prostate cancer with the potential for therapeutic stratification. *Clin. Cancer Res.*, **19** (15), 4058–4066.
55. Kasper, S. and Cookson, M.S. (2006) Mechanisms leading to the development of hormone-resistant prostate cancer. *Urol. Clin. North Am.*, **33** (2), 201–210, vii.
56. Vaz, R.J., Zamora, I., Reiling, S., Shen, J. *et al.* (2010) The challenges of in silico contributions to drug metabolism in lead optimization. *Expert Opin. Drug Metab. Toxicol.*, **6** (7), 851–861.
57. Tannock, I.F., de Wit, R., Berry, W.R., Horti, J. *et al.* (2004) Docetaxel plus prednisone or mitoxantrone plus prednisone for advanced prostate cancer. *N. Engl. J. Med.*, **351** (15), 1502–1512.
58. Grubb, R.L.III, and Kibel, A.S. (2007) Prostate cancer: screening, diagnosis and management in 2007. *Mo. Med.*, **104** (5), 408–413; quiz 413–404.
59. Koutsilieris, M. (1993) Osteoblastic metastasis in advanced prostate cancer. *Anticancer Res.*, **13** (2), 443–449.
60. Logothetis, C.J. and Lin, S.H. (2005) Osteoblasts in prostate cancer metastasis to bone. *Nat. Rev. Cancer*, **5** (1), 21–28.
61. Fong, M.K., Hare, R., and Jarkowski, A. (2012) A new era for castrate resistant prostate cancer: a treatment review and update. *J. Oncol. Pharm. Pract.*, **18** (3), 343–354.
62. Rodrigues, D.N., Butler, L.M., Estelles, D.L., de Bono, J.S. (2013) Molecular pathology and prostate cancer therapeutics: from biology to bedside. *J. Pathol.* **232**(2):178–184.
63. Beltran, H. , and Rubin, M.A. (2013) New strategies in prostate cancer: translating genomics into the clinic. *Clin. Cancer Res.*, **19** (3), 517–523.
64. Baca, S.C., Prandi, D., Lawrence, M.S., Mosquera, J.M. *et al.* (2013) Punctuated evolution of prostate cancer genomes. *Cell*, **153** (3), 666–677.
65. Barbieri, C.E., Baca, S.C., Lawrence, M.S., Demichelis, F. *et al.* (2012) Exome sequencing identifies recurrent SPOP, FOXA1 and MED12 mutations in prostate cancer. *Nat. Genet.*, **44** (6), 685–689.
66. Barbieri, C.E. and Tomline, S.A. (2013) The prostate cancer genome: perspectives and potential. *Urol. Oncol.* **32**(1):53.e15–53.e22.
67. Berger, M.F., Lawrence, M.S., Demichelis, F., Dreir, Y. *et al.* (2011) The genomic complexity of primary human prostate cancer. *Nature*, **470** (7333), 214–220.
68. Grasso, C.S., Wu, Y.M., Robinson, D.R., Cao, X. *et al.* (2012) The mutational landscape of lethal castration-resistant prostate cancer. *Nature*, **487** (7406), 239–243.
69. Kumar, A., White, T.A., MacKenzie, A.P., Clegg, N. *et al.* (2011) Exome sequencing identifies a spectrum of mutation frequencies in advanced and lethal prostate cancers. *Proc. Natl Acad. Sci. USA*, **108** (41), 17087–17092.
70. Lindberg, J., Klevebring, D., Liu, W., Neiman, M. *et al.* (2013) Exome sequencing of prostate cancer supports the hypothesis of independent tumour origins. *Eur. Urol.*, **63** (2), 347–353.
71. Weischenfeldt, J., Simon, R., Feuerbach, L., Schlangen, K. *et al.* (2013) Integrative genomic analyses reveal an androgen-driven somatic alteration landscape in early-onset prostate cancer. *Cancer Cell*, **23** (2), 159–170.
72. Taylor, B.S., Schultz, N., Hieronymus, H., Gopalan, A. *et al.* (2010) Integrative genomic profiling of human prostate cancer. *Cancer Cell*, **18** (1), 11–22.
73. Beroukhim, R., Mermel, C.H., Porter, D., Wei, G. *et al.* (2010) The landscape of somatic copy-number alteration across human cancers. *Nature*, **463** (7283), 899–905.
74. Tomlins, S.A., Rhodes, D.R., Perner, S., Dhanasekaran, S.M. *et al.* (2005) Recurrent fusion of TMPRSS2 and ETS transcription factor genes in prostate cancer. *Science*, **310** (5748), 644–648.
75. Setlur, S.R., Mertz, K.D., Hoshida, Y., Demichellis, F. *et al.* (2008) Estrogen-dependent signaling in a molecularly distinct subclass of aggressive prostate cancer. *J. Natl Cancer Inst.*, **100** (11), 815–825.
76. Gonzalgo, M.L. and Isaacs, W.B. (2003) Molecular pathways to prostate cancer. *J. Urol.*, **170** (6, Pt 1), 2444–2452.
77. Rubin, M.A., Maher, C.A., and Chinnaiyan, A.M. (2011) Common gene rearrangements in prostate cancer. *J. Clin. Oncol.*, **29** (27), 3659–3668.

78. King, J.C., Xu, J., Wongvipat, J., Hieronymus, H. *et al.* (2009) Cooperativity of TMPRSS2-ERG with PI3-kinase pathway activation in prostate oncogenesis. *Nat. Genet.*, **41** (5), 524–526.
79. Shen, M.M. (2013) Chromoplexy: a new category of complex rearrangements in the cancer genome. *Cancer Cell*, **23** (5), 567–569.
80. Forment, J.V., Kaidi, A., and Jackson, S.P. (2012) Chromothripsis and cancer: causes and consequences of chromosome shattering. *Nat. Rev. Cancer*, **12** (10), 663–670.
81. Jones, M.J. and Jallepalli, P.V. (2012) Chromothripsis: chromosomes in crisis. *Dev. Cell*, **23** (5), 908–917.
82. Akfirat, C., Zhang, X., Ventura, A., Berel, D. *et al.* (2013) Tumour cell survival mechanisms in lethal metastatic prostate cancer differ between bone and soft tissue metastases. *J. Pathol.*, **230** (3), 291–297.
83. Barbieri, C.E., Demichelis, F., and Rubin, M.A. (2012) Molecular genetics of prostate cancer: emerging appreciation of genetic complexity. *Histopathology*, **60** (1), 187–198.
84. Haffner, M.C., Aryee, M.J., Toubaji, A., Esopi, D.M. *et al.* (2010) Androgen-induced TOP2B-mediated double-strand breaks and prostate cancer gene rearrangements. *Nat. Genet.*, **42** (8), 668–675.
85. Hieronymus, H. and Sawyers, C.L. (2012) Traversing the genomic landscape of prostate cancer from diagnosis to death. *Nat. Genet.*, **44** (6), 613–614.
86. Ribeiro, F.R., Heringue, R., Hektoen, M., Berg, M. *et al.* (2006) Comparison of chromosomal and array-based comparative genomic hybridization for the detection of genomic imbalances in primary prostate carcinomas. *Mol. Cancer*, **5** 33.
87. Kim, J.H., Dhanasekaran, S.M., Mehra, R., Tomlins, S.A. *et al.* (2007) Integrative analysis of genomic aberrations associated with prostate cancer progression. *Cancer Res.*, **67** (17), 8229–8239.
88. Lapointe, J., Li, C., Giacomini, C.P., Salari, K. *et al.* (2007) Genomic profiling reveals alternative genetic pathways of prostate tumorigenesis. *Cancer Res.*, **67** (18), 8504–8510.
89. Lapointe, J., Li, C., Higgins, J.P., van de Rijn, M. *et al.* (2004) Gene expression profiling identifies clinically relevant subtypes of prostate cancer. *Proc. Natl Acad. Sci. USA*, **101** (3), 811–816.
90. Lieberfarb, M.E., Lin, M., Lechpammer, M., Li, C. *et al.* (2003) Genome-wide loss of heterozygosity analysis from laser capture microdissected prostate cancer using single nucleotide polymorphic allele (SNP) arrays and a novel bioinformatics platform dChipSNP. *Cancer Res.*, **63** (16), 4781–4785.
91. Perner, S., Demechelis, F., Beroukhim, R., Schmidt, F.H. *et al.* (2006) TMPRSS2:ERG fusion-associated deletions provide insight into the heterogeneity of prostate cancer. *Cancer Res.*, **66** (17), 8337–8341.
92. Singh, D., Febbo, P.G., Ross, K., Jackson, D.G. *et al.* (2002) Gene expression correlates of clinical prostate cancer behavior. *Cancer Cell*, **1** (2), 203–209.
93. Xue, W., Kitzing, T., Roessler, S., Zuber, J. *et al.* (2012) A cluster of cooperating tumor-suppressor gene candidates in chromosomal deletions. *Proc. Natl Acad. Sci. USA*, **109** (21), 8212–8217.
94. He, W.W., Sciavolino, P.J., Wing, J., Augustus, M. *et al.* (1997) A novel human prostate-specific, androgen-regulated homeobox gene (NKX3.1) that maps to 8p21, a region frequently deleted in prostate cancer. *Genomics*, **43** (1), 69–77.
95. Robbins, C.M., Tembe, W.A., Baker, A., Sinari, S. *et al.* (2011) Copy number and targeted mutational analysis reveals novel somatic events in metastatic prostate tumors. *Genome Res.*, **21** (1), 47–55.
96. Tomlins, S.A., Rhodes, D.R., Yu, J., Varambally, S. *et al.* (2008) The role of SPINK1 in ETS rearrangement-negative prostate cancers. *Cancer Cell*, **13** (6), 519–528.
97. Ateeq, B., Tomlins, S.A., Laxman, B., Asanqani, I.A. *et al.* (2011) Therapeutic targeting of SPINK1-positive prostate cancer. *Sci. Transl. Med.*, **3** (72), 72ra17.
98. Kan, Z., Jaiswal, B.S., Stinson, J., Janakiraman, V. *et al.* (2010) Diverse somatic mutation patterns and pathway alterations in human cancers. *Nature*, **466** (7308), 869–873.
99. Cairns, P., Okami, K., Halachmi, S., Halachmi, N. *et al.* (1997) Frequent inactivation of PTEN/MMAC1 in primary prostate cancer. *Cancer Res.*, **57** (22), 4997–5000.

100. Eastham, J.A., Stapleton, A.M., Gousse, A.E., Timme, T.L. *et al.* (1995) Association of p53 mutations with metastatic prostate cancer. *Clin. Cancer Res.*, **1** (10), 1111–1118.
101. Nagai, Y., Kojima, T., Muro, Y., Hachiya, T. *et al.* (1997) Identification of a novel nuclear speckle-type protein, SPOP. *FEBS Lett.*, **418** (1-2), 23–26.
102. Zhuang, M., Calabrese, M.F., Liu, J., Waddell, M.B. *et al.* (2009) Structures of SPOP–substrate complexes: insights into molecular architectures of BTB-Cul3 ubiquitin ligases. *Mol. Cell*, **36** (1), 39–50.
103. Furukawa, M., He, Y.J., Borchers, C., and Xiong, Y. (2003) Targeting of protein ubiquitination by BTB-Cullin 3-Roc1 ubiquitin ligases. *Nat. Cell Biol.*, **5** (11), 1001–1007.
104. Geng, C., He, B., Xu, L., Barbieri, C.E. *et al.* (2013) Prostate cancer-associated mutations in speckle-type POZ protein (SPOP) regulate steroid receptor coactivator 3 protein turnover. *Proc. Natl Acad. Sci. USA*, **110** (17), 6997–7002.
105. Vivanco, I. and Sawyers, C.L. (2002) The phosphatidylinositol 3-Kinase AKT pathway in human cancer. *Nat. Rev. Cancer*, **2** (7), 489–501.
106. Engelman, J.A. (2009) Targeting PI3K signalling in cancer: opportunities, challenges and limitations. *Nat. Rev. Cancer*, **9** (8), 550–562.
107. Wong, K.K., Engelman, J.A., and Cantley, L.C. (2010) Targeting the PI3K signaling pathway in cancer. *Curr. Opin. Genet. Dev.*, **20** (1), 87–90.
108. Pourmand, G., Ziaee, A.A., Abedi, A.R., Mehrsai, A. *et al.* (2007) Role of PTEN gene in progression of prostate cancer. *Urol. J.*, **4** (2), 95–100.
109. Choucair, K., Ejdelman, J., Brimo, F., Aprikian, A. *et al.* (2012) PTEN genomic deletion predicts prostate cancer recurrence and is associated with low AR expression and transcriptional activity. *BMC Cancer*, **12** 543.
110. McMenamin, M.E., Soung, P., Perea, S., Kaplan, I. *et al.* (1999) Loss of PTEN expression in paraffin-embedded primary prostate cancer correlates with high Gleason score and advanced stage. *Cancer Res.*, **59** (17), 4291–4296.
111. Reid, A.H., Attard, G., Ambroisine, L., Fisher, G. *et al.* (2010) Molecular characterisation of ERG, ETV1 and PTEN gene loci identifies patients at low and high risk of death from prostate cancer. *Br. J. Cancer*, **102** (4), 678–684.
112. Carver, B.S., Tran, J., Gopalan, A., Chen, Z. *et al.* (2009) Aberrant ERG expression cooperates with loss of PTEN to promote cancer progression in the prostate. *Nat. Genet.*, **41** (5), 619–624.
113. Kim, M.J., Cardiff, R.D., Desai, N., Banach-Petrosky, W.A. *et al.* (2002) Cooperativity of Nkx3.1 and Pten loss of function in a mouse model of prostate carcinogenesis. *Proc. Natl Acad. Sci. USA*, **99** (5), 2884–2889.
114. Di Cristofano, A., Pesce, B., Cordon-Cardo, C., and Pandolfi, P.P. (1998) Pten is essential for embryonic development and tumour suppression. *Nat. Genet.*, **19** (4), 348–355.
115. Podsypanina, K., Ellenson, L.H., Nemes, A., Gu, J. *et al.* (1999) Mutation of Pten/Mmac1 in mice causes neoplasia in multiple organ systems. *Proc. Natl Acad. Sci. USA*, **96** (4), 1563–1568.
116. Gao, T., Furnari, F., and Newton, A.C. (2005) PHLPP: a phosphatase that directly dephosphorylates Akt, promotes apoptosis, and suppresses tumor growth. *Mol. Cell*, **18** (1), 13–24.
117. Gewinner, C., Wang, Z.C., Richardson, A., Teruya-Feldstein, J. *et al.* (2009) Evidence that inositol polyphosphate 4-phosphatase type II is a tumor suppressor that inhibits PI3K signaling. *Cancer Cell*, **16** (2), 115–125.
118. Jaiswal, B.S., Janakiraman, V., Kljavin, N.M., Chaudhuri, S. *et al.* (2009) Somatic mutations in p85alpha promote tumorigenesis through class IA PI3K activation. *Cancer Cell*, **16** (6), 463–474.
119. Ueki, K., Fruman, D.A., Yballe, C.M., Fasshauer, M. *et al.* (2003) Positive and negative roles of p85 alpha and p85 beta regulatory subunits of phosphoinositide 3-kinase in insulin signaling. *J. Biol. Chem.*, **278** (48), 48453–48466.
120. Chen, M., Pratt, C.P., Zeeman, M.E., Schultz, N. *et al.* (2011) Identification of PHLPP1 as a tumor suppressor reveals the role of feedback activation in PTEN-mutant prostate cancer progression. *Cancer Cell*, **20** (2), 173–186.

121. Hodgson, M.C., Shao, L.J., Frolov, A., Li, R. *et al.* (2011) Decreased expression and androgen regulation of the tumor suppressor gene INPP4B in prostate cancer. *Cancer Res.*, **71** (2), 572–582.
122. Sun, X., Huang, J., Homma, T., Kita, D. *et al.* (2009) Genetic alterations in the PI3K pathway in prostate cancer. *Anticancer Res.*, **29** (5), 1739–1743.
123. Carver, B.S., Chapinski, C., Wongvipat, J., Hieronymus, H. *et al.* (2011) Reciprocal feedback regulation of PI3K and androgen receptor signaling in PTEN-deficient prostate cancer. *Cancer Cell*, **19** (5), 575–586.
124. Mangelsdorf, D.J., Thummel, C., Beato, M., Herrlich, P. *et al.* (1995) The nuclear receptor superfamily: the second decade. *Cell*, **83** (6), 835–839.
125. Knudsen, K.E. and Scher, H.I. (2009) Starving the addiction: new opportunities for durable suppression of AR signaling in prostate cancer. *Clin. Cancer Res.*, **15** (15), 4792–4798.
126. Beltran, H., Yelensky, R., Frampton, G.M., Park, K. *et al.* (2013) Targeted next-generation sequencing of advanced prostate cancer identifies potential therapeutic targets and disease heterogeneity. *Eur. Urol.*, **63** (5), 920–926.
127. Liu, W., Laitinen, S., Khan, S., Vihinen, M. *et al.* (2009) Copy number analysis indicates monoclonal origin of lethal metastatic prostate cancer. *Nat. Med.*, **15** (5), 559–565.
128. Holcomb, I.N., Young, J.M., Coleman, I.M., Salari, K. *et al.* (2009) Comparative analyses of chromosome alterations in soft-tissue metastases within and across patients with castration-resistant prostate cancer. *Cancer Res.*, **69** (19), 7793–7802.
129. Lin, C., Yang, L., Tanasa, B., Hutt, K. *et al.* (2009) Nuclear receptor-induced chromosomal proximity and DNA breaks underlie specific translocations in cancer. *Cell*, **139** (6), 1069–1083.
130. Bakin, R.E., Gioeli, D., Sikes, R.A., Bissonette, E.A. *et al.* (2003) Constitutive activation of the Ras/mitogen-activated protein kinase signaling pathway promotes androgen hypersensitivity in LNCaP prostate cancer cells. *Cancer Res.*, **63** (8), 1981–1989.
131. Pearson, H.B., Phesse, T.J., and Clarke, A.R. (2009) K-ras and Wnt signaling synergize to accelerate prostate tumorigenesis in the mouse. *Cancer Res.*, **69** (1), 94–101.
132. Palanisamy, N., Ateeq, B., Kalyana-Sundaram, S., Pflueger, D. *et al.* (2010) Rearrangements of the RAF kinase pathway in prostate cancer, gastric cancer and melanoma. *Nat. Med.*, **16** (7), 793–798.
133. Fizazi, K. (2007) The role of Src in prostate cancer. *Ann. Oncol.*, **18** (11), 1765–1773.
134. Ingley, E. (2012) Functions of the Lyn tyrosine kinase in health and disease. *Cell Commun. Signal.*, **10** (1), 21.
135. Chang, Y.M., Kung, H.J., and Evans, C.P. (2007) Nonreceptor tyrosine kinases in prostate cancer. *Neoplasia*, **9** (2), 90–100.
136. Myoui, A., Nishimura, R., Williams, P.J., Hiraga, T. *et al.* (2003) C-SRC tyrosine kinase activity is associated with tumor colonization in bone and lung in an animal model of human breast cancer metastasis. *Cancer Res.*, **63** (16), 5028–5033.
137. Tatarov, O., Mitchell, T.J., Seywright, M., Leung, H.Y. *et al.* (2009) SRC family kinase activity is up-regulated in hormone-refractory prostate cancer. *Clin. Cancer Res.*, **15** (10), 3540–3549.
138. Martin, G.S. (2001) The hunting of the Src. *Nat. Rev. Mol. Cell Biol.*, **2** (6), 467–475.
139. Summy, J.M. and Gallick, G.E. (2003) Src family kinases in tumor progression and metastasis. *Cancer Metastasis Rev.*, **22** (4), 337–358.
140. Yeatman, T.J. (2004) A renaissance for SRC. *Nat. Rev. Cancer*, **4** (6), 470–480.
141. Migliaccio, A., Castoria, G., Di Domenico, M., de Falco, A. *et al.* (2000) Steroid-induced androgen receptor-oestradiol receptor beta-Src complex triggers prostate cancer cell proliferation. *EMBO J.*, **19** (20), 5406–5417.
142. Lee, L.F., Guan, J., Qiu, Y., and Kung, H.J. (2001) Neuropeptide-induced androgen independence in prostate cancer cells: roles of nonreceptor tyrosine kinases Etk/Bmx, Src, and focal adhesion kinase. *Mol. Cell. Biol.*, **21** (24), 8385–8397.
143. Lee, L.F., Louie, M.C., Desai, S.J., Yang, J. *et al.* (2004) Interleukin-8 confers androgen-independent growth and migration of LNCaP: differential effects of tyrosine kinases Src and FAK. *Oncogene*, **23** (12), 2197–2205.
144. Unni, E., Sun, S., Nan, B., McPhaul, M.J. *et al.* (2004) Changes in androgen receptor

nongenotropic signaling correlate with transition of LNCaP cells to androgen independence. *Cancer Res.*, **64** (19), 7156–7168.
145. Moasser, M.M., Srethapakdi, M., Sachar, K.S., Kraker, A.J. *et al.* (1999) Inhibition of Src kinases by a selective tyrosine kinase inhibitor causes mitotic arrest. *Cancer Res.*, **59** (24), 6145–6152.
146. Nam, S., Kim, D., Cheng, J.Q., Zhang, S. *et al.* (2005) Action of the Src family kinase inhibitor, dasatinib (BMS-354825), on human prostate cancer cells. *Cancer Res.*, **65** (20), 9185–9189.
147. Recchia, I., Rucci, N., Festuccia, C., Bologna, M. *et al.* (2003) Pyrrolopyrimidine c-Src inhibitors reduce growth, adhesion, motility and invasion of prostate cancer cells in vitro. *Eur. J. Cancer*, **39** (13), 1927–1935.
148. Slack, J.K., Adams, R.B., Rovin, J.D., Bissonette, E.A. *et al.* (2001) Alterations in the focal adhesion kinase/Src signal transduction pathway correlate with increased migratory capacity of prostate carcinoma cells. *Oncogene*, **20** (10), 1152–1163.
149. Park, S.I., Zhang, J., Phillips, K.A., Araujo, J.C. *et al.* (2008) Targeting SRC family kinases inhibits growth and lymph node metastases of prostate cancer in an orthotopic nude mouse model. *Cancer Res.*, **68** (9), 3323–3333.
150. Saad, F. (2009) Src as a therapeutic target in men with prostate cancer and bone metastases. *BJU Int.*, **103** (4), 434–440.
151. Sanjay, A., Houghton, A., Neff, L., DiDomenico, E. *et al.* (2001) Cbl associates with Pyk2 and Src to regulate Src kinase activity, alpha(v)beta(3) integrin-mediated signaling, cell adhesion, and osteoclast motility. *J. Cell Biol.*, **152** (1), 181–195.
152. Bolen, J.B., Rowley, R.B., Spana, C., and Tsygankov, A.Y. (1992) The Src family of tyrosine protein kinases in hemopoietic signal transduction. *FASEB J.*, **6** (15), 3403–3409.
153. Goldenberg-Furmanov, M., Stein, I., Pikarsky, E., Rubin, H. *et al.* (2004) Lyn is a target gene for prostate cancer: sequence-based inhibition induces regression of human tumor xenografts. *Cancer Res.*, **64** (3), 1058–1066.
154. Bates, R.C., Edwards, N.S., Burns, G.F., and Fisher, D.E. (2001) A CD44 survival pathway triggers chemoresistance via lyn kinase and phosphoinositide 3-kinase/Akt in colon carcinoma cells. *Cancer Res.*, **61** (13), 5275–5283.
155. Grishin, A.V., Azhipa, O., Semenov, I., and Corey, S.J. (2001) Interaction between growth arrest-DNA damage protein 34 and Src kinase Lyn negatively regulates genotoxic apoptosis. *Proc. Natl Acad. Sci. USA*, **98** (18), 10172–10177.
156. Roginskaya, V., Zuo, S., Caudell, E., Nambudiri, G. *et al.* (1999) Therapeutic targeting of Src-kinase Lyn in myeloid leukemic cell growth. *Leukemia*, **13** (6), 855–861.
157. Sumitomo, M., Shen, R., Walburg, M., Dai, J. *et al.* (2000) Neutral endopeptidase inhibits prostate cancer cell migration by blocking focal adhesion kinase signaling. *J. Clin. Invest.*, **106** (11), 1399–1407.
158. da Silva, H.B., Amaral, E.P., Nolasco, E.L., de Victo, N.C. *et al.* (2013) Dissecting major signaling pathways throughout the development of prostate cancer. *Prostate Cancer*, **2013** 920612.
159. Dhir, R., Ni, Z., Lou, W., DeMiguel, F. *et al.* (2002) Stat3 activation in prostatic carcinomas. *Prostate*, **51** (4), 241–246.
160. Venkitaraman, A.R. (2002) Cancer susceptibility and the functions of BRCA1 and BRCA2. *Cell*, **108** (2), 171–182.
161. Gao, B., Shen, X., Kunos, G., Meng, Q. *et al.* (2001) Constitutive activation of JAK-STAT3 signaling by BRCA1 in human prostate cancer cells. *FEBS Lett.*, **488** (3), 179–184.
162. Aziz, M.H., Manoharan, H.T., Church, D.R., Dreckschmidt, N.E. *et al.* (2007) Protein kinase C epsilon interacts with signal transducers and activators of transcription 3 (Stat3), phosphorylates Stat3Ser727, and regulates its constitutive activation in prostate cancer. *Cancer Res.*, **67** (18), 8828–8838.
163. Hafeez, B.B., Zhong, W., Weichert, J., Dreckschmidt, N.E. *et al.* (2011) Genetic ablation of PKC epsilon inhibits prostate cancer development and metastasis in transgenic mouse model of prostate

adenocarcinoma. *Cancer Res.*, **71** (6), 2318–2327.

164. Vazquez, F., Grossman, S.R., Takahashi, Y., Rokas, M.V. *et al.* (2001) Phosphorylation of the PTEN tail acts as an inhibitory switch by preventing its recruitment into a protein complex. *J. Biol. Chem.*, **276** (52), 48627–48630.

165. Wu, X., Hepner, K., Castelino-Prabhu, S., Do, D. *et al.* (2000) Evidence for regulation of the PTEN tumor suppressor by a membrane-localized multi-PDZ domain containing scaffold protein MAGI-2. *Proc. Natl Acad. Sci. USA*, **97** (8), 4233–4238.

4
Cancer Stem Cells*

Mei Zhang and Jeffrey M. Rosen
University of Pittsburgh, Department of Developmental Biology, A406, 204 Craft Avenue, Pittsburgh, PA 15213, USA

*This chapter has previously been published in: Meyers, R.A. (Ed.) Stem Cells, 2013, 978-3-527-32925-0.

Translational Medicine Cancer: Advances in Molecular Biology and Medicine,
First Edition. Edited by Robert A. Meyers.

Keywords

Cancer stem cell
A cell that can both self-renew, that is, regenerate itself, and differentiate along a particular lineage to give rise to the heterogeneity of cells within a tumor. The putative enrichment of such cells after chemotherapy and radiation treatment has been hypothesized to be responsible for cancer recurrence.

Tumor-initiating cells
Cells with an increased potential to initiate and propagate tumors in comparison to the bulk of the tumor cells with limited proliferative potential.

Wnt/ß-catenin signaling
This is activated through the accumulation of β-catenin in the nucleus, thus leading to an activation of the downstream targets, having been involved in the regulation of both normal and cancer stem cell activities.

Notch signaling
This is activated in many human cancers, and plays an important role in the regulation of asymmetric cell fate decisions in stem cells. It may also be the direct target of the microRNAs that are closely correlated with the invasion and acquisition of stem-like properties of the tumor cells.

Hedgehog signaling
This is activated through the binding of hedgehog ligands to derepress the Gli family of transcription factors and other transcriptional targets. It is involved in the maintenance and regeneration of adult tissues through the regulation of adult stem cells.

Radiation and chemotherapy resistance
An increased resistance to the treatments demonstrated by a subpopulation of cells as compared to other cell populations. It may contribute to relapse and metastasis of the tumors.

Limiting dilution transplantation
An assay commonly used to determine the frequency of self-renewing cells within a normal or tumor tissue after the dissociated cells have been transplanted into recipient animals at decreasing doses.

Sphere assay
An assay developed to study the "stemness" (sphere-forming) ability of both normal and tumor cells by plating the dissociated cells in suspension under serum-free medium conditions.

Cancer stem cells (CSCs) are a subpopulation of cells with the potential to both self-renew – that is, to regenerate themselves – and to differentiate along a particular lineage to give rise to the heterogeneity of cells within tumors. Due to their intrinsic properties, CSCs have reported to be enriched after chemotherapy and radiation treatment, and thus have been hypothesized to be responsible for cancer recurrence. CSCs also may enter the circulatory system and colonize at remote sites, resulting in metastasis. The cell of origin of CSCs remains controversial, as indicated by evidence suggesting an origin from normal stem cells, as well as dedifferentiation from more mature cells. CSCs actively interact with their surrounding microenvironment, the niche. Multiple signaling pathways, many of which are regulated by microRNAs, also have been demonstrated to regulate the activities of both normal and cancer stem cells. Therapeutic targeting of CSCs, as well as the bulk tumor cells, therefore, may be required to eradicate residual disease.

1 Introduction

Cancer stem cells (CSCs) are defined as those cells within a tumor that share properties similar to normal stem cells, such as self-renewal, and the ability to differentiate into the multiple cell types present in a tumor. However, this should not imply that CSCs can only arise from a normal stem cell, or are rare and quiescent. Functionally, CSCs have an increased potential to initiate and propagate tumors – hence, their designation as tumor-initiating cells (TICs) – in comparison to the bulk of the tumor cells, which have limited proliferative potential. CSCs are both experimentally and clinically resistant to radiation and chemotherapy, potentially resulting in residual disease which can lead to recurrence and metastasis [1–4]. Similar to their normal counterparts, CSCs rely on their microenvironment to sustain self-renewal and regulate differentiation [5, 6]. Certain pathways, which have been proven to play roles in controlling normal stem cell properties, also have been shown to be critical for CSCs [7, 8]. Targeting those signaling pathways (as discussed in detail below), as well as the interaction of CSCs with the microenvironment, may provide strategies to eliminate the CSC subpopulation [5, 6, 9–11]. In addition, noncoding RNAs – functional RNA molecules that do not encode proteins such as microRNAs (miRNAs) – have been shown to regulate the CSC properties of self-renewal and differentiation, and thus also provide a potentially new therapeutic approach [12].

2 Identification and Characterization of the CSCs

2.1 Identification

Initially, CSCs were functionally identified in acute myeloid leukemia (AML) by using antibodies which recognized specific cell-surface markers, followed by fluorescence-activated cell sorting (FACS) to isolate a subpopulation of cells capable of tumor formation after transplantation into immunodeficient mice [13–15]. There has, however, been some controversy as to whether there is a similar subpopulation of CSCs within all tumors, and whether the frequencies of these cells is an experimental limitation defined by the compromised microenvironment in which CSCs are engrafted in xenograft models using immunocompromised mice. For example, Kelly *et al.* observed that the majority of cells isolated from relatively homogeneous mouse lymphomas were able to initiate tumor formation [16]. However, by applying the strategy of transplanting FACS-sorted single cells into immunodeficient mice, CSCs have now been identified in many solid tumors, including cancers of the brain [17], breast [18], colon [19], ovary [20], pancreas [21], prostate [22, 23], skin [24, 25], bladder [26], and bone marrow [27, 28]. Thus, the CSC hierarchical model has been validated in a number of specific cases in addition to AML, and may account at least in part for the observed intra-tumoral heterogeneity that appears to be a property of many cancers.

Alternatively, brain tumor stem cells have been identified by culturing tumor cells under non-adherent serum-free conditions that were designed to enrich for the normal neural stem cells. The resulting spheroids, termed *neurospheres*, like their normal counterpart, expressed neural stem cells but lacked the expression of differentiated neuronal-specific markers [17, 29]. An analogous spheroid culture procedure designated as the *mammosphere assay* was developed by Dontu and colleagues to enrich for mammary gland stem cells [30], and this assay has subsequently been used widely as a surrogate for *in vivo* stem cell self-renewal. However, the clonal nature of the spheres has been controversial, as aggregations have been reported [31]. Therefore, disturbance-free culturing conditions, as well as low cell densities, were recommended when using sphere assays as an *in vivo* self-renewal surrogate [32].

2.2 Cell of Origin of CSCs

One fundamental question that remains to be answered is, "From where do CSCs originate?" Are these cells only derived from a normal stem cell, perhaps after the sequential accumulation of mutations, or can they arise from more differentiated cells with an acquired self-renewal property, perhaps in a manner analogous to that observed for the generation of induced pluripotent stem cells (iPSCs)? Direct evidence that normal stem cells can develop into CSCs came from a study using murine brain stem cells, of which overexpression of the nuclear receptor tailless, *Tlx*, only in the neural stem cells of adult mice resulted in a glioblastoma-like lesion; this eventually developed into glioblastomas upon loss of the tumor suppressor p53 [33]. Stem cells with an increased expression of *Tlx* promote blood vessel formation, thus enabling themselves to migrate and grow into a typical coral-like form of glioblastoma. Another study on the cell origin of the CSCs using lineage tracing methodology was performed by Han Clevers and colleagues, in which the deletion of adenomatous polyposis coli (Apc) in long-lived intestinal stem cells in a leucine-rich-repeat containing G-protein-coupled receptor 5 (Lgr5) knockin tamoxifen-inducible Cre recombinase mouse model led to the development of intestinal adenomas. In contrast, the deletion of Apc in short-lived transit-amplifying cells only resulted in the transient appearance of microadenomas, and very rare cases of large adenoma formation even after 30 weeks [34]. This provided evidence that stem cells are the cells of origin of intestinal cancer in this mouse model.

On the other hand, the potential role of de-differentiation has been illustrated by studies in which extracellular signals such as bone morphogenetic proteins (BMPs) and growth factors were able to de-differentiate non-stem cells into cells with stem-like characteristics [35, 36]. Parallel observations have also been reported in brain cancer cells [37]. The perivascular niche had the potential to drive the de-differentiation process, resulting in glioma cells with acquired long-term stem cell characteristics.

Determination of the cell of origin of CSCs should not rely only on histological markers of the tumor cells. While it had been thought that prostate cancers might arise from luminal cells that express prostate-specific antigen, Witte and colleagues have shown that basal cells isolated from primary benign human prostate tissue have an ability to develop into human prostate cancers without a basal phenotype, but with an expansion of luminal cells [38]. Similarly, most BRCA1 breast cancers display a basal-like phenotype. However, the deletion of *Brca1* in mouse mammary luminal progenitors resulted in tumors that phenotypically resembled human BRCA1 and sporadic basal-like tumors. In contrast, the deletion of *Brca1* in basal cells did not result in a similar phenotype, which suggested that there is potential plasticity of stem and progenitor cells and, again, that the tumor phenotype is not always equivalent to the cellular origin of the tumors [39]. Notwithstanding the findings of these studies, in the absence of appropriate lineage tracing experiments, it is not possible to definitively identify the cell of origin for most cancers.

2.3 Are CSCs a Rare Population within a Tumor?

The percentage of CSCs within human breast pleural effusions has been reported to range from 10% to 60%, using $CD44^{+}CD24^{-/low}$ as a marker of CSCs from these tumors [18]. A relatively rare molecular subtype of human breast cancer, the claudin-low type, has been identified and characterized [40, 41]. These tumors uniquely express low levels of tight junction as well as adherens junction genes, including claudin 3 and E-cadherin, and high levels of epithelial-to-mesenchymal transition (EMT) markers. In addition, they display CSC-like features. When compared to luminal A, luminal B, HER2-enriched, and basal-like human breast cancer subtypes, the claudin-low tumors have the highest expression of aldehyde dehydrogenase 1 family, member A1 (ALDH1A1) mRNA, a marker of breast stem/progenitor cells [42]. Such claudin-low tumors have also been identified in a number of distinct mouse models, including a p53-null transplant model with similar characteristics of high gene expression of EMT markers, and low to absent expression of both claudin 3 and E-cadherin [43]. Limiting dilution transplantation analysis has revealed that p53-null claudin-low tumors are highly enriched for CSCs as compared to other adenocarcinomas arising from the same model, which in turn suggests that cancer cells with stem-like properties may not necessarily comprise a rare population in some tumor types.

2.4 Quiescence of the Stem Cells

Mammalian adult stem cells are generally in a quiescent state, and such quiescence is critical for the maintenance and protection of the stem cell pool [44–46]. If CSCs represent "altered adult stem cells" they also might be quiescent, thus allowing them to evade most chemo- and radiotherapies. Tumorigenic $CD24^{+}$ cells with stem properties expressing stem cell-associated genes Nestin, Oct4, Notch 1, and Notch 4 have been suggested to display a quiescent phenotype in primary ovarian tumors [47]. In addition, a human normal mammary stem cell (hNMSC) signature generated from a group of slow-growing, and thus fluorescent label retaining cells (LRCs), was found to correlate with the degree of severity in primary breast tumors, with grade 3 tumors containing more LRCs than grade 1 tumors [48]. These authors were also able to show that breast cancer cells positive for the hNMSC signature were more tumorigenic as xenografts than those cells lacking the hNMSC signature. However, it should be noted that the hNMSC signature identified in these studies does not overlap with several other signatures derived from breast CSCs [49–51].

By using a lipophilic tracer dye, DiI ($C_{59}H_{97}ClN_2O_4$), a population of elongated fibroblast-like LRCs was identified from BxPC-3 and Panc03.27 pancreatic adenocarcinoma cells [52]. These LRCs, comprising approximately 3% of the total cell number, had an increased soft agar colony-forming ability, and displayed an increased invasive potential and xenograft tumor formation as compared to the non-LRCs. These cells also expressed a higher level of the EMT markers, vimentin and Twist, consistent with the EMT features of the stem-like cells. However, these LRCs were enriched, but only partially overlapped with the tentative pancreatic CSC populations identified by $CD24^+CD44^+$, $CD133^+$, and $ALDH1^+$. The functional role of these cell-surface markers in the LRC populations remains to be determined.

The question remains, however, as to whether CSCs always represent a quiescent population of cells in all tumors. Li and Clevers [53] have proposed a theory of coexistence of both quiescent and active adult stem cells in several tissues, based on the observation that both cycling yet long-lived populations of stem cells have been identified [54]. Unlike the more classical model, in which the balance between stem cell self-renewal and differentiation is controlled by asymmetric cell division [55–57], this model proposes that two types of stem cell – quiescent and active – are located in close proximity. Active stem cells play major roles in tissue regeneration, and can regulate the quiescent population through a negative feedback loop [53]. In this case, the "gold standard" limiting dilution transplantation assays, which are most commonly used in the characterization of stem cells from various tissues, might only identify active stem cells, without detecting quiescent stem cells. This may also hold true in the case of CSCs. Therefore, an efficient eradication of the cancer cells responsible for tumorigenesis may require combined strategies targeting different states of the CSCs.

2.5
Cell Plasticity

The stem cell hierarchy is usually thought of as a unidirectional program in which stem cells generate progenitors with a limited self-renewal ability, and then these produce more differentiated lineage cancer cells. However, studies with glioblastoma have identified a population of self-renewing $CD133^-$ cells, which can generate both $CD133^+$ and $CD133^-$ progeny [58]. Similarly, using BxPC-3 and Panc03.27 pancreatic adenocarcinoma cells, Dembinski and Krauss found that not only did LRCs have the capacity to produce non-LRCs, but that the non-LRCs could also produce LRCs. However, the LRCs exhibited a greater potential for tumor formation as xenografts than the non-LRCs, which suggested a dynamic exchange between these two types of cell [52]. By using the H3K4 demethylase, JARID1B, as a marker of slow-cycling cells in melanomas, Herlyn and colleagues [59] have reported that JARID1B-negative cells can become positive, demonstrating the existence of a dynamic switching between JARID1B-positive and -negative cells. Thus, this does not follow a hierarchically organized CSC model. Most recently, studies have shown that basal-like human mammary epithelial cells (HMECs) are able to spontaneously dedifferentiate into stem-like cells, and that transformed non-stem cancer cells can give rise to cells with CSC-like features [60]. These observations suggest there may be exceptions to the simple CSC hierarchical model, and that plasticity should also be considered as a factor which may influence tumor development.

2.6 Circulating Tumor Cells

Circulating tumor cells (CTCs) are extremely rare cells that have become detached from a primary solid tumor and circulate in the patient's blood. Dependent on the isolation techniques used, between one and 1,000 CTC per ml of blood may be detected in the circulation of patients with metastatic disease [61–63]. These cells are capable of initiating a clonal metastatic lesion when colonized, and maintain similar cytomorphologic features and transcriptional profiles when compared to original primary tumor cells [64, 65]. Previously, CTCs have been identified in prostate [66, 67], breast [68–70], and colon cancers [71]. $CD90^+$ cells have been isolated and characterized as CSCs in several hepatocellular carcinoma (HCC) cell lines, in human liver cancer specimens, and in blood samples from patients through serial transplantation into immunodeficient mice. A direct correlation of the number of $CD90^+$ in tumors and in blood has been reported [72]. In addition, a strong correlation between the number of the CTCs and the prognosis of metastatic breast cancer has also been shown [73].

Various methods have been designed to capture and characterize CTCs [74], based on the cells' unique physical properties (e.g., size and density), as well as fiber-optic array-scanning technology (FAST) [75–77], laser-scanning cytometry [78], and surface-enhanced Raman spectroscopy (SERS) [79]; in addition, antibodies against cell-surface antigens have also been used [e.g., anti-epithelial cell adhesion molecule (EpCAM)-coated CTC-chip] [80]. In the latter case, the use of the epithelial marker, EpCAM, may not allow for the isolation of CTCs which have undergone a transient EMT. Notwithstanding, improved methods for the isolation of CTCs are currently under development, and this should facilitate their isolation from the circulation, allowing further characterization of the tumorigenic potential of such cells.

3 Signaling Pathways and Factors Involved in the Regulation of CSCs

A number of signaling pathways, including Wnt, Notch, Hedgehog, and PI3K/Akt/PTEN, are important in the regulation of normal adult stem cells [81]. The abnormal regulation of these pathways stimulates neoplastic proliferation in the same tissues [81, 82]. In addition, different signaling pathways may be active in CSCs of different tumor subtypes, reflecting the roles of these pathways in the cell of origin of various tumors.

3.1 Wnt/β-Catenin Canonical Signaling Pathways

In the absence of Wnt ligands, cytoplasmic β-catenin complexes with Axin, APC, GSK-3, and CK1 to cause phosphorylation and the proteosomal degradation of β-catenin. In the presence of Wnt ligands, receptors such as the lipoprotein receptor-related protein 5/6 (LRP5/6) and Frizzled activate the key regulator Disheveled to inhibit the activity of Axin, APC, GSK-3, and CK1. This leads to a cytoplasmic accumulation of stabilized β-catenin and its translocation to the nucleus. In the nucleus, β-catenin binds with T-cell factor/lymphoid enhancer factor (TCF/LEF) family members and acts as a transcriptional activator to regulate target genes.

This canonical Wnt/β-catenin signaling pathway is well known for its roles in the regulation of stem cell activities, and its activation has been involved in various cancers [83]. For example, perturbation of the Wnt signaling pathway has been shown to lead to hyperplasia and mammary tumor development in mice [84, 85]. In the Pten heterozygous background, the induction of Wnt-1 in the mammary glands of transgenic mice resulted in the formation of tumors containing both myoepithelial and luminal epithelial cells with a common Pten loss mutation, suggesting a common cell of origin within such tumors. In contrast, the induction of Neu, H-Ras, or polyoma middle T antigen diminished the myoepithelial cell population, which suggests that mammary stem cells and/or progenitors may be the targets for Wnt1-induced oncogenesis [86]. Similarly, as noted in Sect. 2.2, intestinal macroadenomas can only be generated by deletion of the Wnt pathway gene Apc in intestinal stem cells, but not in short-lived transit-amplifying cells [34]. In addition, interference of the Wnt signaling pathway in cancers of the skin, prostate, and intestine has resulted in the loss of self-renewal ability and an increased differentiation of the CSCs, as demonstrated by both *in vivo* and *in vitro* assays. This suggests that Wnt signaling plays a role in maintaining the stem cell function of CSCs in many cancers [87–89].

Wnt signaling also plays a role in the interaction with the niche microenvironment to sustain the stem cell pool (also see Sect. 4). For example, Wnt signaling has defined a CSC population in close proximity to the stromal microenvironment in colon cancer. Through regulation of the factors secreted by the microenvironment, β-catenin-dependent transcription was activated, which led to tumor formation originating from CSCs [90].

3.2 Hedgehog (Hh) Signaling Pathway

To date, three mammalian Hedgehog ligands have been identified, of which Sonic Hedgehog is the best studied. In the absence of Hh ligand, its receptor Patched (Ptch) represses Smoothened (Smo). Binding of Hh removes this repression, allowing Smo to activate the Gli family of transcription factors and other transcriptional targets.

Hh signaling is involved in the maintenance and regeneration of adult tissues through the regulation of adult stem cells. Hh signaling also has been implicated in the development of cancers through the regulation of CSCs [91, 92]. Chemoresistance of Bcr-Abl-positive leukemic stem cells (LSCs) has been proposed as the cause of relapse in patients with chronic myeloid leukemia (CML). Hh signaling is active in LSCs through an upregulation of Smo. The conditional deletion or chemical inhibition of Smo results in a reduced number of CML stem cells, a reduced potential of Bcr-Abl-positive leukemias to re-grow after transplantation, and a delayed relapse after treatment cessation [93, 94]. Conversely, constitutively active Smo increased the number of CML stem cells and led to a more rapid disease occurrence [94]. Likewise, addition of the Hh ligand increased the self-renewal ability of a multiple myeloma stem cell population, while inhibition of the Hh activity with the Hh inhibitor, cyclopamine or with a blocking antibody 5E1 greatly reduced the clonal expansion of the multiple myeloma stem cells, and instead promoted their terminal differentiation [95].

The self-renewal potential of glioma CSCs is also regulated by Hh-Gli signaling. The Hh effector Gli was identified initially because it was overexpressed in some primary glioblastoma multiforme (GBM) tumors and GBM cell lines [34, 91, 96]. Blocking the function of the Hh-Gli pathway with cyclopamine or through lentiviral-mediated silencing of Smo prevented the growth of human glioma xenografts and neurosphere formation, which suggested the importance of Hh signaling in brain tumor development.

Hh signaling may also regulate stem cell functions through the interaction with stem cell-related factors, such as Bmi1. A high expression of components of the Hh signaling pathway (Ptch1, Gli1, and Gli2) regulated Bmi-1 in stem/progenitor-enriched mammospheres. Expression changes in Bmi-1 thus affecting Hh signaling also resulted in an altered mammary development in a humanized non-obese diabetic-severe combined immunodeficient (NOD/SCID) mouse model [92].

Stem cells have been reported to express an EMT signature [97, 98]. Activation of the Hh pathway results in an increased EMT phenotype (increased expression of Snail and reduced expression of E-cadherin) [89]. The inhibition of Hh signaling by cyclopamine caused a marked reduction of the EMT phenotype, and a reduction in the invasive potential in pancreatic cancer cell lines, as well as a dramatic inhibition of metastasis in an orthotopic pancreatic xenograft model. The overexpression of Gli1 resulted in an enhanced invasive potential, and almost no expression of E-cadherin, suggesting that Gli1 is involved in the process of EMT [99].

3.3 Notch Signaling Pathway

The Notch receptor is a single-pass transmembrane receptor. In the absence of Notch signaling, target genes are repressed as the DNA-binding protein, CSL (CBF1/RBP-Jk, Suppressor of Hairless, LAG-1) is inhibited by transcriptional repressors. In mammals, binding of the transmembrane Notch ligands, Delta-like, and Jagged, initiates proteosomal cleavages mediated by ADAM (a disintegrin and metalloprotease) family proteases and γ-secretase. This frees the Notch intracellular domain, which then translocates to the nucleus, binds to CSL and to the Notch coactivator, Mastermind, to upregulate transcription of the Hes (Hairy/Enhancer of Split) and Hey (Hairy/Enhancer of Split related with YRPW) gene families.

Notch has been reported to play a key role, in the regulation of asymmetric cell fate decisions in human mammary stem cells [100–104]. The deregulation of Notch signaling can cause cancer; for example, expression of the activated form of the Notch1 receptor, Notch1 receptor intracellular domain (N1IC), in mammary cells of mouse mammary tumor virus (MMTV)/N1IC transgenic mice increased the survival potential of the $CD24^{+}$ $CD29^{high}$ progenitor cells. An expansion of such progenitor cells led to the formation of basal-like ductal tumors through a cyclin D1-dependent pathway [105].

Notch signaling is also activated in many human cancers, with approximately 50% of human breast cancers showing defects in Notch signaling. The loss of microRNA (miR)-200 with an acquired

EMT signature in tumor cells usually correlates with the invasion and acquisition of stem-like properties [106]. In a study using human metastatic prostate cancer cells, miR-200c and miR-141 each directly inhibited expression of the Notch ligand Jag1, thus reducing cell proliferation. The loss of miR-200 in these cells may increase Jag1 activity through regulation of the transcription factor, ZEB1, which further promoted interactions between tumor cells [107]. This hypothesis has been corroborated by studies of aggressive pancreatic adenocarcinomas and the basal type of breast cancer, in which reduced miR-200 expression coordinated the upregulation of ZEB1/Jag1-mediated Notch activation [108].

3.4 PI3K/Akt/PTEN Signaling Pathway

The activation of the PI3K/Akt/PTEN signaling pathway in stem cells has been inferred by the observation of increased expression levels of PI3K p110α- and β-proteins in cells grown under conditions favoring sphere formation. Deletion of the key negative regulator in the Akt pathway, PTEN, by shRNA knockdown in prostate cancer cells resulted in the subsequent activation of the PI3K/Akt pathway, and a higher potential for both *in vitro* sphere formation and *in vivo* tumor growth than in the control cells [109]. Similarly, brain cells isolated from the Pten knockout mice, as well as neurospheres derived from Pten-deficient mice, showed an increased self-renewal capacity as compared to their respective controls [110].

Wicha and colleagues have shown that the PTEN/Akt/β-catenin signaling pathway regulates mammary stem/progenitor self-renewal through the phosphorylation of glycogen synthase kinase 3β (GSK3β) [9]. Akt activation can also directly phosphorylate, and thus activate, β-catenin at Ser552 [8]. A study using a p53-null mouse model of breast cancer also reported the downregulation of PTEN and an increased activity of phospho-AKT in freshly isolated CSCs and CSCs-enriched mammospheres, as compared to non-CSCs and total tumor cells depleted of CSCs [10].

Side population (SP) cells with a higher ability to efflux the DNA-binding Hoechst dye through the ATP-binding cassette sub-family G member 2 (ABCG2) transporter exhibit increased non-adherent sphere formation potential, as well as a higher frequency of tumor formation and a decreased latency after *in vivo* transplantation [111–113]. The PI3K/Akt pathway also regulates the transporter activity of ABCG2 in the stable cell line U87MG, and that the loss of PTEN led to an increased percentage of SP cells [113].

3.5 Tumor Suppressor p53

Self-renewal is an important property of stem cells required to maintain the stem cell pool [114]. Increasing evidence has shown that the tumor suppressor p53 regulates stem cell self-renewal. Compared to p53 wild-type, p53 deficiency promoted stem cell self-renewal, as demonstrated by increased mammosphere formation and a higher tumorigenic potential after transplantation into recipient mice. An increased tumorigenesis due to p53 deficiency was observed in hematopoietic cells, in an ErbB2 transgenic model of breast cancer, and in brain tumors [115–117]. A reduced expression of the cell cycle regulator p21 was detected in p53-null neurospheres [117], which suggested that dysregulated p53 cell cycle control can result in an imbalance between symmetric and asymmetric division. The pharmacological activation of p53 has been reported to induce asymmetric division in CSCs and decreased tumor development, without affecting other cell types [116]. In addition, RNA interference (RNAi) knockdown of p53 has been shown to result in increased self-renewal of $Kras^{G12D}$-expressing myeloid progenitor cells *in vitro* [118]. Further studies on the mechanisms by which p53 regulates stem cell self-renewal have revealed an important role for miRNAs (see Sect. 3.6).

3.6
MicroRNAs (miRNAs)

miRNAs are small noncoding RNA molecules, averaging 22 nucleotides in length, that bind to their target mRNAs to repress gene expression. An intimate relationship among the expression of certain miRNAs, EMT, and stem cell activity has been established in the past few years [12, 108, 119, 120]. For example, let-7 miRNAs and miRNA (miR)-93 are markedly downregulated in a population of self-renewing progenitor cells isolated from a mouse mammary epithelial cell line, Comma-Dβ. The expression of let-7 causes the loss of self-renewal in these progenitors [121]. Likewise, studies using breast cancer lines and primary cancer cells have shown that let-7 miRNAs are decreased in CSCs and increased with differentiation. When let-7 is re-expressed in CSCs using a lentiviral construct, the CSCs displayed a reduced proliferation, a decreased mammosphere formation, and a reduced ability both to form tumors and to develop metastasis in NOD/SCID mice [122], which suggests an important role of let-7 miRNAs in regulating breast CSCs. In addition, three clusters of miRs – miR-200c-141, miR-200b-200a-429, and miR-183-96-182 – were downregulated in human breast CSCs [12]. Among these, miR-200c greatly reduced the tumorigenic potential of the human breast CSCs, as demonstrated by a limiting dilution assay using immunodeficient mice. In another study using MCF-10A cells expressing tamoxifen-inducible ER-Src, a miR-200b-Suz12-cadherin pathway has been proposed, in which miR-200 decreases the expression and binding of polycomb repressor complex 2 (PRC2) subunit Suz12, and promotes CSC formation and growth [123].

The loss of p53 reduces miR-200c expression, thus triggering EMT and an expansion of the mammary stem cell pool. Reactivation of miR-200c was able to block EMT and induce differentiation of the mammary cells [124]. MiR-34a, also a direct transcriptional target of p53, was downregulated in $CD44^+$ prostate CSCs isolated from both xenograft and primary tumors, while the overexpression of miR-34a in such cells blocked tumorigenesis, and metastasis. In this study, CD44 has been identified as a direct target of miR-34a, and CD44 knockdown resembled miR-34a overexpression in blocking prostate tumorigenesis and metastasis [125]. On the other hand, miR-34, through regulation of its other downstream targets Bcl-2 and Notch, is able to partially restore the tumor suppressor function of p53, and thus to regulate p53-deficient human pancreatic CSC self-renewal [126].

Other factors that play roles in the functions of CSCs include the signal transducer and activator of transcription 3 (STAT3) that regulates the self-renewal of glioblastoma stem cells (GBM-SCs) [127]. Inhibitor targeting STAT3 blocked GBM-SC self-renewal potential, as shown by a decreased ability to form neurospheres. The polycomb group transcriptional repressor Bmi-1 plays an important role in the regulation of hematopoietic, neural, mammary, and prostate stem cells, and their malignant transformation [92, 128–130]. In a Pten-null prostate cancer model, Bmi-1 knockdown through siRNA prevented the development of hyperplasia, as well as prostate cancer growth induced by Pten deletion [130].

4
CSCs and the Microenvironment

As discussed above, a reservoir population of adult mammalian stem cells is thought to exist in a quiescent state. The quiescence of adult stem cells is regulated by their specific microenvironment, the niche, which is in close proximity to stem cells, and is critical to maintain and to protect the stem cell pool [44–46, 131]. Similarly, CSCs also rely on the CSC niche to retain their self-renewal and multipotency [132, 133]. For example, stromal/fibroblast cells in cooperation with hematopoietic progenitor cells provide a niche that is required for tumor cell metastasis to the lung [134]. Therefore, tumor growth involves not only the biology of tumor cells themselves, but also the complex interaction between tumor cells and the tumor microenvironment. Moreover, the CSC niche may also contribute to the enhanced therapy resistance by providing a protective role to the CSCs against chemotherapy [135, 136].

The intestinal stem cell niche is one of the most widely studied niche microenvironments. Both the Wnt and BMP pathways play important roles, the latter having been shown to play a suppressive role, involved in maintaining stem cell pool [137–140]. Similar to the normal intestinal stem cell niche, the colon cancer niche is composed of myofibroblasts and mesenchymal stem cells which support tumor growth and metastasis [141]. Consistently, a high Wnt pathway activity defines a colon CSC population, which is regulated by hepatocyte growth factor (HGF) produced by myofibroblasts residing in the tumor microenvironment [90]. A well-established connection between inflammatory processes and colon cancer has also been described in models of colitis-associated cancer, where interleukin 6 produced by the infiltrating immune cells enhances the proliferation of colon cancer-initiating cells [142].

Several factors secreted by the niche cells play a critical role in regulating CSCs. For example, vascular endothelial growth factor (VEGF), pigment epithelium-derived factor, and stem cell factor all serve as niche factors for neural stem cell renewal (symmetric division), and stimulate neurogenesis (asymmetric cell division) [143–147]. Similarly, vascular endothelial cells provide a niche microenvironment for patient-derived brain tumor cells, maintaining the stem-like features of the brain CSCs, as well as promoting tumorigenicity when cotransplanted in immunocompromised mice [132]. A soluble factor, nitric oxide (NO), as secreted by endothelial cells, mediates the CSCs–niche interaction and regulates CSC self-renewal through activated Notch signaling in a platelet-derived growth factor (PDGF)-induced glioma mouse model [37]. The inhibition of Notch signaling in a three-dimensional explant culture model of surgical GBM specimens resulted in loss of CSC attachment to their vascular niche, and an enhanced efficacy of radiation therapy of CSCs [135]. Interestingly, brain CSCs can also promote the formation of blood vessels by secreting VEGF [148], which suggests the existence of a reciprocal association between the CSCs and the microenvironment.

Hypoxic tumor cells are relatively resistant to both chemotherapy and radiation therapy, due to their location and reduced formation of oxygen free radicals that induce DNA damage [149]. Accordingly, GBM CSC maintenance is regulated by changes in oxygen levels in the cellular environment through hypoxia-inducible factors (HIFs), of which HIF2a has been suggested to play an important role for GBM CSC self-renewal and tumorigenic potential [150]. Taken together, these data support the role of the microenvironment/niche in the maintenance CSCs, and their treatment resistance.

5 Radiation and Chemotherapy Resistance of CSCs

Some adult stem cells are thought to be quiescent and resistant to radiation and chemotherapeutic agents. Likewise, CSCs have similar properties and are able to efflux chemotherapeutic drugs. Furthermore, CSCs exhibit increased levels and activation of DNA repair proteins.

The most direct clinical evidence for the chemotherapy resistance of CSCs has been derived from studies with breast cancer. Breast CSCs $CD44^+CD24^{-/low}$ were isolated from paired core needle biopsies of primary breast cancer patients before and after neoadjuvant chemotherapy. The chemotherapy led to an increased percentage of $CD44^+CD24^{-/low}$ cells and an increased self-renewal capacity, as measured using mammosphere formation assays [3]. Additional evidence for the chemoresistance of CSCs has also been obtained from studies using cancer cell lines. For example, after chemotherapeutic treatment, $CD133^+$ and $CD44^+$ cells (both are putative colorectal CSCs), were enriched 16- to 30-fold and twofold, respectively, and these cells displayed a reduced proliferation [151]. Putative HCC CSCs, identified as $CD133^+$ cells, that were isolated from both a human HCC cell line and xenograft mouse models, were also enriched following chemotherapy through activation of the Akt/PKB- and Bcl-2-mediated cell survival signaling pathways [152]. Furthermore, a stem cell-enriched spheroid population isolated from GBM U87MG cells displayed a stronger drug resistance to conventional anticancer drug treatment, as compared to the parental cells. Such spheroids showed both increased gene expression and immunohistochemical staining for MDR1 (multidrug resistance 1) and CD133 (a marker for a series of leukemia and GBM CSCs), suggesting that such stem cell-enriched spheroids may be resistant to current chemotherapy [153].

The cell cycle inhibitor p21 has been shown to prevent LSCs from accumulating DNA damage and to sustain the stemness of the tumor stem cells [154]. In solid tumor studies, CSCs identified from GBM had a higher DNA repair rate, and were, therefore, more resistant to ionizing radiation treatment than were more differentiated glioma cells [148]. Breast CSC/progenitor cells may also be enriched after radiation, because such cells are particularly resistant to radiation [1, 2, 4]. For instance, the percentage of a tumorigenic CSC subpopulation from a p53-null mouse model of breast cancer was increased after radiation treatment [10]. Similarly, radiation treatment of the GBM neurospheres has also led to an enrichment of such stem-like cells, suggesting that such therapy preferentially targets more differentiated neoplastic cells but not the CSCs [96]. In addition, Diehn *et al.* suggested that the CSCs isolated from a genetically engineered MMTV-Wnt-1 mouse model contained a lower level of reactive oxygen species (ROS) than did the bulk of tumor cells, which might contribute to their radioresistance [155]. Furthermore, using a human breast tumor xenograft model, an increased percentage of the ALDH$^+$ breast CSCs was observed after radiation therapy [11], supporting the hypothesis that the treatment resistance of CSCs may lead to tumor relapse.

6
Targeting CSCs

6.1
Targeting CSCs and Their Microenvironment

The existence of CSCs, and the implication that they may contribute to both chemo- and radio- resistance, have led to changes in the methods used not only for drug evaluation but also to monitor progression after treatment, and of strategies to prevent metastasis.

A number of signaling pathways are aberrantly activated in CSCs. Examples of the approaches targeting such pathways are:

- Pharmacologic Hh pathway inhibition with cyclopamine has been shown to reduce the *in vitro* invasiveness of pancreatic cancer cells, and the potential to develop pulmonary micrometastases in a pancreatic cancer xenograft model [99]. Cyclopamine and Smo inhibition also have been used to target chemo-resistant mouse and human CML and to reduce the LSC pool in CML [93, 94].
- The PTEN/Akt/β-catenin signaling pathway regulates both human and mouse mammary stem/progenitor self renewal, either through the phosphorylation of GSK3β or direct phosphorylation and activation of β-catenin at Ser552 [9, 10]. The inhibition of Akt and TORC1 and TORC2, using Perifosine, was able to block pathway activation and reduce tumor growth in both primary breast tumor xenografts and the p53-null mammary tumor model. Due to the complexity of signaling pathways in CSCs, a combinatorial treatment strategy may be needed to efficiently eliminate CSCs.

Various alternatives have been investigated in targeting CSCs, a few of which are listed below. A novel application using nanotechnology with intravenously administered, optically activated gold nanoshells also has sensitized breast CSCs to radiation therapy [11]. Since miRNAs, such as miR-34a, play an important role in the regulation of stem cell activity, the modulation of miRNAs or agents targeting these small noncoding RNAs provides another potential therapeutic resource to suppress CSCs [124, 125]. However, as normal and CSCs may share similar regulatory pathways, toxicity in mice receiving systemically delivered miR-34a will need to be assessed.

Tumor-associated myofibroblasts (TAFs) have been shown to maintain stem cell-like features of colorectal carcinoma cell lines [156], and are a major cellular component of colon cancer stroma. The targeting of TAFs has also been shown to improve the efficacy of chemotherapy [157]. In addition, the inhibition of angiogenesis reduced the number of CSCs, blocked tumor growth [132], and resulted in a higher susceptibility of CSCs to cytotoxic drugs. This suggests that targeting the microenvironment may represent another effective strategy for combating cancer.

6.2
Targeting CSCs while Sparing the Normal Stem Cells

Since many of the regulatory pathways required to target CSCs are also present in normal stem cells, an appropriate technology to selectively inhibit CSCs, such as miRNA-based therapeutic agents, is urgently required. However, it will be critical first to identify differences between these populations that may provide a therapeutic window [158].

For example, cutaneous CSCs rely on Wnt/β-catenin signaling to maintain their stemness, while normal epidermal homeostasis does not [87]. Thus, targeting of the canonical Wnt pathway in this case may provide a potentially selective target to eradicate cutaneous CSCs. The conditional deletion of Pten in adult hematopoietic cells led to the generation of transplantable leukemia-initiating cells, which subsequently could develop into myeloproliferative disease and later progress to leukemias. At the same time, Pten deletion also caused hematopoietic stem cell (HSC) proliferation, and, therefore, failed to maintain long-term multilineage reconstitution ability. The mTOR inhibitor rapamycin was able to deplete leukemia-initiating cells and to restore normal HSC function [159]. High-throughput screens have identified small molecules (e.g., salinomycin) which selectively targeted the EMT pathways in breast CSCs. This was demonstrated by a 20-fold reduction of breast CSCs, $CD44^{high}CD24^{low}$ cells, as compared to the vehicle-treated controls [160]. Gene sets associated with the stem cell properties as previously identified [30, 49, 50] were also reduced in the salinomycin-treated group.

Some tumor cell lines are heterogeneous and contain subpopulations of CSCs. For example, human breast cancer cell lines, MCF7, SUM149, SUM159, SUM1315, MDA.MB.231, and the mouse mammary cell line, CommaD, contain heterogeneous populations, including cells that can self-renew and can therefore reconstitute the parental line. These cell lines also contain subpopulations of cells that are resistant to chemotherapy and radiation, and as such may be useful for high-throughput drug screening studies [1, 160, 161]. The results of these screens will need to be validated in appropriate *in vivo* preclinical models, however.

7
Conclusions

The identification of the CSCs, first in the hematopoietic system and then in solid tumors, represents a major paradigm change influencing the strategies used for future cancer treatments. Despite controversies related to the existence and frequency of CSCs in some tumors (e.g., melanoma [25, 162]), increasing evidence from both the laboratory and the clinic has clearly shown the existence of intratumoral heterogeneity. This heterogeneity appears to be due in part to a hierarchy in which a tumorigenic subpopulation that is more resistant to chemo- and radiation therapies with stem cell properties has the potential to generate the other cell types present within the tumor. However, this hierarchy may not be unidirectional [60], and the role of plasticity in generating CSCs and especially treatment resistance still requires further investigation. The limiting dilution transplantation assay into immunodeficient mice is considered as the "gold standard" to assess various human CSCs populations. However, technical issues, such as the mechanical and enzymatic methods used to dissociate primary solid tumors, and whether the mouse or even the humanized recipient mouse microenvironment play a role in the human xenograft tumor initiation and sustained growth, are important caveats that must be considered when interpreting these experiments. Finally, a more complete understanding of CSCs will also require a more detailed genetic and epigenetic analysis at the single-cell level, because the CSC analyses performed to date have been conducted on heterogeneous cell populations.

References

1. Chen, M.S., Woodward, W.A., Behbod, F., Peddibhotla, S. *et al.* (2007) Wnt/beta-catenin mediates radiation resistance of Sca1+ progenitors in an immortalized mammary gland cell line. *J. Cell Sci.*, **120**, 468–477.
2. Debeb, B.G., Xu, W., and Woodward, W.A. (2009) Radiation resistance of breast cancer stem cells: understanding the clinical framework. *J. Mammary Gland Biol. Neoplasia*, **14**, 11–17.
3. Li, X., Lewis, M.T., Huang, J., Gutierrez, C. *et al.* (2008) Intrinsic resistance of tumorigenic breast cancer cells to chemotherapy. *J. Natl. Cancer Inst.*, **100**, 672–679.
4. Woodward, W.A., Chen, M.S., Behbod, F., Alfaro, M.P. *et al.* (2007) WNT/beta-catenin mediates radiation resistance of mouse mammary progenitor cells. *Proc. Natl Acad. Sci. USA*, **104**, 618–623.
5. Ginestier, C., Liu, S., Diebel, M.E., Korkaya, H. *et al.* (2010) CXCR1 blockade selectively targets human breast cancer stem cells in vitro and in xenografts. *J. Clin. Invest.*, **120**, 485–497.
6. Fillmore, C.M., Gupta, P.B., Rudnick, J.A., Caballero, S. *et al.* (2010) Estrogen expands breast cancer stem-like cells through paracrine FGF/Tbx3 signaling. *Proc. Natl Acad. Sci. USA*, **107** (50), 21737–21742.
7. Zeng, Y.A. and Nusse, R. (2010) Wnt proteins are self-renewal factors for mammary stem cells and promote their long-term expansion in culture. *Cell Stem Cell*, **6**, 568–577.
8. He, X.C., Yin, T., Grindley, J.C., Tian, Q. *et al.* (2007) PTEN-deficient intestinal stem cells initiate intestinal polyposis. *Nat. Genet.*, **39**, 189–198.
9. Korkaya, H., Paulson, A., Charafe-Jauffret, E., Ginestier, C. *et al.* (2009) Regulation of mammary stem/progenitor cells by PTEN/Akt/beta-catenin signaling. *PLoS Biol.*, **7**, e1000121.
10. Zhang, M., Atkinson, R.L., and Rosen, J.M. (2010) Selective targeting of radiation-resistant tumor-initiating cells. *Proc. Natl Acad. Sci. USA*, **107**, 3522–3527.
11. Atkinson, R.L., Zhang, M., Diagaradjane, P., Peddibhotla, S. *et al.* (2010) Thermal enhancement with optically activated gold nanoshells sensitizes breast cancer stem cells to radiation therapy. *Sci. Transl. Med.*, **2**, 55ra79.
12. Shimono, Y., Zabala, M., Cho, R.W., Lobo, N. *et al.* (2009) Downregulation of miRNA-200c links breast cancer stem cells with normal stem cells. *Cell*, **138**, 592–603.
13. Hope, K.J., Jin, L., and Dick, J.E. (2004) Acute myeloid leukemia originates from a hierarchy of leukemic stem cell classes that differ in self-renewal capacity. *Nat. Immunol.*, **5**, 738–743.
14. Lapidot, T., Sirard, C., Vormoor, J., Murdoch, B. *et al.* (1994) A cell initiating human acute myeloid leukaemia after transplantation into SCID mice. *Nature*, **367**, 645–648.
15. Bonnet, D. and Dick, J.E. (1997) Human acute myeloid leukemia is organized as a hierarchy that originates from a primitive hematopoietic cell. *Nat. Med.*, **3**, 730–737.
16. Kelly, P.N., Dakic, A., Adams, J.M., Nutt, S.L. *et al.* (2007) Tumor growth need not be driven by rare cancer stem cells. *Science*, **317**, 337.
17. Singh, S.K., Clarke, I.D., Terasaki, M., Bonn, V.E. *et al.* (2003) Identification of a cancer stem cell in human brain tumors. *Cancer Res.*, **63**, 5821–5828.
18. Al-Hajj, M., Wicha, M.S., Benito-Hernandez, A., Morrison, S.J. *et al.* (2003) Prospective identification of tumorigenic breast cancer cells. *Proc. Natl Acad. Sci. USA*, **100**, 3983–3988.
19. O'Brien, C.A., Pollett, A., Gallinger, S., and Dick, J.E. (2007) A human colon cancer cell capable of initiating tumour growth in immunodeficient mice. *Nature*, **445**, 106–110.
20. Zhang, S., Balch, C., Chan, M.W., Lai, H.C. *et al.* (2008) Identification and characterization of ovarian cancer-initiating cells from primary human tumors. *Cancer Res.*, **68**, 4311–4320.
21. Li, C., Heidt, D.G., Dalerba, P., Burant, C.F. *et al.* (2007) Identification of pancreatic

cancer stem cells. *Cancer Res.*, **67**, 1030–1037.

22. Maitland, N.J. and Collins, A.T. (2008) Prostate cancer stem cells: a new target for therapy. *J. Clin. Oncol.*, **26**, 2862–2870.
23. Lang, S.H., Frame, F.M., and Collins, A.T. (2009) Prostate cancer stem cells. *J. Pathol.*, **217**, 299–306.
24. Schatton, T., Murphy, G.F., Frank, N.Y., Yamaura, K. *et al.* (2008) Identification of cells initiating human melanomas. *Nature*, **451**, 345–349.
25. Boiko, A.D., Razorenova, O.V., van de Rijn, M., Swetter, S.M. *et al.* (2010) Human melanoma-initiating cells express neural crest nerve growth factor receptor CD271. *Nature*, **466**, 133–137.
26. Chan, K.S., Espinosa, I., Chao, M., Wong, D. *et al.* (2009) Identification, molecular characterization, clinical prognosis, and therapeutic targeting of human bladder tumor-initiating cells. *Proc. Natl Acad. Sci. USA*, **106**, 14016–14021.
27. Matsui, W., Huff, C.A., Wang, Q., Malehorn, M.T. *et al.* (2004) Characterization of clonogenic multiple myeloma cells. *Blood*, **103**, 2332–2336.
28. Matsui, W., Wang, Q., Barber, J.P., Brennan, S. *et al.* (2008) Clonogenic multiple myeloma progenitors, stem cell properties, and drug resistance. *Cancer Res.*, **68**, 190–197.
29. Hemmati, H.D., Nakano, I., Lazareff, J.A., Masterman-Smith, M. *et al.* (2003) Cancerous stem cells can arise from pediatric brain tumors. *Proc. Natl Acad. Sci. USA*, **100**, 15178–15183.
30. Dontu, G., Abdallah, W.M., Foley, J.M., Jackson, K.W. *et al.* (2003) In vitro propagation and transcriptional profiling of human mammary stem/progenitor cells. *Genes Dev.*, **17**, 1253–1270.
31. Mori, H., Fujitani, T., Kanemura, Y., Kino-Oka, M. *et al.* (2007) Observational examination of aggregation and migration during early phase of neurosphere culture of mouse neural stem cells. *J. Biosci. Bioeng.*, **104**, 231–234.
32. Coles-Takabe, B.L., Brain, I., Purpura, K.A., Karpowicz, P. *et al.* (2008) Don't look: growing clonal versus nonclonal neural stem cell colonies. *Stem Cells*, **26**, 2938–2944.
33. Liu, H.K., Wang, Y., Belz, T., Bock, D. *et al.* (2010) The nuclear receptor tailless induces long-term neural stem cell expansion and brain tumor initiation. *Genes Dev.*, **24**, 683–695.
34. Barker, N., Ridgway, R.A., van Es, J.H., van de Wetering, M. *et al.* (2009) Crypt stem cells as the cells-of-origin of intestinal cancer. *Nature*, **457**, 608–611.
35. Kondo, T. and Raff, M. (2000) Oligodendrocyte precursor cells reprogrammed to become multipotential CNS stem cells. *Science*, **289**, 1754–1757.
36. Sharif, A., Legendre, P., Prevot, V., Allet, C. *et al.* (2007) Transforming growth factor alpha promotes sequential conversion of mature astrocytes into neural progenitors and stem cells. *Oncogene*, **26**, 2695–2706.
37. Charles, N., Ozawa, T., Squatrito, M., Bleau, A.M. *et al.* (2010) Perivascular nitric oxide activates notch signaling and promotes stem-like character in PDGF-induced glioma cells. *Cell Stem Cell*, **6**, 141–152.
38. Goldstein, A.S., Huang, J., Guo, C., Garraway, I.P. *et al.* (2010) Identification of a cell of origin for human prostate cancer. *Science*, **329**, 568–571.
39. Molyneux, G., Geyer, F.C., Magnay, F.A., McCarthy, A. *et al.* (2010) BRCA1 basal-like breast cancers originate from luminal epithelial progenitors and not from basal stem cells. *Cell Stem Cell*, **7**, 403–417.
40. Herschkowitz, J.I., Simin, K., Weigman, V.J., Mikaelian, I. *et al.* (2007) Identification of conserved gene expression features between murine mammary carcinoma models and human breast tumors. *Genome Biol.*, **8**, R76.
41. Prat, A., Parker, J.S., Karginova, O., Fan, C. *et al.* (2010) Phenotypic and molecular characterization of the claudin-low intrinsic subtype of breast cancer. *Breast Cancer Res.*, **12**, R68.
42. Ginestier, C., Hur, M.H., Charafe-Jauffret, E., Monville, F. *et al.* (2007) ALDH1 is a marker of normal and malignant human mammary stem cells and a predictor of poor clinical outcome. *Cell Stem Cell*, **1**, 555–567.
43. Herschkowitz, J.I., Zhao, W., Zhang, M., Usary, J. *et al.* (2011) Comparative oncogenomics identifies breast tumors enriched in functional tumor-initiating cells. *Proc. Natl Acad. Sci. USA*, **109** (8), 2778–2783.

44. Arai, F., Hirao, A., Ohmura, M., Sato, H. *et al.* (2004) Tie2/angiopoietin-1 signaling regulates hematopoietic stem cell quiescence in the bone marrow niche. *Cell*, **118**, 149–161.
45. Potten, C.S., Booth, C., and Pritchard, D.M. (1997) The intestinal epithelial stem cell: the mucosal governor. *Int. J. Exp. Pathol.*, **78**, 219–243.
46. Yoshihara, H., Arai, F., Hosokawa, K., Hagiwara, T. *et al.* (2007) Thrombopoietin/MPL signaling regulates hematopoietic stem cell quiescence and interaction with the osteoblastic niche. *Cell Stem Cell*, **1**, 685–697.
47. Gao, M.Q., Choi, Y.P., Kang, S., Youn, J.H. *et al.* (2010) CD24 + cells from hierarchically organized ovarian cancer are enriched in cancer stem cells. *Oncogene*, **29**, 2672–2680.
48. Pece, S., Tosoni, D., Confalonieri, S., Mazzarol, G. *et al.* (2010) Biological and molecular heterogeneity of breast cancers correlates with their cancer stem cell content. *Cell*, **140**, 62–73.
49. Liu, R., Wang, X., Chen, G.Y., Dalerba, P. *et al.* (2007) The prognostic role of a gene signature from tumorigenic breast-cancer cells. *N. Engl. J. Med.*, **356**, 217–226.
50. Shipitsin, M., Campbell, L.L., Argani, P., Weremowicz, S. *et al.* (2007) Molecular definition of breast tumor heterogeneity. *Cancer Cell*, **11**, 259–273.
51. Creighton, C.J., Li, X., Landis, M., Dixon, J.M. *et al.* (2009) Residual breast cancers after conventional therapy display mesenchymal as well as tumor-initiating features. *Proc. Natl Acad. Sci. USA*, **106**, 13820–13825.
52. Dembinski, J.L. and Krauss, S. (2009) Characterization and functional analysis of a slow cycling stem cell-like subpopulation in pancreas adenocarcinoma. *Clin. Exp. Metastasis*, **26**, 611–623.
53. Li, L. and Clevers, H. (2010) Coexistence of quiescent and active adult stem cells in mammals. *Science*, **327**, 542–545.
54. Baker, M. (2007) Stem cells by any other name. *Nature*, **449**, 389.
55. Morrison, S.J. and Kimble, J. (2006) Asymmetric and symmetric stem-cell divisions in development and cancer. *Nature*, **441**, 1068–1074.
56. Jan, Y.N. and Jan, L.Y. (2001) Asymmetric cell division in the *Drosophila* nervous system. *Nat. Rev. Neurosci.*, **2**, 772–779.
57. Wang, H., Ouyang, Y., Somers, W.G., Chia, W. *et al.* (2007) Polo inhibits progenitor self-renewal and regulates Numb asymmetry by phosphorylating Pon. *Nature*, **449**, 96–100.
58. Chen, R., Nishimura, M.C., Bumbaca, S.M., Kharbanda, S. *et al.* (2010) A hierarchy of self-renewing tumor-initiating cell types in glioblastoma. *Cancer Cell*, **17**, 362–375.
59. Roesch, A., Fukunaga-Kalabis, M., Schmidt, E.C., Zabierowski, S.E. *et al.* (2010) A temporarily distinct subpopulation of slow-cycling melanoma cells is required for continuous tumor growth. *Cell*, **141**, 583–594.
60. Chaffer, C.L., Brueckmann, I., Scheel, C., Kaestli, A.J. *et al.* (2011) Normal and neoplastic nonstem cells can spontaneously convert to a stem-like state. *Proc. Natl Acad. Sci. USA*, **108** (19), 7950–7955.
61. Cohen, S.J., Alpaugh, R.K., Gross, S., O'Hara, S.M. *et al.* (2006) Isolation and characterization of circulating tumor cells in patients with metastatic colorectal cancer. *Clin. Colorectal Cancer*, **6**, 125–132.
62. Nagrath, S., Sequist, L.V., Maheswaran, S., Bell, D.W. *et al.* (2007) Isolation of rare circulating tumour cells in cancer patients by microchip technology. *Nature*, **450**, 1235–1239.
63. He, W., Wang, H., Hartmann, L.C., Cheng, J.X. *et al.* (2007) In vivo quantitation of rare circulating tumor cells by multiphoton intravital flow cytometry. *Proc. Natl Acad. Sci. USA*, **104**, 11760–11765.
64. Marrinucci, D., Bethel, K., Luttgen, M., Bruce, R.H. *et al.* (2009) Circulating tumor cells from well-differentiated lung adenocarcinoma retain cytomorphologic features of primary tumor type. *Arch. Pathol. Lab. Med.*, **133**, 1468–1471.
65. Helzer, K.T., Barnes, H.E., Day, L., Harvey, J. *et al.* (2009) Circulating tumor cells are transcriptionally similar to the primary tumor in a murine prostate model. *Cancer Res.*, **69**, 7860–7866.
66. Panteleakou, Z., Lembessis, P., Sourla, A., Pissimissis, N. *et al.* (2009) Detection of circulating tumor cells in prostate cancer patients: methodological pitfalls and clinical relevance. *Mol. Med.*, **15**, 101–114.

67. Attard, G., Swennenhuis, J.F., Olmos, D., Reid, A.H. *et al.* (2009) Characterization of ERG, AR and PTEN gene status in circulating tumor cells from patients with castration-resistant prostate cancer. *Cancer Res.*, **69**, 2912–2918.
68. Cristofanilli, M., Budd, G.T., Ellis, M.J., Stopeck, A. *et al.* (2004) Circulating tumor cells, disease progression, and survival in metastatic breast cancer. *N. Engl. J. Med.*, **351**, 781–791.
69. Riethdorf, S., Fritsche, H., Muller, V., Rau, T. *et al.* (2007) Detection of circulating tumor cells in peripheral blood of patients with metastatic breast cancer: a validation study of the CellSearch system. *Clin. Cancer Res.*, **13**, 920–928.
70. Husemann, Y., Geigl, J.B., Schubert, F., Musiani, P. *et al.* (2008) Systemic spread is an early step in breast cancer. *Cancer Cell*, **13**, 58–68.
71. Cohen, S.J., Punt, C.J., Iannotti, N., Saidman, B.H. *et al.* (2008) Relationship of circulating tumor cells to tumor response, progression-free survival, and overall survival in patients with metastatic colorectal cancer. *J. Clin. Oncol.*, **26**, 3213–3221.
72. Yang, Z.F., Ngai, P., Ho, D.W., Yu, W.C. *et al.* (2008) Identification of local and circulating cancer stem cells in human liver cancer. *Hepatology*, **47**, 919–928.
73. Botteri, E., Sandri, M.T., Bagnardi, V., Munzone, E. *et al.* (2010) Modeling the relationship between circulating tumour cells number and prognosis of metastatic breast cancer. *Breast Cancer Res. Treat.*, **122**, 211–217.
74. Yu, M., Stott, S., Toner, M., Maheswaran, S. *et al.* (2011) Circulating tumor cells: approaches to isolation and characterization. *J. Cell Biol.*, **192**, 373–382.
75. Hsieh, H.B., Marrinucci, D., Bethel, K., Curry, D.N. *et al.* (2006) High speed detection of circulating tumor cells. *Biosens. Bioelectron.*, **21**, 1893–1899.
76. Marrinucci, D., Bethel, K., Bruce, R.H., Curry, D.N. *et al.* (2007) Case study of the morphologic variation of circulating tumor cells. *Hum. Pathol.*, **38**, 514–519.
77. Marrinucci, D., Bethel, K., Lazar, D., Fisher, J. *et al.* (2010) Cytomorphology of circulating colorectal tumor cells: a small case series. *J. Oncol.*, **2010**, 861341.
78. Pachmann, K., Clement, J.H., Schneider, C.P., Willen, B. *et al.* (2005) Standardized quantification of circulating peripheral tumor cells from lung and breast cancer. *Clin. Chem. Lab. Med.*, **43**, 617–627.
79. Wang, X., Qian, X., Beitler, J.J., Chen, Z.G. *et al.* (2011) Detection of circulating tumor cells in human peripheral blood using surface-enhanced Raman scattering nanoparticles. *Cancer Res.*, **71** (5), 1526–1532.
80. Sieuwerts, A.M., Kraan, J., Bolt, J., van der Spoel, P. *et al.* (2009) Anti-epithelial cell adhesion molecule antibodies and the detection of circulating normal-like breast tumor cells. *J. Natl Cancer Inst.*, **101**, 61–66.
81. Radtke, F. and Clevers, H. (2005) Self-renewal and cancer of the gut: two sides of a coin. *Science*, **307**, 1904–1909.
82. Taipale, J. and Beachy, P.A. (2001) The Hedgehog and Wnt signalling pathways in cancer. *Nature*, **411**, 349–354.
83. Reya, T. and Clevers, H. (2005) Wnt signalling in stem cells and cancer. *Nature*, **434**, 843–850.
84. Nusse, R., van Ooyen, A., Cox, D., Fung, Y.K.*et al.* (1984) Mode of proviral activation of a putative mammary oncogene (int-1) on mouse chromosome 15. *Nature*, **307**, 131–136.
85. Li, Y., Hively, W.P., and Varmus, H.E. (2000) Use of MMTV-Wnt-1 transgenic mice for studying the genetic basis of breast cancer. *Oncogene*, **19**, 1002–1009.
86. Li, Y., Welm, B., Podsypanina, K., Huang, S. *et al.* (2003) Evidence that transgenes encoding components of the Wnt signaling pathway preferentially induce mammary cancers from progenitor cells. *Proc. Natl Acad. Sci. USA*, **100**, 15853–15858.
87. Malanchi, I., Peinado, H., Kassen, D., Hussenet, T. *et al.* (2008) Cutaneous cancer stem cell maintenance is dependent on beta-catenin signalling. *Nature*, **452**, 650–653.
88. Bisson, I. and Prowse, D.M. (2009) WNT signaling regulates self-renewal and differentiation of prostate cancer cells with stem cell characteristics. *Cell Res.*, **19**, 683–697.
89. Zheng, H., Ying, H., Wiedemeyer, R., Yan, H. *et al.* (2010) PLAGL2 regulates Wnt signaling to impede differentiation in neural

stem cells and gliomas. *Cancer Cell*, **17**, 497–509.

90. Vermeulen, L., De Sousa, E.M.F., van der Heijden, M., Cameron, K. *et al.* (2010) Wnt activity defines colon cancer stem cells and is regulated by the microenvironment. *Nat. Cell Biol.*, **12**, 468–476.
91. Clement, V., Sanchez, P., de Tribolet, N., Radovanovic, I. *et al.* (2007) HEDGEHOG-GLI1 signaling regulates human glioma growth, cancer stem cell self-renewal, and tumorigenicity. *Curr. Biol.*, **17**, 165–172.
92. Liu, S., Dontu, G., Mantle, I.D., Patel, S. *et al.* (2006) Hedgehog signaling and Bmi-1 regulate self-renewal of normal and malignant human mammary stem cells. *Cancer Res.*, **66**, 6063–6071.
93. Dierks, C., Beigi, R., Guo, G.R., Zirlik, K. *et al.* (2008) Expansion of Bcr-Abl-positive leukemic stem cells is dependent on Hedgehog pathway activation. *Cancer Cell*, **14**, 238–249.
94. Zhao, C., Chen, A., Jamieson, C.H., Fereshteh, M. *et al.* (2009) Hedgehog signalling is essential for maintenance of cancer stem cells in myeloid leukaemia. *Nature*, **458**, 776–779.
95. Peacock, C.D., Wang, Q., Gesell, G.S., Corcoran-Schwartz, I.M. *et al.* (2007) Hedgehog signaling maintains a tumor stem cell compartment in multiple myeloma. *Proc. Natl Acad. Sci. USA*, **104**, 4048–4053.
96. Bar, E.E., Chaudhry, A., Lin, A., Fan, X. *et al.* (2007) Cyclopamine-mediated hedgehog pathway inhibition depletes stem-like cancer cells in glioblastoma. *Stem Cells*, **25**, 2524–2533.
97. Mani, S.A., Guo, W., Liao, M.J., Eaton, E.N. *et al.* (2008) The epithelial-mesenchymal transition generates cells with properties of stem cells. *Cell*, **133**, 704–715.
98. Kong, D., Banerjee, S., Ahmad, A., Li, Y. *et al.* (2010) Epithelial to mesenchymal transition is mechanistically linked with stem cell signatures in prostate cancer cells. *PLoS ONE*, **5**, e12445.
99. Feldmann, G., Dhara, S., Fendrich, V., Bedja, D. *et al.* (2007) Blockade of hedgehog signaling inhibits pancreatic cancer invasion and metastases: a new paradigm for combination therapy in solid cancers. *Cancer Res.*, **67**, 2187–2196.
100. Dontu, G., Jackson, K.W., McNicholas, E., Kawamura, M.J. *et al.* (2004) Role of Notch signaling in cell-fate determination of human mammary stem/progenitor cells. *Breast Cancer Res.*, **6**, R605–R615.
101. Dontu, G., El-Ashry, D., and Wicha, M.S. (2004) Breast cancer, stem/progenitor cells and the estrogen receptor. *Trends Endocrinol. Metab.*, **15**, 193–197.
102. Dontu, G., Liu, S., and Wicha, M.S. (2005) Stem cells in mammary development and carcinogenesis: implications for prevention and treatment. *Stem Cell Rev.*, **1**, 207–213.
103. Dontu, G. and Wicha, M.S. (2005) Survival of mammary stem cells in suspension culture: implications for stem cell biology and neoplasia. *J. Mammary Gland Biol. Neoplasia*, **10**, 75–86.
104. Liu, S., Dontu, G., and Wicha, M.S. (2005) Mammary stem cells, self-renewal pathways, and carcinogenesis. *Breast Cancer Res.*, **7**, 86–95.
105. Ling, H., Sylvestre, J.R., and Jolicoeur, P. (2010) Notch1-induced mammary tumor development is cyclin D1-dependent and correlates with expansion of pre-malignant multipotent duct-limited progenitors. *Oncogene*, **29**, 4543–4554.
106. Valastyan, S. and Weinberg, R.A. (2009) MicroRNAs: Crucial multi-tasking components in the complex circuitry of tumor metastasis. *Cell Cycle*, **8**, 3506–3512.
107. Vallejo, D.M., Caparros, E., and Dominguez, M. (2011) Targeting Notch signalling by the conserved miR-8/200 microRNA family in development and cancer cells. *EMBO J.*, **30**, 756–769.
108. Brabletz, S., Bajdak, K., Meidhof, S., Burk, U. *et al.* (2011) The ZEB1/miR-200 feedback loop controls Notch signalling in cancer cells. *EMBO J.*, **30**, 770–782.
109. Dubrovska, A., Kim, S., Salamone, R.J., Walker, J.R. *et al.* (2009) The role of PTEN/Akt/PI3K signaling in the maintenance and viability of prostate cancer stem-like cell populations. *Proc. Natl Acad. Sci. USA*, **106**, 268–273.
110. Groszer, M., Erickson, R., Scripture-Adams, D.D., Dougherty, J.D. *et al.* (2006) PTEN negatively regulates neural stem cell self-renewal by modulating G0-G1 cell cycle entry. *Proc. Natl Acad. Sci. USA*, **103**, 111–116.
111. Goodell, M.A., Brose, K., Paradis, G., Conner, A.S. *et al.* (1996) Isolation and

functional properties of murine hematopoietic stem cells that are replicating in vivo. *J. Exp. Med.*, **183**, 1797–1806.
112. Wu, C., and Alman, B.A. (2008) Side population cells in human cancers. *Cancer Lett.*, **268**, 1–9.
113. Bleau, A.M., Hambardzumyan, D., Ozawa, T., Fomchenko, E.I. *et al.* (2009) PTEN/PI3K/Akt pathway regulates the side population phenotype and ABCG2 activity in glioma tumor stem-like cells. *Cell Stem Cell*, **4**, 226–235.
114. Molofsky, A.V., Pardal, R., and Morrison, S.J. (2004) Diverse mechanisms regulate stem cell self-renewal. *Curr. Opin. Cell Biol.*, **16**, 700–707.
115. TeKippe, M., Harrison, D.E., and Chen, J. (2003) Expansion of hematopoietic stem cell phenotype and activity in Trp53-null mice. *Exp. Hematol.*, **31**, 521–527.
116. Cicalese, A., Bonizzi, G., Pasi, C.E., Faretta, M. *et al.* (2009) The tumor suppressor p53 regulates polarity of self-renewing divisions in mammary stem cells. *Cell*, **138**, 1083–1095.
117. Meletis, K., Wirta, V., Hede, S.M., Nister, M. *et al.* (2006) p53 suppresses the self-renewal of adult neural stem cells. *Development*, **133**, 363–369.
118. Zhao, Z., Zuber, J., Diaz-Flores, E., Lintault, L. *et al.* (2010) p53 loss promotes acute myeloid leukemia by enabling aberrant self-renewal. *Genes Dev.*, **24**, 1389–1402.
119. Gregory, P.A., Bert, A.G., Paterson, E.L., Barry, S.C. *et al.* (2008) The miR-200 family and miR-205 regulate epithelial to mesenchymal transition by targeting ZEB1 and SIP1. *Nat. Cell Biol.*, **10**, 593–601.
120. Polyak, K. and Weinberg, R.A. (2009) Transitions between epithelial and mesenchymal states: acquisition of malignant and stem cell traits. *Nat. Rev. Cancer*, **9**, 265–273.
121. Ibarra, I., Erlich, Y., Muthuswamy, S.K., Sachidanandam, R. *et al.* (2007) A role for microRNAs in maintenance of mouse mammary epithelial progenitor cells. *Genes Dev.*, **21**, 3238–3243.
122. Yu, F., Yao, H., Zhu, P., Zhang, X. *et al.* (2007) let-7 regulates self renewal and tumorigenicity of breast cancer cells. *Cell*, **131**, 1109–1123.
123. Iliopoulos, D., Lindahl-Allen, M., Polytarchou, C., Hirsch, H.A. *et al.* (2010) Loss of miR-200 inhibition of Suz12 leads to polycomb-mediated repression required for the formation and maintenance of cancer stem cells. *Mol. Cell*, **39**, 761–772.
124. Chang, C.J., Chao, C.H., Xia, W., Yang, J.Y. *et al.* (2011) p53 regulates epithelial-mesenchymal transition and stem cell properties through modulating miRNAs. *Nat. Cell Biol.*, **13** (3), 317–323.
125. Liu, C., Kelnar, K., Liu, B., Chen, X. *et al.* (2011) The microRNA miR-34a inhibits prostate cancer stem cells and metastasis by directly repressing CD44. *Nat. Med.*, **17**, 211–215.
126. Ji, Q., Hao, X., Zhang, M., Tang, W. *et al.* (2009) MicroRNA miR-34 inhibits human pancreatic cancer tumor-initiating cells. *PLoS ONE*, **4**, e6816.
127. Sherry, M.M., Reeves, A., Wu, J.K., and Cochran, B.H. (2009) STAT3 is required for proliferation and maintenance of multipotency in glioblastoma stem cells. *Stem Cells*, **27**, 2383–2392.
128. Park, I.K., Qian, D., Kiel, M., Becker, M.W. *et al.* (2003) Bmi-1 is required for maintenance of adult self-renewing haematopoietic stem cells. *Nature*, **423**, 302–305.
129. Molofsky, A.V., Pardal, R., Iwashita, T., Park, I.K. *et al.* (2003) Bmi-1 dependence distinguishes neural stem cell self-renewal from progenitor proliferation. *Nature*, **425**, 962–967.
130. Lukacs, R.U., Memarzadeh, S., Wu, H., and Witte, O.N. (2010) Bmi-1 is a crucial regulator of prostate stem cell self-renewal and malignant transformation. *Cell Stem Cell*, **7**, 682–693.
131. Morrison, S.J. and Spradling, A.C. (2008) Stem cells and niches: mechanisms that promote stem cell maintenance throughout life. *Cell*, **132**, 598–611.
132. Calabrese, C., Poppleton, H., Kocak, M., Hogg, T.L. *et al.* (2007) A perivascular niche for brain tumor stem cells. *Cancer Cell*, **11**, 69–82.
133. Konopleva, M.Y. and Jordan, C.T. (2011) Leukemia stem cells and microenvironment: biology and therapeutic targeting. *J. Clin. Oncol.*, **29**, 591–599.
134. Kaplan, R.N., Riba, R.D., Zacharoulis, S., Bramley, A.H. *et al.* (2005) VEGFR1-positive haematopoietic bone marrow progenitors initiate the pre-metastatic niche. *Nature*, **438**, 820–827.

135. Hovinga, K.E., Shimizu, F., Wang, R., Panagiotakos, G. *et al.* (2010) Inhibition of notch signaling in glioblastoma targets cancer stem cells via an endothelial cell intermediate. *Stem Cells*, **28**, 1019–1029.
136. Folkins, C., Man, S., Xu, P., Shaked, Y. *et al.* (2007) Anticancer therapies combining antiangiogenic and tumor cell cytotoxic effects reduce the tumor stem-like cell fraction in glioma xenograft tumors. *Cancer Res.*, **67**, 3560–3564.
137. Clevers, H. (2006) Wnt/beta-catenin signaling in development and disease. *Cell*, **127**, 469–480.
138. He, X.C., Zhang, J., Tong, W.G., Tawfik, O. *et al.* (2004) BMP signaling inhibits intestinal stem cell self-renewal through suppression of Wnt-beta-catenin signaling. *Nat. Genet.*, **36**, 1117–1121.
139. Kosinski, C., Li, V.S., Chan, A.S., Zhang, J. *et al.* (2007) Gene expression patterns of human colon tops and basal crypts and BMP antagonists as intestinal stem cell niche factors. *Proc. Natl Acad. Sci. USA*, **104**, 15418–15423.
140. Gregorieff, A., Pinto, D., Begthel, H., Destree, O. *et al.* (2005) Expression pattern of Wnt signaling components in the adult intestine. *Gastroenterology*, **129**, 626–638.
141. Shinagawa, K., Kitadai, Y., Tanaka, M., Sumida, T. *et al.* (2010) Mesenchymal stem cells enhance growth and metastasis of colon cancer. *Int. J. Cancer*, **127**, 2323–2333.
142. Grivennikov, S., Karin, E., Terzic, J., Mucida, D. *et al.* (2009) IL-6 and Stat3 are required for survival of intestinal epithelial cells and development of colitis-associated cancer. *Cancer Cell*, **15**, 103–113.
143. Breier, G., Albrecht, U., Sterrer, S., and Risau, W. (1992) Expression of vascular endothelial growth factor during embryonic angiogenesis and endothelial cell differentiation. *Development*, **114**, 521–532.
144. Sun, L., Hui, A.M., Su, Q., Vortmeyer, A. *et al.* (2006) Neuronal and glioma-derived stem cell factor induces angiogenesis within the brain. *Cancer Cell*, **9**, 287–300.
145. Jin, K., Mao, X.O., Sun, Y., Xie, L. *et al.* (2002) Stem cell factor stimulates neurogenesis in vitro and in vivo. *J. Clin. Invest.*, **110**, 311–319.
146. Ramirez-Castillejo, C., Sanchez-Sanchez, F., Andreu-Agullo, C., Ferron, S.R. *et al.* (2006) Pigment epithelium-derived factor is a niche signal for neural stem cell renewal. *Nat. Neurosci.*, **9**, 331–339.
147. Knoblich, J.A. (2008) Mechanisms of asymmetric stem cell division. *Cell*, **132**, 583–597.
148. Bao, S., Wu, Q., Sathornsumetee, S., Hao, Y. *et al.* (2006) Stem cell-like glioma cells promote tumor angiogenesis through vascular endothelial growth factor. *Cancer Res.*, **66**, 7843–7848.
149. Hockel, M. and Vaupel, P. (2001) Tumor hypoxia: definitions and current clinical, biologic, and molecular aspects. *J. Natl Cancer Inst.*, **93**, 266–276.
150. Li, Z., Bao, S., Wu, Q., Wang, H. *et al.* (2009) Hypoxia-inducible factors regulate tumorigenic capacity of glioma stem cells. *Cancer Cell*, **15**, 501–513.
151. Dallas, N.A., Xia, L., Fan, F., Gray, M.J. *et al.* (2009) Chemoresistant colorectal cancer cells, the cancer stem cell phenotype, and increased sensitivity to insulin-like growth factor-I receptor inhibition. *Cancer Res.*, **69**, 1951–1957.
152. Ma, S., Lee, T.K., Zheng, B.J., Chan, K.W. *et al.* (2008) CD133+ HCC cancer stem cells confer chemoresistance by preferential expression of the Akt/PKB survival pathway. *Oncogene*, **27**, 1749–1758.
153. Nakai, E., Park, K., Yawata, T., Chihara, T. *et al.* (2009) Enhanced MDR1 expression and chemoresistance of cancer stem cells derived from glioblastoma. *Cancer Invest.*, **27**, 901–908.
154. Viale, A., De Franco, F., Orleth, A., Cambiaghi, V. *et al.* (2009) Cell-cycle restriction limits DNA damage and maintains self-renewal of leukaemia stem cells. *Nature*, **457**, 51–56.
155. Diehn, M., Cho, R.W., Lobo, N.A., Kalisky, T. *et al.* (2009) Association of reactive oxygen species levels and radioresistance in cancer stem cells. *Nature*, **458**, 780–783.
156. Kirkland, S.C. (2009) Type I collagen inhibits differentiation and promotes a stem cell-like phenotype in human colorectal carcinoma cells. *Br. J. Cancer*, **101**, 320–326.
157. Loeffler, M., Kruger, J.A., Niethammer, A.G., and Reisfeld, R.A. (2006) Targeting tumor-associated fibroblasts improves

cancer chemotherapy by increasing intratumoral drug uptake. *J. Clin. Invest.*, **116**, 1955–1962.

158. Jordan, C.T. (2010) Targeting myeloid leukemia stem cells. *Sci. Transl. Med.*, **2**, 31ps21.

159. Yilmaz, O.H., Valdez, R., Theisen, B.K., Guo, W. *et al.* (2006) Pten dependence distinguishes haematopoietic stem cells from leukaemia-initiating cells. *Nature*, **441**, 475–482.

160. Gupta, P.B., Onder, T.T., Jiang, G., Tao, K. *et al.* (2009) Identification of selective inhibitors of cancer stem cells by high-throughput screening. *Cell*, **138**, 645–659.

161. Fillmore, C.M. and Kuperwasser, C. (2008) Human breast cancer cell lines contain stem-like cells that self-renew, give rise to phenotypically diverse progeny and survive chemotherapy. *Breast Cancer Res.*, **10**, R25.

162. Quintana, E., Shackleton, M., Foster, H.R., Fullen, D.R. *et al.* (2010) Phenotypic heterogeneity among tumorigenic melanoma cells from patients that is reversible and not hierarchically organized. *Cancer Cell*, **18**, 510–523.

5
Chromatin Remodeling in Carcinoma Cells*

Therese M. Becker
University of Sydney, Westmead Institute for Cancer Research at Westmead Millennium Institute, Westmead Hospital, Westmead, Sydney, New South Wales, Australia

*This chapter has previously been published in: Meyers, R.A. (Ed.) Epigenetic Regulation and Epigenomics, 2012, 978-3-527-32682-2.

Translational Medicine Cancer: Advances in Molecular Biology and Medicine,
First Edition. Edited by Robert A. Meyers.

Keywords

Chromatin
An organized structure of eukaryotic DNA.

Histone modification
Covalent changes to histones.

ATP-dependent chromatin remodeling
Changing the histone–DNA interaction to unwind or pack DNA.

SWI/SNF
A specific chromatin-remodeling complex.

Carcinoma
A cancer derived from epithelial cells.

Tumor suppressor
A protein with a function that opposes tumor development and progression.

Senescence
Secure withdrawal from proliferation.

Carcinoma arises from the malignant transformation of epithelial cells, which involves the deregulation of genetic information leading to excessive proliferation, cell death evasion, and invasive potential. The traditional dogma was that the underlying genetic deregulation emerged through an accumulation of individual mutations that inactivate tumor suppressor genes and activate oncogenes. During recent years, it has become evident that epigenetic changes are potently contributing to tumor formation. Epigenetics describes the modulation of genetic information by mechanisms other than sequence alterations of the underlying DNA. A major part of epigenetic gene regulation is accomplished through chromatin remodeling. In eukaryotic cells, huge amounts of DNA are managed through intricate mechanisms that allow DNA condensation into tightly packable protein–DNA units (chromatin), and chromatin-remodeling processes render DNA either accessible or inaccessible to diverse DNA processes such as replication and transcription. The role of chromatin remodeling in carcinoma cells is reviewed in this chapter.

1 Introduction

The malignant transformation of normal cells into cancer cells involves dynamic changes of the cellular genome. Cancer cells have accumulated a range of genetic alterations that render them unresponsive to normal cellular regulatory mechanisms, and thereby they have acquired the main hallmarks of cancer cells:

- *Excessive proliferation* by becoming independent of external stimuli to activate proliferative signaling pathways and by "ignoring" anti-proliferative signals.
- *Avoidance of cell death* by opposing normal pathways of programmed cell death (apoptosis) and developing immortality mechanisms, which means cancer cells have in principal an unlimited life span.
- *Gain of invasive potential* to invade other tissues and form metastasis (as reviewed by Hanahan and Weinberg [1]).

The chance to accumulate genetic alterations that cause tumor development increases with the number of cell divisions that a single cell undergoes during its life span. Thus, it is not surprising that the majority of tumors arise in tissues with a high proliferation rate. Epithelial cells undergo relatively frequent cell divisions to renew epithelial tissue and to ensure its function. Cancers that arise from epithelial cells are termed *carcinomas*, and comprise over 90% of all cancers [2] (see Table 1).

The genetic changes relating to malignant transformation may be direct mutational changes, which will produce hyperactive oncogenes or inactive tumor suppressor genes [1]. Moreover, proliferation, cell death avoidance, and invasive potential may be achieved by more global changes in gene regulation. Often, transformed cells have reverted their overall gene expression pattern from a growth-arrested, differentiated state in which cells function as part of a specialized tissue to a stem cell-like proliferative state [3]. The stem cell state is normally maintained during embryonic development or during later life stages in long-term progenitor cells, which continue to divide and give rise to cells, which will growth arrest and differentiate. Progenitor cells are characteristic for tissues with a high cell renewal, such as colon epithelial tissue or skin tissue. It has been proposed that, in a 60-year-old individual, the colon epithelial cells – which arise from progenitor cells in the crypts of the colon epithelium – have arisen through about 3000 cell divisions, whereas skin keratinocytes – which originate from progenitors in the basal layer of the skin – have arisen from about 300 divisions by this age [4]. Not surprisingly, in these progenitor cells the proliferation genes are highly expressed, while differentiation gene expression is limited. The switch from one gene expression state to the other can be accomplished by expressing or repressing an entire cassette of genes associated with either proliferation or differentiation. In mammalian cells, these gene cassettes are often organized in adjacent genomic regions that allow coordinated regulation. Such regulation can be achieved by chromatin-remodeling processes that either allow or deny access to the transcriptional machinery or transcriptional repressors [5]. Consequently, tumor cells may achieve a more stem cell-like gene expression pattern by targeting chromatin-remodeling processes during cellular transformation, while alterations of chromatin-remodeling functions will contribute to the development of the carcinoma, and its progression.

Some insights into chromatin-remodeling events, and their contribution to carcinoma formation, are provided in this chapter.

Tab. 1 Types of carcinoma.

Carcinoma type	*Commonest tissues*	*Proportion of all cancers (%)*
External epithelia	Skin, large intestine, lung, stomach, cervix	56
Internal epithelia	Breast, prostate, ovary, bladder, pancreas	36

Data derived from Ref. [2].

2 Chromatin

Eukaryotic cells are characterized by their nuclear compartmentation of DNA, and by the arrangement of DNA molecules into highly organized, dynamic DNA–protein structures referred to as *chromatin*. Within chromatin, the DNA (in units of 146 bp) is wrapped around and covalently bound to histone octamer protein complexes; the resulting structures are termed *nucleosomes*. The nucleosome histone octamers consist of H2A, H2B, H3, and H4 or specialized histone variants [6] (Fig. 1). The nucleosomes can be further folded (condensed) with the aid of the H1 linker histone, and may be organized in several hierarchical steps into tightly condensed chromosomes [7] (Fig. 2). Chromatin is a dynamic structure that is maintained by a series of sophisticated chromatin-remodeling procedures that either allow or deny local and global access to the DNA. Thus, chromatin remodeling regulates all cellular aspects related to DNA maintenance, repair, transcription, and replication (for reviews, see Refs [3, 5]).

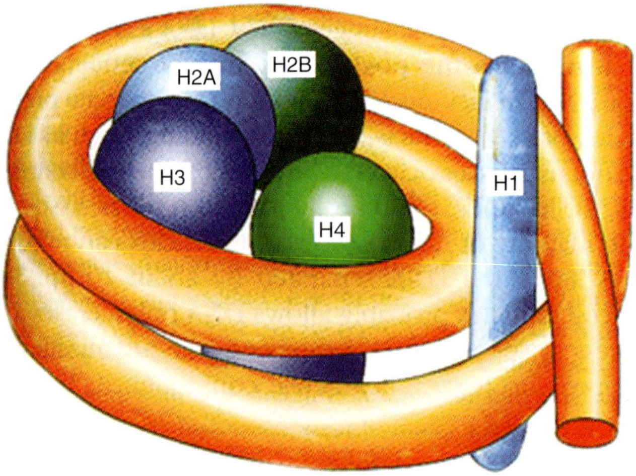

Fig. 1 The nucleosome, the basic unit of chromatin. This schematic shows the winding of the double-stranded DNA molecule around a histone octamer, consisting of H2A, H2B, H3, and H4. The histone linker H1 is also represented. Reproduced with permission from Ref. [7]; © 1996–2011, themededicalbiochemistrypage.org.

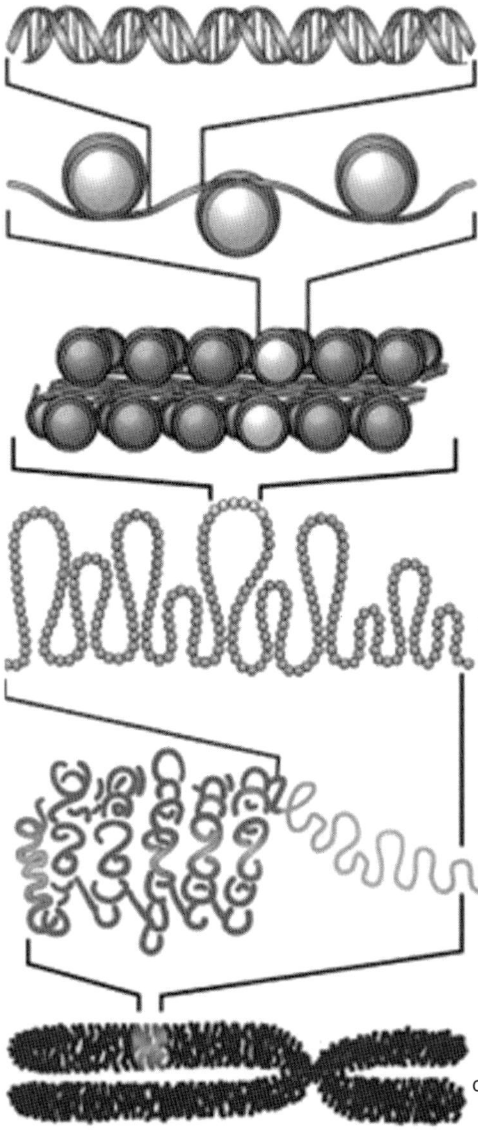

Fig. 2 Chromatin condensation. Schematic demonstrating the formation of higher-order chromatin complexes from the double-stranded DNA molecule over nucleosome formation to condensation into chromosomes. Reproduced with permission from Ref. [7]; © 1996–2011, themedicalbiochemistrypage.org

3
Chromatin Remodeling

In principle, chromatin remodeling describes a series of coordinated cellular mechanisms that alter the nucleosome configuration such that chromatin and DNA become either accessible or inaccessible to proteins that regulate DNA processes such as transcription, replication, telomere maintenance, and DNA repair.

Chromatin remodeling involves two major steps:

- Histone-modifying enzymes covalently add or remove modifications (e.g., acetylations, methylations, or phosphorylations) at amino acid residues of the N-terminal tail of histones which protrude from the globular nucleosome core. These modifications alter the histones' affinity for binding with the DNA and with specific DNA-processing molecules. For example, acetylated histones, which are found preferentially in actively transcribed and regulatory DNA regions (euchromatin), are highly interactive with bromodomains, which are protein domains found in DNA-interacting enzymes such as the ATP-dependent chromatin-remodeling factors BRG1 and BRM.
- Additionally, chromatin remodeling involves the action of ATP-dependent chromatin-remodeling complexes. These complexes utilize the energy of ATP hydrolyses to affect DNA–histone interactions and, thereby, the repositioning of nucleosomes. These functions alter the DNA–histone interface and either allow or inhibit access of transcriptional repressors or the transcriptional machinery, which includes facilitating molecules such as RNA polymerase II (RNApol-II) and the TATTA box binding protein (TBP) and sequence-specific transcription factors [3].

Both of these processes are tightly cooperative. Histone modifications are often required to form docking stations for ATP- dependent chromatin-remodeling complexes, whereas some histone-modifying proteins are known to function while interacting with ATP-dependent chromatin-remodeling complexes [8].

For more detailed information concerning the physical and biochemical mechanisms of chromatin remodeling around histone octamers, the reader is referred to reviews focusing on this area (e.g., Ref. [9]).

3.1 Histone Modifications

3.1.1 Histone Acetyl Transferases (HATs)

Histone acetyl transferase (HAT) complexes transfer acetyl groups to the N-terminal lysine residues of histones. The acetylations of histones are thought to alter the overall charge of lysine residues, and thereby weaken electrostatic histone–DNA interactions such that acetylated DNA regions become less tightly folded. As a result, acetylation makes DNA more accessible to the transcriptional machinery, and ATP-dependent bromodomain-containing chromatin remodelers usually promote gene expression in acetylated chromatin regions [10]. More open and transcriptionally active chromatin with acetylated histones is also referred to as *euchromatin* [11].

3.1.2 Histone Deacetylases (HDACs)

Histone deacetylase (HDAC) complexes are the direct antagonists of HATs, and remove acetyl groups from lysine residues. Deacetylation promotes a tightening of the DNA–histone interactions which, in conjunction with further modifications, leads to chromatin condensation which prevents transcription from occurring [12] (Fig. 2). The deacetylation function of HDACs is essential for the formation of heterochromatin which, in contrast to the transcriptionally active and highly acetylated euchromatin, does not contain histone acetylations and is characterized by the heterochromatin protein HP1 and

Tab. 2 Effects of histone lysine methylations.

Lysine methylation state	*Effect, chromatin type*
H3K4me2/me3	Euchromatin, Active transcription, Usually combined with acetylation
H3K27me1	Euchromatin, Active transcription, Usually combined with acetylation
H3K36me2/me3	Euchromatin, Active transcription, Usually combined with acetylation
H3K9me2/me3	Heterochromatin No transcription Binds heterochromatin protein HP1
H4K20me2/me3	Heterochromatin No transcription
H3K27me3	Facultative heterochromatin No transcription

increased chromatin condensation, thus not allowing access to transcriptional activators [13].

3.1.3 Histone Methyltransferases

Methylations occur on histones at either lysine or arginine residues. Histone methylation affects four different N-terminal histone tail lysines of H3 (K4, K9, K27, and K36), one H3 core lysine (K79), and the H4 lysine K20. Additionally, methylations have been identified for the N-terminal arginines of H3 (R2, R8, R17, and R26), and for H4 on R3 [14]. Three different degrees of methylation have been identified: monomethylations (me1); dimethylations (me2); and trimethylations (me3). The different orders of methylation state produce a diversity of bio chemical characteristics, and create conditions for a range of specific interactions with effector protein complexes [10, 15]. Histone lysine methylations have been associated with both transcriptional activation or repression: H3K4me2/me3, H3K27me1, and H3K36me2/me3 are found at actively transcribed euchromatic regions (usually in combination with acetylation), while H3K9me2/me3 attracts the heterochromatin protein HP1 and is linked to heterochromatin formation. It is found together with H4K20me2/me3 in inactive heterochromatin regions [10]. Thus, the outcomes of adjacent H3K9me2/me3 and H3K4me/2me3 oppose each other and, indeed, H3 is never found methylated at both K4 and K9. H3K27me3 is linked to facultative heterochromatin in euchromatic regions [15, 16]. For example, EZH2 silencing of the cell cycle inhibitor $p16^{INK4a}$ (see Sect. 4.1.1.3) involves H3K27me3 at $p16^{INK4a}$ regulatory sequences [17]. The effects of histone methylation are listed in Table 2.

3.1.4 Histone Demethylases

Although, originally, histone methylation turnover was thought to be negligible [18], today it has become clear that methylation of histones is a dynamic process, with specialized enzymes that catalyze the removal of methyl groups having been identified; these are termed *histone demethylases*. Currently, the importance of histone demethylases in cancer is emerging, together with their discovery [14].

3.1.5 Histone Modifications in Carcinoma

It has been suggested that combinations of specific histone methylation patterns over an extended genetic region are "read" by DNA-interacting proteins as an "epigenetic code" for gene regulation, and that carcinoma cells have evolved their own epigenetic code to promote proliferation, survival, and invasion [19].

It has been shown that the loss of monoacetylation and trimethylation of histone H4 arises early during the development of carcinoma of the breast, lung, and colon [20], while risk-associated patterns of H3 and H4 acetylation and dimethylation have been reported for prostate cancer [21]. In breast cancer, a low or absent H4K16Ac was identified, while moderate acetylation (H3K9Ac, H3K18Ac, H4K12Ac), lysine methylation (H3K4me2, H4K20me3), and arginine methylation (H4R3me2) were each correlated with aggressive subtypes, and high relative levels of global histone acetylation and methylation were associated with a favorable prognosis [22]. A low expression of H3K18Ac and H3K27me3 also correlated with esophageal squamous cell carcinoma [23], whilst in non-small-cell lung cancer (NSCLC) patients H3K4me2, H3K9Ac were shown to be prognostic markers, together with H2AK5Ac [24]. Low H3K4me2 and H3K18Ac were also significant predictors for overall survival in pancreatic adenocarcinoma, which also displays low levels of H3K9me2 [25]. These data are summarized in Table 3.

3.1.6 Histone-Modifying Enzymes and Carcinoma: Acetylation

The initial evidence for an alteration of HATs in cancer derived from viral oncogenes such as E1A, which target the methyltransferases p300 and CBP. Moreover, mutations of p300 and CBP have been associated with colorectal, breast, gastric, pancreatic, and hepatocellular cancers [29, 30]. In the case of HDACs, it appears that overexpression is the most common mode of malignant adaptation, with HDAC1 being overexpressed in prostate [31], gastric [32], colorectal [33, 34], and breast carcinoma [35]. HDAC2 is overexpressed in colon [33, 36], cervical [37], and gastric carcinoma [38], while an increased expression of HDAC3 is seen in colon carcinoma [33] and an overexpression of HDAC6 was identified in breast cancer specimens [39]. The overexpression of HDACs in cancer is proposed to promote facultative heterochromatin, which silences a range of tumor suppressor and anti-apoptotic proteins. Consequently, the specific silencing of HDACs with small interfering RNA (siRNA) molecules reduced tumor cell growth and, most importantly, caused programmed cell death in tumor cells by derepressing the apoptotic pathways. It

Tab. 3 Histone modifications associated with carcinoma.

Carcinoma type	*Histone modification*	*Reference(s)*
Breast cancer	High: H3R17me2 low/absent: H4K16Ac, H3K9Ac, H3K18Ac, H4K12Ac, H3K4me2, H4K20me3, H4R3me2	[20, 22, 26]
Prostate cancer	Low: H3K18Ac, H4K12Ac	[21]
Hepatocellular carcinoma	Histone hyperacetylation	[27]
Oral carcinoma	Histone hyperacetylation	[28]
Esophageal small-cell cancer	Low: H3K18Ac, H3K27me3	[23]
Non-small-cell lung cancer	High: H3K4me2, H4K8Ac, low: H2AK5Ac, H2BK12Ac	[24]
Pancreatic cancer	Low: H3K4me2, H3K18Ac, H3K9me2	[25]

is, therefore, important to highlight that the use of small molecules targeting and inhibiting HDACs (i.e., HDAC inhibitors) has been suggested as an anti-cancer treatment. Subsequent studies conducted both *in vitro* and *in vivo* have shown much promise, and have linked HDAC inhibitors to cell-cycle arrest, the induction of apoptosis, and the repression of angiogenesis and invasion [40].

3.1.7 Histone-Modifying Enzymes and Carcinoma: Methylation

Normally, the histone methyltransferase and polycomb protein EZH2 plays an important role during development and tissue renewal, as it is involved in shutting down the expression of genes that promote growth arrest and differentiation. In proliferating stem cells, EZH2 facilitates the trimethylation of H3K27 and facultative heterochromatin formation. The genes shut down by EZH2 often have a tumor-suppressor function in normal cells, such that an increased EZH2 level leads to excess proliferation and favors tumor formation [41]. The overexpression of EZH2 has been demonstrated in prostate, breast, colorectal, endometrial, hepatocellular, and bladder cancers, as well as in melanoma [42–44], while specific EZH2 silencing using siRNA molecules caused growth arrest in prostate cancer cells [45].

The lysine-specific demethylase 1 (LSD1), which removes methylations from H3K4me2/me3, was found to be downregulated and linked to invasiveness in breast cancer [46], whereas LSD1 was overexpressed and correlated with a poor prognosis in prostate cancer [47]. The histone demethylase JMJD2C, which removes methyl groups from H3K9me3, may destabilize heterochromatin by disallowing HP1 binding; typically, JMJD2C is amplified and thus highly expressed in prostate cancer [48]. JMJD2C is also found amplified in lung sarcomatoid carcinoma, esophageal carcinoma, and squamous cell carcinoma [49, 50]. In contrast, the histone demethylase JMJD3 is specific for the removal of trimethylation from H3K27 in facultative heterochromatin and is, therefore, a pivotal player in the derepression of cell cycle inhibitory genes. JMJD3 has also been linked to keratinocyte growth arrest and differentiation, with one of the derepressed targets being the cell-cycle inhibitor and tumor suppressor $p16^{INK4a}$ (see Sect. 4.1.1.3). It is not surprising, therefore, that JMJD3 inhibition contributes to mouse fibroblast transformation [51, 52]. JARID1B (PLU-1), which specifically demethylates H3K4me3/me2, was found to be overexpressed in both breast and prostate carcinoma [53, 54], while JARID1C was overexpressed in renal cell carcinoma [55]. Of particularly interest is that the expression of JARID1B and JARID1A expression was recently shown to be dynamic within cancer cell populations, and to be associated with drug resistance. For instance, only a small proportion of proliferating melanoma cells expressed JARID1B, and this correlated with a decreased proliferation rate of these cells in culture. However, an elevated expression of JARID1B was associated with an increased resistance against specific BRAF inhibitors targeted against the melanoma-associated, oncogenic $BRAF^{V600E}$ mutation. Consequently, JARID1B-expressing melanoma cells were shown to accumulate in BRAF inhibitor-treated melanoma cell cultures [56]. Similarly, JARID1A, which is also expressed transiently in only a small proportion of NSCLC cells, contributes to resistance to epidermal growth factor receptor (EGFR) tyrosine kinase inhibitors [57]. These data not only highlight a role for histone demethylation in tumor drug resistance, but also emphasize a new challenge related to the dynamic nature of the expression of these demethylases. This, in turn, suggests that a small proportion of tumor cells with adapted chromatin remodeling features may survive drug treatment and have the potential to establish new proliferating subpopulations and, thus, relapse.

Regarding the role of histone methylation and demethylation in cancer, therapeutic targeting is an option. Due to the fact that methylases and demethylases are quite specific with regards to the histone residues that they modify (as opposed to HATs and HDACs, which are quite promiscuous in their acetylation targets), the small molecules required for their inhibition need to be tailored for specific methylases or demethylases (for a review, see Ref. [58]).

3.1.8 Other Histone Modifications

Histones have further been shown to be phosphorylated at serine or tyrosine residues, ubiquitylated or sumoylated at lysine residues, and poly-ADP ribosylated at glutamate residues; they may also be altered by arginine deamination and proline isomerization [10, 59]. Clearly, when these modifications have been more extensively investigated, associations with carcinoma development and progression will emerge. Thus far, histone phosphorylation may have potential links to tumor development, with roles in DNA repair, mitosis, and apoptosis (as reviewed by Wang *et al.* [59]). The H2A.X histone variant is phosphorylated at Ser139, and is linked to DNA double-strand break repair [60], whereas aurora-kinase-B phosphorylates H3 at Ser10 (H3S10), which is crucial for chromosome condensation and appropriate segregation during mitosis [61]. Interestingly, an increased H3S10 phosphorylation has been associated with cell transformation in response to oncogenic RAS signaling [62]. Finally, H2BS14 phosphorylation follows apoptotic stimuli and plays a role in peroxide-induced apoptosis [63, 64].

3.2 ATP-Dependent Chromatin Remodeling

3.2.1 ATP-Dependent Chromatin-Remodeling Complexes

In humans, four families of ATP-dependent chromatin-remodeling complexes have been characterized to date:

- The SWI/SNF (named due to its homology to the yeast "switching defective/sucrose nonfermenting") complexes are thought to bind promoter regions by being attracted to acetylated histones via the bromodomain of their enzymatic cores BRG1 or BRM, and mainly activate gene expression. More recently, this view was challenged when it was shown that only about 12% of BRG1 complexes bound to active promoters while, quite unexpectedly, a higher BRG1 occupancy was found at distant regulatory enhancer or repressor motives. This distant BRG1–DNA interaction is proposed to confer predominantly gene-repressive functions (as reviewed in Ref. [3]). The underlying mechanisms of distant gene regulation remain to be further defined, but will no doubt reveal important chromatin-remodeling insights and have implications on malignant cell transformation.
- The ISWI ("imitation switching defective") complexes have so-called SANT and SLIDE domains, which mediate binding to unmodified histones.
- The NuRD/Mi-2/CHD ("nucleosome remodeling and acetylation"/"chromatin helicase DNA-binding") complexes have

chromodomains that specifically interact with methylated histones.

- The INO80 ("inositol requiring 80") complexes have SPLIT domains for histone interaction.

All four remodeling complexes have been associated with activation and repression of transcription (for a review, see Ref. [8]).

3.2.2 SWI/SNF in Carcinoma

SWI/SNF complexes are the most extensively studied human ATP-dependent chromatin-remodeling complexes, and thus have provided most of the data implicating ATP-dependent chromatin remodeling in malignant transformation and tumor progression. SWI/SNF complexes consist of either BRG1 or BRM as their enzymatic ATPase core. Both core proteins are able to interact via their bromodomain with acetylated histones, and to catalyze ATP hydrolyses to alter DNA–histone interactions. The SWI/SNF are multisubunit complexes that contain, in addition to BRG1 or BRM, seven to eleven subunits termed BRG1-associated factors (BAFs), and are approximately 2 MDa in size. Specific BAF subunit composition is thought to alter enzymatic activity, to facilitate binding to DNA sequence-specific transcription factors, and to target the remodeling complex to DNA and nucleosomes. Protein subunits in human cells include BAF155, BAF170, BAF180, BAF200, BAF270A, BAF270B, BAF53A, BAF53B, BAF57, BAF60a, BAF60b, BAF60c, BRG1, BRM, β-actin, and BAF47 (the BAF-associated number refers to the protein size, in kDa; see Table 4) [65, 66]. The subunit composition of SWI/SNF chromatin-remodeling factors can vary, depending on the biological context. For example, during the development stage, pluripotent, proliferating stem cells express mainly BRG1, which is paired with specific proliferation-associated BAFs, including BAF155 and BAF53A. However, once the

Tab. 4 Human SWI/SNF subunits.

Subunit	*Gene*	*Relevance*
BRG1	*SMARCA4*	Proliferating stem cells; context-dependent potential oncogenic and tumor-suppressor activity
BRM	*SMRACA2*	Differentiated cells; potential tumor suppressor
BAF155	*SMARCC1*	Proliferating stem cells, potential oncogene
BAF47 (hSNF5, INI)	*SMARCB1*	Bona fide tumor suppressor
BAF57	*SMARCE1*	Mediates hormone receptor-dependent transcription
BAF53A	*ACTL6A*	Proliferating stem cells, potential oncogene
BAF53B	*ACTL6B*	Differentiated cells
BAF180	*PBRM1*	Potential tumor suppressor
BAF270A (BAF250A)	*ARID1A*	Potential tumor suppressor
BAF270B (BAF250B)	*ARID1B*	—
BAF170	*SMARCC2*	Differentiated cells
BAF200	*ARID2*	—
BAF60a	*SMARCD1*	Potential tumor suppressor
BAF60b	*SMARCD2*	—
BAF60c	*SMARCD3*	—
β-Actin	*ACTB*	—

cells have differentiated, growth arrest-associated BAFs make up the SWI/SNF complex. In this case, BRG1 is replaced by BRM, while BAF155 is no longer part of the complex which, instead, contains BAF170 and BAF53B rather than BAF53A [3]. This highlights the fact that cells rely on different chromatin-remodeling factor compositions, depending on their proliferative fate and that, during tumorigenesis, cells may revert to a more "stem cell-like" chromatin-remodeling control by switching to a proliferative SWI/SNF composition. Moreover, as the multiple subunits of chromatin-remodeling complexes influence the enzymatic activity and specificity of the entire complex, alterations in any of the subunits may be associated with selective advantages for cancer cells.

3.2.2.1 SWI/SNF Subunits Linked to Cancer

BAF47 Germline mutations of the BAF47 subunit of SWI/SNF, which originally was identified as INI1 and is also known as *hSNF5*, predispose carriers to rhabdoid carcinomas [67, 68]. BAF47 is also frequently found to be somatically impaired in rhabdoid carcinomas [69], as well as in a subset of central primitive neuroectodermal tumors (cPNETs) and medulloblastomas [70]. BAF47 fulfills the criteria of a tumor suppressor gene, as mice heterozygous for BAF47 are tumor-prone (a complete BAF47 knockout is developmentally lethal) [71], while the re-expression of BAF47 in rhabdoid carcinoma cells causes cell-cycle arrest and an onset of senescence [72]. Arrest and senescence are associated with expression of the cell-cycle inhibitor $p16^{INK4a}$ [73, 74]. Importantly, the reintroduction of BAF47 does not affect the SWI/SNF-driven regulation of a range of other target genes [75]; this suggests that BAF47 provides specificity for the $p16^{INK4a}$ promoter and, indeed, it has been shown that BAF47 can bind the $p16^{INK4a}$ promoter [76, 77]. With $p16^{INK4a}$ regulation as a predominant target of BAF47, it is no surprise that BAF47 loss is associated with melanoma progression and chemoresistance [78], since it is known that $p16^{INK4a}$ plays a prominent role in melanoma predisposition and development (for reviews, see Refs [79, 80]) as well as chemoresistance [81]. Interestingly, one report has linked the tumor-suppressor function of BAF47 to the presence of BRG1. In this case, the investigators suggested that, in the absence of BAF47, BRG1 may function as an oncogene and drive tumor formation, whereas the loss of both BAF47 and BRG1 together had little effect [82].

BRG1 The potential role for BRG1 as an oncogene is controversial, although this would be in line with the observation that, in normal proliferating stem cells, the SWI/SNF complexes contain BRG1 rather than BRM together with a proliferation-associated BAF subunit composition, whereas differentiated nonproliferating cells rely on BRM as their SWI/SNF core unit [3]. Yet, there are other more compelling data which suggest that BRG1 may act as an oncogene. For example, an increased BRG1 expression was correlated with oral carcinoma development [83], while in prostate carcinoma an increased BRG1 and a reduced BRM expression was found in malignant versus normal tissue, and was associated with high-grade tumors as well as the invasive potential of prostate carcinoma cells [84]. Further, an elevated BRG1 expression was also identified in gastric carcinoma compared to normal mucosa, and increasing BRG1 levels were correlated with tumor progression [85, 86]. Additionally, mechanistic links have been identified to an oncogenic role of BRG1, based on the results of studies of colorectal carcinoma where BRG1 – but not BRM – was overexpressed. In this case, the *PTEN* gene – which encodes a tumor suppressor and phosphatase, which antagonizes the oncogenic phosphatidylinositol 3-kinase (PI3K)–AKT pathway – was shown to be repressed by the BRG1–SWI/SNF complex in colorectal carcinoma cells [87]. In addition, BRG1 was found to interact with the oncogene β-catenin and to promote the expression of β-catenin target genes [88].

In stark contrast to these reports, mounting evidence has emerged in recent years that BRG1 acts as a tumor suppressor. Such a role for BRG1 is highlighted by the fact that BRG1 hemizygous mice are susceptible to tumors (complete BRG1 knockout is developmentally lethal) [89], whilst a total loss of BRG1 has been identified in tumor tissue and also shown to enhance the development of human lung cancer [90]. The loss of BRG1 expression in lung cancer is also associated with a poor prognosis [91, 92], while more recently acquired data have suggested that a significant proportion of primary lung tumors already carry biallelic inactivated BRG1, placing it at an early stage of tumor development [93]. Moreover, BRG1 is silenced or mutated in many human tumor cell lines derived from breast, pancreatic, ovarian, lung, brain, prostate, and colon carcinomas [94–97]. BRG1 has also been shown to be lost in established neuroendocrine carcinomas and adenocarcinomas of the cervix [98]. The tumor-suppressive function of BRG1 is largely attributed to cell-cycle inhibitory roles (these are discussed more fully in Sect. 3.1.2).

It is important to emphasize that both oncogenic or tumor-suppressive roles for BRG1 are plausible, based on the assumption that the effect of BRG1 depends largely on the subunit composition of the predominant SWI/SNF complex (e.g., the presence of BAF47, as outlined in the section titled "BAF47"), and that this composition is determined by the biological context [3]. Therefore, even seemingly contradictive observations of increased or decreased levels of BRG1 in the same tumor type may be explained by genetic tumor heterogeneity, since not all tumors – even of the same type – necessarily employ the same pathway alterations to achieve malignant features. For example, as outlined above, the overexpression of BRG1 was correlated with a progression of colorectal and prostate carcinoma, which implied an oncogenic role [84, 87], whereas BRG1 was found to be lost or mutated in cell lines derived from these malignancies, which implied a tumor-suppressive role [97]. Moreover, whereas a common loss of BRG1 was identified in melanoma progression, there was no effect of BRG1 silencing on cell-cycle progression in melanoma cells [99]. Nonetheless, other investigators reported increased BRG1 levels associated with melanoma [100], a discrepancy which may be explained by known genomic melanoma heterogeneity being dependent on ethnic background and exposure to ultraviolet sunlight in an analyzed melanoma cohort [101, 102]. The UK-derived sample set of melanoma is likely to have reduced BAF47 leels, as was found in a melanoma cohort investigated by the same laboratory [78]. However, as this is in line with an oncogenic role for BRG1 [82], it would be interesting to determine whether the Australian melanoma cohort expresses BAF47.

BRM In contrast to BRG1, its homolog BRM appears to be rarely mutated in cancer, although recently a "hotspot" BRM mutation was identified in squamous and basal cell carcinomas [103]. BRM expression is more commonly reduced or lost. In prostate carcinoma, a reduced BRM expression was found in malignant versus normal tissue and also in high-grade lesions [84], while targeted BRM silencing in mice caused prostate hyperplasia. This was in line with the observation that BRM loss was associated with an increased proliferation in human prostate carcinoma cells [104]. BRM is also often lost or reduced in cell lines or tumor tissue such as gastric carcinoma and lung tumors, while a reduced expression of BRM in mice was associated with a carcinogen-induced development of lung carcinoma. Moreover, there is evidence that histone modifications play a role in the loss of BRM expression, since drugs that can inhibit HDAC function (i.e., HDAC inhibitors) may also restore BRM expression [86, 105]. Notably, the expression of BRM may play a role in preventing proliferation in oncogenic RAS-expressing fibroblasts, and this BRM function can also be prevented by exposure to HDAC inhibitors. Interestingly, however, the HDAC inhibitory effect is targeted to the acetylation of BRM rather than acetylation of histone tails. Typically, HDAC inhibitor treatment caused an accumulation of acetylated BRM and prevented its growth-inhibitory functions [106].

BAF57 The SWI/SNF subunit BAF57 mediates interaction with the hormone-dependent estrogen and androgen receptors. BAF57 is essential for estrogen-dependent transcriptional activation [107], and a truncating BAF57 mutation has been identified in a breast ductual carcinoma cell line. Although this mutant BAF57 no longer binds the estrogen receptor, it can still bind to the downstream coactivator SRC1a, rendering SRC1a transcriptional activity independent of estrogen [108]. Additionally, in a subset of prostate carcinoma, an increased BAF57 level was shown to be linked to an androgen-dependent proliferation [109].

BAF155 High levels of BAF155, which normally is a subunit found in proliferating stem cells [3], were associated with cervical carcinoma [110] and also identified in prostate carcinoma versus normal differentiated prostate tissue. Such an elevated BAF155 expression was also correlated with tumor progression and metastasis [111].

BAF180 The novel technology of exome sequencing has identified BAF180 as the second most common gene to be altered by truncating mutations in 41% of renal cell carcinoma [112]. Additionally, BAF180 has been shown, together with BRD7, to cooperate with *p53* in target gene expression during oncogene-induced senescence, and was suggested to act as a tumor suppressor since it was found to be mutated in breast cancer [113, 114].

BAF270A The SWI/SNF subunit BAF270A, which is deficient in approximately 30% of renal carcinomas and 10% of breast carcinomas [115], has been shown to be essential for normal cell cycle arrest, as it has been linked to reduced cyclin D1 and increased *p21* expression [116].

3.2.3 Other ATP-Dependent Chromatin-Remodeling Complexes in Cancer

3.2.3.1 ISWI Amplified Rsf-1, which forms part of the ISWI complex, was identified in prostate carcinoma in comparison to normal tissue [117]. It was also related to ovarian cancer and, more importantly, to the resistance of ovarian carcinoma to the drug paclitaxel; in contrast, the silencing of Rsf-1 in ovarian carcinoma cells led to an increased paclitaxel sensitivity [118].

Additionally, the overexpression of Rsf-1 resulted in DNA damage and chromosomal aberration [119].

3.2.3.2 NuRD/Mi-2/CHD CHD5 has been proposed as a tumor suppressor, since heterozygous mice develop spontaneous tumors, including squamous cell carcinoma [120].

MTA1 (metastasis-associated gene1) or its homologs *MTA2* and *MTA3*, form part of the NuRD/Mi-2/CHD complex. *MTA1*, which is a signal transduction target of the breast cancer-associated oncogene HER2, was found to be overexpressed in breast cancer cell lines and to be tightly linked to invasion in esophageal, colorectal, and gastric carcinomas [121]. The expression of *MTA3* is facilitated by estrogen; moreover, *MTA3* downregulates SNAI1 which, in turn, is a transcriptional repressor of the tumor suppressor and cell–cell adhesion molecule, E-cadherin. Estrogen receptor-negative breast carcinomas thus display a reduced *MTA3* level linked to a loss of E-cadherin and invasiveness [122].

The histone demethylase LSD1 is a component of the NuRD/Mi-2/CHD complex, and is involved in the transcriptional regulation of cell proliferation, survival, and epithelial-to-mesenchymal transition. LSD1 also inhibits the invasion of breast cancer cells *in vitro*, and suppresses breast cancer metastatic potential *in vivo*. In line with this latter finding, LSD1 is downregulated in breast carcinomas [46].

ZIP, a transcriptional repressor, recruits the NuRD/Mi-2/CHD complex to repress expression of the EGFR, inhibits cell proliferation, and suppresses breast carcinogenesis [123].

3.2.3.3 INO80 Recent screenings for proteins, which regulate telomere length, have led to the emergence of INO80 complex proteins, and deletion of the Ies3 subunit has resulted in an impaired growth and telomeric instability in yeast [124].

The Tip60 HAT, which complexes with INO80, acetylates p53 and is involved in the p53 activation in response to DNA damage; it also mediates the decision between cell cycle arrest and apoptosis [125]. Hence, Tip60 is thought to be a tumor suppressor, and has been proposed to be critical to oncogene-induced DNA damage response, which is impaired by reduced Tip60 levels [126].

The INO80 interacting proteins reptin (Tip48) and pontin (Tip49) are overexpressed in hepatocellular, colorectal, breast, gastric, bladder, and non-small-cell lung carcinoma. They are required for telomerase assembly, and they cooperate with Tip60 in DNA damage recognition and repair processes [127]. However, whether these functions of Tip60, reptin, and pontin are determined by interaction to the INO80 complex is unclear, as these proteins have HAT (Tip60) or ATPase (reptin and pontin) activities, and have been shown to interact with chromatin and other remodeling complexes [127]. Nevertheless, data from yeast are in line with a role for INO80 in DNA double-strand break repair and telomerase assembly [124, 128].

4
Remodeling the Hallmarks of Cancer

As outlined above, the main hallmarks of transformed malignant cells are excessive proliferation, avoidance of cell death, and invasive potential. A number of tumor-suppressor genes and oncogenes have been identified in pathways that affect these hallmarks, because their activity is frequently altered by mutations, genetic amplifications, or loss [1].

Moreover, most – if not all – oncogenes and tumor-suppressor genes are regulated by chromatin-remodeling events linked to malignant transformation. Some of these connections are reviewed in the following subsections.

4.1 Excessive Proliferation

4.1.1 Chromatin-Remodeling Complexes and Cell Cycle-Inhibiting Tumor Suppressor-Proteins

The main function of a broad range of tumor-suppressor proteins is that of cell cycle inhibition and, thus, an appropriate control of cell proliferation. These tumor suppressors are usually inactivated during malignant transformation.

4.1.1.1 Chromatin Remodeling and p53 The tumor suppressor and transcription factor p53 is often referred to as the "*guardian of the genome*" [129], to describe its essential role in responding to genomic stress (e.g., radiation- or carcinogen-induced DNA damage). p53 is mutated or deleted in more than 55% of all human cancers [130]. Remarkably, p53 responds to DNA damage by either cell-cycle arrest and DNA repair. Alternately, if the damage is too severe for an efficient repair, p53 will induce a tightly regulated cell death program (apoptosis) to prevent malignant transformation through the accumulation of genetic changes [129]. p53-induced cell cycle arrest is mainly facilitated by the transcriptional upregulation of the cell-cycle inhibitor $p21^{Waf1}$, and this arrest allows for DNA repair to be completed before replication can continue [131]. Alternately, if the DNA damage is too severe, then p53 may transcriptionally activate the transcription of several pro-apoptotic proteins to induce cell death [129]. Emerging evidence has demonstrated that p53 employs the chromatin-remodeling machinery as part of its normal function as a transcription factor, and to decide between growth arrest and cell death. For instance, various HATs not only transfer acetyl groups to histone residues to allow p53 access to its target promoters; rather, they also transfer acetyl groups to a number of transcription factors or transcriptional repressors, including p53. The latter is directly acetylated by p300, CBP, or PCAF at multiple carboxy-terminal residues, and this correlates to increased transcriptional activity [132]. Moreover, p53 acetylation has been linked to genomic stress, and is generally increased following DNA-damaging insults [133–135]. It has also been shown that, following DNA damage, the interaction between p53 and the HAT subunit hADA3 increased such that, in the presence of hADA3, the p53 transcriptional activity is strongly enhanced [136] and hADA3 is a target of E6 human papilloma virus antigen in cervical carcinoma [137]. Moreover, the acetylation of p53 competes directly with ubiquitylation by the ubiquitin ligase mdm2, thereby contributing to p53 accumulation, as the p53 levels are normally regulated by rapid ubiquitin-dependent proteasomal degradation [138]. HDAC2 was also shown to affect p53 transcriptional activity, though not by modulating the acetylation of p53 itself; rather, a knockdown of HDAC2 increased the p53-DNA binding activity [139].

p53 has been reported to interact with the SWI/SNF chromatin-remodeling complex, and this was critical for the transcriptional regulation of $p21^{Waf1}$. Interaction with SWI/SNF was proposed to occur via BAF60A, and this correlated with an increased $p21^{Waf1}$ transcription [140]. Remarkable other investigations linked BAF53A to p53 interaction with the SWI/SNF complex [141] which, in contrast, correlated with $p21^{Waf1}$ repression, and this was of course in line with BAF53A being a "proliferative" SWI/SNF subunit [3, 142]. Moreover, an N-terminal-truncated BAF53A caused cell death via p53 [143]. SWI/SNF and p53 have also provided another spin to the controversy between oncogenic and tumor-suppressive roles for BRG1. An analysis of the effects of BRG1 silencing on p53 stability showed that the knockdown of BRG1, but not BRM, reduced the ubiquitylation of p53 and thus led to an enhanced p53 accumulation and transcriptional upregulation of $p21^{Waf1}$. The authors of the study concluded that, in tumor cells which have retained wild-type p53, this tumor suppressor was disabled by BRG1 in cooperation with CBP [144]. The SWI/SNF subunit BAF180, on the other hand, has been shown (together with BRD7) to induce a number of *p53* target genes in normal cells during oncogene-induced senescence [113], and this agrees with evidence showing BAF180 to be critical for $p21^{Waf1}$ regulation [114]. Furthermore, the presence of p53 is directly linked to chromatin relaxing during DNA repair, and p53-null cells have changed the patterns of histone acetylation compared to p53-expressing cells; this suggests a more global effect of p53 on chromatin remodeling (for a review, see Ref. [145]). The main functions of p53, and the contribution of chromatin remodeling, are shown schematically in Fig. 3.

4.1.1.2 Chromatin Remodeling and Retinoblastoma Protein (pRb) The retinoblastoma protein (pRb) is often referred to as the "*master brake*" of cell proliferation, because the main pRb function is to inhibit the E2F transcription factor from expressing a range of proliferation genes [146]. Thus, active pRb causes cell cycle arrest. On the other hand, when cells receive growth factor signaling and are promoted towards cell division, pRb is sequentially phosphorylated by cyclin-dependent kinases (CDKs), which results in pRb inactivation and, consequently, the derepression of proliferation genes. Mutations of pRb have been associated with the childhood cancer, retinoblastoma. Moreover, the cell cycle regulatory pRb pathway is disturbed in over 90% of cancers [1]. During recent years, it has become increasingly clear that pRb relies on chromatin remodeling for transcriptional repression of extended chromatin regions. pRB was shown to recruit HDAC1 to E2F-responsive promoters [147], while HDAC1 assists the pRb-repressive function by the deacetylation of H4 [148]. The phosphorylation of pRb by CDK4, on the other hand, prevents HDAC1–pRb interaction [149, 150]. Furthermore, pRB is thought to engage the SWI/SNF chromatin-remodeling complex for E2F promoter silencing, and this provides an elegant model of the sequential expression of cyclins, which provide activity to CDKs (and thereby pRb inactivation) at specific stages of the cell cycle. The sequential activity of CDK4/cyclin D in G_1 phase of the cell cycle, of CDK2/cyclin E to promote the entry of DNA replication in S-phase, followed by CDK2/cyclin A to complete the DNA replication, requires an initiation by cyclin D, the expression of which is a direct consequence of growth factor signaling. CDK4/cyclin D complexes initiate phosphorylation of the active pRb, which occupies E2F promoters in complex with HDAC1 and SWI/SNF. In turn, this disrupts the pRb–HDAC1 interaction, which then causes derepression of the first E2F target genes, including cyclin E. This leads to active CDK2/cyclin E and further pRb phosphorylation, which dissociates SWI/SNF from the complex and allows for cyclin A expression and

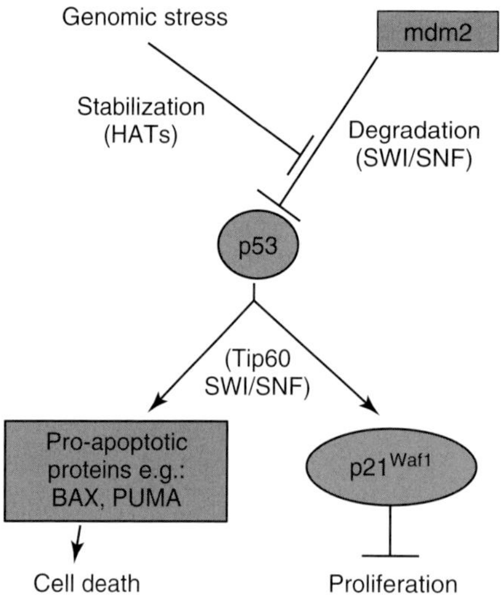

Fig. 3 Chromatin remodeling and p53. p53 has been referred to as the *"guardian of the genome,"* and is stabilized in response to genomic stress. Stabilization involves acetylation by HATs and opposes mdm2-dependent p53 degradation, which involves SWI/SNF. p53 transcriptionally upregulates either $p21^{Waf1}$, causing cell cycle arrest, or pro-apoptotic proteins, leading to cell death. The decision involves Tip60, and transcriptional p53 activities depend on SWI/SNF with specific subunit composition.

traversion through S-phase [150]. There is evidence that both BRG1 and BRM can function in these HDAC1–SWI/SNF–pRb repressive complexes, and it has been proposed that pRb requires BRG1 binding for cell-cycle arrest [151, 152], while some cancer-associated BRG1 mutations abrogate pRB-mediated arrest [153]. However, the cell cycle inhibitory effect created by the reintroduction of BRG1 is only seen convincingly in cells that are deficient in both BRG1 and BRM, such as SW-13 adenocarcinoma cells [99]. It should also be noted that this arrest may not involve the direct interaction between pRb and BRG1, as it could be shown that SW-13 cells would arrest upon ectopic BRG1 expression, even when a BRG1 mutant was expressed that was disabled in pRb interaction. Instead, BRG1 led to an upregulation of the CDK inhibitor $p21^{Waf1}$, leading in turn to a hypophosphorylated active pRb and cell cycle arrest [154]. Evidence was also produced that SWI/SNF is involved in the transcriptional upregulation of the CDK inhibitors, $p16^{INK4a}$ and $p15^{INK4b}$ [73, 154], as well as the repression of cyclin D expression [155], and all of these SWI/SNF targets were shown to promote pRb activity. A loss of the SWI/SNF subunit BAF47 impaired $p16^{INK4a}$ and $p21^{Waf1}$ expression [76] (the relevance of BAF47 loss in carcinoma was discussed above; see the section titled "BAF47"). Details of the pRb pathway, and its cooperation with chromatin remodeling, is shown schematically in Fig. 4. pRb also interacts with heterochromatin-forming proteins to induce senescence (as discussed in Sect. 3.2.1).

4.1.1.3 Chromatin Remodeling and $p16^{INK4a}$

The main CDK inhibitors relevant in carcinoma are $p21^{Waf1}$ and $p16^{INK4a}$. $p21^{Waf1}$ is not mutated in carcinoma, but

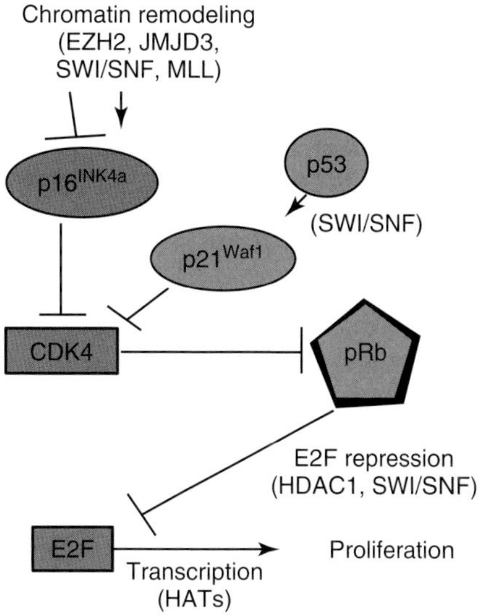

Fig. 4 Chromatin remodeling and pRb. The main function of pRb is to repress E2F from transcriptional activation of proliferation genes, which involves HDAC1 and SWI/SNF. pRb itself can be inhibited by phosphorylation catalyzed by CDKs, while the CDK inhibitors $p21^{Waf1}$ and $p16^{INK4a}$ in turn can keep pRb active by inhibiting the CDKs. Transcriptional regulation of these inhibitors requires SWI/SNF. $p16^{INK4a}$ in particular is strongly regulated by chromatin-remodeling events. The histone methyltransferase EZH2 efficiently represses $p16^{INK4a}$, which is directly opposed by JMJD3, while the histone methyltransferase MLL activates $p16^{INK4a}$ expression.

is deregulated via alterations of p53; in contrast, $p16^{INK4a}$ is one of the most frequently mutated, deleted, or silenced genes in cancer [156]. Moreover, individuals with germline mutations in $p16^{INK4a}$ are at risk of developing pancreatic cancer, and carry a highly increased risk of developing melanoma because these mutations generally produce nonfunctional proteins [80, 157, 158]. The main role of $p16^{INK4a}$ is to inhibit the cyclin D-dependent kinases CDK4 (and CDK6), which places $p16^{INK4a}$ at the forefront of cell-cycle inhibition in early G_0/G_1 phase, where it is involved in the central decision of whether to enter proliferation, or not. In normal cells, $p16^{INK4a}$ causes cell cycle arrest associated with differentiation, and it probably plays an even more important role in senescence (see Sect. 4.1.3). Not surprisingly, $p16^{INK4a}$ is therefore efficiently repressed during development and in proliferating progenitor cell populations, whereas the levels are increased when cells differentiate and senesce [17]. $p16^{INK4a}$ regulation on the level of chromatin modifications plays an important role, and alterations in these mechanisms have been associated with cancer. For example, a reversible repression of $p16^{INK4a}$ is achieved by the formation of facultative heterochromatin, which is recognized by the repressor BMI1, a member of the polycomb group of proteins, a well-characterized repressor of $p16^{INK4a}$. More recently, a critical role for the histone methyltransferase EZH2 was identified in $p16^{INK4a}$ repression. In this case, EZH2 cooperates with BMI1 by the trimethylation of H3K27, which attracts BMI1 binding [17]. Moreover, the histone H3 demethylase

Ndy1 enhances p16^{INK4a} repression by opposing the EZH2 downregulation, usually in association with senescence, and thereby enhances H3K27 trimethylation and BMI1 binding to the p16^{INK4a} promoter. Consequently, Ndy1 overexpression did cause mouse embryonic fibroblast transformation [159]. On the other hand, EZH2 may be downregulated in concert with an upregulation of its opponent, the H3K27 demethylase JMJD3, during HRAS^{G12V}-mediated p16^{INK4a} upregulation and senescence in human fibroblasts. This is in line with the fact that JMJD3 levels are significantly higher in growth-arrested nevi compared to normal skin, while ectopic JMJD3 induces p16^{INK4a} upregulation and senescent features in human fibroblasts [160].

When cells differentiate and reach their finite life span, the SWI/SNF chromatin remodeling complex replaces the BMI1 repressor, which in turn relaxes the chromatin structures around the p16^{INK4a} promoter region. This effect is strictly dependent on the BRG1 and BAF47 subunits, and allows transcription factor access to activate p16^{INK4a} transcription [74].

In contrast to the repressing H3K27me3, H3K4me3 was shown to be required for p16^{INK4a} expression, and was shown to depend on the presence of two components of the H3K4 methyltransferase complex, MLL1 and pRbBP5. Moreover, H3K4me3 status – and thus p16^{INK4a} expression – were reportedly modulated by the pRbBP5 binding partner DDB1, which is a component of the ultraviolet light-damaged DNA-binding protein. DDB1 knockdown reduced H3K4me3 of the p16^{INK4a} promoter and reduced p16^{INK4a} expression. Importantly, in human fibroblasts both MLL1 and DDB1 were crucial in the oncogene-induced upregulation of p16^{INK4a} by H-Ras^{G12V} [161].

4.1.2 Chromatin-Remodeling Complexes and Oncogenic Signaling

The control of oncogenic signaling pathways provides a powerful mechanism to render a cell independent of external growth stimuli, and to ignore anti-proliferative signals as well as to promote resistance to programmed cell death. Hence, such control represents a frequent adaptation to promote malignant transformation.

4.1.2.1 MAPK Pathway The mitogen-activated phosphorylation kinase (MAPK) signaling pathway is frequently aberrantly activated in human cancers, through constitutively activating mutations in RAS or the downstream kinase BRAF, and is thought to confer proliferation and survival mechanisms. There is evidence that MAPK signaling involves the modulation of chromatin. Indeed it has been shown that, upon MAPK signaling, H3S10 is phosphorylated by the H3 kinase MSK1, in association with an increased acetylation of H3 at MAPK target promoters. H1S3 was also found to be phosphorylated in activated RAS expressing mouse fibroblasts, and may destabilize the chromatin structure (for a review, see Ref. [62]). The SWI/SNF core enzyme BRM is proposed to be capable of preventing proliferation in oncogenic RAS-expressing fibroblasts [106]. Consequently, the expression of BRM – but not of BRG1 – is diminished upon the RAS transformation of mouse fibroblasts [162]. A recent search for genes that can cooperate with mutated, oncogenic BRAF^{V600E} in melanoma formation identified SETDB1, which is a histone methyltransferase that facilitates H3K9 trimethylation but opposes BRAF^{V600E}-induced senescence, and thereby accelerates melanoma formation in a zebrafish model. SETDB1 is also frequently overexpressed in human melanomas, and also in NSCLC, small-cell lung cancer (SCLC), as well as ovarian, hepatocellular, and breast carcinoma [163].

4.1.2.2 PI3K-AKT Pathway Another important signaling pathway that very frequently is abnormally activated in cancer is the PI3K–AKT pathway, which facilitates increased proliferation and survival, including resistance against chemotherapeutic drugs [164]. Previously, AKT has been shown to interact with SWI/SNF and to phosphorylate the subunit BAF155 (and potentially also BAF47), though the functional outcome of this modification is unclear [165]. Phosphorylation of the acetyltransferase p300 on Ser1834 by AKT was shown to be critical for the p300-mediated histone acetylation and activation of gene expression [166]. Recently, a link between the histone H3K27 methylase EZH2 and the PI3K–AKT pathway was proposed. In this case, the overexpression of EZH2, which is common in a range of carcinomas (see Sect. 3.1.7), was associated with an upregulation of AKT1 and its active, phosphorylated form. Moreover, AKT1 was critical for the cellular localization of the breast cancer-associated BRCA1 tumor suppressor and genomic stability, as regulated by EZH2 [167].

4.1.3 Senescence

Senescence is a form of stable cell cycle arrest that permanently prevents cells from re-entering the cell cycle in response to physiological stimuli. The onset of senescence is a response to various genomic stress features, including telomere shortening, DNA damage, or oxidative stress. Senescence is seen as a major tumor-suppressive mechanism to prevent cells with deleterious DNA damage from dividing and potentially transmitting any transforming genetic changes to their daughter cells. Not surprisingly, senescence is also activated in response to oncogenes such as B-RAF^{V600E} or N-RAS^{Q61K} [168–170]; this may be a last resort to prevent malignant transformation by these oncogenes, as their signaling potently opposes programmed cell death by altering the balance of pro- and anti-apoptotic proteins [171]. pRb is required for an intact senescence program, and the re-introduction of pRb into SAOS osteosarcoma cells that have lost pRb expression may cause senescence [172]. In contrast, the inactivation of pRb with the viral oncoprotein E1A prevents senescence, whereas a mutant E1A, which is impaired in pRb binding, is unable to prevent senescence [173]. Interestingly, the mutant E1A, which is able to bind pRb but unable to interact with the HDAC p300 and p400, leads to a less-efficient development of senescence [173]. These data highlight the role of chromatin remodeling in cellular senescence programs; indeed, during senescence E2F-responsive genomic promoter regions are stably repressed from transcription by the establishment of heterochromatin regions. These regions are visible as microscopic "senescence-associated heterochromatin foci" (SAHF) when cells are stained with certain DNA-intercalating dyes, such as DAPI. SAHF formation coincides with the recruitment of heterochromatin proteins and pRb to E2F-responsive promoters. A number of heterochromatin-associated histone modifications are characteristic for SAHF; these include a lack of H3K9Ac and H3K4me3, but an accumulation of H3K9me3. The latter attracts HP1 proteins, which are required for heterochromatin assembly [173]. Further, SAHF H3K9 trimethylation is thought to be very stable and to prevent HATs from catalyzing histone acetylations. This contributes to a very secure silencing of E2F-responsive proliferation genes, which is not reversible by physiological stimuli [173]. Importantly, although the p53–p21 pathway was shown to be capable of inducing a number of senescent features – and does so in response to DNA damage, telomere dysfunction and during oncogene-induced senescence [174, 175] – E2F promoter silencing via SAHF formation requires an intact p16–pRb pathway. In fact, it has been argued that SAHF formation is critical for the potency of the senescence program in tumor suppression [170, 173]. As expected, SAHF formation is prevented by cancer-associated $p16^{INK4a}$ mutations [176]. Additionally, there is some evidence that the SWI/SNF complex is required for senescence onset, and the introduction of BRG1 into cells induces senescent features, although this is only convincing in cells that also lack the BRM ATPase [154]. Interestingly, it was found that BRG1 can interact with $p16^{INK4a}$, although the absence of BRG1 neither prevents $p16^{INK4a}$-induced growth arrest nor senescence or SAHF formation. It should be noted that the WMM1175 melanoma cells used in these experiments also express BRM, and so still have functional SWI/SNF complexes [99]. This is in line with a suggested role for BRM in melanocyte senescence, since it was shown that BRM was recruited and required – albeit transiently – by the pRb–HDAC1 complex during the initiation of SAHF [177].

H3K9 trimethylation during SAHF formation is thought to be catalyzed by either the SUV39H1 methyltransferase which, together with HP1 interacts with pRb during E2F promoter silencing in senescence [178], or the RIZ1/PRDM2 H3K9 methyltransferase, which was also shown to cooperate in pRb gene repression and is inactivated in colon, breast, and gastric carcinoma [179]. Furthermore, a search for H3K9me3 interacting proteins to identify proteins involved in the senescence program identified JMJD2C, which is a H3K9- and H3K36-specific demethylase. JMJD2C is a direct antagonist of the SUV39H1 H3K9 methyltransferase, and its overexpression in carcinomas is therefore not surprising [48].

Consistent with gene silencing and decreased acetylation during senescence is the downregulation of the histone acetylases p300/CBP, as observed in melanocytes reaching their finite lifespan [180]. Chromatin remodeling in senescence is shown schematically in Fig. 5.

4.2
Avoidance of Cell Death

Clearly, oncogenic signaling pathways which employ the chromatin-remodeling machinery (Sect. 3.2) can influence apoptotic cell fate. Additionally, as apoptosis involves the fragmentation of DNA it becomes clear that chromatin-remodeling processes must be more intimately involved, to allow the access of endonucleases for DNA cleavage. It is now also recognized that the histone phosphorylations of H2A, H2B, H3, and H4, the dephosphorylation of H1, the acetylation of H3B and H4, and the deacetylations of H4, together with methylations of H3 and H4 and the de-ubiquitylation of H2A, form part of the apoptotic process, and that some of these events are catalyzed by caspases while others involve conventional histone-modifying enzymes (for a review, see Ref. [181]). Furthermore, HDAC inhibitors have shown to have dramatic effects on cultured cells, which includes the onset of apoptosis [40].

There is also some evidence for a contribution of the SWI/SNF complex to the regulation of apoptosis, although this field has not been fully investigated. Although

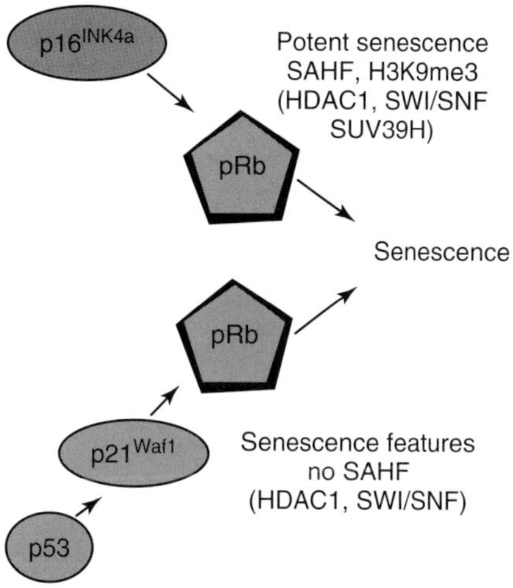

Fig. 5 Chromatin remodeling and senescence. The secure repression of E2F target genes by pRb is central to the cellular senescence program. Regardless of whether this is initiated by $p21^{Waf1}$ or $p16^{INK4a}$, this involves HDAC1 and SWI/SNF action. While p53-$p21^{Waf1}$-induced senescence displays several distinctive features (growth arrest, cell flattening, cell vacuolization, and increased β-galactosidase levels), only $p16^{INK4a}$ action on pRb leads to senescence involving H3K9 trimethylation by SUV39H. This leads to a distinctive formation of SAHF that, arguably, contributes potency to the senescence program.

the knockdown of BAF60A was shown to decrease apoptosis, BAF60A was also proposed to interact with p53 and to enhance p53 transcriptional activity [140]; the BAF57 subunit was suggested to induce apoptosis by an upregulation of the pro-apoptotic tumor suppressor, CYLD [182]. On the other hand, N-terminal-truncated BAF53A caused cell death via p53, which suggested a BAF53A function in survival [143], while others proposed a critical role for SWI/SNF complexes to prevent apoptosis in response to DNA damage, as BRG1 knockdown increased apoptosis and this was regulated by a prolonged p53 activity [183]. Furthermore, the activity of the NuRD/Mi-2/CHD subunit MTA1 was shown to repress p53 activity in prostate carcinoma cells, while the inhibition of MTA1 contributed to p53-driven apoptosis [184].

4.3 Invasive Potential

The most important cell adhesion molecule in the normal homeostasis of epithelial cells is arguably E-cadherin. As a cell–cell adhesion molecule, E-cadherin is pivotal for normal cell–cell interaction so as to form a tight epithelial tissue; indeed, the loss of E-cadherin has been linked with an epithelial to mesenchymal transition of tumor cells. Consequently, the loss of E-cadherin would be critical in allowing cells to dislodge from their normal epithelial tissue environment and to invade other tissues [185]. The expression of E-cadherin also parallels that of $p16^{INK4a}$, in a sense that it is not expressed during development (when cells must migrate to find their appropriate tissue location) yet its level of expression is high during differentiation.

Not surprisingly, E-cadherin is (similar to $p16^{INK4a}$) repressed by facultative heterochromatin formation involving the H3K27 methyltransferase, EZH2 [186]. Furthermore, it was noted above that MTA3 (which is a subunit of the ATP-dependent chromatin-remodeling complex NuRD/Mi-2/CHD) regulates SNAI1, a transcriptional E-cadherin repressor, and also reduces MTA3, which is linked to a loss of E-cadherin and to breast carcinoma invasiveness [122]. The NuRD/Mi-2/CHD subunit LSD1 was also linked to an inhibition of breast cancer invasiveness, and reduced levels of LSD1 have been found in breast carcinomas [46]. The SWI/SNF ATP-dependent chromatin-remodeling complex is also associated with epithelial to mesenchymal transitions. In fact, the introduction of either BRG1 or BRM into cells deficient of both ATPases led to a reversal of CpG island methylation (a regulative modification at the DNA level) and a derepression of the E-cadherin promoter; subsequently, the SWI/SNF complex was found to engage with the E-cadherin promoter to activate transcription [187]. In contrast, it has been shown recently that BRG1 can interact with the ZEB1 protein and aid the repression of E-cadherin. ZEB1–BRG1 colocalization was also observed in invasive colorectal carcinoma cells [188]. Moreover, BRG1 was shown to be critical for matrix metalloproteinase-2 (MM-2) expression, which is correlated with tumor invasion and angiogenesis. BRG1 was shown to facilitate MM-2 expression by recruiting the transcription factors SP1, SP3, and AP2 to the MM-2 promoter [189] and, as noted above, in prostate carcinoma an increased BRG1 and reduced BRM expression was found in malignant versus normal tissue and associated with the invasive potential of the tumor cells [84]. Thus, it is clear that BRG1 possesses both pro- and anti-invasive properties in carcinoma cells, depending on its binding partners.

5
Conclusion

Currently, an overwhelming amount of evidence exists demonstrating the importance of epigenetic alterations of gene regulation in carcinoma. It is recognized that histone modifications are associated with carcinoma progression and resistance to therapeutic intervention, and that they modulate the expression and function of important tumor suppressors and oncogenes. Furthermore, oncogenic signaling pathways engage the functions of ATP-dependent chromatin-remodeling complexes, which is perhaps not surprising as these complexes have important functions in modulating the expression of tumor suppressors. Thus, chromatin remodeling has been linked to all aspects of tumor evolution, and affects the most important hallmarks of cancer, namely proliferation, survival, and invasion. Not surprisingly, chromatin remodeling has been linked to all frequently studied types of carcinoma.

References

1. Hanahan, D. and Weinberg, R.A. (2000) The hallmarks of cancer. *Cell*, **100** (1), 57–70.
2. Cairns, J. (1975) Mutation selection and the natural history of cancer. *Nature*, **255** (5505), 197–200.
3. Hargreaves, D.C. and Crabtree, G.R. (2011) ATP-dependent chromatin remodeling: genetics, genomics and mechanisms. *Cell Res.*, **21** (3), 396–420.
4. Frank, S.A. and Nowak, M.A. (2004) Problems of somatic mutation and cancer. *BioEssays*, **26** (3), 291–299.

5. Davis, P.K. and Brackmann, R.K. (2003) Chromatin remodeling and cancer. *Cancer Biol. Ther.*, **2** (1), 22–29.
6. Luger, K., Mader, A.W., Richmond, R.K., Sargent, D.F., and Richmond, T.J. (1997) Crystal structure of the nucleosome core particle at 2.8 Å resolution. *Nature*, **389** (6648), 251–260.
7. King, M.W. © 1996–2011; *themedicalbiochemistrypage.org/dna.html.*
8. Wang, G.G., Allis, C.D., and Chi, P. (2007) Chromatin remodeling and cancer, Part II: ATP-dependent chromatin remodeling. *Trends Mol. Med.*, **13** (9), 373–380.
9. Saha, A., Wittmeyer, J., and Cairns, B.R. (2006) Chromatin remodelling: the industrial revolution of DNA around histones. *Nat. Rev. Mol. Cell Biol.*, **7** (6), 437–447.
10. Kouzarides, T. (2007) SnapShot: histone-modifying enzymes. *Cell*, **131** (4), 822.
11. Roth, S.Y., Denu, J.M., and Allis, C.D. (2001) Histone acetyltransferases. *Annu. Rev. Biochem.*, **70**, 81–120.
12. Thiagalingam, S., Cheng, K.H., Lee, H.J., Mineva, N., Thiagalingam, A., and Ponte, J.F. (2003) Histone deacetylases: unique players in shaping the epigenetic histone code. *Ann. N. Y. Acad. Sci.*, **983**, 84–100.
13. Taddei, A., Roche, D., Bickmore, W.A., and Almouzni, G. (2005) The effects of histone deacetylase inhibitors on heterochromatin: implications for anticancer therapy? *EMBO Rep.*, **6** (6), 520–524.
14. Agger, K., Christensen, J., Cloos, P.A., and Helin, K. (2008) The emerging functions of histone demethylases. *Curr. Opin. Genet. Dev.*, **18** (2), 159–168.
15. Ng, S.S., Yue, W.W., Oppermann, U., and Klose, R.J. (2009) Dynamic protein methylation in chromatin biology. *Cell. Mol. Life Sci.*, **66** (3), 407–422.
16. Taverna, S.D., Li, H., Ruthenburg, A.J., Allis, C.D., and Patel, D.J. (2007) How chromatin-binding modules interpret histone modifications: lessons from professional pocket pickers. *Nat. Struct. Mol. Biol.*, **14** (11), 1025–1040.
17. Bracken, A.P., Kleine-Kohlbrecher, D., Dietrich, N., Pasini, D., Gargiulo, G., Beekman, C., Theilgaard-Monch, K., Minucci, S., Porse, B.T., Marine, J.C., Hansen, K.H., and Helin, K. (2007) The Polycomb group proteins bind throughout the INK4A-ARF locus and are disassociated in senescent cells. *Genes Dev.*, **21** (5), 525–530.
18. Byvoet, P., Shepherd, G.R., Hardin, J.M., and Noland, B.J. (1972) The distribution and turnover of labeled methyl groups in histone fractions of cultured mammalian cells. *Arch. Biochem. Biophys.*, **148** (2), 558–567.
19. Jenuwein, T. and Allis, C.D. (2001) Translating the histone code. *Science*, **293** (5532), 1074–1080.
20. Fraga, M.F., Ballestar, E., Villar-Garea, A., Boix-Chornet, M., Espada, J., Schotta, G., Bonaldi, T., Haydon, C., Ropero, S., Petrie, K., Iyer, N.G., Perez-Rosado, A., Calvo, E., Lopez, J.A., Cano, A., Calasanz, M.J., Colomer, D., Piris, M.A., Ahn, N., Imhof, A., Caldas, C., Jenuwein, T., and Esteller, M. (2005) Loss of acetylation at Lys16 and trimethylation at Lys20 of histone H4 is a common hallmark of human cancer. *Nat. Genet.*, **37** (4), 391–400.
21. Seligson, D.B., Horvath, S., Shi, T., Yu, H., Tze, S., Grunstein, M., and Kurdistani, S.K. (2005) Global histone modification patterns predict risk of prostate cancer recurrence. *Nature*, **435** (7046), 1262–1266.
22. Elsheikh, S.E., Green, A.R., Rakha, E.A., Powe, D.G., Ahmed, R.A., Collins, H.M., Soria, D., Garibaldi, J.M., Paish, C.E., Ammar, A.A., Grainge, M.J., Ball, G.R., Abdelghany, M.K., Martinez-Pomares, L., Heery, D.M., and Ellis, I.O. (2009) Global histone modifications in breast cancer correlate with tumor phenotypes, prognostic factors, and patient outcome. *Cancer Res.*, **69** (9), 3802–3809.
23. Tzao, C., Tung, H.J., Jin, J.S., Sun, G.H., Hsu, H.S., Chen, B.H., Yu, C.P., and Lee, S.C. (2009) Prognostic significance of global histone modifications in resected squamous cell carcinoma of the esophagus. *Mod. Pathol.*, **22** (2), 252–260.
24. Barlesi, F., Giaccone, G., Gallegos-Ruiz, M.I., Loundou, A., Span, S.W., Lefesvre, P., Kruyt, F.A., and Rodriguez, J.A. (2007) Global histone modifications predict prognosis of resected non small-cell lung cancer. *J. Clin. Oncol.*, **25** (28), 4358–4364.
25. Manuyakorn, A., Paulus, R., Farrell, J., Dawson, N.A., Tze, S., Cheung-Lau, G., Hines, O.J., Reber, H., Seligson, D.B., Horvath, S., Kurdistani, S.K., Guha, C., and Dawson, D.W. (2010) Cellular histone modification patterns predict prognosis and

treatment response in resectable pancreatic adenocarcinoma: results from RTOG 9704. *J. Clin. Oncol.*, **28** (8), 1358–1365.

26. Cheung, N., Chan, L.C., Thompson, A., Cleary, M.L., and So, C.W. (2007) Protein arginine-methyltransferase-dependent oncogenesis. *Nat. Cell Biol.*, **9** (10), 1208–1215.
27. Bai, X., Wu, L., Liang, T., Liu, Z., Li, J., Li, D., Xie, H., Yin, S., Yu, J., Lin, Q., and Zheng, S. (2008) Overexpression of myocyte enhancer factor 2 and histone hyperacetylation in hepatocellular carcinoma. *J. Cancer Res. Clin. Oncol.*, **134** (1), 83–91.
28. Shandilya, J., Swaminathan, V., Gadad, S.S., Choudhari, R., Kodaganur, G.S., and Kundu, T.K. (2009) Acetylated NPM1 localizes in the nucleoplasm and regulates transcriptional activation of genes implicated in oral cancer manifestation. *Mol. Cell. Biol.*, **29** (18), 5115–5127.
29. Iyer, N.G., Ozdag, H., and Caldas, C. (2004) p300/CBP and cancer. *Oncogene*, **23** (24), 4225–4231.
30. Gayther, S.A., Batley, S.J., Linger, L., Bannister, A., Thorpe, K., Chin, S.F., Daigo, Y., Russell, P., Wilson, A., Sowter, H.M., Delhanty, J.D., Ponder, B.A., Kouzarides, T., and Caldas, C. (2000) Mutations truncating the EP300 acetylase in human cancers. *Nat. Genet.*, **24** (3), 300–303.
31. Halkidou, K., Gaughan, L., Cook, S., Leung, H.Y., Neal, D.E., and Robson, C.N. (2004) Upregulation and nuclear recruitment of HDAC1 in hormone refractory prostate cancer. *Prostate*, **59** (2), 177–189.
32. Choi, J.H., Kwon, H.J., Yoon, B.I., Kim, J.H., Han, S.U., Joo, H.J., and Kim, D.Y. (2001) Expression profile of histone deacetylase 1 in gastric cancer tissues. *Jpn. J. Cancer Res.*, **92** (12), 1300–1304.
33. Wilson, A.J., Byun, D.S., Popova, N., Murray, L.B., L'Italien, K., Sowa, Y., Arango, D., Velcich, A., Augenlicht, L.H., and Mariadason, J.M. (2006) Histone deacetylase 3 (HDAC3) and other class I HDACs regulate colon cell maturation and p21 expression and are deregulated in human colon cancer. *J. Biol. Chem.*, **281** (19), 13548–13558.
34. Ishihama, K., Yamakawa, M., Semba, S., Takeda, H., Kawata, S., Kimura, S., and Kimura, W. (2007) Expression of HDAC1 and CBP/p300 in human colorectal carcinomas. *J. Clin. Pathol.*, **60** (11), 1205–1210.
35. Zhang, Z., Yamashita, H., Toyama, T., Sugiura, H., Ando, Y., Mita, K., Hamaguchi, M., Hara, Y., Kobayashi, S., and Iwase, H. (2005) Quantitation of HDAC1 mRNA expression in invasive carcinoma of the breast. *Breast Cancer Res. Treat.*, **94** (1), 11–16.
36. Zhu, P., Huber, E., Kiefer, F., and Gottlicher, M. (2004) Specific and redundant functions of histone deacetylases in regulation of cell cycle and apoptosis. *Cell Cycle*, **3** (10), 1240–1242.
37. Huang, B.H., Laban, M., Leung, C.H., Lee, L., Lee, C.K., Salto-Tellez, M., Raju, G.C., and Hooi, S.C. (2005) Inhibition of histone deacetylase 2 increases apoptosis and p21Cip1/WAF1 expression, independent of histone deacetylase 1. *Cell Death Differ.*, **12** (4), 395–404.
38. Song, J., Noh, J.H., Lee, J.H., Eun, J.W., Ahn, Y.M., Kim, S.Y., Lee, S.H., Park, W.S., Yoo, N.J., Lee, J.Y., and Nam, S.W. (2005) Increased expression of histone deacetylase 2 is found in human gastric cancer. *APMIS*, **113** (4), 264–268.
39. Zhang, Z., Yamashita, H., Toyama, T., Sugiura, H., Omoto, Y., Ando, Y., Mita, K., Hamaguchi, M., Hayashi, S., and Iwase, H. (2004) HDAC6 expression is correlated with better survival in breast cancer. *Clin. Cancer Res.*, **10** (20), 6962–6968.
40. Bolden, J.E., Peart, M.J., and Johnstone, R.W. (2006) Anticancer activities of histone deacetylase inhibitors. *Nat. Rev. Drug Discov.*, **5** (9), 769–784.
41. Vire, E., Brenner, C., Deplus, R., Blanchon, L., Fraga, M., Didelot, C., Morey, L., Van Eynde, A., Bernard, D., Vanderwinden, J.M., Bollen, M., Esteller, M., Di Croce, L., de Launoit, Y., and Fuks, F. (2006) The Polycomb group protein EZH2 directly controls DNA methylation. *Nature*, **439** (7078), 871–874.
42. Moss, T.J. and Wallrath, L.L. (2007) Connections between epigenetic gene silencing and human disease. *Mutat. Res.*, **618** (1–2), 163–174.
43. Kleer, C.G., Cao, Q., Varambally, S., Shen, R., Ota, I., Tomlins, S.A., Ghosh, D., Sewalt, R.G., Otte, A.P., Hayes, D.F., Sabel, M.S., Livant, D., Weiss, S.J., Rubin, M.A., and Chinnaiyan, A.M. (2003) EZH2 is a marker of aggressive breast cancer and promotes

neoplastic transformation of breast epithelial cells. *Proc. Natl Acad. Sci. USA*, **100** (20), 11606–11611.
44. Berezovska, O.P., Glinskii, A.B., Yang, Z., Li, X.M., Hoffman, R.M., and Glinsky, G.V. (2006) Essential role for activation of the Polycomb group (PcG) protein chromatin silencing pathway in metastatic prostate cancer. *Cell Cycle*, **5** (16), 1886–1901.
45. Varambally, S., Dhanasekaran, S.M., Zhou, M., Barrette, T.R., Kumar-Sinha, C., Sanda, M.G., Ghosh, D., Pienta, K.J., Sewalt, R.G., Otte, A.P., Rubin, M.A., and Chinnaiyan, A.M. (2002) The polycomb group protein EZH2 is involved in progression of prostate cancer. *Nature*, **419** (6907), 624–629.
46. Wang, Y., Zhang, H., Chen, Y., Sun, Y., Yang, F., Yu, W., Liang, J., Sun, L., Yang, X., Shi, L., Li, R., Li, Y., Zhang, Y., Li, Q., Yi, X., and Shang, Y. (2009) LSD1 is a subunit of the NuRD complex and targets the metastasis programs in breast cancer. *Cell*, **138** (4), 660–672.
47. Kahl, P., Gullotti, L., Heukamp, L.C., Wolf, S., Friedrichs, N., Vorreuther, R., Solleder, G., Bastian, P.J., Ellinger, J., Metzger, E., Schule, R., and Buettner, R. (2006) Androgen receptor coactivators lysine-specific histone demethylase 1 and four and a half LIM domain protein 2 predict risk of prostate cancer recurrence. *Cancer Res.*, **66** (23), 11341–11347.
48. Cloos, P.A., Christensen, J., Agger, K., Maiolica, A., Rappsilber, J., Antal, T., Hansen, K.H., and Helin, K. (2006) The putative oncogene GASC1 demethylates tri- and dimethylated lysine 9 on histone H3. *Nature*, **442** (7100), 307–311.
49. Italiano, A., Attias, R., Aurias, A., Perot, G., Burel-Vandenbos, F., Otto, J., Venissac, N., and Pedeutour, F. (2006) Molecular cytogenetic characterization of a metastatic lung sarcomatoid carcinoma: 9p23 neocentromere and 9p23-p24 amplification including JAK2 and JMJD2C. *Cancer Genet. Cytogenet.*, **167** (2), 122–130.
50. Yang, Z.Q., Imoto, I., Fukuda, Y., Pimkhaokham, A., Shimada, Y., Imamura, M., Sugano, S., Nakamura, Y., and Inazawa, J. (2000) Identification of a novel gene, GASC1, within an amplicon at 9p23-24 frequently detected in esophageal cancer cell lines. *Cancer Res.*, **60** (17), 4735–4739.
51. Sen, G.L., Webster, D.E., Barragan, D.I., Chang, H.Y., and Khavari, P.A. (2008) Control of differentiation in a self-renewing mammalian tissue by the histone demethylase JMJD3. *Genes Dev.*, **22** (14), 1865–1870.
52. Agger, K., Cloos, P.A., Rudkjaer, L., Williams, K., Andersen, G., Christensen, J., and Helin, K. (2009) The H3K27me3 demethylase JMJD3 contributes to the activation of the INK4A-ARF locus in response to oncogene- and stress-induced senescence. *Genes Dev.*, **23** (10), 1171–1176.
53. Yamane, K., Tateishi, K., Klose, R.J., Fang, J., Fabrizio, L.A., Erdjument-Bromage, H., Taylor-Papadimitriou, J., Tempst, P., and Zhang, Y. (2007) PLU-1 is an H3K4 demethylase involved in transcriptional repression and breast cancer cell proliferation. *Mol. Cell*, **25** (6), 801–812.
54. Scibetta, A.G., Santangelo, S., Coleman, J., Hall, D., Chaplin, T., Copier, J., Catchpole, S., Burchell, J., Taylor-Papadimitriou, J. (2007) Functional analysis of the transcription repressor PLU-1/JARID1B. *Mol. Cell. Biol.*, **27** (20), 7220–7235.
55. Dalgliesh, G.L., Furge, K., Greenman, C., Chen, L., Bignell, G., Butler, A., Davies, H., Edkins, S., Hardy, C., Latimer, C., Teague, J., Andrews, J., Barthorpe, S., Beare, D., Buck, G., Campbell, P.J., Forbes, S., Jia, M., Jones, D., Knott, H., Kok, C.Y., Lau, K.W., Leroy, C., Lin, M.L., McBride, D.J., Maddison, M., Maguire, S., McLay, K., Menzies, A., Mironenko, T., Mulderrig, L., Mudie, L., O'Meara, S., Pleasance, E., Rajasingham, A., Shepherd, R., Smith, R., Stebbings, L., Stephens, P., Tang, G., Tarpey, P.S., Turrell, K., Dykema, K.J., Khoo, S.K., Petillo, D., Wondergem, B., Anema, J., Kahnoski, R.J., Teh, B.T., Stratton, M.R., and Futreal, P.A. (2010) Systematic sequencing of renal carcinoma reveals inactivation of histone modifying genes. *Nature*, **463** (7279), 360–363.
56. Roesch, A., Fukunaga-Kalabis, M., Schmidt, E.C., Zabierowski, S.E., Brafford, P.A., Vultur, A., Basu, D., Gimotty, P., Vogt, T., and Herlyn, M. (2010) A temporarily distinct subpopulation of slow-cycling melanoma cells is required for continuous tumor growth. *Cell*, **141** (4), 583–594.
57. Sharma, S.V., Lee, D.Y., Li, B., Quinlan, M.P., Takahashi, F., Maheswaran, S., McDermott,

U., Azizian, N., Zou, L., Fischbach, M.A., Wong, K.K., Brandstetter, K., Wittner, B., Ramaswamy, S., Classon, M., and Settleman, J. (2010) A chromatin-mediated reversible drug-tolerant state in cancer cell subpopulations. *Cell*, **141** (1), 69–80.
58. Spannhoff, A., Sippl, W., and Jung, M. (2009) Cancer treatment of the future: inhibitors of histone methyltransferases. *Int. J. Biochem. Cell Biol.*, **41** (1), 4–11.
59. Wang, G.G., Allis, C.D., and Chi, P. (2007) Chromatin remodeling and cancer, Part I: Covalent histone modifications. *Trends Mol. Med.*, **13** (9), 363–372.
60. Fernandez-Capetillo, O., Lee, A., Nussenzweig, M., and Nussenzweig, A. (2004) H2AX: the histone guardian of the genome. *DNA Repair (Amst.)*, **3** (8–9), 959–967.
61. Terada, Y. (2006) Aurora-B/AIM-1 regulates the dynamic behavior of HP1alpha at the G_2-M transition. *Mol. Biol. Cell*, **17** (7), 3232–3241.
62. Dunn, K.L. and Davie, J.R. (2005) Stimulation of the Ras-MAPK pathway leads to independent phosphorylation of histone H3 on serine 10 and 28. *Oncogene*, **24** (21), 3492–3502.
63. Ahn, S.H., Cheung, W.L., Hsu, J.Y., Diaz, R.L., Smith, M.M., and Allis, C.D. (2005) Sterile 20 kinase phosphorylates histone H2B at serine 10 during hydrogen peroxide-induced apoptosis in *S. cerevisiae*. *Cell*, **120** (1), 25–36.
64. Cheung, W.L., Ajiro, K., Samejima, K., Kloc, M., Cheung, P., Mizzen, C.A., Beeser, A., Etkin, L.D., Chernoff, J., Earnshaw, W.C., and Allis, C.D. (2003) Apoptotic phosphorylation of histone H2B is mediated by mammalian sterile twenty kinase. *Cell*, **113** (4), 507–517.
65. Halliday, G.M., Bock, V.L., Moloney, F.J., and Lyons, J.G. (2009) SWI/SNF: a chromatin-remodelling complex with a role in carcinogenesis. *Int. J. Biochem. Cell Biol.*, **41** (4), 725–728.
66. Reisman, D., Glaros, S., and Thompson, E.A. (2009) The SWI/SNF complex and cancer. *Oncogene*, **28** (14), 1653–1668.
67. Biegel, J.A., Zhou, J.Y., Rorke, L.B., Stenstrom, C., Wainwright, L.M., and Fogelgren, B. (1999) Germ-line and acquired mutations of INI1 in atypical teratoid and rhabdoid tumors. *Cancer Res.*, **59** (1), 74–79.
68. Sevenet, N., Sheridan, E., Amram, D., Schneider, P., Handgretinger, R., and Delattre, O. (1999) Constitutional mutations of the hSNF5/INI1 gene predispose to a variety of cancers. *Am. J. Hum. Genet.*, **65** (5), 1342–1348.
69. Rousseau-Merck, M.F., Versteege, I., Legrand, I., Couturier, J., Mairal, A., Delattre, O., and Aurias, A. (1999) hSNF5/INI1 inactivation is mainly associated with homozygous deletions and mitotic recombinations in rhabdoid tumors. *Cancer Res.*, **59** (13), 3152–3156.
70. Sevenet, N., Lellouch-Tubiana, A., Schofield, D., Hoang-Xuan, K., Gessler, M., Birnbaum, D., Jeanpierre, C., Jouvet, A., and Delattre, O. (1999) Spectrum of hSNF5/INI1 somatic mutations in human cancer and genotype-phenotype correlations. *Hum. Mol. Genet.*, **8** (13), 2359–2368.
71. Guidi, C.J., Sands, A.T., Zambrowicz, B.P., Turner, T.K., Demers, D.A., Webster, W., Smith, T.W., Imbalzano, A.N., and Jones, S.N. (2001) Disruption of Ini1 leads to peri-implantation lethality and tumorigenesis in mice. *Mol. Cell. Biol.*, **21** (10), 3598–3603.
72. Reincke, B.S., Rosson, G.B., Oswald, B.W., and Wright, C.F. (2003) INI1 expression induces cell cycle arrest and markers of senescence in malignant rhabdoid tumor cells. *J. Cell. Physiol.*, **194** (3), 303–313.
73. Betz, B.L., Strobeck, M.W., Reisman, D.N., Knudsen, E.S., and Weissman, B.E. (2002) Re-expression of hSNF5/INI1/BAF47 in pediatric tumor cells leads to G_1 arrest associated with induction of p16ink4a and activation of RB. *Oncogene*, **21** (34), 5193–5203.
74. Kia, S.K., Gorski, M.M., Giannakopoulos, S., and Verrijzer, C.P. (2008) SWI/SNF mediates polycomb eviction and epigenetic reprogramming of the INK4b-ARF-INK4a locus. *Mol. Cell. Biol.*, **28** (10), 3457–3464.
75. Doan, D.N., Veal, T.M., Yan, Z., Wang, W., Jones, S.N., and Imbalzano, A.N. (2004) Loss of the INI1 tumor suppressor does not impair the expression of multiple BRG1-dependent genes or the assembly of SWI/SNF enzymes. *Oncogene*, **23** (19), 3462–3473.
76. Chai, J., Charboneau, A.L., Betz, B.L., and Weissman, B.E. (2005) Loss of the

hSNF5 gene concomitantly inactivates p21CIP/WAF1 and p16INK4a activity associated with replicative senescence in A204 rhabdoid tumor cells. *Cancer Res.*, **65** (22), 10192–10198.

77. Oruetxebarria, I., Venturini, F., Kekarainen, T., Houweling, A., Zuijderduijn, L.M., Mohd-Sarip, A., Vries, R.G., Hoeben, R.C., and Verrijzer, C.P. (2004) P16INK4a is required for hSNF5 chromatin remodeler-induced cellular senescence in malignant rhabdoid tumor cells. *J. Biol. Chem.*, **279** (5), 3807–3816.
78. Lin, H., Wong, R.P., Martinka, M., and Li, G. (2009) Loss of SNF5 expression correlates with poor patient survival in melanoma. *Clin. Cancer Res.*, **15** (20), 6404–6411.
79. Hayward, N.K. (2003) Genetics of melanoma predisposition. *Oncogene*, **22** (20), 3053–3062.
80. Becker, T.M., Rizos, H., Kefford, R.F., and Mann, G.J. (2001) Functional impairment of melanoma-associated p16(INK4a) mutants in melanoma cells despite retention of cyclin-dependent kinase 4 binding. *Clin. Cancer Res.*, **7** (10), 3282–3288.
81. Gallagher, S.J., Thompson, J.F., Indsto, J., Scurr, L.L., Lett, M., Gao, B.F., Dunleavey, R., Mann, G.J., Kefford, R.F., and Rizos, H. (2008) p16INK4a expression and absence of activated B-RAF are independent predictors of chemosensitivity in melanoma tumors. *Neoplasia*, **10** (11), 1231–1239.
82. Wang, X., Sansam, C.G., Thom, C.S., Metzger, D., Evans, J.A., Nguyen, P.T., and Roberts, C.W. (2009) Oncogenesis caused by loss of the SNF5 tumor suppressor is dependent on activity of BRG1, the ATPase of the SWI/SNF chromatin remodeling complex. *Cancer Res.*, **69** (20), 8094–8101.
83. Gunduz, E., Gunduz, M., Ouchida, M., Nagatsuka, H., Beder, L., Tsujigiwa, H., Fukushima, K., Nishizaki, K., Shimizu, K., and Nagai, N. (2005) Genetic and epigenetic alterations of BRG1 promote oral cancer development. *Int. J. Oncol.*, **26** (1), 201–210.
84. Sun, A., Tawfik, O., Gayed, B., Thrasher, J.B., Hoestje, S., Li, C., and Li, B. (2007) Aberrant expression of SWI/SNF catalytic subunits BRG1/BRM is associated with tumor development and increased invasiveness in prostate cancers. *Prostate*, **67** (2), 203–213.
85. Sentani, K., Oue, N., Kondo, H., Kuraoka, K., Motoshita, J., Ito, R., Yokozaki, H., and Yasui, W. (2001) Increased expression but not genetic alteration of BRG1, a component of the SWI/SNF complex, is associated with the advanced stage of human gastric carcinomas. *Pathobiology*, **69** (6), 315–320.
86. Yamamichi, N., Inada, K., Ichinose, M., Yamamichi-Nishina, M., Mizutani, T., Watanabe, H., Shiogama, K., Fujishiro, M., Okazaki, T., Yahagi, N., Haraguchi, T., Fujita, S., Tsutsumi, Y., Omata, M., and Iba, H. (2007) Frequent loss of Brm expression in gastric cancer correlates with histologic features and differentiation state. *Cancer Res.*, **67** (22), 10727–10735.
87. Watanabe, T., Semba, S., and Yokozaki, H. (2011) Regulation of PTEN expression by the SWI/SNF chromatin-remodelling protein BRG1 in human colorectal carcinoma cells. *Br. J. Cancer*, **104** (1), 146–154.
88. Barker, N., Hurlstone, A., Musisi, H., Miles, A., Bienz, M., and Clevers, H. (2001) The chromatin remodelling factor Brg-1 interacts with beta-catenin to promote target gene activation. *EMBO J.*, **20** (17), 4935–4943.
89. Bultman, S., Gebuhr, T., Yee, D., La Mantia, C., Nicholson, J., Gilliam, A., Randazzo, F., Metzger, D., Chambon, P., Crabtree, G., and Magnuson, T. (2000) A Brg1 null mutation in the mouse reveals functional differences among mammalian SWI/SNF complexes. *Mol. Cell*, **6** (6), 1287–1295.
90. Glaros, S., Cirrincione, G.M., Palanca, A., Metzger, D., and Reisman, D. (2008) Targeted knockout of BRG1 potentiates lung cancer development. *Cancer Res.*, **68** (10), 3689–3696.
91. Fukuoka, J., Fujii, T., Shih, J.H., Dracheva, T., Meerzaman, D., Player, A., Hong, K., Settnek, S., Gupta, A., Buetow, K., Hewitt, S., Travis, W.D., and Jen, J. (2004) Chromatin remodeling factors and BRM/BRG1 expression as prognostic indicators in non-small cell lung cancer. *Clin. Cancer Res.*, **10** (13), 4314–4324.
92. Reisman, D.N., Sciarrotta, J., Wang, W., Funkhouser, W.K., and Weissman, B.E. (2003) Loss of BRG1/BRM in human lung cancer cell lines and primary lung cancers: correlation with poor prognosis. *Cancer Res.*, **63** (3), 560–566.

93. Rodriguez-Nieto, S., Canada, A., Pros, E., Pinto, A.I., Torres-Lanzas, J., Lopez-Rios, F., Sanchez-Verde, L., Pisano, D.G., Sanchez-Cespedes, M. (2011) Massive parallel DNA pyrosequencing analysis of the tumor suppressor BRG1/SMARCA4 in lung primary tumors. *Hum. Mutat.*, **32** (2), E1999–E2017.
94. Muchardt, C. and Yaniv, M. (2001) When the SWI/SNF complex remodels...the cell cycle. *Oncogene*, **20** (24), 3067–3075.
95. Decristofaro, M.F., Betz, B.L., Rorie, C.J., Reisman, D.N., Wang, W., and Weissman, B.E. (2001) Characterization of SWI/SNF protein expression in human breast cancer cell lines and other malignancies. *J. Cell. Physiol.*, **186** (1), 136–145.
96. Medina, P.P., Romero, O.A., Kohno, T., Montuenga, L.M., Pio, R., Yokota, J., Sanchez-Cespedes, M. (2008) Frequent BRG1/SMARCA4-inactivating mutations in human lung cancer cell lines. *Hum. Mutat.*, **29** (5), 617–622.
97. Wong, A.K., Shanahan, F., Chen, Y., Lian, L., Ha, P., Hendricks, K., Ghaffari, S., Iliev, D., Penn, B., Woodland, A.M., Smith, R., Salada, G., Carillo, A., Laity, K., Gupte, J., Swedlund, B., Tavtigian, S.V., Teng, D.H., and Lees, E. (2000) BRG1, a component of the SWI-SNF complex, is mutated in multiple human tumor cell lines. *Cancer Res.*, **60** (21), 6171–6177.
98. Kuo, K.T., Liang, C.W., Hsiao, C.H., Lin, C.H., Chen, C.A., Sheu, B.C., and Lin, M.C. (2006) Downregulation of BRG-1 repressed expression of CD44s in cervical neuroendocrine carcinoma and adenocarcinoma. *Mod. Pathol.*, **19** (12), 1570–1577.
99. Becker, T.M., Haferkamp, S., Dijkstra, M.K., Scurr, L.L., Frausto, M., Diefenbach, E., Scolyer, R.A., Reisman, D.N., Mann, G.J., Kefford, R.F., and Rizos, H. (2009) The chromatin remodelling factor BRG1 is a novel binding partner of the tumor suppressor p16INK4a. *Mol. Cancer*, **8**, 4.
100. Lin, H., Wong, R.P., Martinka, M., and Li, G. (2010) BRG1 expression is increased in human cutaneous melanoma. *Br. J. Dermatol.*, **163** (3), 502–510.
101. Curtin, J.A., Fridlyand, J., Kageshita, T., Patel, H.N., Busam, K.J., Kutzner, H., Cho, K.H., Aiba, S., Brocker, E.B., LeBoit, P.E., Pinkel, D., and Bastian, B.C. (2005) Distinct sets of genetic alterations in melanoma. *N. Engl. J. Med.*, **353** (20), 2135–2147.
102. Eide, M.J., and Weinstock, M.A. (2005) Association of UV index, latitude, and melanoma incidence in nonwhite populations – US Surveillance, Epidemiology, and End Results (SEER) Program, 1992 to 2001. *Arch. Dermatol.*, **141** (4), 477–481.
103. Moloney, F.J., Lyons, J.G., Bock, V.L., Huang, X.X., Bugeja, M.J., and Halliday, G.M. (2009) Hotspot mutation of Brahma in non-melanoma skin cancer. *J. Invest. Dermatol.*, **129** (4), 1012–1015.
104. Shen, H., Powers, N., Saini, N., Comstock, C.E., Sharma, A., Weaver, K., Revelo, M.P., Gerald, W., Williams, E., Jessen, W.J., Aronow, B.J., Rosson, G., Weissman, B., Muchardt, C., Yaniv, M., and Knudsen, K.E. (2008) The SWI/SNF ATPase Brm is a gatekeeper of proliferative control in prostate cancer. *Cancer Res.*, **68** (24), 10154–10162.
105. Glaros, S., Cirrincione, G.M., Muchardt, C., Kleer, C.G., Michael, C.W., and Reisman, D. (2007) The reversible epigenetic silencing of BRM: implications for clinical targeted therapy. *Oncogene*, **26** (49), 7058–7066.
106. Bourachot, B., Yaniv, M., and Muchardt, C. (2003) Growth inhibition by the mammalian SWI-SNF subunit Brm is regulated by acetylation. *EMBO J.*, **22** (24), 6505–6515.
107. Garcia-Pedrero, J.M., Kiskinis, E., Parker, M.G., and Belandia, B. (2006) The SWI/SNF chromatin remodeling subunit BAF57 is a critical regulator of estrogen receptor function in breast cancer cells. *J. Biol. Chem.*, **281** (32), 22656–22664.
108. Kiskinis, E., Garcia-Pedrero, J.M., Villaronga, M.A., Parker, M.G., and Belandia, B. (2006) Identification of BAF57 mutations in human breast cancer cell lines. *Breast Cancer Res. Treat.*, **98** (2), 191–198.
109. Link, K.A., Balasubramaniam, S., Sharma, A., Comstock, C.E., Godoy-Tundidor, S., Powers, N., Cao, K.H., Haelens, A., Claessens, F., Revelo, M.P., and Knudsen, K.E. (2008) Targeting the BAF57 SWI/SNF subunit in prostate cancer: a novel platform to control androgen receptor activity. *Cancer Res.*, **68** (12), 4551–4558.
110. Shadeo, A., Chari, R., Lonergan, K.M., Pusic, A., Miller, D., Ehlen, T., Van Niekerk, D., Matisic, J., Richards-Kortum, R., Follen, M., Guillaud, M., Lam, W.L., and MacAulay, C. (2008) Up regulation in gene expression

of chromatin remodelling factors in cervical intraepithelial neoplasia. *BMC Genomics*, **9**, 64.

111. Heeboll, S., Borre, M., Ottosen, P.D., Andersen, C.L., Mansilla, F., Dyrskjot, L., Orntoft, T.F., and Torring, N. (2008) SMARCC1 expression is upregulated in prostate cancer and positively correlated with tumour recurrence and dedifferentiation. *Histol. Histopathol.*, **23** (9), 1069–1076.
112. Varela, I., Tarpey, P., Raine, K., Huang, D., Ong, C.K., Stephens, P., Davies, H., Jones, D., Lin, M.L., Teague, J., Bignell, G., Butler, A., Cho, J., Dalgliesh, G.L., Galappaththige, D., Greenman, C., Hardy, C., Jia, M., Latimer, C., Lau, K.W., Marshall, J., McLaren, S., Menzies, A., Mudie, L., Stebbings, L., Largaespada, D.A., Wessels, L.F., Richard, S., Kahnoski, R.J., Anema, J., Tuveson, D.A., Perez-Mancera, P.A., Mustonen, V., Fischer, A., Adams, D.J., Rust, A., Chan-on, W., Subimerb, C., Dykema, K., Furge, K., Campbell, P.J., Teh, B.T., Stratton, M.R., and Futreal, P.A. (2011) Exome sequencing identifies frequent mutation of the SWI/SNF complex gene PBRM1 in renal carcinoma. *Nature*, **469** (7331), 539–542.
113. Burrows, A.E., Smogorzewska, A., and Elledge, S.J. (2010) Polybromo-associated BRG1-associated factor components BRD7 and BAF180 are critical regulators of p53 required for induction of replicative senescence. *Proc. Natl Acad. Sci. USA*, **107** (32), 14280–14285.
114. Xia, W., Nagase, S., Montia, A.G., Kalachikov, S.M., Keniry, M., Su, T., Memeo, L., Hibshoosh, H., and Parsons, R. (2008) BAF180 is a critical regulator of p21 induction and a tumor suppressor mutated in breast cancer. *Cancer Res.*, **68** (6), 1667–1674.
115. Wang, X., Nagl, N.G. Jr, Flowers, S., Zweitzig, D., Dallas, P.B., and Moran, E. (2004) Expression of p270 (ARID1A), a component of human SWI/SNF complexes, in human tumors. *Int. J. Cancer*, **112** (4), 636.
116. Nagl, N.G. Jr, Patsialou, A., Haines, D.S., Dallas, P.B., Beck, G.R. Jr, and Moran, E. (2005) The p270 (ARID1A/SMARCF1) subunit of mammalian SWI/SNF-related complexes is essential for normal cell cycle arrest. *Cancer Res.*, **65** (20), 9236–9244.
117. Mohamed, M.A., Greif, P.A., Diamond, J., Sharaf, O., Maxwell, P., Montironi, R., Young, R.A., and Hamilton, P.W. (2007) Epigenetic events, remodelling enzymes and their relationship to chromatin organization in prostatic intraepithelial neoplasia and prostatic adenocarcinoma. *Br. J. Urol. Int.*, **99** (4), 908–915.
118. Choi, J.H., Sheu, J.J., Guan, B., Jinawath, N., Markowski, P., Wang, T.L., Shih Ie, M. (2009) Functional analysis of 11q13.5 amplicon identifies Rsf-1 (HBXAP) as a gene involved in paclitaxel resistance in ovarian cancer. *Cancer Res.*, **69** (4), 1407–1415.
119. Sheu, J.J., Guan, B., Choi, J.H., Lin, A., Lee, C.H., Hsiao, Y.T., Wang, T.L., Tsai, F.J., Shih Ie, M. (2010) Rsf-1, a chromatin remodeling protein, induces DNA damage and promotes genomic instability. *J. Biol. Chem.*, **285** (49), 38260–38269.
120. Bagchi, A., Papazoglu, C., Wu, Y., Capurso, D., Brodt, M., Francis, D., Bredel, M., Vogel, H., and Mills, A.A. (2007) CHD5 is a tumor suppressor at human 1p36. *Cell*, **128** (3), 459–475.
121. Bowen, N.J., Fujita, N., Kajita, M., and Wade, P.A. (2004) Mi-2/NuRD: multiple complexes for many purposes. *Biochim. Biophys. Acta*, **1677** (1–3), 52–57.
122. Fujita, N., Jaye, D.L., Kajita, M., Geigerman, C., Moreno, C.S., and Wade, P.A. (2003) MTA3, a Mi-2/NuRD complex subunit, regulates an invasive growth pathway in breast cancer. *Cell*, **113** (2), 207–219.
123. Li, R., Zhang, H., Yu, W., Chen, Y., Gui, B., Liang, J., Wang, Y., Sun, L., Yang, X., Zhang, Y., Shi, L., Li, Y., and Shang, Y. (2009) ZIP: a novel transcription repressor, represses EGFR oncogene and suppresses breast carcinogenesis. *EMBO J.*, **28** (18), 2763–2776.
124. Yu, E.Y., Steinberg-Neifach, O., Dandjinou, A.T., Kang, F., Morrison, A.J., Shen, X., and Lue, N.F. (2007) Regulation of telomere structure and functions by subunits of the INO80 chromatin remodeling complex. *Mol. Cell. Biol.*, **27** (16), 5639–5649.
125. Tang, Y., Luo, J., Zhang, W., and Gu, W. (2006) Tip60-dependent acetylation of p53 modulates the decision between cell-cycle arrest and apoptosis. *Mol. Cell*, **24** (6), 827–839.

126. Fazzio, T.G., Huff, J.T., and Panning, B. (2008) An RNAi screen of chromatin proteins identifies Tip60-p400 as a regulator of embryonic stem cell identity. *Cell*, **134** (1), 162–174.
127. Grigoletto, A., Lestienne, P., and Rosenbaum, J. (2011) The multifaceted proteins Reptin and Pontin as major players in cancer. *Biochim. Biophys. Acta*, **1815** (2), 147–157.
128. Morrison, A.J., Highland, J., Krogan, N.J., Arbel-Eden, A., Greenblatt, J.F., Haber, J.E., and Shen, X. (2004) INO80 and gamma-H2AX interaction links ATP-dependent chromatin remodeling to DNA damage repair. *Cell*, **119** (6), 767–775.
129. Lane, D.P. (1992) Cancer. p53, guardian of the genome. *Nature*, **358** (6381), 15–16.
130. Greenblatt, M.S., Bennett, W.P., Hollstein, M., and Harris, C.C. (1994) Mutations in the p53 tumor suppressor gene: clues to cancer etiology and molecular pathogenesis. *Cancer Res.*, **54** (18), 4855–4878.
131. el-Deiry, W.S., Tokino, T., Velculescu, V.E., Levy, D.B., Parsons, R., Trent, J.M., Lin, D., Mercer, W.E., Kinzler, K.W., and Vogelstein, B. (1993) WAF1, a potential mediator of p53 tumor suppression. *Cell*, **75** (4), 817–825.
132. Prives, C., and Manley, J.L. (2001) Why is p53 acetylated? *Cell*, **107** (7), 815–818.
133. Ito, A., Lai, C.H., Zhao, X., Saito, S., Hamilton, M.H., Appella, E., and Yao, T.P. (2001) p300/CBP-mediated p53 acetylation is commonly induced by p53-activating agents and inhibited by MDM2. *EMBO J.*, **20** (6), 1331–1340.
134. Liu, L., Scolnick, D.M., Trievel, R.C., Zhang, H.B., Marmorstein, R., Halazonetis, T.D., and Berger, S.L. (1999) p53 sites acetylated in vitro by PCAF and p300 are acetylated in vivo in response to DNA damage. *Mol. Cell. Biol.*, **19** (2), 1202–1209.
135. Sakaguchi, K., Herrera, J.E., Saito, S., Miki, T., Bustin, M., Vassilev, A., Anderson, C.W., and Appella, E. (1998) DNA damage activates p53 through a phosphorylation-acetylation cascade. *Genes Dev.*, **12** (18), 2831–2841.
136. Wang, T., Kobayashi, T., Takimoto, R., Denes, A.E., Snyder, E.L., el-Deiry, W.S., and Brachmann, R.K. (2001) hADA3 is required for p53 activity. *EMBO J.*, **20** (22), 6404–6413.
137. Kumar, A., Zhao, Y., Meng, G., Zeng, M., Srinivasan, S., Delmolino, L.M., Gao, Q., Dimri, G., Weber, G.F., Wazer, D.E., Band, H., and Band, V. (2002) Human papillomavirus oncoprotein E6 inactivates the transcriptional coactivator human ADA3. *Mol. Cell. Biol.*, **22** (16), 5801–5812.
138. Brooks, C.L. and Gu, W. (2003) Ubiquitination, phosphorylation and acetylation: the molecular basis for p53 regulation. *Curr. Opin. Cell Biol.*, **15** (2), 164–171.
139. Harms, K.L. and Chen, X. (2007) Histone deacetylase 2 modulates p53 transcriptional activities through regulation of p53-DNA binding activity. *Cancer Res.*, **67** (7), 3145–3152.
140. Oh, J., Sohn, D.H., Ko, M., Chung, H., Jeon, S.H., and Seong, R.H. (2008) BAF60a interacts with p53 to recruit the SWI/SNF complex. *J. Biol. Chem.*, **283** (18), 11924–11934.
141. Lee, D., Kim, J.W., Seo, T., Hwang, S.G., Choi, E.J., and Choe, J. (2002) SWI/SNF complex interacts with tumor suppressor p53 and is necessary for the activation of p53-mediated transcription. *J. Biol. Chem.*, **277** (25), 22330–22337.
142. Wang, M., Gu, C., Qi, T., Tang, W., Wang, L., Wang, S., and Zeng, X. (2007) BAF53 interacts with p53 and functions in p53-mediated p21-gene transcription. *J. Biochem.*, **142** (5), 613–620.
143. Lee, J.H., Lee, J.Y., Chang, S.H., Kang, M.J., and Kwon, H. (2005) Effects of Ser2 and Tyr6 mutants of BAF53 on cell growth and p53-dependent transcription. *Mol. Cell*, **19** (2), 289–293.
144. Naidu, S.R., Love, I.M., Imbalzano, A.N., Grossman, S.R., and Androphy, E.J. (2009) The SWI/SNF chromatin remodeling subunit BRG1 is a critical regulator of p53 necessary for proliferation of malignant cells. *Oncogene*, **28** (27), 2492–2501.
145. Allison, S.J. and Milner, J. (2004) Remodelling chromatin on a global scale: a novel protective function of p53. *Carcinogenesis*, **25** (9), 1551–1557.
146. Weintraub, S.J., Chow, K.N., Luo, R.X., Zhang, S.H., He, S., and Dean, D.C. (1995) Mechanism of active transcriptional repression by the retinoblastoma protein. *Nature*, **375** (6534), 812–815.
147. Brehm, A., Miska, E.A., McCance, D.J., Reid, J.L., Bannister, A.J., and Kouzarides,

T. (1998) Retinoblastoma protein recruits histone deacetylase to repress transcription. *Nature*, **391** (6667), 597–601.

148. Ferreira, R., Naguibneva, I., Mathieu, M., Ait-Si-Ali, S., Robin, P., Pritchard, L.L., Harel-Bellan, A. (2001) Cell cycle-dependent recruitment of HDAC-1 correlates with deacetylation of histone H4 on an Rb-E2F target promoter. *EMBO Rep.*, **2** (9), 794–799.
149. Takaki, T., Fukasawa, K., Suzuki-Takahashi, I., and Hirai, H. (2004) Cdk-mediated phosphorylation of pRB regulates HDAC binding in vitro. *Biochem. Biophys. Res. Commun.*, **316** (1), 252–255.
150. Zhang, H.S., Gavin, M., Dahiya, A., Postigo, A.A., Ma, D., Luo, R.X., Harbour, J.W., and Dean, D.C. (2000) Exit from G_1 and S phase of the cell cycle is regulated by repressor complexes containing HDAC-Rb-hSWI/SNF and Rb-hSWI/SNF. *Cell*, **101** (1), 79–89.
151. Dunaief, J.L., Strober, B.E., Guha, S., Khavari, P.A., Alin, K., Luban, J., Begemann, M., Crabtree, G.R., and Goff, S.P. (1994) The retinoblastoma protein and BRG1 form a complex and cooperate to induce cell cycle arrest. *Cell*, **79** (1), 119–130.
152. Strobeck, M.W., Knudsen, K.E., Fribourg, A.F., DeCristofaro, M.F., Weissman, B.E., Imbalzano, A.N., and Knudsen, E.S. (2000) BRG-1 is required for RB-mediated cell cycle arrest. *Proc. Natl Acad. Sci. USA*, **97** (14), 7748–7753.
153. Bartlett, C., Stammler, T., Rosson, G.S., and Weissman, B.E. (2010) BRG1 mutations found in human cancer cell lines inactivate Rb-mediated cell cycle arrest. *J. Cell. Physiol.*, **226** (8), 1989–1997.
154. Kang, H., Cui, K., and Zhao, K. (2004) BRG1 controls the activity of the retinoblastoma protein via regulation of p21CIP1/WAF1/SDI. *Mol. Cell. Biol.*, **24** (3), 1188–1199.
155. Rao, M., Casimiro, M.C., Lisanti, M.P., D'Amico, M., Wang, C., Shirley, L.A., Leader, J.E., Liu, M., Stallcup, M., Engel, D.A., Murphy, D.J., and Pestell, R.G. (2008) Inhibition of cyclin D1 gene transcription by Brg-1. *Cell Cycle*, **7** (5), 647–655.
156. Sharpless, N.E. and DePinho, R.A. (1999) The INK4A/ARF locus and its two gene products. *Curr. Opin. Genet. Dev.*, **9** (1), 22–30.
157. Goldstein, A.M., Chan, M., Harland, M., Gillanders, E.M., Hayward, N.K., Avril, M.F., Azizi, E., Bianchi-Scarra, G., Bishop, D.T., Bressac-de Paillerets, B., Bruno, W., Calista, D., Cannon Albright, L.A., Demenais, F., Elder, D.E., Ghiorzo, P., Gruis, N.A., Hansson, J., Hogg, D., Holland, E.A., Kanetsky, P.A., Kefford, R.F., Landi, M.T., Lang, J., Leachman, S.A., Mackie, R.M., Magnusson, V., Mann, G.J., Niendorf, K., Newton Bishop, J., Palmer, J.M., Puig, S., Puig-Butille, J.A., de Snoo, F.A., Stark, M., Tsao, H., Tucker, M.A., Whitaker, L., and Yakobson, E. (2006) High-risk melanoma susceptibility genes and pancreatic cancer, neural system tumors, and uveal melanoma across GenoMEL. *Cancer Res.*, **66** (20), 9818–9828.
158. McKenzie, H.A., Fung, C., Becker, T.M., Irvine, M., Mann, G.J., Kefford, R.F., and Rizos, H. (2010) Predicting functional significance of cancer-associated p16(INK4a) mutations in CDKN2A. *Hum. Mutat.*, **31** (6), 692–701.
159. Tzatsos, A., Pfau, R., Kampranis, S.C., and Tsichlis, P.N. (2009) Ndy1/KDM2B immortalizes mouse embryonic fibroblasts by repressing the Ink4a/Arf locus. *Proc. Natl Acad. Sci. USA*, **106** (8), 2641–2646.
160. Barradas, M., Anderton, E., Acosta, J.C., Li, S., Banito, A., Rodriguez-Niedenfuhr, M., Maertens, G., Banck, M., Zhou, M.M., Walsh, M.J., Peters, G., and Gil, J. (2009) Histone demethylase JMJD3 contributes to epigenetic control of INK4a/ARF by oncogenic RAS. *Genes Dev.*, **23** (10), 1177–1182.
161. Kotake, Y., Zeng, Y., and Xiong, Y. (2009) DDB1-CUL4 and MLL1 mediate oncogene-induced p16INK4a activation. *Cancer Res.*, **69** (5), 1809–1814.
162. Muchardt, C., Bourachot, B., Reyes, J.C., and Yaniv, M. (1998) ras transformation is associated with decreased expression of the brm/SNF2alpha ATPase from the mammalian SWI-SNF complex. *EMBO J.*, **17** (1), 223–231.
163. Ceol, C.J., Houvras, Y., Jane-Valbuena, J., Bilodeau, S., Orlando, D.A., Battisti, V., Fritsch, L., Lin, W.M., Hollmann, T.J., Ferre, F., Bourque, C., Burke, C.J., Turner, L., Uong, A., Johnson, L.A., Beroukhim, R., Mermel, C.H., Loda, M., Ait-Si-Ali, S., Garraway, L.A., Young, R.A., and Zon, L.I. (2011) The histone methyltransferase

SETDB1 is recurrently amplified in melanoma and accelerates its onset. *Nature*, **471** (7339), 513–517.

164. Fresno Vara, J.A., Casado, E., de Castro, J., Cejas, P., Belda-Iniesta, C., and Gonzalez-Baron, M. (2004) PI3K/Akt signalling pathway and cancer. *Cancer Treat. Rev.*, **30** (2), 193–204.
165. Foster, K.S., McCrary, W.J., Ross, J.S., and Wright, C.F. (2006) Members of the hSWI/SNF chromatin remodeling complex associate with and are phosphorylated by protein kinase B/Akt. *Oncogene*, **25** (33), 4605–4612.
166. Huang, W.C. and Chen, C.C. (2005) Akt phosphorylation of p300 at Ser-1834 is essential for its histone acetyltransferase and transcriptional activity. *Mol. Cell. Biol.*, **25** (15), 6592–6602.
167. Gonzalez, M.E., Duprie, M.L., Krueger, H., Merajver, S.D., Ventura, A.C., Toy, K.A., and Kleer, C.G. (2011) Histone methyltransferase EZH2 induces Akt-dependent genomic instability and BRCA1 inhibition in breast cancer. *Cancer Res.*, **71** (6), 2360–2370.
168. Campisi, J. (2005) Senescent cells, tumor suppression, and organismal aging: good citizens, bad neighbors. *Cell*, **120** (4), 513–522.
169. Scurr, L.L., Pupo, G.M., Becker, T.M., Lai, K., Schrama, D., Haferkamp, S., Irvine, M., Scolyer, R.A., Mann, G.J., Becker, J.C., Kefford, R.F., and Rizos, H. (2010) IGFBP7 is not required for B-RAF-induced melanocyte senescence. *Cell*, **141** (4), 717–727.
170. Haferkamp, S., Scurr, L.L., Becker, T.M., Frausto, M., Kefford, R.F., and Rizos, H. (2009) Oncogene-induced senescence does not require the p16(INK4a) or p14ARF melanoma tumor suppressors. *J. Invest. Dermatol.*, **129** (8), 1983–1991.
171. Xue, L., Murray, J.H., and Tolkovsky, A.M. (2000) The Ras/phosphatidylinositol 3-kinase and Ras/ERK pathways function as independent survival modules each of which inhibits a distinct apoptotic signaling pathway in sympathetic neurons. *J. Biol. Chem.*, **275** (12), 8817–8824.
172. Xu, H.J., Zhou, Y., Ji, W., Perng, G.S., Kruzelock, R., Kong, C.T., Bast, R.C., Mills, G.B., Li, J., and Hu, S.X. (1997) Reexpression of the retinoblastoma protein in tumor cells induces senescence and telomerase inhibition. *Oncogene*, **15** (21), 2589–2596.
173. Narita, M., Nunez, S., Heard, E., Lin, A.W., Hearn, S.A., Spector, D.L., Hannon, G.J., and Lowe, S.W. (2003) Rb-mediated heterochromatin formation and silencing of E2F target genes during cellular senescence. *Cell*, **113** (6), 703–716.
174. Beausejour, C.M., Krtolica, A., Galimi, F., Narita, M., Lowe, S.W., Yaswen, P., and Campisi, J. (2003) Reversal of human cellular senescence: roles of the p53 and p16 pathways. *EMBO J.*, **22** (16), 4212–4222.
175. Haferkamp, S., Tran, S.L., Becker, T.M., Scurr, L.L., Kefford, R.F., and Rizos, H. (2009) The relative contributions of the p53 and pRb pathways in oncogene-induced melanocyte senescence. *Aging (Albany N.Y.)*, **1** (6), 542–556.
176. Haferkamp, S., Becker, T.M., Scurr, L.L., Kefford, R.F., and Rizos, H. (2008) p16INK4a-induced senescence is disabled by melanoma-associated mutations. *Aging Cell*, **7** (5), 733–745.
177. Bandyopadhyay, D., Curry, J.L., Lin, Q., Richards, H.W., Chen, D., Hornsby, P.J., Timchenko, N.A., and Medrano, E.E. (2007) Dynamic assembly of chromatin complexes during cellular senescence: implications for the growth arrest of human melanocytic nevi. *Aging Cell*, **6** (4), 577–591.
178. Nielsen, S.J., Schneider, R., Bauer, U.M., Bannister, A.J., Morrison, A., O'Carroll, D., Firestein, R., Cleary, M., Jenuwein, T., Herrera, R.E., and Kouzarides, T. (2001) Rb targets histone H3 methylation and HP1 to promoters. *Nature*, **412** (6846), 561–565.
179. Kim, K.C., Geng, L., and Huang, S. (2003) Inactivation of a histone methyltransferase by mutations in human cancers. *Cancer Res.*, **63** (22), 7619–7623.
180. Bandyopadhyay, D., Okan, N.A., Bales, E., Nascimento, L., Cole, P.A., and Medrano, E.E. (2002) Down-regulation of p300/CBP histone acetyltransferase activates a senescence checkpoint in human melanocytes. *Cancer Res.*, **62** (21), 6231–6239.
181. Fullgrabe, J., Hajji, N., and Joseph, B. (2010) Cracking the death code: apoptosis-related histone modifications. *Cell Death Differ.*, **17** (8), 1238–1243.
182. Wang, L., Baiocchi, R.A., Pal, S., Mosialos, G., Caligiuri, M., and Sif, S. (2005) The

BRG1- and hBRM-associated factor BAF57 induces apoptosis by stimulating expression of the cylindromatosis tumor suppressor gene. *Mol. Cell. Biol.*, **25** (18), 7953–7965.

183. Park, J.H., Park, E.J., Hur, S.K., Kim, S., and Kwon, J. (2009) Mammalian SWI/SNF chromatin remodeling complexes are required to prevent apoptosis after DNA damage. *DNA Repair (Amst.)*, **8** (1), 29–39.

184. Kai, L., Samuel, S.K., and Levenson, A.S. (2010) Resveratrol enhances p53 acetylation and apoptosis in prostate cancer by inhibiting MTA1/NuRD complex. *Int. J. Cancer*, **126** (7), 1538–1548.

185. Kang, Y. and Massague, J. (2004) Epithelial-mesenchymal transitions: twist in development and metastasis. *Cell*, **118** (3), 277–279.

186. Fujii, S. and Ochiai, A. (2008) Enhancer of zeste homolog 2 downregulates E-cadherin by mediating histone H3 methylation in gastric cancer cells. *Cancer Sci.*, **99** (4), 738–746.

187. Banine, F., Bartlett, C., Gunawardena, R., Muchardt, C., Yaniv, M., Knudsen, E.S., Weissman, B.E., and Sherman, L.S. (2005) SWI/SNF chromatin-remodeling factors induce changes in DNA methylation to promote transcriptional activation. *Cancer Res.*, **65** (9), 3542–3547.

188. Sanchez-Tillo, E., Lazaro, A., Torrent, R., Cuatrecasas, M., Vaquero, E.C., Castells, A., Engel, P., and Postigo, A. (2010) ZEB1 represses E-cadherin and induces an EMT by recruiting the SWI/SNF chromatin-remodeling protein BRG1. *Oncogene*, **29** (24), 3490–3500.

189. Ma, Z., Chang, M.J., Shah, R., Adamski, J., Zhao, X., and Benveniste, E.N. (2004) Brg-1 is required for maximal transcription of the human matrix metalloproteinase-2 gene. *J. Biol. Chem.*, **279** (44), 46326–46334.

6
Intracellular Signaling in Cancer

David M. Terrian[1], Brian D. Lehmann[2], and James A. McCubrey[3]
[1]The Brody School of Medicine at East Carolina University, Anatomy and Cell Biology, Brody Medical Sciences Building, 600 Moye Boulevard, Greenville, NC 27834-4354, USA
[2]Vanderbilt-Ingram Cancer Center, Department of Biochemistry, 652 Preston Research Building, Nashville, TN 37232-6307, USA
[3]The Brody School of Medicine at East Carolina University, Microbiology and Immunology, Leo Jenkins Cancer Center, Brody Medical Sciences Building, 600 Moye Boulevard, Greenville NC 27834-4354, USA

Translational Medicine Cancer: Advances in Molecular Biology and Medicine,
First Edition. Edited by Robert A. Meyers.

Keywords

Cancer gene
A mutated gene that has been causally implicated in oncogenesis (i.e., conferring a clonal growth advantage) by at least two independent reports confirming somatic/germline mutations in primary patient specimens.

Chromosomal translocation
Genomic rearrangements that create chimeric genes; the most common mutation in cancer genes, often resulting in fusion proteins that include an oncogenic kinase catalytic domain and regulatory elements of an unrelated protein.

Human kinome
A catalog of the 621 protein kinase genes encoded by the human genome (viewable at: www.kinase.com).

Protein kinase domain
A conserved sequence of amino acids composed of 12 subdomains and responsible for the catalytic transfer of the γ-phosphate of ATP to protein substrates.

Neoplasm
Any new growth of tissue that is both uncontrolled and progressive.

Cancer genes that encode protein kinases are the most commonly mutated class of genes to be causally linked to the progressive outgrowth of malignant neoplasms. Protein kinases operate as important molecular switches that impinge, either directly or indirectly, on an integrated signaling circuitry that often becomes genetically reprogrammed, through successive mutations, to transmit many of the growth-stimulating and survival signals that govern the transformation of normal human cells into a cancerous cell mass. Of the 491 genes presently included in a comprehensive catalog of known human cancer genes, 6.5% (32 distinct genes) encode protein kinase domain sequences. This number far exceeds a random prediction of 2.0% (9.8 genes), and reinforces the hypothesis that a majority of human cancer cells carry genetic defects in the growth-signaling circuitry that enable these cells to grow autonomously and evade cellular senescence and death. In this chapter, a general overview is first provided of a prototypical growth-signaling circuit, together with a detailed analysis of the canonical Ras/Raf/MEK/ERK pathway, because it has a central role in cancer cell signaling. This analysis is followed by a discussion of each of the kinases known to be mutated in cancer; descriptions will be provided of the domain organization and normal function(s) of the protein kinase, its oncogenic alterations in cancer, and how these alter its activity and the development of one or more of the hallmarks of cancer. The tumor/cancer types in which these changes are most common will also be identified.

1 Overview of the Prototypical Growth-Signaling Circuit

1.1 Signaling Pathways in Normal Human Cells

Protein phosphorylation is the most common post-translational protein modification in mammalian cells, and is a fundamental mechanism for the direct or indirect control of physiological processes of crucial importance to cell growth and survival. Protein phosphorylation is mediated by a superfamily of protein kinases comprising approximately 2% of the human genome (i.e., a Human Kinome that presently includes 621 distinct genes inferred from amino acid sequences of the kinase domain; see *www.kinome.com*), and a diverse collection of unrelated substrate proteins, comprising more than 10% of the genome. Collectively, these proteins control the responsiveness of a cell to its environment by coordinating complex functions such as normal embryonic development, cell division, differentiation, adhesion, motility, and metabolic activity.

All eukaryotic protein kinases have conserved a $\sim 250-300$ amino acid "kinase domain," composed of 12 subdomains and 12 relatively ($> 95\%$) invariant residues. The kinase domain uses the γ-phosphate of ATP or GTP to generate phosphate monoesters using two types of acceptor: (i) protein phenolic groups of tyrosine amino acids (i.e., the protein-tyrosine kinases, PTKs); and (ii) protein alcohol groups on serine/threonine (ST) amino acids (i.e., the protein Ser/Thr kinases). A third class, termed dual-specificity protein kinases, is able to phosphorylate either substrate (both Tyr and Ser/Thr). The kinase domain folds into a two-lobed, three-dimensional structure separated by a deep hydrophobic cleft. When in a catalytically active conformation, the kinase domain is able to perform three interdependent functions: (i) binding and orienting the ATP/GTP phosphate donor as complex with divalent cation (Mg^{2+} or Mn^{2+}); (ii) binding and orienting the protein substrate; and (iii) transferring the γ-phosphate from ATP/GTP to the acceptor hydroxyl residue of the protein substrate. The latter reaction occurs within a subdomain referred to as the catalytic loop. In a normal cell, the catalytic activity of most protein kinases is constrained by an intrasteric inhibition (at both the ATP/GTP and substrate peptide binding sites) within the kinase domain. This suppression of catalytic activity is often achieved when a "pseudosubstrate" sequence, located outside of the kinase domain and containing the consensus recognition sequence of true peptide substrate lacking the phosphoryl acceptor, interacts with the kinase domain and thereby blocks or distorts the active site. This form of autoinhibition is common among both the PTK and Ser/Thr kinase families, and serves as just one example of the multiple layers of regulatory constraint normally exerted on a protein kinase in order to maintain a balance between the rate of cell-cycle progression and cell death. Other important examples of mechanisms capable of relieving the autoinhibition of a protein kinase include the role of protein–protein interactions (both homo- and hetero-dimerization), and activating ligands and/or second messengers. Limiting the intensity/duration of changes in protein phosphorylation are protein phosphatases and a diverse array of mechanisms controlling the genesis and half-life of the protein kinase. The present understanding of how multiple protein kinases normally function within a signaling pathway has been derived largely from studies focused on delineating immediate upstream and downstream interactions through gain- and loss-of-function (LOF) manipulations, arranging these interactions into orderly and conditional cascades of molecular events, and ultimately linking cell-surface receptors to an effector that is capable of eliciting a biological response. Analyses of such linear pathways have provided valuable insights into the interactions between components, both within and between certain pathways.

Many of the genes that cause cancer encode receptor PTKs (RTK), a subclass of transmembrane-spanning receptor that responds to certain extracellular cues (ligands) with a disinhibition of their intrinsic PTK activity. The cytoplasmic Ser/Thr protein kinases AKT and Raf function intracellularly as important points for the convergence of many of the signals emanating from these upstream oncogenic RTKs.

A typical graphic representation of such a signaling network is shown in Fig. 1, encompassing AKT, Raf, and several interactive signaling pathways regulating the phosphorylation of its downstream effectors, such as c-Myc, Elk, cAMP response element binding protein (CREB), NF-kappa B, and Forkhead. These phosphorylated proteins, in turn, mediate the effects of AKT and Raf on gene transcription, and on the proliferation, growth, differentiation and survival of the cell.

1.2
Genetic Rewiring of Signaling Circuits in Cancer Cells

Most PTKs that have thus far been implicated in cancer undergo mutations that result in a loss of their normal autoinhibition of catalytic activity. In certain signaling "cascades," such as the mitogen-activated protein (MAP) kinase pathway (Fig. 1), this loss of repression would be sufficient to trigger the constitutive activation of all downstream components, since the activity of MAP kinases is dependent on activation of the upstream kinase. However, the regulation of kinase activity is generally more complex, making it more difficult to predict how the oncogenic deregulation of a given kinase may contribute to cellular transformation. For example, the activation of 3-phosphoinositide-dependent protein

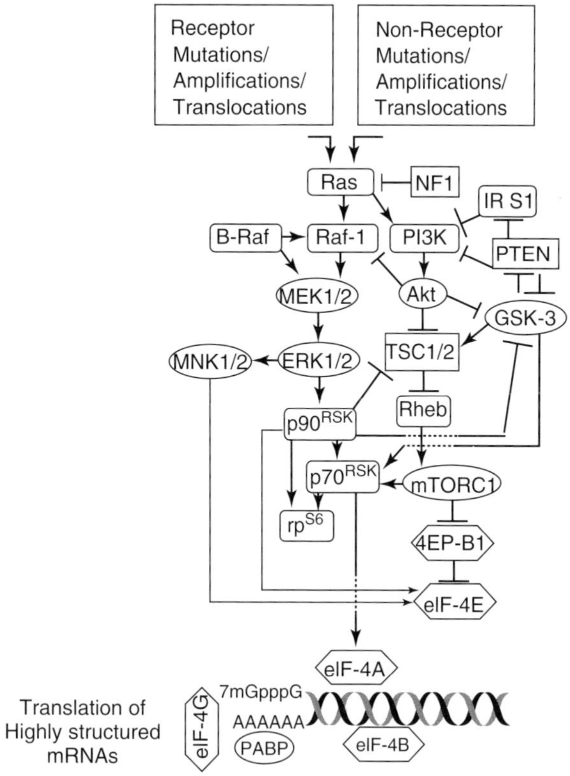

Fig. 1 Activation of the Ras/Raf/MEK/ERK pathway by genetic mutations and interactions with PI3K/PTEN/Akt/mTORC1 pathway. The RAS/Raf/MEK/ERK and PI2K/PTEN/Akt/mTORC1 pathways can be activated by either mutations or deregulation of upstream receptors or mutations and deletions of intrinsic pathway components. Signaling molecules mutated in human cancer are indicated in black symbols. The Ras/Raf/MEK/ERK and PI3K/PTEN/Akt/mTORC1 pathways often interact to cross-regulate each other as well as to control protein translation. In addition, there are other kinases such as glycogen synthetase kinase-3 (GSK-3) which also closely interacts and is regulated by these pathways.

kinase-1 (PDK1) is not sufficient to induce the phosphorylation of certain of its downstream substrates (e.g., p70 ribosomal S6 kinase; p70S6K). Rather, these substrate proteins must first undergo conformational rearrangements that render their PDK1 phosphorylation sites accessible to PDK1. Such modes of substrate-directed regulation are not frequently integrated into a typical signaling diagram. A traditional view of signaling networks provides a valuable and empirical foundation for generating meaningful hypotheses pertaining to the potential downstream effects that a constitutively active (mutant) oncogenic kinase might have within a network such as that shown in Fig. 1. However, these predictions are often intuitive and can be somewhat simplistic when applied to the downstream effectors of a protein kinase such as PDK1.

Systems biology and computational modeling have now entered the field of cell biology and are changing the way in which signaling networks are regarded in complex biological systems. A goal of systems biology is to understand, from a global perspective, the constellation of biological behaviors that can emerge from single/multiple mutations within the components of a cell's extensive network of signaling modules. This approach to interrogating cells, together with high-throughput genomic and proteomic screening approaches, promises to lead to the identification of unexpected "cluster groups" of genes/proteins of similar or related functions that are mutually influenced by experimentally defined mutational events. Though this approach presents a formidable challenge, it constitutes a major new frontier in the study of signaling in cancer, and has gained a significant degree of traction with the recent identification of a limited and well-characterized set of cancer genes encoding signaling molecules.

1.3
Identification of Activating Mutations in Cancer Genes Encoding Signaling Molecules

Since publication of the original census in 2004, the number of known genes causally implicated in cancer has risen by 41% (from 291 to 491), while the number of cancer genes encoding protein kinase domain sequences has risen from 27 to 32 (see Table 1 and Cancer Genome Project: *http://www.sanger.ac.uk/cancergenome/projects/census/*). Consequently, the protein kinase domain remains the most common Pfam domain to be causally linked to cancer. With the identification of each of these mutated kinases, it will now be possible to apply the principles of systems biology to a global analysis of how the introduction of each mutant kinase alters cellular behavior, and to exploit genomic and proteomic strategies to extract responsive cluster groups. With such an expansive inventory of new data, computational modeling can then begin to evaluate whether a specific combination of mutations ("hits") would be capable of transforming a normal cell into a cancer cell. The oncogenic themes likely to arise from such an effort can be expected to vary among cancers and according to the sequence in which genetic mutations are introduced within a given cell.

The successes of the epidermal growth factor receptor (EGFR) kinase inhibitor gefitinib (Iressa®) in the treatment of non-small-cell lung cancer (NSCLC) and of the ABL kinase inhibitor imatinib (Gleevec®) in treating chronic myeloid leukemia (CML), have demonstrated the potential for targeting specific activating mutations in protein kinases that promote proliferative signals in cancer cells. However, it is also now apparent that the effectiveness of such treatments will be limited, and will be critically dependent on identifying the dominant genetic lesion responsible for activating a given kinase in each malignancy, and thereafter developing drugs to target this "Achilles' heel." For example, only about 10–19% of patients with NSCLC respond to treatment with gefitinib, although a majority of these patients express the EGFR. However, a subset of patients, representing about 10% of lung cancers with tumors positive for mutations clustered around the ATP-binding pocket of the tyrosine kinase domain, respond dramatically to gefitinib. Consequently, the patient's response to therapy will be dependent on correctly matching a given kinase inhibitor to the causal oncogenic event in each case. As a prelude to this emerging effort, each of the 32 protein kinases currently known to be capable of functioning as dominant human cancer genes will now be described. In addition, the domain organization and normal function(s) of the protein will be described,

Tab. 1 Summary and classification of protein kinases in the human Cancer Gene Census.

Protein kinase classification	*Symbol*	*Name*	*Cytogenetic band*
Protein-tyrosine kinase family			
Receptor tyrosine kinase subfamily	ALK	Anaplastic lymphoma kinase (Ki-1)	2p23
	EGFR	Epidermal growth factor receptor	7p12
	ERBB2	v-erb-b2 erythroblastic leukemia viral oncogene	17q21
	FGFR1	Fibroblast growth factor receptor 1	8p11
	FGFR2	Fibroblast growth factor receptor 2	10q26
	FGFR3	Fibroblast growth factor receptor 3	4p16
	FLT3	fms-related tyrosine kinase 3	13q12
	KIT	v-kit Hardy-Zuckerman 4 feline sarcoma viral oncogene	4q12
	MET	met proto-oncogene	7q21
	NTRK1	Neurotropic tyrosine kinase, receptor type 1	1q21
	NTRK3	Neurotropic tyrosine kinase, receptor type 3	15q25
	PDGFRA	Platelet-derived growth factor receptor, α receptor	4q11
	PDGFRB	Platelet-derived growth factor receptor, β polypeptide	5q33.1
	RET	ret proto-oncogene	10q11
Cytoplasmic tyrosine kinase subfamily	ABL1	v-abl Abelson murine leukemia viral oncogene homolog 1	9q34.1
	ABL2	v-abl Abelson murine leukemia viral oncogene homolog 2	1q34
	ITK	IL2-inducible T-cell kinase	5q31
	JAK1	Janus kinase 1	1p32.2
	JAK2	Janus kinase 2	9p24
	JAK3	Janus kinase 3	19p13.1
	LCK	Lymphocyte-specific protein tyrosine kinase	1p34
	SYK	Spleen tyrosine kinase	9q22
Dual-specificity kinase family			
Kinase group MEK/STE7	MAP2K4	Mitogen-activated protein kinase kinase 4	17p12
Protein-serine/threonine kinase family			
Kinase group TKL	BRAF	v-raf murine sarcoma viral oncogene homolog B1	7q34

(*continued overleaf*)

Tab. 1 *(Continued)*

Protein kinase classification	*Symbol*	*Name*	*Cytogenetic band*
Kinase group CAMK	PIM-1	pim-1 oncogene	6p12
Kinase group STE	STK11	Serine/threonine kinase 11 gene (LKB1)	19p13.3
Kinase group AGC	AKT1	v-akt murine thymoma viral oncogene homolog 1	14q32
	AKT2	v-akt murine thymoma viral oncogene homolog 1	19q13.1
Kinase group CMGC	CDK4	Cyclin-dependent kinase 4	12q14
	CDK6	Cyclin-dependent kinase 6	7q21
	CDK12	Cyclin-dependent kinase 12	17q12
Phosphatidylinositol kinase (PIK)-related subfamily			
	ATM	Ataxia telangiectasia mutated	11q22

together with its oncogenic alterations in cancer and how this alters its activity. The tumor/cancer types in which these changes are most common will be identified. First, however, an analysis is provided of the canonical Ras/Raf/MEK/ERK pathway that is of such importance in cancer cell signaling.

2
Ras/Raf/MEK/ERK Pathway

Usually, signaling commences upon ligation of a growth factor/cytokine/interleukin (IL)/mitogen (ligand) to its cognate receptor at the cell surface. This event can result in the activation of many downstream signaling cascades, including the Ras/Raf/MEK/ERK pathway. This cascade can further transmit its signals to the nucleus to control gene expression, to the translational apparatus to enhance the translation of "weak" mRNAs, to the apoptotic machinery to regulate apoptosis, or to other events involved in the regulation of cellular proliferation and cell cycle progression. Regulation of the Ras/Raf/MEK/ERK and Ras/PI3K/PTEN/Akt/mTOR pathways is mediated by a series of kinases, phosphatases, GTP:GDP exchange, and scaffolding proteins. There are also many tumor suppressor proteins which interact with these cascades and which frequently serve to fine-tune or limit activity, including RKIP, protein serine/threonine phosphatase 2A (PP2A), DUSP5, and DUSP6. Mutations occur in many of the genes in these pathways, leading to uncontrolled regulation and aberrant signaling. An overview of the effects of mutations and the activation of the Ras/Raf/MEK/ERK pathway and the frequently co-associated Ras/PI3K/PTEN/Akt/mTOR pathway, and how they interact, is presented in Fig. 1. At this point, those genes which are abnormally expressed in human cancer will be identified, the aim being to illustrate the importance of these genes and pathways.

Following the stimulation of a growth factor receptor (GFR), a Src homology 2 domain-containing protein (Shc), an adaptor protein, becomes associated with the C terminus of activated GFRs, such as EGFR, insulin like growth factor-1 receptor (IGF-1R), vascular endothelial growth factor receptor (VEGFR), and many others. *EGFR* mutations can contribute to the transformation of multiple cell lineages, and these alterations are considered driver mutations

Shc recruits the growth factor receptor-bound protein 2 (Grb2) protein and the son of sevenless (SOS) homolog protein (a guanine nucleotide exchange factor; GEF), resulting in a loading of the membrane-bound GDP:GTP exchange protein (GTPase) Ras with GTP. *RAS* is frequently mutated in many diverse human cancers. *RAS* mutations are often driver mutations. GEFs promote Ras activation by displacing GDP from Ras, which leads to GTP binding.

2.1 Ras

Ras activation is suppressed by the GTPase activating proteins (GAPs) that stimulate the GTPase activity of Ras. There are two prominent GAP proteins, p120GAP and NF1. *NF1* is a tumor suppressor gene and has both driver and gatekeeper gene functions. Germline mutations at *NF1* lead to neurofibromatosis.

Ras can also be activated by GFRs, such as insulin receptor (IR), via intermediates such as insulin receptor substrate (IRS) proteins that bind Grb2. *IRS4* has recently been documented to be mutated in melanoma. Ras:GTP then recruits the S/T kinase Raf to the membrane where it becomes activated, likely via a Src-family tyrosine (Y) kinase.

Both, *RAS* and *RAF* are members of multigene families, there being three Ras members (*KRAS*, *NRAS*, and *HRAS*) and three *RAF* members (*BRAF*, *RAF1*; also known as c-Raf, and *ARAF*). *BRAF* is frequently mutated in melanomas and certain other cancers, and these mutations are frequently driver mutations.

2.2 Raf

Raf-1 can be regulated by dephosphorylation by the PP2A and others. PP2A has been reported to both positively and negatively regulate Raf-1. *PP2A* is also considered a tumor suppressor gene and has gatekeeper gene functions.

Raf phosphorylates and activates the mitogen-activated protein kinase kinase-1 (MEK1) (a dual specificity kinase (T/Y) on S/T residues). Other proteins such as kinase suppressor of Ras (KSR) have recently been shown to phosphorylate MEK1. KSR has scaffolding properties and interacts with Raf, MEK, and extracellular signal-regulated kinase (ERK) which regulate ERK activation. KSR can form dimers with various Raf proteins and alter the effects of Raf inhibitors. KSR1 competes with Raf-1 for Raf inhibitor-induced binding to B-Raf, which decreases the normal ERK activation observed after Raf-inhibitor treatment.

2.3 MEK

MEK1 phosphorylates ERK1 and ERK2 at specific T/Y residues. *MEK1* was originally not thought to be mutated frequently in human cancer, but recent large-scale mutation screening studies and investigations aimed at determining mechanisms of resistance to small-molecule inhibitors have indicated that *MEK1* is mutated in certain human cancers and is also mutated in certain inhibitor-resistant cells. *MEK1* is also considered to be a driver oncogene in certain cancers. Rac (Ras-related gene) and p21-activating kinases (PAKs) can also regulate MEK/ERK activation.

2.4 ERK

Activated ERK1 and ERK2 S/T kinases phosphorylate and activate a variety of substrates, including p90 Ribosomal six kinase-1 ($p90^{Rsk1}$), and this pathway has been implicated in cancer progression. ERK1 and 2 are considered by some as gatekeeper genes. ERK also phosphorylates MAPK signal integrating kinases (Mnk1/2), which can in turn phosphorylate eukaryotic translation initiation factor 4E (eIF4E), a key protein involved in the translation of difficult mRNAs. *EIF4E* is considered to be a gatekeeper gene.

$p90^{Rsk1}$ can activate the CREB transcription factor as well as proteins involved in the regulation of protein translation [e.g., Mnk-1, p70S6K, eukaryotic translation initiation factor 4B (eIF4B), and ribosomal protein S6 (rpS6)].

The number of ERK1/2 substrates/targets is easily in the hundreds, and include different types of molecules such as other kinases, phosphatases, GFRs, cytokines, cell-cycle regulator proteins, transcription factors, or proteins involved in mRNA translation or apoptosis. The suppression of MEK and ERK can have profound effects on cell growth, inflammation, and aging. Activated ERK can also phosphorylate "upstream" Raf-1 and MEK1, which alters their activity. Depending upon the site phosphorylated on Raf-1, ERK phosphorylation can either enhance or inhibit Raf-1 activity. In contrast, some studies have indicated that when MEK1 is phosphorylated by ERK, its activity decreases. Other recent studies have indicated that ERK does not display a negative feedback inhibition of B-Raf. ERK also phosphorylates SOS at multiple sites, leading to the dissociation of SOS from GRB2 and preventing Ras activation. ERK can also phosphorylate EGFR and suppress its activity. The dual specificity phosphatases (*DUSPs*) (aka MKPs) are transcriptionally induced by ERK phosphorylation of transcription factors (e.g., Ets). The DUSPs serve as negative feedback regulators to suppress ERK activity. Some of the events induced by ERK phosphorylation are rapid, such as post-translational modification (PTM), while other events require gene transcription and translation (e.g., ERK phosphorylation of Ets which induces transcription of DUSPs). The *DUSPs* are potentially tumor suppressor genes, and DUSP mutations have been detected in various cancers. An overview of the regulatory loops in the Ras/Raf/MEK/ERK pathway, as well as the inhibitors of this pathway, is presented in Fig. 2.

3 Raf/MEK/ERK Inhibitors

3.1 Raf Inhibitors

Among the Raf inhibitors that have been developed, some are being used for therapy while others are being evaluated in clinical trials. Raf inhibitors have in general exhibited greater response rates in clinical trials than MEK inhibitors, though this may be related to the broader therapeutic index of Raf inhibitors that suppress ERK activity in a mutant-allele specific fashion as opposed to MEK inhibitors, which suppress MEK activity in both tumor and normal cells. Some inhibitors (i.e., sorafenib; Bayer) were initially thought to specifically inhibit Raf but have subsequently been shown to have multiple targets (e.g., VEGF-R, Flt-3, c-Kit, PDGF-R). However, that does not preclude their usefulness in cancer therapy. Sorafenib has been approved for the treatment of certain cancers such as renal cell carcinoma

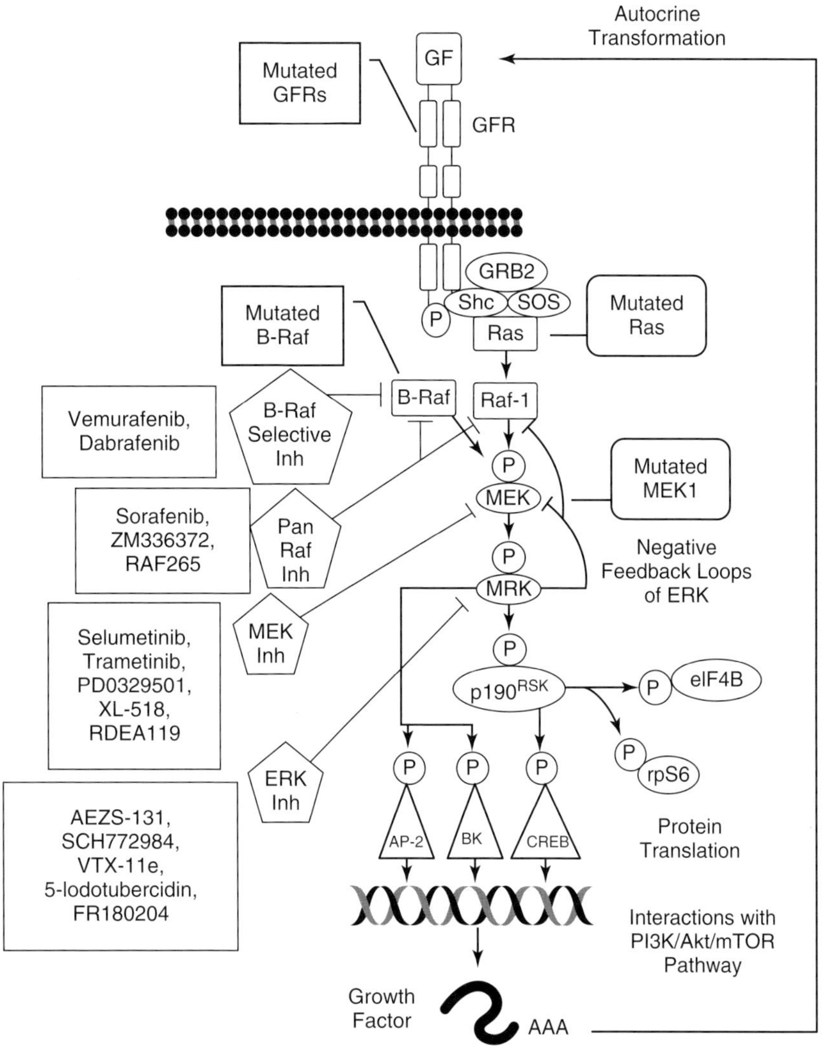

Fig. 2 Common sites of inhibition of the Ras/Raf/MEK/ERK and negative feed-back signaling pathways. Some commonly used inhibitors of this pathway and sites of negative ERK feed-back signaling are indicated. Also indicated are sites where oncogenic mutations feed into the Ras/Raf/MEK/ERK pathway.

(RCC) and unresectable hepatocellular carcinoma (HCC). When sorafenib was evaluated in the Sorafenib HCC Assessment Randomized Protocol (SHARP) trial, the drug was shown to be effective in prolonging median survival and time-to-progression in patients with advanced HCC.

PLX-4032 (Zelboraf®; vemurafenib; Plexxikon/Roche) is a B-Raf inhibitor that has been evaluated in clinical trials. Vemurafenib has been approved by the US Food and Drug Administration (FDA) for the treatment of patients with unresectable or metastatic melanoma carrying the *BRAF*

(V600E) mutation. In order for vemurafenib to be clinically effective, it must suppress downstream ERK activation essentially completely. Vemurafenib has been evaluated clinically in patients with metastatic or unresectable papillary thyroid cancers (PTCs) that have the *BRAF* V600E mutation and are also resistant to radioactive iodine therapy.

GSK2118436 (dabrafenib) is an ATP-competitive inhibitor of mutant B-Raf, WT B-Raf and WT Raf-1, and was developed by GlaxoSmithKline. In clinical trials conducted in patients with melanoma, brain metastases and other solid tumors, dabrafenib – as an active inhibitor of *BRAF* V600E – was shown to be safe and to elicit a good response.

Raf-265, an ATP-competitive pan-Raf inhibitor developed by Novartis, has been evaluated in clinical trials for the treatment of patients with locally advanced or metastatic melanoma.

3.2 MEK Inhibitors

A range of specific inhibitors of MEK have been developed that include PD0325901 (all Pfizer), Selumetinib (ARRY-142886, AZD6244; Astra-Zeneca), MEK162/ARRY-162 (Novartis), XL518 (GDC-0973; Genentech), RDEA119/Refametinib (Ardea Biosciences/Bayer), and Trametinib (GSK112012, GlaxoSmithKline).

MEK inhibitors differ from most other kinase inhibitors in that they do not compete with ATP binding (non-ATP competitive), which confers a high specificity. Most MEK inhibitors are specific and do not inhibit many different protein kinases although, as will be discussed below, certain MEK inhibitors are more specific than others. The crystal structures of MEK1 and MEK2 have been solved as ternary complexes with ATP, while earlier studies with an MEK inhibitor, PD184352, revealed that both MEK1 and MEK2 have unique inhibitor binding sites located on a hydrophobic pocket adjacent to, but not overlapping with, the ATP-binding site. Furthermore, effective targeting of MEK1/MEK2 is highly specific, as ERK1/ERK2 are the only well-described downstream targets. One distinct advantage of inhibiting MEK is that it can be targeted without knowledge of the precise genetic mutation that results in its aberrant activation. This is not the case when targeting Raf, as certain Raf inhibitors will activate Raf; moreover, certain B-Raf-specific inhibitors will not be effective in the presence of *RAS* mutations (as discussed above).

One advantage of targeting MEK is that the Ras/Raf/MEK/ERK pathway is a convergence point where a number of upstream signaling pathways can be blocked with the inhibition of MEK. For example, MEK inhibitors (e.g., selumetinib; AZD6244) have been investigated for the treatment of pancreatic cancers, breast cancers and other cancers such as hematopoietic malignancies, including multiple myeloma.

Selumetinib inhibits MEK1 *in vitro*, with an IC_{50} value of 14.1 ± 0.79 nM; it is specific for MEK1 as it did not appear to inhibit any of the approximately 40 other kinases in the panel tested. Selumetinib is not competitive with ATP. Molecular modeling studies have indicated that selumetinib binds to an allosteric binding site on MEK1/MEK2. The binding sites on MEK1/MEK2 are relatively unique to these kinases and may explain the high specificity of MEK inhibitors. This binding may lock MEK1/2 in an inactivate conformation that enables the binding of ATP and substrate, but prevents the molecular interactions required for catalysis and access to the

ERK activation loop. Selumetinib inhibited downstream ERK1/ERK2 activation in *in vitro* cell line assays with stimulated and unstimulated cells, and also inhibited activation in tumor-transplant models. Selumetinib did not prevent the activation of the related ERK5 that occurs with some older MEK1 inhibitors, which are no longer being investigated in clinical trials. The inhibition of ERK1/2 suppresses their ability to phosphorylate and modulate the activity of Raf-1, B-Raf and MEK1, but not MEK2, as the latter lacks the ERK1/ERK2 phosphorylation site. In essence, by inhibiting ERK1/2 the negative loop of Raf-1 and MEK phosphorylation is suppressed, and hence there will be an accumulation of activated Raf-1 and MEK. This biochemical feedback loop may provide a rationale for combining Raf and MEK inhibitors in certain therapeutic situations.

RDEA119/Refametinib is a more recently described MEK inhibitor developed by Ardea Biosciences. It is a highly selective MEK inhibitor that displays a >100-fold selectivity in kinase inhibition in a panel of 205 kinases. In contrast, in the same kinase specificity analysis, other recently developed MEK inhibitors (e.g., PD0325901) also inhibited the Src and RON kinases.

Trametinib (GSK1120212) is an allosteric MEK inhibitor developed by GlaxoSmithKline. It has been shown to be effective when combined with dabrafenib in certain dabrafenib-resistant *BRAF* V600 melanoma lines that also had mutations at *NRAS* or *MEK*. The combination of trametinib and the PI3K/mTOR dual inhibitor GSK2126458 also enhanced cell growth inhibition in these B-Raf inhibitor-resistant *BRAF* mutant melanoma lines.

GDC-0973 (XL518) is a potent and selective MEK inhibitor developed by Genentech. The effects of combining GDC-0973 and the PI3K inhibitor GDC-0941 on the proliferation of *BRAF* and *KRAS* mutant cancer cells indicated combination efficacy, both *in vitro* and *in vivo*.

3.3 ERK Inhibitors

At least two ERK molecules are regulated by the Raf/MEK/ERK cascade, namely ERK1 and ERK2. Though little is known about the differential *in vivo* targets of ERK1 and ERK2, the development of specific ERK1 and ERK2 inhibitors is ongoing and may be useful in the treatment of certain diseases such as those leukemias where an elevated ERK activation is associated with a poor prognosis, including acute myeloid leukemia and acute lymphoblastic leukemia (ALL). Among other ERK inhibitors that have been described, AEZS-131 (developed by Aeterna Zentaris) has been reported as a highly selective ERK 1/2 inhibitor. Other ERK inhibitors (e.g., SCH772984, VTX-11e, 5-Iodotubercidin, and FR180204) have also been developed and may be useful for the treatment of cancers that have become resistant to MEK inhibitors.

4 Protein Kinases Encoded by Cancer Genes

4.1 Protein-Tyrosine Kinase Family

4.1.1 Receptor Protein-Tyrosine Kinase Subfamily

More than half of the protein kinases included in the current census of human cancer genes are RTKs. Somatic chromosomal translocations that result in the fusion of a given tyrosine kinase domain of one gene with the regulatory/oligomerization region of another unrelated gene are most common. In many instances, a number of different genes will form chimeric transcripts with the same *RTK* gene. Generally, the fusion proteins encoded by these translocated cancer genes phosphorylate their protein substrates in a ligand-independent manner. Such mutations are often found in lymphomas, leukemia, epithelial neoplasms, and solid tumors.

Anaplastic Lymphoma Kinase ALK The anaplastic lymphoma kinase (*ALK*) gene, located on human chromosome 2p23, is expressed only in the small intestine, testis and brain, and encodes a Type Ia membrane protein (carboxy-terminal kinase domain oriented intracellularly) with a glycosylated amino-terminal extracellular region that binds the ligands pleiotrophin and midkine. Ligand-binding specificity is dependent on two cysteine-rich clusters that are adjacent to the major ligand-binding domain. The extracellular domain of ALK also contains a low-density lipoprotein-A module, which suggests that ALK may also interact with low-density lipoproteins. Other motifs present in the extracellular domain include a glycine-rich patch and a pair of Meprin/A-5/protein-tyrosine phosphatase Mu (MAM) domains, which may play a role in mediating certain cell–cell interactions. As with other RTKs (see below), tyrosine residues in the intracellular domain of ALK regulate the enzyme's intrinsic kinase activity, and provide phosphorylation-dependent recruitment sites for the Src homology 2 (SH-2) domains of various adapter proteins and enzymes that propagate intracellular signaling cascades. For ALK, these include the IRS1, Shc protein, and phospholipase C-γ (PLC-γ). The adapter protein Shc forms complexes with GRB2 and the SOS protein to activate the Ras/Raf/MEK/ERK signaling cascade (Fig. 3).

Although ALK-deficient mice do not display any marked developmental defects, the *drosophila* homolog (i.e., *DAlk*) is required for normal development of the gut musculature. A common feature among all ALK chimeric proteins is the presence of an oligomerization domain in the partner protein and oligomerization-mediated kinase activation. Although the intracellular localization of these fusion proteins is determined predominantly by the amino-terminal fusion partner, all fusion proteins result in an aberrant activation of multiple downstream signaling cascades that are responsible for cellular transformation. Approximately half of all non-Hodgkin lymphomas classified as anaplastic large-cell lymphoma (ALCL) are associated with a common somatic chromosomal translocation, t(2;5)(p23;q35). This chimeric gene encodes an oncogenic fusion protein comprised of the complete cytosolic and kinase domains of ALK joined to the amino-terminal 117 amino acid residues of the nucleolar phosphoprotein nucleophosmin (NPM) (Fig. 2). The transforming activity of the fusion protein (NPM-ALK) requires activation of its kinase function, which is a result of the oligomerization mediated by the NPM segment. Although less common, other ALK fusion partners include TPM3, TFGxl, trk fused gene (TFG), ATIC, CLTC, MSN, TPM4, MyH9, RANBP2, ALO17, and CARS. Constitutive ALK signaling activates pathways that are capable of promoting the emergence of at least three distinct hallmarks of cancer, namely enhanced proliferation, survival, and angiogenesis (Fig. 3). Proliferation is enhanced by activation of the MAP kinase pathway through Ras or PLC-γ. PLC-γ-mediated hydrolysis of the membrane phospholipids produces localized increases in inositol triphosphate and diacyglycerol which, in turn, stimulates the release of Ca^{2+} from intracellular stores

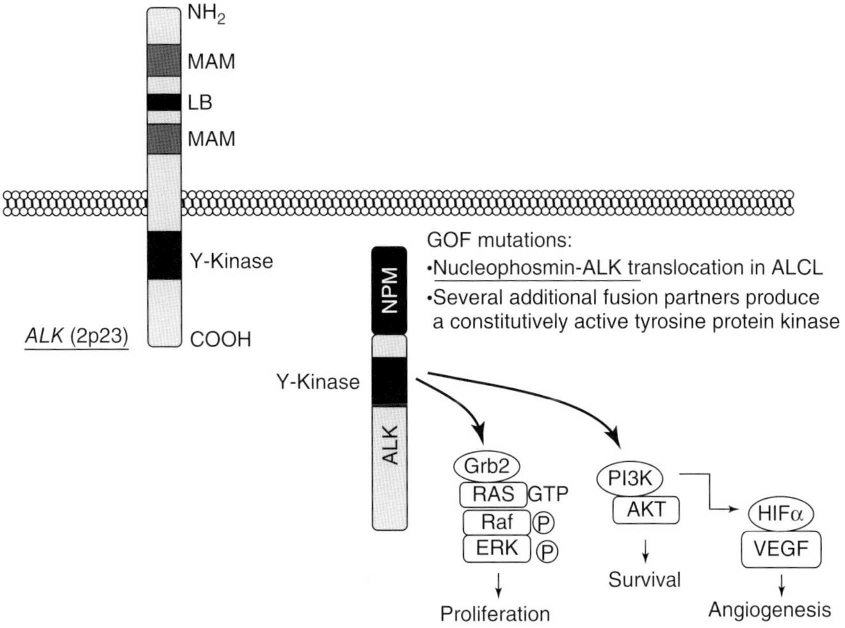

Fig. 3 Domain organization of fused and normal ALK. The nucleophosmin (NPM)-anaplastic lymphoma kinase (ALK) fusion protein places the dimerization domain of NPM amino-terminal to ALK, inducing a gain-of-function (GOF) and constitutive activation of the ALK protein tyrosine kinase domain (Y-kinase) and increased cancer cell proliferation, survival, and tumor angiogenesis. The extracellular ligand-binding (LB) and MAM domains of ALK are shown.

and the activity of the Ser/Thr protein kinase C. Although the precise nature of the interaction between PI3K and ALK is unknown, a potential complex includes NPM–ALK and the regulatory p85 subunit of PI3K with GAB2, Shc, and/or CrkL, an interaction recently identified in the human Karpass 299 cell line.

EGFR and ERBB2 The EGFR subfamily of RTKs encompasses four closely related receptors, EGFR (also known as ErbB-1 or HER1), ErbB-2 (HER2), ErbB-3 (HER3), and ErbB4 (HER4), which drive proliferation, differentiation, survival, and cell motility. Of these, EGF*R* and ErbB-*2*, located on chromosomes 7p12 and 17q21, respectively, are best characterized for their role in human cancers. EGFRs are composed of an extracellular domain (containing two receptor ligand-binding domains and a furin-like domain), a transmembrane region, and the cytoplasmic domain (containing the catalytic PTK domain). EGFR signaling is normally triggered by the binding of its natural ligand, such as EGF, betacellulin, heparin-binding EGF, and transforming growth factor-α. These EGFR ligands are all membrane-bound proteins containing one or more EGF-like motifs that interact with the receptor. Members of the EGFR subfamily are capable of forming as many as 10 different homodimers and heterodimers when occupied by an appropriate ligand, and the affinity of each EGFR combination differs for each ligand. For example, an EGFR homodimer binds EGF, betacellulin and heparin-binding

EGF with about the same affinity, whereas an ErbB2:3 heterodimer preferentially binds the heregulin β isoform versus the heregulin α isoform. Receptor dimerization initiates transphosphorylation of the PTK activation loop, which is required for optimal kinase function. Multiple tyrosine residues in the cytoplasmic tail are also phosphorylated, permitting the recruitment of adapter signaling proteins. Adapter proteins containing SH-2 domains recognize and bind phosphorylated Tyr residues and feed into signaling pathways, including the Ras/Raf/MEK/ERK, PI3K/AKT, and PLCγ/PKC pathways. For example, phosphoTyr1068 binds the adaptor protein GRB2. The result is the potential initiation of at least four hallmarks of cancer, namely enhanced cancer cell proliferation, survival, metastasis, and tumor angiogenesis (Fig. 4).

Biological responses are complex and diverse, depending upon receptor composition, ligand identity, and duration of signaling. The EGF*R* (ErbB-1) and ErbB-*2* genes are commonly amplified, rearranged, or mutated in many human cancers, including head and neck cancer, NSCLC, breast cancer, and gliomas. For example, EGF*R* is overexpressed in 40–80% of NSCLCs. As discussed above (see Sect. 1.3), however, only a subset of these patients respond to the EGFR kinase inhibitor gefitinib. Therefore, overexpression by itself is not sufficient to predict responses, and the evidence suggests that activating mutations are also required. In NSCLC, these activating mutations are somatic and clustered around the ATP-binding pocket of the EGFR catalytic domain. All mutations are heterozygous and consist either of small, in-frame deletions, or amino acid substitutions within exons 18–21. Importantly, these mutations lead to increased responsiveness to EGF and also confer susceptibility to the inhibitor gefitinib. In contrast, rearrangements within the extracellular domain of EGFR are common in gliomas and colorectal cancer, and

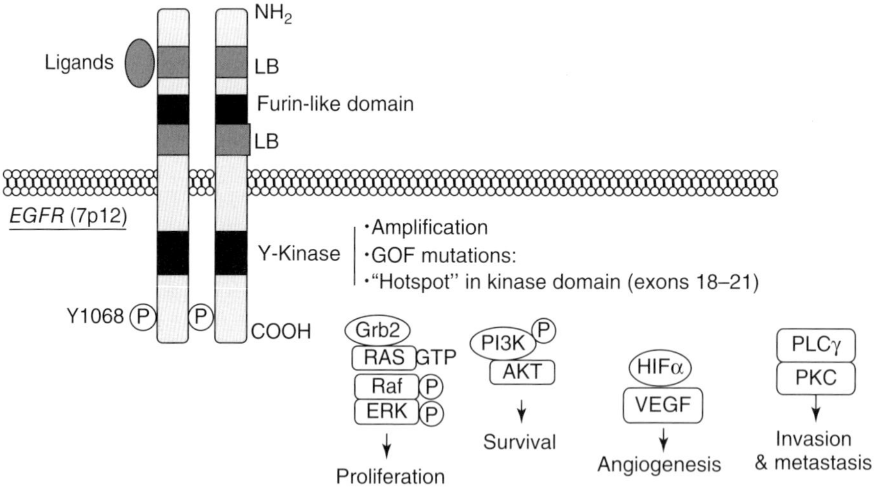

Fig. 4 EGFR amplification and GOF mutations localized to the Y-kinase increase responsiveness to ligand binding and signaling downstream via pathways which may promote cancer cell proliferation, survival, metastasis, and tumor angiogenesis. Positions of the LB domains, furin-like domain and the phospho-Tyr1068 which binds GRB2 adaptor protein are indicated.

are associated with kinase activation. A number of monoclonal antibodies targeting the extracellular ligand-binding region (e.g., cetuximab) have been developed to specifically block such mutant EGFRs. The *erbB-2* gene is amplified in 20–30% of breast and ovarian cancers. ErbB-2 does not directly bind to ligands, and yet its overexpression promotes oncogenesis through decreases in attenuation pathways and signal potentiation. ErbB-2 amplifies signaling indirectly by increasing the ligand affinity of its heterodimeric binding partners (i.e., EGFR, ErbB-3, and ErbB-4) and prolonging the surface expression of these RTKs by slowing their rate of internalization. In addition, ErbB-2 enhances EGFR recycling and reduces lysosomal targeting.

FGFR1, FGFR2, and FGFR3 The fibroblast growth factor receptor (FGFR) family is comprised of four high-affinity RTKs (FGFR1–4) that share several structural features in common, including two or three immunoglobulin-like domains and an acidic box in their amino-terminal extracellular domains, a single transmembrane domain, and an intracellular PTK domain split by a short (14-amino acid) insertion. While three of these RTKs have now been classified as cancer genes (*FGFR1–3*), the fourth (*FGFR4*) is also known to be overexpressed in mammary and ovarian carcinomas. The chromosomal locations of the *FGFR1–FGFR3* genes are 8p11, 10q26, and 4p16, respectively. Distinct FGFR mutations have been shown to mediate a variety of autosomal disorders of bone growth and embryonic development, and a discussion of each would be beyond the scope of the present chapter. However, it is important to note that the majority of mutant FGFRs identified to date possess elevated PTK activity relative to their wild-type counterparts, resulting in ligand-independent activation. While mutations in the extracellular domain of FGFRs have been associated with dwarfism, Pfeiffer syndrome and other developmental abnormalities, the most common mutations associated with human cancer involve a subset of acquired (somatic) translocations that target chromosome band 8p11 (*FGFR1*), and this will serve as an instructive example of how the constitutive activation of FGFRs might contribute to oncogenesis. Crystal structures of an inactive FGFR1 reveal that the activation loop normally occludes substrate tyrosine binding to the catalytic site. In solution, equilibrium exists between this inactive conformation (called *cis* inhibition) and the active loop conformation. The family of fibroblastic growth factors (FGFs) capable of activating the FGFRs includes at least 24 members sharing 30–50% amino acid sequence homology, and are designated FGF-1 through FGF-24. Certain FGF family members are considered to be oncogenic (e.g., FGF-3, FGF-4, FGF-5, and FGF-6), and all have been implicated in the control of a variety of biological processes. In response to ligand occupancy, FGFRs undergo a ligand-induced conformational shift that promotes dimerization and *trans*-phosphorylation of tyrosine residues in the activation loops of the binding partner. This leads to a movement of the activation loop away from the catalytic site that permits substrate access, while *trans*-phosphorylation provides binding sites for the recruitment of signal transducers and activators of transcriptions (STATs) via their SH-2 domains. In acute myelogenous leukemia (AML), chromosomal rearrangements lead to a fusion between the ZNF198 at exon 17 and FGFR1 at exon 9, containing the PTK domain (Fig. 5).

ZNF198 is a widely expressed gene that encodes a 155-kDa protein with

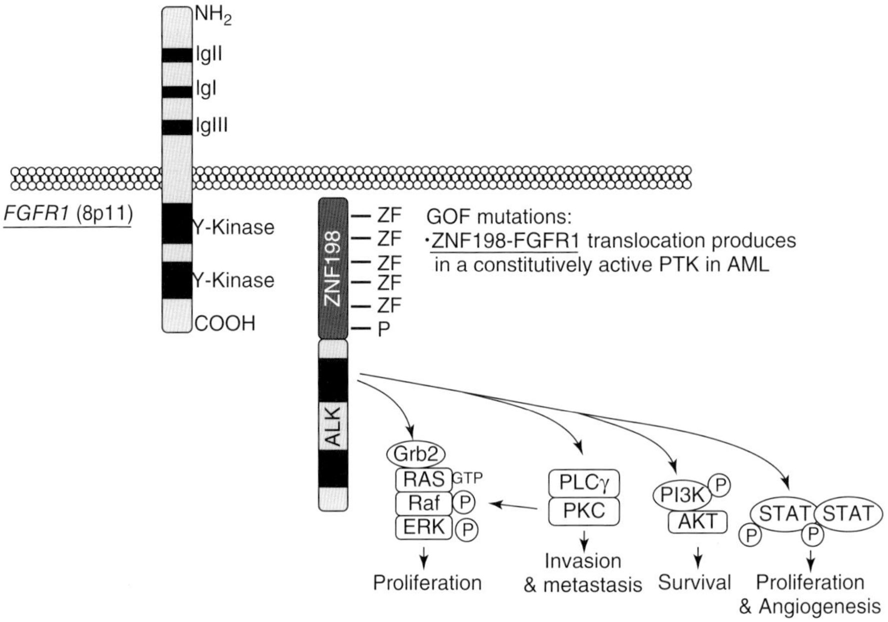

Fig. 5 Schematic structure of fused and normal FGFR1. FGFR1 has two or three extracellular immunoglobulin (Ig)-like domains and a pair of intracellular Y-kinase domains. The Ig domains of FGFR1 are lost in the chromosomal translocation that forms a constitutively active ZNF198–FGFR1 fusion protein. The ZNF198 fusion partner retains its five zinc fingers (ZFs) and a proline-rich domain. FGFR1 phosphorylation of STAT5 and other downstream substrates promote cancer cell proliferation, survival, metastasis, and tumor angiogenesis.

five Zn finger-related motifs (known as MYM domains) that are retained in the ZNF198–FGFR1 chimeric protein. ZNF198–FGFR1 has constitutive PTK activity, and is thought to rely on STAT5 activation in the transformation of hematopoietic cells. FGFR1 phosphorylation of STAT5 promotes dimerization, and dimeric STATs are exported to the nucleus, where these latent transcription factors associate with the promoters of early response genes such as c-*fos*, c-*jun* and c-*myc*, and initiate the transcription of cell-cycle regulatory genes. FGFRs are mitogenic for endothelial cells and play a significant role in angiogenesis by stimulating the production and release of agents that break down basement membranes, thereby creating a microenvironment that is permissive to metastasis. At least two additional hallmarks of cancer that are associated with constitutive FGFR signaling are enhanced cancer cell proliferation and survival.

FLT3 FMS-like tyrosine kinase 3 (FLT3) is a member of the platelet-derived growth factor receptor (PDGFR) family, and is located on human chromosome 13q12. Four of the five members of the PDGFR gene family are regarded as authentic cancer genes; these include KIT (see below), the platelet-derived growth factor receptor type A (PDGFRA) and platelet-derived growth factor receptor type B (PDGFRB), as well as FLT3. The latter contains five extracellular ligand-binding immunoglobulin (Ig)-like domains and a cytoplasmic split tyrosine kinase motif (Fig. 6).

FLT3 is expressed in hematopoietic cells, placenta, gonads, and brain. The ligand for FLT3 is a hematopoietic growth factor termed the Flt3 ligand (FL). The *FL* gene encodes a transmembrane protein, but alternative splicing can result in the generation of a soluble form of the FL. Flt3 binds both the membrane-bound and soluble forms of FL, the binding of which activates tyrosine kinase activity and stimulates proliferation. Although FL does not efficiently induce the proliferation of normal cells in its own right, it does reinforce signaling via other growth factor receptors. Although FLT3-deficient mice develop normally, they possess specific deficiencies in primitive B-lymphoid progenitors, and $FLT^{-/-}$ stem cells are impaired in their ability to reconstitute T cells and myeloid cells when transplanted into bone marrow. Thus, FLT3 signaling is critical for the normal development of stem cells and B cells. FLT3 signaling is mediated by phosphorylation-dependent interactions between its carboxy-terminal cytoplasmic tail and enzymes and adapter proteins, such as the regulatory p85 subunit of PI3K, RAS/GAP, PLC-γ, Shc, and GRB2 (Fig. 6). The activity of FLT3 is normally autoinhibited by intramolecular interactions between the cytosolic juxtamembrane domain and kinase domain, preventing both ATP and substrate from binding to

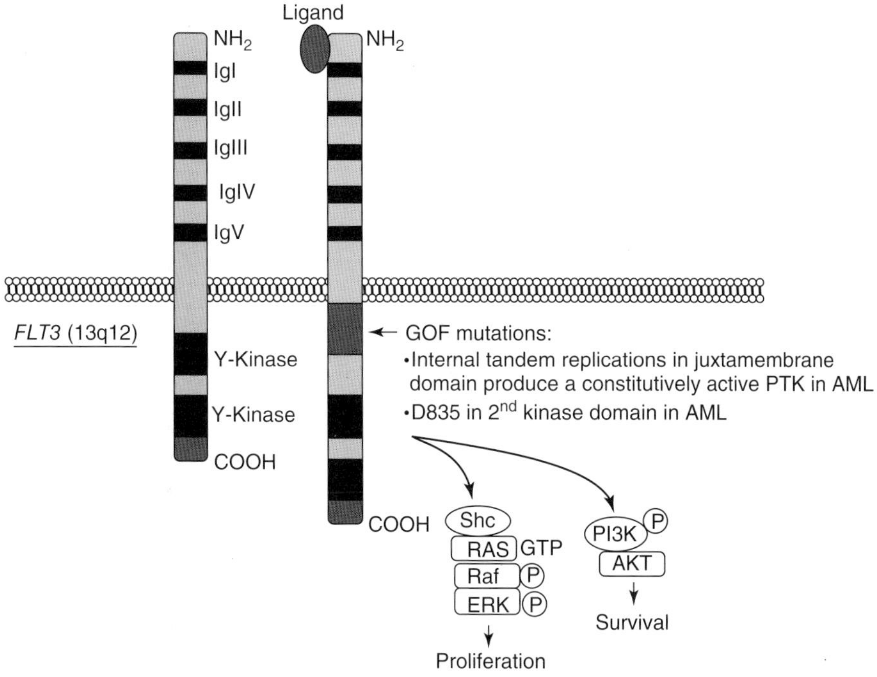

Fig. 6 Schematic structure of fused and normal PDGFR family of PTKs includes a single-pass transmembrane protein containing five extracellular Ig-like motifs and a split intracellular Y-kinase domain. In the case of FLT3 (shown above), tandem repeats within the cytosolic juxtamembrane domain often disrupt autoinhibition of PTK activity in leukemia and are associated with an increased cancer cell proliferation and survival.

the activation loop. Although FLT3 is not normally expressed in mature hematopoietic cells, somatic mutants of the *FLT3* gene are frequently expressed in hematological malignancies. Examples include FLT3 expression in 70% of AMLs, about 30% of ALLs, and some CML cases in lymphoid blast crisis. Although a number of different activating mutations of the *FLT3* gene have been documented, two types are by far the most common. Small tandem duplications of amino acids within the juxtamembrane domain and point mutations within the activation loop occur in 24% and 7% of AML patients, respectively. Although duplications within the juxtamembrane domain are highly variable, each is thought to move this cytosolic domain away from the activation loop, triggering PTK activity. Point mutations of the nucleotides encoding Asp835, within the FLT3 activation loop, similarly disrupt the autoinhibition of kinase activity. Since *FLT3* mutations tend to be associated with a poor prognosis in leukemia, it is thought that mutant *FLT3* may cooperate with other leukemia oncogenes to confer a more aggressive phenotype of this disease (e.g., increased proliferation, survival, and angiogenesis; Fig. 6).

KIT The proto-oncogene *c-kit*, located on human chromosome 4q12, encodes the stem cell growth factor (SCF) receptor KIT/SCFR (stem cell growth factor receptor), a human homolog of the *v-kit* Hardy-Zuckerman 4 feline sarcoma viral oncogene. The domain organization of KIT has been conserved by all members of the PDGFR family of glycosylated, single-pass transmembrane proteins containing five extracellular Ig-like domains and a split intracellular PTK domain (as shown in Fig. 6). The cytosolic juxtamembrane region appears to provide a crucial level of KIT autoinhibition, as mutations within this region are often associated with constitutive, ligand-independent, kinase activity. The importance of this receptor in normal hematopoiesis and melanogenesis was established by the finding that naturally occurring germline LOF mutations result in developmental defects in both processes. In a normal cell, KIT activation stimulates the proliferation and differentiation of cells such as hematopoietic stem cells, mast cells and germ cells through the canonical RTK pathway; that is, SCF binding initiates KIT dimerization and the transphosphorylation of the dimeric KIT subunits. These phosphotyrosine residues enable the docking of SH-2 domain-containing signaling molecules including GRB2, SHP-1 (a protein-tyrosine phosphatase expressed in hematopoietic cells), and members of the Src family of cytosolic tyrosine kinases. The dephosphorylation of KIT by SHP-1 functionally limits the activity of this RTK, and the half-life of the receptor is limited by the adapter protein APS, which promotes a ubiquitin-mediated degradation of KIT through the recruitment of E3 ubiquitin ligase c-Cbl. Thus, both SHP-1 and APS may function as tumor suppressors for KIT-induced malignancies. Downstream of KIT, PI3K-dependent activation of AKT and the phosphorylation/inactivation of the proapoptotic protein Bcl-2-associated death promoter (BAD) appear to be crucial mediators of KIT-induced cellular transformation. More than 30 gain-of-function (GOF) mutations in KIT result in a constitutively active kinase, and these have now been identified in a variety of different human malignancies. These mutations can generally be found in one of two regions. Deletions of a few codons within the autoinhibitory juxtamembrane region, encoded by exon 11, are observed in gastrointestinal stromal tumors, while single amino acid changes in exon 17, in the carboxy-terminal half of the kinase domain, are associated with mast cell leukemia and, to a lesser extent, with T-cell lymphoma and AML. Other GOF point mutations have been identified in exon 13, encoding the amino-terminal half of the KIT kinase domain. In addition to these somatic mutations, GOF mutations of the *c-kit* gene have also been associated with germline tumors such as testicular seminomas and ovarian dysgerminomas. Experimentally, the coexpression of KIT and its ligand SCF has been shown to enhance the autocrine growth of cells and tumorigenesis. Approximately 80% of the patients with activating KIT mutations in exon 11 respond to the EGFR inhibitor imatinib.

MET The human *Met* (hepatocyte growth factor; HGF) receptor gene is located on chromosome 7q21, and is expressed in a variety of organs and cell types. Following its glycosylation, the MET precursor protein (190 kDa) is cleaved into a 50-kDa α-chain and 140-kDa β-chain that are linked via disulfide bonds. After membrane insertion of the β-subunit, these heterodimers form a mature receptor consisting of an extracellular α-subunit joined to a single-pass membrane protein (i.e., β-subunit) with a large ectodomain and intracellular PTK domain (Fig. 7).

Stimulation of MET by its natural ligand, hepatocyte growth factor/scatter factor (HGF/SF), can lead to a broad range of responses in normal cells, such as increased proliferation, scattering, enhanced motility, and angiogenesis (Fig. 6). The downstream effects of MET activation include the stimulation of both the PI3K and MAP kinase pathways (see Figs 1 and 7). A series of autophosphorylation events, including phosphorylation within the β-chain activation loop at Tyr1234 and Tyr1235, and substrate docking sites at Tyr1249 and Tyr1356, stimulate MET catalytic activity, and facilitate the docking of SH-2 domain and MET-binding domain (MBD)-containing proteins including Src, GRB2, PI3K, and PLC-γ. Proteosomal degradation of MET is mediated through interactions between the multisubstrate docking site of MET and c-Cbl. A score of MET-activating mutations, several of which are located in the PTK domain of the β-subunit, have been identified in various solid tumors and metastatic carcinomas. Of particular interest are various germline as well as somatic missense mutations within exons 16–19 of the *Met* gene. These mutations commonly generate amino acid substitutions within the kinase domain. For example, the N1118Y substitution impacts the highly conserved ATP-binding site, while Y1253D affects the regulatory site on MET in squamous cell cancer. MET overexpression has been observed in various tumor specimens, including NSCLC, mesothelioma, pancreatic cancer, osteosarcoma and ovarian carcinoma, as well as metastatic lesions. Studies of small cell lung carcinoma have revealed that interactions between PI3K and MET sustain survival and motility. The direct involvement of MET in the process of oncogenesis is implicated by the presence of germline as well as somatic missense mutations, which characterize hereditary papillary renal carcinoma. MET signaling activates pathways that are capable of promoting the emergence of at least four distinct hallmarks of cancer, namely enhanced proliferation, survival, angiogenesis, and invasiveness (Fig. 7).

NTRK1 and NTRK3 Neurotrophins are a family of four secreted and growth-promoting polypeptides that are known to be widely distributed in both neural and non-neural tissues and include nerve growth factor, brain-derived neurotrophic

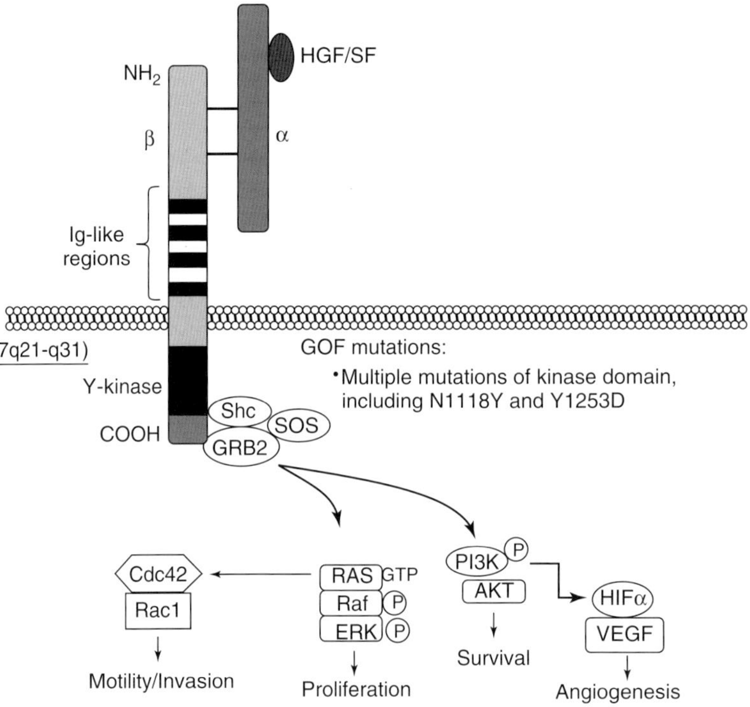

Fig. 7 MET PTK receptor and its domains are represented in this figure. The mature protein has an α-subunit containing a ligand-binding domain for HGF/SF and β-subunit containing four Ig-like motifs and a Y-kinase domain. Many pathways are activated, as indicated downstream of the dimerized receptor.

factor, and neurotrophins (NTs) 3 and 4. Three homologous neurotrophin tyrosine receptor kinases (NTRKs 1–3), are responsive to one or more of the neurotrophins, and both NTRK1 and NTRK3 are known to form oncogenic fusion proteins in soft-tissue tumors. The human *NTRK1* and *NTRK3* genes are located on chromosomes 1q21 and 15q25, respectively. While nerve growth factor is the preferred ligand of NTRK1, NTRK3 binds neurotrophin 3 with high affinity. NTRK receptors share a common structural organization of their amino-terminal extracellular domains, which includes three leucine-rich 24-residue motifs flanked by a pair of juxtamembrane C2-type Ig-like domains (Fig. 8).

The major ligand-binding domains are localized to the membrane-proximal Ig-like domains, which are also required to inhibit a spontaneous dimerization and activation of the receptors in the absence of ligand. The intracellular carboxy-terminus of these receptors contains a PTK domain plus several Tyr-containing motifs. These motifs are involved in regulating kinase activity, and also in the phosphorylation-dependent recruitment of several small G proteins (e.g., Ras, Rap-1, and members of the Cdc-42/Rac/Rho family) and intermediates in the MAP kinase, PI3K, and PLC-γ signaling cascades, ultimately resulting in an activation of gene expression in a cell-dependent manner. Chromosomal translocations that form oncogenic fusion

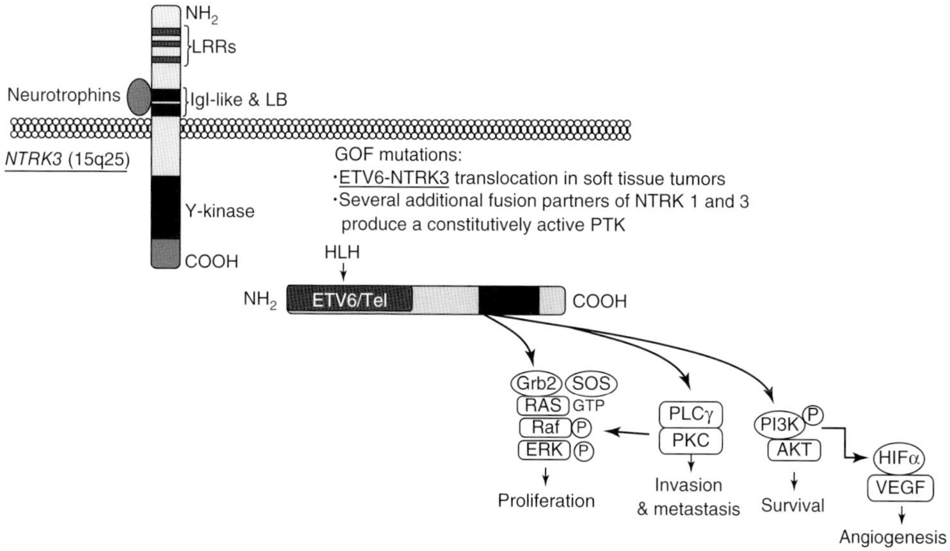

Fig. 8 Domain organization of fused and normal NTRK3. NTRK3 has three extracellular leucine-rich repeats (LRRs) and two immunoglobulin-like (Ig-like) motifs, which are also the major ligand-binding (LB) region of this PTK receptor. NTRKs form a variety of distinct oncogenic fusion proteins, but all retain a carboxy-terminal Y-kinase domain. In the example shown, the ETVS6/Tel fusion partner contains a helix–loop–helix (HLH) domain that mediates ligand-independent dimerization of NTRK3 proteins and promotes cancer cell proliferation, survival, metastasis, and tumor angiogenesis.

proteins containing the kinase domains of either NTRK1 or NTRK3 have been identified in various carcinomas. In the case of NTRK1, oncogenic activity may be conferred by a chromosomal translocation involving the 5′-dimerization domain of a non-muscle tropomyosin gene and the 3-kinase domain of *Ntrk1*, giving rise to a constitutively active PTK. In other cases, rearrangements have been identified involving either the 5-translocated promoter region (TPR) from the *Met* oncogene, or the 5′-region of *TFG*. The *TPR–Ntrk1* translocation forms a fusion protein composed of the amino-terminal portion of TPR and the carboxy-terminal tyrosine kinase domain of NTRK1. Transgenic mice that express human TPR–NTRK1 in the thyroid develop follicular hyperplasia and papillary carcinoma. Chromosomal translocations involving *Ntrk3* often involve the breakpoint regions t(12;15)(p13;q25), and result in a chimera consisting of the NTRK3 kinase domain and the dimerization domain of the ETV6/Tel transcription factor (Fig. 8).

The oncogenic fusion proteins produced by any of the rearrangements discussed above exhibit three common characteristics that are absent from NTRK1 or NTRK3, namely a ubiquitous cytosolic expression, ligand-independent dimerization via coiled-coil domain interactions, and constitutive PTK activity. These fusions have been implicated in the genesis of human colon carcinoma and papillary thyroid carcinoma in the case of NTRK1, and in secretory breast carcinoma and congenital fibrosarcoma in the case of NTRK3. Collectively, these chromosomal rearrangements and constitutive kinase activity are capable of promoting the emergence of at least four distinct hallmarks of cancer, namely enhanced cancer cell proliferation, survival, tumor invasion and metastasis, and tumor angiogenesis (Fig. 8).

PDGFRA and PDGFRB PDGFRA and PDGFRB are encoded by genes located in human chromosomes 4q11 and 5q31, respectively. PDGFRA exhibits a pronounced structural and sequence similarity with members of the PDGFR family, including PDGFRB and FLT3 (see Fig. 6). The main distinction between PDGFRA and PDGFRB is exemplified by their ligand preferences, with PDGFRB selectively interacting with PDGF homodimers containing B-type polypeptide chains (PDGF BB), while PDGFRA interacts more promiscuously with the PDGF AA, BB, and AB dimers. In addition, these receptors exhibit temporally distinct and tissue-specific activities in the mammalian embryo and adult. For example, PDGFRA expression in the nervous system coincides with the development of glial cells, which includes processes of glial cell migration and proliferation. The functional PDGFRA consists of an extracellular domain containing five Ig-like domains punctuated by eight N-linked glycosylation sites, a transmembrane domain, and two cytoplasmic kinase domains (as shown in Fig. 6). The extracellular Ig-like domains are essential for the correct folding of the PDGFRA protein, thereby enabling receptor–ligand interactions. Ligand engagement induces receptor dimerization and autophosphorylation, which in turn generates phosphorylation-dependent recruitment sites for adapter proteins and enzymes capable of activating the PI3K and MAP kinase pathways (e.g., Src, PI3K, PLC-γ, the Crk group of adapter proteins, and SHP-2). PDGFRA is also capable of stimulating c-Jun N-terminal kinase (JNK1) to promote apoptosis, and of inhibiting cell-cycle progression via the cyclin-dependent kinase (CDK) inhibitor p21Cip1. Signaling through the PDGFRA also stimulates membrane ruffling, cytoskeletal rearrangements, and migration in a cell-type-specific manner. PDGFRA activity is limited by ligand-induced Src activation and receptor trafficking/internalization. In cancer, a potentially activating deletion of an 81-amino acid stretch within the fifth Ig-like domain of the extracellular stalk of PDGFRA has been described. This deletion mutant is found amplified in glioblastoma, where it is likely to give rise to an alternatively folded receptor capable of sustaining PDGFRA catalytic activity in the absence of ligand. Recently, a transforming PDGFRA deletion mutant lacking Ig-like domain-encoding exons 8 and 9 has also been identified in gliomas. This mutant protein is capable of autophosphorylation and, therefore, activation in the absence of PDGF. In addition, mutations and deletions within the PDGFRA activation loop (exon 18) and juxtamembrane region (exon 12), that have been detected in gastrointestinal pacemaker cell tumors (i.e., "gastrointestinal stromal tumors"), are also capable of producing ligand-independent activation. PDGFRA and its ligands have also been reported to facilitate tumorigenesis in the absence of genetic mutations, by establishing autocrine signaling loops, as in the case of medulloblastomas where autocrine stimulation promotes the emergence of at least four distinct hallmarks of cancer, namely enhanced proliferation, survival, angiogenesis, and metastasis.

RET The glial cell line-derived neurotrophic factor (GDNF) family, consisting of GDNF, neurturin, artemin and persephin, all transmit signals promoting neuronal survival, proliferation, and differentiation via the transmembrane receptor tyrosine kinase RET (rearranged in transformation) and its coreceptors GFRα-1 to -4. RET is normally expressed throughout the nervous system, within neural crest-derived cells, and in the urogenital system, and is required for neuronal and kidney development. The GFRα-1–4 coreceptors are glycosyl-phosphatidyl-inositol-anchored proteins that selectively recognize and bind certain members of the GDNF family and associate with RET kinase to trigger intracellular signaling. For instance, GDNF preferentially binds to the GFRα-1–RET heterodimer. The *RET* gene has been mapped to human chromosome 10q11, and is known to give rise to a transcript which is alternatively spliced to yield the short, middle, and long isoforms of RET, each containing an extracellular domain with three cadherin-like domains and a cysteine-rich region (Fig. 9). When bound to GDNF, GFRα-1 dimerizes with RET, and stimulates the transphosphorylation of specific Tyr residues within the cytoplasmic domain of RET. The three RET isoforms have a number of Tyr residues in common, including those residing within the kinase domain and carboxy-terminus, though the long RET isoform contains an additional carboxy-terminal Tyr1096 residue (Fig. 9). Once phosphorylated, these residues serve as scaffolding sites for signaling complexes. The conserved Tyr residues bind GRB10, PLC-γ, and Shc, while Tyr1096 binds GRB2 specifically and links RET activation to the cell proliferation and survival pathways controlled by the Ras/Raf/MEK/ERK and PI3K/AKT pathways, respectively (Fig. 9).

Underscoring the importance of Tyr1062, are studies that demonstrate that a point mutation within this codon significantly decreases the transforming activity of the RET protein. RET-mediated transformation results from various activating germline mutations in early-onset multiple endocrine neoplasia types 2A and 2B and late-onset familial medullary thyroid carcinoma. In addition, genetic rearrangements have been identified in papillary thyroid carcinoma. Chromosomal translocations involving the intracellular domain of RET have given rise to 10 known fusion proteins (RET/PTC1–9 and RET/ELKS) prevalent in papillary thyroid carcinoma in a geographically distinct manner. These include translocations between *RET* and *H4* (Fig. 9), *RIα*, *ELE1*, *RFG5*, *hTIF*, *RFG7*, *kinectin*, *RFG9*, and *ELKS*. The most common rearrangements, *RET/PTC1* and *RET/PTC3*, involve inversions of the *H4* inv(10)(q11.2q21) and *ELE1* genes, and account for 90% of all rearrangements in papillary thyroid carcinoma. The results of studies have suggested that RET/PTC becomes ectopically localized to the cytoplasm, where it activates STAT3 by phosphorylation of Tyr705 (Fig. 9). The constitutively active STAT3 promotes cellular transformation via the stimulation of *cyclin-D1*, *D2*, *D3*, and *bcl-XL* transcription. RET/PTC has also been implicated in the transcriptional upregulation of the osteopontin (*OPN*) gene, resulting in an increase in the proliferative and invasive activities of cells that respond to autocrine stimulation by OPN (Fig. 9). While RET/PTC rearrangements predominate in papillary thyroid carcinoma, germline mutations within *RET* are responsible for the onset of multiple endocrine neoplasia types 2A and 2B, and familial medullary thyroid carcinoma. Point mutations or base-pair duplications of the *RET* gene have been reported in patients with multiple endocrine neoplasia 2A and familial medullary thyroid carcinoma. Missense mutations within *RET* have also

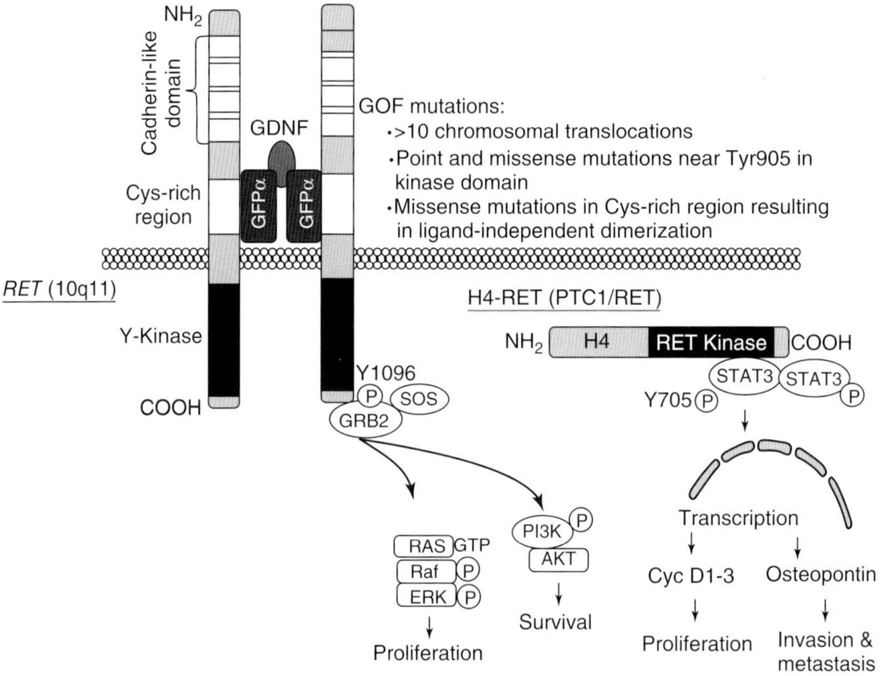

Fig. 9 Domain organization of fused and normal RET kinase. When bound to its ligand, GFRα-RET heterodimers dimerize and transphosphorylate specific Tyr residues within the cytoplasmic domain. Each RET protein also contains three extracellular cadherin-like domains and a cysteine-rich region. A common rearrangement involves a translocation of the *H4* gene, which produces a cytosolic fusion protein (H4–RET) that promotes cancer cell proliferation, survival, and metastasis via STAT3 and osteopontin.

been described in patients with multiple endocrine neoplasia 2A or 2B. Point mutations observed in multiple endocrine neoplasia 2A and familial medullary thyroid carcinoma, target the extracellular cysteine residues that normally contribute to intramolecular disulfide bonds, and promote RET dimerization. The formation of RET homodimers leads to its ligand-independent activation. On the other hand, RET mutations in multiple endocrine neoplasia 2B do not induce dimerization; rather, mutations target the catalytic core and may alter substrate specificity by inducing conformational changes in the region containing Tyr905, a conserved residue which is important for catalysis.

4.1.2 Cytoplasmic Protein-Tyrosine Kinase Subfamily

ABL1 and ABL2 The *ABL1* gene (chromosomal band 9q34.1), along with the closely related *ABL2* or *ARG* (Abl related gene; chromosomal band 1q24), belongs to the Abelson family of genes that encode prototypic nonreceptor tyrosine kinases involved in various cellular processes. The *ABL1* gene product, c-Abl, is a ubiquitously expressed protein that colocalizes with actin-based cytoskeletal structures in the cytoplasm, and binds chromatin when imported to the nucleus. These intermolecular interactions are mediated by carboxy-terminal actin- and DNA-binding domains of c-Abl

(Fig. 10). The amino-terminus of c-Abl contains an SH-3 domain and an SH-2 domain. Autoinhibition of c-Abl is achieved by a novel variation on a familiar theme, and the details of this mechanism only recently became apparent when the crystal structure of the protein became available for examination. As in the case of the closely related Src kinases, the amino-terminus of the protein functions as an autoinhibitory "Cap" region, preventing the substrate from docking with the kinase domain. However, this is not achieved by the binding of its SH-2 domain to a phosphotyrosine residue, as in other Src-like kinases. Rather, an amino-terminal myristate is responsible for binding to a hydrophobic pocket in the kinase domain that stabilizes an autoinhibited conformation (Fig. 10).

It is this distinction in the autoinhibitory mechanism employed by c-Abl, as compared to those of the other Src kinases, that explains the ability of the drug imatinib to selectively inhibit the catalytic activity of c-Abl, but not that of c-Src. The targeted disruption of c-Abl in mice is characterized by lymphopenia, a failure to thrive, and neonatal mortality. In the cytoplasm, c-Abl participates in the propagation of mitogenic signals via the Ras/Raf/MEK/ERK and PI3K/AKT pathways (Fig. 10). This signaling is negatively regulated by the interactions of c-Abl with PSTPIP1, a PEST-type protein Tyr phosphatase that dephosphorylates c-Abl. c-Abl signaling in the nucleus results in a suppression of cell growth via the c-Abl-dependent transactivation of various proapoptotic genes. During the early G_1 phase of the cell cycle, this nuclear activity of c-Abl is repressed due to its sequestration by the pocket protein Rb. Following an oncogenic and reciprocal chromosomal translocation, the control of c-Abl activity is lost in cells, and this generates what has been termed the Philadelphia chromosome. This mutation is present in most cases of CML and in many cases of ALL. Although similar translocations occur in CML and ALL, their c-Abl products are distinct, producing 210-kDa and 180–185-kDa proteins, respectively. Both involve a t(9;22)(q34, q11) translocation of the *ABL* gene to the center of the *BCR* gene, producing a fused transcript that translates into a chimeric BCR-Abl protein that is constitutively active (Fig. 10). C-Abl activity is sustained by the loss of its autoinhibitory (myristoylated) amino-terminal tail and the dimerization triggered by the presence of coiled-coil

Fig. 10 Domain organization of fused and normal ABL1. Translocation of the *ABL1* gene to the center of the *BCR* gene produces chimeric BCR–ABL1. This, in turn, results in a deletion of the autoinhibitory amino-terminal myristoylation site of ABL1 and dimerization of BCR coiled-coil domains triggers ligand-independent kinase activity, and promotes tumor cell proliferation, survival, and angiogenesis by localizing the protein in the cytosol and preventing DNA binding.

domains within the amino-terminus of BCR (Fig. 10). BCR-Abl dimerization, in turn, allows for transphosphorylation of the fusion protein and an activation of oncogenic signaling. The formation of BCR-Abl homo-oligomers and hetero-oligomers with BCR further promotes oncogenesis by localizing the protein in the cytosol and preventing DNA binding. The use of the small-molecule tyrosine kinase inhibitor imatinib in the treatment of CML patients who are BCR-Abl-positive has met with success, but drug resistance has occurred frequently. Mutations of residues in or near the ATP-binding pocket of c-Abl prevent imatinib binding and account for a majority of drug-resistant cases. The homologous *ARG* gene is involved in a similar translocation, but with a different fusion partner, Ets variant gene 6 (*ETV6/Tel*). ETV6/Tel is a member of the Ets family of transcription factors, and is frequently found to be rearranged with a variety of translocation partners in human leukemias. This fusion protein contains the complete SH-3, SH-2 and PTK domains of ARG, along with the helix–loop–helix oligomerization domain of ETV6/Tel (for example, see Fig. 8). The translocation t(1;12)(q25;p13) causes malignancy in adult patients with AML.

ITK and SYK The human spleen tyrosine kinase (*SYK*) gene is located on chromosome 9q22 and encodes a nonreceptor cytoplasmic tyrosine kinase that has a pair of amino-terminal SH-2 domains followed by the PTK domain (Fig. 11). The interleukin-2-inducible T-cell kinase (*ITK*) gene (chromosomal band 5q31) and *SYK* gene form a constitutively active fusion tyrosine kinase ITK-SYK that contains an amino-terminal phosphatidylinositol-3,4,5-triphosphate (PIP3)-binding pleckstrin homology (PH) domain and proline-rich Tec homology (TH) domain from ITK, followed by the SYK carboxy-terminal PTK domain (Fig. 11).

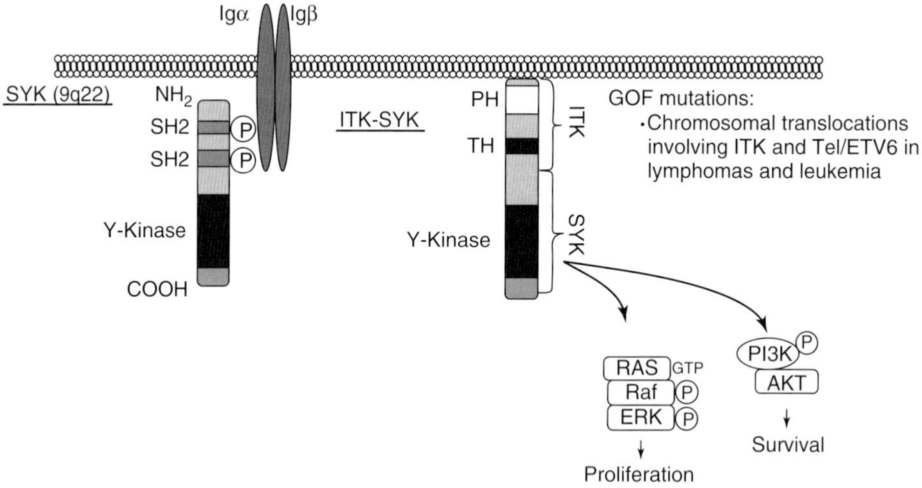

Fig. 11 Schematic representation of SYK kinase and the lymphoma-associated ITK–SYK fusion protein. SYK has dual SH2 protein interaction domains and a carboxy-terminal Y-kinase domain. The chromosomal translocation of the pleckstrin homology (PH) and Tec homology (TH) domain of ITK with the Y-kinase domain of SYK produces a constitutively active kinase that promotes the proliferation and survival of T cells in certain cases of PTCL.

This novel chromosomal translocation is evident in a subset of peripheral T-cell lymphomas (PTCLs) derived from follicular helper T cells. This lymphoma-associated fusion protein is constitutively active, and the PH domain colocalizes this activity with lipid rafts and T-cell receptors (TCRs) in the plasma membrane, where it triggers antigen receptor signaling. SYK signaling has important roles in both B- and T-cell development in association with the activation of the Ras/Raf/ERK signaling cascade and mTOR, possibly via PI3K/AKT (Fig. 11). Although its role in epithelial malignancies remains controversial, the chronic activation of SYK and persistent antigen receptor signaling are capable of inducing oncogenesis in clinically relevant models of PTCL, an aggressive malignancy with a generally poor prognosis. *SYK* is also a target for translocations associated with the *TEL/ETV6* gene on 12p12-13.

JAK-1, JAK-2, and JAK-3 Janus kinase (JAK) is a family of four nonreceptor PTKs that are essential for propagating certain cytokine-induced signals via STAT transcription factors. Three members of this subfamily of PTKs (i.e., *JAK1, 2,* and *3*) are oncogenes; *JAK1* is located on human chromosome 1p32.2, *JAK2* on 9p24, and *JAK3* on chromosome 19p13.1. The JAK proteins range from 120 to 130 kDa and have in common seven Janus Homology (JH) domains (JH1 to JH7), which are sites of interaction between the protein kinase and its associated cytokine receptor. A carboxy-terminal JH1 domain contains the kinase domain responsible for the phosphorylation of its preferred substrate; the Tyr, neutral, basic amino acid tripeptide motif. Amino-terminal to the catalytic domain is a regulatory pseudokinase domain, referred to as the JH2 domain (Fig. 12). Normally, JAK PTK activity is cytokine-dependent. IL-3, IL-5, and/or IL-6

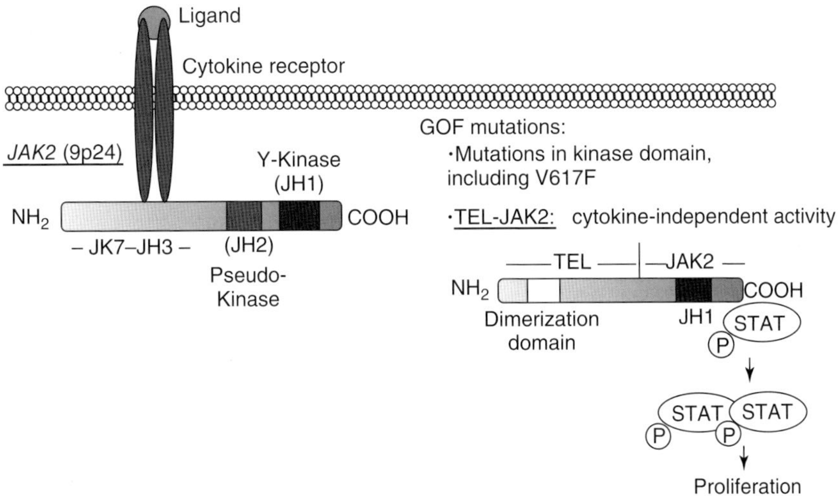

Fig. 12 Schematic structure of fused and normal JAK-2. Jak2 has seven JH domains; the JH3–7 domains interact with cytokine receptors and PTK substrates, JH2 is a regulatory pseudokinase domain, and JH1 contains the carboxy-terminal Y-kinase domain. The dimerization domain of TEL triggers ligand-independent activity of chimeric TEL–JAK-2 and phosphorylation of STAT proteins. Consequently, oncogenic signaling and cancer cell proliferation are increased.

interactions with cytokine receptors initiate receptor dimerization and an activating transphosphorylation of the constitutively associated JAK proteins. Once active, JAKs phosphorylate Tyr residues on the carboxy-terminal region of the cytokine receptors, providing binding sites for the recruitment of STATs via their SH-2 domains. This in turn permits JAKs to phosphorylate STAT proteins, thereby promoting the disassociation of STATs from cytokine receptors, and their dimerization. When exported to the nucleus, dimeric STATs associate with the promoters of early-response genes such as c-*fos*, c-*jun* and c-*myc*, and initiate the transcription of cell-cycle regulatory genes such as *cyclin D1* and cancer cell proliferation (Fig. 12).

Overexpression and activating mutations of JAK1, such as A634D, V658F, R724H, and R879C, have been observed in ALL. A GOF mutation in the pseudokinase domain of JAK2, V617F, produces a constitutively active PTK in myeloproliferative neoplasms and AML. This persistent activation of JAK2 significantly enhances the proliferation of hematopoietic cells. Deregulation of JAK-2 activity may also arise as a result of a somatic chromosomal translocation between the dimerization domain of the ETS-like transcription factor Tel/ETV6 and JAK2 (Fig. 12). The resulting fusion protein (TEL–JAK2) contains the carboxy-terminal protein dimerization domain of Tel and the amino-terminal JAK2 catalytic domain, and is able to form homodimers capable of transphosphorylating its binding partner, thus leading to constitutive activation. The Tel–JAK2 chimera is localized ectopically within cells, and may undergo unusual interactions with members of the early gene induction pathway in a cytokine-independent manner, thereby supporting cellular transformation. The presence of the *Tel–JAK2* translocation and resulting transformation of hematopoietic cells underscore the importance of JAK-2 signaling in hematopoiesis. Examples of the Tel–JAK2-driven transformation include cases of T-cell childhood ALL, AML, ALL, and atypical CML. The finding that JAK2 inhibition by a specific kinase blocker (i.e., AG-490) induces the apoptosis of leukemic cells further supports the direct linkage between constitutive JAK2 activation and malignant transformation. Finally, whilst somatic mutations of JAK3 are best known for their association with severe combined immunoinsufficiency (SCID) disorders, they are also associated with certain leukemias, such as ALL and AML. Point mutations of JAK3 that have been identified in cases of AML are mostly clustered in the pseudokinase (JH2) domain, and JAK3 is also overexpressed in hematopoietic malignancies.

LCK Lymphocyte-specific protein tyrosine kinase (LCK) is a member of the Src tyrosine kinase family and is located on the short arm of human chromosome 1 (1p34). LCK expression predominates in T cells, where its activity is crucial for normal T-cell development and activation. LCK is closely associated with the TCRs CD4 and CD8. LCK is a multidomain protein, typical of the Src kinases, including SH-3, SH-2, tyrosine kinase domains, and a unique domain that is responsible (among other things) for the membrane anchoring of LCK. During the process of acquiring adaptive immunity, T-cell activation is initiated by the binding of CD4 and CD8 to antigen-presenting cells via type I and type II glycoproteins of the major histocompatibility complex. The proximity of LCK to these activated TCRs allows for the transduction of signals responsible for the T-cell response, which includes the production of IL-2 and inositide (1,4,5) triphosphate. LCK activation requires the autophosphorylation of Tyr394 and dephosphorylation of Tyr505, in the carboxy-terminal tail. The protein Tyr phosphatase CD45 regulates LCK activity in a bimodal fashion, not only by mediating Tyr505 dephosphorylation (enhanced activity) but also by dephosphorylation of Tyr394 (suppressed activity). LCK promotes T-cell activation by phosphorylating immunoreceptor Tyr-based activation motifs present within TCR complexes. When phosphorylated, these motifs enable the recruitment of SH-2 domain-containing proteins (e.g., ZAP70) to the TCR. In addition, LCK phosphorylates the membrane–cytoskeleton linker protein ezrin to promote T-cell activation. In T cells, the LKC-dependent phosphorylation of the linker protein Shc also initiates mitogenic signaling via the Ras/Raf/MEK/ERK pathway. Because of the critical position occupied by LCK in the TCR signaling pathway, its necessarily transient activity is highly regulated through ubiquitination under the control of the Cbl ubiquitin ligase. LCK mutations resulting in constitutive autophosphorylation and/or deregulated catalytic function have been described in T-cell leukemia, where they contribute to the oncogenic potential of this kinase. Insertion of the tripeptide QKP between the cytosolic SH-2 and kinase domains may also enhance catalytic function by disrupting the normal repressive interaction of its SH-2 domain with its catalytic site. Three other recorded mutations, resulting in a substitution within the region critical for interaction between LCK and CD4, and two alternative substitutions within the kinase domain, may result in a GOF for LCK. In addition to these mutations, a chromosomal translocation (t(1;7)(p34;q34)) has been observed in T-cell ALL, resulting in an oncogenic fusion protein (i.e., βTCR:LCK) that contains the βTCR enhancer region and the 5′ promoter region of LCK. In the case of βTCR:LCK fusion, a GOF would result from an alleviation of transcriptional repression by the βTCR enhancer region.

4.2 Dual-Specificity Protein-Serine/Threonine and Tyrosine Kinase Family

4.2.1 Kinase Group MEK/STE7

MAP2K4/MEK4 and MEK were discussed in Sect. 2.

4.3 Protein-Serine/Threonine Kinase Family

4.3.1 Kinase Group TKL (Tyrosine Kinase-Like)

BRAF was discussed in Sect.2.

4.3.2 Kinase Group CAMK (Calcium/Calmodulin-Dependent Protein Kinase)

PIM-1 A common site for proviral integration of the Moloney murine leukemia virus (PIM) in mice encodes the *PIM-1* gene. The Ser/Thr kinase PIM-1 is the product of this gene, which is located on the p12 segment of human chromosome 6. In addition to PIM-1, both PIM-2, and PIM-3 make up the PIM family of kinases, which may serve redundant functions as observed in PIM-1-deficient mice where compensation for the loss of *PIM-1* was evident. PIM-1 functions to promote hematopoiesis in response to growth factors such as IL-2 and IL-3, and its expression is confined, throughout fetal development, to hematopoietic sites within the liver and spleen, as well as within the thymus, lymph nodes, and testis. A high expression of PIM-1 has also been observed in myeloid cell and B-cell lines. PIM-1 has been characterized as a 33-kDa cytoplasmic protein, though recent studies have revealed that its nuclear localization is crucial for its antiapoptotic function in Burkitt's lymphoma. The upregulation of PIM-1 has been implicated in BCR/Abl-mediated leukemogenesis, where it acts downstream of STAT5 to confer protection from apoptosis and promote cell-cycle progression. The elevated levels of PIM-1 facilitate phosphorylation and inactivation of the proapoptotic protein BAD, as well as a stabilization of MDM2, an E3 ligase that targets BAD for destruction (Fig. 13). PIM-1 overexpression has also been shown to positively influence a variety of human leukemias, though overexpression alone is not sufficient for cellular transformation, and cooperation between PIM-1 and c-myc is often observed during this process. The identification of a breakpoint region on chromosome 6 has raised the possibility of a translocation between *PIM-1* and the proto-oncogene *c-ABL* (t(6;9)(p21; q33)) in some cases of myeloid leukemia. In response to cytokines and various mitogens, *PIM-1* expression is induced via the JAK/STAT5 and MAP kinase pathways, and its activity may be modulated through a series of autophosphorylation events. With the exception of Etk, which promotes the activation of PIM-1 via the phosphorylation of Tyr218, upstream kinase regulators of PIM-1 have not been identified. Recent studies have shown that phosphorylation of the nuclear mitotic apparatus protein by PIM-1 facilitates interactions between this protein, the heterochromatic protein-1β, and the spindle proteins dynein and dynactin, which are crucial for mitosis. When overexpressed in human prostate epithelial cells, PIM-1 has been shown to disrupt the mitotic spindle checkpoint and induce genomic instability, which is a critical hallmark of cancer. The oncogenic kinase PIM-1 also alleviates transcriptional repression by directly phosphorylating HPI, a member of the transcriptional repression machinery, thereby regulating chromatin structure (Fig. 13). Other putative substrates of PIM-1 include p100, a coactivator of the transcription factor c-myb, the cell-cycle phosphatase Cdc25A, and the CDK inhibitor $p21^{cip1}$. PIM1 phosphorylation of $p21^{cip1}$ promotes the translocation of this cell-cycle inhibitor to the cytoplasm, and its inactivation (Fig. 13).

It is evident that PIM-1 regulates the activity of many proteins that have the potential to promote tumor growth and, as expected, its expression is tightly regulated via transcriptional and translational mechanisms. The presence of dyad symmetry elements enables the *PIM-1* gene to form stem–loop structures, and to block transcriptional elongation in the absence

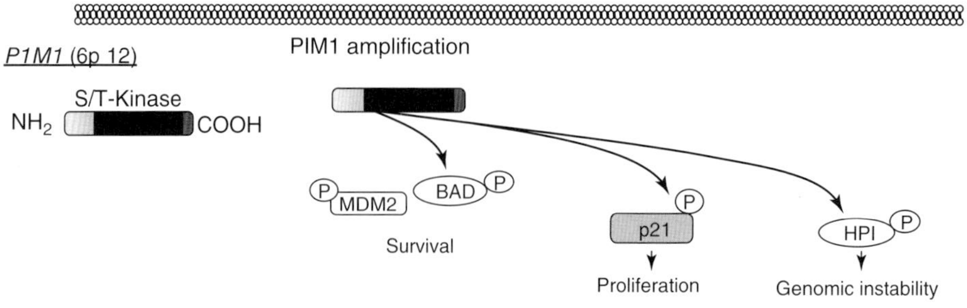

Fig. 13 PIM1 is a cytosolic Ser/Thr protein kinase. The overexpression of PIM1 enhances cell survival, proliferation, and genomic instability by phosphorylating and inactivating BAD, $p21^{Cip1}$, and HPI1, respectively.

of appropriate stimuli. Another regulatory element is present in the form of an A/U-rich region within the 3′-untranslated region (UTR) of PIM-1 mRNA. This region promotes RNA instability, which can be overcome in the presence of certain cytokines and mitogens. Translation of the PIM-1 message is regulated by the presence of a 5′-UTR containing an internal ribosomal entry site (IRES). PIM-1 activity may also be regulated via autophosphorylation, while its turnover rate is dependent upon ubiquitination. In the event of deregulated PIM-1 expression, oncogenesis is most likely facilitated by the PIM-1-dependent stimulation of cell-cycle progression and mitosis.

4.3.3 Kinase Group STE (Homologs of Yeast Sterile 7, Sterile 11, Sterile 20 Kinase)

STK11 Serine–threonine kinase 11 (STK11), also known as LKB1, is a crucial regulator of cellular metabolism and polarity. This cytoplasmic kinase is expressed ubiquitously in human tissues and exhibits catalytic activity only in the presence of its binding partner STRAD (STE20-related adapter). As STRAD lacks intrinsic kinase activity, the association of the pseudokinase domain of STRADα with the kinase domain of STK11 is postulated to result in intermolecular interactions capable of stimulating the STK11 catalytic core. In the cytoplasm, the association of STK11 with STRADα gives rise to a series of STK11 autophosphorylation events and results in the phosphorylation of STRADα. The STRADα and MO25 complex interacts with STK11 to promote its nucleocytoplasmic transport and cytoplasmic retention by blocking access to importin, an important mediator of nuclear import. The induction of cell polarity by STK11 may result from its association with the Ser/Thr kinase PAR1, leading to destabilization and a polarized localization of the microtubules. STK11 also plays a role in axis induction during development by stimulating the Wnt signaling pathway. The tumor suppressor activity of STK11 is mediated by the energy sensor AMP-activated protein kinase (AMPKα). In response to the cellular stress caused by ATP depletion, STK11 directly phosphorylates Thr172 of AMPKα (Fig. 14). This results in a downregulation of ATP-consuming processes, such as the translational activity mediated by mTOR, and upregulation of ATP-producing processes such as glycolysis via increased translocation of the glucose transporter GLUT4.

Therefore, inactivating mutations of STK11, as observed in various carcinomas, offer cells a distinct growth advantage by removing the inhibition that is normally placed on protein synthesis and glucose uptake. The tumor-suppression activity of STK11 makes it a target for inactivating mutations such as those observed in patients with the hereditary Peutz–Jeghers syndrome. Indeed, the *STK11* gene was mapped to human chromosome 19p13.3 – a site susceptible to germline mutations in patients with Peutz–Jeghers syndrome. In addition to the formation of benign intestinal hamartomatous polyps, individuals with Peutz–Jeghers syndrome exhibit a predisposition towards gastrointestinal, pancreatic, ovarian, testicular, uterine, and breast carcinomas. Studies have identified various deviations of the wild-type STK11 transcript; these are caused by translational frameshift insertions or deletions, nonsense mutations, missense mutations, and a disruption of STK11 mRNA splicing through exonal splice site mutations. To date, 145 distinct germline STR11 mutations have been reported. The functional consequence

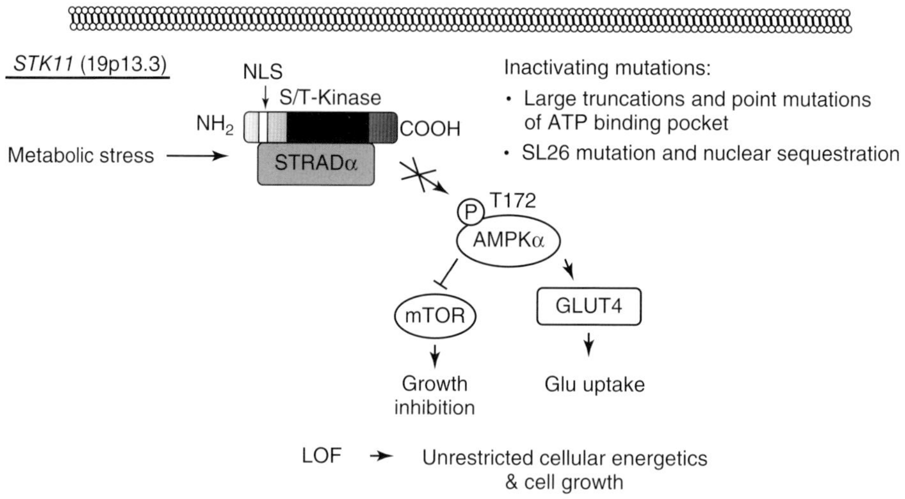

Fig. 14 STK11/LKB1 translocates from the nucleus to cytoplasm when bound to STRADα, where it functions as a tumor suppressor by restricting protein translation by blocking mTOR activation in an AMPK-dependent fashion. Loss-of-function (LOF) STK11/LKB1 mutations lead to deregulated cell growth and increased glycolysis via upregulation of glucose uptake.

of these mutations is the generation of an inactive STK11 protein. Germinal *STK11* mutations are often accompanied by somatic mutations which, in turn, lead to a homozygous deletion of *STK11*. Truncations have also been reported in patients with pancreatic cancer, mucinous minimal deviation adenocarcinoma of the uterine cervix, lung adenocarcinoma, and malignant melanoma. In general, *STK11* mutations affect the biological activity of STK11 by one of two common means. Large truncations or targeted mutations of invariant residues within the STK11 kinase domain destroy its ability to bind ATP and catalyze phosphotransfers. The SL26 mutant present in some cases of Peutz–Jeghers syndrome does not affect kinase activity; rather, it promotes the loss of cytoplasmic retention, which is accompanied by a constitutive nuclear localization of STK11 (Fig. 14). It is hypothesized that such nuclear localization results from an inability of the SL26 mutant to interact with its cytoplasmic substrates. Nuclear localization also leads to a silencing of STK11 signaling because of the lack of substrates and activating factors. Paradoxically, owing to a marked increase in the AMP/ATP ratio, cells with loss of STK11 expression are rendered more susceptible to cell death caused by agents such as the AMP analog AICAR. This response may represent a unique therapeutic opportunity for those patients in whom carcinomas exhibit a loss of STK11 activity.

4.3.4 Kinase Group AGC (Containing PKA, PKG, PKC)

AKT1 and AKT2 The *AKT1* gene (on human chromosome 14q32.33) and *AKT2* gene, located at 9q13.1, produce a phosphoinositide-dependent Ser/Thr kinase, crucial for regulating cellular events including apoptosis, differentiation, proliferation, and metabolism. The gene product of *AKT2*, also referred to as protein kinase B beta (PKB-β), is a broadly expressed protein containing a single phosphoinositide-binding PH domain at its amino terminal. Adjacent to this PH domain is the kinase domain, followed by a carboxy-terminal regulatory hydrophobic domain (HD) (Fig. 15). The AKT isoforms (AKT1 and AKT2) function downstream of various RTKs, and the biogenesis of an active AKT requires the sequential recruitment of the inactive protein to the plasma membrane, followed by the transphosphorylation of a conserved Thr residue within the activation loop (Thr308 in AKT1 and Thr306 in AKT2) and Ser residue within the regulatory HD motif (Ser473 in AKT1 and Ser474 in AKT2) (Fig. 15).

AKT is recruited to the plasma membrane through the binding of its amino-terminal PH domain to PIP3, a lipid product of PI3K. This protein–lipid interaction is tightly controlled by the lipid phosphatase PTEN (phosphatase and tensin homolog), which inhibits AKT and is also a cancer (tumor) suppressor gene that undergoes both somatic and germline inactivating mutations. By anchoring AKT to the membrane lipid bilayer, PIP3 alters the conformation of the AKT kinase domain and exposes the activation loop phosphorylation site to PDK1. Following this "priming" step, the carboxy-terminal HD motif of AKT is phosphorylated by mTOR complex 2 (mTORC2), a multimeric protein complex composed of mTOR, rapamycin-insensitive companion of mTOR (RICTOR), GβL, and mammalian stress-activated protein kinase interacting protein 1 (mSIN1). Once fully activated, AKT is capable of enhancing cellular proliferation and the generation of survival signals through some of the pathways and effector molecules included in Figs 1 and 15. AKT2 was included in the census of cancer genes because of its amplification in ovarian and pancreatic cancers. The region that is amplified in ovarian carcinomas (19q13.1-q13.2) spans the entire *AKT2* gene. While no activating mutations have been documented for AKT2 in cancer, a somatic mutation in the

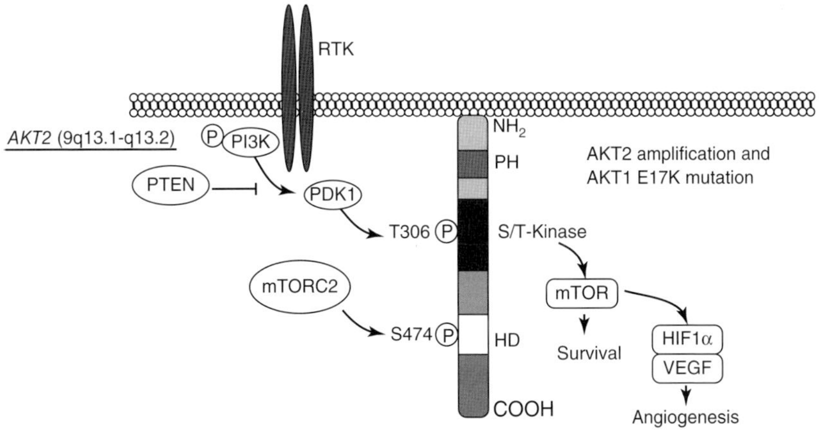

Fig. 15 Schematic structure of AKT2. AKT2 has an amino-terminal pleckstrin homology (PH) domain that is adjacent to a Ser/Thr protein kinase domain, followed by a carboxy-terminal regulatory hydrophobic domain (HD). AKT functions downstream of various RTKs and is activated by phosphorylation of two distinct sites PDK1 and mTORC2. Several details of the downstream signaling events that propagate oncogenic signaling by AKT are presented in Fig. 1.

HD of AKT1 (E17K) was recently reported. Tumors in which AKT2 has been amplified may not be responsive to small-molecule inhibitors of kinase activity. Depending on the genetic profile presented by a given patient, other oncogenes might contribute to the transforming activity of AKT2. For example, the coexistence of an AKT2 amplification and LOF mutation in PTEN would allow an enhancement of AKT2 signaling in the presence of endogenous growth factors. The human prostate epithelial cancer cell line LNCaP harbors such a PTEN mutation and possesses constitutive AKT kinase activity. Another interesting candidate in ovarian cancer is SERTA domain-containing protein 1 (SEI-1), which is located in a genetic region of chromosome 19q13.1-q13.2, neighboring AKT2. This genetic region is coamplified along with AKT2 in a subset of ovarian cancer cell lines. The *SEI-1* gene encodes a CDK4-binding protein, which renders the activity of cyclin D–CDK4 complexes resistant to the inhibitory effect of p16INK4α. Given the proliferation-related functions of AKT2 and SEI-1, it is possible that both gene amplifications at chromosome 19 have the potential to advance the growth of certain ovarian tumors.

4.3.5 Kinase Group CMGC (Containing Cdk, Mapk, Gsk3, Clk)

CDK4, CDK6, and CDK12 and the INK4 Inhibitors

Of the 25 human genes in the CDK family (CDK1–20 and five CDK-like genes), three meet the criteria for inclusion in the current Cancer Gene Census. These are *CDK4* (on chromosome 12q14), *CDK6* (on chromosome 7q21), and *CDK12* (on chromosome 17q12). These protein kinases are best known for their role in regulating a cell's progression through various checkpoints in the cell cycle. For example, the *CDK4* gene encodes a small (303-amino acid) cytosolic Ser/Thr protein kinase essential for coordinating the cell's progression through the early growth (G_1) phase of the cell cycle. CDK4 is activated by the binding of D-type cyclins (which impart basal activity to the kinase) and, by phosphorylation (which fully activates the kinase). When active, CDK4 interferes with the tumor suppressor activity of the retinoblastoma protein Rb and its homologs (Fig. 16).

CDK4 is inactivated by binding of the INK4 family of CDK inhibitors. G_1 progression depends on extracellular mitogenic signals that upregulate cyclin D expression, committing the cell to another round of cell

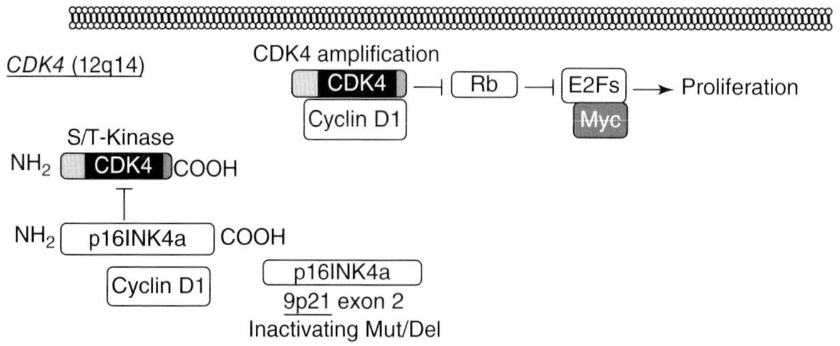

Fig. 16 INK4 proteins interfere with the tumor suppressor activity of Rb by preventing D-type cyclins from activating CDK4 and the subsequent phosphorylation of Rb. INK4 loss-of-function mutations and CDK4 amplification are both capable of hindering the sequestration of E2F transcription factors by Rb, thereby removing the G_1 to S cell-cycle checkpoints.

division. INK4 proteins bind monomeric CDK4, preventing its stable interaction with cyclin D and subsequent phosphorylation of Rb. Underphosphorylated Rb sequesters the E2F family of transcription factors and prevents them inducing progression from G_1 to S phase of the cell cycle. INK4 proteins also bind the cyclin D–CDK4 complex, forming an inactive ternary complex. There are four INK4 proteins (P16INK4a, P15INK4b, P18INK4c, and P19INK4d), all of which possess ankyrin repeats involved in binding CDK4. The binding of these proteins substantially alters the conformation of the CDK4 ATP-binding site and renders the kinase inactive. Mutations leading to a loss of control over the CDK4/INK4/Rb pathway are among the most common in human cancer. CDK4 and CDK6 are often amplified and overexpressed in glioblastoma (50%), and CDK4 is also overexpressed in uterine cervical carcinoma (26%), breast carcinoma (16%), and osteosarcoma (16%). In addition, a germline R24H point mutation of *CDK4* has been reported in some cases of familial melanoma. This mutation contributes to malignant transformation by preventing the interaction between p16INK4a and CDK4, but is rarely encountered in other human neoplasms. More common are germline mutations in the binding region of the *INK4* genes, lowering the affinity of the INK4 proteins for CDK4. In the case of p16INK4a, such mutations occur frequently in leukemia (58%), bladder carcinoma (50%), glioma (35%), nasopharyngeal carcinoma (35%), and pancreatic cancer (21%). Promoter methylation is also a common mechanism of inactivation of the *INK4* genes in cancer. Finally, in some cases of ovarian cancer somatic missense mutations have recently been identified in CDK12.

4.3.6 Phosphatidylinositol Kinase (PIK)-Related Subfamily

ATM The human tumor suppressor gene ataxia-telangiectasia mutated (*ATM*) is located on chromosome 11q22 and is frequently mutated in a rare autosomal recessive disorder causing neuronal degeneration, immunologic deficiency, radiosensitivity, and cancer predisposition. Over one-third of ataxia-telangiectasia patients develop lymphoid cancers that include non-Hodgkin's lymphoma, Hodgkin's lymphoma, and other leukemias. The *ATM* gene encodes a large (3056-amino acid) nuclear protein that is expressed in testis, spleen, and thymus of adult mice and plays a critical role in maintaining genetic stability in response to DNA damage. ATM is a member of the recently identified phosphatidylinositol kinase (PIK)-related subfamily that phosphorylates Ser/Thr residues, rather than lipids. All members contain a conserved carboxy-terminal catalytic domain, referred to as the PI-kinase domain, and a novel amino-terminal FAT (FRAP, ATM, and TRRAP) domain that possibly regulates the conformation of the PI-kinase domain (Fig. 17).

Structurally, ATM is composed of two regions referred to as the head and arm. Three-dimensional reconstructions of ATM bound to DNA suggest that the kinase uses its arm to clamp around the double helix, but whether this protein–DNA interaction is required for the ATM-dependent repair of double-strand breaks is not known. ATM activation in response to DNA double-strand breaks is rapidly triggered following ionizing radiation and functions to arrest the cell-cycle progression of damaged cells in G_1, S, or G_2. The G_1/S checkpoint is activated through the induction of p53 (Fig. 17). ATM positively regulates the p53 pathway by increasing the half-life of the p53 protein,

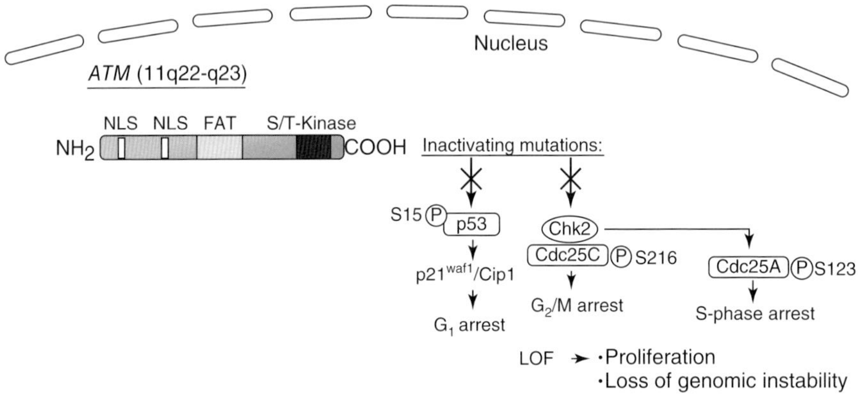

Fig. 17 ATM is a large nuclear protein that contains two amino-terminal nuclear localization signals (NLSs), a FAT (FRAP, ATM, and TRRAP) domain, and a carboxy-terminal Ser/Thr protein kinase domain. Loss-of-function (LOF) mutations prevent ATM from blocking cell cycle progression from G_1 to S phase in response to DNA damage, thereby predisposing cancer cells to genomic instability.

either by directly phosphorylating Ser15, or indirectly through activation of other kinases such as Chk2 and c-Abl. ATM, Chk2, and c-Abl are all capable of phosphorylating p53, thus stabilizing the latter protein by interfering with its binding to the p53 regulatory protein, murine double minute 2. In addition to regulating the G_1/S checkpoint, ATM engages the S and G_2/M checkpoints in response to ionizing radiation. It triggers the G_2/M checkpoint by Chk2-mediated phosphorylation of Ser216 on Cdc25C, thereby promoting the binding of 14-3-3 proteins. The Cdc25C/14-3-3 complex is then exported from the nucleus, preventing Cdc25C from activating the nuclear Cdc2/cyclin B complex required for mitosis (Fig. 17). In a similar fashion, ATM controls the S-phase checkpoint through the Chk2-dependent phosphorylation of Cdc25A Ser123. This leads to Cdc25A degradation. Most of the ATM mutations associated with cancer are derived from large deletions, causing protein truncations that inactivate the kinase. Other inactivating mutations are represented by missense mutations (e.g., L1420F) or small in-frame deletions/insertions. This loss of ATM function is thought to predispose to cancer by increasing genomic instability.

Acknowledgments

Grants from the National Institutes of Health (Grant R01 CA98195 to J.A.M.) and from the DOD Prostate Cancer Research Program (DAMD 17-02-1-0053 to D.M.T.) provided financial support for these studies.

Abbreviations

ALCL	anaplastic large-cell lymphoma
ALK	anaplastic lymphoma kinase
ALL	acute lymphoblastic leukemia
ARG	Abl-related gene
ATM	ataxia telangiectasia mutated kinase
BAD	Bcl-2-associated death promoter
CDK	cyclin-dependent kinase
CML	chronic myelogenous leukemia
DUSP	dual-specificity phosphatases
EGFR	epidermal growth factor receptor

ERK	extracellular signal regulated kinase
eIF4E	eukaryotic translation initiation factor 4E
FGFR	fibroblast growth factor receptor
FLT3	FMS-like tyrosine kinase 3
GAP	GTPase-activating protein
GDNF	glial cell line-derived neurotrophic factor
GEF	guanine nucleotide exchange factor
GOF	gain-of-function
GRB2	growth factor receptor bound protein 2
IGF-1R	insulin-like growth factor-1 receptor
IRS	insulin receptor substrate
JAK	Janus kinase
JNK	c-Jun-N-terminal kinase
LCK	T-lymphocyte protein-tyrosine kinase
MAP	mitogen-activated protein
MEK	MAP kinase-kinase
Mnk	1/2 MAPK signal-integrating kinases
NSCLC	non-small-cell lung cancer
NTRK	neurotrophin tyrosine receptor kinase
PDGFR	platelet-derived growth factor receptor
PDK1	3-phosphoinositide-dependent protein kinase-1
PI3K	phosphoinositide 3-kinase
PIM	proviral integration of the Moloney murine leukemia virus
PLCγ	phospholipase C γ
PTEN	phosphatase and tensin homolog
PTK	protein-tyrosine kinase
RET	rearranged in transformation
RTK	receptor protein-tyrosine kinase
SCF	stem cell growth factor
SGK	serum glucocorticoid regulated kinase
SH	Src homology domain
STAT	signal transducers and activators of transcription
STRAD	STE20-related adapter
STK11	serine-threonine kinase 11
TCR	T-cell receptor
TPR	translocated promoter region

Bibliography

Books and Reviews

Blume-Jensen, P. and Hunter, T. (2001) Oncogenic kinase signaling. *Nature*, **411**, 355–365.

Chappell, W.H., Steelman, L.S., Long, J.M., Kempf, R.C. *et al.* (2011) Ras/Raf/MEK/ERK and PI3K/PTEN/Akt/mTOR inhibitors: rationale and importance to inhibiting these pathways in human health. *Oncotarget*, **2**, 135–164.

Futreal, P.A., Coin, L., Marshall, M., Down, T. *et al.* (2004) A census of human cancer genes. *Nat. Rev.*, **4**, 177–183.

Hanahan, D. and Weinberg, R.A. (2011) Hallmarks of cancer: the next generation. *Cell*, **144**, 646–674.

Hanks, S.K. and Hunter, T. (1995) Protein kinases 6. The eukaryotic protein kinase superfamily: kinase (catalytic) domain structure and classification. *FASEB J.*, **9**, 576–596.

Hubbard, S.R. and Till, J.H. (2000) Protein tyrosine kinase structure and function. *Annu. Rev. Biochem.*, **69**, 373–398.

McCubrey, J.A., Steelman, L.S., Chappell, W.H., Abrams, S.L. *et al.* (2012) Ras/Raf/MEK/ERK and PI3K/PTEN/Akt/mTOR cascade inhibitors: how mutations can result in therapy resistance and how to overcome resistance. *Oncotarget*, **3**, 1068–1111.

McCubrey, J.A., Steelman, L.S., Chappell, W.H., Abrams, S.L. *et al.* (2012) Mutations and deregulation of Ras/Raf/MEK/ERK and PI3K/PTEN/Akt/mTOR cascades which alter therapy response. *Oncotarget*, **3**, 954–987.

McCubrey, J.A., Steelman, L.S., Kempf, C.R., Chappell, W.H. *et al.* (2011) Therapeutic resistance resulting from mutations in Raf/MEK/ERK and PI3K/PTEN/Akt/mTOR signaling pathways. *J. Cell. Physiol.*, **226**, 2762–2781.

Steelman, L.S., Chappell, W.H., Abrams, S.L., Kempf, R.C. *et al.* (2011) Roles of the

Raf/MEK/ERK and PI3K/PTEN/Akt/mTOR pathways in controlling growth and sensitivity to therapy-implications for cancer and aging. *Aging (Albany)*, **3**, 192–222.

Steelman, L.S., Franklin, R.A., Abrams, S.L., Chappell, W. *et al.* (2011) Roles of the Ras/Raf/MEK/ERK pathway in leukemia therapy. *Leukemia*, **25**, 1080–1094.

Steelman, L.S., Pohnert, S.C., Shelton, J.G., Franklin, R.A. *et al.* (2004) JAK/STAT, Raf/MEK/ERK, PI3K/AKT and BCR-ABL in cell cycle progression and leukemogenesis. *Leukemia*, **18**, 189–218.

Terrian, D.M. (ed) (2003) *Cancer Cell Signaling*, Humana Press, Totowa.

Primary Literature

Agnes, F., Shamoon, B., Dina, C., Rosnet, O. *et al.* (1994) Genomic structure of the downstream part of the human FLT3 gene: exon/intron structure conservation among genes encoding receptor tyrosine kinases (RTK) of subclass III. *Gene*, **145**, 283–288.

Alessi, D.R., Andjelkovic, M., Caudwell, B., Cron, P. *et al.* (1996) Mechanism of activation of protein kinase B by insulin and IGF-1. *EMBO J.*, **15**, 6541–6551.

Bai, R.Y., Ouyang, T., Miething, C., Morris, S.W. *et al.* (2000) Nucleophosmin-anaplastic lymphoma kinase associated with anaplastic large-cell lymphoma activates the phosphatidylinositol 3-kinase/AKT antiapoptotic signaling pathway. *Blood*, **96**, 4319–4327.

Bischof, D., Pulford, K., Mason, D.Y., and Morris, S.W. (1997) Role of the nucleophosmin (NPM) portion of the non-Hodgkin's lymphoma-associated NPM-anaplastic lymphoma kinase fusion protein in oncogenesis. *Mol. Cell. Biol.*, **17**, 2312–2325.

Blume-Jensen, P., Janknecht, R., and Hunter, T. (1998) The kit receptor promotes cell survival via activation of PI 3-kinase and subsequent AKT-mediated phosphorylation of BAD on Ser136. *Curr. Biol.*, **8**, 779–782.

Bosch, E., Cherwinski, H., Peterson, D., and McMahon, M. (1997) Mutations of critical amino acids affect the biological and biochemical properties of oncogenic A-Raf and Raf-1. *Oncogene*, **15**, 1021–1033.

Bossotti, R., Isacchi, A., and Sonnhammer, E.L.L. (2000) FAT: a novel domain in PIK-related kinases. *Trends Biochem. Sci.*, **25**, 225–227.

Brown, V.L., Harwood, C.A., Crook, T., Cronin, J.G. *et al.* (2004) p16INK4a and p14ARF tumor suppressor genes are commonly inactivated in cutaneous squamous cell carcinoma. *J. Invest. Dermatol.*, **122**, 1284–1292.

Carow, C.E., Levenstein, M., Kaufmann, S.H., Chen, J. *et al.* (1996) Expression of the hematopoietic growth factor receptor FLT3 (STK-1/Flk2) in human leukemias. *Blood*, **87**, 1089–1096.

Cazzaniga, G., Tosi, S., Aloisi, A., Giudici, G. *et al.* (1999) The tyrosine kinase abl-related gene ARG is fused to ETV6 in an AML-M4Eo patient with a t(1;12)(q25;p13): molecular cloning of both reciprocal transcripts. *Blood*, **94**, 4370–4373.

Chang, F., Lee, J.T., Navolanic, P.M., Steelman, J.G. *et al.* (2003) Involvement of PI3K/AKT pathway in cell cycle progression, apoptosis, and neoplastic transformation: a target for cancer chemotherapy. *Leukemia*, **17**, 590–603.

Chang, F., Steelman, L.S., Lee, J.T., Shelton, J.G. *et al.* (2003) Signal transduction mediated by the Ras/Raf/MEK/ERK pathway from cytokine receptors to transcription factors: potential targeting for therapeutic intervention. *Leukemia*, **17**, 1263–1293.

Clark, S.S., McLaughlin, J., Crist, W.M., Champlin, R. *et al.* (1987) Unique forms of the abl tyrosine kinase distinguish Ph1positive CML from Ph1-positive ALL. *Science*, **235**, 85–88.

Cong, F., Spencer, S., Cote, J.F., Wu, Y. *et al.* (2000) Cytoskeletal protein PSTPIP1 directs the PEST-type protein tyrosine phosphatase to the c-Abl kinase to mediate Abl dephosphorylation. *Mol. Cell*, **6**, 1413–1423.

Crews, C.M., Alessandrini, A., and Erikson, R.L. (1992) The primary structure of MEK, a protein kinase that phosphorylates the ERK gene product. *Science*, **258**, 478–480.

Cunningham, D., Humblet, Y., Siena, S., Khayat, D. *et al.* (2004) Cetuximab monotherapy and cetuximab plus irinotecan in irinotecan-refractory metastatic colorectal cancer. *N. Engl. J. Med.*, **351**, 337–345.

Dong, J., Phelps, R.G., Qiao, R., Yao, S. *et al.* (2003) B Raf oncogenic mutations correlate with progression rather than initiation of human melanoma. *Cancer Res.*, **63**, 3883–3886.

Druker, B.J. (2004) Molecularly targeted therapy: have the floodgates opened? *Oncologist*, **9**, 357–360.

Druker, B.J., Talpaz, M., Testa, D.J., Peng, B. *et al.* (2001) Efficacy and safety of a specific inhibitor of the BCR-ABL tyrosine kinase in chronic myeloid leukemia. *N. Engl. J. Med.*, **344**, 1031–1037.

Eley, G., Frederick, L., Wang, X.Y., Smith, D.I. *et al.* (1998) 3′ end structure and rearrangements of EGFR in glioblastomas. *Genes Chromosomes Cancer*, **23**, 248–254.

Falck, J., Mailand, N., Syljuasen, R.G., Bartek, J. *et al.* (2001) The ATM-Chk2-Cdc25A checkpoint pathway guards against radioresistant DNA synthesis. *Nature*, **410**, 766–767.

Feng, J., Park, J., Cron, P., Hess, D. *et al.* (2004) Identification of a PKB/AKT hydrophobic motif Ser-473 kinase as DNA-dependent protein kinase. *J. Biol. Chem.*, **279**, 41189–41196.

Gatti, R.A., Berkel, I., Boder, E., Braedt, G. *et al.* (1988) Localization of an ataxia-telangiectasia gene to chromosome 11q22–23. *Nature*, **336**, 577–580.

Gilliland, D.G. and Griffin, J.D. (2002) The roles of FLT3 in hematopoiesis and leukemia. *Blood*, **100**, 1532–1542.

Goldman, J.M. and Melo, J.V. (2003) Chronic myeloid leukemia – advances in biology and new approaches to treatment. *N. Engl. J. Med.*, **349**, 1451–1464.

Golub, T.R., Goga, A., Barker, G.F., Afar, D.E. *et al.* (1996) Oligomerization of the ABL tyrosine kinase by the Ets protein TEL in human leukemia. *Mol. Cell. Biol.*, **16**, 4107–4116.

Haeder, M., Rotsch, M., Bepler, G., Hennig, C. *et al.* (1988) Epidermal growth factor receptor expression in human lung cancer cell lines. *Cancer Res.*, **48**, 1132–1136.

Hall, M. and Peters, G. (1996) Genetic alterations of cyclins, cyclin-dependent kinases, and Cdk inhibitors in human cancer. *Adv. Cancer Res.*, **68**, 67–108.

Hantschel, O. and Superti-Furga, G. (2004) Regulation of the c-Abl and Bcr-Abl tyrosine kinases. *Nat. Rev. Mol. Cell Biol.*, **5**, 33–44.

Heath, C. and Cross, N.C. (2004) Critical role of STAT5 activation in transformation mediated by ZNF198-FGFR1. *J. Biol. Chem.*, **279**, 6666–6673.

Huang, W., Kessler, D.S., and Erikson, R.L. (1995) Biochemical and biological analysis of Mek1 phosphorylation site mutants. *Mol. Biol. Cell*, **6**, 237–245.

Iijima, Y., Okuda, K., Tojo, A., Tri, N.K. *et al.* (2002) Transformation of Ba/F3 cells and Rat-1 cells by ETV6/ARG. *Oncogene*, **21**, 4374–4383.

Iwahara, T., Fujimoto, J., Wen, D., Cupples, R. *et al.* (1997) Molecular characterization of ALK, a receptor tyrosine kinase expressed specifically in the nervous system. *Oncogene*, **14**, 439–449.

Jeffrey, P.D., Tong, L., and Pavletich, N.P. (2000) Structural basis of inhibition of CDK-cyclin complexes by INK4 inhibitors. *Genes Dev.*, **14**, 3115–3125.

Johnson, D.E. and Williams, L.T. (1993) Structural and functional diversity in the FGF receptor multigene family. *Adv. Cancer Res.*, **60**, 1–41.

Keung, Y.K., Beaty, M., Steward, W., Jackle, B. *et al.* (2002) Chronic myelocytic leukemia with eosinophilia, t(9;12)(q34;p13), and ETV6-ABL gene rearrangement: case report and review of the literature. *Cancer Genet. Cytogenet.*, **138**, 139–142.

Khosravi, R., Maya, R., Gottlieb, T., Oren, M. *et al.* (1999) Rapid ATM-dependent phosphorylation of MDM2 precedes p53 accumulation in response to DNA damage. *Proc. Natl Acad. Sci. USA*, **96**, 14973–14977.

King, C.R., Kraus, M.H., and Aaronson, S.A. (1985) Amplification of a novel v-erbB-related gene in a human mammary carcinoma. *Science*, **229**, 974–976.

de Klein, A., van Kessel, A.G., Grosveld, G., Bartram, C.R. *et al.* (1982) A cellular oncogene is translocated to the Philadelphia chromosome in chronic myelocytic leukemia. *Nature*, **300**, 765–767.

Kulkarni, S., Reiter, A., Smedley, D., Goldman, J.M. *et al.* (1999) The genomic structure of ZNF198 and location of breakpoints in the t(8;13) myeloproliferative syndrome. *Genomics*, **55**, 118–121.

Lacronique, V., Boureux, A., Monni, R., Dumon, S. *et al.* (2000) Transforming properties of chimeric TEL-JAK proteins in Ba/F3 cells. *Blood*, **95**, 2076–2083.

Leonard, J.H., Kearsley, J.H., Chenevix-Trench, G., and Hayward, N.K. (1991) Analysis of gene amplification in head-and-neck squamous-cell carcinoma. *Int. J. Cancer*, **48**, 511–515.

Llorca, O., Rivera-Calzada, A., Grantham, J., and Willison, K.R. (2003) Electron microscopy and 3D reconstruction reveal that human ATM kinase uses an arm-like domain to clamp around double-stranded DNA. *Oncogene*, **22**, 3867–3874.

Lyman, S.D., Stocking, K., Davison, B., Fletcher, F. *et al.* (1995) Structural analysis of human and

murine flt3 ligand genomic loci. *Oncogene*, **11**, 1165–1172.
Lynch, T.J., Bell, D.W., Sordella, R., Gurubhagavatula, S. *et al.* (2004) Activating mutations in the epidermal growth factor receptor underlying responsiveness of non-small-cell lung cancer to Gefitinib. *N. Engl. J. Med.*, **350**, 2129–2038.
Ma, P.C., Maulik, G., Christensen, J., and Salgia, R. (2003) c-Met: structure, functions and potential for therapeutic inhibition. *Cancer Metastasis Rev.*, **22**, 309–325.
Marais, R., Light, Y., Paterson, H.F., Mason, C.S. *et al.* (1997) Differential regulation of Raf-1, A-Raf, and B-Raf by oncogenic ras and tyrosine kinases. *J. Biol. Chem.*, **272**, 4378–4383.
Maroe, N., Rottapel, R., Rosnet, O., Marchetto, S. *et al.* (1993) Biochemical characterization and analysis of the transforming potential of the FLT3/FLK2 receptor tyrosine kinase. *Oncogene*, **8**, 909–918.
Matsuoka, S., Huang, M., and Elledge, S.J. (1998) Linkage of ATM to cell cycle regulation by the Chk2 protein kinase. *Science*, **282**, 1893–1897.
McWhirter, J.R., Galasso, D.L., and Wang, J.Y. (1993) A coiled-coil oligomerization domain of Bcr is essential for the transforming function of Bcr-Abl oncoproteins. *Mol. Cell. Biol.*, **13**, 7587–7595.
Meydan, N., Grunberger, T., Dadi, H., Shahar, M. *et al.* (1996) Inhibition of acute lymphoblastic leukemia by a Jak-2 inhibitor. *Nature*, **379**, 645–648.
Mora, A., Komander, D., van Aalten, D.M.F., and Alessi, D.R. (2004) PDK1, the master regulator of AGC kinase signal transduction. *Semin. Cell Dev. Biol.*, **15**, 161–170.
Morris, S.W., Kirstein, M.N., Valentine, M.B., Dittmer, K.G. *et al.* (1994) Fusion of a kinase gene, ALK, to a nucleolar protein gene, NPM, in non-Hodgkin's lymphoma. *Science*, **263**, 1281–1284.
Muenke, M., Schell, U., Hehr, A., Robin, N.H. *et al.* (1994) A common mutation in the fibroblast growth factor receptor 1 gene in Pfeiffer syndrome. *Nat. Genet.*, **8**, 269–274.
Neilson, K.M. and Friesel, R. (1996) Ligand-independent activation of fibroblast growth factor receptors by point mutations in the extracellular, transmembrane, and kinase domains. *J. Biol. Chem.*, **271**, 25049–25057.
Passos-Bueno, M.R., Wilcox, W.R., Jabs, E.W., Sertie, A.L. *et al.* (1999) Clinical spectrum of fibroblast growth factor receptor mutations. *Hum. Mutat.*, **14**, 115–125.
Pluk, H., Dorey, K., and Superti-Furga, G. (2002) Autoinhibition of c-Abl. *Cell*, **108**, 247–259.
Roche-Lestienne, C., Soenen-Cornu, V., Grad-Duflos, N., Lai, J.L. *et al.* (2002) Several types of the Abl gene can be found in chronic myeloid leukemia patients resistant to STI571, and they can pre-exist to the onset of treatment. *Blood*, **100**, 1014–1018.
Russell, J.P., Powell, D.J., Cunnane, M., Greco, A. *et al.* (2000) The TRK-T1 fusion protein induces neoplastic transformation of thyroid epithelium. *Oncogene*, **19**, 5729–5735.
Saltzman, A., Stone, M., Franks, C., Searfoss, G. *et al.* (1998) Cloning and characterization of human Jak-2 kinase: high mRNA expression in immune cells and muscle tissue. *Biochem. Biophys. Res. Commun.*, **246**, 627–633.
Shtivelman, E., Lifshitz, B., Gale, R.P., and Cana-ani, E. (1985) Fused transcript of abl and bcr genes in chronic myelogenous leukemia. *Nature*, **315**, 550–554.
Sommer, S.S., Jiang, Z., Feng, J., Buzin, C.H. *et al.* (2003) ATM missense mutations are frequent in patients with breast cancer. *Cancer Genet. Cytogenet.*, **145**, 115–120.
Sozeri, O., Vollmer, K., Liyanage, M., Frith, D. *et al.* (1992) Activation of the c-Raf protein kinase by protein kinase C phosphorylation. *Oncogene*, **7**, 2259–2262.
Tang, T.C.-M., Sham, J.S.T., Xie, D., Fang, Y. *et al.* (2002) Identification of a candidate oncogene SEI-1 within a minimal amplified region at 19q13.1 in ovarian cancer cell lines. *Cancer Res.*, **62**, 7157–7161.
Thompson, F.H., Nelson, M.A., Trent, J.M., Guan, X.Y. *et al.* (1996) Amplification of 19q13.1-q13.2 sequences in ovarian cancer. G-band, FISH, and molecular studies. *Cancer Genet. Cytogenet.*, **87**, 55–62.
Thorstenson, Y.R., Roxas, A., Kroiss, R., Jenkins, M.A. *et al.* (2003) Contributions of ATM mutations to familial breast and ovarian cancer. *Cancer Res.*, **63**, 3325–3333.
Tybulewicz, V.L., Crawford, C.E., Jackson, P.K., Bronson, R.T. *et al.* (1991) Neonatal lethality and lymphopenia in mice with homozygous disruption of the c-abl protooncogene. *Cell*, **65**, 1153–1163.
Van Etten, R.A., Jackson, P., and Baltimore, D. (1989) The mouse type IV c-abl gene product

is a nuclear protein, and activation of transforming ability is associated with cytoplasmic localization. *Cell*, **58**, 669–678.

Weber, C.K., Slupsky, J.R., Herrmann, C., Schuler, M. *et al.* (2000) Mitogenic signaling of Ras is regulated by differential interaction with Raf isozymes. *Oncogene*, **19**, 169–176.

Wolfel, T., Hauer, M., Schneider, J., Serrano, M. *et al.* (1995) A p16INK4a-insensitive CDK4 mutant targeted by cytolytic T lymphocytes in a human melanoma. *Science*, **269**, 1281–1284.

Wong, A.J., Ruppert, J.M., Bigner, S.H., Grzeschik, C.H. *et al.* (1992) Structural alterations of the epidermal growth factor receptor gene in human gliomas. *Proc. Natl Acad. Sci. USA*, **89**, 2965–2969.

Zhang, B.H. and Guan, K.L. (2000) Activation of B-Raf kinase requires phosphorylation of the conserved residues Thr598 and Ser601. *EMBO J.*, **19**, 5429–5439.

Zimmermann, S. and Moelling, K. (1999) Phosphorylation and regulation of Raf by AKT (protein kinase B). *Science*, **286**, 1741–1744.

Zuo, L., Weger, J., Yang, Q., Goldstein, A.M. *et al.* (1996) Germline mutations in the p16INK4a binding domain of CDK4 in familial melanoma. *Nat. Genet.*, **12**, 97–99.

7
Molecular Biology of Liver Cancer

Tulin Ersahin[1], *Mehmet Ozturk*[2], *and Rengul Cetin-Atalay*[1,3]
[1] *Bilkent University, Department of Molecular Biology and Genetics, Faculty of Science, 06800 Ankara, Turkey*
[2] *Dokuz Eylul University, Advanced Biomedical Research Center, Faculty of Medicine, 035340 Izmir, Turkey*
[3] *Middle East Technical University, Department of Health Informatics, Graduate School of Informatics METU, 06800 Ankara, Turkey*

Translational Medicine Cancer: Advances in Molecular Biology and Medicine,
First Edition. Edited by Robert A. Meyers.

Keywords

HBV
A small DNA virus that infects hepatocytes in the liver, causing acute or chronic hepatitis.

HCV
A small RNA virus that infects hepatocytes in the liver, causing acute or chronic hepatitis.

Genome-wide expression
An experimental approach for the identification of disease-specific gene expression profiles.

Gene signatures
Expression pattern of a set of genes associated with a clinical subtype of a disease.

PI3K/AKT pathway
A signaling pathway involved in cell survival that is hyperactivated in various cancers, including liver cancer.

MAPK/ERK pathway
A kinase protein cascade involved in cell survival that is upregulated in various cancers, including liver cancer.

p53
A tumor suppressor protein that is inactivated by mutation in cancers, including liver cancer.

hTERT
Human telomerase reverse transcriptase that is inactivated by mutation in cancers, including liver cancer.

Hepatocellular carcinoma (HCC) is one of the leading causes of cancer-related mortality worldwide. Recent advances in the molecular profiling of HCC emphasize its intratumoral heterogeneity and reveal how cellular pathways are altered in favor of tumor progression. Malignant transformation of primary liver cancer is achieved through the acquisition of cancer hallmark capabilities that promote the uncontrolled proliferation of hepatocytes. In this review, the characteristics and acquired capabilities of human primary liver cancer, based on the HCC-specific genetic and epigenetic alterations, are described and discussed.

1
Introduction

Primary liver cancer is the fifth most frequently diagnosed cancer in men, and the seventh in women. However, due to its aggressive behavior and resistance to conventional therapies, liver cancer is the second most frequent cause of cancer death in men, and the sixth in women [1]. In 2008, a total of 748 300 new liver cancer cases and 695 900 cancer deaths were reported worldwide [2]. Hepatocellular carcinoma (HCC) is the major histological subtype of primary liver cancers, accounting for about 80% of the total liver cancer cases worldwide [3].

The major risk factors for HCC are hepatitis B virus (HBV), hepatitis C virus (HCV) infections, alcohol, and aflatoxin B1 exposure. The development of HCC is a multistep process, where hepatic injury first leads to chronic liver disease, after which continuous inflammation results in cycles of cell death and hepatocyte regeneration. The subsequent expansion of dysplastic nodules, along with telomerase reactivation and increased genomic instability, is followed by malignant transformation [4]. The integration of HBV DNA into the host genome is frequent in HCC, whereas the RNA virus HCV leads to malignant transformation through oxidative stress and a proinflammatory response induced by viral proteins [5–7]. In addition, exposure to exogenous (aflatoxins) or endogenous (toxic metabolites such as steroids and cholesterol and metal ions such as copper, iron, nickel) hepatotoxic factors leads to liver injury and provokes an increase in the proliferating fraction of hepatocytes [8–12]. Therefore, chronic liver regeneration may itself be a source of spontaneous gene mutations leading to HCC [13].

The current classification system (Barcelona Clinic Liver Cancer; BCLC) considers HCC in five stages: very early; early; intermediate; advanced; and end-stage according to the tumor size, nodule number, vascular invasion, metastasis, and liver function [14]. The treatment strategy is determined based on the stage of the patient. Patients with very early and early HCC are treated with resection, liver transplantation or partial hepatectomy, while patients with intermediate HCC are treated with chemoembolization [15]. Unfortunately, however, patients are often diagnosed at an advanced stage, when these treatments cannot be applied. Although chemotherapy is the only treatment option for patients with advanced-stage HCC, patients with the same BCLC stage can still have variable responses due to high genomic heterogeneity of HCC [16, 17]. Therefore, a molecular level classification is necessary for effective diagnosis and personalized treatment options. Currently, genome-wide expression analysis, exome sequencing and microRNA (miRNA) profiling techniques are being used for the efficient molecular classification of HCCs.

Sorafenib (Nexavar; BAY43-9006), a multikinase inhibitor, is the only FDA-approved molecular-targeted agent for the treatment of patients with advanced HCC [18–20]. Sorafenib inhibits Raf, vascular endothelial growth factor receptor (VEGFR) kinase and platelet-derived growth factor receptor (PDGFR) kinase, and thereby suppresses cell proliferation and angiogenesis. In Phase III randomized controlled trials, sorafenib showed an overall survival benefit of three months [15, 21].

Primary liver cancer is a major public health problem, which requires in-depth molecular analysis in order to discover optimal targeted therapeutics. Due to the presence of genomic variations, custom-designed therapies based on the present understanding of the molecular biology of liver cancer will be indispensable for the treatment of HCC in the future.

2
Molecular Hallmarks of Hepatocellular Carcinoma

Malignant transformation requires the cancer cells to acquire several growth-promoting characteristics in order to become tumorigenic and, eventually, malignant. Hepatocellular cancers derive from initially quiescent hepatocytes, the growth of which is tightly controlled. The appearance of tumor in this highly controlled microenvironment suggests that the initial HCC cells acquire phenotypic hallmarks of cancer, as described by Hanahan and Weinberg. In this case, six core and two emerging cancer hallmark capabilities have been suggested, namely: sustaining proliferative signaling; evading growth suppressors; resisting cell death; enabling replicative immortality; inducing angiogenesis; and activating invasion and metastasis, along with deregulating cellular energetics and avoiding immune destruction [22]. The acquisition of these capabilities is facilitated by two enabling characteristics, genome instability and tumor-promoting inflammation during the stepwise progression of HCC.

2.1 Genome Instability and Mutations

Tumorigenesis is a multistep process that initiates from dysplastic lesions which accumulate somatically acquired mutations and genomic instability [23]. Genomic instability, including chromosomal rearrangements and point mutations, is an enabling characteristic of multistep tumorigenesis and underlies the acquisition of hallmark capabilities [22].

HCC harbors chromosomal gains at 1q, 5, 6p, 7, 8q, 17q, and 20, and chromosomal losses at 1p, 4q, 6q, 8p, 13q, 16, 17p, and 21 [24, 25]. The gain of 1q and the loss of 1p and 17p are associated with early stages of HCC, while the gain of 6p, 5q, and 8q and loss of 4q and 8p are associated with advanced-stage HCC [26, 27]. Moreover, the integration of HBV DNA within or upstream of the telomerase reverse transcriptase (*TERT*) gene in the host genome was observed in patients with HBV-related HCCs [28, 29]. DNA-damaging reactive oxygen species (ROS) can also contribute to genomic instability [30, 31].

The *TERT* gene is frequently mutated in HCC, allowing the cells to gain replicative immortality. *TERT* promoter mutations in 44% of 61 HCC patients were shown to be associated with the early stages of HCC, independent of any viral infection, gender, age, and ethnicity [32]. *TP53* is the second most frequently mutated tumor suppressor gene in HCC, and is associated with a poor prognosis [33]. In HCCs, mutation rates are 50% and 20% with or without aflatoxin exposure, respectively [34]. A hot-spot mutation at codon 249 (R249S) is specific to AFB1 exposure in 36% of tumors from Africa and in 32% of tumors from China [35, 36]. Aberrant activation of the Wnt/β-catenin signaling pathway, which plays a key role in hepatocarcinogenesis, is achieved by activating mutations of β-catenin (CTNNB1) and inactivating mutations or loss of heterozygosity (LOH) of AXIN1 [24, 37–39]. Although mutations of the retinoblastoma gene (RB1) are rare, RB is frequently inactivated in HCC through the LOH of 13q, proteosomal degradation, or aberrant cyclin-dependent kinase activity [40–42]. The cyclin-dependent kinase inhibitor 2A gene (p16), which acts as a tumor suppressor downstream of p53 and Rb, is frequently inactivated in HCC [43, 44].

Mutations in the downstream elements of receptor tyrosine kinase (RTK) signaling are also observed in HCC. The inactivating mutations of the tumor suppressor gene phosphatase and tensin homolog (PTEN), which is a negative regulator of the phosphatidylinositol 3-kinase (PI3K)/Akt survival pathway, are observed in HCC,

Tab. 1 Molecular alterations of critical genes in HCC.

Cellular process	*Molecule*	*Alteration in HCC*	*Acquired capability*
Growth factor signaling	EGF/EGFR	Upregulation	Sustaining proliferative signaling
	HGF/MET	Upregulation	
	IGF/IGFR	Overexpression	Resisting cell death
	VEGF/VEGFR	Upregulation	Inducing angiogenesis
	PDGF/PDGFR	Upregulation	
	FGF/FGFR	Upregulation	
Cell cycle regulation	TP53	Inactivating mutation/LOH	Sustaining proliferative signaling Evading growth suppressors
	RB1	Inactivating mutation/LOH	
	c-myc	Overexpression	
	p16 (CDKN2A)	Inactivating mutation/hypermethylation	
	Cyclin D1	Overexpression	
	IRF2	Inactivating mutation	
Ras/RAF pathway	RAS	Activating mutation	Sustaining proliferative signaling Evading growth suppressors
	RPS6KA3	Inactivating mutation	Inducing angiogenesis Activating invasion and metastasis
PI3K/AKT pathway	PI3K-alpha (PIK3CA)	Activating mutation	Sustaining proliferative signaling Evading growth suppressors
	PTEN	Inactivating mutation/LOH	Inducing angiogenesis Activating invasion and metastasis
	AKT	Constitutive activation	Reprogramming energy metabolism
	mTORC1	Upregulation	
JAK/Stat pathway	Stat	Constitutive activation	Tumor-promoting inflammation
	SOCS1, SOCS3	Downregulation	
NF-κB pathway	NF-κB	Constitutive activation	Sustaining proliferative signaling Resisting cell death Tumor-promoting inflammation
Wnt/β-catenin pathway	β-Catenin (CTNNB1)	Activating mutation/overexpression	Sustaining proliferative signaling Tumor-promoting inflammation
	AXIN1, AXIN2	Inactivating mutation/LOH	
	APC	Inactivating mutation	

(*continued overleaf*)

Tab. 1 (*Continued.*)

Cellular process	*Molecule*	*Alteration in HCC*	*Acquired capability*
Hedgehog pathway	SHH	Overexpression	Reprogramming energy metabolism
	SMO	Overexpression	
	HHIP	LOH, hypermethylation	
Histone modification	DNMT1, 3A, 3B	Upregulation	Sustaining proliferative signaling Tumor promoting inflammation
	EZH2	Overexpression	
	ARID1, ARID2	Inactivating mutation	
Apoptosis	Fas	Downregulation	Resisting cell death
	FasL	Upregulation	
	DR5	Downregulation	
Angiogenesis	Angiopoietin	Upregulation	Inducing angiogenesis
	Tie-2	Upregulation	
Immunity	Glypican-3	Upregulation	Evading immune destruction

LOH: Loss of heterozygosity.

but not very frequently [45–49]. Ras proto-oncogenes (H-*ras*, K-*ras*, N-*ras*) are also rarely mutated and activated in HCC [43, 50–52]. Runt-related transcription factor (RUNX) family genes induce a senescence-like growth arrest in response to oncogenic Ras. The expression of one of the members of this family, RUNX3, is decreased in more than 50% of HCC cases [53, 54], its downregulation being associated with escape from apoptosis and sustained growth [55, 56]. Recently, the role of RUNX3 in mediating angiogenesis and epithelial-mesenchymal transition (EMT) was also demonstrated [57]. Thus, the loss of RUNX expression may explain the rarity of Ras mutations in these cancers.

Less-frequent inactivating tumor-suppressing mutations of P14, IGFR2, KLF6, hepatocyte nuclear factor (HNF) 1α, SMAD2, SMAD4 and LKB1/STK11, and less-frequent activating oncogenic mutations of EGFR, Erb2 and PIK3CA have also been reported in HCC [58, 59]. Molecular alterations of critical genes and their contribution to the acquisition of hallmark capabilities are listed in Table 1.

2.2 Sustaining Proliferative Signaling

Constitutive activation of survival pathways, the inactivation of tumor suppressors TP53, Rb and p16, the overexpression of c-myc and cyclin D1, the epigenetic silencing of p16INK4, and the overexpression of E2F family members promote cell cycle progression and a sustained proliferation of HCC cells.

Growth factors and their corresponding activated tyrosine kinase receptors such as EGF-EGFR, IGF-IGFR and HGF-MET transmit the proliferation signal through Ras/Raf/MEK/ERK and PI3K/AKT/mTOR pathways. Constitutive activation of these pathways are maintained through the overexpression of MET and EGFR and the inactivation or downregulation of the negative regulators PTEN and RASSF1A of PI3K/AKT/mTOR and Ras/Raf/MEK/ERK pathways, respectively. The identification of potential driver genes in human liver carcinoma by genome-wide screening revealed genes with specific signaling pathways such as PI3K/Akt/mTOR, AMP-activated protein kinase (AMPK) and EGFR [60]. Cell proliferation in HCV-positive HCC is also associated with Myc and AKT activation [61]. Moreover, it is known that the overexpression of E2F family members, which are responsible for the transcription of genes involved in cell cycle and proliferation, inhibits c-Myc-driven apoptosis through the PI3K/AKT/mTOR pathway [62, 63].

Additionally, large-scale analyses with high numbers of HCC samples revealed proliferation gene signatures. Chen *et al.*, in 2002, showed a high expression of a "proliferation cluster" comprised of genes required for cell cycle progression in 102 primary HCC tumor samples and in 10 HCC cell lines [64]. Ribosomal protein genes were highly expressed, as expected in cells with unlimited cell growth. Upregulated genes were related to DNA replication and G_2/M progression. In another study of 91 human primary HCCs, a proliferation gene signature (containing PCNA, Bub3, MCM2, MCM6, MCM7, cyclinA2, cyclinB1, CKS2, and CDK4) was able to distinguish between two groups of HCC patients with a distinct prognosis [65].

2.3
Evading Growth Suppressors

The inactivation of pRB, p53 and p16, and the overexpression of c-myc and cyclin D1 confer growth advantages in HCC. Activation of the MYC transcription signature is strongly associated with the malignant conversion of preneoplastic liver nodules, while the inactivation of MYC in invasive HCCs leads to a sustained tumor regression as well as proliferation arrest, differentiation and apoptosis of malignant cells [66, 67].

The HCV core protein induces promoter hypermethylation and a downregulation of p16 expression, and subsequently induces Rb phosphorylation that leads to the activation of E2F1 that, in turn, stimulates cell growth [68]. This mechanism is also exploited to overcome stress-induced premature senescence in the presence of HCV-induced oxidative stress [69]. Furthermore, HCV core protein upregulates DNA methyltransferases 1 and 3b, and induces the promoter hypermethylation of retinoic acid receptor-β2 (RAR-β2). This mechanism leads to an escape from RB/E2F-related growth arrest induced by all-*trans* retinoic acid [70].

Upregulation of the oncoprotein gankyrin enhances the transcriptional activity of β-catenin, which in turn transcriptionally activates gankyrin by a positive feedback loop [71]. Gankyrin is highly expressed in HCC, and its overexpression mediates the degradation of the tumor suppressor proteins Rb and p53 and thereby accelerates cell cycle progression.

Growth arrest and DNA damage 45G (GADD45G) is commonly downregulated in oncogene-transformed HCC [72]. The ectopic expression of GADD45G induces senescence in HCC through a repression of the Jak/Stat3 pathway, independently of p53, p16INK4a and Rb. The expression of constitutively activated Stat3 or human telomerase reverse transcriptase (hTERT) reverts GADD45G-induced senescence.

Transforming growth factor-beta (TGF-β) signaling has a growth-suppressive role in the early stages of HCC. Resistance to TGF-β signaling-mediated growth inhibition is a frequent event during the malignant transformation of hepatocytes. In the advanced stages of HCC, TGF-β is secreted by stromal cells in the tumor microenvironment, and therefore induces apoptosis in hepatocytes. However, downregulation of the receptors of TGF-β and upregulation of the EGFR and MEK/ERK pathways confers resistance to TGF-β-induced cell death in liver tumor cells [73–79].

2.4 Resisting Cell Death

The presence of elevated levels of growth factors such as IGF, the upregulation of anti-apoptotic pathways such as NF-κB pathway, the downregulation of death receptors such as DR5 and Fas, and mutations in the tumor suppressor genes such as p53, contribute to an evasion of apoptosis in many cancers, including liver cancer. The IGF/IGFR signaling pathway that is constitutively activated in 20% of HCC via IGF2 or IGFR1 overexpression, regulates proliferation, motility, invasion, and the inhibition of apoptosis. This is correlated with the stage, metastasis and survival of HCC [80–84]. The IGF pathway activates cell-survival pathways PI3K/AKT and RAF/MEK/ERK and provides a mechanism by which to evade apoptosis [80].

Mutations in the tumor suppressor gene *p53* cause a loss of apoptotic response. Large-scale gene expression analysis showed that cell cycle-related genes (*CCNG2, BZAP45*) and cell proliferation-related genes (*SSR1, ANXA2, S100A10,* and *PTMA*) were overexpressed in mutant-*p53* tumors compared to wild-type-*p53* tumors in HCV-related HCC [85].

Tumor necrosis factor (TNF) activates the antiapoptotic NF-κB pathway, the pro-apoptotic caspase-cascade, and JNK kinases in HCC. Inhibition of NF-κB by NEMO results in an upregulation of the death receptor DR5, whose ligand (TRAIL) is predominantly expressed by natural killer (NK) cells and is essentially involved in liver injury in NEMO-deficient hepatocytes [86]. Furthermore, Caspase-8 is frequently inactivated in HCC and therefore interferes with the proapoptotic caspase-cascade [87]. The downregulation of Fas expression, upregulation of its ligand (FasL) expression in hepatocytes, and the elevation of serum soluble Fas levels were also identified as critical players of evasion from immune surveillance, and hepatic carcinogenesis [88].

2.5 Enabling Replicative Immortality

It is known that the reactivation of telomerase maintains telomere length and replicative immortality during cirrhosis, and therefore leads to HCC progression [89, 90]. Not only telomere dysfunction but also oncogene activation, persistent DNA damage and ROS-induced oxidative stress can cause permanent cell cycle arrest, known as senescence [31, 91]. The ability to bypass senescence is a characteristic of liver cancer cells in gaining replicative immortality.

TGF-β, Ras, Raf, Mos, Mek, Myc, E2F, Stat5, Cyclin E and PTEN are all key players of oncogene-induced senescence [92]. The reactivation of telomerase and inactivation of p53, p15, p16 and p21 may play critical roles in the bypass of senescence and maintenance of immortality [91]. Indeed, the expression of human hTERT was able to revert GADD455-induced senescence in HCC [72], while an upregulation of hTERT expression by low-dose cisplatin contributed to cell death resistance in a

HCC cell line; this resistance was reverted by an inhibition of hTERT [93]. Two independent mutations were identified within the core promoter of TERT; these mutations increased transcriptional activity from the TERT promoter by two- to fourfold, and were shown to occur frequently in HCC [94]. A recurrent integration of HBV into the promoter of the TERT gene was correlated with increased TERT expression in HBV-related HCC patient samples [95]. Furthermore, a recent study showed that HCCs expressing "stemness"-related proteins (K19, EpCAM, CD133) have increased telomere length, an increased expression of hTERT and shelterin complex proteins (TRF1, TRF2, TIN2, POT1, TPP1, RAP1), and an increased chromosomal instability compared to HCCs without these markers [96]. Hence, high telomerase activity and long telomeres in HCCs are associated with an aggressive behavior and a poor prognosis [95, 97]. It was reported recently that there is a major shift from senescence-associated gene expression to immortality-associated gene expression during the transition from dysplasia to early HCC lesions. Moreover, a senescence bypass signature was able to differentiate HCC from cirrhosis [98]. Therefore, targeting senescence in liver cancer treatment can be considered as an alternative mechanism in addition to classical chemotherapeutic agents [99].

2.6 Inducing Angiogenesis

Angiogenesis and neovascularization involve interactions between tumor cells, vascular endothelial cells (VECs) and their supporting pericytes in order to supply oxygen and nutrients to the growing tumor [30]. In HCC, the balance between proangiogenic and antiangiogenic factors is disrupted due to an excess secretion of angiogenic factors by endothelial cells and pericytes in the tumor microenvironment [100]. Angiogenic factors, including vascular endothelial growth factor (VEGF), fibroblast growth factor (FGF), platelet-derived growth factor (PDGF) and angiopoietin-2, Tie-2 are upregulated in HCC. This process induces angiogenic signaling through activation of the RAF/MEK/ERK, PI3K/AKT/mTOR, JAK/Stat, and HGF/MET pathways [101–106].

VEGF and its receptors VEGFR-1 and VEGFR-2 are overexpressed in HCC, and are associated with aggressiveness and a poor prognosis [107–111]. The HBV × antigen also upregulates VEGFR-3 [112]. VEGF acts synergistically with FGF, the overexpression of which is correlated with HCC angiogenesis [113, 114].

Hypoxia, which occurs during fibrosis, cirrhosis and malignant transformation, enhances the proliferation, angiogenesis, metastasis, chemoresistance and radioresistance of HCC [100]. Hypoxia-induced factor 1 alpha (HIF-1α) promotes hepatocyte EMT through PI3K/Akt, TGF-β and β-catenin signaling, and this is associated with an enhanced metastatic potential and a poor prognosis in HCC [115–117]. Bone morphogenetic protein 4 (BMP-4) is also induced in hypoxic conditions and promotes vasculogenesis and tumor progression in HCC [118].

2.7 Activating Invasion and Metastasis

Cell detachment is an early step of tumor invasion, requiring alterations to the adhesive properties of cancer cells in general. Therefore, EMT is critical for activating the invasion and metastasis of HCC. The upregulation and activation of Twist, Snail, Slug, Zeb1/2 and Vimentin, and the downregulation of E-cadherin and HNF-4α, frequently occur during EMT and correlate with a poor prognosis in liver cancer [119–122]. Furthermore, p53 regulates EMT through miR-200 family members and miR-192, which targets Zeb1 and Zeb2 [123]. HBx expression was shown to induce EMT by activating the PI3K/Akt/GSK-3β pathway, which stabilizes Snail and mediates integrin $\alpha_6\beta_1$ signaling, thus facilitating tumor invasion and metastasis during HCC progression [124, 125]. In HCV-positive HCC patients, the expression of a four-gene signature, including E-cadherin, inhibitor of DNA binding 2 (ID2), MMP9 and transcription factor 3 (TCF3), is correlated with poor prognosis [126].

Osteopontin is overexpressed in metastatic HBV-related HCC and invasion, and metastases are effectively blocked by an osteopontin-specific antibody both *in vitro* and *in vivo* [127]. This observation suggests that osteopontin can be considered as a diagnostic marker and a potential therapeutic target for HBV-related metastatic HCC [128, 129]. The tumor suppressor DLC1 and cytoskeletal protein RhoA are also involved in the prevention of dissemination and metastasis of human HCC cells in nude mice [130, 131]. Additionally, intratumoral hypoxia triggers the invasion and metastasis of HCC through an oncogenic HGF/MET signaling pathway [132]. A Met-regulated gene expression signature defines an aggressive subtype of HCC with an increased vascular invasion rate, microvessel density, and a decreased mean survival time of HCC patients [103].

2.8 Reprogramming Energy Metabolism

Cancer cells reprogram their energy metabolism so that they can use glucose to supply energy through aerobic glycolysis and glutamine to provide intermediates of the tricarboxylic acid (TCA) cycle [133]. In addition, autophagy enables fast-growing cells to break down cellular organelles, and this results in recycled catabolites that can be used for biosynthesis and energy metabolism [22]. mTOR, an evolutionarily conserved serine/threonine kinase located downstream in the PI3K/AKT pathway, also functions as a nutrition sensor to monitor cellular metabolism [134]. In the presence of sufficient energy and nutrients, active mTOR promotes translation and biosynthesis, and hence suppresses autophagy. In the absence of sufficient energy and nutrients, however, mTOR with downregulated activity leads to a reduction in biosynthesis and promotes autophagy [135]. The PI3K/AKT pathway also stimulates glucose uptake and metabolism for the continued growth and survival of cancer cells [136].

In contrast, glutamine uptake and metabolism is under the control of c-myc [137]. Elevated energy consumption and an addiction to mitochondrial glutaminolysis is dependent on the AMPK-related kinase 5 (ARK5) through oncogenic c-myc expression in HCC cells. ARK5 limits protein synthesis via an inhibition of mTORC1 and maintains the high respiratory capacity required for efficient glutamine metabolism [138]. Therefore, targeting cellular energy homeostasis represents a promising therapeutic strategy for HCC cells with a higher c-myc expression. AMPK, which is activated in response to reduced energy levels, promotes ATP production by increasing catabolism, and conserves ATP by switching off biosynthetic pathways. AMPK was found to be dysfunctional in patients with HCC, and low p-AMPK levels were correlated with aggressiveness and a poor prognosis [139].

Moreover, AMPK is also activated by the p53 targets, Sestrin1 and Sestrin2 [140]. p53 stimulates oxidative phosphorylation and reduces the rate of glycolysis through the upregulation of TP53-induced glycolysis and apoptosis regulator (TIGAR) [141]. Therefore, energy metabolism is shifted from mitochondrial respiration towards glycolysis by the loss of p53 in cancer cells.

During HCC development, metabolic remodeling from mitochondrial oxidation to glycolysis was assessed by a combined transcriptomics and metabolomics study in six subgroups of HCC tissues, as defined by Beyoğlu *et al.* [142]. HCC has lower levels of glucose and other metabolites (glycerol 3-phosphate, glycerol 2-phosphate, malate, alanine, and myoinositol) involved in energy production compared to healthy liver. Moreover, concentrations of certain saturated lipids are reduced in a subgroup of HCC cells associated with high serum alpha-fetoprotein (AFP) levels. This is consistent with the previous observations on the upregulation of lipid catabolism accompanied by elevated AFP expression. In another study, 28 metabolites and 169 genes were identified that were involved in energy metabolism associated with aggressive HCC [143]. Metabolic activities within the HCC microenvironment are also promoted by Hedgehog signaling activation by malignant hepatocytes. Hedgehog ligands produced by these cells stimulate glycolysis in the neighboring myofibroblasts, resulting in the release of myofibroblast-derived lactate that can be used as an energy source by the malignant hepatocytes [144].

Mutations that activate oncogenes (e.g., c-myc, HIF-1α, and PI3K/AKT) or inactivate tumor suppressors (e.g., p53, PTEN, TSC2, and LKB1) have been shown to contribute to metabolic alterations in various types of cancer [145]. Oncogene-altered energy metabolism presents a new class of target molecules for tumor therapy. Indeed, activating oxidative phosphorylation by a pyruvate dehydrogenase kinase inhibitor (dichloroacetate) overcame the sorafenib resistance of HCC, while a combination of sorafenib and dichloroacetate resulted in an elevated tumor regression compared to sorafenib alone [146]. Therefore, tumor bioenergetics can be further exploited for HCC therapy.

2.9 Tumor-Promoting Inflammation

The tissue microenvironment plays a critical role in HCC formation and development [100]. The HCC microenvironment is composed of cancer-associated fibroblasts (CAFs), invading inflammatory cells, endothelial cells, pericytes adjacent to the endothelial cells, hepatic stellate cells (HSCs), macrophages (Kupffer cells), dendritic cells (DCs), and stem/progenitor cells. The extracellular matrix (ECM) components, including collagen, fibronectin, laminin, glycosaminoglycans and proteoglycans, provide a supportive microenvironment for these cells [100, 147]. All of these components of the microenvironment interact with each other and produce growth factors, cytokines, chemokines and free radicals that contribute to liver fibrosis, and therefore also tumor initiation and progression. Overexpression of the highly negatively charged ECM protein osteopontin is associated with large tumor size, advanced tumor stage, capsular infiltration, vascular invasion, lymph node invasion, intrahepatic metastasis, early recurrence, and poor prognosis of HCC [148–154]. Furthermore, plasma osteopontin levels were significantly higher in HCC patients, and thus can be considered as an early HCC marker, together with AFP [151, 155, 156].

During HBV- or HCV-infection-associated chronic liver disease, hepatocyte injury leads to inflammatory cell infiltration, where the host immune cells destroy virus-infected hepatocytes [157, 158]. The continuous inflammation results in a cycle of hepatocyte death and proliferation, leading to an increased genomic instability and mutations [4]. When stimulated with proinflammatory cytokines (IL-1β, TNF-α, and PDGF), Kupffer cells and HSCs produce osteopontin that plays an important role in inflammation, growth, invasion, metastasis, angiogenesis and inhibition of apoptosis [159–162]. TNF-α, produced by Kupffer cells and other immune cells, promotes tumor progression mainly through the NF-κB and Akt pathways [159].

Kupffer cells express and release the proinflammatory cytokine interleukin-6 (IL-6). The latter is one of the major mediators of inflammation, and activates the STAT3 pathway to mediate its signal through the gp130 protein. IL-6 also protects liver cells against apoptosis via the STAT3 pathway following viral infection or chemical ingestion [163, 164]; this mechanism is also exploited during tumor promotion [165]. Constitutively activated Stat3 protein maintains NF-κB activity in tumors by preventing the nuclear export of NF-κB complex through RelA acetylation [166]. RelA, in turn, maintains a persistent activation of STAT3 and IL-6 in the HCC microenvironment [167]. Therefore, a conditional knock-out of STAT3 expression impairs liver regeneration, whereas a tissue-specific knock-out of hepatic STAT3 effects glucose homeostasis and the induction of insulin resistance [168, 169].

Oncogenic β-catenin also triggers an inflammatory response, including activation of the NF-κB pathway in hepatocytes, which promotes aggressiveness of HCC in mice [170]. Elevated IL-6 levels and constitutively activated STAT3 has been frequently detected in HCC patients and in cell lines [171–176]. In addition, the Jak-Stat inhibitors SOCS1 and SOCS3 are downregulated in HCC via promoter hypermethylation [177]. Moreover, somatic gain-of-function mutations in the IL6ST gene (gp130), have been identified in inflammatory hepatocellular adenomas [178]. IL-6/STAT3 further activates several interleukins and growth factors.

A low-grade inflammatory response is induced upon lipid accumulation in obesity, which in turn increases IL-6 and TNF expression by adipose tissue and Kupffer cells [179, 180]. IL-6 and TNF signaling promotes the proliferation of damaged hepatocytes via an activation of JAK/STAT and AKT/ERK, respectively [159, 181].

A unique 17-gene immune response signature of the liver microenvironment could predict venous metastases, recurrence, and prognosis in HCC [182]. A global Th1- to Th2-like cytokine shift is associated with HCC metastasis, which is promoted through a shift toward anti-inflammatory/immune-suppressive responses.

2.10 Evading Immune Destruction

Liver cancer develops usually on top of chronic inflammation during fibrosis and cirrhosis. Yet, even though tumor cells exploit inflammation in favor of their growth, as mentioned above, they still need to escape from immune destruction. HCC cells evade immune destruction by expressing immunosuppressive molecules such as PD-L1 and indolamine 2,3-dioxygenase (IDO), and also by secreting cytokines and chemokines such as IL-6, IL-10, TGF-β and VEGF [183]. The resulting immunosuppressive microenvironment is supported by the induction of regulatory dendritic cells (DCregs), regulatory T cells (Tregs), and tumor-associated macrophages (TAMs) and the suppression of DCs, effector T cells and NK cells.

Macrophages are the major infiltrating leukocytes, and are involved in both innate and adaptive immune responses [184]. During either HBV- or HCV-infection, a fibrotic and cirrhotic liver causes inflammatory cell infiltration due to the necrosis of hepatocytes. Therefore, the status of TAMs and other immune cells in the tumor microenvironment is closely associated with the suppression of antitumor immunity and progression of HCC. Polarizing inflammatory responses toward the preferential recruitment of Th2-type cells and Tregs rather than Th1-type cells promotes tumor immune evasion. While M2-type "alternatively activated" macrophages promote tumor progression, M1-type "classically activated" macrophages can exert antitumor activity by killing the tumor cells. Therefore, shifting the macrophage balance from tumor promotion by innate immunity-driven inflammation towards tumor surveillance by adaptive immune responses can be an effective therapeutic strategy [185].

Dendritic cells, which are important for chronic liver inflammation, express and present antigens to infiltrating cytotoxic T lymphocytes (CTLs). The dendritic cells express Glypican 3 (GPC3), the upregulation of which is associated with a poor prognosis in HCC [186]. Indeed, tumor-induced DCregs with hyperactivate Stat3 can facilitate tumor immune evasion independent of Stat3 hyperactivation status in tumors [187]. Additionally, a CTL-mediated immune response can be impaired by Kupffer cells through programmed death ligand 1 (PD-L1), and regulatory dendritic cells (CD14+ CTLA4+) through CTLA4-dependent IL-10 and IDO production, enabling immune evasion [188, 189].

Cytokine production by NK cells in chronic HCV infection is shifted towards the secretion of Th2-type cytokines, promoting an environment which is more permissive for HCV [190]. In chronic HCV patients, NK cells have a reduced cytotoxicity and interferon (IFN)-γ production, and secrete IL-10 and TGF-β; this results in the induction of Th2 cells and Tregs, maintaining immune evasion. Furthermore, many immune system-related genes, including SATB1, TNFRSF5, CTLA4, GITR, SIRPα, PD-L1 and ICOS, have altered expressions in HCC [85, 191, 192].

It has been shown that the only FDA-approved chemotherapeutic agent, sorafenib, also suppresses the proliferation and activation of NK cells in addition to malignant hepatocytes. Consequently, the reduced cytotoxicity of NK cells handicaps HCC patients during treatment by rendering the host more susceptible to tumor growth and metastasis [193]. Therefore, immunotherapeutic approaches activating NK cells can enhance the efficacy of sorafenib.

In light of above-described cellular mechanisms, the hallmarks of cancer are represented in parallel with the altered genes involved in the development of the multistep progression of liver cancer, in Fig. 1. This overall picture demonstrates that genome instability and mutations, along with sustaining proliferative signaling having the highest number of altered genes,

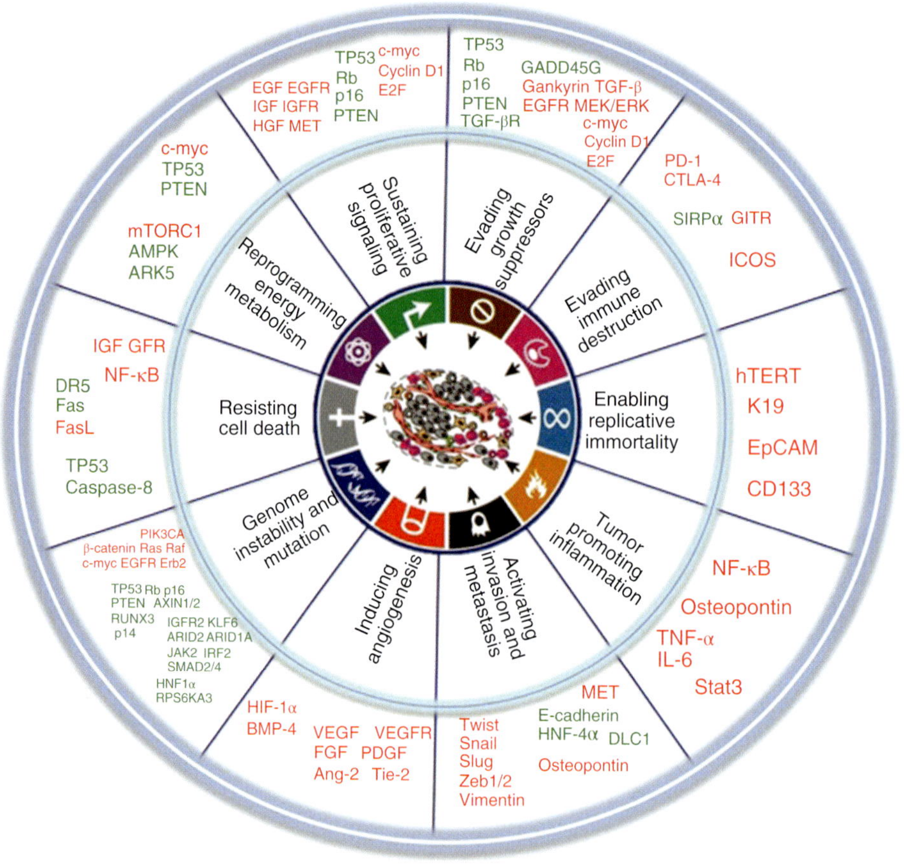

Fig. 1 Molecular hallmarks of hepatocellular carcinoma. Molecular alterations promote the hallmark capabilities through either activation/expression/upregulation are shown in red, while inactivation/loss/downregulation are shown in green. Modified with permission from Ref. [22].

are the most widely studied hallmarks in HCC. Yet, the new emerging capabilities of tumor cells should be further examined in this disease in order to identify novel genes associated with HCC malignancy.

3 Genome-Wide Changes in Hepatocellular Carcinoma

During the past decade, several high-throughput analyses on HCC samples have been reported. Microarray, whole-genome sequencing and exome sequencing studies will enlighten the intra-tumoral heterogeneity of HCCs from various etiologies and histological pathologies. These large-scale studies revealed novel HCC-related pathways or gene signatures that can distinguish subgroups of HCC based on etiology, molecular background, and histopathology. These signatures also predict survival, metastasis, and recurrence. Major findings of recent high-throughput analyses on HCC are presented in Table 2. Some of the novel

Tab. 2 Major findings of recent high-throughput analyses on HCC.

Classification of HCC based on altered gene expression	*Reference*
Highly expressed "proliferation cluster" genes in HCC	[64]
"Proliferation cluster" genes predict survival by 406 gene signature	[65]
HCC classified into two groups based on IFN-regulated and apoptosis-relevant genes	[194]
Identify 240 gene signature for low- to high-grade dysplastic nodules and HCC	[195]
Six subgroups of HCC related to genetic alterations	[43]
G1: IGFR1 activation, AKT activation, developmental imprinting	
G2: PIK3CA and TP53 mutations, AKT activation	
G3: TP53 mutation and overexpression of cell-cycle genes	
G4: a heterogeneous subgroup	
G5: β-catenin mutations, Wnt/β-catenin activation	
G6: G5 with satellite nodules, having higher activation of the Wnt pathway and low E-cadherin expression	
Four neoplastic stages of HCV-positive HCC	[196]
Control vs. cirrhosis: 8 gene signature	
Cirrhosis vs. dysplasia: 24 gene signature	
Dysplasia vs. early HCC: 93 gene signature	
Early vs. advanced HCC: 9 gene signature	
Subclasses of HCC with different genetic backgrounds	[197]
c-Myc induced, 6p/1q-amplified, 17q-amplified	
Five classes of HCV-positive HCC	[111]
CTNNB1 class: β-catenin mutations	
Proliferation class: IGF1R activation and RPS6 phosphorylation	
IFN-related class: a novel class defined by polysomy of chromosome 7	
Three molecular subclasses of HCC	[61]
S1: Wnt/β-catenin activation	
S2: proliferation with Myc and AKT activation and IFN repression	
S3: tumor size, differentiation, and serum AFP levels	
Cirrhosis vs. HCC	[98]
Immortality and senescent signature: 15 immortality gene	
Classification of HCC based on chromosomal imbalance	
Early vs. advanced stages of HCC	[198]
Three HBV-positive HCC subgroups predict survival	[199]
Classification of HCC based on different etiologies	
Invasiveness gene signatures of HBV- or HCV-associated HCC	[200]
HBV- or HCV-positivity	[201]
HBV-positive HCC: 31 genes relating to signal transduction, transcription, metastasis	
HCV-positive HCC: 52 genes related to detoxification and immune response	
HBV-positive HCC vs. non-tumor liver tissues	[202]
44 gene signature	
Early vs. late stages of HBV-positive HCC	[203]
65 gene signature	
Susceptibility to HBV-positive HCC	[204]
1p36.22 locus, and KIF1B-, UBE4B-, PGD-related pathways	
Susceptibility to HCV-positive HCC	[205]
Common variants within the DEPDC5 locus on chromosome 22	

(*continued overleaf*)

Tab. 2 (*Continued.*)

Classification of HCC based on altered gene expression	*Reference*
HCV-positive HCC	[206]
Strong association of a locus in the 5′ flanking region of MICA, which leads to activation of natural killer cells and CD8+ T cells	
Molecular profile of specific gene alterations	
HCV-positive HCC with mutant-p53 vs. wild-type-p53	[85]
Overexpressed: cell cycle- and cell proliferation-related genes	
Underexpressed: immune system-related genes	
Myc signature in cirrhosis vs. nodules vs. HCC	[66]
Met knock-out signature: HGF/Met activation associated with poor survival	[103]
RB knock-out signature: increased proliferation and RB/E2F activity	[207]
TGF-ß knock-out signature: invasive phenotype and increased tumor recurrence	[208]
Identification of genetic alterations	
Somatic substitutions in HCV-positive HCC: T > C/A > G and C > T/G > A 11 731 mutations, including TP53, AXIN1, ADAM22, JAK2, KHDRBS2, NEK8, and TRRAP	[209]
Frequent mutations in HCV-associated HCC, including CTNNB1, TP53, ARID2, DMXL1, and NLP1	[210]
ARID2 as a tumor suppressor gene in HCV-associated HCC	
Cirrhotic vs. non-cirrhotic HCC: G > T and C > A transversions are more frequent in tumors from non-cirrhotic liver	[24]
New recurrent alterations in ARID1A, RPS6KA3, NFE2L2, and IRF2	
Identification of epigenetic alterations	
Predict survival: aberrant DNA methylation signature at promoter sites	[211]
High frequency chromatin regulating gene mutation	[29]
Mutation in ARID2	
Identification of potential tumor driver mutations	
50 potential genes with specific signaling pathways (mTOR, AMPK, and EGFR)	[60]
3 tumor mutations (CCNG1, P62, and an indel/fusion gene) from sequencing of 3 HCC nodules from 1 HBV-positive HCC patient	[212]
13 potential tumor suppressor genes, including XPO4	[213]
Prediction of metastasis from gene signatures	
153 gene signature in HBV-positive HCC	[127]
Osteopontin overexpression	
17 gene stromal tissue signature to predict metastasis	[182]
HBV-positive HCC patients with portal vein tumor thrombosis	[214]
Transversions: C : G > A : T and T : A > A : T	
Mutation in ARID1A	
Prediction of recurrence from gene signatures	
12 gene signature	[215]
Early intrahepatic recurrence	[216]
57 gene signature	[217]
Early recurrence in HBV-positive HCC	[218]
Late recurrence from stromal tissues	[219]
Molecular profile of poorly differentiated cells	
Hepatoblast-like subclass with AP1 activation associated with poor prognosis	[220]
Progenitor-like class with EPCAM and AFP in HBV-positive HCC EPCAM-/AFP+ are associated with poor prognosis	[221]

important genes associated with HCC are also included in the table, along with their respective study.

4 microRNA Profiling of Hepatocellular Carcinoma

Recent findings highlight the importance of microRNAs in mediating the acquired capabilities of HCC. Comprehensive analyses of microRNA expression patterns have revealed a differential expression of miRNAs in metastatic HCC and non-metastatic HCC compared to healthy liver (Table 3). Hence, miRNAs can be utilized as prognostic markers in HCC patients with various clinical phenotypes. miR-26a, the expression of which is reduced in HCC, inhibits angiogenesis by downregulating VEGFA through PIK3C2α/Akt/HIF-1α, and suppresses growth and metastasis through IL-6/Stat3 signaling [222, 223]. miRNA replacement therapy, where miR-26a is administered in a mouse model of HCC, inhibits cancer cell proliferation and induces apoptosis [224]. The delivery of downregulated miRNAs that are highly expressed and therefore tolerated in normal liver inhibits tumorigenicity without toxicity; hence, this approach may be a valuable strategy for miRNA-mediated HCC therapies.

5 Epigenetic Mechanisms

The epigenetic regulation of gene expression involves DNA methylation, post-translational histone modifications, and changes in the expression profiles of chromatin-modifying enzymes, which are highly deregulated in cancers, including HCC [253].

Epigenetic alterations in HCC include global DNA hypomethylation, gene-specific DNA hypermethylation of Rb, E-cadherin, RASSF1A and p16, gene-specific DNA hypomethylation of Vimentin, uPA and CD147, upregulation of DNA methyltransferases DNMT1, DNMT3A, and DNMT3B, and altered histone modification patterns of H3K9 and H3K27 through deregulation of the histone-modifying enzymes HDAC1/2/3, SIRT1, EZH2, and ARID2 [24, 29, 210, 254–260].

The histone methyltransferase EZH2 is overexpressed in HCC and contributes to the epigenetic silencing of target genes that regulate cancer cell growth and survival. Sorafenib was shown to downregulate EZH2 protein levels by accelerating its proteasome-mediated degradation in hepatoma cells, and thereby altering the HCC epigenome by reducing H3K27 trimethylation [261]. The overexpression of EZH2 reverses sorafenib-induced cell cycle arrest, and apoptosis. This epigenetics-based study revealed a novel combinational therapy approach, where an inhibition of EZH2 can be used to increase the sensitivity of HCC cells to chemotherapeutic agents.

All hallmarks of HCC are under the control of epigenetic mechanisms. In return, an altered metabolism in HCC cells can determine the availability of metabolites that are necessary for the functioning of epigenetic modifiers and thereby regulate the cancer epigenome [262]. Aberrant activities of epigenetic regulators and gene-specific methylation alterations in HCC are listed in Tables 4 and 5.

6 Concluding Remarks

HCC, as one of the leading causes of cancer-related death, is mostly a viral

Tab. 3 Alterations in microRNAs and their mechanisms of promoting HCC progression.

microRNA	*Mechanism*	*Alteration in HCC*	*Reference(s)*
miR-101	Targets EZH2 and inhibits HCC progression	Downregulation	[225]
miR-122	Sensitize HCC cells to chemotherapeutic drugs by downregulating MDR related genes	Downregulation	[226]
miR-124	Targets PIK3CA, suppresses PI3K/AKT pathway Suppresses the HCC growth through targeting STAT3 Transient inhibition of HNF4α initiates HCC through a microRNA-inflammatory feedback loop of miR-124, IL6R, STAT3, miR-24, and miR-629	Downregulation	[227–229]
miR-139	Promotes cell proliferation and invasion through the WNT/TCF-4 pathway	Downregulation	[230]
miR-140-5p	Suppresses cell proliferation and metastasis by targeting TGFBR1 and FGF9	Downregulation	[231]
miR-148a	Suppresses EMT and metastasis by targeting Met/Snail Reduces HPIP, represses AKT and ERK and inhibits mTOR through AKT/ERK/FOXO4/ATF5	Downregulation	[232, 233]
miR-155	Targets APC, promotes hepatocyte proliferation and tumorigenesis by activating Wnt signaling	Overexpression	[234]
miR-17	Inhibits cell migration and invasion via suppression of MMP-3 and Akt	Downregulation	[235]
miR-195	Blocks G1/S transition by repressing Rb/E2F signaling	Downregulation	[236]
miR-199a/b-3p	Suppress HCC growth through targeting PAK4 and inhibiting PAK4/Raf/MEK/ERK pathway	Downregulation	[237]
miR-21	Suppresses PTEN, hSulf-1, PDCD4, and RECK and activates EMT via AKT and ERK pathways	Overexpression	[238, 239]
miR-214	Contributes to angiogenesis through activation of HDGF paracrine pathway	Downregulation	[240]
miR-216a/217	Activates the PI3K/Akt and TGF-β pathways by targeting PTEN and SMAD7	Overexpression	[241]
miR-221	Accelerates hepatocyte proliferation during liver regeneration	Overexpression	[242]
miR-222	Promotes metastasis through activating AKT signaling and targeting PPP2R2A	Overexpression	[243]
miR-224	Activates AKT signaling pathway by targeting PPP2R1B	Overexpression	[244]
miR-26a	Inhibits angiogenesis by down-Regulating VEGFA through PIK3C2α/Akt/HIF-1α Suppress HCC growth and metastasis through IL-6-Stat3 signaling	Downregulation	[222, 223]
miR-27a	Reverses drug resistance (MDR) by inhibiting the FZD7/β-catenin pathway	Downregulation	[245]
miR-375	Inhibits autophagy by reducingATG7 expression	Downregulation	[246]

(*continued overleaf*)

Tab. 3 (*Continued.*)

microRNA	*Mechanism*	*Alteration in HCC*	*Reference(s)*
miR-503	Blocks G1/S transition by repressing Rb/E2F signaling	Downregulation	[247]
miR-519d	Targets CDKN1A/p21, PTEN, AKT3, and TIMP2. Promotes cell proliferation, invasion, and impairs apoptosis	Overexpression	[248]
miR-520b	Contributes to escape from growth suppression by targeting MEKK2 and cyclin D1 through JNK and Rb	Downregulation	[249]
miR-612	Suppresses EMT through Akt2	Downregulation	[250]
miR-675	Increases proliferation and inhibits invasiveness by downregulating RB and Twist1	Overexpression	[251]
miR-7	Inhibits HCC cell growth and metastasis by targeting PI3K/AKT pathway	Downregulation	[252]

Tab. 4 Epigenetic deregulations in HCC.

	Alteration in HCC	*Reference(s)*
DNA methyltransferases		
DNMT1, DNMT3A, DNMT3B	Upregulated	[256, 263–266]
Histone deacetylases		
HDAC-1, HDAC-2, HDAC-3	Upregulated	[267, 268]
SIRT1	Upregulated	[269, 270]
SIRT2	Upregulated	[271]
SIRT3	Downregulated	[272]
SIRT6	Downregulated	[273, 274]
Histone methyltransferases		
EZH2	Upregulated	[275, 276]
SUV39H1	Upregulated	[277]
SMYD3	Upregulated	[278]
MMSET (NSD2)	Upregulated	[279]

infection-associated disease, and its etiology is quite well known. Late diagnosis and the paucity of efficient therapeutic interventions are the major reasons why this cancer remains one of the most deadly on a worldwide basis. During the past decade, many of the investigations on liver cancer have focused on molecular classification methods and the identification of novel therapeutic targets and targeted agents. High-throughput analyses of the genomic state and genome-wide expression analyses were the most frequently used approaches for molecular classification, the primary purpose of which is to reduce the heterogeneity of HCC in terms of therapeutic response and patient survival. These studies have allowed the discovery of different molecular subtypes of liver cancer by using so-called "gene

Tab. 5 Hypermethylated and hypomethylated genes in HCC.

Hypermethylated genes	*Description*	*Role in HCC*	*Reference(s)*
RASSF1A	Ras association domain family member 1	Ras signaling pathway	[280–285]
RASSF5 (NORE1B)	Ras association domain family member 5	Ras signaling pathway	[281]
DAB2IP	DAB2 (mitogen-responsive phosphoprotein) interacting protein	Ras GTPase-activating protein	[283–285]
PTEN	Phosphatase and tensin homolog	PI3K/Akt/mTOR pathway	[280, 286]
TP53	Tumor protein p53	Survival, cell death, proliferation, growth	[287]
RB1	Retinoblastoma 1	Cell cycle, proliferation, growth	[288]
CDKN2A (p16)	Cyclin-dependent kinase inhibitor 2A	Cell cycle, proliferation, growth	[283–285, 289–293]
CDKN2B (p15)	Cyclin-dependent kinase inhibitor 2B	Cell cycle, proliferation, growth	[290, 294]
CDKL2	Cyclin-dependent kinase-like 2	Cell cycle, proliferation, growth	[283–285]
DACH1	Dachshund family transcription factor 1	Proliferation, growth	[295]
BMP4	Bone morphogenetic protein 4	Growth, metabolism, angiogenesis	[283–285]
BMP6	Bone morphogenetic protein 6	Growth, metabolism, angiogenesis	[296]
SOCS1	Suppressor of cytokine signaling 1	Jak/Stat pathway	[176, 297, 298]
SOCS3	Suppressor of cytokine signaling 3	Jak/Stat pathway	[174]
SYK	Spleen tyrosine kinase	Immune response	[299]
MAT1A	Methionine adenosyltransferase I, alpha (liver-specific)	Metabolism	[300–302]
GLS2	Glutaminase 2 (liver, mitochondrial)	Metabolism	[303]
GSTP1	Glutathione S-transferase pi 1	Metabolism	[282, 304, 305]
NQO1	NAD(P)H dehydrogenase, quinone 1	Metabolism	[306]
COX-1, COX-2	Cyclooxygenases	Metabolism	[307]
NKX6-2	NK6 homeobox 2	Metabolism	[283–285]
CDH1	Cadherin 1, type 1, E-cadherin	Invasion, metastasis	[308]
SFRPs	Secreted frizzled-related proteins	Wnt/β-catenin pathway	[309–312]

(*continued overleaf*)

Tab. 5 (*Continued.*)

Hypermethylated genes	*Description*	*Role in HCC*	*Reference(s)*
DACT2	Disheveled-binding antagonist of beta-catenin 2	Wnt/β-catenin pathway	[313, 314]
PRDM2 (RIZ1)	PR domain containing 2, with ZNF domain	Epigenetic regulation	[315, 316]
PRDM5	PR domain containing 5	Epigenetic regulation	[317]
CHD5	Chromodomain helicase DNA-binding protein 5	Chromatin remodeling	[318]
DNM3	Dynamin 3	Microtubule dynamics, vesicular transport	[283–285]

Hypomethylated genes	*Description*	*Role in HCC*	*Reference(s)*
MET	Met proto-oncogene	Growth, invasion, metastasis	[319]
AKT3	v-Akt murine thymoma viral oncogene homolog 3	PI3K/Akt/mTOR pathway	[283–285]
CD147	Basigin (ok blood group)	Invasion, metastasis	[257]
VIM	Vimentin	Invasion, metastasis	[260]
TFF3	Frequent trefoil factor 3	Inflammation, immune response	[320]
CCL20	Chemokine (C-C motif) ligand 20	Inflammation, immune response	[283–285]
CD1B	T-cell surface glycoprotein CD1b	Inflammation, immune response	[283–285]
CD1E	T-cell surface glycoprotein CD1e	Inflammation, immune response	[283–285]
CD300E	Immune receptor expressed on myeloid cells 2	Inflammation, immune response	[283–285]
MNDA	Myeloid cell nuclear differentiation antigen	Inflammation, immune response	[283–285]
MAT2A	Methionine adenosyltransferase 2, alpha	Metabolism	[302, 321]
CYP11B1	Cytochrome P450, family 11, subfamily B, polypeptide 1	Drug metabolism	[283–285]
LINE-1	Long interspersed nuclear element 1	Proliferation	[322]

signatures." A gene signature is the expression pattern of a set of genes associated with a clinical subtype of a disease, and in this regard several gene signatures and associated molecular types of HCC have been described. The heterogeneity of liver cancer, in parallel with the multistep evolution of this cancer, is represented in Figure 2. During the progression of HCC, various cellular mechanisms and

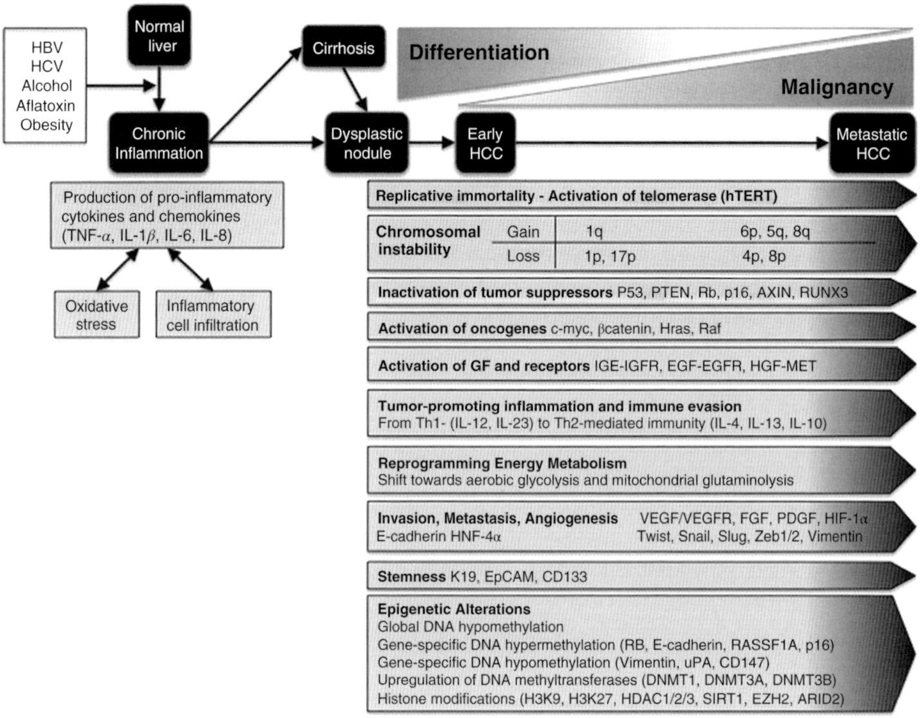

Fig. 2 Multistep evolution of primary liver cancer. The development of HCC is a multistep process, where injured hepatocytes promote chronic inflammation leading to hepatocyte death and regeneration cycles during cirrhosis and enduring liver disease. The subsequent expansion of dysplastic nodules, telomerase reactivation, increased genomic instability, inactivation of tumor suppressors, activation of oncogenes, and increase in growth factor signaling initiates HCC. Chromosomal instability and somatic mutations that favor the uncontrolled growth of HCC cells accumulate as the cancer advances. The acquisition of malignant phenotype is supported by tumor-promoting inflammation, a capability to evade immune destruction, and metabolic alterations that allow the continued growth and survival of cancer cells. Onset of invasive, metastatic, and angiogenetic capabilities promotes progression of carcinoma to the highly malignant metastatic state associated with stemness markers. Molecular alterations throughout the malignant transformation of HCC are regulated at both genetic and epigenetic levels.

their underlying genetic, epigenetic and proteomic alterations enable the acquisition of malignant behavior. These findings demonstrated that HCC is indeed a heterogeneous disease that can be subdivided into more or less homogeneous subclasses. Unfortunately, these findings were of very little assistance to the clinical follow-up of HCC patients, mainly because gene signatures are composed of large sets of genes and are not readily adaptable to routine use.

To date, the outcome of efforts to identify novel targets for HCC treatment has been less than satisfactory, and there are several reasons for this. First, the number of mutant but targetable genes is limited. Second, the pathogenesis of HCC may be related to cellular signaling pathways rather than to specific genes. It is probable that some of the HCC-promoting signaling pathways are activated by mechanisms other than gene mutations. Changes in gene expression or protein networks (such as overall expression, stability, post-translational modifications) may be more critical for HCC pathology than are gene mutations. In HCC, the number of genes with dysregulated expression is extremely high, and consequently a large set of signaling pathways appear to be deregulated in this cancer. The paucity of gene mutations, together with such a high number of genes with expression changes, strongly suggests that the epigenome of HCC is highly affected. DNA methylation changes appear to predominate the HCC epigenome, while a few reported data on histone methylation patterns also indicate that profound changes occur in the organization of HCC nucleosomes and chromatin. Future studies aimed at deciphering the status of HCC epigenome and its effect on HCC proteome may lead to a better understanding of this unusual cancer, and lead in turn to the discovery of novel therapeutic targets.

References

1. Jemal, A., Bray, F., and Ferlay, J. (2011) Global cancer statistics. *CA Cancer J. Clin.*, **61**, 69–90.
2. Ferlay, J., Shin, H.-R., Bray, F., Forman, D., *et al.* (2010) Estimates of worldwide burden of cancer in 2008: GLOBOCAN 2008. *Int. J. Cancer*, **127**, 2893–2917.
3. Perz, J.F., Armstrong, G.L., Farrington, L.A, Hutin, Y.J.F., *et al.* (2006) The contributions of hepatitis B virus and hepatitis C virus infections to cirrhosis and primary liver cancer worldwide. *J. Hepatol.*, **45**, 529–538.
4. Farazi, P.A. and DePinho, R.A. (2006) Hepatocellular carcinoma pathogenesis: from genes to environment. *Nat. Rev. Cancer*, **6**, 674–687.
5. Hino, O., Kajino, K., Umeda, T., and Arakawa, Y. (2002) Understanding the hypercarcinogenic state in chronic hepatitis: a clue to the prevention of human hepatocellular carcinoma. *J. Gastroenterol.*, **37**, 883–887.
6. Bureau, C., Bernad, J., Chaouche, N., Orfila, C., *et al.* (2001) Nonstructural 3 protein of hepatitis C virus triggers an oxidative burst in human monocytes via activation of NADPH oxidase. *J. Biol. Chem.*, **276**, 23077–23083.
7. Gong, G., Waris, G., Tanveer, R., and Siddiqui, A. (2001) Human hepatitis C virus NS5A protein alters intracellular calcium levels, induces oxidative stress, and activates STAT-3 and NF-kappa B. *Proc. Natl Acad. Sci. USA*, **98**, 9599–9604.
8. Kedderis, G.L. (1996) Biochemical basis of hepatocellular injury. *Toxicol. Pathol.*, **24**, 77–83.
9. Aston, N.S., Watt, N., Morton, I.E., Tanner, M.S., *et al.* (2000) Copper toxicity affects proliferation and viability of human hepatoma cells (HepG2 line). *Hum. Exp. Toxicol.*, **19**, 367–376.
10. Ozcelik, D., Ozaras, R., Gurel, Z., Uzun, H., *et al.* (2003) Copper-mediated oxidative stress in rat liver. *Biol. Trace Elem. Res.*, **96**, 209–215.
11. Jung, M., Drapier, J.C., Weidenbach, H., Renia, L., *et al.* (2000) Effects of hepatocellular iron imbalance on nitric oxide and reactive oxygen intermediates production in a model of sepsis. *J. Hepatol.*, **33**, 387–394.
12. Eckers, A., Reimann, K., and Klotz, L.-O. (2009) Nickel and copper ion-induced stress signaling in human hepatoma cells: analysis of phosphoinositide 3′-kinase/Akt signaling. *Biometals*, **22**, 307–316.
13. Fausto, N., Campbell, J.S., and Riehle, K.J. (2006) Liver regeneration. *Hepatology*, **43**, S45–S53.
14. Forner, A., Reig, M.E., de Lope, C.R., and Bruix, J. (2010) Current strategy for staging and treatment: the BCLC update and future prospects. *Semin. Liver Dis.*, **30**, 61–74.
15. Llovet, J.M., Di Bisceglie, A.M., Bruix, J., Kramer, B.S., *et al.* (2008) Design and endpoints of clinical trials in hepatocellular carcinoma. *J. Natl. Cancer Inst.*, **100**, 698–711.
16. El-Serag, H.B. (2011) Hepatocellular carcinoma. *N. Engl. J. Med.*, **365**, 1118–1127.
17. Lee, J.-S., Kim, J.H., Park, Y.-Y., and Mills, G.B. (2011) Systems biology approaches

to decoding the genome of liver cancer. *Cancer Res. Treat.*, **43**, 205–211.
18. Wilhelm, S., Carter, C., Lynch, M., Lowinger, T., *et al.* (2006) Discovery and development of sorafenib: a multikinase inhibitor for treating cancer. *Nat. Rev. Drug Discovery*, **5**, 835–844.
19. Cheng, A.-L., Guan, Z., Chen, Z., Tsao, C.-J., *et al.* (2012) Efficacy and safety of sorafenib in patients with advanced hepatocellular carcinoma according to baseline status: subset analyses of the phase III Sorafenib Asia-Pacific trial. *Eur. J. Cancer*, **48**, 1452–1465.
20. Raoul, J.-L., Bruix, J., Greten, T.F., Sherman, M., *et al.* (2012) Relationship between baseline hepatic status and outcome, and effect of sorafenib on liver function: SHARP trial subanalyses. *J. Hepatol.*, **56**, 1080–1088.
21. Cheng, A.-L., Kang, Y.-K., Chen, Z., Tsao, C.-J., *et al.* (2009) Efficacy and safety of sorafenib in patients in the Asia-Pacific region with advanced hepatocellular carcinoma: a phase III randomised, double-blind, placebo-controlled trial. *Lancet Oncol.*, **10**, 25–34.
22. Hanahan, D. and Weinberg, R.A. (2011) Hallmarks of cancer: the next generation. *Cell*, **144**, 646–674.
23. Stratton, M.R., Campbell, P.J., and Futreal, P.A. (2009) The cancer genome. *Nature*, **458**, 719–724.
24. Guichard, C., Amaddeo, G., Imbeaud, S., Ladeiro, Y., *et al.* (2012) Integrated analysis of somatic mutations and focal copy-number changes identifies key genes and pathways in hepatocellular carcinoma. *Nat. Genet.*, **44**, 694–698.
25. Nault, J.-C. and Zucman-Rossi, J. (2011) Genetics of hepatobiliary carcinogenesis. *Semin. Liver Dis.*, **31**, 173–187.
26. Midorikawa, Y., Yamamoto, S., Tsuji, S., Kamimura, N., *et al.* (2009) Allelic imbalances and homozygous deletion on 8p23.2 for stepwise progression of hepatocarcinogenesis. *Hepatology*, **49**, 513–522.
27. Chochi, Y., Kawauchi, S., Nakao, M., Furuya, T., *et al.* (2009) A copy number gain of the 6p arm is linked with advanced hepatocellular carcinoma : an array-based comparative genomic hybridization study. *J. Pathol.*, **217**, 677–684.
28. Bréchot, C., Gozuacik, D., Murakami, Y., and Paterlini-Bréchot, P. (2000) Molecular bases for the development of hepatitis B virus (HBV)-related hepatocellular carcinoma (HCC). *Semin. Cancer Biol.*, **10**, 211–231.
29. Fujimoto, A., Totoki, Y., Abe, T., Boroevich, K.A, *et al.* (2012) Whole-genome sequencing of liver cancers identifies etiological influences on mutation patterns and recurrent mutations in chromatin regulators. *Nat. Genet.*, **44**, 760–764.
30. Hanahan, D. and Coussens, L.M. (2012) Accessories to the crime: functions of cells recruited to the tumor microenvironment. *Cancer Cell*, **21**, 309–322.
31. Irmak, M.B., Ince, G., Ozturk, M., and Cetin-Atalay, R. (2003) Acquired tolerance of hepatocellular carcinoma cells to selenium deficiency: a selective survival mechanism? *Cancer Res.*, **63**, 6707–6715.
32. Killela, P.J., Reitman, Z.J., Jiao, Y., Bettegowda, C., *et al.* (2013) TERT promoter mutations occur frequently in gliomas and a subset of tumors derived from cells with low rates of self-renewal. *Proc. Natl Acad. Sci. USA*, **110**, 6021–6026.
33. Honda, K., Sbisà, E., Tullo, A, Papeo, P.A, *et al.* (1998) p53 mutation is a poor prognostic indicator for survival in patients with hepatocellular carcinoma undergoing surgical tumour ablation. *Br. J. Cancer*, **77**, 776–782.
34. Hussain, S.P., Schwank, J., Staib, F., Wang, X.W., *et al.* (2007) TP53 mutations and hepatocellular carcinoma: insights into the etiology and pathogenesis of liver cancer. *Oncogene*, **26**, 2166–2176.
35. Bressac, B., Kew, M., Wands, J., and Ozturk, M. (1991) Selective G to T mutations of p53 gene in hepatocellular carcinoma from southern Africa. *Nature*, **350** 429-431.
36. Hsu, I.C., Metcalf, R.A., Sun, T., Welsh, J.A., *et al.* (1991) Mutational hotspot in the p53 gene in human hepatocellular carcinomas. *Nature*, **350**, 427–428.
37. Ishizaki, Y., Ikeda, S., Fujimori, M., Shimizu, Y., *et al.* (2004) Immunohistochemical analysis and mutational analyses of beta-catenin, Axin family and APC genes in hepatocellular carcinomas. *Int. J. Oncol.*, **24**, 1077–1083.
38. Taniguchi, K., Roberts, L.R., Aderca, I.N., Dong, X., *et al.* (2002) Mutational spectrum of beta-catenin, AXIN1, and AXIN2 in

hepatocellular carcinomas and hepatoblastomas. *Oncogene*, **21**, 4863–4871.
39. Laurent-Puig, P. and Zucman-Rossi, J. (2006) Genetics of hepatocellular tumors. *Oncogene*, **25**, 3778–3786.
40. Ozturk, M. (1999) Genetic aspects of hepatocellular carcinogenesis. *Semin. Liver Dis.*, **19**, 235–242.
41. Higashitsuji, H., Itoh, K., Nagao, T., Dawson, S., *et al.* (2000) Reduced stability of retinoblastoma protein by gankyrin, an oncogenic ankyrin-repeat protein overexpressed in hepatomas. *Nat. Med.*, **6**, 96–99.
42. Zhang, X., Xu, H.J., Murakami, Y., Sachse, R., *et al.* (1994) Deletions of chromosome 13q, mutations in Retinoblastoma 1, and retinoblastoma protein state in human hepatocellular carcinoma. *Cancer Res.*, **54**, 4177–4182.
43. Boyault, S., Rickman, D. S., de Reyniès, A., Balabaud, C., *et al.* (2007) Transcriptome classification of HCC is related to gene alterations and to new therapeutic targets. *Hepatology*, **45**, 42–52.
44. Liew, C.T., Li, H.M., Lo, K.W., Leow, C.K., *et al.* (1999) High frequency of p16INK4A gene alterations in hepatocellular carcinoma. *Oncogene*, **18**, 789–795.
45. Bae, J.-J., Rho, J.-W., Lee, T.-J., Yun, S.-S., *et al.* (2007) Loss of heterozygosity on chromosome 10q23 and mutation of the phosphatase and tensin homolog deleted from chromosome 10 tumor suppressor gene in Korean hepatocellular carcinoma patients. *Oncol. Rep.*, **18**, 1007–1013.
46. Fujiwara, Y., Hoon, D.S., Yamada, T., Umeshita, K., *et al.* (2000) PTEN/MMAC1 mutation and frequent loss of heterozygosity identified in chromosome 10q in a subset of hepatocellular carcinomas. *Jpn. J. Cancer Res.*, **91**, 287–292.
47. Kawamura, N., Nagai, H., Bando, K., Koyama, M., *et al.* (1999) PTEN/MMAC1 mutations in hepatocellular carcinomas: somatic inactivation of both alleles in tumors. *Jpn. J. Cancer Res.*, **90**, 413–418.
48. Yao, Y.J., Ping, X.L., Zhang, H., Chen, F.F., *et al.* (1999) PTEN/MMAC1 mutations in hepatocellular carcinomas. *Oncogene*, **18**, 3181–3185.
49. Buontempo, F., Ersahin, T., Missiroli, S., Senturk, S., *et al.* (2011) Inhibition of Akt signaling in hepatoma cells induces apoptotic cell death independent of Akt activation status. *Invest. New Drugs*, **29**, 1303–1313.
50. Challen, C., Guo, K., Collier, J.D., Cavanagh, D., *et al.* (1992) Infrequent point mutations in codons 12 and 61 of ras oncogenes in human hepatocellular carcinomas. *J. Hepatol.*, **14**, 342–346.
51. Takada, S. and Koike, K. (1989) Activated N-ras gene was found in human hepatoma tissue but only in a small fraction of the tumor cells. *Oncogene*, **4**, 189–193.
52. Tsuda, H., Hirohashi, S., Shimosato, Y., Ino, Y., *et al.* (1989) Low incidence of point mutation of c-Ki-ras and N-ras oncogenes in human hepatocellular carcinoma. *Jpn. J. Cancer Res.*, **80**, 196–199.
53. Mori, T., Nomoto, S., Koshikawa, K., Fujii, T., *et al.* (2005) Decreased expression and frequent allelic inactivation of the RUNX3 gene at 1p36 in human hepatocellular carcinoma. *Liver Int.*, **25**, 380–388.
54. Miyagawa, K., Sakakura, C., Nakashima, S., Yoshikawa, T., *et al.* (2006) Down-regulation of RUNX1, RUNX3 and CBFbeta in hepatocellular carcinomas in an early stage of hepatocarcinogenesis. *Anticancer Res.*, **26**, 3633–3643.
55. Nakanishi, Y., Shiraha, H., Nishina, S., Tanaka, S., *et al.* (2011) Loss of runt-related transcription factor 3 expression leads hepatocellular carcinoma cells to escape apoptosis. *BMC Cancer*, **11**, 3.
56. Li, X., Zhang, Y., Zhang, Y., Qiao, T., *et al.* (2008) RUNX3 inhibits growth of HCC cells and HCC xenografts in mice in combination with adriamycin. *Cancer Biol. Ther.*, **7**, 669–676.
57. Tanaka, S., Shiraha, H., Nakanishi, Y., Nishina, S.-I., *et al.* (2012) Runt-related transcription factor 3 reverses epithelial-mesenchymal transition in hepatocellular carcinoma. *Int. J. Cancer*, **131**, 2537–2546.
58. Imbeaud, S., Ladeiro, Y., and Zucman-Rossi, J. (2010) Identification of novel oncogenes and tumor suppressors in hepatocellular carcinoma. *Semin. Liver Dis.*, **30**, 75–86.
59. Villanueva, A., Newell, P., Chiang, D.Y., Friedman, S.L., *et al.* (2007) Genomics and signaling pathways in hepatocellular carcinoma. *Semin. Liver Dis.*, **1**, 55–76.
60. Woo, H.G., Park, E.S., Lee, J.-S., Lee, Y.-H., *et al.* (2009) Identification of potential driver genes in human liver carcinoma by

genomewide screening. *Cancer Res.*, **69**, 4059–4066.
61. Hoshida, Y., Nijman, S.M.B., Kobayashi, M., Chan, J. A., *et al.* (2009) Integrative transcriptome analysis reveals common molecular subclasses of human hepatocellular carcinoma. *Cancer Res.*, **69** 7385–7392.
62. Deng, Q., Wang, Q., Zong, W.-Y., Zheng, D.-L., *et al.* (2010) E2F8 contributes to human hepatocellular carcinoma via regulating cell proliferation. *Cancer Res.*, **70**, 782–791.
63. Ladu, S., Calvisi, D.F., Conner, E.A., Farina, M., *et al.* (2008) E2F1 inhibits c-Myc-driven apoptosis via PIK3CA/Akt/mTOR and COX-2 in a mouse model of human liver cancer. *Gastroenterology*, **135**, 1322–1332.
64. Chen, X., Cheung, S.T., So, S., Fan, T., *et al.* (2002) Gene expression patterns in human liver cancers. *Mol. Biol. Cell* **13**, 1929–1939.
65. Lee, J.-S., Chu, I.-S., Heo, J., Calvisi, D.F., *et al.* (2004) Classification and prediction of survival in hepatocellular carcinoma by gene expression profiling. *Hepatology*, **40**, 667–676.
66. Kaposi-Novak, P., Libbrecht, L., Woo, H.G., Lee, Y.-H., *et al.* (2009) Central role of c-Myc during malignant conversion in human hepatocarcinogenesis. *Cancer Res.*, **69**, 2775–2782.
67. Shachaf, C.M., Kopelman, A.M., Arvanitis, C., Karlsson, A., *et al.* (2004) MYC inactivation uncovers pluripotent differentiation and tumour dormancy in hepatocellular cancer. *Nature* **431**, 1112–1117.
68. Park, S.-H., Lim, J. S., Lim, S.-Y., Tiwari, I., *et al.* (2011) Hepatitis C virus Core protein stimulates cell growth by down-regulating p16 expression via DNA methylation. *Cancer Lett.*, **310**, 61–68.
69. Lim, J.S., Park, S.-H., and Jang, K.L. (2012) Hepatitis C virus Core protein overcomes stress-induced premature senescence by down-regulating p16 expression via DNA methylation. *Cancer Lett.*, **321**, 154–161.
70. Lee, H., Woo, Y.-J., Kim, S.S., Kim, S.-H., *et al.* (2013) Hepatitis C virus Core protein overcomes all-trans retinoic acid-induced cell growth arrest by inhibiting retinoic acid receptor-β2 expression via DNA methylation. *Cancer Lett.*, **335**, 372–379.
71. Dong, L., Yang, G., Pan, Y., Chen, Y., *et al.* (2011). The oncoprotein p28GANK establishes a positive feedback loop in β-catenin signaling. *Cell Res.* **21**, 1248–1261.
72. Zhang, L., Yang, Z., Ma, A., Qu, Y., *et al.* (2014) Growth arrest and DNA damage 45G down-regulation contributes to janus kinase/signal transducer and activator of transcription 3 activation and cellular senescence evasion in hepatocellular carcinoma. *Hepatology*, **59** (1), 178–189.
73. Sugano, Y., Matsuzaki, K., Tahashi, Y., Furukawa, F., *et al.* (2003) Distortion of autocrine transforming growth factor beta signal accelerates malignant potential by enhancing cell growth as well as PAI-1 and VEGF production in human hepatocellular carcinoma cells. *Oncogene*, **22**, 2309–2321.
74. Yamazaki, K., Masugi, Y., and Sakamoto, M. (2011) Molecular pathogenesis of hepatocellular carcinoma: altering transforming growth factor-β signaling in hepatocarcinogenesis. *Dig. Dis.*, **29**, 284–288.
75. Caja, L., Sancho, P., Bertran, E., and Fabregat, I. (2011) Dissecting the effect of targeting the epidermal growth factor receptor on TGF-β-induced-apoptosis in human hepatocellular carcinoma cells. *J. Hepatol.*, **55**, 351–358.
76. Mazzocca, A., Fransvea, E., Dituri, F., Lupo, L., *et al.* (2010) Down-regulation of connective tissue growth factor by inhibition of transforming growth factor beta blocks the tumor-stroma cross-talk and tumor progression in hepatocellular carcinoma. *Hepatology*, **51**, 523–534.
77. Caja, L., Sancho, P., Bertran, E., Iglesias-Serret, D., *et al.* (2009) Overactivation of the MEK/ERK pathway in liver tumor cells confers resistance to TGF-β-induced cell death through impairing up-regulation of the NADPH oxidase NOX4. *Cancer Res.*, **69**, 7595–7602.
78. Van Zijl, F., Mair, M., Csiszar, A., Schneller, D., *et al.* (2009) Hepatic tumor-stroma crosstalk guides epithelial to mesenchymal transition at the tumor edge. *Oncogene*, **28**, 4022–4033.
79. Bierie, B. and Moses, H.L. (2006) Tumour microenvironment: TGFbeta: the molecular Jekyll and Hyde of cancer. *Nat. Rev. Cancer*, **6**, 506–520.
80. Tovar, V., Alsinet, C., Villanueva, A., Hoshida, Y., *et al.* (2010) IGF activation in

a molecular subclass of hepatocellular carcinoma and pre-clinical efficacy of IGF-1R blockage. *J. Hepatol.*, **52**, 550–559.
81. López-Calderero, I., Sánchez Chávez, E., and García-Carbonero, R. (2010) The insulin-like growth factor pathway as a target for cancer therapy. *Clin. Transl. Oncol.*, **12**, 326–338.
82. Nussbaum, T., Samarin, J., Ehemann, V., Bissinger, M., *et al.* (2008) Autocrine insulin-like growth factor-II stimulation of tumor cell migration is a progression step in human hepatocarcinogenesis. *Hepatology*, **48**, 146–156.
83. Pollak, M. (2008) Insulin and insulin-like growth factor signalling in neoplasia. *Nat. Rev. Cancer*, **8**, 915–928.
84. Cells, H.C., Chen, Y., Boyartchuk, V., and Lewis, B.C. (2009) Differential roles of insulin-like growth factor receptor- and insulin receptor-mediated signaling in the phenotypes of hepatocellular carcinoma cells. *Neoplasia*, **11**, 835–845.
85. Okada, T., Iizuka, N., Yamada-Okabe, H., Mori, N., *et al.* (2003) Gene expression profile linked to p53 status in hepatitis C virus-related hepatocellular carcinoma. *FEBS Lett.*, **555**, 583–590.
86. Liedtke, C. and Trautwein, C. (2012) The role of TNF and Fas dependent signaling in animal models of inflammatory liver injury and liver cancer. *Eur. J. Cell Biol.*, **91**, 582–589.
87. Soung, Y.H., Lee, J.W., Kim, S.Y., Sung, Y.J., *et al.* (2005) Caspase-8 gene is frequently inactivated by the frameshift somatic mutation 1225_1226delTG in hepatocellular carcinomas. *Oncogene*, **24**, 141–147.
88. Hammam, O., Mahmoud, O., Zahran, M., Aly, S., *et al.* (2012) The role of fas/fas ligand system in the pathogenesis of liver cirrhosis and hepatocellular carcinoma. *Hepat. Mon.*, **12**, e6132.
89. Oh, B.-K., Jo Chae, K., Park, C., Kim, K., *et al.* (2003) Telomere shortening and telomerase reactivation in dysplastic nodules of human hepatocarcinogenesis. *J. Hepatol.*, **39**, 786–792.
90. Oh, B.-K., Kim, H., Park, Y.N., Yoo, J.E., *et al.* (2008) High telomerase activity and long telomeres in advanced hepatocellular carcinomas with poor prognosis. *Lab. Invest.*, **88**, 144–152.
91. Ozturk, M., Arslan-Ergul, A., Bagislar, S., Senturk, S., *et al.* (2009) Senescence and immortality in hepatocellular carcinoma. *Cancer Lett.*, **286**, 103–113.
92. Di Micco, R., Fumagalli, M., and d'Adda di Fagagna, F. (2007) Breaking news: high-speed race ends in arrest – how oncogenes induce senescence. *Trends Cell Biol.*, **17**, 529–536.
93. Guo, X., Ma, N., Zhou, F., Zhang, L.I., *et al.* (2009) Up-regulation of hTERT expression by low-dose cisplatin contributes to chemotherapy resistance in human hepatocellular cancer cells. *Oncol. Rep.*, **22**, 549–556.
94. Huang, F.W., Hodis, E., Xu, M.J., Kryukov, G.V, *et al.* (2013) Highly recurrent TERT promoter mutations in human melanoma. *Science*, **339**, 957–959.
95. Toh, S.T., Jin, Y., Liu, L., Wang, J., *et al.* (2013) Deep sequencing of the hepatitis B virus in hepatocellular carcinoma patients reveals enriched integration events, structural alterations and sequence variations. *Carcinogenesis*, **34**, 787–798.
96. Kim, H., Yoo, J.E., Cho, J.Y., Oh, B.-K., *et al.* (2013) Telomere length, TERT and shelterin complex proteins in hepatocellular carcinomas expressing "stemness"-related markers. *J. Hepatol.*, **59**, 746–752.
97. Hu, Y., Shen, Y., Ji, B., Yin, S., *et al.* (2011) Liver-specific gene therapy of hepatocellular carcinoma by targeting human telomerase reverse transcriptase with pegylated immuno-lipopolyplexes. *Eur. J. Pharm. Biopharm.*, **78**, 320–325.
98. Yildiz, G., Arslan-Ergul, A., Bagislar, S., Konu, O., *et al.* (2013) Genome-wide transcriptional reorganization associated with senescence-to-immortality switch during human hepatocellular carcinogenesis. *PLoS One*, **8**, e64016.
99. Tuncbilek, M., Guven, E.B., Onder, T., and Cetin-Atalay, R. (2012) Synthesis of novel 6-(4-substituted piperazine-1-yl)-9-(β-D-ribofuranosyl)purine derivatives, which lead to senescence-induced cell death in liver cancer cells. *J. Med. Chem.*, **55**, 3058–3065.
100. Wu, S.-D., Ma, Y.-S., Fang, Y., Liu, L.-L., *et al.* (2012) Role of the microenvironment in hepatocellular carcinoma development and progression. *Cancer Treat. Rev.*, **38**, 218–225.

101. Roberts, L.R. and Gores, G.J. (2005) Hepatocellular carcinoma: molecular pathways and new therapeutic targets. *Semin. Liver Dis.*, **25**, 212–225.
102. Pang, R. and Poon, R.T.P. (2006) Angiogenesis and antiangiogenic therapy in hepatocellular carcinoma. *Cancer Lett.*, **242**, 151–167.
103. Kaposi-novak, P., Lee, J.-S., Gòmez-quiroz, L., Coulouarn, C., *et al.* (2006) Met-regulated expression signature defines a subset of human hepatocellular carcinomas with poor prognosis and aggressive phenotype. *J. Clin. Invest.*, **116** 1582–1595.
104. Mitsuhashi, N., Shimizu, H., Ohtsuka, M., Wakabayashi, Y., *et al.* (2003) Angiopoietins and Tie-2 expression in angiogenesis and proliferation of human hepatocellular carcinoma. *Hepatology*, **37**, 1105–1113.
105. Ueki, T., Fujimoto, J., Suzuki, T., Yamamoto, H., *et al.* (1997) Expression of hepatocyte growth factor and its receptor c-met proto-oncogene in hepatocellular carcinoma. *Hepatology*, **25**, 862–866.
106. Campbell, J.S., Johnson, M.M., Bauer, R.L., Hudkins, K.L., *et al.* (2007) Targeting stromal cells for the treatment of platelet-derived growth factor C-induced hepatocellular carcinogenesis. *Differentiation*, **75**, 843–852.
107. Li, X.M., Tang, Z.Y., Zhou, G., Lui, Y.K., *et al.* (1998) Significance of vascular endothelial growth factor mRNA expression in invasion and metastasis of hepatocellular carcinoma. *J. Exp. Clin. Cancer Res.*, **17**, 13–17.
108. Shimamura, T., Saito, S., Morita, K., Kitamura, T., *et al.* (2000) Detection of vascular endothelial growth factor and its receptor expression in human hepatocellular carcinoma biopsy specimens. *J. Gastroenterol. Hepatol.*, **15**, 640–646.
109. Poon, R.T.P., Ho, J.W.Y., Tong, C.S.W., Lau, C., *et al.* (2004) Prognostic significance of serum vascular endothelial growth factor and endostatin in patients with hepatocellular carcinoma. *Br. J. Surg.*, **91**, 1354–1360.
110. Ng, I.O., Poon, R.T., Lee, J.M., Fan, S.T., *et al.* (2001) Microvessel density, vascular endothelial growth factor and its receptors Flt-1 and Flk-1/KDR in hepatocellular carcinoma. *Am. J. Clin. Pathol.*, **116**, 838–845.
111. Chiang, D.Y., Villanueva, A., Hoshida, Y., Peix, J., *et al.* (2008) Focal gains of VEGFA and molecular classification of hepatocellular carcinoma. *Cancer Res.*, **68**, 6779–6788.
112. Lian, Z., Liu, J., Wu, M., Wang, H.-Y., *et al.* (2007) Hepatitis B x antigen up-regulates vascular endothelial growth factor receptor 3 in hepatocarcinogenesis. *Hepatology*, **45**, 1390–1399.
113. Imura, S., Miyake, H., Izumi, K., Tashiro, S., *et al.* (2004) Correlation of vascular endothelial cell proliferation with microvessel density and expression of vascular endothelial growth factor and basic fibroblast growth factor in hepatocellular carcinoma. *J. Med. Invest.*, **51**, 202–209.
114. Yoshiji, H., Kuriyama, S., Yoshii, J., Ikenaka, Y., *et al.* (2002) Synergistic effect of basic fibroblast growth factor and vascular endothelial growth factor in murine hepatocellular carcinoma. *Hepatology*, **35**, 834–842.
115. Yan, W., Fu, Y., Tian, D., Liao, J., *et al.* (2009) PI3 kinase/Akt signaling mediates epithelial-mesenchymal transition in hypoxic hepatocellular carcinoma cells. *Biochem. Biophys. Res. Commun.*, **382**, 631–636.
116. Copple, B.L. (2010) Hypoxia stimulates hepatocyte epithelial to mesenchymal transition by hypoxia-inducible factor and transforming growth factor-beta-dependent mechanisms. *Liver Int.*, **30**, 669–682.
117. Liu, L., Zhu, X.-D., Wang, W.-Q., Shen, Y., *et al.* (2010) Activation of beta-catenin by hypoxia in hepatocellular carcinoma contributes to enhanced metastatic potential and poor prognosis. *Clin. Cancer Res.*, **16**, 2740–2750.
118. Maegdefrau, U., Amann, T., Winklmeier, A., Braig, S., *et al.* (2009) Bone morphogenetic protein 4 is induced in hepatocellular carcinoma by hypoxia and promotes tumour progression. *J. Pathol.*, **218**, 520–529.
119. Lee, T.K., Poon, R.T.P., Yuen, A.P., Ling, M.T., *et al.* (2006) Twist overexpression correlates with hepatocellular carcinoma metastasis through induction of epithelial-mesenchymal transition. *Clin. Cancer Res.*, **12**, 5369–5376.
120. Niu, R.F., Zhang, L., Xi, G.M., Wei, X.Y., *et al.* (2007) Up-regulation of Twist induces angiogenesis and correlates with metastasis in hepatocellular carcinoma. *J. Exp. Clin. Cancer Res.*, **26**, 385–394.

121. Van Zijl, F., Zulehner, G., Petz, M., Schneller, D., *et al.* (2009) Epithelial-mesenchymal transition in hepatocellular carcinoma. *Future Oncol.*, **5**, 1169–1179.
122. Yang, M.-H., Chen, C.-L., Chau, G.-Y., Chiou, S.-H., *et al.* (2009) Comprehensive analysis of the independent effect of twist and snail in promoting metastasis of hepatocellular carcinoma. *Hepatology*, **50**, 1464–1474.
123. Kim, T., Veronese, A., Pichiorri, F., Lee, T. J., *et al.* (2011) p53 regulates epithelial-mesenchymal transition through microRNAs targeting ZEB1 and ZEB2. *J. Exp. Med.*, **208**, 875–883.
124. Liu, H., Xu, L., He, H., Zhu, Y., *et al.* (2012) Hepatitis B virus X protein promotes hepatoma cell invasion and metastasis by stabilizing Snail protein. *Cancer Sci.*, **103**, 2072–2081.
125. Ke, A.-W., Shi, G.-M., Zhou, J., Huang, X.-Y., *et al.* (2011) CD151 amplifies signaling by integrin $\alpha 6\beta 1$ to PI3K and induces the epithelial-mesenchymal transition in HCC cells. *Gastroenterology*, **140**, 1629–1641.e15.
126. Kim, J., Hong, S.J., Park, J.Y., Park, J.H., *et al.* (2010) Epithelial-mesenchymal transition gene signature to predict clinical outcome of hepatocellular carcinoma. *Cancer Sci.*, **101**, 1521–1528.
127. Ye, Q.-H., Qin, L.-X., Forgues, M., He, P., *et al.* (2003) Predicting hepatitis B virus-positive metastatic hepatocellular carcinomas using gene expression profiling and supervised machine learning. *Nat. Med.*, **9**, 416–423.
128. Sun, B.-S., Dong, Q.-Z., Ye, Q.-H., Sun, H.-J., *et al.* (2008) Lentiviral-mediated miRNA against osteopontin suppresses tumor growth and metastasis of human hepatocellular carcinoma. *Hepatology*, **48**, 1834–1842.
129. Zhao, J., Dong, L., Lu, B., Wu, G., *et al.* (2008) Down-regulation of osteopontin suppresses growth and metastasis of hepatocellular carcinoma via induction of apoptosis. *Gastroenterology*, **135**, 956–968.
130. Xue, W., Krasnitz, A., Lucito, R., Sordella, R., *et al.* (2008) DLC1 is a chromosome 8p tumor suppressor whose loss promotes hepatocellular carcinoma. *Genes Dev.*, **22**, 1439–1444.
131. Zhou, X., Zimonjic, D.B., Park, S.-W., Yang, X.-Y., *et al.* (2008) DLC1 suppresses distant dissemination of human hepatocellular carcinoma cells in nude mice through reduction of RhoA GTPase activity, actin cytoskeletal disruption and down-regulation of genes involved in metastasis. *Int. J. Oncol.*, **32**, 1285–1291.
132. Sullivan, R. and Graham, C.H. (2007) Hypoxia-driven selection of the metastatic phenotype. *Cancer Metastasis Rev.*, **26**, 319–331.
133. Jang, M., Kim, S.S., and Lee, J. (2013) Cancer cell metabolism: implications for therapeutic targets. *Exp. Mol. Med.*, **45**, e45.
134. Guertin, D.A and Sabatini, D.M. (2007) Defining the role of mTOR in cancer. *Cancer Cell*, **12**, 9–22.
135. Xu, X., Ye, L., Araki, K., and Ahmed, R. (2012) mTOR, linking metabolism and immunity. *Semin. Immunol.*, **24**, 429–435.
136. Elstrom, R.L., Bauer, D.E., Buzzai, M., Karnauskas, R., *et al.* (2004) Akt stimulates aerobic glycolysis in cancer cells. *Cancer Res.*, **64**, 3892–3899.
137. Wise, D.R., DeBerardinis, R.J., Mancuso, A., Sayed, N., *et al.* (2008) Myc regulates a transcriptional program that stimulates mitochondrial glutaminolysis and leads to glutamine addiction. *Proc. Natl Acad. Sci. USA*, **105**, 18782–18787.
138. Liu, L., Ulbrich, J., Müller, J., Wüstefeld, T., *et al.* (2012) Deregulated MYC expression induces dependence upon AMPK-related kinase 5. *Nature*, **483**, 608–612.
139. Zheng, L., Yang, W., Wu, F., Wang, C., *et al.* (2013) Prognostic significance of AMPK activation and therapeutic effects of metformin in hepatocellular carcinoma. *Clin. Cancer Res.*, **19**, 5372–5380.
140. Budanov, A.V. and Karin, M. (2008) p53 target genes sestrin1 and sestrin2 connect genotoxic stress and mTOR signaling. *Cell*, **134**, 451–460.
141. Bensaad, K., Tsuruta, A., Selak, M.A, Vidal, M.N.C., *et al.* (2006) TIGAR, a p53-inducible regulator of glycolysis and apoptosis. *Cell*, **126**, 107–120.
142. Beyoğlu, D., Imbeaud, S., Maurhofer, O., Bioulac-Sage, P., *et al.* (2013) Tissue metabolomics of hepatocellular carcinoma: tumor energy metabolism and the role of transcriptomic classification. *Hepatology*, **58**, 229–238.

143. Budhu, A., Roessler, S., Zhao, X., Yu, Z., *et al.* (2013) Integrated metabolite and gene expression profiles identify lipid biomarkers associated with progression of hepatocellular carcinoma and patient outcomes. *Gastroenterology*, **144**, 1066–1075.
144. Chan, I.S., Guy, C.D., Chen, Y., Lu, J., *et al.* (2012) Paracrine Hedgehog signaling drives metabolic changes in hepatocellular carcinoma. *Cancer Res.*, **72**, 6344–6350.
145. Levine, A.J. and Puzio-Kuter, A.M. (2010) The control of the metabolic switch in cancers by oncogenes and tumor suppressor genes. *Science*, **330**, 1340–1345.
146. Shen, Y.-C., Ou, D.-L., Hsu, C., Lin, K.-L., *et al.* (2013) Activating oxidative phosphorylation by a pyruvate dehydrogenase kinase inhibitor overcomes sorafenib resistance of hepatocellular carcinoma. *Br. J. Cancer*, **108**, 72–81.
147. Schrader, J. and Iredale, J.P. (2011) The inflammatory microenvironment of HCC – the plot becomes complex. *J. Hepatol.*, **54**, 853–855.
148. Nagoshi, S. (2014) Osteopontin: versatile modulator of liver diseases. *Hepatol. Res.*, 44 (1), 22–30.
149. Gotoh, M., Sakamoto, M., Kanetaka, K., Chuuma, M., *et al.* (2002) Overexpression of osteopontin in hepatocellular carcinoma. *Pathol. Int.*, **52**, 19–24.
150. Pan, H.-W., Ou, Y.-H., Peng, S.-Y., Liu, S.-H., *et al.* (2003) Overexpression of osteopontin is associated with intrahepatic metastasis, early recurrence, and poorer prognosis of surgically resected hepatocellular carcinoma. *Cancer*, **98**, 119–127.
151. Wu, J.-C., Sun, B.-S., Ren, N., Ye, Q.-H., *et al.* (2010) Genomic aberrations in hepatocellular carcinoma related to osteopontin expression detected by array-CGH. *J. Cancer Res. Clin. Oncol.*, **136**, 595–601.
152. Zhang, H., Ye, Q.-H., Ren, N., Zhao, L., *et al.* (2006) The prognostic significance of preoperative plasma levels of osteopontin in patients with hepatocellular carcinoma. *J. Cancer Res. Clin. Oncol.*, **132**, 709–717.
153. Xie, H., Song, J., Du, R., Liu, K., *et al.* (2007) Prognostic significance of osteopontin in hepatitis B virus-related hepatocellular carcinoma. *Dig. Liver Dis.*, **39**, 167–172.
154. Kim, J., Ki, S.S., Lee, S.D., Han, C.J., *et al.* (2006) Elevated plasma osteopontin levels in patients with hepatocellular carcinoma. *Am. J. Gastroenterol.*, **101**, 2051–2059.
155. Abu El Makarem, M.A., Abdel-Aleem, A., Ali, A., Saber, R., *et al.* (2011) Diagnostic significance of plasma osteopontin in hepatitis C virus-related hepatocellular carcinoma. *Ann. Hepatol.*, **10**, 296–305.
156. Shang, S., Plymoth, A., Ge, S., Feng, Z., *et al.* (2012) Identification of osteopontin as a novel marker for early hepatocellular carcinoma. *Hepatology*, **55**, 483–490.
157. Nakamoto, Y. and Kaneko, S. (2003) Mechanisms of viral hepatitis induced liver injury. *Curr. Mol. Med.*, **3**, 537–544.
158. Herzer, K., Sprinzl, M.F., and Galle, P.R. (2007) Hepatitis viruses: live and let die. *Liver Int.*, **27**, 293–301.
159. Leonardi, G.C., Candido, S., Cervello, M., Nicolosi, D., *et al.* (2012) The tumor microenvironment in hepatocellular carcinoma (review). *Int. J. Oncol.*, **40**, 1733–1747.
160. Ramaiah, S.K. and Rittling, S. (2008) Pathophysiological role of osteopontin in hepatic inflammation, toxicity, and cancer. *Toxicol. Sci.*, **103**, 4–13.
161. Wai, P.Y. and Kuo, P.C. (2008) Osteopontin: regulation in tumor metastasis. *Cancer Metastasis Rev.*, **27**, 103–118.
162. El-Tanani, M.K., Campbell, F.C., Kurisetty, V., Jin, D., *et al.* (2006) The regulation and role of osteopontin in malignant transformation and cancer. *Cytokine Growth Factor Rev.*, **17**, 463–474.
163. Kovalovich, K., DeAngelis, R.A., Li, W., Furth, E.E., *et al.* (2000) Increased toxin-induced liver injury and fibrosis in interleukin-6-deficient mice. *Hepatology*, **31**, 149–159.
164. Haga, S., Terui, K., Zhang, H.Q., Enosawa, S., *et al.* (2003) Stat3 protects against Fas-induced liver injury by redox-dependent and -independent mechanisms. *J. Clin. Invest.*, **112**, 989–998.
165. Yu, H., Pardoll, D., and Jove, R. (2009) STATs in cancer inflammation and immunity: a leading role for STAT3. *Nat. Rev. Cancer*, **9**, 798–809.
166. Lee, H., Herrmann, A., Deng, J.-H., Kujawski, M., *et al.* (2009) Persistently activated Stat3 maintains constitutive NF-kappaB activity in tumors. *Cancer Cell*, **15**, 283–293.

167. He, G. and Karin, M. (2011) NF-κB and STAT3 – key players in liver inflammation and cancer. *Cell Res.*, **21**, 159–168.
168. Li, W., Liang, X., Kellendonk, C., Poli, V., *et al.* (2002) STAT3 contributes to the mitogenic response of hepatocytes during liver regeneration. *J. Biol. Chem.*, **277**, 28411–28417.
169. Inoue, H., Ogawa, W., Ozaki, M., Haga, S., *et al.* (2004) Role of STAT-3 in regulation of hepatic gluconeogenic genes and carbohydrate metabolism in vivo. *Nat. Med.*, **10**, 168–174.
170. Anson, M., Crain-Denoyelle, A.-M., Baud, V., Chereau, F., *et al.* (2012) Oncogenic β-catenin triggers an inflammatory response that determines the aggressiveness of hepatocellular carcinoma in mice. *J. Clin. Invest.*, **122** 586–599.
171. Liu, Y., Fuchs, J., Li, C., and Lin, J. (2010) IL-6, a risk factor for hepatocellular carcinoma: FLLL32 inhibits IL-6-induced STAT3 phosphorylation in human hepatocellular cancer cells. *Cell Cycle*, **9**, 3423–3427.
172. Porta, C., De Amici, M., Quaglini, S., Paglino, C., *et al.* (2008) Circulating interleukin-6 as a tumor marker for hepatocellular carcinoma. *Ann. Oncol.*, **19**, 353–358.
173. Li, W.-C., Ye, S.-L., Sun, R.-X., Liu, Y.-K., *et al.* (2006) Inhibition of growth and metastasis of human hepatocellular carcinoma by antisense oligonucleotide targeting signal transducer and activator of transcription 3. *Clin. Cancer Res.*, **12**, 7140–7148.
174. Niwa, Y., Kanda, H., Shikauchi, Y., Saiura, A., *et al.* (2005) Methylation silencing of SOCS-3 promotes cell growth and migration by enhancing JAK/STAT and FAK signalings in human hepatocellular carcinoma. *Oncogene*, **24**, 6406–6417.
175. Trikha, M., Corringham, R., Klein, B., and Rossi, J.-F. (2003) Targeted anti-interleukin-6. *Clin. Cancer Res.*, **9**, 4653–4665
176. Yoshikawa, H., Matsubara, K., Qian, G.S., Jackson, P., *et al.* (2001) SOCS-1, a negative regulator of the JAK/STAT pathway, is silenced by methylation in human hepatocellular carcinoma and shows growth-suppression activity. *Nat. Genet.*, **28**, 29–35.
177. Calvisi, D.F., Ladu, S., Gorden, A., Farina, M., *et al.* (2006) Ubiquitous activation of Ras and Jak/Stat pathways in human HCC. *Gastroenterology*, **130**, 1117–1128.
178. Rebouissou, S., Amessou, M., Couchy, G., Poussin, K., *et al.* (2009) Frequent in-frame somatic deletions activate gp130 in inflammatory hepatocellular tumours. *Nature*, **457**, 200–204.
179. Park, E.J., Lee, J.H., Yu, G.-Y., He, G., *et al.* (2010) Dietary and genetic obesity promote liver inflammation and tumorigenesis by enhancing IL-6 and TNF expression. *Cell*, **140**, 197–208.
180. Toffanin, S., Friedman, S.L., and Llovet, J.M. (2010) Obesity, inflammatory signaling, and hepatocellular carcinoma – an enlarging link. *Cancer Cell*, **17**, 115–117.
181. Wheelhouse, N.M., Chan, Y.-S., Gillies, S.E., Caldwell, H., *et al.* (2003) TNF-alpha induced DNA damage in primary murine hepatocytes. *Int. J. Mol. Med.*, **12**, 889–894.
182. Budhu, A., Forgues, M., Ye, Q.-H., Jia, H.-L., *et al.* (2006) Prediction of venous metastases, recurrence, and prognosis in hepatocellular carcinoma based on a unique immune response signature of the liver microenvironment. *Cancer Cell*, **10**, 99–111.
183. Yaguchi, T., Sumimoto, H., Kudo-Saito, C., Tsukamoto, N., *et al.* (2011) The mechanisms of cancer immunoescape and development of overcoming strategies. *Int. J. Hematol.*, **93**, 294–300.
184. Lewis, C.E. and Pollard, J.W. (2006) Distinct role of macrophages in different tumor microenvironments. *Cancer Res.*, **66**, 605–612.
185. Allavena, P., Sica, A., Garlanda, C., and Mantovani, A. (2008) The Yin-Yang of tumor-associated macrophages in neoplastic progression and immune surveillance. *Immunol. Rev.*, **222**, 155–161.
186. Shirakawa, H., Suzuki, H., Shimomura, M., Kojima, M., *et al.* (2009) Glypican-3 expression is correlated with poor prognosis in hepatocellular carcinoma. *Cancer Sci.*, **100**, 1403–1407.
187. Cheng, F., Wang, H.-W., Cuenca, A., Huang, M., *et al.* (2003) A critical role for Stat3 signaling in immune tolerance. *Immunity*, **19**, 425–436.
188. Wu, K., Kryczek, I., Chen, L., Zou, W., *et al.* (2009) Kupffer cell suppression of CD8+ T cells in human hepatocellular carcinoma is

mediated by B7-H1/programmed death-1 interactions. *Cancer Res.*, **69**, 8067–8075.

189. Han, Y., Chen, Z., Yang, Y., Jiang, Z., *et al.* (2014) Human CD14(+) CTLA-4(+) regulatory dendritic cells suppress T-cell response by cytotoxic T-lymphocyte antigen-4-dependent IL-10 and indoleamine-2,3-dioxygenase production in hepatocellular carcinoma. *Hepatology*, **59**, 567–579.
190. Brenndörfer, E.D. and Sällberg, M. (2012) Hepatitis C virus-mediated modulation of cellular immunity. *Arch. Immunol. Ther. Exp. (Warsz.)*, **60**, 315–329.
191. Zhao, F., Hoechst, B., Gamrekelashvili, J., Ormandy, L. A, *et al.* (2012) Human CCR4+ CCR6+ Th17 cells suppress autologous CD8+ T cell responses. *J. Immunol.*, **188**, 6055–6062.
192. Pan, Y., Tan, Y., Wang, M., Zhang, J., *et al.* (2013) Signal regulatory protein α is associated with tumor-polarized macrophages phenotype switch and plays a pivotal role in tumor progression. *Hepatology*, **58**, 680–691.
193. Zhang, Q.-B., Sun, H.-C., Zhang, K.-Z., Jia, Q.-A., *et al.* (2013) Suppression of natural killer cells by sorafenib contributes to prometastatic effects in hepatocellular carcinoma. *PLoS One*, **8**, e55945.
194. Breuhahn, K., Vreden, S., Haddad, R., Beckebaum, S., *et al.* (2004) Molecular profiling of human hepatocellular carcinoma defines mutually exclusive interferon regulation and insulin-like growth factor II overexpression. *Cancer Res.*, **64**, 6058–6064.
195. Nam, S.W., Park, J.Y., Ramasamy, A., Shevade, S., *et al.* (2005) Molecular changes from dysplastic nodule to hepatocellular carcinoma through gene expression profiling. *Hepatology*, **42**, 809–818.
196. Wurmbach, E., Chen, Y., Khitrov, G., Zhang, W., *et al.* (2007) Genome-wide molecular profiles of HCV-induced dysplasia and hepatocellular carcinoma. *Hepatology*, **45**, 938–947.
197. Katoh, H., Ojima, H., Kokubu, A., Saito, S., *et al.* (2007) Genetically distinct and clinically relevant classification of hepatocellular carcinoma: putative therapeutic targets. *Gastroenterology*, **133**, 1475–1486.
198. Midorikawa, Y., Tsutsumi, S., Nishimura, K., Kamimura, N., *et al.* (2004) Distinct chromosomal bias of gene expression signatures in the progression of hepatocellular carcinoma. *Cancer Res.*, **64**, 7263–7270.
199. Poon, T.C.W., Wong, N., Lai, P.B.S., Rattray, M., *et al.* (2006) A tumor progression model for hepatocellular carcinoma: bioinformatic analysis of genomic data. *Gastroenterology*, **131**, 1262–1270.
200. Okabe, H., Satoh, S., Kato, T., Kitahara, O., *et al.* (2001) Genome-wide analysis of gene expression in human hepatocellular carcinomas using cDNA microarray: identification of genes involved in viral carcinogenesis and tumor progression. *Cancer Res.*, **61**, 2129–2137.
201. Iizuka, N., Oka, M., Yamada-Okabe, H., Mori, N., *et al.* (2002) Comparison of gene expression profiles between hepatitis B virus- and hepatitis C virus-infected hepatocellular carcinoma by oligonucleotide microarray data on the basis of a supervised learning method. *Cancer Res.*, **62**, 3939–3944.
202. Kim, B.-Y., Lee, J.-G., Park, S., Ahn, J.-Y., *et al.* (2004) Feature genes of hepatitis B virus-positive hepatocellular carcinoma, established by its molecular discrimination approach using prediction analysis of microarray. *Biochim. Biophys. Acta*, **1739**, 50–61.
203. Kim, S.M., Leem, S.-H., Chu, I.-S., Park, Y.-Y., *et al.* (2012) Sixty-five gene-based risk score classifier predicts overall survival in hepatocellular carcinoma. *Hepatology*, **55**, 1443–1452.
204. Zhang, H., Zhai, Y., Hu, Z., Wu, C., *et al.* (2010) Genome-wide association study identifies 1p36.22 as a new susceptibility locus for hepatocellular carcinoma in chronic hepatitis B virus carriers. *Nat. Genet.*, **42**, 755–758.
205. Miki, D., Ochi, H., Hayes, C.N., Abe, H., *et al.* (2011) Variation in the DEPDC5 locus is associated with progression to hepatocellular carcinoma in chronic hepatitis C virus carriers. *Nat. Genet.*, **43**, 797–800.
206. Kumar, V., Kato, N., Urabe, Y., Takahashi, A., *et al.* (2011) Genome-wide association study identifies a susceptibility locus for HCV-induced hepatocellular carcinoma. *Nat. Genet.*, **43**, 455–458.
207. Mayhew, C.N., Carter, S.L., Fox, S.R., Sexton, C.R., *et al.* (2007) RB loss abrogates cell cycle control and genome integrity to

promote liver tumorigenesis. *Gastroenterology*, **133**, 976–984.

208. Coulouarn, C., Factor, V.M., and Thorgeirsson, S.S. (2008) Transforming growth factor-beta gene expression signature in mouse hepatocytes predicts clinical outcome in human cancer. *Hepatology*, **47**, 2059–2067.
209. Totoki, Y., Tatsuno, K., Yamamoto, S., Arai, Y., *et al.* (2011) High-resolution characterization of a hepatocellular carcinoma genome. *Nat. Genet.*, **43**, 464–469.
210. Li, M., Zhao, H., Zhang, X., Wood, L.D., *et al.* (2011) Inactivating mutations of the chromatin remodeling gene ARID2 in hepatocellular carcinoma. *Nat. Genet.*, **43**, 828–829.
211. Hernandez-Vargas, H., Lambert, M.-P., Le Calvez-Kelm, F., Gouysse, G., *et al.* (2010) Hepatocellular carcinoma displays distinct DNA methylation signatures with potential as clinical predictors. *PLoS One*, **5**, e9749.
212. Tao, Y., Ruan, J., Yeh, S.-H., Lu, X., *et al.* (2011) Rapid growth of a hepatocellular carcinoma and the driving mutations revealed by cell-population genetic analysis of whole-genome data. *Proc. Natl Acad. Sci. USA*, **108**, 12042–12047.
213. Zender, L., Xue, W., Zuber, J., Semighini, C.P., *et al.* (2008) An oncogenomics-based in vivo RNAi screen identifies tumor suppressors in liver cancer. *Cell*, **135**, 852–864.
214. Huang, J., Deng, Q., Wang, Q., Li, K.-Y., *et al.* (2012) Exome sequencing of hepatitis B virus-associated hepatocellular carcinoma. *Nat. Genet.*, **44**, 1117–1121.
215. Iizuka, N., Oka, M., Yamada-Okabe, H., Nishida, M., *et al.* (2003) Oligonucleotide microarray for prediction of early intrahepatic recurrence of hepatocellular carcinoma after curative resection. *Lancet*, **361**, 923–929.
216. Kurokawa, Y., Matoba, R., Takemasa, I., Nagano, H., *et al.* (2004) Molecular-based prediction of early recurrence in hepatocellular carcinoma. *J. Hepatol.*, **41**, 284–291.
217. Wang, S.M., Ooi, L.L.P.J., and Hui, K.M. (2007) Identification and validation of a novel gene signature associated with the recurrence of human hepatocellular carcinoma. *Clin. Cancer Res.*, **13**, 6275–6283.
218. Woo, H.G., Park, E.S., Cheon, J.H., Kim, J.H., *et al.* (2008) Gene expression-based recurrence prediction of hepatitis B virus-related human hepatocellular carcinoma. *Clin. Cancer Res.*, **14**, 2056–2064.
219. Hoshida, Y., Villanueva, A., Kobayashi, M., Peix, J., *et al.* (2008) Gene expression in fixed tissues and outcome in hepatocellular carcinoma. *N. Engl. J. Med.*, **359**, 1995–2004.
220. Lee, J.-S., Heo, J., Libbrecht, L., Chu, I.-S., *et al.* (2006) A novel prognostic subtype of human hepatocellular carcinoma derived from hepatic progenitor cells. *Nat. Med.*, **12**, 410–416.
221. Yamashita, T., Forgues, M., Wang, W., Kim, J. W., *et al.* (2008) EpCAM and alpha-fetoprotein expression defines novel prognostic subtypes of hepatocellular carcinoma. *Cancer Res.*, **68**, 1451–1461.
222. Chai, Z., Kong, J., Zhu, X., Zhang, Y., *et al.* (2013) MicroRNA-26a inhibits angiogenesis by down-regulating VEGFA through the PIK3C2α/Akt/HIF-1α pathway in hepatocellular carcinoma. *PLoS One*, **8** (10) (Epub ahead of print).
223. Yang, X., Liang, L., Zhang, X.-F., Jia, H.-L., *et al.* (2013) MicroRNA-26a suppresses tumor growth and metastasis of human hepatocellular carcinoma by targeting interleukin-6-Stat3 pathway. *Hepatology*, **58**, 158–170.
224. Kota, J., Chivukula, R.R., O'Donnell, K.A, Wentzel, E.A, *et al.* (2009) Therapeutic microRNA delivery suppresses tumorigenesis in a murine liver cancer model. *Cell*, **137**, 1005–1017.
225. Xu, L., Beckebaum, S., Lacob, S., Wu, G., *et al.* (2013) MicroRNA-101 inhibits human hepatocellular carcinoma progression through EZH2 downregulation and increased cytostatic drug sensitivity. *J. Hepatol.*, **60**, 590-598.
226. Xu, Y., Xia, F., Ma, L., Shan, J., *et al.* (2011) MicroRNA-122 sensitizes HCC cancer cells to adriamycin and vincristine through modulating expression of MDR and inducing cell cycle arrest. *Cancer Lett.*, **310**, 160–169.
227. Lu, Y., Yue, X., Cui, Y., Zhang, J., *et al.* (2013) MicroRNA-124 suppresses growth of human hepatocellular carcinoma by targeting STAT3. *Biochem. Biophys. Res. Commun.*, **441** (4), 873–879
228. Hatziapostolou, M., Polytarchou, C., Aggelidou, E., Drakaki, A., *et al.* (2011)

An HNF4α-miRNA inflammatory feedback circuit regulates hepatocellular oncogenesis. *Cell*, **147**, 1233–1247.

229. Lang, Q. and Ling, C. (2012) MiR-124 suppresses cell proliferation in hepatocellular carcinoma by targeting PIK3CA. *Biochem. Biophys. Res. Commun.*, **426**, 247–252.

230. Gu, W., Li, X., and Wang, J. (2013) miR-139 regulates the proliferation and invasion of hepatocellular carcinoma through the WNT/TCF-4 pathway. *Oncol. Rep.*, **31**, 397-404.

231. Yang, H., Fang, F., Chang, R., and Yang, L. (2013) MicroRNA-140-5p suppresses tumor growth and metastasis by targeting transforming growth factor β receptor 1 and fibroblast growth factor 9 in hepatocellular carcinoma. *Hepatology*, **58**, 205–217.

232. Zhang, J.-P., Zeng, C., Xu, L., Gong, J., *et al.* (2013) MicroRNA-148a suppresses the epithelial-mesenchymal transition and metastasis of hepatoma cells by targeting Met/Snail signaling. *Oncogene* (Epub ahead of print); doi: 10.1038/onc.2013.369.

233. Xu, X., Fan, Z., Kang, L., Han, J., *et al.* (2013) Hepatitis B virus X protein represses miRNA-148a to enhance tumorigenesis. *J. Clin. Invest.*, **123**, 630–645.

234. Zhang, Y., Wei, W., Cheng, N., Wang, K., *et al.* (2012) Hepatitis C virus-induced up-regulation of microRNA-155 promotes hepatocarcinogenesis by activating Wnt signaling. *Hepatology*, **56**, 1631–1640.

235. Lin, Y.-H., Liao, C.-J., Huang, Y.-H., Wu, M.-H., *et al.* (2013) Thyroid hormone receptor represses miR-17 expression to enhance tumor metastasis in human hepatoma cells. *Oncogene*, **32**, 4509–4518.

236. Xu, T., Zhu, Y., Xiong, Y., Ge, Y.-Y., *et al.* (2009) MicroRNA-195 suppresses tumorigenicity and regulates G1/S transition of human hepatocellular carcinoma cells. *Hepatology*, **50**, 113–121.

237. Hou, J., Lin, L., Zhou, W., Wang, Z., *et al.* (2011) Identification of miRNomes in human liver and hepatocellular carcinoma reveals miR-199a/b-3p as therapeutic target for hepatocellular carcinoma. *Cancer Cell*, **19**, 232–243.

238. Bao, L., Yan, Y., Xu, C., Ji, W., *et al.* (2013) MicroRNA-21 suppresses PTEN and hSulf-1 expression and promotes hepatocellular carcinoma progression through AKT/ERK pathways. *Cancer Lett.*, **337**, 226–236.

239. Liu, C., Yu, J., Yu, S., Lavker, R.M. (2010) MicroRNA-21 acts as an oncomir through multiple targets in human hepatocellular carcinoma. *J. Hepatol.*, **53**, 98–107.

240. Shih, T.-C., Tien, Y.-J., Wen, C.-J., Yeh, T.-S., *et al.* (2012) MicroRNA-214 downregulation contributes to tumor angiogenesis by inducing secretion of the hepatoma-derived growth factor in human hepatoma. *J. Hepatol.*, **57**, 584–591.

241. Xia, H., Ooi, L.L.P.J., and Hui, K.M. (2013) MicroRNA-216a/217-induced epithelial-mesenchymal transition targets PTEN and SMAD7 to promote drug resistance and recurrence of liver cancer. *Hepatology*, **58**, 629–641.

242. Yuan, Q., Loya, K., Rani, B., Möbus, S., *et al.* (2013) MicroRNA-221 overexpression accelerates hepatocyte proliferation during liver regeneration. *Hepatology*, **57**, 299–310.

243. Wong, Q.W.-L., Ching, A.K.-K., Chan, A.W.-H., Choy, K.-W. (2010) MiR-222 overexpression confers cell migratory advantages in hepatocellular carcinoma through enhancing AKT signaling. *Clin. Cancer Res.*, **16**, 867–875.

244. Ma, D., Tao, X., Gao, F., Fan, C., *et al.* (2012) miR-224 functions as an onco-miRNA in hepatocellular carcinoma cells by activating AKT signaling. *Oncol. Lett.*, **4**, 483–488.

245. Chen, Z., Ma, T., Huang, C., Zhang, L., *et al.* (2013) MiR-27a modulates the MDR1/P-glycoprotein expression by inhibiting FZD7/β-catenin pathway in hepatocellular carcinoma cells. *Cell. Signal.*, **25**, 2693–2701.

246. Chang, Y., Yan, W., He, X., Zhang, L., *et al.* (2012) miR-375 inhibits autophagy and reduces viability of hepatocellular carcinoma cells under hypoxic conditions. *Gastroenterology*, **143**, 177–187.e8.

247. Xiao, F., Zhang, W., Chen, L., Chen, F., *et al.* (2013) MicroRNA-503 inhibits the G1/S transition by downregulating cyclin D3 and E2F3 in hepatocellular carcinoma. *J. Transl. Med.*, **11**, 195.

248. Fornari, F., Milazzo, M., Chieco, P., Negrini, M., *et al.* (2012) In hepatocellular carcinoma miR-519d is up-regulated by p53 and DNA hypomethylation and targets CDKN1A/p21, PTEN, AKT3 and TIMP2. *J. Pathol.*, **227**, 275–285.

249. Zhang, W., Kong, G., Zhang, J., Wang, T., *et al.* (2012) MicroRNA-520b inhibits growth of hepatoma cells by targeting MEKK2 and cyclin D1. *PLoS One*, **7**, e31450.
250. Tao, Z.-H., Wan, J.-L., Zeng, L.-Y., Xie, L., *et al.* (2013) miR-612 suppresses the invasive-metastatic cascade in hepatocellular carcinoma. *J. Exp. Med.*, **210**, 789–803.
251. Hernandez, J.M., Elahi, A, Clark, C.W., Wang, J., *et al.* (2013) miR-675 mediates downregulation of Twist1 and Rb in AFP-secreting hepatocellular carcinoma. *Ann. Surg. Oncol.*, 20 (3), 625–635.
252. Fang, Y., Xue, J.-L., Shen, Q., Chen, J., *et al.* (2012) MicroRNA-7 inhibits tumor growth and metastasis by targeting the phosphoinositide 3-kinase/Akt pathway in hepatocellular carcinoma. *Hepatology*, **55**, 1852–1862.
253. Rodríguez-Paredes, M. and Esteller, M. (2011) Cancer epigenetics reaches mainstream oncology. *Nat. Med.*, **17**, 330–339.
254. Zhai, B. (2008) Reduced expression of E-cadherin/catenin complex in hepatocellular carcinomas. *World J. Gastroenterol.*, **14**, 5665.
255. Xu, B., Di, J., Wang, Z., Han, X., *et al.* (2013) Quantitative analysis of RASSF1A promoter methylation in hepatocellular carcinoma and its prognostic implications. *Biochem. Biophys. Res. Commun.*, **438**, 324–328.
256. Saito, Y., Kanai, Y., Nakagawa, T., Sakamoto, M., *et al.* (2003) Increased protein expression of DNA methyltransferase (DNMT) 1 is significantly correlated with the malignant potential and poor prognosis of human hepatocellular carcinomas. *Int. J. Cancer*, **105**, 527–532.
257. Kong, L.-M., Liao, C.-G., Chen, L., Yang, H.-S., *et al.* (2011) Promoter hypomethylation up-regulates CD147 expression through increasing Sp1 binding and associates with poor prognosis in human hepatocellular carcinoma. *J. Cell. Mol. Med.*, **15**, 1415–1428.
258. Chan, C.-F., Yau, T.-O., Jin, D.-Y., Wong, C.-M., *et al.* (2004) Evaluation of nuclear factor-kappaB, urokinase-type plasminogen activator, and HBx and their clinicopathological significance in hepatocellular carcinoma. *Clin. Cancer Res.*, **10**, 4140–4149.
259. Pogribny, I.P. and Rusyn, I. (2012) Role of epigenetic aberrations in the development and progression of human hepatocellular carcinoma. *Cancer Lett.*, 342 (2), 223–230.
260. Kitamura, Y., Shirahata, A., Sakuraba, K., Goto, T., *et al.* (2011) Aberrant methylation of the Vimentin gene in hepatocellular carcinoma. *Anticancer Res.*, **31**, 1289–1291.
261. Wang, S., Zhu, Y., He, H., Liu, J., *et al.* (2013) Sorafenib suppresses growth and survival of hepatoma cells by accelerating degradation of enhancer of zeste homolog 2. *Cancer Sci.*, **104**, 750–759.
262. Puszyk, W.M., Trinh, T.L, Chapple, S.J., and Liu, C. (2013) Linking metabolism and epigenetic regulation in development of hepatocellular carcinoma. *Lab. Invest.*, **93**, 983–990.
263. Choi, M.S., Shim, Y.-H., Hwa, J.Y., Lee, S.K., *et al.* (2003) Expression of DNA methyltransferases in multistep hepatocarcinogenesis. *Hum. Pathol.*, **34**, 11–17.
264. Oh, B.-K., Kim, H., Park, H.-J., Shim, Y.-H., *et al.* (2007) DNA methyltransferase expression and DNA methylation in human hepatocellular carcinoma and their clinicopathological correlation. *Int. J. Mol. Med.*, **20**, 65–73.
265. Park, H.-J., Yu, E., and Shim, Y.-H. (2006) DNA methyltransferase expression and DNA hypermethylation in human hepatocellular carcinoma. *Cancer Lett.*, **233**, 271–278.
266. Nagai, M., Nakamura, A., Makino, R., and Mitamura, K. (2003) Expression of DNA (5-cytosin)-methyltransferases (DNMTs) in hepatocellular carcinomas. *Hepatol. Res.*, **26**, 186–191.
267. Wu, L.-M., Yang, Z., Zhou, L., Zhang, F., *et al.* (2010) Identification of histone deacetylase 3 as a biomarker for tumor recurrence following liver transplantation in HBV-associated hepatocellular carcinoma. *PLoS One*, **5**, e14460.
268. Quint, K., Agaimy, A., Di Fazio, P., Montalbano, R., *et al.* (2011) Clinical significance of histone deacetylases 1, 2, 3, and 7: HDAC2 is an independent predictor of survival in HCC. *Virchows Arch.*, **459**, 129–139.
269. Zhang, B., Chen, J., Cheng, A.S.L., and Ko, B.C.B. (2014) Depletion of sirtuin 1 (SIRT1) leads to epigenetic modifications of

telomerase (TERT) gene in hepatocellular carcinoma cells. *PLoS One*, **9**, e84931.
270. Chen, J., Zhang, B., Wong, N., Lo, A.W.I., *et al.* (2011) Sirtuin 1 is upregulated in a subset of hepatocellular carcinomas where it is essential for telomere maintenance and tumor cell growth. *Cancer Res.*, **71**, 4138–4149.
271. Chen, J., Chan, A.W.H., To, K.-F., Chen, W., *et al.* (2013) SIRT2 overexpression in hepatocellular carcinoma mediates epithelial to mesenchymal transition by protein kinase B/glycogen synthase kinase-3β/β-catenin signaling. *Hepatology*, **57**, 2287–2298.
272. Zhang, C.Z., Liu, L., Cai, M., Pan, Y., *et al.* (2012) Low SIRT3 expression correlates with poor differentiation and unfavorable prognosis in primary hepatocellular carcinoma. *PLoS One*, **7**, e51703.
273. Zhang, Z.-G. and Qin, C.-Y. (2014) Sirt6 suppresses hepatocellular carcinoma cell growth via inhibiting the extracellular signal-regulated kinase signaling pathway. *Mol. Med. Rep.*, **9**, 882–888.
274. Marquardt, J.U., Fischer, K., Baus, K., Kashyap, A., *et al.* (2013) Sirtuin-6-dependent genetic and epigenetic alterations are associated with poor clinical outcome in hepatocellular carcinoma patients. *Hepatology*, **58**, 1054–1064.
275. Au, S.L.-K., Wong, C.C.-L., Lee, J.M.-F., Fan, D.N.-Y., *et al.* (2012) Enhancer of zeste homolog 2 epigenetically silences multiple tumor suppressor microRNAs to promote liver cancer metastasis. *Hepatology*, **56**, 622–631.
276. Kondo, Y., Shen, L., Suzuki, S., Kurokawa, T., *et al.* (2007) Alterations of DNA methylation and histone modifications contribute to gene silencing in hepatocellular carcinomas. *Hepatol. Res.*, **37**, 974–983.
277. Fan, D.N.-Y., Tsang, F.H.-C., Tam, A.H.-K., Au, S.L.-K., *et al.* (2013) Histone lysine methyltransferase, suppressor of variegation 3-9 homolog 1, promotes hepatocellular carcinoma progression and is negatively regulated by microRNA-125b. *Hepatology*, **57**, 637–647.
278. Hamamoto, R., Furukawa, Y., Morita, M., Iimura, Y., *et al.* (2004) SMYD3 encodes a histone methyltransferase involved in the proliferation of cancer cells. *Nat. Cell Biol.*, **6**, 731–740.
279. Zhou, P., Wu, L.-L., Wu, K.-M., Jiang, W., *et al.* (2013) Overexpression of MMSET is correlation with poor prognosis in hepatocellular carcinoma. *Pathol. Oncol. Res.*, **19**, 303–309.
280. Schagdarsurengin, U., Wilkens, L., Steinemann, D., Flemming, P., *et al.* (2003) Frequent epigenetic inactivation of the RASSF1A gene in hepatocellular carcinoma. *Oncogene*, **22**, 1866–1871.
281. Macheiner, D., Heller, G., Kappel, S., Bichler, C., *et al.* (2006) NORE1B, a candidate tumor suppressor, is epigenetically silenced in human hepatocellular carcinoma. *J. Hepatol.*, **45**, 81–89.
282. Lambert, M.-P., Paliwal, A., Vaissière, T., Chemin, I., *et al.* (2011) Aberrant DNA methylation distinguishes hepatocellular carcinoma associated with HBV and HCV infection and alcohol intake. *J. Hepatol.*, **54**, 705–715.
283. Song, M.-A., Tiirikainen, M., Kwee, S., Okimoto, G., *et al.* (2013) Elucidating the landscape of aberrant DNA methylation in hepatocellular carcinoma. *PLoS One*, **8**, e55761.
284. Nishida, N., Kudo, M., Nagasaka, T., Ikai, I., *et al.* (2012) Characteristic patterns of altered DNA methylation predict emergence of human hepatocellular carcinoma. *Hepatology*, **56**, 994–1003.
285. Shen, J., Wang, S., Zhang, Y.-J., Kappil, M., *et al.* (2012) Genome-wide DNA methylation profiles in hepatocellular carcinoma. *Hepatology*, **55**, 1799–1808.
286. Wang, L., Wang, W.-L., Zhang, Y., Guo, S.-P., *et al.* (2007) Epigenetic and genetic alterations of PTEN in hepatocellular carcinoma. *Hepatol. Res.*, **37**, 389–396.
287. Pogribny, I.P. and James, S.J. (2002) Reduction of p53 gene expression in human primary hepatocellular carcinoma is associated with promoter region methylation without coding region mutation. *Cancer Lett.*, **176**, 169–174.
288. Edamoto, Y., Hara, A., Biernat, W., Terracciano, L., *et al.* (2003) Alterations of RB1, p53 and Wnt pathways in hepatocellular carcinomas associated with hepatitis C, hepatitis B and alcoholic liver cirrhosis. *Int. J. Cancer*, **106**, 334–341.
289. Matsuda, Y., Ichida, T., Matsuzawa, J., Sugimura, K., *et al.* (1999) p16(INK4) is

inactivated by extensive CpG methylation in human hepatocellular carcinoma. *Gastroenterology*, **116**, 394–400.

290. Roncalli, M., Bianchi, P., Bruni, B., Laghi, L., *et al.* (2002) Methylation framework of cell cycle gene inhibitors in cirrhosis and associated hepatocellular carcinoma. *Hepatology*, **36**, 427–432.
291. Shim, Y.-H., Yoon, G.-S., Choi, H.-J., Chung, Y. H., *et al.* (2003) p16 Hypermethylation in the early stage of hepatitis B virus-associated hepatocarcinogenesis. *Cancer Lett.*, **190**, 213–219.
292. Li, X., Hui, A.-M., Sun, L., Hasegawa, K. *et al.* (2004) p16INK4A hypermethylation is associated with hepatitis virus infection, age, and gender in hepatocellular carcinoma. *Clin. Cancer Res.*, **10**, 7484–7489.
293. Csepregi, A., Ebert, M.P., Röcken, C., Schneider-Stock, R., *et al.* (2010) Promoter methylation of CDKN2A and lack of p16 expression characterize patients with hepatocellular carcinoma. *BMC Cancer*, **10**, 317.
294. Wong, I.H., Lo, Y.M., Yeo, W., Lau, W.Y., *et al.* (2000) Frequent p15 promoter methylation in tumor and peripheral blood from hepatocellular carcinoma patients. *Clin. Cancer Res.*, **6**, 3516–3521.
295. Zhu, H., Wu, K., Yan, W., Hu, L., *et al.* (2013) Epigenetic silencing of DACH1 induces loss of transforming growth factor-β1 antiproliferative response in human hepatocellular carcinoma. *Hepatology*, **58**, 2012–2022.
296. He, Y., Cui, Y., Xu, B., Gu, J., *et al.* (2014) Hypermethylation leads to bone morphogenetic protein 6 downregulation in hepatocellular carcinoma. *PLoS One*, **9**, e87994.
297. Okochi, O., Hibi, K., Sakai, M., Inoue, S., *et al.* (2003) Methylation-mediated silencing of SOCS-1 gene in hepatocellular carcinoma derived from cirrhosis. *Clin. Cancer Res.*, **9**, 5295–5298.
298. Miyoshi, H., Fujie, H., Moriya, K., Shintani, Y., *et al.* (2004) Methylation status of suppressor of cytokine signaling-1 gene in hepatocellular carcinoma. *J. Gastroenterol.*, **39**, 563–569.
299. Yuan, Y., Wang, J., Li, J., Wang, L., *et al.* (2006) Frequent epigenetic inactivation of spleen tyrosine kinase gene in human hepatocellular carcinoma. *Clin. Cancer Res.*, **12**, 6687–6695.
300. Zhang, J., Gong, C., Bing, Y., Li, T., *et al.* (2013) Hypermethylation-repressed methionine adenosyltransferase 1A as a potential biomarker for hepatocellular carcinoma. *Hepatol. Res.*, **43**, 374–383.
301. Tomasi, M.L., Li, T.W.H., Li, M., Mato, J.M., *et al.* (2012) Inhibition of human methionine adenosyltransferase 1A transcription by coding region methylation. *J. Cell. Physiol.*, **227**, 1583–1591.
302. Frau, M., Tomasi, M.L., Simile, M.M., Demartis, M.I., *et al.* (2012) Role of transcriptional and posttranscriptional regulation of methionine adenosyltransferases in liver cancer progression. *Hepatology*, **56**, 165–175.
303. Zhang, J., Wang, C., Chen, M., Cao, J., *et al.* (2013) Epigenetic silencing of glutaminase 2 in human liver and colon cancers. *BMC Cancer*, **13**, 601.
304. Zhang, Y., Chen, Y., Ahsan, H., Lunn, R.M., *et al.* (2005) Silencing of glutathione S-transferase P1 by promoter hypermethylation and its relationship to environmental chemical carcinogens in hepatocellular carcinoma. *Cancer Lett.*, **221**, 135–143.
305. Zhong, S., Tang, M.W., Yeo, W., Liu, C., *et al.* (2002) Silencing of GSTP1 gene by CpG island DNA hypermethylation in HBV-associated hepatocellular carcinomas. *Clin. Cancer Res.*, **8**, 1087–1092.
306. Tada, M., Yokosuka, O., Fukai, K., Chiba, T., *et al.* (2005) Hypermethylation of NAD(P)H: quinone oxidoreductase 1 (NQO1) gene in human hepatocellular carcinoma. *J. Hepatol.*, **42**, 511–519.
307. Fernández-Alvarez, A., Llorente-Izquierdo, C., Mayoral, R., Agra, N., *et al.* (2012) Evaluation of epigenetic modulation of cyclooxygenase-2 as a prognostic marker for hepatocellular carcinoma. *Oncogenesis*, **1**, e23.
308. Jiang, L., Chan, J.Y.-W., and Fung, K.-P. (2012) Epigenetic loss of CDH1 correlates with multidrug resistance in human hepatocellular carcinoma cells. *Biochem. Biophys. Res. Commun.*, **422**, 739–744.
309. Takagi, H., Sasaki, S., Suzuki, H., Toyota, M., *et al.* (2008) Frequent epigenetic inactivation of SFRP genes in hepatocellular carcinoma. *J. Gastroenterol.*, **43**, 378–389.

310. Quan, H., Zhou, F., Nie, D., Chen, Q., *et al.* (2013) Hepatitis C virus core protein epigenetically silences SFRP1 and enhances HCC aggressiveness by inducing epithelial-mesenchymal transition. *Oncogene*, 33 (22), 2826–2835.
311. Xie, Q., Chen, L., Shan, X., Shan, X., *et al.* (2013) Epigenetic silencing of SFRP1 and SFRP5 by Hepatitis B Virus X protein enhances hepatoma cell tumorigenicity through wnt signaling pathway. *Int. J. Cancer*, 135 (3), 635–646.
312. Shih, Y.-L., Hsieh, C.-B., Yan, M.-D., Tsao, C.-M., *et al.* (2013) Frequent concomitant epigenetic silencing of SOX1 and secreted frizzled-related proteins (SFRPs) in human hepatocellular carcinoma. *J. Gastroenterol. Hepatol.*, **28**, 551–559.
313. Zhang, X., Yang, Y., Liu, X., Herman, J. G., *et al.* (2013) Epigenetic regulation of the Wnt signaling inhibitor DACT2 in human hepatocellular carcinoma. *Epigenetics*, **8** (4) (Epub ahead of print).
314. Gao, S., Yang, Z., Zheng, Z.-Y., Yao, J., *et al.* (2013) Reduced expression of DACT2 promotes hepatocellular carcinoma progression: involvement of methylation-mediated gene silencing. *World J. Surg. Oncol.*, **11**, 57.
315. Zhang, C., Li, H., Wang, Y., Liu, W., *et al.* (2010) Epigenetic inactivation of the tumor suppressor gene RIZ1 in hepatocellular carcinoma involves both DNA methylation and histone modifications. *J. Hepatol.*, **53**, 889–895.
316. Piao, G.H., Piao, W.H., He, Y., Zhang, H.H., *et al.* (2008) Hyper-methylation of RIZ1 tumor suppressor gene is involved in the early tumorigenesis of hepatocellular carcinoma. *Histol. Histopathol.*, **23**, 1171–1175.
317. Shu, X., Geng, H., Li, L., Ying, J., *et al.* (2011) The epigenetic modifier PRDM5 functions as a tumor suppressor through modulating WNT/β-catenin signaling and is frequently silenced in multiple tumors. *PLoS One*, **6**, e27346.
318. Zhao, R., Wang, N., Huang, H., Ma, W., *et al.* (2014) CHD5, a tumor suppressor is epigenetically silenced in hepatocellular carcinoma. *Liver Int.* (Epub ahead of print).
319. Ogunwobi, O.O., Puszyk, W., Dong, H.-J., and Liu, C. (2013) Epigenetic upregulation of HGF and c-Met drives metastasis in hepatocellular carcinoma. *PLoS One*, **8**, e63765.
320. Okada, H., Kimura, M.T., Tan, D., Fujiwara, K., *et al.* (2005) Frequent trefoil factor 3 (TFF3) overexpression and promoter hypomethylation in mouse and human hepatocellular carcinomas. *Int. J. Oncol.*, **26**, 369–377.
321. Yang, H., Huang, Z.Z., Zeng, Z., Chen, C., *et al.* (2001) Role of promoter methylation in increased methionine adenosyltransferase 2A expression in human liver cancer. *Am. J. Physiol. Gastrointest. Liver Physiol.*, **280**, G184–G190.
322. Gao, X., Qu, J., Chang, X., Lu, Y., *et al.* (2014) Hypomethylation of long interspersed nuclear element-1 promoter is associated with poor outcomes for curative resected hepatocellular carcinoma. *Liver Int.*, **34**, 136–146.

8
Prostate Tissue Stem Cells and Prostate Cancer Progression*

Collene R. Jeter[1] *and Dean G. Tang*[1,2,3]
[1] *University of Texas M.D. Anderson Cancer Center, Department of Molecular Carcinogenesis, Science Park, 1808 Park Road 1C, Smithville, TX 78957, USA*
[2] *The Graduate School for Biological Sciences (GSBS), Program in Molecular Carcinogenesis, Houston, TX 77030, USA*
[3] *The University of Texas M.D. Anderson Cancer Center, Centers for Cancer Epigenetics, Stem Cell and Developmental Biology, RNA Interference and Non-coding RNAs, and Molecular Carcinogenesis, Houston, TX 77030, USA*

*This chapter has previously been published in: Meyers, R.A. (Ed.) Stem Cells, 2013, 978-3-527-32925-0.

Translational Medicine Cancer: Advances in Molecular Biology and Medicine,
First Edition. Edited by Robert A. Meyers.

Keywords

Cancer stem cells (CSCs)
A phenotypically distinct subset of tumor cells with tumor-initiating cell capabilities that can undergo self-renewing cell division to give rise to tumorigenic (CSCs) and non-tumorigenic daughter cells. The term is often used synonymously with tumor-initiating cells.

Castration
The removal of androgens by either surgical removal of the testes (decreasing circulating levels of the androgen hormone testosterone) or by pharmacological agents that antagonize androgen signaling. Castration induces prostate regression due to apoptosis of androgen-dependent prostate cells and inhibition of proliferation. Clinically referred to as androgen-deprivation therapy (ADT) in the treatment of prostate cancer.

Castration-resistant prostate cancer (CRPC)
CRPC occurs when prostate tumors become resistant to castration and can continue to grow and metastasize even in the absence of androgen.

HPCa
Human prostate cancer, often refers to primary human prostate tumor specimens.

Self-renewal
Renewing stem cell or CSC cell division, either maintaining at least one daughter cell with stem cell characteristics (asymmetric) or expanding the stem cell population (symmetric).

Transformation
The multistep process of loss of normal cellular growth constraint resulting from genetic mutations (and epigenetic alterations) in oncogenes and tumor suppressors.

Tumor heterogeneity has been ascribed to the presence of cells with stem cell characteristics in a variety of malignancies, and prostate cancer (PCa) is no exception. Phenotypically distinct subsets of tumor cells that are functionally unique in tumor-regenerating potential, termed *cancer stem cells* (CSCs), have been identified in both mouse and human prostate cancer. Whether, or not, CSCs arise from normal tissue stem cells is an important philosophical question with potential clinical implications. Regardless of the origin of the tumor-initiating cell, prostatic CSCs share certain attributes with normal prostate stem cells, including self-renewal, multi-lineage potential, and castration-resistance. In this chapter, cellular hierarchies in both normal prostate and prostate tumors are reviewed, and the cell-of-origin from which prostate tumors may arise is further considered.

1 Introduction

1.1 Conceptual Overview: Stem Cells and Cancer Stem Cells

Somatic tissue stem cells in the adult are critical for cellular homeostasis and replacing damaged cells. Defects in the regulation of the number and/or nature of stem cell division events can disrupt homeostasis, leading to hyperproliferative diseases, including cancer. In a classical hierarchical paradigm, relatively rare and quiescent stem cells undergo asymmetric self-renewing cell divisions to generate the more rapidly proliferating lineage-restricted progenitor cells which, in turn, divide to give rise to the terminally differentiated functional cells that comprise the preponderance of the tissue. The asymmetric nature of these specialized, renewing cell divisions provides the multiple cell lineages required to "recapitulate" the tissue concurrently with maintenance of the stem cell population required for a continued tissue-regenerating activity over the life of the organism. Thus, normal adult stem cells are fundamental to somatic tissue complexity and functional regeneration.

Solid tumors, like the normal tissues from which they arise, are also highly complex. In mimicking normal tissues and organs, a spectrum of cancer cells comprise intratumoral phenotypes that range from primitive stem-like cells to those displaying varying degrees of differentiation. Cell fractionation using cell-surface markers in common with normal stem cells to enrich relatively rare tumorigenic subsets of cells has provided tantalizing evidence that a special subset of tumor cells, termed cancer stem cells (CSCs) may be uniquely capable of tumor regeneration and propagation in a variety of solid tumors, including breast [1, 2], brain [3, 4], colon [5, 6], and prostate [7–9], among many others. However, in order for this to reflect a true hierarchical organization of tumor cells, the CSCs must be able to self-renew and, of paramount importance, give rise to nontumorigenic constituent components. Therefore, full tumor recapitulation following the transplantation of a single candidate CSC, including both tumorigenic and nontumorigenic cancer cells in a similar ratio as the parental tumor, is an ideal biological characterization assay evincing CSC properties.

Tumorigenesis requires the accumulation of numerous genetic mutations (and epigenetic alterations) in oncogenes and tumor suppressor genes in order for the cell undergoing oncogenic transformation to sufficiently manifest the aptly described "hallmarks of cancer" [10, 11]. Conceptually, the prolonged life of the stem cell represents a greater opportunity for the accrual of sufficient molecular alterations to serve as a tumor-initiating cell leading to cancerous properties and phenotypes. In support, transgenic mouse models expressing constitutively active oncogenes or inactivated tumor suppressor genes specifically in somatic tissue stem cells have provided solid evidence that these cells can serve as the cell-of-origin for a variety of cancer types (for a review, see Ref. [12]). Among the best exemplified are transgenic mouse studies in which molecular transformation specifically in renewing intestinal crypt stem cells efficiently promotes intestinal carcinoma, shown by lineage-tracing to be clonally derived and recapitulating tumor heterogeneity [13, 14].

Self-renewal is essential to normal stem cell biology and, in a similar manner, is equally crucial to the CSC archetype. CSCs derived from normal stem cells would be intrinsically imbued with renewal characteristics. However, the derivation of CSCs does not require origination in a normal stem cell; equally possible is the oncogenic transformation of transit-amplifying (i.e., progenitor) cells imparting renewal as part of the neoplastic process. Furthermore, as tumor cells are susceptible to genetic instability, additional molecular alterations may generate tumor-propagating CSCs distinct from the cell of origin in a process referred to as *clonal evolution* [15, 16]. Therefore, the presence of renewing CSCs *per se* is not necessarily indicative of the cell of origin from which the tumor was derived; nevertheless, such cells represent clinically relevant tumor cell subsets driving disease progression. An important aside is that recent observations that unselected human melanoma cell transplantation into certain strains of immunocompromised mice, can exhibit broad efficiency in transplantation with up to a 1 : 4 tumor (re)-initiating cell frequency [17], in an apparently marker independent fashion [18], although this has been called into question [19]. In some tumor types and in a cell-of-origin and oncogenic mutation-dependent fashion, a heterogeneous CSC-based model, clonal evolution or a more homogeneous (or even plastic) model may better describe the cellular architecture comprising the tumor. Regardless, in this chapter evidence will be presented that renewing prostatic stem cells exist in the normal mouse and human prostate, and further, that such cells are susceptible to oncogenic transformation leading to heterogeneous tumors reminiscent of clinical specimens. Thus, regardless of the origin or frequency of such stem-like cells, the definitive identification of a prospectively purified normal stem cell or CSC ideally provides evidence that the cell possesses the archetypical property of self-renewal, as may be evinced by serial sphere formation and/or serial tissue or tumor development.

CSCs and their nontumorigenic daughter cells are not the only cellular components in a tumor, which consists also of endothelial cells, fibroblastic stromal cells, and a variety of immune cells. The stromal cells, in particular, may support the cell nonautonomous renewal, survival, and/or proliferation of subsets of cancer cells, a phenomenon that is especially important in prostate cancer (PCa) (see below).

1.2
Prostate Histology, Cell Lineages, and Morphogenesis

The prostate gland is an androgen-dependent, male secretory gland, the function of which is to generate seminal fluids. In both the mouse and human, the prostate glands exhibit a tubular architecture composed of an epithelial parenchyma surrounded by fibroblastic and contractile smooth muscle mesenchymal cells [20]. The two principal epithelial cell types, distinguished by the expression of basal versus luminal keratins (K) and differentiation markers, such as androgen receptor (AR), NK3 homeobox 1 (NKX3.1), and prostate-specific antigen (PSA), are organized in layers (Fig. 1). The underlying basal cells ($K5/14^+$, $p63^+$, $CD44^+$, integrin $\alpha6^{high}$, $AR^{-/low}$) are in contact with the basement membrane, while the columnar, exocrine luminal cells ($K8/18^+$, $NKX3.1^+$, AR^{high}, PSA^+) above face the lumen [8, 20–25]. Interspersed between these two cell types are other cells that are present in lower frequencies, including the post-mitotic neuroendocrine cells (e.g., synaptophysin$^+$, chromogranin A^+, AR^-) and intermediate transit-amplifying cells that coexpress basal and luminal markers (e.g., $K5^+/K8^+$ double-positive and/or $K19^+$) [26–28]

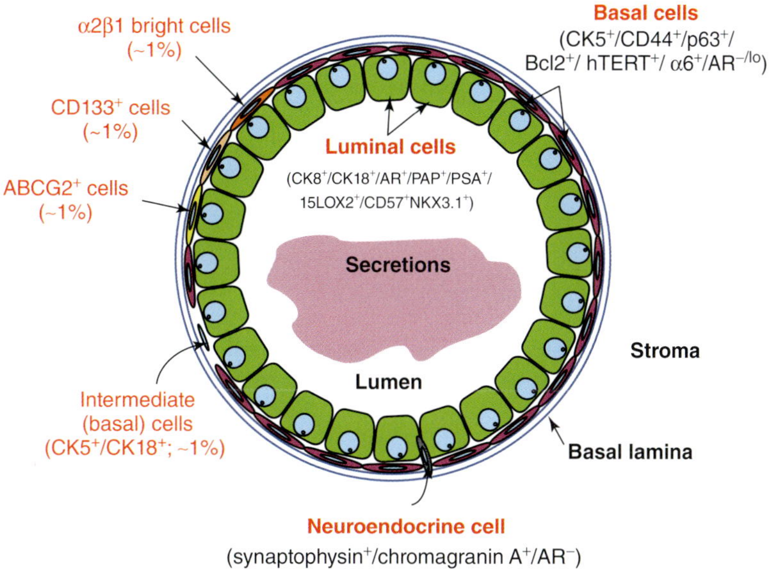

Fig. 1 A schematic of a human prostatic tubule and its cellular composition. Basal cells and luminal cells are largely discriminated on the basis of cytokeratin staining, with keratin 5 (CK5) and 14 (CK14) being highly expressed by basal cells and keratin 8 (CK8) and 18 (CK18) being highly expressed by luminal cells. Other luminal-specific markers include the arachidonic acid-metabolizing enzyme 15-lipoxygenase-2 (15Lox-2) and the homeobox transcription factor NKX3.1. See text for more details.

(Fig. 1). In studies of human prostatic glands, $CD44^+$ basal cells, when cocultured with stromal cells, appeared to be able to generate PSA-producing luminal cells, which suggests that some basal cells might behave like stem cells [29]. Indeed, several minor subsets (generally ~1%) of epithelial cells in the basal layer, which express high levels of the cell surface-proteins integrin $\alpha_2\beta_1$ [30], CD133 [31], or ABCG2 [32], have been reported to possess stem cell properties (Fig. 1). Of particular interest, several distinct multipotent prostatic stem cells have been identified in mouse prostate that are capable of the *functional regeneration* of prostate tissue (see below).

Prostate morphogenesis is an androgen-dependent process. The prostate gland, which initially is an ambisexual primordial organ, develops from the endoderm-derived urogenital sinus epithelium and surrounding embryonic mesenchyme [20]. In response to androgen (at 10 weeks' gestation in humans, and perinatally in mice), the urogenital sinus epithelium begins the process of ductal budding into the surrounding mesenchyme [20, 33]. The prostatic ducts, which initially are solid, begin a process of branching morphogenesis, canalization, and cell-fate determination of basal and luminal secretory cells organized in layers and comprising a pseudostratified epithelium [20, 33]. The mature prostate consists of ducts that branch out from the urethra and terminate into acini. In both human and mouse, prostate epithelial cells and stromal cells express AR and are responsive to androgen. Of particular importance, tissue recombination experiments (see below) using AR^{null} rodent (versus wild-type) epithelial and mesenchymal cells have shown that androgen-stimulated production of AR^{+} *mesenchymal* cell-derived inductive paracrine signals are essential to drive early prostate gland development (for a review, see Ref. [20]).

The prostate gland continues to depend on androgen for its maturation and cellular homeostasis. In the adult rodent, hormone ablation by either surgical or chemical castration induces involution of the gland, predominantly due to apoptosis of androgen-dependent secretory, luminal cells. However, androgen-independent (but androgen-responsive) prostate cells survive as dormant cells in the residual gland, predominantly proximal to the urethra. Subsequent androgen replacement promotes functional regeneration, a process that can be repeated seemingly indefinitely, and this remarkably resilient regenerative capacity is among the early evidence that the prostate harbors renewing somatic tissue stem cells [34, 35]. Extrapolating from these observations, the prostate stem cells must be naturally refractory to androgen deprivation (exhibiting androgen-independent cell survival), although the gland remains responsive (either directly or indirectly) to androgen stimulation.

1.3
Prostate Cancer Pathology

Adenocarcinoma of the prostate is the most commonly diagnosed cancer in men and is the second-leading cause of cancer-related deaths, with 217 730 new cases and 32 050 resultant mortalities reported in the United States in 2010 (The American Cancer Society). Pathological grading of prostate carcinoma is performed routinely according to the Gleason system, which reflects the degree of differentiation of tumor cells; a higher Gleason score (maximum 10) indicates a poorly differentiated tumor and a worse prognosis. Unlike other epithelial tumors, such as breast cancer, histopathologically distinct subtypes of prostate cancer that correspond to a particular prognosis or optimum course of treatment are generally lacking. The predominant diagnosis is acinar adenocarcinomas, with other classifications occurring very rarely, such as neuroendocrine prostate tumors (for a review, see Ref. [21]).

Prostate cancer is also described as multifocal, such that the diseased prostate can contain numerous independent histological foci that are genetically distinct [21]. These observations, taken together with the high frequency of PCa occurrence, suggest that the prostate is particularly susceptible to cancer, and it is tempting to surmise that stem cells in the prostate contribute to the apparent ease with which this gland is subject to oncogenic transformation. One point of contention among PCa experts is how the cell-of-origin relates to a predominance of luminal cells within prostate tumors, often with a complete loss of cells expressing basal markers (e.g., K5, K14, and p63), a diagnostic characteristic of prostate carcinoma [21]. Hypotheses include a luminal stem or progenitor cell-of-origin for PCa or, alternatively, a transformed basal stem cell that gives rise to more differentiated (presumably nontumorigenic) luminal daughter cells [22, 25, 36, 37].

Prostate cancer initially responds very well to conventional therapeutics, including surgical resection (radical prostatectomy), radiation therapy, and androgen-deprivation therapy (ADT). However, subsets of cells in advanced prostate cancer survive hormone ablation, heralding a lethal stage of the disease referred to as castration-resistant prostate cancer (CRPC). Analogous to the stem cells in the normal prostate, stem-like cells in prostate tumors are hypothesized to exhibit androgen independence and account for the recurrence of prostate cancer in castration-resistant disease [22, 25, 38]. A further understanding of the parallels between normal stem cells and CSCs in the prostate may thus provide novel therapeutic avenues for the treatment of CRPC.

2 Mouse Prostate Stem and Progenitor Cells

2.1 Normal Prostate Tissue Reconstituting Cells

Contrasting with the globular human prostate, the mouse prostate is compartmentalized into several lobes, each of which maintains a tubular architecture and adjoins the urethra to release prostate secretions (Fig. 2; [39]). Although the basal and luminal histological characteristics of the prostate are generally conserved between mouse and human prostate epithelial cells, until relatively recently only corollary evidence suggested that mouse prostate stem cells even existed. Over the past five years or so, investigations have revealed multiple populations of mouse prostate cells with stem cell characteristics (see below), although the precise cell lineage hierarchy of the prostate remains unknown. To begin to address this deficiency, ideally it should be possible to discern prostatic stem cells from progenitor cells, including analyses of renewal and *the* differentiation potential of a candidate stem (or progenitor) cell. The scientific literature regarding the evidence for mouse prostatic stem cells is reviewed in the following subsection, and the possible interrelationships between these cells are considered.

One critical question is whether the basal and luminal layers each possess their own stem cells, or do cells in one layer provide progenitors for both? The results of involution–regeneration assays in the mouse prostate have suggested that regenerative prostatic cells could potentially exist in both basal and luminal layers. E. Wilson and colleagues meticulously performed 5′-bromo-deoxyuridine (BrdU) cell labeling prior to chase during 16 cycles of mouse prostate involution (castration/androgen deprivation) and regeneration (androgen supplementation) to reveal slow-cycling, label-retaining cells (LRCs) in *both* basal and luminal cell subsets in prostatic tubules proximal to the urethra (Fig. 2b) [35], a region that is rich in smooth muscle cell-derived quiescence-promoting transforming growth factor-beta (TGF-β) [40]. Of note, the region proximal to the urethra (Fig. 2b), which is proposed to be the prostate stem cell "niche," is enriched in quiescent (G_0) cells expressing the mouse stem cell marker Sca-1 (stem cell antigen-1; [41, 42]). Slow-cycling LRCs have been evinced to possess stem cell properties in a variety of tissues, including the oral mucosa [43], skin [43, 44], mammary [45], and cornea [46], among others. Considering that LRCs exist in both the basal and luminal layers of the mouse prostate following numerous rounds of involution–regeneration, these findings beg the question: Which of these cellular compartments actually possess stem cell activities, or could both?

Tissue regeneration capabilities are of paramount importance when testing prospectively purified stem cells. Considering that prostate epithelial–stromal cell signaling is critical to prostate development, the establishment of tissue recombination assays has proven invaluable in the analysis of the reconstitution abilities

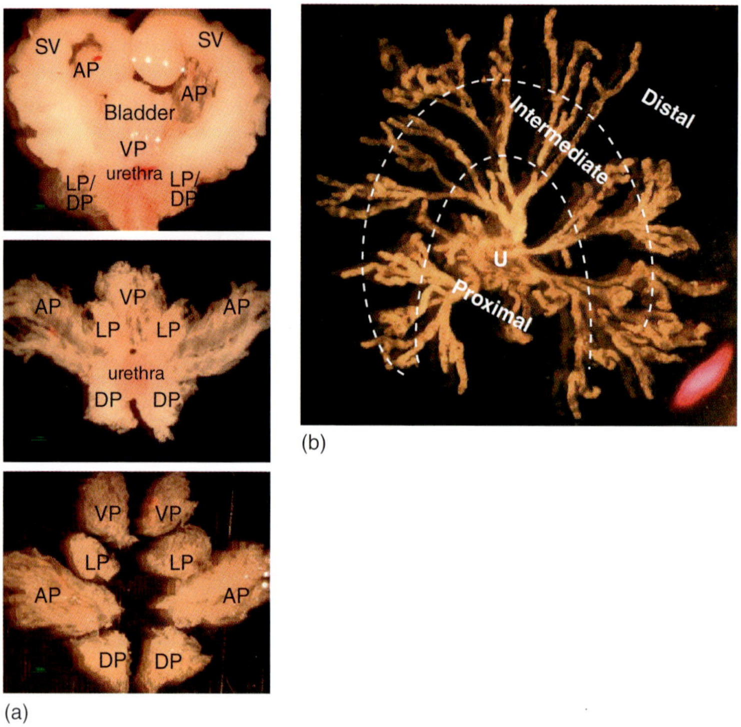

Fig. 2 Anatomy of the mouse prostate gland. (a) The gland is composed of two lobes each of: anterior prostate (AP), ventral prostate (VP), lateral prostate (LP) and dorsal prostate (DP), each surrounding the urethra; (b) A branched microdissected ventral prostate lobe showing the tubular architecture. The region proximal to the urethra (U) has been implicated as harboring stem-like cells, for example, including slow-cycling label-retaining cells. Images kindly provided by M. Suraneni.

of prostate epithelial cells [47–49]. In brief, dissociated prostate epithelial cells are combined with embryonic rat (or mouse) urogenital sinus mesenchyme (UGSM) and transplanted under the kidney capsule in immunocompromised recipient mice. Stromal–epithelial cell signaling induces the growth and differentiation of epithelial cells, permitting an analysis of the proliferative potential and multipotency of candidate prostatic stem cells.

Prostate tissue regeneration of prospectively purified normal mouse prostate epithelial stem cells via distinct cell-surface markers has provided compelling evidence that such cells do, in fact, exist in the prostate. The first such analysis of dissociated subsets of prostate epithelial cells were Sca-1^+ mouse prostate cells. Cell labeling experiments using mixed green fluorescence protein (GFP) tagged (and non-tagged) Sca-1^+ cells demonstrated clonal origins for single prostatic tubules, while immunohistochemical analysis revealed a bipotency of the grafted cells, as the outgrowths contained both $p63^+$ basal and AR^{high} luminal cells [42]. However, Sca-1^+ prostate cells are heterogeneous, including stem and non-stem cells, such that this marker alone could only enrich for cells with reconstitution capabilities.

Combinatorial marker profiling by O. Witte's group has further discriminated and functionally characterized lineage (stromal, endothelial, and hematopoietic) -negative, Sca-1^{+} cells coexpressing CD49f (integrin α_6) as candidate prostate stem cells. These Lin^{-}Sca-1^{+}CD49f^{+} LSC (luminal stem cell) mouse prostate cells constituted approximately 1% of the total prostate epithelial cells, localized predominantly in the basal layer proximal to the urethra and overlapped (ca. 70%) with the basal marker K5. These cells also exhibited an enhanced clonal growth and serial (i.e., renewing) sphere-forming potential *in vitro*, as well as an ability to regenerate prostatic tubules when transplanted as tissue recombinants under the renal capsule [22, 50]. Of particular importance, these tubules were shown to be of clonal origin and to contain both basal (e.g., K5^{+}, p63^{+}) and luminal (K8^{+}) cells. This provided compelling evidence of the existence (and identity) of bipotent normal mouse prostatic basal stem cells.

The defining stem cell functional assay is *single-cell tissue reconstitution*, a technically challenging feat recently accomplished by W. Gao and colleagues to identify a Lin^{-}Sca – 1^{+}CD133^{+}CD44^{+}CD117^{+} multipotent prostate stem cell in the basal compartment [51]. Immunostaining showed that CD117^{+} (c-kit, stem cell factor receptor) was enriched proximal to the urethra, predominantly in the basal layer (2.65% of total basal cells); however, it was noteworthy that about 1% of luminal cells also expressed CD117. The fractionated CD117^{+} cells were shown to possess two important stem cell characteristics, namely multipotency and self-renewal. CD117^{+} cells, when combined with rat UGSM and transplanted under the kidney capsule, regenerated prostate-like structures that comprised cells of the basal (e.g., K14^{+}), luminal (e.g., K18^{+}) and, for the first time, neuroendocrine (e.g., synaptophysin^{+}) lineages, thereby demonstrating tripotency [51]. Renewal was demonstrated via serial transplantation assays at low cell numbers. The single-cell reconstitution assays were focused on CD117-expressing basal cells (Lin^{-}Sca-1^{+}CD133^{+}CD44^{+}CD117^{+}), and remarkably achieved engraftment in 14 of 97 single-cell transplantations (along with rat UGSM). The single-cell-derived outgrowths were also shown to be composed of all three principal prostate cell types (basal, luminal, and neuroendocrine), with luminal cells expressing the critical differentiation-related homeobox protein, Nkx3-1, and the terminal differentiation marker, probasin (Pb); this suggested that the reconstituted tubules possessed functional secretory activities [51].

Some elegant studies conducted by M. Shen and colleagues, using a lineage-tracing methodology, revealed a castration-resistant Nkx3.1 expressing (CARN) *luminal* cell that also demonstrated multipotency and tissue reconstitution capacity [52]. Nkx3.1 is an androgen-regulated homeobox transcription factor that is critical to luminal cell differentiation and is broadly expressed in luminal cells. Castration dramatically reduces the expression level of this luminal cell marker, in part due to apoptosis of the luminal cells. However, in the castrated rodent prostate a low percentage (0.7%) of Nkx3.1-expressing cells persists as quiescent CARNs that are strictly luminal (positive for K18 and negative for the basal marker p63 and the neuroendocrine marker synaptophysin) and further are CD117^{-}; this suggests that this subset of cells is non-overlapping with the reported CD117^{+} basal stem cell subset [52]. By using transgenic mice with a knock-in allele placing tamoxifen-inducible Cre downstream of the Nkx3.1 promotor crossed with R26R-YFP (yellow fluorescent protein) Cre-reporter mice, it was shown that CARNs possessed proliferative potential and bipotency, and were able to give rise to additional luminal cells (e.g., $K18^+$) and, at a low percentage, also basal (e.g., $K5^+$, $p63^+$) cells. The maintenance of a constant CARN percentage following multiple rounds of involution–regeneration, together with BrdU uptake, was taken as evidence of the long-term self-renewal potential of these cells. Furthermore, single-cell, tissue-recombinant reconstitution assays gave rise to prostatic ductal structures that were composed entirely of YFP^+ cells expressing predominantly luminal (e.g., $K18^+$, AR^+) molecular markers, along with a lower frequency of basal (e.g., $K5^+$, $p63^+$) and neuroendocrine (synaptophysin) markers; these critical findings corroborated the claims that CARNs are prostatic stem cells [52].

The interrelationships among these prostatic cell lineages are currently unclear. The $Lin^-Sca\text{-}1^+CD49f^+$ mouse prostate cells were later further fractionated via the epithelial cell adhesion molecule (EpCAM) family member tumor-associated calcium signal transducer 2 (Trop-2), to reveal serial sphere-forming and tripotent cells in the $Trop\text{-}2^{high}$ subset [53], and suggesting that the true stem cells would be better defined as $Lin^-Sca\text{-}1^+CD49f^+Trop\text{-}2^{high}$. The potential overlap between $Lin^-Sca\text{-}1^+CD49f^+Trop\text{-}2^{high}$ and $Lin^-Sca\text{-}1^+CD133^+CD44^+CD117^+$ cells has not been established experimentally. Indeed, it is possible that these two Sca-1 expressing cell populations identify the overlapping, basally localized prostate stem cells. Alternatively, as the single-cell reconstitution of prostatic glands has not been demonstrated for $Lin^-Sca\text{-}1^+CD49f^+Trop\text{-}2^{high}$ cells, the $Lin^-Sca\text{-}1^+CD133^+CD44^+CD117^+$ cells might represent a more primitive stem cell population. Further marker characterization of these basal stem cell subsets, coupled with single-cell functional assays, is required to address whether or not these two prostate stem cells are distinct, or are overlapping with each other.

Whether luminal stem/progenitor cells arise from basal stem cells, or whether luminal stem cells exist independently, is also unclear. Of note, the attenuation of CD117 function with a blocking antibody inhibited prostate regeneration, decreased proliferation, and also increased the ratio of basal to luminal cells, suggesting that CD117 may function in basal cell differentiation into luminal cells [51]. However, a functional role for CD117 in luminal cell differentiation from luminal stem/progenitor cells, or some cell non-autonomous effect of CD117 inhibition (e.g., effects on stromal cells), cannot be eliminated on the basis of these findings. With regards to the candidate LSCs, CARNs could be either true prostate stem cells or, alternatively, a facultative stem cell that only displays full multipotency under castrated conditions. In the intact prostate, it is conceivable that Nkx3.1-expressing subsets could function as either LSCs or luminal progenitor cells. The discrimination between these two hypothetical scenarios could be accomplished by combining Nkx3.1 with other molecular markers to determine specific subsets of Nkx3.1-expressing luminal cells that possess stem cell activity in the non-castrated condition. In summary, as all three candidate stem cells have been shown to possess tripotency and self-renewal, a definitive establishment of lineage hierarchy is not possible until further experiments have been performed to clarify these outstanding questions.

2.2 Tumor Initiation and CSCs in Basal and Luminal Cell Compartments

The etiology of PCa is unknown. Although prostate tumors are comprised predominantly of luminal-like cancer cells, the origin of these cells could be either transformed luminal stem/progenitor cells or, alternatively, a transformed basal stem cell giving rise to luminal (and neuroendocrine) daughter cells. Strangely, experimental evidence is available to support either premise, such that neither has been definitively eliminated. The available evidence will be reviewed, and the implications discussed, at this point.

A histopathological evaluation of transgenic mouse prostate cancer models has provided the opportunity to evaluate the cell-of-origin for PCa. A homozygous deletion of floxed *Pten* (phosphatase and tensin homolog deleted on chromosome 10) via a Pb promoter-driven Cre recombinase (Pb-Cre4) broadly expressed in mouse prostate epithelial cells recapitulated the clinical manifestations of human prostate cancer, including an initially androgen-dependent carcinoma that progressed to metastatic and castration-resistant disease. In these *Pten*-null prostate tumors, the $Lin^{-}Sca\text{-}1^{+}CD49f^{high}$ prostate stem cells are significantly increased concordantly with an overall increase in $K5^{+}$ and $p63^{+}$ basal cells [54, 55]. Furthermore, purified $Lin^{-}Sca\text{-}1^{+}CD49f^{high}$ cells were found to be tumorigenic upon transplantation into immunocompromised mice [55]. Although this clearly demonstrated that $Lin^{-}Sca\text{-}1^{+}CD49f^{high}$ basal stem cells could serve as tumor-initiating cells, the general expansion of the basal population was not consistent with human PCa luminal phenotypes. In comparison, knocking out *Pten* via a luminal cell-specific, PSA-driven Cre-recombinase gave rise to hyperplastic cells expressing luminal markers (K8) and progenitor markers (K19, Sca-1, and Trop-2), but not basal markers (p63). This suggested that tumor-initiating cells in this model would arise in luminal progenitor cells [56].

Other transgenic mouse tumor models are also consistent with a luminal progenitor tumor-initiating cell for PCa. A luminal progenitor phenotype arises from genetic deletions of the well-known tumor suppressors p53 and Rb in the Pb compartment via Pb-Cre4 [57]. In this case, early neoplastic cells expressed luminal markers (K8 and AR) as well as neuroendocrine markers (synaptophysin and chromogranin A). This suggested that tumors originated in a luminal progenitor cell, and more closely approximated human cancer phenotypes in the clinic [57]. As both of these transgenic mouse models [55, 57] utilized the same Pb-Cre4, a promiscuous promoter in the prostate, it is important to bear in mind that distinct lineages of normal cells may exhibit a unique susceptibility to certain transforming genetic alterations. Of interest, prostate tumors in $Pten^{-/-}TP53^{-/-}$ mice (deleted via the same Pb-Cre4 recombinase) demonstrated an increase in $K8^{+}$ cells and K5/K8 double-positive cells [58], thus reinforcing the proposal that the genetics of the alteration may give rise to unique tumor-initiating cells or CSCs, similar to observations in murine lung cancer models [59].

An empirical analysis of the ability of a cell to serve as a cell-of-origin for PCa has required genetic alterations specifically in the subset of cells in question. In this regard, $Lin^{-}Sca\text{-}1^{+}CD49f^{high}$ cells have been shown to be efficient targets of lentiviral-mediated oncogenic transformation (e.g., overexpression of ERG1 or activated AKT1 and AR) upon kidney capsule transplantation in immunocompromised mice relative to the transduced $Lin^{-}Sca\text{-}1^{-}CD49f^{low}$ subset of luminal cells [23]. Of particular interest, histological analysis has demonstrated that the transformation of $Lin^{-}Sca\text{-}1^{+}CD49f^{high}$ cells can lead to outgrowths displaying $K8^{+}$ luminal-like cell expansion, similar to human PCa histopathology [23]. On the other hand, Pten deletion in CARNs, followed by androgen supplementation, was recently shown to rapidly induce carcinoma, thus demonstrating the oncogenic potential of luminal cells [52]. These observations support the premise that malignant transformation may originate in a broad spectrum of cells.

3 Human Prostate Stem Cells

3.1 Normal Human Prostate (NHP) Cells and Immortalization

Although anatomically distinct, normal human prostate (NHP) exocrine glands are reminiscent of mouse prostatic tubules, and the same principal cell layers and cell types comprise the prostate in both species (Figs. 1 and 2). In the human prostatic gland (Fig. 1), several basally localized subpopulations – for example, those expressing an intermediate profile including basal (K5 +) and luminal (K8 +) markers [60], or stem cell markers such as CD133 (prominin-1), $\alpha_2\beta_1$ integrin, or the multidrug-resistance transporter, ABCG2 – have been reported to possess stem/progenitor cell properties [26, 30–32] (for reviews, see Refs [8, 22, 24, 61]). $CD133^{+}$ α_2 β_1^{+} double-positive NHP cells have been reported to possess long-term proliferative potential and to reconstitute prostatic-like acini in tissue reconstitution assays with human stromal cells [31]. Glycosylated-CD133 alone has been shown to be sufficient to enrich for tissue-reconstituting cells *in vivo* that can self-renew and give rise to transit amplifying and neuroendocrine cells upon culturing *in vitro* [9]. However, the NHP epithelial cells have a finite life span in culture, thus precluding the long-term analysis of cell growth characteristics.

Primary NHP epithelial cells isolated from normal/benign human prostates and cultured in serum/androgen-free conditions are $CD44^{+}\alpha2\beta1^{+}p63^{+}AR^{-}PSA^{-}$ progenitor cells [62, 63]. Exogenous human telomerase reverse transcriptase (hTERT) and p16-shRNA (short-hairpin RNA) leads to the immortalization of NHP progenitor cells, which are karyotypically normal with a global stem cell gene expression profile, unlimited proliferative capacity, and multilineage differentiation potential [62]. hTERT immortalization has also been used to generate, from nonmalignant prostate tissues and malignant tumors, prostate epithelial cell lines that also display expression of the stem cell marker CD133 and possess sphere-forming potential [64]. These immortalized cell lines may be useful model systems for the elucidation of malignant transformation and comparative analysis with cancer cell lines (as described below).

3.2 CSC Analysis in Prostate Cancer Cell Lines and Xenografts

Human prostate cancer (hPCa) cell lines and xenografts have been broadly used to characterize putative prostate CSCs, largely due to difficulties in reconstituting primary tumors upon the transplantation of primary PCa cells into immunocompromised mice (see Sect. 3.3). Prospectively purified

CSCs have been identified largely on the basis of cell-surface markers expressed by their corresponding normal stem cell counterparts, and characterized as cells with a higher clonogenic potential *in vitro* and tumor-initiating abilities *in vivo*. Dissociated cells from cell lines or xenografts may be sorted into various subsets that are then tested for their clonal, clonogenic (i.e., sphere-forming), and tumorigenic potential (for reviews, see Refs [25, 65–67]).

Similar to the experiments described above for NHP stem cell studies, the hTERT immortalization of malignant tumor-derived epithelial cell lines and characterization has demonstrated that subsets of these cells exhibit both stem cell and CSC properties. hTERT-immortalized tumor-derived lines form spheres upon anchorage deprivation in serum-free media, and express CD133 and the chemokine receptor CXCR4 [64]. In separate experiments, hTERT-immortalized PCa cells derived from Gleason 8-9 tumors expressed early progenitor markers CD44 and Nestin and, of particular importance, could reconstitute tumors upon transplantation *in vivo* [68]. However, because the genetic modifications of hPCa cells to force the establishment of immortalized lines via hTERT expression could, undoubtedly, modulate the expression CSC genes and characteristics, evidence of endogenous expression and activity of such markers in non-modified cells is desirable.

One of the earliest efforts along this line was to characterize the Hoechst dye-effluxing side population (SP) cells isolated from the androgen-dependent LAPC9 HPCa xenograft tumors (derived from a hPCa bone metastasis; [69]. The SP cells in LAPC9 tumors represented less than 1% of the bulk tumor mass, but were enriched in tumorigenic potential as compared to the non-SP cells [70]. The ABC half-transporter ABCG2 was also assayed as a candidate marker for prostate CSCs, potentially mediating the SP phenotype. However, $ABCG2^+$ and $ABCG2^-$ Du145 cells (an undifferentiated AR^- hPCa cell line and xenograft) were only slightly more tumorigenic, which suggested that ABCG2 might not be the key mediator of the SP phenotype accounting for the increased tumorigenic potential of these cells [70].

Subsequently, attention was turned to the normal and cancer stem/progenitor marker CD44 [1, 7, 8]. Comparative analyses of sorted $CD44^+$ cells versus their $CD44^-$ isogenic counterparts from xenograft tumors, including LAPC4 (lymph node metastasis-derived), LAPC9, and Du145, revealed that the $CD44^+$ PCa cell population was enriched for highly tumorigenic and often metastatic PCa cells [7]. Of particular interest, $CD44^+$ (AR^-) cells from LAPC4 and LAPC9 tumors were shown to possess sphere-forming abilities and multilineage differentiation potential, as evidenced by the ability of these cells to undergo asymmetric cell division to give rise to $CD44^-$ and/or AR^+ cells [7]. In separate studies, it has been shown that $CD44^+$ LAPC-4 sphere cells colocalize with the slow-cycling BrdU-retaining cells and express certain stem cell associated genes and, although they express little AR, they can give rise to AR^+ cells [25]. Most recently, the $CD44^+$ subset of PCa cells was further characterized by combining this marker with integrin $\alpha_2\beta_1$ [8]. The results suggested that hPCa cells are organized as a tumorigenic hierarchy in which tumor-initiating cells are enriched in the SP [70], Lin^-CD44^+ PCa cells [7], and further enriched in $CD44^+$ α_2 β_1^+PCa cells [8]. A comparison of the tumor-initiating cell frequencies of several purified populations of PCa cells suggested that the $CD44^+$ PCa cell subset was heterogeneous, and harbored both primitive CSCs and more mature progenitors [7, 8, 70].

Numerous other research groups have also found evidence that long-termed cultured hPCa cell lines and xenografts contain stem-like cells that may be defined as CSCs. $CD44^+CD24^-$ cells comprising 0.04% of LNCaP and up to 10% of Du145 cells have been shown to spontaneously form spheres in serum-free media, while the addition of serum triggered a reversion to adherent cell morphology and growth in parallel with cell differentiation, thus recapitulating the parental population [71]. $CD133^+$ cells in LNCaP, LAPC-4, and CWR22Rv1 lines represent 1–5% of the total population, but have the ability to clonally recapitulate the parental population, which is suggestive of renewal and multilineage differentiation potential [9].

Most recently, in a study conducted by Scher and colleagues, it was found that subsets of CWR22 cells could form spheres that were composed largely of undifferentiated cells expressing certain stem cell (and CSC) -related markers such as CD44 and integrin α_2 [72]. Although some of these markers enriched for sphere-forming cells, the authors sought novel markers that may further discriminate prostate CSCs. Subsequently, they reported that the human pluripotency stem cell marker TRA-1-60, particularly when combined with CD151 and CD160, could significantly enrich for sphere-forming and tumor-initiating cells exhibiting an increased nuclear factor (NF) $-\kappa$B signaling and the ability to recapitulate parental tumor heterogeneity in serial transplantation experiments. Similar triple-positive cells with increased tumorigenic potential were also identified in other hPCa lines, including Du145, PC3, and VCaP [72].

3.3 Primary Human Prostate Cancer

In comparison to the discoveries using rodent prostate tumor models and/or human xenografts, there is a paucity of information relating primary hPCa to CSCs, largely because of the hurdles presented by the need for stromal cell signaling to drive tumor development [66]. $CD44^+\ \alpha_2\ \beta_1^{hi}CD133^+$ cells with clonal growth potential have been reported to track potential prostate CSCs, although the actual ability of these cells to initiate tumor development was not tested [73]. Immunohistochemistry has been used to identify subsets of primary hPCa cells in clinical tumor specimens *in situ* expressing CD133, Trop-2, and (or) $\alpha_2\beta_1$ as potential markers of CSCs, although this is speculative owing to a lack of functional characterization [74]. Consistent with these findings, benign human prostate basal stem-like cells enriched via Trop-2 and CD49f have been shown to be susceptible to lentiviral-mediated oncogenic manipulation and to initiate tumor development upon transplantation into immunodeficient mice [75]. Similarly, CD133 has been used to fractionate tumorigenic subsets of basal cells from early-passage human xenografts [76]. Finally, TRA-1-$60^+CD151^+CD166^+$ triple-positive cells have also been identified as a rare percentage (ca. 2.5%) in primary PCa samples that was higher than matching benign tissues [72], although the functional importance of this expression is currently unknown. Of note, a technical improvement to methods for functionally assaying the tumorigenicity of HPCa cell subsets may be the inclusion of neonatal seminal vesicle mesenchyme to improve hPCa primary tumor take upon transplantation directly into immunodeficient mice [77].

4 Molecular Mechanisms Regulating Prostate Cell Fate

Although significant progress has been made in delineating cell lineages in the prostate, further understanding of normal mouse and human prostate homeostasis will be essential to elucidate PCa cellular origins and molecular mechanisms. Some of the candidate cellular and molecular mechanisms by which prostate growth is restrained or, from a different perspective, how prostate tumor development may be promoted, are briefly described in the following subsection.

Stromal factors, such as TGF-β, have been shown to be important regulators of prostate epithelial cell homeostasis. For example, TGF-β plays a critical role in maintaining the relative quiescence of Bcl-2-overexpressing, proximal Sca-1^+ cells under physiological conditions [40]. Upon androgen depletion (castration), a reversal in the TGF-β gradient places the distal tips of prostatic tubules in a higher TGF-β concentration, such that these Bcl-2^{low} cells are susceptible to TGF-β-induced apoptosis [40]. Among many other cellular responses to these shifts in important signaling molecules, decreases in TGF-β proximal to the urethra could allow (presumably basally localized) prostate stem cells to hyperproliferate. Other paracrine growth-stimulatory factors generated by the stroma may include FGF10, Sonic Hedgehog, and Wnt/β-catenin [21].

Fibroblastic cells from prostate tumors, termed cancer-associated fibroblasts (CAFs), enhanced the sphere-forming efficiency of $Lin^-Sca\text{-}1^+CD49f^{high}$ prostate tumor epithelial cells from *Pten*-null mice [78]. Of note, this coculture system prevented direct cell–cell contact between the fibroblasts and the CSCs, suggesting that paracrine factors had mediated the observed effect. In tissue/tumor reconstitution assays, CAFs increased the engraftment efficiency, proliferative index, and histopathological phenotypes relative to normal prostate fibroblasts or UGSM, both of which were better than epithelial cells alone [78]. More recently, a method to establish first-passage grafts of hPCa clinical tumor specimens in immunocompromised mice has been developed, using neonatal mouse seminal vesicle mesenchyme [77]. A complete understanding of the optimum stromal cells and stromal factors promoting prostate tumor growth, tumor recapitulation, and prostate tumor-initiating cell activity await further analyses, with implications in the development of therapeutics blocking essential stromal–epithelial cell signaling in patient tumors.

Aberrant self-renewal, whether exhibited by neoplastic stem cells already possessing this hallmark stem cell characteristic, or acquired by transformed progenitor (or otherwise lineage-restricted) cells, has been hypothesized to be among the critical forces driving tumor development [11]. Interestingly, the embryonic stem cell self-renewal gene Nanog is preferentially expressed in $CD44^+$ PCa cells, and plays a functional role in PCa cell clonogenic growth and tumorigenicity [79]. Remarkably, inducible Nanog expression is sufficient to promote prostate tumor growth and castration-resistant tumor development [80]. Nanog expression has also been reported in other PCa cell subsets, including $CD117^+ABCG2^+$ drug-resistant 22RV1 cells [68, 81] and hTERT-immortalized $CD44^+$ and $Nestin^+$ primary human prostate tumor derived cells with tumor reconstitution capabilities [68].

MicroRNAs have recently been shown to regulate the CSC characteristics of prostate tumor cells. Subsets of hPCa cells displaying epithelial–mesenchymal transition (EMT) phenotypes often express stem cell-related genes such as *Nanog*, *Lin28B*, and *Notch1*, along with a decreased expression of miR-200 and/or let-7 family miRNAs. EMT phenotypes in PCa cells were reversed by the exogenous application of miR-200, which also decreased sphere-forming potential and some stem cell-related genes [24]. Recently, an unbiased miRNA expression profiling in PCa stem/progenitor cells (marked by CD44, CD133, and/or $\alpha_2\beta_1$) revealed an underexpression of miR-34a in these cell subsets [82]. An exogenous application of miR-34a to LAPC4, LAPC9, and Du145 tumors decreased the tumor-regenerating capabilities of bulk or purified $CD44^+$ hPCa cells [82]. miR-34a was shown to suppress prostate CSC characteristics, such as serial sphere-formation and serial tumor transplantation. Decreases in migration and metastasis were also observed, while a therapeutic application of miR-34a via tail vein injection was found to extend the survival of prostate tumor-bearing mice relative to controls. Mechanistically, at least one CSC-related downstream target of miR-34a that could account for the observed phenomena appeared to be CD44, which was shown to be reciprocally expressed with miR-34a and contained a miR-34a targeting sequence [82].

5
Outstanding Questions

During recent years, research groups have accrued a significant compilation of evidence supporting the premise that stem cells and CSCs exist in the normal prostate and prostate tumors, respectively, and, that PCa can arise in both basal and luminal cellular compartments. However, how this relates to the cell-of-origin of patient tumors comprised largely of cells displaying a remarkably luminal phenotype is unclear. An efficient tumor formation of fractionated basal stem cell subsets giving rise to PCa recapitulating luminal phenotypes of the human disease provides compelling support for the validity of the CSC paradigm in prostate cancer [23]:

- Do transformed basal stem cells lose their basal characteristics during the transformation process?
- Are the luminal cells observed the nontumorigenic daughter cells of transformed basal stem cells?
- Could transformed basal stem cells give rise to transformed luminal-like daughter cells, both of which possess tumorigenicity?
- Alternatively, may luminal stem/progenitor cells be the cell-of-origin for PCa, at least in some primary hPCa samples, as has been shown by lineage-tracking of transformed CARN cells [52]?
- Or – in a final scenario that has not yet been eliminated – can a reversible plasticity in tumor cellular phenotypes allow for the interconversion of tumorigenic (or androgen-independent or metastatic) and nontumorigenic cell subsets?
- Finally, which cell-intrinsic and cell-nonautonomous molecular mechanisms drive prostate neoplasia and, in particular, the acquisition of CSC properties?

The answers to these questions await future investigations that will undoubtedly illuminate the present understanding of prostate tumorigenesis and create new opportunities for PCa therapeutics for the specific targeting of stem-like cells in prostate tumors.

Acknowledgments

Studies conducted in the Tang laboratory were supported in part by grants from NIH (R01-ES015888 and 1R21CA 150009), Department of Defense (W81XWH-08-1-0472 and W81XWH-11-1-0331), Elsa Pardee Foundation, and CPRIT funding (RP120380), Elsa Pardee Foundation, and the MDACC Center for Cancer Epigenetics and Laura & John Arnold Foundation RNA Center pilot grant, and by two Center Grants (CCSG-5 P30 CA166672 and ES007784).

References

1. Al-Hajj, M., Wicha, M.S., Benito-Hernandez, A., Morrison, S.J. *et al.* (2003) Prospective identification of tumorigenic breast cancer cells. *Proc. Natl Acad. Sci. USA*, **100**, 3983–3988.
2. Zucchi, I., Sanzone, S., Astigiano, S., Pelucchi, P. *et al.* (2007) The properties of a mammary gland cancer stem cell. *Proc. Natl Acad. Sci. USA*, **104** (25), 10476–10481.
3. Singh, S.K., Clarke, I.D., Terasaki, M., Bonn, V.E. *et al.* (2003) Identification of a cancer stem cell in human brain tumors. *Cancer Res.*, **63** (18), 5821–5828.
4. Singh, S.K., Hawkins, C., Clarke, I.D., Squire, J.A. *et al.* (2004) Identification of human brain tumour initiating cells. *Nature*, **432** (7015), 396–401.
5. O'Brien, C.A., Pollett, A., Gallinger, S., and Dick, J.E. (2007) A human colon cancer cell capable of initiating tumour growth in immunodeficient mice. *Nature*, **445** (7123), 106–110.
6. Ricci-Vitiani, L., Lombardi, D.G., Pilozzi, E., Biffoni, M. *et al.* (2007) Identification and expansion of human colon-cancer-initiating cells. *Nature*, **445** (7123), 111–115.
7. Patrawala, L., Calhoun, T., Schneider-Broussard, R., Li, H. *et al.* (2006) Highly purified $CD44^+$ prostate cancer cells from xenograft human tumors are enriched in tumorigenic and metastatic progenitor cells. *Oncogene*, **25** (12), 1696–1708.
8. Patrawala, L., Calhoun-Davis, T., Schneider-Broussard, R., and Tang, D.G. (2007) Hierarchical organization of prostate cancer cells in xenograft tumors: the $CD44^+alpha_2beta_1{}^+$ cell population is enriched in tumor-initiating cells. *Cancer Res.*, **67** (14), 6796–6805.
9. Vander Griend, D.J., Karthaus, W.L., Dalrymple, S., Meeker, A. *et al.* (2008) The role of CD133 in normal human prostate stem cells and malignant cancer-initiating cells. *Cancer Res.*, **68** (23), 9703–9711.
10. Hanahan, D. and Weinberg, R.A. (2000) The hallmarks of cancer. *Cell*, **100** (1), 57–70.
11. Hanahan, D. and Weinberg, R.A. (2011) Hallmarks of cancer: the next generation. *Cell*, **144** (5), 646–674.
12. Visvader, J.E. (2011) Cells of origin in cancer. *Nature*, **469** (7330), 314–322.
13. Barker, N., Ridgway, R.A., van Es, J.H., van de Wetering, M. *et al.* (2009) Crypt stem cells as the cells-of-origin of intestinal cancer. *Nature*, **457** (7229), 608–611.
14. Zhu, L., Gibson, P., Currle, D.S., Tong, Y. *et al.* (2009) Prominin 1 marks intestinal stem cells that are susceptible to neoplastic transformation. *Nature*, **457** (7229), 603–607.
15. Marusyk, A. and Polyak, K. (2010) Tumor heterogeneity: causes and consequences. *Biochim. Biophys. Acta*, **1805** (1), 105–117.
16. Visvader, J.E. and Lindeman, G.J. (2008) Cancer stem cells in solid tumours: accumulating evidence and unresolved questions. *Nat. Rev. Cancer*, **8** (10), 755–768.
17. Quintana, E., Shackleton, M., Sabel, M.S., Fullen, D.R. *et al.* (2008) Efficient tumour formation by single human melanoma cells. *Nature*, **456** (7222), 593–598.
18. Quintana, E., Shackleton, M., Foster, H.R., Fullen, D.R. *et al.* (2010) Phenotypic heterogeneity among tumorigenic melanoma cells from patients that is reversible and not hierarchically organized. *Cancer Cell*, **18** (5), 510–523.
19. Boiko, A.D., Razorenova, O.V., van de Rijn, M., Swetter, S.M. *et al.* (2010) Human melanoma-initiating cells express neural crest nerve growth factor receptor CD271. *Nature*, **466** (7302), 133–137.
20. Cunha, G.R., Ricke, W., Thomson, A., Marker, P.C. *et al.* (2004) Hormonal, cellular, and molecular regulation of normal and

neoplastic prostatic development. *J. Steroid Biochem. Mol. Biol.*, **92** (4), 221–236.
21. Shen, M.M., and Abate-Shen, C. (2010) Molecular genetics of prostate cancer: new prospects for old challenges. *Genes Dev.*, **24** (18), 1967–2000.
22. Lawson, D.A. and Witte, O.N. (2007) Stem cells in prostate cancer initiation and progression. *J. Clin. Invest.*, **117** (8), 2044–2050.
23. Lawson, D.A., Zong, Y., Memarzadeh, S., Xin, L. *et al.* (2010) Basal epithelial stem cells are efficient targets for prostate cancer initiation. *Proc. Natl Acad. Sci. USA*, **107** (6), 2610–2615.
24. Kong, D., Banerjee, S., Ahmad, A., Li, Y. *et al.* (2010) Epithelial to mesenchymal transition is mechanistically linked with stem cell signatures in prostate cancer cells. *PLoS ONE*, **5** (8), e12445.
25. Tang, D.G., Patrawala, L., Calhoun, T., Bhatia, B. *et al.* (2007) Prostate cancer stem/progenitor cells: identification, characterization, and implications. *Mol. Carcinog.*, **46** (1), 1–14.
26. Hudson, D.L., Guy, A.T., Fry, P., O'Hare, M.J. *et al.* (2001) Epithelial cell differentiation pathways in the human prostate: identification of intermediate phenotypes by keratin expression. *J. Histochem. Cytochem.*, **49** (2), 271–278.
27. van Leenders, G., Dijkman, H., Hulsbergen-van de Kaa, C., Ruiter, D. *et al.* (2000) Demonstration of intermediate cells during human prostate epithelial differentiation in situ and in vitro using triple-staining confocal scanning microscopy. *Lab. Invest.*, **80** (8), 1251–1258.
28. Matusik, R.J., Jin, R.J., Sun, Q., Wang, Y. *et al.* (2008) Prostate epithelial cell fate. *Differentiation*, **76** (6), 682–698.
29. Liu, A.Y., True, L.D., LaTray, L., Nelson, P.S. *et al.* (1997) Cell-cell interaction in prostate gene regulation and cytodifferentiation. *Proc. Natl Acad. Sci. USA*, **94** (20), 10705–10710.
30. Collins, A.T., Habib, F.K., Maitland, N.J., and Neal, D.E. (2001) Identification and isolation of human prostate epithelial stem cells based on alpha(2)beta(1)-integrin expression. *J. Cell Sci.*, **114** (Pt 21), 3865–3872.
31. Richardson, G.D., Robson, C.N., Lang, S.H., Neal, D.E. *et al.* (2004) CD133, a novel marker for human prostatic epithelial stem cells. *J. Cell Sci.*, **117** (Pt 16), 3539–3545.
32. Huss, W.J., Gray, D.R., Greenberg, N.M., Mohler, J.L. *et al.* (2005) Breast cancer resistance protein-mediated efflux of androgen in putative benign and malignant prostate stem cells. *Cancer Res.*, **65** (15), 6640–6650.
33. Lang, S.H., Frame, F.M., and Collins, A.T. (2009) Prostate cancer stem cells. *J. Pathol.*, **217** (2), 299–306.
34. English, H.F., Santen, R.J., and Isaacs, J.T. (1987) Response of glandular versus basal rat ventral prostatic epithelial cells to androgen withdrawal and replacement. *Prostate*, **11** (3), 229–242.
35. Tsujimura, A., Koikawa, Y., Salm, S., Takao, T. *et al.* (2002) Proximal location of mouse prostate epithelial stem cells: a model of prostatic homeostasis. *J. Cell Biol.*, **157** (7), 1257–1265.
36. Kasper, S. (2008) Stem cells: the root of prostate cancer? *J. Cell. Physiol.*, **216** (2), 332–336.
37. Wang, Z.A. and Shen, M.M. (2010) Revisiting the concept of cancer stem cells in prostate cancer. *Oncogene*, **30** (11), 1261–1271.
38. Litvinov, I.V., De Marzo, A.M., and Isaacs, J.T. (2003) Is the Achilles' heel for prostate cancer therapy a gain of function in androgen receptor signaling? *J. Clin. Endocrinol. Metab.*, **88** (7), 2972–2982.
39. Abate-Shen, C. and Shen, M.M. (2000) Molecular genetics of prostate cancer. *Genes Dev.*, **14** (19), 2410–2434.
40. Salm, S.N., Burger, P.E., Coetzee, S., Goto, K. *et al.* (2005) TGF-{beta} maintains dormancy of prostatic stem cells in the proximal region of ducts. *J. Cell Biol.*, **170** (1), 81–90.
41. Burger, P.E., Xiong, X., Coetzee, S., Salm, S.N. *et al.* (2005) Sca-1 expression identifies stem cells in the proximal region of prostatic ducts with high capacity to reconstitute prostatic tissue. *Proc. Natl Acad. Sci. USA*, **102** (20), 7180–7185.
42. Xin, L., Lawson, D.A., and Witte, O.N. (2005) The Sca-1 cell surface marker enriches for a prostate-regenerating cell subpopulation that can initiate prostate tumorigenesis. *Proc. Natl Acad. Sci. USA*, **102** (19), 6942–6947.
43. Cotsarelis, G., Sun, T.T., and Lavker, R.M. (1990) Label-retaining cells reside in the bulge area of pilosebaceous unit: implications for follicular stem cells, hair cycle, and skin carcinogenesis. *Cell*, **61** (7), 1329–1337.
44. Tumbar, T., Guasch, G., Greco, V., Blanpain, C. *et al.* (2004) Defining the epithelial stem

cell niche in skin. *Science*, **303** (5656), 359–363.
45. Welm, B.E., Tepera, S.B., Venezia, T., Graubert, T.A. *et al.* (2002) Sca-1(pos) cells in the mouse mammary gland represent an enriched progenitor cell population. *Dev. Biol.*, **245** (1), 42–56.
46. Lavker, R.M., Tseng, S.C., and Sun, T.T. (2004) Corneal epithelial stem cells at the limbus: looking at some old problems from a new angle. *Exp. Eye Res.*, **78** (3), 433–446.
47. Cunha, G.R. and Lung, B. (1978) The possible influence of temporal factors in androgenic responsiveness of urogenital tissue recombinants from wild-type and androgen-insensitive (Tfm) mice. *J. Exp. Zool.*, **205** (2), 181–193.
48. Hayward, S.W., Haughney, P.C., Rosen, M.A., Greulich, K.M. *et al.* (1998) Interactions between adult human prostatic epithelium and rat urogenital sinus mesenchyme in a tissue recombination model. *Differentiation*, **63** (3), 131–140.
49. Xin, L., Ide, H., Kim, Y., Dubey, P. *et al.* (2003) In vivo regeneration of murine prostate from dissociated cell populations of postnatal epithelia and urogenital sinus mesenchyme. *Proc. Natl Acad. Sci. USA*, **100** (Suppl. 1), 11896–11903.
50. Lawson, D.A., Xin, L., Lukacs, R.U., Cheng, D. *et al.* (2007) Isolation and functional characterization of murine prostate stem cells. *Proc. Natl Acad. Sci. USA*, **104** (1), 181–186.
51. Leong, K.G., Wang, B.E., Johnson, L., and Gao, W.Q. (2008) Generation of a prostate from a single adult stem cell. *Nature*, **456** (7223), 804–808.
52. Wang, X., Kruithof-de Julio, M., Economides, K.D., Walker, D. *et al.* (2009) A luminal epithelial stem cell that is a cell of origin for prostate cancer. *Nature*, **461** (7263), 495–500.
53. Goldstein, A.S., Lawson, D.A., Cheng, D., Sun, W. *et al.* (2008) Trop2 identifies a subpopulation of murine and human prostate basal cells with stem cell characteristics. *Proc. Natl Acad. Sci. USA*, **105** (52), 20882–20887.
54. Wang, S., Garcia, A.J., Wu, M., Lawson, D.A. *et al.* (2006) Pten deletion leads to the expansion of a prostatic stem/progenitor cell subpopulation and tumor initiation. *Proc. Natl Acad. Sci. USA*, **103** (5), 1480–1485.
55. Mulholland, D.J., Xin, L., Morim, A., Lawson, D. *et al.* (2009) Lin-Sca-1^{+}CD49^{fhigh} stem/progenitors are tumor-initiating cells in the Pten-null prostate cancer model. *Cancer Res.*, **69** (22), 8555–8562.
56. Korsten, H., Ziel-van der Made, A., Ma, X., van der Kwast, T. *et al.* (2009) Accumulating progenitor cells in the luminal epithelial cell layer are candidate tumor initiating cells in a Pten knockout mouse prostate cancer model. *PLoS ONE*, **4** (5), e5662.
57. Zhou, Z., Flesken-Nikitin, A., and Nikitin, A.Y. (2007) Prostate cancer associated with p53 and Rb deficiency arises from the stem/progenitor cell-enriched proximal region of prostatic ducts. *Cancer Res.*, **67** (12), 5683–5690.
58. Abou-Kheir, W.G., Hynes, P.G., Martin, P.L., Pierce, R. *et al.* (2010) Characterizing the contribution of stem/progenitor cells to tumorigenesis in the $Pten^{-/-}TP53^{-/-}$ prostate cancer model. *Stem Cells*, **28** (12), 2129–2140.
59. Curtis, S.J., Sinkevicius, K.W., Li, D., Lau, A.N. *et al.* (2010) Primary tumor genotype is an important determinant in identification of lung cancer propagating cells. *Cell Stem Cell*, **7** (1), 127–133.
60. Wang, Y., Hayward, S., Cao, M., Thayer, K. *et al.* (2001) Cell differentiation lineage in the prostate. *Differentiation*, **68** (4-5), 270–279.
61. Laffin, B. and Tang, D.G. (2010) An old player on a new playground: bmi-1 as a regulator of prostate stem cells. *Cell Stem Cell*, **7** (6), 639–640.
62. Bhatia, B., Jiang, M., Suraneni, M., Patrawala, L. *et al.* (2008) Critical and distinct roles of p16 and telomerase in regulating the proliferative life span of normal human prostate epithelial progenitor cells. *J. Biol. Chem.*, **283** (41), 27957–27972.
63. Bhatia, B., Tang, S., Yang, P., Doll, A. *et al.* (2005) Cell-autonomous induction of functional tumor suppressor 15-lipoxygenase 2 (15-LOX2) contributes to replicative senescence of human prostate progenitor cells. *Oncogene*, **24** (22), 3583–3595.
64. Miki, J., Furusato, B., Li, H., Gu, Y. *et al.* (2007) Identification of putative stem cell markers, CD133 and CXCR4, in hTERT-immortalized primary nonmalignant and malignant tumor-derived human prostate epithelial cell lines and in prostate cancer specimens. *Cancer Res.*, **67** (7), 3153–3161.

65. Xin, L., Lukacs, R.U., Lawson, D.A., Cheng, D. *et al.* (2007) Self-renewal and multilineage differentiation in vitro from murine prostate stem cells. *Stem Cells*, **25** (11), 2760–2769.
66. Li, H., Jiang, M., Honorio, S., Patrawala, L. *et al.* (2009) Methodologies in assaying prostate cancer stem cells. *Methods Mol. Biol.*, **568**, 85–138.
67. Lukacs, R.U., Goldstein, A.S., Lawson, D.A., Cheng, D. *et al.* (2010) Isolation, cultivation and characterization of adult murine prostate stem cells. *Nat. Protoc.*, **5** (4), 702–713.
68. Gu, G., Yuan, J., Wills, M., and Kasper, S. (2007) Prostate cancer cells with stem cell characteristics reconstitute the original human tumor in vivo. *Cancer Res.*, **67** (10), 4807–4815.
69. Craft, N., Chhor, C., Tran, C., Belldegrun, A. *et al.* (1999) Evidence for clonal outgrowth of androgen-independent prostate cancer cells from androgen-dependent tumors through a two-step process. *Cancer Res.*, **59** (19), 5030–5036.
70. Patrawala, L., Calhoun, T., Schneider-Broussard, R., Zhou, J. *et al.* (2005) Side population is enriched in tumorigenic, stem-like cancer cells, whereas ABCG2+ and ABCG2- cancer cells are similarly tumorigenic. *Cancer Res.*, **65** (14), 6207–6219.
71. Hurt, E.M., Kawasaki, B.T., Klarmann, G.J., Thomas, S.B. *et al.* (2008) CD44+ CD24(-) prostate cells are early cancer progenitor/stem cells that provide a model for patients with poor prognosis. *Br. J. Cancer*, **98** (4), 756–765.
72. Rajasekhar, V.K., Studer, L., Gerald, W., Socci, N.D. *et al.* (2011) Tumour-initiating stem-like cells in human prostate cancer exhibit increased NF-kappaB signalling. *Nat. Commun.*, **2**, 162.
73. Collins, A.T., Berry, P.A., Hyde, C., Stower, M.J. *et al.* (2005) Prospective identification of tumorigenic prostate cancer stem cells. *Cancer Res.*, **65** (23), 10946–10951.
74. Trerotola, M., Rathore, S., Goel, H.L., Li, J. *et al.* (2010) CD133, Trop-2 and alpha2beta1 integrin surface receptors as markers of putative human prostate cancer stem cells. *Am. J. Transl. Res.*, **2** (2), 135–144.
75. Goldstein, A.S., Huang, J., Guo, C., Garraway, I.P. *et al.* (2010) Identification of a cell of origin for human prostate cancer. *Science*, **329** (5991), 568–571.
76. Maitland, N.J., Frame, F.M., Polson, E.S., Lewis, J.L. *et al.* (2011) Prostate cancer stem cells: do they have a basal or luminal phenotype? *Horm. Cancer*, **2** (1), 47–61.
77. Toivanen, R., Berman, D.M., Wang, H., Pedersen, J. *et al.* (2011) Brief report: a bioassay to identify primary human prostate cancer repopulating cells. *Stem Cells*, **29** (8), 1310–1314.
78. Liao, C.P., Adisetiyo, H., Liang, M., and Roy-Burman, P. (2010) Cancer-associated fibroblasts enhance the gland-forming capability of prostate cancer stem cells. *Cancer Res.*, **70** (18), 7294–7303.
79. Jeter, C.R., Badeaux, M., Choy, G., Chandra, D. *et al.* (2009) Functional evidence that the self-renewal gene NANOG regulates human tumor development. *Stem Cells*, **27** (5), 993–1005.
80. Jeter, C.R., Liu, B., Liu, X., Chen, X. *et al.* (2011) NANOG promotes cancer stem cell characteristics and prostate cancer resistance to androgen deprivation. *Oncogene*, **30**, 3833–3845.
81. Liu, T., Xu, F., Du, X., Lai, D. *et al.* (2010) Establishment and characterization of multidrug resistant, prostate carcinoma-initiating stem-like cells from human prostate cancer cell lines 22RV1. *Mol. Cell. Biochem.*, **340** (1-2), 265–273.
82. Liu, C., Kelnar, K., Liu, B., Chen, X. *et al.* (2011) The microRNA miR-34a inhibits prostate cancer stem cells and metastasis by directly repressing CD44. *Nat. Med.*, **17** (2), 211–215.

9
The Stem Cell Niche and Its Role in Self-Renewal, Aging, and Malignancy*

Peter Breslin S.J., Andrew Volk, and Jiwang Zhang
Loyola University Medical Center, Cardinal Bernardin Cancer Center, Department of Pathology, Department of Cellular and Molecular Physiology, 2160 S. First Avenue, Maywood, IL 60153, USA

*This chapter has previously been published in: Meyers, R.A. (Ed.) Stem Cells, 2013, 978-3-527-32925-0.

Translational Medicine Cancer: Advances in Molecular Biology and Medicine,
First Edition. Edited by Robert A. Meyers.

Keywords

Stem cell
Any cell that can maintain long-term tissue homeostasis through controlled self-renewal and differentiation into lineage-specific progeny.

Niche
The cellular and extracellular microenvironment that supports the stem cell.

Quiescence
A reversible cellular state characterized by lack of replication and low levels of metabolism. Quiescent cells are considered to be "resting," but can still respond to stimulation.

Aging
Dynamic process that involves the gradual wearing down of tissue function over time.

Cancer stem cell
A key cell existing in a tumor which can produce the bulk of the tumor cells required for tumor initiation, progression, and relapse.

Cells within tissues are organized in specific patterns supported by constructive materials, such as stromal cells and extracellular matrix (ECM), and are regulated by cell binding and diffused factors. These patterned areas are determined by signals from adjacent cells which form a microenvironment. Stem cells reside in a specialized location of the tissue characterized by a unique microenvironment; this location is referred to as the "*stem cell niche.*"

Signaling in the niche regulates the balance of self-renewal and differentiation in stem cells, and plays an important role in tissue generation during homeostasis and regeneration in response to damage. The niche provides a shelter that protects stem cells against unwanted insults, and also provides the correct factors to support self-renewal and maintains the undifferentiated state of the stem cells. Because of the role that niche dysregulation plays in malignancies and tissue degeneration, the niche and niche signaling can be used as targets to develop drugs to potentially treat these diseases. Alterations in niche signaling are also involved the aging process, as they influence stem cell behavior. Thus, the manipulation of niche signaling may enable the most effective strategies for cancer prevention, for slowing down the natural aging program, and for treating aging-related disorders.

1 The Stem Cell Niche and Its Role in Stem Cell Self-Renewal

1.1 Concept of the Stem Cell

Cells within a tissue are heterogeneous in terms of their morphology, rate of proliferation, and function. Stem cells represent a small subset of these cells at the top of the hierarchy which, when activated, can produce multipotent progenitor cells. The progenitor cells can then give rise to all types of tissue-specific mature cells from the particular tissue in order to maintain tissue homeostasis and to repair inevitable tissue damage (Fig. 1) [1–6]. Previous studies have suggested that the self-renewal of stem cells is controlled by their intrinsic machinery, and regulated by a network of signaling induced by microenvironmental factors generated by the niche.

Modern science has known about stem cells since the mid-1900s, when specialized cells called hematopoietic stem cells (HSCs) were identified in the bone marrow (BM) and shown to be capable of dividing and regenerating the blood tissues. [7]. More recently, however, similar cells have been also characterized and identified in all major tissues of the body [6, 8–10]. These cells share some common features: notably, they remain undifferentiated and quiescent until needed, and when they divide they have the unique ability to regenerate the tissue from which they were isolated. Moreover, the tissue-regenerative activity of stem cells can be maintained for the entire lifetime of the organism [1–6, 8–14].

1.1.1 Stem Cell Self-Renewal

There are three paradigms by which stem cell division can be conceptualized [8, 15] (Fig. 2):

- *Symmetrical self-renewal*, in which the stem cell is stimulated to divide and, as a result, generates two new undifferentiated stem cells. This is commonly seen in situations that call for increased numbers of stem cells. For example, when stem cells are transplanted into immune-depleted irradiated mice, the transplanted stem cells will undergo symmetric expansion until the number of stem cells is adequate to reconstitute effective hematopoiesis in the recipient [7, 8, 15, 16].
- *Symmetric division and differentiation*, where the stem cells are signaled to divide and, as a result, differentiate into two active progenitor cells without generating any new stem cells. This is useful in a tissue should there be proportionally too many stem cells, or if the particular tissue demand for new mature effector cells is exceptionally high. Defects in self-renewal signaling from the niche can result in this condition.
- *Asymmetric division*, in which case the stem cell divides and creates two progeny cells, one of which is a stem cell and the other is a progenitor cell, simultaneously generating progenitor cells while also maintaining the stem cell pool [17–24] (Fig. 2).

Self-renewal is a type of cell division that incorporates the maintenance of the undifferentiated multipotent state. It is a process by which stem cells divide to maintain the stem cell pool through asymmetric division, or to expand the pool by promoting symmetric expansion and repressing differentiation. The balance of asymmetric division and symmetric expansion or differentiation (Fig. 2) is controlled by a cell-autonomous machinery, and is highly regulated by both cellular and extracellular components derived

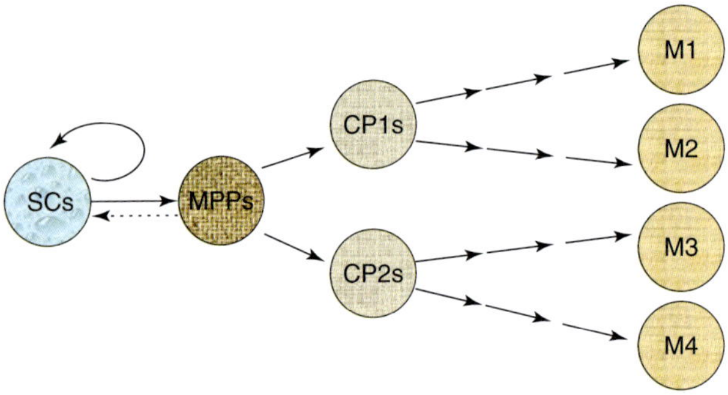

Fig. 1 Hierarchy of stem cells (SCs) and their progeny. MPP, multipotent progenitor cell; CP, committed progenitor cell; M, mature cell.

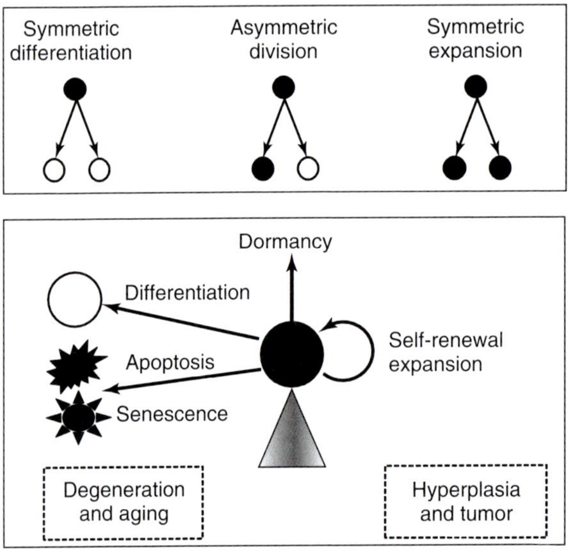

Fig. 2 Stem cell division and fate paradigms.

from the stem cell niche [25–30]. It is of the utmost importance for organisms to retain a stable number of stem cells. If the division of these cells is dominated by symmetric division and differentiation (reduction of self-renewal), the stem cell pool will eventually become depleted and this may lead to tissue-degenerative disorders or a premature aging phenotype in the organism due to stem cell depletion [16, 31–33]. In contrast, if the division of these cells is dominated by symmetric expansion (increase of self-renewal), this may result in tissue hyperplasia or tumor development due to the over-production of stem cells [11, 34–39] (Fig. 2).

1.1.2 Cell-Intrinsic Machinery Controls Stem Cell Self-Renewal

The complete mechanism by which stem cell self-renewal is regulated is not yet fully understood. In pluripotent stem cells – such as embryonic stem (ES) cells and induced pluripotent stem cells (iPSCs)—a key self-renewal-related core transcriptional network has been described which contains pluripotent transcriptional factors such as Oct4, Nanog, and Sox2 [40–43]. This network promotes the self-renewal of pluripotent stem cells mainly by blocking multilineage differentiation and promoting proliferation. However, this self-renewal network has not yet been identified in adult stem cells, although some self-renewal-related factors such as Hoxb4, Bmi-1, β-catenin, Notch, FoxOs, and c-Myc, as well as several tissue-specific transcriptional factors, have been reported in HSCs, neural, and other tissue stem cells; various different factors may contribute differently to self-renewal in different tissue stem cells. In fact, in adult stem cells, self-renewal involves the combined effects of multiple cellular responses that includes differentiation blockage and prevention of cell death (such as apoptosis or senescence), in combination with controlled proliferation. Among these factors, differentiation blockage is also the key for the self-renewal of adult stem cells (Fig. 3) [34, 44–56].

Today, increasing evidence indicates that differentiation is implicated as the default pathway for stem cell division. In order to promote stem cell self-renewal, the inhibition of differentiation is an absolute requirement. For tissues such as muscle and germ cells, in which stem cell differentiation has only one lineage choice, blocking stem cell differentiation may be relatively easy to achieve. Transcriptional regulation induced by a single signal, such as bone morphogenetic protein (BMP)-Bam in *Drosophila* germline stem cells (GSCs) [57–62] and Notch-CSL in satellite muscle stem cells (MuSCs), might be sufficient to maintain them in the undifferentiated state [55, 56, 63–66]. However, in tissues capable of producing multiple lineages of cells, such as hematopoietic and neural tissues, to inhibit the differentiation of one lineage will in most cases lead to an increase in differentiation towards another alternative lineage option. Therefore, in such tissues, a more fundamental gene regulatory machinery might be involved. Examples of such regulatory machinery include Polycomb group protein-mediated chromosomal remodeling and miRNA-mediated panels of gene repression. To date, several Polycomb group proteins, such as Bmi1 and Ezh2, have been shown to be involved in the regulation of self-renewal of neural, hematopoietic, and other stem cells through repressing the expression of differentiation-related genes, maintaining the levels of expression

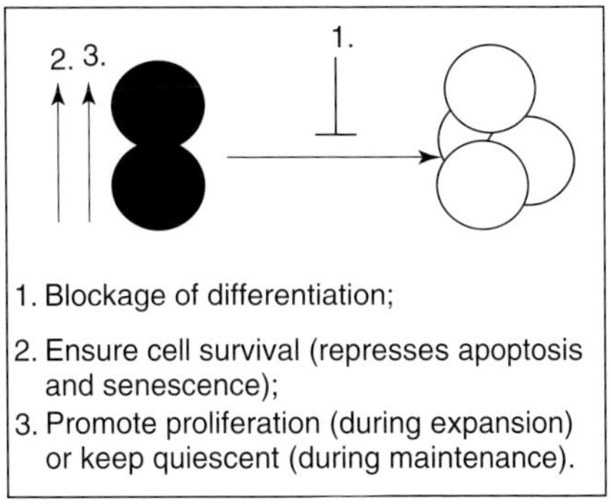

Fig. 3 Cell-intrinsic machinery controls stem cell self-renewal.

of self-renewal related genes, and also inhibiting negative cell-cycle regulators and senescence signals [34, 44–46, 67–77]. An inactivation of multiple downstream targets of Bmi-1, such as p16^{Ink4a}, p19^{Arf}, and Trp53, is needed in order to improve the self-renewal capacity of stem cells, as shown by studying HSCs [39]. In addition, several self-renewal-related miRNAs have been identified in many tissue stem cells [78–87].

Recently, the results of several studies have suggested that the cell cycle must be controlled such that stem cells traverse the cycle slowly in order for the self-renewal regulatory machinery to function correctly. Both, *in vitro* and *in vivo* studies have demonstrated that an over-stimulation of stem cell proliferation will significantly compromise stem cell self-renewal activity, and result in stem cell depletion [88–99]. This might be due to the fact that, with a very limited time to engage the cell cycle, the self-renewal machinery might be unable to complete its task, telomere extension may not be successfully completed, and the DNA repair machinery may fail to correct all of the errors in the duplicated copy of DNA. In support of this notion, most – if not all – tissue stem cells are relatively slow-cycling when they replicate. It has been suggested that the maintenance of a slow-cycling quiescent state is one of the most important mechanisms used by many tissue stem cells to retain their multipotency, as long as the senescence and apoptotic pathways are repressed [9, 44, 73, 95, 96, 100–106]. Normally, the majority of stem cells in adult tissues are maintained in a quiescent state with only a few of them being activated; this allows them to proceed to self-renew or to differentiate in order to regenerate tissues (Fig. 2). This dynamic balance of quiescence versus activation and self-renewal versus differentiation is regulated primarily by signals from the niche (as will be explained in Sect. 1.2.3).

It is known that the major cell death-inducing mechanisms—apoptosis and senescence—are involved in controlling stem cell numbers in almost all tissues studied. The inhibition of pro-survival signaling, promoting pro-apoptotic signaling, or turning on the senescence pathway leads to stem cell depletion and the failure of tissue function, as demonstrated by numerous animal model studies [44, 45, 103, 107–117]. Whereas, transgenic over-expression of several pro-survival genes induces an expansion of the stem cell pool [101, 116, 118], extensive DNA damage – as induced by oxygen stress, mutagens, and/or irradiation – has been reported to be a major cause of stem cell apoptosis and senescence. In addition, DNA damage also induces cell cycle arrest, significantly compromising stem cell self-renewal activity. Maintaining the integrity of the genome is critical for maintenance of the stem cell pool [119–124]. It is well known that maintaining telomere length is one of the most important mechanisms underlying stem cell self-renewal. Significant telomere shortening during cell division is usually recognized as DNA damage by DNA repair systems, which induce cell cycle arrest and/or apoptosis in stem cells [36, 125–129]. This process can be reversed by a transgenic over-activation of telomerase in stem cells, immortalizing them via an enhancement of stem cell self-renewal.

The results of these studies, collectively, suggest that self-renewal is a cell-autonomous characteristic of the stem cell which is controlled by its intrinsic regulatory machinery. However, stem cell self-renewal can only be achieved when the stem cell is seeded into a

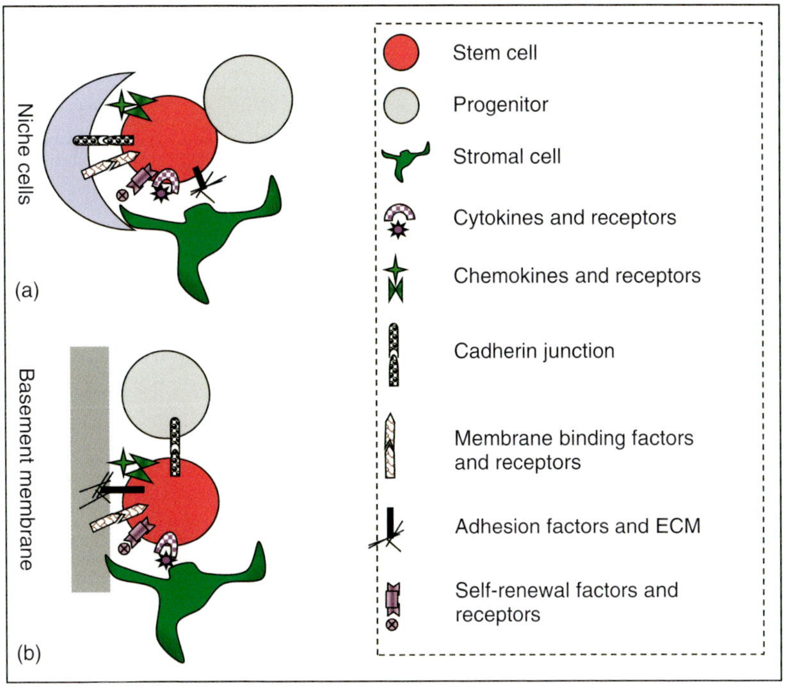

Fig. 4 Proposed model for the cellular and noncellular components of the stem cell niche. The stem cell niche is formed by stationary components which provide a stable physical residence for the stem cells, together with other support cells and extracellular components. Based on the type of stationary component, cellular or noncellular, two types of stem cell niche have been proposed. (a) The stationary components of the cellular niche are specialized niche cells; (b) The stationary component of the noncellular niche is a specialized ECM called the basement membrane.

correct microenvironment. This niche microenvironment-dependent feature of stem cell self-renewal suggests that the cell-intrinsic molecular machinery is most likely subject to regulatory signals emanating from the niche microenvironment.

1.2 Concept of the Stem Cell Niche and Regulation of Stem Cell Self-Renewal

The stem cell niche, which is the home of the stem cell, is composed of the combined contributions of stationary components and support cells, self-renewal factors, a cytokine milieu, and the ECM (Fig. 4). The niche creates an enriching microenvironment where the stem cells are maintained in an undifferentiated multipotent state by promoting self-renewal and by preventing apoptosis and differentiation. Similar to the actual stem cell, there exists a dynamic balance of signaling from the niche, in conjunction with a dynamic balance in the activation state of the stem cells in the niche. The different niche components play different roles in controlling stem cell activities, and the combined effects of these niche signals regulate the dynamics of quiescence versus proliferation, and of self-renewal versus differentiation in these cells [14, 27–29,

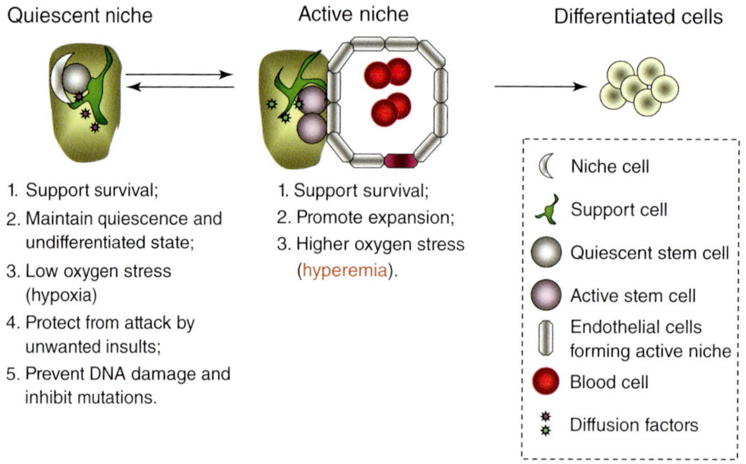

Fig. 5 Niches regulate stem cell self-renewal.

130, 131]. In many tissues, two niches have been described, namely quiescent and active [132–134]. The quiescent niche maintains the stem cell in a quiescent state for long-term maintenance, whereas the active niche directs stem cell proliferation in order to increase numbers and promote differentiation for tissue regeneration purposes (Fig. 5). The dynamic state of stem cells can be regulated by dynamic trafficking of the stem cells back and forth between these two niches (as described in Sect. 1.1.).

1.2.1 **Stationary Components of the Stem Cell Niche**

A primary role of the stem cell niche is to provide a stable physical residence for the stem cell. In order to provide such a residence for stem cell attachment, at least one stationary component is required within the niche to act as a docking apparatus.

The stationary component in some niches is a specialized niche cell that mediates stem cell adhesion via cadherins [29, 104, 131]. In such a niche, the stem cells connect to the niche cells via cadherin-mediated cell adhesion junctions, while niche cells produce factors such as cytokines, chemokines, and self-renewal factors. Most of the self-renewal factors identified are development-related, such as Notch ligands, Wnts, BMPs, and Hedgehog [54, 135–145]. It is important to note that the factors produced by different types of tissue stem cell niche are not exactly the same; also, different types of tissue stem cells utilize different factors to accomplish self-renewal. Interestingly, some of these factors are produced in a cell membrane-binding form which can only be signaled by stem cells when they are physically in contact with niche cells. These contact-dependent signals augment the diffuse signals that form concentration gradients in which the highest levels are immediately localized to niche cells. Such signaling gradients regulate the dynamic activation state of stem cells, and thus serve to maintain the cell in a quiescent state. Should the stem cell detach from the niche, it would lose this juxtacrine signal and find itself in a lower concentration of diffusing niche factors. This would lead to stem cell proliferation and differentiation [135, 139, 146–151].

One example of a stationary niche component is the cap cell/GSC niche in the *Drosophila* ovary, in which cap cells provide the stationary components for the GSC niche. Through E-cadherin-mediated cell–cell adhesion junctions, GSCs connect to the niche cells (cap cells). It has been shown that E-cadherin-mediated cell–cell adhesion junctions are important for keeping the GSCs adjacent to niche signals, so that self-renewal can be accomplished. The niche cells produce a concentration gradient of factors such as Unpaired and BMP, which induce self-renewal signaling in GSCs. Disruption of this junction leads to GSC detachment from cap cells and the migration of stem cells away from the cap cell, which results in stem cell differentiation and depletion [23, 59, 152, 153]. The GSC niche in the *Drosophila* testis is almost identical to that in the ovary [154–157]. Another example of this type of niche is the HSC osteoblastic niche, in which SNO (spindle-shaped N-cadherin$^+$ osteoblastic) cells, through N-cadherin-mediated adhesion junctions, retain HSCs within the niche environment [132, 158–162]. The SNO cells produce factors that include cell membrane-binding forms of stem cell factor (SCF), Jagged, Angiopoietin-1, and also diffuse factors such as stromal-derived factor-1 (SDF-1; also called *CXCL12*), Wnt and BMPs, which restrict proliferation and maintain the "stemness" of stem cells. The disruption of such adhesion junctions or niche signals will significantly compromise HSC activity and most likely result in stem cell depletion [132, 158–161, 163, 164].

In some tissues the stem cell niche does not include a canonical niche cell, and such niches will be referred to as "*noncellular*" niches. In noncellular types of niche the stem cells are attached to the basement membrane (which is a specialized ECM) through integrin-mediated hemidesmosome and focal adhesion junctions (Fig. 4). The self-renewal factors might be concentrated in the basement membrane ECM, where they form a gradient to regulate the dynamics of stem cell activities. Examples of such niches are epithelial stem cell niches in skin and mammary glands [165, 166]. The deletion of integrins from such stem cells leads to a depletion of stem cells in these tissues [165, 167].

In addition to their adhesive function, cadherin- and integrin-mediated attachments of stem cells to niches also act as external mediators to connect the cytoskeleton in stem cells and to control the orientation of the mitotic spindle apparatus and, thus, stem cell polarity. Such polarity has been reported to be implicated in regulating asymmetric division, which is essential for stem cell self-renewal [23, 156, 158, 165].

At this point it is necessary to note that such cadherin- and integrin-mediated attachments have the ability to be repeatedly broken down and reformed, allowing for stem cells to detach from the niche and mobilize, then to follow homing signals and reattach to their niche cells [29, 131].

1.2.2 **Support Cells**

In addition to a stationary niche component, there is also the presence of support cells which accompany the stem cells to form the niche construct. These cells provide vital support by augmenting factors involved in the regulation of self-renewal and survival, while simultaneously effecting the migration or trafficking of stem cells (Figs. 4 and 5). An example of such a component is the CAR (CXCL-12-abundant reticular) cell in the HSC niche [168]. Another example is the presence of escort cells or somatic stem cells in *Drosophila*

GSC niches in either the ovary or testis, respectively [29, 168].

In the hematopoietic system, the CAR cell may be responsible for both the trafficking of HSCs between the quiescent osteoblastic niche and the active vascular niche, as well as secreting additional SDF-1 to augment such secretions from SNO cells [168]. Receptor interaction with SDF-1 causes the HSCs to move towards the source of the ligand, in a process known as *chemotaxis*. The CAR cells also provide survival signals that augment osteoblastic and vascular endothelial niche cell signals. It has been shown that CAR cells secrete SDFs along with interleukin (IL)-6, which has been shown to increase the affinity of HSCs for vascular cell adhesion molecules (VCAMs) and thus the stem cells' affinity for the niche [147, 169–174]. More recently, two other types of support cell – $CD146^+$ osteoprogenitors and $Nestin^+$ mesenchymal stem cells (MSCs) – were identified in BM niches [175, 176]. As with CAR cells, $Nestin^+$ MSCs also accompany HSCs and support HSC activities *in vivo* and *in vitro*. $CD146^+$ osteoprogenitors reside on the sinusoidal wall, and might be involved in the generation of both BM osteoblastic niche cells and vascular niche cells. The relationship among these three types of niche support cells is not yet known, however.

In the *Drosophila* ovary, in addition to GSCs, another type of cell exists, known as somatic stem cells. The cap (niche) cells contact both GSCs and somatic stem cells, this contact being mediated by E-cadherin; such contact supports the self-renewal of both GSCs and somatic stem cells [154, 164, 177, 178]. The somatic stem cells also produce the signaling molecules required for GSC self-renewal [179]. When activated, the GSCs produce progeny cells that migrate towards the opposite end of the ovary, away from the cap cell. Interestingly, somatic stem cells also produce progeny and always accompany the germline cells at different stages of development. In fact, just as is observed in the ovarian niche, the testicular niche is also shared by both GSCs and somatic stem cells [154]. The two types of stem cell are at once competitive for the niche and also are mutually interdependent. In addition, the results of recent studies have suggested that niche cells are also produced by somatic stem cells [154].

1.2.3 Extracellular Components

The next component of the stem cell niche to consider is the extracellular microenvironment, which includes homing molecules responsible for attracting stem cells to the niche, soluble factors responsible for regulating the quiescence or proliferation of stem cells, and ECM components responsible for orientation and location signaling. The extracellular components of the niche are comprised of signals that elicit a response in the stem cell but are not derived from a direct physical contact between the stem cell and the canonical niche cell (Fig. 4). Such signals have several roles to play in stem cell support, including releasing homing molecules, maintaining survival signaling, maintaining a productive attachment to the matrix, and modulating stem cell activation.

Extracellular Matrix (ECM) The ECM consists of materials that occupy the space between the cells of a tissue and which form a network of complex molecules consisting of glycosaminoglycans and hyaluronic acid as well as proteins (including collagens, fibronectins, and laminins), which are produced by both stem cells and niche cells. Stem cells are immersed in the niche ECM through surface receptors such as integrins and CD44 [180–189]. The stem

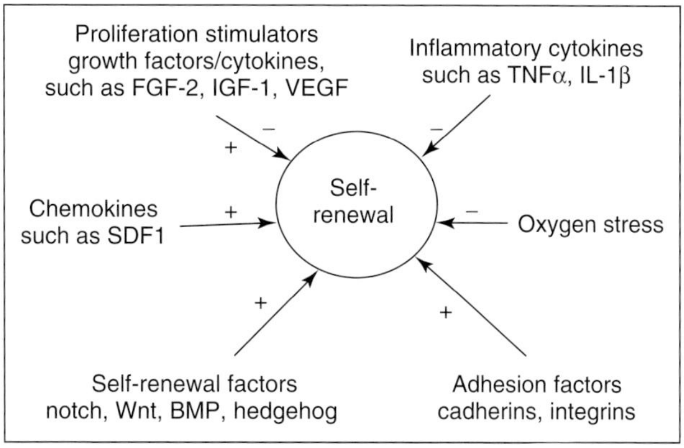

Fig. 6 Niche factors regulate stem cell self-renewal.

cells are connected to the ECM materials, which induce the focal adhesion signaling that is essential for stem cell survival and migration. Integrin-mediated signaling is also implicated in the task of correctly arranging the mitotic spindle during cell division [18, 20, 23], which induces proper cell polarity for asymmetric self-renewal division.

Growth Factors, Cytokines, and Chemokines Growth factors and cytokines induce the dynamic balance of quiescence or proliferation and of self-renewal or differentiation in stem cells by cooperating with other niche factors. Some growth factors and cytokines are required to maintain the quiescent state of stem cells; these include thrombopoietin (TPO), SCF, and SDF1 in the BM HSC niche. Their depletion will increase stem cell proliferation and induce stem cells to be released from their niche environment (this is termed *mobilization*) [139, 147–149], while some other factors – including IL-3 and granulocyte colony-stimulating factor (G-CSF) – in the HSC niche likely induce stem cell proliferation and differentiation [170, 190, 191]. Most of these niche factors also induce survival signaling to prevent stem cell apoptosis or senescence, provoked by inflammatory factors such as interferons and tumor necrosis factor alpha (TNF-α) or other unwanted stress [192–196]. A dynamically balanced signal stimulated by these positive and negative factors regulates the homeostasis of stem cells, and thus the normal functioning of the tissue (Fig. 6).

Stem cells have the capacity to move if suitably stimulated. This provides them with a unique mobility property either within or outside of the niche. Stem cells outside the niche are not as effective or easily maintained as they are inside the niche; therefore, niche cells produce homing factors (chemokines) to attract stem cells to find their home and be retained in their correct niche. For example, in the BM niche, SNO cells, in combination with CAR cells, secrete SDF-1, which acts as a chemoattractant to HSCs [170, 172]. SDF-1 creates a concentration gradient that is strongest in BM niches and becomes weaker farther away from the BM. HSCs express CXCR-4, a receptor with a high affinity for SDF-1, which senses the SDF-1 gradient by activating an intracellular signaling cascade involved in cell mobility [189, 197]. The HSCs use this receptor signal to "follow" the SDF-1 concentration gradient to the BM niche. Once the

HSCs reach the BM niche which has the highest SDF-1 concentration, SDF-1 can then utilize attachment tools to find and connect with a niche within the niche microenvironment.

Finally, many tissue niches are located in an environment that is inherently hypoxic, due to the distance that the stem cell niches are generally located from active vascular diffusion [120, 198, 199]. By limiting the exposure of cells to oxygen, the niche may reduce the effects of DNA damage that are associated with exposure to reactive oxygen species (ROS) [52, 53, 120, 200–203]. Also, since ROS are generated as a natural byproduct of oxidative phosphorylation, the maintenance of stem cells in a quiescent state reduces metabolic demand and may create a beneficial effect of reducing the amount of ROS generated within the stem cell [202, 204, 205]. By incorporating a hypoxic niche with a low metabolic demand, the stem cell niche may limit the stem cell's exposure to ROS, and thereby prevent oxidative stress on the stem cell. It has been shown that a reduced oxidative stress leads to the maintenance of quiescence and "stemness" [198, 199, 205–207]. In neural stem cells (NSCs), which are found in a hypoxic niche environment, the hypoxia-induced transcription factors HIF-1α and HIF-1β are stabilized, leading to an inhibition of apoptosis by p53 and the promotion of self-renewal and proliferation by Cyclin D1 [208]. Thus, in most tissues the hypoxic condition of the niche also plays an essential role in the self-renewal of stem cells.

The entire complement of niche factors execute their roles on stem cells through stimulating signals that regulate the stem cell-intrinsic machinery, by blocking differentiation, preventing cell death, and/or adjusting the speed with which the cells move through the cell cycle. The local niche factors, which are also controlled by systematic regulators such as soluble factors in blood or sympathetic neural stimulation, coordinate stem cells behaviors to fit the systemic requirements of the body [209–212]. Thus, the stem cell niche is equally important to tissue homeostasis and organismal well-being as the stem cell itself.

It is important to mention here that the theory of stem cells and their niches has also been widely studied in other tissues, such as intestine and hair follicles. However, in this chapter attention will be focused on data obtained from studies of murine HSCs and *Drosophila* GSCs to develop the concept of both stem cells and niches. The concepts advanced here should also be applicable to studies of stem cells and niches in other tissues. Those readers interested in a detailed introduction to stem cells and niches in intestinal and hair follicle cells should consult the recent excellent reviews of Linheng Li, Hans Clevers, and Elaine Fuchs [29, 106, 131, 134, 213, 214].

2 The Stem Cell Niche and Its Role in Aging

In many contexts, the concept of things "wearing out" is familiar, and in such cases this is generally associated with a reduced or impaired functionality. In the same way, degenerative changes can occur in tissues of the body. On an organismal level, aging can be understood as a progressive systemic decline in functionality, associated with an increased risk of death over time, and includes corresponding biochemical and histological changes [208]. At the level of individual cells also, changes to cellular structures and altered behaviors occur as a function of the passage of time. The importance of aging effects on stem cells and their niches should not be underestimated, as one of the prime reasons for the existence of stem cells is to counteract the effects of aging.

As organisms age, they need to continuously undergo regeneration in order to maintain their normal functions. The stem cells of tissues address this need by producing new tissue cells to replace old, dysfunctional cells. Another way to consider this is to conceptualize aging as a type of wounding process for which stem cells are the healing response; the aging of stem cells consists of a progressive impairment of this healing response. General aging processes that affect stem cells and niches are often accelerated by systematic damage induced by infections, chemicals, and physical insults (e.g., irradiation) that accrue during day-to-day life. These only serve to enhance the "wear-and tear" on tissue stem cells, retarding further their ability to heal the wounds of aging. As a consequence, during the aging process the functions of tissues will progressively decline due to the compromised ability of stem cells to regenerate the tissue.

2.1 Age-Associated Changes in Stem Cell Behaviors

With the possible exception of adult GSCs [215], most stem cells appear to accumulate some signs of aging [37, 38, 216–222]. These signs are most evident as changes in essential behaviors which, collectively, define stem cells as stem cells. Aging negatively impacts stem cells in several ways, including a reduction in the ability to proliferate, an impairment in the capacity to self-renew, alterations in the process of differentiation, a decreased breadth of developmental potency and plasticity, an increased propensity toward senescence or apoptosis, and a reduction in functional ability (which can be associated with a decrease in cell number, but not necessarily so) [220].

Although the regenerative abilities of all tissues are significantly compromised in aged animals, aging-related changes in stem cells can be significantly different from tissue to tissue. Based on whether their aging-related changes are cell-autonomous or niche-dependent, tissues might be considered as belonging to one of two types:

- In the first type of tissue, represented by BM hematopoietic and neural tissues, aging-related changes observed in stem cells are partly cell-intrinsic and partly extrinsically (niche) caused. The tissue regenerative ability of this type of cell can only be partially induced when seeded into a younger tissue environment [223–226].
- The second type of tissue, which would include MuSCs and liver stem cells, is tissue in which self-renewal and tissue regenerative ability are less affected by aging [211, 227, 228]. The aging-related changes of this type of stem cell are more dependent upon aging-related changes in their niche environment; these types of changes can show rejuvenation if the cells are placed into a younger environment.

2.1.1 Aging in Hematopoietic Stem Cells

The aged hematopoietic system exhibits a reduced capacity to regenerate and return to normal homeostasis after injury or stress. As a consequence, an elevated incidence of age-associated anemia, a decreased competence of the adaptive immune system, and a predisposition to myelogenous diseases are commonly observed in the elderly. Studies have suggested that both cell-intrinsic alterations and defective niche functioning are responsible for these aging-related changes in HSCs.

The hematopoietic reconstitutive capacity of aged HSCs is two- to fourfold poorer than that of their younger counterparts, as shown by competitive repopulation assays following transplantation [229]. It was found that, with respect to reduction in the abilities of self-renewal, homing (finding the way back to BM niches) and lodging (long-term occupancy of BM niches for blood regeneration), multiple cell-intrinsic defects might contribute to functional declines in aged HSCs [221, 222, 230–232]. A reduction of the self-renewal ability of aged HSCs might be a consequence of a reduction in the mechanisms by which senescence and apoptosis are prevented [233] and/or a disruption in cell polarity-related asymmetric division [17]. Impaired homing and lodging abilities might be caused by a compromised response to niche-produced chemoattractive signaling, and a reduction in adhesion molecule-mediated niche attachment, respectively. Indeed, it has been reported that older HSCs express significantly lower levels of adhesion molecules and localize more distantly from BM niche cells compared to younger HSCs [234]. Consistent with these findings, in aged mice an increased frequency of HSCs was observed in the spleen and in the peripheral circulation, which suggested a increased detachment of HSCs from BM niches and subsequent mobilization. Interestingly, this decline in HSC function is not always associated with a decrease in HSC numbers. In fact, the C57BL/6 strain of mice actually shows increased numbers of HSCs with age; these cells are, however, less functional [235]. In addition to these cell-autonomous changes in HSCs, it has been observed that young HSCs are impaired in their homing to aged niches [221]. Furthermore, older BM niches cannot host younger HSCs properly, which leads to the increased mobilization of younger HSCs in older recipients following transplantation [221]. The results of these studies suggest that an increased mobilization of HSCs in old mice is the result of the combined effects of both HSC and niche alterations that result from aging [217, 221].

Increased numbers of granulocyte/monocytes (which are myeloid cells) and a corresponding decrease in both T and B lymphocytes, are consistently observed in older individuals. This suggests a possible myeloid differentiation preference over the lymphoid lineage in older HSCs. This notion was supported by the observation of a systemic increase in the expression of myeloid molecular signatures (such as Pml, Fli1, and Runx1), coupled with a corresponding reduction in the expression of lymphoid cell markers (such as in IL-7 and BAFF) in aged compared to younger HSCs [222]. Interestingly, this skewing in lineage choice in old HSCs seems cell-autonomous because it cannot be reversed by transplantation into young BM [230–232]. Previously, it had been proposed that this age-associated change in HSC lineage potential may be due to an alteration of epigenetic modifications guided by instructive signals emanating from aged systemic and BM niches [76]. However, recent studies have proposed that the situation might be due to the clonal selection of myeloid-biased HSCs. Accordingly, these investigators have proposed that HSCs are heterogeneous populations comprised of lineage-balanced HSCs, myeloid-biased HSCs, and lymphoid-biased HSCs. During the aging process, it is suggested that a gradual increase occurs in myeloid-biased HSCs, together with a reduction in balanced and lymphoid-biased HSCs, and that this leads eventually to the observed predominance of myeloid cells over lymphoid cells [236]. The presumption here, however, is that myeloid-biased HSCs typically have a longer life span than either of the other two types.

2.1.2 Aging in Neural Stem Cells

The development of mitotic quiescence is the most significant change observed in NSCs in the aged animal; such quiescence results in significantly reduced neurogenesis, and in some cases can progress to neurodegenerative disease. It has been shown in various studies that NSCs in both the subventricular zone (SVZ) of the lateral ventricles of the brain and the subgranular zone (SGZ) in the hippocampus become exceedingly quiescent with age [223, 224, 226, 237]. In the young adult mouse SVZ, for example, a high number of proliferating ($MASH^+$) stem cells is observed, but levels of these cells are reduced significantly by middle age; this corresponds to a notably atrophied SVZ seen at this time. Thus, by middle age fewer new neuroblasts are being added to the mouse olfactory bulb. Interestingly, the number of NSCs does not change with aging, as shown by an *in vitro* neurosphere-forming assay; rather, NSCs lose their proliferative capacity and become excessively quiescent. Reduced pERK levels, together with increased pAKT levels in the NSCs of older compared to younger mouse brains, may help account for some of these proliferative changes [226, 238]. Interestingly, it has been shown that when mice become engaged in active physical exercise and/or are raised in an enriched environment, the decline in NSC activity generally seen with aging is reversed [219, 238]. These data suggest that it might be possible to (at least partially) induce quiescent NSCs to become actively neurogenic.

2.1.3 Aging in Muscle Stem Cells

The proliferation and myogenic differentiation of satellite cells, which are stem cells in skeletal muscle tissue (MuSCs), are significantly impacted by aging. Older MuSCs show a higher nucleus-to-cytoplasm ratio than younger cells, and also show a delayed responsiveness to activating stimuli and a reduced proliferative capacity, most likely because of epigenetic stem cell changes resulting from impaired Notch functioning [211, 218]. In addition, the myogenic differentiation of MuSCs is reduced when these cells are older, due to decreased levels of myogenic regulatory factors such as Myf-5, MyoD, and Myf-4, as well as reductions in intracellular levels of the differentiation inducers S100B and RAGE [239, 240].

However, MuSCs typically show practically no change in number with age, except under certain instances. In some strains of mice, for example, females show a reduction in stem cell numbers with age [241, 242], but in most other instances the MuSC numbers do not suffer any notable decline with aging. In fact, it has been shown that their numbers are actually increased if the experimental animals have been exercised regularly [243]. Consistent with these observations, muscle tissue regenerative capacity is not necessarily negatively impacted [55, 244–248].

Both, *in vitro* and *in vivo* studies have shown that Notch and Wnt signaling induce MuSC proliferation and differentiation, whereas transforming growth factor-beta (TGF-β) signaling represses their proliferation [212, 245, 248]. A dynamically balanced network consisting of these positive and negative factors controls the proliferation and differentiation of MuSCs and, in turn, myogenesis [245]. It has been shown that the proliferation of MuSCs in aged animals was significantly impaired due to elevated levels of TGF-β and a reduction of Notch ligand-Delta in niche cells [56, 228, 245, 248]. Interestingly, this aging-related defect can be rescued by the recovery of Notch function, which can re-confer a proliferative capacity similar to that seen in young MuSCs, and suggests a central role for Notch signaling in this network [66, 228].

2.2 Aging-Related Alterations in Cell-Intrinsic Signaling Pathways

The most important and pathogenic-related changes in aged stem cells from many tissues are related to: (i) the gradual but constant accumulation of DNA damage; and (ii) the premature induction of apoptosis and senescence. Together, these two phenomena account for the majority of age-related changes observed to occur in stem cells (Fig. 7).

It has been estimated that each and every cell in the human body incurs several thousand DNA lesions each day. Genetic studies have suggested that accelerating DNA damage, impairing the DNA repair machinery or accelerating telomeric shortening, will lead to premature aging symptoms in many organ systems, and also result in HSC depletion and BM failure [125, 128, 249–251]. As they give rise to all tissue cells, it can be argued that stem cells need to be the most vigorously protected of all cells, and consequently the presence of damaged DNA in stem cells ought to invoke a robust damage repair response. Unfortunately, however, the DNA repair machinery and chromatin remodeling function are both reduced in older HSCs and NSCs. Mice having a reduced DNA repair capacity show an accelerated early aging and reduced stem cell function [119, 221, 252, 253]. To add to this burden, the levels of ROS and inflammatory stresses are increased in the older stem cell niches of both HSCs and NSCs. The FoxO family of proteins, which normally protect against oxidative stress via the transcriptional regulation of manganese superoxide dismutase and catalase gene expression, are reduced in older stem cells,

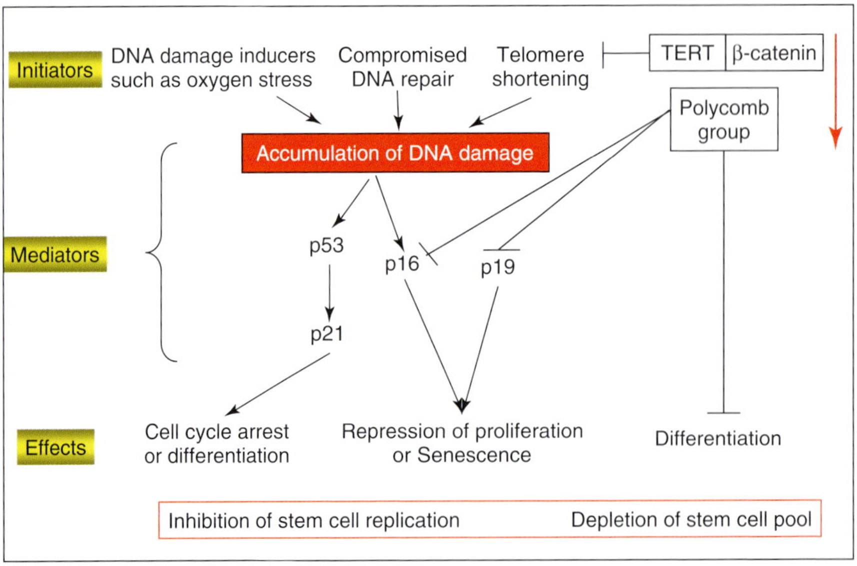

Fig. 7 Signaling pathways related to stem cell aging.

thus likely contributing significantly to the ROS-induced accumulation of damage [52, 53, 254]. In both hematopoietic tissue and brain, increased levels of ROS lead to HSC and NSC depletion with increasing age, attributable to a disruption in PI3 kinase signaling and p38 MAPK signaling [113, 255, 256]. Conversely, a caloric restriction has been shown to actually improve stem cell functioning, possibly because this lowers the metabolic rates in cells so as to reduce ROS levels, leading to less ROS-related toxicity.

It is well known that as cells age, their telomeric chromosomal ends become shorter because telomerase (a unique type of DNA polymerase) is unable to replicate chromosome ends. Such telomeric shortening is likely interpreted as DNA damage by the stem cell. A decline in telomere integrity, therefore, is one of the key stem cell-intrinsic factors leading to symptoms of aging in these cells. Telomere loss and uncapping provokes progressive tissue atrophy, stem cell depletion, organ system failure, and impaired responses to tissue injury [127, 128, 257–259]. To further support this notion, a recent study has demonstrated that aging-related stem cell defects and tissue degeneration can be largely restored by reactivating telomerase [127]. It should be pointed out that not only stem cells but also niche cells are subject to aging-related telomere dysfunction [128, 258].

In response to DNA damage, the cells' protective machinery will become activated, inducing cell-cycle arrest by increasing the expression of negative cell-cycle regulators such as p53-p21^{Waf} [1]. Thus, chronic high levels of p21^{Waf1} are often taken as evidence of aging. This cell-cycle arrest induced by DNA damage allows the DNA repair machinery to fix the problems, lest the damaged cell be eliminated by either apoptotic or senescence mechanisms. This explains why the quiescence, senescence, and apoptosis of stem cells are increased in the elderly [70, 72, 73, 260, 261]. However, increased levels of negative regulators of the cell cycle do not always promote the effects of aging. The role of p21^{Waf1} in the cell is essentially protective, and while it seems to promote aging by impairing stem cell functions in cells with shortened telomeres, in fact the inactivation of p21^{Waf1} in animals showing telomeric dysfunction will extend the survival of HSCs. Actually, the inactivation of p21^{Waf1} in animals showing telomeric dysfunction will extend the survival of HSCs. It may be that increased levels of p53, p21^{Waf1}, and p16^{Ink4a} can delay the signs of stem cell aging, but only if such cells have a robust telomerase activity [262].

In addition to DNA damage, another intracellular sign of aging is the impairment of signaling which should direct the prevention of either apoptosis or senescence. Studies utilizing gene-knockout techniques have shown that Bmi-1, a member of the Polycomb family of proto-oncogenes, is consistently required for the self-renewal of diverse adult stem cells, partly by repressing the expression of tumor suppressors such as p16^{Ink4a} and p19^{Arf}. These tumor suppressors inhibit stem cell regenerative capacity by promoting stem cell death or inducing senescence. Bmi-1 overexpression significantly promotes the self-renewal of HSCs and NSCs to induce hematopoiesis and neurogenesis, respectively, in mice [221]. It was found that Bmi-1 is deficient in older HSCs and NSCs, leading to excessive levels of p16^{Ink4a} and p19^{Arf} in these cells [76, 256]. These studies demonstrate that tumor suppressors such as p16^{Ink4a} and p19^{Arf} induce stem cell senescence and apoptosis, thus playing critical roles in promoting the process of aging [35, 263, 264]. However, Bim-1 also appears to have a significant impact on the BM microenvironment which is independent of $p16^{Ink4a}$ and $p19^{Arf}$ [263]. Likewise, it was found that other Polycomb group proteins such as Ezh2 and Cbx7 would promote self-renewal in several types of stem cells by exactly the same mechanism [74, 76, 265].

It should be understood that DNA damage and apoptosis/senescence signaling are coordinated signaling networks, and always operate together during aging. It has been shown that a loss of Bmi-1 leads to an increased mitochondrial dysfunction, and that old dysfunctional mitochondria generate increased amounts of ROS. This, in turn, can lead to increased p53 activity and hence accelerated apoptosis in these cells, leading to NSC depletion [125].

2.3 The Aging Stem Cell Niche

The stem cell-intrinsic factors described above only partially account for the aging phenotype observed in stem cells [216]. Stem cell niches, like stem cells themselves, are also subject to aging. The major aging-related changes of the stem cell niche are: a reduction in adhesive molecule expression and alterations in the components of the ECM, which compromise stem cell–niche interactions and niche signaling; a reduction in major self-renewal stimulators such as Notch-ligands and Wnt; increased levels of negative regulators of self-renewal, such as TGF-β and IL-β; and a reduction in the activity of molecules that stimulate proliferation, such as fibroblast growth factor-2 (FGF-2), insulin-like growth factor-1 (IGF-1), and vascular endothelial growth factor (VEGF) (see Fig. 6).

The NSC niche, which includes at minimum the SVZ of the lateral ventricles and the SGZ in the hippocampus, undergoes atrophic changes with age, with predictable effects on NSC number and functioning. Several molecular changes have been described that account for impaired niche-NSC signaling with age. Aging induces the activation of microglia and astrocytes in the hippocampus, resulting in increased IL-1β expression [224]. This cytokine is associated with epigenetic modifications, such as increased H3K9 trimethylation, that increase the silencing of gene expression. Niche astrocyte-generated TGF-β1 almost completely blocks the generation of new neurons from NSCs in the brains of aged mice [266]. A poorer neurogenesis also results from a decreased availability of growth factors within the neural niche. Reduced FGF-2 is seen by middle age, as is a decline in IGF-1 functioning [267]. Age-related changes in niche-associated astrocytes result in reduced VEGF and FGF-2 signaling; these changes in turn limit NSC maintenance and contribute to decreased neurogenesis [268]. Perhaps most importantly, in aged stem cells and aged niches, there is reduced production of the key stem cell growth-related factor Notch [17]. If Notch signaling is reduced in the aging brain, NSCs will all respond by differentiating, thus soon depleting the NSC pool [269]. NSCs also have abundant Wnt receptors. Wnt signals from niche cells may decrease with age, reducing neurogenesis from the hippocampus [270].

As would be expected, MuSCs also show aging changes that are related to the increased age of their niche, which is primarily the myofiber itself, with contributions from local fibroblasts, neural elements of the neuromuscular junction, and vascular elements [218]. The aged muscle cell niche is linked to a decrease in the ability of MuSCs to undergo self-renewal. Studies of heterochronic parabionts have shown that older satellite cells, when placed into young niches, show an increased regenerative capacity [256]. This study, in addition to others, suggested that MuSC niche changes are the primary cause of aging in MuSCs, which is regulated by systematic factors present in serum. As noted above for the neural niche, Notch signaling – which is centrally involved in self-renewal behavior in all stem cells – becomes increasingly perturbed in MuSC niches. Dner, which is a Notch activator, is reduced in old compared to young MuSCs. Old MuSCs fail to activate Notch signaling in response to injury due to a lack of concomitant, injury-stimulated Notch ligand Delta upregulation [66]. The reduction in Notch signaling is also accompanied by excessive TGF-β activity, leading to the induction of chronic high levels of CDK inhibitors, thus further impairing self-renewal capacity [245, 271]. Lack of Notch activation, combined with TGF-β upregulation [228, 244], leads to excessive pSmad3-mediated transcription of cell-cycle inhibitors, thereby antagonizing MuSC proliferation, as noted above [33]. Added to this is the observation of increased activity of Wnt inhibitors such as Wif1 and Sfrp2, further compromising self-renewal as MuSCs and their niches age [212, 272].

Signaling from niche to MuSCs is also mediated by various integrins and M-cadherin. Such signaling from the niche changes with age, and damage to (and increased thickness of) the basal lamina with aging is thought to reduce the physical connection between MuSCs and myofibers. The adhesion of MuSCs to their niches is reduced in aging muscle tissue, which may be related to the observation of a thickening of the basal lamina of aged muscle fibers [218].

Aged BM stromal cells are less effective than their younger counterparts in supporting HSC expansion, as shown in *in vitro* studies [273]. The most significant aging-related alteration observed in the BM microenvironment is an increase in the proportion of adipose tissue [274]. It has been reported that adipocytes function as negative regulators of HSC self-renewal [275] and, in BM, are generated by MSCs which are also the parental cells for osteoblasts, fibroblasts, and endothelial cells – the important components of HSC niches. The results of these studies suggested that, as with the lymphoid-myeloid lineage switch in HSCs described above, MSCs in the BM might also shift their lineage differentiation fate during aging. The production of more-negative and less-supportive niche components in aged MSCs might significantly impact the aging of HSCs. Levels of TGF-β and other inflammatory factors are increased in aged animals, and this has been suggested to be the cause of the lineage shift of HSCs [276]. However, whether these factors are also behind the increase in adipose tissue in aged animals merits further study.

3 The Stem Cell Niche and Its Role in Carcinogenesis

3.1 Cancer Stem Cell Concept

During the past 15 years, a growing body of experimental evidence has suggested that the neoplastic cells in many types of tumor are functionally heterogeneous in terms of their ability to initiate tumor growth *in vivo* by transplantation [277, 278]. It was proposed that, as is the case with cells in normal tissues, the neoplastic

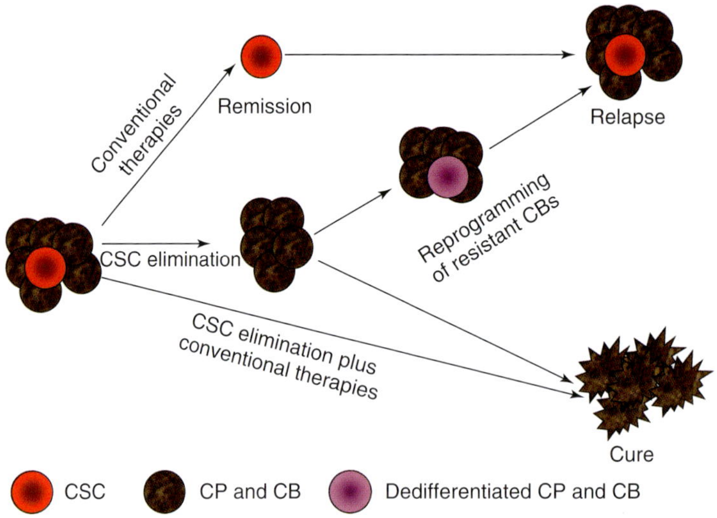

Fig. 8 Model for cancer therapy and cancer stem cell elimination. SC, stem cell; MPP, multipotent progenitor; CP, committed progenitor; M, mature cell; CIC, cancer-initiating cell; mCIC, metastatic cancer-initiating cell; CSC, cancer stem cell; CB, cancer blast.

cells of a tumor are also organized as a hierarchy consisting of stem cells, progenitors, and differentiated cells. Cancer stem cells (CSCs) comprise only a small subset of cancer cells, and are located at the apex of this developmental hierarchy. It has been hypothesized that only CSCs are capable of sustaining tumor formation and growth, giving rise to cancer progenitors and differentiated blasts, which form the bulk of a tumor. The relative resistance of CSCs to standard anti-cancer drugs, which only inhibit the bulk of replicating cancer blasts, has led to the conclusion that CSCs account for the phenomena of drug resistance and disease relapse [279–285]. The notion that only CSCs are able to generate new tumors in distant organs implies that CSCs might be responsible for tumor metastases, the most advanced stage of cancer development [286]. It is now believed that cancers might be cured by completely eliminating all of these carcinogenic "seed cells." Such a suggestion leads to CSC research currently being one of the most important and exciting areas of cancer research, stimulating both clinicians and basic research teams to work in collaboration to seek more efficient and specific methods to kill CSCs. Moreover, it is predicted that a combination of specific CSC killers and more conventional anti-cancer agents will prove to be more efficient in eradicating tumors and lead to improved patient outcomes [287–289] (Fig. 8).

The CSC model includes two important, related concepts [290, 291]:

- Cancer-initiating cells (CICs) or pre-CSCs describe the cell population from which tumors originate; these cells may sustain mutations that can allow them to survive for long enough to accumulate additional mutations, eventually leading to full transformation.
- CSCs describe the existing subpopulation of cancer cells within the tumor which can continuously replenish a tumor's growth during its progression, and can

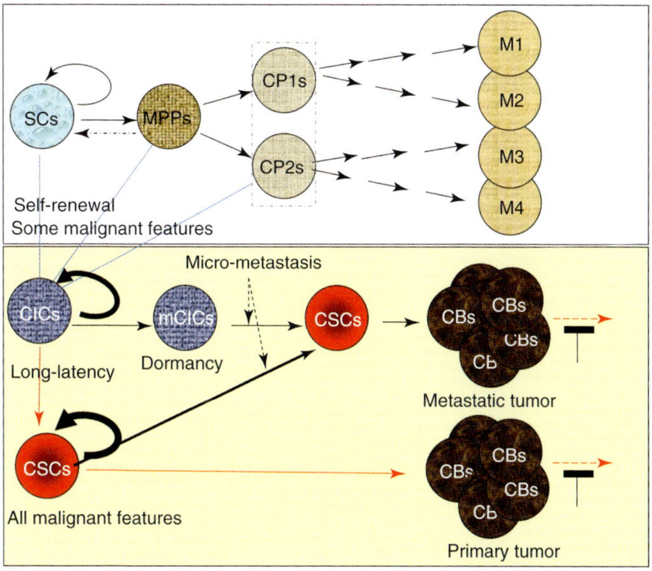

Fig. 9 Model of cancer initiation and incorporation of the cancer stem cell. CSC, cancer stem cell; CB, cancer blast; CP, committed progenitor.

also regenerate tumors during relapse or metastasis.

Although CICs and CSCs can each generate cancers in recipient animals following their transplantation, additional mutations are required in order for CICs to become fully transformed and show uncontrolled growth. Thus, a relatively longer latency is needed for CICs to develop into tumors compared to CSCs (Fig. 9).

3.1.1 Cancer Initiator Cells, the Origin of Cancer

The concept of "CSCs" was actually first conceptualized 150 years ago. The similarity between embryonic tissue and cancerous tumors was noted in early pathology studies, and led to the "embryonal rest" theory of cancer development. The proposal was that cancers are initiated from an activated "embryonic remnant" existing within adult tissues which normally lies dormant [292]. These supposedly deeply sleeping "embryonic remnants" can be reawakened by mutagens or some other such stimulation, and are capable of proliferation in an uncontrolled manner. This theory might have validity when considering some fetal tumors, including teratocarcinomas, Wilms' tumors (nephroblastomas), and neuroblastomas in children [293, 294]. However, modern evidence suggests that most cancers more likely originate from adult tissue cells.

Technical advances in the identification, purification, and phenotypic and functional analyses of adult stem cells in most normal tissues have significantly enhanced the ability to search for the source of cancer cells. The multiple genetic mutations theory of carcinogenesis suggests that tissue stem cells are more likely to be CICs because of their long life span (capacity for self-renewal), which allows them the ability to accumulate all six mutations generally believed to be minimally required for the development of full-blown cancer. Short-lived progenitors would then be eliminated from the tissue before they were able to sustain additional mutations. This notion was supported by experiments which showed that the molecular lesions characteristic of cancer cells existed within cells that showed a normal tissue stem cell phenotype [295–297]. Further support was derived from the fact that CSCs in most tumors express the phenotypic markers of their corresponding normal tissue stem cells. By using *in vivo* labeling and tracking techniques, Barker *et al.* determined that deletion of the *APC* gene, which is linked to the induction of intestinal cancers, originates in crypt stem cells. However, more recent studies have shown that cancers can also originate from progenitors at different stages of maturation. For example, Jamieson *et al.* found that committed myeloid progenitors or granulocyte-macrophage progenitors (GMPs) from Bcr-Abl^+ chronic myeloid leukemia (CML) patients in blast crisis expressed high levels of the active form of β-catenin, and had self-renewal capacity; thus, it was proposed that GMPs could be the CICs in this type of leukemia [296]. This notion has been supported experimentally by studies using murine leukemia models when, by using antibody-based cell-surface marker staining and flow cytometric sorting techniques, different populations of hematopoietic cells – including HSCs, common myeloid progenitors (CMPs), GMPs, and common lymphoid progenitors (CLPs) – were first purified. These purified cells were then infected with a leukemic fusion gene-expressing virus and examined to determine which infected cells could generate leukemia in recipient mice. Through such studies, it was found that all leukemic fusion genes tested could induce the transformation of normal wild-type HSCs to leukemic stem cells (LSCs). However, several leukemic fusion genes, including MLL-AF9 [298], MLL-ENL [299], and MOZ-TIF2 [300], were sufficient to induce the leukemic transformation of differentiated progenitors, even in committed GMPs. The results of these studies suggested that some oncogenes might induce self-renewal signaling in differentiated progenitors, allowing them to sustain additional mutations that would result in a full leukemic transformation. It should be added that this experimental system, in which viral infection is used to overexpress leukemic genes in hematopoietic progenitors, might not reflect the actual level and cell type of leukemic gene expression in human leukemia samples because the promoter used to drive the leukemic genes in viral infections is much stronger than the actual endogenous promoters driving the expression of oncogenes in human leukemic cells. In addition, the actual promoter might have cell-type selection, whereas the artificial viral promoters do not have such selection. Thus, whether human leukemia can actually be initiated from differentiated progenitors still requires considerable further experimental confirmation.

3.1.2 Cancer Stem Cells, Cancer Progression, Metastasis, and Relapse

CSCs represent a subset of cells within a tumor which have the ability to self-renew and can sustain tumor growth. Among all cancer cells, only CSCs are capable of forming tumors upon transplantation, so as to recapitulate the entire cancer cell population of the original tumor. The strategy and techniques used to identify CSCs were exactly the same as those used to identify normal tissue stem cells [277, 278, 301–303]. As is the case with normal tissue stem cells, the number and activities of CSCs can only be determined by functional transplantation assays. Thus, the *in-vivo* transplantation and tumor regeneration method is considered to be the "gold standard" assay to study the tumor initiation capacity of CSCs.

LSCs were the first type of CSC to be described experimentally. By using fluorescence-activated cell sorting (FACS) to separate human acute myeloid leukemia (AML) cells into $CD34^+CD38^-$, $CD34^+CD38^+$, and $CD34^-$ subpopulations, and then transplanting them into non-obese diabetic-severe combined immunodeficient (NOD-SCID) mice (with significantly impaired B- and T-cell development) to study their leukemogenic activity, Dick and colleagues showed that LSCs (also referred to as SCID leukemia-initiating cells; SL-ICs) presented with a $CD34^+CD38^-$ phenotype in the peripheral blood of AML patients [277, 278]. The frequency of these cells was 1/250 000 in patient samples. These data suggested that LSCs were rare, capable of partial differentiation, and were restricted to the immature phenotype associated with HSCs in normal blood differentiation. The finding of LSCs in human leukemic samples was confirmed by several other laboratories [304, 305]. Subsequently, the NOD-SCID mouse-based xenotransplantation system has been used as the "gold standard" assay to functionally examine the frequency and self-renewal capacity of LSCs. By using this assay system, the frequency of LSCs in a leukemia sample was determined by serial dilution and transplantation to determine the smallest number of transplanted cells that could initiate leukemia in recipients. The self-renewal capacity of LSCs was examined by serial transplantation to determine whether the leukemogenic cells could first be expanded in recipient mice, and then be retransplanted into secondary recipients. The use of more highly immunocompromised mouse models, such as the NSG mouse (NOD/SCID, interleukin-2 receptor gamma chain-null, with impaired natural killer (NK) cell development in addition to impaired B- and T-cell development), suggested that the frequency of LSCs ranged from 0.1% to 10%, and was some 200- to 800-fold higher than had previously been proposed [306–308]. Residual immune responses and microenvironmental factors in recipients (e.g., murine cytokines might not be capable of supporting the growth of all human LSCs) might compromise the engraftment capacity of LSCs, yielding an underestimate of the actual frequency of LSCs in patient samples. Nonetheless, the LSC concept of leukemogenesis was strongly supported in almost all of these studies.

By extending this FACS purification-NOD/SCID mouse transplantation system to study CSCs in solid tumors, investigators have subsequently identified CSCs in various human solid tumors, including breast cancer ($CD44^+CD24^{-/low}ESA^+Lineage^-$) [309], brain cancer ($CD133^+$) [310–312], breast and colon cancers ($CD133^+$) [313, 314], pancreatic adenocarcinoma ($CD44^+CD24^+$ ESA^+) [315], and prostate cancer ($CD44^+$) [316]. Moreover, all of these studies demonstrated a unifying concept – the CSC theory – for the development of cancer.

At this point it is important to note that the xenotransplantation technique used in CSC studies has certain significant limitations which could yield unexpected conclusions:

- Sample preparation (e.g., tissue collection and digestion) might cause a loss of the viability, biomarkers, and "stemness" of CSCs.

- Antibody staining and sorting processes (a niche-free procedure) might result in a loss of viability and "stemness" of CSCs.
- Some antibodies used in CSC isolation might influence CSC activities (e.g., engraftment and tumorigenesis) [317, 318].
- Transplantation host and transplantation sites might affect the tumorigenic activity of CSCs [317].

As any or all of these factors could lead to erroneous estimations of the frequency of CSCs [301, 319, 320], their careful consideration is required when interpreting any experimental data obtained.

3.1.3 Cancer Stem Cells are Relatively Chemo- and Radio-Resistant

The discovery of CSCs in leukemia and solid tumors has changed the present view of carcinogenesis and chemotherapy. The current body of evidence suggests that CSCs are generally insensitive to more standard forms of chemotherapy and radiation treatment [283, 321]. These traditional anti-cancer therapies mainly inhibit bulk-replicating cancer cells and induce the temporary remission of cancers; however, CSCs may not be effectively inhibited by these methods, at least not as they are now being utilized. In support of this notion, a significant enrichment of CSCs was observed in tumors following treatment [322], and these residual CSCs – either immediately or after a certain period of dormancy – actively regenerated tumors (termed *tumor relapse*) [323]. Various studies showed that the elevated expression of an LSC "gene signature" is independently associated with adverse outcomes in patients with AML [324, 325]. The frequency and number of CSCs in a tumor will also have a major impact on disease prognosis; for example, the frequency of CSCs at the time of diagnosis correlates closely with the duration of relapse-free survival, such that the higher the CSC frequency in the tumor, the shorter will be the duration of relapse-free survival [308, 326]. It has also been suggested that CSCs can be used to monitor minimal residual disease, and also as an index to evaluate the therapeutic response to specific cancer treatments [324, 327, 328].

CSCs, like normal stem cells, are slow cell cyclers [329] and express higher levels of multidrug-resistance genes (e.g., multifunctional efflux ABC transporters [284, 288, 330]), and thus cannot be targeted by many conventional anti-cancer therapies. In addition, a significantly greater checkpoint activation in response to DNA damage and an increase in DNA repair capacity were found in many CSCs, which might also help to account for the drug and/or radio-resistance of CSCs [322]. Furthermore, the significantly increased levels of NF − κB and/or PI3K/Akt found in many CSCs suggests that CSCs might be protected by these survival signals [283, 331–333]. Therefore, inducing entry into the cell cycle of otherwise quiescent CSCs [282, 329, 334–336], inhibiting the expression of multidrug-resistance genes [284, 330], suppressing DNA damage checkpoint responses (such as a specific inhibitor of Chk1 and Chk2 [322]) and inhibiting PI3K/Akt and NF − κB signaling [332, 337, 338] in CSCs may all factor into the ability to overcome chemo- and/or radio-resistance, and could provide a therapeutic model for more effective treatments for cancers [339]. In addition to these cell-autonomous mechanisms, microenvironmental (niche)-protective activities also contribute to the chemo- and/or radio-resistance of CSCs. Thus, niche signaling molecules could also potentially be used as targets for anti-cancer drugs (see below).

3.1.4 Complexity of Cancer Stem Cells

Most recent studies have suggested that the hierarchy consisting of stem cells – progenitors – differentiated cells in tumors might in fact be considerably more complicated than has been proposed previously [286, 287, 340, 341]:

- First, the size of the CSC compartment or frequency of CSCs in tumors was found to be highly variable in patients. In some types of tumors (e.g., melanoma), single melanoma cells – regardless of phenotype – could reconstitute the disease in an immunocompromised mouse strain, suggesting a high frequency of CSCs in some tumors [319, 342, 343]. Other studies have suggested that CSCs can be induced by increasing the genomic instability of cancer cells, as the increased complexity of genetic lesions in cancer cells will enhance their tumorigenic potential [344]. Thus, an increased frequency of CSCs in tumors might serve as an index of disease progression.
- Second, the biomarkers used when examining CSCs are not always faithfully associated with their tumorigenic activity. Several markers used previously are more likely only to enrich the frequency of CSCs, whereas cancer cells negative for these markers also contain some tumorigenic cells [318, 345–347].
- Third, in some tumors more than two types of CSCs are present [348]. Interestingly, different types of CSC in the same sample can give rise to each other when the correct environment is supplied [348–350]. Most interestingly, significant genetic diversities (such as different panels of gene mutations and copy numbers) were found within neoplastic cells of the same tumor, and CSCs existed in neoplastic cells with different genetic lesions [351, 352].

All of these factors make it very difficult to use CSCs as indicators to monitor disease progression, and also much more difficult to purify CSCs for further biological studies. In addition, the results suggest that CSCs might be more plastic in their phenotypes and genotypes than normal tissue stem cells, owing to alterations in their epigenetic and/or other regulatory machinery [346, 353, 354].

3.2 Cancer Stem Cell Niches

The fact that most cancers could be transplanted into xenographic recipients suggested that cell-autonomous changes in CSCs are the initial pathogenic cause for these diseases [11, 355–357]. However, in order for tumors to develop in recipients, CSCs need to be implanted into a correct microenvironment, combined with specific supporting materials (such as Matrigel®) for their efficient engraftment within the new hosts [343]. The host niche environment not only provides an available space within which CSCs can reside, but also produces the factors required to support CSC growth. In addition, it was found that niches also provide an anti-immune microenvironment to prevent CSCs from an immune-reaction attack [358–360]. Furthermore, donor cell-derived leukemia development in the clinical allogeneic transplantation setting suggested that, in some cases, a lesion in niche cells could be the major cause of the eventual development of cancer [357, 361–363]. However, the architecture and components of CSC niches are not well understood [364, 365].

3.2.1 Contribution of the Niche Microenvironment to Tumorigenesis

The contribution of the microenvironment to tumorigenesis is significantly different from one tumor to another. In most tumors, the cell-intrinsic defects of neoplastic cells are the key factors for tumor formation, while the stromal cell environment functions only in an auxiliary fashion to support tumor cell proliferation and survival. The special gene-expression and protein-secretion profiles of these tumor stromal cells make them more effective than their normal counterparts in supporting tumor cell growth. It has been shown in various studies that mesenchymal stromal cells and endothelial cells isolated from the tumor environment induce a survival signal in cancer cells that protects these cells from the chemotoxic effects of anti-tumor medications and irradiation [366–368]. For example, carcinoma-associated fibroblasts (CAFs) promote $CD44^{+}CD24^{-}$ breast CSC proliferation and expansion through the secretion of SDF-1, which accelerates tumor generation when cotransplanted [369, 370].

However, in some tumors the same molecular lesions in neoplastic cells were also detected in their non-neoplastic supporting cells [11, 355–357]. It was suggested that in these tumors, in addition to the lesion in the neoplastic cells, the molecular lesions observed in stromal cells are absolutely required for tumorigenesis. For example, genetic studies showed that, in addition to homozygous mutations in the *NF1* gene in neoplastic astroglial cells, a heterozygous mutation of the same gene in brain environmental cells was also absolutely necessary for the development of neurofibromas [371–373]. It was also found that Schwann cells, mast cells, and/or microglia with heterozygous *NF1* mutations would produce specific factors, such as hyaluronidase, to promote *Nf1*-null astrocyte growth [374].

Interestingly, in some other tumors the tumor environment plays an instructive role in tumor formation. Experimental data have suggested that certain microenvironmental lesions are sufficient to drive the tumor-like growth of normal tissue stem or progenitor cells. For example, in hematopoietic tissue, allogeneic transplantation studies have suggested that the BM environment in some leukemia patients can induce the same immunophenotypical and molecular features as recipient cancer cells [375]. The specific deletion of the retinoic acid γ receptor (RARγ), the retinoblastoma gene (*Rb*), or *Dicer1* from BM niche cells is sufficient to induce myeloproliferative disorders, and even leukemia development [376–378]. In the intestine, the inactivation of TGF-β signaling in fibroblasts of the lamina propria leads to an abundant expansion of stromal cells and tumor formation, as demonstrated in TGFRII-knockout and Smad4-knockout studies [361, 362]. In the mammary gland, the deregulation of TGF-α in stromal cells leads to the development of breast cancer [379]. The results of these studies suggested that certain oncogenic factors are produced by microenvironmental stromal cells which are normally repressed. These factors most likely are cytokines or inflammatory factors which, when overproduced by niche cells, stimulate an excessive proliferation of normal stem and progenitor cells, resulting ultimately in the development of a tumor. In fact, most cytokines, when overexpressed, can induce hyperplasia or tumor-like growth in their target tissues [334, 380–387].

3.2.2 The Role of the Normal Stem Cell Niche in Tumorigenesis

In theory, because of its repressive activity on proliferation and low oxygen stress feature, the quiescent stem cell niche might inhibit tumorigenesis by restricting DNA damage and the subsequent accumulation of mutations (Fig. 10). In contrast,

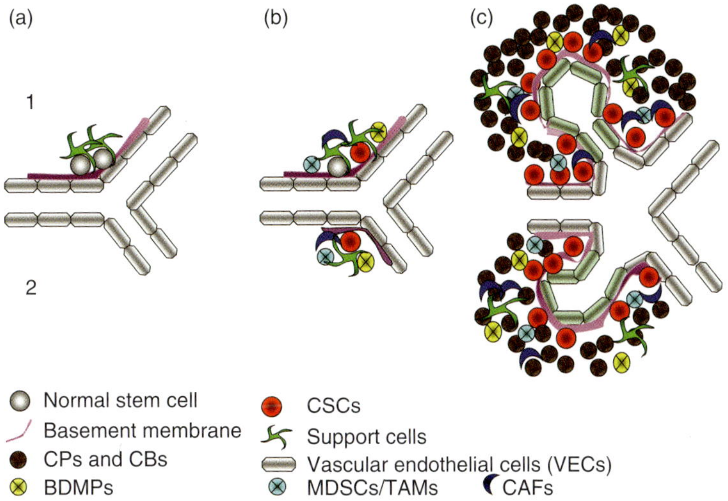

Fig. 10 Cancer stem cell niche model. (a) Normal situation containing a pre-existing niche and normal stem cell; (b) The CSCs either compete with the normal stem cells for space in the existing niche (1) or independently generate a new niche (2); (c) During cancer development, the CSCs induce angiogenesis and create more niche cells to support tumor growth. BDMPs, BM-derived myeloid progenitor cells; MDSCs, myeloid-derived suppressor cells; TAMs, tumor-associated macrophages; CAFs, carcinoma-associated fibroblasts and CSCs. All of these contribute to neovascularization and *de novo* niche formation.

stem cells in a proliferative niche would be subject to a considerable number of mutations, and therefore the proliferative niche that is normally formed by vascular endothelial cells is more likely to provide an environment that is more favorable to support a large expansion of CSCs and, hence, tumor growth. If mutations were to occur in stem cells within a quiescent niche, or stem cells in a proliferative niche were to home back to a quiescent niche after the occurrence of mutations, then these pre-CSCs might remain dormant for relatively long periods of time. Sustaining additional genetic lesions would eventually permit them to overcome the repressive signals emanating from the quiescent niche, or to even convert the quiescent niche to a proliferative niche by inducing neovascularization. Indeed, this might explain the state of dormancy seen in some tumors in clinical patients [388, 389].

Xenotransplantation studies have shown that LSCs may employ the same molecules, and "hijack" the same signaling machinery used by normal HSCs; this would allow them to "home" to where they would compete with normal HSCs to occupy both the endosteal quiescent niche and the endothelial vascular niche [317, 329, 390]. In fact, the BM endosteal niche was found to maintain LSCs in a state of quiescence and to protect them against chemotoxic agent-induced killing. In a colorectal cancer model in which APC deletion is known to be a crucial event, the initiation of cancer was observed from crypt stem cells in the crypt niche [391]. Taken together, the results of these studies have suggested that normal stem cell niches might be required for cancer initiation, but might also play essential roles in disease relapse.

The high frequency of tumor metastases in bone, lung, and liver suggests that certain peculiar qualities must exist regarding these tissues that make them relatively attractive for the CSCs to colonize [392–394]. It was speculated that the normal tissue stem cell niche might also play a central role in a large proportion of tumor cells that become metastatic. For example, the results of studies in animal models have suggested that cancer cells might usurp HSC homing pathways in the mouse to colonize the BM by directly competing with HSCs for occupancy of the HSC niche. Moreover, the manipulation of BM niche size or signaling can have a significant influence on the frequency of bone metastases [395].

The limited size of normal stem cell niches restricts their ability to host a significantly expanded number of CSCs. It is known that tumors do not always form at sites in close proximity to existing normal stem cell niches. Indeed, the presence of LSCs in the BM, spleen, and even in peripheral blood and other tissues during disease progression suggests that they can self-renew and survive independently of normal HSC niches. The results of *in vitro* culture studies have implied that LSCs rely on certain hematopoietic cytokines for their self-renewal and proliferation. It is believed that, *in vivo*, CSCs may induce the proliferation of local cells or attract BM-derived cells to sites where they reside, or are going to occupy. Both local and BM-derived cells form a type of *ad hoc* niche to support CSC proliferation and survival (Fig. 10). This concept was strongly supported by observations of the *de novo* construction of a pre-metastatic niche for incoming CSCs in destination organs during cancer metastasis [396].

3.2.3 **The Potential Components of CSC Niches**

In tumors, in addition to neoplastic cells there are also non-neoplastic components that construct a complex microenvironment to support tumor growth (Fig. 10). The cellular components of this constructed niche include local cells such as stromal fibroblasts, mesenchymal cells, adipocytes, and endothelial cells, as well as BM-derived inflammatory cells. These non-neoplastic cells in the tumor microenvironment impart growth and survival regulatory signals to the neoplastic cells through the elaboration of specific paracrine factors, such as growth factors, chemokines, and ECM proteins. Myofibroblasts (in cancerous tissue, known as cancer-activated fibroblasts, *CAFs*) and α-smooth muscle actin-expressing fibroblasts have been specifically identified as a type of important niche cell for CSCs; such CAFs are significantly increased in most cancers. This type of cell can be induced from local fibroblasts [388] or derived from BM MSCs [389, 397], which actively produce factors such as SDF1 Wnt5a, BMP4, and IL-6 to induce cancer cell expansion. In addition, the blood vasculature has been found to be a critical component of the tumor microenvironment that forms niches for CSCs in some tumors. As an example, a perivascular niche has been described in neuronal tumors which stimulates PI3K/Akt signaling to protect CSCs from apoptosis induced by either radio- or chemo-therapeutic agents [359, 360, 398, 399]. In almost all solid tumors, a significant vascularization was consistently observed due to neoangiogenesis, which is essential for the growth of solid tumors [369, 388, 400]. Transplantation studies have demonstrated that most CSCs preferentially reside in specific zones adjacent to sinusoidal endothelial cells [329, 390, 401, 402], which suggests that neovascularization may provide new niches for the large number of expanded CSCs. Interestingly, both local endothelial cells and BM-derived inflammatory cells contribute significantly to neovascularization in such tumors [359, 403], whereas CAFs promote tumor vascularization by producing VEGF-A [404]. The results of several studies have also suggested that the CSCs also transdifferentiate into endothelial progenitors, and are involved in the build-up of new blood vasculature [401, 405–407].

3.2.4 The Contribution of BM-Derived Inflammatory Cells to the CSC Niche

Inflammatory cells (such as immune cells, mast cells, macrophages, BM-derived myeloid progenitors (BDMPs), and BM-derived endothelial progenitors [408–410]) and mediators such as chemokines, cytokines, and prostaglandins, are present in the microenvironment of most – if not all – tumors (Fig. 4). This pro-tumor inflammatory microenvironment can be established before or after a malignant change occurs, which in turn contributes greatly to tumor progression by enhancing the proliferation, survival, metastasis, and immune surveillance of malignant cells, as well as promoting angiogenesis [403, 411, 412]. BDMPs and myeloid-derived suppressor cells (MDSCs) are the key components of cancer-related inflammation [413–417].

Tumor-secreted and host-secreted inflammatory chemoattractants such as CSF1, IL-1β, IL-6, IL-23, S100A8, S100A9, VEGF-A, TGF-β, lysyl oxidase, and TNF-α in primary tumor sites (*note*: different tumors may produce different chemoattractants) induce the accumulation of MDSCs from BM, an immature population of myeloid cells within tumor sites [380, 384, 403, 410, 413, 415, 416, 418]. These cells downregulate immune surveillance by inhibiting both adaptive and innate anti-tumor immunity. In the tumor microenvironment, hypoxia (via HIF-1α) dramatically alters the function of the MDSCs and redirects their differentiation towards tumor-associated macrophages (TAMs), which have an even higher immune suppressive activity [205]. TAMs can also differentiate from monocytes, which are directly recruited into tumors from the peripheral circulation by tumor-secreted CSF1 [409, 419]. Both MDSCs and TAMs promote the tumor's ability to evade immune attack through a repression of T cell- and NK cell-mediated immune responses. They also enhance tumor progression and metastasis through the secretion of cytokines such as IL-1 β, TNF-α, IL4, and IL6 to promote tumor growth or matrix metalloproteinases (MMPs) to release tumor cells for metastasis [205, 411, 420–424]. Interestingly, MDSCs and $CD11b^+$ myelomonocytic cells can also be incorporated into the tumor endothelium and become directly involved in neovasculogenesis in tumors [425]. Thus, BM-derived inflammatory cells are involved in the construction of a physical niche (blood vasculature) to host CSCs, an immune niche to prevent CSCs from immune attack, and a secretory niche (cytokines and matrix proteins) to promote self-renewal, survival, and motility of CSCs. Moreover, during tumor metastasis, BM-derived myeloid progenitor cells form the key components of the pre-metastatic niche in organs destined for incoming tumor cells. It was observed that primary tumor cells produce factors that upregulate fibronectin in resident fibroblasts or other factors such as collagen IV in destination organs, which direct the $VEGFR1^+CD133^+CD34^+c\text{-}Kit^+$ BDMP cells or $CD11b^+$ myeloid cells to sites of future metastases. Thus, BM-derived inflammatory cells play a critical role in tumor generation, growth, and progression at nearly all stages of tumor development. Reducing the accumulation of MDSCs and other myeloid progenitors by repressing inflammatory mediators could significantly suppress tumorigenesis and limit later tumor progression and metastasis [396, 414, 426–428].

3.2.5 The CSC Niche and Drug Resistance

When the CSCs have found their niche, induced by chemoattractants such as SDF-1, they will become attached to niche cells and the ECM architecture for engraftment. Inside the niche, the cell membrane-binding form of factors such as Notch, SCF, and Sonic Hedgehog, or locally diffused forms of factors such as Wnt, TGF-1β, and SDF-1 can stimulate both self-renewal and survival signals in CSCs [340, 341, 429–432]. The direct interaction of CSCs with the ECM, through adhesion molecules that include integrins or CD44, will induce additional signals to enhance the self-renewal expansion and motility of CSCs [433, 434]. All of these chemical signatures, when stimulated by the niche microenvironment, have been reported to mediate both chemo- and radio-resistance in CSCs [282, 285, 403, 435, 436]. In addition, the cancer-related inflammatory reaction aids in the proliferation and survival of malignant cells by producing cytokines, promoting angiogenesis, and subverting adaptive immune responses, all of which alter the responses of CSCs to hormones and chemotherapeutic agents [357, 385, 403, 437]. It is believed that niche-mediated drug resistance is a transient state, and can be reversed by removing the operative niche signals. Therefore, treatments which combine the inhibition of niche-mediated signaling and of angiogenesis with the repression of cancer-related inflammatory reactions, together with more standard therapeutic strategies, might have significant value in eliminating cancer cells – including the crucial CSCs – from patients. In this way, a dramatic improvement may be created (at least potentially) in the prognosis for this very important class of diseases.

References

1. Shizuru, J.A., Negrin, R.S., and Weissman, I.L. (2005) Hematopoietic stem and progenitor cells: clinical and preclinical regeneration of the hematolymphoid system. *Annu. Rev. Med.*, **56**, 509–538.
2. Orkin, S.H. and Zon, L.I. (2008) Hematopoiesis: an evolving paradigm for stem cell biology. *Cell*, **132**, 631–644.
3. van der Flier, L.G., and Clevers, H. (2009) Stem cells, self-renewal, and differentiation in the intestinal epithelium. *Annu. Rev. Physiol.*, **71**, 241–260.
4. Weissman, I.L. (2002) The road ended up at stem cells. *Immunol. Rev.*, **185**, 159–174.
5. Fuchs, E. (2007) Scratching the surface of skin development. *Nature*, **445**, 834–842.
6. Shi, Y., Sun, G., Zhao, C., and Stewart, R. (2008) Neural stem cell self-renewal. *Crit. Rev. Oncol. Hematol.*, **65**, 43–53.
7. Becker, A.J., McCulloch, E.A., and Till, J.E. (1963) Cytological demonstration of the clonal nature of spleen colonies derived from transplanted mouse marrow cells. *Nature*, **197**, 452–454.
8. Bryder, D., Rossi, D.J., and Weissman, I.L. (2006) Hematopoietic stem cells: the paradigmatic tissue-specific stem cell. *Am. J. Pathol.*, **169**, 338–346.
9. Cotsarelis, G., Cheng, S.Z., Dong, G., Sun, T.T. *et al.* (1989) Existence of slow-cycling limbal epithelial basal cells that can be preferentially stimulated to proliferate: implications on epithelial stem cells. *Cell*, **57**, 201–209.
10. Barker, N., van Es, J.H., Kuipers, J., Kujala, P. *et al.* (2007) Identification of stem cells in small intestine and colon by marker gene Lgr5. *Nature*, **449**, 1003–1007.
11. Reya, T., Morrison, S.J., Clarke, M.F., and Weissman, I.L. (2001) Stem cells, cancer, and cancer stem cells. *Nature*, **414**, 105–111.
12. Weissman, I.L. (2000) Translating stem and progenitor cell biology to the clinic: barriers and opportunities. *Science*, **287**, 1442–1446.
13. Moore, K.A. (2004) Recent advances in defining the hematopoietic stem cell niche. *Curr. Opin. Hematol.*, **11**, 107–111.

14. Morrison, S.J. and Spradling, A.C. (2008) Stem cells and niches: mechanisms that promote stem cell maintenance throughout life. *Cell*, **132**, 598–611.
15. Molofsky, A.V., Pardal, R., and Morrison, S.J. (2004) Diverse mechanisms regulate stem cell self-renewal. *Curr. Opin. Cell Biol.*, **16**, 700–707.
16. Bell, D.R., and Van Zant, G. (2004) Stem cells, aging, and cancer: inevitabilities and outcomes. *Oncogene*, **23**, 7290–7296.
17. Florian, M.C. and Geiger, H. (2010) Concise review: polarity in stem cells, disease, and aging. *Stem Cells*, **28**, 1623–1629.
18. Gonczy, P. (2008) Mechanisms of asymmetric cell division: flies and worms pave the way. *Nat. Rev. Mol. Cell Biol.*, **9**, 355–366.
19. Williams, S.E., Beronja, S., Pasolli, H.A., and Fuchs, E. (2010) Asymmetric cell divisions promote Notch-dependent epidermal differentiation. *Nature*, **470**, 353–358.
20. Lechler, T. and Fuchs, E. (2005) Asymmetric cell divisions promote stratification and differentiation of mammalian skin. *Nature*, **437**, 275–280.
21. Tajbakhsh, S., Rocheteau, P., and Le Roux, I. (2009) Asymmetric cell divisions and asymmetric cell fates. *Annu. Rev. Cell Dev. Biol.*, **25**, 671–699.
22. Wu, M., Kwon, H.Y., Rattis, F., Blum, J. *et al.* (2007) Imaging hematopoietic precursor division in real time. *Cell Stem Cell*, **1**, 541–554.
23. Marthiens, V., Kazanis, I., Moss, L., Long, K. *et al.* (2010) Adhesion molecules in the stem cell niche--more than just staying in shape? *J. Cell Sci.*, **123**, 1613–1622.
24. Yamashita, Y.M., Fuller, M.T., and Jones, D.L. (2005) Signaling in stem cell niches: lessons from the *Drosophila* germline. *J. Cell Sci.*, **118**, 665–672.
25. Conover, J.C. and Notti, R.Q. (2008) The neural stem cell niche. *Cell Tissue Res.*, **331**, 211–224.
26. Suda, T., Arai, F., and Shimmura, S. (2005) Regulation of stem cells in the niche. *Cornea*, **24**, S12–S17.
27. Voog, J. and Jones, D.L. (2010) Stem cells and the niche: a dynamic duo. *Cell Stem Cell*, **6**, 103–115.
28. Zhang, J. and Li, L. (2008) Stem cell niche: microenvironment and beyond. *J. Biol. Chem.*, **283**, 9499–9503.
29. Li, L. and Xie, T. (2005) Stem cell niche: structure and function. *Annu. Rev. Cell Dev. Biol.*, **21**, 605–631.
30. Fuchs, E. (2009) Finding one's niche in the skin. *Cell Stem Cell*, **4**, 499–502.
31. Benitah, S.A., Frye, M., Glogauer, M., and Watt, F.M. (2005) Stem cell depletion through epidermal deletion of Rac1. *Science*, **309**, 933–935.
32. Morrison, S.J., Wandycz, A.M., Akashi, K., Globerson, A., *et al.* (1996) The aging of hematopoietic stem cells. *Nat. Med.*, **2**, 1011–1016.
33. Silva, H. and Conboy, I.M. (2008) *Aging and Stem Cell Renewal*, Harvard Stem Cell Institute, Cambridge, MA.
34. Wang, Y., Guan, Y., Wang, F., Huang, A. *et al.* (2010) Bmi-1 regulates self-renewal, proliferation and senescence of human fetal neural stem cells in vitro. *Neurosci. Lett.*, **476**, 74–78.
35. Nishino, J., Kim, I., Chada, K., and Morrison, S.J. (2008) Hmga2 promotes neural stem cell self-renewal in young but not old mice by reducing p16Ink4a and p19Arf Expression. *Cell*, **135**, 227–239.
36. Lansdorp, P.M. (2005) Role of telomerase in hematopoietic stem cells. *Ann. N. Y. Acad. Sci.*, **1044**, 220–227.
37. Krtolica, A. (2005) Stem cell: balancing aging and cancer. *Int. J. Biochem. Cell Biol.*, **37**, 935–941.
38. Van Zant, G. and Liang, Y. (2003) The role of stem cells in aging. *Exp. Hematol.*, **31**, 659–672.
39. Akala, O.O., Park, I.K., Qian, D., Pihalja, M. *et al.* (2008) Long-term haematopoietic reconstitution by Trp53-/-p16Ink4a-/-p19Arf-/- multipotent progenitors. *Nature*, **453**, 228–232.
40. Chambers, I. and Tomlinson, S.R. (2009) The transcriptional foundation of pluripotency. *Development*, **136**, 2311–2322.
41. Jaenisch, R. and Young, R. (2008) Stem cells, the molecular circuitry of pluripotency and nuclear reprogramming. *Cell*, **132**, 567–582.
42. Ying, Q.L., Wray, J., Nichols, J., Batelle-Morera, L. *et al.* (2008) The ground state of embryonic stem cell self-renewal. *Nature*, **453**, 519–523.
43. Silva, J. and Smith, A. (2008) Capturing pluripotency. *Cell*, **132**, 532–536.

44. Molofsky, A.V., Pardal, R., Iwashita, T., Park, I.K. *et al.* (2003) Bmi-1 dependence distinguishes neural stem cell self-renewal from progenitor proliferation. *Nature*, **425**, 962–967.
45. Park, I.K., Qian, D., Kiel, M., Becker, M.W. *et al.* (2003) Bmi-1 is required for maintenance of adult self-renewing haematopoietic stem cells. *Nature*, **423**, 302–305.
46. Lessard, J. and Sauvageau, G. (2003) Bmi-1 determines the proliferative capacity of normal and leukaemic stem cells. *Nature*, **423**, 255–260.
47. Lowry, W.E., Blanpain, C., Nowak, J.A., Guasch, G. *et al.* (2005) Defining the impact of beta-catenin/Tcf transactivation on epithelial stem cells. *Genes Dev.*, **19**, 1596–1611.
48. Reya, T., Duncan, A.W., Ailles, L., Domen, J. *et al.* (2003) A role for Wnt signalling in self-renewal of haematopoietic stem cells. *Nature*, **423**, 409–414.
49. Satoh, Y., Matsumura, I., Tanaka, H., Ezoe, S. *et al.* (2004) Roles for c-Myc in self-renewal of hematopoietic stem cells. *J. Biol. Chem.*, **279**, 24986–24993.
50. Antonchuk, J., Sauvageau, G., and Humphries, R.K. (2002) HOXB4-induced expansion of adult hematopoietic stem cells ex vivo. *Cell*, **109**, 39–45.
51. Abramovich, C., Pineault, N., Ohta, H., and Humphries, R.K. (2005) Hox genes: from leukemia to hematopoietic stem cell expansion. *Ann. N. Y. Acad. Sci.*, **1044**, 109–116.
52. Tothova, Z., Kollipara, R., Huntly, B.J., Lee, B.H. *et al.* (2007) FoxOs are critical mediators of hematopoietic stem cell resistance to physiologic oxidative stress. *Cell*, **128**, 325–339.
53. Yalcin, S., Zhang, X., Luciano, J.P., Mungamuri, S.K. *et al.* (2008) Foxo3 is essential for the regulation of ataxia telangiectasia mutated and oxidative stress-mediated homeostasis of hematopoietic stem cells. *J. Biol. Chem.*, **283**, 25692–25705.
54. Weber, J.M. and Calvi, L.M. (2010) Notch signaling and the bone marrow hematopoietic stem cell niche. *Bone*, **46**, 281–285.
55. Schubert, C. (2004) Notch, the new muscle booster. *Nat. Med.*, **10**, 24.
56. Conboy, I.M. and Rando, T.A. (2002) The regulation of Notch signaling controls satellite cell activation and cell fate determination in postnatal myogenesis. *Dev. Cell*, **3**, 397–409.
57. Zhang, J. and Li, L. (2005) BMP signaling and stem cell regulation. *Dev. Biol.*, **284**, 1–11.
58. Xi, R., Doan, C., Liu, D., and Xie, T. (2005) Pelota controls self-renewal of germline stem cells by repressing a Bam-independent differentiation pathway. *Development*, **132**, 5365–5374.
59. Song, X., Wong, M.D., Kawase, E., Xi, R. *et al.* (2004) Bmp signals from niche cells directly repress transcription of a differentiation-promoting gene, bag of marbles, in germline stem cells in the *Drosophila* ovary. *Development*, **131**, 1353–1364.
60. Xie, T. and Spradling, A.C. (1998) decapentaplegic is essential for the maintenance and division of germline stem cells in the *Drosophila* ovary. *Cell*, **94**, 251–260.
61. Kai, T. and Spradling, A. (2004) Differentiating germ cells can revert into functional stem cells in *Drosophila melanogaster* ovaries. *Nature*, **428**, 564–569.
62. Pan, L., Chen, S., Weng, C., Call, G. *et al.* (2007) Stem cell aging is controlled both intrinsically and extrinsically in the *Drosophila* ovary. *Cell Stem Cell*, **1**, 458–469.
63. Kitamoto, T. and Hanaoka, K. (2010) Notch3 null mutation in mice causes muscle hyperplasia by repetitive muscle regeneration. *Stem Cells*, **28**, 2205–2216.
64. Mitchell, K.J., Pannerec, A., Cadot, B., Parlakian, A. *et al.* (2010) Identification and characterization of a non-satellite cell muscle resident progenitor during postnatal development. *Nat. Cell Biol.*, **12**, 257–266.
65. Chen, J.F., Tao, Y., Li, J., Deng, Z. *et al.* (2010) microRNA-1 and microRNA-206 regulate skeletal muscle satellite cell proliferation and differentiation by repressing Pax7. *J. Cell Biol.*, **190**, 867–879.
66. Conboy, I.M., Conboy, M.J., Smythe, G.M., and Rando, T.A. (2003) Notch-mediated restoration of regenerative potential to aged muscle. *Science*, **302**, 1575–1577.
67. Iwama, A., Oguro, H., Negishi, M., Kato, Y. *et al.* (2004) Enhanced self-renewal of hematopoietic stem cells mediated by the polycomb gene product Bmi-1. *Immunity*, **21**, 843–851.

68. Godlewski, J., Nowicki, M.O., Bronisz, A., Williams, S. *et al.* (2008) Targeting of the Bmi-1 oncogene/stem cell renewal factor by microRNA-128 inhibits glioma proliferation and self-renewal. *Cancer Res.*, **68**, 9125–9130.
69. Lukacs, R.U., Memarzadeh, S., Wu, H., and Witte, O.N. (2010) Bmi-1 is a crucial regulator of prostate stem cell self-renewal and malignant transformation. *Cell Stem Cell*, **7**, 682–693.
70. Miki, J., Fujimura, Y., Koseki, H., and Kamijo, T. (2007) Polycomb complexes regulate cellular senescence by repression of ARF in cooperation with E2F3. *Genes Cells*, **12**, 1371–1382.
71. Lee, K., Adhikary, G., Balasubramanian, S., Gopalakrishnan, R. *et al.* (2008) Expression of Bmi-1 in epidermis enhances cell survival by altering cell cycle regulatory protein expression and inhibiting apoptosis. *J. Invest. Dermatol.*, **128**, 9–17.
72. Molofsky, A.V., He, S., Bydon, M., Morrison, S.J. *et al.* (2005) Bmi-1 promotes neural stem cell self-renewal and neural development but not mouse growth and survival by repressing the p16Ink4a and p19Arf senescence pathways. *Genes Dev.*, **19**, 1432–1437.
73. Bruggeman, S.W., Valk-Lingbeek, M.E., van der Stoop, P.P., Jacobs, J.J. *et al.* (2005) Ink4a and Arf differentially affect cell proliferation and neural stem cell self-renewal in Bmi1-deficient mice. *Genes Dev.*, **19**, 1438–1443.
74. Kamminga, L.M., Bystrykh, L.V., de Boer, A., Houwer, S. *et al.* (2006) The Polycomb group gene Ezh2 prevents hematopoietic stem cell exhaustion. *Blood*, **107**, 2170–2179.
75. Sander, S., Bullinger, L., Klapproth, K., Fiedler, K. *et al.* (2008) MYC stimulates EZH2 expression by repression of its negative regulator miR-26a. *Blood*, **112**, 4202–4212.
76. De Haan, G. and Gerrits, A. (2007) Epigenetic control of hematopoietic stem cell aging the case of Ezh2. *Ann. N. Y. Acad. Sci.*, **1106**, 233–239.
77. Schuringa, J.J. and Vellenga, E. (2010) Role of the polycomb group gene BMI1 in normal and leukemic hematopoietic stem and progenitor cells. *Curr. Opin. Hematol.*, **17**, 294–299.
78. Petriv, O.I., Kuchenbauer, F., Delaney, A.D., Lecault, V. *et al.* (2010) Comprehensive microRNA expression profiling of the hematopoietic hierarchy. *Proc. Natl Acad. Sci. USA*, **107**, 15443–15448.
79. Garcia, P. and Frampton, J. (2008) Hematopoietic lineage commitment: miRNAs add specificity to a widely expressed transcription factor. *Dev. Cell*, **14**, 815–816.
80. Ferretti, E., De Smaele, E., Miele, E., Laneve, P. *et al.* (2008) Concerted microRNA control of Hedgehog signalling incerebellar neuronal progenitor and tumour cells. *EMBO J.*, **27**, 2616–2627.
81. Han, Y.C., Park, C.Y., Bhagat, G., Zhang, J. *et al.* (2010) microRNA-29a induces aberrant self-renewal capacity in hematopoietic progenitors, biased myeloid development, and acute myeloid leukemia. *J. Exp. Med.*, **207**, 475–489.
82. Arnold, C.P., Tan, R., Zhou, B., Yue, S.B. *et al.* (2011) MicroRNA programs in normal and aberrant stem and progenitor cells. *Genome Res.*, **21** (5), 798–810.
83. Starnes, L.M. and Sorrentino, A. (2010) Regulatory circuitries coordinated by transcription factors and microRNAs at the cornerstone of hematopoietic stem cell self-renewal and differentiation. *Curr. Stem Cell Res. Ther.*, **6** (2), 142–161.
84. Navarro, F. and Lieberman, J. (2010) Small RNAs guide hematopoietic cell differentiation and function. *J. Immunol.*, **184**, 5939–5947.
85. O'Connell, R.M., Chaudhuri, A.A., Rao, D.S., Gibson, W.S. *et al.* (2010) MicroRNAs enriched in hematopoietic stem cells differentially regulate long-term hematopoietic output. *Proc. Natl Acad. Sci. USA*, **107**, 14235–14240.
86. Inui, M., Martello, G., and Piccolo, S. (2010) MicroRNA control of signal transduction. *Nat. Rev. Mol. Cell Biol.*, **11**, 252–263.
87. Herrera-Merchan, A., Cerrato, C., Luengo, G., Dominguez, O. *et al.* (2010) miR-33-mediated downregulation of p53 controls hematopoietic stem cell self-renewal. *Cell Cycle*, **9**, 3277–3285.
88. Hock, H., Hamblen, M.J., Rooke, H.M., Schindler, J.W. *et al.* (2004) Gfi-1 restricts proliferation and preserves functional integrity of haematopoietic stem cells. *Nature*, **431**, 1002–1007.

89. McMahon, K.A., Hiew, S.Y., Hadjur, S., Veiga-Fernandes, H. *et al.* (2007) Mll has a critical role in fetal and adult hematopoietic stem cell self-renewal. *Cell Stem Cell*, **1**, 338–345.
90. Jude, C.D., Climer, L., Xu, D., Artinger, E. *et al.* (2007) Unique and independent roles for MLL in adult hematopoietic stem cells and progenitors. *Cell Stem Cell*, **1**, 324–337.
91. Passegue, E., Wagers, A.J., Giuriato, S., Anderson, W.C. *et al.* (2005) Global analysis of proliferation and cell cycle gene expression in the regulation of hematopoietic stem and progenitor cell fates. *J. Exp. Med.*, **202**, 1599–1611.
92. Cheng, T., Rodrigues, N., Shen, H., Yang, Y. *et al.* (2000) Hematopoietic stem cell quiescence maintained by p21cip1/waf1. *Science*, **287**, 1804–1808.
93. Kippin, T.E., Martens, D.J., and van der Kooy, D. (2005) p21 loss compromises the relative quiescence of forebrain stem cell proliferation leading to exhaustion of their proliferation capacity. *Genes Dev.*, **19**, 756–767.
94. Lacorazza, H.D., Yamada, T., Liu, Y., Miyata, Y. *et al.* (2006) The transcription factor MEF/ELF4 regulates the quiescence of primitive hematopoietic cells. *Cancer Cell*, **9**, 175–187.
95. Wilson, A., Laurenti, E., Oser, G., van der Wath, R.C. *et al.* (2008) Hematopoietic stem cells reversibly switch from dormancy to self-renewal during homeostasis and repair. *Cell*, **135**, 1118–1129.
96. Wilson, A., Laurenti, E., and Trumpp, A. (2009) Balancing dormant and self-renewing hematopoietic stem cells. *Curr. Opin. Genet. Dev.*, **19**, 461–468.
97. Chen, C., Liu, Y., Liu, R., Ikenoue, T. *et al.* (2008) TSC-mTOR maintains quiescence and function of hematopoietic stem cells by repressing mitochondrial biogenesis and reactive oxygen species. *J. Exp. Med.*, **205**, 2397–2408.
98. Zhang, J., Grindley, J.C., Yin, T., Jayasinghe, S. *et al.* (2006) PTEN maintains haematopoietic stem cells and acts in lineage choice and leukaemia prevention. *Nature*, **441** (7092), 518–522.
99. Yilmaz, O.H., Valdez, R., Theisen, B.K., Guo, W. *et al.* (2006) Pten dependence distinguishes haematopoietic stem cells from leukaemia-initiating cells. *Nature*, **441** (7092), 475–482.
100. Foudi, A., Hochedlinger, K., Van Buren, D., Schindler, J.W. *et al.* (2009) Analysis of histone 2B-GFP retention reveals slowly cycling hematopoietic stem cells. *Nat. Biotechnol.*, **27**, 84–90.
101. Domen, J., Cheshier, S.H., and Weissman, I.L. (2000) The role of apoptosis in the regulation of hematopoietic stem cells: overexpression of Bcl-2 increases both their number and repopulation potential. *J. Exp. Med.*, **191**, 253–264.
102. Domen, J. (2001) The role of apoptosis in regulating hematopoietic stem cell numbers. *Apoptosis*, **6**, 239–252.
103. Motoyama, N., Wang, F., Roth, K.A., Sawa, H. *et al.* (1995) Massive cell death of immature hematopoietic cells and neurons in Bcl-x-deficient mice. *Science*, **267**, 1506–1510.
104. Kuang, S., Gillespie, M.A., and Rudnicki, M.A. (2008) Niche regulation of muscle satellite cell self-renewal and differentiation. *Cell Stem Cell*, **2**, 22–31.
105. Hock, H., Meade, E., Medeiros, S., Schindler, J.W. *et al.* (2004) Tel/Etv6 is an essential and selective regulator of adult hematopoietic stem cell survival. *Genes Dev.*, **18**, 2336–2341.
106. Fuchs, E. (2009) The tortoise and the hair: slow-cycling cells in the stem cell race. *Cell*, **137**, 811–819.
107. Chen, J. (2004) Senescence and functional failure in hematopoietic stem cells. *Exp. Hematol.*, **32**, 1025–1032.
108. Smith, K.S., Chanda, S.K., Lingbeek, M., Ross, D.T. *et al.* (2003) Bmi-1 regulation of INK4A-ARF is a downstream requirement for transformation of hematopoietic progenitors by E2a-Pbx1. *Mol. Cell*, **12**, 393–400.
109. Lindsten, T., Zong, W.X., and Thompson, C.B. (2005) Defining the role of the Bcl-2 family of proteins in the nervous system. *Neuroscientist*, **11**, 10–15.
110. Opferman, J.T., Iwasaki, H., Ong, C.C., Suh, H. *et al.* (2005) Obligate role of anti-apoptotic MCL-1 in the survival of hematopoietic stem cells. *Science*, **307**, 1101–1104.
111. Leung, C.G., Xu, Y., Mularski, B., Liu, H. *et al.* (2007) Requirements for survivin in terminal differentiation of erythroid cells

and maintenance of hematopoietic stem and progenitor cells. *J. Exp. Med.*, **204**, 1603–1611.

112. Orelio, C. and Dzierzak, E. (2007) Bcl-2 expression and apoptosis in the regulation of hematopoietic stem cells. *Leuk. Lymphoma*, **48**, 16–24.
113. Ito, K., Hirao, A., Arai, F., Matsuoka, S. *et al.* (2004) Regulation of oxidative stress by ATM is required for self-renewal of haematopoietic stem cells. *Nature*, **431**, 997–1002.
114. Domen, J., Gandy, K.L., and Weissman, I.L. (1998) Systemic overexpression of BCL-2 in the hematopoietic system protects transgenic mice from the consequences of lethal irradiation. *Blood*, **91**, 2272–2282.
115. Peters, R., Leyvraz, S., and Perey, L. (1998) Apoptotic regulation in primitive hematopoietic precursors. *Blood*, **92**, 2041–2052.
116. Ogilvy, S., Metcalf, D., Print, C.G., Bath, M.L. *et al.* (1999) Constitutive Bcl-2 expression throughout the hematopoietic compartment affects multiple lineages and enhances progenitor cell survival. *Proc. Natl Acad. Sci. USA*, **96**, 14943–14948.
117. Tang, M., Wei, X., Guo, Y., Breslin, P. *et al.* (2008) TAK1 is required for the survival of hematopoietic cells and hepatocytes in mice. *J. Exp. Med.*, **205** (7), 1611–1619.
118. Domen, J. and Weissman, I.L. (2000) Hematopoietic stem cells need two signals to prevent apoptosis; BCL-2 can provide one of these, Kitl/c-Kit signaling the other. *J. Exp. Med.*, **192**, 1707–1718.
119. Park, Y. and Gerson, S.L. (2005) DNA repair defects in stem cell function and aging. *Annu. Rev. Med.*, **56**, 495–508.
120. Eliasson, P. and Jonsson, J.I. (2010) The hematopoietic stem cell niche: low in oxygen but a nice place to be. *J. Cell. Physiol.*, **222**, 17–22.
121. Blanpain, C., Mohrin, M., Sotiropoulou, P.A., and Passegue, E. (2011) DNA-damage response in tissue-specific and cancer stem cells. *Cell Stem Cell*, **8**, 16–29.
122. Hasty, P., Campisi, J., Hoeijmakers, J., van Steeg, H. *et al.* (2003) Aging and genome maintenance: lessons from the mouse? *Science*, **299**, 1355–1359.
123. Kenyon, J. and Gerson, S.L. (2007) The role of DNA damage repair in aging of adult stem cells. *Nucleic Acids Res.*, **35**, 7557–7565.
124. Lombard, D.B., Chua, K.F., Mostoslavsky, R., Franco, S. *et al.* (2005) DNA repair, genome stability, and aging. *Cell*, **120**, 497–512.
125. Sahin, E. and Depinho, R.A. (2010) Linking functional decline of telomeres, mitochondria and stem cells during ageing. *Nature*, **464**, 520–528.
126. Flores, I., Benetti, R., and Blasco, M.A. (2006) Telomerase regulation and stem cell behaviour. *Curr. Opin. Cell Biol.*, **18**, 254–260.
127. Jaskelioff, M., Muller, F.L., Paik, J.H., Thomas, E. *et al.* (2011) Telomerase reactivation reverses tissue degeneration in aged telomerase-deficient mice. *Nature*, **469**, 102–106.
128. Ju, Z. and Lenhard Rudolph, K. (2008) Telomere dysfunction and stem cell ageing. *Biochimie*, **90**, 24–32.
129. Shay, J.W. and Wright, W.E. (2010) Telomeres and telomerase in normal and cancer stem cells. *FEBS Lett.*, **584**, 3819–3825.
130. Moore, K.A. and Lemischka, I.R. (2006) Stem cells and their niches. *Science*, **311**, 1880–1885.
131. Xie, T. and Li, L. (2007) Stem cells and their niche: an inseparable relationship. *Development*, **134**, 2001–2006.
132. Zhang, J., Niu, C., Ye, L., Huang, H. *et al.* (2003) Identification of the haematopoietic stem cell niche and control of the niche size. *Nature*, **425**, 836–841.
133. Kiel, M.J., Yilmaz, O.H., Iwashita, T., Yilmaz, O.H. *et al.* (2005) SLAM family receptors distinguish hematopoietic stem and progenitor cells and reveal endothelial niches for stem cells. *Cell*, **121**, 1109–1121.
134. Li, L. and Clevers, H. (2010) Coexistence of quiescent and active adult stem cells in mammals. *Science*, **327**, 542–545.
135. Fleming, H.E., Janzen, V., Celso, C.L., Guo, J. *et al.* (2008) Wnt signaling in the niche enforces hematopoietic stem cell quiescence and is necessary to preserve self-renewal in vivo. *Cell Stem Cell*, **2**, 274–283.
136. Ferraro, F., Celso, C.L., and Scadden, D. (2010) Adult stem cells and their niches. *Adv. Exp. Med. Biol.*, **695**, 155–168.
137. Sato, T., van Es, J.H., Snippert, H.J., Stange, D.E. *et al.* (2011) Paneth cells constitute the niche for Lgr5 stem cells in intestinal crypts. *Nature*, **469**, 415–418.

138. Trowbridge, J.J., Scott, M.P., and Bhatia, M. (2006) Hedgehog modulates cell cycle regulators in stem cells to control hematopoietic regeneration. *Proc. Natl Acad. Sci. USA*, **103**, 14134–14139.
139. Thoren, L.A., Liuba, K., Bryder, D., Nygren, J.M. *et al.* (2008) Kit regulates maintenance of quiescent hematopoietic stem cells. *J. Immunol.*, **180**, 2045–2053.
140. Nie, Y., Han, Y.C., and Zou, Y.R. (2008) CXCR4 is required for the quiescence of primitive hematopoietic cells. *J. Exp. Med.*, **205**, 777–783.
141. Calvi, L.M., Adams, G.B., Weibrecht, K.W., Weber, J.M. *et al.* (2003) Osteoblastic cells regulate the haematopoietic stem cell niche. *Nature*, **425**, 841–846.
142. Rattis, F.M., Voermans, C., and Reya, T. (2004) Wnt signaling in the stem cell niche. *Curr. Opin. Hematol.*, **11**, 88–94.
143. Zeng, Y.A. and Nusse, R. (2010) Wnt proteins are self-renewal factors for mammary stem cells and promote their long-term expansion in culture. *Cell Stem Cell*, **6**, 568–577.
144. Joshi, P.A., Jackson, H.W., Beristain, A.G., Di Grappa, M.A. *et al.* (2010) Progesterone induces adult mammary stem cell expansion. *Nature*, **465**, 803–807.
145. Aguirre, A., Rubio, M.E., and Gallo, V. (2010) Notch and EGFR pathway interaction regulates neural stem cell number and self-renewal. *Nature*, **467**, 323–327.
146. Heissig, B., Hattori, K., Dias, S., Friedrich, M. *et al.* (2002) Recruitment of stem and progenitor cells from the bone marrow niche requires MMP-9 mediated release of kit-ligand. *Cell*, **109**, 625–637.
147. Tzeng, Y.S., Li, H., Kang, Y.L., Chen, W.C. *et al.* (2011) Loss of Cxcl12/Sdf-1 in adult mice decreases the quiescent state of hematopoietic stem/progenitor cells and alters the pattern of hematopoietic regeneration after myelosuppression. *Blood*, **117**, 429–439.
148. Yoshihara, H., Arai, F., Hosokawa, K., Hagiwara, T. *et al.* (2007) Thrombopoietin/MPL signaling regulates hematopoietic stem cell quiescence and interaction with the osteoblastic niche. *Cell Stem Cell*, **1**, 685–697.
149. Qian, H., Buza-Vidas, N., Hyland, C.D., Jensen, C.T. *et al.* (2007) Critical role of thrombopoietin in maintaining adult quiescent hematopoietic stem cells. *Cell Stem Cell*, **1**, 671–684.
150. Stier, S., Ko, Y., Forkert, R., Lutz, C. *et al.* (2005) Osteopontin is a hematopoietic stem cell niche component that negatively regulates stem cell pool size. *J. Exp. Med.*, **201**, 1781–1791.
151. Adams, G.B. and Scadden, D.T. (2006) The hematopoietic stem cell in its place. *Nat. Immunol.*, **7**, 333–337.
152. Song, X. and Xie, T. (2002) DE-cadherin-mediated cell adhesion is essential for maintaining somatic stem cells in the *Drosophila* ovary. *Proc. Natl Acad. Sci. USA*, **99**, 14813–14818.
153. Song, X., Zhu, C.H., Doan, C., and Xie, T. (2002) Germline stem cells anchored by adherens junctions in the *Drosophila* ovary niches. *Science*, **296**, 1855–1857.
154. Voog, J., D'Alterio, C., and Jones, D.L. (2008) Multipotent somatic stem cells contribute to the stem cell niche in the *Drosophila* testis. *Nature*, **454**, 1132–1136.
155. Boyle, M., Wong, C., Rocha, M., and Jones, D.L. (2007) Decline in self-renewal factors contributes to aging of the stem cell niche in the *Drosophila* testis. *Cell Stem Cell*, **1**, 470–478.
156. Yamashita, Y.M., Jones, D.L., and Fuller, M.T. (2003) Orientation of asymmetric stem cell division by the APC tumor suppressor and centrosome. *Science*, **301**, 1547–1550.
157. Hombria, J.C. and Brown, S. (2002) The fertile field of *Drosophila* Jak/STAT signalling. *Curr. Biol.*, **12**, R569–R575.
158. Fumio Arai, A.H., Kentaroh, H., and Toshio, S. (2004) Role of cell adhesion in the maintenance of hematopoietic stem cells in the bone marrow niche. *Blood* (ASH Annual Meeting Abstracts), **104**, 669.
159. Haug, J.S., He, X.C., Grindley, J.C., Wunderlich, J.P. *et al.* (2008) N-cadherin expression level distinguishes reserved versus primed states of hematopoietic stem cells. *Cell Stem Cell*, **2**, 367–379.
160. Hosokawa, K., Arai, F., Yoshihara, H., Iwasaki, H. *et al.* (2010) Knockdown of N-cadherin suppresses the long-term engraftment of hematopoietic stem cells. *Blood*, **116**, 554–563.
161. Wein, F., Pietsch, L., Saffrich, R., Wuchter, P. *et al.* (2010) N-cadherin is expressed on human hematopoietic progenitor cells and mediates interaction with human

mesenchymal stromal cells. *Stem Cell Res.*, **4**, 129–139.
162. Wilson, A., Murphy, M.J., Oskarsson, T., Kaloulis, K. *et al.* (2004) c-Myc controls the balance between hematopoietic stem cell self-renewal and differentiation. *Genes Dev.*, **18**, 2747–2763.
163. Arai, F., Hirao, A., Ohmura, M., Sato, H. *et al.* (2004) Tie2/angiopoietin-1 signaling regulates hematopoietic stem cell quiescence in the bone marrow niche. *Cell*, **118**, 149–161.
164. Xie, T., Song, X., Jin, Z., Pan, L. *et al.* (2008) Interactions between stem cells and their niche in the *Drosophila* ovary. *Cold Spring Harbor Symp. Quant. Biol.*, **73**, 39–47.
165. Taddei, I., Deugnier, M.A., Faraldo, M.M., Petit, V. *et al.* (2008) Beta1 integrin deletion from the basal compartment of the mammary epithelium affects stem cells. *Nat. Cell Biol.*, **10**, 716–722.
166. Tumbar, T., Guasch, G., Greco, V., Blanpain, C. *et al.* (2004) Defining the epithelial stem cell niche in skin. *Science*, **303**, 359–363.
167. Watt, F.M. (2002) Role of integrins in regulating epidermal adhesion, growth and differentiation. *EMBO J.*, **21**, 3919–3926.
168. Sugiyama, T., Kohara, H., Noda, M., and Nagasawa, T. (2006) Maintenance of the hematopoietic stem cell pool by CXCL12-CXCR4 chemokine signaling in bone marrow stromal cell niches. *Immunity*, **25**, 977–988.
169. Jung, Y., Wang, J., Schneider, A., Sun, Y.X. *et al.* (2006) Regulation of SDF-1 (CXCL12) production by osteoblasts; a possible mechanism for stem cell homing. *Bone*, **38**, 497–508.
170. Lapidot, T., Dar, A., and Kollet, O. (2005) How do stem cells find their way home? *Blood*, **106** (6), 1901–1910.
171. Peled, A., Kollet, O., Ponomaryov, T., Petit, I. *et al.* (2000) The chemokine SDF-1 activates the integrins LFA-1, VLA-4, and VLA-5 on immature human CD34(+) cells: role in transendothelial/stromal migration and engraftment of NOD/SCID mice. *Blood*, **95**, 3289–3296.
172. Voermans, C., Anthony, E.C., Mul, E., van der Schoot, E. *et al.* (2001) SDF-1-induced actin polymerization and migration in human hematopoietic progenitor cells. *Exp. Hematol.*, **29**, 1456–1464.
173. Sharma, M., Afrin, F., Satija, N., Tripathi, R.P. *et al.* (2011) Stromal-derived factor-1/CXCR4 signaling: indispensable role in homing and engraftment of hematopoietic stem cells in bone marrow. *Stem Cells Dev.*, **20** (6), 933–946.
174. Omatsu, Y., Sugiyama, T., Kohara, H., Kondoh, G. *et al.* (2010) The essential functions of adipo-osteogenic progenitors as the hematopoietic stem and progenitor cell niche. *Immunity*, **33**, 387–399.
175. Tripodo, C., Di Bernardo, A., Ternullo, M.P., Guarnotta, C. *et al.* (2009) CD146(+) bone marrow osteoprogenitors increase in the advanced stages of primary myelofibrosis. *Haematologica*, **94**, 127–130.
176. Mendez-Ferrer, S., Michurina, T.V., Ferraro, F., Mazloom, A.R. *et al.* (2010) Mesenchymal and haematopoietic stem cells form a unique bone marrow niche. *Nature*, **466**, 829–834.
177. Kirilly, D. and Xie, T. (2007) The *Drosophila* ovary: an active stem cell community. *Cell Res.*, **17**, 15–25.
178. Kai, T. and Spradling, A. (2003) An empty *Drosophila* stem cell niche reactivates the proliferation of ectopic cells. *Proc. Natl Acad. Sci. USA*, **100**, 4633–4638.
179. Lopez-Onieva, L., Fernandez-Minan, A., and Gonzalez-Reyes, A. (2008) Jak/Stat signalling in niche support cells regulates dpp transcription to control germline stem cell maintenance in the *Drosophila* ovary. *Development*, **135**, 533–540.
180. Wilson, A. and Trumpp, A. (2006) Bone-marrow haematopoietic-stem-cell niches. *Nat. Rev. Immunol.*, **6**, 93–106.
181. Ellis, S.J. and Tanentzapf, G. (2010) Integrin-mediated adhesion and stem-cell-niche interactions. *Cell Tissue Res.*, **339**, 121–130.
182. Tanentzapf, G., Devenport, D., Godt, D., and Brown, N.H. (2007) Integrin-dependent anchoring of a stem-cell niche. *Nat. Cell Biol.*, **9**, 1413–1418.
183. Priestley, G.V., Scott, L.M., Ulyanova, T., and Papayannopoulou, T. (2005) Lack of alpha4 integrin expression in stem cells restricts competitive function and self-renewal activity. *Blood*, **107** (7), 2959–2967.
184. Scott, L.M., Priestley, G.V., and Papayannopoulou, T. (2003) Deletion of alpha4 integrins from adult hematopoietic cells reveals roles in homeostasis,

regeneration, and homing. *Mol. Cell. Biol.*, **23**, 9349–9360.
185. Umemoto, T., Yamato, M., Shiratsuchi, Y., Terasawa, M. *et al.* (2006) Expression of integrin beta3 is correlated to the properties of quiescent hemopoietic stem cells possessing the side population phenotype. *J. Immunol.*, **177**, 7733–7739.
186. Campos, L.S., Leone, D.P., Relvas, J.B., Brakebusch, C. *et al.* (2004) beta1 integrins activate a MAPK signalling pathway in neural stem cells that contributes to their maintenance. *Development*, **131**, 3433–3444.
187. Krause, D.S., Lazarides, K., von Andrian, U.H., and Van Etten, R.A. (2006) Requirement for CD44 in homing and engraftment of BCR-ABL-expressing leukemic stem cells. *Nat. Med.*, **12**, 1175–1180.
188. Jin, L., Hope, K.J., Zhai, Q., Smadja-Joffe, F. *et al.* (2006) Targeting of CD44 eradicates human acute myeloid leukemic stem cells. *Nat. Med.*, **12**, 1167–1174.
189. Avigdor, A., Goichberg, P., Shivtiel, S., Dar, A. *et al.* (2004) CD44 and hyaluronic acid cooperate with SDF-1 in the trafficking of human CD34+ stem/progenitor cells to bone marrow. *Blood*, **103**, 2981–2989.
190. Cottler-Fox, M.H., Lapidot, T., Petit, I., Kollet, O. *et al.* (2003) Stem cell mobilization. *Hematology (Am. Soc. Hematol. Educ. Program)*, 419–437.
191. Nervi, B., Link, D.C., and DiPersio, J.F. (2006) Cytokines and hematopoietic stem cell mobilization. *J. Cell. Biochem.*, **99**, 690–705.
192. Baldridge, M.T., King, K.Y., Boles, N.C., Weksberg, D.C. *et al.* (2010) Quiescent haematopoietic stem cells are activated by IFN-gamma in response to chronic infection. *Nature*, **465**, 793–797.
193. Essers, M.A., Offner, S., Blanco-Bose, W.E., Waibler, Z. *et al.* (2009) IFNalpha activates dormant haematopoietic stem cells in vivo. *Nature*, **458**, 904–908.
194. Sato, T., Onai, N., Yoshihara, H., Arai, F. *et al.* (2009) Interferon regulatory factor-2 protects quiescent hematopoietic stem cells from type I interferon-dependent exhaustion. *Nat. Med.*, **15**, 696–700.
195. Mozzetta, C., Consalvi, S., Saccone, V., Forcales, S.V. *et al.* (2010) Selective control of Pax7 expression by TNF-activated p38alpha/polycomb repressive complex 2 (PRC2) signaling during muscle satellite cell differentiation. *Cell Cycle*, **10**, 191–198.
196. Palacios, D., Mozzetta, C., Consalvi, S., Caretti, G. *et al.* (2010) TNF/p38alpha/polycomb signaling to Pax7 locus in satellite cells links inflammation to the epigenetic control of muscle regeneration. *Cell Stem Cell*, **7**, 455–469.
197. Dar, A., Goichberg, P., Shinder, V., Kalinkovich, A. *et al.* (2005) Chemokine receptor CXCR4-dependent internalization and resecretion of functional chemokine SDF-1 by bone marrow endothelial and stromal cells. *Nat. Immunol.*, **6**, 1038–1046.
198. Simsek, T., Kocabas, F., Zheng, J., Deberardinis, R.J. *et al.* (2010) The distinct metabolic profile of hematopoietic stem cells reflects their location in a hypoxic niche. *Cell Stem Cell*, **7**, 380–390.
199. Mohyeldin, A., Garzon-Muvdi, T., and Quinones-Hinojosa, A. (2010) Oxygen in stem cell biology: a critical component of the stem cell niche. *Cell Stem Cell*, **7**, 150–161.
200. Ito, K., Hirao, A., Arai, F., Takubo, K. *et al.* (2006) Reactive oxygen species act through p38 MAPK to limit the lifespan of hematopoietic stem cells. *Nat. Med.*, **12** (4), 446–451.
201. Iiyama, M., Kakihana, K., Kurosu, T., and Miura, O. (2006) Reactive oxygen species generated by hematopoietic cytokines play roles in activation of receptor-mediated signaling and in cell cycle progression. *Cell. Signal.*, **18**, 174–182.
202. Kentaro Hosokawa, F.A., Yoshihara, H., Takubo, K., Ito, K. *et al.* (2006) Reactive oxygen species control hematopoietic stem cell-niche interaction through the regulation of N-cadherin. *Blood* (ASH Annual Meeting Abstracts), **108**, 86.
203. Omori, E., Morioka, S., Matsumoto, K., and Ninomiya-Tsuji, J. (2008) TAK1 regulates reactive oxygen species and cell death in keratinocytes, which is essential for skin integrity. *J. Biol. Chem.*, **283**, 26161–26168.
204. Ceradini, D.J. and Gurtner, G.C. (2005) Homing to hypoxia: HIF-1 as a mediator of progenitor cell recruitment to injured tissue. *Trends Cardiovasc. Med.*, **15**, 57–63.
205. Corzo, C.A., Condamine, T., Lu, L., Cotter, M.J. *et al.* (2010) HIF-1alpha regulates function and differentiation of myeloid-derived

suppressor cells in the tumor microenvironment. *J. Exp. Med.*, **207**, 2439–2453.

206. Gustafsson, M.V., Zheng, X., Pereira, T., Gradin, K. *et al.* (2005) Hypoxia requires notch signaling to maintain the undifferentiated cell state. *Dev. Cell*, **9**, 617–628.

207. Thom, S.R., Bhopale, V.M., Velazquez, O.C., Goldstein, L.J., *et al.* (2006) Stem cell mobilization by hyperbaric oxygen. *Am. J. Physiol. Heart Circ. Physiol.*, **290**, H1378–H1386.

208. Liu, Y., Elf, S.E., Asai, T., Miyata, Y. *et al.* (2009) The p53 tumor suppressor protein is a critical regulator of hematopoietic stem cell behavior. *Cell Cycle*, **8**, 3120–3124.

209. Larsson, J. and Scadden, D. (2006) Nervous activity in a stem cell niche. *Cell*, **124**, 253–255.

210. Katayama, Y., Battista, M., Kao, W.M., Hidalgo, A. *et al.* (2006) Signals from the sympathetic nervous system regulate hematopoietic stem cell egress from bone marrow. *Cell*, **124**, 407–421.

211. Conboy, I.M., Conboy, M.J., Wagers, A.J., Girma, E.R. *et al.* (2005) Rejuvenation of aged progenitor cells by exposure to a young systemic environment. *Nature*, **433**, 760–764.

212. Brack, A.S., Conboy, M.J., Roy, S., Lee, M. *et al.* (2007) Increased Wnt signaling during aging alters muscle stem cell fate and increases fibrosis. *Science*, **317**, 807–810.

213. Simons, B.D. and Clevers, H. (2011) Strategies for homeostatic stem cell self-renewal in adult tissues. *Cell*, **145**, 851–862.

214. Blanpain, C. and Fuchs, E. (2009) Epidermal homeostasis: a balancing act of stem cells in the skin. *Nat. Rev. Mol. Cell Biol.*, **10**, 207–217.

215. Massasa, E., Costa, X.S., and Taylor, H.S. (2010) Failure of the stem cell niche rather than loss of oocyte stem cells in the aging ovary. *Aging (Albany NY)*, **2**, 1–2.

216. Sharpless, N.E. and Schatten, G. (2009) Stem cell aging. *J. Gerontol. A Biol. Sci. Med. Sci.*, **64**, 202–204.

217. Beerman, I., Maloney, W.J., Weissmann, I.L., and Rossi, D.J. (2010) Stem cells and the aging hematopoietic system. *Curr. Opin. Immunol.*, **22**, 500–506.

218. Gopinath, S.D. and Rando, T.A. (2008) Stem cell review series: aging of the skeletal muscle stem cell niche. *Aging Cell*, **7**, 590–598.

219. Rando, T.A. (2006) Stem cells, ageing and the quest for immortality. *Nature*, **441**, 1080–1086.

220. Flores, I. and Blasco, M.A. (2010) The role of telomeres and telomerase in stem cell aging. *FEBS Lett.*, **584**, 3826–3830.

221. Ergen, A.V. and Goodell, M.A. (2010) Mechanisms of hematopoietic stem cell aging. *Exp. Gerontol.*, **45**, 286–290.

222. Rossi, D.J., Bryder, D., Zahn, J.M., Ahlenius, H. *et al.* (2005) Cell intrinsic alterations underlie hematopoietic stem cell aging. *Proc. Natl Acad. Sci. USA*, **102**, 9194–9199.

223. Lazarov, O., Mattson, M.P., Peterson, D.A., Pimplikar, S.W. *et al.* (2010) When neurogenesis encounters aging and disease. *Trends Neurosci.*, **33**, 569–579.

224. Kuzumaki, N., Ikegami, D., Imai, S., Narita, M. *et al.* (2010) Enhanced IL-1beta production in response to the activation of hippocampal glial cells impairs neurogenesis in aged mice. *Synapse*, **64**, 721–728.

225. Maslov, A.Y., Barone, T.A., Plunkett, R.J., and Pruitt, S.C. (2004) Neural stem cell detection, characterization, and age-related changes in the subventricular zone of mice. *J. Neurosci.*, **24**, 1726–1733.

226. Bouab, M., Paliouras, G.N., Aumont, A., Forest-Berard, K. *et al.* (2011) Aging of the subventricular zone neural stem cell niche: evidence for quiescence-associated changes between early and mid-adulthood. *Neuroscience*, **173**, 135–149.

227. Geiger, H., Rennebeck, G., and Van Zant, G. (2005) Regulation of hematopoietic stem cell aging in vivo by a distinct genetic element. *Proc. Natl Acad. Sci. USA*, **102**, 5102–5107.

228. Carlson, M.E., Suetta, C., Conboy, M.J., Aagaard, P. *et al.* (2009) Molecular aging and rejuvenation of human muscle stem cells. *EMBO Mol. Med.*, **1**, 381–391.

229. Geiger, H. and Rudolph, K.L. (2009) Aging in the lympho-hematopoietic stem cell compartment. *Trends Immunol.*, **30**, 360–365.

230. Xing, Z., Ryan, M.A., Daria, D., Nattamai, K.J. *et al.* (2006) Increased hematopoietic stem cell mobilization in aged mice. *Blood*, **108** (7), 2190–2197.

231. Liang, Y., Van Zant, G., and Szilvassy, S.J. (2005) Effects of aging on the homing and engraftment of murine hematopoietic stem and progenitor cells. *Blood*, **106**, 1479–1487.

232. Geiger, H., Koehler, A., and Gunzer, M. (2007) Stem cells, aging, niche, adhesion and Cdc42: a model for changes in cell-cell interactions and hematopoietic stem cell aging. *Cell Cycle*, **6**, 884–887.
233. Janzen, V., Forkert, R., Fleming, H.E., Saito, Y. *et al.* (2006) Stem-cell ageing modified by the cyclin-dependent kinase inhibitor p16INK4a. *Nature*, **443**, 421–426.
234. Kohler, A., Schmithorst, V., Filippi, M.D., Ryan, M.A. *et al.* (2009) Altered cellular dynamics and endosteal location of aged early hematopoietic progenitor cells revealed by time-lapse intravital imaging in long bones. *Blood*, **114**, 290–298.
235. Drummond-Barbosa, D. (2008) Stem cells, their niches and the systemic environment: an aging network. *Genetics*, **180**, 1787–1797.
236. Beerman, I., Bhattacharya, D., Zandi, S., Sigvardsson, M. *et al.* (2010) Functionally distinct hematopoietic stem cells modulate hematopoietic lineage potential during aging by a mechanism of clonal expansion. *Proc. Natl Acad. Sci. USA*, **107**, 5465–5470.
237. Muller-Sieburg, C. and Sieburg, H.B. (2008) Stem cell aging: survival of the laziest? *Cell Cycle*, **7**, 3798–3804.
238. Galvan, V. and Jin, K. (2007) Neurogenesis in the aging brain. *Clin. Intervention Aging*, **2**, 605–610.
239. Beccafico, S., Riuzzi, F., Puglielli, C., Mancinelli, R. *et al.* (2010) Human muscle satellite cells show age-related differential expression of S100B protein and RAGE. *Age (Dordr.)*, **33** (4), 523–541.
240. Corbu, A., Scaramozza, A., Badiali-DeGiorgi, L., Tarantino, L. *et al.* (2010) Satellite cell characterization from aging human muscle. *Neurol. Res.*, **32**, 63–72.
241. Day, K., Shefer, G., Shearer, A., Yablonka-Reuveni, Z. (2010) The depletion of skeletal muscle satellite cells with age is concomitant with reduced capacity of single progenitors to produce reserve progeny. *Dev. Biol.*, **340**, 330–343.
242. Shefer, G., Rauner, G., Yablonka-Reuveni, Z., and Benayahu, D. (2010) Reduced satellite cell numbers and myogenic capacity in aging can be alleviated by endurance exercise. *PLoS ONE*, **5**, e13307.
243. Renault, V., Piron-Hamelin, G., Forestier, C., DiDonna, S. *et al.* (2000) Skeletal muscle regeneration and the mitotic clock. *Exp. Gerontol.*, **35**, 711–719.
244. Carlson, M.E., Hsu, M., and Conboy, I.M. (2008) Imbalance between pSmad3 and Notch induces CDK inhibitors in old muscle stem cells. *Nature*, **454**, 528–532.
245. Carlson, M.E., Conboy, M.J., Hsu, M., Barchas, L. *et al.* (2009) Relative roles of TGF-beta1 and Wnt in the systemic regulation and aging of satellite cell responses. *Aging Cell*, **8**, 676–689.
246. Otto, A., Schmidt, C., Luke, G., Allen, S. *et al.* (2008) Canonical Wnt signalling induces satellite-cell proliferation during adult skeletal muscle regeneration. *J. Cell Sci.*, **121**, 2939–2950.
247. Bernard, F., Krejci, A., Housden, B., Adryan, B. *et al.* (2010) Specificity of Notch pathway activation: twist controls the transcriptional output in adult muscle progenitors. *Development*, **137**, 2633–2642.
248. Brack, A.S., Conboy, I.M., Conboy, M.J., Shen, J. *et al.* (2008) A temporal switch from notch to Wnt signaling in muscle stem cells is necessary for normal adult myogenesis. *Cell Stem Cell*, **2**, 50–59.
249. Liang, Y. and Van Zant, G. (2003) Genetic control of stem-cell properties and stem cells in aging. *Curr. Opin. Hematol.*, **10**, 195–202.
250. Ruzankina, Y., Pinzon-Guzman, C., Asare, A., Ong, T. *et al.* (2007) Deletion of the developmentally essential gene ATR in adult mice leads to age-related phenotypes and stem cell loss. *Cell Stem Cell*, **1**, 113–126.
251. Sharpless, N.E. and DePinho, R.A. (2007) How stem cells age and why this makes us grow old. *Nat. Rev. Mol. Cell Biol.*, **8**, 703–713.
252. Asai, T., Liu, Y., Bae, N., and Nimer, S.D. (2010) The p53 tumor suppressor protein regulates hematopoietic stem cell fate. *J. Cell. Physiol.*, **226** (9), 2215–2221.
253. Takubo, K., Ohmura, M., Azuma, M., Nagamatsu, G. *et al.* (2008) Stem cell defects in ATM-deficient undifferentiated spermatogonia through DNA damage-induced cell-cycle arrest. *Cell Stem Cell*, **2**, 170–182.
254. Moller, C., Alfredsson, J., Engstrom, M., Wootz, H. *et al.* (2005) Stem cell factor promotes mast cell survival via inactivation of FOXO3a-mediated transcriptional induction and MEK-regulated phosphorylation of

the proapoptotic protein Bim. *Blood*, **106**, 1330–1336.

255. Kim, J. and Wong, P.K. (2009) Loss of ATM impairs proliferation of neural stem cells through oxidative stress-mediated p38 MAPK signaling. *Stem Cells*, **27**, 1987–1998.

256. Rossi, D.J., Jamieson, C.H., and Weissman, I.L. (2008) Stems cells and the pathways to aging and cancer. *Cell*, **132**, 681–696.

257. Kamminga, L.M. and de Haan, G. (2006) Cellular memory and hematopoietic stem cell aging. *Stem Cells*, **24**, 1143–1149.

258. Ju, Z., Jiang, H., Jaworski, M., Rathinam, C. *et al.* (2007) Telomere dysfunction induces environmental alterations limiting hematopoietic stem cell function and engraftment. *Nat. Med.*, **13**, 742–747.

259. Ju, Z. and Rudolph, K.L. (2006) Telomeres and telomerase in stem cells during aging and disease. *Genome Dyn.*, **1**, 84–103.

260. Stepanova, L. and Sorrentino, B.P. (2005) A limited role for p16Ink4a and p19Arf in the loss of hematopoietic stem cells during proliferative stress. *Blood*, **106**, 827–832.

261. Sharpless, N.E. (2004) Ink4a/Arf links senescence and aging. *Exp. Gerontol.*, **39**, 1751–1759.

262. Symonds, C.E., Galderisi, U., and Giordano, A. (2009) Aging of the inceptive cellular population: the relationship between stem cells and aging. *Aging (Albany NY)*, **1**, 372–381.

263. Oguro, H., Iwama, A., Morita, Y., Kamijo, T. *et al.* (2006) Differential impact of Ink4a and Arf on hematopoietic stem cells and their bone marrow microenvironment in Bmi1-deficient mice. *J. Exp. Med.*, **203** (10), 2247–2253.

264. Tzatsos, A. and Bardeesy, N. (2008) Ink4a/Arf regulation by let-7b and Hmga2: a genetic pathway governing stem cell aging. *Cell Stem Cell*, **3**, 469–470.

265. Gil, J., Bernard, D., Martinez, D., and Beach, D. (2004) Polycomb CBX7 has a unifying role in cellular lifespan. *Nat. Cell Biol.*, **6**, 67–72.

266. Buckwalter, M.S., Yamane, M., Coleman, B.S., Ormerod, B.K. *et al.* (2006) Chronically increased transforming growth factor-beta1 strongly inhibits hippocampal neurogenesis in aged mice. *Am. J. Pathol.*, **169**, 154–164.

267. Riddle, D.R. and Lichtenwalner, R.J. (2007) *Neurogenesis in the Adult and Aging Brain*, CRC Press, Boca Raton, FL.

268. Bernal, G.M. and Peterson, D.A. (2011) Phenotypic and gene expression modification with normal brain aging in GFAP-positive astrocytes and neural stem cells. *Aging Cell*, **10** (3), 466–482.

269. Imayoshi, I., Sakamoto, M., Yamaguchi, M., Mori, K. *et al.* (2010) Essential roles of Notch signaling in maintenance of neural stem cells in developing and adult brains. *J. Neurosci.*, **30**, 3489–3498.

270. Lie, D.C., Colamarino, S.A., Song, H.J., Desire, L. *et al.* (2005) Wnt signalling regulates adult hippocampal neurogenesis. *Nature*, **437**, 1370–1375.

271. Scime, A., Desrosiers, J., Trensz, F., Palidwor, G.A. *et al.* (2010) Transcriptional profiling of skeletal muscle reveals factors that are necessary to maintain satellite cell integrity during ageing. *Mech. Ageing Dev.*, **131**, 9–20.

272. Wagers, A.J. (2008) Wnt not, waste not. *Cell Stem Cell*, **2**, 6–7.

273. Mauch, P., Botnick, L.E., Hannon, E.C., Obbagy, J. *et al.* (1982) Decline in bone marrow proliferative capacity as a function of age. *Blood*, **60**, 245–252.

274. French, R.A., Broussard, S.R., Meier, W.A., Minshall, C. *et al.* (2002) Age-associated loss of bone marrow hematopoietic cells is reversed by GH and accompanies thymic reconstitution. *Endocrinology*, **143**, 690–699.

275. Naveiras, O., Nardi, V., Wenzel, P.L., Hauschka, P.V. *et al.* (2009) Bone-marrow adipocytes as negative regulators of the haematopoietic microenvironment. *Nature*, **460**, 259–263.

276. Challen, G.A., Boles, N.C., Chambers, S.M., and Goodell, M.A. (2010) Distinct hematopoietic stem cell subtypes are differentially regulated by TGF-beta1. *Cell Stem Cell*, **6**, 265–278.

277. Lapidot, T., Sirard, C., Vormoor, J., Murdoch, B. *et al.* (1994) A cell initiating human acute myeloid leukaemia after transplantation into SCID mice. *Nature*, **367**, 645–648.

278. Bonnet, D. and Dick, J.E. (1997) Human acute myeloid leukemia is organized as a hierarchy that originates from a primitive hematopoietic cell. *Nat. Med.*, **3**, 730–737.

279. Maitland, N.J. and Collins, A.T. (2008) Prostate cancer stem cells: a new target for therapy. *J. Clin. Oncol.*, **26**, 2862–2870.
280. Lee, C.J., Dosch, J., and Simeone, D.M. (2008) Pancreatic cancer stem cells. *J. Clin. Oncol.*, **26**, 2806–2812.
281. Zhao, C., Chen, A., Jamieson, C.H., Fereshteh, M. *et al.* (2009) Hedgehog signalling is essential for maintenance of cancer stem cells in myeloid leukaemia. *Nature*, **458**, 776–779.
282. Eyler, C.E. and Rich, J.N. (2008) Survival of the fittest: cancer stem cells in therapeutic resistance and angiogenesis. *J. Clin. Oncol.*, **26**, 2839–2845.
283. Liu, G., Yuan, X., Zeng, Z., Tunici, P. *et al.* (2006) Analysis of gene expression and chemoresistance of CD133+ cancer stem cells in glioblastoma. *Mol. Cancer*, **5**, 67.
284. Lu, C. and Shervington, A. (2008) Chemoresistance in gliomas. *Mol. Cell. Biochem.*, **312**, 71–80.
285. Dean, M., Fojo, T., and Bates, S. (2005) Tumour stem cells and drug resistance. *Nat. Rev. Cancer*, **5**, 275–284.
286. Dhawan, P., Ahmad, R., Srivastava, A.S., and Singh, A.B. (2011) Cancer stem cells and colorectal cancer: an overview. *Curr. Top. Med. Chem.*, **11** (13), 1592–1598.
287. Takebe, N., Harris, P.J., Warren, R.Q., and Ivy, S.P. (2011) Targeting cancer stem cells by inhibiting Wnt, Notch, and Hedgehog pathways. *Nat. Rev. Clin. Oncol.*, **8**, 97–106.
288. Moore, N. and Lyle, S. (2011) Quiescent, slow-cycling stem cell populations in cancer: a review of the evidence and discussion of significance. *J. Oncol.*, **2011**, pii: 396076.
289. Ito, K., Bernardi, R., Morotti, A., Matsuoka, S. *et al.* (2008) PML targeting eradicates quiescent leukaemia-initiating cells. *Nature*, **453**, 1072–1078.
290. Visvader, J.E. (2011) Cells of origin in cancer. *Nature*, **469**, 314–322.
291. Dick, J.E. (2008) Stem cell concepts renew cancer research. *Blood*, **112**, 4793–4807.
292. Wicha, M.S., Liu, S., and Dontu, G. (2006) Cancer stem cells: an old idea--a paradigm shift. *Cancer Res.*, **66**, 1883–1890; discussion 1895–1886.
293. Sell, S. (2004) Stem cell origin of cancer and differentiation therapy. *Crit. Rev. Oncol. Hematol.*, **51**, 1–28.
294. Sell, S. (2008) Alpha-fetoprotein, stem cells and cancer: how study of the production of alpha-fetoprotein during chemical hepatocarcinogenesis led to reaffirmation of the stem cell theory of cancer. *Tumour Biol.*, **29**, 161–180.
295. Miyamoto, T., Weissman, I.L., and Akashi, K. (2000) AML1/ETO-expressing nonleukemic stem cells in acute myelogenous leukemia with 8;21 chromosomal translocation. *Proc. Natl Acad. Sci. USA*, **97**, 7521–7526.
296. Jamieson, C.H., Ailles, L.E., Dylla, S.J., Muijtjens, M. *et al.* (2004) Granulocyte-macrophage progenitors as candidate leukemic stem cells in blast-crisis CML. *N. Engl. J. Med.*, **351**, 657–667.
297. Levis, M., Murphy, K.M., Pham, R., Kim, K.T. *et al.* (2005) Internal tandem duplications of the FLT3 gene are present in leukemia stem cells. *Blood*, **106**, 673–680.
298. Krivtsov, A.V., Twomey, D., Feng, Z., Stubbs, M.C. *et al.* (2006) Transformation from committed progenitor to leukaemia stem cell initiated by MLL-AF9. *Nature*, **442**, 818–822.
299. Cozzio, A., Passegue, E., Ayton, P.M., Karsunky, H. *et al.* (2003) Similar MLL-associated leukemias arising from self-renewing stem cells and short-lived myeloid progenitors. *Genes Dev.*, **17**, 3029–3035.
300. Huntly, B.J., Shigematsu, H., Deguchi, K., Lee, B.H. *et al.* (2004) MOZ-TIF2, but not BCR-ABL, confers properties of leukemic stem cells to committed murine hematopoietic progenitors. *Cancer Cell*, **6**, 587–596.
301. Clarke, M.F., Dick, J.E., Dirks, P.B., Eaves, C.J. *et al.* (2006) Cancer stem cells–perspectives on current status and future directions: AACR Workshop on cancer stem cells. *Cancer Res.*, **66**, 9339–9344.
302. Wang, J.C. and Dick, J.E. (2005) Cancer stem cells: lessons from leukemia. *Trends Cell Biol.*, **15**, 494–501.
303. Hope, K.J., Jin, L., and Dick, J.E. (2004) Acute myeloid leukemia originates from a hierarchy of leukemic stem cell classes that differ in self-renewal capacity. *Nat. Immunol.*, **5**, 738–743.
304. Jordan, C.T., Upchurch, D., Szilvassy, S.J., Guzman, M.L. *et al.* (2000) The interleukin-3 receptor alpha chain is a unique marker for human acute myelogenous leukemia stem cells. *Leukemia*, **14**, 1777–1784.
305. Gilliland, D.G., Jordan, C.T., and Felix, C.A. (2004) The molecular basis of leukemia.

Hematology (Am. Soc. Hematol. Educ. Program.), 80–97.

306. Lapidot, T., Fajerman, Y., and Kollet, O. (1997) Immune-deficient SCID and NOD/SCID mice models as functional assays for studying normal and malignant human hematopoiesis. *J. Mol. Med.*, **75**, 664–673.
307. Ailles, L.E., Gerhard, B., Kawagoe, H., and Hogge, D.E. (1999) Growth characteristics of acute myelogenous leukemia progenitors that initiate malignant hematopoiesis in nonobese diabetic/severe combined immunodeficient mice. *Blood*, **94**, 1761–1772.
308. Pearce, D.J., Taussig, D., Zibara, K., Smith, L.L. *et al.* (2006) AML engraftment in the NOD/SCID assay reflects the outcome of AML: implications for our understanding of the heterogeneity of AML. *Blood*, **107**, 1166–1173.
309. Al-Hajj, M., Wicha, M.S., Benito-Hernandez, A., Morrison, S.J. *et al.* (2003) Prospective identification of tumorigenic breast cancer cells. *Proc. Natl Acad. Sci. USA*, **100**, 3983–3988.
310. Singh, S.K., Hawkins, C., Clarke, I.D., Squire, J.A. *et al.* (2004) Identification of human brain tumour initiating cells. *Nature*, **432**, 396–401.
311. Galli, R., Binda, E., Orfanelli, U., Cipelletti, B. *et al.* (2004) Isolation and characterization of tumorigenic, stem-like neural precursors from human glioblastoma. *Cancer Res.*, **64**, 7011–7021.
312. Hemmati, H.D., Nakano, I., Lazareff, J.A., Masterman-Smith, M. *et al.* (2003) Cancerous stem cells can arise from pediatric brain tumors. *Proc. Natl Acad. Sci. USA*, **100**, 15178–15183.
313. Ricci-Vitiani, L., Lombardi, D.G., Pilozzi, E., Biffoni, M. *et al.* (2007) Identification and expansion of human colon-cancer-initiating cells. *Nature*, **445**, 111–115.
314. O'Brien, C.A., Pollett, A., Gallinger, S., and Dick, J.E. (2007) A human colon cancer cell capable of initiating tumour growth in immunodeficient mice. *Nature*, **445**, 106–110.
315. Li, C., Heidt, D.G., Dalerba, P., Burant, C.F. *et al.* (2007) Identification of pancreatic cancer stem cells. *Cancer Res.*, **67**, 1030–1037.
316. Patrawala, L., Calhoun-Davis, T., Schneider-Broussard, R., and Tang, D.G. (2007) Hierarchical organization of prostate cancer cells in xenograft tumors: the $CD44^{+}alpha_2beta_{1+}$ cell population is enriched in tumor-initiating cells. *Cancer Res.*, **67**, 6796–6805.
317. Ishikawa, F., Yoshida, S., Saito, Y., Hijikata, A. *et al.* (2007) Chemotherapy-resistant human AML stem cells home to and engraft within the bone-marrow endosteal region. *Nat. Biotechnol.*, **25**, 1315–1321.
318. Taussig, D.C., Miraki-Moud, F., Anjos-Afonso, F., Pearce, D.J. *et al.* (2008) Anti-CD38 antibody-mediated clearance of human repopulating cells masks the heterogeneity of leukemia-initiating cells. *Blood*, **112**, 568–575.
319. Kelly, P.N., Dakic, A., Adams, J.M., Nutt, S.L. *et al.* (2007) Tumor growth need not be driven by rare cancer stem cells. *Science*, **317**, 337.
320. Hill, R.P. (2006) Identifying cancer stem cells in solid tumors: case not proven. *Cancer Res.*, **66**, 1891–1895; discussion 1890.
321. Kang, M.K. and Kang, S.K. (2007) Tumorigenesis of chemotherapeutic drug-resistant cancer stem-like cells in brain glioma. *Stem Cells Dev.*, **16**, 837–847.
322. Bao, S., Wu, Q., McLendon, R.E., Hao, Y. *et al.* (2006) Glioma stem cells promote radioresistance by preferential activation of the DNA damage response. *Nature*, **444**, 756–760.
323. Croker, A.K. and Allan, A.L. (2008) Cancer stem cells: implications for the progression and treatment of metastatic disease. *J. Cell. Mol. Med.*, **12**, 374–390.
324. Gentles, A.J., Plevritis, S.K., Majeti, R., and Alizadeh, A.A. (2010) Association of a leukemic stem cell gene expression signature with clinical outcomes in acute myeloid leukemia. *J. Am. Med. Assoc.*, **304**, 2706–2715.
325. Masuda, S. (2011) Leukemic stem cell gene expression signature and clinical outcomes in acute myeloid leukemia. *J. Am. Med. Assoc.*, **305**, 1094; author reply 1094–1095.
326. van Rhenen, A., van Dongen, G.A., Kelder, A., Rombouts, E.J. *et al.* (2007) The novel AML stem cell associated antigen CLL-1 aids in discrimination between normal and leukemic stem cells. *Blood*, **110**, 2659–2666.
327. Baumann, M., Dubois, W., and Suit, H.D. (1990) Response of human squamous cell

carcinoma xenografts of different sizes to irradiation: relationship of clonogenic cells, cellular radiation sensitivity in vivo, and tumor rescuing units. *Radiat. Res.*, **123**, 325–330.
328. Yaromina, A., Krause, M., Thames, H., Rosner, A. *et al.* (2007) Pre-treatment number of clonogenic cells and their radiosensitivity are major determinants of local tumour control after fractionated irradiation. *Radiother. Oncol.*, **83**, 304–310.
329. Saito, Y., Uchida, N., Tanaka, S., Suzuki, N. *et al.* (2010) Induction of cell cycle entry eliminates human leukemia stem cells in a mouse model of AML. *Nat. Biotechnol.*, **28**, 275–280.
330. Lou, H. and Dean, M. (2007) Targeted therapy for cancer stem cells: the patched pathway and ABC transporters. *Oncogene*, **26**, 1357–1360.
331. Guzman, M.L., Neering, S.J., Upchurch, D., Grimes, B. *et al.* (2001) Nuclear factor-kappaB is constitutively activated in primitive human acute myelogenous leukemia cells. *Blood*, **98**, 2301–2307.
332. Guzman, M.L., Rossi, R.M., Karnischky, L., Li, X. *et al.* (2005) The sesquiterpene lactone parthenolide induces apoptosis of human acute myelogenous leukemia stem and progenitor cells. *Blood*, **105**, 4163–4169.
333. Jordan, C.T. and Guzman, M.L. (2004) Mechanisms controlling pathogenesis and survival of leukemic stem cells. *Oncogene*, **23**, 7178–7187.
334. Essers, M.A. and Trumpp, A. (2010) Targeting leukemic stem cells by breaking their dormancy. *Mol. Oncol.*, **4**, 443–450.
335. Zeng, Z., Shi, Y.X., Samudio, I.J., Wang, R.Y. *et al.* (2009) Targeting the leukemia microenvironment by CXCR4 inhibition overcomes resistance to kinase inhibitors and chemotherapy in AML. *Blood*, **113**, 6215–6224.
336. Jin, L., Tabe, Y., Konoplev, S., Xu, Y. *et al.* (2008) CXCR4 up-regulation by imatinib induces chronic myelogenous leukemia (CML) cell migration to bone marrow stroma and promotes survival of quiescent CML cells. *Mol. Cancer Ther.*, **7**, 48–58.
337. Griffin, J.D. (2001) Leukemia stem cells and constitutive activation of NF-kappaB. *Blood*, **98**, 2291.
338. Hassane, D.C., Sen, S., Minhajuddin, M., Rossi, R.M. *et al.* (2011) Chemical genomic screening reveals synergism between parthenolide and inhibitors of the PI-3 kinase and mTOR pathways. *Blood*, **116**, 5983–5990.
339. Weissman, I. (2005) Stem cell research: paths to cancer therapies and regenerative medicine. *J. Am. Med. Assoc.*, **294**, 1359–1366.
340. Wang, J., Wakeman, T.P., Lathia, J.D., Hjelmeland, A.B. *et al.* (2010) Notch promotes radioresistance of glioma stem cells. *Stem Cells*, **28**, 17–28.
341. Pannuti, A., Foreman, K., Rizzo, P., Osipo, C. *et al.* (2010) Targeting Notch to target cancer stem cells. *Clin. Cancer Res.*, **16**, 3141–3152.
342. Somervaille, T.C. and Cleary, M.L. (2006) Identification and characterization of leukemia stem cells in murine MLL-AF9 acute myeloid leukemia. *Cancer Cell*, **10**, 257–268.
343. Quintana, E., Shackleton, M., Sabel, M.S., Fullen, D.R. *et al.* (2008) Efficient tumour formation by single human melanoma cells. *Nature*, **456**, 593–598.
344. Armanios, M. and Greider, C.W. (2005) Telomerase and cancer stem cells. *Cold Spring Harbor Symp. Quant. Biol.*, **70**, 205–208.
345. Taussig, D.C., Vargaftig, J., Miraki-Moud, F., Griessinger, E. *et al.* (2010) Leukemia-initiating cells from some acute myeloid leukemia patients with mutated nucleophosmin reside in the CD34(-) fraction. *Blood*, **115**, 1976–1984.
346. Sarry, J.E., Murphy, K., Perry, R., Sanchez, P.V. *et al.* (2011) Human acute myelogenous leukemia stem cells are rare and heterogeneous when assayed in NOD/SCID/IL2Rgammac-deficient mice. *J. Clin. Invest.*, **121**, 384–395.
347. Wang, J., Sakariassen, P.O., Tsinkalovsky, O., Immervoll, H. *et al.* (2008) CD133 negative glioma cells form tumors in nude rats and give rise to CD133 positive cells. *Int. J. Cancer*, **122**, 761–768.
348. Goardon, N., Marchi, E., Atzberger, A., Quek, L. *et al.* (2011) Coexistence of LMPP-like and GMP-like leukemia stem cells in acute myeloid leukemia. *Cancer Cell*, **19**, 138–152.

349. Wei, J., Wunderlich, M., Fox, C., Alvarez, S. *et al.* (2008) Microenvironment determines lineage fate in a human model of MLL-AF9 leukemia. *Cancer Cell*, **13**, 483–495.
350. Majeti, R. and Weissman, I.L. (2011) Human acute myelogenous leukemia stem cells revisited: there's more than meets the eye. *Cancer Cell*, **19**, 9–10.
351. Lorico, A. and Rappa, G. (2011) Phenotypic heterogeneity of breast cancer stem cells. *J. Oncol.*, **2011**, 135039.
352. Notta, F., Mullighan, C.G., Wang, J.C., Poeppl, A. *et al.* (2011) Evolution of human BCR-ABL1 lymphoblastic leukaemia-initiating cells. *Nature*, **469**, 362–367.
353. Roesch, A., Fukunaga-Kalabis, M., Schmidt, E.C., Zabierowski, S.E. *et al.* (2010) A temporarily distinct subpopulation of slow-cycling melanoma cells is required for continuous tumor growth. *Cell*, **141**, 583–594.
354. Sarig, R., Rivlin, N., Brosh, R., Bornstein, C. *et al.* (2010) Mutant p53 facilitates somatic cell reprogramming and augments the malignant potential of reprogrammed cells. *J. Exp. Med.*, **207**, 2127–2140.
355. Maitland, N.J. and Collins, A. (2005) A tumour stem cell hypothesis for the origins of prostate cancer. *BJU Int.*, **96**, 1219–1223.
356. Jordan, C.T. (2004) Cancer stem cell biology: from leukemia to solid tumors. *Curr. Opin. Cell Biol.*, **16**, 708–712.
357. Konopleva, M.Y. and Jordan, C.T. (2011) Leukemia stem cells and microenvironment: biology and therapeutic targeting. *J. Clin. Oncol.*, **29**, 591–599.
358. Yen, T.H. and Wright, N.A. (2006) The gastrointestinal tract stem cell niche. *Stem Cell Rev.*, **2**, 203–212.
359. Veeravagu, A., Bababeygy, S.R., Kalani, M.Y., Hou, L.C., and Tse, V. (2008) The cancer stem cell-vascular niche complex in brain tumor formation. *Stem Cells Dev.*, **17**, 859–867.
360. Calabrese, C., Poppleton, H., Kocak, M., Hogg, T.L. *et al.* (2007) A perivascular niche for brain tumor stem cells. *Cancer Cell*, **11**, 69–82.
361. Jacoby, R.F., Schlack, S., Cole, C.E., Skarbek, M. *et al.* (1997) A juvenile polyposis tumor suppressor locus at 10q22 is deleted from nonepithelial cells in the lamina propria. *Gastroenterology*, **112**, 1398–1403.
362. Bhowmick, N.A., Chytil, A., Plieth, D., Gorska, A.E. *et al.* (2004) TGF-beta signaling in fibroblasts modulates the oncogenic potential of adjacent epithelia. *Science*, **303**, 848–851.
363. Niewerth, D., Creutzig, U., Bierings, M.B., and Kaspers, G.J. (2010) A review on allogeneic stem cell transplantation for newly diagnosed pediatric acute myeloid leukemia. *Blood*, **116**, 2205–2214.
364. Lane, S.W., Scadden, D.T., and Gilliland, D.G. (2009) The leukemic stem cell niche: current concepts and therapeutic opportunities. *Blood*, **114**, 1150–1157.
365. Li, L. and Neaves, W.B. (2006) Normal stem cells and cancer stem cells: the niche matters. *Cancer Res.*, **66**, 4553–4557.
366. Sneddon, J.B., Zhen, H.H., Montgomery, K., van de Rijn, M. *et al.* (2006) Bone morphogenetic protein antagonist gremlin 1 is widely expressed by cancer-associated stromal cells and can promote tumor cell proliferation. *Proc. Natl Acad. Sci. USA*, **103**, 14842–14847.
367. Barcellos-Hoff, M.H. and Ravani, S.A. (2000) Irradiated mammary gland stroma promotes the expression of tumorigenic potential by unirradiated epithelial cells. *Cancer Res.*, **60**, 1254–1260.
368. Olumi, A.F., Grossfeld, G.D., Hayward, S.W., Carroll, P.R. *et al.* (1999) Carcinoma-associated fibroblasts direct tumor progression of initiated human prostatic epithelium. *Cancer Res.*, **59**, 5002–5011.
369. Huang, M., Li, Y., Zhang, H., and Nan, F. (2010) Breast cancer stromal fibroblasts promote the generation of $CD44^{+}CD24^{-}$ cells through SDF-1/CXCR4 interaction. *J. Exp. Clin. Cancer Res.*, **29**, 80.
370. Liao, C.P., Adisetiyo, H., Liang, M., Roy-Burman, P. (2010) Cancer-associated fibroblasts enhance the gland-forming capability of prostate cancer stem cells. *Cancer Res.*, **70**, 7294–7303.
371. Staser, K., Yang, F.C., and Clapp, D.W. (2010) Mast cells and the neurofibroma microenvironment. *Blood*, **116**, 157–164.
372. Zhu, Y., Ghosh, P., Charnay, P., Burns, D.K. *et al.* (2002) Neurofibromas in NF1: Schwann cell origin and role of tumor environment. *Science*, **296**, 920–922.
373. Bajenaru, M.L., Hernandez, M.R., Perry, A., Zhu, Y. *et al.* (2003) Optic nerve glioma in

mice requires astrocyte Nf1 gene inactivation and Nf1 brain heterozygosity. *Cancer Res.*, **63**, 8573–8577.

374. Daginakatte, G.C. and Gutmann, D.H. (2007) Neurofibromatosis-1 (Nf1) heterozygous brain microglia elaborate paracrine factors that promote Nf1-deficient astrocyte and glioma growth. *Hum. Mol. Genet.*, **16**, 1098–1112.
375. Flynn, C.M. and Kaufman, D.S. (2007) Donor cell leukemia: insight into cancer stem cells and the stem cell niche. *Blood*, **109**, 2688–2692.
376. Perry, J.M. and Li, L. (2007) Disrupting the stem cell niche: good seeds in bad soil. *Cell*, **129**, 1045–1047.
377. Walkley, C.R., Olsen, G.H., Dworkin, S., Fabb, S.A. *et al.* (2007) A microenvironment-induced myeloproliferative syndrome caused by retinoic acid receptor gamma deficiency. *Cell*, **129**, 1097–1110.
378. Raaijmakers, M.H., Mukherjee, S., Guo, S., Zhang, S. *et al.* (2010) Bone progenitor dysfunction induces myelodysplasia and secondary leukaemia. *Nature*, **464**, 852–857.
379. Chepko, G., Slack, R., Carbott, D., Khan, S. *et al.* (2005) Differential alteration of stem and other cell populations in ducts and lobules of TGFalpha and c-Myc transgenic mouse mammary epithelium. *Tissue Cell*, **37**, 393–412.
380. Lazennec, G. and Richmond, A. (2010) Chemokines and chemokine receptors: new insights into cancer-related inflammation. *Trends Mol. Med.*, **16**, 133–144.
381. Patel, S.A., Heinrich, A.C., Reddy, B.Y., and Rameshwar, P. (2009) Inflammatory mediators: parallels between cancer biology and stem cell therapy. *J. Inflamm. Res.*, **2**, 13–19.
382. Groot Kormelink, T., Abudukelimu, A., and Redegeld, F.A. (2009) Mast cells as target in cancer therapy. *Curr. Pharm. Des.*, **15**, 1868–1878.
383. Gonda, T.A., Tu, S., and Wang, T.C. (2009) Chronic inflammation, the tumor microenvironment and carcinogenesis. *Cell Cycle*, **8**, 2005–2013.
384. Gabrilovich, D.I. and Nagaraj, S. (2009) Myeloid-derived suppressor cells as regulators of the immune system. *Nat. Rev. Immunol.*, **9**, 162–174.
385. Quante, M. and Wang, T.C. (2008) Inflammation and stem cells in gastrointestinal carcinogenesis. *Physiology (Bethesda)*, **23**, 350–359.
386. Aguirre-Ghiso, J.A. (2007) Models, mechanisms and clinical evidence for cancer dormancy. *Nat. Rev. Cancer*, **7**, 834–846.
387. Allan, A.L., Vantyghem, S.A., Tuck, A.B., and Chambers, A.F. (2006) Tumor dormancy and cancer stem cells: implications for the biology and treatment of breast cancer metastasis. *Breast Dis.*, **26**, 87–98.
388. Kojima, Y., Acar, A., Eaton, E.N., Mellody, K.T. *et al.* (2010) Autocrine TGF-beta and stromal cell-derived factor-1 (SDF-1) signaling drives the evolution of tumor-promoting mammary stromal myofibroblasts. *Proc. Natl Acad. Sci. USA*, **107**, 20009–20014.
389. Worthley, D.L., Ruszkiewicz, A., Davies, R., Moore, S. *et al.* (2009) Human gastrointestinal neoplasia-associated myofibroblasts can develop from bone marrow-derived cells following allogeneic stem cell transplantation. *Stem Cells*, **27**, 1463–1468.
390. Sipkins, D.A., Wei, X., Wu, J.W., Runnels, J.M. *et al.* (2005) In vivo imaging of specialized bone marrow endothelial microdomains for tumour engraftment. *Nature*, **435**, 969–973.
391. Barker, N., Ridgway, R.A., van Es, J.H., van de Wetering, M. *et al.* (2009) Crypt stem cells as the cells-of-origin of intestinal cancer. *Nature*, **457**, 608–611.
392. Weber, M.H., Goltzman, D., Kostenuik, P., Rabbani, S. *et al.* (2000) Mechanisms of tumor metastasis to bone. *Crit. Rev. Eukaryot. Gene Expr.*, **10**, 281–302.
393. Hall, F.M. (2004) Bone metastasis. *N. Engl. J. Med.*, **351**, 195–196; author reply 195–196.
394. Zhao, R. and Xi, R. (2010) Stem cell competition for niche occupancy: emerging themes and mechanisms. *Stem Cells Rev.*, **6**, 345–350.
395. Shiozawa, Y., Pedersen, E.A., Havens, A.M., Jung, Y. *et al.* (2011) Human prostate cancer metastases target the hematopoietic stem cell niche to establish footholds in mouse bone marrow. *J. Clin. Invest.*, **121** (4), 1298–1312.
396. Kaplan, R.N., Riba, R.D., Zacharoulis, S., Bramley, A.H. *et al.* (2005) VEGFR1-positive haematopoietic bone marrow

progenitors initiate the pre-metastatic niche. *Nature*, **438**, 820–827.
397. Quante, M., Tu, S.P., Tomita, H., Gonda, T. *et al.* (2011) Bone marrow-derived myofibroblasts contribute to the mesenchymal stem cell niche and promote tumor growth. *Cancer Cell*, **19**, 257–272.
398. Hambardzumyan, D., Becher, O.J., Rosenblum, M.K., Pandolfi, P.P. *et al.* (2008) PI3K pathway regulates survival of cancer stem cells residing in the perivascular niche following radiation in medulloblastoma in vivo. *Genes Dev.*, **22**, 436–448.
399. Krishnamurthy, S., Dong, Z., Vodopyanov, D., Imai, A. *et al.* (2010) Endothelial cell-initiated signaling promotes the survival and self-renewal of cancer stem cells. *Cancer Res.*, **70**, 9969–9978.
400. Orimo, A., Gupta, P.B., Sgroi, D.C., Arenzana-Seisdedos, F. *et al.* (2005) Stromal fibroblasts present in invasive human breast carcinomas promote tumor growth and angiogenesis through elevated SDF-1/CXCL12 secretion. *Cell*, **121**, 335–348.
401. Zhang, L.Z., Zhang, C.Q., Yan, Z.Y., Yang, Q.C. *et al.* (2011) Tumor-initiating cells and tumor vascularization. *Pediatr. Blood Cancer*, **56**, 335–340.
402. Colmone, A., Amorim, M., Pontier, A.L., Wang, S. *et al.* (2008) Leukemic cells create bone marrow niches that disrupt the behavior of normal hematopoietic progenitor cells. *Science*, **322**, 1861–1865.
403. Mantovani, A., Allavena, P., Sica, A., and Balkwill, F. (2008) Cancer-related inflammation. *Nature*, **454**, 436–444.
404. Guo, X., Oshima, H., Kitmura, T., Taketo, M.M. *et al.* (2008) Stromal fibroblasts activated by tumor cells promote angiogenesis in mouse gastric cancer. *J. Biol. Chem.*, **283**, 19864–19871.
405. Wang, R., Chadalavada, K., Wilshire, J., Kowalik, U. *et al.* (2011) Glioblastoma stem-like cells give rise to tumour endothelium. *Nature*, **468**, 829–833.
406. Ricci-Vitiani, L., Pallini, R., Biffoni, M., Todaro, M. *et al.* (2011) Tumour vascularization via endothelial differentiation of glioblastoma stem-like cells. *Nature*, **468**, 824–828.
407. Bautch, V.L. (2010) Cancer: Tumour stem cells switch sides. *Nature*, **468**, 770–771.
408. Pekarek, L.A., Starr, B.A., Toledano, A.Y., and Schreiber, H. (1995) Inhibition of tumor growth by elimination of granulocytes. *J. Exp. Med.*, **181**, 435–440.
409. Kim, R., Emi, M., Tanabe, K., and Arihiro, K. (2006) Tumor-driven evolution of immunosuppressive networks during malignant progression. *Cancer Res.*, **66**, 5527–5536.
410. Kusmartsev, S. and Gabrilovich, D.I. (2006) Role of immature myeloid cells in mechanisms of immune evasion in cancer. *Cancer Immunol. Immunother.*, **55**, 237–245.
411. Coussens, L.M., Tinkle, C.L., Hanahan, D., and Werb, Z. (2000) MMP-9 supplied by bone marrow-derived cells contributes to skin carcinogenesis. *Cell*, **103**, 481–490.
412. Lyden, D., Hattori, K., Dias, S., Costa, C. *et al.* (2001) Impaired recruitment of bone-marrow-derived endothelial and hematopoietic precursor cells blocks tumor angiogenesis and growth. *Nat. Med.*, **7**, 1194–1201.
413. Condamine, T. and Gabrilovich, D.I. (2011) Molecular mechanisms regulating myeloid-derived suppressor cell differentiation and function. *Trends Immunol.*, **32**, 19–25.
414. Bunt, S.K., Yang, L., Sinha, P., Clements, V.K. *et al.* (2007) Reduced inflammation in the tumor microenvironment delays the accumulation of myeloid-derived suppressor cells and limits tumor progression. *Cancer Res.*, **67**, 10019–10026.
415. Ostrand-Rosenberg, S. and Sinha, P. (2009) Myeloid-derived suppressor cells: linking inflammation and cancer. *J. Immunol.*, **182**, 4499–4506.
416. Enioutina, E.Y., Bareyan, D., and Daynes, R.A. (2011) A role for immature myeloid cells in immune senescence. *J. Immunol.*, **186**, 697–707.
417. Hiratsuka, S., Watanabe, A., Aburatani, H., and Maru, Y. (2006) Tumour-mediated upregulation of chemoattractants and recruitment of myeloid cells predetermines lung metastasis. *Nat. Cell Biol.*, **8**, 1369–1375.
418. Tadmor, T., Attias, D., and Polliack, A. (2011) Myeloid-derived suppressor cells – their role in haemato-oncological malignancies and other cancers and possible implications for therapy. *Br. J. Haematol.*, **153** (5), 557–567.
419. Lin, E.Y., Nguyen, A.V., Russell, R.G., and Pollard, J.W. (2001) Colony-stimulating

factor 1 promotes progression of mammary tumors to malignancy. *J. Exp. Med.*, **193**, 727–740.

420. Yang, L., DeBusk, L.M., Fukuda, K., Fingleton, B. *et al.* (2004) Expansion of myeloid immune suppressor Gr^+CD11b^+ cells in tumor-bearing host directly promotes tumor angiogenesis. *Cancer Cell*, **6**, 409–421.
421. Bollrath, J., Phesse, T.J., von Burstin, V.A., Putoczki, T. *et al.* (2009) gp130-mediated Stat3 activation in enterocytes regulates cell survival and cell-cycle progression during colitis-associated tumorigenesis. *Cancer Cell*, **15**, 91–102.
422. Grivennikov, S., Karin, E., Terzic, J., Mucida, D. *et al.* (2009) IL-6 and Stat3 are required for survival of intestinal epithelial cells and development of colitis-associated cancer. *Cancer Cell*, **15**, 103–113.
423. Ahn, G.O. and Brown, J.M. (2008) Matrix metalloproteinase-9 is required for tumor vasculogenesis but not for angiogenesis: role of bone marrow-derived myelomonocytic cells. *Cancer Cell*, **13**, 193–205.
424. Murdoch, C., Muthana, M., Coffelt, S.B., and Lewis, C.E. (2008) The role of myeloid cells in the promotion of tumour angiogenesis. *Nat. Rev. Cancer*, **8**, 618–631.
425. Worthley, D.L., Giraud, A.S., and Wang, T.C. (2010) Stromal fibroblasts in digestive cancer. *Cancer Microenviron.*, **3**, 117–125.
426. Fujita, M., Kohanbash, G., Fellows-Mayle, W., Hamilton, R.L. *et al.* (2011) COX-2 blockade suppresses gliomagenesis by inhibiting myeloid-derived suppressor cells. *Cancer Res.*, **71** (7), 2664–2674.
427. Erler, J.T., Bennewith, K.L., Cox, T.R., Lang, G. *et al.* (2009) Hypoxia-induced lysyl oxidase is a critical mediator of bone marrow cell recruitment to form the premetastatic niche. *Cancer Cell*, **15**, 35–44.
428. Carlesso, N. and Cardoso, A.A. (2010) Stem cell regulatory niches and their role in normal and malignant hematopoiesis. *Curr. Opin. Hematol.*, **17**, 281–286.
429. Meads, M.B., Gatenby, R.A., and Dalton, W.S. (2009) Environment-mediated drug resistance: a major contributor to minimal residual disease. *Nat. Rev. Cancer*, **9**, 665–674.
430. Hazlehurst, L.A., Landowski, T.H., and Dalton, W.S. (2003) Role of the tumor microenvironment in mediating de novo resistance to drugs and physiological mediators of cell death. *Oncogene*, **22**, 7396–7402.
431. Marhaba, R., Klingbeil, P., Nuebel, T., Nazarenko, I. *et al.* (2008) CD44 and EpCAM: cancer-initiating cell markers. *Curr. Mol. Med.*, **8**, 784–804.
432. Hambardzumyan, D., Becher, O.J., and Holland, E.C. (2008) Cancer stem cells and survival pathways. *Cell Cycle*, **7**, 1371–1378.
433. Matsunaga, T., Takemoto, N., Sato, T., Takimoto, R. *et al.* (2003) Interaction between leukemic-cell VLA-4 and stromal fibronectin is a decisive factor for minimal residual disease of acute myelogenous leukemia. *Nat. Med.*, **9**, 1158–1165.
434. Farmer, P., Bonnefoi, H., Anderle, P., Cameron, D. *et al.* (2009) A stroma-related gene signature predicts resistance to neoadjuvant chemotherapy in breast cancer. *Nat. Med.*, **15**, 68–74.
435. Emmenegger, U. and Kerbel, R.S. (2010) Cancer: chemotherapy counteracted. *Nature*, **468**, 637–638.
436. Gilbert, L.A. and Hemann, M.T. (2010) DNA damage-mediated induction of a chemoresistant niche. *Cell*, **143**, 355–366.
437. Konopleva, M., Tabe, Y., Zeng, Z., and Andreeff, M. (2009) Therapeutic targeting of microenvironmental interactions in leukemia: mechanisms and approaches. *Drug Resist. Update*, **12**, 103–113.

10
Tumor Immunology

Sukhmani K. Padda[1], Rupa Narayan[1], Bryan M. Burt[2], and Edgar G. Engleman[3]
[1] Stanford University, Department of Medicine, Division of Hematology/Oncology, 875 Blake Wilbur Drive, Stanford, CA 94305, USA
[2] Baylor College of Medicine, Department of Surgery, One Baylor Plaza, Houston, TX 77030, USA
[3] Stanford University, Stanford Blood Center, 3373 Hillview Avenue, Palo Alto, CA 94304, USA

Translational Medicine Cancer: Advances in Molecular Biology and Medicine,
First Edition. Edited by Robert A. Meyers.

Keywords

Cancer immunoediting
Cancer immunoediting is a conceptual framework that integrates the dual role of the immune system in both host protection from tumor development and in the promotion of tumor growth. There are three distinct phases: elimination, equilibrium, and escape.

Tumor microenvironment (**TME**)
The TME refers to the environment surrounding the tumor that includes diverse cell types such as fibroblasts, endothelial cells, and immune cells (T cells, natural killer cells, T regulatory cells, myeloid derived suppressor cells, dendritic cells, and macrophages). The TME cells, along with their secreted cytokines and chemokines, have an impact on tumor growth and progression.

Tumor-infiltrating lymphocytes (**TILs**)
TILs are immune cells, usually T cells, that are found infiltrating the tumor. Generally, the presence of TILs such as $CD8^+$ cytotoxic T lymphocytes and $CD45RO^+$ memory T cells is associated with an improved prognosis. TILs can also be harvested and expanded for antitumor immunotherapy. This approach has been used most successfully in metastatic melanoma.

Myeloid-derived suppressor cells (**MDSCs**)
MDSCs are immunosuppressive cells of myeloid lineage found in the TME. They exert their immunosuppressive effects by direct mechanisms, using arginase-1 and nitric oxide pathways, and also by indirect mechanisms. The presence of MDSCs in tumors is generally associated with a poor prognosis.

T regulatory cells (Tregs)
Tregs are immunosuppressive cells found in the TME. They can develop *de novo* in the thymus (natural Tregs) or from differentiation of naïve T cells (induced T regs). They suppress immune responses by contact-dependent and -independent mechanisms. The prognostic significance of tumor-associated Tregs varies.

Dendritic cells (DCs)
DCs are the most potent antigen-presenting cells, and bridge the innate and adaptive immune systems. DCs have both immunostimulatory as well as tolerogenic, immunosuppressive populations. There is currently one FDA-approved DC therapy for cancer.

Tumor-associated macrophages (TAMs)
TAMs are generally thought to promote tumor initiation, progression, and metastasis through both their proinflammatory and immunosuppressive actions. They play an important role in tumor cell intravasation (blood vessel invasion) and also stimulate angiogenesis (blood vessel production).

Monoclonal antibodies (mAbs)
MAbs are derived from immunoglobulins (Ig). IgG mAbs typically contain a single Fv receptor that recognizes a single antigenic target. Chimeric mAbs contain murine variable regions and human constant regions; humanized mAbs contain only murine components in the smaller complementarity-determining regions of the variable region; and fully human mAbs contain no murine components. There are multiple FDA approved mAbs for cancer treatment.

Adoptive T-cell therapy, including T-cell receptors (TCRs) and chimeric antigen receptors (CARs)
In general, engineered TCRs are generated by cloning α and β chains of a TCR from tumor-associated antigen-specific T-cell clones and expressing these chains in T cells. An advantage of engineered TCRs is that they behave like natural TCRs and can recognize intracellular antigens that are processed and presented in association with major histocompatibility (MHC) class I or II determinants. CARs are recombinant receptors comprised of an extracellular single-chain variable fragment from an antigen-specific mAb and an intracellular T-cell triggering moiety (i.e., CD3ζ). Advantages of CAR T-cell therapy include recognizing antigens independent of MHC molecules, and the potential to recognize nonpeptides. There are several ongoing clinical trials with both of these therapies.

Tumor immunology is the study of the interplay between cancerous and precancerous tumors and the host's immune system. Tumors utilize a variety of mechanisms, including cancer immunoediting, to promote their growth and escape the host immune response. The tumor microenvironment (TME) is also critical for tumor development and growth, and a variety of host immune cells in the TME, including T regulatory cells (Tregs), myeloid-derived suppressor cells (MDSCs), dendritic cells (DCs) and macrophages, contribute to this process. The overall goal of immunotherapy is to overcome cancer immunoediting and the suppressive TME by directing the power and specificity of the immune system against the tumor and the factors that promote tumor growth. Monoclonal antibodies (including checkpoint-blocking antibodies), adoptive cell therapies (including those using T cells with engineered αβ T-cell receptors and chimeric antigen receptors) and DC vaccines are among the approaches that have shown recent promise in the clinic.

Abbreviations in Text (alphabetical)

AID	Activation-induced cytidine deaminase
AIDS	Acquired immune deficiency syndrome
ADCC	Antibody-dependent cell-mediated cytotoxicity
ALL	Acute lymphoblastic leukemia
APC	Antigen-presenting cell
Arg1	Arginase-1
CA-IX	Carboxy-anhydrase-IX
CAR	Chimeric antigen receptor
CCL-	Chemokine ligand
CCR-	Chemokine receptor
CD	Cluster of differentiation
CDR	Complementarity-determining regions
CEA	Carcinoembryonic antigen
CLL	Chronic lymphocytic leukemia
COX2	Cyclooxygenase-2
CRC	Colorectal cancer
CSF-1 and CSF-1R	Colony-stimulating factor 1 and receptor
CTL	Cytotoxic T lymphocyte
CTLA-4	Cytotoxic T lymphocyte antigen-4
CXCL-	Chemokine (C-X-C motif) ligand
CXCR-	Chemokine (C-X-C motif) receptor
DAMPs	Damage-associated molecular pattern molecules
DC	Dendritic cell
DLBCL	Diffuse large B-cell lymphoma
EGF	Epidermal growth factor
EGFR	Epidermal growth factor receptor
EMT	Epithelial-mesenchymal transition
FDA	Food and Drug Administration
FcR	Fc receptor
FGF2	Basic fibroblast growth factor
FISH	Fluorescence *in situ* hybridization
Flt3	Fms-like tyrosine kinase 3
FOXP3	Forkhead box P3

Fv	Variable fragment
GD2	Disialoganglioside
gp100	Glycoprotein 100
GM-CSF	Granulocyte macrophage colony-stimulating factor
HIF1α	Hypoxia-inducible factor 1-alpha
HLA	Human leukocyte antigen
HMGB1	High-mobility group box 1
HSCT	Hematopoietic stem cell transplant
IAE	Immune-related adverse event
IDC	Immature dendritic cell
IDO	Indoleamine 2,3-dioxygenase
IFN-γ	Interferon-gamma
Ig	Immunoglobulin
IHC	Immunohistochemistry
IL-	Interleukin
iNOS or NOS2	Nitric oxide synthase
KRAS	V-Ki-ras2 Kirsten rat sarcoma viral oncogene homolog
LAG3	Lymphocyte activation gene-3
LAP	Latency-associated protein
L-arg	L-Arginine
MAb	Monoclonal antibody
MAGE	Melanoma-associated antigen
MALT	Mucosa-associated lymphoid tissue
MART-1	Melanoma antigen recognized by T cells 1
mCRPC	Metastatic castrate-resistant prostate cancer
M-CSF	Macrophage colony-stimulating factor
MDSC	Myeloid-derived suppressor cell
MET	Mesenchymal-epithelial transition
MHC	Major histocompatibility complex
MICA	MHC class I polypeptide-related sequence A
MMP	Matrix metalloproteinases
NF-κβ	Nuclear factor of kappa light polypeptide gene-enhancer in B cells
NCI	National Cancer Institute
NHL	Non-Hodgkin's lymphoma
NO	Nitric oxide
NSCLC	Non-small cell lung cancer
NY-ESO-1	Cancer-testis antigen
OS	Overall survival
PAP	Prostatic acid phosphatase
PBMC	Peripheral blood mononuclear cell
PD-1	Programmed death receptor-1
PDGF	Platelet-derived growth factor
PD-L1	Programmed death receptor ligand-1
PD-L2	Programmed death receptor ligand-2

PFS	Progression-free survival
PGE2	Prostaglandin E2
PGF	Placental growth factor
regDC	Regulatory dendritic cell
RNA	Ribonucleic acid
RNI	Reactive nitrogen intermediates
ROS	Reactive oxygen species
SCF-1	Stem cell factor
scFv	Single-chain variable fragment
SDF1α	Stromal-derived factor 1α
STAT	Signal transducer and activator of transcription
TAA	Tumor-associated antigen
TAM	Tumor-associated macrophage
TCR	T-cell receptor
TDM1	Ado-trastuzumab emtansine
TGF-β	Transforming growth factor-β
TIL	Tumor-infiltrating lymphocyte
TIM3	T-cell immunoglobulin domain and mucin domain 3
TLR	Toll-like receptor
TME	Tumor microenvironment
TNF-α	Tumor necrosis factor-α
Treg	Regulatory T-cell
VEGF	Vascular endothelial growth factor
VEGFR-1 and VEGFR-2	Vascular endothelial growth factor receptor 1 and 2
α-FR	Alpha folate receptor

1 Introduction

For several decades, slow but steady progress has been made in the identification of immune cells and factors capable of attacking tumor cells, as well as those that promote the growth of tumors and facilitate their escape from immune effector cells. Until about 15 years ago, clinically effective immunotherapeutic approaches to cancer had remained elusive, but this changed with the development of chimeric, humanized, and fully human monoclonal antibodies (mAbs) targeting tumor-associated antigens (TAAs) or components of the tumor microenvironment (TME). Until very recently, these mAbs were the only immunotherapeutic agents in wide clinical use, and there was little optimism that immune-based therapies would extend beyond this approach. Within the past few years, however, the situation has changed dramatically, starting with the United States Food and Drug Administration (FDA) approving the first therapeutic cancer vaccine in 2010 [1]. Most recently, clinical trials evaluating mAbs that block immune checkpoints [2–5] and adoptive cell therapies with engineered chimeric antigen receptor (CAR) T cells [6, 7] have generated enormous excitement on the basis of significantly greater potency and duration of response than seen with prior immunotherapies.

The role of the immune system in cancer is complex. Although it promotes cancer development, growth and metastases, it can also be harnessed to effectively and adaptively target tumors for killing [8]. In the first part of this chapter, the consequences of the interaction between tumors and the immune system will be discussed, including: (i) the concept of cancer immunoediting [8]; (ii) the role of chronic inflammation in tumor development; and (iii) the role of specific immune cells within the TME, including dendritic cells (DCs), T regulatory cells (Tregs), macrophages, and myeloid-derived suppressor cells (MDSCs) in promoting tumor development and growth. In the second part of the chapter, antitumor immunotherapies will be described, including: (i) mAbs; (ii) T-cell-based therapies, including expanded tumor infiltrating lymphocytes (TILs) and T cells with genetically engineered T-cell receptors (TCRs) such as CARs; and finally (iii) DC-based vaccines.

2 Consequences of the Interaction between Tumors and the Immune System

2.1 Cancer Immunoediting

The capacity of the immune system to control cancer growth has been clearly demonstrated over the past two decades. A wide variety of studies have demonstrated that mice genetically deficient in critical components of the immune system are more susceptible to develop tumors in almost all experimental settings, including models of transplantable, spontaneous, virus-induced, and carcinogen-induced cancers [9].

Cancer immunoediting is a conceptual framework described by Schreiber and colleagues that integrates the dual role of the immune system in both host protection from tumor development and in the promotion of tumor growth [10]. This elegant theory takes into account the often conflicting and opposing roles of the immune system. The immune system not only suppresses tumor growth by destroying cancer cells and inhibiting their outgrowth, but also promotes malignant progression by selecting for tumor cells that are better fit to survive in an immunocompetent host, or by altering the TME in a way that fosters tumor outgrowth. This concept applies the theory of evolution and natural selection on a microscopic scale, and suggests that tumors which become clinically relevant must contain subpopulations of cancer cells that can survive the pressures of the host immune system. Thereby, these tumors have evolved successful strategies to evade host immunosurveillance.

Observations from several classic experiments clarify the basic tenets of immunoediting. For example, immunodeficient mice are more prone to develop tumors in response to treatment with carcinogens than are immunocompetent mice [11]. Further, the tumors that develop in immunodeficient mice are qualitatively different from those that develop in immunocompetent mice. Specifically, tumor cells from carcinogen-treated immunodeficient mice are more likely to be rejected when transplanted into an immunocompetent mouse than are tumor cells grown in carcinogen-treated immunocompetent mice [11]. These findings suggest that tumors grown in immunodeficient mice are "unedited" and have a more immunogenic phenotype that allows their rejection by a competent immune system. Thus, the immune system not only protects the host from tumor formation, but also shapes tumor immunogenicity.

Three distinct phases of the cancer immunoediting process have been described, termed elimination, equilibrium, and escape [10, 12].

2.1.1 Elimination Phase

In the elimination phase, the innate and adaptive immune system work together to detect and destroy developing tumors before they become clinically apparent. The exact mechanisms by which the immune system senses a developing tumor are incompletely understood. Some hypothesized mechanisms include classic danger signals such as interferon-gamma (IFN-γ) produced by innate immune cells [13]; damage-associated molecular pattern (DAMP) molecules directly released from dying tumor cells, such as high-mobility group box 1 (HMGB1) [14]; and stress ligands such as RAE-1 or MICA/B that can be expressed on the surface of tumor cells and recognized by the adaptive immune system [15]. Although effective immunoediting can be achieved in some experimental systems in which only innate immunity is present [16], most systems require the additional expression of tumor antigens that can trigger effector T-cell expansion and activate the adaptive immune system.

This is a fundamental principle in tumor immunology and in immunoediting, where cancer cells express antigens that distinguish them from the normal host and permit recognition and destruction by T cells. The existence of TAAs was first demonstrated in experiments in which mice immunized with chemically – induced tumors were protected from a re-challenge with the same tumor. This phenomenon has now been reproduced in a wide variety of experimentally induced tumors. Molecular studies have demonstrated that some of these TAAs are truly specific to the tumor. Such antigens are often products of mutated cellular genes, normal genes that are aberrantly expressed, or genes that encode viral proteins. Other tumor antigens may be expressed at higher levels by tumor cells, but are not tumor-specific. Immune responses directed against such antigens may affect normal cells in addition to tumor cells. The term TAA, as used in this chapter, encompasses both types of tumor antigen. A number of human TAAs have been discovered and include overexpressed cellular antigens (such as HER-2), differentiation antigens (such as the melanocyte differentiation antigens), oncofetal antigens (such as carcinoembryonic antigen; CEA), and cancer/testis antigens which are expressed in germ cells but are silent in somatic cells (such as melanoma-associated antigen (MAGE) and NY-ESO-1) [17]. As discussed elsewhere in this chapter, the major objective of cancer immunotherapies, such as DC vaccines, CAR T cells, and checkpoint-blocking mAbs, is to induce or direct the immune response against TAAs, resulting in tumor elimination.

2.1.2 Equilibrium Phase

If tumor cells are successfully recognized and destroyed by the immune system, the immunoediting process concludes with the elimination phase. Rare variants of tumor cells, however, may survive the elimination phase and enter the equilibrium phase. In this phase, the adaptive immune system prevents tumor outgrowth by maintaining a functional state of dormancy. Evidence for the equilibrium phase has been demonstrated in experiments in which immunocompetent mice treated with low doses of carcinogen harbored occult tumor cells without clinically evident tumors for long periods of time. However, when T cells (and other components of the adaptive immune system) were depleted in such mice, tumors rapidly appeared at the original carcinogen injection site [18]. This suggests that the adaptive immune system was maintaining animals in a state of equilibrium free of visible tumor. Therefore, while the immune system provides suppression of tumor outgrowth, at the same time it also provides the selective pressures necessary for "successful" tumors to develop qualities that allow them to escape their immunologic restraints.

2.1.3 Escape Phase

Tumors that acquire the ability to evade recognition by the immune system will enter the escape phase and emerge as clinically apparent tumors. Tumor escape can occur through a variety of mechanisms. However, for the most part, either the tumor cell population itself changes in response to the immune system's editing function or the immune system changes to promote a more immunosuppressive state within the TME. Adaptations at the level of the tumor cell include those leading to a decreased recognition of the tumor cell by the immune system. This may include a loss of tumor antigen expression or presentation, or an increased resistance to the cytotoxic effects of immunity via mechanisms such as upregulation of anti-apoptotic pathways [12, 19–21]. Recent studies have additionally shown that tumor cell escape can be dependent on the nature of tumor cell mutations, whereby a T-cell-dependent selection process allows tumor cells that fail to express highly antigenic mutations to escape. An alternative mechanism of immune escape is one in which tumor cells promote an immunosuppressive state within their microenvironment by producing immunoregulatory cytokines and enzymes such as vascular endothelial growth factor (VEGF), transforming growth factor-β (TGF-β), and indoleamine 2,3-dioxygenase (IDO), as well as by recruiting immune cells such as Tregs and MDSCs that can suppress the development of antitumor immunity [12].

2.1.4 Cancer Immunoediting in Humans

Although the majority of mechanistic studies of immunoediting have been performed in mice, there are also data supporting the occurrence of immunoediting in humans. One example of immunoediting in humans is a loss of the MART-1 (melanoma antigen recognized by T cells 1) antigen that is seen in melanoma patients after they are treated with MART-1-specific T cells [22]. The strongest evidence of immunoediting in humans comes from reports that correlate the quality, quantity and microanatomic location of tumor-infiltrating immune cells in human cancers, particularly cytotoxic $CD8^+$ T cells and Treg cells, with long-term clinical outcomes [23]. Other studies have shown that cancer patients can develop potent T-cell and antibody responses to TAAs [24]. Finally, human immunodeficiency has been linked to an increased cancer risk, as occurs in patients with acquired immune deficiency syndrome (AIDS) or in those maintained on chronic immunosuppressive medications following organ transplantation. Although cancers arising in these patient populations are often of a viral etiology – such as certain lymphomas, Kaposi's sarcoma and cervical cancer – there is evidence that these patients are also at higher risk of developing cancers of the lung, colon, pancreas, kidney, head and neck, and endocrine system [12]. It has become generally accepted that clinically evident and progressively growing human tumors have developed effective strategies to evade natural antitumor immune mechanisms. Thus, the development of effective cancer immunotherapies must take into consideration these many factors, and aim to increase the quality or quantity of effector immune cells and/or eliminate tumor-induced immunosuppressive mechanisms.

2.2
The Role of Chronic Inflammation in Tumor Development

Rudolf Virchow first postulated a link between inflammation and cancer in the nineteenth century, when he observed the infiltration of leukocytes into tumors [25]. Since then, an abundance of experimental, clinical and epidemiologic evidence has confirmed a close association between inflammation and tumorigenesis [9, 12, 26–28]. Inflammation normally functions to maintain tissue health in response to insults such as tissue damage or infection. The first line of defense against pathogens and foreign agents are cells of the innate immune system including granulocytes, macrophages, DCs, and innate lymphocytes. These cells lay the foundation for the state of *acute inflammation* in which a number of cytokines, chemokines and matrix remodeling proteins are generated to recruit circulating leukocytes into damaged tissue and eliminate pathogens. In tumors, these normally well-orchestrated series of events fail to resolve and this results in a state of *chronic inflammation*, which has been shown to contribute to tumorigenesis at all stages. This occurs through various mechanisms such as generation of genotoxic stresses, induction of cellular proliferation, and enhancement of angiogenesis (blood vessel formation) and invasion [28]. Based on the continuous cellular renewal and proliferation induced by tumor-associated chronic inflammation, tumors have been referred to as "wounds that do not heal" [29].

In many cases, human tumor development is linked to chronic infections, dietary factors, obesity, inhaled pollutants, tobacco use, or autoimmunity [28]. It was initially believed that these factors directly initiate the formation of cancerous cells. However, it is now apparent that many of these factors indirectly promote tumor formation by activating chronic inflammation. For example, persistent infection with the bacterium *Helicobacter pylori* is associated with the development of gastric cancer and mucosa-associated lymphoid tissue (MALT) lymphoma, and infection with hepatitis B or C viruses increase the risk of hepatocellular carcinoma. Immune dysregulation and autoimmunity is another type of chronic inflammation that precedes tumor development; an example is inflammatory bowel disease, which increases the risk of colorectal cancer (CRC). Moreover, chronic inflammation induced by environmental exposures such as tobacco smoke can precipitate chronic obstructive pulmonary disease (COPD), an inflammatory condition associated with a higher risk of lung cancer.

As noted in Sect. 2.1, there is also an abundance of evidence that the immune system can protect the host against tumor development. Thus, it is likely that cancer immunoediting and tumor-promoting inflammation coexist in the same TME. This coexistence is possible at least in part due to the functional plasticity of myeloid and lymphoid immune cells which, depending on their context and environment, can demonstrate both protumor and antitumor functions (Table 1). The specific tumor-promoting mechanisms used by these cell types are reviewed in Sect. 2.3. The net outcome of tumor-associated inflammation depends on the relative dominance of its tumor-promoting or tumor-suppressive actions. Most (if not

Tab. 1 Immune cell types.

Immune cell type	*Tumor-suppressive functions*	*Tumor-promoting functions*
Macrophages	Produce IL-12 and NO[a)]	Suppress antitumor immune responses, promote angiogenesis, invasiveness, metastases
DCs	Produce IL-12 and IFN-γ	Suppress antitumor immune responses, promote angiogenesis
MDSCs	Activate NK cells	Suppress cytotoxic T cells and NK cells, expand Tregs
Neutrophils	Tumor cell cytotoxicity	Produce immunosuppressive cytokines
$CD8^+$ T cells	Tumor cell cytotoxicity, produce IFN-γ	Produce immunosuppressive cytokines
$CD4^+$ T cells	Assist $CD8^+$ T-cell antitumor responses	Produce immunosuppressive cytokines
T regulatory cells	Suppression of tumor-promoting inflammation	Suppress antitumor immune responses
B cells	Assist T-cell immunity	Promote metastases and immunosuppressive microenvironment

IL = interleukin; NO = nitric oxide; DC = dendritic cell; IFN = interferon; MDSC = myeloid-derived suppressor cell; NK = natural killer cell; Treg = T regulatory cell.
a) NO can also be tumor-promoting.
Adapted from Ref. [9].

all) established solid tumors, however, are associated with an intrinsic chronic inflammatory response that supports a protumorigenic environment.

2.2.1 Tumor Initiation and Inflammation

One way in which an inflammatory TME may foster the initiation of tumors is by promoting genomic instability. Inflammation may increase cellular mutagenesis via reactive oxygen species (ROS) and reactive nitrogen intermediates (RNI). Both ROS and RNI are produced by activated inflammatory cells, or by the upregulation of enzymes that induce genomic instability, such as activation-induced cytidine deaminase (AID) [30]. Alternatively, chronic inflammation releases a variety of cytokines and growth factors that can endow a stem cell-like phenotype upon tumor progenitor cells and stimulate stem cell expansion [31].

2.2.2 Tumor Promotion and Inflammation

Following the initiation of tumor development, chronic inflammation promotes tumor proliferation and survival through a variety of mechanisms. One of these mechanisms involves the production of tumor-promoting cytokines that activate transcription factors advantageous to the survival of premalignant and malignant cells. A number of cytokines produced by tumor-infiltrating immune cells have these effects, including tumor necrosis factor-α (TNF-α) and interleukin (IL)-6, IL-1, IL-17 and IL-23. Among the different transcription factors that are activated in most cancers, NF-$\kappa\beta$ (nuclear factor of kappa light polypeptide gene-enhancer in B cells) and STAT3 (signal transducer and activator of transcription 3) can activate genes that control cell proliferation and survival, as well as angiogenesis, motility and invasiveness [28, 32]. Another major tumor-promoting pathway involving infiltrating immune cells is the suppression of antitumor immune responses by immunosuppressive cells in the TME, such as MDSCs and Tregs. The mechanisms by which these cells mediate their effects are reviewed in Sect. 2.3.

2.2.3 Angiogenesis and Inflammation

The continued growth of large tumors requires angiogenesis, a process which is largely dependent on hypoxia, which has also recently been shown to be regulated by inflammatory cells such as tumor-associated macrophages (TAMs). In response to hypoxia, TAMs upregulate important proangiogenic genes including hypoxia-inducible factor 1-α (HIF1α), VEGF, IL-8, and chemokine (C-X-C motif) ligand (CXCL) 1 [33].

2.2.4 Metastases and Inflammation

The development of metastases is a critical step in tumor evolution and is regulated on multiple levels by immune mediators. An early step in the metastatic process is known as epithelial–mesenchymal transition (EMT), in which cancer cells acquire the characteristics that allow them to increase motility and invade basement membranes. EMT is regulated by immune cell-derived cytokines such as TGF-β [34]. Cancer cell invasion requires proteolysis of the extracellular matrix (ECM), a process that is regulated by matrix metalloproteinase (MMP) enzymes secreted by tumor-infiltrating myeloid cells [35]. Once these cancer cells reach the circulation and target organs, their survival is also regulated by a variety of cytokines and chemokines [28]. The landing zone for these metastatic cells is referred to as the "premetastatic niche," reflecting the view that certain sites within target organs may be already primed by tumor-derived inflammatory signals to facilitate tumor cell entry and survival [36].

In summary, chronic inflammation can influence nearly every aspect of tumor development and progression. A thorough investigation of these various components of the inflammatory response has the potential to lead to the discovery of novel therapeutic targets and new strategies to prevent, as well as treat, cancer.

2.3 Immune Cells in the Tumor Microenvironment (TME)

2.3.1 Introduction

Tumors are complex and heterogeneous groups of abnormally mutated cells. Tumor-derived factors, including growth factors and cytokines, recruit varied cell types to the TME, including fibroblasts, immune cells and endothelial cells. The T-cell responses to TAAs are well-characterized and as discussed below, high densities of TILs correlate with an improved outcome in several types of cancer [23, 37]. Despite the presence of such antitumor T cells, malignant tumors can grow and ultimately kill their host. A major goal in tumor immunology is to better understand how tumors evade antitumor immune responses (Table 2). In addition to immunoediting, this evasion can occur through both a direct inhibition of conventional effector T-cell responses as well as through the modulation of a variety of host immune cells, including Tregs, MDSCs, DCs and macrophages. These latter cells cannot only promote a local immunosuppressive microenvironment but can also have systemic effects that promote tumor growth. The focus of this section will be on the immune cells in the TME, with special attention paid to Tregs, MDSCs, DCs and macrophages.

Tab. 2 Mechanisms of tumor escape.

Tumor or tumor-associated antigen (TAA) mechanisms
Downregulation of MHC-I molecules
Downregulation of genes associated with antigen presentation
Loss of antigenic epitopes
TAA, often self-antigens so may be protected by immune tolerance
Modulation of expression of TAA
Physical barriers preventing access of effector immune cells to tumor cells (such as mucin production)
Expression of coinhibitory molecules such as PD-L1
Secretion of factors causing abnormal myelopoiesis and promoting recruitment, expansion, and function of suppressive immune cells
Mechanisms associated with host immune system
Ignorance: immune system may not recognize low levels of TAAs
Tolerance of T cells to TAAs (anergy or loss of activation to TCR stimulation): can be caused by membrane-bound and soluble factors derived from other cells, including Tregs, MDSCs, DCs, and macrophages, as well as tumor
Suppression of T cells: from tumor-derived factors and factors derived from immunosuppressive myeloid cells and Tregs
T-cell exhaustion
Defects in antigen presentation by APCs
Downregulation of MHC-I and/or MHC-II molecules and positive costimulatory signals
Expression of immunosuppressive cytokines such as TGF-β and IL-10
Expression of enzymes modulating amino acid metabolism affecting T-cell function (arginase and arginine depletion, IDO and tryptophan depletion)
Production of reactive oxygen species and reactive nitrogen species
Promotion of angiogenesis (particularly by tumor-associated macrophages)

MHC = major histocompatibility complex; PD-L1 = programmed death receptor ligand 1; TCR = T-cell receptor; Treg = T regulatory cell; MDSC = myeloid-derived suppressor cell; DC = dendritic cell; APC = antigen-presenting cell; TGF = transforming growth factor; IL = interleukin; IDO = indoleamine 2,3-dioxygenase.
Adapted from Ref. [38].

Mechanisms that promote immune tolerance are required to prevent immune responses to self-antigens that can result in autoimmune disease, regulate immune responses to foreign antigens, and limit injury to normal tissue from inflammation. Interestingly, tumors can promote a tolerogenic environment that facilitates immune evasion. Conventional T-cell tolerance or anergy refers to a lack of response to TCR stimulation by antigen, as reflected by an absence of cytolytic activity and reduced IFN-γ secretion [39–41]. Additionally, tolerized tumor-specific $CD8^+$ T cells can also suppress other $CD8^+$ T cells via alteration in costimulatory and coinhibitory signals in the TME [42, 43]. A better understanding of the mechanisms responsible for TME immunosuppression and tumor immune evasion has led to the development of targeted therapeutic approaches to reduce immune suppression, and will hopefully lead to more effective antitumor immunotherapies [44].

2.3.2 Tumor-Infiltrating Lymphocytes (TILs)

Recent studies have revealed the prognostic value of the type and density of immune cells in the TME. An increased density of TILs is associated with an improved prognosis in several types of cancer [23, 45, 46]. Galon *et al.* correlated clinical outcome with the type, density and location of immune cells in 415 CRC samples [47]. A higher $CD3^+$ T-cell density was associated with longer disease-free survival, while a strong immune reaction at either the center or invasive margin of the tumor correlated with a favorable prognosis, regardless of the local extent of tumor or invasion of regional lymph nodes [47]. There was also an inverse association between high-risk colorectal tumors (perineural or lymphovascular invasion) and the frequency of infiltrating ($CD45RO^+$) memory T cells, with $CD45RO^+$ status emerging as an independent favorable prognostic factor in a multivariate analysis [48]. This suggests that the initial immune response likely affects tumor progression, and a decline in TILs is reflective of immune escape [49]. In another study of ovarian carcinoma samples, patients with tumors containing T cells had a strikingly longer progression-free survival (PFS) than patients whose tumors did not contain T cells (PFS 22.4 months versus 5.8 months, $P < 0.001$) and overall survival (OS; 50.3 months versus 18.0 months, $P < 0.001$) [50]. These studies demonstrate that markers of immune activation, including infiltration of tumors with $CD3^+$ T cells, $CD8^+$ cytotoxic T lymphocytes (CTLs) and $CD45RO^+$ memory T cells, are associated with improved outcomes [23]. Further studies are being conducted to clarify which tumor-specific and host-specific factors lead to a robust lymphocytic infiltrate, how this infiltrate evolves with tumor progression, and whether it plays a role in the response to chemotherapies and immunotherapies.

2.3.3 Myeloid-Derived Suppressor Cells (MDSCs)

T-cell activation requires contact with professional antigen-presenting cells (APCs), which include activated macrophages, DCs and B lymphocytes. APCs present the antigen bound to a major histocompatibility complex (MHC), and also supply the soluble and surface-bound activation stimuli needed for T-cell activation, including cytokines, chemokines and costimulatory molecules [51]. Recently, intense investigations have been conducted of the regulatory immune cell populations that control the activation, expansion and function of T lymphocytes. Such regulatory populations include Tregs (described further below) as well as suppressive cells of nonlymphoid origin, including MDSCs.

MDSC Phenotype MDSCs consist of monocytes and granulocytes in various states of maturation [52]. These cells have immunosuppressive properties and act via antigen-specific and nonspecific mechanisms. MDSCs were first noted to accumulate in lymphoid tissues in tumor-bearing mice, as well as during states of chronic inflammation, infection, sepsis and trauma [53–55]. In tumor-bearing mice, MDSCs are characterized by the expression of CD11b and Gr-1 [51]. More recent subclassifications include mononuclear monocytic MDSCs ($CD11b^+Ly6C^{high}Ly6G^-$) and granulocytic-like polymorphonuclear MDSCs ($CD11b^+Ly6G^+Ly6C^{low/int}$) [52, 56–58]. The classification of human MDSCs has been more challenging due to the lack of a Gr1 homolog and phenotypic heterogeneity; however, recent consensus has described human MDSCs as $CD33^+HLA\text{-}DR^{(low/-)}CD11b^+$ [58].

MDSCs and Cancer In murine tumor models, MDSCs are present in tumors and peripheral blood, and exhibit immunosuppressive properties. In patients with certain cancers, MDSCs also infiltrate tumors and blood and have immunosuppressive properties, as in head and neck cancers where they can suppress IL-2 secretion [59]. The frequency of MDSCs also correlates with worse prognostic features; for example, patients with non-small-cell lung cancer (NSCLC) have higher levels of MDSCs compared to healthy subjects, and MDSCs are also more frequent in patients with advanced-stage disease compared to those with earlier-stage disease [60]. In another study of solid tumors from 106 patients with breast, colon and pancreatic cancer, higher levels of circulating MDSCs correlated with an advanced clinical stage and an increased metastatic tumor burden [61]. There is a growing body of studies suggesting that MDSC levels are increased in patients with cancer, and that they may have a negative prognostic value. However, the predictive value of MDSCs needs to be evaluated in a prospective manner [62].

MDSC Recruitment and Expansion Although MDSCs are more frequent in tumor-bearing mice compared to naïve tumor-free mice, the extent of MDSC expansion varies with the tumor model [52, 56]. This supports the hypothesis that tumor-derived factors are important in MDSC recruitment and expansion, and may also help to explain differences in the subsets of expanded MDSCs between murine tumor models. Factors that affect MDSC recruitment and/or expansion in murine models include granulocyte macrophage colony-stimulating factor (GM-CSF), stem cell factor (SCF) -1, macrophage-colony-stimulating factor (M-CSF), VEGF, prostaglandins, and chemokine receptor (CXCR) 2 ligands, among others [63–68].

MDSC Immunosuppression Mechanisms MDSCs can suppress $CD8^+$ T-cell proliferation and function in tumor-bearing mice [56] via direct mechanisms (arginase-1 (Arg1) and nitric oxide (NO) pathways), and also via indirect mechanisms (modulation of other cells, such as Treg populations) [58]. L-arginine (L-arg) is a semi-essential amino acid that can be converted to L-ornithine and urea by Arg1. L-Arginine can also undergo oxidation to form citrulline and NO via inducible nitric oxide synthase (iNOS; also known as NOS2) [69–71]. MDSCs express high levels of Arg1, which leads to extracellular arginine deprivation and a subsequent reduced expression of the critical T-cell signaling ζ-chain of the CD3 complex [72, 73]. L-Arginine starvation also causes T-cell cycle arrest at the G_0-G_1 phase [74]. MDSCs from patients with renal cell carcinoma have increased Arg1 activity, along with low levels of L-arg and high levels of ornithine in the plasma [75]. High levels of Arg1 are also found in tumor samples from patients with NSCLC [73] and pancreatic cancer [76]. Tumor-derived soluble factors, such as cyclooxygenase-2 (COX-2) and prostaglandin E2 (PGE2), can also induce Arg1 activity in MDSCs [77]. NO has both antitumor and immunosuppressive effects [58, 78]. NO affects IL-2 receptor downstream signaling [79–81], preventing T-cell activation and promoting T-cell apoptosis via Fas-dependent mechanisms [82, 83]. It has also been proposed that the coexpression of Arg1 and NOS2 favors the production of RNI and ROS [58, 84], which can suppress T-cell activity by affecting TCR signaling and inducing apoptosis [76, 85–89]. MDSCs also have indirect immunosuppressive effects. They can act as tolerogenic APCs, as shown in a mouse model of B-cell lymphoma, in which MDSCs had the ability to take up and present antigen to $CD4^+FOXP3^+$(forkhead box P3) Tregs through Arg1-dependent expansion of these cells [90]. In another murine tumor model, MDSCs secreted IFN-γ and IL-10 through iNOS-independent pathways to induce $CD25^+FOXP3^+$ Tregs [91]. In summary, MDSCs are immunosuppressive, and their presence correlates with a poor prognosis in cancer.

2.3.4 T Regulatory Cells (Tregs)

Treg Phenotype Suppressive T cells were first recognized when mice subjected to thymectomy on day 3 after birth were noted to have autoimmune diseases, suggesting the presence of immunosuppressive lymphocytes in the thymus. These were later isolated and termed Tregs [92–95]. Phenotypically, the best-described $CD4^+$ Treg subset is identified by the expression of CD25 (the IL-2 receptor α chain) [96] and the transcription factor FOXP3, which is necessary for the development and functioning of Tregs [97–99]. Expression of CD25 and FOXP3, however, is not limited to Tregs and can be found on other immune cells [100, 101]. Shortly after the identification of $CD4^+CD25^+$-expressing Tregs in mice, several groups identified the presence of $CD4^+CD25^+$ regulatory T cells in humans (for a review, see Ref. [95]). Tregs are a heterogeneous population of T cells and can either develop in the thymus (natural Tregs) or be induced via *de novo* differentiation from $CD4^+CD25^-$ naïve T cells (induced Tregs) [95, 102]. Tregs mediate peripheral tolerance, thereby preventing autoimmunity, and reduce potentially tissue-damaging effects of inflammation after injury or infection [95]. Tregs are capable of suppressing the proliferation of conventional T effector cells [103] by expressing inhibitory cytokines (i.e., IL-10 and TGF-β) and cell-surface molecules (i.e., cytotoxic T lymphocyte antigen-4; CTLA-4) [103–105].

Tregs and Cancer Tregs are present at an increased frequency in human tumor samples, peripheral blood, tumor-draining lymph nodes and ascites [106–108]. The prognostic value of Tregs in patients with cancer appears to vary depending on the underlying tumor type and molecular subtype, the tumor stage and grade, and the particular Treg subtype evaluated [109]. This variability in prognostic outcome has been demonstrated in studies of CRC [110, 111] and ovarian cancer. In studies of ovarian epithelial cancers by Curiel *et al.*, an increased tumor Treg frequency was predictive of a poor survival after controlling for variables such as tumor stage and surgical debulking [106]. However, in another study of ovarian cancer, the presence of tumoral $FOXP3^+$ cells proved to be a positive prognostic factor in a multivariate analysis [112]. The markers used to identify Tregs differed between the two studies. In the latter study, Treg presence was also associated with the presence of CTLs, and a high CTL/Treg ratio had a positive prognostic impact on disease-free survival, making this ratio a potentially important prognostic tool. In attempts to explain the variable impact of Tregs on prognosis, it has been proposed by some groups that Treg function may vary depending on the environment, whereas others have proposed that the immunoinhibitory function of Tregs may actually be favorable in certain environments by suppressing tumor-promoting inflammation (for a review, see Ref. [95]).

Treg Recruitment and Expansion Several studies have been conducted to determine how Tregs are recruited to and expand in tumors. Chemokines secreted by tumor cells and immune cells, such as TAMs and DCs – along with the expression of integrin and CXCRs on Tregs – can impact Treg migration to the TME [95, 113]. For instance, in ovarian cancer samples, Curiel *et al.* found high levels of tumor cell and macrophage secretion of chemokine ligand (CCL) 22 and a correspondingly enhanced Treg expression of CCL2 receptor, CCR4 [106]. In a murine ovarian cancer xenograft model, CCL22 also induced Treg migration into tumors, whereas in a melanoma model this migration was mediated by CCL2 (another ligand for the CCR4 receptor) [114]. Tumor hypoxia may also affect chemokine secretion and Treg migration [115]. Tumor- and TME-derived factors, such as TGF-β and IL-10 [116], can modify Treg expansion and differentiation both directly and indirectly. For instance, TGF-β can convert $CD4^+CD25^-$ naïve T cells into Tregs via the induction of FOXP3 expression [117]. The expression by DCs and tumors of IDO, an enzyme that depletes tryptophan, can result in Treg expansion [118–120] and the suppression of conventional effector T cells [95].

Treg Immunosuppression Mechanisms Tregs suppress T-cell immune responses through contact-dependent and contact-independent mechanisms [95, 121]. Direct contact between Tregs and APCs, as well as conventional T effector cells, can be immunosuppressive. This occurs through modulation of cell-surface inhibitory molecule expression such as CTLA-4, TIM3 (T-cell immunoglobulin domain and mucin domain 3) and LAG3 (lymphocyte activation gene-3) [122–126], and also through the secretion of molecules that result in the apoptosis of conventional T cells such as granzyme and galectin-1 [127, 128]. Additionally, $CD4^+CD25^+FOXP3^+$ Tregs from cancer patients cultured *in vitro* with IL-2 were found to suppress $CD8^+$ T-cell proliferation via Fas-mediated apoptosis [129]. Contact-independent mechanisms of Treg suppression are mediated by cytokines such as TGF-β, latency-associated protein (LAP), IL-35 and IL-10 [121, 130, 131], which can limit proliferation, cytokine production and target cell lysis by conventional T effector cells, as well as polarize DCs towards a tolerogenic phenotype [132]. Conversely, Treg-mediated cytokine depletion and the resultant apoptosis of conventional T effectors is another proposed mechanism [126]. Although multiple mechanisms have been described by which Tregs can mediate immunosuppression *in vitro*, the mechanisms by which these cells affect the immune response *in vivo* are not well understood [121, 131]. Clearly, further studies will help to shed light on the potential therapeutic impact of modulating Treg frequency and activity.

2.3.5 Dendritic Cells (DCs)

DC Phenotype DCs are involved in both innate and adaptive immune responses as a result of their acquiring, processing, and presenting antigens. They are the most potent APCs, and can prime and activate both $CD4^+$ and $CD8^+$ T cells, as well as B cells and NK (natural killer) cells, but they also function in maintaining tolerance to self-antigens [133]. DCs consist of a heterogeneous group of cells, each of which is included in one of two main lineages, plasmacytoid DCs and conventional DCs [134]. Among these two lineages, conventional DCs are believed to be the major antigen-processing and -presenting cells, while plasmacytoid DCs are a major source of type I IFNs. The remainder of the following discussion of DCs will focus on conventional DCs, which can be found in lymphoid (including spleen, thymus, lymph nodes) and nonlymphoid peripheral tissues, and originate from progenitors in the bone marrow. Immature dendritic cells (iDCs) have traditionally been phenotypically characterized by a low surface expression of MHC-II molecules and costimulatory molecules such as CD40, CD80 and CD86, whereas mature DCs have been phenotypically characterized by upregulated cell-surface expression of MHC-II molecules and costimulatory molecules. In their immature state, DCs can take up and process antigen and, once they are activated and achieve functional maturity, they can present antigen. Recent terminology in the literature has also described "mature DCs" as DCs capable of immune stimulation and "iDCs" as cells involved in tolerance; however, this terminology has been found to be problematic [135] since phenotypically "mature DCs" have been found to induce tolerance, and a high expression of costimulatory molecules does not always predict immunogenicity [135, 136]. A functional terminology combined with phenotypic characterization has been suggested in which terms such as "phenotypically immature tolerogenic DC" or "phenotypically mature tolerogenic DC" are used [135]. This functional characterization is logical, since ligands such as CD80 and CD86 can promote T-cell activation when ligated to CD28, but promote tolerance when ligated to an inhibitory ligand such as CTLA-4 [137, 138]. Tesone *et al.* have proposed designating DC populations as pre-DCs, iDCs or mature or classical DCs, based on their relative expression of MHC class II [139]. There is also a population of classical DCs with a regulatory phenotype, known as regulatory or tolerogenic dendritic cells (regDCs), that sometimes can be identified by the presence of inhibitory programmed death receptor ligand 1 (PD-L1) expression ($CD11c^{+}CD11b^{+/-}MHC\text{-}II^{+}CD86^{hi}PD\text{-}L1^{+}$). In their model, common myeloid progenitors can give rise to both MDSCs and pre-DCs, which in turn can give rise to both immunosuppressive iDCs and regDCs.

DCs and Cancer DCs can process and present TAAs from necrotic and apoptotic tumor cells [38]. Both, iDCs and regDCs have been identified within tumors, and are thought to contribute to the immunosuppressive TME. Several studies have revealed a reduction of phenotypically mature DCs in tumor samples and peripheral blood from cancer patients [140–142], and an increased frequency of iDCs in peripheral blood from patients with breast, lung and head/neck cancer [143]. Both, the lack of mature DCs and accumulation of iDCs in the TME may contribute to a reduced capacity to stimulate antitumor T cells [38] as a consequence of defective antigen presentation and a lack of costimulatory signals.

Tumor- and tumor stroma-derived factors are hypothesized to cause defective DC differentiation, maturation and survival [38], as well as to allow DCs to acquire a regulatory phenotype [139, 144, 145]. Specifically, tumor-derived factors are thought to cause abnormal myelopoiesis in the host, leading to at least three outcomes: (i) a reduced number of mature immune stimulatory DCs; (ii) an accumulation of iDCs; and (iii) an overall accumulation of immature myeloid cells, including MDSCs [38]. Tumor-derived factors inhibiting DC maturation include VEGF [143, 146], M-CSF, IL-6 [147, 148], IL-10 [149], TGF-β [150] and gangliosides [151], and targeting these factors can potentially abrogate their effects [147, 148, 152].

Mechanisms of DC Mediated Immunosuppression Regulatory DCs have been shown to directly suppress $CD8^+$ T-cell function by reducing T-cell expansion and cytokine production and promoting T-cell anergy [145]. DCs can also have indirect inhibitory effects through their production of the enzyme IDO which induces Tregs [153, 154] and also immunosuppressive cytokines (i.e., TGF-β and IL-10), as well as increased expression of coinhibitory molecules (i.e., programmed death receptor 1; PD-1, PD-L1, and CD277) [139, 155, 156] and Arg1 [157] in a manner similar to MDSCs and macrophages.

2.3.6 Macrophages

Macrophage Phenotype Macrophages are found in all tissues in adult mammals and have diverse functions due to their extraordinary plasticity [158]. In mice, macrophages are derived from the yolk sac and bone marrow progenitors [159–161], and are phenotypically characterized by $CD11b^+$, $CSF\text{-}1R^+$, and $F4/80^+$, with macrophage colony-stimulating factor 1 receptor (CSF-1R) being one of the key regulators of macrophage differentiation [158]. In humans, tissue-resident macrophages are derived from monocytes, which originate from bone marrow myeloid progenitor cells [162]. Their diverse functions include the maintenance of tissue homeostasis and integrity through mechanisms such as clearance of apoptotic cells and tissue remodeling and repair. Phagocytosis and the delivery of cytokines, proteases and signaling molecules are key to their function [158, 162]; however, they also have important roles in both innate and adaptive immunity. Under certain conditions, macrophages can become highly polarized into one of two inflammatory states. The M1 phenotype includes classically activated macrophages, which are induced upon exposure to IFN-γ and/or Toll like receptor (TLR) engagement. M1 cells upregulate MHC-II expression and NF-κβ and STAT1 signaling, produce ROS, RNI, and cytokines (TNF-α, IL-1, IL-12, and IL-2), and overall promote antimicrobicidal and cellular immunity [158, 162]. M2 cells are induced upon exposure to factors such as IL-4 and IL-13, produce molecules such as Arg1, PD-L2, and IL-10 [158], and have roles in tissue repair and regulation of the duration and magnitude of immune responses [158, 162]. Whereas the M1 and M2 system provides a convenient framework for understanding the diverse functions of macrophages, most macrophages do not strictly conform to M1 or M2.

Macrophages and Cancer In tumors, macrophages are notably increased and comprise a major component of the immune cell infiltrate [163, 164]. Several studies have shown a correlation between macrophage density in human tumors and poor prognosis [163, 165]. Tumor- and stroma-derived chemoattractants such as CCL2, CCL3, CCL4, CCL5 and placental growth factor (PGF) are thought to help recruit monocytes, which then differentiate into TAMs in the TME [164]. Although macrophages can have both antitumor and protumor effects, their net effect is to promote tumor initiation, progression and metastasis via their proinflammatory and immunoregulatory functions [165]. Macrophage production of inflammatory cytokines such as IFN-γ and TNF-α may help produce a state of chronic inflammation that promotes a mutagenic environment [158, 165]. As tumors become established, TAMs are thought to adopt an immunosuppressive phenotype that promotes tumor progression [165, 166]. TAMs downregulate the production of factors such as IL-12, IL-18, and TLR signaling pathway components [165], and secrete immunosuppressive factors such as Argl, PGE2, TGF-β and IL-10 [165]. Macrophages can also recruit immunosuppressive cells, including Tregs, to the TME through their secretion of chemokines such as CCL22 [106].

2.4 The Role of the Immune System in Cancer Metastasis

Metastasis is a poor prognostic sign in all cancers. By definition, the presence of metastases indicates that the tumor has spread from its initial location to other sites, and once this occurs, cures are rare. Here, the important role of TAMs in metastasis development are highlighted. TAMs support many processes required for metastasis development, including tumor cell migration, invasion and intravasation (invasion into bloodstream), and also the stimulation of angiogenesis [164, 165, 167, 168]. Complex signaling allows for TAM involvement in these processes, and examples are provided below. Studies of breast cancer and glioma models showed that tumor-derived CSF-1 stimulates macrophage production of epidermal growth factor (EGF), which leads to tumor cell migration [165]. IL-4 from the tumor or stroma fosters a macrophage phenotype that promotes tumor invasion [165]. Tumors can recruit proangiogenic macrophages through tumor-associated hypoxia [169] and through the secretion of soluble factors such as CSF-1, stromal-derived factor 1α (SDF1α) and VEGF [164, 169]. Specifically, TAMs expressing the angiopoietin receptor TIE2 are essential for tumor angiogenesis in various tumor models [170]. TAMs can also secrete several proangiogenic molecules, including VEGF, platelet-derived growth factor (PDGF) and basic fibroblast growth factor (FGF2) as well as MMPs [164, 171], which remodel the ECM and promote tumor migration and invasion [158]. Myeloid cell populations can also impact lymphatic endothelial cells and produce factors in certain cancer models that promote lymphangiogenesis [172, 173], in which both vasculature and lymphatic vessels are known routes of tumor dissemination [174, 175].

Other elements of the TME support metastatic tumor spread [174]. Primary tumor invasion is thought to involve EMT, during which tumor cells can acquire stem cell-like properties and a migratory phenotype, whereas later metastatic stages may involve mesenchymal-epithelial transitioning (MET) to support tumor growth [176–179]. Stromal-derived products may facilitate these phenotypic changes; for instance, TAM-derived TGF-β was found to promote EMT in a teratocarcinoma model [179]. As previously mentioned, macrophages can help tumor cells intravasate into the circulation by localizing to the peripheral tumor stroma and blood vessels [174]; however, the exact kinetics of tumor entry into the circulation has yet to be elucidated. The formation of metastases itself is an inefficient process, with only a small percentage of intravasated tumor cells being capable of establishing secondary metastatic sites. However, circulating platelets can promote tumor cell extravasation into organs by forming aggregates with tumor cells and anchoring the cells to the endothelium [180]. Tumor cells can also secrete factors that lead to the establishment of pre-metastatic niches, characterized by clusters of bone marrow-derived cells including MDSCs, fibroblasts and associated cytokines [36, 174, 181]. Tumor-derived exosomes (vesicles) may also promote tumorigenic niches and direct bone marrow-derived cells to these areas [174, 182]. Once seeded at secondary organ sites, metastases may either be suppressed by immune-mediated cytolysis or exist as dormant micrometastases [183]. Many factors help tumor cells exit dormancy and develop macrometastases and outgrowth, including immunosuppressive mechanisms, as demonstrated with TAMs in pancreatic and breast cancer [184, 185], along with sustained vascularization [174]. In summary, immune cells play important roles in tumor metastasis through the promotion of angiogenesis and tumor cell migration, as well as the establishment of pre-metastatic niches and support of tumor extravasation and growth at seeded secondary sites.

3 Immunotherapeutic Approaches to Cancer

3.1 Monoclonal Antibodies

3.1.1 Introduction

B cells are immune cells that can differentiate into plasma cells which are responsible for the production of proteins known as immunoglobulins (Igs) or antibodies. The Igs recognize a specific site on an antigen known as an epitope, and promote an immune response via activation of the complement system and binding to Fc receptors on several other types of immune cells [186]. The basic Ig structure is Y-shaped and contains four chains: two light chains (either kappa or lambda restricted), and two heavy chains [187]. There is one amino-terminal variable region per chain and the number of carboxy-terminal constant domains varies, with three to four per heavy chain and one per light chain. The Ig can be digested into an Fc fragment and Fab domain [187]. The Fab domain can be further broken down into the variable fragment (Fv), which is responsible for the antigen specificity of the Ig, and the distal Fc receptor, which is responsible for its effector functions. There are several isotypes of Ig, which are defined by their constant regions and serve different roles in the immune response. IgG is the predominant isotype and includes four subclasses (IgG1–4), of which IgG1 is most efficient at binding Fc receptors on other immune cells [188]. Hereafter, Igs will be referred to as antibodies.

The technologies enabling the manufacturing of mAbs have revolutionized cancer therapy, with multiple FDA-approved mAb therapies (Table 3) and countless mAbs being developed preclinically and undergoing clinical trials. The term monoclonal refers to the homogenous antibody product containing a single Fv receptor that recognizes a single antigenic target. MAbs were first produced with mouse hybridoma technology by Kohler and Milstein in 1975 [189]. However, these mAbs did not achieve clinical relevance because certain nonhuman amino acid sequences were recognized as foreign by the human immune system, and this resulted in the production of neutralizing anti-murine antibodies in recipients [190]. Chimeric mouse-human mAbs, and humanized and fully human mAbs were eventually produced and advanced into the clinic. Chimeric mAbs contain murine variable regions and human constant regions [191]; humanized mAbs contain only murine components in the smaller complementarity-determining regions (CDRs) of the variable region; and fully human mAbs contain no murine components [192]. In general, chimeric,

Tab. 3 FDA-approved monoclonal antibodies for cancer therapy.

Name (Company)	*Structure*	*Target*	*Mechanism of action*[a]	*FDA-approved cancer indication*
Targeting tumor surface antigens				
Rituximab (Genentech, Inc.)	Chimeric IgG1κ (type I)[b]	CD20	ADCC CDC	Diffuse large B-cell lymphoma [193–195] Follicular lymphoma [196] Chronic lymphocytic leukemia [197]
Ofatumumab (GlaxoSmithKline)	Human IgG1κ (type I)[b]	CD20	ADCC CDC	Chronic lymphocytic leukemia [198]
Obinutuzumab (Genentech, Inc.)	Humanized IgG1 (type II)[b]	CD20	ADCC ADCP	Chronic lymphocytic leukemia [199]
Cetuximab (Bristol-Myers Squibb)	Chimeric IgG1κ	EGFR	ADCC Blocks cell signaling	Colorectal cancer (*KRAS* wild-type[c]) [200, 201] Head and neck squamous cell cancer [202–204]
Panitumumab (Amgen, Inc.)	Human IgG2κ	EGFR	Blocks cell signaling	Colorectal cancer (*KRAS* wild-type[c]) [205]
Pertuzumab (Genentech, Inc.)	Humanized IgG1κ	Her2	ADCC Blocks cell signaling	$Her2^+$ breast cancer [206, 207]
Trastuzumab (Genentech, Inc.)	Humanized IgG1κ	Her2	ADCC Blocks cell signaling	$Her2^+$ breast cancer [208] $Her2^+$ gastric cancer [209]
Targeting the tumor microenvironment				
Bevacizumab (Genentech, Inc.)	Humanized IgG1κ	VEGF	Inhibits angiogenesis	Colorectal cancer [210, 211] Non-small-cell lung cancer (non-squamous) [212] Renal cell cancer [213] Glioblastoma multiforme [214] Ovarian Cancer [215] Cervical Cancer [216]
Ziv-Aflibercept (Regeneron Pharmaceuticals, Inc.)	Human IgG1 recombinant fusion protein (VEGFR1/2)	VEGF[d]	Inhibits angiogenesis	Colorectal cancer [217]
Ramucirumab (Eli Lilly & Co.)	Human IgG1κ	VEGFR2	Inhibits angiogenesis	Gastric Cancer [218, 219]

Antibody conjugates (radioimmunotherapy and chemoimmunotherapy)				
Ibritumomab tiuxetan – Y^{90} (Spectrum Pharmaceuticals, Inc.)	Murine IgG1κ linked to yttrium-90	CD20	βeta-emitter radiation	Follicular lymphoma or transformed B-cell non-Hodgkin's lymphoma [220, 221]
Tositumomab-I^{131} (GlaxoSmithKline)	Murine IgG2aλ linked to iodine-131	CD20	Gamma-emitter radiation ADCC CDC	Follicular lymphoma or transformed B-cell non-Hodgkin's lymphoma [222]
Brentuximab vedotin (Seattle Genetics, Inc.)	Chimeric IgG1κ conjugated to anti-microtubule agent monomethyl auristatin E (MMAE)	CD30	Complex internalized and MMAE released → cell cycle arrest + apoptosis	Anaplastic large cell lymphoma [223] Hodgkin's lymphoma [224]
Ado-trastuzumab emtansine (Genentech, Inc.)	Humanized IgG1κ conjugated to anti-microtubule agent DM1	Her2	Complex internalized and DM1 released → cell cycle arrest + apoptosis ADCC	$Her2^+$ breast cancer [225]
Others				
Alemtuzumab (Genzyme)	Humanized IgG1κ	CD52	ADCC	Chronic lymphocytic leukemia [226]
Denosumab (Amgen, Inc.)	Human IgG2κ	RANK ligand	Inhibits ligand and receptor binding on osteoclasts	Bone metastases [227] Giant cell tumor of bone [228]

ADCC = antibody-dependent cell-mediated cytotoxicity; CDC = complement-dependent cytotoxicity; ADCP = antibody-dependent cell-mediated phagocytosis; EGFR = epidermal growth factor receptor; VEGF = vascular endothelial growth factor or VEGF-A; VEGFR1/2 = vascular endothelial growth factor receptors 1 and 2.

a) Mechanism of action supplemented by data from each drug's package insert.
b) Type I refers to non-glycoengineered and type II to glycoengineered.
c) *KRAS* mutated colorectal cancers do not respond to anti-EGFR targeted therapies.
d) This fusion protein targets VEGF-A, VEGF-B, and placental growth factor.

humanized, and fully human mAbs do not induce neutralizing antibodies in humans.

The targets of cancer mAbs that have demonstrated clinical efficacy include: (i) antigens expressed on the tumor cell surface; (ii) antigens that are involved in tumor angiogenesis; and (iii) antigens which act as "brakes" for the immune system (i.e., immune checkpoints) that are expressed on either the tumor surface or immune cells [229]. Additionally, in an attempt to direct cytotoxic therapy specifically to the tumor, mAbs are being conjugated to radioisotopes and chemotherapeutic agents. More creative designs – including bispecific antibodies – are being developed in order to target more than one antigen, including antigens on more than one cell type [230]. Most recently, blinatumomab, a bispecific T-cell engager, has been approved for the treatment of acute lymphoblastic leukemia (ALL) [231]. This antibody works by targeting both CD3 on T cells and CD19 on leukemia cells, forming a cytolytic synapse between the T-cell and the leukemia cell to enhance tumor kill by the immune system [231]. MAbs employ several mechanisms of action to attack cancer, such as inducing antibody-dependent cell-mediated cytotoxicity (ADCC), complement-dependent cytotoxicity, antibody-dependent cellular phagocytosis, alteration of signal transduction, and blockade of receptor–ligand interactions [188].

3.1.2 Targeting Antigens on the Tumor Cell Surface

Rituximab, a chimeric IgG1κ targeting CD20, was the first mAb to be approved for the treatment of cancer. CD20 is an antigen expressed on the surface of B-cell malignancies, including chronic lymphocytic leukemia (CLL) and the majority of non-Hodgkin's lymphomas (NHLs), including diffuse large B-cell lymphoma (DLBCL) and follicular lymphoma. CD20 is not a tumor-specific antigen since it is also expressed on normal pre-B and mature B cells. Rituximab received FDA approval in February 2006 for the first-line treatment of DLBCL [193–195], after three randomized trials conducted in 1855 patients had demonstrated an improvement in event-free survival with the addition of rituximab to standard anthracycline-based chemotherapy. Rituximab has since revolutionized the treatment of B-cell malignancies and also certain autoimmune disorders, in part because it is generally well tolerated. Since its initial FDA approval, rituximab has been approved for the treatment of follicular lymphoma at various stages of disease [196] and, in combination with chemotherapy, in previously treated and untreated CLL [197]. It remains under active clinical investigation [232].

The adverse events related to rituximab are well known. As would be suspected with infusions of any protein-based therapies – especially those with murine components – infusion reactions with rituximab are common (≥25%). Infusion reactions usually occur during the first infusion and generally resolve in less than 24 h. The reactions may be associated with a range of signs and symptoms: fevers, rigors, hives, hypotension, angioedema (swelling of mouth and tongue), bronchospasm, hypoxia, and even shock or death. There are algorithms in place for the prevention and management of infusion-related reactions, including interrupting or slowing the infusion and premedicating with an antihistamine (diphenhydramine) and acetaminophen [233]. These infusion reactions are seen to some degree with all mAbs. Given that CD20 is expressed on normal B cells, on-target B-cell depletion and associated hypogammaglobulinemia are common with rituximab, and there is an increase in susceptibility to infections, but not to the degree that was initially feared. Rare reactivation of the Hepatitis B virus has been seen along with JC virus-induced progressive multifocal leukoencephalopathy [233]. Such on-target B-cell depletion is seen with other mAbs targeting CD20 (as described below), and highlight the fact that mAbs have on-target toxicity in normal tissues or cells.

After rituximab, other mAbs targeting CD20 have been engineered. Ofatumumab, unlike rituximab, is a fully human IgG1κ mAb and targets a different epitope on CD20 [234]. It received accelerated approval by the FDA in October 2009 as a single agent for refractory CLL [198]. Its efficacy as a single agent is notable since rituximab demonstrated no single agent activity in the treatment of a similar group of CLL patients [235]. It is difficult to discern if this difference in activity is attributable to the engineering of the mAbs or to differences in the clinical trials. Despite ofatumumab's fully human structure, infusion reactions are still common (≥40% with the first infusion). Most recently, in November 2013 the FDA approved another anti-CD20 mAb, obinutuzumab, in combination with chemotherapy for CLL [199]. Obinutuzumab was the first drug to be approved by the FDA under a "breakthrough therapy designation." This means that its approval was based on preliminary clinical evidence that the drug may offer significant improvement over currently available therapies. Obinutuzumab is a humanized IgG1 mAb that targets epitopes of CD20 that overlap with those recognized by rituximab. Unlike rituximab and ofatumumab, it is a type II glycoengineered mAb and thus has increased affinity for Fcγ receptor IIIA on effector immune cells, enhanced cell-mediated cytotoxicity, and decreased complement-dependent toxicity [236]. Additional clinical trials should provide further insight into the utility of this newly approved therapy.

The ErbB transmembrane receptor tyrosine kinase family (Her1 or epidermal growth factor receptor (EGFR), Her2, Her3, and Her4) is a rationale mAb target [237]. Most of these receptors require ligand activation of the receptor, except for Her2, which can be independently activated. This growth factor signaling pathway is dysregulated and overexpressed in many epithelial cancers, with resulting increased cell proliferation, migration, angiogenesis, and resistance to apoptosis. For example, EGFR is overexpressed in CRC, head and neck cancer, and lung cancer, and Her2 is overexpressed in a subset of breast cancers [238]. The extent of Her2 expression on tumor cells is a reliable predictor of clinical response to treatment with Her2 mAbs. Her2 expression on tumors can be tested by either assaying protein expression with immunohistochemistry (IHC) or Her2 gene amplification on chromosome 17 with fluorescence *in-situ* hybridization (FISH) [239]. On the other hand, testing for EGFR with either IHC or FISH is not a reliable predictor of response to anti-EGFR mAbs [240].

There are two FDA-approved mAbs that target EGFR, cetuximab (chimeric IgG1κ) and panitumumab (human IgG2κ). Cetuximab is approved for the treatment of metastatic *KRAS* wild-type CRC, either in combination with chemotherapy or as a single agent [200, 201], and for squamous cell head and neck cancer in combination with chemotherapy or radiation, or as a single agent [202–204]. Panitumumab is also approved as a single agent in metastatic *KRAS* wild-type CRC [205]. Cetuximab and panitumumab target two overlapping epitopes on EGFR, but at distinct sites [241], and are often used interchangeably in the management of CRC [242]. After patients have progressed on one anti-EGFR mAb, it is unlikely that they will benefit from the other. A potential advantage of panitumumab over cetuximab is its fully human content, resulting in less severe infusion reactions, and it also can be considered if a patient has had a severe reaction to cetuximab. Of note, EGFR is not a tumor-specific antigen and is found in normal epithelial tissues including skin, hair follicles and the gastrointestinal tract. Not surprisingly, therefore, cetuximab and panitumumab have side effects that include skin rash and diarrhea, of which the former can be severe in ~10–15% of cases [200–205]. Dermatologic toxicity can be managed by limiting exposure to the sun, and treatment with emollients and topical or oral antibiotics.

Two mAbs (both humanized IgG1κ) have been approved for the treatment of Her2-positive ($Her2^+$) breast cancer, namely pertuzumab [206, 207] and trastuzumab [208]. These mAbs have revolutionized the treatment of breast cancer patients in what was previously considered a poor prognostic subgroup. Trastuzumab is also approved in the first-line setting in combination with chemotherapy for a subset of gastric cancer that expresses Her2 [209]. This class of mAb also brought about the first FDA approval of two mAbs for combined use. The combination of pertuzumab, trastuzumab and chemotherapy is approved in the neoadjuvant (preoperative) setting in $Her2^+$ breast cancer, based on an improvement in pathologic complete response at the time of surgery, and also in metastatic $HER2^+$ breast cancer based on an improvement in PFS [206, 207]. This combination of mAbs is especially intriguing, since both bind the same target. The mechanism of synergy is thought to be secondary to cooperation between these mAbs when they are colocalized on Her2 and "clamp" it, potentially inhibiting Her2 dimerization [243]. This suggests that the development of a mAb which binds different epitopes on the same target may result in a synergistic inhibition of the target, and should be a strategy employed in clinical trials moving forward. Anti-Her2 mAbs also have a unique toxicity of cardiomyopathy (which can be reversible), and cardiac function monitoring with echocardiograms is required [244]. Dual Her2 inhibition with trastuzumab and pertuzumab does not seem to increase this risk [245].

3.1.3 **Targeting Antigens Involved in Tumor Angiogenesis**

Solid tumors require angiogenesis to sustain growth, invasion and metastases. Angiogenesis is frequently dysregulated in solid tumors, and it is a validated target for tumor control. VEGF-A or VEGF appears to be one of the most important proteins in regulating both normal and pathologic angiogenesis. VEGF has two receptor tyrosine kinases, including vascular endothelial growth factor receptor 1 (VEGFR-1) and VEGFR2, of which the latter is responsible for VEGF effects. Unlike previously described mAbs, which target an antigen on the tumor cell surface, VEGF is found in the circulation. Targeting VEGF has been shown to decrease tumor blood perfusion and volume, interstitial fluid pressure and microvascular density, and also normalizes the vasculature [246].

There are currently three approved mAbs that target angiogenesis, bevacizumab, ziv-aflibercept, and ramucirumab. Bevacizumab is a humanized IgG1κ anti-VEGF mAb [210]; ramucirumab is a human IgG1κ anti-VEGFR-2 mAb [218, 219]; and ziv-aflibercept is engineered as a recombinant fusion protein with the receptor domains of VEGF receptors, VEGFR-1 and VEGFR-2

[217]. Thus, ziv-aflibercept targets not only VEGF but also other proteins which are important in angiogenesis, such as VEGF-B and PGF. Bevacizumab is approved in combination with standard chemotherapy for the treatment of metastatic CRC [210, 211], non-squamous NSCLC [212], renal cell cancer [213], ovarian cancer [215], and cervical cancer [216], and as a single agent in glioblastoma multiforme [214]. Given that the tumor stroma is the target of bevacizumab rather than an antigen on the tumor itself, it is often used after disease progression in CRC [210]. Conversely, its anti-angiogenic properties do not translate into clinical benefit across all tumor types, as was demonstrated when the FDA revoked the accelerated approval of bevacizumab for breast cancer [247]. Ziv-aflibercept is approved in combination with chemotherapy for metastatic CRC [217]. Ramucirumab is approved as a single agent and in combination with chemotherapy for gastric cancer and gastro-esophageal cancer [218, 219]. Since these mAbs affect angiogenesis, they have a unique side-effect profile that includes hypertension, proteinuria, problems with wound or surgical healing, gastrointestinal perforation, fistula development, arterial thrombotic events (including heart attack and stroke), and bleeding events that can range from minor to severe. Currently, there are multiple ongoing trials with mAbs targeting various components of angiogenesis and the tumor stroma, but these are beyond the scope of this chapter [232].

3.1.4 Monoclonal Antibodies Conjugated to Radioisotopes and Chemotherapeutic Agents

MAbs are also being used to selectively deliver radiotherapy and cytotoxic chemotherapy to tumors while sparing normal tissue. There are two approved radioimmunotherapies that target CD20, ibritumomab tiuxetan-Y^{90}, and tositumumab-I^{131}. Ibritumomab is a murine IgG1κ anti-CD20 mAb that is conjugated to the linker-chelator tiuxetan, which provides a restricted chelation site for binding either indium-111 (In^{111}) or yttrium-90 (Y^{90}). It is approved for the treatment of follicular lymphoma or transformed B-cell NHL, and is dosed in conjunction with rituximab [220, 221]. Tositumomab-I^{131} is a murine IgG2aλ anti-C20 mAb that is covalently linked to iodine-131 (I^{131}), and it was initially approved for the treatment of follicular lymphoma that is refractory to rituximab [222]; however, it is no longer being sold as of February 2014 because of its limited use [248]. Both drugs require a test dose to ensure adequate biodistribution before administering the full dose (i.e., to confirm that radiation therapy is being delivered to the tumor). The choice of radioisotope that is linked to the antibody is important. Y^{90} is a beta-emitter while I^{131} is a gamma-emitter, which translates into Y^{90} having a higher energy delivery, a longer path length, and a shorter half-life [249]. Like all forms of radiation therapy, ibritumomab tiuxetan-Y^{90} and tositumumab-I^{131} can cause myelosuppression along with secondary leukemias and myelodysplasia in about 5–10% of treated patients.

FDA-approved antibody–drug conjugates (ADCs) include anti-CD30 brentuximab vedotin and anti-Her2 trastuzumab emtansine (TDM1). CD30 is a member of the tumor necrosis family, and is found most commonly on anaplastic large-cell lymphoma and Hodgkin's lymphoma, embryonal carcinomas, and activated B and T cells [250]. Despite preclinical promise, unconjugated human or chimeric anti-CD30 mAbs are of limited clinical efficacy [251]. Brentuximab vedotin is a chimeric IgG1κ anti-CD30 mAb conjugated to a cytotoxic anti-microtubule agent, monomethyl auristatin E. Based on a >70% clinical response rate, it is approved for relapsed anaplastic large-cell lymphoma [223] and Hodgkin's lymphoma after failure of autologous hematopoietic stem cell transplant (HSCT) [224]. This is an excellent example of how the clinical activity of a mAb can be enhanced by conjugation to a cytotoxic agent. TDM1 is a humanized IgG1κ anti-Her2 mAb conjugated via a stable linker to the anti-microtubule agent, DM1 or emtansine [225]. In a randomized trial, it demonstrated improved survival in metastatic $Her2^+$ breast cancer patients when compared to a combination of the small-molecule Her2 tyrosine kinase inhibitor, lapatinib, plus chemotherapy. Unlike anti-CD30 mAbs, unconjugated anti-Her2 mAbs such as trastuzumab do have single-agent activity in $Her2^+$ breast cancer [252]. However, the efficacy of this mAb can also potentially be enhanced by conjugating it to a cytotoxic agent, and anti-Her2 mAb is being compared to TDM1 in several clinical trials [232].

3.1.5 **Targeting Antigens Involved in Tumor Immune Evasion (Checkpoint Inhibitors)**

One of the most exciting classes of mAbs that are being actively studied are immune checkpoint inhibitors (Table 4). When an APC comes into contact with a T cell, it presents an antigen via MHC-I or MHC-II to the TCR. Following this primary and necessary interaction, several other coinhibitory and costimulatory signals

Tab. 4 Immune checkpoint inhibitors in clinical trials.

Drug	*Company*	*mAb or fusion protein*	*Target*
Ipilimumab	Bristol-Myers Squibb	Human IgG1κ	CTLA-4[a)]
Tremelimumab	Pfizer, Inc./MedImmune, Inc.	Human IgG2κ	CTLA-4[a)]
Nivolumab/BMS-936558/MDX-1106	Bristol-Myers Squibb	Human IgG4κ	PD-1[a)]
MK-3475/Pembrolizumab	Merck, Inc.	Humanized IgG4κ	PD-1[a)]
Pidilizumab/CT-011	Curetech Ltd.	Humanized IgG1κ	PD-1[a)]
AMP-224	GlaxoSmithKline/ Amplimmune, Inc.	Human Fc B7-DC IgG1 fusion	PD-1[a)]
MPDL3280A	Genentech, Inc./Roche	Human IgG1κ	PD-L1[b)]
BMS-936559	Bristol-Myers Squibb	Human IgG4	PD-L1[b)]
MEDI4736	MedImmune, Inc.	Human IgG1κ	PD-L1[b)]
IMP321	Immutep	Human Fc LAG-3 IgG1 fusion	LAG-3[a)]
Lirilumab/IPH2102/BMS-986015	Innate Pharma/Bristol-Myers Squibb	Human IgG4κ	KIR[a)]
MGA271	MacroGenics, Inc.	Humanized IgG1κ	B7-H3[b)]

mAb = monoclonal antibody; CTLA-4 = cytotoxic T-lymphocyte antigen-4; PD-1 = programmed death receptor 1; B7-DC = programmed death receptor ligand 2; PD-L1 = programmed death receptor ligand 1; LAG3 = lymphocyte activation gene-3; KIR = killer cell immunoglobulin-like receptor.

a) Inhibitory receptors. Nivolumab is approved for use in Japan for unresectable melanoma. Ipilimumab and pembrolizumab are approved in the United States for the treatment of unresectable melanoma.

b) Inhibitory ligands.

Adapted from Ref. [229].

fine-tune the immune response (for a review, see Ref. [229]). Coinhibitory signals are also known as immune checkpoints and dampen the immune response. The most extensively studied checkpoint molecules are CTLA-4 and PD-1 and their ligands, including CD80/B7-1 and CD86/B7-2, and PD-L1 and PD-L2, respectively. The expression of CTLA-4 is restricted to T cells, and it is induced at the time of T-cell activation while PD-1 is expressed on B cells, T cells and NK cells and maintains peripheral tolerance in tissues or tumors. Given the role of CTLA-4 in regulating early T-cell activation, CTLA-4 knockout mice suffer massive inflammation and death, while PD-1 knockout mice usually suffer nonfatal autoimmune conditions.

Ipilimumab, a human IgG1κ anti-CTLA-4 mAb, was the first of the immune checkpoint inhibitors to complete clinical development. It received FDA approval in early 2011 for the treatment of metastatic melanoma after a randomized trial had demonstrated an improvement in OS from 6.4 to 10.0 months ($p < 0.001$) [2]. This was the first time that any therapy in melanoma had demonstrated a survival benefit. Melanoma is thought to be an immunogenic tumor, with high-dose IL-2 previously yielding durable responses in a small fraction of patients [253]. Given the ability of ipilimumab to activate the immune system, it is not surprising that there is a 10–15% rate of severe immune-related adverse events (iAEs), mostly affecting the gastrointestinal tract with diarrhea and colitis and the skin with pruritus and rash [2]. These iAEs, however, are mostly reversible with corticosteroids. Another anti-CTLA-4 inhibitor, tremelimumab (human IgG2 anti-CTLA-4), did not show improvement in OS in untreated metastatic melanoma [254]. The authors concluded that this lack of benefit could have been related to a difference in the patient population (fewer poor-risk patients) and an unintended crossover, since ipilimumab became available during this trial.

There are currently two major classes of mAbs targeting the PD-1 axis: PD-1 inhibitors and PD-L1 inhibitors. PD-1 inhibitors block the interaction of PD-1 and its two ligands, PD-L1 and PD-L2 [4, 5]. PD-L1 inhibitors block the interaction of PD-L1 and its receptors, PD-1 and B7-1 [3, 255]. Nivolumab, a fully human IgG4 anti-PD-1, was the first of these mAbs reported to have clinical activity. Its clinical activity generated significant excitement as durable responses were seen not only in "immunogenic" tumors (i.e., melanoma and renal cell carcinoma) but also in a non-immunogenic tumor, NSCLC. The latter observation was particularly impressive as these were refractory cases, and prior to this study, NSCLC had not been responsive to any immunotherapies. Nivolumab is well tolerated, with only a 6% rate of severe drug-related iAEs; the most concerning of these, pneumonitis (inflammation of the lungs), was associated with three deaths [4]. It is also notable that PD-1 inhibitors do not have the same degree of gastrointestinal adverse events as ipilimumab. Most recently, nivolumab was approved in Japan for the treatment of unresectable melanoma, and pembrolizumab, a humanized IgG4κ anti-PD1 mAb, was FDA approved for the treatment of unresectable melanoma [256]. Of the PD-L1 inhibitors, MPDL3280A is the most advanced in development. MPDL3280A is an anti-PD-L1 human engineered IgG1κ mAb designed to lack the capacity to mediate ADCC and thereby avoid killing activated T cells that express PD-L1 [3]. Like nivolumab, MPDL3280A has demonstrated responses in similar cancer types in a refractory setting, and only a 2% rate of severe drug-related iAEs has been seen in initial trials. Remarkably, the combined use of ipilimumab and nivolumab has demonstrated a synergistic benefit in melanoma with greater efficacy and more prolonged responses, though at the expense of more frequent severe iAEs (>50%) [257].

3.2 Adoptive T-Cell Immunotherapy

3.2.1 Introduction

The most widely practiced form of adoptive cell therapy is allogeneic HSCT, formerly known as bone marrow transplantation, which is used to treat leukemia, lymphoma and myeloma, and has an independent T-cell-mediated antitumor effect [258]. Given their ability to recognize TAAs and lyse tumor cells, expand, and develop memory, T cells are an ideal cell type for use in adoptive therapy. T cells either express αβ or γδ TCRs; more is known about the αβ TCRs, which are expressed on T helper and cytotoxic cells. αβ TCRs recognize specific antigens in association with MHC-I and MHC-II molecules [259]. TILs and genetically engineered T cells will be described in the following sections.

3.2.2 Tumor-Infiltrating Lymphocytes (TILs)

As mentioned above, the presence of TILs in many cancer types is associated with an improved prognosis, and is thought to reflect a specific antitumor immune response. The ability to expand TILs *in vitro* was first described in 1986 [260]; these TILs were directly extracted from the patient's tumor and expanded *ex vivo* for 14–100 days using IL-2 (a cytokine that interacts with its receptor on T cells to induce proliferation) [260]. Addition to the culture media of a CD3 agonist (OKT3; CD3 present on all T cells) and irradiated allogeneic feeder cells from donor peripheral blood mononuclear cells (PBMCs) has improved cell yields [261, 262]. With this methodology, TILs that were preselected for potency and diverse antigen recognition could be expanded in a 14-day protocol from an average of 3.4×10^7 cells to 4.1×10^{10} cells. Despite the rapid expansion protocol, the entire TIL generation process from culture manipulation to infusion required 6–8 weeks. Another important discovery related to TIL therapy was identifying the need for lymphodepleting chemotherapy prior to TIL infusion. In a murine model, lymphodepleting chemotherapy with cyclophosphamide, followed by the administration of autologous TILs and IL-2, resulted in a complete regression of MC-38 colon adenocarcinoma hepatic metastases [263]. Lymphodepletion correlates with an improved trafficking of TILs to tumor sites [264], and is thought to destroy Tregs [265], MDSCs and other tolerogenic cells to allow for the persistence and expansion of TILs [266].

Although TILs have been studied in ovarian cancer, renal cell carcinoma, lung cancer and gastric cancer, most trials have been conducted in melanoma and have taken place at the National Cancer Institute (NCI) Surgery Branch. In 1988, the first human trial with TILs showed a 55% response rate (11 of 20) in metastatic melanoma patients [267], though these responses were only short-lived due to the limited persistence of TILs [268]. Hypotheses to explain the limited TIL persistence have included inadequate lymphodepletion, repeated *in vitro* stimulations, and infusions restricted to $CD8^+$ T cells. In 2002, the NCI group reported significantly better results with heterogeneous $CD4^+$ and $CD8^+$ TILs, which had undergone only a single round of *in vitro* stimulation, and were administered after fludarabine and cyclophosphamide chemotherapy. The TILs expanded *in vivo*, expressed antigen-specific lytic activity and secreted cytokines *in vivo*, trafficked to tumor sites, and persisted for at least four months [269]. Most importantly, there was a 51% (n = 18) clinical response rate [270]. On-target toxicity associated with melanocyte differentiation antigens expressed in the skin and eyes manifested as autoimmune vitiligo and uveitis, respectively [269]. TIL persistence was associated with a better response [271]; for example, in responders at 1–2 months, persistence was 26% (range 2–96%) compared to only 5% (range <1–9%) in nonresponders. Attempts at myeloablative lymphodepleting chemotherapy with the addition of total body lymphoid irradiation, followed by HSCT rescue, seemed to increase response rates at the higher radiation dose but with worse toxicity [272]. Durability of responses from these three trials in 93 patients was impressive [273], with an overall response rate of 56% (n = 52/93) and 20% (n = 19/93) achieving a complete response lasting for three or more years. Such responses have not been seen with other current FDA-approved therapies for melanoma, and could represent a cure in a subset of patients. Despite impressive clinical activity in melanoma, the widespread use of TILs has not occurred because: (i) pre-existing tumor-specific lymphocytes must be sampled and expanded *ex vivo* (~45% of melanoma patients) [273]; (ii) TIL generation requires six weeks; (iii) there is a potential need for myeloablative conditioning [273]; and (iv) importantly, the cost and technical complexity of the entire process are daunting.

3.2.3 Adoptive Therapies with T Cells Containing Engineered TCRs and CARs

Given the cumbersome requirements associated with TIL therapy, significant effort has been undertaken to engineer T cells that can target tumors in a more efficient fashion. The two most common mechanisms of engineering include: (i) genetically engineering the TCR by cloning α and β chains from TAA-specific T-cell clones [274]; and (ii) creating new recombinant receptors comprised of both an extracellular single-chain variable fragment (scFv) from an antigen-specific mAb and an intracellular T cell triggering moiety (i.e., CD3ζ) [275]. The main advantage of CAR T-cell therapy over T cells with engineered TCRs is the ability of CAR T cells to recognize antigen independent of MHC. Downregulation of the MHC complex is a well-known mechanism of tumor immune evasion, and MHC independence also allows for therapy to be applicable to the entire patient population, without the need for human leukocyte antigen (HLA) type restriction. CAR T cells can also potentially target non-peptides such as carbohydrates. CAR T cells cannot target intracellular antigens, in contrast to T cells with engineered TCRs, but retain the antigen specificity and activation and effector functions of conventional T cells [276]. For both types of adoptive T-cell therapies, it is important to choose target antigens expressed mainly or exclusively on tumor cells to minimize off-target effects on normal tissues. Potential TAA targets, as previously reviewed, include: (i) tissue-specific differentiation antigens whose off-target effect on normal tissue is tolerable, such as prostate-specific antigen (PSA) in prostate cancer or glycoprotein 100 (gp100) or MART-1 in melanoma; (ii) cancer-testis antigens such as MAGE and NY-ESO-1; (iii) overexpressed proteins such as Her2 in breast cancer; (iv) mutated antigens; and (v) viral antigens such as human papilloma virus in cervical cancer [17].

Tab. 5 Published engineered T-cell receptor (TCR) trials.

Target	*Tumor type* (n = *total number of patients)*	*Responses*	*Toxicity*	*Reference(s)*
MART-1	Melanoma ($n = 31$)	4	None[a)]	[274, 277]
MART-1 gp100 (murine)	Melanoma ($n = 36$)	MART-1: 6 gp100: 3	Autoimmune melanocyte destruction[b)]: vitiligo, uveitis, hearing loss, vertigo	[277]
CEA	Colorectal cancer ($n = 3$)	1	Autoimmune colitis[b)]	[278]
NY-ESO-1	Melanoma and synovial cell sarcoma (SCC) ($n = 17$)	Melanoma: 5 SCC: 4	None	[279]
MAGE-A3	Melanoma (majority) ($n = 9$)	4	Neurologic toxicity[c)]	[280]

MART-1 and gp100 = melanocyte differentiation antigens; CEA = carcinoembryonic antigen; NY-ESO-1 and MAGE-A3 = cancer-testis antigens.

a) This trial used a low-affinity TCR, which most likely accounts for its lack of toxicity.

b) Expected on-target toxicities.

c) This was an unexpected but on-target toxicity due to this TCR targeting MAGE-A12, which was presumably present in a subset of the patients' brains. There was also another MAGE-A3 TCR, which caused severe cardiac toxicity and death in two patients by cross-reacting with a protein found in cardiac muscle, Titin.

Engineered αβ TCRs T cells with engineered TCRs were initially studied in 17 refractory metastatic melanoma patients [274] (Table 5). PBMCs were collected from patients by leukapheresis and transduced with a retroviral vector containing genes encoding the α and β chains of a TCR that recognized MART-1 in a HLA-A-*0201 restricted fashion (known as DMF4). The average gene transfer efficiency was approximately 42% (range 17–67%) and T cells from all patients were biologically active, as indicated by their ability to secrete proinflammatory IFN-γ after exposure to MART-1. *In vitro* expansion protocols were similar to those described with TIL therapy, with a culture period of 6–9 days [262]. In eight patients with available samples at >50 days, these T cells with engineered TCRs persisted (>17%). Two patients had an objective clinical response, and this correlated with T-cell expansion *in vivo* of 30- to 1400-fold. There were no toxicities seen with this therapy, but the response rate was lower than that previously seen with TIL therapy (~50%).

Because DMF4 cells had a limited ability to recognize small amounts of antigen, a high-affinity TCR for HLA-A*0201-restricted MART-1 known as DMF5 was designed [277]. Testing the affinity of the TCR for the MHC complex is possible *in vitro* [281], but it is more difficult to determine its affinity *in vivo*. The association between TCR affinity and T-cell activity is also difficult to predict, but a surrogate of *in vivo* TCR affinity can be derived from the use of MHC tetramers, which consist of multiple MHC molecules and the antigen of interest [282]. When a T cell binds to an MHC tetramer complex, a signal is generated that can be measured using flow cytometry. CD8 is another molecule found on the surface of T cells that helps to stabilize the TCR/antigen interaction, and the determination of CD8 independence during tetramer staining indicates a higher-affinity TCR. DMF5 bound MHC tetramers more efficiently than DMF4 (30–60% versus 0–2%, respectively), and binding did not depend on CD8 coreceptor signaling. DMF5 cells recognized 100-fold lower peptide concentrations, secreted higher levels of IFN-γ, and killed the target more efficiently. In 36 melanoma patients, 20 were treated with DMF5 and 16 with T cells expressing a gp100-engineered murine TCR [277]. At this point, attention will be focused on DMF5 to illustrate the differences between T cells with TCRs recognizing the same target with different affinities. The transduction efficiency was 71%, and all patients had persisting T cells at 1 month. There were six responses, in which no variables related to the T cells seemed to correlate with response. Unlike DMF4 cells, the higher-affinity DMF5 cells had more toxicity, including a reversible diffuse erythematous skin rash and eventual destruction of epidermal melanocytes resulting in vitiligo, reversible uveitis, and reversible hearing loss and dizziness. Although it is difficult to compare across trials, especially with such limited numbers, the response rate was lower and the on-target toxicity was higher than TIL therapy in melanoma [273].

T cells expressing other engineered TCRs have been used in clinical trials and have taught valuable lessons (Table 5). First, on-target toxicity can be significant. T cells expressing a TCR targeting CEA – a protein that is present in CRC and is also expressed on normal epithelial cells in the gastrointestinal tract [283] – induced an inflammatory colitis in two of three patients [278]. Second, targeting certain TAAs can result in clinical activity across tumor types. In a 17-patient trial with a TCR targeting NY-ESO-1, there was notable clinical activity in both melanoma and synovial sarcoma [279]. The third and final lesson was that toxicity may come as a surprise, especially if there is a cross-reactive target [284]. In a nine-patient study of T cells expressing a murine TCR targeting MAGE-A3, there was severe neurologic toxicity in three patients that ranged from seizures and stroke to coma. This TCR also targeted MAGE-A12, MAGE-A2, and MAGE-A6 [285], and the authors hypothesized that this toxicity was due to recognition of MAGE-A12 protein in the brain. In another MAGE-A3 targeted engineered TCR trial, cross-reactivity with the peptide Titin led to rapid cardiac death in two patients [284].

CAR T Cells As mentioned, CARs are constructed by attaching an external scFv from a mAb to a T-cell intracellular signaling chain such as TCRζ or FcRγ [275]. There have been three generations of CARs: the first generation CAR contained only the intracellular TCRζ or FcRγ domain; the second generation CAR included additional intracellular costimulatory domains such as CD28 or CD137 (4-1BB); and the third generation CAR had two or more costimulatory signaling domains. These first-generation CAR T cells were tested with various targets in a variety of cancer types but did not have significant clinical activity (Table 6) [286–292]. The majority of these trials showed that the *in vivo* persistence of first-generation CAR T cells was limited; for example, in the ovarian cancer trial, α-folate receptor (α-FR) CAR T cells represented >1% of circulating lymphocytes after infusion but had declined by day 5 and were no longer detectable in the majority of patients by week 3 [286]. There were several hypotheses to explain this lack of CAR

Tab. 6 Published chimeric antigen receptor (CAR) T-cell trials.

CAR construct	*Target*	*Cancer (n = number of patients)*[a]	*Responses*	*Toxicity*[a]	*Reference(s)*
FcRγ	α-Folate receptor	Ovarian ($n = 14$)	No responses	Gr 1-2 nausea, fatigue Gr 3-4 dyspnea, hypotension	[286]
CD3ζ	L1-CAM (CD171)	Neuroblastoma ($n = 6$)	1	Decreased blood counts, bacteremia, interstitial pneumonitis	[287]
CD3ζ	CD20	B-cell non-Hodgkin's lymphoma and mantle cell lymphoma ($n = 7$)	1 CR (prior to CAR T) 1 PR (after CAR T)	None related to CAR T cells	[290]
CD3ζ	CD19 CD20	Diffuse large B-cell lymphoma and follicular lymphoma after autologous transplant ($n = 4$)	N/A	Fever, rigors Gr 2 hepatitis, anemia Gr 3 lymphopenia, eosinophilia	[291]
FcRγ	CA-IX	Renal cell cancer ($n = 12$)	0	Gr 2–4 transaminitis[b]	[292]
CD3ζ	GD2	Neuroblastoma ($n = 19$)	3 out of 11	Gr 1–3 localized pain	[288]
CD3ζ-CD28	CD19	B-cell malignancies($n = 8$)	6 achieved remission	Significant cytokine storm[c], B-cell depletion[b], Hypogammaglobulinemia[b]	[294, 295]
CD3ζ-4-1BB	CD19	Chronic lymphocytic leukemia ($n = 3$)	1 CR 2 PR	Cytokine symptoms[c], B-cell depletion[b]	[6, 296]

CD3ζ-CD28	CD19	Chronic lymphocytic leukemia (CLL) ($n = 8$)	CLL: 1	Gr 5 hypotension, renal failure; Gr 4 neutropenia; Gr 3 febrile neutropenia; Gr 3 hypotension	[297]
		Acute lymphoblastic leukemia (ALL) ($n = 1$)	ALL: persistent B-cell depletion		
CD3ζ and CD3ζ-CD28	CD19	Non-Hodgkin's lymphoma ($n = 6$)	0	None	[298]
CD3ζ-CD28	CD19	Acute lymphoblastic leukemia ($n = 5$)	5 CR by minimal residual disease	Cytokine symptoms[c)]	[7]
CD3ζ-4-1BB	CD19	Acute lymphoblastic leukemia ($n = 30$)	27 CR (22 by minimal residual disease)	Cytokine release syndrome ($n = 8$ severe)[c)]; encephalopathy ($n = 13$); and B-cell aplasia[b)]	[301]
CD3ζ-CD28-4-1BB	ERBB2	Colorectal cancer ($n = 1$)	Not evaluable	Respiratory distress and death	[299]
CD3ζ-CD28-4-1BB	CD20	Mantle cell lymphoma ($n = 3$) and follicular lymphoma ($n = 1$)	1 PR	Gr 3 Hypoxemia, Gr 1–2 orthostatic hypotension, fevers, chills, no B-cell aplasia	[300]

FcR = Fc receptor; gr = grade; L1-CAM = L1-cell adhesion molecule; CR = complete response; PR = partial response; CAR T = chimeric antigen receptor T cells; CA-IX = carboxy-anhydrase-IX; GD2 = disialoganglioside.

a) Number of patients represent those treated with at least one dose of CAR T therapy. Grade of adverse event given if indicated (Grade 1 = mild; Grade 2 = moderate; Grade 3 = severe; Grade 4 = life- threatening; Grade 5 = death).
b) On-target toxicity.
c) Cytokine symptoms may include fevers, rigors, dyspnea, hypotension, transient cardiac dysfunction.

T-cell persistence, including the generation of human immune responses against the murine portion of the CAR construct [286] or to the transgene and retroviral vectors [291] along with prolonged *ex vivo* culture time producing anergic exhausted CAR T cells. The α-FR CAR T cells in the ovarian cancer trial also did not appear to traffic to tumors and persisted mainly in the lung, liver and spleen [286]. Like T cells with engineered TCRs, the first-generation CAR T cells demonstrated on-target effects in normal tissue. For example, carboxy-anhydrase-IX (CA-IX) is expressed in normal biliary epithelium, and CA-IX CAR T cells resulted in hepatitis [292]. Although the primary objective of these Phase I trials was to examine safety, the CAR T cells did not show any significant antitumor activity, which was most likely related to their lack of persistence [271]. Currently, several methods are being developed to prolong the *in vivo* persistence of CAR T cells, mostly through enhancing costimulatory signals [293]. Although the GD2 (disialoganglioside) CAR T cells were considered first generation, they persisted for ≥90 weeks and this was correlated with a predominant central memory phenotype ($CD45RO^+CD62L^+$) CAR T and higher numbers of $CD4^+$ cells [288]. This indicates that the phenotype of the infused T cells may also influence *in vivo* persistence and efficacy.

Second-generation CAR T cells directed at CD19 are mainly being developed for B-cell malignancies, such as CLL [6] and acute lymphoblastic leukemia (ALL) [7], since CD19 is expressed in these cancer types (Table 6). The second-generation CAR T cells contain costimulatory molecules, which are frequently lacking on tumor cell targets. Compared to first-generation CAR T cells, these cells have more antitumor activity, traffic more efficiently to tumor sites, expand, persist, and develop memory. However, their use is associated with an elevated risk of treatment-related cytokine toxicities. In the first case report in a poor risk, multiply pretreated, p53-deficient CLL patient, autologous T cells were genetically transduced with a lentivirus containing a costimulatory domain CD137 (4-1BB) in addition to the CD3ζ domain. The results were promising, and an update with two additional patients was published [296]. These CD19 CAR T cells expanded *in vivo* by 1000- to 10 000-fold [6], and at peak expansion comprised 10% to >95% of circulating lymphocytes. This peak coincided with leukemia cell killing and subsequent delayed tumor lysis syndrome (metabolic complications secondary to tumor breakdown) and cytokine release symptoms between days 7 and 21. These CD19 CAR T cells also tracked efficiently to the sites of disease and, despite having a murine scFv domain, persisted and retained their function in the blood and bone marrow for 3–6 months. Moreover, the $CD4^+$ CAR T cells developed a central memory phenotype, while the $CD8^+$ CAR T cells developed an effector memory phenotype. Antitumor activity was notable, with two complete responses and one partial response lasting for eight months or more. The overall ratio of infused CAR T cells to CLL targets was as high as 1 : 93 000. Given that CD19 is a B-cell-lineage marker, there was on-target toxicity consisting of prolonged B-cell depletion and hypogammaglobulinemia. In two subsequent Phase I trials with CD19-CD3ζ-CD28 CAR T cells in CLL ($n = 8$ treated) and B-cell ALL ($n = 1$ treated), CAR T cell persistence was associated with a low disease burden and exposure to lymphodepleting chemotherapy [297]. The *in vivo* expansion and persistence of second-generation CAR T cells was also confirmed in a trial in which patients served as their own controls, initially receiving first-generation CD19 CAR T cells (CD3ζ) followed by second-generation CAR T cells (CD3ζ-CD28) [298]. Most recently, in an exciting CAR T trial, all five relapsed refractory ALL patients achieved a complete response with no minimal residual disease. Similar to the CLL trial, there was substantial *in vivo* CAR T expansion and persistence; however, cytokine toxicities occurred earlier at days 3–5. This toxicity appeared to be related to the tumor burden, with patients having higher tumor burdens experiencing worse symptoms. This argues for a maximum tumor reduction prior to CAR T-cell infusion, especially as corticosteroid therapy to treat cytokine storm can lead to decreased CAR T-cell persistence [7, 296]. All but one of the ALL patients went on to an allogeneic HSCT, which is the only hope for cure in many of these patients. These results were confirmed in a larger 30 patient ALL clinical trial, with a 90% complete hematologic remission rate and a 6-month event-free survival of 67% (95% CI 51–88) [301].

Third-generation CARs have multiple costimulatory domains. In a pilot trial with the third-generation CD20 CAR T cells (CD3ζ-CD28-4-1BB) in NHL, CAR expression was limited due to transduction with naked plasmid DNA. However, the *in vivo* persistence of these CAR T cells was noted, with peak levels of 1.0–3.2% found immediately after the infusion and remaining detectable for 9–12 months [300]. These CAR T cells were also found to traffic to tumor sites that included malignant lymph nodes and bone marrow at 24–48 h after the third infusion. Despite the murine scFv variable fragment in the CAR construct, there was no detectable anti-CAR humoral or cellular immune responses. However, third-generation CAR T cells come with a cautionary tale, as demonstrated with an ERBB2-CAR (CD3ζ-CD28-4-1BB) T-cell, which led to respiratory distress and death in a patient, presumably due to recognition of low levels of ERBB-2 in the lungs and subsequent cytokine storm [299].

CAR T cells represent a novel and promising therapy. Many variables remain to be evaluated and optimized, including the choice of method for transducing the CAR (i.e., lentivirus or retrovirus vector versus nonviral vectors such as the *sleeping beauty* transposon system), the optimal dose and dosing schedule of CAR T cells, and strategies to combat any unintended adverse events related to the CAR T cells, such as integrating suicide genes to enable destruction of the cells [302]. There is also concern regarding virally transduced CAR T cells, given the potential of insertional mutagenesis. However, the long-term monitoring of gamma-retroviral transduced CAR T cells in three HIV clinical trials (>500 patient–years) demonstrated that, despite the persistence of these cells for more than a decade, there was no evidence of transformation or integration at sites near growth control sites within the genome [303].

3.3 Dendritic (DC) Therapy

As mentioned above, DCs are the most potent APCs and are an important bridge between the innate and adaptive immune systems, not only eliciting immune responses but also playing a role in central and peripheral tolerance. Given their critical role in the immune system, there has been significant interest in harnessing the potential of DCs to induce antitumor immunity.

Conceptually, there are several ways to utilize DCs in the treatment of cancer (for a review, see Ref. [304] and Table 7). The first

Tab. 7 Types of dendritic cell (DC) therapy.

Types of dendritic cell therapy	*Methodology*
Unloaded DCs	DCs not stimulated either *in vivo* or *ex vivo* against a TAA
DCs loaded with TAA *ex vivo*	Autologous cancer cell lysate
	Allogeneic cancer cell lysate
	Recombinant TAAs
	Whole-tumor RNA
	TAA coding RNA
	TAA expressing gene vectors
DCs loaded *in vivo*	TAA fused to monoclonal antibody
	Injection of tumors with DCs or DC activators or chemoattractants
	Injection of DC growth factors followed by activation and loading of these cells *in situ*
	DC targeted immunoliposomes
DC-derived exosomes	Vesicles (non-cellular) secreted from DCs with MHC and costimulatory molecules

TAA = tumor-associated antigen; RNA = ribonucleic acid; MHC = major histocompatibility complex.

method to be developed clinically involves the loading and activation of DCs *ex vivo*, followed by the delivery of these cells systemically by, for example, intravenous or subcutaneous injection. The present authors' group performed the first clinical study with TAA-pulsed DCs [305], and carried out additional trials which demonstrated both the safety of the approach and its potential to induce clinically meaningful responses in patients with advanced cancers [306, 307]. In this approach, DCs isolated from the blood of patients are loaded with TAAs in the form of recombinant proteins or peptides, autologous or allogeneic cancer cell lysates, transfected whole-tumor RNA, TAA-coding RNAs or gene vectors, and "dendritomes" (DCs fused to inactivated cancer cells). Compared to unloaded DCs, the loaded DCs have the potential to deliver a coordinated TAA-specific immune response. Another method involves the direct delivery of unloaded DCs or molecules that attract DCs into tumors, where the DCs can potentially acquire TAAs and present the processed antigens to T cells either in the tumor itself or, more likely, in the tumor-draining lymph nodes [308, 309]. A third approach is to load and activate DCs *in vivo*. For example, this can be achieved by inducing the release of large numbers of iDCs from bone marrow using the Fms-like tyrosine kinase 3 (Flt3) ligand, followed by local delivery (e.g., into the skin) of TAA and a DC-activating agent [310, 311], or delivering TAA fused to a DC-targeting mAb [312] or TAA in immunoliposomes [313]. The fourth method is to use DC-derived exosomes [314], which are antigen-presenting vesicles secreted by DCs that retain functional MHC molecules and costimulatory molecules.

Today, many methods are being explored to enhance DC therapy (for a review, see Ref. [315]). An enhanced polarization of monocyte-derived DCs in culture has improved the clinical efficacy, with properties such as maturity, responsiveness to chemokines, and IL-12p70-secreting ability emerging as the most important [316]. Changes in cytokine culture cocktails are also being explored [317], along with shorter culture times [318]. Overall, there is a significant effort to standardize the design and manufacturing of DCs [319]. There is also ongoing debate about which DC subtype is best and the potential of improved efficacy from the use of naturally occurring DCs (plasmacytoid or myeloid) over *ex vivo* DCs derived from monocytes and $CD34^+$ progenitors [320]. There are many variables that can be optimized, including the route of administration, dose and dosing schedule [315], along with evaluating the effects of coadministered cytokines [321].

Sipuleucel-T became the first FDA-approved DC vaccine in 2010 [1], when it was approved for the treatment of metastatic castrate-resistant prostate cancer (mCRPC). The vaccine is composed of autologous PBMCs, including APCs that are loaded and activated *ex vivo* with a recombinant fusion protein, PA2024, for 36–44 h. PA2024 is composed of prostatic acid phosphatase (PAP) fused to GM-CSF. PAP is found on prostate cancer cells, and normal prostate and GM-CSF is known to activate DCs [322]. mCRPC patients ($n = 512$) were randomized to receive either Sipuleucel-T or placebo, and were scheduled for leukapheresis at weeks 0, 2, and 4 to collect PBMCs, followed three days later by an infusion of Sipuleucel-T or placebo. In the treatment arm, each dose of Sipuleucel-T contained at least 40 million PBMCs expressing the costimulatory molecule, CD54, which is a marker of activated DCs [323]. After a median follow-up of 34.1 months, Sipuleucel-T resulted in an improved median OS of 25.8 months compared to 21.7 months ($p = 0.032$) [1]. It also induced a humoral and cell-mediated immune response, which is hypothesized to account for this survival benefit [324]. Antibody titers of >400 to PA2024 and PAP were observed at a higher rate in the Sipuleucel-T group than the placebo group (66.2% versus 2.9% and 28.5% versus 1.4%, respectively). T-cell proliferation to PA2024 and PAP was higher in the Sipuleucel-T group (73% versus 12.1% and 27.3% versus 8%, respectively). The development of high antibody titers (>400) was predictive of an improved survival. Adverse events were reported more frequently in the Sipuleucel-T group and included fevers, chills, flu-like illness, muscle aches, hypertension, hyperhidrosis (excessive sweating) and groin pain. However, these were short-lived, usually occurred within the first 24 h, were resolved within 1–2 days, and were generally not severe. As of July 2012, there were 120 clinical trials ongoing with DC vaccines (see Ref. [304]). Some of these trials may lead to additional FDA-approved DC therapies, and it is possible that DC therapy may prove most effective in a combination strategy, for example, with chemotherapies [325] or immune checkpoint blockade [326].

4 Conclusions

As a result of remarkable advances in the present understanding of the immune system, and of the mechanisms that enable tumors to escape the immune response, novel – and increasingly effective – forms of tumor immunotherapy have been developed during recent years. Indeed, immunotherapy is becoming a core component of cancer therapy and a renewed enthusiasm is permeating the field. Knowledge of the structure of Igs and the identification of appropriate TAAs has facilitated the construction, optimization and antitumor potency of human mAbs. Likewise, knowledge of the structure of TCRs and the molecular signaling pathways that lead to a T-cell response has allowed the creation of T cells with CARs directed against antigens on tumors. Understanding the mechanism by which DCs induce antitumor immunity has enabled the development of the first effective therapeutic cancer vaccine. The discovery of immune checkpoints has led to development of checkpoint inhibitors which have demonstrated efficacy in patients with non-immunogenic, as well as immunogenic, tumors. The combined use of these immunotherapies could potentially lead to cures of advanced cancers which previously were thought to be incurable.

Acknowledgments

This work was supported by a TL1 Clinical Research Training Program of the Stanford Clinical and Translational Science Award to Spectrum (NIH TL1 TR 001084), (RN&SKP), a Stanford Cancer Institute Fellowship Award (SKP), and NIH grant 5U01CA141468 (EGE).

References

1. Kantoff, P.W., Higano, C.S., Shore, N.D., Berger, E.R. *et al.* (2010) Sipuleucel-T immunotherapy for castration-resistant prostate cancer. *N. Engl. J. Med.*, **363**, 411–422.
2. Hodi, F.S., O'Day, S.J., McDermott, D.F., Weber, R.W. *et al.* (2010) Improved survival with ipilimumab in patients with metastatic melanoma. *N. Engl. J. Med.*, **363**, 711–723.
3. Herbst, R.S., Gordon, M.S., Fine, G.D., Sosman, J.A. *et al.* (2013) A study of MPDL3280A, an engineered PD-L1 antibody in patients with locally advanced or metastatic tumors. *J. Clin. Oncol.*, **31**(Suppl.), Abstract 3000.
4. Topalian, S.L., Hodi, F.S., Brahmer, J.R., Gettinger, S.N. *et al.* (2012) Safety, activity, and immune correlates of anti-PD-1 antibody in cancer. *N. Engl. J. Med.*, **366**, 2443–2454.
5. Hamid, O., Robert, C., Daud, A., Hodi, F.S. *et al.* (2013) Safety and tumor responses with lambrolizumab (anti-PD-1) in melanoma. *N. Engl. J. Med.*, **369**, 134–144.
6. Porter, D.L., Levine, B.L., Kalos, M., Bagg, A. *et al.* (2011) Chimeric antigen receptor-modified T cells in chronic lymphoid leukemia. *N. Engl. J. Med.*, **365**, 725–733.
7. Brentjens, R.J., Davila, M.L., Riviere, I., Park, J. *et al.* (2013) CD19-targeted T cells rapidly induce molecular remissions in adults with chemotherapy-refractory acute lymphoblastic leukemia. *Sci. Transl. Med.*, **5**, 177ra138.
8. Dunn, G.P., Old, L.J., and Schreiber, R.D. (2004) The three Es of cancer immunoediting. *Annu. Rev. Immunol.*, **22**, 329–360.
9. Chow, M.T., Moller, A., and Smyth, M.J. (2012) Inflammation and immune surveillance in cancer. *Semin. Cancer Biol.*, **22**, 23–32.
10. Schreiber, R.D., Old, L.J., and Smyth, M.J. (2011) Cancer immunoediting: integrating immunity's roles in cancer suppression and promotion. *Science*, **331**, 1565–1570.
11. Shankaran, V., Ikeda, H., Bruce, A.T., White, J.M. *et al.* (2001) IFNgamma and lymphocytes prevent primary tumour development and shape tumour immunogenicity. *Nature*, **410**, 1107–1111.
12. Vesely, M.D., Kershaw, M.H., Schreiber, R.D., and Smyth, M.J. (2011) Natural innate and adaptive immunity to cancer. *Annu. Rev. Immunol.*, **29**, 235–271.
13. Matzinger, P. (1994) Tolerance, danger, and the extended family. *Annu. Rev. Immunol.*, **12**, 991–1045.
14. Sims, G.P., Rowe, D.C., Rietdijk, S.T., Herbst, R. *et al.* (2010) HMGB1 and RAGE in inflammation and cancer. *Annu. Rev. Immunol.*, **28**, 367–388.
15. Guerra, N., Tan, Y.X., Joncker, N.T., Choy, A. *et al.* (2008) NKG2D-deficient mice are defective in tumor surveillance in models of spontaneous malignancy. *Immunity*, **28**, 571–580.

16. O'Sullivan, T., Saddawi-Konefka, R., Vermi, W., Koebel, C.M. *et al.* (2012) Cancer immunoediting by the innate immune system in the absence of adaptive immunity. *J. Exp. Med.*, **209**, 1869–1882.
17. Cheever, M.A., Allison, J.P., Ferris, A.S., Finn, O.J. *et al.* (2009) The prioritization of cancer antigens: a national cancer institute pilot project for the acceleration of translational research. *Clin. Cancer Res.*, **15**, 5323–5337.
18. Koebel, C.M., Vermi, W., Swann, J.B., Zerafa, N. *et al.* (2007) Adaptive immunity maintains occult cancer in an equilibrium state. *Nature*, **450**, 903–907.
19. DuPage, M., Mazumdar, C., Schmidt, L.M., Cheung, A.F. *et al.* (2012) Expression of tumour-specific antigens underlies cancer immunoediting. *Nature*, **482**, 405–409.
20. Zitvogel, L., Tesniere, A., and Kroemer, G. (2006) Cancer despite immunosurveillance: immunoselection and immunosubversion. *Nat. Rev. Immunol.*, **6**, 715–727.
21. Khong, H.T. and Restifo, N.P. (2002) Natural selection of tumor variants in the generation of "tumor escape" phenotypes. *Nat. Immunol.*, **3**, 999–1005.
22. Maeurer, M.J., Gollin, S.M., Storkus, W.J., Swaney, W. *et al.* (1996) Tumor escape from immune recognition: loss of HLA-A2 melanoma cell surface expression is associated with a complex rearrangement of the short arm of chromosome 6. *Clin. Cancer Res.*, **2**, 641–652.
23. Galon, J., Angell, H.K., Bedognetti, D., and Marincola, F.M. (2013) The continuum of cancer immunosurveillance: prognostic, predictive, and mechanistic signatures. *Immunity*, **39**, 11–26.
24. Dougan, M. and Dranoff, G. (2009) Immune therapy for cancer. *Annu. Rev. Immunol.*, **27**, 83–117.
25. Virchow, R. (1881) An address on the value of pathological experiments. *Br. Med. J.*, **2**, 198–203.
26. Coussens, L.M., Zitvogel, L., and Palucka, A.K. (2013) Neutralizing tumor-promoting chronic inflammation: a magic bullet? *Science*, **339**, 286–291.
27. Elinav, E., Nowarski, R., Thaiss, C.A., Hu, B. *et al.* (2013) Inflammation-induced cancer: crosstalk between tumours, immune cells and microorganisms. *Nat. Rev. Cancer*, **13**, 759–771.
28. Grivennikov, S.I., Greten, F.R., and Karin, M. (2010) Immunity, inflammation, and cancer. *Cell*, **140**, 883–899.
29. Dvorak, H.F. (1986) Tumors: wounds that do not heal. Similarities between tumor stroma generation and wound healing. *N. Engl. J. Med.*, **315**, 1650–1659.
30. Okazaki, I.M., Kotani, A., and Honjo, T. (2007) Role of AID in tumorigenesis. *Adv. Immunol.*, **94**, 245–273.
31. Chen, X., Xu, H., Yuan, P., Fang, F. *et al.* (2008) Integration of external signaling pathways with the core transcriptional network in embryonic stem cells. *Cell*, **133**, 1106–1117.
32. Yu, H., Kortylewski, M., and Pardoll, D. (2007) Crosstalk between cancer and immune cells: role of STAT3 in the tumour microenvironment. *Nat. Rev. Immunol.*, **7**, 41–51.
33. Kujawski, M., Kortylewski, M., Lee, H., Herrmann, A. *et al.* (2008) Stat3 mediates myeloid cell-dependent tumor angiogenesis in mice. *J. Clin. Invest.*, **118**, 3367–3377.
34. Yang, J. and Weinberg, R.A. (2008) Epithelial-mesenchymal transition: at the crossroads of development and tumor metastasis. *Dev. Cell*, **14**, 818–829.
35. Kitamura, T., Kometani, K., Hashida, H., Matsunaga, A. *et al.* (2007) SMAD4-deficient intestinal tumors recruit CCR1+ myeloid cells that promote invasion. *Nat. Genet.*, **39**, 467–475.
36. Kaplan, R.N., Riba, R.D., Zacharoulis, S., Bramley, A.H. *et al.* (2005) VEGFR1-positive haematopoietic bone marrow progenitors initiate the pre-metastatic niche. *Nature*, **438**, 820–827.
37. Mlecnik, B., Bindea, G., Pages, F., and Galon, J. (2011) Tumor immunosurveillance in human cancers. *Cancer Metastasis Rev.*, **30**, 5–12.
38. Gabrilovich, D. (2004) Mechanisms and functional significance of tumour-induced dendritic-cell defects. *Nat. Rev. Immunol.*, **4**, 941–952.
39. Ganss, R. and Hanahan, D. (1998) Tumor microenvironment can restrict the effectiveness of activated antitumor lymphocytes. *Cancer Res.*, **58**, 4673–4681.
40. Anderson, M.J., Shafer-Weaver, K., Greenberg, N.M., and Hurwitz, A.A. (2007) Tolerization of tumor-specific T cells despite efficient initial priming in a

primary murine model of prostate cancer. *J. Immunol.*, **178**, 1268–1276.
41. Hurwitz, A.A. and Watkins, S.K. (2012) Immune suppression in the tumor microenvironment: a role for dendritic cell-mediated tolerization of T cells. *Cancer Immunol. Immunother.*, **61**, 289–293.
42. Shafer-Weaver, K.A., Anderson, M.J., Stagliano, K., Malyguine, A. *et al.* (2009) Cutting edge: tumor-specific CD8+ T cells infiltrating prostatic tumors are induced to become suppressor cells. *J. Immunol.*, **183**, 4848–4852.
43. Shafer-Weaver, K.A., Watkins, S.K., Anderson, M.J., Draper, L.J. *et al.* (2009) Immunity to murine prostatic tumors: continuous provision of T-cell help prevents CD8 T-cell tolerance and activates tumor-infiltrating dendritic cells. *Cancer Res.*, **69**, 6256–6264.
44. Devaud, C., John, L.B., Westwood, J.A., Darcy, P.K. *et al.* (2013) Immune modulation of the tumor microenvironment for enhancing cancer immunotherapy. *Oncoimmunology*, **2**, e25961.
45. Angell, H. and Galon, J. (2013) From the immune contexture to the Immunoscore: the role of prognostic and predictive immune markers in cancer. *Curr. Opin. Immunol.*, **25**, 261–267.
46. Fridman, W.H., Pages, F., Sautes-Fridman, C., and Galon, J. (2012) The immune contexture in human tumours: impact on clinical outcome. *Nat. Rev. Cancer*, **12**, 298–306.
47. Galon, J., Costes, A., Sanchez-Cabo, F., Kirilovsky, A. *et al.* (2006) Type, density, and location of immune cells within human colorectal tumors predict clinical outcome. *Science*, **313**, 1960–1964.
48. Pages, F., Berger, A., Camus, M., Sanchez-Cabo, F. *et al.* (2005) Effector memory T cells, early metastasis, and survival in colorectal cancer. *N. Engl. J. Med.*, **353**, 2654–2666.
49. Mlecnik, B., Tosolini, M., Kirilovsky, A., Berger, A. *et al.* (2011) Histopathologic-based prognostic factors of colorectal cancers are associated with the state of the local immune reaction. *J. Clin. Oncol.*, **29**, 610–618.
50. Zhang, L., Conejo-Garcia, J.R., Katsaros, D., Gimotty, P.A. *et al.* (2003) Intratumoral T cells, recurrence, and survival in epithelial ovarian cancer. *N. Engl. J. Med.*, **348**, 203–213.
51. Bronte, V., Apolloni, E., Cabrelle, A., Ronca, R. *et al.* (2000) Identification of a CD11b(+)/Gr-1(+)/CD31(+) myeloid progenitor capable of activating or suppressing CD8(+) T cells. *Blood*, **96**, 3838–3846.
52. Movahedi, K., Guilliams, M., Van den Bossche, J., Van den Bergh, R. *et al.* (2008) Identification of discrete tumor-induced myeloid-derived suppressor cell subpopulations with distinct T cell-suppressive activity. *Blood*, **111**, 4233–4244.
53. Bunt, S.K., Sinha, P., Clements, V.K., Leips, J. *et al.* (2006) Inflammation induces myeloid-derived suppressor cells that facilitate tumor progression. *J. Immunol.*, **176**, 284–290.
54. Cuenca, A.G., Delano, M.J., Kelly-Scumpia, K.M., Moreno, C. *et al.* (2011) A paradoxical role for myeloid-derived suppressor cells in sepsis and trauma. *Mol. Med.*, **17**, 281–292.
55. Makarenkova, V.P., Bansal, V., Matta, B.M., Perez, L.A. *et al.* (2006) CD11b+/Gr-1+ myeloid suppressor cells cause T cell dysfunction after traumatic stress. *J. Immunol.*, **176**, 2085–2094.
56. Youn, J.I., Nagaraj, S., Collazo, M., and Gabrilovich, D.I. (2008) Subsets of myeloid-derived suppressor cells in tumor-bearing mice. *J. Immunol.*, **181**, 5791–5802.
57. Peranzoni, E., Zilio, S., Marigo, I., Dolcetti, L. *et al.* (2010) Myeloid-derived suppressor cell heterogeneity and subset definition. *Curr. Opin. Immunol.*, **22**, 238–244.
58. Serafini, P. (2013) Myeloid derived suppressor cells in physiological and pathological conditions: the good, the bad, and the ugly. *Immunol. Res.*, **57**, 172–184.
59. Pak, A.S., Wright, M.A., Matthews, J.P., Collins, S.L. *et al.* (1995) Mechanisms of immune suppression in patients with head and neck cancer: presence of CD34(+) cells which suppress immune functions within cancers that secrete granulocyte-macrophage colony-stimulating factor. *Clin. Cancer Res.*, **1**, 95–103.
60. Wang, S., Fu, Y., Ma, K., Liu, C. *et al.* (2014) The significant increase and dynamic changes of the myeloid-derived suppressor cells percentage with chemotherapy in

advanced NSCLC patients. *Clin. Transl. Oncol.*, **16**(7), 616–622.
61. Diaz-Montero, C.M., Salem, M.L., Nishimura, M.I., Garrett-Mayer, E. *et al.* (2009) Increased circulating myeloid-derived suppressor cells correlate with clinical cancer stage, metastatic tumor burden, and doxorubicin-cyclophosphamide chemotherapy. *Cancer Immunol. Immunother.*, **58**, 49–59.
62. Montero, A.J., Diaz-Montero, C.M., Kyriakopoulos, C.E., Bronte, V. *et al.* (2012) Myeloid-derived suppressor cells in cancer patients: a clinical perspective. *J. Immunother.*, **35**, 107–115.
63. Serafini, P., Carbley, R., Noonan, K.A., Tan, G. *et al.* (2004) High-dose granulocyte-macrophage colony-stimulating factor-producing vaccines impair the immune response through the recruitment of myeloid suppressor cells. *Cancer Res.*, **64**, 6337–6343.
64. Pan, P.Y., Wang, G.X., Yin, B., Ozao, J. *et al.* (2008) Reversion of immune tolerance in advanced malignancy: modulation of myeloid-derived suppressor cell development by blockade of stem-cell factor function. *Blood*, **111**, 219–228.
65. Gabrilovich, D., Ishida, T., Oyama, T., Ran, S. *et al.* (1998) Vascular endothelial growth factor inhibits the development of dendritic cells and dramatically affects the differentiation of multiple hematopoietic lineages in vivo. *Blood*, **92**, 4150–4166.
66. Sinha, P., Clements, V.K., Fulton, A.M., and Ostrand-Rosenberg, S. (2007) Prostaglandin E2 promotes tumor progression by inducing myeloid-derived suppressor cells. *Cancer Res.*, **67**, 4507–4513.
67. Talmadge, J.E., Hood, K.C., Zobel, L.C., Shafer, L.R. *et al.* (2007) Chemoprevention by cyclooxygenase-2 inhibition reduces immature myeloid suppressor cell expansion. *Int. Immunopharmacol.*, **7**, 140–151.
68. Sawanobori, Y., Ueha, S., Kurachi, M., Shimaoka, T. *et al.* (2008) Chemokine-mediated rapid turnover of myeloid-derived suppressor cells in tumor-bearing mice. *Blood*, **111**, 5457–5466.
69. Wu, G., Morris, S.M., Jr, (1998) Arginine metabolism: nitric oxide and beyond. *Biochem. J.*, **336** (Pt. 1), 1–17.
70. Bogdan, C. (2001) Nitric oxide and the immune response. *Nat. Immunol.*, **2**, 907–916.
71. Raber, P., Ochoa, A.C., and Rodriguez, P.C. (2012) Metabolism of L-arginine by myeloid-derived suppressor cells in cancer: mechanisms of T cell suppression and therapeutic perspectives. *Immunol. Invest.*, **41**, 614–634.
72. Rodriguez, P.C., Zea, A.H., Culotta, K.S., Zabaleta, J. *et al.* (2002) Regulation of T cell receptor CD3zeta chain expression by L-arginine. *J. Biol. Chem.*, **277**, 21123–21129.
73. Rodriguez, P.C., Quiceno, D.G., Zabaleta, J., Ortiz, B. *et al.* (2004) Arginase I production in the tumor microenvironment by mature myeloid cells inhibits T-cell receptor expression and antigen-specific T-cell responses. *Cancer Res.*, **64**, 5839–5849.
74. Rodriguez, P.C., Quiceno, D.G., and Ochoa, A.C. (2007) L-arginine availability regulates T-lymphocyte cell-cycle progression. *Blood*, **109**, 1568–1573.
75. Zea, A.H., Rodriguez, P.C., Atkins, M.B., Hernandez, C. *et al.* (2005) Arginase-producing myeloid suppressor cells in renal cell carcinoma patients: a mechanism of tumor evasion. *Cancer Res.*, **65**, 3044–3048.
76. Schmielau, J. and Finn, O.J. (2001) Activated granulocytes and granulocyte-derived hydrogen peroxide are the underlying mechanism of suppression of t-cell function in advanced cancer patients. *Cancer Res.*, **61**, 4756–4760.
77. Rodriguez, P.C., Hernandez, C.P., Quiceno, D., Dubinett, S.M. *et al.* (2005) Arginase I in myeloid suppressor cells is induced by COX-2 in lung carcinoma. *J. Exp. Med.*, **202**, 931–939.
78. Fukumura, D., Kashiwagi, S., and Jain, R.K. (2006) The role of nitric oxide in tumour progression. *Nat. Rev. Cancer*, **6**, 521–534.
79. Duhe, R.J., Evans, G.A., Erwin, R.A., Kirken, R.A. *et al.* (1998) Nitric oxide and thiol redox regulation of Janus kinase activity. *Proc. Natl Acad. Sci. USA*, **95**, 126–131.
80. Bingisser, R.M., Tilbrook, P.A., Holt, P.G., and Kees, U.R. (1998) Macrophage-derived nitric oxide regulates T cell activation via reversible disruption of the Jak3/STAT5 signaling pathway. *J. Immunol.*, **160**, 5729–5734.

81. Mazzoni, A., Bronte, V., Visintin, A., Spitzer, J.H. *et al.* (2002) Myeloid suppressor lines inhibit T cell responses by an NO-dependent mechanism. *J. Immunol.*, **168**, 689–695.
82. Saio, M., Radoja, S., Marino, M., and Frey, A.B. (2001) Tumor-infiltrating macrophages induce apoptosis in activated CD8(+) T cells by a mechanism requiring cell contact and mediated by both the cell-associated form of TNF and nitric oxide. *J. Immunol.*, **167**, 5583–5593.
83. Macphail, S.E., Gibney, C.A., Brooks, B.M., Booth, C.G. *et al.* (2003) Nitric oxide regulation of human peripheral blood mononuclear cells: critical time dependence and selectivity for cytokine versus chemokine expression. *J. Immunol.*, **171**, 4809–4815.
84. Xia, Y. and Zweier, J.L. (1997) Superoxide and peroxynitrite generation from inducible nitric oxide synthase in macrophages. *Proc. Natl Acad. Sci. USA*, **94**, 6954–6958.
85. Brito, C., Naviliat, M., Tiscornia, A.C., Vuillier, F. *et al.* (1999) Peroxynitrite inhibits T lymphocyte activation and proliferation by promoting impairment of tyrosine phosphorylation and peroxynitrite-driven apoptotic death. *J. Immunol.*, **162**, 3356–3366.
86. Alvarez, B. and Radi, R. (2003) Peroxynitrite reactivity with amino acids and proteins. *Amino Acids*, **25**, 295–311.
87. Nagaraj, S., Gupta, K., Pisarev, V., Kinarsky, L. *et al.* (2007) Altered recognition of antigen is a mechanism of CD8+ T cell tolerance in cancer. *Nat. Med.*, **13**, 828–835.
88. Otsuji, M., Kimura, Y., Aoe, T., Okamoto, Y. *et al.* (1996) Oxidative stress by tumor-derived macrophages suppresses the expression of CD3 zeta chain of T-cell receptor complex and antigen-specific T-cell responses. *Proc. Natl Acad. Sci. USA*, **93**, 13119–13124.
89. Kono, K., Salazar-Onfray, F., Petersson, M., Hansson, J. *et al.* (1996) Hydrogen peroxide secreted by tumor-derived macrophages down-modulates signal-transducing zeta molecules and inhibits tumor-specific T cell-and natural killer cell-mediated cytotoxicity. *Eur. J. Immunol.*, **26**, 1308–1313.
90. Serafini, P., Mgebroff, S., Noonan, K., and Borrello, I. (2008) Myeloid-derived suppressor cells promote cross-tolerance in B-cell lymphoma by expanding regulatory T cells. *Cancer Res.*, **68**, 5439–5449.
91. Huang, B., Pan, P.Y., Li, Q., Sato, A.I. *et al.* (2006) Gr-1 + CD115+ immature myeloid suppressor cells mediate the development of tumor-induced T regulatory cells and T-cell anergy in tumor-bearing host. *Cancer Res.*, **66**, 1123–1131.
92. Nishizuka, Y. and Sakakura, T. (1969) Thymus and reproduction: sex-linked dysgenesia of the gonad after neonatal thymectomy in mice. *Science*, **166**, 753–755.
93. Sakaguchi, S., Takahashi, T., and Nishizuka, Y. (1982) Study on cellular events in post-thymectomy autoimmune oophoritis in mice. I. Requirement of Lyt-1 effector cells for oocytes damage after adoptive transfer. *J. Exp. Med.*, **156**, 1565–1576.
94. Sakaguchi, S., Fukuma, K., Kuribayashi, K., and Masuda, T. (1985) Organ-specific autoimmune diseases induced in mice by elimination of T cell subset. I. Evidence for the active participation of T cells in natural self-tolerance; deficit of a T cell subset as a possible cause of autoimmune disease. *J. Exp. Med.*, **161**, 72–87.
95. Banerjee, A., Vasanthakumar, A., and Grigoriadis, G. (2013) Modulating T regulatory cells in cancer: how close are we? *Immunol. Cell Biol.*, **91**, 340–349.
96. Sakaguchi, S., Sakaguchi, N., Asano, M., Itoh, M. *et al.* (1995) Immunologic self-tolerance maintained by activated T cells expressing IL-2 receptor alpha-chains (CD25). Breakdown of a single mechanism of self-tolerance causes various autoimmune diseases. *J. Immunol.*, **155**, 1151–1164.
97. Hori, S., Nomura, T., and Sakaguchi, S. (2003) Control of regulatory T cell development by the transcription factor Foxp3. *Science*, **299**, 1057–1061.
98. Fontenot, J.D., Gavin, M.A., and Rudensky, A.Y. (2003) Foxp3 programs the development and function of CD4 + CD25+ regulatory T cells. *Nat. Immunol.*, **4**, 330–336.
99. Khattri, R., Cox, T., Yasayko, S.A., and Ramsdell, F. (2003) An essential role for Scurfin in CD4 + CD25+ T regulatory cells. *Nat. Immunol.*, **4**, 337–342.
100. Bisikirska, B., Colgan, J., Luban, J., Bluestone, J.A. *et al.* (2005) TCR stimulation with modified anti-CD3 mAb

expands CD8+ T cell population and induces CD8 + CD25+ Tregs. *J. Clin. Invest.*, **115**, 2904–2913.
101. Roncarolo, M.G. and Gregori, S. (2008) Is FOXP3 a bona fide marker for human regulatory T cells? *Eur. J. Immunol.*, **38**, 925–927.
102. Josefowicz, S.Z., Lu, L.F., and Rudensky, A.Y. (2012) Regulatory T cells: mechanisms of differentiation and function. *Annu. Rev. Immunol.*, **30**, 531–564.
103. Levings, M.K., Sangregorio, R., and Roncarolo, M.G. (2001) Human cd25(+)cd4(+) t regulatory cells suppress naive and memory T cell proliferation and can be expanded in vitro without loss of function. *J. Exp. Med.*, **193**, 1295–1302.
104. Rubtsov, Y.P., Rasmussen, J.P., Chi, E.Y., Fontenot, J. *et al.* (2008) Regulatory T cell-derived interleukin-10 limits inflammation at environmental interfaces. *Immunity*, **28**, 546–558.
105. Nakamura, K., Kitani, A., and Strober, W. (2001) Cell contact-dependent immunosuppression by CD4(+)CD25(+) regulatory T cells is mediated by cell surface-bound transforming growth factor beta. *J. Exp. Med.*, **194**, 629–644.
106. Curiel, T.J., Coukos, G., Zou, L., Alvarez, X. *et al.* (2004) Specific recruitment of regulatory T cells in ovarian carcinoma fosters immune privilege and predicts reduced survival. *Nat. Med.*, **10**, 942–949.
107. Woo, E.Y., Chu, C.S., Goletz, T.J., Schlienger, K. *et al.* (2001) Regulatory CD4(+)CD25(+) T cells in tumors from patients with early-stage non-small cell lung cancer and late-stage ovarian cancer. *Cancer Res.*, **61**, 4766–4772.
108. Woo, E.Y., Yeh, H., Chu, C.S., Schlienger, K. *et al.* (2002) Cutting edge: regulatory T cells from lung cancer patients directly inhibit autologous T cell proliferation. *J. Immunol.*, **168**, 4272–4276.
109. deLeeuw, R.J., Kost, S.E., Kakal, J.A., and Nelson, B.H. (2012) The prognostic value of FoxP3+ tumor-infiltrating lymphocytes in cancer: a critical review of the literature. *Clin. Cancer Res.*, **18**, 3022–3029.
110. Sinicrope, F.A., Rego, R.L., Ansell, S.M., Knutson, K.L. *et al.* (2009) Intraepithelial effector (CD3+)/regulatory (FoxP3+) T-cell ratio predicts a clinical outcome of human colon carcinoma. *Gastroenterology*, **137**, 1270–1279.
111. Salama, P., Phillips, M., Grieu, F., Morris, M. *et al.* (2009) Tumor-infiltrating FOXP3+ T regulatory cells show strong prognostic significance in colorectal cancer. *J. Clin. Oncol.*, **27**, 186–192.
112. Leffers, N., Gooden, M.J., de Jong, R.A., Hoogeboom, B.N. *et al.* (2009) Prognostic significance of tumor-infiltrating T-lymphocytes in primary and metastatic lesions of advanced stage ovarian cancer. *Cancer Immunol. Immunother.*, **58**, 449–459.
113. Jacobs, J.F., Nierkens, S., Figdor, C.G., de Vries, I.J. *et al.* (2012) Regulatory T cells in melanoma: the final hurdle towards effective immunotherapy? *Lancet Oncol.*, **13**, e32–e42.
114. Kimpfler, S., Sevko, A., Ring, S., Falk, C. *et al.* (2009) Skin melanoma development in ret transgenic mice despite the depletion of CD25 + Foxp3+ regulatory T cells in lymphoid organs. *J. Immunol.*, **183**, 6330–6337.
115. Facciabene, A., Peng, X., Hagemann, I.S., Balint, K. *et al.* (2011) Tumour hypoxia promotes tolerance and angiogenesis via CCL28 and T(reg) cells. *Nature*, **475**, 226–230.
116. Seo, N., Hayakawa, S., Takigawa, M., and Tokura, Y. (2001) Interleukin-10 expressed at early tumour sites induces subsequent generation of CD4(+) T-regulatory cells and systemic collapse of antitumour immunity. *Immunology*, **103**, 449–457.
117. Chen, W., Jin, W., Hardegen, N., Lei, K.J. *et al.* (2003) Conversion of peripheral CD4 + CD25- naive T cells to CD4 + CD25+ regulatory T cells by TGF-beta induction of transcription factor Foxp3. *J. Exp. Med.*, **198**, 1875–1886.
118. Fallarino, F., Vacca, C., Orabona, C., Belladonna, M.L. *et al.* (2002) Functional expression of indoleamine 2,3-dioxygenase by murine CD8 alpha(+) dendritic cells. *Int. Immunol.*, **14**, 65–68.
119. Chung, D.J., Rossi, M., Romano, E., Ghith, J. *et al.* (2009) Indoleamine 2,3-dioxygenase-expressing mature human monocyte-derived dendritic cells expand potent autologous regulatory T cells. *Blood*, **114**, 555–563.

120. Fallarino, F., Grohmann, U., You, S., McGrath, B.C. *et al.* (2006) The combined effects of tryptophan starvation and tryptophan catabolites down-regulate T cell receptor zeta-chain and induce a regulatory phenotype in naive T cells. *J. Immunol.*, **176**, 6752–6761.

121. Sakaguchi, S., Miyara, M., Costantino, C.M., and Hafler, D.A. (2010) FOXP3+ regulatory T cells in the human immune system. *Nat. Rev. Immunol.*, **10**, 490–500.

122. Wing, K., Onishi, Y., Prieto-Martin, P., Yamaguchi, T. *et al.* (2008) CTLA-4 control over Foxp3+ regulatory T cell function. *Science*, **322**, 271–275.

123. Huang, C.T., Workman, C.J., Flies, D., Pan, X. *et al.* (2004) Role of LAG-3 in regulatory T cells. *Immunity*, **21**, 503–513.

124. Liang, B., Workman, C., Lee, J., Chew, C. *et al.* (2008) Regulatory T cells inhibit dendritic cells by lymphocyte activation gene-3 engagement of MHC class II. *J. Immunol.*, **180**, 5916–5926.

125. Sakuishi, K., Ngiow, S.F., Sullivan, J.M., Teng, M.W. *et al.* (2013) TIM3FOXP3 regulatory T cells are tissue-specific promoters of T-cell dysfunction in cancer. *Oncoimmunology*, **2**, e23849.

126. Dobrzanski, M.J. (2013) Expanding roles for CD4 T cells and their subpopulations in tumor immunity and therapy. *Front. Oncol.*, **3**, 63.

127. Gondek, D.C., Lu, L.F., Quezada, S.A., Sakaguchi, S. *et al.* (2005) Cutting edge: contact-mediated suppression by CD4 + CD25+ regulatory cells involves a granzyme B-dependent, perforin-independent mechanism. *J. Immunol.*, **174**, 1783–1786.

128. Garin, M.I., Chu, C.C., Golshayan, D., Cernuda-Morollon, E. *et al.* (2007) Galectin-1: a key effector of regulation mediated by CD4 + CD25+ T cells. *Blood*, **109**, 2058–2065.

129. Strauss, L., Bergmann, C., and Whiteside, T.L. (2009) Human circulating CD4 + CD25highFoxp3+ regulatory T cells kill autologous CD8+ but not CD4+ responder cells by Fas-mediated apoptosis. *J. Immunol.*, **182**, 1469–1480.

130. Andersson, J., Tran, D.Q., Pesu, M., Davidson, T.S. *et al.* (2008) CD4+ FoxP3+ regulatory T cells confer infectious tolerance in a TGF-beta-dependent manner. *J. Exp. Med.*, **205**, 1975–1981.

131. Belkaid, Y. (2007) Regulatory T cells and infection: a dangerous necessity. *Nat. Rev. Immunol.*, **7**, 875–888.

132. Facciabene, A., Motz, G.T., and Coukos, G. (2012) T-regulatory cells: key players in tumor immune escape and angiogenesis. *Cancer Res.*, **72**, 2162–2171.

133. Watowich, S.S. and Liu, Y.J. (2010) Mechanisms regulating dendritic cell specification and development. *Immunol. Rev.*, **238**, 76–92.

134. Villadangos, J.A. and Schnorrer, P. (2007) Intrinsic and cooperative antigen-presenting functions of dendritic-cell subsets in vivo. *Nat. Rev. Immunol.*, **7**, 543–555.

135. Reis e Sousa, C. (2006) Dendritic cells in a mature age. *Nat. Rev. Immunol.*, **6**, 476–483.

136. Albert, M.L., Jegathesan, M., and Darnell, R.B. (2001) Dendritic cell maturation is required for the cross-tolerization of CD8+ T cells. *Nat. Immunol.*, **2**, 1010–1017.

137. Perez, V.L., Van Parijs, L., Biuckians, A., Zheng, X.X. *et al.* (1997) Induction of peripheral T cell tolerance in vivo requires CTLA-4 engagement. *Immunity*, **6**, 411–417.

138. Probst, H.C., McCoy, K., Okazaki, T., Honjo, T. *et al.* (2005) Resting dendritic cells induce peripheral CD8+ T cell tolerance through PD-1 and CTLA-4. *Nat. Immunol.*, **6**, 280–286.

139. Tesone, A.J., Svoronos, N., Allegrezza, M.J., and Conejo-Garcia, J.R. (2013) Pathological mobilization and activities of dendritic cells in tumor-bearing hosts: challenges and opportunities for immunotherapy of cancer. *Front. Immunol.*, **4**, 435.

140. Hoffmann, T.K., Muller-Berghaus, J., Ferris, R.L., Johnson, J.T. *et al.* (2002) Alterations in the frequency of dendritic cell subsets in the peripheral circulation of patients with squamous cell carcinomas of the head and neck. *Clin. Cancer Res.*, **8**, 1787–1793.

141. Troy, A.J., Summers, K.L., Davidson, P.J., Atkinson, C.H. *et al.* (1998) Minimal recruitment and activation of dendritic cells within renal cell carcinoma. *Clin. Cancer Res.*, **4**, 585–593.

142. Della Bella, S., Gennaro, M., Vaccari, M., Ferraris, C. *et al.* (2003) Altered maturation of peripheral blood dendritic cells in patients with breast cancer. *Br. J. Cancer*, **89**, 1463–1472.
143. Almand, B., Resser, J. R., Lindman, B., Nadaf, S. *et al.* (2000) Clinical significance of defective dendritic cell differentiation in cancer. *Clin. Cancer Res.*, **6**, 1755–1766.
144. Scarlett, U.K., Rutkowski, M.R., Rauwerdink, A.M., Fields, J. *et al.* (2012) Ovarian cancer progression is controlled by phenotypic changes in dendritic cells. *J. Exp. Med.*, **209**, 495–506.
145. Norian, L.A., Rodriguez, P.C., O'Mara, L.A., Zabaleta, J. *et al.* (2009) Tumor-infiltrating regulatory dendritic cells inhibit CD8+ T cell function via L-arginine metabolism. *Cancer Res.*, **69**, 3086–3094.
146. Gabrilovich, D.I., Ishida, T., Nadaf, S., Ohm, J.E. *et al.* (1999) Antibodies to vascular endothelial growth factor enhance the efficacy of cancer immunotherapy by improving endogenous dendritic cell function. *Clin. Cancer Res.*, **5**, 2963–2970.
147. Menetrier-Caux, C., Montmain, G., Dieu, M.C., Bain, C. *et al.* (1998) Inhibition of the differentiation of dendritic cells from CD34(+) progenitors by tumor cells: role of interleukin-6 and macrophage colony-stimulating factor. *Blood*, **92**, 4778–4791.
148. Ratta, M., Fagnoni, F., Curti, A., Vescovini, R. *et al.* (2002) Dendritic cells are functionally defective in multiple myeloma: the role of interleukin-6. *Blood*, **100**, 230–237.
149. Sharma, S., Stolina, M., Lin, Y., Gardner, B. *et al.* (1999) T cell-derived IL-10 promotes lung cancer growth by suppressing both T cell and APC function. *J. Immunol.*, **163**, 5020–5028.
150. Flavell, R.A., Sanjabi, S., Wrzesinski, S.H., and Licona-Limon, P. (2010) The polarization of immune cells in the tumour environment by TGFbeta. *Nat. Rev. Immunol.*, **10**, 554–567.
151. Shurin, G.V., Shurin, M.R., Bykovskaia, S., Shogan, J. *et al.* (2001) Neuroblastoma-derived gangliosides inhibit dendritic cell generation and function. *Cancer Res.*, **61**, 363–369.
152. Engleman, E.G., Brody, J., and Soares, L. (2004) Using signaling pathways to overcome immune tolerance to tumors. *Sci. STKE*, **2004**, pe28.
153. Curti, A., Trabanelli, S., Onofri, C., Aluigi, M. *et al.* (2010) Indoleamine 2,3-dioxygenase-expressing leukemic dendritic cells impair a leukemia-specific immune response by inducing potent T regulatory cells. *Haematologica*, **95**, 2022–2030.
154. Belladonna, M.L., Volpi, C., Bianchi, R., Vacca, C. *et al.* (2008) Cutting edge: autocrine TGF-beta sustains default tolerogenesis by IDO-competent dendritic cells. *J. Immunol.*, **181**, 5194–5198.
155. Krempski, J., Karyampudi, L., Behrens, M.D., Erskine, C.L. *et al.* (2011) Tumor-infiltrating programmed death receptor-1+ dendritic cells mediate immune suppression in ovarian cancer. *J. Immunol.*, **186**, 6905–6913.
156. Frigola, X., Inman, B.A., Krco, C.J., Liu, X. *et al.* (2012) Soluble B7-H1: differences in production between dendritic cells and T cells. *Immunol. Lett.*, **142**, 78–82.
157. Narita, Y., Kitamura, H., Wakita, D., Sumida, K. *et al.* (2013) The key role of IL-6-arginase cascade for inducing dendritic cell-dependent CD4(+) T cell dysfunction in tumor-bearing mice. *J. Immunol.*, **190**, 812–820.
158. Wynn, T.A., Chawla, A., and Pollard, J.W. (2013) Macrophage biology in development, homeostasis and disease. *Nature*, **496**, 445–455.
159. Schulz, C., Gomez Perdiguero, E., Chorro, L., Szabo-Rogers, H. *et al.* (2012) A lineage of myeloid cells independent of Myb and hematopoietic stem cells. *Science*, **336**, 86–90.
160. Ginhoux, F., Greter, M., Leboeuf, M., Nandi, S. *et al.* (2010) Fate mapping analysis reveals that adult microglia derive from primitive macrophages. *Science*, **330**, 841–845.
161. Hoeffel, G., Wang, Y., Greter, M., See, P. *et al.* (2012) Adult Langerhans cells derive predominantly from embryonic fetal liver monocytes with a minor contribution of yolk sac-derived macrophages. *J. Exp. Med.*, **209**, 1167–1181.
162. Gordon, S. and Taylor, P.R. (2005) Monocyte and macrophage heterogeneity. *Nat. Rev. Immunol.*, **5**, 953–964.
163. Bingle, L., Brown, N.J., and Lewis, C.E. (2002) The role of tumour-associated

macrophages in tumour progression: implications for new anticancer therapies. *J. Pathol.*, **196**, 254–265.

164. Murdoch, C., Muthana, M., Coffelt, S.B., and Lewis, C.E. (2008) The role of myeloid cells in the promotion of tumour angiogenesis. *Nat. Rev. Cancer*, **8**, 618–631.
165. Qian, B.Z. and Pollard, J.W. (2010) Macrophage diversity enhances tumor progression and metastasis. *Cell*, **141**, 39–51.
166. Saccani, A., Schioppa, T., Porta, C., Biswas, S.K. *et al.* (2006) p50 nuclear factor-kappaB overexpression in tumor-associated macrophages inhibits M1 inflammatory responses and antitumor resistance. *Cancer Res.*, **66**, 11432–11440.
167. DeNardo, D.G., Andreu, P., and Coussens, L.M. (2010) Interactions between lymphocytes and myeloid cells regulate pro- versus anti-tumor immunity. *Cancer Metastasis Rev.*, **29**, 309–316.
168. Hanahan, D. and Coussens, L.M. (2012) Accessories to the crime: functions of cells recruited to the tumor microenvironment. *Cancer Cell*, **21**, 309–322.
169. Du, R., Lu, K.V., Petritsch, C., Liu, P. *et al.* (2008) HIF1alpha induces the recruitment of bone marrow-derived vascular modulatory cells to regulate tumor angiogenesis and invasion. *Cancer Cell*, **13**, 206–220.
170. Mazzieri, R., Pucci, F., Moi, D., Zonari, E. *et al.* (2011) Targeting the ANG2/TIE2 axis inhibits tumor growth and metastasis by impairing angiogenesis and disabling rebounds of proangiogenic myeloid cells. *Cancer Cell*, **19**, 512–526.
171. Lin, E.Y., Li, J.F., Bricard, G., Wang, W. *et al.* (2007) Vascular endothelial growth factor restores delayed tumor progression in tumors depleted of macrophages. *Mol. Oncol.*, **1**, 288–302.
172. Schoppmann, S.F., Birner, P., Stockl, J., Kalt, R. *et al.* (2002) Tumor-associated macrophages express lymphatic endothelial growth factors and are related to peritumoral lymphangiogenesis. *Am. J. Pathol.*, **161**, 947–956.
173. Zumsteg, A., Baeriswyl, V., Imaizumi, N., Schwendener, R. *et al.* (2009) Myeloid cells contribute to tumor lymphangiogenesis. *PLoS One*, **4**, e7067.
174. Quail, D.F. and Joyce, J.A. (2013) Microenvironmental regulation of tumor progression and metastasis. *Nat. Med.*, **19**, 1423–1437.
175. Alitalo, A. and Detmar, M. (2012) Interaction of tumor cells and lymphatic vessels in cancer progression. *Oncogene*, **31**, 4499–4508.
176. Mani, S.A., Guo, W., Liao, M.J., Eaton, E.N. *et al.* (2008) The epithelial-mesenchymal transition generates cells with properties of stem cells. *Cell*, **133**, 704–715.
177. Thiery, J.P., Acloque, H., Huang, R.Y., and Nieto, M.A. (2009) Epithelial-mesenchymal transitions in development and disease. *Cell*, **139**, 871–890.
178. Gao, D., Vahdat, L.T., Wong, S., Chang, J.C. *et al.* (2012) Microenvironmental regulation of epithelial-mesenchymal transitions in cancer. *Cancer Res.*, **72**, 4883–4889.
179. Bonde, A.K., Tischler, V., Kumar, S., Soltermann, A. *et al.* (2012) Intratumoral macrophages contribute to epithelial-mesenchymal transition in solid tumors. *BMC Cancer*, **12**, 35.
180. Gay, L.J. and Felding-Habermann, B. (2011) Contribution of platelets to tumour metastasis. *Nat. Rev. Cancer*, **11**, 123–134.
181. Psaila, B. and Lyden, D. (2009) The metastatic niche: adapting the foreign soil. *Nat. Rev. Cancer*, **9**, 285–293.
182. Peinado, H., Aleckovic, M., Lavotshkin, S., Matei, I. *et al.* (2012) Melanoma exosomes educate bone marrow progenitor cells toward a pro-metastatic phenotype through MET. *Nat. Med.*, **18**, 883–891.
183. Aguirre-Ghiso, J.A. (2007) Models, mechanisms and clinical evidence for cancer dormancy. *Nat. Rev. Cancer*, **7**, 834–846.
184. Yoshikawa, K., Mitsunaga, S., Kinoshita, T., Konishi, M. *et al.* (2012) Impact of tumor-associated macrophages on invasive ductal carcinoma of the pancreas head. *Cancer Sci.*, **103**, 2012–2020.
185. Qian, B., Deng, Y., Im, J. H., Muschel, R.J. *et al.* (2009) A distinct macrophage population mediates metastatic breast cancer cell extravasation, establishment and growth. *PLoS One*, **4**, e6562.
186. Langman, R.E. (2000) The specificity of immunological reactions. *Mol. Immunol.*, **37**, 555–561.
187. Schroeder, H.W., Jr, and Cavacini, L. (2010) Structure and function of immunoglobulins. *J. Allergy Clin. Immunol.*, **125**, S41–S52.

188. Sliwkowski, M.X. and Mellman, I. (2013) Antibody therapeutics in cancer. *Science*, **341**, 1192–1198.
189. Kohler, G. and Milstein, C. (1975) Continuous cultures of fused cells secreting antibody of predefined specificity. *Nature*, **256**, 495–497.
190. Shawler, D.L., Bartholomew, R.M., Smith, L.M., and Dillman, R.O. (1985) Human immune response to multiple injections of murine monoclonal IgG. *J. Immunol.*, **135**, 1530–1535.
191. Boulianne, G.L., Hozumi, N., and Shulman, M.J. (1984) Production of functional chimaeric mouse/human antibody. *Nature*, **312**, 643–646.
192. Jones, P.T., Dear, P.H., Foote, J., Neuberger, M.S. *et al.* (1986) Replacing the complementarity-determining regions in a human antibody with those from a mouse. *Nature*, **321**, 522–525.
193. Habermann, T.M., Weller, E.A., Morrison, V.A., Gascoyne, R.D. *et al.* (2006) Rituximab-CHOP versus CHOP alone or with maintenance rituximab in older patients with diffuse large B-cell lymphoma. *J. Clin. Oncol.*, **24**, 3121–3127.
194. Coiffier, B., Lepage, E., Briere, J., Herbrecht, R. *et al.* (2002) CHOP chemotherapy plus rituximab compared with CHOP alone in elderly patients with diffuse large-B-cell lymphoma. *N. Engl. J. Med.*, **346**, 235–242.
195. Pfreundschuh, M., Trumper, L., Osterborg, A., Pettengell, R. *et al.* (2006) CHOP-like chemotherapy plus rituximab versus CHOP-like chemotherapy alone in young patients with good-prognosis diffuse large-B-cell lymphoma: a randomised controlled trial by the MabThera International Trial (MInT) Group. *Lancet Oncol.*, **7**, 379–391.
196. Hiddemann, W., Kneba, M., Dreyling, M., Schmitz, N. *et al.* (2005) Frontline therapy with rituximab added to the combination of cyclophosphamide, doxorubicin, vincristine, and prednisone (CHOP) significantly improves the outcome for patients with advanced-stage follicular lymphoma compared with therapy with CHOP alone: results of a prospective randomized study of the German Low-Grade Lymphoma Study Group. *Blood*, **106**, 3725–3732.
197. Hallek, M., Fischer, K., Fingerle-Rowson, G., Fink, A.M. *et al.* (2010) Addition of rituximab to fludarabine and cyclophosphamide in patients with chronic lymphocytic leukaemia: a randomised, open-label, phase 3 trial. *Lancet*, **376**, 1164–1174.
198. Wierda, W.G., Kipps, T.J., Mayer, J., Stilgenbauer, S. *et al.* (2010) Ofatumumab as single-agent CD20 immunotherapy in fludarabine-refractory chronic lymphocytic leukemia. *J. Clin. Oncol.*, **28**, 1749–1755.
199. Goede, V., Fischer, K., Humphrey, K., Asikanius, E. *et al.* (2013) Obinutuzumab (GA101) plus chlorambucil (Clb) or rituximab (R) plus Clb versus Clb alone in patients with chronic lymphocytic leukemia (CLL) and preexisting medical conditions (comorbidities): final stage 1 results of the CLL11 (BO21004) phase III trial. *J. Clin. Oncol.*, **31** (Suppl.), Abstract 7004).
200. Van Cutsem, E., Kohne, C.H., Hitre, E., Zaluski, J. *et al.* (2009) Cetuximab and chemotherapy as initial treatment for metastatic colorectal cancer. *N. Engl. J. Med.*, **360**, 1408–1417.
201. Jonker, D.J., O'Callaghan, C.J., Karapetis, C.S., Zalcberg, J.R. *et al.* (2007) Cetuximab for the treatment of colorectal cancer. *N. Engl. J. Med.*, **357**, 2040–2048.
202. Bonner, J.A., Harari, P.M., Giralt, J., Azarnia, N. *et al.* (2006) Radiotherapy plus cetuximab for squamous-cell carcinoma of the head and neck. *N. Engl. J. Med.*, **354**, 567–578.
203. Vermorken, J.B., Trigo, J., Hitt, R., Koralewski, P. *et al.* (2007) Open-label, uncontrolled, multicenter phase II study to evaluate the efficacy and toxicity of cetuximab as a single agent in patients with recurrent and/or metastatic squamous cell carcinoma of the head and neck who failed to respond to platinum-based therapy. *J. Clin. Oncol.*, **25**, 2171–2177.
204. Vermorken, J.B., Mesia, R., Rivera, F., Remenar, E. *et al.* (2008) Platinum-based chemotherapy plus cetuximab in head and neck cancer. *N. Engl. J. Med.*, **359**, 1116–1127.
205. Van Cutsem, E., Peeters, M., Siena, S., Humblet, Y. *et al.* (2007) Open-label phase III trial of panitumumab plus best supportive care compared with best supportive care alone in patients with chemotherapy-refractory metastatic colorectal cancer. *J. Clin. Oncol.*, **25**, 1658–1664.
206. Gianni, L., Pienkowski, T., Im, Y. H., Roman, L. *et al.* (2012) Efficacy and

safety of neoadjuvant pertuzumab and trastuzumab in women with locally advanced, inflammatory, or early HER2-positive breast cancer (NeoSphere): a randomised multicentre, open-label, phase 2 trial. *Lancet Oncol.*, **13**, 25–32.

207. Baselga, J., Cortes, J., Kim, S.B., Im, S.A. *et al.* (2012) Pertuzumab plus trastuzumab plus docetaxel for metastatic breast cancer. *N. Engl. J. Med.*, **366**, 109–119.

208. Perez, E.A., Romond, E.H., Suman, V.J., Jeong, J.H. *et al.* (2011) Four-year follow-up of trastuzumab plus adjuvant chemotherapy for operable human epidermal growth factor receptor 2-positive breast cancer: joint analysis of data from NCCTG N9831 and NSABP B-31. *J. Clin. Oncol.*, **29**, 3366–3373.

209. Bang, Y.J., Van Cutsem, E., Feyereislova, A., Chung, H.C. *et al.* (2010) Trastuzumab in combination with chemotherapy versus chemotherapy alone for treatment of HER2-positive advanced gastric or gastro-oesophageal junction cancer (ToGA): a phase 3, open-label, randomised controlled trial. *Lancet*, **376**, 687–697.

210. Bennouna, J., Sastre, J., Arnold, D., Osterlund, P. *et al.* (2013) Continuation of bevacizumab after first progression in metastatic colorectal cancer (ML18147): a randomised phase 3 trial. *Lancet Oncol.*, **14**, 29–37.

211. Hurwitz, H., Fehrenbacher, L., Novotny, W., Cartwright, T. *et al.* (2004) Bevacizumab plus irinotecan, fluorouracil, and leucovorin for metastatic colorectal cancer. *N. Engl. J. Med.*, **350**, 2335–2342.

212. Sandler, A., Gray, R., Perry, M.C., Brahmer, J. *et al.* (2006) Paclitaxel-carboplatin alone or with bevacizumab for non-small-cell lung cancer. *N. Engl. J. Med.*, **355**, 2542–2550.

213. Escudier, B., Pluzanska, A., Koralewski, P., Ravaud, A. *et al.* (2007) Bevacizumab plus interferon alfa-2a for treatment of metastatic renal cell carcinoma: a randomised, double-blind phase III trial. *Lancet*, **370**, 2103–2111.

214. Friedman, H.S., Prados, M.D., Wen, P.Y., Mikkelsen, T. *et al.* (2009) Bevacizumab alone and in combination with irinotecan in recurrent glioblastoma. *J. Clin. Oncol.*, **27**, 4733–4740.

215. Pujade-Lauraine, E., Hilpert, F., Weber, B., Reuss, A. *et al.* (2014) Bevacizumab combined with chemotherapy for platinum-resistant recurrent ovarian cancer: The AURELIA open-label randomized phase III trial. *J. Clin. Oncol.*, **32**(13), 1302–1308.

216. Tewari, K.S., Sill, M.W., Long, H.J. 3rd, Penson, R.T. (2014) Improved survival with bevacizumab in advanced cervical cancer. *N. Engl. J. Med.*, **370**(8), 734–743.

217. Van Cutsem, E., Tabernero, J., Lakomy, R., Prenen, H. *et al.* (2012) Addition of aflibercept to fluorouracil, leucovorin, and irinotecan improves survival in a phase III randomized trial in patients with metastatic colorectal cancer previously treated with an oxaliplatin-based regimen. *J. Clin. Oncol.*, **30**, 3499–3506.

218. Wilke, H., Muro, K., Van Cutsem, E., Oh, S.C. *et al.* (2014) RAINBOW Study Group. Ramucirumab plus paclitaxel versus placebo plus paclitaxel in patients with previously treated advanced gastric or gastro-oesophageal junction adenocarcinoma (RAINBOW): a double-blind, randomised phase 3 trial. *Lancet Oncol.* **15** (11), 1224–1235.

219. Fuchs, C.S., Tomasek, J., Yong, C.J., Dumitru, F. *et al.* (2014) REGARD Trial Investigators. Ramucirumab monotherapy for previously treated advanced gastric or gastro-oesophageal junction adenocarcinoma REGARD): an international, randomised, multicentre, placebo-controlled, phase 3 trial. *Lancet.* **383**(9911), 31–39.

220. Morschhauser, F., Radford, J., Van Hoof, A., Botto, B. *et al.* (2013) 90Yttrium-ibritumomab tiuxetan consolidation of first remission in advanced-stage follicular non-Hodgkin lymphoma: updated results after a median follow-up of 7.3 years from the International, Randomized, Phase III First-Line Indolent trial. *J. Clin. Oncol.*, **31**, 1977–1983.

221. Witzig, T.E., Gordon, L.I., Cabanillas, F., Czuczman, M.S. *et al.* (2002) Randomized controlled trial of yttrium-90-labeled ibritumomab tiuxetan radioimmunotherapy versus rituximab immunotherapy for patients with relapsed or refractory low-grade, follicular, or transformed B-cell non-Hodgkin's lymphoma. *J. Clin. Oncol.*, **20**, 2453–2463.

222. Kaminski, M.S., Zelenetz, A.D., Press, O.W., Saleh, M. *et al.* (2001) Pivotal

study of iodine I 131 tositumomab for chemotherapy-refractory low-grade or transformed low-grade B-cell non-Hodgkin's lymphomas. *J. Clin. Oncol.*, **19**, 3918–3928.

223. Pro, B., Advani, R., Brice, P., Bartlett, N. L. *et al.* (2012) Brentuximab vedotin (SGN-35) in patients with relapsed or refractory systemic anaplastic large-cell lymphoma: results of a phase II study. *J. Clin. Oncol.*, **30**, 2190–2196.
224. Younes, A., Gopal, A.K., Smith, S.E., Ansell, S.M. *et al.* (2012) Results of a pivotal phase II study of brentuximab vedotin for patients with relapsed or refractory Hodgkin's lymphoma. *J. Clin. Oncol.*, **30**, 2183–2189.
225. Verma, S., Miles, D., Gianni, L., Krop, I.E. *et al.* (2012) Trastuzumab emtansine for HER2-positive advanced breast cancer. *N. Engl. J. Med.*, **367**, 1783–1791.
226. Hillmen, P., Skotnicki, A.B., Robak, T., Jaksic, B. *et al.* (2007) Alemtuzumab compared with chlorambucil as first-line therapy for chronic lymphocytic leukemia. *J. Clin. Oncol.*, **25**, 5616–5623.
227. Henry, D.H., Costa, L., Goldwasser, F., Hirsh, V. *et al.* (2011) Randomized, double-blind study of denosumab versus zoledronic acid in the treatment of bone metastases in patients with advanced cancer (excluding breast and prostate cancer) or multiple myeloma. *J. Clin. Oncol.*, **29**, 1125–1132.
228. Chawla, S., Henshaw, R., Seeger, L., Choy, E. *et al.* (2013) Safety and efficacy of denosumab for adults and skeletally mature adolescents with giant cell tumour of bone: interim analysis of an open-label, parallel-group, phase 2 study. *Lancet Oncol.*, **14**, 901–908.
229. Pardoll, D.M. (2012) The blockade of immune checkpoints in cancer immunotherapy. *Nat. Rev. Cancer*, **12**, 252–264.
230. Bargou, R., Leo, E., Zugmaier, G., Klinger, M. *et al.* (2008) Tumor regression in cancer patients by very low doses of a T cell-engaging antibody. *Science*, **321**, 974–977.
231. Topp, M.S., Gökbuget, N., Zugmaier, G., Klappers, P. *et al.* (2014) Phase II Trial of the Anti-CD19 Bispecific T Cell-Engager Blinatumomab Shows Hematologic and Molecular Remissions in Patients with Relapsed or efractory B-Precursor Acute Lymphoblastic Leukemia. *J. Clin. Oncol.*
232. US National Institutes of Health ClinicalTrials.gov, *http://clinicaltrials.gov/* (accessed 8 December 2014).
233. (2013) Rituximab [package insert]. Genentech, South San Francisco, CA.
234. Du, J., Yang, H., Guo, Y., and Ding, J. (2009) Structure of the Fab fragment of therapeutic antibody Ofatumumab provides insights into the recognition mechanism with CD20. *Mol. Immunol.*, **46**, 2419–2423.
235. Itala, M., Geisler, C.H., Kimby, E., Juvonen, E. *et al.* (2002) Standard-dose anti-CD20 antibody rituximab has efficacy in chronic lymphocytic leukaemia: results from a Nordic multicentre study. *Eur. J. Haematol.*, **69**, 129–134.
236. Mossner, E., Brunker, P., Moser, S., Puntener, U. *et al.* (2010) Increasing the efficacy of CD20 antibody therapy through the engineering of a new type II anti-CD20 antibody with enhanced direct and immune effector cell-mediated B-cell cytotoxicity. *Blood*, **115**, 4393–4402.
237. Yarden, Y. and Sliwkowski, M.X. (2001) Untangling the ErbB signalling network. *Nat. Rev. Mol. Cell Biol.*, **2**, 127–137.
238. Burstein, H.J. (2005) The distinctive nature of HER2-positive breast cancers. *N. Engl. J. Med.*, **353**, 1652–1654.
239. Dixon, J.M., Wilson, V., Verrill, M., and Symmans, W.F. (2012) HER2 testing in patients with breast cancer. *Br. Med. J.*, **344**, e3958.
240. Hecht, J.R., Mitchell, E., Neubauer, M.A., Burris, H.A., III, *et al.* (2010) Lack of correlation between epidermal growth factor receptor status and response to Panitumumab monotherapy in metastatic colorectal cancer. *Clin. Cancer Res.*, **16**, 2205–2213.
241. Voigt, M., Braig, F., Gothel, M., Schulte, A. *et al.* (2012) Functional dissection of the epidermal growth factor receptor epitopes targeted by panitumumab and cetuximab. *Neoplasia*, **14**, 1023–1031.
242. Kim, R. (2009) Cetuximab and panitumumab: are they interchangeable? *Lancet Oncol.*, **10**, 1140–1141.
243. Fuentes, G., Scaltriti, M., Baselga, J., and Verma, C.S. (2011) Synergy between trastuzumab and pertuzumab for human epidermal growth factor 2 (Her2) from colocalization: an in silico based mechanism. *Breast Cancer Res.*, **13**, R54.

244. Suter, T.M., Procter, M., van Veldhuisen, D.J., Muscholl, M. *et al.* (2007) Trastuzumab-associated cardiac adverse effects in the herceptin adjuvant trial. *J. Clin. Oncol.*, **25**, 3859–3865.
245. Swain, S.M., Ewer, M.S., Cortes, J., Amadori, D. *et al.* (2013) Cardiac tolerability of pertuzumab plus trastuzumab plus docetaxel in patients with HER2-positive metastatic breast cancer in CLEOPATRA: a randomized, double-blind, placebo-controlled phase III study. *Oncologist*, **18**, 257–264.
246. Willett, C.G., Boucher, Y., di Tomaso, E., Duda, D.G. *et al.* (2004) Direct evidence that the VEGF-specific antibody bevacizumab has antivascular effects in human rectal cancer. *Nat. Med.*, **10**, 145–147.
247. Tanne, J.H. (2011) FDA cancels approval for bevacizumab in advanced breast cancer. *Br. Med. J.*, **343**, d7684.
248. GSK Press Release (2014) GSK to discontinue manufacture and sale of the BEXXAR® therapeutic regimen (tositumomab and iodine I 131 tositumomab). GSK, Aug. 07 2013 (Retrieved 26 January 2014). Available at: http://us.gsk.com/html/media-news/pressreleases/2013/GSK-to-discontinue-manufacture-and-sale-of-the-BEXXAR-therapeutic-regimen.html.
249. Cheson, B.D. (2003) Radioimmunotherapy of non-Hodgkin lymphomas. *Blood*, **101**, 391–398.
250. Falini, B., Pileri, S., Pizzolo, G., Durkop, H. *et al.* (1995) CD30 (Ki-1) molecule: a new cytokine receptor of the tumor necrosis factor receptor superfamily as a tool for diagnosis and immunotherapy. *Blood*, **85**, 1–14.
251. Ansell, S.M., Horwitz, S.M., Engert, A., Khan, K.D. *et al.* (2007) Phase I/II study of an anti-CD30 monoclonal antibody (MDX-060) in Hodgkin's lymphoma and anaplastic large-cell lymphoma. *J. Clin. Oncol.*, **25**, 2764–2769.
252. Vogel, C.L., Cobleigh, M.A., Tripathy, D., Gutheil, J.C. *et al.* (2002) Efficacy and safety of trastuzumab as a single agent in first-line treatment of HER2-overexpressing metastatic breast cancer. *J. Clin. Oncol.*, **20**, 719–726.
253. Atkins, M.B., Lotze, M.T., Dutcher, J.P., Fisher, R.I. *et al.* (1999) High-dose recombinant interleukin 2 therapy for patients with metastatic melanoma: analysis of 270 patients treated between 1985 and 1993. *J. Clin. Oncol.*, **17**, 2105–2116.
254. Ribas, A., Kefford, R., Marshall, M.A., Punt, C.J. *et al.* (2013) Phase III randomized clinical trial comparing tremelimumab with standard-of-care chemotherapy in patients with advanced melanoma. *J. Clin. Oncol.*, **31**, 616–622.
255. Brahmer, J.R., Tykodi, S.S., Chow, L.Q., Hwu, W.J. *et al.* (2012) Safety and activity of anti-PD-L1 antibody in patients with advanced cancer. *N. Engl. J. Med.*, **366**, 2455–2465.
256. (2014) PD-1 Inhibitor Approved for Melanoma. *Cancer Discov.*, **4**(11), 1249.
257. Wolchok, J.D., Kluger, H., Callahan, M.K., Postow, M.A. *et al.* (2013) Nivolumab plus ipilimumab in advanced melanoma. *N. Engl. J. Med.*, **369**, 122–133.
258. Weiden, P.L., Flournoy, N., Thomas, E.D., Prentice, R. *et al.* (1979) Antileukemic effect of graft-versus-host disease in human recipients of allogeneic-marrow grafts. *N. Engl. J. Med.*, **300**, 1068–1073.
259. Davis, M.M., Boniface, J.J., Reich, Z., Lyons, D. *et al.* (1998) Ligand recognition by alpha beta T cell receptors. *Annu. Rev. Immunol.*, **16**, 523–544.
260. Topalian, S.L., Muul, L.M., Solomon, D., and Rosenberg, S.A. (1987) Expansion of human tumor infiltrating lymphocytes for use in immunotherapy trials. *J. Immunol. Methods*, **102**, 127–141.
261. Dudley, M.E., Wunderlich, J.R., Shelton, T.E., Even, J. *et al.* (2003) Generation of tumor-infiltrating lymphocyte cultures for use in adoptive transfer therapy for melanoma patients. *J. Immunother.*, **26**, 332–342.
262. Riddell, S.R. and Greenberg, P.D. (1990) The use of anti-CD3 and anti-CD28 monoclonal antibodies to clone and expand human antigen-specific T cells. *J. Immunol. Methods*, **128**, 189–201.
263. Rosenberg, S.A., Spiess, P., and Lafreniere, R. (1986) A new approach to the adoptive immunotherapy of cancer with tumor-infiltrating lymphocytes. *Science*, **233**, 1318–1321.

264. Pockaj, B.A., Sherry, R.M., Wei, J.P., Yannelli, J.R. *et al.* (1994) Localization of 111indium-labeled tumor infiltrating lymphocytes to tumor in patients receiving adoptive immunotherapy. Augmentation with cyclophosphamide and correlation with response. *Cancer*, **73**, 1731–1737.
265. North, R. J. (1982) Cyclophosphamide-facilitated adoptive immunotherapy of an established tumor depends on elimination of tumor-induced suppressor T cells. *J. Exp. Med.*, **155**, 1063–1074.
266. Gattinoni, L., Finkelstein, S.E., Klebanoff, C.A., Antony, P.A. *et al.* (2005) Removal of homeostatic cytokine sinks by lymphodepletion enhances the efficacy of adoptively transferred tumor-specific CD8+ T cells. *J. Exp. Med.*, **202**, 907–912.
267. Rosenberg, S.A., Packard, B.S., Aebersold, P.M., Solomon, D. *et al.* (1988) Use of tumor-infiltrating lymphocytes and interleukin-2 in the immunotherapy of patients with metastatic melanoma. A preliminary report. *N. Engl. J. Med.*, **319**, 1676–1680.
268. Dudley, M.E., Wunderlich, J., Nishimura, M.I., Yu, D. *et al.* (2001) Adoptive transfer of cloned melanoma-reactive T lymphocytes for the treatment of patients with metastatic melanoma. *J. Immunother.*, **24**, 363–373.
269. Dudley, M.E., Wunderlich, J.R., Robbins, P.F., Yang, J.C. *et al.* (2002) Cancer regression and autoimmunity in patients after clonal repopulation with antitumor lymphocytes. *Science*, **298**, 850–854.
270. Dudley, M.E., Wunderlich, J.R., Yang, J.C., Sherry, R.M. *et al.* (2005) Adoptive cell transfer therapy following non-myeloablative but lymphodepleting chemotherapy for the treatment of patients with refractory metastatic melanoma. *J. Clin. Oncol.*, **23**, 2346–2357.
271. Robbins, P.F., Dudley, M.E., Wunderlich, J., El-Gamil, M. *et al.* (2004) Cutting edge: persistence of transferred lymphocyte clonotypes correlates with cancer regression in patients receiving cell transfer therapy. *J. Immunol.*, **173**, 7125–7130.
272. Dudley, M.E., Yang, J.C., Sherry, R., Hughes, M.S. *et al.* (2008) Adoptive cell therapy for patients with metastatic melanoma: evaluation of intensive myeloablative chemoradiation preparative regimens. *J. Clin. Oncol.*, **26**, 5233–5239.
273. Rosenberg, S.A., Yang, J.C., Sherry, R.M., Kammula, U.S. *et al.* (2011) Durable complete responses in heavily pretreated patients with metastatic melanoma using T-cell transfer immunotherapy. *Clin. Cancer Res.*, **17**, 4550–4557.
274. Morgan, R.A., Dudley, M.E., Wunderlich, J.R., Hughes, M.S. *et al.* (2006) Cancer regression in patients after transfer of genetically engineered lymphocytes. *Science*, **314**, 126–129.
275. Eshhar, Z. (2008) The T-body approach: redirecting T cells with antibody specificity. *Handb. Exp. Pharmacol.*, **329-342**.
276. Gross, G., Waks, T., and Eshhar, Z. (1989) Expression of immunoglobulin-T-cell receptor chimeric molecules as functional receptors with antibody-type specificity. *Proc. Natl Acad. Sci. USA*, **86**, 10024–10028.
277. Johnson, L.A., Morgan, R.A., Dudley, M.E., Cassard, L. *et al.* (2009) Gene therapy with human and mouse T-cell receptors mediates cancer regression and targets normal tissues expressing cognate antigen. *Blood*, **114**, 535–546.
278. Parkhurst, M.R., Yang, J.C., Langan, R.C., Dudley, M.E. *et al.* (2011) T cells targeting carcinoembryonic antigen can mediate regression of metastatic colorectal cancer but induce severe transient colitis. *Mol. Ther.*, **19**, 620–626.
279. Robbins, P.F., Morgan, R.A., Feldman, S.A., Yang, J.C. *et al.* (2011) Tumor regression in patients with metastatic synovial cell sarcoma and melanoma using genetically engineered lymphocytes reactive with NY-ESO-1. *J. Clin. Oncol.*, **29**, 917–924.
280. Morgan, R.A., Chinnasamy, N., Abate-Daga, D., Gros, A. *et al.* (2013) Cancer regression and neurological toxicity following anti-MAGE-A3 TCR gene therapy. *J. Immunother.*, **36**, 133–151.
281. Corr, M., Slanetz, A.E., Boyd, L.F., Jelonek, M.T. *et al.* (1994) T cell receptor-MHC class I peptide interactions: affinity, kinetics, and specificity. *Science*, **265**, 946–949.
282. Choi, E.M., Chen, J.L., Wooldridge, L., Salio, M. *et al.* (2003) High avidity antigen-specific CTL identified by CD8-independent tetramer staining. *J. Immunol.*, **171**, 5116–5123.
283. Hammarstrom, S. (1999) The carcinoembryonic antigen (CEA) family: structures,

suggested functions and expression in normal and malignant tissues. *Semin. Cancer Biol.*, **9**, 67–81.

284. Cameron, B.J., Gerry, A.B., Dukes, J., Harper, J.V. *et al.* (2013) Identification of a Titin-derived HLA-A1-presented peptide as a cross-reactive target for engineered MAGE A3-directed T cells. *Sci. Transl. Med.*, **5**, 197ra103.
285. Chinnasamy, N., Wargo, J.A., Yu, Z., Rao, M. *et al.* (2011) A TCR targeting the HLA-A*0201-restricted epitope of MAGE-A3 recognizes multiple epitopes of the MAGE-A antigen superfamily in several types of cancer. *J. Immunol.*, **186**, 685–696.
286. Kershaw, M.H., Westwood, J.A., Parker, L.L., Wang, G. *et al.* (2006) A phase I study on adoptive immunotherapy using gene-modified T cells for ovarian cancer. *Clin. Cancer Res.*, **12**, 6106–6115.
287. Park, J.R., Digiusto, D.L., Slovak, M., Wright, C. *et al.* (2007) Adoptive transfer of chimeric antigen receptor re-directed cytolytic T lymphocyte clones in patients with neuroblastoma. *Mol. Ther.*, **15**, 825–833.
288. Louis, C.U., Savoldo, B., Dotti, G., Pule, M. *et al.* (2011) Antitumor activity and long-term fate of chimeric antigen receptor-positive T cells in patients with neuroblastoma. *Blood*, **118**, 6050–6056.
289. Pule, M.A., Savoldo, B., Myers, G.D., Rossig, C. *et al.* (2008) Virus-specific T cells engineered to coexpress tumor-specific receptors: persistence and antitumor activity in individuals with neuroblastoma. *Nat. Med.*, **14**, 1264–1270.
290. Till, B.G., Jensen, M.C., Wang, J., Chen, E.Y. *et al.* (2008) Adoptive immunotherapy for indolent non-Hodgkin lymphoma and mantle cell lymphoma using genetically modified autologous CD20-specific T cells. *Blood*, **112**, 2261–2271.
291. Jensen, M.C., Popplewell, L., Cooper, L.J., DiGiusto, D. *et al.* (2010) Antitransgene rejection responses contribute to attenuated persistence of adoptively transferred CD20/CD19-specific chimeric antigen receptor redirected T cells in humans. *Biol. Blood Marrow Transplant.*, **16**, 1245–1256.
292. Lamers, C.H., Sleijfer, S., van Steenbergen, S., van Elzakker, P. *et al.* (2013) Treatment of metastatic renal cell carcinoma with CAIX CAR-engineered T cells: clinical evaluation and management of on-target toxicity. *Mol. Ther.*, **21**, 904–912.
293. Kowolik, C.M., Topp, M.S., Gonzalez, S., Pfeiffer, T. *et al.* (2006) CD28 costimulation provided through a CD19-specific chimeric antigen receptor enhances in vivo persistence and antitumor efficacy of adoptively transferred T cells. *Cancer Res.*, **66**, 10995–11004.
294. Kochenderfer, J.N., Dudley, M.E., Feldman, S.A., Wilson, W.H. *et al.* (2012) B-cell depletion and remissions of malignancy along with cytokine-associated toxicity in a clinical trial of anti-CD19 chimeric-antigen-receptor-transduced T cells. *Blood*, **119**, 2709–2720.
295. Kochenderfer, J.N., Wilson, W.H., Janik, J.E., Dudley, M.E. *et al.* (2010) Eradication of B-lineage cells and regression of lymphoma in a patient treated with autologous T cells genetically engineered to recognize CD19. *Blood*, **116**, 4099–4102.
296. Kalos, M., Levine, B.L., Porter, D.L., Katz, S. *et al.* (2011) T cells with chimeric antigen receptors have potent antitumor effects and can establish memory in patients with advanced leukemia. *Sci. Transl. Med.*, **3**, 95ra73.
297. Brentjens, R.J., Riviere, I., Park, J.H., Davila, M.L. *et al.* (2011) Safety and persistence of adoptively transferred autologous CD19-targeted T cells in patients with relapsed or chemotherapy refractory B-cell leukemias. *Blood*, **118**, 4817–4828.
298. Savoldo, B., Ramos, C.A., Liu, E., Mims, M.P. *et al.* (2011) CD28 costimulation improves expansion and persistence of chimeric antigen receptor-modified T cells in lymphoma patients. *J. Clin. Invest.*, **121**, 1822–1826.
299. Morgan, R.A., Yang, J.C., Kitano, M., Dudley, M.E. *et al.* (2010) Case report of a serious adverse event following the administration of T cells transduced with a chimeric antigen receptor recognizing ERBB2. *Mol. Ther.*, **18**, 843–851.
300. Till, B.G., Jensen, M.C., Wang, J., Qian, X. *et al.* (2012) CD20-specific adoptive immunotherapy for lymphoma using a chimeric antigen receptor with both CD28

and 4-1BB domains: pilot clinical trial results. *Blood*, **119**, 3940–3950.
301. Maude, S.L., Frey, N., Shaw, P.A., Aplenc, R. *et al.* (2014) Chimeric antigen receptor T cells for sustained remissions in leukemia. *N. Engl. J. Med.* **371**(16), 1507–1517.
302. Budde, L.E., Berger, C., Lin, Y., Wang, J. *et al.* (2013) Combining a CD20 chimeric antigen receptor and an inducible caspase 9 suicide switch to improve the efficacy and safety of T cell adoptive immunotherapy for lymphoma. *PLoS One*, **8**, e82742.
303. Scholler, J., Brady, T.L., Binder-Scholl, G., Hwang, W.T. *et al.* (2012) Decade-long safety and function of retroviral-modified chimeric antigen receptor T cells. *Sci. Transl. Med.*, **4**, 132ra153.
304. Vacchelli, E., Vitale, I., Eggermont, A., Fridman, W.H. *et al.* (2013) Trial watch: dendritic cell-based interventions for cancer therapy. *Oncoimmunology*, **2**, e25771.
305. Hsu, F.J., Benike, C., Fagnoni, F., Liles, T.M. *et al.* (1996) Vaccination of patients with B-cell lymphoma using autologous antigen-pulsed dendritic cells. *Nat. Med.*, **2**, 52–58.
306. Fong, L., Brockstedt, D., Benike, C., Wu, L. *et al.* (2001) Dendritic cells injected via different routes induce immunity in cancer patients. *J. Immunol.*, **166**, 4254–4259.
307. Fong, L., Hou, Y., Rivas, A., Benike, C. *et al.* (2001) Altered peptide ligand vaccination with Flt3 ligand expanded dendritic cells for tumor immunotherapy. *Proc. Natl Acad. Sci. USA*, **98**, 8809–8814.
308. Furumoto, K., Soares, L., Engleman, E.G., and Merad, M. (2004) Induction of potent antitumor immunity by in situ targeting of intratumoral DCs. *J. Clin. Invest.*, **113**, 774–783.
309. Saji, H., Song, W., Furumoto, K., Kato, H. *et al.* (2006) Systemic antitumor effect of intratumoral injection of dendritic cells in combination with local photodynamic therapy. *Clin. Cancer Res.*, **12**, 2568–2574.
310. Merad, M., Sugie, T., Engleman, E.G., and Fong, L. (2002) In vivo manipulation of dendritic cells to induce therapeutic immunity. *Blood*, **99**, 1676–1682.
311. Okano, F., Merad, M., Furumoto, K., and Engleman, E.G. (2005) In vivo manipulation of dendritic cells overcomes tolerance to unmodified tumor-associated self antigens and induces potent antitumor immunity. *J. Immunol.*, **174**, 2645–2652.
312. Bonifaz, L.C., Bonnyay, D.P., Charalambous, A., Darguste, D.I. *et al.* (2004) In vivo targeting of antigens to maturing dendritic cells via the DEC-205 receptor improves T cell vaccination. *J. Exp. Med.*, **199**, 815–824.
313. van Broekhoven, C.L., Parish, C.R., Demangel, C., Britton, W.J. *et al.* (2004) Targeting dendritic cells with antigen-containing liposomes: a highly effective procedure for induction of antitumor immunity and for tumor immunotherapy. *Cancer Res.*, **64**, 4357–4365.
314. Viaud, S., Thery, C., Ploix, S., Tursz, T. *et al.* (2010) Dendritic cell-derived exosomes for cancer immunotherapy: what's next? *Cancer Res.*, **70**, 1281–1285.
315. Butterfield, L.H. (2013) Dendritic cells in cancer immunotherapy clinical trials: are we making progress? *Front. Immunol.*, **4**, 454.
316. Mailliard, R.B., Wankowicz-Kalinska, A., Cai, Q., Wesa, A. *et al.* (2004) alpha-type-1 polarized dendritic cells: a novel immunization tool with optimized CTL-inducing activity. *Cancer Res.*, **64**, 5934–5937.
317. Palucka, K. and Banchereau, J. (2012) Cancer immunotherapy via dendritic cells. *Nat. Rev. Cancer*, **12**, 265–277.
318. Kvistborg, P., Boegh, M., Pedersen, A.W., Claesson, M.H. *et al.* (2009) Fast generation of dendritic cells. *Cell. Immunol.*, **260**, 56–62.
319. Nicolette, C.A., Healey, D., Tcherepanova, I., Whelton, P. *et al.* (2007) Dendritic cells for active immunotherapy: optimizing design and manufacture in order to develop commercially and clinically viable products. *Vaccine*, **25** (Suppl. 2), B47–B60.
320. Bol, K.F., Tel, J., de Vries, I.J., and Figdor, C.G. (2013) Naturally circulating dendritic cells to vaccinate cancer patients. *Oncoimmunology*, **2**, e23431.
321. Berntsen, A., Brimnes, M.K., Thor Straten, P., and Svane, I.M. (2010) Increase of circulating CD4 + CD25highFoxp3+ regulatory T cells in patients with metastatic renal cell carcinoma during treatment with dendritic cell vaccination and low-dose interleukin-2. *J. Immunother.*, **33**, 425–434.
322. Markowicz, S. and Engleman, E.G. (1990) Granulocyte-macrophage colony-stimulating factor promotes differentiation and survival of human peripheral blood

dendritic cells in vitro. *J. Clin. Invest.*, **85**, 955–961.

323. Sheikh, N.A. and Jones, L.A. (2008) CD54 is a surrogate marker of antigen presenting cell activation. *Cancer Immunol. Immunother.*, **57**, 1381–1390.

324. Sheikh, N.A., Petrylak, D., Kantoff, P.W., Dela Rosa, C. *et al.* (2013) Sipuleucel-T immune parameters correlate with survival: an analysis of the randomized phase 3 clinical trials in men with castration-resistant prostate cancer. *Cancer Immunol. Immunother.*, **62**, 137–147.

325. Ridolfi, L., Petrini, M., Fiammenghi, L., Granato, A.M. *et al.* (2010) Unexpected high response rate to traditional therapy after dendritic cell-based vaccine in advanced melanoma: update of clinical outcome and subgroup analysis. *Clin. Dev. Immunol.*, **2010**, 504979.

326. Ribas, A., Comin-Anduix, B., Chmielowski, B., Jalil, J. *et al.* (2009) Dendritic cell vaccination combined with CTLA4 blockade in patients with metastatic melanoma. *Clin. Cancer Res.*, **15**, 6267–6276.

Part II
Molecular Diagnostics and Imaging

Translational Medicine Cancer: Advances in Molecular Biology and Medicine,
First Edition. Edited by Robert A. Meyers.

11
DNA in Neoplastic Disease Diagnosis

Eric Q. Konnick[1,2], Cecilia C.S. Yeung[2,3], and David Wu[1]
[1] University of Washington Medical Center, Department of Laboratory Medicine, Box 357110, 1959 NE Pacific Street, Seattle, WA 98195-7110, USA
[2] University of Washington Medical Center, Department of Pathology, Box 357470, 1959 NE Pacific Street, Seattle, WA 98195-7470, USA
[3] Fred Hutchinson Cancer Research Center, Clinical Research Division, G7910, P.O. Box 19024, Seattle, WA 98109-1024, USA

Translational Medicine Cancer: Advances in Molecular Biology and Medicine,
First Edition. Edited by Robert A. Meyers.

Keywords

Benign tumor
A neoplasm that is capable of dysregulated growth, but is not capable of invasion of adjacent tissues or metastatic spread. May cause local mass effect and/or organ compromise that can lead to systemic consequences, including death.

Cytogenetics
The study of chromosomes and chromosome abnormalities.

Deoxyribonucleic acid (DNA)
A polymer of monosaccharides with a phosphate group and one of several nitrogen-containing, complementary nucleobases. The monosaccharides polymerize to form long chains, and are paired with complementary polymer strands. DNA is transcribed to messenger RNA by RNA polymerases. Messenger RNA is translated into proteins by ribosomes.

DNA microarray
A high-throughput multiplex assay where numerous spots of DNA probes adherent to a solid substrate are hybridized to a sample nucleic acid and read in parallel to generate massive amounts of data through various detection methods.

DNA sequencing
The process used to identify the precise order of nucleotides within a DNA molecule.

Epigenetics
Heritable changes in gene expression that are not caused by changes in the DNA sequence, often due to the addition or removal of specific chemical signals from the DNA backbone, such as methylation or acetylation.

Karyotype
Description of the number and appearance of chromosomes.

Malignancy
Descriptive term for neoplasia which exhibits invasion, metastasis, or compromise of other organ functions, that may lead to death.

Metastasis
The spread of a neoplasm from one organ or part to another non-adjacent organ or part of the body. Metastasis can occur via the lymphatic or hematogenous (blood) spread of viable, neoplastic cells, or can occur through seeding of the fluid within the visceral spaces (abdominal or pleural cavities).

Neoplasia
Abnormal growth or division of cells, which is not coordinated with the surrounding tissues.

Polymerase chain reaction (PCR)
A method for amplifying specific sequences of DNA. It requires short oligonucleotide primers, a thermostable DNA polymerase, and deoxynucleotide triphosphates (dNTPs), which are combined in a buffer and placed in an instrument that automates the sequential heating and cooling of the reaction mixture at precise temperatures. With each cycle, under optimal conditions, the amount of target DNA will be doubled. The amplified DNA sequence can then be identified using one of several methods.

RNA
Ribonucleic acid; transcribed from DNA, with a similar structure, but with a different sugar–backbone structure. RNA molecules are used for many different functions, including messenger RNA (mRNA), noncoding RNA, transfer RNA (tRNA), ribosomal RNA (rRNA), and microRNA.

Diagnosis of neoplasia has undergone a transformation during the past 50 years with the discovery of DNA as the carrier of genetic information and insight into the central role of DNA in many neoplastic processes. Techniques developed to aid in the diagnostic process have become key tools for the clinical evaluation of specific neoplastic entities. As much as previous generations of diagnostic methods have transformed pathology, new tools and techniques are rapidly making their way into clinical laboratories and show promise to reshape the role of diagnostic medicine.

1 History of DNA Detection in Neoplastic Diagnosis

Neoplasia, both malignant and benign, has been recognized for thousands of years, although the underlying process has been a mystery for most of human history. In fact, the origin of the word cancer is the Greek *karkinos* (Crab), describing the infiltrating and tenacious nature of this process [1, 2]. Associations have long been made showing that individual families can have a predisposition to specific diseases (e.g., hemophilia) [3, 4], and that some cancers appear to have an inheritable component [5]. Other observations showed that some unique cancers were associated with specific professions, such as scrotal squamous cell carcinoma in young chimney sweeps [6]. These observations suggested that there were both environmental and, occasionally, inheritable factors which influenced the development of cancers, but the underlying causes were still unknown.

DNA was established as the carrier of genetic material in the 1940s [7], and the inheritance theory devised by Mendel [8] allowed for an understanding of inheritance in genetic quanta. Even before the discovery of staining methods to identify specific chromosomal alterations, some pioneers had noted that abnormal mitoses were present in cancers [9], suggesting a role in neoplasia. The identification of the structure of DNA [10] and the subsequent determination of the genetic code [11] allowed scientists to understand how changes at the nucleotide level could result in protein changes, though the association with neoplasia was not clear.

During the mid-twentieth century, the identification of a specific translocation between chromosomes 9 and 22 (the Philadelphia chromosome) and its association with chronic myelogenous leukemia (CML) demonstrated that specific chromosomal alterations were associated with specific neoplastic conditions [12]. The study and enumeration of the genetic complement using stained metaphase chromosomes became known as the karyotype, and the field of cytogenetics was formed around the evaluation of such genetic methods. Chromosome visualization technology continued to advance, and new staining methods [13, 14] allowed for other recurrent chromosomal abnormalities to be associated with other neoplastic conditions [15, 16], as well as suggesting possible prognostic implications of some of these rearrangements [17]. How the underlying changes resulted in neoplasia was still largely unknown.

1.1
Beginning of the DNA Era

Discoveries made during the early 1970s allowed scientists to use restriction enzymes from bacteria to cleave DNA in a reproducible manner that allowed the identification of changes in the DNA sequence if the change occurred at one of these sites [18]. This discovery not only allowed interrogation of genomes but also laid the groundwork for future technologies, including DNA cloning, DNA sequencing, and *in-situ* hybridization. The identification of recurrent translocations in many neoplasms offered the opportunity for the identification of these genetic structural rearrangements by molecular methods. Initially, karyotyping was the test of choice, but was time-consuming, laborious, and required viable neoplastic cells. The development of molecular cloning allowed for the introduction of *in-situ* hybridization methods, first with radioactive elements [19] and later fluorescence-based [20], which allowed for a rapid and specific identification of relatively precise chromosomal alterations. During the same time period, techniques to determine the sequence of DNA bases were developed by Frederick Sanger and colleagues, using radiolabeled nucleotides [21–23]. This technology allowed the determination of each subsequent base in a sequence by incorporating nucleotides – specifically di-deoxynucleotides – that prevented any further elongation of the DNA molecule and could be visualized on an autoradiographic polyacrylamide gel after size separation. This process was laborious and time-consuming, but was an important

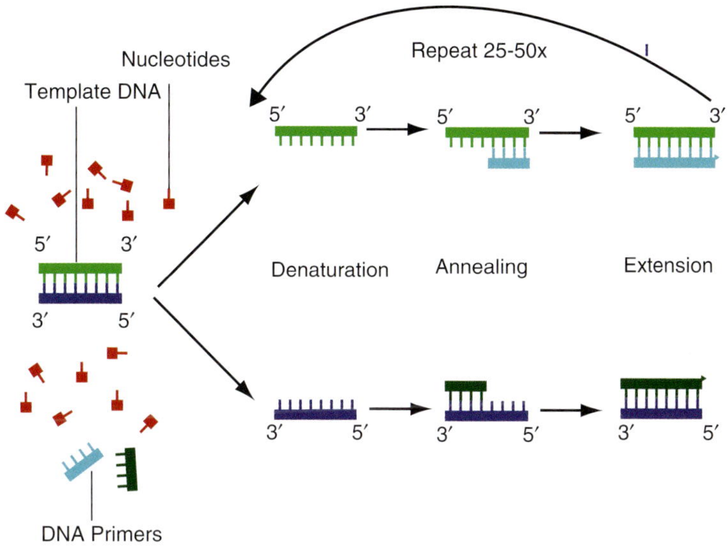

Fig. 1 Schematic showing the general principle of the polymerase chain reaction. Template DNA is incubated with synthetic oligonucleotide primers specific for the sequence of interest, with deoxynucleotide triphosphates (dNTPs), and a thermostable *Thermus aquaticus* (*taq*) polymerase. Template DNA is denatured using a high temperature, and the reaction mixture is cooled to allow annealing of the oligonucleotide primers. Primers are extended by *taq* polymerase. The process is repeated, with the amount of template DNA effectively doubling for each reaction cycle.

step in understanding the sequence and structure of genes. Other advances were soon to follow. The polymerase chain reaction (PCR) was developed during the mid-1980s by Kary B. Mullis (Fig. 1) [24], with clinical uses published almost immediately [25], and the use of thermostable enzymes described soon after [26, 27]. The PCR allowed for a rapid amplification of specific nucleotide sequences and, when combined with restriction enzyme technology (restriction fragment length polymorphism; RFLP), offered a relatively quick and simple way to identify point mutations in neoplastic tissue [28]. During the 1980s and 1990s, improvements in PCR technology, Sanger sequencing methods (Fig. 2), improved computing power, and ancillary techniques allowed for an increased understanding of the role of genes in disease. During this time period, numerous genes were associated with specific neoplastic conditions [29, 30]. These technological advancements led to a rapid implementation of PCR-based assays in clinical laboratories for well-known genetic predispositions [31], infectious diseases associated with cancers [32, 33], and characteristic alterations of specific neoplasms [34, 35]. PCR-based assays were also described as a possible method of residual disease detection when the genetic abnormalities were characteristic for a given disease [36].

1.2 Human Genome Project

During the 1990s and early 2000s, Sanger sequencing continued its rapid advancement, with the transition from slab-gel to capillary-gel electrophoresis-based methods [37]. The drive to complete the Human

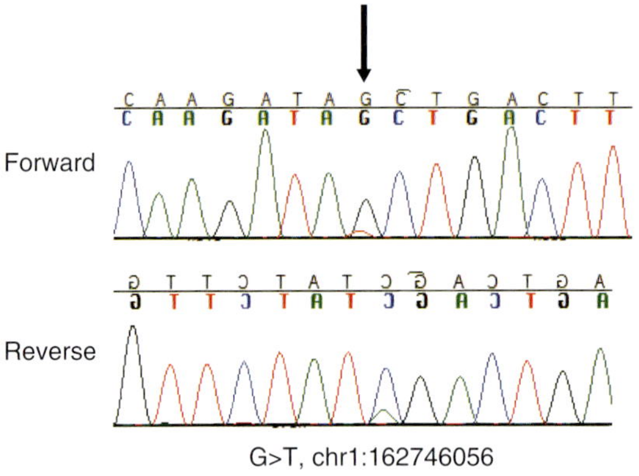

Fig. 2 Screen capture of Sanger sequencing using fluorescently labeled dideoxynucleotide terminators. In this method, a single reaction tube is required for each sequencing reaction, which includes dideoxynucleotide terminators labeled with different colored fluorophores. After the sequencing reaction, the sequencing product is processed and placed on an automated instrument which separates the differently sized DNA fragments using an electric field applied to a polyacrylamide gel or a sequencing capillary. A laser and fluorescent detector are positioned at a fixed point along the electrophoresis apparatus, and DNA bases are interrogated for fluorescence as the DNA fragment passes the laser/detector. These data are represented by the electropherogram of peaks of different colors. Calling of individual bases can be automated to increase throughput. A nucleotide G-to-T substitution is identified in a minority of the sequencing reads (~15–20%) at the highlighted position.

Genome Project [38] resulted in innovations in robotics, automation, and informatics [39] that were soon incorporated into clinical laboratories. During this time, DNA sequencing technology improved and DNA sequences of 500–1000 bp became possible using Sanger chemistry. The scope of this project was massive, using highly robotized workflows at multiple centers across the world. During the late 1990s, the government-run effort was challenged by a private consortium using a novel approach of "shotgun sequencing," which sheared DNA into random fragments and then used techniques to attach primers and sequence the intervening DNA [40, 41]. This technique relied heavily on the ability of computers to reassemble the DNA sequence by identifying regions that overlapped with one another. This situation led to a competition between the groups, which evolved into a grand collaboration resulting in the first draft of the human genome release in 2001 [42, 43].

2 Evolution of Massively Parallel Sequencing

The first descriptions of massively parallel sequencing were published in the late 1990s and mid 2000s [44, 45]. Although the specific implementations were different, what they held in common were the ability to sequence millions of short DNA sequence reads, in parallel, without requiring knowledge of the sequence, a priori. This technology has been termed massively parallel or "next-generation sequencing" (NGS). These new techniques relied on fragmentation of the DNA to be sequenced and the ligation of a common

DNA sequence which would allow amplification and enrichment of the sequences after a single DNA molecule isolation step (analogous to shotgun sequencing). The sequencing was then performed in a parallel fashion, with millions of simultaneous, short DNA segments that could be sequenced simultaneously, with the process generating massive amounts of data. These data then required assembly and computational alignment against the reference human genome, using powerful computer algorithms. Since the initial description, the technology has evolved and multiple different methods have been introduced and commercialized. There has been a similar rapid adoption of this technology in research and diagnostic laboratories, which will likely continue as the cost of the instruments and reagents continues rapidly to decline.

While there are numerous permutations of the massively parallel sequencing model, several methods have enjoyed early success in research and clinical laboratories. These individual methods include sequencing by synthesis, pyrosequencing, sequencing by ligation, and semiconductor sequencing [46, 47]. This list is likely to change with time and will certainly evolve, as evidenced by the advancements of third-generation technologies such as Pacific Biosystems and nanopore technologies. Their descriptions offer insight into the overall technique, and highlight some strengths and weaknesses of the methods. Descriptions of the dominant technologies in this field are included below; however, the aim of this chapter is to present a survey of the technology as of early 2015, and not an exhaustive evaluation of all approaches that are, or may be, available in the near future (Table 1). In general, these techniques have the following steps in common: isolation of genetic material; shearing of the genetic material with subsequent ligation of adapter molecules with known sequence ("barcodes"); enrichment of the sequence of interest (hybrid-capture and/or PCR); sequencing; data analysis; and variant identification. The incorporation of sequence-tagged primers, allowing for the unique identification of multiplexed samples, was an innovation that dramatically reduced the cost of obtaining sequence information compared to previous technologies.

A key component of NGS is the bioinformatics required to analyze the sequencing data and annotate the observed variants. These data analyses pipelines consist of computer programs, which de-convolute the multiplexed sequences, assemble short sequences into longer composites, align the fragments of DNA sequence to a template (Fig. 3), identify variants from the reference sequence, and detect single nucleotide variants, insertion/deletions, and other structural and copy number changes [51]. Although the individual pipeline components are constantly evolving and changing, a major factor is the computing and storage infrastructure necessary to support NGS operation and data interpretation [52].

The raw data generated from individual experiments can currently exceed a terabyte in storage capacity, and processed files can consume hundreds of gigabytes. The long-term storage of these data files can challenge storage networks and can consume resources in excess of generating the experimental data [53]. These factors require thoughtful evaluation to optimize data analysis and storage, and will require new training considerations as molecular pathologists with the aid of bioinformaticians provide clinical-grade NGS tests [54, 55]. The depth and complexity of these data may further require new information systems to facilitate clinical utility and

Tab. 1 Comparison of nucleic acid sequencing technologies.

Method	*Manufacturer*	*Read length (bp)*	*Accuracy (%)*	*Time per run*	*Cost per 1 million bases (in US$)*	*Advantages*	*Disadvantages*
Single-molecule real-time sequencing	Pacific Biosciences	10 000–15 000	85.0	30 min to 4 h	$0.13–$0.60	Longest read length and fast	Low accuracy, moderate throughput, and expensive equipment
Ion semiconductor	Ion Torrent	~400	98.0	2 h	$1	Less expensive equipment and fast	Homopolymer errors
Pyrosequencing	454	700	99.9	24 h	$10	Long read size and fast	Homopolymer errors and cost
Sequencing by synthesis	Illumina	100–300	98.0	Varies (1–10 d)	$0.07	Potential for high sequence yield, depending upon sequencer model and desired application	Expensive equipment and short read lengths
Sequencing by ligation	SOLiD	50 + 35 or 50 + 50	99.9	1–2 wk	$0.13	Accuracy, low cost per base	Short read assembly, difficulties with palindromes, slow
Chain termination	Sanger	Up to 1 000	99.9	20 min to 3 h	$2400	Long individual reads and useful for many applications	High cost, low throughput
Nanopore	Oxford Nanopore	Up to 6 000	92–95	Hours	Unknown	Compact instrumentation, long read lengths	Lower accuracy, unknown cost

This table represents a best estimate of the listed parameters at the time of submission. The stated parameters are estimates based on published data [48–50] and the personal experience of the authors.

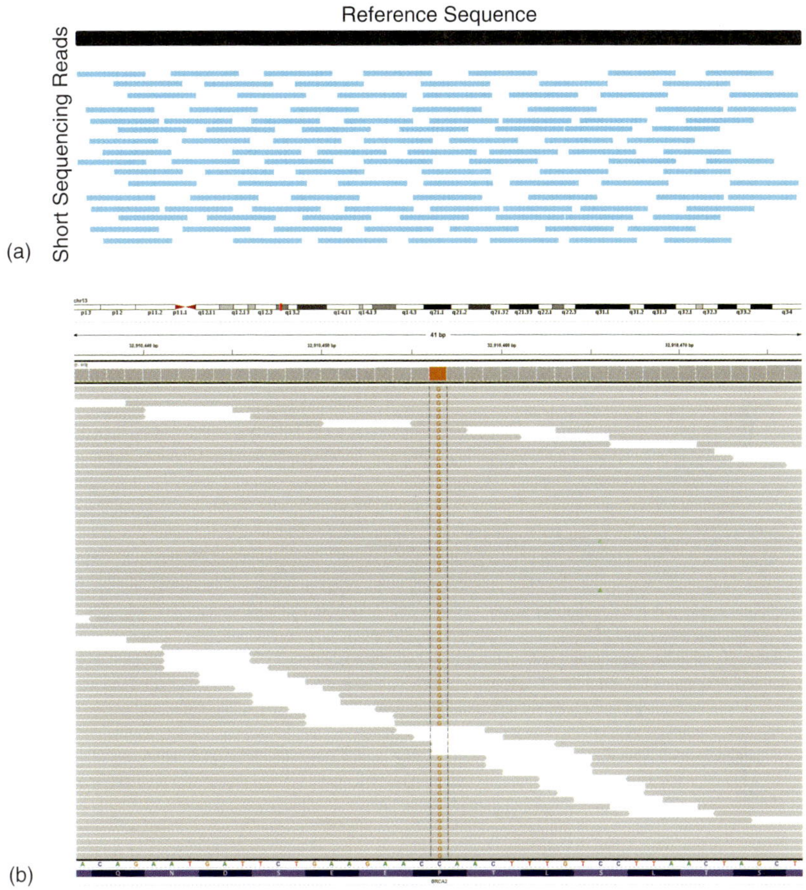

Fig. 3 (a) Schematic showing the alignment of numerous short DNA sequencing reads to a template. Note the overlap in sequence from multiple short-read length fragments. When numerous fragments overlap at a given position, low-frequency variants may be identified in a small subset of the reads. Similarly, when there is low coverage of a given region of DNA, rare variants may not be detected because they represent the minor population of the sample; (b) Screen capture of clinical next-generation sequence data as displayed using the Integrated Genomics Viewer (Broad Institute, Cambridge, MA, USA). The highlighted base is a variant that is identified in the *BRCA2* gene.

reporting [56]. It is also important to realize that the physicians who are most likely to use NGS technology in the setting of neoplasia have a wide level of prior knowledge. Even at tertiary-care medical centers, decision support systems are a necessary application to aid in the interpretation of genomic tests results and treatment decisions [57, 58].

2.1 Descriptions of Second-Generation (Next-Generation) Sequencing (NGS) Approaches

2.1.1 Sequencing by Synthesis (Illumina)

The sequencing by synthesis method is a hybrid between fluorescent di-deoxy Sanger sequencing and solid-surface sequencing by synthesis, which was a novel approach to allow for massively parallel sequencing [59, 60]. The key innovation in this technology was the use of fluorescently labeled, reversible di-deoxy terminators [61] which allows for a decoupling of the chemical reaction and the acquisition of base identity. In this approach, DNA fragments with ligated adapters are attached to a glass substrate ("flow cell") such that individual molecules are spatially separated. The attached sequences are then amplified in place such that products of the reaction are attached to the substrate, forming a clonal cluster of identical DNA fragments. Fluorescently tagged, reversible terminators are added, and a digital camera is used to determine the identity of the incorporated base in each DNA cluster, based on the incorporated fluorescent signal. The reversible terminator is then cleaved, and the process is repeated. One key advantage of this technology was the decoupling of the enzymatic reaction from the image acquisition, which could allow for large arrays of DNA to be interrogated by a single camera. This method has also been suggested to have the highest throughput per run and lowest error rates [62]. Hardware, chemistry, and fluidics factors have been described as limiting the read length and accuracy of the method [63], but continuous hardware and software updates have improved performance over time. This method also shows decreased coverage with increased GC content [64–66] and may be susceptible to bias introduced at the PCR stage [67].

2.1.2 Semiconductor-Based (Ion Torrent)

In this method, minute reaction wells with a semiconductor at the base are flooded with droplets containing single strands of DNA [68]. The reaction plate is sequentially flooded with a single type of nucleotide. If a nucleotide is incorporated into a nascent strand of DNA, a hydrogen ion is released upon nucleotide addition, and the highly sensitive ion sensor in the instrument can detect this ion. In the case of a homopolymer repeat stretch, multiple nucleotides will be incorporated in a single cycle. This leads to a corresponding number of released hydrogens and a proportionally higher electronic signal. This feature of the method can result in a large number of false-positive insertion/deletion (indel) calls caused by the high frequency of homopolymer sequencing errors. Multiple methods have been devised to decrease the rate of false-positive calls by the bioinformatics pipelines commonly used, which may be beneficial in the setting of clinical testing [62, 69, 70]. Semiconductor NGS methods have been compared to other platforms and found to have a higher throughput than other chemistries [62].

2.1.3 Long-Read, Single-Molecule Sequencing Approaches ("Third-Generation Sequencing")

The majority of NGS sequencing technologies require amplification of the nucleic acid sequences of interest, which can result in sequencing artifacts and bias of PCR reactions which can result in incorrect or incomplete data. The short lengths of NGS sequencing reads in the currently most commonly used methods also makes interpretation of structural rearrangements and pseudogenes difficult, with complicated computational algorithms required to infer the presence of such alterations [71–73]. Several methods which are able to sequence individual nucleic acids to read lengths in the kilobase range or greater have been developed. Fewer numbers of long reads at lower coverage have been demonstrated to improved assemblies compared to short-read sequencing at higher depths [74, 75].

Pacific Biosciences RS (PacBio) Platform Despite the great gains provided by second-generation sequencing methods and instrumentation, they have several drawbacks. First, they require amplification of source DNA before sequencing, leading to amplification artifacts and biased coverage of the genome related to the chemical–physical properties of the DNA. Second, current technologies produce relatively short reads, with median lengths of 100 bp for Illumina and ~700 bp for 454. Short reads make assembly and related analyses difficult, especially in regions of high homology on in genes that have pseudogenes [76, 77]. Pacific Biosciences (PacBio) has developed a single-molecule sequencing method that is carried out in wells on a chip, which is called a single-molecule, real-time (SMRT) cell [78]. The technology utilized is based on the concept of sequencing by synthesis in a zero-mode waveguide [79]. Zero-mode waveguides are small wells with an unmodified polymerase attached to the bottom the well, and with the associated detectors also located at the well bottom. Fluorescently labeled nucleotides are free in solution, and as nucleotides are added the fluorescence at the point of incorporation is detected. After incorporation, the fluorescent label is removed and the process is repeated.

The PacBio method does not require amplification and reduces compositional bias [80], and has also been reported to produce nucleotide sequences in excess of tens of thousands of base pairs [81] with a rapid time to sequence data from extracted DNA [82]. These reported performance specifications have promise in resolving complex structural rearrangements and large deletions [83, 84], as well as resolving areas within the genome that have proved structurally or diagnostically problematic, such as pseudogenes [85]. Such techniques may also be useful in determining if a genetic alterations are in cis or trans, which could be useful in the setting of neoplasia and multiple mutations present within a single gene.

While the long read lengths offer many potential advantages, the reported accuracy of the sequencing reads is currently much less than that of other technologies. The reported nucleotide call accuracy is only ~85% [81, 82], with uniformly distributed errors dominated by point insertions and deletions. Accuracy can be improved by combining data from different methods in algorithmic approaches [78, 86], but such approaches require multiple technologies and increased costs. The accuracy of the sequencing data can also be improved using an approach called "circular consensus sequencing" (CCS), which uses multiple passes across an insert molecule to assemble a more accurate sequence, but at the expense of read length [87, 88]. Method development is ongoing.

Nanopore Technologies Nanopore sequencing of DNA is a technique of interrogating single molecules, and has the promise of minimal sample preparation, long sequencing reads, low cost, and high speed with compact instrumentation [89]. These methods are based on the techniques that pass DNA molecules through artificial nanoscale pores composed of organic or nonorganic molecules, and can be detected based on electrical or other signals that are generated during passage [90–93]. In most methods, DNA passing through the nanopore changes its ion current, with multiple parameters impacting the measured current, and each nucleotide impacting the ion flow in such a way that the sequence can be determined [94].

The current nanopore sequencing schemes in development differ primarily in the construction of the pores and the technology needed to detect the nucleotide sequences. The two general classes of pore technology are solid-state- and protein-based nanopore sequencing. Protein nanopore sequencing utilizes membrane protein complexes alpha-hemolysin and *Mycobacterium smegmatis* porin A (MspA) [90, 91, 94], which show great promise given their ability to distinguish between individual and groups of nucleotides. Solid-state nanopore sequencing utilizes synthetic materials, which offers the possibility of custom pore configurations for different applications as well as stable mechanical, thermal, and chemical properties. The competing nanopore technologies offer different performance characteristics and are still undergoing active development by different groups.

The concept of nanopore sequencing originated from the idea that single-stranded DNA or RNA molecules can be driven in a strict linear sequence through a biological pore by electrophoresis or other mechanisms, and can be detected by a change in ionic current while moving through the pore. Pores are constructed to contain a detection region that is able recognize different bases, with each base generating specific signals corresponding to the sequence. A precise control of nucleic acid transport through the pore is critical for accurate measurement, and numerous enzymes such as exonucleases and polymerases have been employed for this purpose [91].

The read lengths for some applications have been shown to be ~5–6 kb [95], but error rates for the technology have been shown to range between 5% and 8% for insertions deletions, and substitutions for some methods [48]. Some evaluations of early instruments and chemistries have indicated that only a fraction of the sequencing reads are able to be mapped to the reference sequence of a known sequencing target [96]. These factors limit the current standalone, direct utility of the technologies in the diagnostic setting, but may be used as a method to generate a scaffold sequence that can be combined with other sequencing methods. Some analyses have suggested that accuracy can be improved through increasing the read coverage of individual sequenced molecules [97]. The methods may also require bioinformatics tools specific to nanopore sequencing [98].

Third-Generation Sequencing Summary The current technologies for single-molecule, long-read nucleic acid sequencing promise several features which are attractive to diagnostic applications in neoplastic disease. The ability for sequencing reads up to tens of kilobases in length would be useful for the determination of structural rearrangements such as insertions, deletions, and translocations that are common in neoplasia, and could resolve the phase of compound mutations in a gene of interest. The current use of the techniques is limited by the high error rates across multiple types of variation, with resulting mapping difficulties. Techniques that combine different sequencing methods have use in applications such as the *de-novo* assembly of genomes and neoplasia research, but are currently cumbersome for routine diagnostic use. Strategies have been developed to increase the read depth of individually sequenced molecules, often at the expense of read length, which dilutes one of the main advantages of these techniques. Current methods are undergoing continuous improvement and refinement which will likely address shortcomings precluding their widespread diagnostic use.

2.1.4 Other Technologies Described in the Literature

Several technologies were prominent during the early years of NGS introduction, but have become less common as other techniques have evolved.

Pyrosequencing Sequencing by Synthesis (454) The pyrosequencing (454) method amplifies DNA inside water droplets in an oil solution (emulsion PCR) [99], with each droplet containing a single DNA template attached to a single primer-coated bead that then forms a clonal colony. The sequencing instrument contains many picoliter-volume wells each containing a single bead and sequencing enzymes. Pyrosequencing uses luciferase to generate light for detection when individual nucleotides are added to the nascent DNA, and the combined data are used to generate sequence read-outs [100]. This technology has been noted to produce read lengths intermediate between those of Sanger and other NGS platforms [46, 63]. The pyrosequencing method is capable of producing large amounts of long, high-quality sequencing reads but, similar to the semiconductor-based method, the homopolymers are sequenced in one cycle, which results in homopolymer uncertainty and incorporation asynchronization. Computational methods have been proposed to decrease the false-positive rates called in common bioinformatics pipelines [62, 69, 70]. This chemistry has been suggested to produce the longest reads (up to 700 bases) and most contiguous assemblies, but compared to other similar systems it has been described as having the lower throughput [49, 62] and a higher cost per base [49]. Beginning in 2016, this technology will no longer be supported by the manufacturer in its current form.

Sequencing by Oligonucleotide Ligation and Detection (SOLiD) In sequencing by oligonucleotide ligation and detection (SOLiD), the nucleic acid is isolated and used to prepare magnetic beads with a single target sequence per bead. DNA is amplified using emulsion PCR to increase the number of copies of the unique target DNA bound to each bead, with a known adapter sequence incorporated. The PCR products are then covalently bound to a glass slide. The captured DNA is sequenced by using primers to the adapter sequence and a set of four fluorescently labeled, two-base probes compete for ligation to the sequencing primer, which obtains specificity by interrogating every first and second base in each ligation reaction [101]. Multiple cycles of ligation, detection and cleavage are performed, with the number of cycles determining the eventual read length. Following a series of ligation cycles, the extension product is removed and the template is reset with a primer complementary to the $(n - 1)$ position for a second round of ligation cycles. This process is repeated for each adapter, allowing for each base to be interrogated in two independent ligation reactions by two different primers [102]. The sequencing by ligation method has been reported to have difficulty with palindromic sequences [103]. This method has been described to have intermediate throughput and costs, as compared to other NGS methods [49].

2.2 Scale of Sequencing

Each of these sequencing-based methods is capable of being used on multiple scales. Briefly, the scales of NGS are whole-genome, exome, transcriptome, and targeted panels. Applications used extensively by research groups include sequencing an entire genome for the identification of alterations with potential clinical importance [104]. This involves obtaining enough DNA from a tissue to proceed through the various preparation stages, although a common approach has been to use a relatively low depth of coverage to maximize the use of reagents and instruments. Exome-level sequencing requires the isolation of relevant coding regions, generally accomplished through a hybrid-capture technique utilizing antisense oligonucleotides to select the sequences of interest [105, 106]. When used in the setting of neoplasia, whole-genome and whole-exome sequencing are often performed on both neoplastic and non-neoplastic tissue to allow for benign, germline variants to be excluded from consideration as "drivers" of neoplasia. A transcriptome analysis ("RNA-Seq") can be performed by capturing RNA from a given sample source, and creating and then sequencing the cDNA [47]. Targeted panels can utilize either hybrid capture techniques [107, 108] or PCR-based amplification techniques [109–112]. The potential advantages of hybrid capture methods include the possibility of detecting structural rearrangements such as translocations, inversions, and large insertions or deletions, as well as the identification of copy number changes. The determination of these types of rearrangements has been accomplished using software that can be incorporated into the data analysis pipeline after sequence acquisition [71, 73, 113–115].

2.3 Microarray Testing

2.3.1 Evolution of Microarray Testing

During the 1990s and 2000s, comparable technologic progression resulted in the development of microarray approaches for surveying the genome. Parallel with the evolution of DNA sequencing, there was a rapid change from first "dot-blot" assays, where DNA was manually adhered to membranes and subsequently hybridized to a test sample, to the earliest renditions of modern microarrays in which high-density DNA "dots" were generated by robotics and could generate massive amounts of data within a relatively short time [116, 117]. DNA microarrays are developed by binding homogeneous, small pieces of known DNA sequences onto the surface of a solid substrate, thus creating a microscopic spot. DNA or RNA (newer generations, which include protein/nucleic acid complexes) are then hybridized to these micro spots, so as to create a read signal that can be captured by a device that conveys the information to a computer for translation.

First-generation microarray analysis was performed by comparing differentially labeled DNA from a known reference sample and a test sample. The products are hybridized to a solid substrate "chip" and, in cases where there was a 1 : 1 ratio of normal to test sample DNA, this was interpreted as normal; in contrast, gains and losses were identified when the ratio of the two differentially labeled samples was higher or lower than 1 (Fig. 4). This technique was further refined when microarray chips could capture higher numbers of data points and no longer needed comparative samples. Following an improved understanding of the human genome, microarray chips with copy number alteration (CNA) probes were able to identify gains and deletions of chromosomes with higher resolution, enabling the visualization of microdeletion and gains. The subsequent incorporation of single-nucleotide polymorphism (SNP) probes allowed for the capture of copy neutral loss of heterozygosity (cnLOH) that

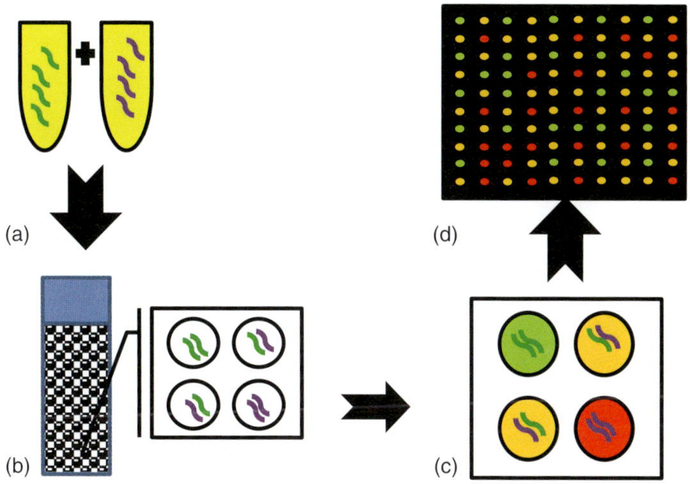

Fig. 4 A simplified schematic of newer-generation high-resolution DNA microarray chips, many of which feature copy number alteration (CNA) and single nucleotide polymorphism (SNP) probes. (a) Sample DNA is processed through a number of preparative steps, usually including amplification; (b) The prepared sample is hybridized onto a microarray "chip," where the sample DNA would bind specific spots with a prelabeled and known DNA sequence; (c) A higher-magnification view of the individual spots on the microarray chip: when DNA is bound it results in a signal of varying intensity depending on the amount of bound DNA; (d) A simplified diagram of how the specific probes may be arranged on a given microarray chip. In this schematic, the CNA probes are placed in the lower three rows, while the SNP probes are placed in the other rows. The corresponding signals are seen with differential amounts of DNA binding.

would have been previously missed when applying conventional cytogenetics.

2.3.2 **Current State of Testing**

Newer DNA microarrays utilizing CNA and SNP probes provide a high-resolution, virtual karyotype with added cnLOH data that enables surveillance of the entire human genome for chromosomal aberrations (Figs 5 and 6). Two major limitations should be considered when using microarrays for making clinical decisions. First, microarrays have a lower sensitivity than some other methods, requiring purity of the target population of cells to exceed 20%, in most cases. Second, DNA microarrays are not generally able to detect balanced translocations, unless it is a custom array built specifically to capture translocation sequences, or specific modifications have been developed [118].

There are a plethora of microarrays that interrogate different starting materials, ranging from DNA, messenger RNA (mRNA) -based expression arrays, microRNA, and protein–DNA. Specifically pertaining to DNA microarrays, there are a wide variety of assays with different utilities. These range from custom arrays that can target a specific pathway, sequencing arrays that provide single base-pair information to a specific region of the genome arrays that combine both CNA and SNP data, whole-genome arrays that provide varying levels of resolution, and epigenetic microarrays that can provide methylation data.

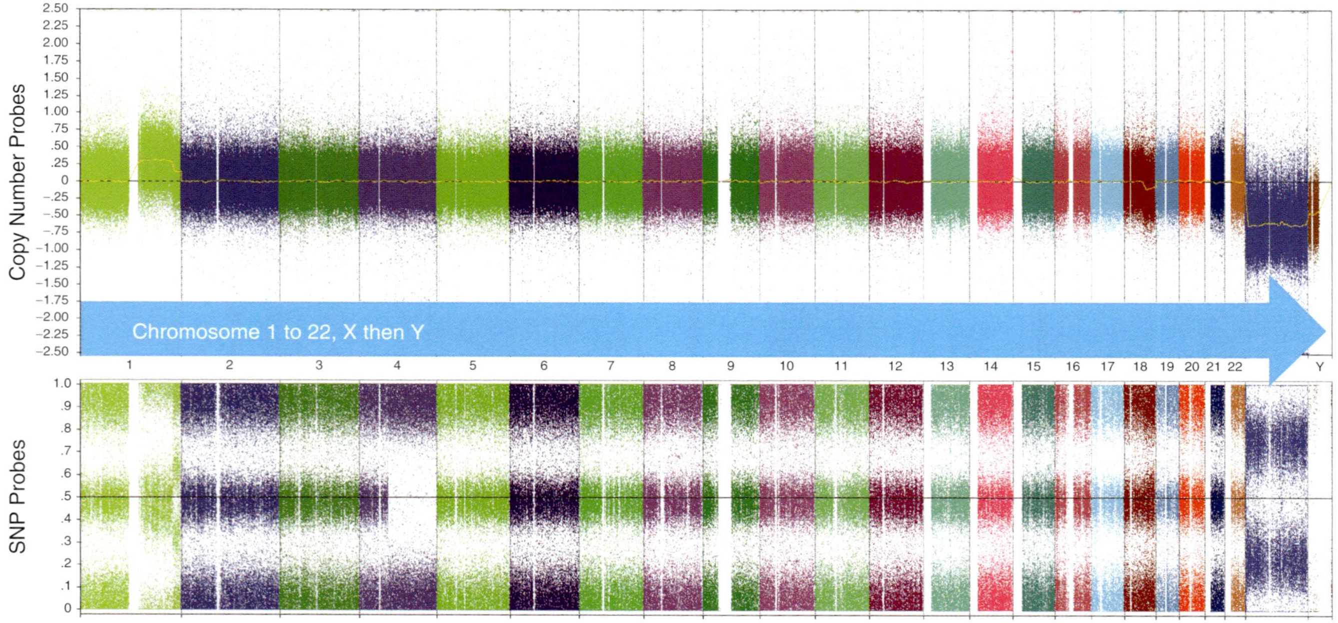

Fig. 5 Example of data from a high-resolution genomic microarray assay with both copy number alteration (CNA) and single nucleotide polymorphism (SNP) probes. This view allows for visualization of the entire genome from chromosome 1 on the left to the gender chromosomes on the far right. There are two rows of data: the top row of data are generated from the spots with copy number probes; the lower row are generated from data with the SNP probes.

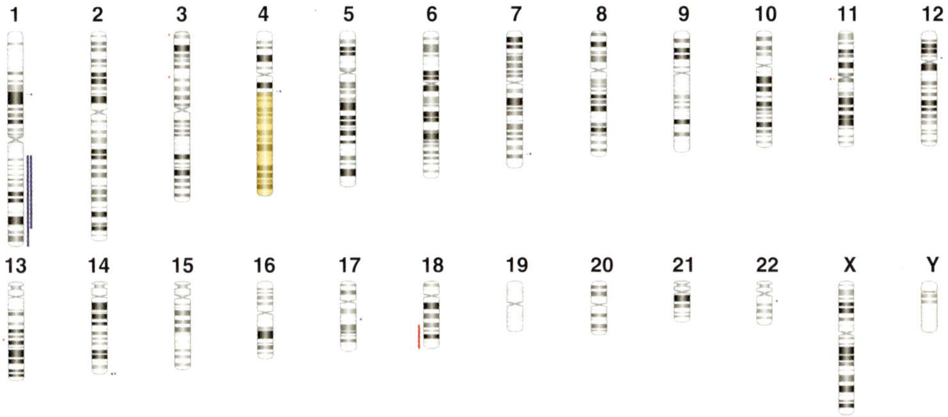

Fig. 6 Example of virtual karyotype generated from high-resolution genomic microarray assay with both copy number alteration (CNA) and single nucleotide polymorphism (SNP) probes. In this software, copy gains are highlighted in blue and shown to the right of the respective chromosome (1q), copy losses are shown in red and shown to the left of the respective chromosome (18q), and copy neutral loss of heterozygosity (cnLOH) are highlighted in gold on top of the chromosome (4q).

3 Examples of Individual Uses of Molecular Techniques in Neoplasia

As molecular and therapeutic modalities have improved over time, the use of molecular diagnostics has become commonplace in most areas of laboratory medicine. Currently, molecular diagnostics are employed in the diagnosis, prognosis, aiding in treatment decision-making, disease monitoring, and resistance testing. Some examples are included in the following sections to describe how molecular techniques have been employed in the various facets noted above. This is not meant to be an exhaustive or comprehensive list, but rather a highlight of several different applications of molecular diagnostic techniques in the clinical realm.

3.1 Diagnosis

Traditional pathologic diagnosis of neoplastic tissue requires a dual understanding of the clinical presentation and the histology through microscopic examination. Tools such as immunohistochemistry (IHC) can be used to help identify protein expression profiles of neoplastic cells, which may aid in classification. When these parameters clearly differentiate an entity, the diagnosis of a clinical pathologic sample is not difficult. However, when there are spectrums of diseases that are widely overlapping the exact neoplastic entity can be difficult to determine, and additional tools such as karyotype, fluorescence *in-situ* hybridization (FISH), PCR, or DNA sequencing can be utilized. There are many examples in which key molecular or cytogenetic abnormalities of DNA are now routinely used to define a clinical diagnosis (Table 2). Examples of some are described in the following sections.

3.1.1 Chronic Myelogenous Leukemia (CML)

As noted previously (see Sect. 1.1), CML was discovered to harbor the t(9;22) breakpoint cluster region (*BCR*)/*ABL*

Tab. 2 Examples of molecular or cytogenetic aberrations in tumors useful pathologic classification, prognostication, or clinical targeting.

	Molecular/cytogenetic aberrations
Acute myeloid leukemia	*CBFB-MYH11, DEK-NUP214, PML-RARA, inv(16), RUNX1/RUNX1T1, FLT3, IDH1, IDH2, KIT, MLL-MLLT3, NPM1, CEBPA, KIT, DNMT3A, TP53, WT1, RPN1-EVI1, RBM15-MKL1, PTEN, PHF6, TP53, TET2, RAS, ASXL1, GATA2*
Acute lymphoid leukemia	*BCR-ABL, MLL, CRLF2, JAK2, JAK1, JAK3, IL7R, IKZF1, NOTCH1, FBXW7, IDH2, DNMT3A, ETV6, NF1*
Bladder carcinoma	*FGFR3-TACC3, ERBB2, TSC1, EGFR, PIK3CA, RAS, VEGF*
Brain cancer, glioblastoma multiforme, and others	*1p19q, ACVR1, BRAF, EGFR, IDH1, IDH2, CDKN2A, PPM1D, PTEN, PDGFR, NF1, MGMT, FGFR, SHH, TACC, TP53*
Breast cancers	*ERBB2, BRCA1, BRCA2, CDH1, AKT1, PIK3CA, PTEN, ESR1, FGFR1, FGFR2, PALB2*
Colorectal carcinoma	*KRAS, BRAF, EGFR, APC, MUTYH, POLD1, POLE, TP53, MLH1, MSH2, AKT1, NRAS, PIK3CA, PTEN, SMAD4*
Esophageal adenocarcinoma	*EGFR, MET, ERBB2, TP53, SYNE1, ARID1A, HGF, FGFR4, ESRRB*
Gastric carcinoma	*MET, EGFR, CDH1, CD274, ERBB2, JAK2, PIK3CA, PDCD1LG2, RHOA*
Head and neck neoplasms and thyroid	*BRAF, HRAS, KRAS, NRAS, RET*
Lung neoplasms	*ALK-EML4, RET, ROS1, EGFR, NRTK, KRAS, ERBB2, MET, DDR2, NKX2.1, PIK3CA, FGFR1, AKT1, BRAF, CCND1, MAP2K1, STK11, NRAS*
Pancreatic adenocarcinoma	*KRAS, CDKN2A, TP53, SMAD4, DPC4, EGFR, ERBB2, TGFB, VEGF, MAPK, PIK3CA, NFKB*
Prostate neoplasms	*TMPRSS2-ERG, PTEN, AR, MSR, EGFR, RB, MYC, SPOP, ATRX*
Non-Hodgkin B-cell lymphomas	*IGH, IGK, ID1, BCL2-IGH, MYC-IGH, CCNDA-IGH, BRAF, MYD88, MALT1, trisomy 12, del 13q, TP53*
Mature T-cell lymphomas	*TRB, TRG, STAT3, STAT5B, IL2RG, JAK1, JAK3, TET2, IDH2, DNMT3A, PLCG1, RHOA, TRA-TCL1, IDH2, NPM1-ALK, TLX1, TLX3, TAL1, LMO2, i(7q)(q10), 9q gain*
Melanoma and basal cell carcinoma	*BRAF, NRAS, PTCH1, CTNNB1, GNA11, GNAQ, KIT, MEK1*
Myeloproliferative neoplasms	*ASXL1, BCR-ABL, EZH2, IDH1, IDH2, JAK2, MPL, CALR, CSF3R, DNMT3A, TET2, PDGFRB*
Myelodysplastic syndromes	*DNMT3A, TET2, IDH1/2, SF3B1, U2AF1, SRSF2, ZRSR2, EZH2, ASXL1, JAK2, KRAS, CBL, EVI, RUNX1, GATA2, Monosomy 5, 7,* complex karyotype
Ovarian and endometrial cancers	*AKT1, BRCA1, BRCA2, PTEN, SMARCA4, TP53, BRAF, KRAS, PIK3CA, PTEN, CTNNB1, FBXW7, HRAS, NRAS, CDKN2A, FGFR2, FGFR3, FOXL2, PALB2 PPP2R1A, PIK3R1, PIK3R2, CCNE1, CDK4, CCND1, MCL1*
Sarcomas	*KIT, SSX2, SSX4, ALK, EWSR1-FLI1, COL1A1-PDGFB, MED12 PDGRA, PAX3-FKHR, PAX7-FKH4, JAZF1-JJAZ1, ASPSCR1-TFE3, EWSR1-DD1T3, SDHB*

Given the rapidity of change in this field, this represents at best an incomplete list of known and recently discovered molecular genetic aberrations present in the listed neoplasms.

translocation [12, 119], which was a very useful cytogenetic finding. As investigations continued into the development and pathogenesis of many other neoplasms, it became obvious that this particular chromosomal rearrangement seemed unique to CML. As newer techniques were developed and applied to the diagnostic paradigm of CML, assays using FISH, PCR, microarray and NGS have all been developed. The quantification of *BCR/ABL* after treatment with tyrosine kinase inhibitors (TKIs) is now used routinely to predict the outcome of patients, especially in those who achieve a complete response based on an absence of *BCR/ABL* transcript by sensitive molecular techniques [120].

3.1.2 Hairy Cell Leukemia

Large-scale investigations into other neoplasms have been undertaken to identify recurrent alterations that may be pathogenic, characteristic, prognostic, and/or possibly to guide therapy. One such example is in the case of hairy cell leukemia (HCL), which was shown to harbor a *BRAF* (B-Raf proto-oncogene, serine/threonine kinase) p.V600E mutation in the majority of tested samples [121]. While HCL displays a characteristic morphology and immunophenotype in many cases, the observation of *BRAF* p.V600E mutations being common in this entity may allow for the resolution of difficult or equivocal cases [122]. More recently, clinicians have targeted this specific mutation in *BRAF* through the use of *BRAF* inhibitors for the treatment of refractory HCL [123].

3.1.3 Acute Promyelocytic Leukemia

A characteristic translocation involving the *PML* (promyelocytic leukemia protein) and retinoic acid receptor (*RARA*) genes on chromosomes 15 and 17, respectively, is observed in the majority of acute promyelocytic leukemia (APL) cases. APL is unique because the translocation protein product creates a block preventing the maturation of promyelocytes to neutrophils, and treatment with retinoic acid (vitamin A) allows the maturation to progress [124]. Another unique feature of this disease is that chemotherapeutic induction therapy is associated with a significant risk of disseminated intravascular coagulation, with associated morbidity and mortality [125]. Thus, it is critical that APL cases be rapidly and accurately identified to allow for the care team to initiate the correct therapies and monitor for adverse events [126]. While routine cytogenetics (karyotype) can be used to diagnose t(15 : 17) in APL, FISH has been demonstrated to have superior sensitivity [127]. Although the routine testing of patient samples for t(15;17) may be best performed using FISH, the identification of a chimeric RNA by reverse-transcription polymerase chain reaction (RT-PCR) has been described as the "gold standard" for the identification of a translocation transcript [128]. Although RT-PCR may be more sensitive than other methods in most cases, it can be limited by technical factors such as degraded RNA or large distances between the primer binding sites. There are also cases of masked [129] or so-called "cryptic" translocations identified by PCR [130] and NGS techniques [131] which were not identified with FISH techniques. In these rare cases, techniques such as microarray have been shown to be useful in establishing the diagnosis [132].

3.1.4 NUT Midline Carcinoma

NUT midline carcinoma (NMC) is a rare cancer that occurs across the age spectrum, is typically present in the midline, shows variable degrees of squamous differentiation, and has characteristic fusions between *NUTM1* and multiple partners, most commonly *BRD4* [133, 134]. These carcinomas most commonly arise in the head, neck, mediastinum, or aerodigestive tract [134–137]. The observed translocations are diagnostic and often are often the only chromosomal abnormality [138], but many cases of this type of carcinoma likely are not recognized because of variation in the clinical and morphologic appearance [135]. The identification of this type of carcinoma is important because these patients typically have a poor prognosis [134]. Within this group of carcinomas, translocations that involve *NUTM1*, but not *BRD4*, may have a different clinical course compared to patients with the *NUTM1–BRD4* translocations [139].

3.1.5 Fibrolamellar Carcinoma

Fibrolamellar carcinoma (FLC) is a rare type of liver cancer that has epidemiology, etiology, and prognosis that is distinct from other forms of hepatic carcinomas. FLC usually occurs in younger individuals and in the absence of other factors such as infection or inflammation that are associated with other forms of liver cancer such as hepatocellular carcinoma (HCC). The prognosis for these patients is generally more favorable compared to HCC, with survivals of ~40% versus 25%, respectively [140]. The gross and histologic appearances of these neoplasms can be characteristic, with the neoplastic cells resembling well-differentiated hepatocytes with an eosinophilic and granular cytoplasm and a characteristic fibrotic stroma [141]. The etiology of FLC was unknown until a recurrent *DNAJB1–PRKACA* fusion was identified in all of 15 FLC cases analyzed [142]. This fusion has been studied in FLC and other forms of liver neoplasia and appears to be specific for this neoplasm [143]. Although, many FLC cases appear to have a classic histomorphology, some studies have suggested that interobserver agreements of fibrous stroma may be suboptimal [144], suggesting that the use of RT-PCR or FISH techniques to identify the *DNAJB1–PRKACA* fusion may be diagnostically valuable.

3.2 Prognosis

The discovery of recurrent cytogenetic or molecular abnormalities in many different types of malignancies has led clinicians and investigators to determine if these molecular changes have differential effects on patient prognosis or outcome. Examples of where molecular or cytogenetic prognostication is used in the clinical realm are included in Table 1. A brief description of some of these are included in the following sections.

3.2.1 Acute Myeloid Leukemia

Investigations of acute myeloid leukemia (AML), a blood cancer that results in the proliferation of immature cells, have revealed that there were populations that had recurrent morphologic similarities which correlated with cytogenetic abnormalities. One example was the discovery of subtypes of AML that affect core binding factor (CBF) -related alterations. This group of diseases shows distinct and recurrent rearrangements involving components of the CBF heterodimeric transcription factor [(*CBFβ*) in inv16 and *RUNX1/RUNX1T1* in t(8;21)]. This observation was important because this subset of AML patients generally had a better prognosis, with a greater response to chemotherapy than other types of AML patients, and improved survival. While the prognosis for most patients with CBF AML is improved compared to other types of AML, various studies have revealed that a subset of these patients relapsed at a higher rate than most patients in this category. It was subsequently shown that alterations in several exons of the *KIT* gene corresponded with a higher risk of relapse after chemotherapy [145]. Mutations in *KIT* in CBF-positive, adult AML patients appeared to decrease the overall survival and increase the relapse rate compared to wild-type, and there was an increased five-year relapse rate of CBF AML with *KIT* mutations in exon 17 compared to wild-type *KIT*. *KIT* mutations in pediatric AML cases did not show a difference in overall survival, disease-free survival, or relapse compared to patients lacking such mutations [145, 146]. As a result of these findings, it has been recommended that individuals with CBF-positive AML be evaluated for alterations in *KIT* to help direct therapy and monitoring.

AML with recurrent genetic abnormalities has been noted to have differences in prognosis related to the observed mutations [147]. For example, AML with mutated nucleophosmin 1 (*NPM1*) shows a good response to induction chemotherapy, and these patients have a favorable prognosis when the AML blasts show a normal karyotype [148]. When mutations in fms-related tyrosine kinase 3 (*FLT3*) are observed, whether in the presence of *NPM1* or not, the prognosis is poorer in these patients [149, 150]. In the setting of normal karyotype, *CEBPA* (CCAAT/enhancer-binding protein alpha) mutations have been associated with a favorable response [151]. In these examples, it is important to determine the impact of prognostic factors to help guide the clinician with regards to therapeutic decisions that include specific chemotherapy regimens, and whether bone marrow transplantation should be considered earlier in the treatment course [152].

3.2.2 Inherited Cancer Syndromes

Molecular diagnostic techniques are also commonly employed in the evaluation of individuals for inherited genetic defects, which may predispose them to the development of malignancies. Alterations of genes such as *BRCA1*, *BRCA2*, *APC*, *TP53*, *PIK3CA*, *RET* and many other genes have been associated with inherited syndromes that predispose individuals to malignancy. The identification of families and individuals with inherited mutations that put them at risk for the development of malignancy allows for a more intensive screening (e.g., Lynch and inherited breast cancer) of organs where malignancies are atypical in affected patients (e.g., *APC*), or screening for prophylactic procedures, such as colectomy [153–155]. The approach for many years has consisted either of a step-wise testing or a combination of methods to identify not only missense or nonsense mutations but also structural rearrangements. The implementation of screening assays using NGS offers the possibility for identifying additional individuals who may be at increased risk for inherited disease, with decreased effort and increased sensitivity [156, 157]. While special techniques will likely continue to be required for technical reasons under certain circumstances [76, 158], the implementation of appropriately constructed panels may allow for an efficient and timely identification of individuals at risk for cancers due to inherited genes.

3.3 Treatment Guidance

Although the identification of the Philadelphia chromosome as a recurrent abnormality in CML was a boon to diagnosis, initially it was unknown how this alteration was related to the neoplasm.

Extensive studies revealed that a fusion protein resulted when the *ABL1* (ABL proto-oncogene 1, non-receptor tyrosine kinase) gene on chromosome 9 (region q34) was translocated to the *BCR* gene on chromosome 22 (region q11). Further studies showed that this rearrangement resulted in a constitutive transcription of the chimeric gene, with hyperactivation of the *ABL1* tyrosine kinase. This observation led to the development of TKIs that targeted the active site of the chimeric protein [159, 160]. The treatment of CML is necessary because, after a relatively indolent phase, the majority of patients will show transformation to AML or acute lymphoblastic leukemia (ALL), with increased numbers of blasts and a near-uniform mortality rate. Treatment with oral TKIs has resulted in dramatic decreases in numbers of neoplastic cells, with many people showing a prolonged response over years with increased survival [161]. Similar therapies to other aberrantly expressed or chimeric proteins have been introduced, such as *ALK–EML4* fusions observed in some forms of lung cancer [162]. These cases highlight situations where specific chromosomal alterations resulting in aberrant chimeric proteins are required in order for a therapy to be effective, thus mandating selection of the appropriate patients through the use of correct diagnostic techniques.

Examples also exist of mutations that are less-striking than large, structural rearrangements which have been associated with responses to other therapies. An example of a mutation that requires identification prior to initiation of therapy is *BRAF* p.V600E in the setting of melanoma. This mutation is present in a large proportion of melanomas, and leads to an increased activation of *BRAF* tyrosine kinase [163]. Inhibitors have been developed to this molecule that have shown an impressive short-term resolution of widespread disease [164, 165]. It is important that only patients who will benefit from these drugs be identified prior to treatment, because one of the side effects of the medication is an induction of cutaneous squamous cell carcinomas in individuals taking the medication [166]. Similar examples have been demonstrated in other malignancies, including lung cancer, where activating mutations in specific proteins are amenable to inhibition by small molecules that target these constitutively active proteins [167, 168].

It is not surprising that, while some mutations confer susceptibility to a given therapy, other mutations confer resistance to different medications. A well-known example of a mutation which results in a lack of response to therapy is in the setting of colorectal carcinoma (CRC). The expression of epidermal growth factor receptor (*EGFR*) on CRC cells drove the development of treatments targeting this molecule. Disappointing responses in clinical trials for *EGFR*-targeted therapy allowed for further studies to identify mutations in the *KRAS* (Kirsten rat sarcoma viral oncogene homolog) oncogene that effectively led to resistance against these therapies in a large proportion of patients. Mutations in *KRAS* are most common in codon 12 or 13, and the observed mutations result in an impaired hydrolysis of GTP bound to *KRAS*, thereby impairing its inactivation. The identification of mutations in *KRAS* codon 12 or 13 predicts resistance to anti-*EFGR* therapies, and consequently the testing of patients who are being considered for anti-*EGFR* therapies is mandatory prior to the initiation of therapy [169, 170]. *BRAF* p.V600E is also observed in a subset of *KRAS*-negative CRCs [171], with a decreased overall survival compared to patients with wild-type *BRAF* when treated with anti-EFGR therapy [172]. In this example, identification of the mutations resulted in patients not being prescribed a drug that would not benefit them, thus preventing unnecessary costs and side effects.

3.4 Microarrays in Clinical Use

The first clinical microarray tests approved by the Food and Drug Administration (FDA) included a microarray chip to identify cytochrome P450 (CYP450) for drug efficacy and toxicity in individual patients [173], and another to prognosticate invasive breast cancer [174]. DNA microarray tests, including the aforementioned FDA-approved tests, have shown significant clinical utility in terms of disease diagnosis/classification, prognostication, and therapy guidance.

Microarrays are increasingly used in clinical cancer diagnostics. Several large studies with AML and myelodysplastic syndromes (MDSs) have shown that microarray testing increased the discovery of genomic aberrations and, in addition, provided pertinent data in prognostication [175–177]. In breast cancer, microarray testing aids in the classification of tumors, which is closely associated with the prognosis [178, 179]. Gene expression arrays provide data for improved classification/prognosis and therapy guidance in several cancers, including diffuse large B-cell lymphoma [180], breast [181], lung [182], kidney [183], sarcoma [184], and others.

In addition to their clear utility in the setting of neoplasia, microarray techniques have shown broad clinical applicability in the setting of prenatal testing [185], constitutional genetic testing [186], and microbiology [187]. Because of the widely customizable format of microarrays, there is an increasing use in clinical testing as a large amount of data pertaining not only to diagnosis and prognosis but also therapeutic guidance can be achieved from a single, possibly more economical assay (when compared to NGS or a panel of more than five FISH tests), and with a relatively faster turnaround-time (compared to NGS).

3.5 Additional Applications of Molecular Diagnostics

An early use of PCR-based assays was the detection of minimal residual disease (MRD) in the post-therapy setting via specific and detectable mutations [36], and this continues to be important to the present day [188, 189]. Methods used for this type of monitoring must be exquisitely sensitive and specific, especially if clinical decisions will be made based on the results. While PCR has been used in many settings, specificity has been achieved through the detection of targets that are not present in the germline sequence, such as translocations or somatic mutations. Many translocations have regular but inconsistent breakpoints which may require a complex design of PCR primers or probes [190, 191]. Despite the technical challenges in patients with structural rearrangements that are detectable by PCR, this method offers a robust and reliable method of MRD detection compared to FISH [192, 193], because many FISH protocols require 5–10% of the total cells present to be positive for the abnormality to distinguish a true positive case from background signals [194]. NGS offers the potential to simplify the process of MRD detection because the specific mutation or alteration does not necessarily need to be known at the initiation of testing. Studies comparing NGS- and PCR-based MRD assays have shown similar sensitivity [195], with some studies of deep sequencing showing increased sensitivity [196]. These methods will likely be increasingly utilized in the future as laboratories become comfortable with the techniques and their limitations.

It has long been observed that cells with abnormalities can be detected in circulating blood, even in otherwise healthy individuals [197]. The detection of these abnormal cells or even cell-free DNA offers numerous diagnostic possibilities, from prenatal diagnosis [198] to neoplastic diagnosis and evaluation [199]. One limitation has been the assay sensitivity, though when NGS is used to a high depth of coverage these methods can identify such aberrant cells with high sensitivity and specificity [200, 201]. This type of non-invasive testing is very attractive from multiple standpoints and, despite the technical challenges, will likely become a more prominent diagnostic method in the future. Success identifying DNA from neoplasia in the peripheral blood has sparked other groups to attempt using other unconventional sources of DNA that may be useful in screening, such as stool in the setting of CRC [202]. It is likely that innovations in this area will continue.

Resistance to specific therapies develops during the course of targeted therapy as a result of genetic mutations, and occasionally is observed *de novo*. As additional drugs have become available for targeting specific genetic alterations [203, 204], molecular methods may be needed to evaluate specific neoplasms for resistant mutations and possible alternative therapeutic strategies [205, 206]. This will be an important aspect of diagnostic testing, especially as more medications become available to treat malignancies that have developed resistance to prior therapies.

NGS has revealed a huge spectrum of possible genetic alterations that may not only be associated with neoplasia but may also offer targets for therapeutic intervention. While some genetic alterations are specific to a given neoplasm, many other alterations seem to have a wide distribution across neoplasms [207]. Common mutations that can be targeted by drugs in a specific neoplasm most likely have already been identified; furthermore, clinical trials have been designed and initiated in many cases. In the future, the identification of individuals to participate in clinical trials may be achieved not on their specific tumor type but rather on the underlying mutations observed, with many different types of malignancy being included in clinical trials based on their genetic characteristics. Whilst this potential change in the selection process will certainly complicate clinical trial design, it may be necessary to identify the populations that could potentially benefit from new medication and therapeutic strategies [208–210].

Currently, routine clinical testing in the setting of neoplasia is primarily accomplished with DNA, although other targets such as RNA and methylated DNA are also utilized. As the current understanding of neoplasia increases, it is likely that the scope of targets that may be beneficial in diagnosis, prognosis, and/or therapy will expand further. Possible future targets include epigenetic changes, transcriptome analysis, or changes within the regulatory molecules, such as microRNA. The use of other techniques within the clinical setting will be driven by findings currently underway in large-scale research studies. The constant advancement of the field requires that practitioners be fluent with the latest research and findings.

4 Assay Selection for Specific Clinical Applications

Targeted testing has been used for many years when a simple rapid test is able to differentiate patients into categories necessary for treatment determination. Some examples of this are *PML–RARA* translocation testing evaluating for APL [211] and *BRAF* testing in melanoma [212], where the results of testing can be rapidly generated to direct targeted therapy. These types of assay can provide quick turnaround-times, providing important information to clinicians. More recently, tests to evaluate treatment efficacy have been introduced for CML using quantitative assays for *BCR–ABL* [213] which can provide guidance for therapy administration. The applications for these types of rapid test have several characteristics in common:

- The targets for the tests tend to be relatively common in a given type of tumor, with a predictable range of variants that are easily identifiable.
- The results of the test will be used in the short term to direct clinical decisions.
- The laboratory methods selected have a rapid turnaround-time to facilitate clinical decision making.

Many of the laboratory methods can provide rapid testing, but specific limitations of these methodologies must be considered when implementing or using them. One very important consideration is to evaluate the target for less common but equally important variants, such as the *BRAF* p.V600K mutation in melanoma [214]. If designed and used correctly, single-target assays can be extremely useful in directing patient care. Targeted assays often take the form of single-target PCR assays or Sanger sequencing assays. While targeted PCR assays can detect a mutation at 1% or less of a mixture using special techniques [215], Sanger sequencing is typically only able to detect a minor population which is greater than 10–15% of the total [216]. Another caveat to single-target testing is in the cases of low-frequency alterations that are associated with targeted therapy, such as *ALK* rearrangements and targeted therapy in lung cancer. Economic analyses have suggested that this type of testing and therapy may not be cost-effective [217], which may limit its implementation in settings where resource management is prioritized.

Small panels of tests have been developed when there are a relatively few significant mutations that may direct immediate therapy or future therapeutic decisions. Often, these types of test were constructed with decision tree support for laboratories to allow them to appropriately manage their workflow. Examples of this type of testing include the sequential testing of *EGFR* and *ALK* rearrangements in lung cancer [218, 219], multiple FISH assays for hematopoietic neoplasms [220, 221], sequential testing of *KIT* exons which will result in changes in the management of gastrointestinal stromal tumor (GIST) and AML [145, 222], or the sequential testing of cancer risk genes [223, 224]. These types of panel tend to increase in cost over single tests, be more complex for the laboratory to manage, and have increased turnaround-times compared to single-analyte tests. Likewise, these types of test may not be cost-effective, given the ease of use and robust information that NGS-based assays are likely to provide [225].

Limited panel-based NGS tests are likely to take the place of small- to medium-sized panel assays. NGS assays have reliably identified point mutations down to a minor frequency of 1% in mixing studies [226], which is comparable to many sensitive, targeted PCR-based assays. Amplification-based NGS can quickly generate information on positional alterations of tens to hundreds of genes, with effort similar that required for multi-part panel tests. Amplification-based NGS gene panel tests have compared favorably to single-target assays [227, 228], and can be accomplished in a similar timeframe to FISH and multiple PCR assays. One limitation of most NGS panel tests is the difficulty of assessing structural or copy number variants [107, 108]. Until this technical limitation is overcome in a format that is amenable to wide implementation, NGS panel implementations may be restricted to point mutation assessment in most cases.

There is a general interest in the medical community over the possible applications of whole-genome and whole-exome sequencing in routine medical care [229] and cancer [230]. The use of exome [231] and whole-genome [131, 232] sequencing analysis has been described in the context of patient care and treatment decision in oncology testing, but the time, expense and complexity may preclude its routine use for at least the next few years. At the time of writing, studies of whole-genome sequencing have been suggested to have technical and throughput limitations even in the more-straightforward analysis of inherited disorders [233], which suggests that the routine analysis of more complex neoplastic tissue is unlikely in the near term. However, if it can be shown that there is some benefit to exome-level or whole-genome-level analysis over targeted or panel testing, then this approach may become commonplace, despite the current limitations.

5 Summary

The analysis of DNA has played a key role in the diagnosis of neoplasia for many years. Advances over the past 40 years have demonstrated the clear benefits of using DNA-based methods to allow for a more accurate diagnosis, prognosis, and recommendations of treatments in the setting of neoplasia. New technologies are expanding the current insight into the role of DNA in neoplastic disease, and will likely result in new therapies directed at molecular aberrations that have their roots in disturbances at the DNA level. Many of the techniques that have been developed have been continually refined over time, and will continue to play a role in the care of patients with neoplastic diseases. However, careful consideration must be given to the strength, weaknesses and limitations of the different methods under consideration for any particular application. While much information has been discovered, a huge amount remains unknown or is poorly understood. Additional investigations will lead to a better understanding of epigenetics, RNA editing, and gene regulation through microRNAs, and may in turn lead to new diagnostic opportunities.

References

1. Mukherjee, S. (2011) *The Emperor of All Maladies: A Biography of Cancer*, Scribner, New York: xviii, **573**, 12 p., 8 p. of plates.
2. Weinberg, R.A. (2014) *The Biology of Cancer*. Garland Science, New York, 875 pp.
3. Lannoy, N. and Hermans, C. (2010) The 'royal disease' – haemophilia A or B? A haematological mystery is finally solved. *Haemophilia*, **16**, 843–847.
4. Stevens, R.F. (1999) The history of haemophilia in the royal families of Europe. *Br. J. Haematol.*, **105**, 25–32.

5. Abramson, D.H. (2005) Retinoblastoma in the 20th century: past success and future challenges the Weisenfeld lecture. *Invest. Ophthalmol. Visual Sci.*, **46**, 2683–2691.
6. Waldron, H.A. (1983) A brief history of scrotal cancer. *Br. J. Ind. Med.*, **40**, 390–401.
7. Avery, O.T., Macleod, C.M., and McCarty, M. (1944) Studies on the chemical nature of the substance inducing transformation of pneumococcal types: induction of transformation by a desoxyribonucleic acid fraction isolated from *Pneumococcus* type III. *J. Exp. Med.*, **79**, 137–158.
8. Sorsby, A. (1965) Gregor Mendel. *Br. Med. J.*, **1**, 333–338.
9. Boveri, T. (2008) Concerning the origin of malignant tumours by Theodor Boveri. Translated and annotated by Henry Harris. *J. Cell Sci.*, **121** Suppl. 1, 1–84.
10. Watson, J.D. and Crick, F.H. (1953) Molecular structure of nucleic acids; a structure for deoxyribose nucleic acid. *Nature*, **171**, 737–738.
11. Nirenberg, M.W. and Matthaei, J.H. (1961) The dependence of cell-free protein synthesis in E. coli upon naturally occurring or synthetic polyribonucleotides. *Proc. Natl Acad. Sci. USA*, **47**, 1588–1602.
12. Nowell, P.C. and Hungerford, D.A. (1961) Chromosome studies in human leukemia. II. Chronic granulocytic leukemia. *J. Natl Cancer Inst.*, **27**, 1013–1035.
13. Sakurai, M., Hayata, I., and Sandberg, A.A. (1976) Prognostic value of chromosomal findings in Ph1-positive chronic myelocytic leukemia. *Cancer Res.*, **36**, 313–318.
14. Caspersson, T., Farber, S., Foley, G.E., Kudynowski, J. *et al.* (1968) Chemical differentiation along metaphase chromosomes. *Exp. Cell. Res.*, **49**, 219–222.
15. Lawler, S.D., Pentycross, C.R., and Reeves, B.R. (1968) Chromosomes and transformation of lymphocytes in lymphoproliferative disorders. *Br. Med. J.*, **4**, 213–219.
16. Moorhead, P.S. (1965) Human tumor cell line with a quasi-diploid karyotype (RPMI 2650). *Exp. Cell. Res.*, **39**, 190–196.
17. Rowley, J.D. (1981) Association of specific chromosome abnormalities with type of acute leukemia and with patient age. *Cancer Res.*, **41**, 3407–3410.
18. Smith, H.O. and Wilcox, K.W. (1970) A restriction enzyme from *Hemophilus influenzae*. I. Purification and general properties. *J. Mol. Biol.*, **51**, 379–391.
19. Price, P.M. and Hirschhorn, K. (1975) In situ hybridization of chromosome loci. *Fed. Proc.*, **34**, 2227–2232.
20. Pinkel, D., Straume, T., and Gray, J.W. (1986) Cytogenetic analysis using quantitative, high-sensitivity, fluorescence hybridization. *Proc. Natl Acad. Sci. USA*, **83**, 2934–2938.
21. Sanger, F. and Coulson, A.R. (1975) A rapid method for determining sequences in DNA by primed synthesis with DNA polymerase. *J. Mol. Biol.*, **94**, 441–448.
22. Sanger, F., Nicklen, S., and Coulson, A.R. (1977) DNA sequencing with chain-terminating inhibitors. *Proc. Natl Acad. Sci. USA*, **74**, 5463–5467.
23. Maxam, A.M. and Gilbert, W. (1977) A new method for sequencing DNA. *Proc. Natl Acad. Sci. USA*, **74**, 560–564.
24. Bartlett, J.M. and Stirling, D. (2003) A short history of the polymerase chain reaction. *Methods Mol. Biol.*, **226**, 3–6.
25. Saiki, R.K., Bugawan, T.L., Horn, G.T., Mullis, K.B., and Erlich, H.A. (1986) Analysis of enzymatically amplified beta-globin and HLA-DQ alpha DNA with allele-specific oligonucleotide probes. *Nature*, **324**, 163–166.
26. Lawyer, F.C., Stoffel, S., Saiki, R.K., Myambo, K. *et al.* (1989) Isolation, characterization, and expression in *Escherichia coli* of the DNA polymerase gene from *Thermus aquaticus*. *J. Biol. Chem.*, **264**, 6427–6437.
27. Saiki, R.K., Gelfand, D.H., Stoffel, S., Scharf, S.J. *et al.* (1988) Primer-directed enzymatic amplification of DNA with a thermostable DNA polymerase. *Science*, **239**, 487–491.
28. Dryja, T.P., Rapaport, J.M., Joyce, J.M., and Petersen, R.A. (1986) Molecular detection of deletions involving band q14 of chromosome 13 in retinoblastomas. *Proc. Natl Acad. Sci. USA*, **83**, 7391–7394.
29. Lübbert, M., Mirro, J., Miller, C.W., Kahan, J. *et al.* (1990) N-ras gene point mutations in childhood acute lymphocytic leukemia correlate with a poor prognosis. *Blood*, **75**, 1163–1169.
30. Iggo, R., Gatter, K., Bartek, J., Lane, D., and Harris, A.L. (1990) Increased expression of mutant forms of p53 oncogene in primary lung cancer. *Lancet*, **335**, 675–679.

31. Komminoth, P., Kunz, E.K., Matias-Guiu, X., Hiort, O. *et al.* (1995) Analysis of RET protooncogene point mutations distinguishes heritable from nonheritable medullary thyroid carcinomas. *Cancer*, **76**, 479–489.
32. Claas, E.C., Melchers, W.J., van der Linden, H.C., Lindeman, J., and Quint, W.G. (1989) Human papillomavirus detection in paraffin-embedded cervical carcinomas and metastases of the carcinomas by the polymerase chain reaction. *Am. J. Pathol.*, **135**, 703–709.
33. Feinmesser, R., Miyazaki, I., Cheung, R., Freeman, J.L. *et al.* (1992) Diagnosis of nasopharyngeal carcinoma by DNA amplification of tissue obtained by fine-needle aspiration. *N. Engl. J. Med.*, **326**, 17–21.
34. Melo, J.V., Gordon, D.E., Cross, N.C., and Goldman, J.M. (1993) The ABL-BCR fusion gene is expressed in chronic myeloid leukemia. *Blood*, **81**, 158–165.
35. Dobrovic, A., Trainor, K.J., and Morley, A.A. (1988) Detection of the molecular abnormality in chronic myeloid leukemia by use of the polymerase chain reaction. *Blood*, **72**, 2063–2065.
36. Thompson, J.D., Brodsky, I., and Yunis, J.J. (1992) Molecular quantification of residual disease in chronic myelogenous leukemia after bone marrow transplantation. *Blood*, **79**, 1629–1635.
37. Huang, X.C., Quesada, M.A., and Mathies, R.A. (1992) DNA sequencing using capillary array electrophoresis. *Anal. Chem.*, **64**, 2149–2154.
38. Hood, L.E., Hunkapiller, M.W., and Smith, L.M. (1987) Automated DNA sequencing and analysis of the human genome. *Genomics*, **1**, 201–212.
39. Zimmermann, J., Voss, H., Schwager, C., Stegemann, J., and Ansorge, W. (1988) Automated Sanger dideoxy sequencing reaction protocol. *FEBS Lett.*, **233**, 432–436.
40. Fleischmann, R.D., Adams, M.D., White, O., Clayton, R.A. *et al.* (1995) Whole-genome random sequencing and assembly of *Haemophilus influenzae* Rd. *Science*, **269**, 496–512.
41. Venter, J.C., Adams, M.D., Sutton, G.G., Kerlavage, A.R. *et al.* (1998) Shotgun sequencing of the human genome. *Science*, **280**, 1540–1542.
42. Venter, J.C., Adams, M.D., Myers, E.W., Li, P.W. *et al.* (2001) The sequence of the human genome. *Science*, **291**, 1304–1351.
43. Lander, E.S., Linton, L.M., Birren, B., Nusbaum, C. *et al.* (2001) Initial sequencing and analysis of the human genome. *Nature*, **409**, 860–921.
44. Shendure, J., Porreca, G.J., Reppas, N.B., Lin, X. *et al.* (2005) Accurate multiplex polony sequencing of an evolved bacterial genome. *Science*, **309**, 1728–1732.
45. Brenner, S., Johnson, M., Bridgham, J., Golda, G. *et al.* (2000) Gene expression analysis by massively parallel signature sequencing (MPSS) on microbead arrays. *Nat. Biotechnol.*, **18**, 630–634.
46. Mardis, E.R. (2008) Next-generation DNA sequencing methods. *Annu. Rev. Genomics Hum. Genet.*, **9**, 387–402.
47. Pareek, C.S., Smoczynski, R., and Tretyn, A. (2011) Sequencing technologies and genome sequencing. *J. Appl. Genet.*, **52**, 413–435.
48. Jain, M., Fiddes, I.T., Miga, K.H., Olsen, H.E. *et al.* (2015) Improved data analysis for the MinION nanopore sequencer. *Nat. Methods*, **12** (4), 351–356.
49. Liu, L., Li, Y., Li, S., Hu, N. *et al.* (2012) Comparison of next-generation sequencing systems. *J. Biomed. Biotechnol.*, **2012**, 251364.
50. Quail, M.A., Smith, M., Coupland, P., Otto, T.D. *et al.* (2012) A tale of three next generation sequencing platforms: comparison of Ion Torrent, Pacific Biosciences and Illumina MiSeq sequencers. *BMC Genomics*, **13**, 341.
51. Gullapalli, R.R., Lyons-Weiler, M., Petrosko, P., Dhir, R. *et al.* (2012) Clinical integration of next-generation sequencing technology. *Clin. Lab. Med.*, **32**, 585–599.
52. Richter, B.G. and Sexton, D.P. (2009) Managing and analyzing next-generation sequence data. *PLoS Comput. Biol.*, **5**, e1000369.
53. Batley, J. and Edwards, D. (2009) Genome sequence data: management, storage, and visualization. *Biotechniques*, **46**, 333–334, 336.
54. Sinard, J.H., Powell, S.Z., and Karcher, D.S. (2014) Pathology training in informatics: evolving to meet a growing need. *Arch. Pathol. Lab. Med.*, **138**, 505–511.
55. Haspel, R.L., Olsen, R.J., Berry, A., Hill, C.E. *et al.* (2014) Progress and potential:

training in genomic pathology. *Arch. Pathol. Lab. Med.*, **138**, 498–504.
56. Starren, J., Williams, M.S., and Bottinger, E.P. (2013) Crossing the omic chasm: a time for omic ancillary systems. *JAMA*, **309**, 1237–1238.
57. Overby, C.L., Kohane, I., Kannry, J.L., Williams, M.S. *et al.* (2013) Opportunities for genomic clinical decision support interventions. *Genet. Med.*, **15**, 817–823.
58. Gray, S.W., Hicks-Courant, K., Cronin, A., Rollins, B.J., and Weeks, J.C. (2014) Physicians' attitudes about multiplex tumor genomic testing. *J. Clin. Oncol.*, **32** (13), 1317–1323.
59. Guo, J., Xu, N., Li, Z., Zhang, S. *et al.* (2008) Four-color DNA sequencing with 3'-O-modified nucleotide reversible terminators and chemically cleavable fluorescent dideoxynucleotides. *Proc. Natl Acad. Sci. USA*, **105**, 9145–9150.
60. Ju, J., Kim, D.H., Bi, L., Meng, Q. *et al.* (2006) Four-color DNA sequencing by synthesis using cleavable fluorescent nucleotide reversible terminators. *Proc. Natl Acad. Sci. USA*, **103**, 19635–19640.
61. Bentley, D.R., Balasubramanian, S., Swerdlow, H.P., Smith, G.P. *et al.* (2008) Accurate whole human genome sequencing using reversible terminator chemistry. *Nature*, **456**, 53–59.
62. Loman, N.J., Misra, R.V., Dallman, T.J., Constantinidou, C. *et al.* (2012) Performance comparison of benchtop high-throughput sequencing platforms. *Nat. Biotechnol.*, **30**, 434–439.
63. Kircher, M., Heyn, P., and Kelso, J. (2011) Addressing challenges in the production and analysis of illumina sequencing data. *BMC Genomics*, **12**, 382.
64. Dohm, J.C., Lottaz, C., Borodina, T., and Himmelbauer, H. (2008) Substantial biases in ultra-short read data sets from high-throughput DNA sequencing. *Nucleic Acids Res.*, **36**, e105.
65. Benjamini, Y. and Speed, T.P. (2012) Summarizing and correcting the GC content bias in high-throughput sequencing. *Nucleic Acids Res.*, **40**, e72.
66. Ross, M.G., Russ, C., Costello, M., Hollinger, A. *et al.* (2013) Characterizing and measuring bias in sequence data. *Genome Biol.*, **14**, R51.
67. Aird, D., Ross, M.G., Chen, W.S., Danielsson, M. *et al.* (2011) Analyzing and minimizing PCR amplification bias in Illumina sequencing libraries. *Genome Biol.* **12**, R18.
68. Rothberg, J.M., Hinz, W., Rearick, T.M., Schultz, J. *et al.* (2011) An integrated semiconductor device enabling non-optical genome sequencing. *Nature*, **475**, 348–352.
69. Yeo, Z.X., Chan, M., Yap, Y.S., Ang, P. *et al.* (2012) Improving indel detection specificity of the Ion Torrent PGM benchtop sequencer. *PLoS One*, **7**, e45798.
70. Zeng, F., Jiang, R., and Chen, T. (2013) PyroHMMsnp: an SNP caller for Ion Torrent and 454 sequencing data. *Nucleic Acids Res.*, **41**, e136.
71. Chen, K., Wallis, J.W., McLellan, M.D., Larson, D.E. *et al.* (2009) BreakDancer: an algorithm for high-resolution mapping of genomic structural variation. *Nat. Methods*, **6**, 677–681.
72. Abel, H.J., Al-Kateb, H., Cottrell, C.E., Bredemeyer, A.J. *et al.* (2014) Detection of gene rearrangements in targeted clinical next-generation sequencing. *J. Mol. Diagn.*, **16**, 405–417.
73. Wang, J., Mullighan, C.G., Easton, J., Roberts, S. *et al.* (2011) CREST maps somatic structural variation in cancer genomes with base-pair resolution. *Nat. Methods*, **8**, 652–654.
74. Gnerre, S., Maccallum, I., Przybylski, D., Ribeiro, F.J. *et al.* (2011) High-quality draft assemblies of mammalian genomes from massively parallel sequence data. *Proc. Natl Acad. Sci. USA*, **108**, 1513–1518.
75. Schatz, M.C., Delcher, A.L., and Salzberg, S.L. (2010) Assembly of large genomes using second-generation sequencing. *Genome Res.*, **20**, 1165–1173.
76. Vaughn, C.P., Robles, J., Swensen, J.J., Miller, C.E. *et al.* (2010) Clinical analysis of PMS2: mutation detection and avoidance of pseudogenes. *Hum. Mutat.*, **31**, 588–593.
77. Hansen, M.F., Neckmann, U., Lavik, L.A., Vold, T. *et al.* (2014) A massive parallel sequencing workflow for diagnostic genetic testing of mismatch repair genes. *Mol. Genet. Genomic Med.*, **2**, 186–200.
78. Au, K.F., Underwood, J.G., Lee, L., and Wong, W.H. (2012) Improving PacBio long read accuracy by short read alignment. *PLoS One*, **7**, e46679.

79. Eid, J., Fehr, A., Gray, J., Luong, K. *et al.* (2009) Real-time DNA sequencing from single polymerase molecules. *Science*, **323**, 133–138.
80. Schadt, E.E., Turner, S., and Kasarskis, A. (2010) A window into third-generation sequencing. *Hum. Mol. Genet.*, **19**, R227–R240.
81. Rasko, D.A., Webster, D.R., Sahl, J.W., Bashir, A. *et al.* (2011) Origins of the *E. coli* strain causing an outbreak of hemolytic-uremic syndrome in Germany. *N. Engl. J. Med.*, **365**, 709–717.
82. Chin, C.S., Sorenson, J., Harris, J.B., Robins, W.P. *et al.* (2011) The origin of the Haitian cholera outbreak strain. *N. Engl. J. Med.*, **364**, 33–42.
83. Okoniewski, M.J., Meienberg, J., Patrignani, A., Szabelska, A. *et al.* (2013) Precise breakpoint localization of large genomic deletions using PacBio and Illumina next-generation sequencers. *Biotechniques*, **54**, 98–100.
84. Ritz, A., Bashir, A., Sindi, S., Hsu, D. *et al.* (2014) Characterization of structural variants with single molecule and hybrid sequencing approaches. *Bioinformatics*, **30**, 3458–3466.
85. Huddleston, J., Ranade, S., Malig, M., Antonacci, F. *et al.* (2014) Reconstructing complex regions of genomes using long-read sequencing technology. *Genome Res.*, **24**, 688–696.
86. Koren, S., Schatz, M.C., Walenz, B.P., Martin, J. *et al.* (2012) Hybrid error correction and de novo assembly of single-molecule sequencing reads. *Nat. Biotechnol.*, **30**, 693–700.
87. Archer, J., Weber, J., Henry, K., Winner, D. *et al.* (2012) Use of four next-generation sequencing platforms to determine HIV-1 coreceptor tropism. *PLoS One*, **7**, e49602.
88. Jiao, X., Zheng, X., Ma, L., Kutty, G. *et al.* (2013) A benchmark study on error assessment and quality control of CCS reads derived from the PacBio RS. *J. Data Min. Genomics Proteomics*, **4**, 1–5.
89. Laszlo, A.H., Derrington, I.M., Ross, B.C., Brinkerhoff, H. *et al.* (2014) Decoding long nanopore sequencing reads of natural DNA. *Nat. Biotechnol.*, **32**, 829–833.
90. Stoddart, D., Heron, A.J., Klingelhoefer, J., Mikhailova, E. *et al.* (2010) Nucleobase recognition in ssDNA at the central constriction of the alpha-hemolysin pore. *Nano Lett.*, **10**, 3633–3637.
91. Manrao, E.A., Derrington, I.M., Laszlo, A.H., Langford, K.W. *et al.* (2012) Reading DNA at single-nucleotide resolution with a mutant MspA nanopore and phi29 DNA polymerase. *Nat. Biotechnol.*, **30**, 349–353.
92. Tsutsui, M., He, Y., Furuhashi, M., Rahong, S. *et al.* (2012) Transverse electric field dragging of DNA in a nanochannel. *Sci. Rep.*, **2**, 394.
93. Singer, A., McNally, B., Torre, R.D., and Meller, A. (2012) DNA sequencing by nanopore-induced photon emission. *Methods Mol. Biol.*, **870**, 99–114.
94. Derrington, I.M., Butler, T.Z., Collins, M.D., Manrao, E. *et al.* (2010) Nanopore DNA sequencing with MspA. *Proc. Natl Acad. Sci. USA*, **107**, 16060–16065.
95. Ashton, P.M., Nair, S., Dallman, T., Rubino, S. *et al.* (2014) MinION nanopore sequencing identifies the position and structure of a bacterial antibiotic resistance island. *Nat. Biotechnol.*, **33**, 296–300.
96. Mikheyev, A.S. and Tin, M.M. (2014) A first look at the Oxford Nanopore MinION sequencer. *Mol. Ecol. Resour.*, **14**, 1097–1102.
97. O'Donnell, C.R., Wang, H., and Dunbar, W.B. (2013) Error analysis of idealized nanopore sequencing. *Electrophoresis*, **34**, 2137–2144.
98. Loman, N.J., and Quinlan, A.R. (2014) Poretools: a toolkit for analyzing nanopore sequence data. *Bioinformatics*, **30**, 3399–3401.
99. Dressman, D., Yan, H., Traverso, G., Kinzler, K.W., and Vogelstein, B. (2003) Transforming single DNA molecules into fluorescent magnetic particles for detection and enumeration of genetic variations. *Proc. Natl Acad. Sci. USA*, **100**, 8817–8822.
100. Margulies, M., Egholm, M., Altman, W.E., Attiya, S. *et al.* (2005) Genome sequencing in microfabricated high-density picolitre reactors. *Nature*, **437**, 376–380.
101. Valouev, A., Ichikawa, J., Tonthat, T., Stuart, J. *et al.* (2008) A high-resolution, nucleosome position map of *C. elegans* reveals a lack of universal sequence-dictated positioning. *Genome Res.*, **18**, 1051–1063.

102. Thiede, C., Koch, S., Creutzig, E., Steudel, C. *et al.* (2006) Prevalence and prognostic impact of NPM1 mutations in 1485 adult patients with acute myeloid leukemia (AML). *Blood*, **107**, 4011–4020.
103. Huang, Y.F., Chen, S.C., Chiang, Y.S., Chen, T.H., and Chiu, K.P. (2012) Palindromic sequence impedes sequencing-by-ligation mechanism. *BMC Syst. Biol.*, **6** Suppl. 2, S10.
104. Tuna, M. and Amos, C.I. (2013) Genomic sequencing in cancer. *Cancer Lett.*, **340**, 161–170.
105. Clark, M.J., Chen, R., Lam, H.Y., Karczewski, K.J. *et al.* (2011) Performance comparison of exome DNA sequencing technologies. *Nat. Biotechnol.*, **29**, 908–914.
106. Wooderchak-Donahue, W.L., O'Fallon, B., Furtado, L.V., Durtschi, J.D. *et al.* (2012) A direct comparison of next generation sequencing enrichment methods using an aortopathy gene panel-clinical diagnostics perspective. *BMC Med. Genomics*, **5**, 50.
107. Pritchard, C.C., Salipante, S.J., Koehler, K., Smith, C. *et al.* (2013) Validation and implementation of targeted capture and sequencing for the detection of actionable mutation, copy number variation, and gene rearrangement in clinical cancer specimens. *J. Mol. Diagn.*, **16**, 56–67.
108. Cottrell, C.E., Al-Kateb, H., Bredemeyer, A.J., Duncavage, E.J. *et al.* (2014) Validation of a next-generation sequencing assay for clinical molecular oncology. *J. Mol. Diagn.*, **16**, 89–105.
109. Tsongalis, G.J., Peterson, J.D., de Abreu, F.B., Tunkey, C.D. *et al.* (2014) Routine use of the Ion Torrent AmpliSeq™ Cancer Hotspot Panel for identification of clinically actionable somatic mutations. *Clin. Chem. Lab. Med.*, **52**, 707–714.
110. Singh, R.R., Patel, K.P., Routbort, M.J., Reddy, N.G. *et al.* (2013) Clinical validation of a next-generation sequencing screen for mutational hotspots in 46 cancer-related genes. *J. Mol. Diagn.*, **15**, 607–622.
111. Beadling, C., Neff, T.L., Heinrich, M.C., Rhodes, K. *et al.* (2013) Combining highly multiplexed PCR with semiconductor-based sequencing for rapid cancer genotyping. *J. Mol. Diagn.*, **15**, 171–176.
112. Hadd, A.G., Houghton, J., Choudhary, A., Sah, S. *et al.* (2013) Targeted, high-depth, next-generation sequencing of cancer genes in formalin-fixed, paraffin-embedded and fine-needle aspiration tumor specimens. *J. Mol. Diagn.*, **15**, 234–247.
113. Escaramís, G., Tornador, C., Bassaganyas, L., Rabionet, R. *et al.* (2013) PeSV-Fisher: identification of somatic and non-somatic structural variants using next generation sequencing data. *PLoS One*, **8**, e63377.
114. Ye, K., Schulz, M.H., Long, Q., Apweiler, R., and Ning, Z. (2009) Pindel: a pattern growth approach to detect break points of large deletions and medium sized insertions from paired-end short reads. *Bioinformatics*, **25**, 2865–2871.
115. Krumm, N., Sudmant, P.H., Ko, A., O'Roak, B.J. *et al.* (2012) Copy number variation detection and genotyping from exome sequence data. *Genome Res.*, **22**, 1525–1532.
116. Southern, E.M. (2001) DNA microarrays. History and overview. *Methods Mol. Biol.*, **170**, 1–15.
117. Hoheisel, J.D. (1994) Application of hybridization techniques to genome mapping and sequencing. *Trends Genet.*, **10**, 79–83.
118. Greisman, H.A., Hoffman, N.G., and Yi, H.S. (2011) Rapid high-resolution mapping of balanced chromosomal rearrangements on tiling CGH arrays. *J. Mol. Diagn.*, **13**, 621–633.
119. Tough, I.M., Court Brown, W.M., Baikie, A.G., Buckton, K.E. *et al.* (1961) Cytogenetic studies in chronic myeloid leukaemia and acute leukaemia associated with monogolism. *Lancet*, **1**, 411–417.
120. Hughes, T.P., Hochhaus, A., Branford, S., Müller, M.C. *et al.* (2010) Long-term prognostic significance of early molecular response to imatinib in newly diagnosed chronic myeloid leukemia: an analysis from the International Randomized Study of Interferon and STI571 (IRIS). *Blood*, **116**, 3758–3765.
121. Tiacci, E., Trifonov, V., Schiavoni, G., Holmes, A. *et al.* (2011) BRAF mutations in hairy-cell leukemia. *N. Engl. J. Med.*, **364**, 2305–2315.
122. Rider, T., Powell, R., Gover, R., Ansell, R. *et al.* (2013) Molecular detection of BRAF-V600E is superior to flow cytometry for disease evaluation in hairy cell leukaemia. *Hematol. Oncol.*, **32**, 158–161.

123. Dietrich, S., Glimm, H., Andrulis, M., von Kalle, C. *et al.* (2012) BRAF inhibition in refractory hairy-cell leukemia. *N. Engl. J. Med.*, **366**, 2038–2040.
124. Licht, J.D. (2009) Acute promyelocytic leukemia – weapons of mass differentiation. *N. Engl. J. Med.*, **360**, 928–930.
125. Choudhary, A. and DeLoughery, T.G. (2012) Bleeding and thrombosis in acute promyelocytic leukemia. *Am. J. Hematol.*, **87**, 596–603.
126. Mi, J.Q., Li, J.M., Shen, Z.X., Chen, S.J., and Chen, Z. (2012) How to manage acute promyelocytic leukemia. *Leukemia*, **26**, 1743–1751.
127. Brockman, S.R., Paternoster, S.F., Ketterling, R.P., and Dewald, G.W. (2003) New highly sensitive fluorescence in situ hybridization method to detect PML/RARA fusion in acute promyelocytic leukemia. *Cancer Genet. Cytogenet.*, **145**,144–151.
128. Sanz, M.A., Grimwade, D., Tallman, M.S., Lowenberg, B. *et al.* (2009) Management of acute promyelocytic leukemia: recommendations from an expert panel on behalf of the European LeukemiaNet. *Blood*, **113**, 1875–1891.
129. Campbell, L.J., Oei, P., Brookwell, R., Shortt, J. *et al.* (2013) FISH detection of PML-RARA fusion in ins(15;17) acute promyelocytic leukaemia depends on probe size. *Biomed. Res. Int.*, **2013**, 164501.
130. Kim, M.J., Cho, S.Y., Kim, M.H., Lee, J.J. *et al.* (2010) FISH-negative cryptic PML-RARA rearrangement detected by long-distance polymerase chain reaction and sequencing analyses: a case study and review of the literature. *Cancer Genet. Cytogenet.*, **203**, 278–283.
131. Welch, J.S., Westervelt, P., Ding, L., Larson, D.E. *et al.* (2011) Use of whole-genome sequencing to diagnose a cryptic fusion oncogene. *JAMA*, **305**, 1577–1584.
132. Koshy, J., Qian, Y.W., Bhagwath, G., Willis, M. *et al.* (2012) Microarray, gene sequencing, and reverse transcriptase-polymerase chain reaction analyses of a cryptic PML-RARA translocation. *Cancer Genet.*, **205**, 537–540.
133. French, C.A. (2010) Demystified molecular pathology of NUT midline carcinomas. *J. Clin. Pathol.*, **63**, 492–496.
134. Bauer, D.E., Mitchell, C.M., Strait, K.M., Lathan, C.S. *et al.* (2012) Clinicopathologic features and long-term outcomes of NUT midline carcinoma. *Clin. Cancer Res.*, **18**, 5773–5779.
135. Stelow, E.B., Bellizzi, A.M., Taneja, K., Mills, S.E. *et al.* (2008) NUT rearrangement in undifferentiated carcinomas of the upper aerodigestive tract. *Am. J. Surg. Pathol.*, **32**, 828–834.
136. Ziai, J., French, C.A., and Zambrano, E. (2010) NUT gene rearrangement in a poorly-differentiated carcinoma of the submandibular gland. *Head Neck Pathol.*, **4**, 163–168.
137. Bishop, J.A. and Westra, W.H. (2012) NUT midline carcinomas of the sinonasal tract. *Am. J. Surg. Pathol.*, **36**, 1216–1221.
138. French, C.A. (2013) The importance of diagnosing NUT midline carcinoma. *Head Neck Pathol.*, **7**, 11–16.
139. French, C.A., Kutok, J.L., Faquin, W.C., Toretsky, J.A. *et al.* (2004) Midline carcinoma of children and young adults with NUT rearrangement. *J. Clin. Oncol.*, **22**, 4135–4139.
140. Eggert, T., McGlynn, K.A., Duffy, A., Manns, M.P. *et al.* (2013) Fibrolamellar hepatocellular carcinoma in the USA, 2000–2010: a detailed report on frequency, treatment and outcome based on the Surveillance, Epidemiology, and End Results database. *United Eur. Gastroenterol. J.*, **1**, 351–357.
141. Klein, W.M., Molmenti, E.P., Colombani, P.M., Grover, D.S. *et al.* (2005) Primary liver carcinoma arising in people younger than 30 years. *Am. J. Clin. Pathol.*, **124**, 512–518.
142. Honeyman, J.N., Simon, E.P., Robine, N., Chiaroni-Clarke, R. *et al.* (2014) Detection of a recurrent DNAJB1–PRKACA chimeric transcript in fibrolamellar hepatocellular carcinoma. *Science*, **343**, 1010–1014.
143. Graham, R.P., Jin, L., Knutson, D.L., Kloft-Nelson, S.M. *et al.* (2015) DNAJB1–PRKACA is specific for fibrolamellar carcinoma. *Mod. Pathol.*, **28** (6), 822–829.
144. Malouf, G., Falissard, B., Azoulay, D., Callea, F. *et al.* (2009) Is histological diagnosis of primary liver carcinomas with fibrous stroma reproducible among experts? *J. Clin. Pathol.*, **62**, 519–524.
145. Paschka, P., Marcucci, G., Ruppert, A.S., Mrózek, K. *et al.* (2006) Adverse prognostic significance of KIT mutations in adult acute

myeloid leukemia with inv(16) and t(8;21): a Cancer and Leukemia Group B Study. *J. Clin. Oncol.*, **24**, 3904–3911.
146. Pollard, J.A., Alonzo, T.A., Gerbing, R.B., Ho, P.A. *et al.* (2010) Prevalence and prognostic significance of KIT mutations in pediatric patients with core binding factor AML enrolled on serial pediatric cooperative trials for de novo AML. *Blood*, **115**, 2372–2379.
147. Ferrara, F., Palmieri, S., and Leoni, F. (2008) Clinically useful prognostic factors in acute myeloid leukemia. *Crit. Rev. Oncol. Hematol.*, **66**, 181–193.
148. Falini, B., Mecucci, C., Tiacci, E., Alcalay, M. *et al.* (2005) Cytoplasmic nucleophosmin in acute myelogenous leukemia with a normal karyotype. *N. Engl. J. Med.*, **352**, 254–266.
149. Whitman, S.P., Ruppert, A.S., Radmacher, M.D., Mrózek, K. *et al.* (2008) FLT3 D835/I836 mutations are associated with poor disease-free survival and a distinct gene-expression signature among younger adults with de novo cytogenetically normal acute myeloid leukemia lacking FLT3 internal tandem duplications. *Blood*, **111**, 1552–1559.
150. Whitman, S.P., Archer, K.J., Feng, L., Baldus, C. *et al.* (2001) Absence of the wild-type allele predicts poor prognosis in adult *de novo* acute myeloid leukemia with normal cytogenetics and the internal tandem duplication of FLT3: a cancer and leukemia group B study. *Cancer Res.*, **61**, 7233–7239.
151. Preudhomme, C., Sagot, C., Boissel, N., Cayuela, J.M. *et al.* (2002) Favorable prognostic significance of CEBPA mutations in patients with *de novo* acute myeloid leukemia: a study from the Acute Leukemia French Association (ALFA). *Blood*, **100**, 2717–2723.
152. Döhner, K., Schlenk, R.F., Habdank, M., Scholl, C. *et al.* (2005) Mutant nucleophosmin (NPM1) predicts favorable prognosis in younger adults with acute myeloid leukemia and normal cytogenetics: interaction with other gene mutations. *Blood*, **106**, 3740–3746.
153. Kim, T.M., Lee, S.H., and Chung, Y.J. (2013) Clinical applications of next-generation sequencing in colorectal cancers. *World J. Gastroenterol.*, **19**, 6784–6793.
154. Pritchard, C.C., Smith, C., Salipante, S.J., Lee, M.K. *et al.* (2012) ColoSeq provides comprehensive lynch and polyposis syndrome mutational analysis using massively parallel sequencing. *J. Mol. Diagn.*, **14**, 357–366.
155. Shirts, B.H., Salipante, S.J., Casadei, S., Ryan, S. *et al.* (2014) Deep sequencing with intronic capture enables identification of an APC exon 10 inversion in a patient with polyposis. *Genet. Med.*, **16** (10), 783–786.
156. Stadler, Z.K., Schrader, K.A., Vijai, J., Robson, M.E., and Offit, K. (2014) Cancer genomics and inherited risk. *J. Clin. Oncol.*, **32**, 687–698.
157. Walsh, T., Lee, M.K., Casadei, S., Thornton, A.M. *et al.* (2010) Detection of inherited mutations for breast and ovarian cancer using genomic capture and massively parallel sequencing. *Proc. Natl Acad. Sci. USA*, **107**, 12629–12633.
158. Vaughn, C.P., Baker, C.L., Samowitz, W.S., and Swensen, J.J. (2013) The frequency of previously undetectable deletions involving 3' exons of the PMS2 gene. *Genes Chromosomes Cancer*, **52**, 107–112.
159. Deininger, M.W., Goldman, J.M., Lydon, N., and Melo, J.V. (1997) The tyrosine kinase inhibitor CGP57148B selectively inhibits the growth of BCR-ABL-positive cells. *Blood*, **90**, 3691–3698.
160. Druker, B.J., Tamura, S., Buchdunger, E., Ohno, S. *et al.* (1996) Effects of a selective inhibitor of the Abl tyrosine kinase on the growth of Bcr-Abl positive cells. *Nat. Med.*, **2**, 561–566.
161. Huang, X., Cortes, J., and Kantarjian, H. (2012) Estimations of the increasing prevalence and plateau prevalence of chronic myeloid leukemia in the era of tyrosine kinase inhibitor therapy. *Cancer*, **118**, 3123–3127.
162. Kwak, E.L., Bang, Y.J., Camidge, D.R., Shaw, A.T. *et al.* (2010) Anaplastic lymphoma kinase inhibition in non-small-cell lung cancer. *N. Engl. J. Med.*, **363**, 1693–1703.
163. Bucheit, A.D., Syklawer, E., Jakob, J.A., Bassett, R.L. *et al.* (2013) Clinical characteristics and outcomes with specific BRAF and NRAS mutations in patients with metastatic melanoma. *Cancer*, **119**, 3821–3829.
164. Kim, T., Kim, J., and Lee, M.G. (2010) Inhibition of mutated BRAF in melanoma. *N. Engl. J. Med.*, **363**, 2261; author reply 2262.

165. Chapman, P.B., Hauschild, A., Robert, C., Haanen, J.B. *et al.* (2011) Improved survival with vemurafenib in melanoma with BRAF V600E mutation. *N. Engl. J. Med.*, **364**, 2507–2516.
166. Oberholzer, P.A., Kee, D., Dziunycz, P., Sucker, A. *et al.* (2012) RAS mutations are associated with the development of cutaneous squamous cell tumors in patients treated with RAF inhibitors. *J. Clin. Oncol.*, **30**, 316–321.
167. Hammerman, P.S., Sos, M.L., Ramos, A.H., Xu, C. *et al.* (2011) Mutations in the DDR2 kinase gene identify a novel therapeutic target in squamous cell lung cancer. *Cancer Discovery*, **1**, 78–89.
168. Rosell, R., Bivona, T.G., and Karachaliou, N. (2013) Genetics and biomarkers in personalisation of lung cancer treatment. *Lancet*, **382**, 720–731.
169. Van Cutsem, E., Peeters, M., Siena, S., Humblet, Y. *et al.* (2007) Open-label phase III trial of panitumumab plus best supportive care compared with best supportive care alone in patients with chemotherapy-refractory metastatic colorectal cancer. *J. Clin. Oncol.*, **25**, 1658–1664.
170. Amado, R.G., Wolf, M., Peeters, M., Van Cutsem, E. *et al.* (2008) Wild-type KRAS is required for panitumumab efficacy in patients with metastatic colorectal cancer. *J. Clin. Oncol.*, **26**, 1626–1634.
171. Rajagopalan, H., Bardelli, A., Lengauer, C., Kinzler, K.W. *et al.* (2002) Tumorigenesis: RAF/RAS oncogenes and mismatch-repair status. *Nature*, **418**, 934.
172. Roth, A.D., Tejpar, S., Delorenzi, M., Yan, P. *et al.* (2010) Prognostic role of KRAS and BRAF in stage II and III resected colon cancer: results of the translational study on the PETACC-3, EORTC 40993, SAKK 60-00 trial. *J. Clin. Oncol.*, **28**, 466–474.
173. Heller, T., Kirchheiner, J., Armstrong, V.W., Luthe, H. *et al.* (2006) AmpliChip CYP450 GeneChip: a new gene chip that allows rapid and accurate CYP2D6 genotyping. *Ther. Drug Monit.*, **28**, 673–677.
174. Jonsdottir, K., Assmus, J., Slewa, A., Gudlaugsson, E. *et al.* (2014) Prognostic value of gene signatures and proliferation in lymph-node-negative breast cancer. *PLoS One*, **9**, e90642.
175. Gondek, L.P., Tiu, R., O'Keefe, C.L., Sekeres, M.A. *et al.* (2008) Chromosomal lesions and uniparental disomy detected by SNP arrays in MDS, MDS/MPD, and MDS-derived AML. *Blood*, **111**, 1534–1542.
176. Tiu, R.V., Gondek, L.P., O'Keefe, C.L., Elson, P. *et al.* (2011) Prognostic impact of SNP array karyotyping in myelodysplastic syndromes and related myeloid malignancies. *Blood*, **117**, 4552–4560.
177. Rucker, F.G., Bullinger, L., Schwaenen, C., Lipka, D.B. *et al.* (2006) Disclosure of candidate genes in acute myeloid leukemia with complex karyotypes using microarray-based molecular characterization. *J. Clin. Oncol.*, **24**, 3887–3894.
178. Choi, H., Jung, C., Sohn, S.K., Kim, S. *et al.* (2013) Genome-wide genotype-based risk model for survival in acute myeloid leukaemia patients with normal karyotype. *Br. J. Haematol.*, **163**, 62–71.
179. Cusumano, P.G., Generali, D., Ciruelos, E., Manso, L. *et al.* (2014) European inter-institutional impact study of MammaPrint. *Breast*, **23**, 423–428.
180. Bea, S., Zettl, A., Wright, G., Salaverria, I. *et al.* (2005) Diffuse large B-cell lymphoma subgroups have distinct genetic profiles that influence tumor biology and improve gene-expression-based survival prediction. *Blood*, **106**, 3183–3190.
181. Andre, F., Bachelot, T., Commo, F., Campone, M. *et al.* (2014) Comparative genomic hybridisation array and DNA sequencing to direct treatment of metastatic breast cancer: a multicentre, prospective trial (SAFIR01/UNICANCER). *Lancet Oncol.*, **15**, 267–274.
182. Craddock, K.J., Lam, W.L., and Tsao, M.S. (2013) Applications of array-CGH for lung cancer. *Methods Mol. Biol.*, **973**, 297–324.
183. Chen, M., Ye, Y., Yang, H., Tamboli, P. *et al.* (2009) Genome-wide profiling of chromosomal alterations in renal cell carcinoma using high-density single nucleotide polymorphism arrays. *Int. J. Cancer*, **125**, 2342–2348.
184. Silveira, S.M., Villacis, R.A., Marchi, F.A., Barros Filho M. de, C. *et al.* (2013) Genomic signatures predict poor outcome in undifferentiated pleomorphic sarcomas and leiomyosarcomas. *PLoS One*, **8**, e67643.
185. Wapner, R.J., Martin, C.L., Levy, B., Ballif, B.C. *et al.* (2012) Chromosomal microarray versus karyotyping for prenatal diagnosis. *N. Engl. J. Med.*, **367**, 2175–2184.

186. Ellison, J.W., Ravnan, J.B., Rosenfeld, J.A., Morton, S.A. *et al.* (2012) Clinical utility of chromosomal microarray analysis. *Pediatrics*, **130**, e1085–e1095.
187. Miller, M.B. and Tang, Y.W. (2009) Basic concepts of microarrays and potential applications in clinical microbiology. *Clin. Microbiol. Rev.*, **22**, 611–633.
188. Hourigan, C.S. and Karp, J.E. (2013) Minimal residual disease in acute myeloid leukaemia. *Nat. Rev. Clin. Oncol.*, **10**, 460–471.
189. Yeung, D.T., Parker, W.T., and Branford, S. (2011) Molecular methods in diagnosis and monitoring of haematological malignancies. *Pathology*, **43**, 566–579.
190. Belaud-Rotureau, M.A., Parrens, M., Dubus, P., Garroste, J.C. *et al.* (2002) A comparative analysis of FISH, RT-PCR, PCR, and immunohistochemistry for the diagnosis of mantle cell lymphomas. *Mod. Pathol.*, **15**, 517–525.
191. Einerson, R.R., Kurtin, P.J., Dayharsh, G.A., Kimlinger, T.K, and Remstein, E.D. (2005) FISH is superior to PCR in detecting t(14;18)(q32;q21)-IgH/bcl-2 in follicular lymphoma using paraffin-embedded tissue samples. *Am. J. Clin. Pathol.*, **124**, 421–429.
192. Kim, Y.J., Kim, D.W., Lee, S., Kim, H.J. *et al.* (2002) Comprehensive comparison of FISH, RT-PCR, and RQ-PCR for monitoring the BCR-ABL gene after hematopoietic stem cell transplantation in CML. *Eur. J. Haematol.*, **68**, 272–280.
193. Polampalli, S., Choughule, A., Prabhash, K., Amare, P. *et al.* (2011) Role of RT-PCR and FISH in diagnosis and monitoring of acute promyelocytic leukemia. *Indian J. Cancer*, **48**, 60–67.
194. Tubbs, R.R. and Stoler, M.H. (2009) *Cell-and Tissue Based Molecular Pathology.* Churchill Livingstone/Elsevier, Philadelphia, 1st edn, 441 pp.
195. Ladetto, M., Brüggemann, M., Monitillo, L., Ferrero, S. *et al.* (2013) Next-generation sequencing and real-time quantitative PCR for minimal residual disease detection in B-cell disorders. *Leukemia*, **28**, 1299–1307.
196. Faham, M., Zheng, J., Moorhead, M., Carlton, V.E. *et al.* (2012) Deep-sequencing approach for minimal residual disease detection in acute lymphoblastic leukemia. *Blood*, **120**, 5173–5180.
197. Limpens, J., Stad, R., Vos, C., de Vlaam, C. *et al.* (1995) Lymphoma-associated translocation t(14;18) in blood B cells of normal individuals. *Blood*, **85**, 2528–2536.
198. Kitzman, J.O., Snyder, M.W., Ventura, M., Lewis, A.P. *et al.* (2012) Noninvasive whole-genome sequencing of a human fetus. *Sci. Transl. Med.*, **4**, 137ra76.
199. Breitenbuecher, F., Hoffarth, S., Worm, K., Cortes-Incio, D. *et al.* (2014) Development of a highly sensitive and specific method for detection of circulating tumor cells harboring somatic mutations in non-small-cell lung cancer patients. *PLoS One*, **9**, e85350.
200. Leary, R.J., Sausen, M., Kinde, I., Papadopoulos, N. *et al.* (2012) Detection of chromosomal alterations in the circulation of cancer patients with whole-genome sequencing. *Sci. Transl. Med.*, **4**, 162ra54.
201. Lowes, L.E., and Allan, A.L. (2014) Recent advances in the molecular characterization of circulating tumor cells. *Cancers (Basel)*, **6**, 595–624.
202. Imperiale, T.F., Ransohoff, D.F., Itzkowitz, S.H., Levin, T.R. *et al.* (2014) Multitarget stool DNA testing for colorectal-cancer screening. *N. Engl. J. Med.*, **370**, 1287–1297.
203. Lee, H.K., Kim, H.W., Lee, I.Y., Lee, J. *et al.* (2014) G-749, a novel FLT3 kinase inhibitor can overcome drug resistance for the treatment of acute myeloid leukemia. *Blood*, **123** (14), 2209–2219.
204. Shaw, A.T., Kim, D.W., Mehra, R., Tan, D.S. *et al.* (2014) Ceritinib in ALK-rearranged non-small-cell lung cancer. *N. Engl. J. Med.*, **370**, 1189–1197.
205. Soverini, S., Hochhaus, A., Nicolini, F.E., Gruber, F. *et al.* (2011) BCR-ABL kinase domain mutation analysis in chronic myeloid leukemia patients treated with tyrosine kinase inhibitors: recommendations from an expert panel on behalf of European LeukemiaNet. *Blood*, **118**, 1208–1215.
206. Jabbour, E., Branford, S., Saglio, G., Jones, D. *et al.* (2011) Practical advice for determining the role of BCR-ABL mutations in guiding tyrosine kinase inhibitor therapy in patients with chronic myeloid leukemia. *Cancer*, **117**, 1800–1811.
207. The Cancer Genome Atlas Research Network (2013) Genomic and epigenomic landscapes of adult *de novo* acute myeloid leukemia. *N. Engl. J. Med.*, **368**, 2059–2074.

208. Simon, R. and Roychowdhury, S. (2013) Implementing personalized cancer genomics in clinical trials. *Nat. Rev. Drug Discovery*, **12**, 358–369.

209. Schilsky, R.L. (2014) Implementing personalized cancer care. *Nat. Rev. Clin. Oncol.*, **11** (7), 432–438.

210. Simon, R. (2008) The use of genomics in clinical trial design. *Clin. Cancer Res.*, **14**, 5984–5993.

211. Dimov, N.D., Medeiros, L.J., Kantarjian, H.M., Cortes, J.E. *et al.* (2010) Rapid and reliable confirmation of acute promyelocytic leukemia by immunofluorescence staining with an antipromyelocytic leukemia antibody: the M.D. Anderson Cancer Center experience of 349 patients. *Cancer*, **116**, 369–376.

212. Hay, R., MacRae, E., Barber, D., Khalil, M., and Demetrick, D.J. (2007) BRAF mutations in melanocytic lesions and papillary thyroid carcinoma samples identified using melting curve analysis of polymerase chain reaction products. *Arch. Pathol. Lab. Med.*, **131**, 1361–1367.

213. Dufresne, S.D., Belloni, D.R., Levy, N.B., and Tsongalis, G.J. (2007) Quantitative assessment of the BCR-ABL transcript using the Cepheid Xpert BCR-ABL Monitor assay. *Arch. Pathol. Lab. Med.*, **131**, 947–950.

214. Anderson, S., Bloom, K.J., Vallera, D.U., Rueschoff, J. *et al.* (2012) Multisite analytic performance studies of a real-time polymerase chain reaction assay for the detection of BRAF V600E mutations in formalin-fixed, paraffin-embedded tissue specimens of malignant melanoma. *Arch. Pathol. Lab. Med.*, **136**, 1385–1391.

215. Pritchard, C.C., Akagi, L., Reddy, P.L., Joseph, L., and Tait, J.F. (2010) COLD-PCR enhanced melting curve analysis improves diagnostic accuracy for KRAS mutations in colorectal carcinoma. *BMC Clin. Pathol. England*, **10**, 6.

216. Dijkstra, J.R., Heideman, D.A., Meijer, G.A., Boers, J.E. *et al.* (2013) KRAS mutation analysis on low percentage of colon cancer cells: the importance of quality assurance. *Virchows Arch.*, **462**, 39–46.

217. Djalalov, S., Beca, J., Hoch, J.S., Krahn, M. *et al.* (2014) Cost effectiveness of EML4-ALK fusion testing and first-line crizotinib treatment for patients with advanced ALK-positive non-small-cell lung cancer. *J. Clin. Oncol.*, **32**, 1012–1019.

218. Rosell, R., Moran, T., Queralt, C., Porta, R. *et al.* (2009) Screening for epidermal growth factor receptor mutations in lung cancer. *N. Engl. J. Med.*, **361**, 958–967.

219. Dacic, S. (2013) Molecular genetic testing for lung adenocarcinomas: a practical approach to clinically relevant mutations and translocations. *J. Clin. Pathol.*, **66**, 870–874.

220. Lu, G., Muddasani, R., Orlowski, R.Z., Abruzzo, L.V. *et al.* (2013) Plasma cell enrichment enhances detection of high-risk cytogenomic abnormalities by fluorescence *in situ* hybridization and improves risk stratification of patients with plasma cell neoplasms. *Arch. Pathol. Lab. Med.*, **137**, 625–631.

221. Coleman, J.F., Theil, K.S., Tubbs, R.R., and Cook, J.R. (2011) Diagnostic yield of bone marrow and peripheral blood FISH panel testing in clinically suspected myelodysplastic syndromes and/or acute myeloid leukemia: a prospective analysis of 433 cases. *Am. J. Clin. Pathol.*, **135**, 915–920.

222. Lasota, J. and Miettinen, M. (2006) KIT and PDGFRA mutations in gastrointestinal stromal tumors (GISTs). *Semin. Diagn. Pathol.*, **23**, 91–102.

223. Lastra, E., García-González, M., Llorente, B., Bernuy, C. *et al.* (2012) Lynch syndrome diagnostics: decision-making process for germ-line testing. *Clin. Transl. Oncol.*, **14**, 254–262.

224. Hegde, M., Ferber, M., Mao, R., Samowitz, W. *et al.* (2014) ACMG technical standards and guidelines for genetic testing for inherited colorectal cancer (Lynch syndrome, familial adenomatous polyposis, and MYH-associated polyposis). *Genet. Med.*, **16**, 101–116.

225. Kurian, A.W., Hare, E.E., Mills, M.A., Kingham, K.E. *et al.* (2014) Clinical evaluation of a multiple-gene sequencing panel for hereditary cancer risk assessment. *J. Clin. Oncol.*, **32** (19), 2001–2009.

226. de Biase, D., Visani, M., Malapelle, U., Simonato, F. *et al.* (2013) Next-generation sequencing of lung cancer EGFR exons 18-21 allows effective molecular diagnosis of small routine samples (cytology and biopsy). *PLoS One*, **8**, e83607.

227. McCourt, C.M., McArt, D.G., Mills, K., Catherwood, M.A. *et al.* (2013) Validation of next generation sequencing technologies in comparison to current diagnostic gold standards for BRAF, EGFR and KRAS mutational analysis. *PLoS One*, **8**, e69604.

228. Tuononen, K., Mäki-Nevala, S., Sarhadi, V.K., Wirtanen, A. *et al.* (2013) Comparison of targeted next-generation sequencing (NGS) and real-time PCR in the detection of EGFR, KRAS, and BRAF mutations on formalin-fixed, paraffin-embedded tumor material of non-small cell lung carcinoma-superiority of NGS. *Genes Chromosomes Cancer*, **52**, 503–511.

229. Johansen Taber, K.A., Dickinson, B.D., and Wilson, M. (2014) The promise and challenges of next-generation genome sequencing for clinical care. *JAMA Intern. Med.*, **174**, 275–280.

230. Daniels, M., Goh, F., Wright, C.M., Sriram, K.B. *et al.* (2012) Whole-genome sequencing for lung cancer. *J. Thorac. Dis.*, **4**, 155–163.

231. Tenedini, E., Bernardis, I., Artusi, V., Artuso, L. *et al.* (2013) Targeted cancer exome sequencing reveals recurrent mutations in myeloproliferative neoplasms. *Leukemia*, 28, 1052–1059.

232. Mardis, E.R., Ding, L., Dooling, D.J., Larson, D.E. *et al.* (2009) Recurring mutations found by sequencing an acute myeloid leukemia genome. *N. Engl. J. Med.*, **361**, 1058–1066.

233. Dewey, F.E., Grove, M.E., Pan, C., Goldstein, B.A. *et al.* (2014) Clinical interpretation and implications of whole-genome sequencing. *JAMA*, **311**, 1035–1045.

12
Metabolomic Imaging for Cancer Detection

Emily A. Decelle and Leo L. Cheng
Massachusetts General Hospital, Harvard Medical School, Departments of Radiology and Pathology, Boston, Massachusetts 02114, USA

Keywords

Cancer
A broad group of diseases involving unregulated cell growth.

Translational Medicine Cancer: Advances in Molecular Biology and Medicine,
First Edition. Edited by Robert A. Meyers.

Metabolomics
The scientific study of global variations of metabolites.

Metabolomic imaging
The mapping of cancer-specific biomolecular profile values onto anatomic structures to guide disease localization diagnosis and characterization.

Magnetic resonance spectroscopy
A research technique that exploits the magnetic properties of certain atomic nuclei to determine the physical and chemical properties of molecules.

Mass spectrometry
An analytical technique that identifies atoms and molecules by generating charged molecules and measuring their mass-to-charge ratios.

The field of metabolomics for cancer diagnosis and characterization is relatively new, and especially the application of metabolomics for *in-vivo* cancer imaging is still in its infancy. Cancer metabolomics involves the study of global variations of metabolites, with which malignancy conditions can be evaluated by profiling the entire measurable metabolome, instead of focusing only on certain metabolites or isolated metabolic pathways. At present, the study of cancer metabolomics is mainly accomplished utilizing magnetic resonance spectroscopy (MRS) and mass spectrometry (MS). These studies aim to uncover disease-specific metabolomic profiles in order to better understand variations in the involved pathways of cancer metabolism, and to utilize these profiles to establish metabolomic criteria for cancer detection, characterization, and patient prognostication and monitoring. The ultimate goal for cancer metabolomics is its implementation in the clinic to improve clinical abilities for cancer diagnosis and characterization, and through the development of *in-vivo* metabolomic imaging these clinical goals can be achieved noninvasively.

1
Introduction to Metabolomics

The molecular biological science of "-omics" evolves from the suffix "-ome," denoting "…all of the specified constituents of a cell, considered collectively or in total," according to the *Oxford English Dictionary* (Third Edition). Within this scientific topic, the earlier-established disciplines of genomics and proteomics paved the way for the introduction and development of a newer "-omics" family member, metabolomics. Metabolomics is the study of global variations of metabolites, the small-molecule intermediates and products of cellular metabolism, from all measurable metabolic pathways. More specifically, in relation to human oncology, cancer metabolomics can be applied to investigate global metabolite variations in relation to physiological and pathological conditions arising due to oncological development and progression [1].

1.1 Background

During the past half-century – a time when hypothesis-driven research and evidence-based medicine have arisen as the dominant mechanisms for medical research and clinical decision-making – the conception of the study of "-omics" presented somewhat of a deviation from this classic research paradigm. After technological innovation had allowed for the completion of the Human Genome Project in 2003, the field of genomics blossomed, in contrast to classical genetics – that is, the study of individual genes – to reflect the study of correlations between large arrays of genome data and human physiological conditions. The inherent exploratory nature of searching for these correlations, often involving genes or proteins with unknown functions, was a departure from the standard hypothesis-based research structure. However, the introduction of this "-omics" branch to the study of genes allowed for a greater exploration into unknown realms of molecular biology, without sacrificing scientific accuracy, but with the understanding that future efforts would be necessary to fully comprehend the findings that involved the unknown building blocks of life.

While the terms *genome* andproteome refer to the complete set of genes and proteins, respectively, but only to a certain extent as determined by measurement capabilities, the metabolome refers to the complete set of all small and identifiable molecules (<1500 Da), termed metabolites, measurable under certain experimental conditions. The Human Metabolome Database (HMDB) was first published in 2007, and is currently the most complete and comprehensive collection of human metabolites, including records of nearly 21 000 detected metabolites in the human body [2]. As opposed to genomics and proteomics, which often uncover the involvement of unknown genes and proteins, very few new metabolites may be discovered from metabolomic studies with currently employed analytical methodologies. Rather, at present, the main goal of metabolomics is to interpret the global alterations of the vast amount of known metabolites represented by various metabolic profiles of numerous known metabolic pathways. In this way, metabolomics may be considered less complicated than proteomics and genomics in terms of the observation of new individual elements, as is seen in genomics with the development of gene microarrays, but as complex as genomics and proteomics with regard to evaluations of their connections with physiology and pathology.

The cornerstone of the principle of metabolomics, and of the other "-omics" disciplines, is the widely accepted scientific concept that all processes of human biology are interconnected. Therefore, alterations in one metabolic pathway almost certainly affect the others, and so studying an overall profile of metabolic changes is more representative and informative than studying individual metabolites or isolated pathways. Additionally, changes occurring "upstream" in genes and proteins lead to changes in cellular metabolism. Hence, a metabolomic study of physiological changes will give a "downstream" or terminal view of the biological system, encompassing all alterations in genetic regulation, enzyme activity, and metabolic reactions, ending with the final manifestations in histology [3]. The flow diagram in Fig. 1 not only presents the conceptual evolution of a disease, but also characterizes the temporal progression. In this context, studying metabolomics might have certain advantages over genomics and proteomics, specifically for defining

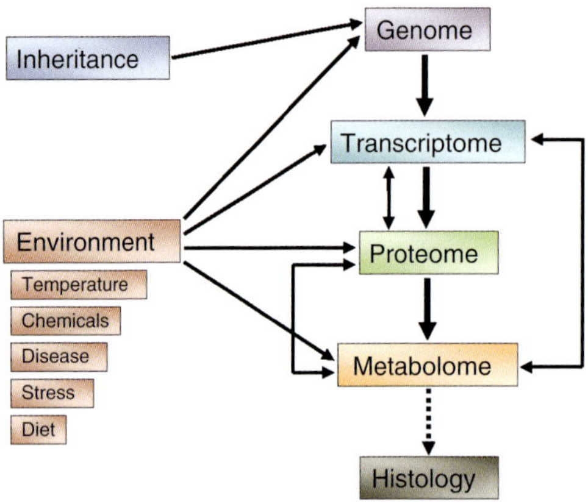

Fig. 1 The biological organization of the "-omes." The classical view of biological organization is to consider the flow of information from the genome, originally a result of inheritance, to the transcriptome, the proteome, and the metabolome, and then eventually resulting in the histological display of tissue. However, each tier of organization depends on the other, so a perturbation in one network can affect another. Furthermore, multiple environmental factors have a crucial impact on not only expression and concentrations of transcripts, proteins, and metabolites, but also on the genome by selecting for adaptive changes in subpopulations of cells within a tumor. Metabolomics can potentially probe much more than classical metabolism and metabolic disorders – it can also be used to monitor changes in the genome or to measure the effects of downregulation or upregulation of a specific gene transcript. Adapted with permission from Ref. [3], with modifications.

changes in phenotype, physiological and pathological status and, therefore, function.

1.2 Applications to Cancer

Current disease management for oncology can be classified into the following stages: screening; detection; diagnosis; prognostication; treatment election; and therapy response. The current "gold standard" in cancer diagnosis continues to be histopathology, where tissue samples are examined microscopically to visually evaluate disease status through the characterizations of cellular morphological changes. Histopathology, established throughout the past century using collections of medical evidence from symptomatic patient populations, has functioned as an invaluable decision-making tool in cases where a patient displays symptoms of a clinically significant disease. However, in the era of cancer screening, histopathology has shown its inadequacy and limitations for diagnosing early-stage, asymptomatic disease, and further for the prognostication of these screening-identified patients, due to the inability to delineate probable disease courses based only on morphological appearance. Therefore, the realization of these current limitations has led to proposals and investigations into cancer biomarkers that possess the ability to reflect the biological activities of disease and reveal insight into the biological processes of cancer development and progression.

Typically, biomarkers have been sought in terms of individual genes, proteins, or metabolites, which are indicative of the possible existence of certain pathological statuses to assist cancer screening, and may even be used for prognostic or predictive purposes. Current well-established examples include *Her2* gene status for breast cancer [4] and the prostate-specific antigen (PSA) for prostate cancer (PCa) [5]. While these biomarkers have improved screening, diagnostic and prognostic capabilities for cancer, more accurate disease markers are being actively investigated and have been reported in the format of genomics and proteomics, particularly for those that are able to rectify the limitations of current histopathology for detecting cancer at earlier stages and predicting disease aggressiveness for these patients.

Along this line of reasoning, and especially considering the "downstream" status of metabolomics towards cancer presentations visible from histology, the study of cancer metabolomics investigates possible cancer metabolic markers in the form of metabolic profiles under the governing assumption that cancer cells possess a unique and characteristic metabolic phenotype. In fact, cancer – more than any other type of disease – can most effectively be studied with metabolomics because of the existence of an inherent division between diseased and healthy (i.e., benign as identified by histology) tissue, in the form of a cancer lesion or tumor. At present, the tumor metabolome remains in the early stages of characterization, and there are numerous gaps in knowledge, especially taking into account the specific characteristics of tumors such as type, grade, and stage. In general, it is understood that tumors have a high glucose uptake, elevated phospholipid levels, and increased glycolytic activity; however, specific tumor types often differ in the variations of multiple metabolites including alanine, citrate, glycine, lactate, cholines, and lipids [6]. Thus, metabolomics becomes an important player in further characterizing and probing these metabolic changes because of the interconnectedness of all metabolic pathways in the human body. Metabolic processes before, during, or after malignant transformation can occur in parallel, and with multiple feedback loops. The goals of cancer metabolomic studies are twofold:

- To uncover disease-specific metabolomic profiles, in conjunction with traditional histopathology and patient clinical outcomes, for better understanding the contributions of individual and all involved pathways of cancer metabolism.
- To utilize these profiles to establish metabolomic criteria for cancer detection, disease characterization, and patient prognostication, applicable through metabolomic imaging.

2 Methodologies

At present, studies of cancer metabolomics are conducted utilizing two major spectroscopic techniques: (i) nuclear magnetic resonance (NMR) spectroscopy, which now is more commonly referred to as magnetic resonance spectroscopy (MRS) in the medical context; and (ii) mass spectrometry (MS) [7]. The relevant MS techniques can be further divided into either gas chromatography (GC)-based or liquid chromatography (LC)-based MS.

The MRS and MS techniques each have distinct strengths and weaknesses. The MS format is highly sensitive and has an ability to identify metabolites over a large range of molecular weights at low concentrations, whereas MRS detects mostly unbound metabolites in the cytoplasm that are of low molecular weight at the level of a single amino acid. Notably, MRS requires metabolite concentrations at the micromolar level (or mmol kg^{-1}), which is at least 10^3- to 10^6-fold higher than the detection limits of MS. One positive aspect of MRS techniques is that – at least in in principle – they do not require destruction of the tissue samples when used *ex vivo*, and this allows histopathological analyses of the tissue to be performed following spectroscopy. Furthermore, whilst MRS presents the strongest potential for translation into clinical use in the form of noninvasive *in-vivo* MRS, or magnetic resonance spectroscopy imaging (MRSI), MS – by definition – is limited to *ex-vivo* or *in-vitro* use, to the best of present knowledge and imagination. The details of each of these spectroscopic methods will be discussed in the following sections, to further elucidate their contributions to the study of cancer metabolomics and, especially, of cancer metabolomic imaging.

2.1 Nuclear Magnetic Resonance (NMR) Spectroscopy

When placed in a magnetic field, the energy status of certain nuclei can split, or degenerate, into different levels, and the energy differences between and among these levels depends on the strength of the magnet field (usually reported in Tesla, T), the type of nucleus, and its surrounding electronic structure (chemical environment). By absorbing and re-emitting electromagnetic radiation energy equivalent to the energy difference among the degenerated levels in the radiofrequency range, the nuclei can transfer themselves between these levels. NMR spectroscopy exploits this quantum mechanical phenomenon by resonating the transition of the nuclei in the magnetic field with accurately measured amounts of re-emitted energy to identify molecules based on their nucleus types and chemical structures. First predicted according to the theory of quantum mechanics by Rabi (who received the Nobel Prize in Physics, in 1944), and then experimentally observed during the mid-1940s by Bloch and Purcell (Nobel Prize in Physics, 1952), NMR spectroscopy was mostly used in physics and chemistry studies and not applied to medicine until the 1970s. In 1971, the proton (^{1}H) relaxation time was first used for tumor detection [8], and signals from multiple compounds in a suspension of red blood cells were recorded in 1977 [9]. Studies aimed at correlating the variations in the NMR-measureable metabolites with cancer pathologies thus began. With the invention of magnetic resonance imaging (MRI) by Lauterbur and Mansfield (Nobel Prize in Physiology or Medicine, 2003), the study of human malignancy with *in-vivo* MRS has projected itself as the eventual goal for clinical translation. To achieve this objective, the accurate characterization of individual cancer metabolites and the tumor metabolome is necessary. To date far, this task has been carried out via high-resolution *ex vivo* measurements of tissue samples at high magnetic field strength.

Originally, *ex-vivo* metabolic measurements of cancer tissues were conducted using conventional liquid NMR methods, designed to analyze aqueous solutions mainly for organic chemistry. In those studies, the intact tissues were immersed in a solution media and measured with low spectral resolution due to their semi-solid nature. To improve spectral resolution for identification of individual metabolite peaks, chemical extractions of the water (solvent) soluble metabolites from the tissue were also employed.

Both of these approaches presented serious shortcomings, however. The low spectral resolution achieved with intact tissues prevented the identification of many metabolites, while the extraction methods used not only destroyed the tissue architecture (preventing subsequent histopathology) but the detected metabolites also depended on the extraction method employed and the completeness of the procedures. In the study of a notoriously heterogeneous disease such as cancer, the destruction of tissue samples was especially problematic because histopathology could not be confirmed either before or following spectroscopy. The solution to these problems was the introduction of intact tissue NMR using the high-resolution magic angle spinning (HRMAS) method, which has since provided significant advances in the field of cancer metabolomics.

2.1.1 HRMAS MRS

Introduced for intact biological tissue analysis in 1996, the HRMAS MRS method applies mechanical rotation of the sample at an angle of 54.7° from the direction of the magnetic field; high-resolution spectra of intact tissue are obtained, allowing for the observation of individual metabolites [10]. In early studies, it was established that the HRMAS MRS method could produce high-resolution spectra of specimens of malignant lymph nodes [10], Pick's disease brain tissue [11], and human breast ductal carcinomas [12]. The most critical and advantageous feature of HRMAS MRS for cancer studies is that the method preserves the tissue's histological structure, allowing for histopathological evaluations of the same specimens following spectroscopic analysis [13]. For the establishment of cancer metabolic markers at present – either in the form of single metabolites or with metabolic profiles as metabolomics – and before tissue spectroscopic results can be evaluated against patient outcomes with longitudinal observations, traditional histopathology continues to function as the best reference available for correlating metabolic variations with tumor characteristics. Therefore, because of the well-recognized heterogeneity of human cancer distributions, it is necessary to use both methods (MRS and histopathology) in parallel, on the same sample, to uncover and expose potential cancer biomarkers.

2.1.2 Individual Metabolite Correlations versus Metabolic Profiling

Metabolomics studies for numerous types of cancer have utilized the HRMAS MRS method since 1996 [10]. In general, metabolomics-based cancer studies can follow one of two different formats. Many research studies involve the broad-spectrum metabolic "profiling" of samples by dividing the entire meaningful spectra into a large number of bins, and then seeking correlations between the set of individual metabolites, either as concentrations or as metabolite ratios, represented by certain bins, and different cancer characteristics. These studies are valuable in illuminating the most relevant metabolic "players" in cancer development and progression. Early findings included linear correlations between the concentrations of lactate and lipids, and the percentage of tumor necrosis in human glioblastoma multiforme specimens, as well as linear correlations between metabolic ratios of phosphorylcholine to choline and the percentage of malignant glioma cells [14]. A more recent study of brain tumors found differences in the concentrations of 20 individual metabolites between different tumor types, including astrocytoma grade II, grade III gliomas, glioblastomas, metastases, meningiomas, and lymphomas [15].

Subsequently, in a study of PCa, the concentrations of specific metabolites, including phosphocholine, glycerophosphocholine, lactate and alanine were found to be higher in PCa tissue than in benign stromal tissue, while the concentrations of citrate and polyamines were high in benign epithelial prostate tissue [16].

One of the earliest cancer studies using HRMAS involved an investigation into the metabolic changes occurring in multiple types of brain tumor [13]. In this study, when the concentrations of 11 individual metabolites and the ratios of 15 metabolites to creatine were measured, the metabolic ratios were seen to present a greater sensitivity in differentiating tumors from normal brain tissue than individual metabolites. This discovery was an early precursor to the second type of research format employed in studies of cancer metabolomics, namely the creation of metabolomic profiles that summarize variations in multiple metabolites. In the same way that metabolic ratios were shown to be stronger biomarkers for differentiating cancer characteristics than individual metabolites, metabolomic profiles can provide the greatest sensitivity for identifying and classifying cancer characteristics, while providing an even more accurate picture of the metabolic changes occurring in malignancy.

2.1.3 Improved Characterization with Metabolomic Profiles

In the search for more robust biomarkers for cancer detection and characterization, a study of PCa metabolomics using HRMAS MRS revealed the benefit of utilizing metabolomic profiles, in place of individual metabolites or metabolite ratios. In 2005, the metabolomic profiles obtained from 199 tissue samples taken from 82 PCa patients were reported to be able to differentiate cancerous from histologically benign samples from the same patient with 98.2% accuracy (Fig. 2). Notably, the ability of three variables – principal component 13 (PC13), phosphocholine and choline – to differentiate between these groups was compared. While each variable/metabolite was able to differentiate the groups on their own, the significance was greatly enhanced when the variables were combined, demonstrating the advantage of applying a metabolomic profile constituting multiple metabolic changes, over employing changes in only individual metabolites or ratios, for the detection of cancer.

The metabolomic profiles calculated from samples that contained only histologically benign tissue (179 of the total 199 samples), collected from cancerous prostate glands, were also used to differentiate the cancer pathological stage, to predict tumor perineural invasion, and to delineate a subset of less aggressive tumors with a Gleason score (GS) of 6 and a pathological tumor stage (pT) of 2 [17]. Since the strength of metabolomic profiles for both detecting and characterizing PCa were first observed, further investigations have been made into the abilities of metabolomic profiles. Additional studies have investigated the ability of profiles to differentiate PCa aggressiveness (based on GS) in the hopes that, by establishing profiles for malignant potential, the prognostic biomarkers could be translated into clinical applications [18]. Another indicator of malignant potential in PCa is the probability of biochemical recurrence following radical prostatectomy. A study evaluated 16 patients with PCa recurrence, clinically and pathologically matched with 16 patients without recurrence, and found that the calculated metabolomic profiles were able to predict cancer recurrence with an overall accuracy of 78% [19].

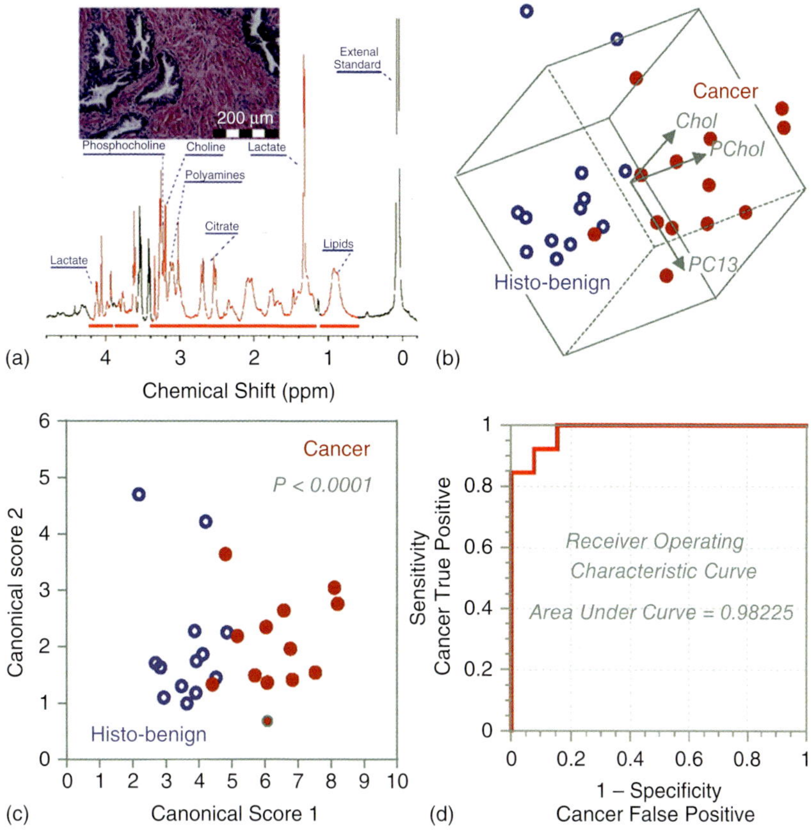

Fig. 2 Differentiating prostate cancer from histologically benign samples. (a) High-resolution magic angle spinning ^{1}H MR spectrum of intact tissue obtained from the removed prostate of a 61-year-old patient with GS 6 T2b tumors. Histopathology analysis of the tissue sample (inset) after spectroscopy measurements showed that the sample contained 40% histopathologically defined benign epithelium and 60% stromal structures, with no identifiable cancerous glands. Cellular metabolites are labeled on the spectrum. The 36 most intense resonance peaks or metabolite groups above the horizontal bars were selected for analyses, whereas the other regions were excluded from calculation, partly due to surgery-related alcohol contamination; (b) Three-dimensional plot of principal component 13 (PC13 correlates linearly with percent volume of cancer cells in tissue samples) versus phosphocholine versus choline. Cancerous and histologically benign (histo-benign) tissue samples from 13 patients can be visually separated in observation plane. The paired Student's *t*-test results (cancer versus histo-benign from the same patients) for PC13, phosphocholine, and choline are 0.012, 0.004, and 0.001, respectively; (c) Canonical plot resulting from discriminant analysis of the three variables in panel (b) presents the maximum separation between the two groups; (d) The resulting receiver operating characteristic curves indicates the accuracy of using the three variables in panel (b) to positively identify cancer samples. Adapted with permission from Ref. [17].

After demonstrating the great potential of metabolomic profiles for cancer detection and characterization, studies using HRMAS MRS to investigate metabolic biomarkers began adopting the profile approach in favor of individual metabolites in the evaluation of other types of cancer. However, an additional benefit gained from the use of metabolomic profiles is an ability to probe the profiles for the most influential or significant contributing metabolites. In this way, the sensitivity of the profiles for cancer classification can be obtained, as can the ability to elucidate details regarding the variations that occur in specific metabolites or metabolic pathways. In a study of childhood neuroblastomas, the metabolomic profiles were able to differentiate neuroblastoma and healthy adrenal tissue sample groups, and also revealed the significance of changes to a number of different metabolic processes based on the contributions to the profiles from creatine, glutamine, acetate, glycine, myoinositol, glutathione, glutamate, serine, and ascorbic acid [20]. Metabolomics have also been pursued in an effort to further characterize certain types of brain tumors, including an improved grading of malignancy for low- and high-grade oligodendrogliomas [21] and the prediction of aggressiveness for recurrent low-grade gliomas [22].

With the improved capability of computed tomography (CT) in observing early, small lesions of lung cancer, but a concern for overutilization of the test due to its radiation hazard, there has been an increased need for new diagnostic measures for lung cancer given the lack of general screening methods, besides CT, and the poor prognosis at late detection. By using HRMAS MRS, metabolomics have been investigated in lung cancer tissue, and calculated profiles have been shown to differentiate tumor tissue from noninvolved adjacent tissues [23]. Specifically, the most discriminating metabolites were found to be lactate, phosphocholine and glycophosphocholine, the levels of which were elevated in tumor tissues, while the levels of glucose, myoinositol, inosine/adenosine and acetate were reduced. An additional domain which is currently being investigated, utilizing metabolomic analysis and HRMAS MRS, is the possibility of a serum screening test for lung cancer diagnosis. In a recent study, the metabolomic profiles calculated from paired tissue and serum samples from 14 lung cancer patients were shown capable of differentiating two different lung cancer types – adenocarcinoma and squamous cell carcinoma – as well as differentiating themselves from sera obtained from seven healthy controls (Fig. 3) [24]. If serum metabolomic profiles could be established that might indicate human lung cancer at asymptomatic stages and patients subsequently triaged to other more-specific testing (e.g., CT), then treatment could be moved to earlier stages and the prognosis for many patients would be improved.

The search for breast cancer biomarkers has also been actively pursued using metabolomics and HRMAS MRS. Various research groups have investigated the ability of metabolomic profiles to predict tumor size, histological grade, axillary lymph node status, *HER2* status, estrogen and progesterone receptor status, and triple negativity in breast cancer patients [25–27]. Among a population of patients with estrogen receptor-positive breast cancer, metabolomic profiles were also able to predict five-year survival [28]. The results obtained showed that the metabolomic profiles of breast cancer would have the potential to provide supplemental information for clinical decision making, especially in terms of determining

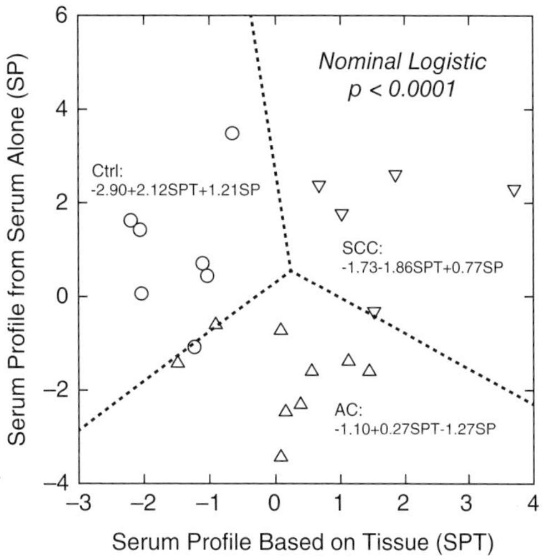

Fig. 3 Differentiations among three tested serum groups (adenocarcinoma, AC; squamous cell carcinoma, SCC; and control). Differentiations were based on nominal logistic and linear discriminant analyses of serum metabolomic profile presented as canonical scores and calculated based on profile coefficients obtained from tissue analyses (Serum profile tissue; SPT) and using serum spectral data independently (Serum profile; SP). Δ denote ACs, ∇ denote SCCs, ○ denote control sera. Nominal logistic regression equations for each group are presented in the figure. Adapted with permission from Ref. [24].

the prognosis for individual patients. Most recently, metabolomics was investigated as a potential tool for breast cancer surgery, where tumor borders could be immediately delineated by MRS to prevent reoperation or local recurrence. In this study, the metabolomic profiles could be used to differentiate between tumor and noninvolved adjacent tissues with high sensitivity and specificity [29]. An additional study demonstrated the ability of metabolomic profiles to predict pathologic response to neoadjuvant chemotherapy in patients with locally advanced breast cancer, whereas individual metabolites and metabolite ratios were unable to differentiate pathologic response groups, further reinforcing the strong support for metabolomic profiling [30].

Although brain, prostate, lung and breast cancer are the most commonly studied cancers with metabolomics using HRMAS MRS, other types of cancer studies have also been pursued. Metabolomic profiles from urine were shown to be capable of differentiating patients with small bowel neuroendrocrine neoplasms from those with pancreatic neuroendrocrine neoplasms. The profiles were also capable of differentiating between functional and nonfunctional neuroendrocrine neoplasms, as well as identifying patients with metastases [31]. In a study of human rectal adenocarcinoma, the metabolomic profiles were able to discriminate between malignant and benign biopsies, and also to correlate with tissue pathologies [32]. By using fine-needle aspiration biopsies for papillary thyroid carcinoma diagnosis, the calculated metabolomic profiles could be used to differentiate three different histological cancer groups, including papillary thyroid carcinoma, follicular adenomas, and normal thyroid samples [33]. Finally, by using traditional NMR spectroscopy techniques with serum samples, metabolomic profiles were capable of distinguishing low-grade urinary bladder cancer from high-grade cancer, and accurately identifying women with epithelial ovarian cancer, with benign ovarian cysts, and those in good health [34, 35].

2.2 Mass Spectrometry (MS)

Whereas, NMR-based measurements normally can be used to detect between 20 and 200 metabolites (depending on whether intact tissue, tissue extracts, blood, or urine are involved), the more sensitive analyses with MS can detect about 2000 metabolites. Moreover, MS is an especially valuable tool for detecting metabolites at low concentrations.

2.2.1 Principle of MS Analysis

A classical MS analysis involves an initial separation of molecules by liquid chromatography (LC-MS), or the high-resolution approach of gas chromatography (GC-MS), based on mass. The output of separated molecules from the chromatographic process is subsequently ionized and detected based on the principle of MS (the measurement of mass-to-charge ratio). This includes all possible fragments of molecules produced during the ionization procedures, to achieve molecular identifications. Although other separation methods, including capillary electrophoresis (CE)-MS and Fourier transform (FT)-MS have been described, these are less frequently used than chromatography and their details are beyond the scope of this review. From the above description, it is clear that classical MS can be performed neither on an intact specimen, nor *in vivo*. Rather, the sample must be converted into a physical state that can be passed through the chromatography columns, and this inevitably results in the destruction of all tissue histological structures during the process. Nevertheless, interesting metabolite data have been acquired – and metabolomic profiles generated – from MS-based analyses of human cancer samples.

In an effort to overcome the limitations of MS in terms of sample destruction, new protocols have been proposed to preserve the histological structures of intact tissue samples. In addition to MS imaging (see Sect. 3.2), a method known as Molecular Preservation by Extraction and Fixation (mPREF) has been described in which methanol fixation was used to preserve various human biopsy tissues (including prostate and kidney) that could then be chemically stained and examined immunohistochemically. The methanolic fixative solution would then be analyzed for small-molecule biomarkers, using MS [36, 37].

2.2.2 Application of MS Metabolomics in Cancer

Reports have been made of extensive studies with PCa utilizing the MS method for metabolite identification. In these cases, individual metabolic variations have been investigated that occur during androgen deprivation therapy (ADT) for PCa, as measured from serum samples [38], and the development of PCa bone metastases [39]. In similar fashion to metabolomic profiling performed with NMR-based technologies, many MS "metabolomics" studies have focused on the discovery of individual metabolites as biomarkers for cancer. One such study, conducted in 2009, used combined high-throughput LC-MS and GC-MS, and reported the detection of 626 metabolites from 42 prostate tissue samples. These findings included 60 metabolites detected in localized or metastatic PCa, but not in benign prostate tissue, as well as the upregulation of 50 metabolites and downregulation of 37 metabolites, in PCa (Fig. 4). To summarize, these results have revealed an increase in amino acid metabolism and nitrogen breakdown pathways during PCa metastatic progression [40]. While the global metabolic variations described in this study provided the most robust view of the changes that occur during PCa development, the research groups identified the individual metabolite sarcosine as being an important indicator of PCa progression to metastasis that could be measured in the urine. A later study, however, found no significant differences in urinary sarcosine levels between PCa patients and healthy controls, but did identify a profile of nine metabolic markers that could differentiate the PCa group [41]. These results further support the strength of using metabolomics profiles, established with MS, for cancer diagnosis and characterization over individual metabolic biomarkers.

Metabolomic analysis with MS has also been successful in characterizing ovarian cancer. In one study, metabolite profiling was achieved using GC-MS, and the uncovered metabolomic profiles were capable of differentiating borderline ovarian tumors (which have an indolent clinical course) from invasive ovarian carcinomas, with 88% separation accuracy (Fig. 5) [42]. Additionally, a combination of LC-MS and GC-MS identified metabolomic profiles that could distinguish primary epithelial ovarian cancer from metastatic tumors resulting from ovarian cancer [43]. Of interest, both of these studies reported the identification of novel metabolites in the ovary and/or metabolites, the role of which in ovarian physiology is still unknown. These findings not only enhance the present understanding of human ovarian biochemistry but also reveal the potential of MS for exposing a multitude of unknown metabolites that may be clinically relevant for diagnosis or management.

Similar to NMR-based methods, MS has also been used to investigate the potential for a serum screening test to detect early-stage lung cancer, before it becomes clinically significant. In a recent study, GC-MS and LC-MS were both used to uncover the metabolic signatures in the plasma of 31 stage I human lung adenocarcinoma patients and 28 healthy controls. The established metabolomic profiles were used to differentiate between the cancer and control groups, and revealed an interesting variation in the sex hormone metabolic pathway that could be used as a metabolic biomarker of lung cancer development [44]. Breast cancer has also been studied using plasma, where 375 metabolites were identified and 78 were found to differ between breast cancer cases and healthy controls. In this study, the research group performed a pathway-based statistical analysis for the creation of profiles that could characterize breast cancer, and found that the metabolomic profiles could make a significant discrimination between breast cancer patients and healthy controls, and also identify differences in patients with triple-negative breast cancer [45]. Global metabolic variations in serum/plasma have also been studied in renal cell carcinoma, gastric and colorectal carcinoma, [46–49].

2.2.3 Combining MS and NMR

MS- and NMR-based methods have also been utilized in conjunction to probe for possible biomarkers for several types of cancer. In one such study seeking metabolic variations that could discriminate hepatocellular carcinoma from liver cirrhosis, the decision was taken

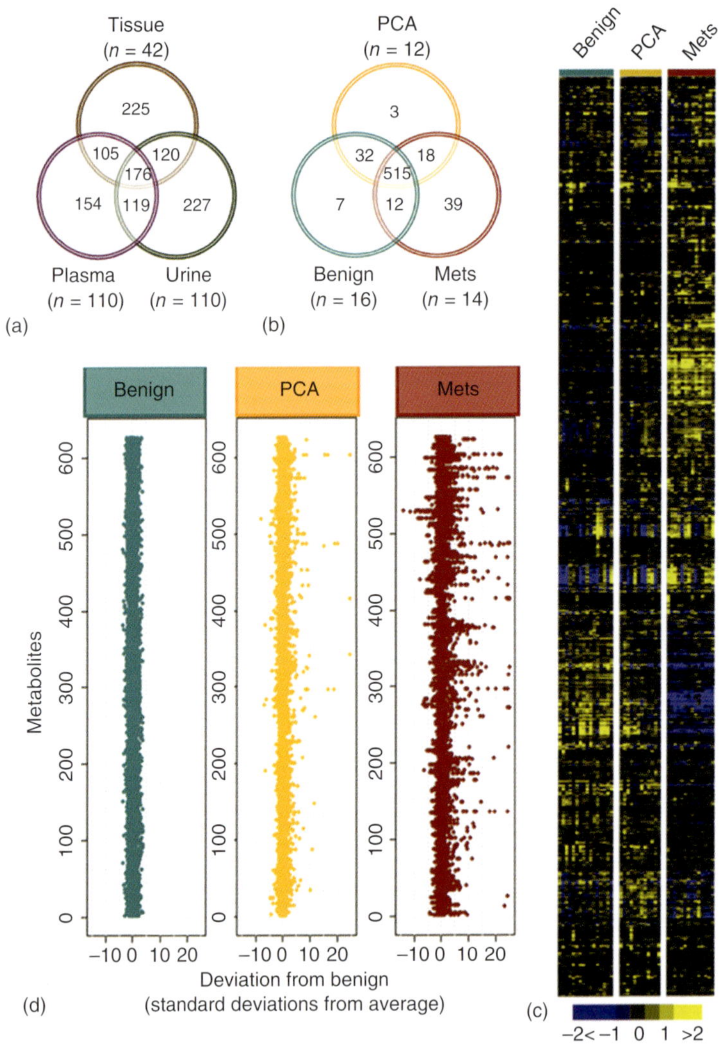

Fig. 4 Metabolomic profiling of prostate cancer. (a) Venn diagram of the total metabolites detected across 42 prostate-related tissues and 110 matched plasma and urine samples; (b) Venn diagram of 626 metabolites in tissues measured across 16 benign adjacent prostate, 12 clinically localized prostate cancers (PCa), and 14 metastatic prostate cancers (Mets); (c) Heat-map representation of unsupervised hierarchical clustering of panel (b) (rows) grouped by sample type (columns). Shades of yellow and blue represent elevation and decrease of a metabolite, respectively, relative to the median metabolite levels (see color scale); (d) The number of standard deviations from the mean, or z-scores, were plotted for each of the 626 metabolites, where the plots were normalized to the mean of the benign prostate samples. The plots revealed robust metabolic alterations in metastatic tumors (z-score range: −13.6 to 81.9) compared to fewer changes in clinically localized prostate cancer samples (z-score range: −7.7 to 45.8). The max SD has been truncated at 25 for clarity. Adapted with permission from Ref. [40].

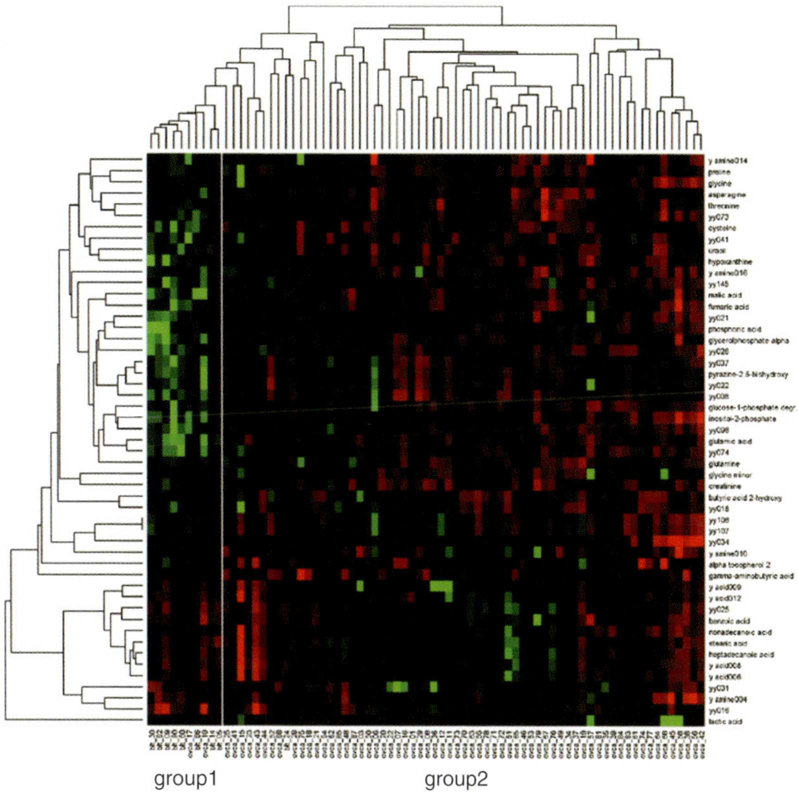

Fig. 5 Supervised clustering. Hierarchical clustering was made to visualize the differences in metabolic signatures between borderline tumors and ovarian carcinomas (ovcas), and led to the separation of these tumors by clustering for 51 significantly ($p < 0.01$) regulated metabolites. Two groups of tumors are distinguished. Group 1 contains most of the borderline tumors (as well as two carcinomas), whereas group 2 contains most of the carcinomas as well as one borderline tumor. blt, borderline tumor. Adapted with permission from Ref. 42.

to use the two techniques in tandem in order to enhance visualization of the metabolites and achieve the most comprehensive metabolomic profile possible. The metabolomic profiles acquired were capable of differentiating patients with hepatocellular carcinoma from those with liver cirrhosis, as well as from healthy volunteers. The profiles not only revealed a total of 32 potential biomarkers, but also showed that the metabolic alterations occurring in hepatocellular carcinoma included changes in the synthesis of ketone bodies, the citrate cycle, phospholipid metabolism, sphingolipid metabolism, fatty acid metabolism, amino acid catabolism, and bile acid metabolism [50]. In another study, the MRS and MS metabolic profiles of paired bone marrow and peripheral blood samples from children with acute lymphoblastic leukemia were examined to investigate the metabolomics of the tumor microenvironment. Significant changes in lipid metabolism in untreated cases, and a strong correlation between cases in remission and amino acid metabolites, were uncovered [51].

2.3 Data Analyses

Metabolomic profiling involves the global assessment of a large amount of metabolites measured with either MRS or MS. However, no matter from which modality these data are acquired, multivariate analysis becomes necessary to extract meaningful information from the data sets. Multiple steps are involved in the multivariate analyses, including pattern recognition, metabolite identification, profile construction, and biomarker validation. Pattern recognition employs statistical methods to orient the data sets into recognizable groups, such as cancer versus benign, or aggressive versus nonaggressive. In order to reach these specific data orientations, both supervised and unsupervised pattern recognition statistical methods can be employed. The unsupervised pattern recognition method – such as hierarchical cluster analysis and principal component analysis – measures the innate variation in a data set and then tests the results for their potential correlations with these groups. Contrastingly, supervised pattern recognition uses prior information to generate the patterns, with examples including partial least-squares discriminant analysis, canonical analysis, and neural networks. Following these pattern recognition procedures, the specific factors that are responsible for the group divisions, with weighted contributions, can be identified as spectral regions and related metabolites. The combinations of information from recognition and identification of group patterns will lead to the proposal and generation of metabolomic profiles specific to the evaluated groups or conditions (often referred to as a "training cohort"), as the working hypotheses. Finally, the hypothetical cancer metabolomic profiles ("markers") need to be validated by independent patient groups (often designated as a "testing cohort"), not including those involved in the generation of these profiles. The evaluation of metabolomic profile association with a particular cancer characteristic, such as tumor grade, metastatic status or response to treatment, can then be achieved by using a standard Student's *t*-test, analysis of variance (ANOVA), or a similar statistical process. Unfortunately, when the analytic procedures deal only with certain known cancer metabolites, the above-described metabolomic approach may be simplified by proposing these individual metabolites as the major biomarkers for the condition, as illustrated in Fig. 6 [6].

To further emphasize, it is extremely important that analytic steps and safeguards should be considered to avoid false-positive results. Metabolomic profiles, biomarkers or classifiers that can differentiate between disease conditions and are obtained from a "training cohort" should only be presented as conclusive when they have been verified with an independent "testing cohort," and the sensitivity and specificity for differentiating the testing cohort be evaluated by receiver operating characteristic (ROC) curves [52, 53].

At this point it is worth noting that cancer metabolomic analyses are conceptually different from cancer metabolic analyses, although both approaches use the same data set, measured with either MRS or MS. Classic cancer metabolic analyses evaluate the relationship between individual metabolites, or metabolite ratios, and disease conditions (as noted above). Consequently, it might be difficult – or

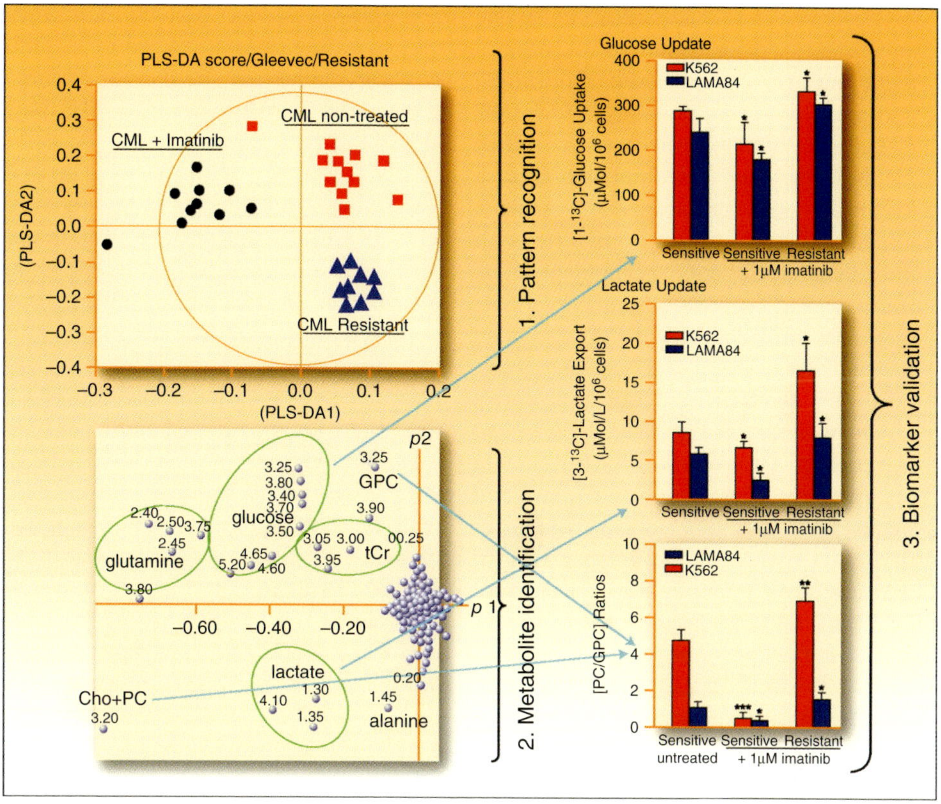

Fig. 6 Simplified steps of metabolomics analysis for certain known cancer metabolites. The example is given for imatinib treatment in chronic myeloid leukemia cells. Step 1: ^{1}H-NMR spectra of cell extracts followed by principal component analysis for pattern recognition. In this step, spectral data sets are entered into statistical programs that facilitate pattern recognition, or group clustering, such as cancer versus noncancer. Step 2: Metabolite identification resulting in a biomarker. In step 2, the spectral regions that are responsible for group clustering in step 1 are identified and linked to a specific metabolite based on known NMR chemical shifts. Step 3: Metabolite quantification and validation. In the final step, a standard Student's *t*-test or ANOVA can be used to associate the identified biomarkers with specific characteristics, such as response to therapy. Adapted with permission from Ref. [6].

even conflicting – to use a single metabolite to interpret two or more cancer-related clinical parameters. However, the above-described data processing mechanism for generating cancer metabolomic profiles shows clearly that, even when using a single set of cancer spectroscopic data, numerous profiles can be formed to represent different disease conditions by variations in the combinations of all metabolites in the data set.

3
Metabolomic Imaging

At present, human cancer metabolomics has mainly been studied *ex vivo* in an effort to describe and characterize metabolic alterations during malignant development and progression. Results from these endeavors have demonstrated the proof-of-concept of the existence of cancer metabolomics and the improved disease diagnosis and characterization abilities of metabolomics when compared with individual metabolites. In principle, this concept and approach can be developed and translated into metabolomic imaging for clinical use, as is always the ultimate goal for any biomedical research endeavor. Research toward such translations has been witnessed with the technological innovation and advancements both in MRSI, and in the form of mass spectrometry imaging (MSI), with the understanding that the ultimate target of MR-based metabolomic imaging is noninvasive *in-vivo* examinations, while MS-based imaging would likely be restricted to *ex-vivo* evaluations of human specimens.

3.1 MR-Based Metabolomic Imaging

MRSI utilizes the technology employed in MR scanners for imaging in humans. While the more commonly used MRI measures water signals to investigate anatomy and function, MRS can also be acquired by suppressing water signals and measuring cellular metabolites noninvasively. The use of MRI localization techniques with MRS measurements of different anatomic locations (often referred to as "voxels") within a region of interest can capture the spatial distributions of metabolic intensities to form the MRS image. In contrast to the implementation of standard MRSI, where the intensities of single metabolites are used to map the spatial distributions of individual metabolites, metabolomic imaging utilizes the multivoxel spectroscopy from MRSI but maps the anatomic image with values of disease condition-specific metabolomic profiles calculated from MRS of each voxel.

3.1.1 Examples of Standard MRSI

MRSI of the brain has the longest history of investigation, with studies being traced back to the early 1980s [54]. A leading reason for this is the ease of MRSI measurements in the brain to produce high spectral resolution; this benefit is due to a lack of internal motion of the brain as opposed to the abdomen, as well as the ability to stabilize the head with tight-fitting coils during the examination. The study of a patient with an oligodendroglioma represents a good example of the value of MRSI for brain tumor diagnosis and analysis. By overlaying the MRS results with the MRI images acquired from the brain using a 3 T MR scanner, metabolic maps are produced that reflect localized disease-related metabolic activities (Fig. 7) [55]. In this case, the research group specifically examined the metabolite maps of the choline/*N*-acetylaspartate ratio, lipids and water, for which the metabolite changes are observed to coincide with the lesions reflected by hyperintensities on the MRI. However, this standard procedure for *in-vivo* MRSI interpretation involves the simultaneous evaluation of multiple maps of individual metabolites by the observer, and requires the observer to generate an image of tumor involvement through his/her mental conclusion. The ultimate goal for metabolomic imaging is to generate a series of distribution maps of disease conditions using a single set of multivoxel MRS, independent of the ability of the observers to analyze the acquired raw spectroscopic data.

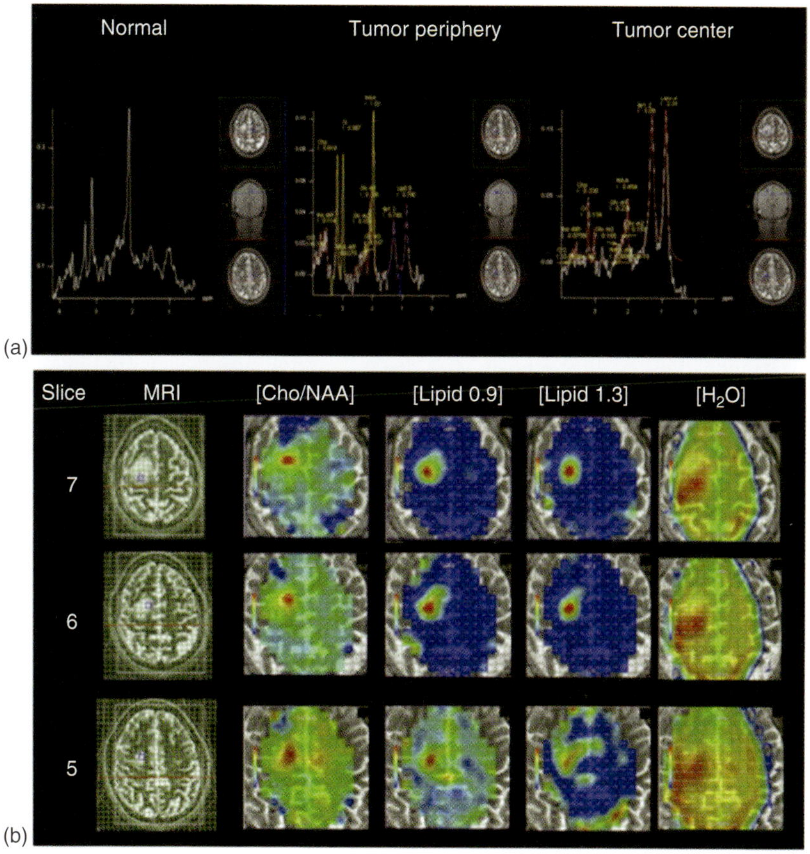

Fig. 7 3D metabolite mapping in a patient with a brain tumor at 3 T. (a) Localized spectra in the periphery of the tumor show elevated choline and lipid peaks. Decreased creatine and *N*-acetylaspartate (NAA), and strongly elevated lipid peaks are measured in the center of the tumor; (b) T_2-weighted MRI, metabolite maps, and water reference scan demonstrating spatial heterogeneity of the lesion. Data were acquired in 10 min, including water reference scan, using 32 × 32 × 8 spatial matrix and 7 mm isotropic voxel dimensions (0.34 ml voxel size). Figure adapted with permission from Ref. [55].

Beyond being able to coregister an MRI image of the brain with the individual metabolic changes occurring in tumor lesions for diagnostic purposes, *in-vivo* MRSI holds the critical potential to go beyond diagnosis (which can often be accomplished with MRI alone) and provide information from *in-vivo* analyses with regards to tumor type, grade, and prognosis. Currently, tumor type and grade is determined from an histopathological analysis of surgically removed brain tumor tissue. Consequently, many of the early brain tumor metabolomics studies focused on comparisons between *ex-vivo* HRMAS MRS and *in-vivo* MRSI to help elucidate the relevant metabolites being observed *in vivo* by applying the high-resolution spectra obtained *ex vivo*. These studies have been performed with several different types of brain tumor, including three histological subtypes of meningiomas (meningothelial, fibrous, and oncocytic), pediatric brain tumors, astrocytomas, and glioblastomas [56–58]. Overall, a good agreement was observed between the *ex-vivo* and *in-vivo* analyses, such that the metabolites measured using HRMAS could serve to improve the analysis and interpretation of MRS performed *in vivo*, thereby strengthening its applicability in the cancer clinic.

Recently, the ability of MRSI alone to characterize brain tumors, in terms of tumor type and prognosis, has been investigated. In one laboratory, when the potential for both of these characterizations in pediatric brain tumors was examined, it was found that single-voxel MRSI would allow discrimination between different pediatric tumor types including medullablastomas, ependymomas, and pilocytic astrocytomas [59]. Additionally, single-voxel MRSI allowed for the prediction of survival for pediatric brain tumor patients. Although individual metabolites were able to predict survival, the powers of prediction were enhanced to an accuracy level similar to that of histological grading, when a multivariate model based on three MRS biomarkers was used [60]. Similarly, in adults with glioblastoma multiforme who were imaged prior to and following radiation treatment, the MRSI parameters were more successful in predicting survival than were anatomic, diffusion, or perfusion imaging [61].

The possible application of MRSI to PCa detection and characterization is especially important, due to the currently limited ability to image PCa for diagnosis. Unlike the case of brain tumors, the use of MRI alone is insufficient to identify PCa lesions; rather, random biopsies are executed to sample the entire prostate in order to uncover possible cancer lesions for patients at risk for PCa as indicated by PSA screening. Metabolomic imaging has the potential to rectify these difficulties, and if MRSI were to be adopted in the PCa clinic for the detection and evaluation of tumors *in vivo*, patients could avoid the invasive nature of a biopsy procedure. Early studies have shown that, when combined, MRI and MRSI were able to improve PCa localization, assessment of the spatial extent, measurement of therapeutic response, and prediction of aggressiveness [62]. Recently, by using a combination of MRSI and diffusion-weighted imaging (DWI), the presence of PCa – including tumor volume and aggressiveness, based on GS – has been identified [63].

Currently, a full understanding of the molecular biology reflected in prostate MRSI, and the ability to interpret the *in-vivo* spectroscopy, is still in its early stages, and therefore – similar to the above-discussed studies of brain tumors – *in-vivo* MRSI is currently being performed in parallel with *ex-vivo* HRMAS analysis to elucidate PCa-specific metabolic profiles. Specifically, investigations have been undertaken to assess the ability of MRI/MRSI in guiding prostate biopsies, while also performing HRMAS MRS studies on the biopsy samples obtained. In one such study, MRI/MRSI at 1.5 T was able to discriminate healthy from cancerous prostate tissue with an overall accuracy of 81%, while the high-resolution spectra acquired from the subsequent *ex-vivo* analysis revealed metabolic biomarkers that could aid in the future clinical interpretation of *in-vivo* MRSI [64]. *Ex-vivo* HRMAS has also been used to assess the ability of *in-vivo* MRSI to determine GS, this being a potentially valuable approach that would allow the prediction of cancer aggressiveness without a need for invasive procedures [65].

While MRI is increasingly being used for breast cancer staging and screening

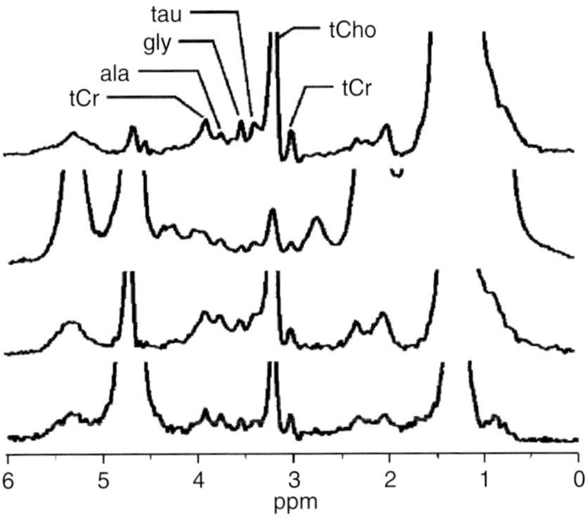

Fig. 8 Examples of high-quality 4 T breast spectra showing additional metabolites. These spectra all show a clear tCho resonance as well as several resonances previously reported only in *ex-vivo* studies. Considering chemical shift and relative signal amplitudes, resonances were identified as tCr = total creatine CH_3 (3.03 ppm) and CH_2 (3.92 ppm), tCho = total choline-containing compounds (3.22 ppm), tau = taurine (3.42 ppm), gly = glycine (3.55 ppm), and ala = alanine (3.78 ppm). Adapted with permission from Ref. [66].

for high-risk patients, MRSI is also being explored for the characterization of breast cancer. The most frequent investigations involve the single resonance at ~3.2 ppm, which is well known to be associated with malignancy. Unfortunately, this single resonance contains other resonances contributed by several metabolites that include phosphocholine, glycerophosphocholine and free choline, which has led to the entire peak being referred to as total choline-containing compounds (tCho). The aim of most of these MRSI studies is focused on the measurement of tCho as an indicator of breast cancer, as opposed to the examination of a metabolomic profile as a means of characterizing malignancy; this is due to a lack of sufficient spectral resolution, caused in part by a low clinical field strength as well as strong lipid resonances. The application of higher field strengths provides a potential means for metabolomics to become relevant to breast cancer studies, as the sensitivity and higher spectral resolution allows for the detection of additional metabolites (Fig. 8) [66]. In the past, *ex-vivo* metabolomic analyses have identified a number of biomarkers that can differentiate between breast cancer and normal tissue [67], and consequently the goal must be to establish these profiles *ex vivo* and then to apply them in the clinical setting. On a related note, MRSI has not been studied extensively in human lung cancer, due not only to artifacts caused by strong respiratory motions but also to the well-demonstrated efficacy of CT for the diagnosis and characterization of lung cancer.

3.1.2 Illustration of MRS-Based Metabolomic Imaging

The ultimate goal of MRS-based metabolomics in the cancer clinic is to construct disease-specific metabolomic profiles that can be presented as simplified maps, not only to locate tumors but also to provide an accurate assessment of the disease. A preliminary study in PCa patients revealed such an ability in the *ex-vivo* analysis of five whole prostate glands that had been removed via prostatectomy of biopsy-proven PCa. The prostate glands were measured using multivoxel MRS on a 7 T whole-body MR scanner, followed by whole-mount histology. Due to an absence of PCa parameters for metabolomic profile constructions determined specifically for 7 T measurements, the research group applied the PCa metabolomic profiles obtained with intact tissue analysis using HRMAS at 14 T to the multivoxel MRS results, and calculated a malignancy index that could be mapped across the prostate glands. Ultimately, when the imaging results were compared with the whole-mount prostate histology (Fig. 9), the calculated metabolomic profiles, or "metabolic index maps," were shown to be capable of discriminating between PCa-positive and PCa-negative regions in the prostate tissues, with an overall accuracy of 93–97% [68].

The creation of metabolomic maps for MRSI analyses of cancer resolves many of the challenges facing clinical translation, including complications in the interpretation of MRSI results. The collaborative efforts with *ex-vivo* analysis (especially HRMAS MRS) have the potential to establish metabolomics maps that are more sensitive than any single metabolites, such that they can be used first for cancer detection but, more importantly, they can ultimately provide a simple and noninvasive means for diagnosing and characterizing the malignant potential of multiple diseases.

3.2 MS Imaging

Whilst it is difficult to envision the *in-vivo* implementation of MS-based imaging in humans, as opposed to that promised by MRS-based metabolomic imaging, MSI may prove to be a powerful tool for cancer diagnosis and characterization in the *ex-vivo* pathology setting. Innovation in MS technologies permits molecules on the surface of a biological specimen to be MS-detected after being bombarded and ionized by tiny laser or ion beams, with precise control over both the size of the beam and the location at which it comes into contact with the tissue surface. This allows for the generation of a complete MS profile of hundreds of diverse biomolecules in human tissues, including metabolites, peptides and proteins, and their respective locations in each tissue sample. Although somewhat time-consuming, by moving the beam across the two-dimensional surface of the tissue and reducing the size of the beam to a microscopic scale, a set of MS-detected molecules of the entire surface with high spatial sensitivity, at the single-cell level, can be obtained [69]. From this set of MS data, the intensities of any single molecule – or metabolomic profiles – can be extracted and mapped onto the surface of the histological details. Therefore, the development and maturation of MSI has the potential to rectify the destructive nature of MS for *ex-vivo* tissues, and to augment and enhance current clinical pathology practices into molecular-level analyses – an idea first proposed when protein expression profiles obtained from MSI were seen to discriminate among lung cancer subtypes [70].

Whilst earlier reports of MSI have mostly focused on proteomics, with the majority of studies having evaluated protein expression

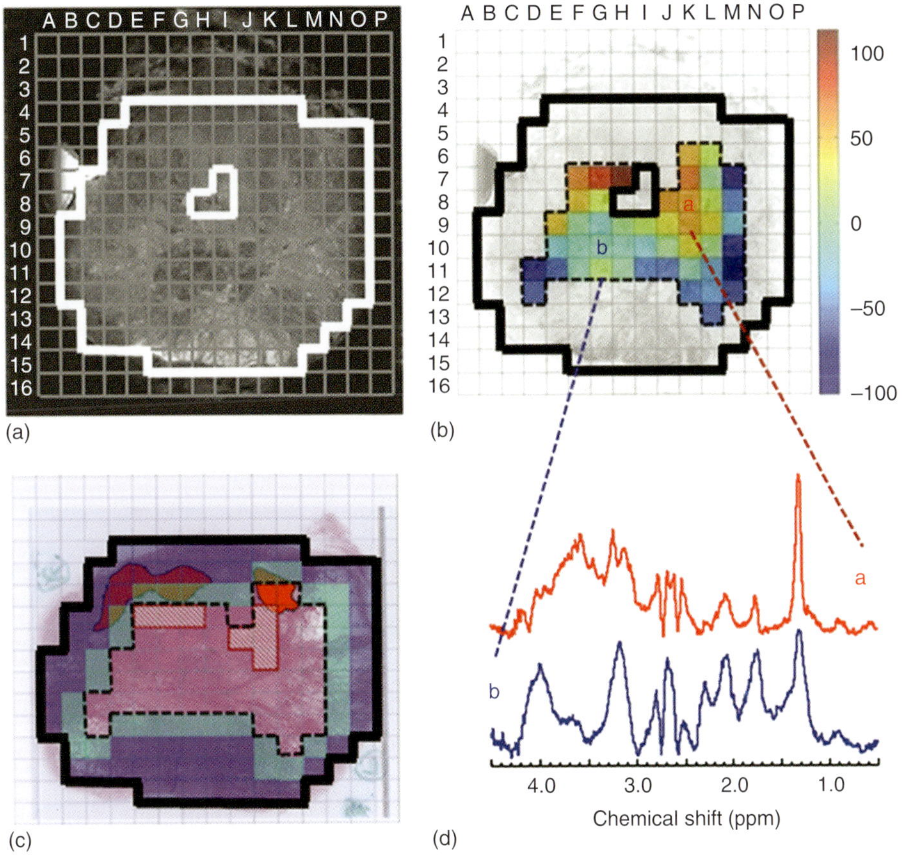

Fig. 9 Metabolomic imaging of cancer from excised human prostate gland. (a) MRI scan of a prostate axial cross-section (from a 47-year-old with tumor stage pT2cNxMx patient) overlaid with a grid to indicate the locations of 16×16 voxels for which multivoxel MR spectra were acquired. The outer white border delineates the outline of the prostate cross-section; the inner white border circles the urethra. Spectra in voxels outside the outer border and inside the inner border were not included in the analyses; (b) Because of the magnetic susceptibility interference at the tissue–air interface, spectra in voxels between solid and dashed lines were eliminated from further analysis. The values of the metabolomic profile for the remaining voxels were calculated with the overall loading factors identified from *ex-vivo* prostate tissue analysis at 14 T for all remaining voxels; the values were then mapped onto the MR image with a color range calibrated to −100 and +100; (c) Histologically identified cancer regions are circled in red and plotted onto a whole-mounted prostate histology image at approximately the same prostate cross-sectional level as in panel (b). Metabolomic profile regions having at least two connected voxels with profile values greater than M + SD are plotted in shaded red. Partial tumor sizes are estimated by excluding their overlaps with purple voxels; here, the "partial tumor size" for the right lesion on the top right of the histological slide equals the total tumor size. For the cancer region on the top left of the histological slide, it is less than half the tumor size. Geometric centers are used to calculate the center of profile-elevated regions and to estimate the center of partial tumor size, excluding the discarded voxels; (d) Representative spectra from voxels with elevated [a in panel (b)] and nonelevated [b in panel (b)] profile values are plotted. Adapted with permission from Ref. [68].

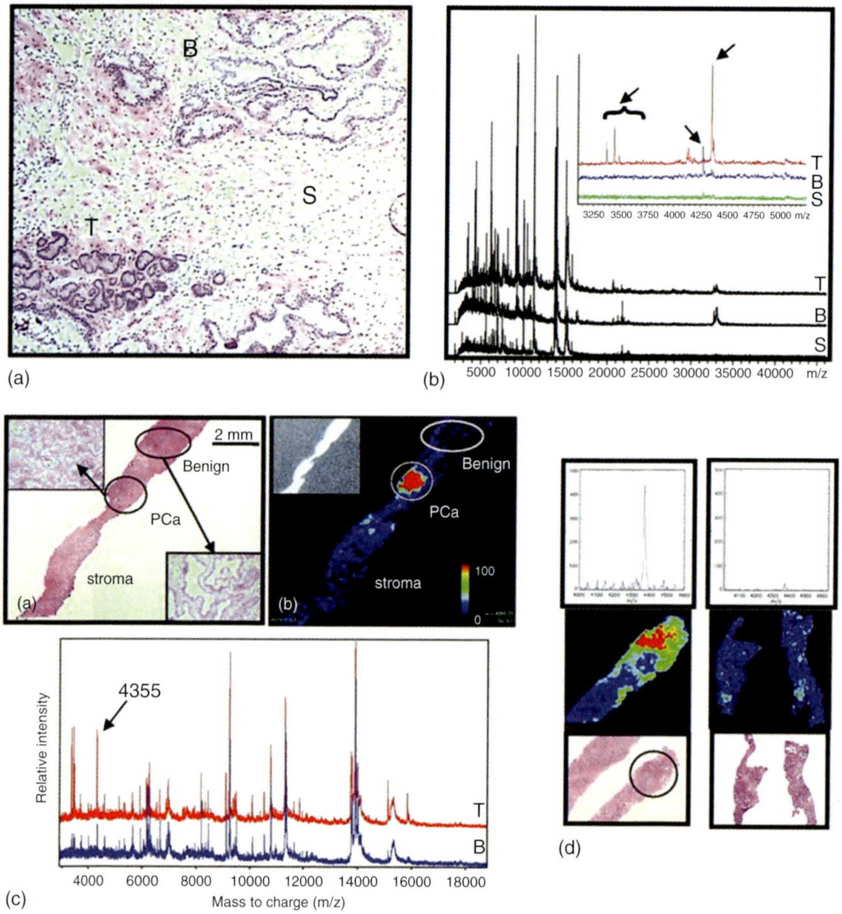

Fig. 10 Direct tissue mass spectrometric analysis of human prostate tissue reveals cell-specific profiles. Frozen prostate tissue was subjected to MALDI-MSI as described. (a) Representative histologic image of hematoxylin and eosin (H&E)-stained prostate tissue, showing areas of prostate adenocarcinoma (T), benign prostate glands (B), and benign stroma (S); (b) Resulting average spectra acquired from each region, showing characteristic profiles for different cell types. The inset is an expanded view of the mass range *m/z* 3000–5300, showing differences in the profiles for each cell type; (c) MALDI-MSI of a single prostate tissue containing tumor and uninvolved regions. Panel a: H&E-stained image of a tissue specimen containing a defined area of PCa glands and benign glands. The insets are magnified (×10) views of each cell type. Panel b: Two-dimensional ion density map showing high expression of a peak at *m/z* 4355 (red) in the PCa area (the inset is a scan of the tissue after matrix deposition). Panel c: Average spectra from a single spot obtained in the PCa (T) region and from the benign adjacent glandular (B) area displaying differential expression of the ion at *m/z* 4355; (d) MALDI-MSI two-dimensional ion density maps of a PCa-containing tissue and a benign prostate tissue. Circled: Pathology-defined regions of PCa. Areas in red from the resulting MALDI-MSI indicate high expression of *m/z* 4355. Top: Spectra exported from representative regions of each tissue are shown in the *m/z* range of 4000–4600 and display the *m/z* 4355 peak profile. Adapted with permission from Ref. [71].

differences, they have not yet extended into metabolomics. Multiple studies have uncovered protein expression patterns that were able to discriminate between PCa and benign tissue (Fig. 10) [72, 71], and similar findings were revealed for the differentiation of myxofibrosarcoma nodules from noncancerous tissue [73].

Recently, further technological advancements have allowed for a better profiling of metabolites using MSI [74]. Indeed, studies have been reported of investigations into metabolomic variations in animal models, including healthy rat brain and mice bearing drug-sensitive tumor xenografts [68, 75]. In a study of breast tumor mouse xenografts, three-dimensional MSI was used to measure the metabolic response of tumor sections to chemotherapy (Fig. 11) [68]. These findings, including both the differential protein expression patterns and the more recent metabolomic variations, indicate a potentially important role for MSI in the discovery of cancer biomarkers, as well as the identification of malignancy or monitoring of treatment in the clinic, with the expectation of providing novel information of clinical significance.

4 Limitations and Future Directions

Today, biomarkers are being widely investigated and have now begun to be used in clinical medicine for disease diagnosis and patient prognostication. The novel discipline of metabolomics has the potential to

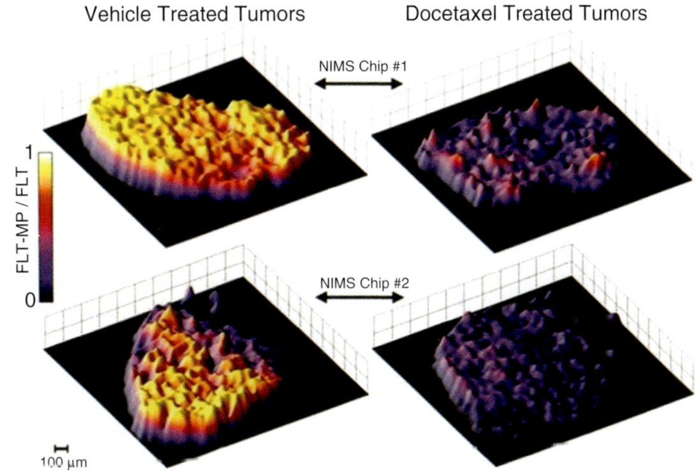

Fig. 11 Nanostructure-initiator mass spectrometry (NIMS) FLT-MP/FLT ratiometric imaging of docetaxel- and vehicle-treated breast tumor mouse xenografts. Three-dimensional NIMS images of tumor sections from fluorothymidine (FLT)-treated mice shows a significantly higher conversion of FLT to FLT-MP (30-deoxy-30-fluorothymidine monophosphate) in vehicle-treated (left) versus docetaxel-treated mice (right). Tumor sections from one set of vehicle- and docetaxel-treated mice were placed in close proximity on a single NIMS chip and imaged using NIMS (Chip #1, top). As further validation, tumor sections from a second set of vehicle- and docetaxel-treated mice were placed in close proximity on a second NIMS chip and imaged using NIMS (Chip #2, bottom). A total of four independent tumors was imaged: two from docetaxel-treated mice and two from vehicle-treated mice. The FLT-MP/FLT ratios were normalized to a maximum value of 1 for each NIMS chip. Adapted with permission from Ref. [76].

provide more robust biomarkers to enhance even further the interpretations of disease status, based on holistic evaluations of the existing biomolecular information related to physiological and pathological conditions. In order for metabolomics to become a valuable clinical tool for cancer screening, diagnosis, characterization, prognosis, and monitoring, several obstacles must first be overcome, however. To date, the majority of cancer metabolomic studies have been carried out *ex vivo*, and consequently the application of metabolomic imaging must first be translated from the successful *ex-vivo* profiles and protocols into an *in-vivo* setting for MR-based imaging, or for *ex-vivo* MSI. For instance, MRS-based translations include the consideration of differences in magnetic field strengths between the *ex-vivo* high field at 7 T or above, and the clinical reality of 3 T, or even 1.5 T. Although metabolomic studies should be incorporated into both preclinical and clinical research, the validation of profiles may depend on the integration of MRS- and MS-observable metabolites, the establishment of metabolomic databases, and the combination of metabolomic findings with related genomic and/or proteomic analyses.

A new focus for future metabolomics studies – termed metabolomic fields – may be of particular value or interest in the study of certain cancers. As most MRS-visible metabolites are of low molecular weight and reside unbound in the cytoplasm, they are not prevented from migrating away from the cancer lesions; hence, it is likely that a cancer metabolomic "lesion" may expand the histological lesion. This observed phenomenon of cancer metabolite delocalization beyond the histological lesion has been implied previously [17], and may be beneficial for the diagnosis of specific cancer types, such as rectifying the acknowledged histological sampling errors that occur during the diagnosis of PCa. It is through these metabolomic fields that metabolomic profiles, determined from histologically benign regions of PCa, may indicate a patient's clinical and pathological status and may even be able to predict the aggressive potential of the cancer. In this way, metabolomics would be of great clinical value for increasing the power of *ex-vivo* pathological analyses.

The ultimate goal of metabolomic imaging in the cancer clinic – with either MRS or MS – is to improve the abilities of both radiological and pathological analyses in the detection, identification and evaluation of cancer, and the monitoring of its therapy. Based on their similarities to histology – in the sense of conceptual evaluation and temporal development – metabolomics and metabolomic imaging will surely have the ability to identify and even predict the bimolecular changes that occur during cancer development and progression, thereby strengthening the clinical possibility of implementing "personalized" medicine.

References

1. Cheng, L.L. and Pohl, U. (2007) The Role of NMR-Based Metabolomics in Cancer, in: Lindon, J.C., Nicholls, J.K., Holmes, E. (Eds), *The Handbook of Metabonomics and Metabolomics*. Amsterdam, Elsevier, pp. 345–374.
2. Wishart, D.S., Jewison, T., Guo, A.C., Wilson, M., *et al.* (2013) HMDB 3.0 – the human metabolome database in 2013. *Nucleic Acids Res.*, **41** (Database issue), D801-D807.
3. Griffin, J.L. and Shockcor, J.P. (2004) Metabolic profiles of cancer cells. *Nat. Rev. Cancer*, **4**(7), 551–561.
4. Slamon, D.J., Clark, G.M., Wong, S.G., Levin, W.J., Ullrich, A., and McGuire, W.L. (1987) Human breast cancer: correlation of relapse and survival with amplification of the

HER-2/neu oncogene. *Science*, **235**(4785), 177–182.
5. Stamey, T.A., Yang, N., Hay, A.R., McNeal, J.E., Freiha, F.S., and Redwine, E. (1987) Prostate-specific antigen as a serum marker for adenocarcinoma of the prostate. *N. Engl. J. Med.*, **317**(15), 909–916.
6. Spratlin, J.L., Serkova, N.J., and Eckhardt, S.G. (2009) Clinical applications of metabolomics in oncology: a review. *Clin. Cancer Res.*, **15**(2), 431–440.
7. Dunn, W.B., Bailey, N.J., and Johnson, H.E. (2005) Measuring the metabolome: current analytical technologies. *Analyst*, **130**(5), 606–625.
8. Damadian, R. (1971) Tumor detection by nuclear magnetic resonance. *Science*, **171**(976), 1151–1153.
9. Brown, F.F., Campbell, I.D., Kuchel, P.W., and Rabenstein, D.C. (1977) Human erythrocyte metabolism studies by 1H spin echo NMR. *FEBS Lett.*, **82**(1), 12–16.
10. Cheng, L.L., Lean, C.L., Bogdanova, A., Wright, S.C. *et al.* (1996) Enhanced resolution of proton NMR spectra of malignant lymph nodes using magic-angle spinning. *Magn. Reson. Med.*, **36**(5), 653–658.
11. Cheng, L.L., Ma, M.J., Becerra, L., Ptak, T. *et al.* (1997) Quantitative neuropathology by high resolution magic angle spinning proton magnetic resonance spectroscopy. *Proc. Natl Acad. Sci. USA*, **94**(12), 6408–6413.
12. Cheng, L.L., Chang, I.W., Smith, B.L., and Gonzalez, R.G. (1998) Evaluating human breast ductal carcinomas with high-resolution magic-angle spinning proton magnetic resonance spectroscopy. *J. Magn. Reson.*, **135**(1), 194–202.
13. Cheng, L.L., Chang, I.W., Louis, D.N., and Gonzalez, R.G. (1998) Correlation of high-resolution magic angle spinning proton magnetic resonance spectroscopy with histopathology of intact human brain tumor specimens. *Cancer Res.*, **58**(9), 1825–1832.
14. Cheng, L.L., Anthony, D.C., Comite, A.R., Black, P.M., Tzika, A.A., and Gonzalez, R.G. (2000) Quantification of microheterogeneity in glioblastoma multiforme with ex vivo high-resolution magic-angle spinning (HRMAS) proton magnetic resonance spectroscopy. *Neuro Oncol.*, **2**(2), 87–95.
15. Wright, A.J., Fellows, G.A., Griffiths, J.R., Wilson, M., Bell, B.A., and Howe, F.A. (2010) Ex-vivo HRMAS of adult brain tumours: metabolite quantification and assignment of tumour biomarkers. *Mol. Cancer*, **9**, 66.
16. Swanson, M.G., Zektzer, A.S., Tabatabai, Z.L., Simko, J. *et al.* (2006) Quantitative analysis of prostate metabolites using 1H HR-MAS spectroscopy. *Magn. Reson. Med.*, **55**(6), 1257–1264.
17. Cheng, L.L., Burns, M.A., Taylor, J.L., He, W. *et al.* (2005) Metabolic characterization of human prostate cancer with tissue magnetic resonance spectroscopy. *Cancer Res.*, **65**(8), 3030–3034.
18. Giskeodegard, G.F., Bertilsson, H., Selnaes, K.M., Wright, A.J. *et al.* (2013) Spermine and citrate as metabolic biomarkers for assessing prostate cancer aggressiveness. *PLoS ONE*, **8**(4), e62375.
19. Maxeiner, A., Adkins, C.B., Zhang, Y., Taupitz M., *et al.* (2010) Retrospective analysis of prostate cancer recurrence potential with tissue metabolomic profiles. *Prostate*, **70**(7), 710–717.
20. Imperiale, A., Elbayed, K., Moussallieh, F.M., Neuville, A., *et al.* (2011) Metabolomic pattern of childhood neuroblastoma obtained by (1)H-high-resolution magic angle spinning (HRMAS) NMR spectroscopy. *Pediatr. Blood Cancer*, **56**(1), 24–34.
21. Erb, G., Elbayed, K., Piotto, M., Raya, J. *et al.* (2008) Toward improved grading of malignancy in oligodendrogliomas using metabolomics. *Magn. Reson. Med.*, **59**(5), 959–965.
22. Constantin, A., Elkhaled, A., Jalbert, L., Srinivasan, R. *et al.* (2012) Identifying malignant transformations in recurrent low grade gliomas using high resolution magic angle spinning spectroscopy. *Artif. Intell. Med.*, **55**(1), 61–70.
23. Rocha, C.M., Barros, A.S., Gil, A.M., Goodfellow, B.J. *et al.* (2010) Metabolic profiling of human lung cancer tissue by 1H high resolution magic angle spinning (HRMAS) NMR spectroscopy. *J. Proteome Res.*, **9**(1), 319–332.
24. Jordan, K.W., Adkins, C.B., Su, L., Halpren, E.F. *et al.* (2010) Comparison of squamous cell carcinoma and adenocarcinoma of the lung by metabolomic analysis of tissue-serum pairs. *Lung Cancer*, **68**(1), 44–50.
25. Bathen, T.F., Jensen, L.R., Sitter, B., Fjösne, H.E., *et al.* (2007) MR-determined metabolic phenotype of breast cancer in prediction of

lymphatic spread, grade, and hormone status. *Breast Cancer Res. Treat.*, **104**(2), 181–189.

26. Giskeodegard, G.F., Grinde, M.T., Sitter, B., Axelson, D.E. *et al.* (2010) Multivariate modeling and prediction of breast cancer prognostic factors using MR metabolomics. *J. Proteome Res.*, **9**(2), 972–979.
27. Choi, J.S., Baek, H.M., Kim, S., Kim, M.J. *et al.* (2012) HR-MAS MR spectroscopy of breast cancer tissue obtained with core needle biopsy: correlation with prognostic factors. *PLoS ONE*, **7**(12), e51712.
28. Giskeodegard, G.F., Lundgren, S., Sitter, B., Fjøsne, H.E., *et al.* (2012) Lactate and glycine-potential MR biomarkers of prognosis in estrogen receptor-positive breast cancers. *NMR Biomed.*, **25**(11), 1271–1279.
29. Bathen, T.F., Geurts, B., Sitter, B., Fjøsne, H.E. *et al.* (2013) Feasibility of MR metabolomics for immediate analysis of resection margins during breast cancer surgery. *PLoS ONE*, **8**(4), e61578.
30. Choi, J.S., Baek, H.M., Kim, S., Kim, M.J. *et al.* (2013) Magnetic resonance metabolic profiling of breast cancer tissue obtained with core needle biopsy for predicting pathologic response to neoadjuvant chemotherapy. *PLoS ONE*, **8**(12), e83866.
31. Kinross, J.M., Drymousis, P., Jimenez, B., and Frilling A. (2013) Metabonomic profiling: a novel approach in neuroendocrine neoplasias. *Surgery*, **154**(6), 1185–1192; discussion 92–93.
32. Jordan, K.W., Nordenstam, J., Lauwers, G.Y., Rothenberger, D.A., *et al.* (2009) Metabolomic characterization of human rectal adenocarcinoma with intact tissue magnetic resonance spectroscopy. *Dis. Colon Rectum*, **52**(3), 520–525.
33. Jordan, K.W., Adkins, C.B., Cheng, L.L., and Faquin, W.C. (2011) Application of magnetic-resonance-spectroscopy-based metabolomics to the fine-needle aspiration diagnosis of papillary thyroid carcinoma. *Acta Cytol.*, **55**(6), 584–589.
34. Bansal, N., Gupta, A., Mitash, N., Shakya, P.S. *et al.* (2013) Low- and high-grade bladder cancer determination via human serum-based metabolomics approach. *J. Proteome Res.*, **12**(12), 5839–5850.
35. Odunsi, K., Wollman, R.M., Ambrosone, C.B., Hutson, A. *et al.* (2005) Detection of epithelial ovarian cancer using 1H-NMR-based metabonomics. *Int. J. Cancer*, **113**(5), 782–788.
36. Shuster, J.R., Lance, R.S., and Troyer, D.A. (2011) Molecular preservation by extraction and fixation, mPREF: a method for small molecule biomarker analysis and histology on exactly the same tissue. *BMC Clin. Pathol.*, **11**, 14.
37. Brown, M.V., McDunn, J.E., Gunst, P.R., Smith, E.M. *et al.* (2012) Cancer detection and biopsy classification using concurrent histopathological and metabolomic analysis of core biopsies. *Genome Med.*, **4**(4), 33.
38. Saylor, P.J., Karoly, E.D., and Smith, M.R. (2012) Prospective study of changes in the metabolomic profiles of men during their first three months of androgen deprivation therapy for prostate cancer. *Clin. Cancer Res.*, **18**(13), 3677–3685.
39. Thysell, E., Surowiec, I., Hornberg, E., Crnalic, S., *et al.* (2010) Metabolomic characterization of human prostate cancer bone metastases reveals increased levels of cholesterol. *PLoS ONE*, **5**(12), e14175.
40. Sreekumar, A., Poisson, L.M., Rajendiran, T.M., Khan, A.P. *et al.* (2009) Metabolomic profiles delineate potential role for sarcosine in prostate cancer progression. *Nature*, **457**(7231), 910–914.
41. Wu, H., Liu, T., Ma, C., Xue, R. *et al.* (2011) GC/MS-based metabolomic approach to validate the role of urinary sarcosine and target biomarkers for human prostate cancer by microwave-assisted derivatization. *Anal. Bioanal. Chem.*, **401**(2), 635–646.
42. Denkert, C., Budczies, J., Kind, T., Weichert, W. *et al.* (2006) Mass spectrometry-based metabolic profiling reveals different metabolite patterns in invasive ovarian carcinomas and ovarian borderline tumors. *Cancer Res.*, **66**(22), 10795–10804.
43. Fong, M.Y., McDunn, J., and Kakar, S.S. (2011) Identification of metabolites in the normal ovary and their transformation in primary and metastatic ovarian cancer. *PLoS ONE*, **6**(5), e19963.
44. Wen, T., Gao, L., Wen, Z., Wu, C. *et al.* (2013) Exploratory investigation of plasma metabolomics in human lung adenocarcinoma. *Mol. Biosyst.*, **9**(9), 2370–2378.
45. Shen, J., Yan, L., Liu, S., Ambrosone, C.B., and Zhao H. (2013) Plasma metabolomic

profiles in breast cancer patients and healthy controls: by race and tumor receptor subtypes. *Transl. Oncol.*, **6**(6), 757–765.

46. Miyagi, Y., Higashiyama, M., Gochi, A., Akaike, M.. *et al.* (2011) Plasma free amino acid profiling of five types of cancer patients and its application for early detection. *PLoS ONE*, **6**(9), e24143.
47. Catchpole, G., Platzer, A., Weikert, C., Kempkensteffen, C., *et al.* (2011) Metabolic profiling reveals key metabolic features of renal cell carcinoma. *J. Cell. Mol. Med.*, **15**(1), 109–118.
48. Tan, B., Qiu, Y., Zou, X., Chen, T. *et al.* (2013) Metabonomics identifies serum metabolite markers of colorectal cancer. *J. Proteome Res.*, **12**(6), 3000–3009.
49. Kim, K., Aronov, P., Zakharkin, S.O., Anderson, D. *et al.* (2009) Urine metabolomics analysis for kidney cancer detection and biomarker discovery. *Mol. Cell. Proteomics*, **8**(3), 558–570.
50. Liu, Y., Hong, Z., Tan, G., Dong, X. *et al.* (2014) NMR and LC/MS-based global metabolomics to identify serum biomarkers differentiating hepatocellular carcinoma from liver cirrhosis. *Int. J. Cancer*, 135(3), 658–668.
51. Tiziani, S., Kang, Y., Harjanto, R., Axelrod, J., *et al.* (2013) Metabolomics of the tumor microenvironment in pediatric acute lymphoblastic leukemia. *PLoS ONE*, **8**(12), e82859.
52. McNeil, B.J., Keller, E., and Adelstein, S.J. (1975) Primer on certain elements of medical decision making. *N. Engl. J. Med.*, **293**(5), 211–215.
53. McNeil, B.J., Varady, P.D., Burrows, B.A., and Adelstein, S.J. (1975) Measures of clinical efficacy. Cost-effectiveness calculations in the diagnosis and treatment of hypertensive renovascular disease. *N. Engl. J. Med.*, **293**(5), 216–221.
54. Cady, E.B., Costello, A.M., Dawson, M.J., Delpy, D.T., *et al.* (1983) Non-invasive investigation of cerebral metabolism in newborn infants by phosphorus nuclear magnetic resonance spectroscopy. *Lancet*, **1**(8333), 1059–1062.
55. Posse, S., Otazo, R., Dager, S.R., and Alger, J. (2012) MR spectroscopic imaging: principles and recent advances. *J. Magn. Reson. Imaging*, **37**(6), 1301–1325.
56. Tugnoli, V., Schenetti, L., Mucci, A., Parenti, F., *et al.* (2006) Ex vivo HR-MAS MRS of human meningiomas: a comparison with in vivo 1H MR spectra. *Int. J. Mol. Med.*, **18**(5), 859–869.
57. Wilson, M., Davies, N.P., Grundy, R.G., and Peet, A.C. (2009) A quantitative comparison of metabolite signals as detected by in vivo MRS with ex vivo 1H HR-MAS for childhood brain tumours. *NMR Biomed.*, **22**(2), 213–219.
58. Opstad, K.S., Wright, A.J., Bell, B.A., Griffiths, J.R., and Howe, F.A. (2010) Correlations between in vivo (1)H MRS and ex vivo (1)H HRMAS metabolite measurements in adult human gliomas. *J. Magn. Reson. Imaging*, **31**(2), 289–297.
59. Vicente, J., Fuster-Garcia, E., Tortajada, S., García-Gómez, J.M., *et al.* (2013) Accurate classification of childhood brain tumours by in vivo (1)H MRS – a multi-centre study. *Eur. J. Cancer*, **49**(3), 658–667.
60. Wilson, M., Cummins, C.L., Macpherson, L., Sun, Y. *et al.* (2013) Magnetic resonance spectroscopy metabolite profiles predict survival in paediatric brain tumours. *Eur. J. Cancer*, **49**(2), 457–464.
61. Li, Y., Lupo, J.M., Parvataneni, R., Lamborn, K.R. *et al.* (2013) Survival analysis in patients with newly diagnosed glioblastoma using pre- and postradiotherapy MR spectroscopic imaging. *Neuro Oncol.*, **15**(5), 607–617.
62. Kurhanewicz, J., Swanson, M.G., Nelson, S.J., and Vigneron, D.B. (2002) Combined magnetic resonance imaging and spectroscopic imaging approach to molecular imaging of prostate cancer. *J. Magn. Reson. Imaging*, **16**(4), 451–463.
63. Nagarajan, R., Margolis, D., Raman, S., Sarma, M.K. *et al.* (2012) MR spectroscopic imaging and diffusion-weighted imaging of prostate cancer with Gleason scores. *J. Magn. Reson. Imaging*, **36**(3), 697–703.
64. Swanson, M.G., Vigneron, D.B., Tabatabai, Z.L., Males, R.G. *et al.* (2003) Proton HR-MAS spectroscopy and quantitative pathologic analysis of MRI/3D-MRSI-targeted postsurgical prostate tissues. *Magn. Reson. Med.*, **50**(5), 944–954.
65. Selnaes, K.M., Gribbestad, I.S., Bertilsson, H., Wright, A. *et al.* (2012) Spatially matched in vivo and ex vivo MR metabolic profiles

of prostate cancer – investigation of a correlation with Gleason score. *NMR Biomed.*, **26**(5), 600–606.

66. Haddadin, I.S., McIntosh, A., Meisamy, S., Corum, C. *et al.* (2009) Metabolite quantification and high-field MRS in breast cancer. *NMR Biomed.*, **22**(1), 65–76.
67. Budczies, J., Denkert, C., Muller, B.M., Brockmöller, S.F., *et al.* (2012) Remodeling of central metabolism in invasive breast cancer compared to normal breast tissue – a GC-TOFMS based metabolomics study. *BMC Genomics*, **13**, 334.
68. Wu, C.L., Jordan, K.W., Ratai, E.M., Sheng, J. *et al.* (2010) Metabolomic imaging for human prostate cancer detection. *Sci. Transl. Med.*, **2**, 16ra8.
69. Zavalin, A., Todd, E.M., Rawhouser, P.D., Yang, J., Norris, J.L., and Caprioli, R.M. (2012) Direct imaging of single cells and tissue at sub-cellular spatial resolution using transmission geometry MALDI MS. *J. Mass Spectrom.*, **47**(11), 1473–1481.
70. Yanagisawa, K., Shyr, Y., Xu, B.J., Larsen, P.H. *et al.* (2003) Proteomic patterns of tumour subsets in non-small-cell lung cancer. *Lancet*, **362**(9382), 433–439.
71. Cazares, L.H., Troyer, D., Mendrinos, S., Lance, R.A. *et al.* (2009) Imaging mass spectrometry of a specific fragment of mitogen-activated protein kinase/extracellular signal-regulated kinase kinase kinase 2 discriminates cancer from uninvolved prostate tissue. *Clin. Cancer Res.*, **15**(17), 5541–5551.
72. Schwamborn, K., Krieg, R.C., Reska, M., Jakse, G., Knuechel, R., and Wellmann, A. (2007) Identifying prostate carcinoma by MALDI-Imaging. *Int. J. Mol. Med.*, **20**(2), 155–159.
73. Jones, E.A., van Remoortere, A., van Zeijl, R.J., Hogendoorn, P.C., *et al.* (2011) Multiple statistical analysis techniques corroborate intratumor heterogeneity in imaging mass spectrometry datasets of myxofibrosarcoma. *PLoS ONE*, **6**(9), e24913.
74. Shanta, S.R., Kim, T.Y., Hong, J.H., Lee, J.H. *et al.* (2012) A new combination MALDI matrix for small molecule analysis: application to imaging mass spectrometry for drugs and metabolites. *Analyst*, **137**(24), 5757–5762.
75. Benabdellah, F., Touboul, D., Brunelle, A., and Laprevote, O. (2009) In situ primary metabolites localization on a rat brain section by chemical mass spectrometry imaging. *Anal. Chem.*, **81**(13), 5557–5560.
76. O'Brien, P.J., Lee, M., Spilker, M.E., Zhang, C.C. *et al.* (2013) Monitoring metabolic responses to chemotherapy in single cells and tumors using nanostructure-initiator mass spectrometry (NIMS) imaging. *Cancer Metab.*, **1**(1), 1–4.

13
Microfluidics in Nanomedicine

YongTae Kim[1] and Robert Langer[2]
[1]Georgia Institute of Technology, George W. Woodruff School of Mechanical Engineering, Wallace H. Coulter Department of Biomedical Engineering, Institute for Electronics and Nanotechnology, Parker H. Petit Institute for Bioengineering and Bioscience, 345 Ferst Drive, Atlanta, GA 30318, USA
[2]Massachusetts Institute of Technology, Department of Chemical Engineering, Harvard-MIT Division of Health Sciences and Technology, David H. Koch Institute for Integrative Cancer Research, 500 Main Street, Cambridge, MA 02139, USA

Translational Medicine Cancer: Advances in Molecular Biology and Medicine,
First Edition. Edited by Robert A. Meyers.

Keywords

Microfluidics
The science and technology that involves the manipulation of nanoscale amounts of fluids in microscale fluidic channels for applications that include chemical synthesis, and biological analysis and engineering.

Nanotechnology
The manipulation of matter on atomic and molecular scales.

Nanomedicine
The medical application of nanotechnology for the advanced diagnosis, treatment and prevention of a number of diseases.

Biomimetic microsystem
A microscale device that mimics biological systems and is used to probe complex human problems.

Clinical translation
Clinical translation involves the application of discoveries made in the laboratory to diagnostic tools, medicines, procedures, policies and education, in order to improve the health of individuals and the community.

Nanomedicine is the medical application of nanotechnology for the treatment and prevention of major ailments, including cancer and cardiovascular diseases. Despite the progress and potential of nanomedicines, many such materials fail to reach clinical trials due to critical challenges that include poor reproducibility in high-volume production that have led to failure in animal studies and clinical trials. Recent approaches using microfluidic technology have provided emerging platforms with great potential to accelerate the clinical translation of nanomedicine. Microfluidic technologies for nanomedicine development are reviewed in this chapter, together with a detailed discussion of microfluidic assembly, characterization and evaluation of nanomedicine, and a description of current challenges and future prospects.

1 Introduction

Nanomedicine is the medical application of nanotechnology that uses engineered nanomaterials for the robust delivery of therapeutic and diagnostic agents in the advanced treatment of many diseases, including cancer [1–3], atherosclerosis [4–6], diabetes [7–9], pulmonary diseases [10, 11] and disorders of the central nervous system [12, 13]. One key advantage of nanomedicine is the ability to deliver poorly water-soluble drugs [14–16] or plasma-sensitive nucleic acids (e.g., small interfering (si)RNA [17, 18]) into the circulation with enhanced stability. Nanomedicine is also capable of providing contrast agents for different imaging modalities and the targeting of specific sites for the delivery of drugs and/or genes [19–23]. Engineered nanomaterials, developed as particulates that are widely referred to as nanoparticles (NPs), have been formulated using a variety of materials that includes lipids, polymers, inorganic nanocrystals, carbon nanotubes, proteins, and DNA origami [24–36]. The ultimate goal of nanomedicine is to achieve a robust, targeted delivery of complex assemblies that contain sufficient amounts of multiple therapeutic and diagnostic agents for highly localized drug release, but with no adverse side effects [37, 38], and a reliable detection of any site-specific therapeutic response [39, 40].

1.1 Nanomedicine Development

Typical nanomedicine development processes for the clinical translation include benchtop syntheses, characterizations, *in-vitro* evaluations, *in-vivo* evaluations with animal models, and scaled-up production in readiness for clinical trials. Although, previously, several NPs have been reported as superior platforms, many are still far from their first stages of patient clinical trials due to several critical challenges [41, 42]. Such challenges mainly result from batch-to-batch variations of NPs produced in the benchtop synthesis process, and from insignificant outcomes in the *in-vitro* evaluation process under physiologically irrelevant conditions. These limitations ultimately lead to highly variable results in the *in-vivo* evaluation, or to failure in clinical trials. In order to address these challenges, the following methodologies need to be established in the nanomedicine development process:

- Nanomedicine needs to be continuously produced in a high-throughput fashion. The large-scale, continuous production of nanomedicines will allow a robust supply of highly reproducible materials for the *in-vitro* and *in-vivo* evaluation stages and clinical trials, ultimately increasing the success rate in clinical trials.
- Nanomedicines synthesized using large-scale, continuous production methods also need to be characterized in a high-throughput manner. Rapid characterization will create an efficient production cycle for an optimized nanomedicine via feedback loops between the synthesis and characterization stages.
- The *in-vitro* evaluation of nanomedicine must be conducted in more physiologically relevant environments. Highly repeatable results obtained from these biomimetic conditions will allow the obviation of a number of simple screening experiments in animal studies, not only saving costly animal models but also accelerating the clinical translation.

1.2 Microfluidics Technology

Microfluidics technology provides highly compatible platforms to create a new nanomedicine development pipelines that include the required methodologies introduced above. Basically, microfluidics presents a number of useful capabilities to manipulate very small quantities of samples, and to detect substances with a high resolution for a wide range of applications, including chemical syntheses [43, 44] and biological analysis [45, 46]. More importantly, the adaptability of microfluidics allows its integration with many other technologies, such as micro/nanofabrication, electronics, and feedback control systems [47–52]. Recently, microfluidic platforms integrated with control systems and advanced microfabrication technologies have been used to address the critical challenges in nanomedicine [53–57]. For example, the continuous synthesis of NPs in microfluidics has demonstrated a versatility to produce a variety of NPs with different sizes, shapes, and surface compositions [58, 59]. Several advances have recently been made in the label-free detection, characterization and identification of single NPs [60]. The confluence of microfluidics and biomimetic design has enabled the creation of physiologically relevant microenvironments for the evaluation of drug candidates [61–63]. The key microfluidic technologies in nanomedicine, including microfluidic assembly, and the characterization and evaluation of nanomedicines, are discussed in the following sections (see Fig. 1), and their current challenges and future research directions are highlighted.

2 Microfluidic Assembly of Nanomedicines

The bulk synthesis of NPs typically has strong dependencies on nonstandard multistep processes which are time-consuming, difficult to scale up, and depend heavily on specific synthetic conditions in

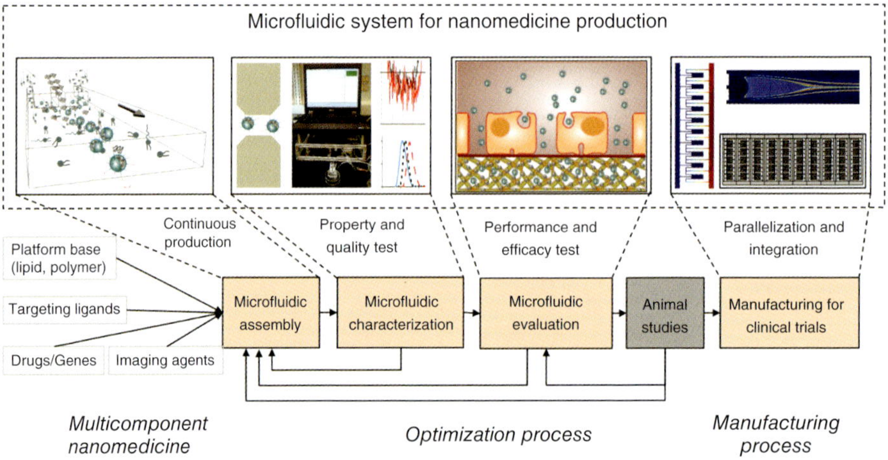

Fig. 1 A new nanomedicine development pipeline using microfluidic systems. First, a designed nanomedicine with multiple precursor components is continuously assembled through controlled strong mixing patterns in the Microfluidic Assembly stage, and the properties of the nanomedicine produced are identified at the Microfluidic Characterization stage. Only if those properties meet the nanomedicine design criteria will the performance and efficacy of a selected nanomedicine be evaluated in *in-vitro* biomimetic microsystems that recapitulate the structure and function of human organs in the Microfluidic Evaluation stage. If the targeting, therapeutic, and imaging efficacies are satisfactory in the *in-vitro* model system, the nanomedicine will then be validated with animal models. All nanomedicine candidates that are unsuccessful at the above stage will be reformulated in the Microfluidic Assembly stage and go through the iterative processes. If successful in animal models, the selected nanomedicine will then be manufactured through parallelized microfluidic platforms. The pressure and flow patterns in the integrated microfluidic system are regulated by high-precision control systems.

the laboratory. This reliance of NPs on such nonstandard multistep processes inevitably causes high batch-to-batch variations in their physico-chemical properties [64–69]. Batch size is also subject to custom protocols that vary among laboratories, leading to difficulties in screening and identifying optimal NP physico-chemical characteristics for enhanced drug delivery. Furthermore, the introduction and combination of multiple materials for creating multicomponent NPs compromises the expected functionality of the individual elements. This is largely because of an inability to precisely control the continuous assembly process in various conventional bulk syntheses that involve the macroscopic mixing of precursor solutions [58, 70]. As the micrometer- and nanometer-scale interactions of precursors will direct the characteristics of NPs, it is essential that their composition is fine-tuned in order to attain the anticipated functionalities of multicomponent NP assemblies. In general, the central challenge for the synthesis of multicomponent NPs is to establish large-scale and continuous manufacturing methodologies with high reproducibility.

Amphiphilic blocks self-assemble spontaneously into NPs through size-dependent formation mechanisms and on timescales governed by diffusion [71]. The physico-chemical properties of NPs are, at least in part, determined by the timescales at which the multiple solutions mix in the system [72], as well as the thermodynamic characteristics of the block polymers [73]. Thus, a mixing timescale that is longer than the characteristic time for chemical chain formation will result in an uncontrolled aggregation due to incomplete solvent change. Conversely, a complete solvent change through shorter mixing times in rapid precipitations can result in stable assembly kinetics that lead to the production of homogeneous NPs [74]. One critical difference between conventional bulk synthesis and microfluidic assembly is the mixing time, which occurs on the order of seconds in bulk synthesis and contrasts with those in the millisecond and microsecond range in microfluidic assembly [75]. This shorter mixing time results in more homogeneous NPs by reducing the aggregation of precursors, leading to high reproducibility, which in turn prevents the subsequent thermal and mechanical agitation needed in conventional bulk synthesis for NP homogenization. Therefore, a precise control of microfluidic flow patterns with tunable characteristic mixing times will offer a better understanding of the effect of the mixing time on NP reproducibility and homogeneity.

Microfluidic technologies have demonstrated a better control over effective mixing of the precursor solutions for assembling a range of NP types (Fig. 2a) when compared to conventional bulk methods, due to the larger contact surface areas given per unit volume of fluid in microfluidics [58, 70]. For example, typical laminar flows in microfluidics enabled the controlled syntheses of several NPs (Fig. 2b), including liposomes [76–78] and polymeric NPs [75, 79, 80], with a narrower size distribution compared to those of conventional bulk synthesis. Under laminar flow conditions at a low Reynolds number ($\sim$1),[1] mixing occurs only through diffusion across the interface between two miscible fluids moving next to each other in viscous flows. Unfortunately, NP synthesis by diffusive mixing does not allow for the development

[1] The Reynold's number is a dimensionless number that provides a ratio of inertial to viscous forces to quantify the relative importance of these two types of force for given flow conditions.

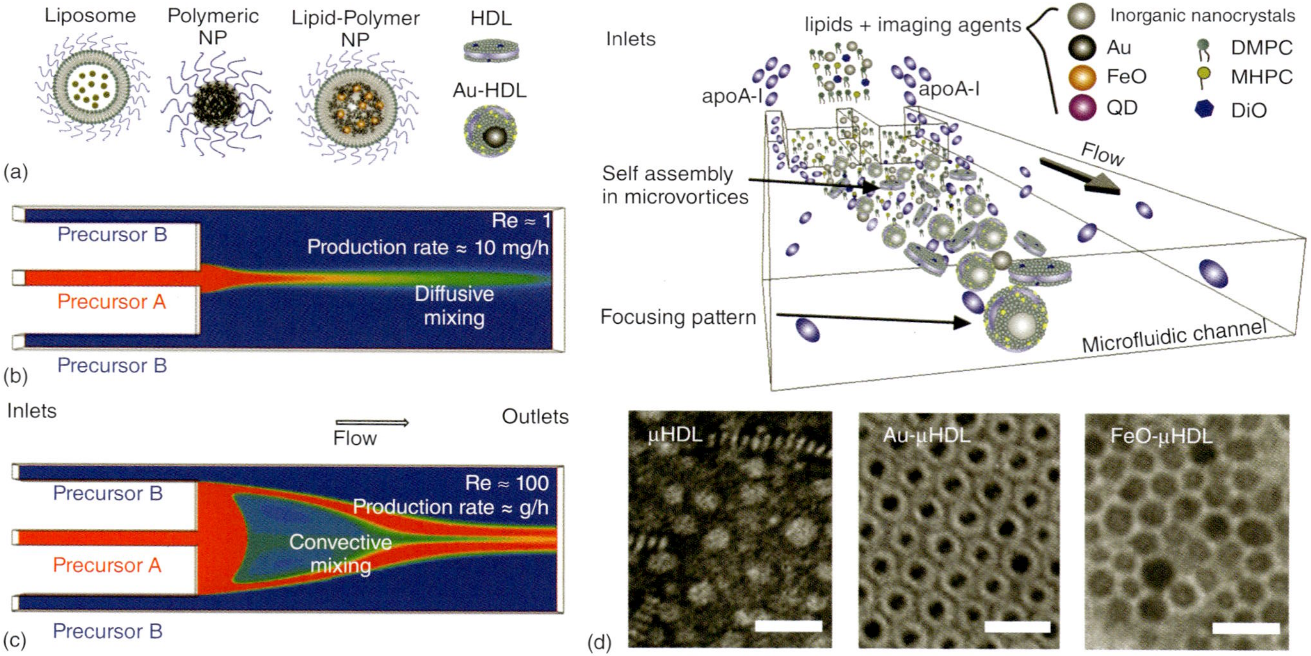

Fig. 2 (a) Representative nanomedicine types (liposome, polymeric NP, lipid-polymer NP, and high-density lipoprotein (HDL)) that have been assembled by microfluidic technology; (b) A microfluidic channel that creates diffusive mixing across laminar flow interfaces at Re ~ 1; (c) A microfluidic channel that creates convective mixing between two precursors at Re ~ 100; (d) Microfluidic reconstitution of HDL (referred to as ■ μHDL) with inorganic nanocrystals, such as gold (Au) and iron oxide (FeO), using controlled microvortices. The transmission electron microscopy images show μHDL, Au-μHDL, and FeO-μHDL (scale bars = 20 nm). Reproduced with permission from Refs [100].

of materials such as lipid–polymer hybrid NPs [81], which require a strong mixing of solutions in the aqueous and organic phases. Lipid–polymer hybrid NPs have shown a higher drug loading within the polymer core and a slower drug release due to the lipid shell when compared to pure polymeric NPs [82, 83]. Furthermore, the diffusive mixing required is difficult to scale up and ultimately leads to a limited controllability of the precursor mixing time, thereby restricting NP homogeneity and leading to a low-throughput production of NPs.

One approach to facilitate precursor mixing (i.e., shortening the mixing times) is to use convective mixing, thereby increasing the interfacial surface area between fluids and reducing the diffusion length scales. Whereas, conventional microfluidic systems exploit easy-to-control flow patterns, which are strictly laminar at low Re values ($\sim$1), an increase in Re ($10 < Re < 30$) generates complex flow patterns under a variety of geometric conditions of the microfluidic channel, such as local microvortices and flow separation due to an increase in inertial forces [84–86]. In order to implement convective mixing in microfluidic devices, microfluidic platforms have been designed for the rapid mixing of fluids using relatively higher inertial forces in localized regions with moderate Re values ($10 < Re < 100$) [87–93]. Furthermore, microvortices have demonstrated the ability to rapidly manipulate, sort, and excite particles in microfluidics [94–97]. Recently, a new generation of three-dimensional (3D) focusing patterns in a simple, single-layer microfluidic channel has allowed the development of a pattern-controllable microvortex platform (Fig. 2c). This device has been used for the highly reproducible synthesis of lipid–polymer hybrid NPs with multiple drugs and imaging agents [98, 99], and multifunctional high-density lipoprotein -derived NPs [100] with high productivity (up to $1\,g\,h^{-1}$) (Fig. 2d).

3
Microfluidic Characterization of Nanomedicines

The most important properties of NPs to be characterized before probing their interaction with biological systems are size, shape, surface chemistry/charge, and stability. The development of novel NP characterization tools will impact heavily on nanomedicine, as the lack of characterization standards and quality control tools for NPs has inhibited their clinical adoption to date. One practical obstacle to the clinical-scale commercialization of NPs is an inability to certify the stability of formulations, as even small property variations will have significant effects on *in-vivo* distribution, causing unpredictable therapy outcomes. Recently, several studies have been conducted on NP quality evaluation in microfluidics. For example, a rapid liposome quality assessment in microfluidics allows for quantitative results on liposome formulation composition and stability using dielectric spectroscopy and multivariate data analysis methods [101]. Instantaneous immobilization by ultra-rapid cooling in microfluidics reveals the formation of nonequilibrium liposomes in detail [102]. Yet, in spite of recent advances in the label-free characterization of single NPs [60], it remains difficult to effectively integrate a high-throughput microfluidic technology capable of detecting NPs (or their motion) with currently available characterization equipment that includes dynamic light scattering (DLS) [103], transmission electron microscopy (TEM) [104], atomic force microscopy (AFM) [105], Auger electron spectroscopy (AES) [106], nuclear magnetic resonance (NMR) [107], and flow cytometry [108].

Meanwhile, recent advances in nanofabrication and microfluidics have allowed for the development of high-throughput devices capable of characterizing NP properties, including size and surface charge. The most common electrical technique is to probe impedance changes in nanowire-embedded microfluidic and nanofluidic

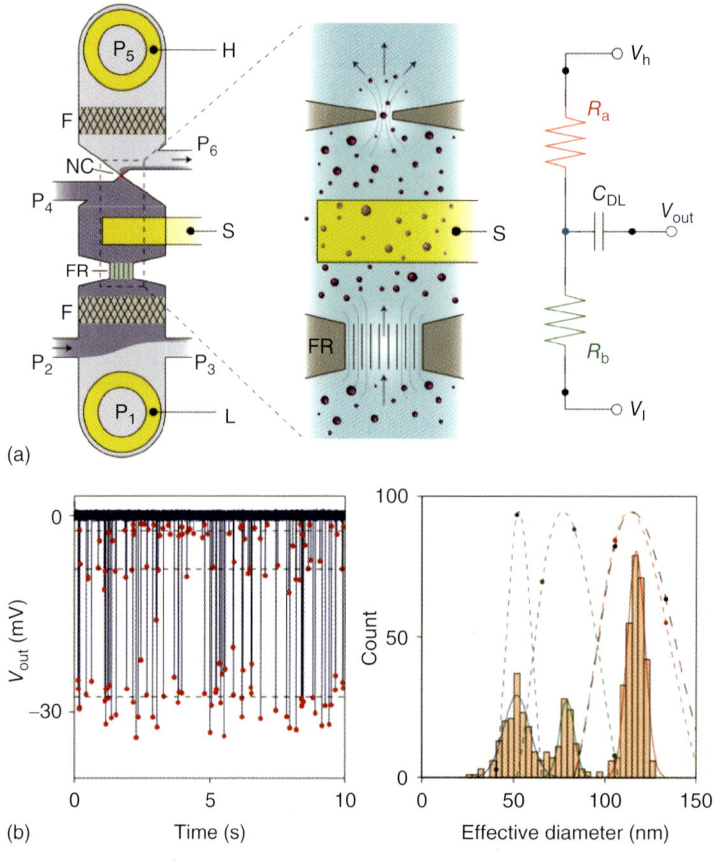

Fig. 3 (a) Overall device layout (left) depicting the electrical and fluidic components: external voltage bias electrodes (H, L) and sensing electrode (S); embedded filters (F); fluid resistor (FR); nanoconstriction (NC); pressure-regulated fluidic ports (P_1~P_6). The nanoparticle (NP) suspension enters at P2 and exits at P6. A detailed image of the dashed box area in (a) (middle) shows the key sensing parts. While NPs flow in the direction of the arrows, changes in the electrical potential in the NC are detected by the electrode S. Electrical circuit expression of the device (right): a constant bias voltage V_h (V_l); Resistors R_a and R_b represent the resistance of the nanoconstriction and the fluidic resistor, respectively; (b) An example analysis of a NP mixture with polydispersity. Left: Output voltage over time for a mixture of NPs of different diameters. Events marked with red circles cluster around three values of V (horizontal dashed black lines). Right: Histogram of effective diameters (40 s measurements). Reproduced with permission from Ref. [117].

channels, and to detect any perturbation in the local electrical properties of the nanowires in response to disturbance by NP solutions. For example, nanowire field-effect transistors allowed for the real-time sensitive detection of label-free molecules [109–111]. A combination of these technologies with nanoscale mechanical systems offers real-time, high-precision, single-molecule/NP/cell detectors, such as advanced mass spectrometry [112] and microfluidic and nanofluidic channel resonators [113–115].

With recent advances in the fundamental physical chemistry of nanoscale pore sensors, several pore-based sensors have been developed in the nanoscale range, offering a rapid and specific, yet simple, biosensing strategy with an improved measurement sensitivity over a wide particle size range [116]. An example of this is a high-throughput microfluidic analyzer that has been developed to detect and characterize unlabeled NPs in a multicomponent mixture at a rate of 500 000 particles per second (Fig. 3) [117]. In this case, a real-time single-nucleotide detection of a model G487A mutation (which is responsible for glucose-6-phosphate dehydrogenase deficiency) was achieved by leveraging the *in-situ* reaction-monitoring capability of the nanopore platform [118]. Tunable pore sensors, which can elastically adjust the size of their pore, have been used to count and detect the size and concentration of smaller NPs compared to other techniques such as flow cytometry [119]. This platform also allows for a simultaneous extraction of the size and zeta-potential of NPs from their charge density, under electrophoretic forces [120]. NP translocation was also detected using a pressure-reversal technique through a cone-shaped nanopore membrane [121]. Today, these approaches, all of which employ size-tunable pore sensors, are starting to provide a better understanding of the fundamental behavior of NPs, as well as a high-throughput characterization of their properties.

4
Microfluidic Evaluation of Nanomedicines

Nanomedicines needs to be nontoxic, biodegradable, sufficiently stable to be delivered to targeted sites, and to have a superior therapeutic advantage over the free drug [122, 123]. Conventionally, nanomedicine evaluation has been made in static cell culture plates, but unfortunately this neglects the important effects of flowing conditions and subsequent transport phenomena on the microenvironment. In contrast to static conditions, flowing conditions assist in the homogeneous distribution of NPs with no gravitational sedimentation, which is similar to the physiological conditions encountered *in vivo*. For example, microfluidic approaches have been used to measure the cytotoxicity of quantum dots (QDs) in a flowing condition [124, 125] and to examine the stability of multicomponent NPs across a laminar flow interface [126]. Compared to tests conducted in conventional static plates, these approaches have provided a more accurate approximation of nanomedicine performance *in vivo*. Microfluidic approaches were also used to evaluate the selective binding of NPs to cells while varying the fluid shear stress, the targeting ligand concentration, receptor expression on target cells, and NP size [127, 128]. The targeted delivery of a nanomedicine represents a powerful technology for the development of safer and more effective therapeutics compared to systemic delivery by nontargeted formulations [83]. Indeed, such approaches show that the targeting performance of an NP can be examined under more physiologically relevant conditions, which is preferential to examining a wide array of cell–particle interactions prior to *in-vivo* experiments. The accurate detection of biomarkers also holds significant promise for "personalized" cancer diagnostics, with more physiologically relevant 3D platforms having been developed for identifying and validating ubiquitous biomarkers [129].

4.1 Mimicking Physiological Environments

Today, an increasing number of engineered NPs requires a reliable high-throughput screening methodology with more physiologically accurate conditions. Whereas, microfluidic approaches have demonstrated the potential to closely approximate physiological environments, current preclinical studies on drug candidates mostly rely on costly and highly variable animal models, mainly because existing cell culture models fail to recapitulate the organ-level pathophysiology of humans. This lack of accurate predictive models highlights the need for better approaches to mimic the structure and function of cells, tissues and organs, as well as the dynamically changing environments *in vivo*. Recently, the evolution of microfluidics has witnessed the integration of *in-vitro* cellular approaches onto chips, which allow real-time, *in-vitro* microscopic observations to be made as well as an evaluation of cell function [130]. But, in order to probe the targeting, therapeutic and diagnostic efficacy of NPs in spatially and temporally regulated environments, it is important first to examine how the NPs interact with cells, tissues and organisms under more physiologically realistic conditions [131, 132]. Consequently, the microfluidic approaches for replicating organ-level structure and function will be discussed at this point (Fig. 4a) [133], and current applications and potentials for the *in-vitro* evaluation of nanomedicines highlighted.

4.2 Endothelial Cell Systems

The vascular endothelium is a crucial target for therapeutic intervention in pathological processes that include inflammation, atherosclerosis, and thrombosis. Endothelial cells exist under dynamically changing mechanical stresses that are generated by blood flow patterns.

Fig. 4 (a) Schematic of microengineered biomimetic systems with spatiotemporal control over physiological effectors, including mechanical cues, chemical factors, electrical signals, multilayered platform with 3D scaffold; (b) Schematic depiction of a biomimetic model that mimics the function of the blood–brain barrier. b.END3 brain endothelial cells and C8-D1A astrocyte cells are cultured on either side of a porous membrane between two microfluidic flow chambers; (c) Schematic of a microfluidic model that mimics the permeable endothelium in artery-surrounding microvessels. Permeability is detected using microelectrodes embedded in the chip. The fluorescent image shows the disrupted endothelial connections. Adherens junctions are shown in green, and nuclei in blue (scale bar = 20 μm). Schematic and TEM image of nanoparticles used for NP translocation studies in the chip (scale bar = 100 nm); (d) Schematic of a lung-on-a-chip device showing IL-2-induced pulmonary edema (scale bar in contrast image = 200 μm). The graph shows barrier permeability in response to IL-2, with and without cyclic strain. Error bars indicate SEM. The fluorescent images show that immunostaining of epithelial occulidin (green) and vascular endothelial cadherin (VE-cadherin; red) with 10% strain with and without IL-2 (scale bars = 30 μm). Reproduced with permission from Refs [133, 143, 149, and 152].

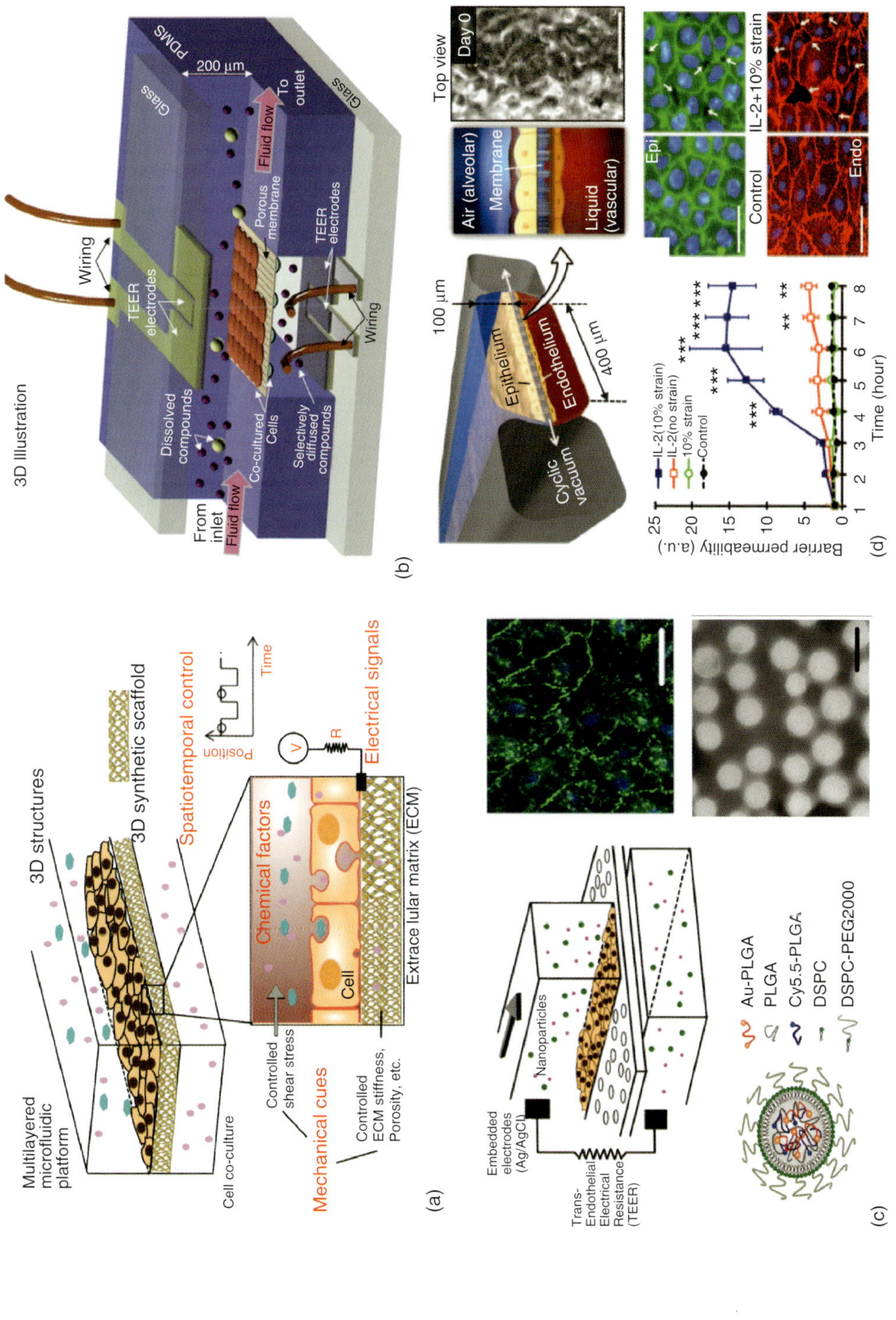
Multilayered microfluidic platform
3D structures
3D synthetic scaffold
Cell co-culture
Spatiotemporal control
Position
Time
Chemical factors
Cell
Controlled shear stress
Mechanical cues
Controlled ECM stiffness, Porosity, etc.
Extracellular matrix (ECM)
Electrical signals
V
R
(a)
3D Illustration
From inlet
Fluid flow
Dissolved compounds
Co-cultured Cells
Selectively diffused compounds
TEER electrodes
Wiring
Porous membrane
200 μm
PDMS
Glass
To outlet
(b)
Trans-Endothelial Electrical Resistance (TEER)
Embedded electrodes (Ag/AgCl)
Nanoparticles
Au-PLGA
PLGA
Cy5.5-PLGA
DSPC
DSPC-PEG2000
(c)
100 μm
Cyclic vacuum
Epithelium
Endothelium
400 μm
Air (alveolar)
Membrane
Liquid (vascular)
Top view
Day 0
IL-2(10% strain)
IL-2(no strain)
10% strain
Control
Barrier permeability (a.u.)
Time (hour)
Epi
Endo
IL-2+10% strain
(d)

Yet, endothelial cell monolayers cultured in conventional multiwell plates fail to reproduce the complex architecture of a vascular network *in vivo* and thus fail to capture the relationship between shear stress experienced by the cells and the local concentration of the drug used. Recent developments in microengineered vascular systems have shown the potential for evaluating nanomedicines under physiologically realistic conditions. For example, replicating the structure and function of blood vessels *in vitro* can be helpful for investigating NP behavior and interaction in and around the targeted sites [134–138]. An accurate reconstitution of the geometric configuration of natural blood vessels is also important, as the interactive effects between blood flow and drug concentration were not captured by a rectangular channel coated with endothelial cells. Rather, a branching network with tubular channels was constructed in order to reproduce these effects [139]. In addition, microvessels supported by the extracellular matrix (ECM), when patterned in a tubular structure, establish the endothelial monolayer, maintain permeability, and are not prone to delamination (which was relatively common in the rectangular channels). The rectangular channels were also very susceptible to delamination that disrupted local permeability, due mainly to the poor connections between the endothelial cells and ECM at the sharp corners [140, 141].

Several additional methods to reproduce microvessels accommodating multiple cells (endothelial cells, pericytes, and astrocytes) have allowed the development of microfluidic models of the blood–brain barrier (BBB) (Fig. 4b) [142–148], as well as for endothelial dysfunction and permeability control in atherosclerosis (Fig. 4c) [149–151].

4.3 "Organ-On-A-Chip"

In combining microfabrication techniques with tissue engineering, the "lung-on-a-chip" device offered a novel *in-vitro* approach to drug screening by mimicking the mechanical and biochemical activities of the human lung (Fig. 4d) [152]. For example, a recent study using this device revealed that mechanical strains associated with physiological breathing movements play an essential role in the development of the increased vascular leakage which leads to pulmonary edema, and that circulating immune cells are not necessary for this disease to develop. The same studies also led to the identification of potential new therapeutics, including angiopoietin-1 (Ang-1) and a new transient receptor potential vanilloid 4 (TRPV4) ion channel inhibitor (GSK2193874), which might prevent the severe toxicity associated with interleukin (IL)-2 therapy. An *in-vitro* model of the intestine has also been developed, together with its crucial microbial symbionts [153, 154]. Whereas, previous *in-vitro* models of intestinal function depend on the use of epithelial cell lines (e.g., Caco-2 cells) which create polarized epithelial monolayers but fail to mimic human intestinal functions for drug development. The recent development of a "gut-on-a-chip" device recreated the gut microenvironment with low shear stress (0.02 dyne cm^{-2}) with cyclic strain (10%; 0.15 Hz) that mimicked physiological peristaltic motions. This precise regulation allowed for an increased exposure of the intestinal surface area and a robust 3D intestinal villi morphogenesis, which mimicked the enhanced cytochrome isoform-based drug-metabolizing activity and the absorptive efficiency of the human intestine.

4.4 Renal Toxicity and Hepatotoxicity

A further use of biomimetic microfluidic platforms for nanomedicine is to evaluate renal toxicity and hepatotoxicity in preclinical studies. Renal excretion represents a clearance pathway for the removal of molecules from vascular compartments, during which time the circulating NPs enter the glomerular capillary and undergo a size-dependent filtration. Those NPs smaller than the pore size of glomerular filtration (~5 nm) can be filtered to enter the proximal tubule, where the brush border of the epithelial cells is negatively charged. As a consequence, positively charged NPs are readily resorbed from the luminal space compared to the negatively charged NPs. The recent development of a microfluidic device lined by human kidney epithelial cells that can be exposed to a shear rate demonstrated a significant increase in albumin transport, glucose reabsorption and brush border alkaline phosphatase activity, all of which are crucial functions of the human kidney proximal tubule [155]. This approach also confirmed that cisplatin toxicity and Pgp efflux transporter activity detected on-chip more closely mimicked the *in-vivo* responses than those obtained with cells maintained on conventional culture plates.

It should be noted that any NPs which are not cleared via the kidney are excreted via the hepatobiliary system. The hepatocytes, which are referred to as potential sites for toxicity, play an important role in liver clearance through endocytosis and the enzymatic breakdown of NPs. As NPs between 10 and 20 nm in size are efficiently eliminated via the liver, any NPs designed in this size range must be modified in order to avoid their prolonged retention in the liver as they undergo excretion. Recently, microfluidic devices have been developed that reconstitute the function of the hepatocytes; for example, a 3D hepatocyte chip has been fabricated for *in-vitro* drug toxicity examinations aimed at predicting drug hepatotoxicity *in-vivo* [156]. This device allowed for the controlled delivery of multiple drug doses to functional primary hepatocytes, while an incorporated concentration gradient generator created *in vitro* dose-dependent drug responses in order to predict *in-vivo* hepatotoxicity.

4.5 Live Tissue Explants

The introduction of *ex-vivo* live tissue explants into microfluidics may provide additional physiological conditions to be investigated. For example, embryonic tissues excised from live frog embryos were used to examine dynamic responses to time-varying chemical stimuli (Fig. 5a) [157]. Carcinoma tumor biopsies were also introduced into a reproducible glass microfluidics system to study the tumor environment, thus offering a preclinical model for the creation of "personalized" treatment regimens [158]. The culturing of brain tissue slices on transistor arrays fabricated on silicon chips may also become a novel platform for neurophysiological and pharmacological studies. Typical microfluidics and semiconductor technologies can be integrated to produce high-resolution, planar transistor arrays for mimicking neuronal structures in long-term studies of topographic mapping [159], and also for mapping evoked extracellular field potentials in organotypic brain slices of rat hippocampus [160].

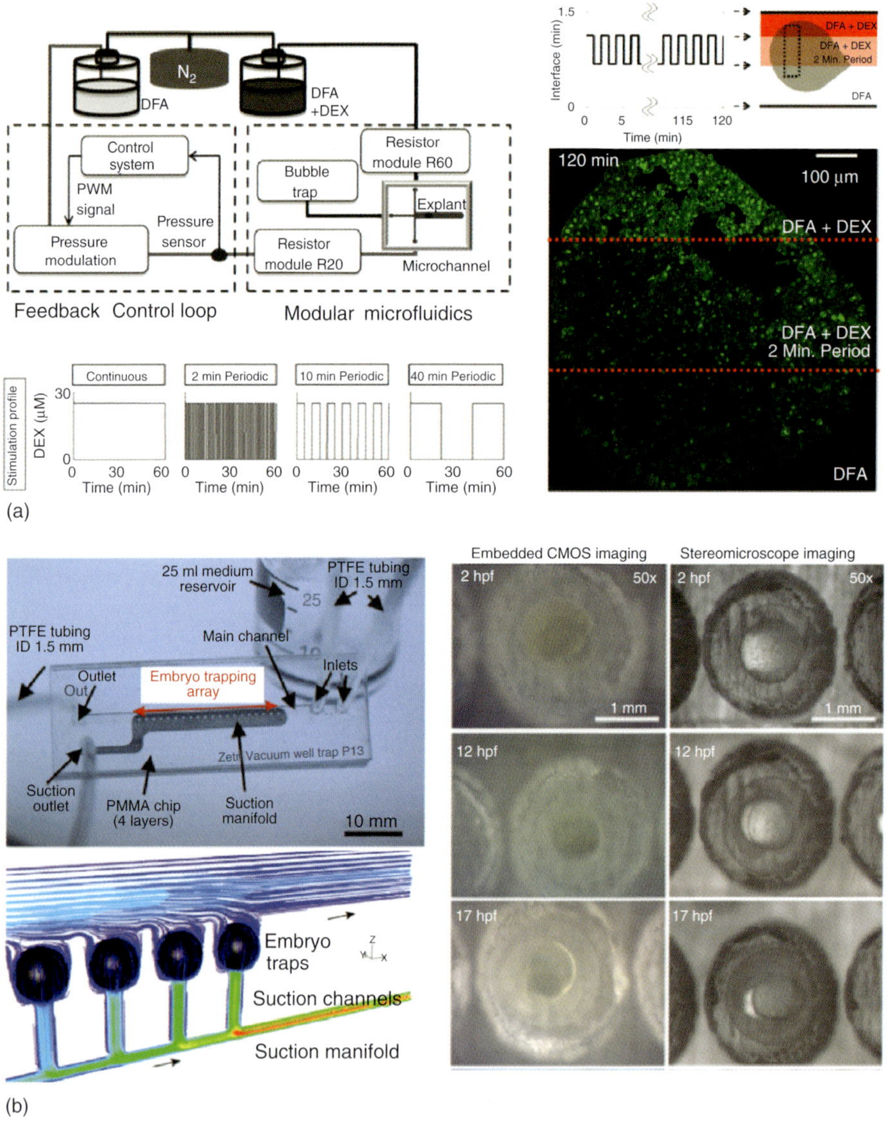

Fig. 5 (a) Schematic of feedback control system that allows for long-term culture of an embryonic tissue excised from a live frog embryo within a microfluidic channel to examine dynamic responses to spatially and temporally varying chemical stimuli. The fluorescent image show the distinct localized responses of an embryonic tissue to dynamic stimuli in a single tissue explant for 2 h, showing that stimulation with localized bursts versus continuous stimulation can result in highly distinct responses. This platform can be used to investigate the cell intercommunication in response to localized drug stimulation for toxicity tests; (b) Schematic of microfluidics-based chip integrating embedded electronic interfaces, and CFD simulation for the flow stream line prediction. Embryos cultured in the device that allowed for immobilization, culture, and treatment of developing zebrafish embryos for toxicity tests. Reproduced with permission from Refs [157] and [167].

4.6 Intact Organisms

Small multicellular organisms such as nematodes, fruit flies, clawed frogs and zebrafish, allow for toxicological screening in the normal physiological environments of intact organisms, providing substantial advantages over cell lines and extracted tissues [161–163]. While the fully automated analysis of these model systems in a high-throughput manner remains challenging, the application of microfluidics to these model organisms has demonstrated the ability to handle multicellular organisms in an efficient manner and to precisely manipulate the local conditions to allow for the assessment and imaging of these small organisms [163–166]. For example, manipulating small organisms, such as the worm *Caenorhabditis elegans*, allowed the observation of neuronal responses in order to correlate the activity of sensory neurons with the worm's behavior *in vivo* [161]. Moreover, the integration of embedded electronic interfaces with microfluidic chip-based technologies allowed for the automatic immobilization, culture, and treatment of developing zebrafish embryos during fish embryo toxicity (FET) biotests (Fig. 5b) [167].

To capture the interactions between multiple organs on microfluidic chips would potentially enable a more accurate model of how organs function and interact with one another for potential drug development applications [168, 169]. For example, the combination of a mathematical model and a multiorgan approach provided a novel platform with improved predictability for testing the toxicity of an anticancer drug, 5-fluorouracil, in a pharmacokinetics-based manner [170]. These approaches can help to achieve a better insight into the mechanisms of action of drug candidates, perhaps leading to patient-specific therapies in the future [171]. While multiorgan chips have the potential to simulate human body functions for patient-specific point-of-care devices and therapies, the inherent complexity of each organ itself hinders the development of reliable "human-on-a-chip" model systems. For example, the practical challenges include an optimization of organ size, the control of fluid volumes, the maintenance of coupled organ systems, and the development of a universal blood substitute [172]. The key question for building multiorgan systems is how to simplify the organ complexity without losing physiological accuracy.

5 Challenges and Opportunities

A new nanomedicine development pipeline using microfluidics technologies includes microfluidic assembly, characterization, evaluation, and the manufacture of nanomedicines. These technologies will allow the robust supply of highly reproducible nanomedicines to the entire development process and thereby increase the success rates in clinical trials. In addition to the stages discussed above, microfluidics technology for nanomedicine manufacture is key to the successful translation of a nanomedicine from the laboratory to the clinic. A long-term vision for the manufacture of nanomedicines is to create reliable, continuous and scalable assembly methodologies for a variety of multifunctional NPs with high reproducibility, yield, and homogeneity [173]. The development of these assembly methods requires microfluidic approaches to allow for an efficient and strong mixing of precursors, modular methods for incorporating multicomponents (e.g., therapeutic compounds, imaging agents, targeting ligands, etc.) into multifunctional NPs, and automated control systems for the large-scale integration and parallelization of microfluidic modules [117, 174]. The ability to integrate microfluidics with dynamics, control, and more complex microfabrication techniques opens the door for high-throughput, automated manufacturing.

A current challenge of nanomedicine manufacture using microfluidics is to optimize and maximize microfluidic devices with tunable mixing flow patterns that are applicable to the synthesis of a wide range of nanomedicine types, without losing the physico-chemical properties of the designed nanomedicine [175]. The key technologies required for the development of these nanomedicine manufacturing techniques are computational fluid dynamics to allow simulation of mixing flow patterns in microfluidic devices, highly reliable microfabrication techniques capable of integrating microscale pumps, valves and detecting sensors [176], and high-precision control systems that regulate parallelization and automation capability [177–180].

While the quantities of NPs synthesized by microfluidic devices are often in the microgram to milligram range, the parallelization of microfluidic channels has the potential to scale-up the synthesis by several orders of magnitude to a clinical scale of grams to kilograms. With parallel and stackable microfluidic systems, gram to kilogram scales of NPs could be prepared with the same properties as those prepared at the bench scale, as long as a precise control of either flow rates or pressure in the microfluidic platform is achieved. Pressure control is far better than flow control for controlling the flow rate into a microfluidic network because the flow rate is proportional to the inlet pressure, which can be easily measured for high bandwidth feedback control [181]. To maintain a precise control over fluid pressure with the potential to scale-up production, it is necessary to isolate the pressure-regulating mechanism from the fluid reservoir and the microfluidic device, so that larger reservoir volumes and diverse microfluidic devices can be used independently and integrated as needed. Three important features should be considered for robust and reliable parallelization:

- Fouled modules must be easily replaced or disconnected from other systems.
- Unexpected disturbances due to air bubble formation in microfluidic devices need to be compensated, as this increases the hydraulic resistance between neighboring devices in the parallelized network, leading to an imprecise regulation of the entire system.
- Bridging or networking channels that connect microfluidic modules need to be well designed with minimal secondary flows, which may affect the main bulk flow streams, leading to chip-to-chip variations.

While the microfluidic assembly of nanomedicine has demonstrated much progress using several platforms and various mixing patterns, practical development has been significantly constrained due to a lack of tools capable of detecting, characterizing, and analyzing NPs in a high-throughput manner. Although single NP detectors and characterization tools using microfluidics and nanofabrication technologies have been demonstrated, these are still limited to specific solutions and NP types and need to be generalized in order to function with multicomponent NPs. In addition, there is a need for a technique allowing the easy preparation of highly concentrated NP samples to be developed, including the purification of toxic solvents via either separation or filtration. Furthermore, while many approaches have shown the promise of combining optical systems with microfluidics [60], the precise and reliable control of light delivery to targeted areas in microfluidic platforms remains an active topic of research and engineering. A combination of microfluidics and optics – termed optofluidics – has demonstrated a synergistic effect for new capabilities in several applications including lens, colloidal suspensions, and flow cytometry [182–185]. In addition, optofluidics technology could be employed to incorporate the microfluidic characterization capabilities into NP production platforms in a high-throughput fashion.

Reliability of biomimetic microsystems that mimic the structure and function of human organs for nanomedicine evaluation is crucial. Current challenges include the reliability of long-term cultivations of multiple cell types [186], as well as real-time monitoring of nanomedicines, cellular response to nanomedicines, and critical chemical cues (e.g., reactive oxygen stress) in 3D microenvironments [187–189]. In addition, the development of synthetic biomaterials remains a critical topic of research for physiological accuracy and niches for specific cells, tissues and organs. Furthermore, the development of *in-vitro* model systems that can accurately replicate the structure and function of *in-vivo* systems necessitates a precise 3D control of dynamically changing properties, such as mechanical properties of the ECM, at a scale comparable to human cells, tissues and organs [132, 172, 190].

6 Concluding Remarks

Microfluidics in nanomedicine has demonstrated the ability to overcome critical issues with conventional approaches used for nanomedicine development. When combined with advanced nanofabrication, synthetic biomaterials and high-precision control systems, microfluidic technologies constitute a novel platform capable of replacing the entire nanomedicine production process in a scalable manner. Although microfluidics as applied to nanomedicine is still in its infancy, it will surely continue to expand to provide innovative systems at industrially relevant scales in the near future.

Acknowledgments

The authors thank the members of their laboratories for participating in stimulating discussions of these investigations.

References

1. Peer, D., Karp, J.M., Hong, S., Farokhzad, O.C., *et al.* (2007) Nanocarriers as an emerging platform for cancer therapy. *Nat. Nanotechnol.* **2**(12), 751–760.
2. Namiki, Y., Fuchigami, T., Tada, N., Kawamura, R., *et al.* (2011) Nanomedicine for cancer: lipid-based nanostructures for drug delivery and monitoring. *Acc. Chem. Res.* **44**(10), 1080–1093.
3. Seigneuric, R., Markey, L., Nuyten, D.S., Dubernet, C., *et al.* (2010) From nanotechnology to nanomedicine: applications to cancer research. *Curr. Mol. Med.* **10**(7), 640–652.
4. Lobatto, M.E., Fuster, V., Fayad, Z.A., and Mulder, W.J. (2011) Perspectives and opportunities for nanomedicine in the management of atherosclerosis. *Nat. Rev. Drug Discovery* **10**(11), 835–852.

5. Psarros, C., Lee, R., Margaritis, M., and Antoniades, C. (2012) Nanomedicine for the prevention, treatment and imaging of atherosclerosis. *Nanomedicine* **8** Suppl. 1, S59–S68.
6. Schiener, M., Hossann, M., Viola, J.R., Ortega-Gomez, A., *et al.* (2014) Nanomedicine-based strategies for treatment of atherosclerosis. *Trends Mol. Med.* **20**(5), 271–281.
7. Sung, H.W., Sonaje, K., and Feng, S.S. (2011) Nanomedicine for diabetes treatment. *Nanomedicine* **6**(8), 1297–1300.
8. Pickup, J.C., Zhi, Z.L., Khan, F., Saxl, T., *et al.* (2008) Nanomedicine and its potential in diabetes research and practice. *Diabetes Metab. Res. Rev.* **24**(8), 604–610.
9. Sonaje, K., Lin, K.J., Wey, S.P., Lin, C.K., *et al.* (2010) Biodistribution, pharmacodynamics and pharmacokinetics of insulin analogues in a rat model: oral delivery using pH-responsive nanoparticles vs. subcutaneous injection. *Biomaterials* **31**(26), 6849–6858.
10. Taratula, O., Kuzmov, A., Shah, M., Garbuzenko, O.B., *et al.* (2013) Nanostructured lipid carriers as multifunctional nanomedicine platform for pulmonary co-delivery of anticancer drugs and siRNA. *J. Controlled Release* **171**(3), 349–357.
11. Mansour, H.M., Rhee, Y.S., and Wu, X. (2009) Nanomedicine in pulmonary delivery. *Int. J. Nanomed.* **4**, 299–319.
12. Muldoon, L.L., Tratnyek, P.G., Jacobs, P.M., Doolittle, N.D., *et al.* (2006) Imaging and nanomedicine for diagnosis and therapy in the central nervous system: report of the 11th Annual Blood-Brain Barrier Disruption Consortium meeting. *Am. J. Neuroradiol.* **27**(3), 715–721.
13. Sharma, H.S. and Sharma, A. (2011) New strategies for CNS injury and repair using stem cells, nanomedicine, neurotrophic factors and novel neuroprotective agents. *Expert Rev. Neurother.* **11**(8), 1121–1124.
14. Morgen, M., Lu, G.W., Du, D., Stehle, R., *et al.* Targeted delivery of a poorly water-soluble compound to hair follicles using polymeric nanoparticle suspensions. *Int. J. Pharm.* 2011, **416**(1), 314–322.
15. Sigfridsson, K., Bjorkman, J.A., Skantze, P., and Zachrisson, H. (2011) Usefulness of a nanoparticle formulation to investigate some hemodynamic parameters of a poorly soluble compound. *J. Pharm. Sci.* **100**(6), 2194–2202.
16. Jia, L. (2005) Nanoparticle formulation increases oral bioavailability of poorly soluble drugs: approaches experimental evidences and theory. *Curr. Nanosci.* **1**(3), 237–243.
17. Gao, W., Xiao, Z., Radovic-Moreno, A., Shi, J., *et al.* (2010) Progress in siRNA delivery using multifunctional nanoparticles. *Methods Mol. Biol.* **629**, 53–67.
18. Leuschner, F., Dutta, P., Gorbatov, R., Novobrantseva, T.I., *et al.* (2011) Therapeutic siRNA silencing in inflammatory monocytes in mice. *Nat. Biotechnol.* **29**(11), 1005–1010.
19. Soppimath, K.S., Aminabhavi, T.M., Kulkarni, A.R., and Rudzinski, W.E. (2001) Biodegradable polymeric nanoparticles as drug delivery devices. *J. Controlled Release* **70**(1-2), 1–20.
20. Chen, X. and Schluesener, H.J. (2008) Nanosilver: a nanoproduct in medical application. *Toxicol. Lett.* **176**(1), 1–12.
21. Huang, X., Jain, P.K., El-Sayed, I.H., and El-Sayed, M.A. (2007) Gold nanoparticles: interesting optical properties and recent applications in cancer diagnostics and therapy. *Nanomedicine* **2**(5), 681–693.
22. Sanvicens, N. and Marco, M.P. (2008) Multifunctional nanoparticles–properties and prospects for their use in human medicine. *Trends Biotechnol.* **26**(8), 425–433.
23. Jain, K.K. (2003) Nanodiagnostics: application of nanotechnology in molecular diagnostics. *Expert Rev. Mol. Diagn.* **3**(2), 153–161.
24. Torchilin, V.P. (2005) Recent advances with liposomes as pharmaceutical carriers. *Nat. Rev. Drug Discovery* **4**(2), 145–160.
25. Gu, F., Zhang, L., Teply, B.A., Mann, N., *et al.* (2008) Precise engineering of targeted nanoparticles by using self-assembled biointegrated block copolymers. *Proc. Natl Acad. Sci. USA* **105**(7), 2586–2591.
26. Salvador-Morales, C., Zhang, L., Langer, R., and Farokhzad, O.C. (2010) Immunocompatibility properties of lipid-polymer hybrid nanoparticles with heterogeneous surface functional groups. *Biomaterials* **30**(12), 2231–2240.
27. Shi, J., Xiao, Z., Votruba, A.R., Vilos, C., *et al.* (2011) Differentially charged hollow core/shell lipid-polymer-lipid hybrid

nanoparticles for small interfering RNA delivery. *Angew. Chem. Int. Ed.* **50**(31), 7027–7031.
28. Bianco, A., Kostarelos, K., and Prato, M. (2005) Applications of carbon nanotubes in drug delivery. *Curr. Opin. Chem. Biol.* **9**(6), 674–679.
29. Huang, X., El-Sayed, I.H., Qian, W., and El-Sayed, M.A. (2006) Cancer cell imaging and photothermal therapy in the near-infrared region by using gold nanorods. *J. Am. Chem. Soc.* **128**(6), 2115–2120.
30. Gupta, A.K. and Gupta, M. (2005) Synthesis and surface engineering of iron oxide nanoparticles for biomedical applications. *Biomaterials* **26**(18), 3995–4021.
31. Mulder, W.J., Koole, R., Brandwijk, R.J., Storm, G., *et al.* (2006) Quantum dots with a paramagnetic coating as a bimodal molecular imaging probe. *Nano Lett.* **6**(1), 1–6.
32. Kratz, F. (2008) Albumin as a drug carrier: design of prodrugs, drug conjugates and nanoparticles. *J. Controlled Release* **132**(3), 171–183.
33. Skajaa, T., Cormode, D.P., Jarzyna, P.A., Delshad, A., *et al.* (2011) The biological properties of iron oxide core high-density lipoprotein in experimental atherosclerosis. *Biomaterials* **32**(1), 206–213.
34. Basta, T., Wu, H.J., Morphew, M.K., Lee, J., *et al.* (2014) Self-assembled lipid and membrane protein polyhedral nanoparticles. *Proc. Natl Acad. Sci. USA* **111**(2), 670–674.
35. Maune, H.T., Han, S.P., Barish, R.D., Bockrath, M., *et al.* (2010) Self-assembly of carbon nanotubes into two-dimensional geometries using DNA origami templates. *Nat. Nanotechnol.* **5**(1), 61–66.
36. Pal, S., Deng, Z., Ding, B., Yan, H., *et al.* (2010) DNA-origami-directed self-assembly of discrete silver-nanoparticle architectures. *Angew. Chem. Int. Ed.* **49**(15), 2700–2704.
37. Farokhzad, O.C., Cheng, J., Teply, B.A., Sherifi, I., *et al.* (2006) Targeted nanoparticle-aptamer bioconjugates for cancer chemotherapy in vivo. *Proc. Natl Acad. Sci. USA* **103**(16), 6315–6320.
38. Xiao, Z., Levy-Nissenbaum, E., Alexis, F., Luptak, A., *et al.* (2012) Engineering of targeted nanoparticles for cancer therapy using internalizing aptamers isolated by cell-uptake selection. *ACS Nano* **6**(1), 696–704.
39. Gianella, A., Jarzyna, P.A., Mani, V., Ramachandran, S., *et al.* (2011) Multifunctional nanoemulsion platform for imaging guided therapy evaluated in experimental cancer. *ACS Nano* **5**(6), 4422–4433.
40. Lobatto, M.E., Fayad, Z.A., Silvera, S., Vucic, E., *et al.* (2010) Multimodal clinical imaging to longitudinally assess a nanomedical anti-inflammatory treatment in experimental atherosclerosis. *Mol. Pharm.* **7**(6), 2020–2029.
41. Venditto, V.J. and Szoka, F.C., Jr, (2013) Cancer nanomedicines: so many papers and so few drugs! *Adv. Drug Delivery Rev.* **65**(1), 80–88.
42. Duncan, R. and Gaspar, R. (2011) Nanomedicine(s) under the microscope. *Mol. Pharm.* **8**(6), 2101–2141.
43. Olofsson, J., Bridle, H., Sinclair, J., Granfeldt, D., *et al.* (2005) A chemical waveform synthesizer. *Proc. Natl Acad. Sci. USA* **102**(23), 8097–8102.
44. Duraiswamy, S. and Khan, S.A. (2010) Plasmonic nanoshell synthesis in microfluidic composite foams. *Nano Lett.* **10**(9), 3757–3763.
45. El-Ali, J., Sorger, P.K., and Jensen, K.F. (2006) Cells on chips. *Nature* **442**(7101), 403–411.
46. Schimek, K., Busek, M., Brincker, S., Groth, B., *et al.* (2013) Integrating biological vasculature into a multi-organ-chip microsystem. *Lab Chip* **13**(18), 3588–3598.
47. Andersson, H. and van den Berg, A. (2004) Microfabrication and microfluidics for tissue engineering: state of the art and future opportunities. *Lab Chip* **4**(2), 98–103.
48. Minteer, S.D. and Moore, C.M. (2006) Overview of advances in microfluidics and microfabrication, in: *Methods in Molecular Biology*, vol. **321**, Humana Press, pp. 1 2.
49. Ozaydin-Ince, G., Coclite, A.M., and Gleason, K.K. (2012) CVD of polymeric thin films: applications in sensors, biotechnology, microelectronics/organic electronics, microfluidics, MEMS, composites and membranes. *Rep. Progr. Phys. Phys. Soc.* **75**(1), 016501.
50. Shih, S.C., Fobel, R., Kumar, P., and Wheeler, A.R. (2011) A feedback control system for high-fidelity digital microfluidics. *Lab Chip* **11**(3), 535–540.
51. Welch, D. and Christen, J.B. (2014) Real-time feedback control of pH within

microfluidics using integrated sensing and actuation. *Lab Chip* **14**(6), 1191–1197.
52. Prohm, C. and Stark, H. (2014) Feedback control of inertial microfluidics using axial control forces. *Lab Chip* **14**(12), 2115–2123.
53. Valencia, P.M., Farokhzad, O.C., Karnik, R., and Langer, R. (2012) Microfluidic technologies for accelerating the clinical translation of nanoparticles. *Nat. Nanotechnol.* 7(10), 623–629.
54. Capretto, L., Carugo, D., Mazzitelli, S., and Nastruzzi, C. (2013) Microfluidic and lab-on-a-chip preparation routes for organic nanoparticles and vesicular systems for nanomedicine applications. *Adv. Drug Delivery Rev.* **65**(11-12), 1496–1532.
55. Hashimoto, M., Tong, R., and Kohane, D.S. (2013) Microdevices for nanomedicine. *Mol. Pharm.* **10**(6), 2127–2144.
56. Bhise, N.S., Ribas, J., Manoharan, V., Zhang, Y.S., *et al.* (2014) Organ-on-a-chip platforms for studying drug delivery systems. *J. Controlled Release* **190**, 82–93
57. Lee, J.B. and Sung, J.H. (2013) Organ-on-a-chip technology and microfluidic whole-body models for pharmacokinetic drug toxicity screening. *Biotechnol. J.* **8**(11), 1258–1266.
58. Song, Y., Hormes, J., Kumar, C.S. Microfluidic synthesis of nanomaterials. *Small*, 2008 **4**(6), 698–711.
59. Marre, S. and Jensen, K.F. (2010) Synthesis of micro and nanostructures in microfluidic systems. *Chem. Soc. Rev.* **39**(3), 1183–1202.
60. Yurt, A., Daaboul, G.G., Connor, J.H., Goldberg, B.B., and Ünlü, M.S. (2012) Single nanoparticle detectors for biological applications. *Nanoscale* **4**(3), 715–726.
61. Neeves, K.B., Onasoga, A.A., and Wufsus, A.R. (2013) The use of microfluidics in hemostasis: clinical diagnostics and biomimetic models of vascular injury. *Curr. Opin. Hematol.* **20**(5), 417–423.
62. Kuo, C.T., Chiang, C.L., Chang, C.H., Liu, H.K., *et al.* (2014) Modeling of cancer metastasis and drug resistance via biomimetic nano-cilia and microfluidics. *Biomaterials* **35**(5), 1562–1571.
63. Domachuk, P., Tsioris, K., Omenetto, F.G., and Kaplan, D.L. (2010) Bio-microfluidics: biomaterials and biomimetic designs. *Adv. Mater.* **22**(2), 249–260.
64. Jonas, A. (1986) Synthetic substrates of lecithin: cholesterol acyltransferase. *J. Lipid Res.* **27**(7), 689–698.
65. Mieszawska, A.J., Mulder, W.J., Fayad, Z.A., and Cormode, D.P. (2013) Multifunctional gold nanoparticles for diagnosis and therapy of disease. *Mol. Pharm.* **10**(3), 831–847.
66. Cormode, D.P., Skajaa, T., van Schooneveld, M.M., Koole, R., *et al.* (2008) Nanocrystal core high-density lipoproteins: a multimodality contrast agent platform. *Nano Lett.* **8**(11), 3715–3723.
67. Chorny, M., Fishbein, I., Danenberg, H.D., and Golomb, G. (2002) Lipophilic drug loaded nanospheres prepared by nanoprecipitation: effect of formulation variables on size, drug recovery and release kinetics. *J. Controlled Release* **83**(3), 389–400.
68. Boehm, A.L., Martinon, I., Zerrouk, R., Rump, E., *et al.* (2003) Nanoprecipitation technique for the encapsulation of agrochemical active ingredients. *J. Microencapsul.* **20**(4), 433–441.
69. Betancourt, T., Brown, B., and Brannon-Peppas, L. (2007) Doxorubicin-loaded PLGA nanoparticles by nanoprecipitation: preparation, characterization and in vitro evaluation. *Nanomedicine* **2**(2), 219–232.
70. Medina-Sanchez, M., Miserere, S., and Merkoci, A. (2012) Nanomaterials and lab-on-a-chip technologies. *Lab Chip* **12**(11), 1932–1943.
71. Johnson, B.K. and Prud'homme, R.K. (2003) Mechanism for rapid self-assembly of block copolymer nanoparticles. *Phys. Rev. Lett.* **91**(11), 118302.
72. Capretto, L., Cheng, W., Carugo, D., Katsamenis, O.L., *et al.* (2012) Mechanism of co-nanoprecipitation of organic actives and block copolymers in a microfluidic environment. *Nanotechnology* **23**(37), 375602.
73. Zhu, Z. (2013) Effects of amphiphilic diblock copolymer on drug nanoparticle formation and stability. *Biomaterials* **34**(38), 10238–10248.
74. Chen, T., Hynninen, A.P., Prud'homme, R.K., Kevrekidis, I.G., *et al.* (2008) Coarse-grained simulations of rapid assembly kinetics for polystyrene-b-poly(ethylene oxide) copolymers in aqueous solutions. *J. Phys. Chem. B* **112**(51), 16357–16366.

75. Karnik, R., Gu, F., Basto, P., Cannizzaro, C., *et al.* (2008) Microfluidic platform for controlled synthesis of polymeric nanoparticles. *Nano Lett.* **8**(9), 2906–2912.
76. van Swaay, D. and deMello, A. (2013) Microfluidic methods for forming liposomes. *Lab Chip* **13**(5), 752–767.
77. Jahn, A., Stavis, S.M., Hong, J.S., Vreeland, W.N., DeVoe, D.L., Gaitan, M. Microfluidic mixing and the formation of nanoscale lipid vesicles. *ACS Nano* 2010 **4**(4), 2077–2087.
78. Hong, J.S., Stavis, S.M., DePaoli Lacerda, S.H., Locascio, L.E. *et al.*, (2010) Microfluidic directed self-assembly of liposome-hydrogel hybrid nanoparticles. *Langmuir* **26**(13), 11581–11588.
79. Kolishetti, N., Dhar, S., Valencia, P.M., Lin, L.Q., *et al.* (2010) Engineering of self-assembled nanoparticle platform for precisely controlled combination drug therapy. *Proc. Natl Acad. Sci. USA* **107**(42), 17939–17944.
80. Rhee, M., Valencia, P.M., Rodriguez, M.I., Langer, R., *et al.*, (2011) Synthesis of size-tunable polymeric nanoparticles enabled by 3D hydrodynamic flow focusing in single-layer microchannels. *Adv. Mater.* **23**(12), H79–H83.
81. Tan, S., Li, X., Guo, Y., and Zhang, Z. (2013) Lipid-enveloped hybrid nanoparticles for drug delivery. *Nanoscale* **5**(3), 860–872.
82. Zhang, L., Chan, J.M., Gu, F.X., Rhee, J.-W., *et al.* (2008) Self-assembled lipid-polymer hybrid nanoparticles: a robust drug delivery platform. *ACS Nano* **2**(8), 1696–1702.
83. Shi, J., Xiao, Z., Kamaly, N., and Farokhzad, O.C. (2011) Self-assembled targeted nanoparticles: evolution of technologies and bench to bedside translation. *Acc. Chem. Res.* **44**(10), 1123–1134.
84. Cheng, C.M., Kim, Y., Yang, J.M., Leuba, S.H., *et al.*, (2009) Dynamics of individual polymers using microfluidic based microcurvilinear flow. *Lab Chip* **9**(16), 2339–2347.
85. Kim, Y., Joshi, S.D., Davidson, L.A., LeDuc, P.R., *et al.*, (2011) Dynamic control of 3D chemical profiles with a single 2D microfluidic platform. *Lab Chip* **11**(13), 2182–2188.
86. Kim, Y., Pekkan, K., Messner, W.C., and Leduc, P.R. (2010) Three-dimensional chemical profile manipulation using two-dimensional autonomous microfluidic control. *J. Am. Chem. Soc.* **132**(4), 1339–1347.
87. deMello, A.J. (2006) Control and detection of chemical reactions in microfluidic systems. *Nature* **442**(7101), 394–402.
88. Lee, M.G., Choi, S., and Park, J.K. (2009)Three-dimensional hydrodynamic focusing with a single sheath flow in a single-layer microfluidic device. *Lab Chip* **9**(21), 3155–3160.
89. Mao, X.L., Lin, S.C.S., Dong, C., and Huang, T.J. (2009) Single-layer planar on-chip flow cytometer using microfluidic drifting based three-dimensional (3D) hydrodynamic focusing. *Lab Chip* **9**(11), 1583–1589.
90. Chang, C.C., Huang, Z.X., and Yang, R.J. (2007) Three-dimensional hydrodynamic focusing in two-layer polydimethylsiloxane (PDMS) microchannels. *J. Micromech. Microeng.* **17**(8), 1479–1486.
91. Nguyen, N.T. and Wu, Z.G. (2005) Micromixers – a review. *J. Micromech. Microeng.* **15**(2), R1–R16.
92. Valencia, P.M., Basto, P.A., Zhang, L., Rhee, M., *et al.* (2010) Single-step assembly of homogenous lipid-polymeric and lipid-quantum dot nanoparticles enabled by microfluidic rapid mixing. *ACS Nano* **4**(3), 1671–1679.
93. Fang, R.H., Chen, K.N., Aryal, S., Hu, C.M., *et al.* (2012) Large-scale synthesis of lipid-polymer hybrid nanoparticles using a multi-inlet vortex reactor. *Langmuir* **28**(39), 13824–13829.
94. Liu, S.J., Wei, H.H., Hwang, S.H., and Chang, H.C. (2010) Dynamic particle trapping, release, and sorting by microvortices on a substrate. *Phys. Rev. E Stat. Nonlin. Soft Matter Phys.* **82**(2 Pt 2), 026308.
95. Stott, S.L., Hsu, C.H., Tsukrov, D.I., Yu, M., *et al.* (2010) Isolation of circulating tumor cells using a microvortex-generating herringbone-chip. *Proc. Natl Acad. Sci. USA* **107**(43), 18392–18397.
96. Hsu, C.H., Di Carlo, D., Chen, C., Irimia, D., *et al.*, (2008) Microvortex for focusing, guiding and sorting of particles. *Lab Chip* **8**(12), 2128–2134.
97. Shelby, J.P., Lim, D.S.W., Kuo, J.S., and Chiu, D.T. (2003) High radial acceleration in microvortices. *Nature* **425**(6953), 38.

98. Kim, Y., Lee, C.B., Ma, M., Mulder, W.J., *et al.* (2012) Mass production and size control of lipid-polymer hybrid nanoparticles through controlled microvortices. *Nano Lett.* **12**(7), 3587–3591.
99. Mieszawska, A.J., Kim, Y., Gianella, A., van Rooy, I., *et al.* (2013) Synthesis of polymer-lipid nanoparticles for image-guided delivery of dual modality therapy. *Bioconjugate Chem.* **24**(9), 1429–1434.
100. Kim, Y., Fay, F., Cormode, D.P., Sanchez-Gaytan, B.L., *et al.* (2013) Single step reconstitution of multifunctional high-density lipoprotein-derived nanomaterials using microfluidics. *ACS Nano* **7**(11), 9975–9983.
101. Birnbaumer, G., Kupcu, S., Jungreuthmayer, C., Richter, L., *et al.* (2011) Rapid liposome quality assessment using a lab-on-a-chip. *Lab Chip* **11**(16), 2753–2762.
102. Jahn, A., Lucas, F., Wepf, R.A., and Dittrich, P.S. (2013) Freezing continuous-flow self-assembly in a microfluidic device: toward imaging of liposome formation. *Langmuir* **29**(5), 1717–1723.
103. Hinterwirth, H., Wiedmer, S.K., Moilanen, M., Lehner, A., *et al.* (2013) Comparative method evaluation for size and size-distribution analysis of gold nanoparticles. *J. Sep. Sci.* **36**(17), 2952–2961.
104. Ungureanu, C., Kroes, R., Petersen, W., Groothuis, T.A., *et al.* (2011) Light interactions with gold nanorods and cells: implications for photothermal nanotherapeutics. *Nano Lett.* **11**(5), 1887–1894.
105. Ramachandran, S. and Lal, R. (2010) Scope of atomic force microscopy in the advancement of nanomedicine. *Indian J. Exp. Biol.* **48**(10), 1020–1036.
106. Wang, Y., Wu, Q., Sui, K., Chen, X.X., *et al.* (2013) A quantitative study of exocytosis of titanium dioxide nanoparticles from neural stem cells. *Nanoscale* **5**(11), 4737–4743.
107. Vasanthakumar, S., Ahamed, H.N., and Saha, R.N. (2014) Nanomedicine I: in vitro and in vivo evaluation of paclitaxel loaded poly-(epsilon-caprolactone), poly (DL-lactide-*co*-glycolide) and poly (DL-lactic acid) matrix nanoparticles in Wistar rats. *Eur. J. Drug Metab. Pharmacokinet.*, in press.
108. Zhu, G., Zheng, J., Song, E., Donovan, M., *et al.* (2013) Self-assembled, aptamer-tethered DNA nanotrains for targeted transport of molecular drugs in cancer theranostics. *Proc. Natl Acad. Sci. USA* **110**(20), 7998–8003.
109. Stern, E., Wagner, R., Sigworth, F.J., Breaker, R., *et al.*, (2007) Importance of the Debye screening length on nanowire field effect transistor sensors. *Nano Lett.* **7**(11), 3405–3409.
110. Patolsky, F., Zheng, G., Hayden, O., Lakadamyali, M., *et al.*, (2004) Electrical detection of single viruses. *Proc. Natl Acad. Sci. USA* **101**(39), 14017–14022.
111. Sridhar, M., Xu, D., Kang, Y., Hmelo, A.B., *et al.* (2008) Experimental characterization of a metal-oxide-semiconductor field-effect transistor-based Coulter counter. *J. Appl. Phys.* **103**(10), 104701–10470110.
112. Naik, A.K., Hanay, M.S., Hiebert, W.K., Feng, X.L., *et al.*, (2009) Towards single-molecule nanomechanical mass spectrometry. *Nat. Nanotechnol.* **4**(7), 445–450.
113. Burg, T.P., Godin, M., Knudsen, S.M., Shen, W., *et al.* (2007) Weighing of biomolecules, single cells and single nanoparticles in fluid. *Nature* **446**(7139), 1066–1069.
114. Lee, J., Shen, W., Payer, K., Burg, T.P., *et al.* (2010) Toward attogram mass measurements in solution with suspended nanochannel resonators. *Nano Lett.* **10**(7), 2537–2542.
115. Lee, J., Chunara, R., Shen, W., Payer, K., *et al.* (2011) Suspended microchannel resonators with piezoresistive sensors. *Lab Chip* **11**(4), 645–651.
116. Kozak, D., Anderson, W., Vogel, R., and Trau, M. (2011) Advances in resistive pulse sensors: devices bridging the void between molecular and microscopic detection. *Nano Today* **6**(5), 531–545.
117. Fraikin, J.L., Teesalu, T., McKenney, C.M., Ruoslahti, E., *et al.* (2011) A high-throughput label-free nanoparticle analyser. *Nat. Nanotechnol.* **6**(5), 308–313.
118. Ang, Y.S. and Yung, L.Y. (2012) Rapid and label-free single-nucleotide discrimination via an integrative nanoparticle-nanopore approach. *ACS Nano* **6**(10), 8815–8823.
119. Roberts, G.S., Yu, S., Zeng, Q., Chan, L.C., *et al.* (2012) Tunable pores for measuring concentrations of synthetic and biological nanoparticle dispersions. *Biosens. Bioelectron.* **31**(1), 17–25.
120. Kozak, D., Anderson, W., Vogel, R., Chen, S., *et al.*, (2012) Simultaneous size and

zeta-potential measurements of individual nanoparticles in dispersion using size-tunable pore sensors. *ACS Nano* **6**(8), 6990–6997.

121. Lan, W.J. and White, H.S. (2012) Diffusional motion of a particle translocating through a nanopore. *ACS Nano* **6**(2), 1757–1765.
122. Love, S.A., Maurer-Jones, M.A., Thompson, J.W., Lin, Y.S., *et al.* (2012) Assessing nanoparticle toxicity. *Annu. Rev. Anal. Chem.* **5**, 181–205.
123. Kim, S.T., Saha, K., Kim, C., and Rotello, V.M. (2013) The role of surface functionality in determining nanoparticle cytotoxicity. *Acc. Chem. Res.* **46**(3), 681–691.
124. Wu, J., Chen, Q., Liu, W., and Zhang, Y. (2012) Cytotoxicity of quantum dots assay on a microfluidic 3D-culture device based on modeling diffusion process between blood vessels and tissues. *Lab Chip* **12**(18), 3474–3480.
125. Mahto, S.K., Yoon, T.H., and Rhee, S.W. (2010) A new perspective on in vitro assessment method for evaluating quantum dot toxicity by using microfluidics technology. *Biomicrofluidics* **4**(3), 034111.
126. Ozturk, S., Hassan, Y.A., and Ugaz, V.M. (2012) A simple microfluidic probe of nanoparticle suspension stability. *Lab Chip* **12**(18), 3467–3473.
127. Farokhzad, O.C., Khademhosseini, A., Jon, S., Hermmann, A., *et al.* (2005) Microfluidic system for studying the interaction of nanoparticles and microparticles with cells. *Anal. Chem.* **77**(17), 5453–5459.
128. Kusunose, J., Zhang, H., Gagnon, M.K., Pan, T., *et al.*, (2013) Microfluidic system for facilitated quantification of nanoparticle accumulation to cells under laminar flow. *Ann. Biomed. Eng.* **41**(1), 89–99.
129. Lai, Y., Asthana, A., and Kisaalita, W.S. (2011) Biomarkers for simplifying HTS 3D cell culture platforms for drug discovery: the case for cytokines. *Drug Discovery Today* **16**(7-8), 293–297.
130. Hsieh, C.C., Huang, S.B., Wu, P.C., Shieh, D.B., *et al.* (2009) A microfluidic cell culture platform for real-time cellular imaging. *Biomed. Microdevices* **11**(4), 903–913.
131. Zhang, X.Q., Xu, X., Bertrand, N., Pridgen, E., *et al.* (2012) Interactions of nanomaterials and biological systems: implications to personalized nanomedicine. *Adv. Drug Delivery Rev.* **64**(13), 1363–1384.
132. Huh, D., Hamilton, G.A., and Ingber, D.E. (2011) From 3D cell culture to organs-on-chips. *Trends Cell Biol.* **21**(12), 745–754.
133. Sei, Y., Justus, K., LeDuc, P., and Kim, Y. (2014) Engineering living systems on chips: from cells to human on chips. *Microfluid. Nanofluid.* **16**(5), 907–920.
134. Zheng, Y., Chen, J., Craven, M., Choi, N.W., *et al.* In vitro microvessels for the study of angiogenesis and thrombosis. *Proc. Natl Acad. Sci. USA* 2012 **109**(24), 9342–9347.
135. Borenstein, J.T., Tupper, M.M., Mack, P.J., Weinberg, E.J., *et al.* (2010) Functional endothelialized microvascular networks with circular cross-sections in a tissue culture substrate. *Biomed. Microdevices* **12**(1), 71–79.
136. Srigunapalan, S., Lam, C., Wheeler, A.R., and Simmons, C.A. (2011) A microfluidic membrane device to mimic critical components of the vascular microenvironment. *Biomicrofluidics* **5**(1), 13409.
137. Shin, Y., Jeon, J.S., Han, S., Jung, G.-S., *et al.* (2011) In vitro 3D collective sprouting angiogenesis under orchestrated ANG-1 and VEGF gradients. *Lab Chip* **11**(13), 2175–2181.
138. Gunther, A., Yasotharan, S., Vagaon, A., Lochovsky, C., *et al.* (2010) A microfluidic platform for probing small artery structure and function. *Lab Chip* **10**(18), 2341–2349.
139. Zhang, B., Peticone, C., Murthy, S.K., and Radisic, M. (2013) A standalone perfusion platform for drug testing and target validation in micro-vessel networks. *Biomicrofluidics* **7**(4), 44125.
140. Esch, M.B., Post, D.J., Shuler, M.L., and Stokol, T. (2011) Characterization of in vitro endothelial linings grown within microfluidic channels. *Tissue Eng. Part A* **17**(23-24), 2965–2971.
141. Wong, K.H., Truslow, J.G., Khankhel, A.H., Chan, K.L., *et al.* (2013) Artificial lymphatic drainage systems for vascularized microfluidic scaffolds. *J. Biomed. Mater. Res. Part A* **101**(8), 2181–2190.
142. Cucullo, L., Marchi, N., Hossain, M., and Janigro, D. (2011) A dynamic in vitro BBB model for the study of immune cell trafficking into the central nervous system. *J. Cereb. Blood Flow Metab.* **31**(2), 767–777.
143. Booth, R. and Kim, H. (2012) Characterization of a microfluidic in vitro model of

the blood-brain barrier (muBBB). *Lab Chip* **12**(10), 1784–1792.
144. Culot, M., Lundquist, S., Vanuxeem, D., Nion, S., *et al.* (2008) An in vitro blood-brain barrier model for high throughput (HTS) toxicological screening. *Toxicol. in Vitro* **22**(3), 799–811.
145. Cucullo, L., McAllister, M.S., Kight, K., Krizanac-Bengez, L., *et al.* (2002) A new dynamic in vitro model for the multidimensional study of astrocyte-endothelial cell interactions at the blood-brain barrier. *Brain Res.* **951**(2), 243–254.
146. Cucullo, L., Hossain, M., Tierney, W., and Janigro, D. (2013) A new dynamic in vitro modular capillaries-venules modular system: cerebrovascular physiology in a box. *BMC Neurosci.* **14**, 18.
147. Prabhakarpandian, B., Shen, M.C., Nichols, J.B., Mills, I.R., *et al.* (2013) SyM-BBB: a microfluidic blood brain barrier model. *Lab Chip* **13**(6), 1093–1101.
148. Parkinson, F.E., Friesen, J., Krizanac-Bengez, L., and Janigro, D. (2003) Use of a three-dimensional in vitro model of the rat blood-brain barrier to assay nucleoside efflux from brain. *Brain Res.* **980**(2), 233–241.
149. Kim, Y., Lobatto, M.E., Kawahara, T., Lee Chung, B., *et al.* (2014) Probing nanoparticle translocation across the permeable endothelium in experimental atherosclerosis. *Proc. Natl Acad. Sci. USA* **111**(3), 1078–1083.
150. Estrada, R., Giridharan, G.A., Nguyen, M.D., Prabhu, S.D., *et al.* (2011) Microfluidic endothelial cell culture model to replicate disturbed flow conditions seen in atherosclerosis susceptible regions. *Biomicrofluidics* **5**(3), 32006–3200611.
151. Polk, B.J. Stelzenmuller, A., Mijares, G., MacCrehan, W., and Gaitan, M. (2006) Ag/AgCl microelectrodes with improved stability for microfluidics. *Sens. Actuators, B* **114**, 239–247.
152. Huh, D., Leslie, D.C., Matthews, B.D., Fraser, J.P., *et al.* (2012) A human disease model of drug toxicity-induced pulmonary edema in a lung-on-a-chip microdevice. *Sci. Transl. Med.* **4**(159), 159ra47.
153. Kim, H.J. and Ingber, D.E. (2013) Gut-on-a-chip microenvironment induces human intestinal cells to undergo villus differentiation. *Integr. Biol.* **5**(9), 1130–1140.
154. Kim, H.J., Huh, D., Hamilton, G., and Ingber, D.E. (2012) Human gut-on-a-chip inhabited by microbial flora that experiences intestinal peristalsis-like motions and flow. *Lab Chip* **12**(12), 2165–2174.
155. Jang, K.J., Mehr, A.P., Hamilton, G.A., McPartlin, L.A., *et al.* (2013) Human kidney proximal tubule-on-a-chip for drug transport and nephrotoxicity assessment. *Integr. Biol.* **5**(9), 1119–1129.
156. Toh, Y.C., Lim, T.C., Tai, D., Xiao, G., *et al.* (2009) A microfluidic 3D hepatocyte chip for drug toxicity testing. *Lab Chip* **9**(14), 2026–2035.
157. Kim, Y., Joshi, S.D., Messner, W.C., and LeDuc, P.R. (2011) Detection of dynamic spatiotemporal response to periodic chemical stimulation in a *Xenopus* embryonic tissue. *PLoS ONE* **6**(1), e14624.
158. Hattersley, S.M., Sylvester, D.C., Dyer, C.E., Stafford, N.D., *et al.* (2012) A microfluidic system for testing the responses of head and neck squamous cell carcinoma tissue biopsies to treatment with chemotherapy drugs. *Ann. Biomed. Eng.* **40**(6), 1277–1288.
159. Hutzler, M. and Fromherz, P. (2004) Silicon chip with capacitors and transistors for interfacing organotypic brain slice of rat hippocampus. *Eur. J. Neurosci.* **19**(8), 2231–2238.
160. Besl, B. and Fromherz, P. (2002) Transistor array with an organotypic brain slice: field potential records and synaptic currents. *Eur. J. Neurosci.* **15**(6), 999–1005.
161. Chronis, N., Zimmer, M., and Bargmann, C.I. (2007) Microfluidics for in vivo imaging of neuronal and behavioral activity in *Caenorhabditis elegans. Nat. Methods* **4**(9), 727–731.
162. George, S., Xia, T., Rallo, R., Zhao, Y., *et al.* (2011) Use of a high-throughput screening approach coupled with in vivo zebrafish embryo screening to develop hazard ranking for engineered nanomaterials. *ACS Nano* **5**(3), 1805–1817.
163. Jimenez, A.M., Roche, M., Pinot, M., and Panizza, P. (2011) Towards high throughput production of artificial egg oocytes using microfluidics. *Lab Chip* **11**(3), 429–434.
164. Crane, M.M., Chung, K., Stirman, J., and Lu, H. (2010) Microfluidics-enabled phenotyping, imaging, and screening of multicellular organisms. *Lab Chip* **10**(12), 1509–1517.

165. Wlodkowic, D., Khoshmanesh, K., Akagi, J., Williams, D.E., *et al.* (2011) Wormometry-on-a-chip: innovative technologies for in situ analysis of small multicellular organisms. *Cytometry, Part A* **79**(10), 799–813.
166. Shi, W., Wen, H., Lin, B., and Qin, J. (2011) Microfluidic platform for the study of *Caenorhabditis elegans*. *Top. Curr. Chem.* **304**, 323–338.
167. Wang, K.I., Salcic, Z., Yeh, J., Akagi, J., *et al.* (2013) Toward embedded laboratory automation for smart lab-on-a-chip embryo arrays. *Biosens. Bioelectron.* **48**, 188–196.
168. Moraes, C., Mehta, G., Lesher-Perez, S.C., and Takayama, S. (2012) Organs-on-a-chip: a focus on compartmentalized microdevices. *Ann. Biomed. Eng.* **40**(6), 1211–1227.
169. Sung, J.H., Esch, M.B., Prot, J.M., Long, C.J., *et al.* (2013) Microfabricated mammalian organ systems and their integration into models of whole animals and humans. *Lab Chip* **13**(7), 1201–1212.
170. Sung, J.H., Kam, C., and Shuler, M.L. (2011) A microfluidic device for a pharmacokinetic-pharmacodynamic (PK-PD) model on a chip. *Lab Chip* **10**(4), 446–455.
171. Williamson, A., Singh, S., Fernekorn, U., and Schober, A. (2013) The future of the patient-specific body-on-a-chip. *Lab Chip* **13**(18), 3471–3480.
172. Wikswo, J.P., Block III,, F.E., Cliffel, D.E., Goodwin, C.R., *et al.* (2013) Engineering challenges for instrumenting and controlling integrated organ-on-chip systems. *IEEE Trans. Biomed. Eng.* **60**(3), 682–690.
173. Olcum, S., Cermak, N., Wasserman, S.C., Christine, K.S., *et al.* (2014) Weighing nanoparticles in solution at the attogram scale. *Proc. Natl Acad. Sci. USA* **111**(4), 1310–1315.
174. Yang, M., Sun, S., Kostov, Y., and Rasooly, A. (2011) A simple 96 well microfluidic chip combined with visual and densitometry detection for resource-poor point of care testing. *Sens. Actuators, B* **153**(1), 176–181.
175. Biswas, S., Miller, J.T., Li, Y., and Nandakumar, K. (2012) Developing a millifluidic platform for the synthesis of ultrasmall nanoclusters: ultrasmall copper nanoclusters as a case study. *Small* **8**(5), 687–698.
176. Hong, J.W. and Quake, S.R. (2005) Integrated nanoliter systems. *Nat. Biotechnol.* **21**(10), 1179–1183.
177. Kobayashi, I., Mukataka, S., and Nakajima, M. (2005) Novel asymmetric through-hole array microfabricated on a silicon plate for formulating monodisperse emulsions. *Langmuir* **21**(17), 7629–7632.
178. Kobayashi, I., Mukataka, S., and Nakajima, M. (2005) Effects of type and physical properties of oil phase on oil-in-water emulsion droplet formation in straight-through microchannel emulsification, experimental and CFD studies. *Langmuir* **21**(13), 5722–5730.
179. Nisisako, T. and Torii, T. (2008) Microfluidic large-scale integration on a chip for mass production of monodisperse droplets and particles. *Lab Chip* **8**(2), 287–293.
180. Li, W., Greener, J., Voicu, D., and Kumacheva, E. (2009) Multiple modular microfluidic (M3) reactors for the synthesis of polymer particles. *Lab Chip* **9**(18), 2715–2721.
181. Kim, Y., LeDuc, P., and Messner, W. (2013) Modeling and control of a nonlinear mechanism for high performance microfluidic systems. *IEEE Trans. Control Syst. Technol.* **21**(1), 203–211.
182. Psaltis, D., Quake, S.R., and Yang, C.H. (2006) Developing optofluidic technology through the fusion of microfluidics and optics. *Nature* **442**(7101), 381–386.
183. Mao, X., Waldeisen, J.R., Juluri, B.K., and Huang, T.J. (2007) Hydrodynamically tunable optofluidic cylindrical microlens. *Lab Chip* **7**(10), 1303–1308.
184. Yang, A.H. and Erickson, D. (2010) Optofluidic ring resonator switch for optical particle transport. *Lab Chip* **10**(6), 769–774.
185. Song, C.L., Luong, T.D., Kong, T.F., and Nguyen, N.T. (2011) Disposable flow cytometer with high efficiency in particle counting and sizing using an optofluidic lens. *Opt. Lett.* **36**(5), 657–659.
186. Ziolkowska, K., Stelmachowska, A., Kwapiszewski, R., and Chudy, M. (2013) Long-term three-dimensional cell culture and anticancer drug activity evaluation in a microfluidic chip. *Biosens. Bioelectron.* **40**(1), 68–74.

187. Lii, J., Hsu, W.J., Parsa, H., and Das, A. (2008) Real-time microfluidic system for studying mammalian cells in 3D microenvironments. *Anal. Chem.* **80**(10), 3640–3647.

188. Cheah, L.T., Dou, Y.H., Seymour, A.M., Dyer, C.E., *et al.* (2010) Microfluidic perfusion system for maintaining viable heart tissue with real-time electrochemical monitoring of reactive oxygen species. *Lab Chip* **10**(20), 2720–2726.

189. Richter, L., Charwat, V., Jungreuthmayer, C., and Bellutti, F. (2011) Monitoring cellular stress responses to nanoparticles using a lab-on-a-chip. *Lab Chip* **11**(15), 2551–2560.

190. Ahmad, A.A., Wang, Y., Gracz, A.D., and Sims, C.E. (2014) Optimization of 3-D organotypic primary colonic cultures for organ-on-chip applications. *J. Biol. Eng.* **8**, 9.

Translational Medicine Cancer

Edited by
Robert A. Meyers

Related Titles

Tsimberidou, A., Kurzrock, R., Anderson, K.C. (eds.)

Targeted Therapy in Translational Cancer Research

2015
ISBN: 978-1-118-46857-9,
also available in digital formats

Tobias, J.S., Hochhauser, D.

Cancer and its Management
7th Edition

2015
ISBN: 978-1-118-46873-9,
also available in digital formats

Brenner, M.K., Hung, M. (eds.)

Cancer Gene Therapy by Viral and Non-viral Vectors

2014
ISBN: 978-1-118-50162-7,
also available in digital formats

Tiwari, A., Choi, J., Patra, H.K. (eds.)

Advanced Theranostics Materials

2015
ISBN: 978-1-118-99829-8,
also available in digital formats

Translational Medicine Cancer

Advances in Molecular Biology and Medicine

Edited by
Robert A. Meyers

Volume 2

WILEY-VCH
Verlag GmbH & Co. KGaA

The Editor

Dr. Robert A. Meyers
Editor-in-Chief
Ramtech Limited
122, Escalle Lane
Larkspur, CA 94939
USA

Cover

Figure with kind permission from Brigitte L. Thériault, Helen Dimaras, Brenda L Gallie and Timothy W Corson (2014): The genomic landscape of retinoblastoma: a review. Clinical & Experimental Ophthalmology, Volume 42, Issue 1: 33–52.

Library of Congress Card No.: applied for

British Library Cataloguing-in-Publication Data
A catalogue record for this book is available from the British Library.

Bibliographic information published by the Deutsche Nationalbibliothek
The Deutsche Nationalbibliothek lists this publication in the Deutsche Nationalbibliografie; detailed bibliographic data are available on the Internet at <http://dnb.d-nb.de>.

Print ISBN: 978-3-527-33569-5
ePDF ISBN: 978-3-527-68186-0
ePub ISBN: 978-3-527-68187-7
Mobi ISBN: 978-3-527-68185-3

Cover Design Adam Design, Weinheim, Germany

Typesetting SPi Global, Chennai, India

Printing and Binding Markono Print Media Pte Ltd, Singapore

Printed on acid-free paper.

Contents

Preface

In 2015, there will be more than 14,000,000 cases of cancer diagnosed worldwide and an estimated 1,658,370 new cancer cases diagnosed and 589,430 cancer deaths in the United States alone. But, cancer prevention and the latest research has resulted in tremendous gains in both longer survival and quality of life such that cancer death rates in the United States have declined 20% from their peak in 1991. Today, there are a record 14.5 million cancer survivors alive in the United States. There is now a race to both prevent and cure cancer once and for all. For the next 10 years, it is forecast that cancer research efforts will focus on areas such as (i) tumor biology and especially cancer stem cells, exploring the theory that cancer stem cells are biologically different from regular cancer cells, have the ability to self-renew, and are more resistant to chemotherapy leading to therapy that attacks these cells, (ii) faster and less expensive next-generation sequencing, liquid biopsies to more accurately detect circulating tumor DNA and microRNA and metabolomics in blood and determine effective treatment, and cancer prevention vaccines, (iii) immunotherapies, some of which reprogram the body's own immune cells to attack cancer, (iv) development of new drugs and methods to carry these drugs to cancerous tumors using very new approaches including nanotechnology, and also (v) cancer etiology, which informs prevention. Given the explosion in understanding how genetics, lifestyle, individual behaviors, social environment, policy, and health-care interventions can reduce the burden of cancer, new insights and integration of knowledge can help shape prevention strategies and lead to reductions in cancer incidence. All of these approaches are leading toward precision personalized medicine and eventual reduction or even elimination of cancer as a fatal disease. Our cancer translational medicine compendium covers the latest developments in all of these areas in detail.

Our compendium is written for university undergraduates, graduate students, faculty, and investigators at research institutes. Our Board of 11 Nobel Laureates approved our overall approach and content and our selection of articles was validated and enhanced by reviewers from major research institutions and each manuscript was then reviewed by scientists from other universities or research institutes. There is a glossary of keywords with definitions at the beginning of each article for the benefit of students. The result is that there are 28 peer reviewed articles at a length of over 750 pages and as such is the largest in depth, up-to-date treatment presently available.

Our compendium is organized into three sections: **Cancer Biology**; **Molecular Diagnostics and Imaging**; and **Cancer Therapy**. The section on **Cancer Biology** begins with an article on cancer etiology and an article on cancer metastasis theory and articles defining the biology of cancer stem cells and their niches and then intracellular signaling and also tumor immunology. An article on chromatin remodeling brings in the concept of epigenetics where the traditional dogma was that the underlying genetic deregulation emerged through an accumulation of individual mutations that inactivate tumor suppressor genes and activate oncogenes. During recent years, it has become evident that epigenetic changes are potently contributing to tumor formation. And there are articles on specific cancers to illustrate the above-mentioned concepts. The section on **Molecular diagnostics and Imaging** covers biomarkers that can be used to detect cancer in different stages and also the relatively new field of metabolomic imaging for cancer detection, which involves the global variations of metabolites, with which malignancy conditions can be evaluated by profiling the entire measurable metabolome and the application of metabolomics for *in vivo* cancer imaging. Also, this section covers the emerging field of microfluidics in nanomedicine, which is capable of serving as contrast agents for different imaging modalities as well as targeting specific sites for delivery of drugs/genes. The section on **Cancer Therapy** includes strategies for targeting cancer stem cells and therapeutics against cancer stem cells; immunologic therapies such as immune checkpoint inhibitors and also immunotherapy with autologous cells; very new and experimental approaches such as carbon nanotubes for enhanced biopharmaceutical delivery and gold and iron oxide nanoparticles with antibody guides to find and destroy cancer cells; RNAi in cancer therapy; role of the microbiota in cancer therapy; radioactive microbes; and surveys of broadly applicable therapeutic approaches such as targeted therapy: genomic approaches; cancer chemotherapy; and cancer pharmacogenomics.

Our participating authors are from some of the most active cancer research institutions including the National Cancer Institute, MD Anderson Cancer Center, Harvard University, Massachusetts General Hospital, the Massachusetts Institute of Technology, Stanford University, University of Oxford, Washington University School of Medicine, Baylor College of Medicine, Pennsylvania State University, Albert Einstein College of Medicine, Cornell University, University of Michigan, Inserm, Fox Chase Cancer Center, and the University of California.

Our team hopes that you, the reader, will benefit from our hard work – finding the content useful in your research and educational. We thank our Managing Editor, Sarah Mellor as well as our Executive Editors, Gregor Ciccetti and Frank Otmar Weinreich for both their advice and hard work in the course of this project.

Larkspur, CA, USA
January 2016

Robert A. Meyers
RAMTECH Limited

Part III
Cancer Therapies

Translational Medicine Cancer: Advances in Molecular Biology and Medicine,
First Edition. Edited by Robert A. Meyers.

14
Theoretical Foundations of Cancer Chemotherapy

Fatemeh Ghari and Nicholas B. La Thangue
University of Oxford, Department of Oncology, Old Road Campus Research Building, Off Roosevelt Drive, Headington, Oxford, OX3 7DQ, UK

Translational Medicine Cancer: Advances in Molecular Biology and Medicine,
First Edition. Edited by Robert A. Meyers.

Keywords

Biomarker
Medical signs whose objective measurement reproducibly correlates with a given physiological or pathological state.

ErbB
The epidermal growth factor receptor.

Epigenetics
Chemical modifications of DNA and DNA-associated proteins (histones).

Apoptosis
Process of programmed cell death.

Synthetic lethality
Lethal combination of loss of two genes, where loss of either alone is permissible to cell viability.

Warburg effect
Metabolic switch of most malignant cells to anaerobic metabolism for ATP production.

Angiogenesis
Process of new blood vessel growth.

Ubiquitin-Proteasome-Pathway
Destruction of substrate proteins marked by a polyubiquitin chain by the proteasome complex.

Hsp90
A ubiquitously expressed protein that acts to protect cells from thermal stress.

The massive research efforts that have been applied to understanding the workings of the cancer cell have provided a wealth of detail on key mechanisms that become abnormal during tumorigenesis. Translating and exploiting this large body of cancer research information in the design and implementation of new drugs should yield better medicines for the cancer patient than those that are currently available. It is an implicit assumption in this objective that mechanism-based agents which target discrete abnormalities in tumor cells will provide improved treatments over conventional therapies. In this respect, an emerging clinical efficacy is beginning to be witnessed in a new generation of anticancer drugs designed to target key mechanisms in tumor cells.

1 Cancer Therapy

Cancer remains largely a clinically unmet disease. Current statistics indicate that cancer affects one person in three, and in the Western world it is the cause of about one-quarter of all mortalities [1]. Worldwide, cancer is the second largest cause of death, and the World Health Organization estimates that, by 2030, there will be 13 million deaths associated with cancer every year [2]. Cancer is not a single disease, as there are over 200 different types of disease that fall into this category. However, four cancers dominate, namely breast, colon, lung, and prostate, and these account for over half of all new cases [3].

Most current cancer therapies employ one of three approaches: surgery; radiotherapy; or chemotherapy. Although surgery and radiotherapy are frequently used to treat localized primary tumors, they have a limited application in disseminated disease in which the tumor is not localized, usually as a result of metastasis [4]. Needless to say, disseminated disease is the cause of most cancer deaths [2]. Current cancer chemotherapy is dominated by treatments that employ cytotoxic agents or hormone-based therapy; typical examples are breast and prostate cancer. Furthermore, while advances have been made in all three approaches to treatment, the impact on mortality rates has been modest. As an example, lung cancer – the leading cause of cancer death worldwide – has a five-year survival rate of less than 10%, and this has improved only marginally from the survival rate that prevailed 30 years ago [5]. There are, however, some notable improvements. Today, cures are achievable in certain childhood leukemias and testicular cancer, with several of the common cancer types (including breast and prostate cancer) exhibiting a five-year survival rate in excess of 50% [5]. Overall, however, conventional chemotherapy is relatively ineffective in many cancers.

This situation reflects the mechanism of action of most current cancer therapy regimes, which are dominated by cytotoxic chemotherapies [4] (Table 1). In general, these chemotherapies target the mechanisms employed by most dividing cells, rather than any specific genetic abnormalities associated with tumor cells. Because of this, the tumor cells' normal healthy counterparts are affected almost as severely as the tumor cells themselves, and this leads to the highly debilitating side effects seen with many therapies, such as myelosuppression, hair loss and gastrointestinal toxicity [4]. This problem, combined with a poor quality of life and diminishing effectiveness in the patient (usually because of the emergence of drug resistance), means that many late-stage cancer treatments provide a minimal survival advantage.

2 The New Era of Mechanism-Based Drug Design

Because of the limited therapeutic value of many current treatments, major efforts are being applied towards identifying new cancer therapies that target specific mechanisms in tumor cells [6] (Table 2), in the hope that this approach will yield drugs with increased specificities for cancer. In this chapter, the progress of this rapidly advancing field is reviewed, and a snapshot is provided of the current situation. Selected examples of mechanism-based drug design programs, illustrating the guiding principles behind the new era of cancer drug development, are also discussed. The success of this approach is clear, and it

Tab. 1 Examples of conventional cancer chemotherapeutic agents [4].

Category	*Example*	*Mechanism of action*
Alkylating agents	Nitrosoureas	Form covalent adducts with DNA (esp. N7 of guanine) to interfere with cell proliferation
Anthracyclines	Doxorubicin	Several mechanisms, including intercalation between DNA, inhibition of topoisomerase II and iron-mediated generation of free radicals
Antimetabolites	5-Fluorouracil	Compete with normal metabolites for incorporation into DNA/RNA
Antitumor antibiotics	Bleomycin	Intercalate DNA strands, to induce spontaneous oxidation, free radical formation, and DNA strand breakage
Topoisomerase I/II inhibitors	Camptothecin	Generate single-/double-strand breaks that activate the checkpoint pathways, leading to cell-cycle arrest/apoptosis
Taxanes	Paclitaxel	Promote microtubule assembly and stability to arrest cells in mitosis
Vinca alkaloids	Vincristine	Block microtubule polymerization, impairing mitotic spindle formation and inducing mitotic checkpoints

Tab. 2 Examples of mechanism-based targeted therapies.

Drug	*Target*
Imatinib	Bcr-*abl* kinase
Trastuzumab	ErbB2/Her2 receptor
Gefitinib	EGFR tyrosine kinase
Olaparib	Poly (ADP-ribose) polymerase 1
Vorinostat	Histone deacetylase
Bevacizumab	Vascular endothelial growth factor
Bortezomib	Proteasome

is anticipated that this situation will be repeated in continuing efforts to deliver tailored therapies to the cancer clinic.

3
Personalized Cancer Medicine and Predictive Biomarkers

During recent years, advances in understanding the biology of cancer have aided the rational design and development of molecularly targeted mechanism-based drugs. An equally crucial venture has been the individualization of these therapies in accordance with the unique genetic and molecular profile of the disease type. The complexity and heterogeneity of various cancer types, along with their inherent interindividual variability, have necessitated the move toward personalized or precision cancer therapy. Accordingly, there has been a clinical need to stratify patients and to target specific therapies to those subpopulations which are most likely to benefit, while avoiding unnecessary harm and toxicity to those who are predicted to be nonresponsive [7].

The realization of personalized cancer medicine requires both an in-depth molecular understanding of the cancer type, as well as validated and robust diagnostic tests to profile the disease and tease out the relevant molecular information. Biomarkers in general terms refer to those characteristics of which objective measurements in the body reproducibly correlates with a

Tab. 3 Examples of predictive biomarkers [9].

Drug	*Cancer type*	*Biomarker*
Gefitinib	Non-small-cell lung cancer	EGFR mutations
Crizotinib	Non-small-cell lung cancer	Rearrangements in ALK
Cetuximab	Metastatic colorectal cancer	Wild-type KRAS
Vemurafenib	Melanoma	V600E BRAF mutation
Trastuzumab	Breast cancer	ErbB2/Her2 receptor overexpression
Olaparib	Breast cancer	BRCA 1/2 mutations

given physiological or pathological state. Biomarkers can serve prognostic, diagnostic or predictive purposes. Predictive biomarkers, which help to predict and assess the probability of a disease's response to a given therapy, are particularly valuable to the application of personalized medicine [8]. The identification and validation of a defined biomarker for a given cancer treatment proves crucial to stratifying patients according to the presence or absence of the biomarker, and hence the likelihood of a therapeutic response.

Currently, a number of validated predictive cancer biomarkers have been identified in the clinic, as outlined in Table 3 [9]. With an exponential rise in the number of mechanism-based therapies in the clinic and under development, and with ongoing issues regarding patient toxicity and a lack of efficacy to therapies, the identification and employment of relevant biomarkers during the coming years will become as important as developing new cancer therapies. Increasing collaborations between academia and the industrial sector, and the utilization of innovative techniques and methodologies, together with a wealth of information generated from global genomics and proteomics studies and patient health records and biobanks, will undoubtedly prove crucial to the success of personalized medicines.

4
Lessons from Gleevec

Imatinib (Gleevec) provides a paradigm for a successfully developed, rationally designed and mechanism-based therapy for the treatment of a specific cancer [10]. In chronic myelogenous leukemia (CML), the consistent chromosomal abnormality, the Philadelphia (Ph) chromosome, is a translocation event between chromosome 9 and 22 that results in the bcr-*abl* fusion product derived from the juxtaposition of c-*abl* oncogene and breakpoint cluster region (bcr) [11] (Fig. 1). bcr-*abl* is a constitutively active tyrosine kinase with oncogenic activity, and can exist in three different forms, namely p185, p220, and p230 [12].

Originally, imatinib was identified by high-throughput screening for compounds with kinase inhibitory activity, and was subsequently optimized using rational drug design procedures for activity against Abl tyrosine kinase in CML cells [13]. A successful Phase I study indicated remarkable single-agent activity during the CML blast crisis, and in Ph-positive acute lymphoblastic leukemia (ALL), although the responses did not appear to be long term in nature. Nevertheless, the success achieved in Phase I trials prompted a further Phase II clinical analysis, followed by US Food

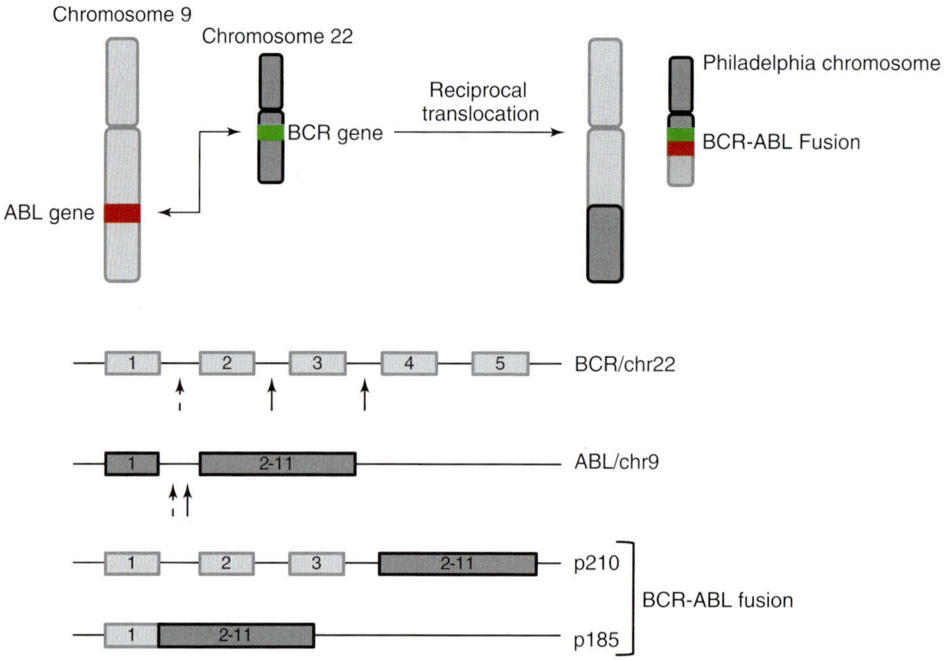

Fig. 1 Chromosomal breakpoints in chronic myelogenous leukemia (CML) and acute lymphoblastic leukemia (ALL). In CML and ALL, a breakpoint occurs between the first exon of c-abl on chromosome 9, translocating most of c-abl to c-bcr on chromosome 22. In CML, the translocations in c-bcr occur after exon 2 or 3, to yield the p210 fusion protein, and in Ph-positive ALL, they occur after exon 1 to create p185. Solid arrows indicate CML breakpoints; broken arrows indicate ALL breakpoints.

and Drug Administration (FDA) approval [13]. Imatinib also demonstrated activity against the c-kit tyrosine kinase, for which gain-of-function mutations are found in gastrointestinal stromal tumors (GISTs). Following encouraging outcomes from clinical studies, imatinib also received FDA approval for use in patients with KIT-positive unresectable and/or metastatic GIST [14].

The appearance of resistance to imatinib was a surprising clinical outcome, as the majority of CML patients that relapse after an initial positive response have a reactivated bcr-*abl* kinase. Although part of the relapse could be attributed to an amplification of the BCR-ABL gene, molecular analyses have indicated that in many cases the resistance was due to point mutations in the *abl* kinase gene that rendered the kinase less sensitive to imatinib [15]. These mutations were found to fall critically at positions that abrogate the kinase interaction with imatinib. One frequently observed mutation is that of the gatekeeper residue, threonine 315 to isoleucine (T315I), located at the periphery of the nucleotide binding site of the kinase, which abolishes an essential H-bond interaction between imatinib and *abl* [16]. These results suggest that, while rationally designed drugs such as imatinib may exhibit a promising clinical efficacy, the appearance of resistance could be a significant problem.

Subsequent to the inadequacies of imatinib, it has been possible with crystal structure studies and rational drug design, to develop second-generation bcr-*abl* kinase inhibitors, including nilotinib (Tasigna) and dasatinib (Sprycel). Both inhibitors show enhanced potency and binding to most imatinib-resistant mutant forms of bcr-*abl* kinase, except for the gatekeeper mutant T315I [17, 18]. Nilotinib, which also has an increased selectivity against *abl* kinase and binds to the inactive conformation of the enzyme, has been approved for use in chronic phase Philadelphia-positive CML [19]. Dasatinib, which binds to the active conformation of the enzyme, has additional activity against src kinases, platelet-derived growth factor receptor (PDGFR), KIT, and ephrin receptor tyrosine kinases, and has been approved for the treatment of all phases of CML with resistance or intolerance to Gleevec [20]. These inhibitors showed that, by improving complementary hydrogen-bonding and hydrophobic interactions with the *abl* kinase, it is possible to augment their binding affinity for the kinase ATP binding site, and thus overcome a large number of mutations, both distal and in the active site, of Bcr-*abl* kinase [21].

Kinases remain the most extensively studied class of drug targets, with involvement in virtually all signal transduction pathways [21]. Oncogenic mutations in particular kinases have been reported across a wide range of tumors, with the resultant oncogene-addiction rendering the tumor selectively susceptible to kinase inhibition. The success of Gleevec, as one of the earliest mechanism-based therapies, is an example of this. At the same time, however, a reminder is provided of the issue of drug resistance as perhaps an inevitable outcome [22]. It is therefore important to continue assessing and evaluating drugs post-approval and, where acquired resistance has developed, attempt to understand the mechanisms involved to develop appropriate alternatives. A rational combination of several drug agents could also be an important strategy to minimizing the likelihood of resistance.

5
The ErbB Family and Kinase Inhibitors

The epidermal growth factor (EGF) family of tyrosine kinase receptors (the Erb receptors) plays a crucial role in regulating proliferation and survival, through downstream signaling to phosphatidylinositide 3-kinase (PI3K) and rat sarcoma (RAS) cascades [23] (Fig. 2). In tumor cells, a variety of mutagenic events give rise to increased ErbB activity, including gene amplification and mutations that alter protein stability and/or activity. Excessive signaling by the ErbB receptor family is a strong oncogenic event that has been associated with the development and progression of many tumor types [23, 24]. These receptors have thus been extensively investigated as potential drug candidates [25], with various humanized monoclonal antibodies and small-molecule inhibitors developed against ErbB1 (also known as EGF receptor and HER1) and ErbB2 (also known as HER2 and Neu).

ErbB2/Her2/neu-positive breast cancers account for 15–25% of such lesions, and are generally associated with a poorer prognosis and an increased tumor aggressiveness [26]. Trastuzumab (Herceptin®) has received regulatory approval as a single agent for the treatment of metastatic breast cancers expressing ErbB2 [27], while cetuximab (Erbitux), an alternative chimeric monoclonal antibody directed against ErbB1/Her1 has been approved for use in

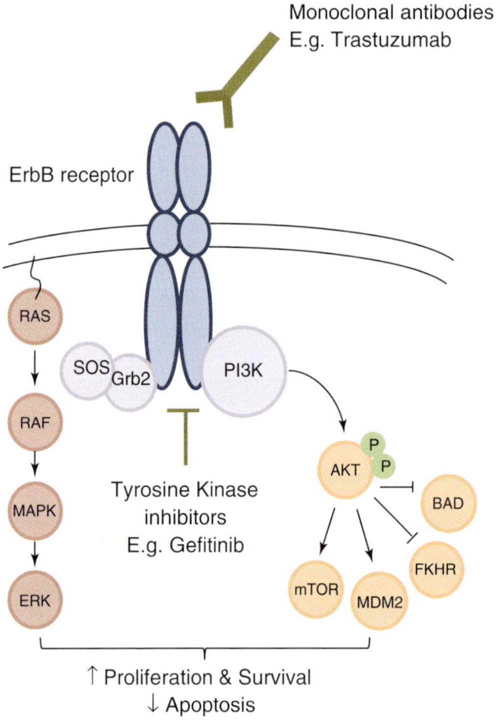

Fig. 2 ErbB receptor signaling. ErbB receptor dimerization upon ligand binding induces receptor autophosphorylation on several tyrosine residues on the cytoplasmic side. This creates a docking site for several phosphotyrosine-binding, SH2-containing adaptor proteins, including Grb2 and PI3K. As a result, a number of signaling events such as the RAS and AKT cascades are activated, which propagate the signal to many other factors to stimulate cell survival and proliferation. Gain of function mutations in KRAS have been reported to mediate resistance to ErbB1 receptor inhibitors, including the monoclonal antibody cetuximab, by overriding the dependence on the upstream ErbB receptor signaling event. Consequently, the KRAS mutation status is used as a predictive biomarker for response to ceruximab in colorectal cancer.

metastatic colorectal cancer (mCRC) [28]. Significantly, the clinical response to cetuximab depends on the Kirsten rat sarcoma (KRAS) status, with tumors harboring wild-type-KRAS responding favorably but those with mutant KRAS not responding [29]. Consequently, the genetic testing of patients for KRAS status has been enforced by the FDA as a predictive biomarker for prescribing cetuximab.

Despite the notable success of trastuzumab in the management of Her2-positive breast cancer, retrospective studies have shown intrinsic or acquired resistance to Herceptin in up to 70% of the cases treated. Various mechanisms for resistance have been described, such as the expression of a truncated p95Her2 fragment, which lacks the extracellular domain and hence the trastuzumab-binding epitope, aberrant activation of the PI3K/Akt pathway, which overrides the tumor addiction to Her2, and the activation of compensatory signaling pathways, particularly the insulin-like growth factor 1 receptor (IGF1R) pathway [30]. Small-molecule tyrosine kinase inhibitors (TKIs) offer an alternative mechanistic approach for inhibiting the ErbB receptor family, with several such compounds either being used in the clinic or undergoing trials. Lapatinib (Tyverb®) is a reversible TKI of ErbB1 and ErbB2, with antitumorigenic activity against Her2-positive breast cancer cell lines, including a subset of p95Her2 expressing trastuzumab-resistant breast tumors [31]. This drug has been approved for use in Her2-positive metastatic breast cancer.

Gefitinib (Iressa®) is an optimized small-molecule ATP competitive antagonist of the ErbB1/epidermal growth factor receptor (EGFR) tyrosine kinase, which was originally identified by high-throughput screening. As ErbB1/EGFR is overexpressed in lung tumors [32], it was predicted that small-molecule inhibitors such as gefitinib could have a therapeutic value. However, an evaluation of gefitinib in a randomized Phase II study for activity against advanced non-small-cell lung cancer (NSCLC) found that it did not provide improvement in survival when added to standard platinum-based chemotherapy [33]. An important achievement following these initially disappointing trials was the subsequent identification of a predictive biomarker for response to gefitinib [34]. Several clinical trials, such as the IRESSA pan-Asian study (IPASS), showed that EGFR mutation-positive tumors are significantly more responsive to gefitinib than EGFR mutation-negative tumors [35]. Accordingly, EGFR mutations, which are present in approximately 30–40% of Asian patients and 10–20% of non-Asian patients with advanced NSCLC, have been used as a predictive biomarker for gefitinib. This drug has been approved and marketed in many countries, including Europe, for the treatment locally advanced or metastatic NSCLC.

The ErbB family of receptors provides another example of the extensive involvement of kinases in cancer, in this instance as a receptor tyrosine kinase. There are several other important examples of kinase inhibitors employed for the management of specific cancer types, including vemurafenib for melanoma and crizotinib for NSCLC. Interestingly, the use of both of these targeted therapies is coupled to testing for their relevant biomarker.

Vemurafenib is a BRAF kinase inhibitor, whose aberrant activity as part of the RAF-mitogen activated protein kinase (MAPK)-extracellular signal regulated kinase (ERK) signaling cascade is a potent driver of uncontrolled cellular proliferation [36]. This compound shows activity against melanoma tumors harboring the BRAF mutation valine 600 to glutamic acid (V600E) [37], the presence of which is tested for as a biomarker for response. The V600E mutation, which accounts for 90% of reported BRAF mutations, is a phosphomimetic mutation that alters the conformation of BRAF to a constitutively active status [36]. Active BRAF drives deregulated cell growth and proliferation and self-sufficiency from external growth signals by activating the downstream MAPK-ERK signaling cascade. BRAF has been reported to be mutated in 40–70% of malignant melanomas, with V600E being by far the most frequent mutation. Intriguingly, although BRAF inhibitors have been found to potently block the proliferation of tumor cells harboring activating mutations in BRAF, paradoxically they augment MAPK signaling in cells with mutant-KRAS [38]. These unexpected findings therefore underscore the need to appropriately identify patients with BRAF mutations prior to commencing therapy with vemurafenib.

Crizotinib is an ATP-competitive inhibitor of the anaplastic lymphoma kinase (ALK), c-Met/hepatocyte growth factor receptor (HGFR), and ROS receptor tyrosine kinases. The ALK protein is normally expressed in the central nervous system, though oncogenic chromosomal translocations for it have been reported in a number of malignancies [39]. The fusion between ALK and echinoderm microtubule associated protein-like 4 (EML4) has been found to be a potent oncogenic driver event in 2–9% of NSCLC tumors [40]. In Phase I clinical trials, crizotinib was found to meaningfully improve overall survival in those patients who carried ALK translocations. These preliminary findings were then successfully extended into later-phase trials with a larger cohort of NSCLC patients with the ALK-EML4 fusion [41]. Crizotinib subsequently gained FDA approval for use in patients with advanced and/or metastatic NSCLC with the ALK-EML4 translocation.

With over 500 kinases in the human genome [21], and improvements in the ability to screen these enzymes and their respective effects, it is hopeful that more examples of cancer-relevant kinase inhibitors will be identified in future years.

6
Targeting Chromatin Control in Cancer

Recent advances have highlighted a new and exciting area for therapeutic intervention, namely that of epigenetic targeting of cancer cells [42]. Chromatin is the DNA–proteinaceous material in which chromosomal DNA resides in the nucleus. Many chromatin proteins are composed of histones that assemble into nucleosomes, and thereby assist in DNA packaging and transcriptional control. The basic structure of the nucleosome consists of 146 bp of DNA wrapped around an octamer of four core histones. The histone tail is subject to a variety of enzymatic modifications, including phosphorylation, acetylation and methylation [43]. Many of these histone modifications, as part of a heritable, self-perpetuating system, form the basis of the epi-(above)-genome. Epigenetic events can direct phenotypically distinct outcomes from an identical underlying genetic material (DNA), and thus can be thought of as non-genetic cellular memory [44].

It has emerged in recent years that cancer cells can exhibit altered epigenetic modifications, such as the hypermethylation of various tumor suppression genes, and hence exhibit aberrant gene silencing. Such "epimutations" justify targeting the enzymes and readers of these epigenetic events as a means of selectively targeting the cancer cells [45].

Histone acetylation on lysine residues is regulated by two groups of enzymes, histone acetyl transferases (HATs), which utilize acetyl-CoA to transfer an acetyl group to the ε-amino site of lysine residues, and histone deacetylases (HDACs), which remove the acetyl group [46]. From the perspective of cancer therapy, HDACs have gained recognition as an important target which, in part, reflects the identification of proteins other than histones that are subject to acetylation control [47]. The activity of cell-cycle regulators, including E2F, p53 and the retinoblastoma protein (pRb), is influenced by acetylation, and pRb controls E2F activity through the recruitment of chromatin-modifying enzymes, such as HDACs [48, 49]. Several oncogenic proteins have altered the recruitment of HDACs, which leads to aberrant gene transcription. This is exemplified by the fusion protein PML-RARα, which occurs through a chromosomal translocation in acute promyelocytic leukemia (APL). The PML-RARα fusion protein recruits HDAC and represses transcription, causing a block to differentiation and promoting the oncogenic phenotype in APL [50]. Not surprisingly, therefore, compounds that inhibit HDAC activity cause potent cell-cycle effects and frequently induce apoptosis [51]. Currently, the two approved HDAC inhibitors vorinostat (Zolinza®) and romidepsin (Istodax®) are being used for the treatment of advanced or refractory cutaneous T-cell lymphoma [52]. PXD101 (Belinostat®) is an alternative highly potent HDAC inhibitor that blocks the proliferation of tumor cells and has recently entered clinical trials. Results from Phase II clinical trials with PXD101 have shown very encouraging activity against T-cell lymphoma [53, 54].

DNA methylation on cytosine (carbon-5) at CpG islands was one of the earliest DNA modifications to be reported. Although cancer cells often exhibit a global hypomethylation state, aberrant CpG hypermethylation is among the most frequently reported epigenetic alterations in cancer. In fact, it is estimated that up to 5–10% of normally unmethylated promoters become hypermethylated in cancer. Though largely dependent on the spatial arrangement of the methylation event, in many instances promoter hypermethylation coincides with transcriptional silencing, a situation observed with several tumor suppressor genes [55]. 5-Azacytidine (Vidaza®) and decitabine (Dacogen®) are cytosine analogs which, following metabolic activation, become incorporated into the DNA, from where they can form a covalent adduct with DNA methyltransferases (DNMTs) and reduce the available pool of the enzyme. Both drugs have been approved for use in myelodysplastic syndromes, which were previously considered largely refractory to therapy [42, 44, 45, 55].

Despite their success, the mechanistic details regarding the mode of action of DNMT inhibitors – and in particular a proven causal link between inhibitor-induced demethylation and observed clinical response – is lacking. A further issue is that of safety, since the pleiotropic and global effects exerted by DNMT inhibitors may be expected to extend beyond cancer cells. In particular, there are concerns that DNA demethylation in normal cells may interfere with the expression of differentiation-associated genes as well as global hypomethylation, potentially leading to genomic instability and even cancer [45]. Consequently, more studies must be conducted to achieve an understanding of these drugs and the full extent of their benefit and adverse effects. Such findings perhaps may also reinforce the need to shift attention away from the global inhibition of epigenetic events towards identifying and targeting more specific cancer-associated changes.

In line with this, the recent development of specific inhibitors of histone modification reading modules has been an exciting achievement, particularly the small-molecule inhibitors of the acetyl-lysine recognition domain, or bromodomains [56]. The recently described compound, JQ1, has been reported to inhibit the bromodomain and extra-terminal (BET) family member BRD4 with high potency [57]. Recurrent oncogenic translocations of BRD4 have been found in several subtypes of human squamous carcinoma, and accordingly the binding of JQ1 to BRD4 was shown to induce strong antiproliferative effects in xenograft models. Since then, JQ1 has been shown to have promising activity against NUT (nuclear protein of the testis) midline carcinoma, several hematopoietic cancers, and glioblastoma tumors [56]. These findings illustrate the possibility of exploiting specific epigenetic readers as reasonable drug targets, and lead the way for many more such compounds during the coming years.

7 Reactivating Cancer Cell Death Pathways

While programmed cell death is an integrated aspect of normal physiological development and tissue homeostasis, it is also frequently deregulated in various pathological conditions. In cancer, the evasion of cell death mechanisms has been identified as one of the critical hallmarks of malignant transformation [58]. With the acquisition and accumulation of oncogenic aberrations, cancer cells also evolve means to suppress cell death pathways, to thus ensure their sustained survival in spite of the accumulating cellular insults and abnormalities. Conventional chemotherapies generally exert their cytotoxicity by activating cell death pathways secondary to excessive damage to the cell. Frequently, however, cancer cells acquire resistance to these therapies by circumventing death mechanisms. It has therefore been argued that an attempted reactivation of cancer cell death pathways and its molecular players could be a potentially effective approach to inducing cancer regression [59].

Several modes of cell death have been defined, including apoptosis, necroptosis, cell death by autophagy and mitotic catastrophe [60]. Apoptosis, as the classical type of programmed cell death, is morphologically characterized by cell shrinkage, nuclear condensation and fragmentation and membrane blebbing. Depending on the origin of the death signal, apoptosis can proceed via the extrinsic or intrinsic pathway, with the former induced by external death ligands of the tumor necrosis factor (TNF) superfamily, and the latter by intracellular stimuli, such as DNA damage and oxidative stress [61].

Extrinsic apoptosis is triggered as the death ligands bind their cognate cell-surface receptors to facilitate the formation of the death inducing signaling complex (DISC), and thus relay the death signal to either direct an activation of the effector caspases or an induction of mitochondrial outer membrane permeabilization (MOMP) [62, 63]. The intrinsic pathway is regulated by a complex interplay between various pro-apoptotic and anti-apoptotic members of the B-cell lymphoma 2 (Bcl-2) family, with a shift in favor of the pro-apoptotic members facilitating MOMP and activation of the caspase cascade. MOMP is a highly regulated process, through which various death-inducing mitochondrial proteins, including cytochrome c (cyt c), are released. In the cytoplasm, cyt c forms the apotosome complex with Apaf-1 and caspase-9, which subsequently activates the effector caspases-3, -6, and/or -7 [64, 65]. The caspase enzymes are a family of cysteinyl aspartate proteinases, which cleave a range of cellular targets to mediate cell death (Fig. 3).

The "BH3-mimetic" drug, ABT-263 (navitoclax), has been shown to inhibit the anti-apoptotic Bcl-2 family members, Bcl-X_L, Bcl-w, and Bcl-2, to thus liberate the pro-apoptotic members Bax and Bak and induce the intrinsic apoptotic pathway [66]. ABT-263 exhibited a limited benefit in a recent Phase II clinical trial for the treatment of small-cell lung cancer (SCLC) as a single agent [67]. However, with preclinical models suggesting this compound to improve tumor sensitivity to standard chemotherapies, it is perhaps reasonable

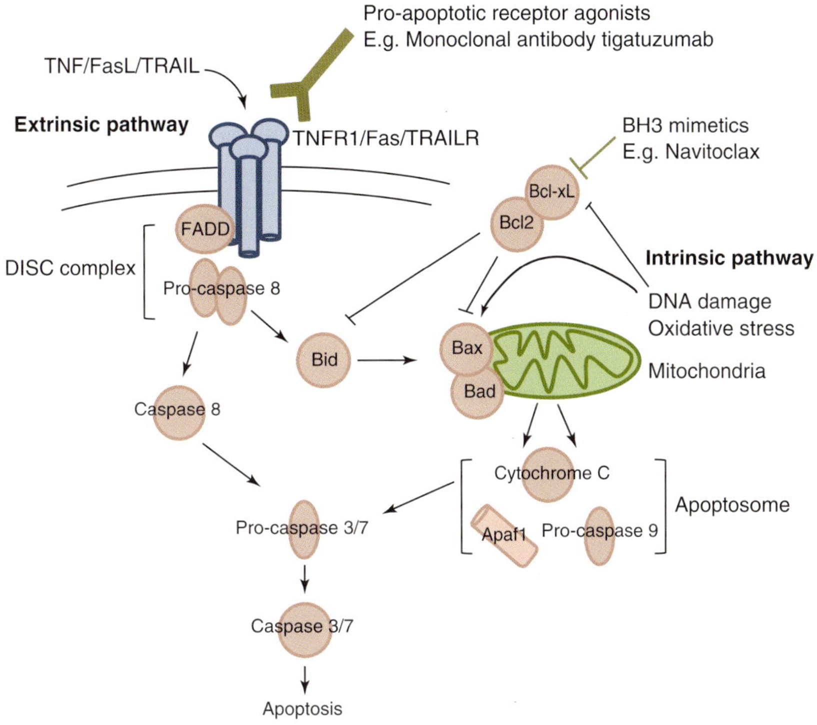

Fig. 3 The extrinsic versus the intrinsic apoptotic pathway. Binding of external factors, such as TNF, Fas ligand and TRAIL, to their corresponding death ligands activates the extrinsic apoptotic pathway. Following oligomerization of the death receptors, the death-inducing signaling complex (DISC) is formed, which propagates the apoptotic signal to direct activation of the effector caspases 3 and 7, or stimulates mitochondrial outer membrane permeabilization (MOMP). The intrinsic pathway is induced by internal insults, including DNA damage. As a result, the balance between the pro-apoptotic and anti-apoptotic members of the BCL-2 family shifts in favor of the former, to then promote MOMP, apoptosome formation, and activation of the effector caspases.

to expect its near future assessment in combination therapies [68].

The pro-apoptotic receptor agonists (PARAs) target the extrinsic apoptotic pathway, among which the TNF-related apoptosis-inducing ligand (TRAIL) system has been the most extensively studied for cancer [69]. Perhaps surprisingly, TRAIL has been found to induce apoptosis in malignant cells with minimal effect on normal cells, and unlike other TNF family members, to only weakly activate the pro-survival NF-κB pathway, hence making it a potentially attractive cancer target [66, 70]. Several agonists and monoclonal antibodies targeted to the TRAIL death receptors DR4 and DR5 are currently being investigated in clinical trials. The murine agonist monoclonal antibody to DR5, tigatuzumab, has been found to be well tolerated with reasonable antitumor activity in recent Phase I trials [71]. In contrast, a Phase II trial of the recombinant human fragment of the TRAIL ligand, Dulanermin, failed to show improvement in patients with NSCLC in combination with other therapies [72].

p53, the most frequently mutated gene in cancer, is a stress-responsive transcription factor, which becomes activated in response to various insults including ionizing radiation, ultraviolet light and many cancer chemotherapeutic drugs [73]. Activating the p53 pathway via pharmacological means has been particularly explored in tumor types where p53 is not mutated but is functionally inactivated [74]. MDM2 and MDMX are the main negative regulators of p53 [75]. The binding of MDM2 to the N-terminal domain of p53 masks its transactivation domain to suppress its activity, and furthermore ubiquitinates p53 to direct it for proteasomal degradation. Particularly in tumors where MDM2 is overexpressed or aberrantly regulated, there is the rationale to stabilize and activate p53 by disrupting its interaction with MDM2 [75]. The small-molecule MDM2 antagonists nutlin-3a and MI-219, have been shown to stabilize p53 by competing with it for interaction with the hydrophobic N-terminal cleft of MDM2 [75]. A derivative of nutlin-3a developed by Roche, RG7112, has improved pharmacological properties and enhanced potency, and has shown promising activity against various solid tumors in preclinical studies, as well as a particularly high activity in ALL xenografts. RG7112 is currently being enrolled into clinical trials [76, 77].

Despite the plethora of molecular targets which are being investigated for the direct targeting of cancer cell death mechanisms, it is perhaps disappointing to see a modest or lack of effect for many of these agents in clinical trials. Given the highly encouraging antitumor activity of many of these therapies in preclinical studies, it may perhaps be that a lack of effective biomarkers and hence poor patient stratification, as well as minimal data on effective drug combinations, could be the reason for a lack of performance observed thus far. Future studies will help to determine whether the observed *in-vitro* effects of these therapies can translate reliably into the clinic.

8
Targeting DNA Repair Mechanisms

Classical chemotherapy and radiotherapy exert their cytotoxic effects by inducing DNA damage, and as such the DNA repair mechanisms which are activated by these agents can in many instances result in resistance to these therapies. The DNA damage response (DDR) is an extensive signaling network which senses various forms of DNA damage to mediate a range of outcomes, from cell cycle arrest, to senescence and apoptosis [78]. At the core of the DDR are the two kinases ataxia telangiectasia mutated (ATM) and ataxia telangiectasia and Rad3 (ATR) -related protein, the former of which is critical for detecting and signaling double-strand DNA breaks and the latter single-strand breaks. These kinases transmit and amplify signals through a plethora of effector molecules, including the cell-cycle checkpoint kinases Chk1 and Chk2, to facilitate the DNA repair process (Fig. 4). Accordingly, various compound inhibitors of ATM and ATR are being investigated in preclinical studies as potential chemo- and radiosensitizers [79, 80].

Chk1 inhibitors have been found to be promising anticancer compounds, particularly against a background of mutated *TP53*, with several such compounds in early-phase clinical trials [81]. DNA damage insults during the G_2/M phase of the cell cycle activate Chk1 through ATR-mediated phosphorylation. Activated Chk1 subsequently phosphorylates the CDC25 family of phosphatases to suppress the activation of cyclin B/cyclin-dependent kinase 1 (CDK1). Consequently, the cells

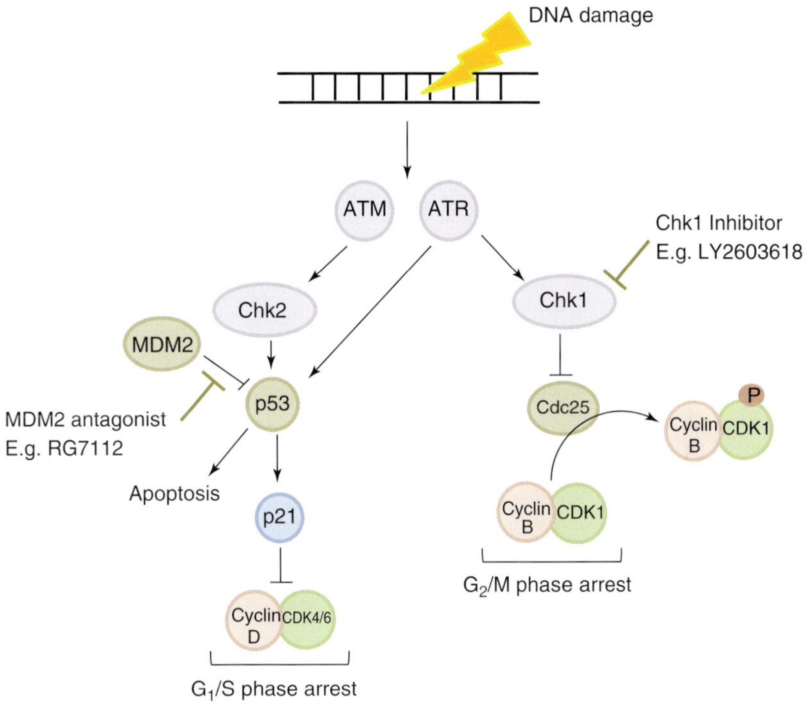

Fig. 4 ATM/ATR signaling to Chk1 and Chk2 kinases. Upon DNA damage, the checkpoint kinases ATM and ATR are activated by double-strand and single-strand breaks, respectively. ATM- and ATR-mediated phosphorylation of Chk1 and Chk2 kinases propagates the DNA damage response. During the G_1/S phase of the cell cycle, p53 becomes activated, to induce cell cycle arrest through upregulation of the CDK inhibitor p21. During G_2 phase, Chk1/2-mediated phosphorylation of cdc25 phosphatase prevents activation of the cdk1/cyclin B complex to cause G_2/M phase arrest.

arrest at G_2/M to allow DNA repair to take place before proceeding into mitosis. The inhibition of Chk1 was found to push the damaged cells into mitosis and reduce their apoptotic threshold [82]. Consequently, Chk1 inhibitors were found to increase the sensitivity of TP53 mutant cancer cells to various chemotherapeutics and radiation. The most advanced Chk1 inhibitor, LY2603618, developed by Eli Lilly, is currently in Phase II clinical trials for advanced NSCLC, in combination with pemetrexed [80].

Synthetic lethality has also offered new opportunities to target cancer cells, particularly taking advantage of situations where a tumor suppressor gene is mutated and to regain its function is not feasible. Two genes are said to be in a "synthetic lethal relationship" if mutation or loss of either alone does not affect cell viability, whereas the loss of both severely compromises cell viability and survival [83]. This relationship has been successfully demonstrated between dysfunctional BRCA1/2 and the loss of the DNA repair enzyme poly-(ADP-ribose) polymerase 1 (PARP1) [84]. Accordingly, the small-molecule PARP1 inhibitor, Olaparib, has entered Phase II clinical trials for use in patients harboring mutations in BRCA1 or 2 [85, 86]. Although this particular example has provided proof-of-concept for the potential success of synthetic lethality, the emergence of resistance can be seen as with many other forms of therapies. Particularly in this instance gain of function mutations in the *BRCA2* gene in cells treated with PARP1 inhibitors has been reported [87]. Nonetheless, synthetic lethality can potentially provide many more novel targets, and offer biochemical specificity based on the deficiencies that the cancer cell harbors. This accordingly relies on an extensive delineation of the genetic networks underlying cancer and the multitude of functions associated with the various proteins, in identifying the most promising lethal interactions.

9
Regulating the Cancer Cell Cycle

The mammalian cell cycle is a highly ordered progression through a series of phases, which culminates in the duplication of the genetic material and division of the cell (Fig. 5). A multitude of kinases are involved in the timely regulation of these processes, a number of which have been explored as potential cancer targets in light of the abnormal growth and division of tumor cells [88]. The cooperative action of the cyclin-dependent kinases (CDKs) and their associated cyclins, is required for sequential phosphorylation of the pRB and release of the E2F transcription factors, the activity of which coordinates the expression of a large body of genes required for cell-cycle progression [89, 90]. In parallel with DNA replication during S-phase, centrosome duplication begins as the cells prepare for mitosis, during which the mitotic spindle forms and chromosomal separation into daughter cells occurs [91]. Several other kinases, in addition to the mitotic CDK/cyclins, are important for the regulation of mitosis and its various aspects, including the Aurora kinases (AKs), Polo-like kinases (PLKs), and the spindle assembly checkpoint (SAC) kinases.

The CDKs are frequently deregulated and overactive in various malignancies, which by contributing to the uncontrolled proliferation of transformed cells, could rationalize their selective inhibition as an antitumor strategy [92]. Depending on their involvement in cell-cycle regulation, the CDK family can be subclassified into the interphase CDKs (CDK-1, -2, -4, and -6), with cell-cycle-regulating functions, and the cell-cycle-unrelated CDKs (CDK-5, -7, -8, -9, -10, -11), which have other specific functions, such as the regulation of transcription. The first-generation pan-CDK inhibitors, such as flavopiridol, olomoucine and R-roscovitine, generally exhibited poor clinical activity and toxicity but paved way for more selective second-generation CDK inhibitors [93]. Several such newer agents are being evaluated in clinical trials, including the aminothiazole SNS-032, which has shown promising potential in Phase I trials to radiosensitize NSCLC cells to ionizing radiation [94]. The synthetic flavone, P276-00, which has increased CDK selectivity for CDK4-cyclinD1, CDK1-cyclinB and CDK9-cyclinT1, has also shown encouraging results in Phase I–II trials for multiple myeloma, mantle cell lymphoma, and melanoma with overexpression of cyclin D [95].

Despite the great length of interest which CDK inhibitors have received as potential anticancer agents, none so far has received clinical approval. Mounting genetic evidence from mouse knockout experiments have shown a redundant role

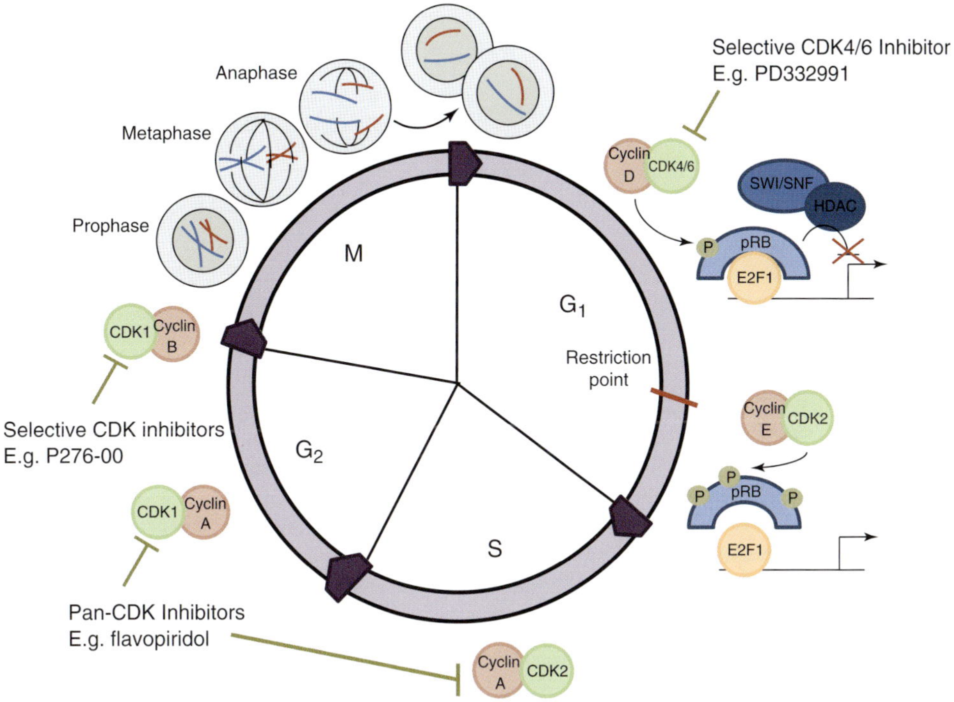

Fig. 5 The cell cycle and the interphase CDK-cyclin kinases. The four phases of the cell cycle are indicated, together with the nature and timing of the cyclin-CDK complexes involved in cell-cycle progression. The restriction point is the point at which cells become committed to cell-cycle progression and no longer require growth factors. In normal cells, the pRb binds to E2F to inactivate E2F activity. As pRb becomes phosphorylated (indicated by P) by CDK complexes, E2F is released leading to the activation of target genes required for cell-cycle progression. Normal control of pRb activity is frequently lost in human tumor cells, either through mutation in pRb or altered levels of its upstream regulators, including CDK complexes.

for the interphase CDKs in control of the cell cycle, except for certain tissues where a particular CDK may have specific functions. In fact, it was found that CDK1 can compensate for a lack of all other interphase CDKs in most mammalian cell lineages [96]. Extending these findings to tumors, it has been suggested that sensitivity to CDK inhibitors may depend on the tumor type, with particular malignancies requiring specific interphase CDKs for proliferation [97]. Accordingly, identifying tumors with a particular reliance on a given CDK, and coupling such information with the relevant biomarkers to appropriately stratify patients, could improve the therapeutic efficacy of selective CDK inhibitors. For instance, combining the selective CDK4/6 inhibitor PD332991 with endocrine therapy in advanced estrogen receptor (ER)-alpha-dependent breast cancer was found to meaningfully improve the patients' overall response, as ER-alpha was reported to activate the CDK4/E2F-dependent transcriptional program to mediate resistance to endocrine therapy [98, 99].

A number of conventional chemotherapies, namely the taxanes and the vinca alkaloids, induce cell death by interfering with the microtubules during mitosis and hence activating the mitotic checkpoint. These antimitotics are highly potent cytotoxics, which also exhibit extensive adverse side effects due to off-target interference with the vital functions of microtubules in nondividing cells [4]. Consequently, efforts have been made to identify mitotic targets which are more selective for highly proliferating cancer cells, with minimal adverse effects on resting cells. Among these, the AK family (-A, -B, and -C) of enzymes, which control chromosome alignment and segregation during mitosis, has been successfully explored as potential targets [100]. The inhibition of Aurora Kinase A (AKA) was shown to cause disorganized mitotic tubule assembly, SAC-induced mitotic arrest, and apoptosis. In contrast, the inhibition of Aurora Kinase B (AKB) was found to abrogate SAC and also to prohibit cytokinesis, leading cells to accumulate their chromosome content and eventually to undergo apoptosis as a result of an enlarged polyploid nuclei.

A number of selective as well as pan-AK inhibitors are currently undergoing clinical trials [101]. For instance, the AKB inhibitor AZD1152, currently in Phase I-II trials for acute myeloid leukemia (AML) and advanced solid tumors, has so far shown promising apoptotic activity, albeit with a dose-limiting toxicity as neutropenia [102]. Other examples include the AKA inhibitor MLN8237, currently in Phase II trials for the treatment of various platinum-resistant/refractory solid tumors [103]. Although none of these agents has yet received approval it will be interesting to see in future years how they perform clinically.

The PLKs are a family of four functionally distinct enzymes, of which PLK1 is the best-characterized member. A number of critical functions have been assigned to PLK1, including centrosome maturation, spindle formation, chromosome segregation and cytokinesis. Furthermore, PLK1 has been reported to be required for full activation of the mitotic CDK1-cyclinB at the G_2/M interphase [104]. Consequently, the inhibition of PLK1 was shown to interfere with mitosis and induce a prolonged mitotic arrest that culminated in cell death [101]. BI2536, a PLK1/PLK2 inhibitor currently in Phase II trials for AML and solid tumors [105], demonstrated potent antitumor activity in xenograft models, though it remains to be seen how well this translates into the clinical setting.

Overall, although the cell cycle and its various aspects of regulation has presented many potential targets for chemotherapy, none of these agents has – rather disappointedly – yet received approval, though this perhaps reflects the complex interplay and redundancy between many of these factors. Furthermore, information regarding the tumor type and stage specifics for a particular cell-cycle regulator is often uncoupled to the clinical evaluation of these inhibitors, raising hopes that an improved understanding of these drugs and their targets might in the future optimize their efficacy.

10
Targeting Cancer Metabolism

Cancer cells show reprogramming of their metabolic pathways in favor of "aerobic glycolysis" to support their enhanced and deregulated growth and proliferation [106]. Famously known as the Warburg effect, the metabolic switch from mitochondrial oxidative phosphorylation to glycolysis, despite the evidently lower yield of ATP, provides the malignant cell with various critical intermediates which can be fed into the anabolic pathways needed to support rapid cell proliferation [107]. Such metabolic reprogramming is associated with the altered expression, mutation and post-translational inactivation of enzymes, and/or switching to alternative enzyme isoforms. These changes collectively – in relation to the importance that metabolic adaptations serve to the survival of cancer cells – have rendered "cancer metabolism" an attractive target for cancer therapy [108]. For this purpose, both direct and indirect targets have been investigated. Indirect targets include those pathways and networks whose aberrant activity contributes to metabolic changes; such targets include protein kinase B (PKB/AKT), MAPK and hypoxia inducible factor 1 (HIF1) [108].

Direct targets range from the glucose transporter GLUT1 to the metabolic enzymes. GLUT1, a transcriptional target of HIF and AKT, is upregulated in many cancer types. The compound Phloretin has been shown to target the GLUT1 transporter, with evidence of cytotoxic activity against melanoma cell lines *in vitro* [109]. As yet, however, no compound that directly targets the GLUT1 transporter has reached the clinic. On the other hand, hexokinase – the first enzyme in the glycolytic pathway, which converts glucose to glucose-6-phosphate – has achieved more success as a target, with two inhibitors of this enzyme (lonidamine and the glucose mimetic 2-deoxyglucose) currently in clinical trials in combination with other chemotherapies for the treatment of various solid tumors [110–113].

Pyruvate kinase (PK) is the final enzyme in the glycolytic pathway, and converts phosphoenolpyruvate to pyruvate. This enzyme is regulated largely through isoform selection, with PKM1 being expressed mostly in normal tissues and PKM2 in rapidly proliferating cells, including tumors [114]. TLN-232 is a seven-amino acid peptide, which selectively targets and inhibits PKM2, and is currently in Phase II clinical trials for various solid tumors [108]. Finally, prior to being excreted from the cell, pyruvate is converted into lactate by the two enzymes lactate dehydrogenase and pyruvate dehydrogenase kinase (PDK). Of these two enzymes, PDK in particular has been investigated as a target, with the PDK1 inhibitor dichloroacetate being entered into early-phase clinical trials [115].

Tumor cells excrete the final product lactate, together with protons, into the extracellular matrix (ECM) by using the monocarboxylate transporter 4 (MCT4). Whilst this helps to maintain a normal intracellular pH it also acidifies the ECM, and this has been associated with an increased mobility and metastatic potential of cancer cells. One strategy for targeting tumor acidification is to inhibit members of the carbonic anhydrase family, particularly the hypoxia-associated carbonic anhydrase IX (CAIX). One such inhibitor, indisulam, is currently in Phase II clinical trials for the treatment of various tumors [116–118].

It is clear that an understanding of the various hallmarks of cancer, such as a deregulated metabolism, can offer many novel and promising targets. Accordingly, a surge of new therapies has been witnessed in clinical evaluations targeting various aspects of cancer metabolism. It is also anticipated that current and future metabolomics screens could be applied to identify other critical cancer pathways and, potentially, new drug targets.

11
Blocking Angiogenesis

Hypoxia, a feature of most locally advanced tumors, develops as the rapidly growing mass of tumor cells fails to become adequately perfused by the abnormal and disorganized tumor microvasculature. As a result of the oxygen diffusion limitations, hypoxic and anoxic regions develop in the tumor [119]. Tumor hypoxia is associated with resistance to therapy and poor prognosis, and coincides with major changes in the gene expression pattern that supports an altered metabolism, the suppression of apoptosis, angiogenesis, epithelial-to-mesenchymal transition, invasiveness, and metastasis [120]. The major pathway activated in hypoxic cells, which helps to coordinate this adaptive response, is that of the transcription factor HIF1 [121].

In cells, HIF1 exists as a heterodimer composed of HIF1α and HIF1β; HIF1β is constitutively expressed, whereas HIF1α is maintained at low levels in most cells under normoxic conditions. The oxygen-dependent regulation of HIF1α is critically controlled by the hydroxylation of specific residues by a series of prolyl and asparaginyl hydroxylases. The modification of two prolyl residues (P402 and P564) mediates binding of the von Hippel–Lindau (VHL) protein, which is the targeting component of an E3 ubiquitin ligase that confers degradation on HIF1 through the proteasomal pathway [122]. A second hydroxylation event occurs in the C-terminal transcriptional activation domain, where an asparagine residue (N803) becomes hydroxylated to block the interaction with the p300 coactivator, thus downregulating HIF activity [123]. Hence, collectively, as these oxygen-dependent post-translational events become rate-limiting under hypoxia, HIF1α is allowed to accumulate and become transcriptionally active.

Angiogenesis is the process of new vessel growth, which is observed during normal growth and development, including wound healing. In tumors also, and in response to hypoxia, angiogenesis is induced by the malignant cells to maintain the supply of nutrients and oxygen to the expanding mass of cells. HIF1 plays an important role in regulating angiogenesis by controlling the expression of several pro-angiogenic factors, including the vascular endothelial growth factor (VEGF) [124–126].

The process of angiogenesis is tightly regulated by the concerted action of many pro- and anti-angiogenic factors. The HIF1 target gene *VEGF* is a crucial regulator of the process, promoting the growth and survival of endothelial cells, and directing both vessel enlargement and branching. VEGF produced by tumor cells acts in a paracrine manner on the nearby endothelial cells to promote angiogenesis. There are three VEGF receptors, VEGFR1–3, of which VEGFR2 in particular seems indispensable for endothelial cell growth and survival. The binding of VEGF to its cognate receptor results in receptor homo-/hetero-dimerization and activating autophosphorylation. The VEGFRs exert their pro-survival and pro-migratory effects by activating downstream MAPK-ERK, PI3K-AKT, T-cell-specific adaptor (TSAd) and p38 MAPK signaling cascades [58, 124, 127]. There are several antitumorigenic therapies which exploit VEGF as a therapeutic target; these include various monoclonal antibodies directed against VEGF ligand(s) or VEGF receptor(s), as well as small-molecule inhibitors of VEGFR tyrosine kinase activity [128].

The therapeutic antibody, bevacizumab (Avastin®), designed to inhibit VEGF activity, provided a 50% increase in patient survival chance compared to patients receiving chemotherapy alone. Bevacizumab has been approved for use in metastatic colon cancer, glioblastoma, advanced non-squamous NSCLC and metastatic kidney cancer [129, 130]. An alternative fully humanized monoclonal antibody, Ramucirumab® [131, 132], directed against VEGFR-2, is being evaluated in late-phase clinical trials for various solid tumors. The small molecules sunitinib (Sutent®), sorafenib (Nexavar®) and pazobanib (Votrient®) are multi-kinase inhibitors which, through targeting VEGFR and PDGFR, also inhibit angiogenesis [133]. These drugs have received regulatory approval for various malignancies, particularly renal cell carcinoma [134].

The unexpected benefit which has emerged from the use of anti-angiogenic therapies is largely indebted to their potential to normalize the tumor vasculature and therefore improve the delivery of other chemotherapeutic agents. In support of this, it is important to emphasize that the anti-angiogenic therapies currently available are poor treatments as monotherapy, yet in combination with other chemotherapies they show markedly improved outcomes [124]. However, despite the success of anti-VEGF therapies – particularly in combination regimens – they are not devoid of problems. In addition to the issue of acquired tumor resistance through redundant switching to other angiogenic factors other than VEGF [135], there are concerns that anti-VEGF therapies, through forcing vessel regression, may exacerbate tumor hypoxia and therefore select for more aggressive and metastatic tumor phenotypes [136]. Hence, future therapies aimed at tumor angiogenesis need to be comprehensively assessed not only in terms of their benefits but also of any potentially serious adverse effects.

12 Regulating Protein Turnover

Increasing evidence suggests that drugs which influence protein degradation may prove to be beneficial in cancer treatment [137]. Protein degradation plays a crucial role in regulating cell viability, not only in the control of damaged proteins but also in controlling key regulatory processes. The turnover of many cell-cycle regulatory proteins is required for normal cell-cycle control, for example, the CDK inhibitor p27, the degradation of which is enhanced in certain tumor cells [138]. Similarly, signaling to the NF-κB transcription factor is dependent upon protein turnover of its specific inhibitor IκB, allowing NF-κB to activate a variety of genes including cytokines (such as TNF) and anti-apoptotic proteins [139].

The proteasome plays a central role in regulating protein turnover. The 26S proteasome is a large multiprotein complex composed of a 20S catalytic core together with two 19S regulatory subunits [140]. This complex degrades proteins covalently marked with a polyubiquitin chain on their lysine residues to short peptides, releasing free ubiquitin in the process (Fig. 6). It has become increasingly clear that compounds which block proteasome function possess potential as anticancer agents [141]. The first compounds to be identified exhibited inappropriate pharmacokinetic properties, but ultimately this led to a generation of compounds with improved properties, including bortezomib (Velcade®). Bortezomib has been shown to be preferentially toxic to cancer cells compared to normal cells, with activity across a wide range of tumor types. For instance, myeloma cells undergo apoptosis when treated with bortezomib; this is in contrast to normal bone marrow stromal cells, which exhibit

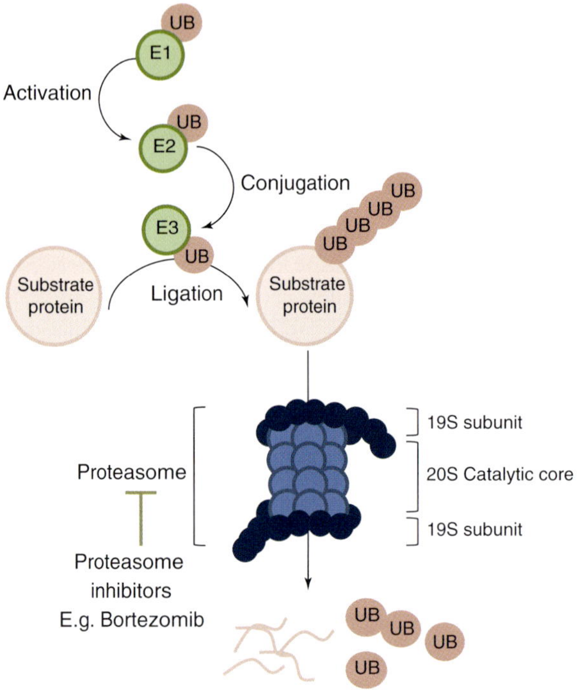

Fig. 6 Protein degradation by the ubiquitin proteasome pathway. Target proteins for degradation are ubiquitinated (UB) at lysine residues through a process involving a ubiquitin-activating enzyme (E1), the transfer of ubiquitin to a conjugating enzyme (E2), which is then transferred to a target protein by an E3 protein ligase. Polyubiquitinated proteins are then recognized and degraded by the 26S proteasome.

a much reduced sensitivity profile [139, 142].

Various mechanisms have been proposed for the mechanism of action of bortezomib, including inhibition of NF-κB activation, defective apoptotic regulation, and the accumulation of misfolded, unfolded and polyubiquitinated proteins. In particular, the latter is thought to overwhelm the clearance of proteins by the endoplasmic reticulum-associated degradation (ERAD) pathway, and therefore induce further endoplasmic reticulum stress and cell death [143]. Bortezomib gained an accelerated FDA approval based on encouraging results obtained from the Phase III VISTA trial, which showed that the addition of bortezomib to the current standard-of-care treatment could improve overall outcome in patients [144]. Other proteasome inhibitors have since entered trials with improved potency and safety profiles, namely Carfilzomib® and NPI-0052 [145].

NEDD8 belongs to the larger family of small ubiquitin-like proteins (UBLs), and shares 60% homology to ubiquitin. Similar to ubiquitin, NEDD8 engages in a cascade involving three dedicated NEDDylation enzymes, E1, E2, and E3, to undergo activation, conjugation and, finally, ligation onto its target substrate [146]. The main substrates identified for NEDD8 are the cullins, which are a component of the multisubunit cullin-RING ligases

(CRLs), a subset of ubiquitin E3 ligases [147, 148]. Cullin NEDDylation has been shown to be an important regulatory event, required for the critical conformational changes that enable recruitment of the ubiquitin E2 enzyme, and hence overall function. CRLs have been found to regulate the degradation of many proteins with cancer-related functions, such as p27, p53, phosphorylated inhibitor of NF-κB alpha (pIκBα), cyclin E and mTOR, among others. This in turn has implicated the NEDDylation pathway as a potential target in cancer, as the deregulation of this process has been shown to contribute to carcinogenesis by promoting the aberrant degradation of various proteins with tumor-suppressive functions [149, 150]. In further support of this, the transcription factor E2F1, endowed with both cell cycle-regulating and apoptotic activities, has been shown to be a direct target for NEDDylation. The outcome of this was found to be increased E2F1 degradation and a correspondingly reduced E2F1-mediated apoptosis [151].

MLN4924 is a small-molecule inhibitor of the NEDD8 activating enzyme (NAE), the proximal regulator of the NEDDylation cascade. Preclinical studies found that MLN4924 could induce apoptosis in a range of cancer cells and effectively disrupted CRL-mediated protein degradation [152]. These studies have highlighted the multiple mechanisms of cell death induced by MLN4924, including the stabilization of CDT1 which promotes DNA rereplication and hence activation of the DDR, induction of apoptotic pathways, and the inhibition of NF-κB and elevated oxidative stress [149]. Given the encouraging preclinical results obtained, MLN4924 has been enrolled into early-phase clinical trials, with initial reports describing promising activity against AML [153].

13
Regulating Hsp90

The expression of the heat-shock proteins (Hsps), a family of highly conserved members, is induced by various environmental and physiological stresses to aid cell survival upon these insults. Heat shock protein 90 (Hsp90), an ATP-dependent member of the Hsp family, is a molecular chaperone that regulates the activity of a variety of cellular proteins, enabling protein folding and preventing the denaturation – and perhaps aggregation – of misfolded proteins [154]. Hsp90 is gaining recognition as a cancer target, in part because it is overexpressed in a variety of tumor cells and associates with oncoproteins, including the tyrosine kinase v-Src, the serine/threonine kinase Raf1, HIF1, and mutated p53 [155]. In light of the cytoprotective nature of Hsp90, and its intricate involvement in regulating many cancer-associated client proteins, it is perceived that a disruption of Hsp90 could in effect interfere with many oncogenic signaling pathways and hence cancer progression [156].

Hsp90 contains three domains: an N-terminal adenine binding domain; a middle linker region for binding to co-chaperones and client proteins; and a C-terminal dimerization domain. Hsp90 can adopt several distinct conformational states, particularly depending on the nucleotide status and bound co-chaperones. In particular, the binding of ATP and its subsequent hydrolysis has been found to promote Hsp90 dimerization, and the adoption of a closed conformation that favors client protein recruitment. Co-chaperones also appear to play an important role in this dynamic structural cycle. More than 20 different co-chaperones have been identified, with a diverse involvement in the activation or inhibition of Hsp90 ATPase activity as well as the recruitment of different client proteins [154, 157, 158].

A small-molecule inhibitor of Hsp90, geldanamycin, blocks the interaction of Hsp90 with client proteins by interacting with the hydrophobic ATP-binding pocket in the N-terminal region and inhibiting the Hsp90 ATPase activity [159]. This, in turn, has been found to disrupt the Hsp90 chaperone complex, leading to the destabilization, ubiquitination and degradation of several of its client proteins, such as Src and ERK. Geldanamycin, and other similar compounds that target the ATPase activity of Hsp90, have demonstrated potent and selective cytotoxic activity, particularly as the Hsp90 complex isolated from tumor cells has been shown to bind these inhibitors with a much higher affinity than the complex isolated from normal cells [156].

In preclinical studies, Hsp90 inhibitors have shown promising activity, with several natural and synthetic compounds in development. NVP-AUY922, a derivative of the natural antifungal antibiotic Radicicol®, binds to the ATP-binding pocket of Hsp90 with greater affinity than does ATP, and therefore impairs Hsp90 function. *In vitro*, the compound has shown potent antitumor activity and is currently being evaluated in Phase II clinical trials [160–162]. While geldanamycin exhibits levels of hepatotoxicity that are too high for clinical application, its derivative, 17-allylaminogeldanamycin (17-AAG), exhibits improved efficacy and has now reached Phase II clinical trials [163, 164]. Despite its potency, 17-AAG suffers from poor solubility and a lack of oral bioavailability; however, the *N,N*-dimethylethylamino analog of 17-AAG (17-DMAG) has demonstrated an improved pharmacokinetic profile and has entered Phase I clinical trials for various solid and hematological malignancies [156].

An interesting property of Hsp90 inhibitors observed in preclinical studies is the induction of cytostasis or apoptosis, depending on the cell type. This may reflect the nature of the Hsp90 chaperone complex and the variation in target client proteins present in cells. Some evidence suggests that the integrity of pRb may dictate sensitivity to Hsp90 inhibition, as cells lacking pRb appear to be quite sensitive to inhibitor-induced apoptosis [165]. In the clinical setting, if cytostasis were to be the primary outcome of treating a patient with an Hsp90 inhibitor, then sustained chronic administration of the drug may not be possible because of the associated toxicity profile. Conversely, if Hsp90 inhibition causes apoptosis upon short-term administration, then efficacy with limited toxicity may be achievable. This situation may be solved eventually by profiling the patient's tumor for the nature and expression level of Hsp90 client proteins. Combined with further information on the appropriate clinical schedule for Hsp90 inhibitors, it may be possible to identify and treat those tumors that will preferentially enter apoptosis upon inhibition of Hsp90 activity. This general principle could be important for other agents, such as HDAC inhibitors.

14
Conclusions and Perspectives

Today, subtle hints are beginning to be seen that the exploitation of mechanism-based approaches will deliver anticancer medicines which have improved benefits over many of the currently available chemotherapeutics. An expanding knowledge of the molecular and genetic defects that underlie cancer has led to the identification of many pathways and targets which can be – and have been – exploited to develop potential mechanism-based anti-cancer medicines. The examples mentioned in this short review, ranging from Gleevec and its derivatives to Velcade, bear witness to the clinical efficacy of such therapies, and encourage further research in the field of drug discovery to complement and parallel the advances made so far in understanding cancer.

Despite the success of many of these mechanism-based therapies, emergent and inherent resistance to these medicines seems to be a recurrent theme. In some cases, such as Gleevec, a molecular understanding of the mechanism of resistance has facilitated the design of next-generation inhibitors capable of targeting resistant populations. This provides a reminder of the importance to understand not only the detailed mode of action of these compounds but also the mechanisms of resistance, as this could sustain further research into the identification of alternative therapies. Furthermore, if drug resistance does appear to be a clinical hurdle it may be possible to modulate its progression by altering drug concentration and exposure via a combination therapy approach.

The genetic complexity of cancer has meant that many of the targeted therapies discussed in this chapter are inadequate as monotherapy, and that a combination of multiple agents might be necessary for a sustained antitumorigenic response [166]. The use of a rational cocktail of molecularly targeted agents could not only improve tumor inhibition through the complementary targeting of various pathways critical for cancer growth, but may also reduce the emergence of resistance. However, most often the combinations to be used are laboriously drawn out from relevant clinical trials, with no systematic methodology currently in place to establish the most appropriate regimen. The combined efforts of research groups in academia and industry, along with the experiences of clinicians, will prove crucial in future years to overcoming these challenges.

One inter-related question concerns the scheduling regime for administering the drug, particularly in combination therapies, which will in turn reflect knowledge of the critical targets through which the inhibitor regulates cancer cell growth, and the expression of these targets in a particular cancer type. In this respect, the different properties of Hsp90 inhibitors in apoptosis and cell cycle arrest have been discussed, and how these properties might impact on the clinical regimen, with patient stratification perhaps required for maximum clinical benefit.

Clearly, whilst many questions remain to be answered, the present era is nevertheless extremely exciting in terms of cancer drug discovery, with specific, newly developed drugs beginning to replace conventional cytotoxic approaches. The views and anticipations outlined in this chapter are shared by many others, and reflect the gathering momentum that the translation of knowledge of cancer cells will in the future yield improved and more efficacious drugs for the treatment of cancer patients.

References

1. Cancer Research UK. Available at: *http://www.cancerresearchuk.org/cancer-info/cancerstats/* (accessed 11 August 2013).
2. World Health Organization. Available at: *http://www.who.int/mediacentre/factsheets/fs297/en/* (accessed 11 August 2013).
3. Cancer Research UK. Available at: *http://www.cancerresearchuk.org/cancer-info/cancerstats/keyfacts/Allcancers combined/* (accessed 11 August 2013).
4. Baba, A.I. and Câtoi, C. (2007) Principles of Anticancer Therapy, in: *Comparative Oncology*, The Publishing House of the Romanian Academy, Bucharest, Ch. 19.

5. Cancer Research UK. Available at: *http://www.cancerresearchuk.org/cancer-info/cancerstats/survival/common-cancers/* (accessed 11 August 2013).
6. Chabner, B.A. and Roberts, T.G. (2005) Timeline: chemotherapy and the war on cancer. *Nat. Rev. Cancer*, **5** (1), 65–72.
7. La Thangue, N.B. and Kerr, D.J. (2011) Predictive biomarkers: a paradigm shift towards personalized cancer medicine. *Nat. Rev. Clin. Oncol.*, **8** (10), 587–596.
8. de Gonzalez, C.D., Clarke, P.A., Al-Lazikani, B., Workman, P. (2013) Personalized cancer medicine: molecular diagnostics, predictive biomarkers, and drug resistance. *Clin. Pharmacol. Ther.*, **93** (3), 252–259.
9. Kumar, S., Mohan, A., and Guleria, R. (2006) Biomarkers in cancer screening, research and detection: present and future: a review. *Biomarkers*, **11** (5), 385–405.
10. Druker, B.J. and Lydon, N.B. (2000) Lessons learned from the development of an abl tyrosine kinase inhibitor for chronic myelogenous leukemia. *J. Clin. Invest.*, **105** (1), 3–7.
11. Deininger, M.W., Goldman, J.M., and Melo, J.V. (2000) The molecular biology of chronic myeloid leukemia. *Blood*, **96** (10), 3343–3356.
12. Advani, A.S. and Pendergast, A.M. (2002) Bcr-Abl variants: biological and clinical aspects. *Leuk. Res.*, **26** (8), 713–720.
13. Capdeville, R., Buchdunger, E., Zimmermann, J., Matter, A. (2002) Glivec (STI571, imatinib), a rationally developed, targeted anticancer drug. *Nat. Rev. Drug Discov.*, **1** (7), 493–502.
14. Siehl, J. and Thiel, E. (2007) C-kit, GIST, and imatinib. *Recent Results Cancer Res.*, **176**, 145–151.
15. Gorre, M.E., Mohammed, M., Ellwood, K., Hsu, N. *et al.* (2001) Clinical resistance to STI-571 cancer therapy caused by BCR-ABL gene mutation or amplification. *Science*, **293** (5531), 876–880.
16. Mian, A.A., Schüll, M., Zhao, Z., Oancea, C. *et al.* (2009) The gatekeeper mutation T315I confers resistance against small molecules by increasing or restoring the ABL-kinase activity accompanied by aberrant transphosphorylation of endogenous BCR, even in loss-of-function mutants of BCR/ABL. *Leukemia*, **23** (9), 1614–1621.
17. Weisberg, E., Manley, P.W., Breitenstein, W., Brüggen, W. *et al.* (2005) Characterization of AMN107, a selective inhibitor of native and mutant Bcr-Abl. *Cancer Cell*, **7** (2), 129–141.
18. Shah, N.P., Lee, F.Y., Luo, R., Jiang, Y. *et al.* (2006) Dasatinib (BMS-354825) inhibits KITD816V, an imatinib-resistant activating mutation that triggers neoplastic growth in most patients with systemic mastocytosis. *Blood*, **108** (1), 286–291.
19. Weisberg, E., Manley, P., Mestan, J., Cowan-Jacob, S. *et al.* (2006) AMN107 (nilotinib): a novel and selective inhibitor of BCR-ABL. *Br. J. Cancer*, **94** (12), 1765–1769.
20. Talpaz, M., Shah, N.P., Kantarijan, H., Donato, N. *et al.* (2006) Dasatinib in imatinib-resistant Philadelphia chromosome-positive leukemias. *N. Engl. J. Med.*, **354** (24), 2531–2541.
21. Zhang, J., Yang, P.L., and Gray, N.S. (2009) Targeting cancer with small molecule kinase inhibitors. *Nat. Rev. Cancer*, **9** (1), 28–39.
22. Barouch-Bentov, R. and Sauer, K. (2011) Mechanisms of drug resistance in kinases. *Expert Opin. Investig. Drugs*, **20** (2), 153–208.
23. Hynes, N.E. and MacDonald, G. (2009) ErbB receptors and signaling pathways in cancer. *Curr. Opin. Cell Biol.*, **21** (2), 177–184.
24. Zhang, H., Berezow, A., Wang, Q., Zhang, G. *et al.* (2007) ErbB receptors: from oncogenes to targeted cancer therapies. *J. Clin. Invest.*, **117** (8), 2051–2058.
25. Ciardiello, F., and Tortora, G. (2001) A novel approach in the treatment of cancer: targeting the epidermal growth factor receptor. *Clin. Cancer Res.*, **7** (10), 2958–2970.
26. Macmillan. Available at: *http://www.macmillan.org.uk/Cancerinformation* (accessed 11 August 2013).
27. Incorvati, J.A., Shah, S., Mu, Y., and Lu, J. (2013) Targeted therapy for HER2-positive breast cancer. *J. Hematol. Oncol.*, **6**, 38.
28. Baselga, J. (2001) The EGFR as a target for anticancer therapy – focus on cetuximab. *Eur. J. Cancer*, **37** (Suppl. 4, S16–S22.
29. Lièvre, A., Bachet, J.B., Le Corre, D., Boige, V. *et al.* (2006) KRAS mutation status is predictive of response to cetuximab therapy

in colorectal cancer. *Cancer Res.*, **66** (8), 3992–3995.

30. Pohlmann, P.R., Mayer, I.A., and Mernaugh, R. (2009) Resistance to trastuzumab in breast cancer. *Clin. Cancer Res.*, **15** (24), 7479–7491.
31. Konecny, G.E., Pegram, M.D., Venkatesan, N., Finn, R. *et al.* (2006) Activity of the dual kinase inhibitor lapatinib (GW572016) against HER-2-overexpressing and trastuzumab-treated breast cancer cells. *Cancer Res.*, **66** (3), 1630–1639.
32. Hirsch, F.R., Varella-Garcia, M., Bunn, P.A., Jr, Di Maria, M.V. *et al.* (2003) Epidermal growth factor receptor in non-small-cell lung carcinomas: correlation between gene copy number and protein expression and impact on prognosis. *J. Clin. Oncol.*, **21** (20), 3798–3807.
33. Manegold, C. (2003) Gefitinib (Iressa, ZD 1839) for non-small cell lung cancer (NSCLC): recent results and further strategies. *Adv. Exp. Med. Biol.*, **532**, 247–252.
34. Maemondo, M., Inoue, A., Kobayashi, K., Sugawara, S. *et al.* (2010) Gefitinib or chemotherapy for non-small-cell lung cancer with mutated EGFR. *N. Engl. J. Med.*, **362** (25), 2380–2388.
35. Mitsudomi, T., Morita, S., Yatabe, Y., Negoro, S. *et al.* (2010) Gefitinib versus cisplatin plus docetaxel in patients with non-small-cell lung cancer harbouring mutations of the epidermal growth factor receptor (WJTOG3405): an open label, randomised phase 3 trial. *Lancet Oncol.*, **11** (2), 121–128.
36. Davies, H., Bignell, G.R., Cox, C., Stephens, P. *et al.* (2002) Mutations of the BRAF gene in human cancer. *Nature*, **417** (6892), 949–954.
37. Chapman, P.B., Hauschild, A., Robert, C., Haanen, J.B., *et al.*, Improved survival with vemurafenib in melanoma with BRAF V600E mutation. *N. Engl. J. Med.*, 2011. **364**(26): 2507–2516.
38. Heidorn, S.J., Milagre, C., Whittaker, S., Noury, A. *et al.* (2010) Kinase-dead BRAF and oncogenic RAS cooperate to drive tumor progression through CRAF. *Cell*, **140** (2), 209–221.
39. Chiarle, R., Voena, C., Ambrogio, C., Piva, R. *et al.* (2008) The anaplastic lymphoma kinase in the pathogenesis of cancer. *Nat. Rev. Cancer*, **8** (1), 11–23.
40. Soda, M., Choi, Y.L., Enomoto, M., Takada, S. *et al.* (2007) Identification of the transforming EML4-ALK fusion gene in non-small-cell lung cancer. *Nature*, **448** (7153), 561–566.
41. Shaw, A.T., Yeap, B.Y., Solomon, B.J., Riety, G.J. *et al.* (2011) Effect of crizotinib on overall survival in patients with advanced non-small-cell lung cancer harbouring ALK gene rearrangement: a retrospective analysis. *Lancet Oncol.*, **12** (11), 1004–1012.
42. Rius, M. and Lyko, F. (2012) Epigenetic cancer therapy: rationales, targets and drugs. *Oncogene*, **31** (39), 4257–4265.
43. Crea, F. (2012) Histone code, human growth and cancer. *Oncotarget*, **3** (1), 1–2.
44. Ellis, L., Atadja, P.W., and Johnstone, R.W. (2009) Epigenetics in cancer: targeting chromatin modifications. *Mol. Cancer Ther.*, **8** (6), 1409–1420.
45. Dawson, M.A. and Kouzarides, T. (2012) Cancer epigenetics: from mechanism to therapy. *Cell*, **150** (1), 12–27.
46. Chan, H.M. and La Thangue, N.B. (2001) p300/CBP proteins: HATs for transcriptional bridges and scaffolds. *J. Cell Sci.*, **114** (Pt 13), 2363–2373.
47. Gallinari, P., Di Marco, S., Jones, P., Pallaoro, M. *et al.* (2007) HDACs, histone deacetylation and gene transcription: from molecular biology to cancer therapeutics. *Cell Res.*, **17** (3), 195–211.
48. Chellappan, S.P., Hiebert, S., Mudryi, M., Horowitz, J.M. *et al.* (1991) The E2F transcription factor is a cellular target for the RB protein. *Cell*, **65** (6), 1053–1061.
49. Classon, M. and Harlow, E. (2002) The retinoblastoma tumour suppressor in development and cancer. *Nat. Rev. Cancer*, **2** (12), 910–917.
50. Lin, R.J., Nagy, L., Inoue, S., Shao, W. *et al.* (1998) Role of the histone deacetylase complex in acute promyelocytic leukaemia. *Nature*, **391** (6669), 811–814.
51. Khan, O. and La Thangue, N.B. (2012) HDAC inhibitors in cancer biology: emerging mechanisms and clinical applications. *Immunol. Cell Biol.*, **90** (1), 85–94.
52. Kim, H.J. and Bae, S.C. (2011) Histone deacetylase inhibitors: molecular mechanisms of action and clinical trials as anti-cancer drugs. *Am. J. Trans. Res.*, **3** (2), 166–179.

53. Molife, L.R., and de Bono, J.S. (2011) Belinostat: clinical applications in solid tumors and lymphoma. *Expert Opin. Investig. Drugs*, **20** (12), 1723–1732.
54. Plumb, J.A., Finn, P.W., Williams, R.J., Bandara, M.J. *et al.* (2003) Pharmacodynamic response and inhibition of growth of human tumor xenografts by the novel histone deacetylase inhibitor PXD101. *Mol. Cancer Ther.*, **2** (8), 721–728.
55. Issa, J.P. (2007) DNA methylation as a therapeutic target in cancer. *Clin. Cancer Res.*, **13** (6), 1634–1637.
56. Barbieri, I., Cannizzaro, E., and Dawson, M.A. (2013) Bromodomains as therapeutic targets in cancer. *Brief. Funct. Genomics*, **12** (3), 219–230.
57. Filippakopoulos, P., Qi, J., Picaud, S., Shen, Y. *et al.* (2010) Selective inhibition of BET bromodomains. *Nature*, **468** (7327), 1067–1073.
58. Hanahan, D. and Weinberg, R.A. (2011) Hallmarks of cancer: the next generation. *Cell*, **144** (5), 646–674.
59. Reed, J.C. (2006) Drug insight: cancer therapy strategies based on restoration of endogenous cell death mechanisms. *Nat. Clin. Pract. Oncol.*, **3** (7), 388–398.
60. Degterev, A. and Yuan, J. (2008) Expansion and evolution of cell death programmes. *Nat. Rev. Mol. Cell Biol.*, **9** (5), 378–390.
61. Fulda, S. and Debatin, K.M. (2004) Targeting apoptosis pathways in cancer therapy. *Curr. Cancer Drug Targets*, **4** (7), 569–576.
62. Ashkenazi, A. (2008) Targeting the extrinsic apoptosis pathway in cancer. *Cytokine Growth Factor Rev.*, **19** (3-4), 325–331.
63. Verbrugge, I., Johnstone, R.W., and Smyth, M.J. (2010) SnapShot: extrinsic apoptosis pathways. *Cell*, **143** (7), 1192, 1192.e1–e2.
64. Fulda, S. and Debatin, K.M. (2006) Extrinsic versus intrinsic apoptosis pathways in anticancer chemotherapy. *Oncogene*, **25** (34), 4798–4811.
65. Jendrossek, V. (2012) The intrinsic apoptosis pathways as a target in anticancer therapy. *Curr. Pharm. Biotechnol.*, **13** (8), 1426–1438.
66. Long, J.S. and Ryan, K.M. (2012) New frontiers in promoting tumour cell death: targeting apoptosis, necroptosis and autophagy. *Oncogene*, **31** (49), 5045–5060.
67. Rudin, C.M., Hann, C.L., Garon, E.B., Ribeiro de Oliveira, M. *et al.* (2012) Phase II study of single-agent navitoclax (ABT-263) and biomarker correlates in patients with relapsed small cell lung cancer. *Clin. Cancer Res.*, **18** (11), 3163–3169.
68. Wong, M., Tan, N., Zha, J., Peale, F.V. *et al.* (2012) Navitoclax (ABT-263) reduces Bcl-x(L)-mediated chemoresistance in ovarian cancer models. *Mol. Cancer Ther.*, **11** (4), 1026–1035.
69. Ashkenazi, A. (2008) Directing cancer cells to self-destruct with pro-apoptotic receptor agonists. *Nat. Rev. Drug Discov.*, **7** (12), 1001–1012.
70. Kruyt, F.A. (2008) TRAIL and cancer therapy. *Cancer Lett.*, **263** (1), 14–25.
71. Forero-Torres, A., Shah, J., Wood, T., Posey, J. *et al.* (2010) Phase I trial of weekly tigatuzumab, an agonistic humanized monoclonal antibody targeting death receptor 5 (DR5). *Cancer Biother. Radiopharm.*, **25** (1), 13–19.
72. Soria, J.C., Márk, Z., Zatloukal, P., Szima, B. *et al.* (2011) Randomized phase II study of dulanermin in combination with paclitaxel, carboplatin, and bevacizumab in advanced non-small-cell lung cancer. *J. Clin. Oncol.*, **29** (33), 4442–4451.
73. Kruse, J.P. and Gu, W. (2009) Modes of p53 regulation. *Cell*, **137** (4), 609–622.
74. Brown, C.J., Lain, S., Verma, C.S., Fersht, A.R. *et al.* (2009) Awakening guardian angels: drugging the p53 pathway. *Nat. Rev. Cancer*, **9** (12), 862–873.
75. Wade, M., Li, Y.C., and Wahl, G.M. (2013) MDM2, MDMX and p53 in oncogenesis and cancer therapy. *Nat. Rev. Cancer*, **13** (2), 83–96.
76. Tovar, C., Graves, B., Packman, K., Filipovic, Z. *et al.* (2013) MDM2 small-molecule antagonist RG7112 activates p53 signaling and regresses human tumors in preclinical cancer models. *Cancer Res.*, **73** (8), 2587–2597.
77. Carol, H., Raynolds, C.P., Kang, M.H., Keir, S.T. *et al.* (2013) Initial testing of the MDM2 inhibitor RG7112 by the pediatric preclinical testing program. *Pediatr. Blood Cancer*, **60** (4), 633–641.
78. Jackson, S.P. and Bartek, J. (2009) The DNA-damage response in human biology and disease. *Nature*, **461** (7267), 1071–1078.
79. Fokas, E., Prevo, R., Hammond, E.M., Brunner, T.B. *et al.* (2013) Targeting ATR

in DNA damage response and cancer therapeutics. *Cancer Treat. Rev.*, **8**, 1451–1459.
80. Furgason, J.M. and Bahassi, el M. (2013) Targeting DNA repair mechanisms in cancer. *Pharmacol. Ther.*, **137** (3), 298–308.
81. Lord, C.J. and Ashworth, A. (2012) The DNA damage response and cancer therapy. *Nature*, **481** (7381), 287–294.
82. Ma, C.X., Janetka, J.W. and Piwnica-Worms, H. (2011) Death by releasing the breaks: CHK1 inhibitors as cancer therapeutics. *Trends Mol. Med.*, **17** (2), 88–96.
83. Kaelin, W.G. (2005) The concept of synthetic lethality in the context of anticancer therapy. *Nat. Rev. Cancer*, **5** (9), 689–698.
84. Turner, N.C., Lord, C.J., Iorns, E., Brough, R. *et al.* (2008) A synthetic lethal siRNA screen identifying genes mediating sensitivity to a PARP inhibitor. *EMBO J.*, **27** (9), 1368–1377.
85. Hutchinson, L. (2010) Targeted therapies: PARP inhibitor olaparib is safe and effective in patients with BRCA1 and BRCA2 mutations. *Nat. Rev. Clin. Oncol.*, **7** (10), 549.
86. Tutt, A., Robson, M., Garber, J.E., Domchek, S.M. *et al.* (2010) Oral poly(ADP-ribose) polymerase inhibitor olaparib in patients with BRCA1 or BRCA2 mutations and advanced breast cancer: a proof-of-concept trial. *Lancet*, **376** (9737), 235–244.
87. Barber, L.J., Sandhu, S., Chen, L., Campbell, J. *et al.* (2013) Secondary mutations in BRCA2 associated with clinical resistance to a PARP inhibitor. *J. Pathol.*, **229** (3), 422–429.
88. de Cárcer, G., Pérez de Castro, I., and Malumbres, M. (2007) Targeting cell cycle kinases for cancer therapy. *Curr. Med. Chem.*, **14** (9), 969–985.
89. DeGregori, J. and Johnson, D.G. (2006) Distinct and overlapping roles for E2F family members in transcription, proliferation and apoptosis, *Curr. Mol. Med.*, **6** (7), 739–748.
90. Stevaux, O. and Dyson, N.J. (2002) A revised picture of the E2F transcriptional network and RB function. *Curr. Opin. Cell Biol.*, **14** (6), 684–691.
91. Satyanarayana, A. and Kaldis, P. (2009) Mammalian cell-cycle regulation: several Cdks, numerous cyclins and diverse com pensatory mechanisms. *Oncogene*, **28** (33), 2925–2939.
92. Aarts, M., Linardopoulos, S. and Turner, N.C. (2013) Tumour selective targeting of cell cycle kinases for cancer treatment. *Curr. Opin. Pharmacol.*, **13**, 529–535.
93. Dickson, M.A. and Schwartz, G.K. (2009) Development of cell-cycle inhibitors for cancer therapy. *Curr. Oncol.*, **16** (2), 36–43.
94. Kodym, E., Kodym, R., Reis, A.E., Habib, A.A. *et al.* (2009) The small-molecule CDK inhibitor, SNS-032, enhances cellular radiosensitivity in quiescent and hypoxic non-small cell lung cancer cells. *Lung Cancer*, **66** (1), 37–47.
95. Shirsath, N.P., Manohar, S.M., and Joshi, K.S. (2012) P276-00, a cyclin-dependent kinase inhibitor, modulates cell cycle and induces apoptosis in vitro and in vivo in mantle cell lymphoma cell lines. *Mol. Cancer*, **11**, 77.
96. Malumbres, M. and Barbacid, M. (2009) Cell cycle, CDKs and cancer: a changing paradigm. *Nat. Rev. Cancer*, **9** (3), 153–166.
97. Lapenna, S. and Giordano, A. (2009) Cell cycle kinases as therapeutic targets for cancer. *Nat. Rev. Drug Discov.*, **8** (7), 547–566.
98. Finn, R.S., Dering, J., Conklin, D., Kalous, O. *et al.* (2009) PD 0332991, a selective cyclin D kinase 4/6 inhibitor, preferentially inhibits proliferation of luminal estrogen receptor-positive human breast cancer cell lines in vitro. *Breast Cancer Res.*, **11** (5), R77.
99. Lange, C.A. and Yee, D. (2011) Killing the second messenger: targeting loss of cell cycle control in endocrine-resistant breast cancer. *Endocr. Relat. Cancer*, **18** (4), C19–C24.
100. Kitzen, J.J., de Jonge, M.J., and Verweij, J. (2010) Aurora kinase inhibitors. *Crit. Rev. Oncol. Hematol.*, **73** (2), 99–110.
101. Pérez de Castro, I., de Cárcer, G., Montoya, G., Malumbres, M. (2008) Emerging cancer therapeutic opportunities by inhibiting mitotic kinases. *Curr. Opin. Pharmacol.*, **8** (4), 375–383.
102. Boss, D.S., Witteveen, P.O., van der Sar, J., Lolkema, M.P. *et al.* (2011) Clinical evaluation of AZD1152, an i.v. inhibitor of Aurora B kinase, in patients with solid malignant tumors. *Ann. Oncol.*, **22** (2), 431–437.
103. Matulonis, U.A., Sharma, S., Ghamande, S., Gordon, S. *et al.* (2012) Phase II study of MLN8237 (alisertib), an investigational

Aurora A kinase inhibitor, in patients with platinum-resistant or -refractory epithelial ovarian, fallopian tube, or primary peritoneal carcinoma. *Gynecol. Oncol.*, **127** (1), 63–69.

104. van de Weerdt, B.C. and Medema, R.H. (2006) Polo-like kinases: a team in control of the division. *Cell Cycle*, **5** (8), 853–864.
105. Steegmaier, M., Hoffmann, M., Baum, A., Lénárt, P. *et al.* (2007) BI 2536, a potent and selective inhibitor of polo-like kinase 1, inhibits tumor growth in vivo. *Curr. Biol.*, **17** (4), 316–322.
106. Tennant, D.A., Durán, R.V., Boulahbel, H., and Gottlieb, E. (2009) Metabolic transformation in cancer. *Carcinogenesis*, **30** (8), 1269–1280.
107. Hsu, P.P. and Sabatini, D.M. (2008) Cancer cell metabolism: Warburg and beyond. *Cell*, **134** (5), 703–707.
108. Tennant, D.A., Durán, R.V., and Gottlieb, E. (2010) Targeting metabolic transformation for cancer therapy. *Nat. Rev. Cancer*, **10** (4), 267–277.
109. Kobori, M., Shinmoto, H., Tsushida, T., and Shinohara, K. (1997) Phloretin-induced apoptosis in B16 melanoma 4A5 cells by inhibition of glucose transmembrane transport. *Cancer Lett.*, **119** (2), 207–212.
110. De Lena, M., Lorusso, V., Latorre, A., Fanizza, G. *et al.* (2001) Paclitaxel, cisplatin and lonidamine in advanced ovarian cancer. A phase II study. *Eur. J. Cancer*, **37** (3), 364–368.
111. Di Cosimo, S., Ferretti, G., Papaldo, P., Carlini, P. *et al.* (2003) Lonidamine: efficacy and safety in clinical trials for the treatment of solid tumors. *Drugs Today (Barc.)*, **39** (3), 157–174.
112. Oudard, S., Carpentier, A., Banu, E., Fauchon, F. *et al.* (2003) Phase II study of lonidamine and diazepam in the treatment of recurrent glioblastoma multiforme. *J. Neurooncol.*, **63** (1), 81–86.
113. Maschek, G., Savari, N., Priebe, W., Braunschweiger, P. *et al.* (2004) 2-deoxy-D-glucose increases the efficacy of adriamycin and paclitaxel in human osteosarcoma and non-small cell lung cancers in vivo. *Cancer Res.*, **64** (1), 31–34.
114. Christofk, H.R., Vander Heiden, M.G., Harris, M.H., Ramanathan, A. *et al.* (2008) The M2 splice isoform of pyruvate kinase is important for cancer metabolism and tumour growth. *Nature*, **452** (7184), 230–233.
115. Sun, R.C., Fadia, M., Dahlstrom, J.E., Parish, C.R. *et al.* (2010) Reversal of the glycolytic phenotype by dichloroacetate inhibits metastatic breast cancer cell growth in vitro and in vivo. *Breast Cancer Res. Treat.*, **120** (1), 253–260.
116. Dittrich, C., Zandvliet, A.S., Gneist, M., Huitema, A.D. *et al.* (2007) A phase I and pharmacokinetic study of indisulam in combination with carboplatin. *Br. J. Cancer*, **96** (4), 559–566.
117. Husain, A. and Madhesia, D. (2012) Heterocyclic compounds as carbonic anhydrase inhibitor. *J. Enzyme Inhib. Med. Chem.*, **27** (6), 773–783.
118. Talbot, D.C., von Pawel, J., Cattell, E., Yule, S.M. *et al.* (2007) A randomized phase II pharmacokinetic and pharmacodynamic study of indisulam as second-line therapy in patients with advanced non-small cell lung cancer. *Clin. Cancer Res.*, **13** (6), 1816–1822.
119. Höckel, M. and Vaupel, P. (2001) Tumor hypoxia: definitions and current clinical, biologic, and molecular aspects. *J. Natl Cancer Inst.*, **93** (4), 266–276.
120. Denko, N.C., Fontana, L.A., Hudson, K.M., Sulphin, P.D. *et al.* (2003) Investigating hypoxic tumor physiology through gene expression patterns. *Oncogene*, **22** (37), 5907–5914.
121. Bárdos, J.I. and Ashcroft, M. (2004) Hypoxia-inducible factor-1 and oncogenic signalling. *BioEssays*, **26** (3), 262–269.
122. Ivan, M., Kondo, K., Yang, H., Kim, W. *et al.* (2001) HIFalpha targeted for VHL-mediated destruction by proline hydroxylation: implications for O_2 sensing. *Science*, **292** (5516), 464–468.
123. Lando, D., Peet, D.J., Whelan, D.A., Gorman, J.J. *et al.* (2002) Asparagine hydroxylation of the HIF transactivation domain a hypoxic switch. *Science*, **295** (5556), 858–861.
124. Carmeliet, P. and Jain, R.K. (2011) Molecular mechanisms and clinical applications of angiogenesis. *Nature*, **473** (7347), 298–307.
125. Dorević, G., Matusan-Ilijas, K., Babarović, E., Hadzisejdi, I. *et al.* (2009) Hypoxia inducible factor-1alpha correlates with vascular endothelial growth factor A and C indicating worse prognosis in clear cell

renal cell carcinoma. *J. Exp. Clin. Cancer Res.*, **28**, 40.
126. Theodoropoulos, V.E., Lazaris, A.Ch., Sofras, F., Gerzellis, I. *et al.* (2004) Hypoxia-inducible factor 1 alpha expression correlates with angiogenesis and unfavorable prognosis in bladder cancer. *Eur. Urol.*, **46** (2), 200–208.
127. Weis, S.M. and Cheresh, D.A. (2011) Tumor angiogenesis: molecular pathways and therapeutic targets. *Nat. Med.*, **17** (11), 1359–1370.
128. Abdelrahim, M., Konduri, S., Basha, R., Philip, P.A. *et al.* (2010) Angiogenesis: an update and potential drug approaches (review). *Int. J. Oncol.*, **36** (1), 5–18.
129. Sandler, A., Gray, R., Perry, M.C., Brahmer, J. *et al.* (2006) Paclitaxel-carboplatin alone or with bevacizumab for non-small-cell lung cancer. *N. Engl. J. Med.*, **355** (24), 2542–2550.
130. Jain, R.K., Duda, D.G., Clark, J.W., and Loeffler, J.S. (2006) Lessons from phase III clinical trials on anti-VEGF therapy for cancer. *Nat. Clin. Pract. Oncol.*, **3** (1), 24–40.
131. Clarke, J.M. and Hurwitz, H.I. (2013) Targeted inhibition of VEGF receptor 2: an update on ramucirumab. *Expert Opin. Biol. Ther.*, **13** (8), 1187–1196.
132. Garon, E.B., Cao, D., Alexandris, E., John, W.J. *et al.* (2012) A randomized, double-blind, phase III study of Docetaxel and Ramucirumab versus Docetaxel and placebo in the treatment of stage IV non-small-cell lung cancer after disease progression after 1 previous platinum-based therapy (REVEL): treatment rationale and study design. *Clin. Lung Cancer*, **13** (6), 505–509.
133. Gotink, K.J. and Verheul, H.M. (2010) Anti-angiogenic tyrosine kinase inhibitors: what is their mechanism of action? *Angiogenesis*, **13** (1), 1–14.
134. Park, S.J., Lee, J.L., Park, I., Park, K. *et al.* (2012) Comparative efficacy of sunitinib versus sorafenib as first-line treatment for patients with metastatic renal cell carcinoma. *Chemotherapy*, **58** (6), 468–474.
135. Bergers, G. and Hanahan, D. (2008) Modes of resistance to anti-angiogenic therapy. *Nat. Rev. Cancer*, **8** (8), 592–603.
136. Pàez-Ribes, M., Allen, E., Hudock, J., Takeda, T. *et al.* (2009) Antiangiogenic therapy elicits malignant progression of tumors to increased local invasion and distant metastasis. *Cancer Cell*, **15** (3), 220–231.
137. Hoeller, D. and Dikic, I. (2009) Targeting the ubiquitin system in cancer therapy. *Nature*, **458** (7237), 438–444.
138. Chiarle, R., Budel, L.M., Skolnik, J., Frizzera, G. *et al.* (2000) Increased proteasome degradation of cyclin-dependent kinase inhibitor p27 is associated with a decreased overall survival in mantle cell lymphoma. *Blood*, **95** (2), 619–626.
139. Ma, M.H., Yang, H.H., Parker, K., Manyak, S. *et al.* (2003) The proteasome inhibitor PS-341 markedly enhances sensitivity of multiple myeloma tumor cells to chemotherapeutic agents. *Clin. Cancer Res.*, **9** (3), 1136–1144.
140. Lander, G.C., Ershin, E., Matyskiela, M.E., Bashore, C. *et al.* (2012) Complete subunit architecture of the proteasome regulatory particle. *Nature*, **482** (7384), 186–191.
141. Driscoll, J.J. and Dechowdhury, R. (2010) Therapeutically targeting the SUMOylation, Ubiquitination and Proteasome pathways as a novel anticancer strategy. *Target. Oncol.*, **5** (4), 281–289.
142. Hideshima, T., Richardson, P., Chauhan, D., Palombella, V.J. *et al.* (2001) The proteasome inhibitor PS-341 inhibits growth, induces apoptosis, and overcomes drug resistance in human multiple myeloma cells. *Cancer Res.*, **61** (7), 3071–3076.
143. Piperdi, B., Ling, Y.H., Liebes, L., Muggia, F. *et al.* (2011) Bortezomib: understanding the mechanism of action. *Mol. Cancer Ther.*, **10** (11), 2029–2030.
144. Spicka, I., Mateos, M.V., Redman, K., Dimopoulos, M.A. *et al.* (2011) An overview of the VISTA trial: newly diagnosed, untreated patients with multiple myeloma ineligible for stem cell transplantation. *Immunotherapy*, **3** (9), 1033–1040.
145. Redic, K. (2013) Carfilzomib: a novel agent for multiple myeloma. *J. Pharm. Pharmacol.*, **65** (8), 1095–1106.
146. Watson, I.R., Irwin, M.S., and Ohh, M. (2011) NEDD8 pathways in cancer, Sine Quibus Non. *Cancer Cell*, **19** (2), 168–176.
147. Morimoto, M., Nishida, T., Nagayama, Y., and Yasuda, H. (2003) Nedd8-modification of Cul1 is promoted by Roc1 as a Nedd8-E3 ligase and regulates its stability. *Biochem. Biophys. Res. Commun.*, **301** (2), 392–398.

148. Ohh, M., Kim, W.Y., Moslehi, J.J., Chen, Y. *et al.* (2002) An intact NEDD8 pathway is required for Cullin-dependent ubiquitylation in mammalian cells. *EMBO Rep.*, **3** (2), 177–182.
149. Soucy, T.A., Smith, P.G., and Rolfe, M. (2009) Targeting NEDD8-activated cullin-RING ligases for the treatment of cancer. *Clin. Cancer Res.*, **15** (12), 3912–3916.
150. Zhao, Y. and Sun, Y. (2013) Cullin-RING Ligases as attractive anti-cancer targets. *Curr. Pharm. Des.*, **19** (18), 3215–3225.
151. Loftus, S.J., Liu, G., Carr, S.M., Munro, S. *et al.* (2012) NEDDylation regulates E2F-1-dependent transcription. *EMBO Rep.*, **13** (9), 811–818.
152. Blank, J.L., Cosmopoulos, K., Bouck, D.C., Garcia, K. *et al.* (2013) Novel DNA damage checkpoints mediating cell death induced by the NEDD8-activating enzyme inhibitor MLN4924. *Cancer Res.*, **73** (1), 225–234.
153. Nawrocki, S.T., Griffin, P., Kelly, K.R., and Carew, J.S. (2012) MLN4924: a novel first-in-class inhibitor of NEDD8-activating enzyme for cancer therapy. *Expert Opin. Investig. Drugs*, **21** (10), 1563–1573.
154. Neckers, L. (2007) Heat shock protein 90: the cancer chaperone. *J. Biosci.*, **32** (3), 517–530.
155. An, W.G., Schulte, T.W., and Neckers, L.M. (2000) The heat shock protein 90 antagonist geldanamycin alters chaperone association with p210bcr-abl and v-src proteins before their degradation by the proteasome. *Cell Growth Differ.*, **11** (7), 355–360.
156. Jego, G., Hazoumé, A., Seigneuric, R., and Garrido, C. (2013) Targeting heat shock proteins in cancer. *Cancer Lett.*, **332** (2), 275–285.
157. Li, J., Soroka, J., and Buchner, J. (2012) The Hsp90 chaperone machinery: conformational dynamics and regulation by co-chaperones. *Biochim. Biophys. Acta*, **1823** (3), 624–635.
158. Zuehlke, A. and Johnson, J.L. (2010) Hsp90 and co-chaperones twist the functions of diverse client proteins. *Biopolymers*, **93** (3), 211–217.
159. Roe, S.M., Prodromou, S., O'Brien, R., Ladbury, J.E. *et al.* (1999) Structural basis for inhibition of the Hsp90 molecular chaperone by the antitumor antibiotics radicicol and geldanamycin. *J. Med. Chem.*, **42** (2), 260–266.
160. Gandhi, N., Wild, A.T., Chettiar, S.T., Aziz, K. *et al.* (2013) Novel Hsp90 inhibitor NVP-AUY922 radiosensitizes prostate cancer cells. *Cancer Biol. Ther.*, **14** (4), 347–356.
161. Garon, E.B., Finn, R.S., Hamidi, H., Dering, J. *et al.* (2013) The HSP90 inhibitor NVP-AUY922 potently inhibits non-small cell lung cancer growth. *Mol. Cancer Ther.*, **12** (6), 890–900.
162. Wainberg, Z.A., Anghel, A., Rogers, A.M., Desai, A.J. *et al.* (2013) Inhibition of HSP90 with AUY922 induces synergy in HER2-amplified trastuzumab-resistant breast and gastric cancer. *Mol. Cancer Ther.*, **12** (4), 509–519.
163. Usmani, S.Z., Bona, R., and Li, Z. (2009) 17 AAG for HSP90 inhibition in cancer – from bench to bedside. *Curr. Mol. Med.*, **9** (5), 654–664.
164. Hostein, I., Robertson, D., DiStefano, F., Workman, P. *et al.* (2001) Inhibition of signal transduction by the Hsp90 inhibitor 17-allylamino-17-demethoxygeldanamycin results in cytostasis and apoptosis. *Cancer Res.*, **61** (10), 4003–4009.
165. Srethapakdi, M., Liu, F., Tavorath, R., and Rosen, N. (2000) Inhibition of Hsp90 function by ansamycins causes retinoblastoma gene product-dependent G1 arrest. *Cancer Res.*, **60** (14), 3940–3946.
166. Li, F., Zhao, C., and Wang, L. (2014) Molecular-targeted agents combination therapy for cancer: developments and potentials. *Int. J. Cancer*, **134** (6) 1257–1269.

15
Cancer Pharmacogenomics

Daniel L. Hertz[1] *and Howard L. McLeod*[2,3]
[1] *University of Michigan College of Pharmacy, Department of Clinical, Social, and Administrative Sciences, 428 Church Street, Ann Arbor, MI 48109, USA*
[2] *Moffitt Cancer Center, Division of Population Sciences, 12902 Magnolia Drive MRC-CANCONT, Tampa, FL 33612, USA*
[3] *DeBartolo Family Personalized Medicine Institute, Tampa, FL 33612, USA*

Translational Medicine Cancer: Advances in Molecular Biology and Medicine,
First Edition. Edited by Robert A. Meyers.

Keywords

Constitutive
A state of constant activation regardless of the presence of an activating stimulus.

Fusion
The combination of two distinct proteins caused by a genetic translocation.

Germline genome
The genome inherited at childbirth that exists in all cells of the patient.

Kinase
Proteins that propagate signals through phosphorylation.

Oncogenic
Causing or giving rise to tumors.

Pharmacogenetics
The use of genetic information to guide pharmacotherapy decisions.

Phenotype
An observable trait, such as protein activity.

Polymorphism
A genetic variant commonly found within the population.

Somatic genome
The genome of the tumor that has acquired oncogenic variation.

The use of genetic information to guide pharmacotherapy decisions, referred to as pharmacogenetics, is particularly important in cancer, which is a disease of the genome. The genetic aberrations that most often cause cancer have been cataloged, initiating research programs to develop drugs to counteract these oncogenic mechanisms. Treatment decisions are now made based on knowledge of the genetic driver within an individual tumor. Aside from the tumor (somatic) genome, treatment decisions can be influenced by the patient (germline) genome. The germline genome dictates patient sensitivity to drug-induced toxicity and determines the patient's exposure to treatment.

Clinicians can use both the somatic and germline genome to select the appropriate dose of the correct drug to maximize the benefit and minimize the toxicity of cancer therapy for an individual patient.

1 Introduction

The present understanding of the mechanisms underlying cancer pathogenesis and awareness of potential strategies to treat cancer have evolved tremendously over the past quarter-century. Not long ago, tumors were classified by their anatomical location and clinical features, and the systemic treatment options were equally toxic to healthy and cancer cells [1]. Whilst the classification of tumors based on their organ of origin endures, tumors from a given organ are no longer treated uniformly. The first stratification of tumors was based on receptor overexpression, such as the estrogen receptor in breast cancer [2]. Many tumors are highly sensitive to drugs that target overexpressed receptors [3]. It is now recognized that receptor overexpression, and other oncogenic drivers, are consequences of aberrations within the tumor genome. Cataloging of the common oncogenic variation in the somatic genome has led to the development and approval of drugs for use only in tumors that harbor a specific genetic aberration. Therapeutic decisions are now frequently made based on the presence of an aberration in the tumor (or somatic) genome.

The somatic genome is not the only genome of interest in cancer pharmacogenetics. The patient's (or germline) genome is also useful for making treatment decisions (Fig. 1) [4]. The germline genome can be used to avoid unsafe drugs and optimize drug dosing. There are several examples in which germline genetics is being used to tailor cancer therapy for individual patients, and many others that are currently under investigation for eventual translation into clinical practice.

2 Somatic Genetics

Oncogenic conversion takes place at the genetic level with the introduction of a mutation, amplification or rearrangement that transforms a healthy cell into a cancer cell. The genomic interrogation of thousands of cancers through The Cancer Genome Atlas (TCGA) has enabled characterization of the oncogenic genetic variations within and across tumor types [5, 6] (Table 1). Dozens of pathways, hundreds of genes, and thousands of individual variants have been described, each representing a potential target for a therapeutic agent. Constitutively active receptors and kinases that initiate and propagate signals, respectively, have been the focus of intense drug development research efforts. For many of these pharmacogenetic examples the drug is only approved for use in tumors that have been verified to harbor the variant of interest, requiring tumor genetic analysis prior to therapy selection (Table 2).

2.1 Receptors

2.1.1 HER2

The first receptors to be utilized to guide cancer treatment were the estrogen and progesterone receptors in breast cancer [2]. The survival benefit of tamoxifen treatment

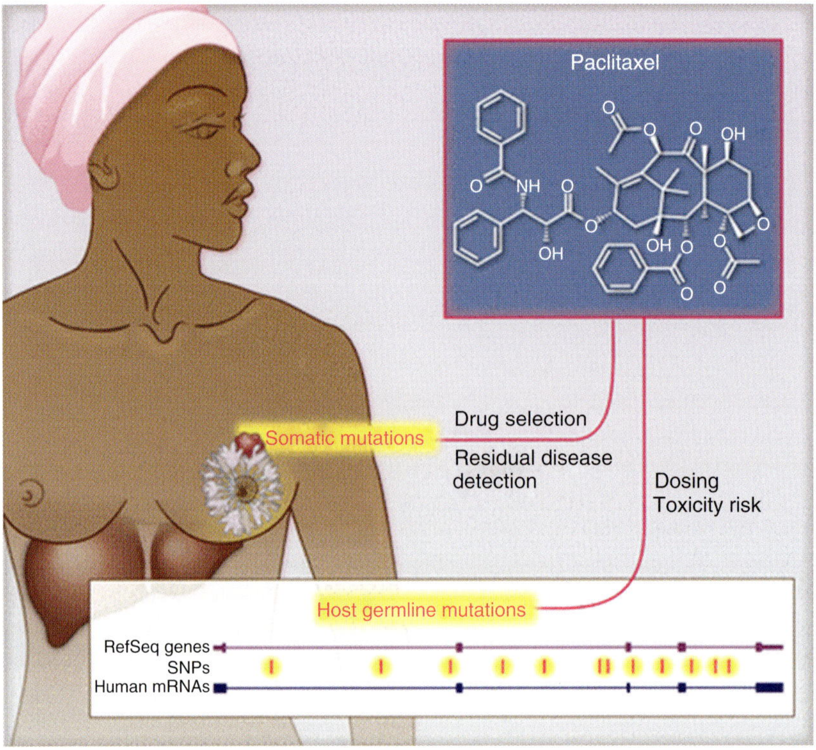

Fig. 1 The tumor (somatic) genome is essential for predicting which drug will be most efficacious, while the patient (germline) genome dictates sensitivity to toxicity and appropriate dose selection. Reproduced with permission from Ref. [4]; © 2013, The American Association for the Advancement of Science.

in patients with tumors expressing these receptors is well established [15]. The Human Epidermal Growth Factor Receptor 2 (EGFR2; HER2) is a third receptor that is overexpressed in 15–20% of breast tumors and is a marker of aggressive disease. Amplification of the *ERBB2* gene leads to HER2 overexpression (HER2+) and an amplified activity of several canonical oncogenic pathways, including the Ras and PI3K kinase signaling cascades (Fig. 2). Various methods of verifying HER2+ tumors exist, and only some of them directly screen the *ERBB2* gene [16], blurring the distinction between receptor-targeted and pharmacogenetic-guided treatment paradigms.

Trastuzumab, a humanized anti-HER2 monoclonal antibody (mAb) showed promising efficacy in preclinical models of HER2+ cancer [17], and was eventually approved for use in patients with HER2+ breast cancer [18]. Following the success of trastuzumab, several other agents were developed and approved for use in HER2+ patients. Trastuzumab has been utilized as a carrier of the highly toxic emtansine, and the resulting antibody–drug conjugate has proven efficacy in trastuzumab-refractory patients [19]. Pertuzumab, another mAb that blocks HER2 heterodimerization, and lapatinib, a small-molecule inhibitor of

Tab. 1 Common gene aberrations and key signaling pathways as reported in *The Cancer Genome Atlas* (TCGA) analyses.

Tumor type	*Amplified genes*	*Deleted genes*	*Mutated genes*	*Key signaling pathways*	*Reference*
Glioblastoma	*EGFR, CDK4, PDGFRA*	*CDKN2A/B, PTEN, CDKN2C*	*TP53, PTEN, NF1*	Ras, PI(3)K, p53, and retinoblastoma	[7]
Ovarian carcinoma	*CCNE1, MYC, MECOM*	*PTEN, RB1, NF1*	*TP53, BRCA1/2*	Ras, PI(3)K, NOTCH, and retinoblastoma	[8]
Breast tumors	*PIK3CA, EGFR, FOXA1*	*MLL3, PTEN, RB1*	*PIK3CA, TP53, GATA3*	PI(3)K, p53, and retinoblastoma	[9]
Squamous cell lung cancer	*SOX2, PDGFRA, KIT*	*PTPRD, CSMD1, LRP1B*	*TP53, CDKN2A, PTEN*	Ras, PI(3)K, oxidative stress response, and squamous differentiation	[10]
Colon and rectal cancer	*ERBB2, MYC, IGF2*	*FHIT, A2BP1, MACROD2*	*ACVR2A, APC, TP53*	WNT, TGF-β, PI(3)K, and p53	[11]
Adult *de novo* acute myeloid leukemia	Few copy-number alterations		*FLT3, NPM1, DNMT3A*	Not reported	[12]
Endometrial carcinoma	*ERBB2, CCNE1, DDAH1*	*19p13.3, CLEC3A/WWOX, PTEN*	*PTEN, PIK3CA, ARID1A*	Ras, β-catenin, and PI(3)K	[13]

HER2 signaling, both improve treatment outcomes when added to trastuzumab [20, 21]. One other tyrosine kinase inhibitor (TKI), neratinib, has shown activity in tumors with HER2-activating mutations [22], as opposed to amplifications, and is currently in Phase II clinical trials for tumors that harbor this rare genetic aberration.

2.1.2 EGFR

The EGFR is a Class I receptor tyrosine kinase (RTK) similar to HER2. Tumors that harbor activating deletions and rearrangements in the *EGFR* gene are typically sensitive to EGFR inhibitor treatment [23]. Because of the frequency of EGFR mutations in non-small cell lung cancer (NSCLC), the EGFR inhibitor erlotinib is superior to chemotherapy in unselected NSCLC patients [24], but patients with EGFR-activating mutations experience the greatest benefit [25]. Gefitinib is another first-generation EGFR inhibitor that benefits patients with EGFR mutations [26, 27]. Resistance to these first-generation inhibitors is typically conferred by the T790M mutation [28]. Afatinib, a second-line EGFR inhibitor, is approved for use in metastatic NSCLC that harbors specific EGFR mutations [29]. Second-line agents may be effective against the T790M resistance mechanism [30], and are being tested in patients who had previously

Tab. 2 Somatic variation used to select cancer treatment.

Protein function	*Genetic marker*	*Variation type*	*Common tumor types*	*Approved agent*	*Resistance mechanism*	*Experimental agent*
Receptor	HER2	Amplification (or mutation)	Breast cancer (found in ovarian, stomach, and uterine cancer also)	Trastuzumab; trastuzumab – emtansine; pertuzumab; lapatinib	Multiple [14]	Neratinib (for HER2 mutation)
	EGFR	Mutations and rearrangements	NSCLC, anal cancer, glioblastoma multiforme	Erlotinib; gefitinib; afatinib	EGFR T790M mutation	—
				Cetuximab; panitumumab	KRAS mutations	—
	PDGFRB	Fusions	CML, myeloproliferative disorders	Imatinib	—	—
	c-KIT	Mutations and amplifications	Gastrointestinal stromal tumors, testicular seminoma, melanoma, AML	Imatinib; dasatinib; sunitinib; nilotinib; sorafenib	Additional KIT mutations	—
	FLT3	Internal tandem duplication	AML	Sorafenib	—	Midostaurin lestauritinib
Kinase	BRAF	Mutations	Melanoma, lymphoma, colorectal cancer, NSCLC, hairy cell leukemia	Vermurafenib; dabrafenib	PDGFRB or EGFR amplification, NRAS mutations	—
	BCR-ABL	Fusion	CML, AML, acute leukocytic leukemia	Imatinib; nilotinib; dasatinib; bosutinib ponatinib	Abl mutation (T315I) or amplification	—
	EML4-ALK	Fusion	NSCLC, neuroblastoma	Crizotinib	ALK mutation (L1196M), EGFR, or c-KIT activation	LDK378 AF802
Other	PML-RAR-α	Fusion	Acute promyelocytic leukemia	ATRA	—	—
	BRCA	Germline mutation	Breast, ovarian, stomach, prostate, colon cancer, and others	None	—	Olaparib veliparib

AML: acute myelogenous leukemia; CML: chronic myelogenous leukemia; NSCLC: non-small-cell lung cancer.

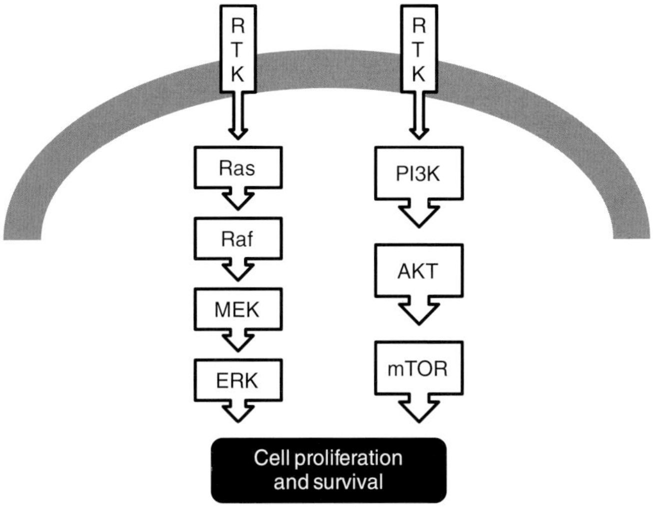

Fig. 2 The receptor tyrosine kinases (RTKs), including EGFR, HER2, PDGFRB, and FLT3, transmit extracellular signals through the Ras and PI3K canonical pathways. Ultimately, this activates gene transcription that regulates cellular proliferation. Amplifications, fusions or mutations that cause constitutively active receptor tyrosine or intracellular kinases cause unchecked cellular replication, the hallmark of cancer. Tumors driven by a given genetic aberration can be very sensitive to inhibition of the constitutively active kinase.

progressed while receiving EGFR inhibitor treatment [31].

Similar to HER2, EGFR has been targeted through the use of mAbs in addition to TKIs. Two EGFR mAbs have been approved for use in metastatic colorectal cancer, cetuximab and panitumumab. Interestingly, these drugs seem to be efficacious regardless of EGFR amplification or mutation [32, 33], and so are used without regard to EGFR status. However, these mAbs do not benefit all metastatic colorectal cancer patients. Tumors that harbor a *KRAS* mutation are inherently resistant to both cetuximab [34, 35] and panitumumab [36, 37].

2.1.3 FLT3

The FMS-like tyrosine kinase receptor-3 gene (FLT3) is a member of the Class III RTK family. An internal tandem duplication in exon 14 of the *FLT3* gene (FLT3-ITD) causes a constitutive initiation of signals that propagate through the same oncogenic pathways as HER2 and EGFR (Fig. 2) [38, 39]. FLT3-ITD confers both a poor prognosis and treatment resistance in acute myeloid leukemia (AML) [40, 41]. Sorafenib is a potent inhibitor of cells bearing FLT3-ITD [42], and the results of early clinical trials in FLT3-ITD AML have been promising [43], although some resistance development has been reported [44]. Several other FLT3-targeted agents are late in clinical development, including midostaurin (which is in Phase III trials) and lestaurtinib, which has already failed a randomized trial, perhaps due to inadequate FLT3 signaling inhibition [45].

2.1.4 PDGFRB

Platelet-derived growth factor receptor-β (PDGFRB), another Class III RTK, is a common member in oncogenic fusion proteins. Several fusion partners have been identified, and in all cases the resultant protein maintains a constitutively active PDGFRB tyrosine kinase domain [46]. The most common partner is the *ETV6/TEL* gene (t(5;12)(q33p13)), which is a transcription factor involved in hematopoiesis [47]. *PDGFRB* gene fusions cause several myeloproliferative disorders that include eosinophilia and chronic myelomonocytic leukemia [48]. Imatinib, which is used in *BCR-ABL* rearrangement (see Sect. 2.2.2), has inhibitory activity towards PDGFRB [49] and is FDA-approved for use in *PDGFRB*-rearranged myelodysplastic/myeloproliferative diseases [50, 51].

2.1.5 c-KIT

Another Class III RTK, c-KIT (CD117, KIT) has common mutations in exon 11 that cause constitutive activity. About three-quarters of gastrointestinal stromal tumors (GISTs) have mutated c-KIT, a marker for poor prognosis [52, 53]. c-KIT mutations are also predictive of response to imatinib [54]. There is structural similarity between c-KIT and the kinase domain that imatinib was designed to inhibit, which explains its potency in c-KIT-mutant GIST cell lines [55]. The efficacy of imatinib mesylate in a Phase III study of unselected GIST patients led to FDA approval [56], though efficacy seemed to depend on which activating mutation was present [57] and imatinib may lack efficacy in tumors with KIT amplification [58]. Imatinib resistance often develops through the introduction of additional KIT mutations [59]. Other KIT inhibitors such as sunitinib, nilotinib and sorafenib have been approved for use in imatinib-resistant disease, and sorafenib (a multikinase inhibitor) has shown particular promise in these tumors [60].

2.2 Kinases

2.2.1 Ras Pathway

The Ras pathway is a complex network of signals that regulate cellular replication, differentiation and apoptosis. The signals originate at cellular receptors (some of which were described in Sect. 2.1) and propagate through multiple levels of kinase phosphorylation, including Ras, Raf, MEK and ERK (Fig. 2) [61]. Approximately one in three cancers have aberrant MAPK pathway signaling [62] due to constitutive activation or amplification somewhere in the network [63]. Currently, no agents have been approved for use specifically in *Ras* mutant tumors, but MEK inhibitors – including selumetinib and MEK162 (Novartis International AG, Basel, Switzerland) – have demonstrated promising efficacy in these tumors [64, 65].

BRAF is a member of the Raf serine/threonine kinases that is often mutated in cancer [66]. Vemurafenib, designed specifically to inhibit BRAF, was found to be particularly selective for the most common activating mutation, V600E [67, 68]. A Phase I clinical study demonstrated exceptional activity in V600E+ metastatic melanoma patients [69], leading to concurrent Phase II and III clinical trials and FDA approval [70]. However, vemurafenib is much less effective in V600E+ tumors that originate from other organs (including metastatic colorectal cancer), seemingly due to compensatory feedback through EGFR [71]. A second BRAF inhibitor, dabrafenib, is also approved in V600E-mutated metastatic melanoma [72]. Dabrafenib efficacy is further enhanced when combined with the MEK inhibitor trametinib [73] which itself is approved for use in V600E or

V600K melanomas that have not had prior treatment with a BRAF or MEK inhibitor.

Somatic MEK mutations in primary tumors are considerably less common than Ras or Raf mutations [74, 75], but are a common mechanism for resistance development. MEK inhibitors, including trametinib and selumetinib, have demonstrated disappointing efficacy as single agents [76], and are primarily being developed for use in combination with other targeted agents or in tumors that have acquired resistance through MEK mutation.

2.2.2 BCR-ABL

Chronic myelogenous leukemia (CML) is characterized by a reciprocal translocation between chromosomes 9 and 22 (t(9;22)) that fuses the breakpoint cluster (*BCR*) and the Abelson tyrosine kinase (*ABL*) genes. Imatinib, a selective inhibitor of BCR-ABL+ cells [77] revolutionized CML treatment [78]. Despite a dramatic initial response, many patients eventually develop resistance to imatinib [79], necessitating the development of newer generations of BCR-ABL inhibitors. The second-generation inhibitors – nilotinib, dasatinib and bosutinib – are more potent than imatinib, have efficacy in tumors that failed imatinib therapy [80, 81], and are superior to imatinib in head-to-head studies of newly diagnosed CML [82, 83]. Unfortunately, none of these second-generation inhibitors is active against the most common resistance mechanism, a mutation in the ABL kinase domain (T315I) [84]. Ponatinib, a third-generation inhibitor that is active against the T315I mutation [85], has received FDA approval in CML and in BCR-ABL+ acute leukocytic leukemia (ALL) [86].

2.2.3 EML4-ALK

Approximately 4% of NSCLCs – more often those from light or nonsmokers – harbor a gene fusion between *EML4* and *ALK* caused by an inversion on chromosome 2p [87]. This fusion results in constitutive kinase signaling and can cause spontaneous lung tumor development in animals [88]. Crizotinib, originally designed for c-MET inhibition, was found to inhibit EML4-ALK signaling [89]. Early clinical studies in EML4-ALK+ NSCLC patients demonstrated impressive efficacy [90] which led to accelerated FDA approval. Several mechanisms of resistance have been identified, including an ALK mutation (L1196M) and the activation of EGFR or c-KIT [91]. Among second-generation ALK inhibitors currently in clinical trials is LDK378 (Novartis, Basel, Switzerland), for which Phase III trials are planned in both treatment-naïve and crizotinib-treated patients. The results of early clinical studies have suggested that LDK378 is effective in patients resistant to crizotinib. Additionally, preclinical data for AF802 (Chugai Pharmaceutical, Tokyo, Japan), another second-generation ALK inhibitor, suggests potential activity in cells harboring the L1196M resistance mutation [92].

2.3 Other Targets

2.3.1 PML-RAR-α

The *PML-RAR*-α reciprocal translocation (t(15;17)(q22;q12)) is the hallmark gene fusion in acute promyelocytic leukemia (APL), a subtype representing <10% of AML. PML-RAR-α interferes with homeostatic retinoic acid signaling. In a healthy cell, retinoic acid binds to RAR-α and regulates the transcription of genes necessary for hematopoietic maturation. The aberrantly fused RAR-α protein is able to bind DNA normally but transcription is repressed, interfering with cellular differentiation [93]. The administration of exogenous all-*trans* retinoic acid (ATRA) can overcome this repression and initiate differentiation of the immature cells [94]. ATRA proved to be practice-changing in APL, and was another example of a targeted therapy that transformed a disease with a terrible prognosis into one that can be effectively treated in most patients [95]. The identification of driver mutations in other forms of leukemia, such as the activating mutation in the *NPM1* gene, may lead to the development of highly effective therapies for patients with other AML subtypes [96].

2.3.2 BRCA

Unlike the targets discussed above, the genetic variants in *BRCA1* and *BRCA2* exist in the germline DNA and increase the patient's risk of breast, ovarian, prostate and other cancers [97]. BRCA is a member of a protein complex that repairs DNA double-strand breaks by homologous recombination (HR) [98]. Poly(ADP-ribose) polymerase (PARP) is an enzyme involved in a complementary DNA repair pathway, termed base-excision repair (BER). A

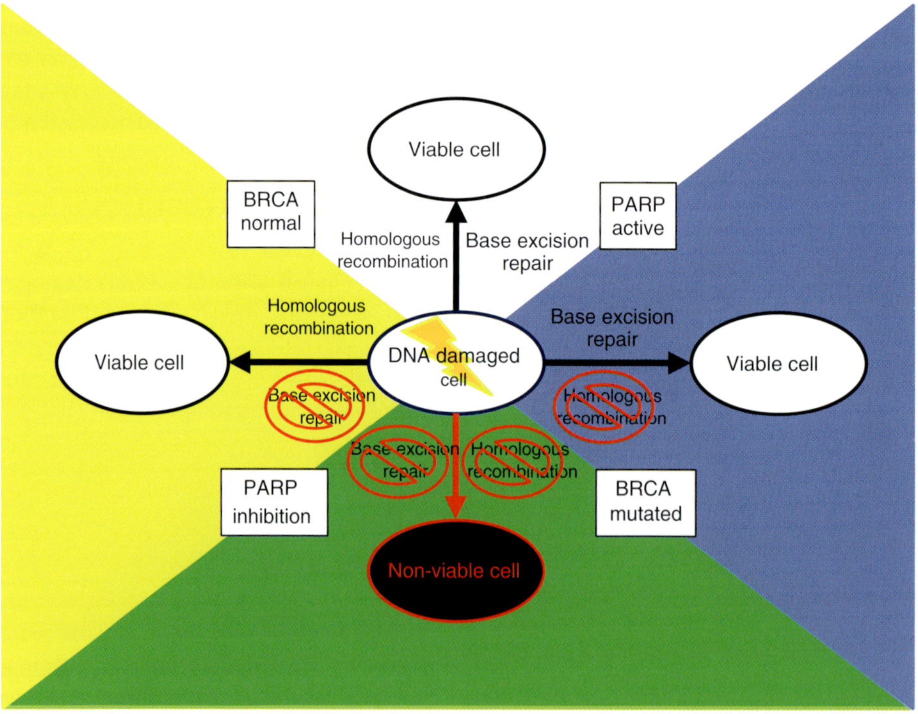

Fig. 3 Normal cells have two complementary mechanisms for DNA repair, namely base-excision repair (BER) and homologous recombination (HR). PARP and BRCA are proteins that are necessary to perform BER and HR, respectively. A cell with BRCA mutation or PARP inhibition is unable to perform one or the other mechanism of DNA repair, but maintains viability. A cell with BRCA mutation that is treated with PARP inhibition is unable to perform either mechanism of DNA repair and dies, a phenomenon known as "synthetic lethality."

cell that can perform either HR or BER is viable, but a cell that loses both DNA repair mechanisms is not viable. Thus, the administration of PARP inhibitors to BRCA-deficient cancer cells causes synthetic lethality (Fig. 3). Based on this hypothesis, and impressive preclinical data [99], clinical development has been pursued for several PARP inhibitors for use in patients with *BRCA1* or *BRCA2* mutations. Phase II results for the PARP inhibitor olaparib in *BRCA1/2* mutant ovarian and breast cancers were impressive [100, 101], and Phase III studies in ovarian cancer patients have been initiated. Initially, iniparib was being developed for use in metastatic breast cancer and also showed promising Phase II results [102]; however, when it failed in a Phase III study this led to conjecture that iniparib does not actually inhibit PARP [103]. Other PARP inhibitors, such as veliparib, are in earlier phases of clinical development.

3 Germline Genetics

The patient's (or germline) genome is also relevant to cancer pharmacogenetics. Variation in the germline genome dictates the expression and function of proteins throughout the patient, and this could influence treatment outcomes either directly, by influencing sensitivity to toxicity, or indirectly through an effect on drug exposure. The validation and translation of germline pharmacogenetic associations has been a challenge because of the difficulty in demonstrating clinical utility [104]. Therefore, only common variants with a dramatic influence on patient outcome have been adopted in clinical practice. Over time, the number of germline pharmacogenetic examples used clinically will undoubtedly expand and the use of germline genetic screening will eventually become a standard of care (Table 3).

Tab. 3 Germline variation that may influence cancer treatment outcomes.

Category	*Drug*	*Gene*	*Phenotype*
Sensitivity to toxicity	Lapatinib	*HLA*	Drug-induced liver injury
	Paclitaxel	*FGD4*	Sensory neuropathy
	Aromatase inhibitors	*TCL1A*	Musculoskeletal adverse events
	Nilotinib	*UGT1A1*	Hyperbilirubinemia
Dose selection	Irinotecan	*UGT1A1*	Neutropenia and diarrhea
	Capecitabine and 5-fluorouracil	*DPYD*	Myelotoxicity
	6-Mercaptopurine and 6-thioguanine	*TPMT*	Myelotoxicity
	Tamoxifen	*CYP2D6*	Efficacy
Supportive care	Codeine and tramadol	*CYP2D6*	Efficacy and toxicity
	Ondansetron and tropisetron		Efficacy
	Tricyclic and selective serotonin reuptake inhibitor antidepressants		Efficacy and toxicity
		CYP2C19	Efficacy and toxicity
	Rasburicase	*G6PD*	Hemolytic anemia

3.1 Sensitivity to Toxicity

3.1.1 HLA-Mediated Hypersensitivity

One of the most promising areas for pharmacogenetics is in the prediction of severe idiosyncratic toxicities. Genome-wide discovery methods have been used to discover human leukocyte antigen (HLA) genotypes that dramatically increase a patient's risk of toxicity. HLA variants have been implicated in drug-induced liver injury, lupus erythematosus, hypersensitivity reaction, Stevens–Johnson syndrome and agranulocytosis, among others [105]. Because these toxicities are severe and unpredictable, the clinical uptake of these findings can be extremely rapid [106]. One example that is specifically relevant to cancer is the finding that lapatinib-induced liver injury is associated with HLA DRB1*0701 [107]. Idiosyncratic toxicities from other cancer drugs may eventually be avoidable by widespread HLA screening.

3.1.2 Pharmacodynamic Sensitivity

For most toxicities there is no single genetic factor that dictates a severe reaction; rather, there are many factors that have marginal effects on the patient's risk of toxicity. These marginal effects are difficult to verify and may never have a role in therapeutic decision-making; nevertheless, they often provide information about the biology, pharmacology and etiology of toxicities about which little is known. One such example is the discovery in a genome-wide association study that a polymorphism in *FGD4* modulates a patient's risk of paclitaxel-induced neuropathy [108]. *FGD4* mutations cause the hereditary neuropathy condition Charcot–Marie–Tooth (subtype 4H) disease [109]. It is possible that these patients have undiagnosed subclinical neuropathy that is exacerbated by paclitaxel treatment.

The T-cell leukemia 1A (TCL1A) gene was discovered in a genome-wide screen of aromatase inhibitor-associated musculoskeletal adverse events, including arthralgias and myalgias [110]. Functional studies revealed that an estrogen response element modulates *TCL1A* expression in a polymorphism-dependent fashion, and alters the expression of interleukins (including IL-17 and IL-12), providing a mechanistic explanation for the discovery [111].

One final example is the sensitivity to nilotinib-induced hyperbilirubinemia in patients with *UGT1A1* variants. Patients with diminished UGT1A1 activity have Craigler–Najjer syndrome type 1, which is characterized by hyperbilirubinemia [112]. When these patients are treated with nilotinib, a *UGT1A1* inhibitor, they are at dramatic risk of hyperbilirubinemia [113]. It would be prudent to avoid nilotinib in patients known to have hyperbilirubinemia, but the use of preemptive *UGT1A1* pharmacogenetic screening is unlikely to become standard of care in this context.

3.2 Dose Selection

In addition to influencing toxicity sensitivity, the germline genome partially dictates drug exposure. Exposure to the active compound – whether the parent compound or a metabolite – can determine if the patient will experience efficacy or toxicity. The formation, distribution and elimination of the active compound are controlled by various enzymes and transporters. For some drugs, the activity of a single protein can explain a substantial portion of the exposure. Genetic variation can either increase or decrease the activity of the protein of interest; for example, gene

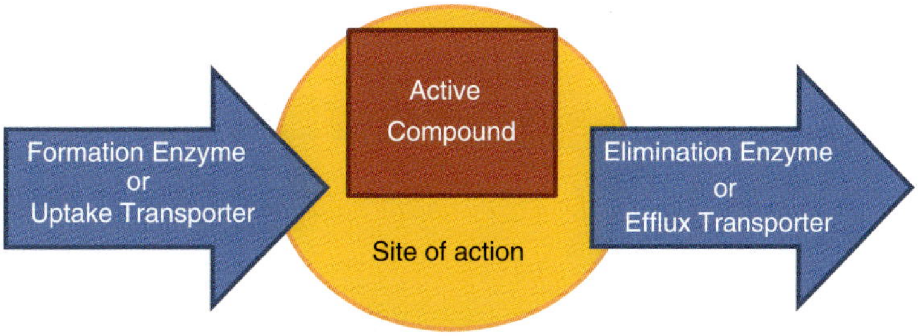

Scenario	Protein of interest	Variant activity	Predicted exposure	Appropriate dose change
1	Formation Enzyme or Uptake Transporter	High	High	Decrease
2	Formation Enzyme or Uptake Transporter	Low	Low	Increase
3	Elimination Enzyme or Efflux Transporter	High	Low	Increase
4	Elimination Enzyme or Efflux Transporter	Low	High	Decrease

Fig. 4 Whether the dose should be increased or decreased in a hypothetical situation depends on several factors. An enzyme can either produce or eliminate the active compound, and a transporter can deliver or remove the active compound from the site of action. A genetically varied enzyme or transporter can be either low or high activity compared to the wild-type. This creates four possible hypothetical scenarios for a variant enzyme or transporter. Each scenario can be predicted to have high or low active drug exposure, meaning that the appropriate dose modification would be to decrease or increase the dose, respectively.

duplications would increase activity while deletions would decrease activity. Meanwhile, mutations and insertions/deletions can have either or no effect.

It is not always straightforward to predict the appropriate dose modification when a variant that modifies drug exposure is present. This requires knowing the active compound, the interaction between protein and compound, and the influence of any genetic variation on protein activity (Fig. 4). Expert guidelines have been published with dosing recommendations for validated gene-drug associations by the Clinical Pharmacogenetics Implementation Consortium (CPIC) [114] and the Pharmacogenetics Working Group of the Royal Dutch Association for the Advancement of Pharmacy (PWG) [115] (Table 4).

3.2.1 UGT1A1 and Irinotecan

Irinotecan is a topoisomerase 1 inhibitor used primarily in the treatment of metastatic colorectal cancer. Its efficacy and toxicity (notably neutropenia and diarrhea) are determined by exposure to the active metabolite, SN-38. Carboxylesterases metabolize irinotecan to SN-38, which is in turn inactivated by UGT1A1 glucuronidation (Fig. 5). *UGT1A1* is highly polymorphic, with more than 130 reported variants, including the UGT1A1*28 promoter polymorphism [118]. UGT1A1*28 has decreased protein expression and enzymatic activity [119, 120], causing an

Tab. 4 Expert guidelines for the use of germline genetics to guide cancer treatment decisions.

Drug	*Gene*	*Variants*	*Phenotype classification*	*Starting dose recommendation*	*Guideline references*
Irinotecan	*UGT1A1*	*28	*28/*28	Decrease dose by 30% for high dose therapy (>250 mg m^{-2})	PWG [115]
5-FU and capecitabine	*DPYD*	*2A, *13, rs67376798	Heterozygote	Reduce dose by 50%	CPIC [116]
			Homozygous variant	Select alternative drug	
		*2A, *3, *7, *8, *9B, *10–*13, 496A>G, IVS10–15T>C, 1156G>T, 1845G>T	IM: 1 variant allele	Reduce dose by 50% or select alternative drug	PWG [115]
			PM: 2 variant alleles	Select alternative drug	
6-MP	*TPMT*	*2, *3A, *3B, *3C, *4	Heterozygote	Reduce dose 30–70%	CPIC [117]
			Homozygous variant	Reduce dose 10-fold and frequency to 3 doses/week	
		*2, *3A–*3D, *4–*18	IM: 1 variant allele	Select alternative drug or reduce dose by 50%	PWG [115]
			PM: 2 variant alleles	Select alternative drug or reduce dose by 90%	
6-TG		*2, *3A, *3B, *3C, *4	Heterozygote	Reduce dose 50–70%	CPIC [117]
			Homozygous variant	Reduce dose 10-fold and frequency to 3 doses per week	
		*2, *3A–*3D, *4–*18	IM: 1 variant allele	Select alternative drug	PWG [115]
			PM: 2 variant alleles	Select alternative drug	

5-FU: 5-fluorouracil, 6-MP: 6-mercaptopurine, 6-TG: 6-thioguanine, IM: intermediate metabolizer, PM: poor metabolizer, PWG: Pharmacogenetics Working Group of the Royal Dutch Association for the Advancement of Pharmacy, and CPIC: Clinical Pharmacogenetics Implementation Consortium.

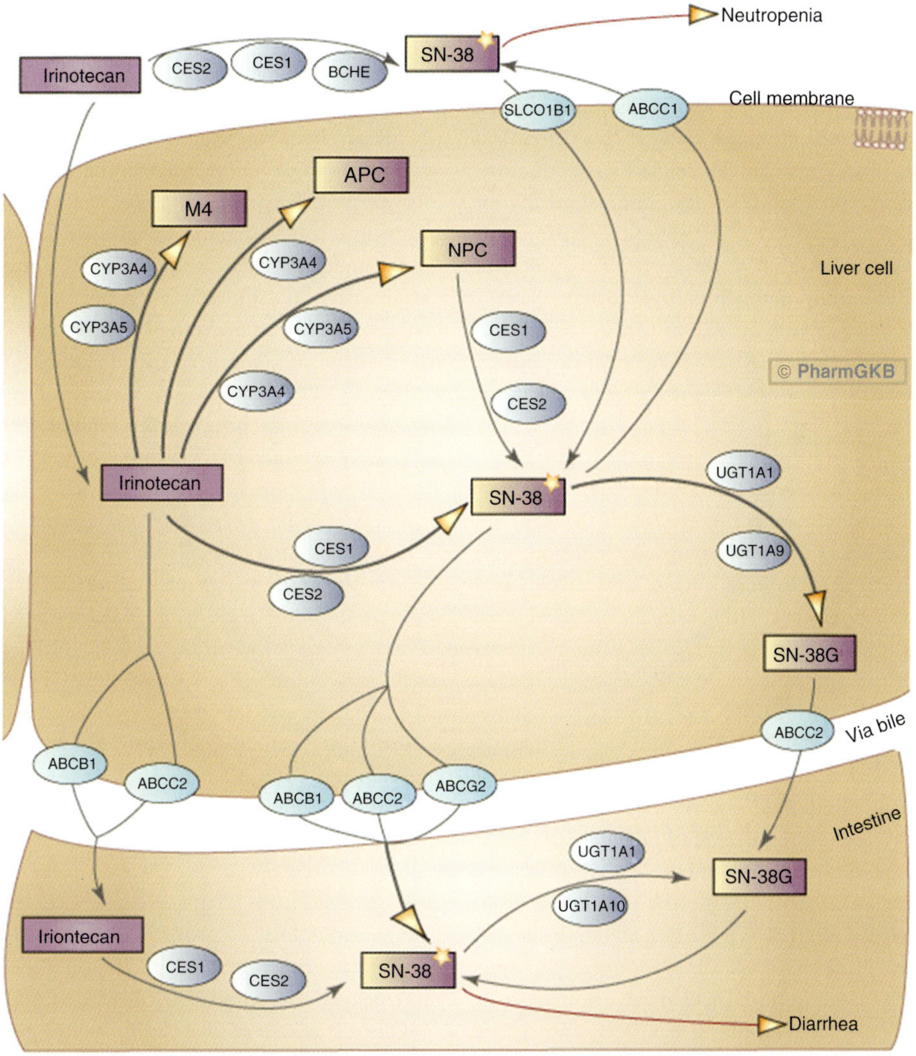

Fig. 5 Irinotecan is metabolized to the active compound SN-38 by carboxylesterases. SN-38 systemic or intestinal exposure causes neutropenia and diarrhea, respectively. Exposure to SN-38 is primarily driven by the hepatic activity of UGT1A1 which is responsible for glucuronidation and inactivation. Genetic variation in the *UGT1A1* promoter, specifically the *28 allele, causes decreased enzyme expression, increased SN-38 exposure, and increased toxicity. Reproduced with permission from Ref. [126]; © 2012, PharmGKB and Stanford University. The original can be accessed at: *http://www.pharmgkb.org/pathway/PA2001*.

increase in SN-38 exposure [121] and a corresponding increase in irinotecan treatment toxicity [122–124], particularly with high-dose therapy [125].

The PWG recommends a 30% dose decrease for UGT1A1*28 homozygous patients receiving high-dose therapy (>250 mg m^{-2}) [115]. Alternatively, the

EGAPP working group did not recommend adjusting irinotecan dosing in these patients, citing a concern that decreasing toxicity may come at the expense of decreased efficacy [127]. The intention of ongoing clinical trials is to demonstrate prospectively the clinical utility of genotype-directed therapy. Ultimately, algorithms that include UGT haplotypes [128] and polymorphisms in other relevant genes such as carboxylesterases [129] and transporters [130–132] may be developed to estimate the ideal irinotecan dose for each patient.

3.2.2 DPYD and Fluoropyrimidines

Fluoropyrimidines, including 5-fluorouracil (5-FU) and its prodrug capecitabine, are used in colorectal, head and neck, and breast cancer. The rate-limiting step in the clearance of 5-FU is catabolism by dihydropyrimidine dehydrogenase (DPYD), which is encoded by the *DPYD* gene. DPYD deficiency owing to *DPYD* genetic variation causes extreme 5-FU exposure [133], and was long ago recognized to cause 5-FU toxicity [134]. The most common polymorphism is a splice-site variant (DPYD*2A, IVS14 +1G → A) which interferes with the transcription of exon 14; there are other non-synonymous exonic variants that also have low-activity phenotypes [135]. The PWG and CPIC guidelines recommend that patients with the intermediate metabolizer phenotype are dose-decreased, while poor metabolizers should not be treated with fluoropyrimidines due to excessive toxicity risk [115, 116].

3.2.3 TPMT and Thiopurines

The thiopurines 6-mercaptopurine and 6-thioguanine used in ALL treatment are inactivated by thiopurine *S*-methyltransferase (TPMT). There is an inverse correlation between TPMT activity and exposure to 6-thioguanine nucleotides that are produced by an alternate metabolic pathway [136, 137]. Most of the low-activity variants in TPMT are non-synonymous polymorphisms that cause low protein expression [138], although important splice variants also exist [139]. Additionally, approximately 1–2% of Caucasians carry a trinucleotide repeat in the promoter region that confers an ultra-rapid metabolizer phenotype [140]. Both, the CPIC and PWG have recommended reducing doses or switching agents for patients that carry low-activity variants [115, 117]. A potentially clinically relevant association between TPMT variation and cisplatin-induced ototoxicity has also been reported, though the mechanism of this interaction or the metabolite responsible for ototoxicity have not been fully elucidated [141, 142].

3.2.4 CYP2D6 and Tamoxifen

Continuous tamoxifen therapy for five or more years is associated with an almost 50% improvement in relapse-free survival in estrogen receptor-positive breast cancer patients, although approximately one in four patients will relapse despite therapy [143]. Tamoxifen is a prodrug that requires metabolic activation by CYP2D6 (and other enzymes) to endoxifen (Fig. 6) [144].

More than 100 CYP2D6 genotypes have been curated by the CYP Alleles Nomenclature Committee, many with functional consequences [146]. Combining a patient's two alleles enables phenotype classifications as a poor, intermediate, extensive or ultra-rapid metabolizer, and this is correlated with endoxifen exposure [147, 148]. However, the relationship between endoxifen exposure and tamoxifen treatment outcome has not been well characterized [149]. More than 20 reports have been made of studies investigating

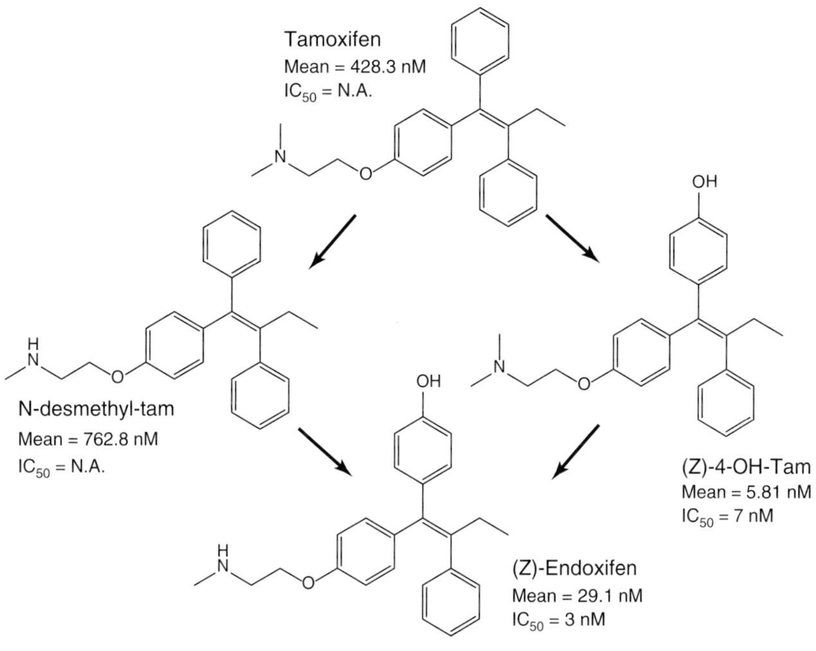

Fig. 6 Tamoxifen is an active pro-drug that is converted to the active metabolite *Z*-endoxifen. Comparing mean concentrations and *in-vitro* IC_{50}-values strongly suggests that endoxifen is responsible for the efficacy of tamoxifen treatment. Many enzymes are involved in the metabolism of tamoxifen, but CYP2D6 is the primary enzyme responsible for the conversion to endoxifen through these two parallel pathways. Reproduced with permission from Ref. [145]; © 2012, *The Oncologist*.

the potential association between CYP2D6 genotype and tamoxifen efficacy, with highly inconsistent results [150–153]. Heterogeneity among the studies, including an inadequate coverage of CYP2D6 alleles, has been suggested as a possible explanation for these discrepant results [145, 154], though it remains possible that the positive studies have merely represented false-positives and publication bias [155].

Prospective genotype-guided studies have shown that doubling the tamoxifen dose in intermediate metabolizers normalizes their endoxifen exposure with that of extensive metabolizers [156], but the clinical utility of this approach is untested. Both, the FDA and CPIC recommend that patients receiving tamoxifen avoid concomitant therapy with CYP2D6 inhibitors, such as the selective-serotonin reuptake inhibitor (SSRI) antidepressants which have themselves been shown to diminish endoxifen exposure [157].

3.3
Supportive Care

Cancer therapy is not limited to treating the tumor; it includes management of the morbidity associated with the disease and with treatment. Oncology supportive care covers many indications and drugs, and is paramount to improving a patient's quality of life [158]. The supportive care issues encountered in a given patient depend on the tumor type, treatment, and the individual. Common elements of supportive care include pain, nausea/vomiting (emesis) and depression/anxiety [159], each of which has data supporting the relevance of pharmacogenomics, particularly with *CYP2D6* or *CYP2C19*.

Narcotic analgesics are commonly used to treat mild to moderate pain, including pain from the tumor or metastases, or treatment-related pain such as that caused by surgery. The activation of codeine to the more potent morphine requires CYP2D6 metabolism. There are established relationships between CYP2D6 phenotype, the metabolic activation of codeine [160], and the efficacy and toxicity of codeine therapy [161]. Similar results have also been reported for the non-narcotic tramadol [162] which requires CYP2D6 activation, but the data are less compelling for other narcotics [163].

Antidepressant use – to treat depression as well as other issues – is common in cancer patients. Although the use of tricyclic antidepressants (TCAs) to manage depression has diminished somewhat, they are still used in some patients to treat depression or neuropathic pain [164]. TCAs are deactivated by CYP2D6 metabolism, and genetic variation influences the exposure of some TCAs [165]. CYP2C19, which is responsible for the interconversion between TCAs, has far less genetic variation than CYP2D6. There are only two loss of activity alleles (*2 and *3) and one gain of activity allele (*17) that are commonly found in patients, though these variants are found in high prevalence leading to substantial variation in CYP2C19 activity within populations [166]. The CPIC has released guidelines on the use of CYP2D6 and CYP2C19 genotype information to guide the dosing of TCAs, making recommendations for patients being treated for depression, but not for neuropathic pain which utilizes lower TCA doses [167].

More recently, the SSRIs have replaced TCAs as front-line therapy for depression. Whilst SSRIs are also metabolized by CYP2D6, the clinical relevance of *CYP2D6* genotyping has not been adequately demonstrated [168]. Some data are available supporting the influence of the *CYP2D6* genotype on venlafaxine efficacy [169, 170] and the *CYP2C19* genotype on sertraline efficacy [169, 171]. The PWG recommends decreasing the sertraline dose by 50% in poor metabolizers [115], despite there being little evidence that this improves treatment outcomes.

Chemotherapy-induced nausea and vomiting is a major problem in patient care, and 5-hydroxytryptamine 3 receptor antagonists (5-HT3RAs) represent a frontline anti-emetic treatment option. Most 5-HT3RAs are metabolized by CYP2D6; this includes ondansetron and tropisetron, which have been shown in some retrospective cohorts to be less effective in ultra-rapid metabolizers [172, 173]. It may be prudent to switch to a non-CYP2D6-dependent agent (e.g., granisetron) in patients who are known ultra-rapid metabolizers, but the benefit of this approach has not been adequately assessed in patients.

Glucose-6-phosphate dehydrogenase (G6PD) is responsible for forming NADPH which protects hemoglobin, and thus red blood cells, from oxidative damage caused by hydrogen peroxide (H_2O_2). Although very rare in Caucasians, G6PD deficiency is common worldwide [174]. Patients who are G6PD-deficient are highly sensitive to drug-induced hemolysis from treatment with drugs that produce H_2O_2 as a byproduct of their mechanism of action. One cancer-relevant example is rasburicase, which is used to prevent and treat tumor lysis syndrome in patients with hematologic cancers. Although this subject has not been investigated comprehensively, there are multiple case reports of patients demonstrating severe toxicity to rasburicase treatment who were eventually determined to be G6PD-deficient [175, 176]. The FDA label contraindicates rasburicase treatment in patients with known G6PD deficiency, and the CPIC has released treatment guidelines for this association [177].

4 Conclusions

Genetics has fundamentally changed the ways in which cancer is described, classified and treated. Both the somatic and germline genomes play a role in selecting the appropriate therapy for an individual patient. In current clinical practice, tumors are assessed genetically for specific oncogenic drivers that are validated targets for approved therapies; however, in the near future this pharmacogenetics approach is likely to give way to a comprehensive assessment of genomics, transcriptomics and epigenomics allowing the oncogenic driver of every tumor to be identified [178]. This, in turn, will enable selection of the drug that is most likely to be efficacious [179]. Analysis of the somatic genome requires concurrent profiling of the germline genome, creating a situation in which patient genetic information is readily available to guide decisions on treatment safety and dosing. While the application of pharmacogenetics to oncology has already made tremendous progress, the era of "personalized" cancer therapy has only just begun.

References

1. Edge, S., Byrd, D., Compton, C., Fritz, A., Greene, F., and Trotti, A. (2010) *y*, 7th edn, Springer, New York.
2. Fisher, B., Redmond, C., Brown, A., Wolmark, N. *et al.* (1981) Treatment of primary breast cancer with chemotherapy and tamoxifen. *N. Engl. J. Med.*, **305** (1), 1–6.
3. Coiffier, B., Lepage, E., Briere, J., Herbrecht, R. *et al.* (2002) CHOP chemotherapy plus rituximab compared with CHOP alone in elderly patients with diffuse large-B-cell lymphoma. *N. Engl. J. Med.*, **346** (4), 235–242.
4. McLeod, H.L. (2013) Cancer pharmacogenomics: early promise, but concerted effort needed. *Science*, **339** (6127), 1563–1566.
5. Collins, F.S. and Barker, A.D. (2007) Mapping the cancer genome. Pinpointing the genes involved in cancer will help chart a new course across the complex landscape of human malignancies. *Sci. Am.*, **296** (3), 50–57.
6. Kandoth, C., McLellan, M.D., Vandin, F., Ye, K. *et al.* (2013) Mutational landscape and significance across 12 major cancer types. *Nature*, **502** (7471), 333–339.
7. Cancer Genome Atlas Research Network (2008) Comprehensive genomic characterization defines human glioblastoma genes and core pathways. *Nature*, **455** (7216), 1061–1068.
8. Cancer Genome Atlas Research Network (2011) Integrated genomic analyses of ovarian carcinoma. *Nature*, **474** (7353), 609–615.
9. Cancer Genome Atlas Network (2012) Comprehensive molecular portraits of human breast tumours. *Nature*, **490** (7418), 61–70.
10. Cancer Genome Atlas Research Network (2012) Comprehensive genomic characterization of squamous cell lung cancers. *Nature*, **489** (7417), 519–525.
11. Cancer Genome Atlas Network (2012) Comprehensive molecular characterization of human colon and rectal cancer. *Nature*, **487** (7407), 330–337.
12. Cancer Genome Atlas Research Network (2013) Genomic and epigenomic landscapes of adult de novo acute myeloid leukemia. *N. Engl. J. Med.*, **368** (22), 2059–2074.

13. Cancer Genome Atlas Research Network, Kandoth, C., Schultz, N., Cherniack, A.D., Akbani, R., Liu, Y. *et al.* (2013) Integrated genomic characterization of endometrial carcinoma. *Nature*, **497** (7447), 67–73.
14. Puglisi, F., Minisini, A.M., De Angelis, C., and Arpino, G. (2012) Overcoming treatment resistance in HER2-positive breast cancer: potential strategies. *Drugs*, **72** (9), 1175–1193.
15. Early Breast Cancer Trialists' Collaborative Group (2011) Relevance of breast cancer hormone receptors and other factors to the efficacy of adjuvant tamoxifen: patient-level meta-analysis of randomised trials. *Lancet*, **378** (9793), 771–784.
16. Wolff, A.C., Hammond, M.E., Schwartz, J.N., Hagerty, K.L. *et al.* (2006) American Society of Clinical Oncology/College of American Pathologists Guideline Recommendations for Human Epidermal Growth Factor receptor 2 testing in breast cancer. *J. Clin. Oncol.*, **25** (1), 118–145.
17. Baselga, J., Norton, L., Albanell, J., Kim, Y., and Mendelsohn, J. (1998) Recombinant humanized anti-HER2 antibody (Herceptin™) enhances the antitumor activity of paclitaxel and doxorubicin against HER2/neu overexpressing human breast cancer xenografts. *Cancer Res.*, **58** (13), 2825–2831.
18. Slamon, D.J., Leyland-Jones, B., Shak, S., Fuchs, H. *et al.* (2001) Use of chemotherapy plus a monoclonal antibody against HER2 for metastatic breast cancer that overexpresses HER2. *N. Engl. J. Med.*, **344** (11), 783–792.
19. Verma, S., Miles, D., Gianni, L., Krop, I.E. *et al.* (2012) Trastuzumab emtansine for HER2-positive advanced breast cancer. *N. Engl. J. Med.*, **367** (19), 1783–1791.
20. Blackwell, K.L., Burstein, H.J., Storniolo, A.M., Rugo, H. *et al.* (2010) Randomized study of lapatinib alone or in combination with trastuzumab in women with ErbB2-positive, trastuzumab-refractory metastatic breast cancer. *J. Clin. Oncol.*, **28** (7), 1124–1130.
21. Baselga, J., Cortes, J., Kim, S.B., Im, S.A. *et al.* (2012) Pertuzumab plus trastuzumab plus docetaxel for metastatic breast cancer. *N. Engl. J. Med.*, **366** (2), 109–119.
22. Bose, R., Kavuri, S.M., Searleman, A.C., Shen, W. *et al.* (2013) Activating HER2 mutations in HER2 gene amplification negative breast cancer. *Cancer Discovery*, **3** (2), 224–237.
23. Paez, J.G., Janne, P.A., Lee, J.C., Tracy, S. *et al.* (2004) EGFR mutations in lung cancer: Correlation with clinical response to gefitinib therapy. *Science*, **304** (5676), 1497–1500.
24. Shepherd, F.A., Rodrigues Pereira, J., Ciuleanu, T., Tan, E.H. *et al.* (2005) Erlotinib in previously treated non-small-cell lung cancer. *N. Engl. J. Med.*, **353** (2), 123–132.
25. Rosell, R., Moran, T., Queralt, C., Porta, R, *et al.* (2009) Screening for epidermal growth factor receptor mutations in lung cancer. *N. Engl. J. Med.*, **361** (10), 958–967.
26. Maemondo, M., Inoue, A., Kobayashi, K., Sugawara, S. *et al.* (2010) Gefitinib or chemotherapy for non-small-cell lung cancer with mutated EGFR. *N. Engl. J. Med.*, **362** (25), 2380–2388.
27. Mok, T.S., Wu, Y.L., Thongprasert, S., Yang, C.H. *et al.* (2009) Gefitinib or carboplatin-paclitaxel in pulmonary adenocarcinoma. *N. Engl. J. Med.*, **361** (10), 947–957.
28. Pao, W., Miller, V.A., Politi, K.A., Riely, G.J. *et al.* (2005) Acquired resistance of lung adenocarcinomas to gefitinib or erlotinib is associated with a second mutation in the EGFR kinase domain. *PLoS Med.*, **2** (3), e73.
29. Sequist, L.V., Yang, J.C., Yamamoto, N., O'Byrne, K. *et al.* (2013) Phase III study of afatinib or cisplatin plus pemetrexed in patients with metastatic lung adenocarcinoma with EGFR mutations. *J. Clin. Oncol.*, **31** (27), 3327–3334.
30. Regales, L., Gong, Y., Shen, R., de Stanchina, E. *et al.* (2009) Dual targeting of EGFR can overcome a major drug resistance mutation in mouse models of EGFR mutant lung cancer. *J. Clin. Invest.*, **119** (10), 3000–3010.
31. Miller, V.A., Hirsh, V., Cadranel, J., Chen, Y.M. *et al.* (2012) Afatinib versus placebo for patients with advanced, metastatic non-small-cell lung cancer after failure of erlotinib, gefitinib, or both, and one or two lines of chemotherapy (LUX-lung 1): a phase 2b/3 randomised trial. *Lancet Oncol.*, **13** (5), 528–538.
32. Chung, K.Y., Shia, J., Kemeny, N.E., Shah, M. *et al.* (2005) Cetuximab shows activity in colorectal cancer patients with tumors that

do not express the epidermal growth factor receptor by immunohistochemistry. *J. Clin. Oncol.*, **23** (9), 1803–1810.
33. Moroni, M., Sartore-Bianchi, A., Benvenuti, S., Artale, S. *et al.* (2005) Somatic mutation of EGFR catalytic domain and treatment with gefitinib in colorectal cancer. *Ann. Oncol.*, **16** (11), 1848–1849.
34. Van Cutsem, E., Kohne, C.H., Hitre, E., Zaluski, J. *et al.* (2009) Cetuximab and chemotherapy as initial treatment for metastatic colorectal cancer. *N. Engl. J. Med.*, **360** (14), 1408–1417.
35. Lievre, A., Bachet, J.B., Le Corre, D., Boige, V. *et al.* (2006) KRAS mutation status is predictive of response to cetuximab therapy in colorectal cancer. *Cancer Res.*, **66** (8), 3992–3995.
36. Amado, R.G., Wolf, M., Peeters, M., Van Cutsem, E. *et al.* (2008) Wild-type KRAS is required for panitumumab efficacy in patients with metastatic colorectal cancer. *J. Clin. Oncol.*, **26** (10), 1626–1634.
37. Douillard, J., Oliner, K.S., Siena, S., Tabernero, J. *et al.* (2013) Panitumumab–FOLFOX4 treatment and RAS mutations in colorectal cancer. *N. Engl. J. Med.*, **369** (11), 1023–1034.
38. Kiyoi, H., Ohno, R., Ueda, R., Saito, H., and Naoe, T. (2002) Mechanism of constitutive activation of FLT3 with internal tandem duplication in the juxtamembrane domain. *Oncogene*, **21** (16), 2555–2563.
39. Choudhary, C., Schwable, J., Brandts, C., Tickenbrock., L. *et al.* (2005) AML-associated Flt3 kinase domain mutations show signal transduction differences compared with Flt3 ITD mutations. *Blood*, **106** (1), 265–273.
40. Kayser, S., Schlenk, R.F., Londono, M.C., Breitenbuecher, F. *et al.* (2009) Insertion of FLT3 internal tandem duplication in the tyrosine kinase domain-1 is associated with resistance to chemotherapy and inferior outcome. *Blood*, **114** (12), 2386–2392.
41. Smith, C.C., Wang, Q., Chin, C.S., Salerno, S. *et al.* (2012) Validation of ITD mutations in FLT3 as a therapeutic target in human acute myeloid leukaemia. *Nature*, **485** (7397), 260–263.
42. Zhang, W., Konopleva, M., Shi, Y.X., McQueen, T. *et al.* (2008) Mutant FLT3: a direct target of sorafenib in acute myelogenous leukemia. *J. Natl Cancer Inst.*, **100** (3), 184–198.
43. Ravandi, F., Cortes, J.E., Jones, D., Faderl, S. *et al.* (2010) Phase I/II study of combination therapy with sorafenib, idarubicin, and cytarabine in younger patients with acute myeloid leukemia. *J. Clin. Oncol.*, **28** (11), 1856–1862.
44. Moore, A.S., Faisal, A., Gonzalez de Castro, D., Bavetsias, V. *et al.* (2012) Selective FLT3 inhibition of FLT3-ITD+ acute myeloid leukaemia resulting in secondary D835Y mutation: a model for emerging clinical resistance patterns. *Leukemia*, **26** (7), 1462–1470.
45. Kottaridis, P.D., Gale, R.E., Frew, M.E., Harrison, G. *et al.* (2001) The presence of a FLT3 internal tandem duplication in patients with acute myeloid leukemia (AML) adds important prognostic information to cytogenetic risk group and response to the first cycle of chemotherapy: Analysis of 854 patients from the United Kingdom Medical Research Council AML 10 and 12 trials. *Blood*, **98** (6), 1752–1759.
46. Steer, E.J. and Cross, N.C (2002) Myeloproliferative disorders with translocations of chromosome 5q31-35: role of the platelet-derived growth factor receptor beta. *Acta Haematol.*, **107** (2), 113–122.
47. Wang, L.C., Swat, W., Fujiwara, Y., Davidson, L. *et al.* (1998) The TEL/ETV6 gene is required specifically for hematopoiesis in the bone marrow. *Genes Dev.*, **12** (15), 2392–2402.
48. Keene, P., Mendelow, B., Pinto, M.R., Bezwoda, W. *et al.* (1987) Abnormalities of chromosome 12p13 and malignant proliferation of eosinophils: a nonrandom association. *Br. J. Haematol.*, **67** (1), 25–31.
49. Carroll, M., Ohno-Jones, S., Tamura, S., Buchdunger, E. *et al.* (1997) CGP 57148, a tyrosine kinase inhibitor, inhibits the growth of cells expressing BCR-ABL, TEL-ABL, and TEL-PDGFR fusion proteins. *Blood*, **90** (12), 4947–4952.
50. Baccarani, M., Cilloni, D., Rondoni, M., Ottaviani, E. *et al.* (2007) The efficacy of imatinib mesylate in patients with FIP1L1-PDGFRalpha-positive hypereosinophilic syndrome. Results of a multicenter prospective study. *Haematologica*, **92** (9), 1173–1179.

51. Metzgeroth, G., Walz, C., Erben, P., Popp, H. *et al.* (2008) Safety and efficacy of imatinib in chronic eosinophilic leukaemia and hypereosinophilic syndrome: a phase-II study. *Br. J. Haematol.*, **143** (5), 707–715.
52. Singer, S., Rubin, B.P., Lux, M.L., Chen, C.J. *et al.* (2002) Prognostic value of KIT mutation type, mitotic activity, and histologic subtype in gastrointestinal stromal tumors. *J. Clin. Oncol.*, **20** (18), 3898–3905.
53. Hirota, S., Isozaki, K., Moriyama, Y., Hashimoto, K. *et al.* (1998) Gain-of-function mutations of c-kit in human gastrointestinal stromal tumors. *Science*, **279** (5350), 577–580.
54. Gronchi, A. (2013) Risk stratification models and mutational analysis: Keys to optimising adjuvant therapy in patients with gastrointestinal stromal tumour. *Eur. J. Cancer*, **49** (4), 884–892.
55. Heinrich, M.C., Griffith, D.J., Druker, B.J., Wait, C.L. *et al.* (2000) Inhibition of c-kit receptor tyrosine kinase activity by STI 571, a selective tyrosine kinase inhibitor. *Blood*, **96** (3), 925–932.
56. Demetri, G.D., von Mehren, M., Blanke, C.D., Van den Abbeele, A.D. *et al.* (2002) Efficacy and safety of imatinib mesylate in advanced gastrointestinal stromal tumors. *N. Engl. J. Med.*, **347** (7), 472–480.
57. Heinrich, M.C., Corless, C.L., Demetri, G.D., Blanke, C.D. *et al.* (2003) Kinase mutations and imatinib response in patients with metastatic gastrointestinal stromal tumor. *J. Clin. Oncol.*, **21** (23), 4342–4349.
58. Hodi, F.S., Corless, C.L., Giobbie-Hurder, A., Fletcher, J.A. *et al.* (2013) Imatinib for melanomas harboring mutationally activated or amplified KIT arising on mucosal, acral, and chronically sun-damaged skin. *J. Clin. Oncol.*, **31** (26), 3182–3190.
59. Wardelmann, E., Thomas, N., Merkelbach-Bruse, S., Pauls, K. *et al.* (2005) Acquired resistance to imatinib in gastrointestinal stromal tumours caused by multiple KIT mutations. *Lancet Oncol.*, **6** (4), 249–251.
60. Montemurro, M., Gelderblom, H., Bitz, U., Schutte, J. *et al.* (2013) Sorafenib as third- or fourth-line treatment of advanced gastrointestinal stromal tumour and pretreatment including both imatinib and sunitinib, and nilotinib: a retrospective analysis. *Eur. J. Cancer*, **49** (5), 1027–1031.
61. Drew, B.A., Burow, M.E., and Beckman, B.S. (2012) MEK5/ERK5 pathway: the first fifteen years. *Biochim. Biophys. Acta*, **1825** (1), 37–48.
62. Martin, G.S. (2003) Cell signaling and cancer. *Cancer Cell*, **4** (3),167–174.
63. Kohno, M., Tanimura, S., and Ozaki, K. (2011) Targeting the extracellular signal-regulated kinase pathway in cancer therapy. *Biol. Pharm. Bull.*, **34** (12), 1781–1784.
64. Janne, P.A., Shaw, A.T., Pereira, J.R., Jeannin, G. *et al.* (2013) Selumetinib plus docetaxel for KRAS-mutant advanced non-small-cell lung cancer: a randomised, multicentre, placebo-controlled, phase 2 study. *Lancet Oncol.*, **14** (1), 38–47.
65. Ascierto, P.A., Schadendorf, D., Berking, C., Agarwala, S.S. *et al.* (2013) MEK162 for patients with advanced melanoma harbouring NRAS or Val600 BRAF mutations: a non-randomised, open-label phase 2 study. *Lancet Oncol.*, **14** (3), 249–256.
66. Solit, D.B., Garraway, L.A., Pratilas, C.A., Sawai, A. *et al.* (2006) BRAF mutation predicts sensitivity to MEK inhibition. *Nature*, **439** (7074), 358–362.
67. Tsai, J., Lee, J.T., Wang, W., Zhang, J. *et al.* (2008) Discovery of a selective inhibitor of oncogenic B-raf kinase with potent antimelanoma activity. *Proc. Natl Acad. Sci. USA*, **105** (8), 3041–3046.
68. Joseph, E.W., Pratilas, C.A., Poulikakos, P.I., Tadi, M. *et al.* (2010) The RAF inhibitor PLX4032 inhibits ERK signaling and tumor cell proliferation in a V600E BRAF-selective manner. *Proc. Natl Acad. Sci. USA*, **107** (33), 14903–14908.
69. Flaherty, K.T., Puzanov, I., Kim, K.B., Ribas, A. *et al.* (2010) Inhibition of mutated, activated BRAF in metastatic melanoma. *N. Engl. J. Med.*, **363** (9), 809–819.
70. Chapman, P.B., Hauschild, A., Robert, C., Haanen, J.B. *et al.* (2011) Improved survival with vemurafenib in melanoma with BRAF V600E mutation. *N. Engl. J. Med.*, **364** (26), 2507–2516.
71. Prahallad, A., Sun, C., Huang, S., Di Nicolantonio, F. *et al.* (2012) Unresponsiveness of colon cancer to BRAF(V600E) inhibition through feedback activation of EGFR. *Nature*, **483** (7387), 100–103.
72. Hauschild, A., Grob, J.J., Demidov, L.V., Jouary, T. *et al.* (2012) Dabrafenib in BRAF-mutated metastatic melanoma:

a multicentre, open-label, phase 3 randomised controlled trial. *Lancet*, **380** (9839), 358–365.
73. Flaherty, K.T., Infante, J.R., Daud, A., Gonzalez, R. *et al.* (2012) Combined BRAF and MEK inhibition in melanoma with BRAF V600 mutations. *N. Engl. J. Med.*, **367** (18), 1694–1703.
74. Murugan, A.K., Dong, J., Xie, J., and Xing, M. (2009) MEK1 mutations, but not ERK2 mutations, occur in melanomas and colon carcinomas, but none in thyroid carcinomas. *Cell Cycle*, **8** (13), 2122–2124.
75. Bromberg-White, J.L., Andersen, N.J., and Duesbery, N.S. (2012) MEK genomics in development and disease. *Brief. Funct. Genomics*, **11** (4), 300–310.
76. O'Neil, B.H., Goff, L.W., Kauh, J.S., Strosberg, J.R. *et al.* (2011) Phase II study of the mitogen-activated protein kinase 1/2 inhibitor selumetinib in patients with advanced hepatocellular carcinoma. *J. Clin. Oncol.*, **29** (17), 2350–2356.
77. Druker, B.J., Tamura, S., Buchdunger, E., Ohno, S. *et al.* (1996) Effects of a selective inhibitor of the abl tyrosine kinase on the growth of bcr-abl positive cells. *Nat. Med.*, **2** (5), 561–566.
78. O'Brien, S.G., Guilhot, F., Larson, R.A., Gathmann, I. *et al.* (2003) Imatinib compared with interferon and low-dose cytarabine for newly diagnosed chronic-phase chronic myeloid leukemia. *N. Engl. J. Med.*, **348** (11), 994–1004.
79. Gorre, M.E., Mohammed, M., Ellwood, K., Hsu, N. *et al.* (2001) Clinical resistance to STI-571 cancer therapy caused by BCR-ABL gene mutation or amplification. *Science*, **293** (5531), 876–880.
80. Kantarjian, H., Giles, F., Wunderle, L., Bhalla, K. *et al.* (2006) Nilotinib in imatinib-resistant CML and Philadelphia chromosome-positive ALL. *N. Engl. J. Med.*, **354** (24), 2542–2551.
81. Talpaz, M., Shah, N.P., Kantarjian, H., Donato, N. *et al.* (2006) Dasatinib in imatinib-resistant Philadelphia chromosome-positive leukemias. *N. Engl. J. Med.*, **354** (24), 2531–2541.
82. Saglio, G., Kim, D.W., Issaragrisil, S., le Coutre, P. *et al.* (2010) Nilotinib versus imatinib for newly diagnosed chronic myeloid leukemia. *N. Engl. J. Med.*, **362** (24), 2251–2259.
83. Kantarjian, H., Shah, N.P., Hochhaus, A., Cortes, J. *et al.* (2010) Dasatinib versus imatinib in newly diagnosed chronic-phase chronic myeloid leukemia. *N. Engl. J. Med.*, **362** (24), 2260–2270.
84. Skaggs, B.J., Gorre, M.E., Ryvkin, A., Burgess, M.R. *et al.* (2006) Phosphorylation of the ATP-binding loop directs oncogenicity of drug-resistant BCR-ABL mutants. *Proc. Natl Acad. Sci. USA*, **103** (51), 19466–19471.
85. O'Hare, T., Shakespeare, W.C., Zhu, X., Eide, C.A. *et al.* (2009) AP24534, a pan-BCR-ABL inhibitor for chronic myeloid leukemia, potently inhibits the T315I mutant and overcomes mutation-based resistance. *Cancer Cell*, **16** (5), 401–412.
86. Cortes, J.E., Kantarjian, H., Shah, N.P., Bixby, D. *et al.* (2012) Ponatinib in refractory Philadelphia chromosome-positive leukemias. *N. Engl. J. Med.*, **367** (22), 2075–2088.
87. Soda, M., Choi, Y.L., Enomoto, M., Takada, S. *et al.* (2007) Identification of the transforming EML4-ALK fusion gene in non-small-cell lung cancer. *Nature*, **448** (7153), 561–566.
88. Soda, M., Takada, S., Takeuchi, K., Choi, Y.L. *et al.* (2008) A mouse model for EML4-ALK-positive lung cancer. *Proc. Natl Acad. Sci. USA*, **105** (50), 19893–19897.
89. Cui, J.J., Tran-Dube, M., Shen, H., Nambu, M. *et al.* (2011) Structure-based drug design of crizotinib (PF-02341066), a potent and selective dual inhibitor of mesenchymal-epithelial transition factor (c-MET) kinase and anaplastic lymphoma kinase (ALK). *J. Med. Chem.*, **54** (18), 6342–6363.
90. Kwak, E.L., Bang, Y.J., Camidge, D.R., Shaw, A.T. *et al.* (2010) Anaplastic lymphoma kinase inhibition in non-small-cell lung cancer. *N. Engl. J. Med.*, **363** (18), 1693–1703.
91. Choi, Y.L., Soda, M., Yamashita, Y., Ueno, T. *et al.* (2010) EML4-ALK mutations in lung cancer that confer resistance to ALK inhibitors. *N. Engl. J. Med.*, **363** (18), 1734–1739.
92. Sakamoto, H., Tsukaguchi, T., Hiroshima, S., Kodama, T. *et al.* (2011) CH5424802, a selective ALK inhibitor capable of blocking the resistant gatekeeper mutant. *Cancer Cell*, **19** (5), 679–690.

93. Di Croce, L., Raker, V.A., Corsaro, M., Fazi, F. *et al.* (2002) Methyltransferase recruitment and DNA hypermethylation of target promoters by an oncogenic transcription factor. *Science*, **295** (5557), 1079–1082.
94. Pandolfi, P.P. (2001) In vivo analysis of the molecular genetics of acute promyelocytic leukemia. *Oncogene*, **20** (40), 5726–5735.
95. Huang, M.E., Ye, Y.C., Chen, S.R., Chai, J.R. *et al.* (1988) Use of all-trans retinoic acid in the treatment of acute promyelocytic leukemia. *Blood*, **72** (2), 567–572.
96. Falini, B., Gionfriddo, I., Cecchetti, F., Ballanti, S. *et al.* (2011) Acute myeloid leukemia with mutated nucleophosmin (NPM1): any hope for a targeted therapy? *Blood Rev.*, **25** (6), 247–254.
97. Mohamad, H.B. and Apffelstaedt, J.P. (2008) Counseling for male BRCA mutation carriers: a review. *Breast*, **17** (5), 441–450.
98. Patel, A.G., Sarkaria, J.N., and Kaufmann, S.H. (2011) Nonhomologous end joining drives poly(ADP-ribose) polymerase (PARP) inhibitor lethality in homologous recombination-deficient cells. *Proc. Natl Acad. Sci. USA*, **108** (8), 3406–3411.
99. Bryant, H.E., Schultz, N., Thomas, H.D., Parker, K.M. *et al.* (2005) Specific killing of BRCA2-deficient tumours with inhibitors of poly(ADP-ribose) polymerase. *Nature*, **434** (7035), 913–917.
100. Tutt, A., Robson, M., Garber, J.E., Domchek, S.M. *et al.* (2010) Oral poly(ADP-ribose) polymerase inhibitor olaparib in patients with BRCA1 or BRCA2 mutations and advanced breast cancer: a proof-of-concept trial. *Lancet*, **376** (9737), 235–244.
101. Audeh, M.W., Carmichael, J., Penson, R.T., Friedlander, M. *et al.* (2010) Oral poly(ADP-ribose) polymerase inhibitor olaparib in patients with BRCA1 or BRCA2 mutations and recurrent ovarian cancer: a proof-of-concept trial. *Lancet*, **376** (9737), 245–251.
102. O'Shaughnessy, J., Osborne, C., Pippen, J.E., Yoffe, M. *et al.* (2011) Iniparib plus chemotherapy in metastatic triple-negative breast cancer. *N. Engl. J. Med.*, **364** (3), 205–214.
103. Patel, A.G., De Lorenzo, S.B., Flatten, K.S., Poirier, G.G., and Kaufmann, S.H. (2012) Failure of iniparib to inhibit poly(ADP-ribose) polymerase in vitro. *Clin. Cancer Res.*, **18** (6), 1655–1662.
104. Relling, M.V., Altman, R.B., Goetz, M.P., and Evans, W.E. (2010) Clinical implementation of pharmacogenomics: overcoming genetic exceptionalism. *Lancet Oncol.*, **11** (6), 507–509.
105. Pavlos, R., Mallal, S., and Phillips, E. (2012) HLA and pharmacogenetics of drug hypersensitivity. *Pharmacogenomics*, **13** (11), 1285–1306.
106. Lai-Goldman, M. and Faruki, H. (2008) Abacavir hypersensitivity: a model system for pharmacogenetic test adoption. *Genet. Med.*, **10** (12), 874–878.
107. Spraggs, C.F., Budde, L.R., Briley, L.P., Bing, N. *et al.* (2011) HLA-DQA1*02:01 is a major risk factor for lapatinib-induced hepatotoxicity in women with advanced breast cancer. *J. Clin. Oncol.*, **29** (6), 667–673.
108. Baldwin, R.M., Owzar, K., Zembutsu, H., Chhibber, A. *et al.* (2012) A genome-wide association study identifies novel loci for paclitaxel-induced sensory peripheral neuropathy in CALGB 40101. *Clin. Cancer Res.*, **18** (18), 5099–5109.
109. Delague, V., Jacquier, A., Hamadouche, T., Poitelon, Y. *et al.* (2007) Mutations in FGD4 encoding the rho GDP/GTP exchange factor FRABIN cause autosomal recessive Charcot–Marie–Tooth type 4H. *Am. J. Hum. Genet.*, **81** (1), 1–16.
110. Ingle, J.N., Schaid, D.J., Goss, P.E., Liu, M. *et al.* (2010) Genome-wide associations and functional genomic studies of musculoskeletal adverse events in women receiving aromatase inhibitors. *J. Clin. Oncol.*, **28** (31), 4674–4682.
111. Liu, M., Wang, L., Bongartz, T., Hawse, J.R. *et al.* (2012) Aromatase inhibitors, estrogens and musculoskeletal pain: estrogen-dependent T-cell leukemia 1A (TCL1A) gene-mediated regulation of cytokine expression. *Breast Cancer Res.*, **14** (2), R41.
112. Bosma, P.J., Seppen, J., Goldhoorn, B., Bakker, C. *et al.* (1994) Bilirubin UDP-glucuronosyltransferase 1 is the only relevant bilirubin glucuronidating isoform in man. *J. Biol. Chem.*, **269** (27), 17960–17964.
113. Singer, J.B., Shou, Y., Giles, F., Kantarjian, H.M. *et al.* (2007) UGT1A1 promoter

polymorphism increases risk of nilotinib-induced hyperbilirubinemia. *Leukemia*, **21** (11), 2311–2315.

114. Relling, M.V. and Klein, T.E. (2011) CPIC: clinical pharmacogenetics implementation consortium of the pharmacogenomics research network. *Clin. Pharmacol. Ther.*, **89** (3), 464–467.
115. Swen, J.J., Nijenhuis, M., de Boer, A., Grandia, L. *et al.* (2011) Pharmacogenetics: from bench to byte an update of guidelines. *Clin. Pharmacol. Ther.*, **89** (5), 662–673.
116. Caudle, K.E., Thorn, C.F., Klein, T.E., Swen, J.J. *et al.* (2013) Clinical pharmacogenetics implementation consortium (CPIC) guidelines for dihydropyrimidine dehydrogenase genotype and fluoropyrimidine dosing. *Clin. Pharmacol. Ther.*, **94** (6), 640–645.
117. Relling, M.V., Gardner, E.E., Sandborn, W.J., Schmiegelow, K. *et al.* (2011) Clinical pharmacogenetics implementation consortium guidelines for thiopurine methyltransferase genotype and thiopurine dosing. *Clin. Pharmacol. Ther.*, **89** (3), 387–391.
118. UGT Nomenclature Committee (2009) UGT Alleles Nomenclature Home Page [Internet], [updated June 2005, cited May 16, 2013]; *http://www.pharmacogenomics.pha.ulaval.ca/cms/ugt_alleles/* (accessed 15 April 2014).
119. Bosma, P.J., Chowdhury, J.R., Bakker, C., Gantla, S. *et al.* (1995) The genetic basis of the reduced expression of bilirubin UDP-glucuronosyltransferase 1 in gilbert's syndrome. *N. Engl. J. Med.*, **333** (18), 1171–1175.
120. Beutler, E., Gelbart, T., and Demina, A. (1998) Racial variability in the UDP-glucuronosyltransferase 1 (UGT1A1) promoter: a balanced polymorphism for regulation of bilirubin metabolism? *Proc. Natl Acad. Sci. USA*, **95** (14), 8170–8174.
121. Iyer, L., Hall, D., Das, S., Mortell, M.A. *et al.* (1999) Phenotype-genotype correlation of *in vitro* SN-38 (active metabolite of irinotecan) and bilirubin glucuronidation in human liver tissue with UGT1A1 promoter polymorphism. *Clin. Pharmacol. Ther.*, **65** (5), 576–582.
122. Innocenti, F., Undevia, S.D., Iyer, L., Chen, P.X. *et al.* (2004) Genetic variants in the UDP-glucuronosyltransferase 1A1 gene predict the risk of severe neutropenia of irinotecan. *J. Clin. Oncol.*, **22** (8), 1382–1388.
123. Iyer, L., Das, S., Janisch, L., Wen, M. *et al.* (2002) UGT1A1*28 polymorphism as a determinant of irinotecan disposition and toxicity. *Pharmacogenomics J.*, **2** (1), 43–47.
124. Toffoli, G., Cecchin, E., Corona, G., Russo, A. *et al.* (2006) The role of UGT1A1*28 polymorphism in the pharmacodynamics and pharmacokinetics of irinotecan in patients with metastatic colorectal cancer. *J. Clin. Oncol.*, **24** (19), 3061–3068.
125. Hoskins, J.M., Goldberg, R.M., Qu, P., Ibrahim, J.G., and McLeod, H.L. (2007) UGT1A1*28 genotype and irinotecan-induced neutropenia: dose matters. *J. Natl Cancer Inst.*, **99** (17), 1290–1295.
126. Whirl-Carrillo, M., McDonagh, E.M., Hebert, J.M., Gong, L. *et al.* (2012) Pharmacogenomics knowledge for personalized medicine. *Clin. Pharmacol. Ther.*, **92** (4), 414–417.
127. Evaluation of Genomic Applications in Practice and Prevention (EGAPP) Working Group (2009) Recommendations from the EGAPP working group: can tumor gene expression profiling improve outcomes in patients with breast cancer? *Genet. Med.*, **11** (1), 66–73.
128. Cecchin, E., Innocenti, F., D'Andrea, M., Corona, G. *et al.* (2009) Predictive role of the UGT1A1, UGT1A7, and UGT1A9 genetic variants and their haplotypes on the outcome of metastatic colorectal cancer patients treated with fluorouracil, leucovorin, and irinotecan. *J. Clin. Oncol.*, **27** (15), 2457–2465.
129. Sai, K., Saito, Y., Tatewaki, N., Hosokawa, M. *et al.* (2010) Association of carboxylesterase 1A genotypes with irinotecan pharmacokinetics in Japanese cancer patients. *Br. J. Clin. Pharmacol.*, **70** (2), 222–233.
130. Xiang, X., Jada, S.R., Li, H.H., Fan, L. *et al.* (2006) Pharmacogenetics of SLCO1B1 gene and the impact of *1b and *15 haplotypes on irinotecan disposition in Asian cancer patients. *Pharmacogenet. Genom.*, **16** (9), 683–691.
131. Di Martino, M.T., Arbitrio, M., Leone, E., Guzzi, P.H. *et al.* (2011) Single nucleotide polymorphisms of ABCC5 and ABCG1 transporter genes correlate to irinotecan-associated gastrointestinal toxicity in

colorectal cancer patients: a DMET microarray profiling study. *Cancer Biol. Ther.*, **12** (9), 780–787.

132. Mathijssen, R.H., Marsh, S., Karlsson, M.O., Xie, R. *et al.* (2003) Irinotecan pathway genotype analysis to predict pharmacokinetics. *Clin. Cancer Res.*, **9** (9), 3246–3253.

133. Diasio, R.B., Beavers, T.L., and Carpenter, J.T. (1988) Familial deficiency of dihydropyrimidine dehydrogenase. biochemical basis for familial pyrimidinemia and severe 5-fluorouracil-induced toxicity. *J. Clin. Invest.*, **81** (1), 47–51.

134. Tuchman, M., Stoeckeler, J.S., Kiang, D.T., O'Dea, R.F. *et al.* (1985) Familial pyrimidinemia and pyrimidinuria associated with severe fluorouracil toxicity. *N. Engl. J. Med.*, **313** (4), 245–249.

135. Mattison, L.K., Soong, R., and Diasio, R.B. (2002) Implications of dihydropyrimidine dehydrogenase on 5-fluorouracil pharmacogenetics and pharmacogenomics. *Pharmacogenomics*, **3** (4), 485–492.

136. Lennard, L., Van Loon, J.A., and Weinshilboum, R.M. (1989) Pharmacogenetics of acute azathioprine toxicity: relationship to thiopurine methyltransferase genetic polymorphism. *Clin. Pharmacol. Ther.*, **46** (2), 149–154.

137. Lennard, L., Lilleyman, J.S., Van Loon, J., and Weinshilboum, R.M. (1990) Genetic variation in response to 6-mercaptopurine for childhood acute lymphoblastic leukaemia. *Lancet*, **336** (8709), 225–229.

138. Tai, H.L., Krynetski, E.Y., Schuetz, E.G., Yanishevski, Y., and Evans, W.E. (1997) Enhanced proteolysis of thiopurine S-methyltransferase (TPMT) encoded by mutant alleles in humans (TPMT*3A, TPMT*2): mechanisms for the genetic polymorphism of TPMT activity. *Proc. Natl Acad. Sci. USA*, **94** (12), 6444–6449.

139. Wang, L., Pelleymounter, L., Weinshilboum, R., Johnson, J.A. *et al.* (2010) Very important pharmacogene summary: thiopurine S-methyltransferase. *Pharmacogenet. Genom.*, **20** (6), 401–405.

140. Roberts, R.L., Gearry, R.B., Bland, M.V., Sies, C.W. *et al.* (2008) Trinucleotide repeat variants in the promoter of the thiopurine S-methyltransferase gene of patients exhibiting ultra-high enzyme activity. *Pharmacogenet. Genom.*, **18** (5), 434–438.

141. Ross, C.J., Katzov-Eckert, H., Dube, M.P., Brooks, B. *et al.* (2009) Genetic variants in TPMT and COMT are associated with hearing loss in children receiving cisplatin chemotherapy. *Nat. Genet.*, **41** (12), 1345–1349.

142. Pussegoda, K., Ross, C.J., Visscher, H., Yazdanpanah, M. *et al.* (2013) Replication of TPMT and ABCC3 genetic variants highly associated with cisplatin-induced hearing loss in children. *Clin. Pharmacol. Ther.*, **94** (2), 243–251.

143. Hackshaw, A., Roughton, M., Forsyth, S., Monson, K. *et al.* (2011) Long-term benefits of 5 years of tamoxifen: 10-year follow-up of a large randomized trial in women at least 50 years of age with early breast cancer. *J. Clin. Oncol.*, **29** (13), 1657–1663.

144. Murdter, T.E., Schroth, W., Bacchus-Gerybadze, L., Winter, S. *et al.* (2011) Activity levels of tamoxifen metabolites at the estrogen receptor and the impact of genetic polymorphisms of phase I and II enzymes on their concentration levels in plasma. *Clin. Pharmacol. Ther.*, **89** (5), 708–717.

145. Hertz, D.L., McLeod, H.L., and Irvin, W.J. (2012) Tamoxifen and CYP2D6: a contradiction of data. *Oncologist*, **17** (5), 620–630.

146. Ingelman-Sundberg, M. (2008) Home Page of the Human Cytochrome P450 (CYP) Allele Nomenclature Database [Internet] [updated 4 September 2008]. Available at: *http://www.cypalleles.ki.se* (accessed 15 April 2014).

147. Borges, S., Desta, Z., Jin, Y., Faouzi, A. *et al.* (2010) Composite functional genetic and comedication CYP2D6 activity score in predicting tamoxifen drug exposure among breast cancer patients. *J. Clin. Pharmacol.*, **50** (4), 450–458.

148. Zanger, U., Raimundo, S., and Eichelbaum, M. (2004) Cytochrome P450 2D6: overview and update on pharmacology, genetics, biochemistry. *Naunyn Schmiedeberg's Arch. Pharmacol.*, **369** (1), 23–37.

149. Madlensky, L., Natarajan, L., Tchu, S., Pu, M. *et al.* (2011) Tamoxifen metabolite concentrations, CYP2D6 genotype, and breast cancer outcomes. *Clin. Pharmacol. Ther.*, **89** (5), 718–725.

150. Rae, J.M., Drury, S., Hayes, D.F., Stearns, V. *et al.* (2012) CYP2D6 and UGT2B7 genotype and risk of recurrence in

tamoxifen-treated breast cancer patients. *J. Natl Cancer Inst.*, **104** (6), 452–460.

151. Regan, M.M., Leyland-Jones, B., Bouzyk, M., Pagani, O. *et al.* (2012) CYP2D6 genotype and tamoxifen response in postmenopausal women with endocrine-responsive breast cancer: the Breast International Group 1-98 trial. *J. Natl Cancer Inst.*, **104** (6), 441–451.
152. Goetz, M.P., Suman, V.J., Hoskin, T.L., Gnant, M. *et al.* (2013) CYP2D6 metabolism and patient outcome in the Austrian Breast and Colorectal Cancer Study Group trial (ABCSG) 8. *Clin. Cancer Res.*, **19** (2), 500–507.
153. Schroth, W., Goetz, M.P., Hamann, U., Fasching, P.A. *et al.* (2009) Association between CYP2D6 polymorphisms and outcomes among women with early stage breast cancer treated with tamoxifen. *J. Am. Med. Assoc.*, **302** (13), 1429–1436.
154. Kiyotani, K., Mushiroda, T., Zembutsu, H., and Nakamura, Y. (2013) Important and critical scientific aspects in pharmacogenomics analysis: lessons from controversial results of tamoxifen and CYP2D6 studies. *J. Hum. Genet.*, **58** (6), 327–333.
155. Lash, T.L., Lien, E.A., Sørensen, H.T., and Hamilton-Dutoit, S. (2009) Genotype-guided tamoxifen therapy: time to pause for reflection? *Lancet Oncol.*, **10** (8), 825–833.
156. Irvin, W.J., Walko, C.M., Weck, K.E., Ibrahim, J.G. *et al.* (2011) Genotype-guided tamoxifen dosing increases active metabolite exposure in women with reduced CYP2D6 metabolism: a multicenter study. *J. Clin. Oncol..*, **29** (24), 3232–3239.
157. Stearns, V., Johnson, M.D., Rae, J.M., Morocho, A. *et al.* (2003) Active tamoxifen metabolite plasma concentrations after coadministration of tamoxifen and the selective serotonin reuptake inhibitor paroxetine. *J. Natl Cancer Inst..*, **95** (23), 1758–1764.
158. Ferrell, B., Paice, J., and Koczywas, M. (2008) New standards and implications for improving the quality of supportive oncology practice. *J. Clin. Oncol.*, **26** (23), 3824–3831.
159. Potter, J., Hami, F., Bryan, T., and Quigley, C. (2003) Symptoms in 400 patients referred to palliative care services: prevalence and patterns. *Palliat. Med.*, **17** (4), 310–314.
160. Kirchheiner, J., Schmidt, H., Tzvetkov, M., Keulen, J.T. *et al.* (2007) Pharmacokinetics of codeine and its metabolite morphine in ultra-rapid metabolizers due to CYP2D6 duplication. *Pharmacogenomics J.*, **7** (4), 257–265.
161. Lotsch, J., Rohrbacher, M., Schmidt, H., Doehring, A., Brockmoller, J., and Geisslinger, G. (2009) Can extremely low or high morphine formation from codeine be predicted prior to therapy initiation? *Pain*, **144** (1-2), 119–124.
162. Stamer, U.M., Musshoff, F., Kobilay, M., Madea, B., Hoeft, A., and Stuber, F. (2007) Concentrations of tramadol and O-desmethyltramadol enantiomers in different CYP2D6 genotypes. *Clin. Pharmacol. Ther.*, **82** (1), 41–47.
163. Crews, K.R., Gaedigk, A., Dunnenberger, H.M., Klein, T.E. *et al.* (2012) Clinical Pharmacogenetics Implementation Consortium (CPIC) guidelines for codeine therapy in the context of cytochrome P450 2D6 (CYP2D6) genotype. *Clin. Pharmacol. Ther.*, **91** (2), 321–326.
164. Laird, B., Colvin, L., and Fallon, M. (2008) Management of cancer pain: basic principles and neuropathic cancer pain. *Eur. J. Cancer*, **44** (8), 1078–1082.
165. Dalen, P., Dahl, M.L., Bernal Ruiz, M.L., Nordin, J., and Bertilsson, L. (1998) 10-hydroxylation of nortriptyline in white persons with 0, 1, 2, 3, and 13 functional CYP2D6 genes. *Clin. Pharmacol. Ther.*, **63** (4), 444–452.
166. Lee, S.J. (2012) Clinical application of CYP2C19 pharmacogenetics toward more personalized medicine. *Front. Genet.*, **3**, 318.
167. Hicks, J.K., Swen, J.J., Thorn, C.F., Sangkuhl, K. *et al.* (2013) Clinical pharmacogenetics implementation consortium guideline for CYP2D6 and CYP2C19 genotypes and dosing of tricyclic antidepressants. *Clin. Pharmacol. Ther.*, **93** (5), 402–408.
168. Thomas, K.L. and Ellingrod, V.L. (2009) Pharmacogenetics of selective serotonin reuptake inhibitors and associated adverse drug reactions. *Pharmacotherapy*, **29** (7), 822–831.
169. Brandl, E.J., Tiwari, A.K., Zhou, X., Deluce, J. *et al.* (2014) Influence of CYP2D6 and CYP2C19 gene variants on antidepressant response in obsessive-compulsive disorder. *Pharmacogenomics J.*, **14** (2), 176–181.

170. Lobello, K.W., Preskorn, S.H., Guico-Pabia, C.J., Jiang, Q. *et al.* (2010) Cytochrome P450 2D6 phenotype predicts antidepressant efficacy of venlafaxine: a secondary analysis of 4 studies in major depressive disorder. *J. Clin. Psychiatry*, **71** (11), 1482–1487.
171. Wang, J.H., Liu, Z.Q., Wang, W., Chen, X.P. *et al.* (2001) Pharmacokinetics of sertraline in relation to genetic polymorphism of CYP2C19. *Clin. Pharmacol. Ther.*, **70** (1), 42–47.
172. Candiotti, K.A., Birnbach, D.J., Lubarsky, D.A., Nhuch, F. *et al.* (2005) The impact of pharmacogenomics on postoperative nausea and vomiting: do CYP2D6 allele copy number and polymorphisms affect the success or failure of ondansetron prophylaxis? *Anesthesiology*, **102** (3), 543–549.
173. Kaiser, R., Sezer, O., Papies, A., Bauer, S. *et al.* (2002) Patient-tailored antiemetic treatment with 5-hydroxytryptamine type 3 receptor antagonists according to cytochrome P-450 2D6 genotypes. *J. Clin. Oncol.*, **20** (12), 2805–2811.
174. Mason, P.J., Bautista, J.M., and Gilsanz, F. (2007) G6PD deficiency: the genotype-phenotype association. *Blood Rev.*, **21** (5), 267–283.
175. Browning, L.A. and Kruse, J.A. (2005) Hemolysis and methemoglobinemia secondary to rasburicase administration. *Ann. Pharmacother.*, **39** (11), 1932–1935.
176. Ng, J.S., Edwards, E.M., and Egelund, T.A. (2012) Methemoglobinemia induced by rasburicase in a pediatric patient: a case report and literature review. *J. Oncol. Pharm. Pract.*, **18** (4), 425–431.
177. McDonagh, E.M., Thorn, C.F., Bautista, J.M., Youngster, I., Altman, R.B. and Klein, T.E. (2012) PharmGKB summary: very important pharmacogene information for G6PD. *Pharmacogenet. Genom.*, **22** (3), 219–228.
178. Roychowdhury, S., Iyer, M.K., Robinson, D.R., Lonigro, R.J. *et al.* (2011) Personalized oncology through integrative high-throughput sequencing: a pilot study. *Sci. Transl. Med.*, **3** (111), 111ra121.
179. Tsimberidou, A.M., Iskander, N.G., Hong, D.S., Wheler, J.J. *et al.* (2012) Personalized medicine in a phase I clinical trials program: the MD Anderson Cancer Center Initiative. *Clin. Cancer Res.*, **18** (22), 6373–6383.

16
Carbon Nanotubes for Enhanced Biopharmaceutical Delivery

Harikrishna Rallapalli[1,2] *and Bryan Ronain Smith*[2]
[1] *New York University Langone Medical Center, Center for Biomedical Imaging, 660 1st Avenue, Room 420, New York, NY 10016, USA*
[2] *Stanford University, Clark Center, 318 Campus Drive, E-150, Stanford, CA 94305, USA*

Keywords

Carbon nanotubes:
Cylindrically arranged carbon atoms formed by rolled sheets of graphene with all sp^2 bonds. A member of the fullerene family.

Translational Medicine Cancer: Advances in Molecular Biology and Medicine,
First Edition. Edited by Robert A. Meyers.

Biopharmaceuticals
Therapeutic compounds extracted from biological sources, or synthesized using biological methods.

Functionalization
The process of modifying nanoparticles by means of chemical reaction or surface adhesion, including adding functional groups, to produce a desired effect.

SWNTs
Single-walled carbon nanotubes.

Delivery
The transport of a compound to its target site within the body.

Carbon nanotubes (CNTs) can be applied as versatile biopharmaceutical delivery systems due to their high drug-loading capacity, excellent cell-penetrating ability, and customizable surface chemistry. Biopharmaceuticals are potent therapeutics when applied correctly, but are challenged by significant drawbacks. The coupling of biopharmaceuticals to CNTs can mitigate these issues, thus additively or synergistically improving their performance. Improved therapeutic effects, enhanced (targeted) delivery to specific tissue types, highly controlled release profiles, and even the visualization of *in-vivo* biodistribution comprise just a few examples of the benefits of such complexes. Furthermore, novel complexes have been tailored to combat cancer and other difficult-to-treat diseases, with promising success. In this chapter, the application of CNTs for the delivery of biopharmaceuticals to disease sites *in vitro* and *in vivo* will be discussed.

1 Introduction

Since their discovery in 1991 by Sumio Iijima, carbon nanotubes (CNTs) have been applied as versatile tools for diverse applications ranging from efficient electronics and high-strength materials for construction to many biological and medical uses [1]. In this chapter, the application of CNTs for the delivery of biopharmaceuticals to disease sites *in vitro* and *in vivo* will be discussed. Although, the early stages of CNT-mediated biopharmaceutical delivery are still being witnessed, ample investigations have suggested that this approach has tremendous potential due to the unique biological and modular properties displayed by CNTs. In many ways, CNTs are analogous to a "tactical rifle" – modifications can be made to both tools according to the needs of the user and the task – making them more lethal to their targets.

Biopharmaceuticals are therapeutic compounds extracted from biological sources or synthesized using biological methods. They are in many ways completely distinct from artificially synthesized pharmaceuticals, and often come with many added benefits. While traditional pharmaceuticals are screened from a list of potential compounds with desired effects, biopharmaceuticals are generally tailored to treat specific biological defects. Quite often, biopharmaceutical mechanisms interact with the same pathways as "typical" disease function, making them highly specific treatments. Their high specificity reduces the risk of side effects, making biopharmaceuticals typically safer than traditional pharmaceuticals [2, 3].

The combination of CNTs and biopharmaceuticals consistently outperforms separate biopharmaceutical treatment or nanoparticle treatment methods, for several reasons. CNT functionalization makes the nanoparticles more biocompatible, allows targeted delivery [4, 5], and eases delivery into the cells. Intrinsically, CNTs are able to enter cells [6, 7] through multiple pathways, and are able to carry a high density of drug to the target [8, 9]. This is accomplished by attachment to the surface, attachment to an encapsulating polymer, or interior loading of the CNT. In contrast to the delivery of biopharmaceuticals alone, conjugation to CNTs reduces some of the inherent drawbacks of certain biopharmaceuticals. For example, whilst small interfering RNA (siRNA) can serve as an excellent anticancer therapeutic, pure siRNA alone easily cannot easily cross the cell membranes due to electrostatic interactions, as both siRNA and the lipid bilayer are negatively charged. The conjugation of siRNA to functionalized carbon nanotubes (f-CNTs) significantly increases their delivery into cells due to CNT uptake through mechanisms other than endocytosis [10]. On the organism scale, CNTs can also act as a vehicle to help overcome various biological barriers in the body to reach target sites better than therapeutic molecules and biopharmaceuticals alone.

2 Nanotube Background

Two distinct varieties of CNT exist today: single-walled nanotubes (SWNTs) and multi-walled nanotubes (MWNTs). SWNTs are composed of single sheets of graphene rolled into a cylinder, while MWNTs are formed from multiple concentric SWNTs. The SWNTs and MWNTs may vary widely in terms of their lengths and diameter, as well as their functionalization (discussed in detail in Sect. 2.3). These physical differences have led to very different biological responses being observed in living subjects [11–14]. For any therapeutic application, it is, of course, necessary that the base nanoparticle is biocompatible. While this point is beyond the scope of this chapter, many recent reviews have explored links between the chemistry, length and other physico-chemical parameters and CNT toxicity [13–15]. Moreover, the distinctive properties of both SWNTs and MWNTs are important not only for biocompatibility, but they have also been exploited for therapeutic applications and biopharmaceutical delivery.

In the past, CNTs were primarily produced in small batches but with high defect rates. The CNT size was difficult to control, and early experiments were difficult to reproduce due to batch-to-batch inconsistencies in physical properties. Today, the diameter, length and mechanical properties can each be adjusted precisely during synthesis, leading to an ability to create biocompatible CNT batches for medical applications. In addition, batches are no longer volume-limited, allowing single, large batches to be used across experiments. The ability to control CNT formation has increased in the past 20 years, producing safer, easier-to-handle CNTs for both *in-vitro* and *in-vivo* experimentation [16].

Many synthesis methods have been developed for SWNTs and MWNTs, each generating a unique variant of CNT with several tunable properties that can be adjusted based on the intended application. After synthesis, the CNT can be functionalized for a variety of purposes.

2.1 Carbon Nanotube Synthesis

CNTs were initially produced by arc-discharge evaporation [1]. This method requires a high voltage to be applied across two graphite electrodes in an inert gas atmosphere, and SWNTs and MWNTs can be selectively synthesized based on the electrode properties. If the electrodes are made from pure graphite rods, then MWNTs will accumulate on the cathode. However, if a metal catalyst is applied in addition to the pure graphite, then SWNTs can be collected in the residue from the chamber. Arc-discharge synthesis methods are limited in the quantity of product they can produce with respect to the initial amount of graphite [17]. Early, large scale arc-discharge methods could produce CNTs ranging from 2 to 20 nm in diameter, with a nanotube yield of 25% [18]. In addition, multiple purification steps are necessary in order to remove any unwanted synthesis byproducts and excess catalyst. Although more high-throughput synthesis methods now exist, many groups have made recent advancements in the performance and safety of arc-discharge synthesis [19]. Today, fully automated arc-discharge methods have been developed to manufacture MWNTs without the need for metal catalysts [20].

Laser ablation synthesis methods have several similar limitations to the arc-discharge synthesis, but can produce more pure products. Laser ablation involves a high-intensity laser impinging upon a carbon target that is held in a fixed 1200 °C oven while released carbon atoms are accelerated towards a copper collection plate. This method can be used to produce both types of CNT: MWNTs are produced simply by the impinging radiation, while SWNTs are produced in the presence of a metal catalyst [17]. Several factors affect the characteristics and quantity of CNTs produced by laser ablation, including the amount/type of metal catalyst, the type of inert gas, and the power/wavelength of the laser. CNTs produced by this method are 90% more pure than those produced by arc-discharge, and have diameter variances of only 1.0–1.6 nm [21, 22]. However, like arc-discharge synthesis, this method is limited in terms of production volume; indeed, the need to scale-up production of CNTs has motivated the development of synthesis methods with minimal physical synthesis limitations.

Modern CNT synthesis methods typically employ chemical vapor deposition (CVD). Essentially, the controlled decomposition of a gas-phase carbon source has the potential to produce CNTs with little impurity and easy-to-control growth [23]. In this process, a metal catalyst "stage" is placed into a controlled-atmosphere chamber and heated to approximately 700 °C. A gas-phase carbon source and a mixed gas are then introduced into the chamber, and carbon is deposited on the catalyst as the gas comes into contact with the plate [24]. The CNT diameters are correlated with the metal catalyst particle size, and the yield is dependent on the metal support compounds [23]. Although this method does not share the same limitations in production volume, all three methods discussed here require the use of metal catalysts – some of which are cytotoxic – to produce SWNTs. The implications of this feature for medical applications will be discussed in Sect. 2.2.

2.2 Cytotoxicity of Carbon Nanotubes

In order for CNTs to be useful for biopharmaceutical delivery, it is critical that their cytotoxicity and long-term toxicity be understood. For that reason – and the fact that pristine CNTs are significantly more cytotoxic than many other nanoparticles – some recent insights are provided into the debate on CNT toxicity. Pristine CNTs are cytotoxic primarily because of their exquisite ability to enter cells [25] and their subsequent potential interactions with DNA and the cytoskeleton. Currently, three mechanisms have been proposed for cell entry, namely nanopenetration, endocytosis, and phagocytosis:

- CNT nanopenetration is a passive process by which CNTs diffuse across cell membranes, most likely by "spearing" them with their very small cross-sectional diameter (1–30 nm).
- Endocytosis is an active process by which CNTs are incorporated into a vesicle for internalization after interaction with the cell surface [25, 26].
- Specialized immune cells known as phagocytes may take up CNTs through a process termed phagocytosis.

All three pathways are represented schematically in Fig. 1 [27]. Recent discoveries regarding the energy dependence of CNT uptake have confirmed that endocytosis is the primary mechanism for transport across cell membranes [26].

After internalization, CNTs may cause toxicity in cells by activating several pathways simultaneously, with the more significant effects involving DNA damage [25]. Changes in cell cycle stage, the generation of apoptotic signals and DNA strand breaks have all been reported in mesothelial cells exposed to pristine SWNTs at approximately 25 μg cm^{-2} [28]. In addition, certain cell types are more susceptible to damage as the dose increases. A correlation between increased dose and increased oxidative stress has been reported *in vivo* in the liver and lung, but interestingly no significant correlation was found in the spleen [29]. Intriguingly, mesothelial cancer cells generally experience an increased phosphorylation of extracellular signal-regulated kinases (ERK) and p38-kinase with respect to their normal counterpart, but are less susceptible to DNA damage than the latter [30].

The physical dimensions of CNTs affect their cell internalization dynamics. A negative correlation between CNT length and macrophage uptake has been reported both *in vivo* and *in vitro* [25]. Short CNTs (<0.22 μm) were found in macrophages four weeks after their subcutaneous injection into rats, whereas longer CNTs (>0.80 μm) were still free-floating in the extracellular milieu. As free CNTs can cause inflammation and local cell death, the toxicity effects of longer CNTs must be considered before their use in medical applications [25].

CNT diameter also affects cellular uptake, and is a significant factor when considering drug loading and surface modification. Jin *et al.* have reported that the endocytosis rate of SWNTs is governed by thermodynamic entropic contributions, and that the rate reaches a maximum when SWNTs display a radius of 25–30 nm (i.e., when clusters at the surface of the cell reach such radii) based on *in-vitro* experiments and computational models; larger or smaller diameters result in significant decreases in simulated uptake [11]. While computational models are not now widely used in the field of medical nanotechnology, such models will become increasingly important as the numbers of nanoparticles rapidly proliferate. The choice and optimization of nanoparticles will then likely eventually be subject to *in-silico* models for major transitions, for example, *in vitro* to preclinical animal models, and preclinical models to patient use in the clinic.

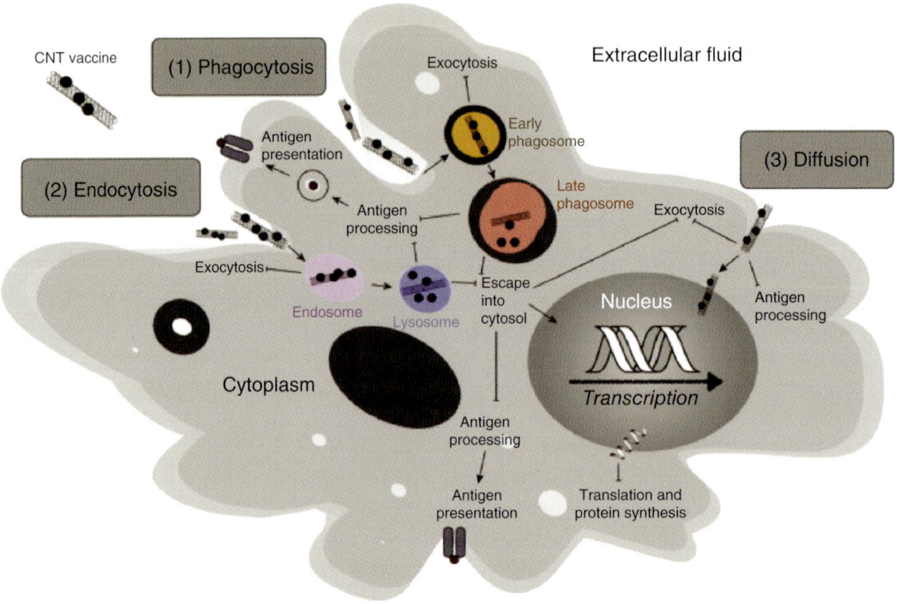

Fig. 1 Mechanisms of carbon nanotube (CNT) uptake by cells. (1) Phagocytes may recognize foreign bodies conjugated to the CNT surface and take up the complex. (2) Functional groups on the CNT surface may interact with cell-surface proteins, resulting in endocytosis. (3) The nanopenetrating ability of CNTs allows them to enter the cell through passive diffusion. (Reproduced with permission from Ref. [27]; © 2014, Elsevier.)

Metal catalysts are commonly used in SWNT synthesis, and are required for the CVD synthesis of any CNT type. They are also incorporated into CNT products at concentrations as high as 40% of the total CNT mass. Catalysts often contain metals such as iron, cobalt, nickel, yttrium [31], and arsenic, some of which can produce free radicals *in vivo*. Small iron oxide nanoparticles, which are commonly used as a catalyst, are known to be biocompatible based on a long history of iron oxide use in humans [25, 32]. However, with other catalysts, free radical production may be a concern, especially during CNT uptake by macrophages [25]. Transition metals can come into contact with superoxide species produced when the enzyme NADPH-oxidase is activated in macrophages, and produce highly reactive hydroxyl groups that have the potential to bind to proteins, lipids and DNA, causing significant damage to the cell [25]. For medical applications of CNTs, purification protocols have been developed and refined to remove any catalyst material from the CNT product by several orders of magnitude, though not without losses in CNT yield [33, 34]. Although these purification procedures leave only trace amounts of metal catalysts, this most likely leads (at least in part) to the minimal toxicity observed in various animal models for optimized CNT formulations [35–37].

Whilst the toxicity of these trace compounds has not been proven in humans, animal studies have provided critical insights. Indeed, many small animal studies have demonstrated a lack of toxicity of correctly functionalized f-CNTs of appropriate length and diameter [35–37]. However, if not correctly modified, the CNTs may have unavoidable structural defects that can have negative effects on cell health [38]. For example, charge density abnormalities caused by five- or seven-membered ring defects in the CNT surface chemistry can cause undesirable nucleic acid binding, or may weaken the bonds between nucleic acid functional groups and the CNT surface. When the CNTs are in cells, such active chemistries can lead to deleterious interactions and interference with DNA [25]. Both, the arc-discharge and laser ablation methods leave the CNT ends "closed," whereas CVD synthesis produces "open" ends. Consequently, CNTs produced using CVD may have "dangling" highly reactive species at each end, which can lead to oxidative stress without further modification [25].

The most common routes for introducing CNTs into the body are intravenous injection, inhalation, and transdermal injection. In general, inhalation and transdermal injections of CNTs result in a distinct biodistribution and pathological effects [39, 40]. Many descriptions have been made of the effects of inhalation of both SWNTs and MWNTs. Based on rodent studies, the consensus is that the inhalation of these structures generally causes adverse effects, but that the results are heavily dependent on CNT functionalization and length [41]. In one study, Ma-Hock *et al.* described the long-term effects of inhaled nanotube exposure in rats [42] when, after 90 days, the inhaled MWNTs had not produced any adverse systemic effects but had caused an increased lung weight, local inflammation, and blood neutrophilia in the case of high inhaled NT concentrations (2.5 mg m^{-3}). Previous conflicting reports of lung toxicity in early studies of long-term CNT inhalation were also summarized [42]. For example, in 2001 Huczko *et al.* reported no significant lung toxicity after four weeks of exposure in guinea pigs [43], but Lam *et al.* subsequently (in 2004) disputed this claim, reporting inflammation and granuloma formation after single-dose exposure [44]. In 2008, Shvedova *et al.* showed that inhaled SWNTs would cause oxidative stress, collagen deposition and fibrosis in mice [45].These apparently conflicting results were most likely due to differences in the manufacture, functionalization and/or dose levels across studies, as well as differences in the species tested. Recent advances in CNT aerosolization testing have led to many studies being conducted in quantitative toxicity measurements [46].

Similar aerosolization advancements have begun to resolve issues with experimental consistency across studies [47]. In a provocative study, Poland *et al.* reported that the pathological effects of CNTs introduced into the abdominal cavity may be similar to those of asbestos, although others have reported that asbestos-like effects are due almost solely to the length/high aspect ratio, chemical functionalization, and bundling activity of CNTs [48–50]. When toxic, MWNTs are reported to produce local inflammation and granuloma in mice, and may have the capacity to cause mesothelioma after intraperitoneal injection. Yet, when correctly functionalized and at the appropriate lengths, minimal toxicity is observed [48–50]. Due to an ability to control the dose precisely, the intravenous injection of CNTs is preferred over inhalation for whole-body, drug-delivery applications; nevertheless, and particularly for studies of toxicity in small animals, inhalation has historically been the more common approach for introducing CNTs into the body [15, 51]. The safe clinical use of CNTs through either inhalation or intravenous injection has been discussed in detail by Saito *et al.* [15], with intravenous injection, using CNTs of small to medium length and appropriate functionalization (e.g., with amphiphilic polymer coating [35]) having been shown to be acutely and chronically safe (for at least three to four months) in rodent models [35, 52]. For these reasons, the intravenous route is the preferred method for nanomedicine applications, including drug delivery.

2.3 Functionalization of Carbon Nanotubes

By design, the surface functionalization of CNTs typically results in an increased biocompatibility, increased circulation times, increased cell- or receptor-specific targeting, and an increased drug-carrying capacity depending on the nature of the attached functional groups. This functionalization is key for medical applications such as biopharmaceutical delivery.

Functionalization can be subdivided into either noncovalent (nondestructive) or covalent (destructive) methods, and further classified based on application. While Mittal has provided an extensive treatment of CNT surface modification methods, the present chapter will detail functionalization methods that are particularly applicable to biopharmaceutical delivery [53]. Currently, oxidation, cycloaddition, and the noncovalent attachment of polymers are three of the most commonly used approaches to produce CNTs for medical applications in living subjects:

- *Oxidation of CNTs:* Oxidation is a common method of functionalization as it is used to help covalently link molecules of interest to CNT sidewalls and end caps. Oxidizing agents, such as nitric acid, are used to produce carboxylic acid groups on defect sites on the CNT sidewall and opened caps. Although oxidized CNTs are generally water-soluble, their aggregation characteristics prevent them from being used *in vivo* directly. To improve the biocompatibility of oxidized CNTs, linker molecules such as polyethylene glycol (PEG) can be covalently attached to oxidized sites [54].
- *Cycloaddition of CNTs:* Cycloaddition is the process of creating a chemical "ring" at the CNT surface, without producing defects on the surface or carboxyl groups at the nanotube ends. The ring provides residual groups, which can be replaced with PEG, proteins, drugs, or certain positively charged moieties. Both, oxidation and cycloaddition are covalent methods of functionalization which can disrupt optical properties of CNTs, thereby limiting certain therapeutic and diagnostic mechanisms of pristine SWNTs [54].
- *Polymer attachment:* Pristine CNTs are not water-soluble and therefore cannot be injected. Moreover, they are cytotoxic unless modified with amphiphilic moieties on the pure graphitic surface. Noncovalent methods to attach water-soluble compounds to CNT surfaces have been developed to easily suspend CNTs in aqueous solutions. Analogous to many phospholipid bilayers, the hydrophobic CNT surface interacts with a hydrophobic segment of a polymer, which is connected to a hydrophilic segment of the polymer that is thermodynamically most stable when presented on the outermost layer of the system in aqueous solvents. Weak van der Waals forces and hydrophobic effects hold the polymer to the CNT surface, preventing nonspecific binding of proteins and other biomolecules to the CNT. Such polymers are commercially available and have been demonstrated to perform well

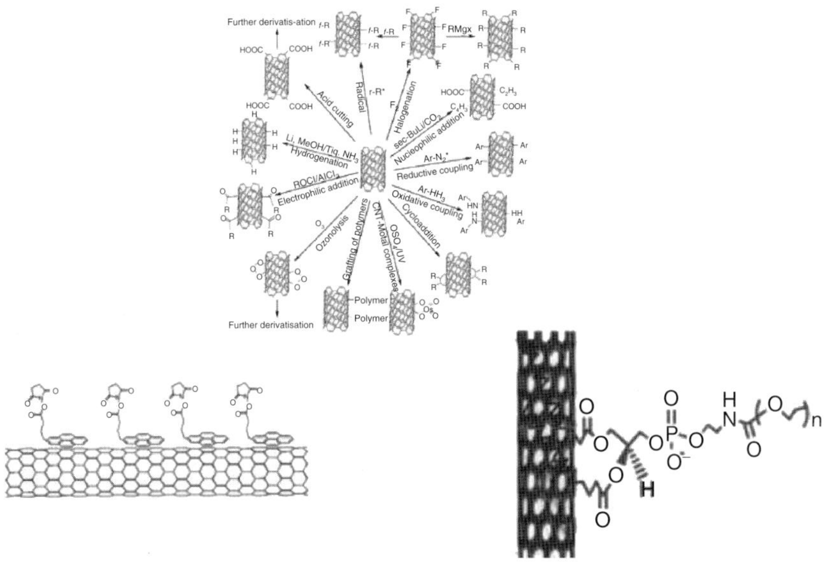

Fig. 2 CNT surface functionalization methods. (a) Examples of direct covalent addition of functional groups and general mechanism for addition of chemical groups. (Reproduced with permission from Ref. [55]; © 2010, Royal Society of Chemistry); (b) Noncovalent functionalization method primarily applied for peptide immobilization. 1-Pyrenebutanoic acid, succinimidyl ester is irreversibly adsorbed onto the hydrophobic surface of CNTs in organic solvents. Proteins then form amide bonds with the exposed succinimidyl ester groups. (Reproduced with permission from Ref. [56]; © 2001, American Chemical Society); (c) Distearoylphosphatidylethanolamine (DSPE) copolymer-based addition of polyethylene glycol (PEG), a molecule commonly used to enhance CNT solubility in water and increase blood circulation time. This approach applies noncovalent hydrophobic interactions of the lipid group to link to the CNT and the PEG group for aqueous solubility. (Reproduced with permission from Ref. [57]; © 2008, Nature Biotechnology.)

in biosensing, suspension stability, and *in-vivo* applications [54].

Liu *et al.* have listed the critical characteristics of noncovalent functionalization coatings for CNTs. The coating should be biocompatible and nontoxic, resist detachment from the CNT surface, have a low critical micelle concentration (CMC) in order to create a stable coating on the CNT surface after application, and have functional groups capable of interacting with biologically active molecules or application-specific compounds [54]. These points are all critical for biopharmaceutical applications of CNTs, and the biopharmaceutical-specific functionalization chemistries will be discussed as each particular chemistry is presented. A general overview of CNT surface functionalization is provided in Fig. 2.

2.4 Pharmacokinetics of Carbon Nanotubes

Although the *in-vivo* kinetics of f-CNTs are still not completely understood, significant efforts have recently been made to understand their short- and long-term biodistributions. This is due in part to the fact that so many different lengths, diameters and functionalization distributions have been tested by different groups [58]. Nevertheless, several characteristics are common across most studies: (i) the CNT's physical properties, particularly functionalization, play a critical role in biodistribution across all time points; (ii) clearance via the urine is comparable to the clearance of larger-diameter nanoparticles (most likely due in part to the large diameter of CNT aggregates); (iii) accumulation is apparent in the liver, spleen, and often the bladder; and (iv) covalent functionalization often promotes a higher stability of CNTs *in vivo* with respect to noncovalent functionalization. Yang *et al.* have provided a detailed overview of the *in-vivo* kinetics of f-CNTs [58].

3
Biopharmaceutical Delivery Using Carbon Nanotubes

Biopharmaceuticals can be broadly categorized into three groups, namely antibodies, nucleic acids, and peptides/proteins (Table 1). Antibodies and peptides/proteins are separated because antibodies are specific to particular antigen epitopes and tend to be larger, whereas peptides/proteins are a general class of macromolecule. The common functionalization methods, recent applications in living subjects, and advantages for each group are discussed in the following sections.

3.1
Peptides/Proteins

Gene expression and cell signaling are commonly accomplished through proteins. The cell membrane can house a variety of proteins serving several purposes, including chemical channels, signaling systems, or pumping systems to build an electrochemical gradient. For therapeutics, membrane-bound proteins that specifically bind to small amino acid sequences (that are related to a particular disease) are of interest. These proteins can be exploited as targets for drug-delivery systems, or they can be blocked by means of a blocker/inhibitor. Conjugating proteins to CNTs can serve both purposes: (i) a drug can be carried by the CNT with a protein-targeting agent; or (ii) high densities of inhibitor (which in many cases is also a protein) can be loaded onto the CNT for delivery [59]. In many instances, the protein conjugated to CNTs can serve as both a targeting agent and as an inhibitor for the membrane-bound protein.

Tab. 1 Biopharmaceutical delivery using carbon nanotubes: an overview.

Biopharmaceutical	*Function*	*Section*
Peptide/protein		3.1
	Targeted drug delivery	3.1.1
	Adjuvant	3.1.2
Antibodies		3.2
	Targeted drug delivery	3.2.1
	Targeting imaging	3.2.2
Nucleic acids		3.3
	Gene silencing	3.3.1

Many functionalization methods for peptide–CNT systems exist, including nonspecific adsorption [60], noncovalent functionalization [61], and covalent functionalization (see Sect. 2.3). Nonspecific adsorption is possible due to the interaction of hydrophobic groups on the protein or polymer with the hydrophobic CNT surface or other hydrophobic scaffolds [62]. Noncovalent functionalization involves the addition of linker molecules with hydrophobic and hydrophilic ends to the CNT surface; the hydrophobic end interacts with the CNT surface, while the hydrophilic end is free to bind protein and provide solubility in an aqueous environment [63].

Noncovalent functionalization is similar to nonspecific adsorption, with the caveat that noncovalent functionalization contains a linker molecule between the nanotube and the biomolecule.

Recent applications of peptides in biopharmaceutical delivery systems can be broadly classified into either targeting or adjuvant approaches. In practice, peptides are more commonly used as targeting agents for CNTs, most likely due to their considerably smaller size. This is logical considering that proteins are often 3–5 nm or more in hydrodynamic diameter; as SWNTs are on the order of 2 nm in diameter, molecules the size of proteins could in many cases adversely modulate some of the properties that make SWNTs and other CNTs so useful in biopharmaceutical delivery.

3.1.1 Peptides and Proteins as Targeting Agents

Peptide- and protein-f-CNTs are efficient drug-delivery systems and imaging agents, mainly because of their targeting ability and high loading capacity due to their extremely high surface area-to-volume ratio.

The use of Arginine-Glycine-Aspartic Acid (RGD) functionalized SWNTs, or other short peptide sequences for receptor-specific targeting, has been explored *in vitro* [64] and *in vivo* [65]. RGD selectively binds the $\alpha_v ß_3$ integrin, a receptor protein that is upregulated on many tumor cell types (e.g., U87MG cells), downregulated on other cells (e.g., MCF-7 cells), and overexpressed on tumor blood vessels [5, 65]. In practice, an RGD ring – cyclic RGD – is typically used for its increased $\alpha_v ß_3$ integrin affinity. Overexpression can be exploited by conjugating RGD to SWNTs to increase homing and efficacy at the target site. Once targeted, therapy can be achieved by incorporating doxorubicin (DOX), a well-characterized chemotherapeutic agent, into RGD–SWNTs to significantly increase binding to U87MG cells with respect to $\alpha_v ß_3$ integrin-deficient MCF-7 cells [64].

A way to selectively target particular subsets of immune cells has long been sought [66]. In a novel application of SWNTs in immune nanomedicine, SWNTs were shown to be taken up with very high selectivity into a monocyte subset [67]; the SWNTs could then be used to deliver molecules to modulate the function of these monocytes. A novel tumor "homing" mechanism was also characterized, showing how RGD-conjugation of SWNTs increased monocyte trafficking into tumor regions after the monocytes had engulfed the SWNTs. In particular, RGD–SWNT-laden circulating monocytes extravasated into tumor regions more frequently than plain or nonspecific control groups [67]. These results suggest that certain peptides can be conjugated to drug carriers for the specific delivery of not only nanoparticles, but also cells carrying such nanoparticles into tumors.

Peptide-conjugated SWNTs can be used as tools for targeted photothermal therapy. SWNTs are inherently excellent absorbers of near-infrared (NIR) excitation, but they must be targeted in order to produce heating at the appropriate site. In order to prevent the irradiation of a nonspecifically targeted region, SWNTs are coupled with a repeating amino acid Lys-Phe-Lys-Ala (KFKA) structure via β-sheet wrapping. This polymer is engineered for water-solubility, binding to tumor tissue by electrostatic interaction, and the ability to be further modified as a carrier moiety for drug delivery. Functionalization was confirmed using atomic force microscopy, transmission electron microscopy, and molecular modeling [68]. The system was tested both *in vitro* and *in vivo* in mice. A high-efficiency, concentration-correlated heat production resulted from the NIR irradiation of peptide-carrying SWNTs in water. Pure peptide was found to cause no cell death below 10 μg ml^{-1} concentrations. *In-vitro* testing confirmed that NIR excitation caused localized cell death. SWNT-$(KFKA)_7$ (a system wherein SWNTs are conjugated with a KFKA sequence which repeats seven times) were injected directly into tumor tissue and subjected to NIR irradiation for 30 s. Localized heating was confirmed using infrared thermometry, with very small temperature increases outside of SWNT localization. A statistically significant suppression of tumor growth was induced by SWNT introduction and ablation [68]. In the next steps, this system should be tested using systemic administration and in a larger animal model more representative of human disease, as this result is promising if it can be successfully administered via a more clinically applicable route.

In addition to radiative therapy methods, peptide f-CNTs can be used to specifically deliver therapeutics to a target site. For example, a multifunctional CNT system which employs asparagine-glycine-arginine (NGR) peptide as a targeting agent has been validated *in vivo*. This system is then loaded with tamoxifen (TMX), an anticancer drug, exploiting the unparalleled loading capacity of CNTs [69]. In addition to chemical therapeutics, the SWNT–peptide retains the useful optical properties of pristine SWNTs. This allows photothermal ablation therapy, which can be used in combination with TMX. In a recent study, peptide-laden SWNTs were taken up into tumors more readily than their pristine counterparts. Although the full NGR/TMX–SWNT system showed cytotoxic effects alone, the combined effects of TMX and NIR irradiation increased the *in vitro* cytotoxicity 1.5-fold. *In-vivo* biodistribution assays confirmed a specific uptake of complete NGR/TMX-SWNT system by tumors, whereas other variants accumulated in the clearance organs. In addition, significant reductions in tumor volume were apparent at 14 days post-injection of NGR/TMX–SWNTs due to the excellent delivery and uptake characteristics of the system [69].

Extracellular receptors may be relatively easy targets for biopharmaceutical delivery systems, but intracellular targets are much more difficult to reach. A method to repair damage to mitochondrial DNA by means of peptide-conjugated CNTs has recently been developed. Mitochondrial targeting can be accomplished via known endogenous peptide markers conjugated to nanoparticles; this is known as the mitochondrial targeting system (MTS). *In vitro*, peptide-conjugated CNTs were significantly more effective in penetrating cell membranes than the peptide alone. Also *in vitro*, MTS–MWNTs aggregate around mitochondrial membranes, which is due to an interaction with the translocator inner membrane/translocator outer membrane protein complex. Importantly, the MTS–MWNT conjugates were also found not to be cytotoxic. The findings of these studies suggest that a need exists to use targeting peptides when delivering therapeutic agents to mitochondria, rather than nonspecific delivery into the cytosol [70].

3.1.2 **Peptides and Proteins as Adjuvants**

CNT-bound peptides can also serve as adjuvants, which are molecules that produce a localized immune response in the host. A resumé of recent progress in this area of technology is now presented.

SWNTs have been employed as peptide carriers to boost weak immune

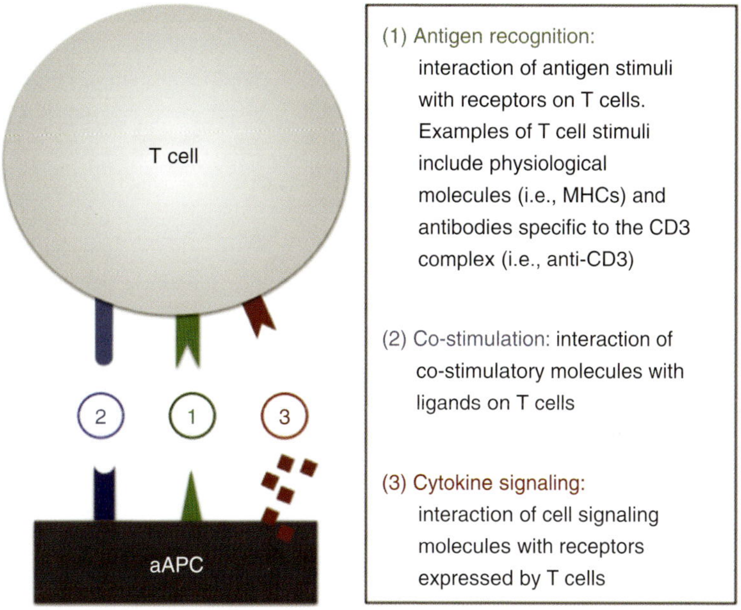

Fig. 3 CNTs can be applied as artificial antigen-presenting cells (aAPCs) in order to produce or boost an immune response to an antigen. (Reproduced with permission from Ref. [71]; © 2011, Elsevier.)

responses. In a recent study, it was found that peptide-laden SWNTs were actively engulfed by immune cells *in vitro*, producing a marked increase in immune response over the pure peptide antigen. In this paradigm, tumor-specific peptide antigens could theoretically "prime" the immune system against tumor subtypes, potentially serving as quasi-vaccines, as shown in Fig. 3 [71].

It is also now possible to induce an immune response to haptens, which are low-molecular-weight compounds that are too small to be presented on major histocompatibility complex (MHC) II molecules. Other methods of immunization against haptens involve conjugation to nanoparticles, and the introduction of these systems into the body. Both, SWNTs and MWNTs were conjugated with azoxystrobin, an antifungal with the broadest known spectrum of activity. When these systems were injected into mice and rabbits for *in-vivo* immunization, the data acquired showed the CNTs to be extremely efficient antigen carriers and to produce an immune response to the hapten complex at a 10-fold lower concentration than did the pure complex alone. Moreover, when CNTs of varying length were tested, the results showed that shorter CNTs would produce a significantly increased immune response in comparison to longer CNTs of similar diameter. To further illustrate the amplification of immune response to CNT antigen carriers, the total CNT–protein complex dose was diluted to one-tenth of the pure protein complex dose. Subsequently, whilst the pure complex produced no significant immune response, the CNT–protein complex, when inoculated, produced significant increases in blood immunoglobulin (Ig) G concentrations [72].

The delivery in this application was to cells of the immune system (e.g., T cells) present in the circulation rather than to a particular region of the body. Consequently, a major feature of this delivery system would be for the presented molecules to circulate for as long as possible in order to enable their maximal exposure to the immune cells.

3.2 Antibodies

Antibodies are a subset of proteins that can serve as versatile biomolecules generated by B cells and are capable of recognizing a wide variety of unique chemical signals or epitopes present on the surface of foreign bodies. By definition, antibodies (or Igs) are large, Y-shaped proteins deployed by the immune system to detect and neutralize antigens. Variable portions at the tips of the structure serve as chemical complements to certain antigens, allowing them to tag the foreign body for phagocytosis, degranulation, or digestion by cytotoxic natural killer (NK) cells. Because of their exquisite sensitivity and specificity for unique antigens, as well as their high binding affinity, antibodies have applications as immunoassay tools, immunization promoters, and targeting agents. Furthermore, it emerges that many antibodies can in fact have therapeutic effects, as is the case with the monoclonal antibody Herceptin® [73]. This is in addition to any drug or other molecules that the CNT can carry into the cell once it is targeted (as will be discussed later). Therefore, the conjugation of antibodies to CNTs as a targeting agent (as well as a potential drug) has great potential to amplify their therapeutic effects.

Recent advances in CNT chemistry have led to the synthesis of antibody f-CNTs primarily for use as targeted drug-delivery vehicles. The covalent functionalization of antibodies to CNTs may be optimal, as the potential loss of the targeting agent would lead to nonspecific drug delivery. To date, three such covalent strategies have been reported:

- A simple oxidation reaction where carboxylic acid groups are appended to defect sites and "open" ends. The antibodies are coupled to the carboxyl groups by amino groups on an amino acid side chain, such as lysine, available on the antibody [74].
- A strategy which also follows standard oxidation techniques – the oxidized CNT is attached to pyrrolidine or groups at the defect sites, and the antibody binds the ring through an amidation reaction.
- A strategy which requires thiolated antibodies and maleimide f-CNT. The complete system consists of a pyrrolidine ring coupled to a maleimide group, which binds the thiolated antibody.

All three methods have been well-characterized and have been validated both *in vitro* and *in vivo* [74].

Currently, two major categories of CNT immunotherapy exist, each with its own benefits. Passive immunotherapy involves the introduction of antibodies or complements as targeting agents for drug delivery vehicles that can then treat the disease site. Active systems boost the host immune response to a specific antigen by introducing pathogen fragments or other adjuvants. Though active systems do not house antibodies themselves, they are designed to drive antibodies and primary immune cells to antigen sites, which has been previously discussed (see Sect. 3.1.2). The targeting and inhibitory action of antibody f-CNTs will be discussed in the following section, beginning with recent *in-vitro* applications and leading to the delivery of these systems in living subjects.

3.2.1 Antibodies as Targeting Agents for Drug Delivery

Multidrug resistance (MDR) is a major problem in clinical oncology today. In time, MDR essentially nullifies many treatment approaches because cancer cells adapt to resist the effects of previously administered drugs. Fortunately, targeted drug delivery can counteract MDR effects by enhancing the local drug concentration at the target, and particularly by enabling the delivery of multiple drugs simultaneously (this is a key advantage of nanoparticle delivery systems, such as CNTs). As an example, P-glycoprotein (P-gp) antibody-functionalized, water-soluble SWNTs were used to target the membrane-bound P-gp on MDR cells. Anti-P-gp was covalently linked to oxidized SWNTs, and the SWNTs were loaded with DOX. In addition to passive diffusion of the drug, it is possible to irradiate the *in-vitro* culture of MDR cells and SWNTs with NIR light in order to boost drug release. The proposed mechanism for attachment (corroborated by fluorescence imaging) is an initial interaction between anti-P-gp functionalized on the SWNT and membrane-bound P-gp, followed by endocytosis by the MDR cells where the drug is released. This enhances the time of contact between the drug and the target cells. This method has been confirmed to produce a 2.4-fold increase in targeted cytotoxicity with respect to free DOX *in vitro* [75], and indicates that the antibodies in this case were effective targeting moieties. It also enabled the delivery of both the antibody and the loaded drug to overcome the drug resistance of the cells, allowing them to be destroyed.

3.2.2 Antibodies as Targeting Agents for Imaging

f-CNTs are useful not only as imaging agents but also as drug carriers. Targeted imaging agents have the potential to serve also as drug carriers in the future, provided that the drug and imaging agent share similar steric and chemical properties; these types of agent are known as "theranostics." It is also important to recognize the prevalence and distribution of antibody targets on the cell of interest, in order to understand the binding and internalization mechanisms and thereby provide a better design of the CNT vehicle for optimal uptake [76]. In the context of imaging agents, antibodies bound to f-CNTs can increase accumulation at a target site, or increase cellular uptake.

Imaging agents and drug-delivery systems are intrinsically connected when working with f-CNTs. When specific surface proteins are upregulated – as often occurs in cancer cells compared to healthy cells – these proteins can be ideal targets for specific drug-delivery vehicles. Her2+ breast cancers are particularly aggressive and are generally less responsive to gene therapy, but display increased Her2 receptor expression on the cell surface. A SWNT–antibody complex which was produced for targeting Her2-positive breast cancer cells *in vitro* [76] consisted of a biotinylated SWNT functionalized to bind a chromophore, a secondary antibody, and primary antibody to the Her2 receptor (Fig. 4). The full complex was seen to bind to the Her2 receptor on BT-474 cells with remarkable specificity, as quantified by Raman imaging and confirmed in control immunofluorescence studies [76].

It is also possible to slow the rate of endocytosis by cooling the cells to 15 °C, and then to obtain confocal Raman spectroscopy images in three dimensions to image the delivery process. The "Raman effect" is an apparent shift in the frequency of monochromatic incident light that inelastically scatters after contact with a

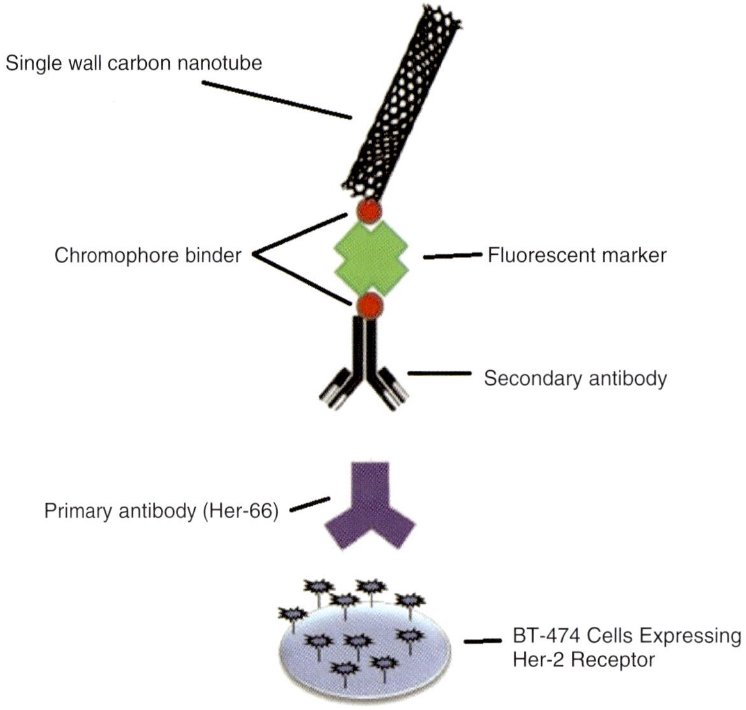

Fig. 4 Schematic diagram of SWNT–antibody complex for Her-2 receptor targeting to the cell surface *in vitro*. The Her-2 distribution on BT-474 cells was evaluated using Her-66 anti-Her-2 monoclonal antibodies. The biotinylated secondary antibody binds the primary antibody and is complexed with a fluorescent marker (FITC) and SWNTs via a chromophore binder (NeutrAvidin-FITC bridge). (Reproduced with permission from Ref. [76]; © 2014, Royal Society of Chemistry.)

target. The frequency shift is associated with the vibrational and rotational energy states of the target molecules and their bonds, which in turn makes it possible to determine the contents of the target by analyzing the frequency shifts of the scattered light [76]. Precision Raman imaging is possible with CNTs due to the unique chemical structure of the C–C bonds within the SWNTs, which makes them easily and very specifically detectable by Raman imaging in a background of carbon (i.e., the milieu of a living subject such as a mouse) [77]. The imaging results showed that SWNTs are localized at the cell membrane, suggesting the start of endocytosis. These results, in combination with data from controls, suggested that antibody-mediated cell binding caused a significant increase in SWNT binding to the cell membrane, which in turn potentiated SWNT internalization by the cell [76].

Antibodies do not need to be directly attached to CNTs in order to serve as targeting agents. A novel two-step system was shown to guide SWNTs to tumor regions by "priming" sites of interest with antibodies, as shown in Fig. 5. The SWNTs were conjugated to oligonucleotide sequences and fluorophores [78], antibodies were employed as targeting agents for specific tumor cell receptors, and complementary

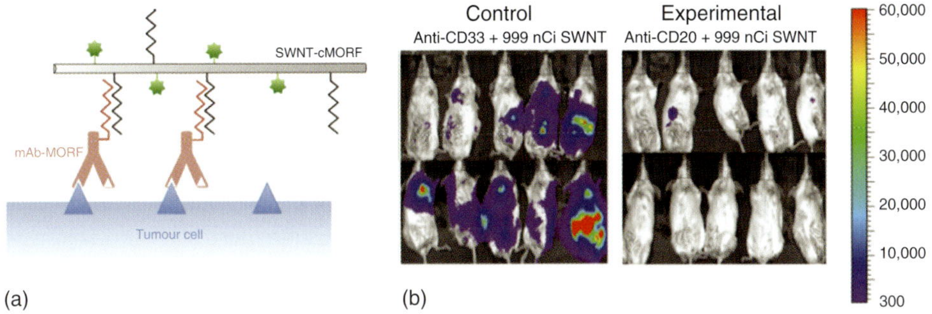

Fig. 5 Imaging agents can be coupled to biopharmaceutical-conjugated CNTs to monitor their *in-vivo* kinetics and therapeutic effect. (a) Schematic representation of two-step SWNT delivery system. Tumor-specific antigens are represented as blue triangles. Antibodies (mAb) with morpholino (MORF) oligonucleotide sequence are shown in red. SWNTs with bound complementary morpholino (cMORF) oligonucleotide sequences and radiotherapeutic agents are shown in gray; (b) Bioluminescence images demonstrating specific $CD20^+$ tumor volume reduction using antibody-conjugated CNTs. Control and experimental conditions are anti-CD33 (Ab for cell-surface markers are not present on this tumor cell type) and anti-CD20 (Ab for cell-surface markers are overexpressed on this tumor cell type) respectively. Both conditions have the same radiotherapeutic dose of 999 nCi. (Reproduced with permission from Ref. [78]; © 2013, Nature Nanotechnology.)

oligonucleotides were covalently bound to the constant regions of the antibody. The SWNT system was loaded with alpha- and gamma-radiation-emitting moieties to serve as a therapeutic agent, and with fluorophores for *in-vivo* tracking. The antibody system was injected intravenously 24 h before the SWNT system was introduced, in order to allow binding to the tumor sites. The combined Ab–SWNT system can be considered as self-assembling because the binding occurs on the target, not during synthesis. This method showed a five- to 10-fold reduction in toxicity over preconjugated systems, which was an amazing achievement. The imaging results indicated dose-dependent therapeutic effects, motivating further exploration to use this method for drug delivery that applies SWNT–Ab systems in a "sandwich assay" technique that is more commonly associated with immunoassays [78].

A MWNT–antibody system to measure levels of prostate stem cell antigen (PSCA) has recently been developed. The purpose of this system is to target cancers that upregulate PSCA, and also to be used as a targeted ultrasound contrast agent *in vitro* and *in vivo* [79]. The hyperechogenicity of MWNTs which was only recently described is somewhat surprising, given the typical size range of many echogenic materials for ultrasonic applications (often $\geq 1\,\mu m$ in diameter, such as microbubbles). Ultrasonic imaging using smaller materials such as MWNTs has great potential, as these agents can reach sites that are inaccessible to larger materials such as microbubbles, including extravascular spaces and cells. Moreover, the high drug-loading capacity of MWNTs will allow for future anticancer drug delivery and real-time tracking. This DOX-loaded system did not cause any adverse hemolytic effects over a large range of drug concentrations ($0–400\,\mu g\,ml^{-1}$) either *in vitro* or *in vivo*, due to complexation with the CNTs [79]. Neither were any organ systems significantly affected after long-term exposure to the MWNT system. However, after an 18-day exposure period the rate of tumor growth was significantly decreased when DOX and PSCA monoclonal antibodies were conjugated to MWNTs with respect to other system variants (i.e., controls) [79].

3.3 Nucleic Acids

Nucleic acids are large macromolecules which store genetic information and are composed of nucleotide bases and a ribose sugar backbone. Ribosomes translate messenger RNA (mRNA) into proteins, which are then packed and processed. This mechanism can be exploited to express exogenous genes or to halt unwanted gene expression by means of engineered nucleic acids delivered through the cell membrane using f-CNTs. The CNTs are able to traverse cell membranes more readily than pure sequences of nucleic acids, and can provide nucleic acid stability against endogenous enzymes, making them a key part of an ideal transport mechanism.

Three overarching methods for nucleotide functionalization of CNTs have been employed in recent studies;

1) Polyethyleneimine (PEI) functionalization was first demonstrated by Liu *et al.* as the first noncovalent method to immobilize DNA on CNTs. PEI is composed of a high density of amines, which readily bind to nucleic acids. The polymer is grafted onto CNTs by a polymerization reaction in the presence of amine-f-CNTs [80].
2) Amino-functionalization of CNTs by the cycloaddition of a pyrrolidine ring bearing a terminal amine group is commonly used to bind plasmid DNA. Pantarotto *et al.* have reported that the addition of this group markedly increases CNT solubility in aqueous solution, as well as enabling covalent immobilization of the plasmid DNA [81].
3) Carboxyl-functionalization is another form of covalent attachment of nucleotides to CNTs; however, this method requires the amino-modification of nucleic acids to overcome electrostatic repulsion forces between the negatively charged nucleic acids and cell membranes [82].

3.3.1 Nucleic Acids as Gene-Silencing Agents

A major focus in the field of nucleic acid delivery has involved the enhancement of siRNA delivery into the cytoplasm [83]. The conjugation of siRNA to CNTs results in four critical benefits when used for treatments in living subjects: (i) ease of entering the cytoplasm; (ii) ability to add targeting ligands for intracellular organelles; (iii) potential therapeutic loading; and (iv) exploitable optical properties for simultaneous cytotoxic and silencing effects. While siRNA has been featured, it is likely that clustered regularly interspersed short palindromic repeats (CRISPR) systems, which only recently have displayed an incredible ability to perform precision gene editing in mammalian cells, will require delivery systems such as can be provided by nanoparticles (including nanotube-based approaches) in the near future.

In the field of cancer therapy, siRNA has been applied to silence the gastrin-releasing peptide receptor (*GRP-R*) gene, which produces a cell-surface receptor (GRP-R) that is commonly upregulated on neuroblastomas. GRP-R is known to promote the growth of epithelial cells and the formation of abnormal tissue growths [83]. The assay was tested *in vitro* and the results confirmed by using NIR fluorescence imaging of siRNA-treated cells and Western blotting for GRP-R. *In-vivo* testing entailed the subcutaneous injection of 1.7 μM SWNT–siRNA into a murine xenograft model, whereby such treatment was found to be effective in limiting the number of mitotic cells, leading to a decreased tumor growth [84]. Conjugation to CNTs thus improves the delivery of siRNA to intracellular targets, facilitating therapy in living subjects.

Although siRNA alone is a useful anticancer therapeutic, a combination of gene silencing and photothermal irradiation can lead to a significant further reduction in tumor volume [85]. In this case, PEI–SWNTs were conjugated with siRNA and NGR, a peptide sequence that can selectively recognize the tumor neovasculature [85]. The system was tested both *in vitro* and *in vivo*. An analysis of cell growth inhibition and transfection efficiency after CNT system incubation with cells and exposure to NIR irradiation occurred *in vitro*. The antitumor activities of fluorescein isothiocyanate (FITC)-labeled SWNT–PEI, SWNT–PEI/NGR, SWNT–PEI/siRNA, and SWNT–PEI/siRNA/NGR systems were tested in human prostate cancer cell line (PC-3) tumor-laden mice. Cellular uptake into the cytoplasm was confirmed using fluorescence microscopy, imaging the FITC label [85]. Nanotube–siRNA complexes were red fluorescence-labeled and cultured with FITC-labeled cells, and imaged at 1, 2, and 4 h time points. Whilst the siRNA-conjugated SWNT–PEI did not completely enter the cell after 4 h, the NGR-conjugated SWNT–PEI showed complete cell entry within 4 h, suggesting that the use of targeting peptides may be useful in nucleic acid delivery. Cell viability was significantly decreased following the NIR irradiation of PC-3 cells exposed to SWNT– PEI/siRNA/NGR.

CdSe quantum dots were used to fluorescently label SWNTs for an *in-vivo* biodistribution analysis, though it should be noted that the size of these labels has a greater likelihood of affecting the unlabeled SWNT biodistribution compared to smaller molecular fluorophores. Both, SWNT–PEI/NGR/quantum dot (QD) and SWNT–PEI/siRNA/QD showed fluorescence activity in tumor regions, with SWNT–PEI/NGR/QD producing a significantly higher signal than SWNT–PEI/QD, suggesting that NGR is important for an enhanced cellular uptake of SWNTs. The targeting peptide may increase the interaction time between CNTs and the cell membrane, thus promoting endocytosis of the complex. Significant reductions in tumor size were noted in SWNT–PEI/siRNA and SWNT–PEI/siRNA/NGR conditions, but not in SWNT–PEI, SWNT–PEI/NGR, or saline conditions. This suggests that siRNA played a critical role in tumor size reduction. Subsequent laser irradiation of all SWNT–PEI/X conditions showed significant reductions in tumor size; the most significant difference being noted in the SWNT–PEI/NGR condition, and again suggesting the power of peptide conjugation as a critical piece in an integrated targeted drug-delivery system [85].

In addition to the innate high-loading capacity of CNTs, it is possible to add scaffolds to CNT surfaces to increase the delivery of siRNA. The effects of branched siRNA scaffold moieties conjugated to MWNTs have been characterized. Branched polymers, or dendrons, can multiply drug-loading capacities but may also lead to steric hindrance and possibly impact SWNT biodistribution. Hence, trade-offs must be considered and typically tested empirically, depending on the particular application. Multiple conjugation methods, including novel "click chemistry" protocols have been used for appending

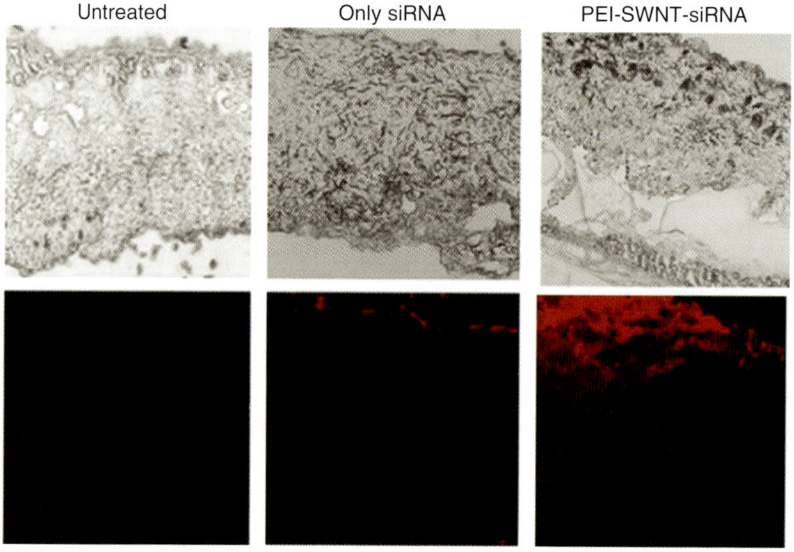

Fig. 6 Topical delivery of siRNA into skin in a murine model. Top row: Bright-field microscopy images of cryosectioned skin samples. Bottom row: Fluorescence images of cryosectioned skin samples. siRNA is Cy-3 fluorescence-labeled, and each condition is tested 4 h after siRNA solution application. The untreated condition shows no siRNA signal, as expected. The siRNA-only condition shows minimal siRNA signal at the surface of the skin. The siRNA–SWNT complex shows both a deeper skin penetration and an increased overall fluorescence signal. (Reproduced with permission from Ref. [87]; © 2014, Elsevier.)

dendrons and siRNA. Several branching chemistry methods were also employed, and the results characterized *in vitro* [86]. The ability to complex siRNA was inversely proportional to the degree of dendron branching in click chemistry-produced branching [86]. This effect may be due to a nonspecific binding of the long alkyl chains with the CNT surface, which hinders siRNA binding to the chain. For example, long chains may wrap around the CNT, producing unwanted conformations and reducing the exposed siRNA binding sites on the branch. However, the loading capacity can be increased by increasing the charge ratio (nitrogen/phosphate), partially mitigating this effect. Two variations of this complex are significantly more effective at delivering siRNA into cells than siRNA alone. The dendron-conjugated MWNTs produce longer-lasting silencing effects than cationic liposomes, though the liposomes have better gene-silencing abilities at early time points [86]. This may be due to faster interaction mechanisms between liposomes and cell membranes than peptide-conjugated CNTs, though MWNT–siRNA complexes display a high stability and an improved siRNA delivery at much longer time points than liposomes. Therefore, MWNT–siRNA exhibit ideal characteristics for controlled, steady-state delivery of the siRNA therapeutic into cells.

The topical delivery of biopharmaceuticals can serve as a highly efficient method for introducing therapeutics into superficial tumor sites such as skin (melanoma is the most common and invasive). Yet, these therapeutics typically require a carrier for adequate delivery. A topical application method for siRNA delivery into melanoma using SWNTs was recently developed. Although modified SWNTs have been successfully applied as transdermal drug delivery vehicles in the past, siRNA delivery was previously hindered by an insufficient drug loading to achieve significant effects. In order to mitigate this issue, siRNA was bound noncovalently to CNTs via specialized scaffolding proteins. Modified PEI-functionalized SWNTs for noncovalent binding to siRNA and an enhanced loading capacity were employed [87]. Binding of siRNA to the PEI–SWNT complex was confirmed using nuclear magnetic resonance (NMR) spectroscopy. Results from murine model experiments showed significant decreases in tumor size as early as 15 days post-injection, an increased siRNA tissue penetration (see Fig. 6), and a consistently small tumor size and weight up to 25 days post application [87]. The study results indicated that a topical application of siRNA-conjugated SWNTs may be a viable treatment mechanism for superficial tumors, due to the unique ability of SWNTs to interact with cell membranes and deliver their cargoes.

4 Conclusions

CNTs are versatile platforms for drug delivery, especially when conjugated to biopharmaceuticals. These versatile nanocarriers support efficient biopharmaceutical delivery through mechanisms distinct from other nanoparticles. Such mechanisms are made possible by the unique, quasi-one-dimensional structure, ultra-high surface area-to-volume ratio and facile chemistries of the CNTs, which allows them to be more efficient and specific to their target sites. The innate ability of CNTs to penetrate cells and to bind a wide range of compounds, in addition to their optical properties, makes them ideal tools for applications as diverse as immune system boosting, disease detection, and drug treatment for cancer and other diseases.

References

1. Iijima, S. (1991) Helical microtubules of graphitic carbon. *Nature*, **354** (6348), 56–58.
2. Zhu, J. (2012) Mammalian cell protein expression for biopharmaceutical production. *Biotechnol. Adv*, **30** (5), 1158–1170.
3. Steinberg, F.M. and Raso, J. (1998) Biotech pharmaceuticals and biotherapy: an overview. *J. Pharm. Pharm. Sci*, **1** (2), 48–59.
4. Smith, B.R., Kempen, P., Bouley, D., Xu, A. *et al.* (2012) Shape matters: intravital microscopy reveals surprising geometrical dependence for nanoparticles in tumor models of extravasation. *Nano Lett*, **12** (7), 3369–3377.
5. Smith, B.R., Zavaleta, C., Rosenberg, J., Tong, R. *et al.* (2013) High-resolution, serial intravital microscopic imaging of nanoparticle delivery and targeting in a small animal tumor model. *Nano Today*, **8** (2), 126–137.
6. Cai, D., Mataraza, J.M., Qin, Z.H., Huang, Z. *et al.* (2005) Highly efficient molecular delivery into mammalian cells using carbon nanotube spearing. *Nat. Methods*, **2** (6), 449–454.
7. Kam, N.W. and Dai, H. (2005) Carbon nanotubes as intracellular protein transporters: generality and biological functionality. *J. Am. Chem. Soc*, **127** (16), 6021–6026.
8. Liu, Z., Chen, K., Davis, C., Sherlock, S. *et al.* (2008) Drug delivery with carbon nanotubes for in vivo cancer treatment. *Cancer Res*, **68** (16), 6652–6660.
9. Madani, S.Y., Naderi, N., Dissanayake, O., Tan, A. *et al.* (2011) A new era of cancer treatment: carbon nanotubes as drug delivery tools. *Int. J. Nanomedicine*, **6**, 2963–2979.
10. Varkouhi, A.K., Foillard, S., Lammers, T., Schiffelers, R.M. *et al.* (2011) SiRNA delivery with functionalized carbon nanotubes. *Int. J. Pharm*, **416** (2), 419–425.
11. Jin, H., Heller, D.A., Sharma, R., and Strano, M.S. (2009) Size-dependent cellular uptake

and expulsion of single-walled carbon nanotubes: single particle tracking and a generic uptake model for nanoparticles. *ACS Nano*, **3** (1), 149–158.
12. Sato, Y., Yokoyama, A., Shibata, K., Akimoto, Y. *et al.* (2005) Influence of length on cytotoxicity of multi-walled carbon nanotubes against human acute monocytic leukemia cell line THP-1 in vitro and subcutaneous tissue of rats in vivo. *Mol. Biosyst*, **1** (2), 176–182.
13. Kostarelos, K., Bianco, A., and Prato, M. (2009) Promises, facts and challenges for carbon nanotubes in imaging and therapeutics. *Nat. Nanotechnol*, **4** (10), 627–633.
14. Kostarelos, K. (2008) The long and short of carbon nanotube toxicity. *Nat. Biotechnol*, **26** (7), 774–776.
15. Saito, N., Haniu, H., Usui, Y., Aoki, K. *et al.* (2014) Safe clinical use of carbon nanotubes as innovative biomaterials. *Chem. Rev*, **114** (11), 6040–6079.
16. Collins, P.G. (2009) Defects and disorder in carbon nanotubes, in *Oxford Handbook of Nanoscience and Technology: Frontiers and Advances*, Oxford University Press, Oxford, pp. 31–81.
17. Thostenson, E.T., Ren, Z., and Chou, T.-W. (2001) Advances in the science and technology of carbon nanotubes and their composites: a review. *Compos. Sci. Technol*, **61** (13), 1899–1912.
18. Ebbesen, T.W. and Ajayan, P.M. (1992) Large-scale synthesis of carbon nanotubes. *Nature*, **358** (6383), 220–222.
19. Hedmer, M., Isaxon, C., Nilsson, P.T., Ludvigsson, L. *et al.* (2014) Exposure and emission measurements during production, purification, and functionalization of arc-discharge-produced multi-walled carbon nanotubes. *Ann. Occup. Hyg*, **58** (3), 355–379.
20. Yousef, S., Khattab, A., Osman, T.A., and Zaki, M. (2013) Effects of increasing electrodes on CNTs yield synthesized by using arc-discharge technique. *J. Nanomater*, **2013**, 9.
21. Guo, T., Nikolaev, P., Thess, A., Colbert, D.T. *et al.* (1995) Catalytic growth of single-walled nanotubes by laser vaporization. *Chem. Phys. Lett*, **243** (1-2), 49–54.
22. Thess, A., Lee, R., Nikolaev, P., Dai, H. *et al.* (1996) Crystalline ropes of metallic carbon nanotubes. *Science*, **273** (5274), 483–487.
23. Eftekhari, A., Jafarkhani, P., and Moztarzadeh, F. (2006) High-yield synthesis of carbon nanotubes using a water-soluble catalyst support in catalytic chemical vapor deposition. *Carbon*, **44** (7), 1343–1345.
24. Sayangdev, N. and Ishwar, K.P. (2008) A model for catalytic growth of carbon nanotubes. *J. Phys. D Appl. Phys*, **41** (6), 065304.
25. Firme, C.P. and Bandaru, P.R. III, (2010) Toxicity issues in the application of carbon nanotubes to biological systems. *Nanomedicine*, **6** (2), 245–256.
26. Yaron, P.N., Holt, B.D., Short, P.A., Lösche, M. *et al.* (2011) Single wall carbon nanotubes enter cells by endocytosis and not membrane penetration. *J. Nanobiotechnol*, **9**, 45.
27. Fadel, T.R. and Fahmy, T.M. (2014) Immunotherapy applications of carbon nanotubes: from design to safe applications. *Trends Biotechnol*, **32** (4), 198–209.
28. Jacobsen, N.R., Pojana, G., White, P., Møller, P. *et al.* (2008) Genotoxicity, cytotoxicity, and reactive oxygen species induced by single-walled carbon nanotubes and C(60) fullerenes in the FE1-Muta™ mouse lung epithelial cells. *Environ. Mol. Mutagen*, **49** (6), 476–487.
29. Lacerda, L., Singh, R., Al-Jamal, K.T., Turton, J., *et al.* (2008) Dynamic imaging of functionalized multi-walled carbon nanotube systemic circulation and urinary excretion. *Adv. Mater*, **20** (2), 225–230.
30. Pacurari, M., Yin, X.J., Zhao, J., Ding, M. *et al.* (2008) Raw single-wall carbon nanotubes induce oxidative stress and activate MAPKs, AP-1, NF-kappaB, and Akt in normal and malignant human mesothelial cells. *Environ. Health Perspect*, **116** (9), 1211–1217.
31. Jakubek, L.M., Marangoudakis, S., Raingo, J., Liu, X. *et al.* (2009) The inhibition of neuronal calcium ion channels by trace levels of yttrium released from carbon nanotubes. *Biomaterials*, **30** (31), 6351–6357.
32. Gupta, A.K. and Gupta, M. (2005) Synthesis and surface engineering of iron oxide nanoparticles for biomedical applications. *Biomaterials*, **26** (18), 3995–4021.
33. Xu, Y.-Q., Peng, H., Hauge, R.H., and Smalley, R.E. (2004) Controlled multistep purification of single-walled carbon nanotubes. *Nano Lett*, **5** (1), 163–168.

34. Ebbesen, T.W., Ajayan, P.M., Hiura, H., and Tanigaki, K. (1994) Purification of nanotubes. *Nature*, **367** (6463), 519.
35. Schipper, M.L., Nakayama-Ratchford, N., Davis, C.R., Kam, N.W. *et al.* (2008) A pilot toxicology study of single-walled carbon nanotubes in a small sample of mice. *Nat. Nanotechnol*, **3** (4), 216–221.
36. Kayat, J., Gajbhiye, V., Tekade, R.K., and Jain, N.K. (2011) Pulmonary toxicity of carbon nanotubes: a systematic report. *Nanomed. Nanotechnol. Biol. Med*, **7** (1), 40–49.
37. Kolosnjaj-Tabi, J., Hartman, K.B., Boudjemaa, S., Ananta, J.S. *et al.* (2010) In vivo behavior of large doses of ultrashort and full-length single-walled carbon nanotubes after oral and intraperitoneal administration to Swiss mice. *ACS Nano*, **4** (3), 1481–1492.
38. Bussy, C., Pinault, M., Cambedouzou, J., Landry, M.J. *et al.* (2012) Critical role of surface chemical modifications induced by length shortening on multi-walled carbon nanotubes-induced toxicity. *Part. Fibre Toxicol*, **9** (1), 46–60.
39. Liu, Y., Zhao, Y., Sun, B., and Chen, C. (2012) Understanding the toxicity of carbon nanotubes. *Acc. Chem. Res*, **46** (3), 702–713.
40. Mostafalou, S., Mohammadi, H., Ramazani, A., and Abdollahi, M. (2013) Different biokinetics of nanomedicines linking to their toxicity; an overview. *Daru*, **21** (1), 14.
41. Zhang, Y., Deng, J., Zhang, Y., Guo, F. *et al.* (2013) Functionalized single-walled carbon nanotubes cause reversible acute lung injury and induce fibrosis in mice. *J. Mol. Med. (Berl.)*, **91** (1), 117–128.
42. Ma-Hock, L., Treumann, S., Strauss, V., Brill, S. *et al.* (2009) Inhalation toxicity of multiwall carbon nanotubes in rats exposed for 3 months. *Toxicol. Sci*, **112** (2), 468–481.
43. Huczko, A., Lange, H., Całko, E., Grubek-Jaworska, H. *et al.* (2001) Physiological testing of carbon nanotubes: are they asbestos-like? *Fullerene Sci. Technol*, **9** (2), 251–254.
44. Lam, C.W., James, J.T., McCluskey, R., and Hunter, R.L. (2004) Pulmonary toxicity of single-wall carbon nanotubes in mice 7 and 90 days after intratracheal instillation. *Toxicol. Sci*, **77** (1), 126–134.
45. Shvedova, A.A., Kisin, E., Murray, A.R., Johnson, V.J. *et al.* (2008) Inhalation vs. aspiration of single-walled carbon nanotubes in C57BL/6 mice: inflammation, fibrosis, oxidative stress, and mutagenesis. *Am. J. Physiol. Lung Cell. Mol. Physiol*, **295** (4), L552–L565.
46. Shaughnessy, P., Adamcakova-Dodd, A., Altmaier, R., and Thorne, P.S. (2014) Assessment of the aerosol generation and toxicity of carbon nanotubes. *Nanomaterials*, **4** (2), 439–453.
47. Kasai, T., Umeda, Y., Ohnishi, M., Kondo, H. *et al.* (2015) Thirteen-week study of toxicity of fiber-like multi-walled carbon nanotubes with whole-body inhalation exposure in rats. *Nanotoxicology*, **9** (4), 413–422.
48. Ali-Boucetta, H., Nunes, A., Sainz, R., Herrero, M.A. *et al.* (2013) Asbestos-like pathogenicity of long carbon nanotubes alleviated by chemical functionalization. *Angew. Chem. Int. Ed*, **52** (8), 2274–2278.
49. Donaldson, K., Murphy, F.A., Duffin, R., and Poland, C.A. (2010) Asbestos, carbon nanotubes and the pleural mesothelium: a review of the hypothesis regarding the role of long fibre retention in the parietal pleura, inflammation and mesothelioma. *Part. Fibre Toxicol*, **7**, 5.
50. Poland, C.A., Duffin, R., Kinloch, I., Maynard, A. *et al.* (2008) Carbon nanotubes introduced into the abdominal cavity of mice show asbestos-like pathogenicity in a pilot study. *Nat. Nanotechnol*, **3** (7), 423–428.
51. Madani, S.Y., Mandel, A., Seifalian, A.M. (2013) A concise review of carbon nanotube's toxicology. *Nano Rev.*, **4**, 4. doi: 10.3402/nano.v4i0.21521
52. Liu, Z., Davis, C., Cai, W., He, L. *et al.* (2008) Circulation and long-term fate of functionalized, biocompatible single-walled carbon nanotubes in mice probed by Raman spectroscopy. *Proc. Natl Acad. Sci. USA*, **105** (5), 1410–1415.
53. Mittal, V. (2011) Carbon nanotubes surface modifications: An overview, in *Surface Modification of Nanotube Fillers*, Wiley-VCH Verlag GmbH & Co. KGaA, Weinheim, pp. 1–23.
54. Liu, Z., Tabakman, S., Welsher, K., and Dai, H. (2009) Carbon nanotubes in biology and medicine: in vitro and in vivo detection, imaging and drug delivery. *Nano Res*, **2** (2), 85–120.
55. Wu, H.-C., Chang, X., Liu, L., Zhao, F. *et al.* (2010) Chemistry of carbon nanotubes in biomedical applications. *J. Mater. Chem*, **20** (6), 1036–1052.

56. Chen, R.J., Zhang, Y.G., Wang, D.W., and Dai, H.J. (2001) Noncovalent sidewall functionalization of single-walled carbon nanotubes for protein immobilization. *J. Am. Chem. Soc*, **123** (16), 3838–3839.
57. Chen, Z., Tabakman, S.M., Goodwin, A.P., Kattah, M.G. *et al.* (2008) Protein microarrays with carbon nanotubes as multicolor Raman labels. *Nat. Biotechnol*, **26** (11), 1285–1292.
58. Yang, S.T., Luo, J., Zhou, Q., and Wang, H. (2012) Pharmacokinetics, metabolism and toxicity of carbon nanotubes for biomedical purposes. *Theranostics*, **2** (3), 271–282.
59. Moore, T.L., Grimes, S.W., Lewis, R.L., and Alexis, F. (2014) Multilayered polymer-coated carbon nanotubes to deliver dasatinib. *Mol. Pharm*, **11** (1), 276–282.
60. Malmsten, M. (2013) Inorganic nanomaterials as delivery systems for proteins, peptides, DNA, and siRNA. *Curr. Opin. Colloid Interface Sci*, **18** (5), 468–480.
61. Witus, L.S., Rocha, J.-D.R., Yuwono, V.M., Paramonov, S.E. *et al.* (2007) Peptides that non-covalently functionalize single-walled carbon nanotubes to give controlled solubility characteristics. *J. Mater. Chem*, **17** (19), 1909–1915.
62. Balavoine, F., Schultz, P., Richard, C., Mallouh, V. *et al.* (1999) Helical crystallization of proteins on carbon nanotubes: a first step towards the development of new biosensors. *Angew. Chem. Int. Ed*, **38** (13-14), 1912–1915.
63. Chen, R.J., Bangsaruntip, S., Drouvalakis, K.A., Kam, N.W. *et al.* (2003) Noncovalent functionalization of carbon nanotubes for highly specific electronic biosensors. *Proc. Natl Acad. Sci. USA*, **100** (9), 4984–4989.
64. Sun, H., She, P., Lu, G., Xu, K. *et al.* (2014) Recent advances in the development of functionalized carbon nanotubes: a versatile vector for drug delivery. *J. Mater. Sci*, **49** (20), 6845–6854.
65. Smith, B.R., Cheng, Z., De, A., Koh, A.L. *et al.* (2008) Real-time intravital imaging of RGD-quantum dot binding to luminal endothelium in mouse tumor neovasculature. *Nano Lett*, **8** (9), 2599–2606.
66. Weissleder, R. and Pittet, M.J. (2008) Imaging in the era of molecular oncology. *Nature*, **452** (7187), 580–589.
67. Smith, B.R., Ghosn, E.E., Rallapalli, H., Prescher, J.A. *et al.* (2014) Selective uptake of single-walled carbon nanotubes by circulating monocytes for enhanced tumour delivery. *Nat. Nanotechnol*, **9** (6), 481–487.
68. Hashida, Y., Tanaka, H., Zhou, S., Kawakami, S. *et al.* (2014) Photothermal ablation of tumor cells using a single-walled carbon nanotube-peptide composite. *J. Control. Release*, **173**, 59–66.
69. Chen, C., Hou, L., Zhang, H., Zhu, L. *et al.* (2013) Single-walled carbon nanotubes mediated targeted tamoxifen delivery system using asparagine-glycine-arginine peptide. *J. Drug Target*, **21** (9), 809–821.
70. Battigelli, A., Russier, J., Venturelli, E., Fabbro, C. *et al.* (2013) Peptide-based carbon nanotubes for mitochondrial targeting. *Nanoscale*, **5** (19), 9110–9117.
71. Villa, C.H., Dao, T., Ahearn, I., Fehrenbacher, N. *et al.* (2011) Single-walled carbon nanotubes deliver peptide antigen into dendritic cells and enhance IgG responses to tumor-associated antigens. *ACS Nano*, **5** (7), 5300–5311.
72. Parra, J., Abad-Somovilla, A., Mercader, J.V., Taton, T.A. *et al.* (2013) Carbon nanotube-protein carriers enhance size-dependent self-adjuvant antibody response to haptens. *J. Control. Release*, **170** (2), 242–251.
73. Izumi, Y., Xu, L., di Tomaso, E., Fukumura, D. *et al.* (2002) Tumour biology: herceptin acts as an anti-angiogenic cocktail. *Nature*, **416** (6878), 279–280.
74. Venturelli, E., Fabbro, C., Chaloin, O., Ménard-Moyon, C. *et al.* (2011) Antibody covalent immobilization on carbon nanotubes and assessment of antigen binding. *Small*, **7** (15), 2179–2187.
75. Li, R., Wu, R., Zhao, L., Wu, M. *et al.* (2010) P-glycoprotein antibody functionalized carbon nanotube overcomes the multidrug resistance of human leukemia cells. *ACS Nano*, **4** (3), 1399–1408.
76. Bajaj, P., Mikoryak, C., Wang, R., Bushdiecker, D.K. II, *et al.* (2014) A carbon nanotube-based Raman-imaging immunoassay for evaluating tumor targeting ligands. *Analyst*, **139** (12), 3069–3076.
77. Zavaleta, C., de la Zerda, A., Liu, Z., Keren, S. *et al.* (2008) Noninvasive Raman spectroscopy in living mice for evaluation of tumor targeting with carbon nanotubes. *Nano Lett*, **8** (9), 2800–2805.
78. Mulvey, J.J., Villa, C.H., McDevitt, M.R., Escorcia, F.E. *et al.* (2013) Self-assembly

of carbon nanotubes and antibodies on tumours for targeted amplified delivery. *Nat. Nanotechnol*, **8** (10), 763–771.

79. Wu, H., Shi, H., Zhang, H., Wang, X. *et al.* (2014) Prostate stem cell antigen antibody-conjugated multiwalled carbon nanotubes for targeted ultrasound imaging and drug delivery. *Biomaterials*, **35** (20), 5369–5380.
80. Liu, Y., Wu, D.C., Zhang, W.D., Jiang, X. *et al.* (2005) Polyethylenimine-grafted multiwalled carbon nanotubes for secure noncovalent immobilization and efficient delivery of DNA. *Angew. Chem. Int. Ed*, **44** (30), 4782–4785.
81. Pantarotto, D., Singh, R., McCarthy, D., Erhardt, M. *et al.* (2004) Functionalized carbon nanotubes for plasmid DNA gene delivery. *Angew. Chem. Int. Ed*, **43** (39), 5242–5246.
82. Bates, K. and Kostarelos, K. (2013) Carbon nanotubes as vectors for gene therapy: past achievements, present challenges and future goals. *Adv. Drug Deliv. Rev*, **65** (15), 2023–2033.
83. Benya, R.V., Kusui, T., Pradhan, T.K., Battey, J.F. *et al.* (1995) Expression and characterization of cloned human bombesin receptors. *Mol. Pharmacol*, **47** (1), 10–20.
84. Qiao, J., Hong, T., Guo, H., Xu, Y.Q. *et al.* (2013) Single-walled carbon nanotube-mediated small interfering RNA delivery for gastrin-releasing peptide receptor silencing in human neuroblastoma, in *NanoBiotechnology Protocols* (eds S.J. Rosenthal and D.W. Wright), Humana Press, pp. 137–147.
85. Wang, L., Shi, J., Zhang, H., Li, H. *et al.* (2013) Synergistic anticancer effect of RNAi and photothermal therapy mediated by functionalized single-walled carbon nanotubes. *Biomaterials*, **34** (1), 262–274.
86. Battigelli, A., Wang, J.T., Russier, J., Da Ros, T. *et al.* (2013) Ammonium and guanidinium dendron-carbon nanotubes by amidation and click chemistry and their use for siRNA delivery. *Small*, **9** (21), 3610–3619.
87. Siu, K.S., Chen, D., Zheng, X., Zhang, X. *et al.* (2014) Non-covalently functionalized single-walled carbon nanotube for topical siRNA delivery into melanoma. *Biomaterials*, **35** (10), 3435–3442.

17
Role of the Microbiota in Cancer Therapy

Phillip J. Daschner
National Cancer Institute, Division of Cancer Biology, 9609 Medical Center Drive, Bethesda, MD 20892, USA

Translational Medicine Cancer: Advances in Molecular Biology and Medicine,
First Edition. Edited by Robert A. Meyers.

Keywords

Microbiome
The combined biogenic material of all microorganisms in a particular environment (e.g., the human gastrointestinal tract).

Dysbiosis
The condition of microbial population imbalances within a host that can promote pathogenesis through cell damage, chronic inflammatory conditions, epigenetic changes, or the production of carcinogenic metabolites.

Probiotic
Live microorganisms which, when administered in adequate amounts, confer a health benefit on the host.

Prebiotic
A non-viable food component that confers a health benefit on the host, associated with modulation of the microbiota.

Synbiotic
A combination of probiotics and prebiotics that beneficially affect the host by improving the survival and implantation of a probiotic.

Endobiotic and xenobiotic
Compounds formed either within or outside of a host cell or tissue.

Microbial metabolites
The byproducts of microbial enzymatic activity derived from drug, dietary, and cellular substrates. These metabolites have a variety of host signaling effects that may have either carcinogenic or protective effects.

Drug metabolism
The enzymatically catalyzed biochemical modification of a therapeutic agent. In humans, this process is carried out by both host and microbial enzymes and can either activate or inhibit drug activity.

Cancer therapeutic
A pharmacological agent administered to treat tumors.

Recent advances in knowledge of the human microbiome's composition and function have generated a flood of new information, technologies, and capabilities that have increased the present understanding of the impact that the microflora has on the initiation, progression, and treatment of cancer. Revealing the molecular mechanisms by which the gastrointestinal microflora can affect carcinogenesis will help to identify both potential new therapeutic targets and microflora–drug interactions that may either disrupt or improve therapeutic efficacy. Clinical translation of this new information is in its infancy, and will be critical to the development of novel strategies for improved therapies based on the rational targeting and manipulation of microbiome function for the treatment of human tumors.

1 Introduction

The human gastrointestinal (GI) tract is colonized by a complex microbial community comprising thousands of species and strains whose individuals total an order of magnitude greater in number than the somatic cells in the human body. These microbes are primarily bacterial, though fungi, archaea and viruses also comprise the community [1, 2]. In addition, the collective genetic material of the microflora, is estimated to encode more than 100-fold the number of genes in the human genome [3]. Humans therefore carry a vastly understudied and complex GI ecosystem composed of their own unique mucosal tissue interacting dynamically with trillions of foreign organisms and their metabolic products.

A growing list of GI resident bacterial species have been associated with carcinogenesis in both humans and in animal models, including *Helicobacter pylori*, enterotoxigenic *Bacteroides fragilis* (ETBF), *Escherichia coli* (NC101), *Fusobacteria nucleatum*, *Salmonella enterica*, and *Streptococcus mitis*. However, treatment strategies designed to inhibit bacterial activities linked to carcinogenesis must go beyond the scope of classical models of acute infectious disease, where the presence of a single biological agent is linked with the onset of disease, and its elimination is curative.

Currently, an important concept emerging from the field of microbiome studies is that gut-resident bacteria interact with their mammalian hosts not only at the level of individual species but also in concert, at the community level, and in the context of a rich and variable nutrient environment. Thus, chronic states of dysbiosis – the condition of microbial population imbalances within the host – can promote cell damage, chronic inflammatory conditions, epigenetic changes, as well as the production of carcinogenic metabolites, leading to neoplasia in the GI tract and other tissues [4–10]. The impact of these host–microbe interactions may be both local and systemic, and can occur through direct cell contact or via the production of secreted factors and metabolites, or by induced signaling through host immune and metabolic pathways [11, 12].

Emerging data have defined a perturbation → dysbiosis → host insult → neoplastic transformation signaling cascade that reveals several targets for intervention strategies based on the disruption of specific points along this pathogenic pathway. In this chapter, evidence will be discussed

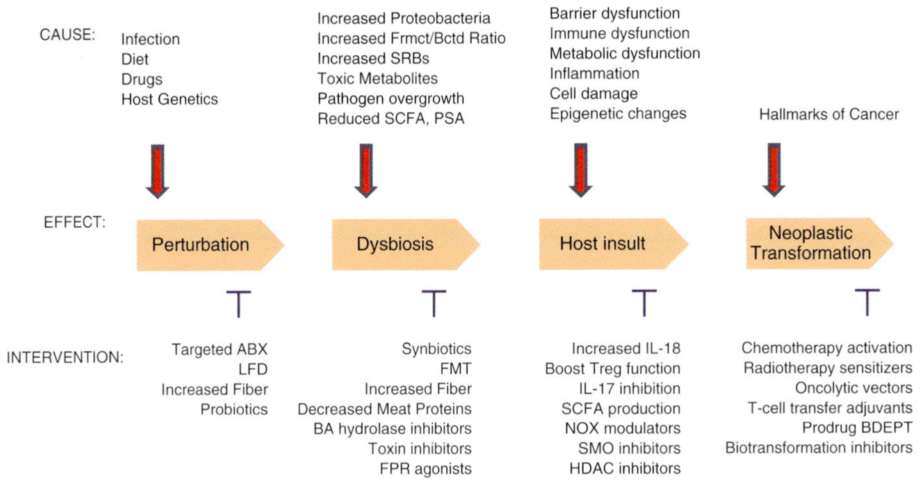

Fig. 1 Schematic of how the human microbiome promotes carcinogenesis through several mechanisms. Examples of cause-and-effect relationships are indicated by arrows. Interventions that target microbial interactions, signals and functions are depicted by T-bars, below arrows. These therapies may prove to be more tractable and effective than targeting host tissue alone, particularly when used in combination with conventional cancer therapies. Abbreviations: ABX, antibiotics; LFD, low-fat diet; Frmct/Bctd, Firmicutes/Bacteroidetes; SRBs, sulfate-reducing bacteria; SCFA, short-chain fatty acids; PSA, *B. fragilis* polysaccharide A; FMT, fecal microbial transplant; BAs, bile acids; FPR, formyl peptide receptor; NOX, NADPH oxidase; SMO, spermine oxidase; HDAC, histone deacetylase; BDEPT, bacterially directed enzyme prodrug therapy.

that identifies several novel therapeutic targets presented by this model (Fig. 1), and potential strategies will be examined to: (i) rationally modify the outcome of host–microbiome interactions to prevent dysbiotic shifts caused by diet, infection, or antibiotic and chemotherapy treatment; and (ii) ameliorate the effects of drug–microflora interactions that may compromise therapeutic efficacy.

2
Understanding Physiologic Effects of GI Microbiota

The human intestinal microbiota affects a broad range of host physiological functions that are important in health and disease. These functions include intestinal epithelial cell (IEC) homeostasis and barrier function, efficient metabolism and energy harvest from dietary sources, vitamin biosynthesis, pathogen protection, and the development and maintenance of innate and adaptive mucosal immune responses [13–16]. The composition of the bacterial community that coregulates these critical host functions can be influenced by several important factors, including host genetics and immune responses, diet, inflammation, and antibiotic (and other drug) treatments [17–24]. Metagenomic studies performed in humans and animal models have established that alterations of the GI microflora community occur during the development of neoplastic and pre-neoplastic disease, and in tumor-bearing versus healthy individuals [25–27]. However, it is currently poorly understood how changes in the microbiome may affect cancer progression and therapeutic interventions [28]. An understanding of the impact of human host–microflora interactions on the initiation, progression and treatment of cancer, and the molecular mechanisms that govern the outcomes of these interactions, will provide new therapeutic strategies and new targets for the treatment of many human tumors, particularly with microbial interactions.

2.1 Alterations of Barrier Function

The human GI tract is lined with a protective mucosal barrier comprised of an epithelial cell membrane that is overlaid with a richly glycosylated mucus layer and underlaid by the dynamic mucosal immune components of the lamina propria and the muscular mucosa. The dense and diverse colonization of intestinal and colonic niches by commensal bacteria in healthy individuals regulates GI tissue development, provides protection against invading pathogenic species, and also functions as an important regulator of GI epithelial, metabolic, and immune homeostasis [29–32], all of which have important influence on host antitumor responses and tolerance to therapy.

2.1.1 Direct Cell Interactions

Compromise of the junctional integrity of gut epithelial cells by viral or bacterial pathogens, cytotoxic chemotherapy agents or developing adenocarcinomas leads to barrier hyperpermeability and exposes the lamina propria to luminal antigens, thereby provoking compensatory immune and wound-repair responses [33, 34]. These responses may become dysregulated in susceptible hosts, resulting in a progression of chronic inflammatory and proliferative disease states. For example, mice carrying a Crohn's disease susceptibility mutation, Atg16L1, exhibit Paneth cell defects, and do not properly resolve inflammation induced by murine norovirus infection. This results in an increased exposure of the submucosal compartment to bacterial antigens and to the development of intestinal lymphoid aggregates, submucosal fibrosis, and epithelial hyperplasia\break [35, 36]. It has also been shown that regulation of the GI epithelial cell cycle and stem cell activation in response to barrier disruption (wound repair) is influenced by specific host–microbiome interactions [37]. Several commensal species of *Lactobacillus* are known to induce NADPH oxidase (NOX)-mediated generation of reactive oxygen species (ROS)-dependent proliferation signals in gut epithelial cells, a function that is markedly impaired in germ-free animal models [38, 39]. These epithelial proliferation signaling pathways are regulated by cell-surface formyl peptide receptors (FPRs) that recognize bacterial cell wall components with taxa-specific formyl group modifications, which may form the basis for a differential activation of NOX-mediated signaling that is essential for barrier restitution following cytotoxic insult [40]. Thus, peptides or other FPR agonists that are capable of enhancing GI barrier function could reduce the effects of chronic mucosal inflammation caused by GI barrier disruption [41–43].

2.1.2 Bacterial Toxins

Several members of the human GI microbial community that are strongly associated with carcinogenic activity (e.g., *H. pylori*, *Salmonella typhimurium*, *E. coli* NC101, *Bacteroides fragilis* ETBF) share the common trait of altering mucosal barrier function through the actions of secreted toxins that target epithelial cell junctions [44–47]. While specific bacterial toxins affect host mucosal cells via distinct signaling mechanisms [48, 49], general strategies to disrupt toxin expression or specific toxin/host cell receptor signal pathways could be effective in treating carcinogenic inflammatory and wound repair responses to enterotoxigenic bacteria.

2.2 Alterations in Metabolism

Microbiota and host enzymatic activities contribute to the co-metabolism of many bioactive foods and cellular compounds, including the activation/deactivation of multiple carcinogens (e.g., *N*-nitroso compounds and heterocyclic amines), drugs (e.g., irinotecan and cyclophosphamide; CTX), and hormones (e.g., estrogens and phytoestrogens) [50–52]. Some of the key classes of bacterial enzymes involved in these metabolic pathways include β-glycosidases, β-glucuronidases, sulfotransferases, arylsulfatases, alcohol dehydrogenases, nitroreductases (NTRs), and lyases [53, 54].

Metagenomic surveys and bioinformatic reconstruction of GI tract microbial community metabolism has indicated that there are stable collective metabolic networks in common among healthy individuals, which suggests that a core functional capacity exists within an individual's heterologous microbial consortia [55]. High-dimensional systems biology approaches have been used to determine functional changes that occur in gut bacterial community metabolic capacity during pathogenesis [56, 57], providing research groups with new approaches to decipher the complex and networked interactions and co-metabolism of drugs and dietary factors within both host and microbial biosynthetic pathways. Identifying differences in microbiome metabolic capacity and activity could help to explain individual variations in the metabolism of drugs and food that underlies differences in both treatment and dietary outcomes [10, 58]. Examples of how microbial metabolic activity affects carcinogenesis are briefly described in the following section, whilst examples of how microbial metabolic activity directly influences treatment outcomes are described in Sect. 3.2.1.

2.2.1 Dietary (Xenobiotic) Metabolites

There is strong evidence that diet plays a critical role in regulating both transient changes and long-term "eubiotic" states of the GI tract microbiome, and thus in turn the enormous and diverse metabolic capacity and output of the gut microbiota, including the activation and detoxification of a variety of ingested drugs and xenobiotics [59–61]. Hence, diet may be considered the primary modulator of human microbiota metabolic function. Bioactive food components metabolized by gut microbes may play either a tumor-initiating or a protective role, whether by facilitating the production of carcinogenic diet-derived metabolites or by promoting the growth of probiotic bacteria and providing the substrates for microbial metabolites with anticancer properties. Consequently, dietary changes that target microbial populations present an opportunity for therapeutic manipulation. As the effects on carcinogenesis of several dietary metabolites have been recently reviewed [61], examples will be only briefly summarized at this point.

Carcinogenic Metabolites It has been demonstrated in both human and animal studies that the gut bacteria can produce a variety of diet-derived genotoxic and inflammatory metabolites that have been linked to carcinogenesis. These compounds are formed through the actions of several classes of widely expressed microbial enzymes, such as azo- and nitroreductases, and β-glucuronidases and β-glucosidases [62, 63], and may also act as substrates for host epithelial cell enzymes. For example, "Western" diets with a high protein content from red and processed meats have been linked to an increased risk of colorectal cancer (CRC), and also with elevated levels of the sulfur amino acids cysteine and methionine. These amino acids are substrates for the production of hydrogen sulfide, a genotoxic gas formed by the metabolic action of a consortium of sulfur-reducing bacteria (SRBs) [64], which intriguingly includes the CRC-associated bacterium *Fusobacterium nucleatum*. This contrasts with the well-established chemoprotective effects of diets high in sulfur from plant-derived indoles and isothiocyanates [65, 66], raising the possibility that levels of amino acid-derived sulfur from red meat – but not glucosinolate and indole-derived sulfur – alters the abundance and function of SRBs to regulate host pathways central to carcinogenesis.

Another class of potentially harmful bacterial metabolites resulting from high-protein diets includes ammonia and *N*-nitroso compounds derived from the microbial fermentation of peptides in the colon [67]. NTRs and nitrate reductases that are expressed primarily by species of Proteobacteria (*Escherichia*, *Pseudomonas*, *Proteus*, *Klebsiella*, and *Neisseria*) have been shown to contribute to the nitrosation of heterocyclic amines and the production of carcinogenic *N*-nitroso compounds. In rats fed a diet with a high meat content, the abundance and activity of NTRs was found to be increased [9, 68, 69]. Thus, strategies that limit the abundance or metabolic activities of SRBs and relevant NTR-producing Proteobacteria may be effective in reducing colonic tumors. Taken together, the results of these studies have clearly established the existence of a microbial component in the production of carcinogenic metabolites from diets with high red meat contents, and offer the opportunity for therapeutic interventions based on disrupting their production.

Protective Effects Several dietary components are significantly metabolized by intestinal bacteria to produce compounds with demonstrated potent anticancer characteristics. These include metabolites of soy isoflavonoids, plant fibers, plant lignans, and polyphenols found in fruits, wine, and tea [70–72]. The polyphenol class of metabolites, and particularly the high-molecular-weight polyphenols that are abundant in several dietary sources, have pleiotropic effects on gut physiology and function, including antioxidant, anti-inflammatory, and other cancer-protective activities [73]. Polymeric polyphenols are poorly absorbed in the human gut and thus become substrates for microbial enzyme activity in the colon (e.g., esterase, glucosidase, demethylase, and hydroxylase) [74]. Several polyphenol metabolites have been associated with lowered risks of colon and prostate cancer in humans, and have been shown to impact on carcinogenic pathways in experimental models [75, 76]. Although the mechanism of action related to a lower cancer risk from a polyphenol-rich diet remains unclear, such diets clearly impact on the composition and structure of gut bacterial communities, such as promoting the growth of probiotic *Lactobacillus* sp. and *Bifidobacterium* sp., and inhibiting disease-associated Enterobacteriaceae strains [77].

For example, the isoflavonoid daidzein is abundant in soy and is metabolized to the byproducts equol and *O*-desmethylangolensin (DMA) by GI bacteria [78]. Equol can inhibit the growth of prostate and breast tumor cells *in vitro* at concentrations that are physiologically relevant in humans [77, 79]. Although it is likely that a consortia of commensal intestinal bacteria contribute to equol and DMA production *in vivo*, specific daidzein-metabolizing species have been identified, including members from the genera *Lactobacillus, Slackia*, and *Eubacteria*, that appear to be key contributors to the process [80, 81]. In addition, a novel three-enzyme daidzein metabolic pathway has been recently determined [82]. Interestingly, only 25–30% of the adult population of Western countries produce equol when consuming a soy-rich diet, compared to 50–60% of adults from Asia. This difference may account for the variable cancer-protective effects of soy that have been reported for different populations [78], and highlights the role that differences in microbial metabolic activity may play in the outcome of diet and drug intervention strategies. Additional studies on ellagic acid, a plant polyphenol that is metabolized by the human colonic microflora to yield urolithins A and B (UROA&B), have shown these compounds to have potent anti-inflammatory activity in the colonic mucosa [83]. Similar to equol, production levels of urolithins depend on the composition of the gut microflora, and this may account for the large interindividual variability observed in human supplementation studies [84].

Butyrate and other short-chain fatty acids (SCFAs) produced by the fermentation of complex plant carbohydrates in the colon are able to promote intestinal barrier integrity, inhibit cell growth, and induce differentiation in colon cancer cells [85], at least partially through its capacity to inhibit histone deacetylases [86], although this effect is dependent on cellular energy status [87]. In human studies, surveys of the gut microbiota from CRC patients have revealed consistent reductions in the abundance of butyrate producers (*Lactobacilli, Bifidobacteria, Enterococci, Streptococci, and Eubacteria*), and reductions in microbial carbohydrate degrading enzymes, such as glycoside-hydrolases, that break down indigestible dietary fiber to SCFAs [88, 89]. A novel strategy for the therapeutic delivery of butyrate *in vivo* has been developed using a diet supplemented with butyrylated maize starch to increase butyrate levels in the colon of AOM treated rats, which resulted in increased apoptosis in DNA-damaged colonocytes and improved barrier function [90]. Although these data indicate a protective effect for butyrate production from the wide array of glycoside-hydrolases that have been identified in the human gut microbiome [24, 91], recent findings demonstrate that butyrate production can also act to promote CRC in mouse models that carry a predisposing DNA repair defect [92].

2.2.2 Host (Endobiotic) Metabolites

In addition to their effects on xenobiotic metabolism, resident microbes also contribute to the GI tract milieu by generating metabolites from host-derived compounds (endobiotics) such as hormones, bile acids, glycans, and histidines through fermentation, reduction, hydrolysis, esterification, and conjugation/deconjugation reactions, or by altering the expression and activity of host genes involved in these metabolic pathways. The metabolic byproducts of endobiotic compounds may have either carcinogenic or protective effects, and potentially offer novel druggable targets for cancer treatment.

Bile Acid Metabolism Primary bile acids such as cholic acid and chenodeoxycholic are synthesized in hepatocytes directly from cholesterol, and are important signaling molecules that regulate biosynthetic and metabolic pathways in the gut and liver which, in turn, affect the pharmacokinetics of chemotherapies [93]. Gastrointestinal bacteria are responsible for the bioconversion of liver-conjugated primary bile acids such as glycocholic or taurocholic acids to secondary bile acids such as deoxycholic acid (DCA) and lithocholic acid, which are likely carcinogens that are strongly associated with increased risk of colon, liver, pancreatic, and gastric cancer in humans [94]. In fact, DCA was shown recently to drive hepatocellular carcinoma development in rodent models by inducing a senescence-associated secretory phenotype in liver stellate cells, a process that was reversed by inhibiting DCA production with either difructose anhydride III or by antibiotic treatment [95].

Two key enzyme classes produced by bacteria that convert primary to secondary bile acids include: (i) deconjugating hydrolases, which are expressed by a broad spectrum of anaerobes; and (ii) 7-alpha-dehydroxlyase, which is primarily expressed by strains of *Lactobacillus* and *Clostridium* [96, 97]. Studies comparing germ-free mice with conventionally housed mice revealed marked differences in both the abundance and content of the total bile acid pools, and indicated that the gut microbiota not only directly deconjugate bile acids but also inhibit their conjugation in the liver by reducing the levels of conjugating substrates such as tauro-β-muricholic acid [98]. Conversely, bile acid levels can alter gut microbial composition [99], thereby affecting secondary bile acid production and enterohepatic circulation. In rodents fed high-cholic acid diets, the prevalence of Firmicutes was overwhelmingly expanded to about 95% of the microbiota, compared to 39% in controls. This expansion was accompanied by an increase in 7-alpha-dehydroxlyase activity and an increased production of DCA [100].

The activity of microbial enzymes that modify bile acids and salts differ among species with respect to substrate specificity, pH optima, enzyme kinetics, and GI tract location [101], and these differences may be exploited for therapeutic purposes. For example, acidification decreases 7-α-dehydroxylation activity and therefore the production of secondary bile acids [102]. A lowered colonic pH can be achieved by increasing the production of SCFA in the colon, which suggests that one mechanism by which probiotic SCFA producers can lower GI cancer risk may be through alterations of the bile acid pool [103]. Other strategies for altering the bile acid pool content include targeting bacterial deconjugating hydrolases and 7-α-dehydroxylation via pharmacologic inhibitors, or by sequestering bile acids biologically in probiotic bacteria, to increase their excretion from the host GI tract. Inhibiting hydrolases not only reduces the formation of harmful secondary bile acids, but may also reduce the formation of genotoxic hydrogen sulfide, a byproduct of the reduction of the sulfur containing taurine moiety of taurocholic acid [104].

Hormone Metabolites Hormones are metabolized by the gut bacteria to yield bioactive compounds that can profoundly affect tumor progression. For example, intestinal bacteria that express methylases, glucuronidases and sulfonases contribute to the proportion of estrogen metabolites that are excreted, absorbed into the blood, or recirculated via the enterohepatic pathway [105]. High systemic estrogen levels contribute to breast, endometrial and other tumors, and the hepatic conjugation of catechol estrogens (e.g., methylation, glucuronidation, and sulfonation) regulates their recirculation and bioavailability. The aggregate of bacterial genes that modify host estrogens has been termed the "estrobolome" [51], to distinguish these activities from host estrogen metabolism. Central to this metabolic activity are bacterial taxa such as Clostridia and Ruminococcaceae that produce β-glucuronides and β-glucuronidases, enzymes which are important in the estrogen conjugation/deconjugation reactions that control hormone activity [106]. Antibiotics have been shown to decrease the enterohepatic circulation of estrogens and to increase the excretion of conjugated estrogens [107], and thus affect the serum levels of active estrogens. This suggests that interventions designed to target the microbial hydroxylation and deconjugation of estrogens could help to reduce estrogen-related cancers.

The gut bacteria also play important roles in regulating hormones involved in energy metabolism and glucose homeostasis. For example, the production of glucagon-like peptide-1 (GLP-1), an incretin hormone that inhibits GI transit and glucagon secretion, and induces insulin production from pancreatic islet cells, is regulated by the microbial production of SCFAs [108]. In germ-free and antibiotic-treated mice, serum levels of GLP-1 are increased, whereas in untreated, conventionally housed mice GLP-1 levels are inversely correlated with SCFA levels. These results suggest that, in the absence of a SCFA energy source in the gut, GLP-1 expression is upregulated to signal a delay in gastric chyme transit that allows for an improved energy harvest in germ-free and antibiotic-treated mice.

2.3 Alterations in Immune Response

The mechanisms by which microflora colonize the GI tract, establish tolerance, and effect host immune system development and inflammatory responses has been reviewed elsewhere [21, 109]. Examples of specific effects that host–microbe interactions have on antitumor immunotherapy and chemotherapy are discussed in Sect. 3.2.2.

3 Microbiota in Cancer Therapy

Tumor therapies using bacterial products have been used at least as far back as Coley's toxins during the late nineteenth century [110]. Modern therapies that include live microorganisms are classified by the US Food and Drug Administration (FDA) as live biotherapeutic products (LBPs), and several examples of these are currently in various stages of development [111]; these include using a consortium of *Clostridium* strains to treat preneoplastic conditions such as irritable bowel disease (IBD) [112]. Other recent efforts to develop antitumor LBPs have included the preclinical characterization of oncolytic bacteria such as *Clostridium novyi* and *Salmonella typhimurium* [113, 114], both of which feature selective antitumor properties based on improving the host antitumor immune responses, or on the homing of tumoricidal bacteria to hypoxic tissues within solid tumors. The only currently approved LBP for cancer therapy remains the use of live, attenuated strains of *Mycobacterium bovis* (*Bacillus*–Calmette–Guerin; BCG) for the treatment of superficial bladder tumors [115]. The development of novel antitumor LBPs is now focused on targeting specific microflora functions and host–microbe interactions that affect a range of carcinogenic signal pathways and therapy responses [116–118]. It has been noted that a detailed knowledge of each cancer patient's unique microbiome is likely to have high translational value to clinical practice, as this information could be exploited for the purposes of optimizing individual therapeutic responses, whether by altering microbial signals to change host metabolic regulation or by developing new metrics for patient stratification based on matching chemotherapy agents to an individual's microbial drug metabolism profile [119].

3.1 Targeting the Microbiota

Human/microbial coevolution has strongly selected for coupling structure-specific substrates with bacterial strain-specific functions in order to promote the colonization and growth of a homeostatic and resilient microbiota within the healthy GI tract [120, 121]. Observations of an altered gut microbiome in patients with IBD and CRC have each prompted treatment strategies directed at changing the gut bacterial communities to re-establish homeostasis in the GI tract, either by selectively removing pathogenic species or by a restoration of commensal species and function. As microbial genetics and function can be readily manipulated both *in vivo* or *ex vivo* by using standard molecular biology techniques or by drug and dietary interventions, the microbiota offers many potentially attractive targets for therapeutic manipulation and engineering. Strategies include eliminating pro-inflammatory taxa, replacing or supplementing deficient microbial metabolic or signaling functions, or the addition of bacterial vectors for the delivery of specific antitumor treatments. Consequently, several therapeutic strategies that either broadly target the microbiome or directly target specific components of the gut microflora community or its functional byproducts, either alone or as adjuvants, have been developed for the treatment of malignancies and preneoplastic conditions [122, 123].

3.1.1 Antibiotics and Bacteriophage

Although therapeutic antitumor antibiotics such as aminoglycosides and anthracyclines have become standard treatment for certain types of cancer, their anticancer activity is primarily attributed to direct effects on tumor cell replication (e.g., DNA synthesis), and their effectiveness is limited by severe cardiotoxicity [124]. Additional studies have found that aminoglycoside antibiotics may be effective in treating colon tumors that carry specific point mutations by inducing translation readthrough of premature stop codons in mutated adenomatous polyposis coli (APC) genes [125].

Perhaps the most direct example of effective antitumor therapy that targets tumor-associated microflora is the treatment of established gastric mucosa-associated lymphoid tissue (MALT) lymphomas, with a triple antibiotic cocktail consisting of clarithromycin, amoxicillin, and lansoprazole [126]. This treatment regime clears the tumor-associated *H. pylori* infection and regresses gastric MALT lymphomas in about 90% of cases [127]. This achievement suggests that as other key "alpha members" of the human microflora that drive carcinogenesis are isolated and identified [25], new antibiotic treatment regimens may be developed to successfully treat developing tumors. However, *H. pylori* remains the only bacterium recognized as a Class I carcinogen [128], and given the unintended pathogenic sequela of many broad-spectrum antibiotic treatments [129–131], it is necessary to develop new antimicrobials with specific and narrow activity spectrums that can also avoid development of multiple drug resistance phenotypes.

Cationic antimicrobial peptides (CAMPs) and peptidoglycan recognition proteins (PGRPs) are among possible solutions. These are naturally occurring amphiphilic compounds that are ubiquitous components of the innate immune defense system and possess mechanisms of action that resist the development of multiple drug resistance [132]. CAMPs can accumulate on the membrane surface of prokaryotes to form high-density clusters that coat the cell surface, much like a carpet [133], analogous to the attachment and cell coating of host immunoglobulin (Ig) A. However, naturally occurring CAMPs have a nonspecific activity profile, killing both Gram-positive and Gram-negative bacteria. In addition, they are subject to proteolytic degradation in the gut and have potentially high immunogenicity and toxicity.

A new class of cyclic D&L-alpha CAMPs has been described that possess several attractive therapeutic features that include proteolytic stability, ease of synthesis, and a lack of natural resistance mechanisms; moreover, these CAMPs can be derived from a potentially vast array of membrane-active sequence space [134]. Lastly, recent efforts to develop synthetic CAMPS with selective antimicrobial activity presents the tantalizing prospect that peptide nanotubes can be designed with narrow-spectrum selectively to target and lyse specific bacterial membranes, or even block the viral cell entry of carcinogenic microbes [135–137].

Another promising method to specifically target potentially carcinogenic members of the microbiota is through phage therapy, a decades-old biocontrol technology that has recently gained renewed clinical interest as a means to combat antibiotic-resistant bacteria. Although there are currently no phage-based therapeutics approved for use in North America and Europe, several recent Phase I trials and extensive clinical experience in Russia has established that phage therapies are generally safe and effective for treating bacterial infections [138, 139].

Most strains of lytic bacteriophage have evolved a highly selective host range, which can be used to advantage for targeting pathogenic bacteria colonizing a biological niche (e.g., the human GI tract) without harming host tissue or beneficial microbes. The narrow host range of bacteriophage is primarily due to specific interactions between virion tail fiber assemblies and bacterial surface membrane proteins that are needed for phage attachment and subsequent transfer of their genome into host cells. This property can be exploited to aid in phage enrichment and isolation from a wide variety of environmental samples [140]. Indeed, if the target host bacterium is culturable, advanced screening platforms can readily isolate and amplify phage stocks for further characterization [141]. The narrow host spectrum of most phage can limit their antibacterial effects, as targeted bacterial populations can downregulate cognate receptor ligands. To address both the limited host range of any single phage and to overcome potential host resistance mechanisms, most phage therapies are administered as a cocktail of related phage types to increase the efficacy of pathogen clearance [142].

To date, more than 6000 types of bacteriophage have been isolated, and other phage are known to infect archaea, fungi, and protists. Newly identified phages can be rapidly characterized and developed to combat clinically significant infectious agents. For example, a panel of *Burkholderia cepacia*-specific phage was recently isolated for development as antimicrobial agents against potentially fatal respiratory infections common in cystic fibrosis patients [143]. In addition, purified phage can be genetically modified to express a number of desirable traits such as enhanced host range, biofilm-degrading enzymes, or specific targeting of antibiotic-resistant bacteria [144]. The continued development of phage therapy is a potentially important tool for combating the growing problem of drug-resistant pathogens, and offers the potential for precise elimination or restructuring of specific components of a dsybiotic microbial community.

3.1.2 Prebiotic/Probiotic Effects

Probiotic bacteria such as *Lactobacillus*, *Bifidobacterium* and *Faecalibcterium* have been officially defined as "... live microorganisms which, when administered in adequate amounts confer a health benefit on the host" [145]. Despite the increasing popularity and use of non-prescribed prebiotic/probiotic products, fundamental knowledge gaps remain regarding their health benefits [146, 147]. These gaps include an understanding of their underlying molecular mechanisms of action, their interactions with resident microbes, and their long-term effects on host physiology and metabolism. However, preclinical studies and several large clinical trials involving patients with inflammatory bowel diseases have shown that probiotic treatments can be more effective and durable than standard therapies using aminosalicylates, steroids, and of other immune suppressants designed to alleviate chronic bowel inflammation [148].

Probiotics produce a variety of endogenous molecules that impact upon the host intestinal biology via several mechanisms, including a modification of bacterial community signaling (e.g., bacteriocins and quorum sensing factors), reinforcement of barrier function, regulation of immune responses, and suppressing the expression of virulence genes [149–151]. Probiotic strains signal their hosts through several classes of low-molecular-weight (LMW) bioactive molecules that include cell wall components (e.g., polysaccharides, lipoproteins, peptidoglycans, teichoic acids, and glycoproteins), secreted proteins, dietary metabolites, vitamins, plasmalogens, secondary bile acids, and biofilms [152, 153]. This suggests that the detection and characterization of LMW microbial byproducts in human physiologic fluids may have diagnostic values and could provide key information that would be useful in the design of specific treatment regimes.

Clinical trials have also demonstrated the cancer-preventive effects of probiotic interventions and reduced complications compared to conventional surgical and chemotherapies [154–157]. Likewise, in preclinical cancer models, probiotic and prebiotic modulations of the microflora improved responses to chemotherapy and radiotherapy treatments [158, 159]. Despite these promising results, however, the efficacy of probiotics as a primary or adjuvant treatment for established tumors has not been clearly demonstrated in human clinical trials [160, 161]. Thoughtful guidelines for the design and evaluation of probiotics in clinical trials (controlling for strain, dosage, treatment duration, and concomitant treatments) and their fit within the US FDA Investigational New Drug Application regulatory framework have been published [147, 162].

Because the *in-vivo* production of bioactive small molecules of microbial origin is often the result of prebiotic (i.e., dietary) secondary metabolism, there is increasing interest in the rational design of specific prebiotic/probiotic combinations – termed "synbiotics" – as preventive and therapeutic treatments that may be used to target specific host–microbe functional interactions. For example, inulin and other resistant starches (RS) that avoid gastric and small bowel breakdown are fermentable carbohydrates that stimulate the growth of human colonic probiotics such as Lactobacilli and Bifidobacteria. These increase the production of SCFAs, which can influence host gene expression and cell differentiation and also inhibit the growth of colon tumor cells, at least in part through their capacity to inhibit histone deacetylases [163]. In a rat model of colon cancer, RS given in combination with *Bifidobacterium lactis* were found to potently enhance antitumor effects when compared to treatment with the probiotic alone [164]. These results were consistent with human dietary studies, which showed increased levels of both butyrate producers and SCFAs in populations that consume high levels of RS and that have a low incidence of colon cancer [165].

Studies of model human symbionts will provide a more complete understanding of gut microflora biology that is needed to develop tailored synbiotic combinations which utilize preferred polysaccharide or other energy sources to provide growth and colonization advantages to administered probiotics. For example, preclinical studies have shown that polysaccharide metabolism in *Bacteroides thetaiotaomicron* (*B. theta*) can be manipulated to derive energy from alternate sugar sources. Expression from a single polysaccharide utilization locus dedicated to dietary fructan metabolism in *B. theta* can be induced or repressed by modulating exposure to fructose-based sugars. Genetic ablation of the proteins involved in the multistep process of *B. theta*'s fructan metabolism variably inhibits the utilization of these polysaccharides, depending on the metabolic step that is compromised [166]. The results of these studies demonstrated natural functional differences in fructan utilization between *Bacteroides* species, and also showed how synbiotic formulations could be used to manipulate a specific component of a patient's microbiota for therapeutic benefit.

Although several clinical intervention studies have shown that probiotic therapies are generally safe, a lack of detailed understanding of many host–microbe interactions and the long-term effects of probiotics warrants some caution be taken to avoid potentially deleterious effects. Some studies have indicated that the administration of probiotic formulations could cause a depletion in resident protective species [167], or provide a growth advantage to opportunistic pathogens by providing a secondary energy substrate [168]. For example, *B. theta* is an early life colonizer of the human gut that metabolizes a variety of polysaccharides, including the complex carbohydrates found in mother's milk. This mutualistic activity is essential for both regulating commensal bacteria colonization and growth, and for the digestion of nutrients used by the neonate [169]. However, *B. theta* polysaccharide foraging of host mucins during dietary deprivation may also provide essential nutrients for pathogen overgrowth by liberating fucose and sialic acid residues as energy sources, especially following antibiotic treatment [168]. Another consideration affecting the efficacy of orally administered probiotics is the loss of viability during gastric and duodenal transit. The use of artificial cell microcapsules may protect delivery of the probiotic bacteria through the gut while concurrently allowing the diffusion of nutrients and metabolites [170].

Taken together, the results of these studies suggest approaches for optimizing synbiotic treatments by using metagenomic and metabolomic analyses of microbial metabolic capacity and activity to identify suitable (perhaps synthetic) carbohydrate supplements and encapsulation methods to selectively adjust the gut community's composition and metabolic output. This type of analysis could also be used to predict how an individual's resident microflora may respond to alterations in the community and changes in the nutrient environment. Such approaches would facilitate safe and personalized therapeutic manipulations of the microbiota [171, 172].

3.1.3 Fecal Microbial Transplantation

While probiotic regimens generally involve treatments using a single or small consortia of species, and produce transient effects on specific microbial populations or functions [173], fecal microbial transplantation (FMT) seeks to permanently replace or restore a dysbiotic community structure via the wholesale transplantation and succession of a healthy microbial community. As FMT involves the transfer of uncharacterized enteric material, even greater (than for probiotic therapy) gaps in knowledge exist regarding its safety, efficacy, mechanisms of action, and long-term effects on host physiology and metabolism. Despite this, FMT may have advantages over more targeted synbiotic treatments that use food-derived or *ex-vivo* cultures instead of human intestine-adapted species that can readily colonize a recipient and persist for extended periods [174].

The dramatic clinical benefits that FMT has realized for patients with recurrent *Clostridium difficile*-mediated infection [175] has raised hope that similar results may be achieved by treating other inflammatory bowel syndromes such as ulcerative colitis (UC) and Crohn's disease, both of which are known to be at high risk for developing CRC, and perhaps for treating colitis-associated adenocarcinomas of the GI tract, though this has yet to be demonstrated in the clinic. Altered gut microbiomes observed in patients with IBD or CRC have led to treatment strategies directed at disrupting microbiota-regulated chronic inflammation. Although it is not clear if the dysbiosis seen in IBD and CRC patients is related to the underlying cause of disease, or is instead a result of mucosal inflammation, intriguing reports have indicated that specific changes in the microbiota, such as an enrichment of *Fusobacteria nucleatum*, are in fact causative [176] and thus amenable to FMT treatment.

Both, pilot and case studies have indicated that FMT may be effective at reducing inflammation in IBD patients [177, 178], in whom dysbiosis is generally similar to that observed in CRC patients. These pathogenic changes in the microflora are characterized by variably lowered GI species diversity that includes fewer Firmicutes and Bacteroidetes, and greater numbers of Enterobacteria and Proteobacteria than in age-matched healthy controls [179]. Long-term follow-up studies showed that FMT produced durable increases in Bacteroidetes and decreases in Proteobacteria [180], while in patients with UC durable changes were detected in the relative abundance of the families Enterobacteriaceae and Lachnospiraceae, which may provide useful diagnostic and clinical response indications after FMT. These studies showed that the colonization of donor microbiota following FMT can be highly variable, and indicate a gradual and sequential process of the establishment of individual taxa that is likely highly influenced by recipient genetics [181]. This is consistent with previous reports of the slow and host-specific recovery of gut communities after antibiotic-induced perturbation [182]. These results also indicate that transferred fecal microbiota may be difficult to maintain stably in recipients with active UC, in contrast to the stable transfer of microbiota to treat recurrent *C. difficile* infections. In addition, it was reported previously that the treatment of *C. difficile*-mediated infection by FMT is usually curative after a single infusion, whereas the treatment of UC often requires serial transplants [177, 183]. This indicates that for different indications, FMT targets different components of the microbiome via different mechanisms of action [184].

The microbiota's established central role in IBD and CRC provides a compelling rationale for the continued development and testing of FMT protocols to treat these conditions. However, given the uncharacterized biotherapeutic agents involved and the nonspecific targeting of FMT treatments, careful and stringent practice guidelines and response metrics must be developed to demonstrate safety and efficacy and to reduce adverse treatment events [185].

3.1.4 Introduction of Engineered Bacteria: "Biotherapeutic Vectors"

The use of live, genetically engineered bacteria to treat human tumors, either as direct oncolytic agents, or as vectors for delivering gene therapy or cytotoxic payloads, or by inducing local innate antitumor immune responses, continues to emerge as a viable experimental therapeutic strategy [186]. The significant and related parallel development of viral vectors for cancer gene therapy has been recently reviewed elsewhere [187] and will not be covered in this bacteria-focused review. However, it is important to note that engineered viral and bacterial therapeutic vectors can both be viewed as specific subsets of "Probiotic" organisms, albeit with very distinct features and functions.

Early clinical trials have proven that bacterial vectors are generally safe and well tolerated [188–190], even though some of these vectors were attenuated strains of pathogenic species. Examples of different bacteria that have shown efficacy as antitumor vectors include *Salmonella* sp., *Bifidobacterium* sp., *Shigella* sp., *Listeria* sp., and *Clostridium* sp. [191]. Ideal candidates for clinical strains should include some common safety features, such as: (i) an ability to selectively colonize and replicate in both large (hypoxic) tumors and small metastatic lesions; (ii) an attenuation of virulence factors to reduce toxicity and vector induction of systemic inflammatory responses; and (iii) the retention of antibiotic sensitivity. Additional safety features have been engineered into the most commonly used bacterial vectors. For example, bacterial cell wall lipids (lipopolysaccharides; LPS) are antigenic and often involved in promoting inflammation and pathogenesis. Disruption of the *msbB* gene, which is involved in the terminal myristoylation of lipid A in *E. coli* and *Salmonella*, resulted in stable mutants with low virulence and toxicity profiles. However, rendering these bacteria nontoxic also reduced their antitumor efficacy [192].

The innate tumor-homing ability of several facultative and obligate anaerobic bacteria has been ascribed to hypoxic environments within most solid tumors. However, the results of recent studies have shown that several other salient features of the tumor microenvironment are also likely involved in bacterial chemotaxis.

Such features include the release of cell nutrients such as amino acids and purines from necrotic tumor tissue, aberrant angiogenesis, and tumor vasculature, and the suppression of immune activity in tumor stroma that may act to inhibit bacterial clearance from tumor tissues [193, 194]. These unique features of the tumor microenvironment can be exploited to optimize the efficacy of engineered bacterial vectors. For example, purine auxotrophs of *Salmonella* have been developed for selective colonization and growth within solid tumors [195], while other bacterial vectors have demonstrated an ability to home to the vasculature of healing wounds, which closely resembles neoangiogenesis in solid tumors. This suggests that bacteria may be induced to home to sites of tumor-specific neoangiogenesis, a process that is often upregulated by oxidative stress signals within an expanding tumor [196].

The natural antineoplastic effects of bacterial tumor colonization may occur through several different mechanisms, or combinations of mechanisms, including intracellular invasion and replication, the expression of cytolytic toxins, or the stimulation of localized cytotoxic T lymphocyte (CTL) and natural killer (NK) antitumor immunity. More commonly, bacterial vectors are genetically modified to attenuate their virulence and replication capacity, toxin production and immunogenicity to improve their safety profile, while still allowing for the efficient delivery of cytotoxic drugs, proteins or genes that produce tumor-cytotoxic effects. Some therapeutic strategies utilize bacterial vectors with intrinsic immune-stimulating effects (e.g., *C. novyi*-NT and *Listeria monocytogenes*-based vaccines), while other species such as *Bifidobacterium* and *Salmonella* have been engineered to induce antitumor immune responses through the expression of tumor antigens. Other strategies deliver apoptosis-signaling proteins such as tumor necrosis factor-related apoptosis-inducing ligand (TRAIL) or apoptin. However, the most widely adapted bacterial vectors have been modified to deliver pro-drug-converting enzymes that can selectively activate cytotoxic agents coadministered with the vector [197–201].

Depending on the vector strain and therapeutic strategy selected, modified bacteria can deliver their therapeutic payloads either intratumorally (e.g., *Bifidobacterium* sp., *Clostridium* sp.) or intracellularly (e.g., *Salmonella* sp., *Shigella* sp., *Listeria* sp.). The term "bactofection" has been used to describe a direct bacterial DNA transfer into target host cells; this is in contrast to the viral vector transfer of DNA by transduction [202]. Each vector and strategy choice has inherent strengths and weaknesses. For example, tumor immunotherapy strategies that express their payload within the vector can bypass DNA entry and tumor cell expression controls, whereas gene therapy strategies that deliver DNA directly inside tumor cells can be targeted more precisely and are less likely to stimulate toxic systemic inflammation.

A significant development in gene-directed enzyme prodrug therapy (GDEPT) strategies has been the adaptation and successful demonstration of using bacterial vectors to deliver and activate anticancer prodrugs into targeted tumor tissues, a process termed bacterially directed enzyme prodrug therapy (BDEPT) [186, 203]. Prodrug formulations are activated by cleavage by either specific endogenous or exogenous vector-expressed enzymes [204, 205]. For example, pyrrolo[2,1-c][1,4]-benzodiazepine-β-glucuronide and BMY 46633 (doxycycline; C-DOX) are prodrug derivatives of doxorubicin that are activated by β-glucuronidase and β-lactamase, respectively. Each of these agents has demonstrated a reduced toxicity profile in treating colorectal tumors when compared to its parent drug [206, 207].

Other BDEPT strategies have successfully employed *S. typhimurium* and *C. beijerinckii* vectors to deliver and express cytosine deaminase in tumor cells, to convert the prodrug 5-fluorocytosine (5-FC) to the cytotoxic chemotherapy agent 5-fluorouracil (5-FU). This therapy has been shown to effectively reduce toxic side effects and improve the therapeutic profile of 5-FU in both animal models and clinical trials [192, 208]. As the cytotoxic effects of 5-FU are largely exerted following its conversion to 5-dFUMP, the *S. typhimurium* vector system has been further optimized to coexpress cytosine deaminase with uracil phosphoribosyltransferase (UPRT) in an effort to increase its cytotoxic and radiosensitizing effects [209].

Gemcitabine is another widely used chemotherapeutic agent that has been adapted for BDEPT. Gemcitabine is a cytidine analog that is phosphorylated in tumor cells to a mononucleotide form (dFdCMP) by the action of deoxycytidine kinase (DCK), and subsequently by other cellular nucleotide kinases to its active metabolites, dFdCDP and dFdCTP, which block DNA synthesis. Resistance to gemcitabine can occur through the inactivation of DCK, while bacterial vectors that express a DCK chimeric protein have been shown to increase tumor sensitivity to gemcitabine in preclinical models.

While most current BDEPT and GDEPT regimens transgenically express eukaryotic converting enzymes, the vast array of bacterial xenobiotic-metabolizing enzymes holds great potential for the screening and discovery of many additional prokaryote drug-metabolizing activities that may markedly expand and improve the spectrum of GDEPT indications. For example, a bacterial NTR was recently used to replace endogenous NAD(P)H:quinone oxidoreductase-1 (NQO1) in the development of a novel GDEPT that combines the bacterial NTR and the prodrug CB1954. The NTR/CB1954 combined therapy (Tretazicar) was found to have improved efficacy against prostate cancer in a clinical trial [210].

3.2 Microbiota Effects on Therapeutic Agents

Recent studies have demonstrated the important concept that when gut bacterial communities are compromised, standard chemotherapy regimens may lose their efficacy [52, 211]. Depending on the selected treatment, microbial effects are mediated through different mechanisms, such as: (i) by direct enzymatic activity on the drug; (ii) by an inhibition of therapy-induced antitumor signaling pathways; or (iii) by altering host enzyme drug metabolism. The results of several studies have shown that drug metabolism is altered during infection [212, 213]. For example, animal models of sepsis induced by bowel puncture have shown hepatic cytochrome P450 (CYP) expression and activity to be significantly reduced [214], indicating that a dysbiotic bacterial community or altered barrier function can affect drug metabolism and activity *in vivo*. These results are also consistent with the finding that intraperitoneally administered LPS inhibits the expression and activity of hepatic drug-metabolizing enzymes [215].

As discussed in Sect. 2.2, a mutualistic system for the co-metabolism of food and xenobiotic compounds has evolved to benefit both host cells and microbes in the human GI niche. Although several host genetic polymorphisms that affect antitumor drug bioavailability, metabolism and toxicity are well characterized, and are currently being screened prior to their administration [216], microbial factors that affect drug metabolism are currently understudied, and are likely to contribute to observed clinical variations in response to both chemotherapy and immunotherapy.

An understanding of the contributions from the gut microflora to drug metabolism, and to the regulation of host genes that mediate antitumor drug responses, offers great potential for accelerating progress in pharmacogenetic testing and personalized medicine by the inclusion of metagenotyping in treatment selection. This may enable a safer and more efficacious use of antineoplastic agents. Specific examples of how microbial activity can affect the efficacy of chemotherapeutic agents are highlighted in the following sections.

3.2.1 Chemical Biotransformation

The direct metabolism of drugs such as acetaminophen (paracetamol) and digoxin by the gut microflora is well documented [58, 217]. Microbial metabolism also affects many chemotherapeutic agents such as CPT-11, 5-aminosalicylic acid (5-ASA), doxorubicin and CB1954, all of which undergo a complex co-metabolism that is carried out by enzymes encoded by both host and commensal gut bacteria [53, 205]. Catabolic hydrolysis, nitroso-reductions and conjugation/deconjugation reactions are among the many chemical transformations catalyzed by gut bacteria that can profoundly affect the activity of an administered chemotherapeutic agent [218]. For example, gut microbial metabolism nonselectively activates the anti-inflammatory prodrug sulfasalazine (which is used to treat precancerous conditions such as IBD and Crohn's disease) into its constituent metabolites 5-ASA and sulfapyridine via cleavage of the compound by bacterial azoreductases [219]. Azoreductases are widely expressed by commensal bacteria representing several phyla of GI tract residents. As the sulfapyridine byproduct is associated with GI toxicity, newer formulations of 5-ASA prodrugs have been developed using alternate azo-bonded groups, as well as a more active positional isomer, 4-ASA, that allows for lower dosing [220]. 5-ASA also undergoes an inactivating acetylation not only in the liver but also by bacterial enzymes at its active site in the GI tract [221]. The bacterial contribution to 5-ASA acetylation was shown to derive from the actions of arylamine N-acetyltransferases that are expressed by species of gut-resident Eubacteria [222].

Activation of the chemotherapeutic agent 5-(aziridin-1-yl)-2,4-dinitrobenzamide (CB1954) and its diol derivative (SN26634) occurs via the catalyzed reduction of its 4-nitro group to a cytotoxic 4-hydroxylamine metabolite. The reduction reaction is carried out in mammalian cells by detoxifying enzymes NQO1 and NQO2, the inhibition of which in the host leads to CB1954 resistance [223]. During its development, CB1954 was found to be only poorly cytotoxic in human tumor cell lines when compared to rodent models, due to a single Tyr104-Gln amino acid change in the NQO1 active site. The expression of a Gln104-Tyr mutant restored NQO1 activity levels in human cells to that of the rat enzyme however [224], and highlighted the need to carefully characterize enzyme activity in animal model systems versus human orthologs. The finding that CB1954 was efficiently activated by bacterial (*E. coli*) NTRs led to improved solid tumor treatments, employing GDEPT strategies that combined CB1954 and NTR (Tretazicar) [225]. Unfortunately, as NTR activity is widely expressed by gut-resident bacteria, concerns have been raised regarding the observed toxic drug activation in non-target tissues such as the GI tract [226]. Together, these results indicate that strategies which selectively inhibit bacterial NTRs may be helpful in reducing GI toxicity to orally administered CB1954 and its analogs.

The degradation of doxorubicin (Dox) into a nontoxic, aglycone intermediate (7-deoxydoxorubicinolone) is catalyzed by the activity of mitochondrial complex I (NADH dehydrogenase), an enzyme complex that includes three core proteins (NuoE, F, and G) that are overexpressed in many tumor types. Therefore, it is proposed that a targeted inhibition of NADH dehydrogenase could sensitize tumor cells to dox's DNA alkylation and crosslinking activity. In addition, the process of reductive deglycosylation of doxorubicin ultimately forms reactive oxygen species (ROS) within host tissues that is likely responsible for the dose-limiting cardiotoxicity observed from this therapy [227]. Importantly, the production of 7-deoxydoxorubicinolone can also be achieved by bacterial NADH dehydrogenase, which is expressed by many human gut bacteria, especially of the phylum Proteobacteria [228], and was also recently isolated from soil-borne *Streptomyces* sp. [229]. The targeted expression of this bacterial enzyme activity offers possibilities for a novel application in drug detoxification, whereby the NADH dehydrogenase-mediated inactivation of doxorubicin in nontumor tissues could potentially reduce toxic side effects in patients undergoing chemotherapy, without compromising drug efficacy in the tumor.

Other microbial metabolic activity results in the reactivation of toxic metabolites that can affect a drug's dose-limiting toxicity. For example, the reactivation of CPT-11 by bacterial β-glucuronidases encoded by *E. coli* and other commensals has been shown to increase toxic side effects in animal models [122]. CPT-11 is a prodrug of irinotecan, a derivative of the topoisomerase inhibitor camptothecin that is widely used to treat a variety of solid tumors. It is activated by nonspecific host esterases to form SN-38, the bioactive compound that binds cellular topoisomerase I and induces apoptosis. SN-38 is glucuronidated to SN-38G in the liver and excreted via the biliary tract to the intestines as an inactive conjugate. Antibiotic cotherapy was shown to reduce the GI toxicity of irinotecan and to improve its efficacy [230], presumably by reducing bacterial β-glucuronidase activity. This interpretation was supported by experiments that specifically inhibited bacterial β-glucuronidases (in contrast to host), causing a significant reduction in both GI toxicity symptoms and inflammation in several tumor animal models [231].

As discussed in Sect. 3.1.4, several well-characterized bacterial enzymes have already been harnessed for use in BDEPT to improve antitumor efficacy in targeted tissues. However, it is likely that many more bacterial enzymes remain to be discovered and characterized; indeed, as knowledge of the microbiota's metabolic capacity improves, many additional BDEPT strategies will surely be developed using novel chimeric compounds delivered by newly modified bacterial vectors. Other bacterial enzymes, that either render antitumor drugs inactive or reactivate an agent in a nontarget tissue to cause adverse reactions and dose-limiting toxicity, are also attractive targets for adjuvant small-molecule inhibitors. These inhibitors can be designed to selectively alter the metabolism of chemotherapeutics and improve clinical outcomes by reducing drug toxicity, improving efficacy, or reducing dose frequency [232].

3.2.2 Modulation of Antitumor Immune Response

The antitumor effects of radiotherapy, GDEPT, and several cytotoxic chemotherapeutic drugs such as CTX, oxaliplatin and CpG oligodeoxynucleotides (CpG-ODN), are achieved in part by an immune-mediated bystander effect that requires the recruitment and activation of an intense inflammatory infiltrate to regress tumors [233–235]. Both, innate and adaptive immune responses participate in the therapy-induced antitumor effects, which generates inflammatory cytokines such as interleukin (IL)-1, tumor necrosis factor (TNF), and IL-17 that are required for efficient tumor killing. These immune-mediated effects are significantly impacted by the gut microbiome, which plays an important role in the development and function of the human immune system and affects a wide range of host immune functions. Indeed, the immunomodulatory effect of bacteria was postulated by Metchnikoff over 100 years ago [236]. Especially important is the microflora's regulation of the unique and complex intestinal mucosal immune system, which has evolved to become well equipped for dealing with a large variety of microbial and food antigens, and which plays an important role in regulating key systemic immune functions such as tolerance, pathogen protection, tumor surveillance, and tumor control, by modulating inflammatory responses, immune cell chemokine signaling, immune homeostasis, and tissue repair processes [21, 22, 237].

Innate immune responses of the mucosal immune system are primarily controlled by two families of molecular pattern recognition receptors, the membrane-bound Toll-like receptors (TLRs) and the cytoplasmic Nod-like receptors (NLRs), that are expressed in immune and APCs as well as IECs [238, 239]. Both receptor families can signal through NF-κB, a key transcriptional regulator of diverse inflammatory responses that is frequently overexpressed in tumors. NF-κB signaling is complex and can exert opposing effects on epithelial and immune cells during tumor development, for example, by promoting a feed-forward loop whereby the activation of NF-κB in immune cells induces the production of cytokines that signal through NF-κB in cancer epithelial cells to induce chemokines that attract more inflammatory cells into the tumor [240]. As this inflammatory response can have both pro- and anti-tumorigenic effects, by inducing either proliferation or apoptosis, a more detailed knowledge of the specific upstream regulation of NF-κB signaling is needed to advance NF-κB-targeted therapeutics towards clinical applications against cancer [241, 242].

Mucosal adaptive immune responses to microbial signals are driven primarily by antigen interactions that polarize and activate different subsets of effector and regulatory T cells. The role of T-cell responses in inhibiting versus promoting tumor progression appears to depend on the qualitative type of host inflammatory response to microbiome signals. Thus, while the transient expression of inflammatory cytokines such as TNF and IL-17 are critical for host control of pathogenic infections, and also mediate host tissue response to cytotoxic drugs such as CTX, oxaliplatin and CpG-ODN, their chronic and dysregulated activity appears to be a major contributor to tumor progression [243, 244]. The tumor-inducing effect of T helper 17 (Th-17)-mediated inflammatory responses has been demonstrated in several infection-induced models of colon, lung and gastric cancer [46, 245], indicating the global importance of this inflammatory signal pathway in pathogenesis.

Several bacterial factors that regulate Th-17 activation and function have been identified. For example, both the bacterial product polysaccharide A (PSA) from *B. fragilis*, or direct IEC contact with *Clostridium* sp. can induce a regulatory T cell (T-reg) phenotype and suppress Th-17 responses elicited by pro-carcinogenic bacteria (e.g., enterotoxigenic *B. fragilis, H. pylori*), thus providing important specific examples of adaptive immunomodulatory effects of commensal bacteria [246, 247]. Other studies have shown that segmented filamentous bacteria can induce the host cell expression of serum amyloid A (SAA), which signals CD4 + T cells to differentiate and express a STAT3-dependent Th-17 proinflammatory phenotype that has been implicated in tumorigenesis in several tissues [248–250]. In addition, the FadA adhesion protein of *Fusobacterium nucleatum* mediates attachment to tumor epithelial cells and activates B-catenin signaling that leads to an increased expression of inflammatory cytokines (IL-6 and TNF), which in turn drives tumor proliferation and progression. Colon tumors colonized with *F. nucleatum* also showed an increased recruitment of tumor-associated myeloid-derived suppressor cells (MDSC), which are tumor-permissive cells that suppress antitumor immunity functions of CD4+ T cells [176, 251]. Thus, strategies that boost anti-inflammatory signals, such as *B. fragilis* PSA, or that block the induction of pro-inflammatory signals, such as SAA expression or *F. nucleatum* FadA activity, may be effective antitumor treatments.

Given the large number of bacterial antigens and metabolites present in the GI lumen, and the network of cellular pattern recognition receptors that has evolved to regulate host responses to these signals, there are undoubtedly many more unidentified bacterial signaling factors that directly and indirectly influence tumor innate and adaptive mucosal immune functions. This complex immune regulatory milieu must be carefully considered when translating to clinical practice any knowledge of an individual's microbiome immune signaling disposition and capacity [252].

The tumoricidal activity of platinum-based drugs such as cisplatin and oxaliplatin relies on the induction of ROS, which are critical mediators of tumor cell DNA damage and apoptosis [234, 253]. The impairment of oxaliplatin-induced generation of ROS, whether by NOX genetic deficiency, by pharmacologic inhibition with *N*-acetylcysteine (NAC), or by depletion of NOX-inducing *Lactobacillus* sp. by antibiotics, results in a reduced efficacy in mouse tumor models [211]. Similarly, antibiotic treatment with vancomycin or a combination treatment of colistin, ampicillin and streptomycin dramatically diminished the effect of CTX and CpG-ODN treatments, due to an inhibition of the inflammatory cytokines TNF, IL-1, and IL-17 [52, 211]. While CTX treatment alone alters the gut microflora composition and induces an inflammatory bacterial translocation into lymph tissue, the inhibition of drug efficacy and IL-17 expression by antibiotics was correlated with decreases in the translocation of Gram-positive species such as *Lactobacillus johnsonii*, *Lactobacillus murinus*, and *Enterococcus hirae*. Moreover, antibiotic treatments that interfered with TNF production and CpG-ODN efficacy correlated with specific decreases in the abundance of *Alistipes* sp. and increases in both *Lactobacillus* and *Clostridium* sp., indicating that commensal bacteria differentially affect drug-induced inflammatory responses. Importantly, orally administered supplements of *Alistipes, L. johnsonii* and *E. hirae* were able to restore TNF and IL-17 expression respectively, and antitumor drug efficacy in treated mice, although the mechanisms regulating this host response have yet to be determined.

Dramatic advances in tumor immunotherapy have highlighted the central role of tumor stroma, especially of CTLs and NK cells, in host tumor surveillance and control. New clinical approaches using the blockade of T-cell-mediated immune signaling checkpoints or adoptive transfer of either autologous tumor-infiltrating lymphocytes (TILs) or genetically reprogrammed T cells carrying chimeric antigen receptors, have demonstrated strong therapeutic efficacy [254, 255]. Further intensive efforts are aimed at developing engineered T cells that can self-renew and persist *in vivo* after adoptive transfer to yield an effective long-term therapeutic response [256].

There is emerging evidence that the efficacy of adoptively transferred T cells is affected by their interactions with resident microflora and their metabolites. For example, specific taxa of resident GI bacteria play an important role in either promoting (e.g., *F. nucleatum*) or blocking (e.g., *Clostridium* sp. in strain clusters IV, XIVa, and XVIII) the production of mucosal inflammatory cytokines and chemokines that signal the recruitment of tumor-infiltrating inflammatory stroma [tumor-associated macrophages (TAMs) and TILs] [113, 155, 257]. Moreover, it has been shown that specific species of GI bacteria differentially function in promoting or blocking the activity of tumor-infiltrating inflammatory stroma, including the activity of adoptively transferred T cells [258]. Taken together, these findings suggest several types of novel approach whereby transferred or engineered antitumor T-cell activity could be modulated by specific manipulations of interacting bacterial antigens and metabolites.

It is not only tumor stromal cells that participate in regulating host antitumor immunity. Interactions between microbes and the innate immune receptors of IECs are also important in signaling functional activities of immune cells that either promote or protect against cancer. For example, inflammasome-derived IL-18 from IECs can ameliorate the development of intestinal inflammation in animal models of IBD, and impair the development of IBD-associated CRC [29, 259]. Recent studies have shown that IL-18 provides protection against colitis-associated cancers through the activation of cellular repair programs and the induction of anti-microbial defenses that reduce inflammation derived from interactions between pathobiont microbiota and damaged mucosal barriers [260]. These data suggest that modifying the intestinal flora with bacteria engineered to induce the expression and secretion of IL-18 from IECs may provide an effective therapeutic strategy against inflammation-induced tumors [261]. Based on the findings of these and other exciting studies, it can be imagined that future strategies for the chemoprevention or treatment of CRC could involve pharmacologics or the introduction of live bacteria that, locally, would either inhibit deleterious signals or activate desired host cell signaling, such as that mediated by the inflammasome activation of IL-18.

In summary, several microbial factors that normally regulate host inflammatory responses to commensal and pathogenic organisms have been identified (e.g., PSA, FadA, CagA). Antibiotic treatments in mouse models have further established a nonspecific correlation between microflora signaling and the efficacy of therapeutic agents, the mechanism of action of which relies on activating antitumor immune responses (e.g., CTX, oxilaplatin, CpG-ODN, and adoptively transferred T cells). These data indicate that many additional drug–microbe–host interactions are likely to be identified that will help to explain the clinically observed variation in patient response to immunotherapy and chemotherapy. Due to the known interindividual variation among microflora communities, it will be important to identify taxa-independent functional features of microbes that regulate drug–microbe and host–microbe interactions. For example, the common expression of key drug-metabolizing enzymes, or shared biosynthetic pathways that produce SCFAs, which were recently discovered to regulate protective colonic T-reg functions [257].

3.2.3 Vaccine Responses

Emerging data indicate that the microbiota affect adaptive immune responses that are critical to the efficacy of antitumor vaccines. Studies of microbial pathogens have identified several mechanisms by which bacteria can subvert adaptive immune responses; these include blocking MHC class II antigen presentation, restricting effector T-cell expansion, reducing antigen uptake by APCs, and eliminating APCs by inducing apoptosis [262–264]. Moreover, vaccination programs in developing countries have revealed a decreased efficacy of vaccines for polio, rotavirus, cholera, and typhoid that has been linked to dysbiotic intestinal microbial composition [265–267].

It has been established in recent studies that the presence and activity of a T-cell-infiltrated tumor microenvironment is a key positive indicator for response to therapeutic cancer vaccines [268, 269]. For example, clinical trials of antitumor vaccines targeting the tumor antigen MAGE-3 have demonstrated a potent induction of effector T-cell response in responsive tumors. Significantly, responses were markedly improved when the vaccine was combined with adjuvants that are either TLR4 or TLR9 agonists, indicating that bacterial pattern recognition receptors are important regulators of cell-mediated vaccine response [270]. Indeed, the administration of probiotic strains of *Bifidobacterium* and *Lactobacillus* was able to affect specific antibody responses to an oral cholera vaccine [271].

These data infer that microflora signals can influence responses to antitumor vaccines, and although this notion awaits experimental confirmation, it provides many opportunities to formulate and test specific hypotheses regarding the mechanisms by which the microbiota may enhance or inhibit vaccine response [272]. For example, it was shown recently that colorectal tumors are frequently associated with enriched populations of *F. nucleatum* [273, 274]. Subsequently, *F. nucleatum* has been shown to promote mucosal inflammatory cytokine production and increase chemokine signaling that recruits tumor-infiltrating inflammatory stroma (TAMs and TILs) [176, 251]. Other studies have shown that *Clostridium* sp. from strain clusters IV, XIVa can induce T-reg functions that interfere with effector T-cell infiltration and activity [113]. Taken together, these data suggest that CRCs with microenvironments enriched for *F. nucleatum* could be especially strong responders to cell-mediated antitumor vaccines, or that conjugate vaccines against *F. nucleatum* could prove effective adjuvant therapy for CRC. Moreover, the targeting of certain strains of *Clostridium* that block inflammatory T-cell activity may also improve responses to anticancer vaccines, especially those that target tumors in the GI tract. Undoubtedly, a better understanding of the relationship between the microbiome and immune function may lead to new approaches to improve vaccine efficacy and identify strategies for the advancement of vaccine design.

3.3 Mining the Human Microbiome

There is a long history of the discovery of antitumor agents isolated as natural products from microbial sources [275]. For example, the anticancer agents dactinomycin, bleomycin doxorubicin, rapamycin, salinosporamide A and bryostatin are all derived from microbial sources [276, 277]. The staggering structural diversity of bioactive compounds produced by microbes is only just starting to be realized, as the biosynthetic pathways for their production is revealed through metagenomic sequencing performed on increasing numbers of microbial samples from mundane and extreme environments [278, 279].

The spectrum of small molecules produced by living organisms has been termed the "parvome" [280], and it therefore follows that the human metabolome and its attendant microbial metabolome are considered a key component of the parvome that contains high clinical relevance. The diverse array of biosynthetic pathways encoded by human gut bacteria provides a potentially rich and readily accessed vein of natural products for which human host cells have presumably co-evolved metabolic pathways to process, reducing the likelihood that these compounds would cause adverse toxic and inflammatory responses. This contrasts with the microbial compounds discovered and isolated from soil organisms and other complex environments, whose natural production and evolution often occurs in environments characterized by sparse nutrition, slow growth, and low cell density [281]. Although modern pharmaceutical processes for the production of microbial natural products can in some ways mimic the high-density, nutrient-rich growth conditions found in the human GI tract, the industrial isolation, purification, fermentation and extraction of microbial compounds does not recreate natural growth and metabolic conditions for soil bacteria. In fact, pharmaceutical-scale processes are designed to narrow the diversity of metabolic byproducts produced in culture, and therefore limit the capacity for the discovery and characterization of unused biosynthetic pathways and novel compounds that may be masked in purified environmental samples [280].

The emerging fields of parvonomics and microbiomics have only recently developed robust bioinformatics pipelines, such as QIIME, Lefse, MG-RAST, and HUMAnN, to identify and characterize different classes of microbial taxa, and their proteins and metabolites, that have potential for use in human drug development. While some currently available methods have been recently reviewed [172], it is expected that new methods and technologies for microbiome analysis will continue rapidly to evolve. This will likely include the integrated analysis of multiple types of "omic" datasets, and the adaptation of current soil microbial ecology techniques such as stable isotope probing, selective enrichment methods, and substrate-induced gene expression [282–284]. The isolation and characterization of microbiome-derived compounds will allow their categorization into several useful functional classes, such as antimicrobials, immunomodulators, gene regulation, and novel xenobiotic-metabolizing enzymes. These compounds, or their byproducts, could in turn find use as biomarkers to define states of health and disease, or predict response, to chemical and biological interventions. Examples of some microbially derived factors with therapeutic potential are highlighted in the following sections.

3.3.1 Novel Drug-Metabolizing Enzymes

The development of novel enzyme-activated prodrug combinations for use in BDEPT and GDEPT strategies (see Sect. 3.1.4) depends on the identification and characterization of specific drug–enzyme interactions, and the bond chemistry of the prodrug. Screening for specific types of enzymatic activity from the vast array of bacterial xenobiotic-metabolizing enzymes holds exciting potential for the discovery of novel drug-metabolizing activities that can be harnessed to markedly expand and improve the spectrum of BDEPT and GDEPT treatments, or used to develop small-molecule inhibitors to avoid drug inactivation or toxic reactivation events. Such discovery may be accomplished through either functional enzyme screens, which offer an unbiased method to mine microflora diversity, or *in-silico* technologies such as virtual sequence homology screens, which compare sequenced open reading frames to the sequences of enzymes libraries [285]. Unfortunately, the former approach has a limited throughput capacity, while the latter technique limits discovery to the identification of enzymes that are closely related to those already in annotated protein libraries. The adaptation of metagenomic analysis approaches will advance the discovery process and allow investigators to bypass culture-dependent methods to identify novel enzyme biosynthesis pathways [286]. For example, recent studies of the *E. coli* polyketide synthase (*pks*+) gene cluster that encodes the colibactin biosynthesis pathway revealed that colibactin is produced as a protoxin and subsequently activated prior to secretion by a periplasmic D-amino peptidase [287]. This suggests that inhibition of the toxin-activating peptidase may inhibit the tumorigenic properties of *pks* + *E. coli* [47].

3.3.2 Antimicrobials

Despite the fact that the human microbiota is contained in densely populated and easily accessed environments such as the skin and urogenital and GI tracts, exploration of the full antibiotic-producing potential of the thousands of species therein has only just begun. Two factors that portend tantalizing opportunities for the further discovery of life-saving drugs with antibiotic and anticancer activity are (i) the past discovery and development of antitumor antibiotics such as anthracyclines and aminoglycosides from bacterial sources; and (ii) the still unknown biosynthetic capacity of the human microbiome that is now accessible through metagenomic analysis. Host epithelial cells too, which are in constant contact with signals from transient and resident microbes, produce a variety of antimicrobial molecules such as defensins, cathelicidins and C-type lectins, and these also may be developed for therapeutic purposes. As the development of these host-derived factors has recently been reviewed [288], the topic will not be discussed at this point.

Metagenomic sequence data from complex environments such as the human GI tract can be exploited by new culture-independent methods for drug discovery and exploration of the biosynthetic capacities [289]. For example, the analysis of small-molecule biosynthesis gene clusters (BGCs) identified from the Human

Microbiome Project (HMP) sequence data have revealed the widespread capacity of human-associated bacteria to produce the thiopeptide class of antibiotics. One such BGC was used to query the HMP reference genome dataset, which identified a common vaginal community member, *Lactobacillus gasseri*, with thiopeptide-producing capacity. This strain was subsequently purified and used in experiments to produce, isolate and structurally characterize the new antibiotic "lactocillin" [290]. Most of the BGCs analyzed in this study were found in bacteria that are abundant members of the human microbiome, such as genera from *Bacteroides*, *Parabacteroides*, *Corynebacterium*, *Rothia*, and *Ruminococcus*, giving further indication that human-associated taxa should be rich sources for experimental mining efforts. Although new antibiotics such as lactocillin may be developed as part of an antimicrobial drug arsenal, it should be noted that at low concentrations many antibiotics act as transcriptional regulators that play a role in bacterial signaling to both host and bacterial cells, and these physiological functions must also be considered during any drug development [291, 292].

3.3.3 Microbes as Biomarkers

Well-characterized and validated biomarkers of disease can be used for cancer detection and diagnosis, or to measure patient response to therapeutics, and may also provide a rationale for choice of therapy. It is notable that the first microbiome-based diagnostic patent was issued in 2012 to a small US company, Metabiomics, by the European Patent Office for a metagenomic profiling system to detect IBD-related disorders. This detection system is currently undergoing clinical trials. In addition, advanced *in-silico* techniques have been devised to analyze individual metagenomic variance profiles as a molecular biomarker that may identify pathogenic or drug-resistance collective phenotypes [293].

Depending on the clinical application, microbiome-based biomarkers may be developed by examining various features and readouts, either alone or in combination with existing biomarkers. For example, strategies to associate specific tumor/microbe interactions may range from an analysis of whole-organism presence/abundance, to the detection/quantification of microbial metabolites, to molecular signatures of host responses to dysbiotic microbiomes. Thus, colonic hyperpermeability and pro-inflammatory cytokine profiles that are associated with specific bacterial taxa could be used to identify individuals at risk for disease progression or poor therapeutic response. Microbiome biomarker development is an active area of research that has already yielded intriguing results that have not only associated microbial population changes with oral, pancreatic and colon cancer, but that may also yield clues regarding the molecular mechanisms linking microbial interactions with these and other types of tumors [294, 295].

Investigations of the oral microbiome demonstrated that *N. elongate* and *S. mitis* can act as highly sensitive and specific (96.4% and 82.1%, respectively) biomarkers that distinguish patients with pancreatic cancer from healthy individuals [296]. While a descriptive study comparing the oral microbiome of healthy and oral squamous cell carcinoma (OSCC) patients found that combined elevated levels of *C. gingivalis*, *P. melaninogenica* and *S. mitis* could predict OSCC with 80% and 83% sensitivity and specificity. A landmark study which compared bacterial taxa abundance in fecal samples from healthy, adenoma and carcinoma patient cohorts, revealed that a specific and sensitive detection of cancerous and precancerous colonic lesions could be achieved by combining metagenomic data with classic biomarkers of CRC such as age, race and body mass index [297]. The analysis revealed an enrichment in samples from patients with carcinoma of specific operational taxonomic units (OTUs) from genera of *Porphyromanas*, Lachnospiraceae, Fusobacteriaceae and Enterobacteriaceae, and a decrease in the abundance of Bacteroides and Clostridiales. These findings comport well with previous studies of microbial immunomodulatory and metabolic effects, in that the cancer-enriched OTUs detected are pro-inflammatory and saccharolytic, whereas the cancer-depleted OTUs comprise complex polysaccharide fermenters that can produce beneficial SCFAs. Importantly, this study also demonstrated that metagenomic analyses can be used to detect the presence of early-stage adenomas, the treatment of which can dramatically improve survival rates [298].

4 Conclusions

A rapidly increasing understanding of the effects of host–microbiome interactions on pathogenesis will have a direct impact on future treatment strategies for many tumor types. The microbiome mediates its effects on cancer treatment through its regulation of barrier function, immune response and metabolism. A mechanistic understanding of how the microflora mediates both oncogenic and protective effects in its host is needed to develop specific and effective treatment strategies that account for the microflora's effects on systemically administered cancer therapeutics. A useful framework for designing and testing such treatment strategies may be to place the intervention in the context of the perturbation → dysbiosis → host insult → neoplastic transformation model pathway (Fig. 1).

Today, many challenges remain before preclinical microbiome studies can be translated to clinical practice. For example, the ability to induce or inhibit biologically relevant microbial functions with small molecules is currently limited by the diversity of existing chemical libraries, which historically are built around eukaryotic targets and mechanisms. New information on bacterial enzyme structure, function and molecular interactions will undoubtedly bolster the development of small-molecule inhibitors that may allow for a significant pharmacologic control of drug metabolism. Moreover, in order to define the molecular mechanisms used by microbiota to regulate treatment response, it will be critical to identify common signaling and metabolic factors from distinct bacterial taxa that are shown to affect host cell biology. Combined experimental approaches will be critical in generating the data needed to accomplish this endeavor, and the continued rapid advancement of technical and bioinformatic methods used to analyze and integrate large datasets will need to be carefully tested and validated in clinical samples.

Despite these challenges, a clinical translation of the evolving understanding of microbiome function is potentially very close. Just as a decade ago, when an improved understanding of tumor vasculature and microenvironment led to new therapeutic strategies, the practical application of new fundamental knowledge of the human microbiome's function has enormous potential for prompting conceptual shifts in the molecular factors and pathways that are therapeutically targeted to treat cancer. For example, future microbiome-focused treatments could involve blocking or replacing the more tractable carcinogenic signals and functions of resident microflora, rather than by directly targeting host tissues and functions, and these novel treatments could be delivered via a rational modification of the microbiome itself.

Several potentially important therapeutic strategies that target microbiome functions have been designed and tested, some with promising results. These include the administration of synbiotics, oncolytic bacterial vectors, targeted immunomodulation, antibiotics, microflora transplantation, small-molecule enzyme inhibitors, and novel prodrug activators. These and other strategies may, for the first time, allow clinical investigators to treat developing tumors not only in the context of host pathophysiology but also as an outcome of multiple, complex and networked interactions with resident microbiota that offer several novel points for therapeutic intervention.

Finally, an appreciation for microbiota variation between individuals and across patient populations, and correlating this variability with specific microbial functions, will continue to emerge as a component of personalized medicine. Differences in gut microbiota metabolic capacity between populations, and the concomitant effects on drug metabolism, will therefore necessitate patient-stratified drug trials to examine effectiveness and toxicity. A better understanding of the complex interplay of microbiome, host genetics, environmental exposures and host immunity will likely guide and improve future cancer treatment strategies.

Acknowledgments

The author wishes to thank Ms Robin Brown for help in creating the graphics for Fig. 1, and Drs Kevin Howcroft and Elizabeth Read-Connole for their critical reading of the manuscript.

References

1. Turnbaugh, P.J., Ley, R.E., Hamady, M., Fraser-Liggett, C.M. *et al.* (2007) *Nature*, **449**, 804.
2. Hooper, L.V. and Gordon, J.I. (2001) *Science*, **292**, 1115.
3. Eckburg, P.B., Bik, E.M., Bernstein, C.N., Purdom, E. *et al.* (2005) *Science*, **308**, 1635.
4. Garrett, W.S., Gordon, J.I., and Glimcher, L.H. (2010) *Cell*, **140**, 859.
5. Garrett, W.S., Gallini, C.A., Yatsunenko, T., Michaud, M. *et al.* (2010) *Cell Host Microbe*, **8**, 292.
6. Segata, N., Izard, J., Waldron, L., Gevers, D. *et al.* (2011) *Genome Biol.*, **12**, R60.
7. Arthur, J.C. and Jobin, C. (2011) *Inflamm. Bowel Dis.*, **17**, 396.
8. Shen, X.J., Rawls, J.F., Randall, T., Burcal, L. *et al.* (2010) *Gut Microbes*, **1**, 138.
9. Huycke, M.M. and Gaskins, H.R. (2004) *Exp. Biol. Med.*, **229**, 586.
10. Turnbaugh, P.J., Ley, R.E., Mahowald, M.A., Magrini, V. *et al.* (2006) *Nature*, **444**, 1027.
11. Karin, M., Lawrence, T., and Nizet, V. (2006) *Cell*, **124**, 823.
12. Martin, F.P., Dumas, M.E., Wang, Y., Quigley, C.L. *et al.* (2007) *Mol. Syst. Biol.*, 3,112.
13. Flint, H.J., Scott, K.P., Louis, P., and Duncan, S.H. (2012) *Nat. Rev. Gastroenterol. Hepatol.*, **9**, 577.
14. Fukata, M. and Abreu, M.T. (2007) *Biochem. Soc. Trans.*, **35**, 1473.
15. Backhed, F., Ley, R.E., Sonnenburg, J.L., Peterson, D.A. *et al.* (2005) *Science*, **307**, 1915.
16. Artis, D. (2008) *Nat. Rev. Immunol.*, **8**, 411.
17. Karrasch, T., Kim, J.S., Muhlbauer, M., Magness, S.T. *et al.* (2007) *J. Immunol.*, **178**, 6522.
18. Chu, F.F., Esworthy, R.S., Chu, P.G., Longmate, J.A. *et al.* (2008) *Cancer Res.*, **64**, 962.
19. Cash, H.L., Whitham, C.V., Behrendt, C.L., and Hooper, L.V. (2006) *Science*, **313**, 1126.
20. Sonnenburg, J.L., Angenent, L.T., and Gordon, J.I. (2004) *Nat. Immunol.*, **5**, 569.

21. Medzhitov, R. (2007) *Nature*, **449**, 819.
22. Honda, K. and Takeda, K. (2009) *Mucosal Immunol.*, **2**, 187.
23. Mai, V., Colbert, L.H., Perkins, S.N., Schatzkin, A. *et al.* (2007) *Mol. Carcinogen.*, **46**, 42.
24. O'Keefe, S.J. (2008) *Curr. Opin. Gastroenterol.*, **24**, 51.
25. Sears, C.L. and Pardoll, D.M. (2011) *J. Infect. Dis.*, **203**, 306.
26. Fox, J.G., Feng, Y., Theve, E.J., Raczynski, A.R. *et al.* (2010) *Gut*, **59**, 88.
27. Zackular, J.P., Baxter, N.T., Iverson, K.D., and Sadler, W.D. (2013) *mBio*, **4**, 1.
28. Collins, D., Hogan, A.M., and Winter, D.C. (2011) *Lancet*, **12**, 504.
29. Peterson, L.W. and Artis, D. (2014) *Nat. Rev. Immunol.*, **14**, 114.
30. Sommer, F. and Backhed, F. (2013) *Nat. Rev. Microbiol.*, **11**, 5189.
31. Buffie, C.G. and Pamer, E.G. (2013) *Nat. Rev. Immunol.*, 13,790.
32. Kamada, N., Chen, G.Y., Inohara, N., and Nunez, G. (2013) *Nat. Immunol.*, **14**, 685.
33. Turner, J.R. (2009) *Nat. Rev. Immunol.*, 9,799.
34. Grivennikov, S.I., Wang, K., Mucida, D., Stewart, C.A. *et al.* (2012) *Nature*, **491**, 254.
35. Caldwell, K., Patel, K.K., Maloney, N.S., Liu, T.C. *et al.* (2010) *Cell*, **141**, 1135.
36. Peterson, D.A. and Turnbaugh, P.J. (2010) *Sci. Transl. Med.*, **2**, 1.
37. Pull, S.L., Doherty, J.M., Mills, J.C., Gordon, J.I. *et al.* (2005) *Proc. Natl Acad. Sci. USA*, **102**, 99.
38. Swanson, P.A., Kumar, A., Samarin, S., Vijay-Kumar, M. *et al.* (2011) *Proc. Natl Acad. Sci. USA*, **108**, 8803.
39. Jones, R.M., Luo, L., Ardita, C.S., Richardson, A.N. *et al.* (2013) *EMBO J.*, **32**, 3017.
40. Alam, A., Leoni, G., Wentworth, C.C., Kwal, J.M. *et al.* (2014) *Mucosal Immunol.*, **7**, 645.
41. Arrieta, M.C., Madsen, K., Doyle, J., and Meddings, J. (2009) *Gut*, **58**, 41.
42. Wentworth, C.C., Alam, A., Jones, R.M., Nusrat, A. *et al.* (2011) *J. Biol. Chem.*, **286**, 38448.
43. Sharma, A., Singla, D., Rashid, M., and Raghavaet, G.P. (2014) *BMC Bioinf.*, **15**, 282.
44. Dutta, U., Garg, P.K., Kumar, R., and Tandon, R.K. (2000) *Am. J. Gastroenterol.*, **95**, 784.
45. Wroblewski, L.E. and Peek, R.M. (2011) *Cell Commun. Signal.*, **9**, 29.
46. Wu, S., Rhee, K.J., Albesiano, E., Rabizadeh, S. *et al.* (2009) *Nat. Med.*, **15**, 1016.
47. Arthur, J.C., Perez-Chanona, E., Mühlbauer, M., Tomkovich, S. *et al.* (2012) *Science*, **338**, 120.
48. Sears, C.L. (2009) *Clin. Microbiol. Rev.*, **22**, 349.
49. Cuevas-Ramos, G., Petit, C.R., Marcq, I., Boury, M. *et al.* (2010) *Proc. Natl Acad. Sci. USA*, **107**, 11537.
50. Mai, V. and Draganov, P.V. (2009) *World J. Gastroenterol.*, **15**, 81.
51. Plottel, C.S. and Blaser, M.J. (2011) *Cell Host Microbe*, **10**, 324.
52. Viaud, S., Saccheri, F., Mignot, G., Yamazaki, T. *et al.* (2013) *Science*, **342**, 971.
53. Haiser, H.J. and Turnbaugh, P.J. (2013) *Pharmacol. Res.*, **69**, 21.
54. Craciun, S. and Balskus, E.P. (2012) *Proc. Natl Acad. Sci. USA*, **109**, 21307.
55. The Human Microbiome Project Consortium (2012) *Nature*, **486**, 207.
56. Borenstein, E., Kupiec, M., Feldman, M.W., and Ruppin, E. (2008) *Proc. Natl Acad. Sci. USA*, **105**, 14482.
57. Peterson, S.N., Meissner, T., Su, A.I., Snesrud, E. *et al.* (2014) *Front. Cell. Infect. Microbiol.*, **14**, 108.
58. Clayton, T.A., Baker, D., Lindon, J.C., Everett, J.R. *et al.* (2009) *Proc. Natl Acad. Sci. USA*, **106**, 14728.
59. Walker, A.W., Ince, J., Duncan, S.H., Webster, L.M. *et al.* (2011) *ISME J.*, **5**, 220.
60. Wu, G.D., Chen, J., Hoffmann, C., Bittinger, K. *et al.* (2011) *Science*, **334**, 105.
61. Louis, P., Hold, G.L., and Flint, H.J. (2014) *Nat. Rev. Microbiol.*, **12**, 661.
62. Rowland, I.R. (2009) *Curr. Pharm. Des.*, **15**, 1524.
63. Rooks, M.G., Veiga, P., Wardwell-Scott, L.H., Tickle, T. *et al.* (2014) *ISME J.*, **8**, 1403.
64. Carbonero, F., Benefiel, A.C., and Gaskins, H.R. (2012) *Nat. Rev. Gastroenterol. Hepatol.*, **9**, 504.
65. Hayes, J.D., Kelleher, M.O., and Eggleston, I.M. (2008) *Eur. J. Nutr.*, **47**, 73.
66. Murillo, G. and Mehta, R.G. (2001) *Nutr. Cancer*, **41**, 17.

67. Russell, W.R., Gratz, S.W., Duncan, S.H., Holtrop, G. *et al.* (2011) *Am. J. Clin. Nutr.*, **93**, 1062.
68. Calmels, S., Ohshima, H., and Bartsch, H. (1988) *J. Gen. Microbiol.*, **134**, 221.
69. Kuhnle, G.G.C., Story, W., Reda, T., Mani, A.R. *et al.* (2007) *Free Radical Biol. Med.*, **43**, 1040.
70. Davis, C.D. and Milner, J.A. (2009) *J. Nutr. Biochem.*, **20**, 743.
71. Holmes, E. (2008) *Nature*, **453**, 396.
72. Blaut, M. and Clavel, T. (2007) *J. Nutr.*, **137**, 751S.
73. Landete, J.M. (2012) *Crit. Rev. Food Sci. Nutr.*, **52**, 936.
74. Selma, M.V., Espin, J.C., and Tomas-Barberan, F.A. (2009) *J. Agric. Food Chem.*, **57**, 6485.
75. Sharma, M., Li, L., Celver, J., Killian, C. *et al.* (2010) *J. Agric. Food Chem.*, **58**, 3965.
76. Hedlund, T.E., van Bokhoven, A., Johannes, W.U., Nordeen, S.K. *et al.* (2006) *Prostate*, **66**, 557.
77. Lee, H.C., Jenner, A.M., Low, C.S., and Lee, Y.K. (2006) *Res. Microbiol.*, **157**, 876.
78. Atkinson, C., Frankenfeld, C.L., and Lampe, J.W. (2005) *Exp. Biol. Med.*, **230**, 155.
79. Raschke, M., Wahala, K., and Pool-Zobel, B.L. (2006) *Br. J. Nutr.*, **96**, 426.
80. Shimada, Y., Yasuda, S., Takahashi, M., Havashi, T. *et al.* (2010) *Appl. Environ. Microbiol.*, **76**, 5892.
81. Matthies, A., Blaut, M., and Braune, A. (2009) *Appl. Environ. Microbiol.*, **75**, 1740.
82. Shimada, Y., Takahashi, M., Miyazawa, N., Abiru, Y. *et al.* (2012) *Appl. Environ. Microbiol.*, **78**, 4902.
83. Larrosa, M., Gonzalez-Sarrfas, M.A., and Yanez-Gascon, M.J. (2010) *J. Nutr. Biochem.*, **21**, 717.
84. Cerda, B., Tomas-Barberan, F.A., and Espin, J.C. (2005) *J. Agric. Food Chem.*, **53**, 227.
85. Fukuda, S., Toh, H., Hase, K., Oshima, K. *et al.* (2011) *Nature*, **469**, 543.
86. Blouin, J.M., Penot, G., Collinet, M., Nacfer, M. *et al.* (2011) *Int. J. Cancer*, **128**, 2591.
87. Donohoe, D.R., Collins, L.B., Wali, A., Bigler, R. *et al.* (2012) *Mol. Cell*, **48**, 612.
88. O'Keefe, S.J., Chung, D., Mahmoud, N., Sepulveda, A.R. *et al.* (2007) *J. Nutr.*, **137**, 175S.
89. Wang, T., Cai, G., Qiu, Y., Fei, N. *et al.* (2012) *ISME J.*, **6**, 320.
90. Clarke, J.M., Young, G.P., Topping, D.L., Bird, A.R. *et al.* (2012) *Carcinogenesis*, 33,197.
91. Moore, A.M., Munck, C., Sommer, M.O., and Dantas, G. (2011) *Front. Microbiol.*, **2**, 188.
92. Belcheva, A., Irrazabal, T., Robertson, S.J., Streutker, C. *et al.* (2014) *Cell*, **158**, 288.
93. Roberts, M.S., Magnusson, B.M., Burczynski, F.J., and Weiss, M. (2002) *Clin. Pharmacokinet.*, **41**, 751.
94. Bernstein, H., Bernstein, C., Payne, C.M., Dvorakova, K. *et al.* (2005) *Mutat. Res.*, **589**, 47.
95. Yoshimoto, S., Loo, T.M., Atarashi, K., Kanda, H. *et al.* (2013) *Nature*, **499**, 97.
96. Gorbach, S.L. and Goldin, B.R. (1990) *Rev. Infect. Dis.*, **12**, S252.
97. Tannock, G.W., Tangerman, A., Van Schaik, A., and McConnell, M.A. (1994) *Appl. Environ. Microbiol.*, **60**, 3419.
98. Ridlon, J.M., Kang, D.J., and Hylemon, P.B. (2006) *J. Lipid Res.*, **47**, 241.
99. Sayin, S.I., Wahlstrom, A., Felin, J., Jantti, S. *et al.* (2013) *Cell Metab.*, **17**, 225.
100. Islam, K.B., Fukiya, S., Hagio, M., Fujii, N. *et al.* (2011) *Gastroenterology*, **141**, 1773.
101. Ridlon, J.M., Alves, J.M., Hylemon, P.B., and Bajaj, J.S. (2013) *Gut Microbes*, **4**, 382.
102. Hill, M.J. (1990) *Mutat. Res.*, **238**, 313.
103. Zampa, A., Silva, S., Fabiani, R., Morozzi, G. *et al.* (2004) *Anaerobe*, **10**, 19.
104. Laue, H., Friedrich, M., Ruff, J., and Cook, A.M. (2001) *J. Bacteriol.*, **183**, 1727.
105. Dabek, M., McCrae, S.I., Stevens, V.J., Duncan, S.H. *et al.* (2008) *FEMS Microbiol. Ecol.*, **66**, 487.
106. Flores, R., Shi, J., Fuhrman, B., Xu, X. *et al.* (2012) *J. Transl. Med.*, **10**, 253.
107. Zhanel, G.G., Siemens, S., Slayter, K., and Mandell, L. (1999) *Can. J. Infect. Dis.*, **10**, 429.
108. Wichmann, A., Allahyar, A., Greiner, T.U., Plovier, H. *et al.* (2013) *Cell Host Microbe*, **14**, 582.
109. Fung, T.C., Artis, D., and Sonnenberg, G.F. (2014) *Immunol. Rev.*, **260**, 35.
110. Coley, W.B. (1893) *Am. J. Med. Sci.*, **105**, 487.
111. Olle, B. (2013) *Nature*, **4**, 309.
112. Atarashi, K., Tanoue, T., Oshima, K., Suda, W. *et al.* (2014) *Nature*, **500**, 232.
113. Barbé, S., Van Mellaert, L., and Anné, J. (2006) *J. Appl. Microbiol.*, **101**, 571.

114. Bermudes, D., Zheng, L., and King, I.C. (2002) *Curr. Opin. Drug Discov. Dev.*, **5**, 194.
115. Redelman-Sidi, G., Glickman, M.S., and Bochner, B.H. (2014) *Nat. Rev. Urol.*, **11**, 153.
116. Elinav, E., Nowarski, R., Thaiss, C.A., Hu, B. *et al.* (2013) *Nat. Rev. Cancer*, **13**, 759.
117. Chen, Z., Guo, L., Zhang, Y., Walzem, R.L. *et al.* (2014) *J. Clin. Invest.*, **124**, 3391.
118. Haiser, H.J. and Turnbaugh, P.J. (2012) *Science*, **336**, 1253.
119. Holmes, E., Li, J.V., Marchesi, J.R., and Nicholson, J.K. (2012) *Cell Metab.*, **16**, 559.
120. Dethlefsen, L., McFall-Ngai, M., and Relman, D.A. (2007) *Nature*, **449**, 811.
121. Sommer, F. and Bäckhed, F. (2013) *Nat. Rev. Microbiol.*, **11**, 227.
122. Wallace, B.D., Wang, H., Lane, D.T., Scott, J.E. *et al.* (2010) *Science*, **330**, 831.
123. Palm, N.W., de Zoete, M.R., Cullen, T.W., Barry, N.A. *et al.* (2014) *Cell*, **158**, 1000.
124. Takimoto, C.H. and Calvo, E. (2008) in *Cancer Management: A Multidisciplinary Approach*, 11th edn (eds R. Pazdur, L.D. Wagman, K.A. Camphausen, and W.J. Hoskins), The Oncology Group, New York, p. 23.
125. Zilberberg, A., Lahav, L., and Rosin-Arbesfeld, R. (2010) *Gut*, **59**, 496.
126. Thiede, C., Wundisch, T., Neubauer, B., Alpen, B. *et al.* (2000) *Cancer Res.*, **156**, 125.
127. Nakamura, T., Inagaki, H., Seto, M., and Nakamura, S. (2003) *J. Gastroenterol.*, **38**, 921.
128. Working Group on the Evaluation of Carcinogenic Risks to Humans (1994) *IARC Monogr. Eval. Carcinog. Risks Hum.*, **61**, 1–241.
129. McFarland, L.V. (1998) *Dig. Dis.*, **16**, 292.
130. Sjölund, M., Wreiber, K., Andersson, D.I., Blaser, M.J. *et al.* (2003) *Ann. Intern. Med.*, **139**, 483.
131. Flanagan, J.L., Brodie, E.L., Weng, L., Lynch, S.V. *et al.* (2007) *J. Clin. Microbiol.*, **45**, 1954.
132. Zasloff, M. (2002) *Nature*, **415**, 389.
133. Shai, Y. and Oren, Z. (2001) *Peptides*, **22**, 1629.
134. Fernandez-Lopez, S., Kim, H.S., Choi, E.C., Delgado, M. *et al.* (2001) *Nature*, **412**, 452.
135. Fletcher, J.T., Finlay, J.A., Callow, M.E., Callow, J.A. *et al.* (2007) *Chemistry*, **13**, 4008.
136. Montero, A., Gastaminza, P., Law, M., Cheng, G. *et al.* (2011) *Chem. Biol.*, **18**, 1453.
137. Lohan, S. and Bisht, G.S. (2013) *Mini-Rev. Med. Chem.*, **13**, 1073.
138. Lu, T.K. and Koeris, M.S. (2011) *Curr. Opin. Microbiol.*, **14**, 524.
139. Harper, D.R. and Enright, M.C. (2011) *J. Appl. Microbiol.*, **111**, 1.
140. Gill, J.J. and Hyman, P. (2010) *Curr. Pharm. Biotech.*, **11**, 2.
141. Singh, A., Arutyunov, D., Szymanski, C.M., and Evoy, S. (2012) *Analyst*, **137**, 3405.
142. Wittebole, X., DeRoock, S., and Opal, S.M. (2014) *Virulence*, **5**, 226.
143. Lynch, K.H., Dennis, J.J. (2012) *Can. J. Microbiol.*, **58**, 221.
144. Lu, T.K. and Collins, J.J. (2009) *Proc. Natl Acad. Sci. USA*, **106**, 4629.
145. Hill, C., Guarner, F., Reid, G., Gibson, G.R. *et al.* (2014) *Nat. Rev. Gastroentrol. Hepatol.*, **11**, 506.
146. Capurso, G., Marignani, M., and Delle Fave, G. (2006) *Dig. Liver Dis.*, **38**, S277.
147. Hoffmann, D.E., Fraser, C.M., Palumbo, F.B., Ravel, J. *et al.* (2013) *Science*, **342**, 314.
148. Meijer, B.J. and Dieleman, L.A. (2011) *J. Clin. Gastroenterol.*, **4**, S319.
149. Liu, C.I., Liu, G.Y., Song, Y., Yin, F., and Hensler, M.E. (2008) *Science*, **319**, 1391.
150. Ashraf, R. and Shah, N.P. (2014) *Crit. Rev. Food Sci. Nutr.*, **54**, 938.
151. Wang, H., Bastian, S.P., Cheah, K.Y., Lawrence, A. *et al.* (2014) *Cancer Biol. Ther.*, **15**, 560.
152. Yan, F., Cao, H., Cover, T.L., Whitehead, R. *et al.* (2007) *Gastroenterology*, **132**, 562.
153. Khazaie, K., Zadeh, M., Khan, M.W., Bere, P. *et al.* (2012) *Proc. Natl Acad. Sci. USA*, **109**, 10462.
154. Rafter, J., Bennett, M., Caderni, G., Clune, Y. *et al.* (2007) *Am. J. Clin. Nutr.*, **85**, 488.
155. Hatakka, K., Holma, R., El-Nezami, H., Suomalainen, T. *et al.* (2008) *Int. J. Food Microbiol.*, **128**, 406.
156. Gianotti, L., Morelli, L., Galbiati, F., Rocchetti, S. *et al.* (2010) *World J. Gastroenterol.*, **16**, 167.
157. Osterlund, P., Ruotsalainen, T., Korpela, R., Saxelin, M. *et al.* (2007) *Br. J. Cancer*, **97**, 1028.

158. Taper, H.S. and Roberfroid, M.B. (2005) *In Vivo*, **19**, 201.
159. Baldwin, C., Millette, M., Oth, D., Ruiz, M.T. *et al.* (2010) *Nutr. Cancer*, **62**, 371.
160. Zhu, Y., Luo, T.M., Jobin, C., and Young, H.A. (2011) *Cancer Lett.*, **309**, 119.
161. Kahouli, I., Tomaro-Duchesneau, C., and Prakash, S. (2013) *J. Med. Microbiol.*, **62**, 1107.
162. FAO (2002) WHO/FAO Working Group Report.
163. Waldecker, M., Kautenburger, T., Daumann, H., Busch, C. *et al.* (2008) *J. Nutr. Biochem.*, **19**, 587.
164. LeLeu, R.K., Hu, Y., Brown, I.L., Woodman, R.J. *et al.* (2010) *Carcinogenesis*, **31**, 246.
165. O'Keefe, S.J., Ou, J., Aufreiter, S., O'Connor, D. *et al.* (2009) *J. Nutr.*, **139**, 2044.
166. Sonnenburg, J.L., Xu, J., Leip, D.D., Chen, C.H. *et al.* (2005) *Science*, **307**, 1955.
167. Arthur, J.C., Gharaibeh, R.Z., Uronis, J.M., Perez-Chanona, E. *et al.* (2013) *Sci. Rep.*, **3**, 2868.
168. Ng, K.M., Ferreyra, J.A., Higginbottom, S.K., Lynch, J.B. *et al.* (2013) *Nature*, **502**, 96.
169. Marcobal, A., Barboza, M., Sonnenburg, E.D., Pudlo, N. *et al.* (2011) *Cell Host Microbe*, **10**, 507.
170. Tomaro-Duchesneau, C., Saha, S., Malhotra, M., Kahouli, I. *et al.* (2013) *J. Pharmaceut.*, **2013** (1), 2041–2052.
171. Panesar, P.S., Kumari, S., and Panesar, R. (2013) *Crit. Rev. Biotechnol.*, **33**, 345.
172. McHardy, I.H., Goudarzi, M., Tong, M., Ruegger, P.M. *et al.* (2013) *Microbiome*, **1**, 17.
173. Sartor, R.B. (2004) *Gastroenterology*, **126**, 1620.
174. Grehan, M.J., Borody, T.J., Leis, S.M., Campbell, J. *et al.* (2010) *J. Clin. Gastroenterol.*, **44**, 551.
175. Gough, E., Shaikh, H., and Manges, A.R. (2011) *Clin. Infect. Dis.*, **53**, 994.
176. Rubinstein, M.R., Wang, X., Liu, W., Hao, Y. *et al.* (2013) *Cell Host Microbe*, **14**, 195.
177. Borody, T.J., Warren, E.F., Leis, S., Surace, R. *et al.* (2003) *J. Clin. Gastroenterol.*, **37**, 42.
178. Angelberger, S., Reinisch, W., Makristathis, A., Lichtenberger, C. *et al.* (2013) *Am. J. Gastroenterol.*, **108**, 1620.
179. Frank, D.N., St Amand, A.L., Feldman, R.A., Boedeker, E.C. *et al.* (2007) *Proc. Natl Acad. Sci. USA*, **104**, 13780.
180. Seekatz, A.M., Aas, J., Gessert, C.E., Rubin, T.A. *et al.* (2014) *mBio*, **5**, 1.
181. Kunde, S., Pham, A., Bonczyk, S., Crumb, T. *et al.* (2013) *J. Pediatr. Gastroenterol. Nutr.*, **56**, 597.
182. Dethlefsen, L. and Relman, D.A. (2011) *Proc. Natl Acad. Sci. USA*, **108**, 4554.
183. Bennet, J.D. and Brinkman, M. (1989) *Lancet*, 1,164.
184. Borody, T.J. and Khoruts, A. (2012) *Nat. Rev. Gastroenterol. Hepatol.*, **9**, 88.
185. Bakken, J.S., Borody, T., Brandt, L.J., Brill, J.V. *et al.* (2011) *Clin. Gastroenterol. Hepatol.*, **9**, 1044.
186. Bernardes, N., Chakrabarty, A.M., and Fialho, A.M. (2013) *Appl. Microbiol. Biotechnol.*, **97**, 5189.
187. Waehler, R., Russell, S.J., and Curiel, D.T. (2007) *Nat. Rev. Genet.*, **8**, 573.
188. Wood, L.M., Guirnalda, P.D., Seavey, M.M., and Paterson, Y. (2008) *Immunol. Res.*, **42**, 233.
189. Li, Z., Lang, E., Runyan, J., Luo, X. *et al.* (2001) *Cancer Res.*, **42**, 687.
190. Toso, J.F., Gill, V.J., Hwu, P., Marincola, F.M. *et al.* (2002) *J. Clin. Oncol.*, **20**, 142.
191. Wei, M.Q., Ellem, K.A., Dunn, P., and West, M.J. (2007) *Eur. J. Cancer*, **43**, 490.
192. Nemunaitis, J., Cunningham, C., Senzer, N., Kuhn, N.J. *et al.* (2003) *Cancer Gene Ther.*, **10**, 737.
193. Kasinskas, R.W. and Forbes, N.S. (2007) *Cancer Res.*, **67**, 3201.
194. Yu, A.Y., Shabahang, S., Timiryasova, T.M., and Zhang, Q. (2004) *Nat. Biotechnol.*, **22**, 313.
195. Mellaert, F.V., Barbe, S., and Anne, J. (2006) *Trends Microbiol.*, **14**, 190.
196. Okuno, Y., Nakamura-Ishizu, A., Otsu, K., Suda, T. *et al.* (2012) *Nat. Med.*, **18**, 1208.
197. Agrawal, N., Bettegowda, C., Cheong, I., Geschwind, J.F. *et al.* (2004) *Proc. Natl Acad. Sci. USA*, **101**, 15172.
198. Wood, L.M. and Paterson, Y. (2014) *Front. Cell. Infect. Microbiol.*, **4**, 51.
199. Ganai, S., Arenas, R.B., and Forbes, N.S. (2009) *Br. J. Cancer*, **101**, 1683.
200. Guan, G.F., Zhao, M., Liu, L.M., Jin, C.S. *et al.* (2013) *Int. J. Med. Sci.*, **10**, 1639.
201. King, I., Itterson, M., and Bermudes, D. (2009) *Methods Mol. Biol.*, **542**, 649.
202. Palffy, R., Gardlik, R., Hodosy, J., Behuliak, M. *et al.* (2006) *Gene Ther.*, **13**, 101.

203. Lehouritis, P., Springer, C., and Tangney, M. (2013) *J. Controlled Release*, **170**, 120.
204. de Graaf, M., Boven, E., Scheeren, H.W., Haisma, H.J. *et al.* (2002) *Curr. Pharm. Des.*, **8**, 1391.
205. Rooseboom, M., Commandeur, J.N.M., and Vermeulen, N.P.E. (2004) *Pharmacol. Rev.*, **56**, 102.
206. Kamal, A., Tekumalla, V., Raju, P., Naidu, V.G. *et al.* (2008) *Bioorg. Med. Chem. Lett.*, **18**, 3769.
207. Svensson, H.P., Vrudhula, V.M., Emswiler, J.E., MacMaster, J.F. *et al.* (1995) *Cancer Res.*, **55**, 2357.
208. Fox, M.E., Lemmon, M.J., Mauchline, M.L., Davis, T.O. *et al.* (1996) *Gene Ther.*, **3**, 173.
209. Xing, L., Sun, X., Deng, X., Kotedia, K. *et al.* (2009) *Radiother. Oncol.*, **92**, 345.
210. Bhaumik, S., Sekar, T.V., Depuy, J., Klimash, J. *et al.* (2012) *Gene Ther.*, **19**, 295.
211. Iida, N., Dzutsev, A., Stewart, C.A., Smith, L. *et al.* (2013) *Science*, **342**, 967.
212. Aitken, A.E., Richardson, T.A., and Morgan, E.T. (2006) *Annu. Rev. Pharmacol. Toxicol.*, **46**, 123.
213. Gandhi, A., Moorthy, B., and Ghose, R. (2012) *Curr. Drug Metab.*, **13**, 1327.
214. Godellas, C.V., Williams, J.F., and Fabri, P.J. (1995) *J. Surg. Res.*, **59**, 783.
215. Ghose, R., White, D., Guo, T., Vallejo, J. *et al.* (2008) *Drug Metab. Dispos.*, **36**, 95.
216. Wilkinson, G.R. (2005) *N. Engl. J. Med.*, **352**, 2211.
217. Haiser, H.J., Gootenberg, D.B., Chatman, K., Sirasani, G. *et al.* (2013) *Science*, **341**, 295.
218. Sousa, T., Paterson, R., Moore, V., Carlsson, A. *et al.* (2008) *Int. J. Pharmacol.*, **363**, 1.
219. Wadworth, A.N. and Fitton, A. (1991) *Drugs*, **41**, 647.
220. Dhaneshwar, S.S. (2014) *World J. Gastroenterol.*, **20**, 3564.
221. Dull, B.J., Salata, K., and Goldman, P. (1987) *Biochem. Pharmacol.*, **36**, 3772.
222. Deloménie, C., Fouix, S., Longuemaux, S., Brahimi, N. *et al.* (2001) *J. Bacteriol.*, **183**, 3417.
223. Chen, S., Wu, K., and Knox, R. (2000) *Free Radical Biol. Med.*, **29**, 276.
224. Wu, K., Eng, E., Knox, R., and Chen, S. (2001) *Arch. Biochem. Biophys.*, **385**, 203.
225. Denny, W.A. (2002) *Curr. Pharm. Des.*, **8**, 1349.
226. Braybrooke, J.P., Slade, A., Deplanque, G., Harrop, R. *et al.* (2005) *Clin. Cancer Res.*, **11**, 1512.
227. Gille, L. and Nohl, H. (1997) *Free Radical Biol. Med.*, **23**, 775.
228. Kerscher, S., Dröse, S., Zickermann, V., and Brandt, U. (2008) in *Bioenergetics: Energy Conservation and Conversion*, vol. **45** (eds G. Schäfer and H.S. Penefsky), Springer-Verlag, Berlin, Heidelberg, p. 185.
229. Westman, E.L., Canova, M.J., Radhi, I.J., Koteva, K. *et al.* (2012) *Chem. Biol.*, **19**, 1255.
230. Stringer, A.M., Gibson, R.J., Logan, R.M., Bowen, J.M. *et al.* (2008) *Cancer Biol. Ther.*, **7**, 1919.
231. Flieger, D., Klassert, C., Hainke, S., Keller, R. *et al.* (2007) *Oncology*, **72**, 10.
232. Wallace, B.D. and Redinbo, M.R. (2013) *Curr. Opin. Chem. Biol.*, **17**, 379.
233. Kuriyama, S., Tsujinoue, H., and Yoshiji, H. (2004) *Methods Mol. Med.*, **90**, 353.
234. Ghiringhelli, F., Apetoh, L., Tesniere, A., Aymeric, L. *et al.* (2009) *Nat. Med.*, **15**, 1170.
235. Viaud, S., Daillère, R., Boneca, I.G., Lepage, P. *et al.* (2014) *Cancer Res.*, **74**, 4217.
236. Anukam, K.C. and Reid, G. (2007) in *Communicating Current Research and Educational Topics and Trends in Applied Microbiology* (ed A. Mendez-Vilas), Formatex.org, Spain, p. 466.
237. Kamada, N. and Núñez, G. (2014) *Gastroenterology*, **146**, 1477.
238. Chen, R., Alvero, A.B., Silasi, D.A., Steffensen, K.D. *et al.* (2008) *Oncogene*, **27**, 225.
239. Clarke, T.B., Davis, K.M., Lysenko, E.S., Zhou, A.Y. *et al.* (2010) *Nat. Med.*, **16**, 228.
240. Grivennikov, G.I., Greten, F.R., and Karin, M. (2010) *Cell*, **140**, 883.
241. Yan, F. and Polk, D.B. (2010) *Gut*, **59**, 421.
242. Sethi, G., Sung, B., and Aggarwal, B.B. (2008) *Exp. Biol. Med.*, **233**, 21.
243. Korn, T., Bettelli, E., Oukka, M., and Kuchroo, V.K. (2009) *Annu. Rev. Immunol.*, **27**, 485.
244. Ciofani, M., Madar, A., Galan, C., Sellars, M. *et al.* (2012) *Cell*, **151**, 289.
245. Pinchuk, I.V., Morris, K.T., Nofchissey, R.A., Earley, R.B. *et al.* (2013) *PLoS One*, **8**, e53798.
246. Atarashi, K., Tanoue, T., Shima, T., Imaoka, S.K. *et al.* (2011) *Science*, **331**, 337.

247. Mazmanian, S.K., Round, J.L., and Kasper, D.L. (2008) *Nature*, **453**, 620.
248. Ivanov, I., Atarashi, K., Manel, N., Brodie, E.L. *et al.* (2009) *Cell*, **139**, 485.
249. Bronte-Tinkew, D.M., Terebiznik, M., Franco, A., Ang, M. *et al.* (2009) *Cancer Res.*, **69**, 632.
250. Fukuda, A., Wang, S.C., Morris, J.P. IV, Folias, A.E. *et al.* (2011) *Cancer Cell*, **19**, 441.
251. Kostic, A.D., Chun, E., Robertson, L., Glickman, J.N. *et al.* (2013) *Cell Host Microbe*, **14**, 207.
252. Karin, M. (2014) *Nat. Med.*, **20**, 126.
253. Ozben, T. (2007) *J. Pharm. Sci.*, **96**, 2181.
254. Pardoll, D.M. (2012) *Nat. Rev. Cancer*, **12**, 252.
255. Grupp, S.A., Kalos, M., Barrett, D., Aplenc, R. *et al.* (2013) *N. Engl. J. Med.*, **368**, 1509.
256. Song, D.G., Ye, Q., Carpenito, C., Poussin, M. *et al.* (2011) *Cancer Res.*, **71**, 4617.
257. Smith, P.M., Howitt, M.R., Panikov, N., Michaud, M. *et al.* (2013) *Science*, **341**, 569.
258. Elinav, E., Nowarski, R., Thaiss, C.A., Hu, B. *et al.* (2013) *Nat. Rev. Cancer*, **13**, 759.
259. Hu, B., Elinav, E., Huber, S., Strowig, T. *et al.* (2013) *Proc. Natl Acad. Sci. USA*, **110**, 9862.
260. Chen, G.Y., Liu, M., Wang, F., Bertin, J. *et al.* (2011) *J. Immunol.*, **186**, 7187.
261. Chen, G.Y. and Nunez, G. (2011) *Gastroenterology*, **141**, 1986.
262. Cheminay, C., Mohlenbrink, A., and Hensel, M. (2005) *J. Immunol.*, **174**, 2892.
263. Gao, L.Y. and Kwaik, Y.A. (2000) *Microbes Infect.*, **2**, 1705.
264. Autenrieth, S.E.I., Soldanova, R., Rösemann, D., Gunst, D. *et al.* (2007) *Cell. Microbiol.*, **9**, 425.
265. Cooper, P.J., Chico, M.E., Losonsky, G., Sandoval, C. *et al.* (2000) *J. Infect. Dis.*, **182**, 1199.
266. Levine, M.M. (2010) *BMC Biol.*, **8**, 129.
267. Serazin, A.C., Shackelton, L.A., Wilson, C., and Bhan, M.K. (2010) *Nat. Immunol.*, **11**, 769.
268. Gajewski, T.F., Schreiber, H., and Fu, Y. (2013) *Nat. Immunol.*, **14**, 1014.
269. Gajewski, T.F., Louahed, J., and Brichard, V.G. (2010) *Cancer J.*, **16**, 399.
270. Brichard, V.G. and Lejeune, D. (2007) *Vaccine*, **25**, B61.
271. Mahalanabis, D., Lopez, A., Sur, D., Deen, J. *et al.* (2008) *PLoS One*, **3**, e2323.
272. Rosana, B.R., Ferreira, L., Caetano, M., Antunes, B. *et al.* (2010) *PLoS Pathog.*, **6**, 1.
273. Castellarin, M., Warren, R.L., Freeman, J.D., Dreolini, L. *et al.* (2011) *Genome Res.*, **22**, 299.
274. Kostic, A.D., Gevers, D., Pedamallu, C.S., Michaud, M. *et al.* (2011) *Genome Res.*, **22**, 292.
275. Craig, G.M., Kingston, D.G., and Newman, D.J. (eds) (2012) *Anticancer Agents from Natural Products*, CRC Press, Boca Raton, FL.
276. Newman, D.J. and Cragg, G.M. (2009) *Curr. Opin. Invest. Drugs*, **10**, 1280.
277. da Rocha, A.B., Lopes, R.M., and Schwartsmann, G. (2001) *Curr. Opin. Pharmacol.*, **1**, 364.
278. Woappi, Y., Gabani, P., and Singh, O.V. (2013) *Br. Microbiol. Res. J.*, **3**, 280.
279. Davies, J. and Ryan, K.S. (2012) *Chem. Biol.*, **7**, 252.
280. Davies, J. (2009) *EMBO Rep.*, **10**, 805.
281. Wang, J.D. and Levin, P.A. (2009) *Nat. Rev. Microbiol.*, **7**, 822.
282. Radajewski, S., Ineson, P., Parekh, N.R., and Murrell, J.C. (2000) *Nature*, **403**, 646.
283. Dantas, G., Sommer, M.O., Oluwasegun, R.D., and Church, G.M. (2008) *Science*, **320**, 100.
284. Uchiyama, T. and Watanabe, K. (2008) *Nat. Protoc.*, **3**, 1202.
285. Ferrer, M., Golyshina, O., Beloqui, A., and Golyshin, P.N. (2007) *Curr. Opin. Microbiol.*, **10**, 207.
286. Guazzaroni, M.E., Beloqui, A., Vieites, J.M., Al-ramahi, Y. *et al.* (2010) in *Handbook of Hydrocarbon and Lipid Microbiology* (ed K.N. Timmis), Springer-Verlag, Berlin, Heidelberg, p. 2911.
287. Brotherton, A.C. and Balskus, E.P. (2013) *J. Am. Chem. Soc.*, **135**, 3359.
288. Gallo, R.L. and Hooper, L.V. (2012) *Nat. Rev. Immunol.*, **12**, 503.
289. Wilson, M.C. and Piel, J. (2013) *Chem. Biol.*, **20**, 636.
290. Donia, M.S., Cimermancic, P., Schulze, C.J., Wieland-Brown, L.C. *et al.* (2014) *Cell*, **158**, 1402.
291. Goh, E.B., Yim, G., Tsui, W., McClure, J. *et al.* (2002) *Proc. Natl Acad. Sci. USA*, **99**, 17025.
292. Linares, J.F., Gustafsson, I., Baquero, F., and Martinez, J.L. (2006) *Proc. Natl Acad. Sci. USA*, **103**, 19484.

293. Schloissnig, S., Arumugam, M., Sunagawa, S., Mitreva, M. *et al.* (2013) *Nature*, **493**, 45.

294. Mager, D.L., Haffajee, A.D., Devlin, P.M., Norris, C.M. *et al.* (2005) *J. Transl. Med.*, **3**, 27.

295. Khan, A.A., Shrivastava, A., and Khurshid, M. (2012) *BBA Rev. Cancer*, **1826**, 331.

296. Farrell, J.J., Zhang, L., Zhou, H., Chia, D. *et al.* (2012) *Gut*, **61**, 582.

297. Zackluar, J.P., Rogers, M.A., Ruffin, M.T., and Schloss, P.D. (2014) *Cancer Prev. Res.*, **7**, 1112.

298. Whitlock, E.P., Lin, J.S., Liles, E., Beil, T.L. *et al.* (2008) *Ann. Intern. Med.*, **149**, 638.

18
Gold and Iron Oxide Nanoparticles with Antibody Guides to Find and Destroy Cancer Cells

Stephanie A. Parker[1], *Isabel A. Soto*[1], *Dickson K. Kirui*[3], *Cameron L. Bardliving*[2], *and Carl A. Batt*[2]
[1]*Cornell University, Department of Biomedical Engineering, 101 Weill Hall, Ithaca, New York 14853, USA*
[2]*Cornell University, Department of Food Science, 351 Stocking Hall, Ithaca, New York 14853, USA*
[3]*Houston Methodist Research Institute, Department of NanoMedicine, 6670 Bertner Ave, Houston, Texas 77030, USA*

Translational Medicine Cancer: Advances in Molecular Biology and Medicine,
First Edition. Edited by Robert A. Meyers.

Keywords

Bioconjugation
A chemical strategy that forms a stable covalent link between two biomolecules.

Theranostics
A combination of diagnostic and therapeutic capabilities into a single agent, in this instance multimodal nanoparticles used for cancer imaging and treatment.

Gold nanoparticles (AuNPs)
Gold particles with diameters between 1 to 100 nm in size.

Superparamagnetic iron oxide nanoparticles (SPIONs)
Iron oxide particles that have a core diameter of less than 20 nm and exhibit superparamagnetic properties at room temperature.

Iron oxide nanoparticles (IONPs)
Iron oxide particles with diameters of between 1 and 100 nm.

Gold and iron oxide hybrid nanoparticles (HNPs)
Metal particles bearing multiple components that are structurally very close to the actual catalysts, with diameters of between 1 and 100 nm. Here, gold and iron oxide composites nanoparticles are reviewed.

Nanoparticles represent an increasingly important field of theranostic agents which can not only sense but also treat disease. Gold nanoparticles (AuNPs) and iron oxide nanoparticles (IONPs) functionalized with antibodies (Abs) specific to tumor antigens can detect and treat cancer at the cellular level. These particles are being investigated for clinical applications, including imaging contrast, thermal tumor therapy, drug delivery, and radiosensitizing agents. In this chapter, a review is provided of the unique properties of these particles which, when combined with targeting moieties, make them suitable for cancer imaging and therapeutics, as evidenced by the recent preclinical, *in-vivo* and *in-vitro* studies, conducted with AuNP and IONP Ab conjugates. Results from these preclinical studies provide evidence for the potential clinical applications, which will advance the diagnosis and treatment of cancer.

1 Introduction

Despite significant advances in cancer research, treatment and diagnosis over past two decades, cancer remains a major health problem in the United States and many other parts of the world. Cancer is the second leading cause of death in the US, accounting for 23% of all deaths in 2009, and is the leading cause of death for both men and women aged 40–79 years [1, 2]. The current standard treatment strategies include surgical resection, chemotherapy and radiotherapy, but these are indiscriminate and unnecessarily lead to the damage of healthy tissues. The limitations of treatments such as chemotherapy and radiotherapy can be attributed to resistance to chemotherapeutics [3], physiological barriers, and poor bioavailability [3, 4]. Advances in nanomedicine have led to the development of immunotargeted nanomaterials that have the potential to address the limitations of conventional cancer therapeutics by targeting cancer at the molecular level. Gold nanoparticles (AuNPs) and iron oxide nanoparticles (IONPs), when conjugated to antibodies, are of particular interest for cancer theranostic applications (Fig. 1). Both, AuNPs and IONPs possess unique optical and magnetic properties that can be exploited for simultaneous diagnostic imaging and therapy, including thermal and controlled drug release for cancer therapy [5] (Fig. 2), which can lead to improved treatments. Recent *in-vitro* and *in-vivo* studies conducted with immunotargeted nanoparticles have shown that these conjugates exhibit a targeted specificity, improved pharmacokinetics, and can even deliver drugs via a leaky vasculature [6].

The synthesis and composition of AuNPs, IONPs and hybrid gold and IO nanoparticles (HNPs), in addition to the clinically relevant properties of these nanomaterials, will first be reviewed. An overview will then be provided of the functionalization strategies used to couple nanoparticles with cancer-targeting antibodies, and some of the targeting moieties currently available will be outlined. Preclinical studies focused on the development of antibody-conjugated AuNPs, IONPs and HNPs for cancer imaging and therapeutics will then be examined, including hyperthermia, targeted drug delivery, and radiation therapy.

2 Composition

During recent years, AuNPs, IONPs and HNPs have been extensively studied and synthesized using several methods that may be applied for biomedical purposes. The unique characteristics of nanoparticles that have spurred interest in them for clinical applications include: (i) their large surface-to-volume ratio; (ii) the physical and chemical properties that can be tuned depending on their size, composition, and shape; (iii) their enhanced penetration and preferential accumulation in tumor tissues; (iv) their high absorption band in the near-infrared (NIR) region [7]; and (v) their ease of surface functionalization with biomolecules [8]. The nanoscale dimensions of nanoparticles also make them suited for targeting cancer at the cellular level [9].

2.1 Gold Nanoparticles

The most common method used for AuNP synthesis is the chemical or electrochemical reduction of a gold(III) precursor [10].

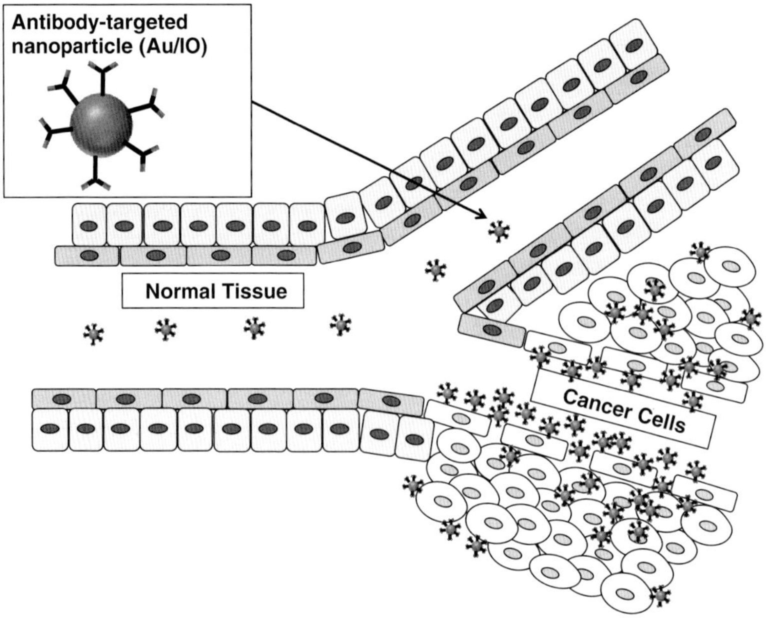

Fig. 1 Antibody-targeted gold (Au)/iron oxide (IO)/hybrid nanoparticles moving through the vascular system of the body to selectively search for and destroy cancerous cells. Normal cells are portrayed as rectangles, while cancer cells are portrayed as oval. Antibody-targeted nanoparticles are able to penetrate the epithelial layer, and antibodies on the particles bind to tumor-specific antigens on the cancer cells, resulting in an accumulation of nanoparticles at the site of disease.

A variety of shapes and structures can be synthesized, with the most prevalent AuNP structures including nanorods, nanospheres and nanoshells, as well as composite particles where gold is combined with other elements such as silica and iron (see Sect. 2.3) [10]. The geometry and composition of AuNPs both affect the nanoparticles' optical properties, including absorption wavelength [11], as well as their interactions with biological molecules, including enzyme catalysis [12]. Hence, AuNPs can be tuned for specific imaging and thermal therapy applications; the varying properties of the nanoparticles are described more fully in Refs [11, 13–15].

AuNPs possess several distinct properties that make them attractive for clinical applications, including an easily modified surface chemistry and size-dependent electrical and optical properties [16–18]. AuNPs possess intense surface plasmon resonance (SPR) scattering, which results in an enhanced electromagnetic field (EMF) at the nanoparticle surface [19]. Scattered light from AuNPs produces an array of colors under dark-field and confocal microscopy imaging applications [20–22]. Due to their high atomic mass AuNPs also absorb significantly more X-ray radiation than do soft tissue cells, and consequently they can be used to enhance cancer radiation therapy and provide an increased contrast in diagnostic computed tomography (CT) scanning [23].

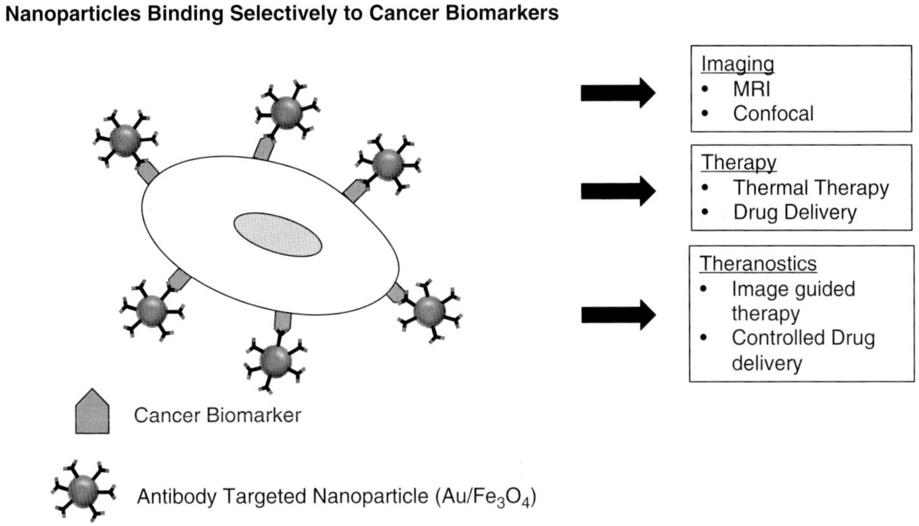

Fig. 2 Antibody-targeted nanoparticles gold (Au), iron oxide (IO), hybrid (Au and IO) nanoparticles binding selectively to cancer biomarkers located on the surface of a cancer cell. This selective and specific binding and the unique optical and magnetic properties of the nanoparticles allows for the imaging and therapy of cancer cells.

2.2 Iron Oxide Nanoparticles

IONPs can be synthesized with a precise size, morphology, surface treatment and magnetic properties [24, 25]. The common methods of synthesis include coprecipitation, microemulsions, and the high-temperature decomposition of organic precursors [25]. Maghemite (Fe_2O_3) and magnetite (Fe_3O_4) iron oxide can each be prepared as monodispersed surface-derivatized nanoparticles [24, 26], and ferric magnetic materials such as iron oxide also possess unique paramagnetic properties. When the size domain of the nanoparticle is reduced below that of a single magnetic domain [27] the magnetic properties are altered, and this results in superparamagnetic behavior [25]. IONPs with a core diameter <20 nm exhibit superparamagnetic properties at room temperature, and are termed superparamagnetic iron oxide nanoparticles (SPIONs).

SPIONs have a significant ability to reduce the magnetic resonance imaging (MRI) signal, thus acting as negative contrast agents and causing a darkening of the MR image [28]. Superparamagnetic iron oxide (SPIO) agents, such as ferumoxides (Feridex IV; Berlex Laboratories. Inc. in the USA and Endorem, Guerbet S.A in Europe), have been approved for clinical applications in the USA [29]. Feridex SPIONs have been approved for intravenous administration and are used as contrast agents to identify liver tumors [30]. These SPIONs serve as attractive platforms for targeted drug delivery, magnetic hyperthermia, immunotherapy, tumor visualization, and the detection of circulating tumor cells [31]. Preclinical studies in animal models have shown that antibody-conjugated SPIONs retain their specificity for antigen-expressing tissues and are also effective contrast-enhancing agents for MRI.

2.3 Hybrid Nanoparticles

The potential benefits of integrating nanomaterials with disparate properties (i.e., magnetism and energy absorption) into HNPs have spurred great interest in the development of various synthetic strategies. HNPs retain the optical and magnetic properties of AuNPs and IONPs, respectively [32–36], and can be fabricated by the aqueous-phase [37] and high-temperature organic phase thermal decomposition of their respective precursors [37–40]. The addition of a gold coating to IONPs has been reported to enhance their chemical stability by protecting the core against oxidation and corrosion [41, 42]. HNPs with gold cores encapsulated within hollow iron oxide shells and gold decorated with iron oxide have been developed for dual-imaging agents, MRI and X-ray detection [43], and multifunctional imaging and thermal applications, respectively [42, 44, 45].

3 Nanoparticle Surface Functionalization

Following their synthesis, several strategies have been employed to functionalize AuNPs and IONPs by allowing the addition of targeting moieties – that is, cancer-specific antibodies. A successful functionalization strategy will not alter the desired properties of the nanoparticles, and will preserve the targeting ability of the immobilized antibodies. First, however, the nanoparticles must be adequately coated, as bare or surfactant-stabilized AuNPs and IONPs are unstable and cytotoxic under physiological conditions [46–48]. These unmodified nanoparticles are also prone to aggregation, and would be rapidly cleared from the body via the reticuloendothelial system (RES) [49]. The resultant shortened blood half-life reduces the efficacy of AuNP- and IONP-based theranostic platforms, where longer a circulation time will increase the probability that the particles can find, attach to and accumulate in, targeted cancer cells.

Fortunately, facile chemistries provide methods for stabilizing the nanoparticles and conjugating antibodies to their surfaces [50]. The most common methods of nanoparticle functionalization include ligand exchange [51–64], coating with amphiphilic polymers [65–78], coating with silica [79–83], polymer coating, and encapsulation (Fig. 3) [84–88]. Following surface coating, AuNPs and IONPs have been shown to retain the desired relativities, biocompatibility properties, native optical properties, and can easily be additionally functionalized with biomolecules, notably monoclonal antibodies (mAbs) and single-chain variable fragments (ScFvs) [89].

3.1 Antibody Immobilization and Orientation on Nanoparticles

Following stabilization coating, nanoparticles can then be functionalized with antibodies for targeted cancer therapeutic applications. Five classes of antibodies have been identified, with immunoglobulin G (IgG) being the most abundant in human serum, and thus the most widely used for biomedical applications [88]. mAbs consist of four polypeptide chains with a molecular weight of 180 kDa, which complicates their conjugation to nanoparticles. mAb and antibody fragments including ScFvs and

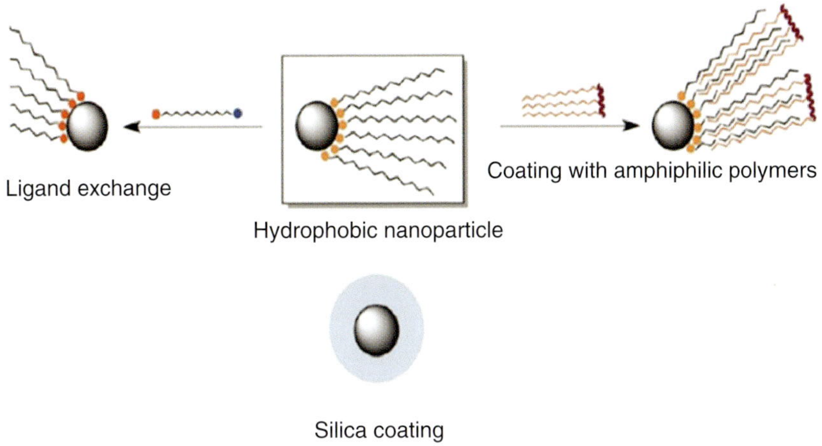

Fig. 3 Primary coating strategies for engineering the surfaces of nanoparticles. The diagram shows ligand exchange, hydrophobic interactions, coating with amphiphilic polymers, and silica coating. Reprinted with permission from Ref. [88]; © 2014, American Chemical Society.

fragment antigen-binding (Fab) antibodies have been investigated in preclinical studies for tumor-targeting nanoparticle treatment applications. The strategies used to produce functional nanoparticles with antibodies include: (i) physical adsorption [90]; (ii) covalent binding via amine groups or carbohydrate groups on antibodies [91]; (iii) the use of adapter biomolecules (streptavidin–biotin, Protein G); and (iv) ionic adsorption. However, a drawback to streptavidin–biotin is that streptavidin is a bacterial protein without a mammalian analog and may cause an immunogenic response in patients [92, 93]. The choice of coupling method depends on the type and amount of functional groups present on the surface-coating layer of the nanoparticles.

It is also important to note that the immobilization of biologically active molecules requires a correct steric availability of the active binding sites in order to avoid reducing or completely eliminating the binding capacity. The orientation of antibodies on the nanoparticle surface must be such that antigen-binding sites are well exposed, without compromising the function of the protein. The steric hindrance of antigen-recognition sites due to randomly oriented antibodies can result in a 100-fold decrease in binding affinity [94]. End-on or flat-on orientations have been found to yield higher antigen-binding capacities compared to random (head-on and sideways) orientations [88]. Typically, to retain full biological activity of the immobilized antibody, the recognition sites – the Fab regions – should be oriented away from surface of the nanoparticle. A directed immobilization of antibody fragments, Fabs and ScFvs, has been achieved via the addition of functionalities to the terminal regions of ScFvs [95] and Fabs [96]. Modification of the internal linker regions of ScFvs has also been reported for site-specific immobilization on the surface of nanoparticles [97].

4
Cancer-Targeting Antibodies

Antibodies are an example of targeting molecules that can allow the nanoparticles

Tab. 1 FDA-approved unmodified monoclonal antibodies for solid tumor and hematological malignancies.

Therapeutic antibody	*Target*	*FDA-approved indication*	*Mechanism of action*
Solid tumor malignancies			
Trastuzumab (Herceptin; Genentech); humanized IgG1	HER2/neu	HER2-overexpressing breast cancer	Inhibition of HER2/neu signaling and ADCC
		HER2-overexpressing metastatic gastric or gastroesophageal junction adenocarcinoma	
Cetuximab (Erbitux; Bristol-Myers Squibb); chimeric human-murine IgG1	EGFR	Colorectal cancer that has metastasized	Inhibition of EGFR signaling and ADCC
		Squamous cell carcinoma of the head and neck	
Bevacizumab (Avastin; Genetech/Roche); humanized IgG1	VEGF	Colorectal cancer that has metastasized	Inhibition of VEGF signaling
		Glioblastoma in patients not responding to other treatments	
		Non-small-cell lung that is locally advance, cannot be surgically removed, has metastasized, or has recurred	
		Renal cell cancer that has metastasized	
Panitumumab (Vectibix; Amgen); human IgG2	EGFR	Colorectal cancer (certain types) in patients whose disease has not responded with other chemotherapy and has metastasized	Inhibition of EGFR signaling
Ipilimumab (Yervoy; Bristol-Myers Squibb); IgG1	CTLA4	Melanoma that cannot be surgically removed or has metastasized	Inhibition of CTLA4 signaling
Pertuzumab (Perjeta; Genentech); humanized IgG1	HER2/neu	Neoadjuvant treatment of HER2-positive breast cancer in combination with trastuzumab and docetaxel (chemotherapeutic)	Inhibition of HER2/neu signaling
		HER2-overexpressing metastatic breast cancer	
Hematological malignancies			
Rituximab (Mabthera; Roche); chimeric human-murine IgG1	CD20	For treatment of CD20-positive B-cell non-Hodgkin's lymphoma CLL	ADCC, induction of apoptosis and CDC
Alemtuzumab (Campath; Genzyme); humanized IgG1	CD52	Approved for single-agent treatment of B-cell CLL.	Induction of apoptosis and CDC
Ofatumumab (Arzerra; Genmab); human IgG1	CD20	For treatment of patients with CLL no longer controlled by other forms of chemotherapy	ADCC and CDC
Obinutuzumab (Gazya; Genetech/Roche); humanized IgG1	CD20	For patients with previously untreated CLL	ADCC and CDC

ADCC, antibody-dependent cellular cytotoxicity; CDC, complement-dependent cytotoxicity; CLL, chronic lymphocytic leukemia; EGFR, epidermal growth factor receptor.

to locate cancerous cells and identify them at a cellular level. On contacting the cancerous cell, the NPs can be used to provide treatment on a highly region-specific basis. Tumor stromal and vascular cells express specific antigens that distinguish them from normal tissue [98], and the identification of these antigens has led to the clinical development of therapeutic mAbs that recognize cancer cells as compared with normal tissues [99, 100]. These cancer antigen-specific antibodies are among the most successful targeted cancer therapeutic strategies for patients with hematological malignancies and solid tumors [98, 101, 102]. Currently, 15 unmodified antibodies have been approved by the Food and Drug Administration for cancer therapy [103, 104]. Tumor-associated antigens recognized by therapeutic antibodies are cell-surface differentiation antigens. Hematopoietic differentiation antigens are generally associated with antibodies that target cluster of differentiation (CD) groups, including CD20, CD20, CD33, and CD52. Cell-surface antigens are glycoproteins and carbohydrates found on the surface of both normal and tumor cells, but they are overexpressed in tumor cell populations. The unmodified mAbs currently approved for solid tumor and hematological malignancies are detailed in Table 1.

ScFv, Ab fragments, have also been explored for use as targeting ligands for cancer therapeutics [105, 106]. ScFvs are recombinant proteins generated from mAbs, and their fragments contain the highly specific variable domains of mAbs, though they are smaller in size. Cancer-targeting ScFvs conjugated to nanoparticles have been used for site-specific therapeutic delivery and imaging [107]. This reduction in size also facilitates a direct and dense immobilization of the antibodies on the surface of nanoparticles [108].

5 AuNPs and IONPs with Antibody Conjugates for Cancer Imaging and Detection

The first step in the process of treating cancer with an antibody-conjugated AuNP and/or IONP is to identify the cancer. Immunotargeted AuNPs and IONPs are of particular interest as these nanoparticles, when selectively targeted to diseased sites, increase the signal-to-noise ratio of traditional imaging technologies, thus making it possible to visualize antigen-expressing cancer cells [109]. The ability to image cancer at the cellular level would provide a significant improvement over traditional methods by helping to diagnose cancers earlier than is currently feasible, and with unaided imaging techniques.

The current standard optical technologies for cancer detection are optical coherence tomography (OCT) and reflectance confocal microscopy (RCM). These technologies image the micro-anatomic features of diseased tissue, and cannot distinguish among cancers based specifically on molecular biomarkers associated with carcinogenesis [110]. The image is generated via reflected light from the endogenous chromophores present in the tissue [111]. The resulting optical signal is weak, and results in poor intracellular contrast and subtle spectral differences between malignant and benign cells. To overcome these limitations, exogenous chemicals have been explored to function as contrast agents; unfortunately, however, organic dyes are subject to rapid photobleaching, biological and chemical degradation, and their signal quickly decreases upon exposure to an excitation light source [112, 113].

AuNPs, IONPs and HNPs would be an improvement over dyes as they are photostable [114, 115] and can carry fluorescent dyes for NIR imaging [113, 116]. Preclinical studies have employed several imaging techniques to examine the capacity of antibody-conjugated AuNPs, IONPs and HNPs for cancer imaging, including NIR, confocal reflectance [117], dark-field, optoacoustic microscopy [118], X-ray microtomography (micro-CT) [119], OCT [120], and MRI [76, 121] contrast agents [122]. Some recent antibody-targeted imaging studies conducted with AuNPs, IONPs and HNPs are outlined in Table 2. The objective of preclinical studies has been to demonstrate the selectivity and quantify the improved imaging as a result of using antibody-conjugated nanoparticles, thereby demonstrating their feasibility for clinical studies. Highlights from recent *in-vivo* imaging studies with AuNPs and IONPs conjugated with cancer-targeting antibodies are provided in the following section. The variety of targeting agents and the subsequent imaging modalities demonstrate the robust nature of these approaches for the detection of a variety of cancers.

5.1 Imaging Studies

Specific examples of imaging studies using antibody-directed nanoparticles demonstrate the utility of the approach. For example, in a recent study by Karmani *et al.* [126], AuNPs conjugated with the anti-CD105 mAb, specific to tumor angiogenesis and tumor progression, were investigated for theranostic, dual-imaging and therapeutic applications in an *in-vivo* mouse model The antibodies were radiolabeled with ^{89}Zr to facilitate positron emission tomography (PET) imaging before their conjugation to 5 nm AuNPs. Male mice bearing melanoma xenografts were intravenously injected with the radiolabeled, antibody-conjugated AuNPs, as well as radiolabeled CD-105 antibody alone and nontargeted AuNPs for comparison. Whole-body PET images and an inductively coupled plasma mass spectrometry (ICP-MS) analysis of the blood and tumors were used to assess the *in-vivo* distribution of the nanoparticles. A high tumor contrast and selective

Tab. 2 Cancer imaging studies conducted with antibody-functionalized gold and iron oxide nanoparticles.

Nanoparticle type	*Cancer type(s)*	*Antibody conjugate*	*Imaging technique (Reference)*
AuNP, GNR	Lung cancer, solid tumor malignancies	Anti-EGFR	Confocal [123], dark-field microscopy [124]
GNR	Breast cancer	Herceptin (anti-HER2/neu)	Micro CT [119]
GNR	Prostate cancer	Anti-PSMA	Optoacoustic imaging [118]
AuNP	Colorectal cancer	Anti-β-catenin and anti-E-cadherin	Confocal [111]
AuNP, HNP	Solid tumor malignancies	Anti-CD105	PET imaging [18], MRI [76]
IONP	Breast cancer	Anti-HER2/neu	MRI [19, 125]

AuNP, gold nanoparticle; GNR, gold nanorod; HNP, hybrid nanoparticle; IONP, iron oxide nanoparticle; MRI, magnetic resonance imaging; Micro CT, X-ray microtomography; PET, positron emission tomography.

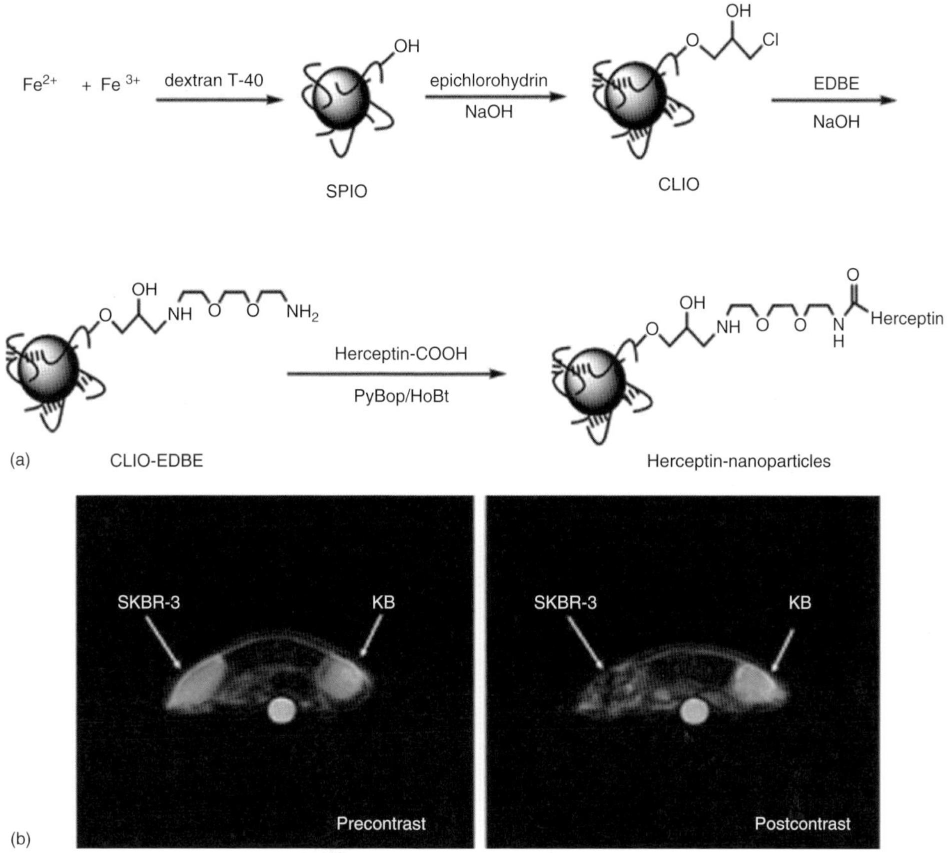

Fig. 4 SPION conjugation scheme and enhanced MRI images. (a) The synthetic scheme of herceptin–nanoparticles. SPIO, superparamagnetic iron oxide; CLIO, cross-linked iron oxide, 2,2′-(ethylenedioxy)bisethylamine; PyBop, (benzotriazol-1-yloxy) tripyrrolidinophosphonium hexafluorophosphate; HoBt, 1-hydroxybenzotriazole; (b) T_2-weighted MRI before (left) and after (right) injection of herceptin–iron oxide nanoparticles. The cell lines used here are SKBR-3 HER2/neu-expressing breast cancer cells and KB non-HER2/neu-expressing epidermal carcinoma cells. Reprinted with permission from Ref. [124]; © 2009, John Wiley & Sons, Ltd.

tumor targeting were observed with the immunotargeted AuNPs, as compared to the antibody–target radioisotope alone or the nontargeted AuNPs. Yet, the major finding of this study was that conjugation of the anti-CD105 antibody with AuNPs did not inhibit tumor uptake, nor the efficacy of tumor-targeting of the antibody.

In another recent study, nanoparticles targeting HER2/neu demonstrated an effective MRI imaging of HER2-expressing cells in xenograft mice [127] (Fig. 4). In this case, SPIONs were first surface-modified with dextran, followed by the addition of primary amine functional groups, and then functionalized with Herceptin in the presence of coupling agents. The targeting of Herceptin–SPIONs to HER2/neu receptor-positive breast cancer cell lines was detected via *in-vitro* and *in-vivo* MRI.

The *in-vitro* MRI signal enhancement increased proportionally as a function of the relative HER2/neu expression level. For *in-vivo* studies, the Herceptin–SPIONs were administered intravenously to tumor-bearing mice prepared by the subcutaneous injection of KB cells (human nasopharyngeal epidermal carcinoma cells lacking HER2/neu receptors) and SKBR-3 (HER2/neu-expressing breast cancer) cells. The tumor site was detected via T_2-weighted MRI, and the administration of Herceptin–SPIONs produced a 45% signal enhancement in the HER2/neu-positive tumor compared to controls, indicating a high level of accumulation within the tumor (Fig. 4). The results of this study demonstrated the potential of SPIONs to be used as MRI contrast agents for the detection of HER2/neu-expressing breast cancer.

6 Antibody-Conjugated AuNPs and IONPs for Cancer Therapeutics

Nanoparticles can serve as theranostic agents, allowing the clinician to initially image and diagnose the cancer and then to follow-up with a target-specific therapy. The advantage is the simplicity of the coupled approach and the ability to initiate treatment only when the diagnostic indicates a need for treatment. Therapeutic strategies for the treatment of cancers employing antibody-conjugated AuNPs and IONPs include hyperthermia, targeted drug delivery, and the enhancement of radiation therapies. These therapeutic methods can be combined with imaging modalities for theranostics, including image-guided hyperthermia and drug delivery. The advantage of the theranostic approach would be that, before the administration of therapeutics, a physician would be able to track the particles and direct treatment at the site of disease. This technology presents a powerful tool that could be used to discover cancer cells and provide treatment immediately, in a potential combined screening and treatment strategy. A brief overview of recent developments in the area of AuNPs and IONPs for cancer therapeutics is provided in the following section.

6.1 Targeted Hyperthermia

One example of a targeted therapeutic approach that utilizes nanoparticles is the exogenous application of energy which is then concentrated by the nanoparticles, resulting in a region-specific increase in temperature. Hyperthermia is the elevation of temperature up to 42–46 °C, usually by photon flux of 5–48 W cm^{-2} [128], for an extended period of time to damage and kill cancer cells by DNA denaturation [129–131]. Hyperthermia has been widely used for cancer treatment [132] to enhance cancer treatment by directly causing cell death, sensitizing cells to chemotherapy [128] and radiotherapy [133], and by promoting tumor reoxygenation [133, 135]. The major technical impedances of traditional hyperthermia treatment are a difficulty of heating the local tumor region without damaging normal tissue [138], and an inability to create hyperthermia uniformly throughout the tumor, leaving cancerous cells and resulting in regrowth of tumor [137].

Targeted AuNP- and IONP-mediated hyperthermia has been explored to enhance treatment specificity and increase heating uniformity [138], and can be initiated via magnetic or light induction. Magnetic hyperthermia is induced by exposing magnetic nanoparticles to an alternating magnetic field (AMF) [31, 139]. Light-induced hyperthermia, photothermal therapy (PTT), is the use of electromagnetic radiation (infrared wavelengths) to convert NIR light energy into heat [140, 141]. The objective of preclinical studies in this area has been to determine the feasibility of administering antibody-conjugated AuNPs and IONPs directed to the disease site by tumor antigen-targeting antibodies, followed by the initiation of thermal activation by either magnetic or light induction. Overviews of some recent *in-vitro* and *in-vivo* studies in this area are detailed in the following sections.

6.1.1 *In-Vitro* Hyperthermia Studies

Nanoparticle-mediated hyperthermia with bioconjugated antigen-targeted photothermal therapeutic agents have been of particular interest for breast cancer theranostic applications [142]. Loo *et al.* reported dual gold nanoshells conjugated with anti-HER2 mAb via attachment with a poly(ethylene glycol) (PEG) linker for imaging and PTT of breast cancer cells. The gold nanoshells were incubated with SKBR3 breast cancer cells for binding studies. The results of the study showed a significantly increased scatter based on optical contrast due to nanoshell selective binding to breast cancer cells as compared to control cell groups [143]. The anti-HER2 nanoshells also exhibited a targeted photothermal effect, with cell death occurring as a result of treatment with a NIR laser after exposure to anti-HER2 nanoshells. The effect was not seen with nanoshells labeled with a nonspecific control antibody and unconjugated nanoshells. The study demonstrated the bioimaging and cancer therapeutic potential of mAb-conjugated gold nanoshells with a clinically relevant biomarker, HER2.

HNPs have also been investigated for use in MRI-guided PTT. Melancon *et al.* reported HNPs for use in MRI-guided laser ablation of head and neck cancer as a minimally invasive alternative for patients who are not candidates for surgery [34]. In this study, a SPIO core covered in amorphous silica was coated with a gold nanoshell (SPIO@Au). The particles were then conjugated with anti-epidermal growth factor receptor (EGFR) Cetuximab (C225) and tested for their capacity to selectively ablate human squamous carcinoma cell lines overexpressing EGFR *in vitro.* The results of the study showed that the C225–SPIO@Au conjugates could produce a temperature in excess of 65 °C, and cell death was only apparent with cells incubated with C225–SPIO@Au nanoshells after exposure to NIR laser light for 3 min. MRI detection was carried out using cells inoculated with the HNP conjugates and then suspended in agar. Cells inoculated with C225–SPIO@Au nanoshells showed darker contrast than the nontargeted control samples and the antibody-blocked control. The biodistribution of these nanoshell conjugates showed a selective uptake into tumors in comparison to PEG-coated SPIO@Au. However, a majority of the conjugates localized to the liver, spleen and kidneys. In addition, the authors did not demonstrate any *in-vivo* ablation of tumors using the C225 functionalized nanoshells.

6.1.2 *In-Vivo* Hyperthermia Studies

Preclinical *in-vivo* studies for targeted hyperthermia have investigated the potential for antibody-conjugated nanoparticles to improve hyperthermia treatment. In one study, multifunctional gold–iron oxide HNPs targeted with humanized single-chain antibody conjugates (A33ScFv) were studied in SW1222 cells (antigen-expressing

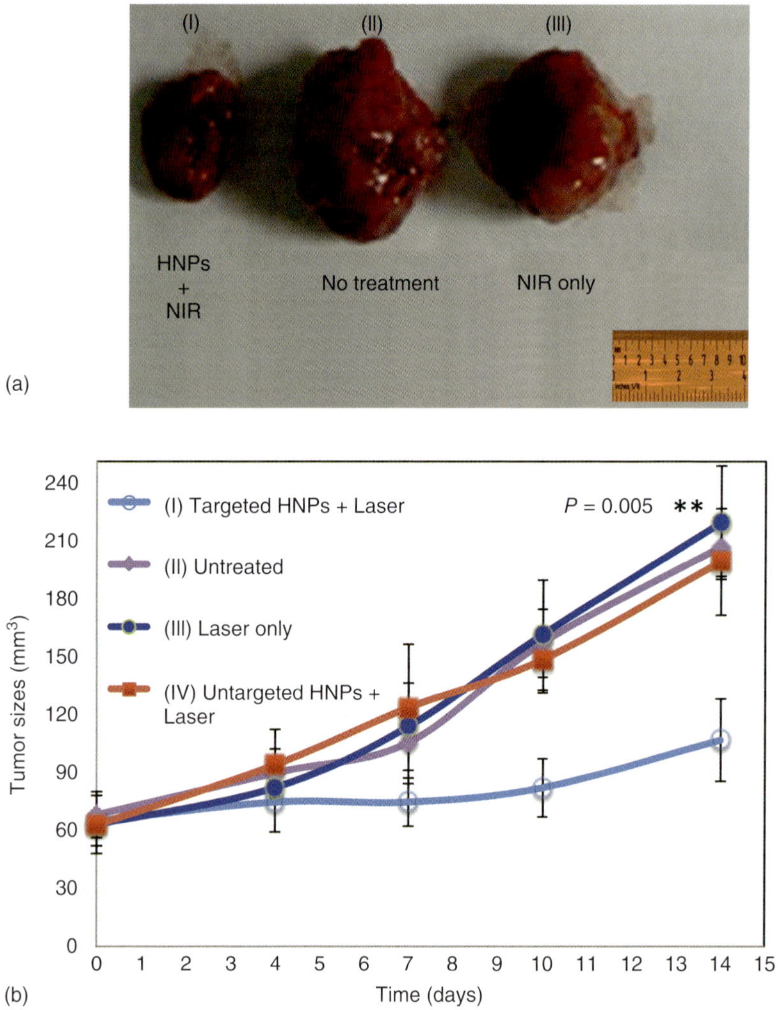

Fig. 5 Photothermal therapy (PTT) response on subcutaneous colorectal tumors. (a) Tumor masses harvested after treatment with HNPs and NIR laser (group I), untreated control (group II), or PBS plus NIR laser (group III); (b) Tumor growth curves after PTT. Tumor size calculated as $a \times b^2/2$ (a = length, b = width), measured 24 h after irradiation and the duration of therapy, (n = 5 mice per group). Reprinted with permission from Ref. [144]; © 2013, Elsevier.

colorectal cancer cell line) for use in laser-assisted PTT using NIR irradiation and MRI contrast agents *in vivo*. The HNPs in this study were effective MRI contrast agents, reducing the post-contrast phase T_2 value by 50%. The HNPs showed localization to SW1222 tumor xenografts within 12 h of systemic injection. The efficacy of photothermal treatment was confirmed by a significant reduction in tumor volume in samples treated with target HNPs exposed to a NIR laser (Fig. 5). Histological analysis further confirmed degradation of the extracellular matrix, cell necrosis, and nucleus damage/shrinkage in the treatment group. PTT resulted in necrosis in approximately 65% of tumor tissues [144].

6.2 Targeted Drug Delivery

Current cancer drug therapies are limited by inefficient delivery due to insolubility, toxicity, and non-specificity [145], but using nanoparticles as the delivery platform for many of these therapies has the potential to overcome these limitations. Nanoparticles provide a large surface area-to-volume ratio, which allows them to bind, adsorb, and carry a large variety of chemical compounds [147]. Coupling chemical compounds with nanoparticles has been shown to increase the aqueous stability of many insoluble drugs and provide specificity, which leads to lower dosage requirements, thereby reducing harmful side effects [145, 147].

The use of AuNPs and IONPs as a drug-delivery system has been shown to increase the effectiveness of several anticancer drugs. Several mechanisms are attributed to this increase in efficiency, including an increased solubility of the active therapy, nanoparticle-mediated cellular uptake, and prolonged circulation and receptor wrapping time [148]. For example, AuNPs have been shown to increase the efficacy of the chemotherapy drugs doxorubicin (DOX) and gemcitabine (Gemzar) [149, 150]. The stability in solution and cytotoxicity of the originally hydrophobic chemotherapy drug, paclitaxel, was also significantly increased after its conjugation to AuNPs [145, 146]. Optimization of this system using PEGylated AuNPs with biotin receptors has further enhanced the cytotoxic effects of paclitaxel [146]. IONPs loaded with chemotherapeutics have been shown to increase the cellular uptake – and thus the potency – of the drugs, including in multidrug-resistant cancer cells [151, 152]. Methotrexate, an antimetabolite used in the treatment of various cancers has been successfully conjugated with IONPs, resulting in an increased cytotoxicity towards target cells due to an improvement in cellular uptake [147].

AuNPs and IONPs can not only carry drugs to target locations, but can also be used to produce a controlled release of the drugs. Triggered drug release from AuNPs and IONPs has been achieved with various activators, including X-rays, glutathione and changes in the environmental pH [67, 148, 153–155]. Notably, the local irradiation of gold nanorods bound to a therapeutic agent via a thiol group has been shown to result in a detachment of the drug in the local area, thus reducing the nonspecific dispersion of stand-alone drugs, which can in turn lead to toxicity in non-targeted organs [158]. Preliminary studies have also shown that radiofrequency electromagnetic field (RF-EMF) activation can also be an effective trigger for the independent release of multiple drugs on a single nanoparticle, using nucleic acid tethers [157]. Some details of recent *in-vitro* and *in-vivo* preclinical studies in this area are outlined in the following sections.

6.2.1 *In-Vitro* Targeted Drug-Delivery Studies

Bisker *et al.* conducted a controlled-release study using AuNPs conjugated with rituximab, an anti-CD20 mAb, for the treatment of leukemia (Fig. 6). [158]. AuNPs of 20 nm diameter were conjugated with PEGylated anti-CD20 via thiol chemistry, and the functionalized AuNPs were then incubated with a human CD-20-positive B-cell line for controlled-release studies. The controlled release of anti-CD20 was activated upon irradiation by a femtosecond pulse train, the wavelength of which was tuned to match

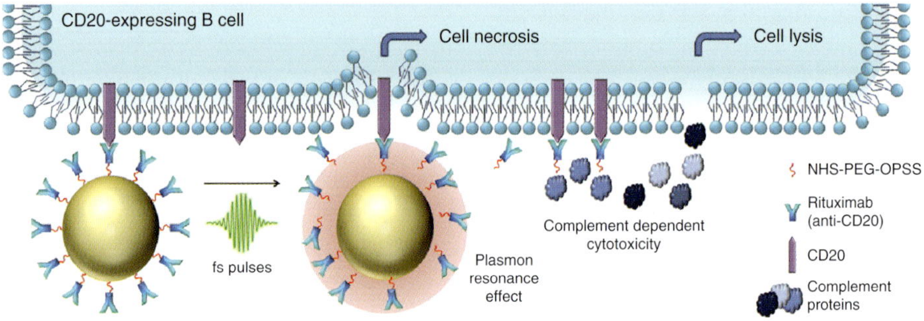

Fig. 6 Controlled release of rituximab from gold nanoparticles. Targeted nanoparticles bind to CD20-expressing cells and are laser-irradiated to release rituximab, which recruits complement proteins to induce complement-dependent cytotoxicity (CDC)-mediated cell death. Reprinted with permission from Ref. [158]; © 2012, Elsevier.

the plasmon resonance of the AuNPs. The released anti-CD20 retained its functionality, inducing a complement-dependent cytotoxicity, where the complement proteins in blood serum acted to induce necrosis of cells marked by rituximab [158]. In addition, plasmonic shock waves emanating from the AuNPs, as a result of irradiation, initiated cell death. The advantages of this system included a high specificity and efficiency, low toxicity, and high levels of repeatability and consistency due to the morphological stability of the nanoparticles.

Dilnawaz *et al.* conducted an *in-vitro* study with HER2 antibody-conjugated glycerol monooleate (GMO) -coated IONPs loaded with different anticancer chemotherapeutic drugs (paclitaxel, rapamycin either alone or combination) and for use as active cancer therapeutics (Fig. 7) [159]. Approximately 95% drug entrapment

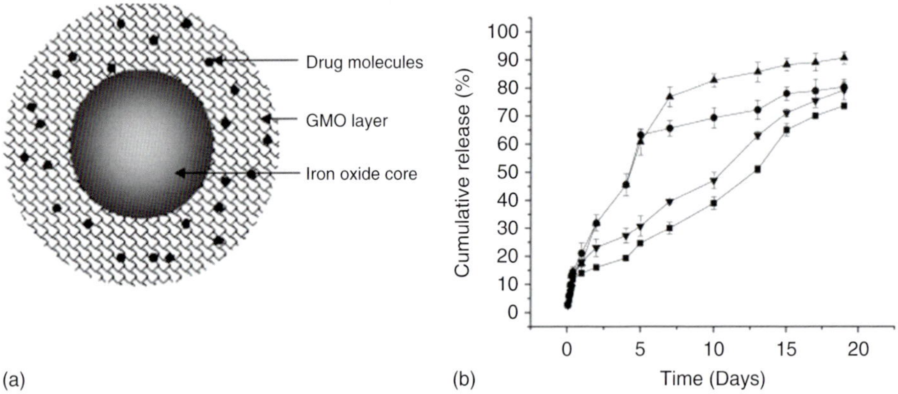

Fig. 7 Drug adsorption and drug release of GMO-coated drug-loaded magnetic iron oxide nanoparticles (IONPs). (a) Schematic representation of drug adsorption in the GMO coating surrounding the IONP core; (b) Release of rapamycin from rapa-GMO-IONPs (•), paclitaxel (*) from pac-GMO-IONPs, rapamycin (▲) from combo-GMO-IONPs, and paclitaxel (▼) from combo-GMO-IONPs under *in-vitro* conditions. Reprinted with permission from Ref. [159]; © 2010, Elsevier.

was achieved via absorption in the GMO coating, and sustained release occurred over a two-week period. Conjugation to the HER2 antibody increased the cellular uptake of GMO–IONPs by human breast cancer carcinoma cell lines (MCF-7) cells threefold when compared to non-targeted GMO–IONPs. The targeted, drug-loaded GMO–IONPs were 55-fold more effective than the native drug, and sevenfold more effective than the unconjugated NPs, which indicated the potential antiproliferation effect in MCF-7 [159]. These findings demonstrated the potential of drug-loaded, antibody-conjugated IONPs as potential drug carriers for improved drug efficiency, while providing selectivity that would reduce the adverse side effects caused by damage to normal healthy tissues.

6.2.2 *In-Vivo* Imaging and Targeted Drug Delivery

Multifunctional IONPs have been developed for combined targeted imaging and drug delivery. Of particular interest are IONPs as nanocarriers for MRI imaging and drug delivery, conjugated with anti-HER2 trastuzumab and DOX [125]. In a recent *in-vivo* study conducted by Zolata *et al.*, polymer-coated IONPs were ^{11}In-labeled and conjugated with trastuzumab–DOX for tumor targeting, drug delivery, controlled release, and dual-mode tumor imaging [160]. The theranostic effects of the immunotargeted and drug-loaded IONPs were evaluated in an HER2-positive breast tumor model bearing BALB/c mice, using biodistribution, molecular imaging, and tumor morphology measurements. Antibody-conjugated nanoparticles from this study accumulated at very high concentrations in the tumors due to their targeting specificity, when compared to non-targeted NP controls. Results obtained using single photon emission computed tomography (SPECT) and MRI showed the tumor region to be clearly contrasted by the antibody-conjugated IONPs. Additionally, the tumor volumes were measured at 12, 24, and 48 h after treatment and were 50%, 25%, and 23% lower, respectively, for treated animals compared to the control group, indicating a significant therapeutic effect. These findings supported the use of multifunctional modified SPIONs for cancer theranostics, namely imaging and drug-delivery applications.

6.3 AuNPs for the Enhancement of Externally Delivered Radiation Therapy

The two major limitations to the efficacy of radiation therapy are the dose limitations that are required to spare healthy surrounding tissues, and the hypoxia-associated reduction in the radiosensitivity of solid tumor cells [161, 162]. AuNPs have been used to address these limitations as radiosensitizers, as they possess a significantly high X-ray absorption coefficient [163, 164]. Both, *in-vitro* and *in-vivo* models have provided evidence that pretreatment with AuNPs has the potential to improve the imaging and radiosenstitization of a tumor, while minimizing the dose to normal tissues [165–171]. AuNPs are expected to increase the effects of radiation treatment due to lethal DNA double-strand breaks (DSBs), when the tumor cells are exposed to radiation [172–174]. Preclinical studies in this area have relied heavily on the leaky tumor microenvironment and intratumoral injection to localize AuNPs to tumor tissues [175]. However, the introduction of targeting moieties on AuNP-based radiosensitizers has the potential to improve tumor specificity and internalization [173, 174, 176].

Thus, antibody-conjugated AuNPs are among the next generation of targeted AuNP radiosensitizers [177]. In a recent study evaluating the effect of Herceptin (trastuzumab)-conjugated AuNPs (Au-T) on HER2-overexpressing MDA-MB-361 on a human breast cancer cell culture and a tumor xenograft mouse model showed that the clonogenic survival of cells exposed to Au-T and X-rays was significantly lower than that of cells exposed to radiation alone, resulting in a dose enhancement of 1.6. *In vivo,* the combination of Au-T and X-irradiation resulted in a regression of MDA-MB-361 tumors by 46% compared to X-irradiation treatment alone [172]. This study was one of the first to demonstrate the potential for nanoparticles conjugated with cancer-targeting antibodies in enhancing externally delivered radiation therapy.

7
Conclusions

AuNPs, IONPs and HNPs functionalized with cancer-specific antibodies provide attractive platforms for potent next-generation cancer theranostics, which couple diagnostics with therapies. Preclinical studies with immunotargeted nanoparticles have demonstrated more specific therapies with the potential to result in reduced adverse side effects in comparison to convectional treatment strategies. Further, treatment at the nanoscale allows for cellular level cancer detection, providing a means to eliminate cancer at the cellular level and thus reduce cancer recurrence. Thus, AuNPs and IONPs, when conjugated with antibodies, enhance the optical signals and provide a sensitive strategy for the detection and destruction of cancer cells which could, potentially, lead to a paradigm shift in cancer treatment strategies.

References

1. Siegel, R., Naishadham, D., and Jemal, A. (2013) Cancer statistics, 2013. CA. *Cancer J. Clin.*, **63** (1), 11–30.
2. Kochanek, K.D., Xu, J., Murphy, S.L., Minino, A.M., and Kung, H.C. (2011) Deaths: final data for 2009. *Natl. Vital Stat. Rep.*, **60**, 1–117.
3. Holohan, C., Van Schaeybroeck, S., Longley, D.B., and Johnston, P.G. (2013) Cancer drug resistance: an evolving paradigm. *Nat. Rev. Cancer*, **13** (10), 714–726.
4. Hull, L.C., Farrell, D., and Grodzinski, P. (2014) Highlights of recent developments and trends in cancer nanotechnology research – view from NCI Alliance for Nanotechnology in Cancer. *Biotechnol. Adv.*, **32** (4), 666–678.
5. Curry, T., Kopelman, R., Shilo, M., and Popovtzer, R. (2014) Multifunctional theranostic gold nanoparticles for targeted CT imaging and photothermal therapy. *Contrast Media Mol. Imaging*, **9** (1), 53–61.
6. Julien, D.C., Behnke, S., Wang, G., Murdoch, G.K. *et al.* (2011) Utilization of monoclonal antibody-targeted nanomaterials in the treatment of cancer. *MAbs*, **3** (5), 467–478.
7. He, X., Wang, K., and Cheng, Z. (2010) In vivo near-infrared fluorescence imaging of cancer with nanoparticle-based probes. *Wiley Interdiscip. Rev. Nanomed. Nanobiotechnol.*, **2** (4), 349–366.
8. Chow, E.K.-H. and D. Ho (2013) Cancer nanomedicine: from drug delivery to imaging. *Sci. Transl. Med.*, **5** (216), 1–12.
9. Alivisatos, P. (2004) The use of nanocrystals in biological detection. *Nat. Biotechnol.*, **22** (1), 47–52.
10. Thakor, A.S., Jokerst, J., Zavaleta, C., Massoud, T.F. *et al.* (2011) Gold nanoparticles: a revival in precious metal administration to patients. *Nano Lett.*, **11** (10), 4029–4036.
11. Cai, W., Gao, T., Hong, H., and Sun, J. (2008) Applications of gold nanoparticles in cancer nanotechnology. *Nanotechnol. Sci. Appl.*, **1**, 17–32.

12. Heddle, J. (2013) Gold nanoparticle–biological molecule interactions and catalysis. *Catalysts*, **3** (3), 683–708.
13. Huang, X.H., Jain, P.K., El-Sayed, I.H., and El-Sayed, M.A. (2007) Gold nanoparticles: interesting optical properties and recent applications in cancer diagnostic and therapy. *Nanomedicine*, **2** (5), 681–693.
14. Huang, X. and El-Sayed, M.A. (2010) Gold nanoparticles: optical properties and implementations in cancer diagnosis and photothermal therapy. *J. Adv. Res.*, **1** (1), 13–28.
15. Ku, K.H., Yang, H., Shin, J.M., and Kim, B.J. (2015) Aspect ratio effect of nanorod surfactants on the shape and internal morphology of block copolymer particles. *J. Polym. Sci. Part A - Polym. Chem.*, **53** (2), 188–192.
16. Cao-Milan, R. and Liz-Marzan, L.M. (2014) Gold nanoparticle conjugates: recent advances toward clinical applications. Expert Opin. *Drug Delivery*, **11** (5), 741–752.
17. Zhou, J., Ralston, J., Sedev, R., and Beattie, D.A. (2009) Functionalized gold nanoparticles: synthesis, structure and colloid stability. *J. Colloid Interface Sci.*, **331** (2), 251–262.
18. Fratoddi, I., Venditti, I., Cametti, C., and Russo, M.V. (2014) Gold nanoparticles and gold nanoparticle-conjugates for delivery of therapeutic molecules. Progress and challenges. J. Mater. *Chem. B*, **2** (27), 4204–4220.
19. Dykman, L. and Khlebtsov, N. (2012) Gold nanoparticles in biomedical applications: recent advances and perspectives. *Chem. Soc. Rev.*, **41** (6), 2256–2282.
20. Klein, S., S. Petersen, U. Taylor, D. Rath. Quantitative visualization of colloidal and intracellular gold nanoparticles by confocal microscopy. *J. Biomed. Optics*, 2010. **15** (3), 036015–036015-11.
21. van Dijk, M.A., Tchebotareva, A.L., Orrit, M., and Lippitz, M. (2006) Absorption and scattering microscopy of single metal nanoparticles. *Phys. Chem. Chem. Phys.*, **8** (30), 3486–3495.
22. Tcherniak, A., Ha, J.W., Dominguez-Medina, S., Slaughter, L.S. *et al.* (2010) Probing a century old prediction one plasmonic particle at a time. *Nano Lett.*, **10** (4), 1398–1404.
23. Dreaden, E.C., Alkilany, A.M., Huang, X.H., Murphy, C.J. *et al.* (2012) The golden age: gold nanoparticles for biomedicine. *Chem. Soc. Rev.*, **41** (7), 2740–2779.
24. Cho, S.J., Idrobo, J.C., Olamit, J., Liu, K. *et al.* (2005) Growth mechanisms and oxidation resistance of gold-coated iron nanoparticles. *Chem. Mater.*, **17** (12), 3181–3186.
25. Teja, A.S. and Koh, P.-Y. (2009) Synthesis, properties, and applications of magnetic iron oxide nanoparticles. *Prog. Cryst. Growth Charact. Mater.*, **55** (1-2), 22–45.
26. Wang, D., He, J., Rosenzweig, N., and Rosenzweig, Z. (2004) Superparamagnetic Fe_2O_3 beads–CdSe/ZnS quantum dots core–shell nanocomposite particles for cell separation. *Nano Lett.*, **4** (3), 409–413.
27. Rosen, J.E., Chan, L., Shieh, D.-B., and Gu, F.X. (2012) Iron oxide nanoparticles for targeted cancer imaging and diagnostics. *Nanomed. Nanotechnol. Biol. Med.*, **8** (3), 275–290.
28. Wang, Y.-X., Hussain, S., and Krestin, G. (2001) Superparamagnetic iron oxide contrast agents: physicochemical characteristics and applications in MR imaging. *Eur. Radiol.*, **11** (11), 2319–2331.
29. Wang, Y.X. (2011) Superparamagnetic iron oxide based MRI contrast agents: current status of clinical application. *Quant. Imaging Med. Surg.*, **1** (1), 35–40.
30. Clement, O., Frija, G., Chambon, C., Schouman-Clayes, E. *et al.* (1991) Liver tumors in cirrhosis: experimental study with SPIO-enhanced MR imaging. *Radiology*, **180** (1), 31–36.
31. Deryl, L.T. and Stefan, H.B. (2011) *Utilization of Magnetic Nanoparticles for Cancer Therapy, in Nanomedicine and Cancer*, Science Publishers, pp. 84–104.
32. Kim, J., Park, S., Lee, J.E., Jin, S.M. *et al.* (2006) Designed fabrication of multifunctional magnetic gold nanoshells and their application to magnetic resonance imaging and photothermal therapy. *Angew. Chem.*, **118** (46), 7918–7922.
33. Wang, L.Y., Luo, J., Fan, Q., Suzuki, M. *et al.* (2005) Monodispersed core–shell Fe_3O_4@Au nanoparticles. *J. Phys. Chem. B*, **109** (46), 21593–21601.
34. Melancon, M.P., Lu, W., Zhong, M., Zhou, M. *et al.* (2011) Targeted multifunctional gold-based nanoshells for magnetic

resonance-guided laser ablation of head and neck cancer. *Biomaterials*, **32** (30), 7600–7608.
35. Li, L., Mak, K.Y., Leung, C.W., Leung, C.H. *et al.* (2014) Synthesis and morphology control of gold/iron oxide magnetic nanocomposites via a simple aqueous method. *IEEE Trans. Magn.*, **50** (1), 1–5.
36. Li, L., Du, Y.M., Mak, K.Y., Leung, C.W. *et al.* (2014) Novel hybrid Au/Fe_3O_4 magnetic octahedron-like nanoparticles with tunable size. *IEEE Trans. Magn.*, **50** (1), 1–5.
37. Xu, Z., Hou, Y., and Sun, S. (2007) Magnetic core/shell Fe_3O_4/Au and Fe_3O_4/Au/Ag nanoparticles with tunable plasmonic properties. *J. Am. Chem. Soc.*, **129** (28), 8698–8699.
38. Lyon, J.L., Fleming, D.A., Stone, M.B., Schiffer, P. *et al.* (2004) Synthesis of Fe oxide core/Au shell nanoparticles by iterative hydroxylamine seeding. *Nano Lett.*, **4** (4), 719–723.
39. Yu, H., Chen, M., Rice, P.M., Wang, S.X. *et al.* (2005) Dumbbell-like bifunctional Au–Fe_3O_4 nanoparticles. *Nano Lett.*, **5** (2), 379–382.
40. Kirui, D.K., Rey, D.A., and Batt, C.A. (2010) Gold hybrid nanoparticles for targeted phototherapy and cancer imaging. *Nanotechnology*, **21** (10), 105105.
41. Barnett, C.M., Gueorguieva, M., Lees, M.R., McGarvey, D.J. *et al.* (2013) Physical stability, biocompatibility and potential use of hybrid iron oxide–gold nanoparticles as drug carriers. *J. Nanopart. Res.*, **15** (6), 1689.
42. Mezni, A., Balti, I., Mlayah, A., Jouini, N. *et al.* (2013) Hybrid Au–Fe_3O_4 nanoparticles: plasmonic, surface enhanced Raman scattering, and phase transition properties. *J. Phys. Chem. C*, **117** (31), 16166–16174.
43. Shevchenko, E.V., Bodnarchuk, M.I., Kovalenko, M.V., Talapin, D.V. *et al.* (2008) Gold/iron oxide core/hollow–shell nanoparticles. *Adv. Mater.*, **20** (22), 4323–4329.
44. Park, H.Y., Schadt, M.J., Wang, L., Lim, I.I.S. *et al.* (2007) Fabrication of magnetic core–shell Fe oxide@Au nanoparticles for interfacial bioactivity and bio-separation. *Langmuir*, **23** (17), 9050–9056.
45. Gautier, J., Allard-Vannier, E., Munnier, E., Souce, M. *et al.* (2013) Recent advances in theranostic nanocarriers of doxorubicin based on iron oxide and gold nanoparticles. *J. Controlled Release*, **169** (1-2), 48–61.
46. Zavisova, V., Koneracka, M., Kovac, J., Kubovcikova, M. *et al.* (2015) The cytotoxicity of iron oxide nanoparticles with different modifications evaluated in vitro. *J. Magn. Magn. Mater.*, **380**, 85–89.
47. Tiwari, P.M., Eroglu, E., Bawage, S.S., Vig, K. *et al.* (2014) Enhanced intracellular translocation and biodistribution of gold nanoparticles functionalized with a cell-penetrating peptide (VG-21) from vesicular stomatitis virus. *Biomaterials*, **35** (35), 9484–9494.
48. Bogdanov, A.A., Gupta, S., Koshkina, N., Corr, S.J. *et al.* (2015) Gold nanoparticles stabilized with MPEG-grafted poly(L-lysine): in vitro and in vivo evaluation of a potential theranostic agent. *Bioconjugate Chem.*, **26** (1), 39–50.
49. Choi, Y.-W., Lee, H., Song, Y., and Sohn, D. (2015) Colloidal stability of iron oxide nanoparticles with multivalent polymer surfactants. *J. Colloid Interface Sci.*, **443**, 8–12.
50. Maldonado, C.R., Salassa, L., Gomez-Blanco, N., and Mareque-Rivas, J.C. (2013) Nano-functionalization of metal complexes for molecular imaging and anticancer therapy. *Coord. Chem. Rev.*, **257** (19–20), 2668–2688.
51. Aryal, S., Remant, B.K.C., Dharmaraj, N., Bhattarai, N. *et al.* (2006) Spectroscopic identification of S–Au interaction in cysteine-capped gold nanoparticles. Spectrochim. *Acta, Part A*, **63** (1), 160–163.
52. Gupta, R.K., Srinivasan, M.P., and Dharmarajan, R. (2011) Synthesis of short chain thiol capped gold nanoparticles, their stabilization and immobilization on silicon surface. *Colloids Surf., A*, **390** (1–3), 149–156.
53. Gupta, R.K., Srinivasan, M.P., and Dharmarajan, R. (2012) Synthesis of 16-mercaptohexadecanoic acid-capped gold nanoparticles and their immobilization on a substrate. *Mater. Lett.*, **67** (1), 315–319.
54. Lavenn, C., Albrieux, F., Tuel, A., and Demessence, A. (2014) Synthesis, characterization and optical properties of an amino-functionalized gold thiolate cluster: Au-10(SPh-pNH(2))(10). *J. Colloid Interface Sci.*, **418**, 234–239.

55. Lin, S.Y., Tsai, Y.T., Chen, C.C., Lin, C.M. *et al.* (2004) Two-step functionalization of neutral and positively charged thiols onto citrate-stabilized Au nanoparticles. *J. Phys. Chem. B*, **108** (7), 2134–2139.
56. Zhou, J.F., Beattie, D.A., Ralston, J., and Sedev, R. (2007) Colloid stability of thymine-functionalized gold nanoparticles. *Langmuir*, **23** (24), 12096–12103.
57. Tan, H., Zhan, T., and Fan, W.Y. (2006) Direct functionalization of the hydroxyl group of the 6-mercapto-1-hexanol (MCH) ligand attached to gold nanoclusters. J. Phys. *Chem. B*, **110** (43), 21690–21693.
58. Yoo, C.I., Seo, D., Chung, B.H., Chung, I.S. *et al.* (2009) A facile one-pot synthesis of hydroxyl-functionalized gold polyhedrons by a surface regulating copolymer. *Chem. Mater.*, **21** (5), 939–944.
59. Goren, M., Galley, N., and Lennox, R.B. (2006) Adsorption of alkylthiol-capped gold nanoparticles onto alkylthiol self-assembled monolayers: an SPR study. *Langmuir*, **22** (3), 1048–1054.
60. Hofmann, A., Schmiel, P., Stein, B., and Graf, C. (2011) Controlled formation of gold nanoparticle dimers using multivalent thiol ligands. *Langmuir*, **27** (24), 15165–15175.
61. Shichibu, Y., Negishi, Y., Tsukuda, T., and Teranishi, T. (2005) Large-scale synthesis of thiolated Au-25 clusters via ligand exchange reactions of phosphine-stabilized Au-11 clusters. *J. Am. Chem. Soc.*, **127** (39), 13464–13465.
62. Woehrle, G.H., Brown, L.O., and Hutchison, J.E. (2005) Thiol-functionalized, 1.5-nm gold nanoparticles through ligand exchange reactions: scope and mechanism of ligand exchange. *J. Am. Chem. Soc.*, **127** (7), 2172–2183.
63. Davis, K., Qi, B., Witmer, M., Kitchens, C.L. *et al.* (2014) Quantitative measurement of ligand exchange on iron oxides via radiolabeled oleic acid. *Langmuir*, **30** (36), 10918–10925.
64. Hofmann, A., Thierbach, S., Semisch, A., Hartwig, A. *et al.* (2010) Highly monodisperse water-dispersable iron oxide nanoparticles for biomedical applications. *J. Mater. Chem.*, **20** (36), 7842–7853.
65. Shem, P.M., Sardar, R., and Shumaker-Parry, J.S. (2009) One-step synthesis of phosphine-stabilized gold nanoparticles using the mild reducing agent 9-BBN. *Langmuir*, **25** (23), 13279–13283.
66. Scaravelli, R.C.B., Dazzi, R.L., Giacomelli, F.C., Machado, G. *et al.* (2013) Direct synthesis of coated gold nanoparticles mediated by polymers with amino groups. *J. Colloid Interface Sci.*, **397**, 114–121.
67. Wang, X.G., Kawanami, H., Islam, N.M., Chattergee, M. *et al.* (2008) Amphiphilic block copolymer-stabilized gold nanoparticles for aerobic oxidation of alcohols in aqueous solution. *Chem. Commun.*, **37**, 4442–4444.
68. Rey, D.A., Strickland, A.D., Kirui, D., Niamsiri, N. *et al.* (2010) In vitro self-assembly of gold nanoparticle-coated poly(3-hydroxybutyrate) granules exhibiting plasmon-induced thermo-optical enhancements. *ACS Appl. Mater. Interfaces*, **2** (7), 1804–1810.
69. Xu, S.Y., Tu, G.L., Peng, B., and Han, X.Z. (2006) Self-assembling gold nanoparticles on thiol-functionalized poly(styrene-*co*-acrylic acid) nanospheres for fabrication of a mediatorless biosensor. *Anal. Chim. Acta*, **570** (2), 151–157.
70. Fuentes, M., Mateo, C., Guisán, J.M., and Fernández-Lafuente, R. (2005) Preparation of inert magnetic nano-particles for the directed immobilization of antibodies. *Biosens. Bioelectron.*, **20** (7), 1380–1387.
71. Almeida, J.P.M., Figueroa, E.R., and Drezek, R.A. (2014) Gold nanoparticle mediated cancer immunotherapy. Nanomed. Nanotechnol. *Biol. Med.*, **10** (3), 503–514.
72. Huang, G., Zhang, C., Li, S., Khemtong, C. *et al.* (2009) A novel strategy for surface modification of superparamagnetic iron oxide nanoparticles for lung cancer imaging. *J. Mater. Chem.*, **19**, 6367–6372.
73. Chen, H.W., Zou, H., Paholak, H.J., Ito, M. *et al.* (2014) Thiol-reactive amphiphilic block copolymer for coating gold nanoparticles with neutral and functionable surfaces. *Polym. Chem.*, **5** (8), 2768–2773.
74. Bloemen, M., Brullot, W., Luong, T., Geukens, N. *et al.* (2012) Improved functionalization of oleic acid-coated iron oxide nanoparticles for biomedical applications. *J. Nanopart. Res.*, **14** (9), 1–10.
75. von Maltzahn, G., Park, J.-H., Agrawal, A., Bandaru, N.K. *et al.* (2009) Computationally guided photothermal tumor therapy using

long-circulating gold nanorod antennas. *Cancer Res.*, **69** (9), 3892–3900.

76. Zhang, S., Gong, M., Zhang, D., Yang, H. *et al.* (2014) Thiol-PEG-carboxyl-stabilized Fe_2O_3/Au nanoparticles targeted to CD105: synthesis, characterization and application in MR imaging of tumor angiogenesis. *Eur. J. Radiol.*, **83** (7), 1190–1198.
77. Tassa, C., Shaw, S.Y., and Weissleder, R. (2011) Dextran-coated iron oxide nanoparticles: a versatile platform for targeted molecular imaging, molecular diagnostics, and therapy. *Acc. Chem. Res.*, **44** (10), 842–852.
78. Li, S., Liu, H., and He, N. (2010) Covalent binding of streptavidin on gold magnetic nanoparticles for bead array fabrication. *J. Nanosci. Nanotechnol.*, **10** (8), 4875–4882.
79. Giaume, D., Poggi, M., Casanova, D., Mialon, G. *et al.* (2008) Organic functionalization of luminescent oxide nanoparticles toward their application as biological probes. *Langmuir*, **24** (19), 11018–11026.
80. Erathodiyil, N. and Ying, J.Y. (2011) Functionalization of inorganic nanoparticles for bioimaging applications. *Acc. Chem. Res.*, **44** (10), 925–935.
81. Graf, C., Vossen, D.L.J., Imhof, A., and van Blaaderen, A. (2003) A general method to coat colloidal particles with silica. *Langmuir*, **19** (17), 6693–6700.
82. Liberman, A., Mendez, N., Trogler, W.C., and Kummel, A.C. (2014) Synthesis and surface functionalization of silica nanoparticles for nanomedicine. *Surf. Sci. Rep.*, **69** (2-3), 132–158.
83. Guerrero-Martínez, A., Pérez-Juste, J., and Liz-Marzán, L.M. (2010) Recent progress on silica coating of nanoparticles and related nanomaterials. *Adv. Mater.*, **22** (11), 1182–1195.
84. Ladj, R., Bitar, A., Eissa, M.M., Fessi, H. *et al.* (2013) Polymer encapsulation of inorganic nanoparticles for biomedical applications. *Int. J. Pharm.*, **458** (1), 230–241.
85. Gun, S., Edirisinghe, M., and Stride, E. (2013) Encapsulation of superparamagnetic iron oxide nanoparticles in poly-(lactide-*co*-glycolic acid) microspheres for biomedical applications. Mater. *Sci. Eng., C*, **33** (6), 3129–3137.
86. Boisselier, E., Diallo, A.K., Salmon, L., Ornelas, C. *et al.* (2010) Encapsulation and stabilization of gold nanoparticles with "click" polyethyleneglycol dendrimers. *J. Am. Chem. Soc.*, **132** (8), 2729–2742.
87. Li, K. and Liu, B. (2012) Polymer encapsulated conjugated polymer nanoparticles for fluorescence bioimaging. *J. Mater. Chem.*, **22** (4), 1257–1264.
88. Fratila, R.M., Mitchell, S.G., del Pino, P., Grazu, V. *et al.* (2014) Strategies for the biofunctionalization of gold and iron oxide nanoparticles. *Langmuir*, **30** (50), 15057–15071.
89. Sperling, R.A. and Parak, W.J. (2010) Surface modification, functionalization and bioconjugation of colloidal inorganic nanoparticles. *Philos. Trans. R. Soc. A*, **368** (1915), 1333–1383.
90. Rayavarapu, R.G., Petersen, W., Ungureanu, C., Post, J.N. *et al.* (2007) Synthesis and bioconjugation of gold nanoparticles as potential molecular probes for light-based imaging techniques. *Int. J. Biomed. Imaging*, **2007**, 10.
91. Puertas, S., Batalla, P., Moros, M., Polo, E. *et al.* (2011) Taking advantage of unspecific interactions to produce highly active magnetic nanoparticle–antibody conjugates. *ACS Nano*, **5** (6), 4521–4528.
92. Chinol, M., Casalini, P., Maggiolo, M., Canevari, S. *et al.* (1998) Biochemical modifications of avidin improve pharmacokinetics and biodistribution, and reduce immunogenicity. *Br. J. Cancer*, **78** (2), 189–197.
93. Cheung, N.K., Modak, S., Lin, Y., Guo, H. *et al.* (2004) Single-chain Fv-streptavidin substantially improved therapeutic index in multistep targeting directed at disialoganglioside GD2. *J. Nucl. Med.*, **45** (5), 867–877.
94. Tajima, N., Takai, M., and Ishihara, K. (2011) Significance of antibody orientation unraveled: well-oriented antibodies recorded high binding affinity. *Anal. Chem.*, **83** (6), 1969–1976.
95. Colombo, M., Sommaruga, S., Mazzucchelli, S., Polito, L. *et al.* (2012) Site-specific conjugation of ScFvs antibodies to nanoparticles by bioorthogonal strain-promoted alkyne–nitrone cycloaddition. *Angew. Chem. Int. Ed.*, **51** (2), 496–499.
96. Hutchins, B.M., Kazane, S.A., Staflin, K., Forsyth, J.S. *et al.* (2011) Site-specific coupling and sterically controlled formation

of multimeric antibody fab fragments with unnatural amino acids. *J. Mol. Biol.*, **406** (4), 595–603.

97. Ackerson, C.J., Jadzinsky, P.D., Jensen, G.J., and Kornberg, R.D. (2006) Rigid, specific, and discrete gold nanoparticle/antibody conjugates. *J. Am. Chem. Soc.*, **128** (8), 2635–2640.
98. Scott, A.M. (2001) Specific targeting, biodistribution and lack of immunogenicity of chimeric anti-GD3 monoclonal antibody KM871 in patients with metastatic melanoma – results of a Phase I trial. *J. Clin. Oncol.*, **19**, 3976–3987.
99. Nelson, A.L. (2010) Development trends for human monoclonal antibody therapeutics. *Nat. Rev. Drug Discov.*, **9**, 767–774.
100. Reichert, J.M. (2008) Monoclonal antibodies as innovative therapeutics. *Curr. Pharm. Biotechnol.*, **9** (6), 423–430.
101. Weiner, L.M., Surana, R., and Wang, S. (2010) Monoclonal antibodies: versatile platforms for cancer immunotherapy. *Nat. Rev. Immunol.*, **10**, 317–327.
102. Alderson, K.L. and Sondel, P.M. (2011) Clinical cancer therapy by NK cells via antibody-dependent cell-mediated cytotoxicity. *J. Biomed. Biotechnol.*, doi:10.1155/2011/379123.
103. Bronte, G., Sortino, G., Passiglia, F., Rizzo, S. *et al.* (2015) Monoclonal antibodies for the treatment of non-haematological tumours: update of an expanding scenario. *Expert Opin. Biol. Ther.*, **15** (1), 45–59.
104. Ribatti, D. (2014) From the discovery of monoclonal antibodies to their therapeutic application: An historical reappraisal. *Immunol. Lett.*, **161** (1), 96–99.
105. Davis, M.E., Chen, Z., and Shin, D.M. (2008) Nanoparticle therapeutics: an emerging treatment modality for cancer. *Nat. Rev. Drug Discov.*, **7** (9), 771–782.
106. Parker, S.A., Diaz, I.L., Anderson, K.A., and Batt, C.A. (2013) Design, production, and characterization of a single-chain variable fragment (ScFv) derived from the prostate specific membrane antigen (PSMA) monoclonal antibody J591. *Protein Expr. Purif.*, **89** (2), 136–145.
107. Huang, X., Yi, C., Fan, Y., Zhang, Y. *et al.* (2014) Magnetic Fe_3O_4 nanoparticles grafted with single-chain antibody (scFv) and docetaxel-loaded β-cyclodextrin potential for ovarian cancer dual-targeting therapy. *Mater. Sci. Eng., C*, **42**, 325–332.
108. Liu, Y., Liu, Y., Mernaugh, R.L., and Zeng, X. (2009) Single chain fragment variable recombinant antibody functionalized gold nanoparticles for a highly sensitive colorimetric immunoassay. *Biosens. Bioelectron.*, **24** (9), 2853–2857.
109. Lee, N., Kim, H., Choi, S.H., Park, M. *et al.* (2011) Magnetosome-like ferrimagnetic iron oxide nanocubes for highly sensitive MRI of single cells and transplanted pancreatic islets. *Proc. Natl Acad. Sci. USA*, **108** (7), 2662–2667.
110. Jain, P.K., El-Sayed, I.H., and El-Sayed, M.A. (2007) Au nanoparticles target cancer. *Nano Today*, **2** (1), 18–29.
111. Lima, K.M.G., Junior, R.F.A., Araujo, A.A., Oliveira, A.L.C.S.L. *et al.* (2014) Environmentally compatible bioconjugated gold nanoparticles as efficient contrast agents for colorectal cancer cell imaging. *Sens. Actuators, B*, **196**, 306–313.
112. Luo, S., Zhang, E., Su, Y., Cheng, T. *et al.* (2011) A review of NIR dyes in cancer targeting and imaging. *Biomaterials*, **32** (29), 7127–7138.
113. Altinoglu, E.I. and Adair, J.H. (2010) Near infrared imaging with nanoparticles. *Wiley Interdiscip. Rev. Nanomed. Nanobiotechnol.*, **2** (5), 461–477.
114. Murphy, C.J., Gole, A.M., Stone, J.W., Sisco, P.N. *et al.* (2008) Gold nanoparticles in biology: beyond toxicity to cellular imaging. *Acc. Chem. Res.*, **41** (12), 1721–1730.
115. Lee, C.N., Wang, Y.M., Lai, W.F., Chen, T.J. *et al.* (2012) Super-paramagnetic iron oxide nanoparticles for use in extrapulmonary tuberculosis diagnosis. *Clin. Microbiol. Infect.*, **18** (6), E149–E157.
116. Skaat, H., Corem-Slakmon, E., Grinberg, I., Last, D. *et al.* (2013) Antibody-conjugated, dual-modal, near-infrared fluorescent iron oxide nanoparticles for antiamyloidgenic activity and specific detection of amyloid-beta fibrils. *Int. J. Nanomed.*, **8**, 4063–4076.
117. Sokolov, K., Follen, M., Aaron, J., Pavlova, I. *et al.* (2003) Real-time vital optical imaging of precancer using anti-epidermal growth factor receptor antibodies conjugated to gold nanoparticles. *Cancer Res.*, **63** (9), 1999–2004.
118. Schol, D., Fleron, M., Greisch, J.F., Jaeger, M. *et al.* (2013) Anti-PSMA

antibody-coupled gold nanorods detection by optical and electron microscopies. *Micron*, **50**, 68–74.

119. Hainfeld, J.F., O'Connor, M.J., Dilmanian, F.A., Slatkin, D.N. *et al.* (2011) Micro-CT enables microlocalisation and quantification of Her2-targeted gold nanoparticles within tumour regions. *Br. J. Radiol.*, **84** (1002), 526–533.
120. Coughlin, A.J., Ananta, J.S., Deng, N., Larina, I.V. *et al.* (2014) Gadolinium-conjugated gold nanoshells for multimodal diagnostic imaging and photothermal cancer therapy. *Small*, **10** (3), 556–565.
121. Lee, N., Choi, Y., Lee, Y., Park, M. *et al.* (2012) Water-dispersible ferrimagnetic iron oxide nanocubes with extremely high r_2 relaxivity for highly sensitive in vivo MRI of tumors. *Nano Lett.*, **12** (6), 3127–3131.
122. Serda, R.E., Adolphi, N.L., Bisoffi, M., and Sillerud, L.O. (2007) Targeting and cellular trafficking of magnetic nanoparticles for prostate cancer imaging. *Mol. Imaging*, **6** (4), 277–288.
123. Kao, H.W., Lin, Y.Y., Chen, C.C., Chi, K.H. *et al.* (2013) Evaluation of EGFR-targeted radioimmuno-gold-nanoparticles as a theranostic agent in a tumor animal model. *Bioorg. Med. Chem. Lett.*, **23** (11), 3180–3185.
124. Gong, T., Olivo, M., Dinish, U.S., Goh, D. *et al.* (2013) Engineering bioconjugated gold nanospheres and gold nanorods as label-free plasmon scattering probes for ultrasensitive multiplex dark-field imaging of cancer cells. *J. Biomed. Nanotechnol.*, **9** (6), 985–991.
125. Choi, W.I., Lee, J.H., Kim, J.Y., Heo, S.U. *et al.* (2015) Targeted antitumor efficacy and imaging via multifunctional nano-carrier conjugated with anti-HER2 trastuzumab. *Nanomedicine*, **11** (2), 359–368.
126. Karmani, L., Bouchat, V., Bouzin, C., Leveque, P. *et al.* (2014) Zr-89-labeled anti-endoglin antibody-targeted gold nanoparticles for imaging cancer: implications for future cancer therapy. *Nanomedicine*, **9** (13), 1923–1937.
127. Chen, T.J., Cheng, T.H., Chen, C.Y., Hsu, S.C. *et al.* (2009) Targeted herceptin-dextran iron oxide nanoparticles for noninvasive imaging of HER2/neu receptors using MRI. *J. Biol. Inorg. Chem.*, **14** (2), 253–260.
128. Vertrees, R.A., Das, G.C., Popov, V.L., Coscio, A.M. *et al.* (2005) Synergistic interaction of hyperthermia and gemcitabine in lung cancer. *Cancer Biol. Ther.*, **4** (10), 1144–1153.
129. Hildebrandt, B., Wust, P., Ahlers, O., Dieing, A. *et al.* (2002) The cellular and molecular basis of hyperthermia. *Crit. Rev. Oncol. Hematol.*, **43** (1), 33–56.
130. Ito, A., Honda, H., and Kobayashi, T. (2006) Cancer immunotherapy based on intracellular hyperthermia using magnetite nanoparticles: a novel concept of 'heat-controlled necrosis' with heat shock protein expression. Cancer Immunol. *Immunother*, **55** (3), 320–328.
131. Chae, S.Y., Kim, Y.-S., Park, M.J., Yang, J. *et al.* (2014) High-intensity focused ultrasound-induced, localized mild hyperthermia to enhance anti-cancer efficacy of systemic doxorubicin: an experimental study. *Ultrasound Med. Biol.*, **40** (7), 1554–1563.
132. Overgaard, J., Gonzalez-Gonzalez, D., Hulshof, M.C., Arcangeli, G. *et al.* (1995) Randomised trial of hyperthermia as adjuvant to radiotherapy for recurrent or metastatic malignant melanoma. *European Society for Hyperthermic Oncology. Lancet*, **345** (8949), 540–543.
133. Roa, W., Zhang, X., Guo, L., Shaw, A. *et al.* (2009) Gold nanoparticle sensitize radiotherapy of prostate cancer cells by regulation of the cell cycle. *Nanotechnology*, **20** (37), 375101.
134. Atkinson, R.L., Zhang, M., Diagaradjane, P., Peddibhotla, S. *et al.* (2010) Thermal enhancement with optically activated gold nanoshells sensitizes breast cancer stem cells to radiation therapy. *Sci. Transl. Med.*, **2** (55), 3001447.
135. Kampinga, H.H. (2006) Cell biological effects of hyperthermia alone or combined with radiation or drugs: a short introduction to newcomers in the field. *Int. J. Hyperthermia*, **22** (3), 191–196.
136. Cui, Z.-G., Piao, J.-L., Kondo, T., Ogawa, R. *et al.* (2014) Molecular mechanisms of hyperthermia-induced apoptosis enhanced by docosahexaenoic acid: implication for cancer therapy. *Chem.- Biol. Interact.*, **215**, 46–53.

137. Saniei, N. (2009) Hyperthermia and cancer treatment. *Heat Transfer Eng.*, **30** (12), 915–917.
138. Sawdon, A., Weydemeyer, E., and Peng, C.A. (2014) Antitumor therapy using nanomaterial-mediated thermolysis. *J. Biomed. Nanotechnol.*, **10** (9), 1894–1917.
139. Dennis, C.L., Jackson, A.J., Borchers, J.A., Hoopes, P.J. *et al.* (2009) Nearly complete regression of tumors via collective behavior of magnetic nanoparticles in hyperthermia. *Nanotechnology*, **20** (39), 395103.
140. Song, X., Gong, H., Yin, S., Cheng, L. *et al.* (2014) Ultra-small iron oxide doped polypyrrole nanoparticles for in vivo multimodal imaging guided photothermal therapy. *Adv. Funct. Mater.*, **24** (9), 1194–1201.
141. Huang, P., Lin, J., Wang, S., Zhou, Z. *et al.* (2013) Photosensitizer-conjugated silica-coated gold nanoclusters for fluorescence imaging-guided photodynamic therapy. *Biomaterials*, **34** (19), 4643–4654.
142. Lee, J., Chatterjee, D.K., Lee, M.H., and Krishnan, S. (2014) Gold nanoparticles in breast cancer treatment: promise and potential pitfalls. *Cancer Lett.*, **347** (1), 46–53.
143. Loo, C., Lowery, A., Halas, N., West, J. *et al.* (2005) Immunotargeted nanoshells for integrated cancer imaging and therapy. *Nano Lett.*, **5** (4), 709–711.
144. Kirui, D.K., Khalidov, I., Wang, Y., and Batt, C.A. (2013) Targeted near-IR hybrid magnetic nanoparticles for in vivo cancer therapy and imaging. *Nanomed. Nanotechnol. Biol. Med.*, **9** (5), 702–711.
145. Gibson, J.D., Khanal, B.P., and Zubarev, E.R. (2007) Paclitaxel-functionalized gold nanoparticles. *J. Am. Chem. Soc.*, **129** (37), 11653–11661.
146. Ding, Y., Zhou, Y.-Y., Chen, H., Geng, D.-D. *et al.* (2013) The performance of thiol-terminated PEG-paclitaxel-conjugated gold nanoparticles. *Biomaterials*, **34** (38), 10217–10227.
147. Sun, C., C. Fang, Z. Stephen, O. Veiseh *et al.* (2008) Tumor-targeted drug delivery and MRI contrast enhancement by chlorotoxin-conjugated iron oxide nanoparticles. *Nanomed.* (London, Engl.), **3** (4), 495–505.
148. Wang, F., Wang, Y.C., Dou, S., Xiong, M.H. *et al.* (2011) Doxorubicin-tethered responsive gold nanoparticles facilitate intracellular drug delivery for overcoming multidrug resistance in cancer cells. *ACS Nano*, **5** (5), 3679–3692.
149. Venkatesan, R., Pichaimani, A., Hari, K., Balasubramanian, P.K. *et al.* (2013) Doxorubicin conjugated gold nanorods: a sustained drug delivery carrier for improved anticancer therapy. *J. Mater. Chem. B*, **1** (7), 1010–1018.
150. Patra, C.R., Bhattacharya, R., Mukhopadhyay, D., and Mukherjee, P. (2010) Fabrication of gold nanoparticles for targeted therapy in pancreatic cancer. *Adv. Drug Delivery Rev.*, **62** (3), 346–361.
151. Laurent, S. and Mahmoudi, M. (2011) Superparamagnetic iron oxide nanoparticles: promises for diagnosis and treatment of cancer. *Int. J. Mol. Epidemiol. Genet.*, **2** (4), 367–390.
152. Kapse-Mistry, S., Govender, T., Srivastava, R., and Yergeri, M. (2014) Nanodrug delivery in reversing multidrug resistance in cancer cells. *Front. Pharmacol*, **5**, 159.
153. Starkewolf, Z.B., L. Miyachi, J. Wong, T. Guo. X-ray triggered release of doxorubicin from nanoparticle drug carriers for cancer therapy. *Chem. Commun.* (Cambridge, Engl.), 2013. **49** (25), 2545–2547.
154. Madhusudhan, A., Reddy, G.B., Venkatesham, M., Veerabhadram, G. *et al.* (2014) Efficient pH dependent drug delivery to target cancer cells by gold nanoparticles capped with carboxymethyl chitosan. *Int. J. Mol. Sci.*, **15** (5), 8216–8234.
155. Derfus, A.M., von Maltzahn, G., Harris, T.J., Duza, T. *et al.* (2007) Remotely triggered release from magnetic nanoparticles. *Adv. Mater.*, **19** (22), 3932–3936.
156. Zhang, X.-D., Wu, D., Shen, X., Chen, J. *et al.* (2012) Size-dependent radiosensitization of PEG-coated gold nanoparticles for cancer radiation therapy. *Biomaterials*, **33** (27), 6408–6419.
157. Dani, R.K., Schumann, C., and Taratula, O. (2014) Temperature-tunable iron oxide nanoparticles for remote-controlled drug release. *AAPS Pharm. Sci. Technol.*, **15** (4), 963–972.
158. Bisker, G., Yeheskely-Hayon, D., Minai, L., and Yelin, D. (2012) Controlled release of Rituximab from gold nanoparticles for phototherapy of malignant cells. *J. Control Release*, **162** (2), 303–309.

159. Dilnawaz, F., Singh, A., Mohanty, C., and Sahoo, S.K. (2010) Dual drug loaded superparamagnetic iron oxide nanoparticles for targeted cancer therapy. *Biomaterials*, **31** (13), 3694–3706.
160. Zolata, H., Abbasi Davani, F., and Afarideh, H. (2015) Synthesis, characterization and theranostic evaluation of Indium-111 labeled multifunctional superparamagnetic iron oxide nanoparticles. *Nucl. Med. Biol.*, **42** (2), 164–170.
161. Lacombe, S. and Sech, C.L. (2009) Advances in radiation biology: radiosensitization in DNA and living cells. *Surf. Sci.*, **603** (10-12), 1953–1960.
162. Harrison, L.B., Chadha, M., Hill, R.J., Hu, K. *et al.* (2002) Impact of tumor hypoxia and anemia on radiation therapy outcomes. *Oncologist*, **7** (6), 492–508.
163. Xi, D., Dong, S., Meng, X., Lu, Q. *et al.* (2012) Gold nanoparticles as computerized tomography (CT) contrast agents. *RSC Adv.*, **2** (33), 12515–12524.
164. Silvestri, A., Polito, L., Bellani, G., Zambelli, V. *et al.* (2015) Gold nanoparticles obtained by aqueous digestive ripening: their application as X-ray contrast agents. *J. Colloid Interface Sci.*, **439**, 28–33.
165. Chithrani, D.B., Jelveh, S., Jalali, F., van Prooijen, M. *et al.* (2010) Gold nanoparticles as radiation sensitizers in cancer therapy. *Radiat. Res.*, **173** (6), 719–728.
166. Geng, F., Xing, J.Z., Chen, J., Yang, R. *et al.* (2014) Pegylated glucose gold nanoparticles for improved in-vivo bio-distribution and enhanced radiotherapy on cervical cancer. *J. Biomed. Nanotechnol.*, **10** (7), 1205–1216.
167. Khoshgard, K., Hashemi, B., Arbabi, A., Rasaee, M.J. *et al.* (2014) Radiosensitization effect of folate-conjugated gold nanoparticles on HeLa cancer cells under orthovoltage superficial radiotherapy techniques. *Phys. Med. Biol.*, **59** (9), 2249.
168. Zhang, X.-D., Luo, Z., Chen, J., Song, S. *et al.* (2015) Ultrasmall glutathione-protected gold nanoclusters as next generation radiotherapy sensitizers with high tumor uptake and high renal clearance. *Sci. Rep.*, **5**, 8669.
169. Yasui, H., Takeuchi, R., Nagane, M., Meike, S. *et al.* (2014) Radiosensitization of tumor cells through endoplasmic reticulum stress induced by PEGylated nanogel containing gold nanoparticles. *Cancer Lett.*, **347** (1), 151–158.
170. Wolfe, T., Chatterjee, D., Lee, J., Grant, J.D. *et al.* (2015) Targeted gold nanoparticles enhance sensitization of prostate tumors to megavoltage radiation therapy in vivo. *Nanomedicine*, **11** (5), 1277–1283.
171. Joh, D.Y., Kao, G.D., Murty, S., Stangl, M. *et al.* (2013) Theranostic gold nanoparticles modified for durable systemic circulation effectively and safely enhance the radiation therapy of human sarcoma cells and tumors. *Transl. Oncol.*, **6** (6), 722–732.
172. Chattopadhyay, N., Cai, Z., Kwon, Y., Lechtman, E. *et al.* (2013) Molecularly targeted gold nanoparticles enhance the radiation response of breast cancer cells and tumor xenografts to X-radiation. *Breast Cancer Res. Treat.*, **137** (1), 81–91.
173. Herold, D.M., Das, I.J., Stobbe, C.C., Iyer, R.V. *et al.* (2000) Gold microspheres: a selective technique for producing biologically effective dose enhancement. *Int. J. Radiat. Biol.*, **76** (10), 1357–1364.
174. Hainfeld, J.F., Slatkin, D.N., and Smilowitz, H.M. (2004) The use of gold nanoparticles to enhance radiotherapy in mice. *Phys. Med. Biol.*, **49** (18), N309–N315.
175. Hainfeld, J.F., Lin, L., Slatkin, D.N., Avraham Dilmanian, F. *et al.* (2014) Gold nanoparticle hyperthermia reduces radiotherapy dose. *Nanomed. Nanotechnol. Biol. Med.*, **10** (8), 1609–1617.
176. Burger, N., Biswas, A., Barzan, D., Kirchner, A. *et al.* (2014) A method for the efficient cellular uptake and retention of small modified gold nanoparticles for the radiosensitization of cells. *Nanomed. Nanotechnol. Biol. Med.*, **10** (6), 1365–1373.
177. Chattopadhyay, N., Cai, Z.L., Pignol, J.P., Keller, B. *et al.* (2010) Design and characterization of HER-2-targeted gold nanoparticles for enhanced X-radiation treatment of locally advanced breast cancer. *Mol. Pharmaceutics*, **7** (6), 2194–2206.

19
Immune Checkpoint Inhibitors

Laura Mansi[1,2], *Franck Pagès*[3], *and Olivier Adotévi*[1,2]
[1] *University of Franche-Comte, INSERM Unit 1098, 30 Bd Fleming, 25030 Besançon Cedex, France*
[2] *University Hospital of Besançon, Department of Medical Oncology, 30 Bd Fleming, 25030 Besançon Cedex, France*
[3] *AP-HP, Paris-Descartes University, Laboratory of Immunology, Immunomonitoring Platform of the Georges Pompidou European Hospital, 20 rue Leblanc, 75908 Paris, France*

Translational Medicine Cancer: Advances in Molecular Biology and Medicine,
First Edition. Edited by Robert A. Meyers.

Keywords

T cell
A type of white blood cell that is of key importance in the immune system. A core component of the adaptive immune system.

Cancer
Cancer or malignant tumor defines a group of diseases involving abnormal proliferative cells with the capacity for migration, neoangiogenesis, and no sensitivity to apoptotic signals.

CTLA-4
Cytotoxic T-lymphocyte-associated protein 4 (also known as cluster of differentiation 152; CD152) is a protein receptor that downregulates the immune system.

Immunotherapy
Treatment that uses certain features of a person's immune system to fight diseases such as cancer.

PD-1/PD-L1
The programmed death 1 (PD-1 or CD279) pathway is a key immune checkpoint. Expression of PD-1 on the T-cell surface exerts an immunoregulatory role in the activation of T cells. PD-ligand 1 (PD-L1) is a transmembrane protein that binds to its receptor, PD-1.

Emerging clinical data suggest that cancer immunotherapy is likely to become a key part of the clinical management of cancer. Knowledge of the basic mechanisms of the immune system as they relate to cancer has been increasing rapidly, and such developments have accelerated the translation of novel immunotherapy techniques into medical breakthroughs for many cancer patients. The development of checkpoint-blocking antibodies, against cytotoxic T-lymphocyte antigen 4 (CTLA-4) and the programmed death 1 receptor (PD-1), has recently demonstrated significant promise in the treatment of malignancies. Ipilimumab (to block CTLA-4) and pembrolizumab (to block PD-1) have been approved by the US Food and Drug Administration for the treatment of advanced melanoma, and additional regulatory approvals are expected for a variety of other agents such as nivolumab (to block PD-1). To combine checkpoint inhibitors approaches with other therapies such as immunomodulators (cytokines, indoleamine 2,3-dioxygenase inhibitors), cytotoxic chemotherapy, radiation therapy or molecularly targeted therapies, may hold the key to the true potential of immunotherapy in the future management of cancer patients.

1 Introduction

Emerging clinical data suggest that cancer immunotherapy is likely to become a key part of the clinical management of cancer. Knowledge of the basic mechanisms of the immune system as they relate to cancer has been increasing rapidly, with recent developments having accelerated the translation of novel immunotherapy techniques into medical breakthroughs for many cancer patients.

Genetic and cellular alterations in tumors enable the immune system to generate T-cell responses that recognize and eradicate cancer cells [1]. The destruction of cancer by T cells is only one step of the cancer immunity cycle, which manages the delicate balance between the recognition of nonself and the prevention of autoimmunity [2]. However, tumors can escape from an equilibrium state previously held in check by the immune system to become significant, as suggested by the theory of immunoediting. Tumors may utilize a variety of mechanisms to evade the immune system, in particular against T-cell attack [1]. Thus, the identification of cancer cell/T-cell inhibitory signals such as cytotoxic T-lymphocyte antigen 4 (CTLA-4) and programmed death 1 (PD-1)/PD-L1, has prompted the development of a new class of immunotherapy.

Preexisting anticancer immune responses can be potentially reinvigorated and possibly expanded by immunotherapy, as the latter specifically hinders immune effector inhibition. A wide variety of approaches exist to elicit an antitumor immune response, involving therapeutic cancer vaccines, adoptive T-cell therapy, antitumor antibodies, and the emerging immune checkpoint blockade.

Moreover, the presence of suppressive factors in the tumor microenvironment (TME) may explain the limited activity observed with previous immune-based therapies. To combine these approaches with other therapies such as immunomodulators (cytokines, indoleamine 2,3-dioxygenase (IDO) inhibitors), cytotoxic chemotherapy, radiation therapy or molecularly targeted therapies, may hold the key to the true potential of immunotherapy in the future management of cancer patients [2, 3].

1.1 Regulation of T-Cell Response: Role of the Costimulatory and Inhibitory Receptor

The adaptive immune response is tightly regulated by multiple costimulatory and coinhibitory pathways. Naïve T cells are activated when the T-cell receptor (TCR) binds to the major histocompatibility complex (MHC)–antigen complex (signal 1) but, unfortunately, signal 1 is insufficient to generate and maintain an adaptive immune response. Rather, the full activation of a T cell also requires the simultaneous engagement of positive costimulatory molecules present on activated antigen-presenting cells (APCs); this is known

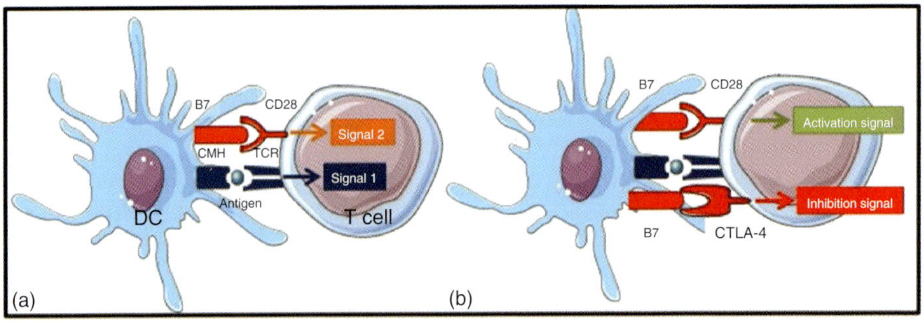

Fig. 1 Immune checkpoints regulate the evolution of immune response. (a) Stimulation of the TCR by MHC/peptide on dendritic cell (DC) complexes delivers signal 1, while interactions between costimulatory ligands on the antigen-presenting cell (APC) and CD28 on the T cell provide signal 2; (b) Activation signal is initiated by binding of B7 molecules on the APC (DC) to cluster of differentiation 28 (CD28) receptors on the T cell. Cytotoxic T-lymphocyte antigen 4 (CTLA-4), expressed on the T cell, binds B7 molecules, resulting in T-cell inactivation. Adapted from Ref. [4].

as signal 2 (Fig. 1). These costimulatory molecules are not present on quiescent APCs, tumor cells or normal host cells. A classic example of a costimulatory pathway is the interaction between B7 expressed on an APC and CD28 expressed on T cells [1] (Figs 1 and 2). When both the MHC–antigen complex and the TCR (signal 1), and the costimulatory pathway (signal 2) are activated, T cells proliferate and are activated. As both costimulatory and coinhibitory molecules may be present at the same time, signal 2 is perhaps more appropriately conceptualized as the sum of both costimulatory signals and coinhibitory signals that determine the T-cell phenotype. After activation, T cells express coinhibitory receptors such as CTLA-4, PD-1, TIM-3 and LAG-3; these compensatory coinhibitors attenuate the immune response and are often co-opted by tumors to evade the host's natural antitumor immune response. Tumor cells and macrophages within the tumor environment can also express coinhibitory receptors and promote immune tolerance (Fig. 3).

2 CTLA-4 Blockade

2.1 Biology of Receptor Ligands

CTLA-4 (CD152) was the first molecule to be identified as being coinhibitory, and is the counterpart of the costimulatory CD28 axis. It is mainly expressed on T cells, where it primarily regulates the amplitude of the early stages of T-cell activation by dendritic cells and other APCs [6–8].

CD28 and CTLA-4 share identical ligands, namely CD80 (also known as B7.1) and CD86 (also known as B7.2) [9–11]. Following activation, the T cells upregulate the surface expression of CTLA-4 that binds B7 molecules with a higher avidity, and thus counteracts CD28's positive costimulatory signal (Fig. 4). This dominance of negative signals results in a reduced T-cell proliferation and decreased interleukin 2 (IL-2) production [1]. Thus, the CTLA-4 inhibitory pathway can modulate an immune response and also avert autoimmunity [12].

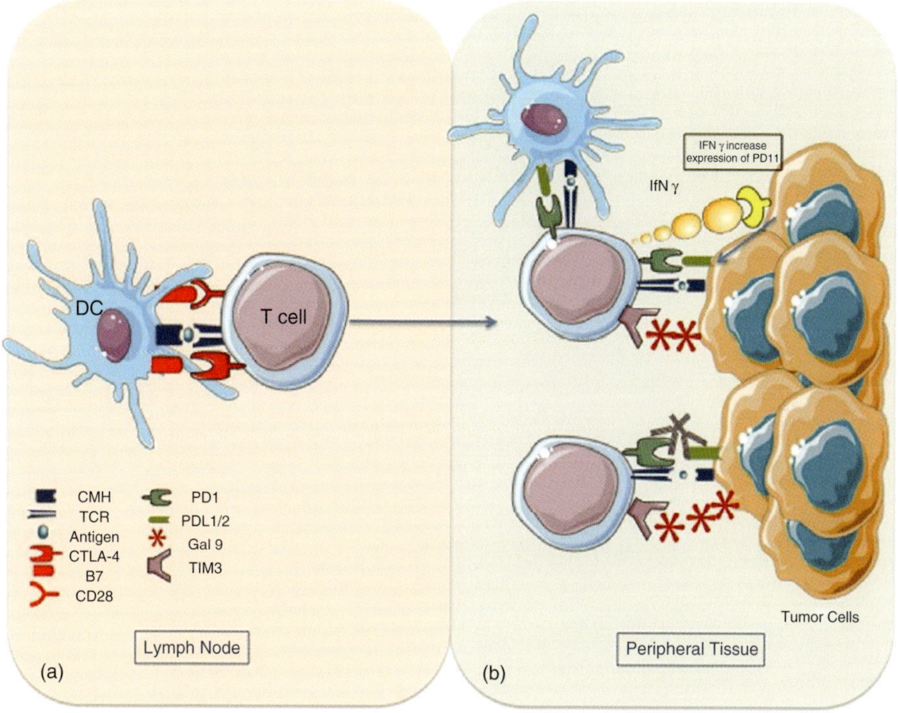

Fig. 2 Targets of antibody immune modulators. (a) The interaction of T-cell receptor (TCR) with a major histocompatibility complex (MHC) molecule expressed by antigen-presenting cells (DC). The interaction of the CD28 receptor on T cell with B7 costimulatory molecules (B7-1 and B7-2) on the DC is necessary to complete T-cell activation. This phase occurs primarily within the lymph nodes. To prevent inappropriate T-cell activation, negative regulators of T-cell immunity, including CTLA-4 and PD-1, are required. CTLA-4 competes with CD28 for the interaction with B7, and it is upregulated after T-cell activation; (b) In the peripheral tissues other negative regulators participate in T-cell inactivation. Moreover, two general mechanisms of expression of immune-checkpoint ligands on tumor cells are identified. (i) Innate immune resistance: Constitutive oncogenic signaling can upregulate PDL1 expression on all tumor cells, independently of inflammatory signals in the tumor microenvironment; (ii) Adaptive immune resistance: PDL1 is induced in response to inflammatory signals that are produced by an active antitumor immune response. For example, the secretion of interferon-γ by activated T cells increases the expression of PDL1 on tumor cells. Adapted from Ref. [1].

The specific signaling pathways by which CTLA-4 blocks T-cell activation are still under investigation, although it has been suggested in a number of studies that activation of the protein phosphatases, the SH2-domain containing tyrosine phosphatase (SHP2; also known as PTPN11) and PP2A, are important in counteracting kinase signals that are induced by TCRs and CD28 [7]. However, CTLA-4 also confers T-cell inhibition through the sequestration of CD80 and CD86 from CD28 engagement, as well as an active removal of CD80 and CD86 from the APC surface [13]. The central role of CTLA-4 for keeping T-cell activation in check is demonstrated by the lethal

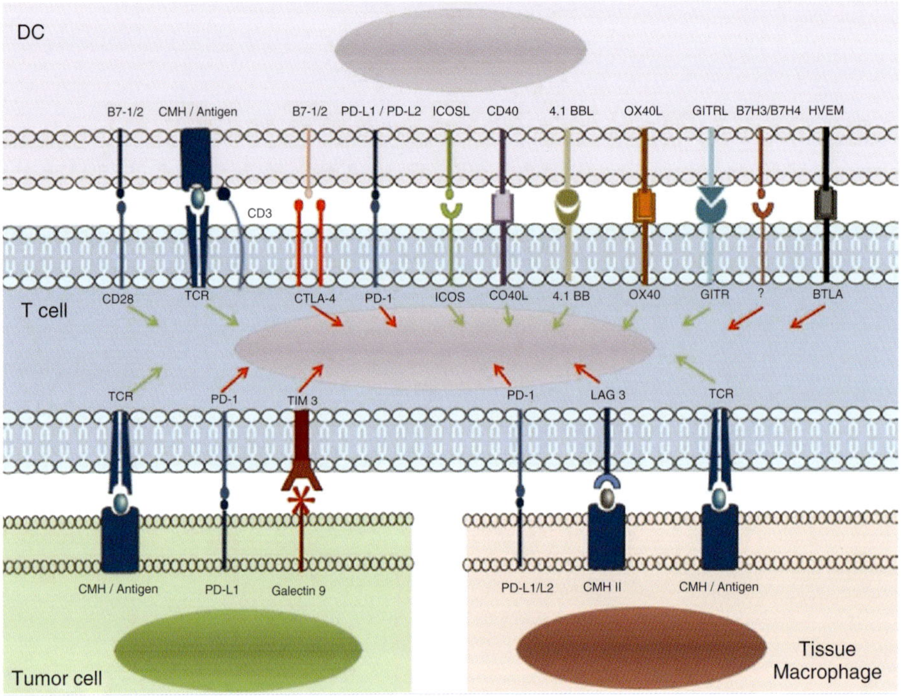

Fig. 3 Multiple costimulatory and coinhibitory ligand–receptor interaction between T cell and dendritic cell (DC), tumor cell, and tissue macrophage in the tumor microenvironment. CTLA-4: cytotoxic T-lymphocyte-associated antigen 4; GAL9, galectin 9; HVEM, herpesvirus entry mediator; ICOS, inducible T cell costimulator; IL, interleukin; LAG3, lymphocyte activation gene 3; PD1, programmed cell death protein 1; PDL, PD1 ligand; TCR, T-cell receptor; TIM3, T-cell membrane protein 3. Adapted from Refs [4, 5].

systemic immune hyperactivation phenotype of *CTLA-4*-knockout mice [14–16].

CTLA-4 is also expressed by regulatory T cells (T-reg) and memory CD4 T cells. These cells may also be targeted by CTLA-4 blockade, and both the enhancement of effector CD4 T cell activity and the inhibition of cell-dependent immunosuppression are probably important factors [17, 18].

2.2 Anti-CTLA-4 Antibodies from Mice to Clinics

The blocking CTLA-4 was initially questioned, as expression of the CTLA-4 ligands is not tumor-specific and the dramatic lethal autoimmune phenotype of CTLA-4-knockout mice predicted a high degree of immune toxicity associated with blockade of this receptor. However, the first preclinical models showed that a therapeutic window could be achieved when CTLA-4 was partially blocked with antibodies [19]. The initial studies demonstrated significant antitumor responses without overt immune toxicities when mice bearing partially immunogenic tumors were treated with CTLA-4 antibodies as single agents. Poorly immunogenic tumors did not respond to anti-CTLA-4 as a single agent, but did respond when anti-CTLA-4 was combined with a granulocyte–macrophage

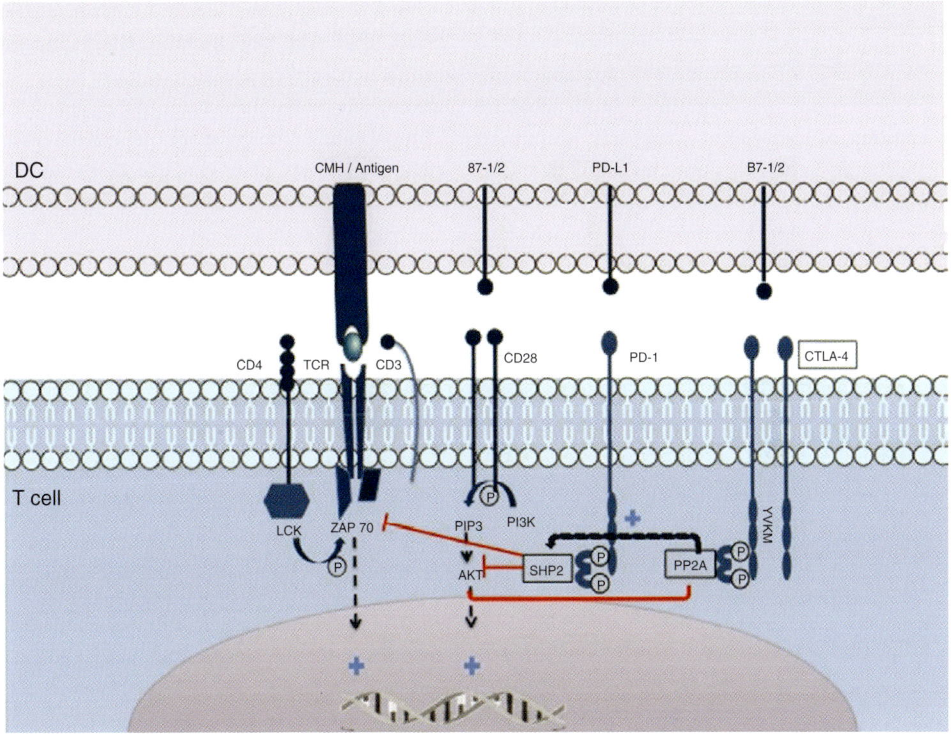

Fig. 4 Cytotoxic T-lymphocyte-associated antigen 4 (CTLA-4) pathways. CTLA-4 recruits the phosphatases SHP2 and PP2A via the YVKM motif in its cytoplasmic domain. SHP2 recruitment results in an attenuation of TCR signaling by dephosphorylating the CD3ζ chain. PP2A recruitment results in downstream dephosphorylation of AKT, further dampening the T-cell activation pathway. Adapted from Refs. [7, 9, 11].

colony-stimulating factor (GM-CSF)-transduced cellular vaccine [20]. These findings suggested that, if there is an endogenous antitumor immune response in the animals after tumor implantation, then CTLA-4 blockade could enhance that endogenous response which, ultimately, could induce tumor regression. In the case of poorly immunogenic tumors, which do not induce substantial endogenous immune responses, the combination of a vaccine and a CTLA-4 antibody could induce a strong enough immune response to slow tumor growth and, in some cases, eliminate established tumors.

On the basis of several preclinical murine models showing improved tumor control after CTLA-4 blockade [19], human monoclonal antibodies that block CTLA-4 have been developed and are currently being tested. Two such CTLA-4-targeting antibodies – ipilimumab and tremelimumab – have been tested in two Phase III clinical studies [5].

Response rates were between 11% with ipilimumab alone in previously treated advanced patients with melanoma, and 15% with ipilimumab plus dacarbazine in treatment-naïve patients [21, 22]. Nevertheless, an improved overall survival with ipilimumab was clearly demonstrated in both of these trials, with a decrease in the probability of death by 34% (compared to vaccine) and 28% (compared to dacarbazine alone). When ipilimumab was combined with dacarbazine, the median duration of best overall response was 19.3 months compared to 8.1 months with dacarbazine monotherapy. The consistent separation of the survival curves by approximately 10% until almost four years of follow-up suggests that the impact on overall survival with ipilimumab is mainly driven by the longlasting clinical benefit achieved in a small proportion of patients. Notably, tremelimumab was also shown to induce durable objective responses in patients with metastatic melanoma; in a Phase III trial, the median response duration was almost threefold longer than the response duration with chemotherapy (35.8 versus 13.7 months; P = 0.0011) [23, 24]. However, it should be stressed that tremelimumab is not yet approved for clinical application.

A long-term follow-up was recently reported on 177 patients with advanced melanoma who participated in three of the earliest ipilimumab trials conducted at the Surgery Branch of the National Cancer Institute [25–29]. In these earlier studies, the median duration of objective responses was up to 88 months (Table 1), with complete responses not being achieved on average until after 30 months of treatment; in fact, one patient received treatment for six years until a complete response was achieved. The durability of complete responses in the long-term analysis is impressive: 14 out of 15 complete responses were ongoing at data cutoff (all six complete responses in patients who were treated with concurrent ipilimumab and IL-2, and eight of nine complete responses in patients who were treated with ipilimumab and gp100 vaccine). The durability of complete responses observed even in the trials that did not include IL-2 led the authors to speculate that a cure may be achievable with this drug. The long-term clinical benefit in a subset of patients with melanoma was also confirmed in earlier Phase II ipilimumab studies, documenting flat survival curves between the four- and five-year marks [30–32].

To date, more than 100 ongoing trials with anti-CTLA-4 drugs for several cancers are recorded on the site of *www.clinicaltrials.gov*.

2.3 Immunological Explanation of the Kinetic and Durable Objective Clinical Responses after Anti-CTLA-4

An important feature of the anti-CTLA-4 clinical responses is their kinetics. Although responses to chemotherapies and tyrosine kinase inhibitors (TKIs) commonly occur within weeks of the initial administration, the response to immune-checkpoint blockers is slower and, in many patients, delayed (for up to six months after treatment initiation) [5]. Recently, at the 2013 European Cancer Congress, Hodi and colleagues presented some updated long-term results from ipilimumab on 1861 melanoma patients. An analysis of these patients showed the median overall survival to be 11.4 months, and 22% of the patients were still alive after three years. There were no deaths among patients who survived beyond seven years, at which time the overall survival rate was 17%. The longest overall survival follow-up in the database was 9.9 years. The analysis showed a plateau of 22% survival at seven years (the notion of "plateau" is a new paradigm to evaluate efficacy in immunotherapy, and its ultimate goal [33]).

Tumor-infiltrating lymphocytes (TILs), including CD8 and CD4 cells, along with

Tab. 1 Responses with CTLA-4 antibodies in clinical trials.

Drug	*Disease*	*Trial (reference)*	*No. of patients*	*Median response duration (months)*	*RR(%)*	*CR(%)*	*PR(%)*	*SD(%)*	*Median OS (months)*
Ipilimumab	Melanoma	Phase I							
		Ipi + gp100 [29]	56	88	13	7	6	NR	14
		Ipi + IL-2 [29]	36	79	25	17	8	NR	16
		Ipi DE ± gp100 [29]	85	42	20	6	14	NR	13
		Phase II							
		Ipi 10 mg kg^{-1} [30]	155	NR	5.8	0	5.8	21	10.2
		Ipi 10 mg kg^{-1} [31]	71	NR	11.1	2.7	8.3	18.1	11.4
		Ipi 3 mg kg^{-1} [31]	71	NR	4.2	0	4.2	22	8.7
		Ipi 10 mg kg^{-1} + budenoside [34]	58	NR	12	2	10	19	17.7
		Ipi 10 mg kg^{-1} + placebo [34]	57	NR	10	0	16	19	19.3
		Phase III							
		Ipi + dacarbazine [22]	250	19.3	15.2	1.6	13.6	18	11.2
		Ipi + gp100 [21]	403	11.5	5.7	0.2	5.5	14.4	10
		Ipi + placebo [21]	137	NR	11	1.5	9.5	17.5	10.1
	Prostate	Phase I [35]							
		Ipi	24	NR	NR	NR	NR	12	NR
		Phase I/II [36]							
		Ipi 10 mg kg^{-1} ± radiotherapy	50	NR	18	2	16	57	11.2
		Phase III [37]							
		Ipi 10 mg kg^{-1} versus placebo	799	NR	NR	NR	NR	NR	NR
	Pancreas	Phase I/II [38]							
		Ipi	27	NR	0	0	0	1	NR
Tremelimumab	**Melanoma**	Phase II [23]							
		Treme 10–15 mg kg^{-1}	84	NR	9.5	2.4	7.1	31	10
		Phase III							
		Treme 15 mg kg^{-1} [24]	328	35.8	11	3	8	NR	12.6
	Prostate	Phase I [39]							
		Treme 15 mg kg^{-1}	11	NR	NR	NR	NR	NR	NR
	Pancreas	Phase I [40]							
		Treme 15 mg kg^{-1}	34	NR	2	0	2	NR	NR

Ipi, ipilimumab; Treme, tremelimumab; DE, deticene; RR, response rate; CR, complete response; PR, partial response; SD, stable disease; NR, not reported; OS, overall survival.

tumor necrosis have been clearly identified in biopsies of melanoma metastases obtained after treatment with ipilimumab [41, 42]. In melanoma patients treated with ipilimumab, the NY-ESO-1 specific antibody and CD8 T responses correlated with clinical benefit [43]. However the implication of these endogenous tumor-reactive T cells remains the subject of investigation.

Immunologically, such delay likely reflects the interval during which the pre-existing, tumor-specific T cells are being activated after uncoupling from the CTLA-4-mediated inhibitory signal, and expanding and infiltrating into the tumor. The observation of an ongoing or even improving tumor response many months or years after the last dose of ipilimumab suggests that the anti-CTLA-4 blockade allows for sufficient activation and expansion of tumor-reactive T cells to control the tumor [5]. Given the multitude of immune escape mechanisms that may impede tumor responses to checkpoint blockade, it is quite remarkable that CTLA-4 blockade as monotherapy induces partial to complete tumor responses that can persist for many years. This result suggests that CTLA-4 is likely one of the predominant inhibitory drivers of pre-existing, endogenous tumor-reactive T cells in these responding patients. The observation that a subset of patients showed delays before achieving a best overall response underlies the characteristic response kinetics associated with ipilimumab. These findings require the re-evaluation of response criteria for immunotherapeutics away from the conventional time-to-progression or Response Evaluation Criteria in Solid Tumors (RECIST) objective response criteria, which were developed on the basis of experiences with chemotherapeutic agents and as the primary measure of drug efficacy [4, 44].

2.4 Anti-CTLA-4 Antibodies Efficacy and Fc-Dependent Depletion of Tumor-Infiltrating Regulatory T Cells

Murine models have demonstrated that deficient CTLA-4 T-reg have an impaired suppressive ability. Moreover, recent preclinical reports have shown that anti-CTLA-4 antibodies have a dual mechanism activity that involves blockade of the inhibitory activity of CTLA-4 on the T-cell effector and also on T-reg. The activity of the antibody on T-reg is mediated via a depletion of T-reg within tumor lesions, and this depletion is dependent on the presence of Fcγ receptor-expressing macrophages within the TME [45–47]. Although, ipilimumab is a human IgG1 antibody, which binds best to most human Fc receptors [48], T-reg depletion has not been yet demonstrated in clinical trials. However, few data are currently available on the effect of ipilimumab on intratumoral T-reg.

3 Targeting the PD-1 PD-L1 Pathway

The PD-1 (CD279) pathway is also a key immune checkpoint and is rapidly attracting a growing interest as a therapeutic target in cancer. The expression of PD-1 on the T-cell surface is a negative regulator of T-cell activity [49].

3.1 Receptor Ligand Biology

3.1.1 PD-1 Expression and Functions on Immune Cells

In contrast to CTLA-4, the major action of PD-1 is to limit autoimmunity and T-cell effector activity in the peripheral tissues

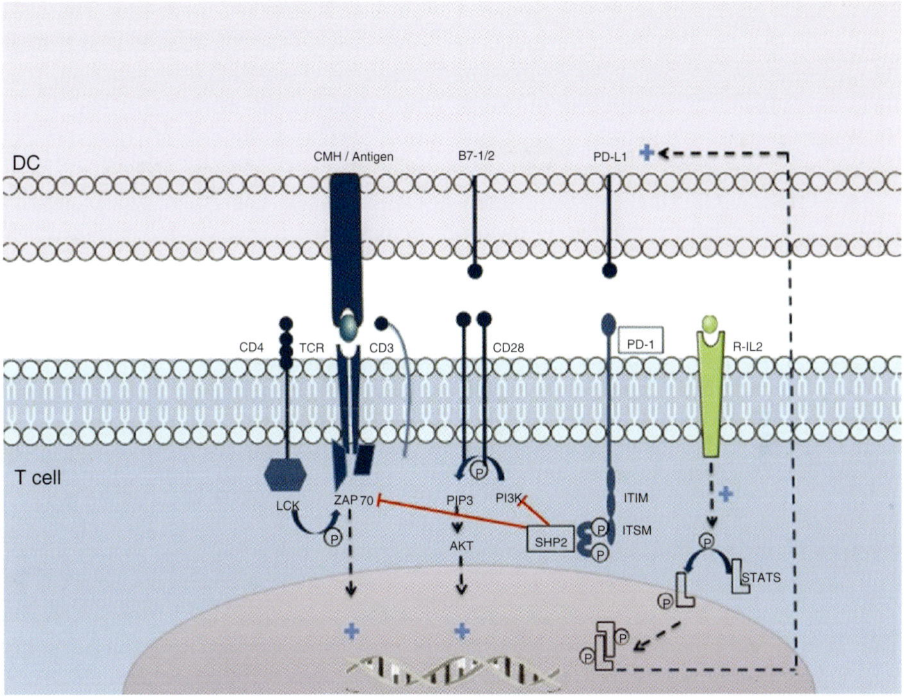

Fig. 5 Program death 1 (PD1) pathways. PD-1 contains in the cytoplasmic domain both an immune-receptor tyrosine-based inhibitory motif (ITIM) and an immune-receptor tyrosine-based switch motif (ITSM). The negative signal of PD-1 is based on the recruitment of SH2-domain containing tyrosine phosphatase (SHP2) through ITSM. SHP2 inhibits phosphatidylinositide 3-kinase (PI3K) phosphorylation, a ZAP 70 pathway that results in T-cell exhaustion. IL-2 signaling via the interleukin-2 receptor (R-IL2) induces signal transducer and activator of transcription-5 phosphorylation (pSTAT-5). Dimerization and nuclear translocation of phosphorylated STAT-5 leads to PD-1 upregulation.

at the time of inflammatory response [4, 50–55] (Fig. 2). The basis of this physiology is that the ligands for PD-1 are upregulated in response to inflammation.

PD-1 is a surface molecule of the immunoglobulin superfamily, and is expressed on activated T cells [51, 52, 56]. The interaction between PD-1 and its ligands, leads to the deactivation of signaling events under CD3-TCR and B28, resulting in a reduced T-cell activity and a decreased duration of T-cell–APC or T-cell–target contact [4, 56–58]. In the cytoplasmic domain, PD-1 contains both an immune-receptor tyrosine-based inhibitory motif (ITIM) and an immunoreceptor tyrosine-based switch motif (ITSM). The negative signal of PD-1 is based on the recruitment of SHP2 through ITSM [56, 59, 60]. T-cell exhaustion results from an inhibition of SHP2 phosphatidylinositide 3-kinase (PI3K)-mediated phosphorylation (Fig. 5) [61].

PD-1 expression is induced by the activation of T cells, and so by the interaction between MHC–antigen and TCR–CD3 [62–64]. The activation of T cells by first signal leads to a nuclear translocation of

NFAT, while NFAT2 binds to the promoter region of the *PD-1* gene, thus inducing its transcription [64, 65]. PD-1 expression is downregulated when there is no more TCR signaling. More precisely, T-bet binds upstream of the *PD-1* gene and represses its transcription [65]. During chronic infection, lymphocytes are persistently exposed to their antigens, and the continuous TCR ligation leads to a downregulation of T-bet, thus maintaining PD-1 transcription [65, 66]. In viral models, the chronicity of viral infection induces a high level of PD-1, which leads to a state of exhaustion, or anergy-specific activated T cells [4, 64].

Like CTLA-4, PD-1 is also highly expressed on T-reg cells, where it may enhance their proliferation in the presence of a ligand [67]. As many tumors are highly infiltrated with T-reg cells, which probably further suppresses effector immune responses, the blockade PD-1 pathway could also enhance antitumor response by reducing both the number and the suppressive activity of intra-tumoral T-reg cells [67].

PD-1 is also more widely expressed than CTLA-4; notably, it is expressed on other activated immune cells such as B cells, monocytes and natural killer (NK) cells which limits their lytic activity [64, 68, 69]. Hence, although PD-1 blockade is typically viewed as enhancing the activity of effector T cells in tissues and in the TME, it also probably enhances NK cell activity in tumors and tissues, and may also enhance antibody production either indirectly or through direct effects on PD-1+ B cells [70].

3.1.2 PD-1 Ligands

PD-1 interacts with its ligands, PD-L1 (B7-H1 or CD274) and PD-L2 (B7-DC or CD273), both of which are members of the B7 family [4, 71]. The interaction of PD-1/PD-L1 occurs predominantly in peripheral tissues, including the TME, and leads to apoptosis and a downregulation of T-cell effector function. Thus, the engagement of PD-1/PD-L1 interaction decreases the risk of collateral tissue damage by T cells [53, 72–74]. PD-L1 is expressed on many cell types, including hematopoietic, endothelial and epithelial, in response to proinflammatory cytokines, notably interferon (IFN) gamma, while B7-DC/PD-L2 is restricted largely to APCs on dendritic cells and macrophages in response to different proinflammatory cytokines [75, 76]. Recently, a molecular interaction between PD-L1 and CD80 was identified [71], whereby CD80 expressed on T cells (and possibly on APCs) could potentially behave as a receptor rather than as a ligand by delivering inhibitory signals when engaged by PD-L1 [77]. The relevance of this interaction in tumor immune resistance has not yet been determined.

3.1.3 PD-1/PDL-1 Regulation in Cancers: Innate versus Adaptive Immune Resistance

Much like chronic viral infections, cancers represent an example of continuous antigen exposure, and consequently it might be reasoned that tumor-specific lymphocytes would express multiple immune checkpoints.

PD-1 is expressed on a large proportion of TILs from many different tumor types. Some of the enhanced PD-1 expression among CD4+ TILs reflects a generally high level of PD-1 expression on T-reg cells which, as noted above, can represent a large proportion of intratumoral CD4+ T cells. An increased PD-1 expression on CD8+ TILs may either reflect an anergic or an exhausted state, as has been suggested by the decreased cytokine production by PD-1+ compared to PD-1− TILs from melanomas [78]. Thus, the expression of PD-1 on TILs is one of the mechanisms of tumor use to evade an active immune response.

Accordingly, the PD-1 ligands are upregulated on the tumor cell surfaces of many different human cancers [78]. PD-L1 is the major PD-1 ligand to be expressed on cells from solid tumors such as melanoma, lung, hepatocellular carcinoma, glioblastoma, kidney, breast, ovarian, pancreatic, and esophageal cancers. In addition to tumor cells, PD-L1 is commonly expressed on myeloid cells within the TME [78].

Although an initial report in renal cancer demonstrated that the expression of PD-L1 on tumor cells has been associated with a poor prognosis, other studies performed on a variety of cancers have suggested that PD-L1 status can either correlate with a worse prognosis, a better prognosis, or show no correlation with clinical outcome [4, 78]. The most probable factors contributing to the wide range of reported outcomes among different patient cohorts are variations in immunohistochemistry technique, cancer type, stage of cancer analyzed, and the treatment history [4, 79–81].

PD-L2 has also been reported to be upregulated in various tumors. PD-L2 is highly upregulated on certain B-cell lymphomas, such as primary mediastinal B-cell lymphoma, follicular cell B-cell lymphoma and Hodgkin's disease [82]. The upregulation of PD-L2 on these lymphomas is commonly associated with gene amplification or rearrangement with the class II MHC transactivator (CIITA) locus, which is highly transcriptionally active in B-cell lymphomas [83].

Two general mechanisms for the regulation of PD-L1 by tumor cells have emerged, namely innate immune resistance and adaptive immune resistance. For some tumors (e.g., glioblastomas) it has been shown that PD-L1 expression is driven by constitutive oncogenic signaling pathways in the tumor cell (this is termed constitutive or innate immune resistance). PD-L1 expression on glioblastomas is enhanced by the deletion or silencing of phosphatase and tensin homolog (PTEN), which involves the PI3K–AKT pathway [84]. Similarly, constitutive anaplastic lymphoma kinase (ALK) signaling, which has been observed in certain lymphomas and occasionally in lung cancer, has been reported to drive PD-L1 expression through signal transducer and activator of transcription 3 (STAT3) signaling [85].

A second alternative mechanism for PD-L1 regulation on tumor cells reflects their adaptation to endogenous tumor-specific immune responses, a process termed adaptive immune resistance [86]. In this model, tumor cells use the natural physiology of PD-1 ligand induction that normally occurs to protect a tissue against immune-mediated damage, in order to protect itself from an antitumor immune response. The expression of PD-L1 represents an adaptive response to endogenous antitumor immunity, and can occur because PD-L1 is induced on most tumor cells in response to IFNs, notably IFN-γ, which also occurs in epithelial and stromal cells in normal tissues [87–89]. Various preclinical and clinical studies support the adaptive immune resistance hypothesis. A strong correlation between cell-surface PD-L1 expression on tumor cells and both lymphocytic infiltration and intratumoral IFN-γ expression has been reported in melanoma. This correlation was seen not only among tumors but within individual PD-L1+ tumors at the regional level, in which regions of lymphocyte infiltration were also located exactly where PD-L1 was expressed on both tumor cells and TILs [86, 90]. These findings suggest the occurrence of a negative feedback loop whereby IFN-γ induces PD-L1 expression, which in turn suppresses the activity of PD-1+ T cells. The mechanism of adaptive immune resistance intrinsically implies that immune surveillance exists even in advanced cancers. However, the tumor ultimately resists immune elimination by upregulating the expression of ligands for inhibitory receptors on tumor-specific lymphocytes, which consequently inhibit antitumor immune responses in the TME [4].

A significant inverse correlation was recently found between PD-L1 expression and intraepithelial CD8 T lymphocyte count, indicating that PD-L1 on tumor cells inhibit the host-tumor immunity and facilitate tumor invasion [91–93]. These data suggest that PD-L1 expression might be a reliable prognostic parameter.

3.2 Clinical Implication PD-1/PD-L1 Blockade

This pathway appears to play an important function in modulating T-cell activity in peripheral tissues. The importance of this pathway in T-cell activity was confirmed in experimental models, which showed that mice deprived of the *PD-1* gene present an enhanced immunity [54]. In support of these findings, knockout models showed that this blockade would result in less immune toxicity than can be explained by the peripheral expression of PD-1 [4, 94].

Data acquired from mice and a human model of chronic viral infections suggested that this exhaustion state is partially reversible by PD-1 pathway blockade [4, 49]. The administration of anti-PD-L1 to mice chronically infected by lymphocytic choriomeningitis virus (LCMV) caused an increased T-CD8-specific response and induced viral control [49, 64]. Recently, humanized murine models with HIV demonstrated excellent results with the PD-L1 antibody, whereby the antibody suppressed HIV viremia and restored CD4 counts [95]. The PD-1 pathway has also been implicated in fungal and parasitic infections.

Many murine tumors models validate the efficacy of blockade PD-1 pathway (PD-1 or its ligands) by enhancing antitumor immunity. Currently, several PD-1 and PD-L1 inhibitors are at early or late stages of clinical development for the treatment of a wide variety of tumor models.

3.2.1 PD-1 Inhibitors

Nivolumab (BMS-936558/MDK-1106) is a fully human IgG4 monoclonal antibody targeting PD-1. In an initial Phase I clinical trial, the safety of nivolumab was evaluated in 39 patients treated for solid cancer [79, 96], while a larger Phase I study (296 patients) demonstrated not only the drug's safety but also data of its efficacy, with a good objective response rate. Some 65% of the patients also had a durable response [97, 98].

The preliminary results of a Phase II trial were presented recently at the 2014 American Society of Clinical Oncology (ASCO) annual meeting, whereby nivolumab as monotherapy in patients with advanced renal cancer achieved a 20% response rate and a median overall survival of more than two years [99]. Moreover, a recent Phase III study evaluating nivolumab versus dacarbazine in patients with previously untreated BRAF wild-type advanced melanoma was stopped early because of a superior overall survival in patients receiving nivolumab compared to the control arm. At one year, the overall survival rate was 72.9% in the nivolumab group, compared to 42.1% in the dacarbazine group (hazard ratio for death, 0.42; 99.79% confidence interval 0.25–0.73; $p < 0.001$) [100]. Nivolumab

Tab. 2 Clinical trials on going with anti-PD-1 and anti-PD-L1 antibodies.

	Clinical trial number[a]	*Comparative arm*
Anti-PD-1		
Nivolumab		
Melanoma	NCT01721772	DTIC
Melanoma BRAF V600	NCT02224781	Sequential ipilimumab/dabrafenib and trametinib
NSCLC	NCT01673867	Docetaxel
	NCT02066636	No
	NCT02041533	Chemotherapy investigator choice
RCC	NCT02231749	Combination with ipilimumab versus sunitinib
Head and neck cancer	NCT02105636	Treatment investigator choice
Pembrolizumab		
NSCLC	NCT02142738	Platinum-based chemotherapy
	NCT02220894	Platinum-based chemotherapy, PDL1 + tumors
Head and neck cancer	NCT02252042	Chemotherapy investigator choice
Urothelial cancer	NCT02256436	Taxane regimen, vinflumine
Anti PD-L1		
MPDL3280 A		
NSCLC	NCT02008227	Docetaxel
Urothelial	NCT02302807	Chemotherapy investigator choice
MEDI4736		
NSCLC	NCT02273375	Placebo

[a]Trials actually recruiting Phase III only (ClinicalTrials.gov.); NSCLC, non-small-cell lung cancer; RCC, renal cell carcinoma.

also demonstrated a substantial efficacy in a Phase I study of patients who had been heavily treated for relapsed Hodgkin's lymphoma, with an objective response rate of 87% being reported, including 17% of patients with complete response [101].

Pembrolizumab (MK-3475) is a highly selective, humanized IgG4 kappa monoclonal antibody targeting PD-1. This novel monoclonal antibody is designed to block PD-1 interaction with its ligands PD-L1 and PD-L2, and shows no cytotoxic [antibody-dependent cell-mediated cytotoxicity (ADCC)/complement-dependent cytotoxicity (CDC)] activity. Pembrolizumab is currently being investigated in a large Phase 1B trial for patients with several solid tumors, such as advanced non-small-cell lung cancer (NSCLC), melanoma, gastric cancer, urothelial cancer, and head and neck carcinoma. The preliminary results have shown promising clinical activity and tolerability with this novel monoclonal antibody. Pembrolizumab was approved by the US Food and Drug Administration in September 2014 for the treatment of patients with unresectable or metastatic melanoma and disease progression following ipilimumab treatment.

Pizilimumab (CT-011) is also a humanized IgG1 monoclonal antibody targeting PD-1, for which a Phase I trial in 17 patients showed clinical benefit in 33% of cases [102].

Currently, several clinical trials are ongoing with these antibodies (only Phase III trials are listed in Table 2).

3.2.2 PD-L1 Inhibitors

Clinical activity has also been observed with anti-PD-L1 inhibitor antibodies, which also inhibit the interaction between PD-L1 and CD80. To date, however, the therapeutic impact of these interactions remains unclear.

BMS-936559 (MDX-1105) is a fully human IgG4 monoclonal antibody, and was the first PD-L1 antibody to demonstrate objective tumor responses. The first results obtained from a Phase I trial of BMS-936559 in 207 patients with advanced solid cancers demonstrated a durable response, with an objective response rate of between 6% and 17%, but stable disease was also prolonged (12–41% at 24 weeks) [103]. Currently, several trials (Phases I–III) are ongoing with additional anti-PD-L1 agents (Table 2).

MPDL3280A is an engineered human monoclonal antibody for which a correlation was demonstrated between response rate and PD-L1-expressing tumors in a Phase I clinical trial; the overall response rate was 39%, and 12% of the patients had a progressive disease, whereas patients with PD-L1-negative tumors had an overall response rate of 13% and 59% had a progressive disease [104, 105]. Encouraging results were also reported at the 2014 ASCO annual meeting for this antibody in a Phase I trial in patients with urinary bladder cancer. In this case, Powles and colleagues described preliminary but promising data in patients with urinary bladder metastatic cancer, with 43% of those who had been treated with the antibody obtaining a positive response in terms of tumor size [106].

MSB0010718C is a fully human IgG1 monoclonal antibody for which Heery *et al.* have presented Phase I trial data in advanced solid malignancies. Notably, it was shown that MSB0010718C can be safely administered intravenously at doses of up to 20 mg kg^{-1} every two weeks until progression [107].

4 Others Checkpoint Inhibitors

Lymphocyte activation gene 3 (LAG3), T-cell immunoglobulin and mucin domain 3 (TIM-3), B- and T-lymphocyte attenuator (BTLA), B7-H3 (CD276), and the family of killer inhibitory receptors (KIRs) have also been associated with the inhibition of lymphocyte activation and an induced lymphocyte anergy.

4.1 TIM-3

(TIM-3) was discovered in 2002 and first investigated in the setting of autoimmunity. In a study conducted by Monney *et al.*, the anti-TIM-3 antibody was shown to exacerbate experimental autoimmune encephalomyelitis [108].

Further studies are needed to confirm that TIM-3 is another inhibitory receptor that specifically causes activated T cell (Th1 and CD8) exhaustion [109–111]. For example, the upregulation of Tim-3 and PD-1 expression is associated with a tumor antigen-specific CD8+ T-cell dysfunction in melanoma patients [112]. Remarkably, the coblockade of Tim-3 and PD-1 has been shown to restore tumor-specific T-cell functions and to promote an effective antitumor immunity in a wide variety of cancers [109, 111, 112]. The C-type lectin galectin-9 (Gal9) is the best known ligand of TIM-3. An interaction between TIM-3 and Gal 9 induces phosphorylation of the Tim-3 cytoplasmic tail at Tyr256 and Tyr263, releasing Bat3 from the TIM-3 cytoplasmic tail and thus causing an accumulation of inactive LcK [113].

Furthermore, the expression of TIM-3 on T cells has been shown to promote myeloid-derived suppressor cells (MDSCs) via Gal9 interaction [114]. Recent studies

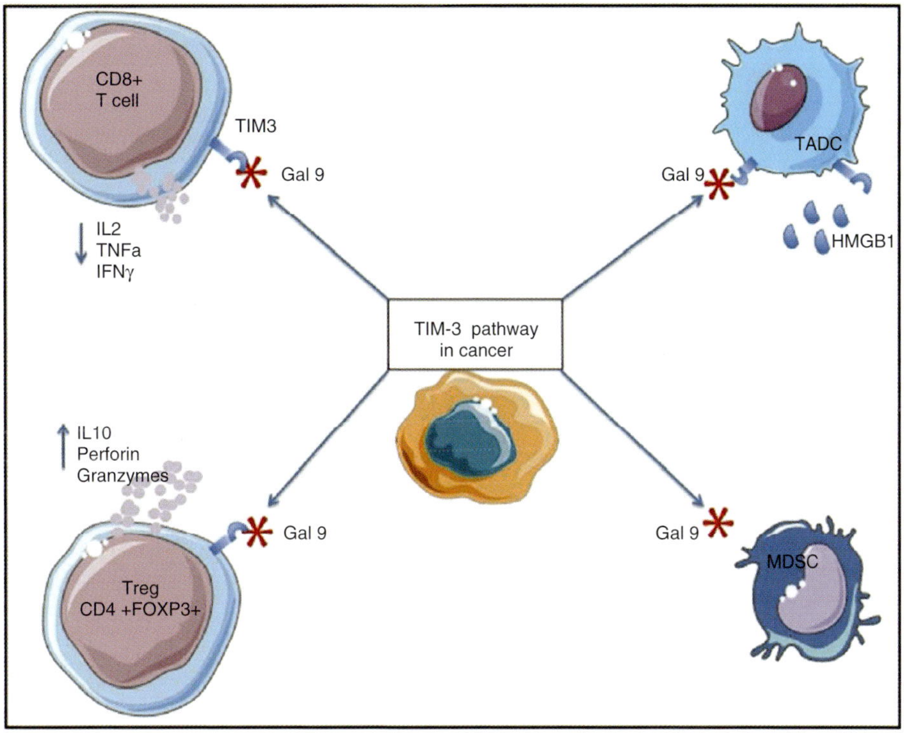

Fig. 6 TIM-3 pathways and immune cells implication. TIM-3 is expressed on CD8 T cell exhausted. TIM-3+ T regulator cells are potent suppressors of immune responses in tumor tissue, and may promote the development of dysfunctional phenotype in intratumoral CD8 T cells. Tumor-associated dendritic cells (TADCs) express TIM-3 and lead to a decrease in free HMGB1 available to bind the nucleic acids released by dying tumor cells. TIM-3 expression also promotes the recruitment of myeloid-derived suppressor cells (MDSCs). IFNγ, interferon-gamma; IL-2, interleukin-2; IL-10, interleukin-10; Gal9, Galectin 9. Adapted from Ref. [113]

have also shown that Tim-3+ T-reg is more suppressive than TIM-3-T-reg in the TME [110]. The results of other studies have also suggested that TIM-3 is upregulated on tumor-associated dendritic cells (TADCs) [115]. Collectively, Tim-3 can function as a determinant of antitumor immunity and thus may emerge as an attractive target in the immunotherapy landscape [1, 116] (Fig. 6).

A better understanding of the complexity of the TIM-3 pathway supports the use of TIM-3-targeted therapies alone, and also in combination with other checkpoint immune inhibitors. However, no clinical data are yet available with anti TIM-3 antibodies.

4.2 LAG3, BTLA

Although the LAG3 (CD223) was previously discovered as a CD4 homolog [117], its role in the immune checkpoint has only recently been defined [118]. LAG3 was shown to enhance the suppressive function of T-reg cells and to impact on CD8+ T-cell function [119, 120]. Like TIM-3,

LAG3 expression is also found on anergic or exhausted T cells, such as those infiltrating the TME [119, 121]. The ligands of LAG3 are MHC class II molecules that are expressed on tumor-infiltrating APCs and which are upregulated on some epithelial cancers (mainly in response to IFN-γ, as does PDL-1). LAG3 binds to MHC class II molecules with a higher avidity than CD4, and this is known to transduce inhibitory signals [122]. Recent preclinical models using antibody blockade LAG3 for cancer treatment have shown an enhanced activation of specific T cells at the tumor site [119, 123], suggesting that LAG3 blockade might be a potential cancer treatment.

The inhibitory function of BTLA on T cells was identified with a BTLA-knockout mouse model [124]. BTLA is constitutively expressed by naïve CD4+ and CD8+ T cells, and is upregulated when T cells are activated [124–126]. Herpes virus entry mediator (HVEM), a member of the tumor necrosis factor receptor (TNFR) superfamily, is the only known ligand of BTLA [127, 128]. BTLA–HVEM interactions lead to $BTLA^{hi}$ T-cell inhibition, and may be involved in tumor progression and resistance to the specific immune response [112, 129].

5 Safety Aspects

The toxicity of checkpoint inhibitors is a limiting problem for the development of this new cancer immunotherapy (Table 3). The major toxicities are immune-related adverse events (IRAEs), which are related to the infiltration of activated CD4 and CD8 T cells, and the increase in inflammatory cytokine production in normal tissues [57, 130]. Both, preclinical models and clinical trials have provided evidence of the role of CTLA-4 blockade in breaking the tolerance to human cancer antigens and also self-antigens [19, 20, 57]. Various studies have been conducted to evaluate the infiltration of depigmentation lesions, revealing the presence of high levels of polymorphonuclear cells and the deposition of antibodies. Major immune toxicities can be explained by the central role of CTLA-4 for keeping T-cell activation in check. The most common IRAEs involve the skin, gastrointestinal tract, liver and endocrine system [105]. In clinical trials, up to 60% of patients treated with ipilimumab reported toxicity, with the toxic effects being severe (grade 3 or 4) in 10–15% of cases [44]. Clinical trials in Phases II and III have shown that the early administration of corticosteroids is essential to manage IRAEs, and to prevent their progression to a more serious toxicity [30, 32, 44, 105].

The most frequent adverse events with agents targeting PD-1 pathway were asthenia, anorexia, diarrhea, nausea, dyspnea, rash, and headache [32, 57, 97]. In clinical trials, 14% of patients treated with nivolumab had serious adverse events (grade 3 or 4).

A better understanding of the mechanism of action of these drugs alone, or in combination, might be useful to improve the toxicity profile. Last, but not least, recent data have suggested a possible correlation between toxicity and clinical efficacy [132, 133].

6 Future Direction: Biomarkers and Combination

6.1 Biomarkers

Due to the cost of immunotherapy, the chance of adverse events, and the

Tab. 3 Check-point inhibitors and safety.

SAE (Grade 3–4)	*CTLA-4 blockade [22, 98, 130, 131]*		*PD-1 blockade [96, 97, 102]*	*PDL-1 blockade [104, 106]*	
	Ipilimumab (%)	*Tremelimumab (%)*	*Nivolumab, Pembrolizumab (%)*	*BMS-936559 (%)*	*MPDL 3280A (%)*
Dermatologic					
Rash	3–4	2.5–18	1–4	<1	<1
Gastrointestinal					
Colitis	2–21	2.1–18	2	0	<1
Diarrhea	4–5.3	5–21	1–3	<1	<1
Nausea	<5	8–12.8	0	0	<1
Endocrine					
Hypothyroidism	0	1	0	<1	<1
Hypophysitis	<1	2	1	0	0
Hypopituitarism	<1	1	0	0	0
Adrenal insufficiency	1.5	1	0	0	0
Hepatic					
Elevation in ALAT	1.5–22	NA	1–7	2	0
Elevation in ASAT	0.8–19	NA	1–6	1-6	0
Hepatitis	<3	1	NA	3	2
Asthenia	6–10	2–13	2	2	2
Pneumonitis	NA	1	1–3	1–3	NA

SAE, serious adverse events; ALAT, alanine aminotransferase; ASAT, aspartate aminotransferase; NA, non-available.

heterogeneity of individual tumors it is important for clinicians and research groups to be able to predict a patient's likelihood of response and adverse events. One method of achieving this is to utilize biomarkers to predict the likelihood of a response so that patients can be stratified prior to therapy. To date, however, no such pretreatment biomarkers have been validated to a point at which they could be applied as part of standard-of-care therapeutic decision-making.

In the case of anti-CTLA-4 therapy, biomarkers such as absolute lymphocyte count, lactate-dehydrogenase (LDH) and C-reactive protein (CRP) are the most frequently identified markers correlating with improved long-term outcome upon treatment with ipilimumab [5, 134, 135]. Interestingly, most of these correlations have not (or to a lesser extent) been found to be predictive in the clinical trials, emphasizing the need for a meta-analysis of all currently available patient data. As these markers are also known to be prognostic markers, their predictive value remains unclear so far. A recent report illustrated the importance of tumor genetics in defining the basis of the clinical benefit from CTLA-4 blockade. The sequences of the whole genome have revealed mutations that lead to neoantigens, which may be immunologically relevant in responses to immune checkpoint blockade. A high mutational load is associated with a benefit from CTLA-4 blockade, but is not sufficient to impart a clinical benefit. These findings provide a rationale to consider examining the exomes of patients treated by CTLA-4 inhibitors [136].

Reports on the anti-PD-1 nivolumab have shown that patients with PD-L1-positive tumors had a greater chance to respond to the therapy than did those with PD-L1-negative tumors [97, 105, 137]. These findings suggest that PD-L1, when detected in biopsy samples by immunohistochemistry, might be capable of predicting the activity of nivolumab in advanced cancer. In addition, in a recent study it was noted that PD-L1 tumor expression and T-cell gene signature correlated with responses to the anti-PD-L1 therapy MPDL3280A [106, 138]. Recent data have also reported objective responses to nivolumab in patients with PD-L1-negative tumors [139], though this may have been due to a heterogeneous tumor expression of PD-1, and a single negative biopsy would be insufficient to determine if a tumor were truly PD-L1-negative. PD-L1 expression may also undergo changes driven by alterations in the TME, for example, infiltrations with immune cells, potentially limiting its use as a clear-cut predictive biomarker as opposed to, for example, BRAFV600 mutational status in melanoma patients [137]. Clearly, further randomized clinical trials using validated immunohistochemical assessments of PD-L1 expression are needed.

As many current immune-based therapies, including immune checkpoint inhibitors, depend on augmenting pre-existing antitumor T-cell immunity in cancer patients, there is a strong rationale to evaluate parameters associated with tumor-specific T-cell responses as potential biomarkers. NY-ESO-1 antibody and CD8 T-cell responses to the cancer/testis antigen NY-ESO-1 before treatment were associated with clinical benefit after ipilimumab treatment [43, 140]. Recently, Meyer *et al.* reported that the frequencies of circulating MDSCs correlated with the clinical outcome of melanoma patients treated with ipilimumab [141].

The use of biomarkers to profile immune cells that have infiltrated a tumor may also prove valuable to clinicians [2, 142]. These profiling tools could be used to understand the dynamic state of the immune system in the individual tumor, and to tailor the immunotherapy selection accordingly.

Large studies have shown that tumor-immune infiltrates in solid tumors such as colorectal cancer can serve as significant prognostic biomarkers, even after adjusting for the tumor–node–metastasis (TNM) classification and tumor molecular biomarkers [143, 144]. A potential clinical translation of these observations is the establishment of an Immunoscore, which is based on the numeration of two lymphocyte populations CD3 and CD8 in the core, and the invasive margin of the tumors [145]. Considering the important role of the host immune reaction in controlling tumor progression, it is now important to initiate the incorporation of an immune parameter such as the Immunoscore as a component of cancer classification and a prognostic tool. Whether the immune infiltrate of the primary tumor evaluated by a robust technology adapted to clinical practice such as the Immunoscore could predict therapeutic responses to immune checkpoint inhibitors is important in patient clinical management. The theranostic value of the Immunoscore is currently under investigation in research laboratories. Recently, Tumeh *et al.* showed that tumor regression after therapeutic PD-1 blockade requires pre-existing CD8 T cells that are negatively regulated by PD-1/PD-L1-mediated adaptive immune resistance [146].

6.2 Combination

Many preclinical models have suggested synergistic effects when checkpoint

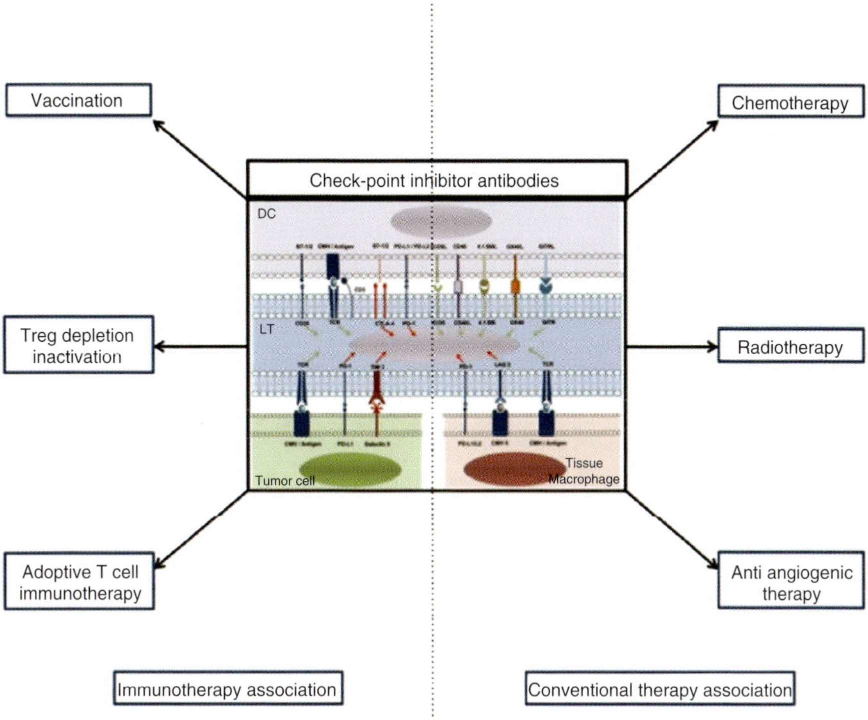

Fig. 7 Check-point inhibitor antibodies and combination currently under development. Adapted from Ref. [56].

inhibitor antibodies were used in combinations (Fig. 7), and the CTLA-4 and PD-1 antibodies association was the first such combination to be investigated. This dual blockade showed promising results of antitumor synergy [121, 147]. For example, in a Phase I clinical trial evaluating the association of ipilimumab with ascending doses of nivolumab, more than half of the patients experienced more than 80% durable tumor reductions. Notably, grade 3 or 4 adverse events related to therapy occurred in 53% of those patients in the combination regimen group [147]. This profile of toxicity suggested a probable switch in upcoming trials to using sequential rather than concomitant immunotherapies. PD-1 blockade has also shown signs of enhanced antitumor efficacy when combined with monoclonal antibody-blocking LAG-3 [123] or TIM-3 [110], which exert only modest effects as monotherapies.

Preclinical and clinical results have also demonstrated synergistic effects of combination checkpoint inhibitor antibodies and other immunotherapy, such as therapeutic vaccines, cytokines, indolamine-2,3-dioxygenase inhibitors, agonists of costimulatory receptors, or adoptive T-cell transfer [137, 145, 148–153].

Finally, synergistic effects have also been demonstrated in preclinical models with conventional anticancer therapy such chemotherapy, radiotherapy, antiangiogenic, and molecular targeted therapies [56]. On the basis of promising results with radiation association, several clinical trials are ongoing to assess the efficacy of checkpoint inhibitor antibodies with radiotherapy [148, 154]. The early clinical data obtained suggest that combined CTLA-4 blockade and vascular endothelial growth factor (VEGF) inhibition can induce durable responses in patients with melanoma [131, 155].

Furthermore, there is emerging evidence for favorable effects of BRAF/MEK pathway inhibition on the endogenous tumor immune response, as well as for potential synergies between mitogen-activated protein kinase (MAPK)-targeted therapy. In fact, immunotherapy clinical trials combining MAPK pathway inhibition with CTLA-4 and PD-L1 blockade in patients with melanoma are currently under way [156–158].

7 Conclusion

The role of checkpoint molecules in regulating immune responses has been demonstrated in preclinical models of chronic infection, self-tolerance and tumor tolerance. Trials with checkpoint inhibitor (CTLA-4, PD-1, and PD-L1) antibodies have demonstrated real evidence of clinical activity in many malignancy types, even in cancers that are not traditionally viewed as amenable to immunotherapy, such as gastric tumors. These encouraging data support the future development of checkpoint antibodies, combinatorial approaches with conventional therapies, and also other immunotherapies and predictive biomarker discovery.

References

1. Vesely, M.D., Kershaw, M.H., Schreiber, R.D., and Smyth, M.J. (2011) Natural innate and adaptive immunity to cancer. *Annu. Rev. Immunol.*, **29**, 235–271.
2. Chen, D.S. and Mellman, I. (2013) Oncology meets immunology: the cancer-immunity cycle. *Immunity*, **39**, 1–10.
3. Motz, G.T. and Coukos, G. (2013) Deciphering and reversing tumor immune suppression. *Immunity*, **39**, 61–73.
4. Pardoll, D.M. (2012) The blockade of immune checkpoints in cancer immunotherapy. *Nat. Rev. Cancer*, **12**, 252–264.
5. Ott, P.A., Hodi, F.S., and Robert, C. (2013) CTLA-4 and PD-1/PD-L1 blockade: new immunotherapeutic modalities with durable clinical benefit in melanoma patients. *Clin. Cancer Res.*, **19**, 5300–5309.
6. Lenschow, D.J., Walunas, T.L., and Bluestone, J.A. (1996) CD28/B7 system of T cell costimulation. *Annu. Rev. Immunol.*, **14**, 233–258.
7. Rudd, C.E., Taylor, A., and Schneider, H. (2009) CD28 and CTLA-4 coreceptor expression and signal transduction. *Immunol. Rev.*, **229**, 12–26.
8. Schwartz, R.H. (1992) Costimulation of T lymphocytes: the role of CD28, CTLA-4, and B7/BB1 in interleukin-2 production and immunotherapy. *Cell*, **71**, 1065–1068.
9. Azuma, M., Ito, D., Yagita, H., Okumura, K. *et al.* (1993) B70 antigen is a second ligand for CTLA-4 and CD28. *Nature*, **366**, 76–79.
10. Freeman, G.J., Gribben, J.G., Boussiotis, V.A., Ng, J.W. *et al.* (1993) Cloning of B7-2: a CTLA-4 counter-receptor that costimulates human T cell proliferation. *Science*, **262**, 909–911.
11. Hathcock, K.S., Laszlo, G., Dickler, H.B., Bradshaw, J. *et al.* (1993) Identification of an alternative CTLA-4 ligand costimulatory for T cell activation. *Science*, **262**, 905–907.
12. Schneider, H., Downey, J., Smith, A., Zinselmeyer, B.H. *et al.* (2006) Reversal of the TCR stop signal by CTLA-4. *Science*, **313**, 1972–1975.
13. Qureshi, O.S., Zheng, Y., Nakamura, K., Attridge, K. *et al.* (2011) Trans-endocytosis of CD80 and CD86: a molecular basis for the cell-extrinsic function of CTLA-4. *Science*, **332**, 600–603.
14. Kuehn, H.S., Ouyang, W., Lo, B., Deenick, E.K. *et al.* (2014) Immune dysregulation in human subjects with heterozygous

germline mutations in CTLA4. *Science*, **345**, 1623–1627.
15. Tivol, E.A., Borriello, F., Schweitzer, A.N., Lynch, W.P. *et al.* (1995) Loss of CTLA-4 leads to massive lymphoproliferation and fatal multiorgan tissue destruction, revealing a critical negative regulatory role of CTLA-4. *Immunity*, **3**, 541–547.
16. Waterhouse, P., Penninger, J.M., Timms, E., Wakeham, A. *et al.* (1995) Lymphoproliferative disorders with early lethality in mice deficient in Ctla-4. *Science*, **270**, 985–988.
17. Hill, J.A., Feuerer, M., Tash, K., Haxhinasto, S. *et al.* (2007) Foxp3 transcription-factor-dependent and -independent regulation of the regulatory T cell transcriptional signature. *Immunity*, **27**, 786–800.
18. Hori, S., Nomura, T., and Sakaguchi, S. (2003) Control of regulatory T cell development by the transcription factor Foxp3. *Science*, **299**, 1057–1061.
19. Leach, D.R., Krummel, M.F., and Allison, J.P. (1996) Enhancement of antitumor immunity by CTLA-4 blockade. *Science*, **271**, 1734–1736.
20. Van Elsas, A., Hurwitz, A.A., and Allison, J.P. (1999) Combination immunotherapy of B16 melanoma using anti-cytotoxic T lymphocyte-associated antigen 4 (CTLA-4) and granulocyte/macrophage colony-stimulating factor (GM-CSF)-producing vaccines induces rejection of subcutaneous and metastatic tumors accompanied by autoimmune depigmentation. *J. Exp. Med.*, **190**, 355–366.
21. Hodi, F.S., O'Day, S.J., McDermott, D.F., Weber, R.W. *et al.* (2010) Improved survival with ipilimumab in patients with metastatic melanoma. *N. Engl. J. Med.*, **363**, 711–723.
22. Robert, C., Thomas, L., Bondarenko, I., O'Day , S. *et al.* (2011) Ipilimumab plus dacarbazine for previously untreated metastatic melanoma. *N. Engl. J. Med.*, **364**, 2517–2526.
23. Camacho, L.H., Antonia, S., Sosman, J., Kirkwood, J.M. *et al.* (2009) Phase I/II trial of tremelimumab in patients with metastatic melanoma. *J. Clin. Oncol.*, **27**, 1075–1081.
24. Ribas, A., Kefford, R., Marshall, M.A., Punt, C.J. *et al.* (2013) Phase III randomized clinical trial comparing tremelimumab with standard-of-care chemotherapy in patients with advanced melanoma. *J. Clin. Oncol.*, **31**, 616–622.
25. Attia, P., Phan, G.Q., Maker, A.V., Robinson, M.R. *et al.* (2005) Autoimmunity correlates with tumor regression in patients with metastatic melanoma treated with anti-cytotoxic T-lymphocyte antigen-4. *J. Clin. Oncol.*, **23**, 6043–6053.
26. Maker, A.V., Phan, G.Q., Attia, P., Yang, J.C. *et al.* (2005) Tumor regression and autoimmunity in patients treated with cytotoxic T lymphocyte-associated antigen 4 blockade and interleukin 2: a phase I/II study. *Ann. Surg. Oncol.*, **12**, 1005–1016.
27. Maker, A.V., Yang, J.C., Sherry, R.M., Topalian, S.L. *et al.* (2006) Intrapatient dose escalation of anti-CTLA-4 antibody in patients with metastatic melanoma. *J. Immunother.*, **29**, 455–463.
28. Phan, G.Q., Yang, J.C., Sherry, R.M., Hwu, P. *et al.* (2003) Cancer regression and autoimmunity induced by cytotoxic T lymphocyte-associated antigen 4 blockade in patients with metastatic melanoma. *Proc. Natl Acad. Sci. USA*, **100**, 8372–8377.
29. Prieto, P.A., Yang, J.C., Sherry, R.M., Hughes, M.S. *et al.* (2012) CTLA-4 blockade with ipilimumab: long-term follow-up of 177 patients with metastatic melanoma. *Clin. Cancer Res.*, **18**, 2039–2047.
30. O'Day, S.J., Maio, M., Chiarion-Sileni, V., Gajewski, T.F. *et al.* (2010) Efficacy and safety of ipilimumab monotherapy in patients with pretreated advanced melanoma: a multicenter single-arm phase II study. *Ann. Oncol.*, **21**, 1712–1717.
31. Wolchok, J.D., Neyns, B., Linette, G., Negrier, S. *et al.* (2010) Ipilimumab monotherapy in patients with pretreated advanced melanoma: a randomised, double-blind, multicentre, phase 2, dose-ranging study. *Lancet Oncol.*, **11**, 155–164.
32. Wolchok, J.D., Weber, J.S., Maio, M., Neyns, B. *et al.* (2013) Four-year survival rates for patients with metastatic melanoma who received ipilimumab in phase II clinical trials. *Ann. Oncol.*, **24**, 2174–2180.
33. McDermott, D., Lebbé, C., Hodi, F.S., Maio, M. *et al.* (2014) Durable benefit and the potential for long-term survival with immunotherapy in advanced melanoma. *Cancer Treat. Rev.*, **40**, 1056–1064.
34. Weber, J., Thompson, J.A., Hamid, O., Minor, D. *et al.* (2009) A randomized, double-blind, placebo-controlled, phase II study comparing the tolerability and efficacy

of ipilimumab administered with or without prophylactic budesonide in patients with unresectable stage III or IV melanoma. *Clin. Cancer Res.*, **15**, 5591–5598.
35. Madan, R.A., Mohebtash, M., Arlen, P.M., Vergati, M. *et al.* (2012) Ipilimumab and a poxviral vaccine targeting prostate-specific antigen in metastatic castration-resistant prostate cancer: a phase 1 dose-escalation trial. *Lancet Oncol.*, **13**, 501–508.
36. Slovin, S.F., Higano, C.S., Hamid, O., Tejwani, S. *et al.* (2013) Ipilimumab alone or in combination with radiotherapy in metastatic castration-resistant prostate cancer: results from an open-label, multicenter phase I/II study. *Ann. Oncol.*, **24**, 1813–1821.
37. Kwon, E.D., Drake, C.G., Scher, H.I., Fizazi, K. *et al.* (2014) Ipilimumab versus placebo after radiotherapy in patients with metastatic castration-resistant prostate cancer that had progressed after docetaxel chemotherapy (CA184-043): a multicentre, randomised, double-blind, phase 3 trial. *Lancet Oncol.*, **15**, 700–712.
38. Royal, R.E., Levy, C., Turner, K., Mathur, A. *et al.* (2010) Phase 2 trial of single agent Ipilimumab (anti-CTLA-4) for locally advanced or metastatic pancreatic adenocarcinoma. *J. Immunother.*, **33**, 828–833.
39. McNeel, D.G., Smith, H.A., Eickhoff, J.C., Lang, J.M. *et al.* (2012) Phase I trial of tremelimumab in combination with short-term androgen deprivation in patients with PSA-recurrent prostate cancer. *Cancer Immunol. Immunother.*, **61**, 1137–1147.
40. Aglietta, M., Barone, C., Sawyer, M.B., Moore, M.J. *et al.* (2014) A phase I dose escalation trial of tremelimumab (CP-675,206) in combination with gemcitabine in chemotherapy-naive patients with metastatic pancreatic cancer. *Ann. Oncol.*, **25**, 1750–1755.
41. Hodi, F.S., Mihm, M.C., Soiffer, R.J., Haluska, F.G. *et al.* (2003) Biologic activity of cytotoxic T lymphocyte-associated antigen 4 antibody blockade in previously vaccinated metastatic melanoma and ovarian carcinoma patients. *Proc. Natl Acad. Sci. USA*, **100**, 4712–4717.
42. Hodi, F.S., Butler, M., Oble, D.A., Seiden, M.V. *et al.* (2008) Immunologic and clinical effects of antibody blockade of cytotoxic T lymphocyte-associated antigen 4 in previously vaccinated cancer patients. *Proc. Natl Acad. Sci. USA*, **105**, 3005–3010.
43. Yuan, J., Adamow, M., Ginsberg, B.A., Rasalan, T.S. *et al.* (2011) Integrated NY-ESO-1 antibody and CD8+ T-cell responses correlate with clinical benefit in advanced melanoma patients treated with ipilimumab. *Proc. Natl Acad. Sci. USA*, **108**, 16723–16728.
44. Hoos, A., Eggermont, A.M.M., Janetzki, S., Hodi, F.S. *et al.* (2010) Improved endpoints for cancer immunotherapy trials. *J. Natl Cancer Inst.*, **102**, 1388–1397.
45. Bulliard, Y., Jolicoeur, R., Windman, M., Rue, S.M. *et al.* (2013) Activating Fc γ receptors contribute to the antitumor activities of immunoregulatory receptor-targeting antibodies. *J. Exp. Med.*, **210**, 1685–1693.
46. Selby, M.J., Engelhardt, J.J., Quigley, M., Henning, K.A. *et al.* (2013) Anti-CTLA-4 antibodies of IgG2a isotype enhance antitumor activity through reduction of intratumoral regulatory T cells. *Cancer Immunol. Res.*, **1**, 32–42.
47. Simpson, T.R., Li, F., Montalvo-Ortiz, W., Sepulveda, M.A. *et al.* (2013) Fc-dependent depletion of tumor-infiltrating regulatory T cells co-defines the efficacy of anti-CTLA-4 therapy against melanoma. *J. Exp. Med.*, **210**, 1695–1710.
48. Bruhns, P., Iannascoli, B., England, P., Mancardi, D.A. *et al.* (2009) Specificity and affinity of human Fcgamma receptors and their polymorphic variants for human IgG subclasses. *Blood*, **113**, 3716–3725.
49. Barber, D.L., Wherry, E.J., Masopust, D., Zhu , B. *et al.* (2006) Restoring function in exhausted CD8 T cells during chronic viral infection. *Nature*, **439**, 682–687.
50. Freeman, G.J., Long, A.J., Iwai, Y., Bourque, K. *et al.* (2000) Engagement of the PD-1 immunoinhibitory receptor by a novel B7 family member leads to negative regulation of lymphocyte activation. *J. Exp. Med.*, **192**, 1027–1034.
51. Ishida, Y., Agata, Y., Shibahara, K., and Honjo, T. (1992) Induced expression of PD-1, a novel member of the immunoglobulin gene superfamily, upon programmed cell death. *EMBO J.*, **11**, 3887–3895.
52. Keir, M.E., Liang, S.C., Guleria, I., Latchman, Y.E. *et al.* (2006) Tissue expression of PD-L1 mediates peripheral T cell tolerance. *J. Exp. Med.*, **203**, 883–895.

53. Keir, M.E., Butte, M.J., Freeman, G.J., and Sharpe, A.H. (2008) PD-1 and its ligands in tolerance and immunity. *Annu. Rev. Immunol.*, **26**, 677–704.
54. Nishimura, H., Nose, M., Hiai, H., Minato, N. *et al.* (1999) Development of lupus-like autoimmune diseases by disruption of the PD-1 gene encoding an ITIM motif-carrying immunoreceptor. *Immunity*, **11**, 141–151.
55. Okazaki, T. and Honjo, T. (2007) PD-1 and PD-1 ligands: from discovery to clinical application. *Int. Immunol.*, **19**, 813–824.
56. Perez-Gracia, J.L., Labiano, S., Rodriguez-Ruiz, M.E., Sanmamed, M.F. *et al.* (2014) Orchestrating immune check-point blockade for cancer immunotherapy in combinations. *Curr. Opin. Immunol.*, **27**, 89–97.
57. Gelao, L., Criscitiello, C., Esposito, A., Goldhirsch, A. *et al.* (2014) Immune check-point blockade in cancer treatment: a double-edged sword cross-targeting the host as an "innocent bystander". *Toxins*, **6**, 914–933.
58. Zinselmeyer, B.H., Heydari, S., Sacristán, C., Nayak, D. *et al.* (2013) PD-1 promotes immune exhaustion by inducing antiviral T cell motility paralysis. *J. Exp. Med.*, **210**, 757–774.
59. Okazaki, T., Maeda, A., Nishimura, H., Kurosaki, T. *et al.* (2001) PD-1 immunoreceptor inhibits B cell receptor-mediated signaling by recruiting src homology 2-domain-containing tyrosine phosphatase 2 to phosphotyrosine. *Proc. Natl Acad. Sci. USA*, **98**, 13866–13871.
60. Saunders, P.A., Hendrycks, V.R., Lidinsky, W.A., and Woods, M.L. (2005) PD-L2:PD-1 involvement in T cell proliferation, cytokine production, and integrin-mediated adhesion. *Eur. J. Immunol.*, **35**, 3561–3569.
61. Nirschl, C.J. and Drake, C.G. (2013) Molecular pathways: coexpression of immune checkpoint molecules: signaling pathways and implications for cancer immunotherapy. *Clin. Cancer Res.*, **19**, 4917–4924.
62. Agata, Y., Kawasaki, A., Nishimura, H., Ishida, Y. *et al.* (1996) Expression of the PD-1 antigen on the surface of stimulated mouse T and B lymphocytes. *Int. Immunol.*, **8**, 765–772.
63. Chemnitz, J.M., Parry, R.V., Nichols, K.E., June, C.H. *et al.* (2004) SHP-1 and SHP-2 associate with immunoreceptor tyrosine-based switch motif of programmed death 1 upon primary human T cell stimulation, but only receptor ligation prevents T cell activation. *J. Immunol.*, **173**, 945–954.
64. Kamphorst, A.O. and Ahmed, R. (2013) Manipulating the PD-1 pathway to improve immunity. *Curr. Opin. Immunol.*, **25**, 381–388.
65. Oestreich, K.J., Yoon, H., Ahmed, R., and Boss, J.M. (2008) NFATc1 regulates PD-1 expression upon T cell activation. *J. Immunol.*, **181**, 4832–4839.
66. Kao, C., Oestreich, K.J., Paley, M.A., Crawford, A. *et al.* (2011) Transcription factor T-bet represses expression of the inhibitory receptor PD-1 and sustains virus-specific CD8+ T cell responses during chronic infection. *Nat. Immunol.*, **12**, 663–671.
67. Francisco, L.M., Salinas, V.H., Brown, K.E., Vanguri, V.K. *et al.* (2009) PD-L1 regulates the development, maintenance, and function of induced regulatory T cells. *J. Exp. Med.*, **206**, 3015–3029.
68. Fanoni, D., Tavecchio, S., Recalcati, S., Balice, Y. *et al.* (2011) New monoclonal antibodies against B-cell antigens: possible new strategies for diagnosis of primary cutaneous B-cell lymphomas. *Immunol. Lett.*, **134**, 157–160.
69. Terme, M., Ullrich, E., Aymeric, L., Meinhardt, K. *et al.* (2011) IL-18 induces PD-1-dependent immunosuppression in cancer. *Cancer Res.*, **71**, 5393–5399.
70. Velu, V., Titanji, K., Zhu, B., Husain, S. *et al.* (2009) Enhancing SIV-specific immunity in vivo by PD-1 blockade. *Nature*, **458**, 206–210.
71. Butte, M.J., Keir, M.E., Phamduy, T.B., Sharpe, A.H. *et al.* (2007) Programmed death-1 ligand 1 interacts specifically with the B7-1 costimulatory molecule to inhibit T cell responses. *Immunity*, **27**, 111–122.
72. Carreno, B.M. and Collins, M. (2002) The B7 family of ligands and its receptors: new pathways for costimulation and inhibition of immune responses. *Annu. Rev. Immunol.*, **20**, 29–53.
73. Greenwald, R.J., Freeman, G.J., and Sharpe, A.H. (2005) The B7 family revisited. *Annu. Rev. Immunol.*, **23**, 515–548.
74. Pentcheva-Hoang, T., Corse, E., and Allison, J.P. (2009) Negative regulators of T-cell

activation: potential targets for therapeutic intervention in cancer, autoimmune disease, and persistent infections. *Immunol. Rev.*, **229**, 67–87.
75. Latchman, Y., Wood, C.R., Chernova, T., Chaudhary, D. *et al.* (2001) PD-L2 is a second ligand for PD-1 and inhibits T cell activation. *Nat. Immunol.*, **2**, 261–268.
76. Tseng, S.Y., Otsuji, M., Gorski, K., Huang, X. *et al.* (2001) B7-DC, a new dendritic cell molecule with potent costimulatory properties for T cells. *J. Exp. Med.*, **193**, 839–846.
77. Paterson, A.M., Brown, K.E., Keir, M.E., Vanguri, V.K. *et al.* (2011) The programmed death-1 ligand 1:B7-1 pathway restrains diabetogenic effector T cells in vivo. *J. Immunol.*, **187**, 1097–1105.
78. Gros, A., Robbins, P.F., Yao, X., Li, Y.F. *et al.* (2014) PD-1 identifies the patient-specific $CD8^+$ tumor-reactive repertoire infiltrating human tumors. *J. Clin. Invest.*, **124**, 2246–2259.
79. Dolan, D.E. and Gupta, S. (2014) PD-1 pathway inhibitors: changing the landscape of cancer immunotherapy. *Cancer Control J. Moffitt Cancer Cent.*, **21**, 231–237.
80. Mellman, I. (2001) Setting logical priorities. *Nature*, **410**, 1026.
81. Zitvogel, L. and Kroemer, G. (2012) Targeting PD-1/PD-L1 interactions for cancer immunotherapy. *Oncoimmunology*, **1**, 1223–1225.
82. Rosenwald, A., Wright, G., Leroy, K., Yu, X. *et al.* (2003) Molecular diagnosis of primary mediastinal B cell lymphoma identifies a clinically favorable subgroup of diffuse large B cell lymphoma related to Hodgkin lymphoma. *J. Exp. Med.*, **198**, 851–862.
83. Steidl, C., Shah, S.P., Woolcock, B.W., Rui, L. *et al.* (2011) MHC class II transactivator CIITA is a recurrent gene fusion partner in lymphoid cancers. *Nature*, **471**, 377–381.
84. Parsa, A.T., Waldron, J.S., Panner, A., Crane, C.A. *et al.* (2007) Loss of tumor suppressor PTEN function increases B7-H1 expression and immunoresistance in glioma. *Nat. Med.*, **13**, 84–88.
85. Marzec, M., Zhang, Q., Goradia, A., Raghunath, P.N. *et al.* (2008) Oncogenic kinase NPM/ALK induces through STAT3 expression of immunosuppressive protein CD274 (PD-L1, B7-H1). *Proc. Natl Acad. Sci. USA*, **105**, 20852–20857.
86. Taube, J.M., Anders, R.A., Young, G.D., Xu, H. *et al.* (2012) Colocalization of inflammatory response with B7-h1 expression in human melanocytic lesions supports an adaptive resistance mechanism of immune escape. *Sci. Transl. Med.*, **4**, 127ra37.
87. Kim, J., Myers, A.C., Chen, L., Pardoll, D.M. *et al.* (2005) Constitutive and inducible expression of b7 family of ligands by human airway epithelial cells. *Am. J. Respir. Cell Mol. Biol.*, **33**, 280–289.
88. Lee, S.-K., Seo, S.-H., Kim, B.-S., Kim, C.D. *et al.* (2005) IFN-gamma regulates the expression of B7-H1 in dermal fibroblast cells. *J. Dermatol. Sci.*, **40**, 95–103.
89. Wilke, C.M., Wei, S., Wang, L., Kryczek, I. *et al.* (2011) Dual biological effects of the cytokines interleukin-10 and interferon-γ. *Cancer Immunol. Immunother.*, **60**, 1529–1541.
90. Gajewski, T.F., Louahed, J., and Brichard, V.G. (2010) Gene signature in melanoma associated with clinical activity: a potential clue to unlock cancer immunotherapy. *Cancer J. Sudbury Mass.*, **16**, 399–403.
91. Hamanishi, J., Mandai, M., Iwasaki, M., Okazaki, T. *et al.* (2007) Programmed cell death 1 ligand 1 and tumor-infiltrating CD8+ T lymphocytes are prognostic factors of human ovarian cancer. *Proc. Natl Acad. Sci. USA*, **104**, 3360–3365.
92. Mahnke, Y.D., Devevre, E., Baumgaertner, P., Matter, M. *et al.* (2012) Human melanoma-specific CD8(+) T-cells from metastases are capable of antigen-specific degranulation and cytolysis directly ex vivo. *Oncoimmunology*, **1**, 467–530.
93. Matsuzaki, J., Gnjatic, S., Mhawech-Fauceglia, P., Beck, A. *et al.* (2010) Tumor-infiltrating NY-ESO-1-specific CD8+ T cells are negatively regulated by LAG-3 and PD-1 in human ovarian cancer. *Proc. Natl Acad. Sci. USA*, **107**, 7875–7880.
94. Dong, H., Strome, S.E., Salomao, D.R., Tamura, H. *et al.* (2002) Tumor-associated B7-H1 promotes T-cell apoptosis: a potential mechanism of immune evasion. *Nat. Med.*, **8**, 793–800.
95. Palmer, B.E., Neff, C.P., Lecureux, J., Ehler, A. *et al.* (2013) In vivo blockade of the PD-1 receptor suppresses HIV-1 viral loads and improves CD4+ T cell levels in humanized mice. *J. Immunol.*, **190**, 211–219.

96. Brahmer, J.R., Drake, C.G., Wollner, I., Powderly, J.D. *et al.* (2010) Phase I study of single-agent anti-programmed death-1 (MDX-1106) in refractory solid tumors: safety, clinical activity, pharmacodynamics, and immunologic correlates. *J. Clin. Oncol.*, **28**, 3167–3175.
97. Topalian, S.L., Hodi, F.S., Brahmer, J.R., Gettinger, S.N. *et al.* (2012) Safety, activity, and immune correlates of anti-PD-1 antibody in cancer. *N. Engl. J. Med.*, **366**, 2443–2454.
98. Topalian, S.L., Sznol, M., McDermott, D.F., Kluger, H.M. *et al.* (2014) Survival, durable tumor remission, and long-term safety in patients with advanced melanoma receiving nivolumab. *J. Clin. Oncol.*, **32**, 1020–1030.
99. Motzer, R.J., Rini, B.I., McDermott, D.F., Redman, B.G. *et al.* (2014) Nivolumab for metastatic renal cell carcinoma (mRCC): results of a randomized, dose-ranging phase II trial. *J. Clin. Oncol.*, **32**, 5s.
100. Robert, C., Long, G.V., Brady, B., Dutriaux, C. *et al.* (2015) Nivolumab in previously untreated melanoma without BRAF mutation. *N. Engl. J. Med.*, **372**, 320–330.
101. Ansell, S.M., Lesokhin, A.M., Borrello, I., Halwani , A. *et al.* (2015) PD-1 blockade with nivolumab in relapsed or refractory Hodgkin's lymphoma. *N. Engl. J. Med.*, **372**, 311–319.
102. Berger, R., Rotem-Yehudar, R., Slama, G., Landes, S. *et al.* (2008) Phase I safety and pharmacokinetic study of CT-011, a humanized antibody interacting with PD-1, in patients with advanced hematologic malignancies. *Clin. Cancer Res.*, **14**, 3044–3051.
103. Lu, J., Lee-Gabel, L., Nadeau, M.C., Ferencz, T.M. *et al.* (2014) Clinical evaluation of compounds targeting PD-1/PD-L1 pathway for cancer immunotherapy. *J. Oncol. Pharm. Pract.* doi: 10.1177/1078155214538087
104. Herbst, R.S., Gordon, M.S., Fine, G.D., Sosman, J.A. *et al.* (2013) A study of MPDL3280A, an engineered PD-L1 antibody in patients with locally advanced or metastatic tumors. *J. Clin. Oncol.*, **31** (31 suppl.), abstract 3000.
105. Weber, J.S., Kudchadkar, R.R., Yu, B., Gallenstein, D. *et al.* (2013) Safety, efficacy, and biomarkers of nivolumab with vaccine in ipilimumab-refractory or -naive melanoma. *J. Clin. Oncol.*, **31**, 4311–4318.
106. Powles, T., Eder, J.P., Fine, G.D., Braiteh, F.S. *et al.* (2014) MPDL3280A (anti-PD-L1) treatment leads to clinical activity in metastatic bladder cancer. *Nature*, **515**, 558–562.
107. Heery, C.R., Coyne, G.H.O., Madan, R.A. *et al.* (2014) Phase I open-label, multiple ascending dose trial of MSB0010718C, an anti-PD-L1 monoclonal antibody, in advanced solid malignancies. *J. Clin. Oncol.*, **32**, 5s.
108. Monney, L., Sabatos, C.A., Gaglia, J.L., Ryu, A. *et al.* (2002) Th1-specific cell surface protein Tim-3 regulates macrophage activation and severity of an autoimmune disease. *Nature*, **415**, 536–541.
109. Baitsch, L., Baumgaertner, P., Devêvre, E., Raghav, S.K. *et al.* (2011) Exhaustion of tumor-specific $CD8^+$ T cells in metastases from melanoma patients. *J. Clin. Invest.*, **121**, 2350–2360.
110. Sakuishi, K., Apetoh, L., Sullivan, J.M., Blazar, B.R. *et al.* (2010) Targeting Tim-3 and PD-1 pathways to reverse T cell exhaustion and restore anti-tumor immunity. *J. Exp. Med.*, **207**, 2187–2194.
111. Zhou, Q., Munger, M.E., Veenstra, R.G., Weigel, B.J. *et al.* (2011) Coexpression of Tim-3 and PD-1 identifies a CD8+ T-cell exhaustion phenotype in mice with disseminated acute myelogenous leukemia. *Blood*, **117**, 4501–4510.
112. Fourcade, J., Sun, Z., Benallaoua, M., Guillaume, P. *et al.* (2010) Upregulation of Tim-3 and PD-1 expression is associated with tumor antigen-specific CD8+ T cell dysfunction in melanoma patients. *J. Exp. Med.*, **207**, 2175–2186.
113. Hastings, W.D., Anderson, D.E., Kassam, N., Koguchi, K. *et al.* (2009) TIM-3 is expressed on activated human CD4+ T cells and regulates Th1 and Th17 cytokines. *Eur. J. Immunol.*, **39**, 2492–2501.
114. Dardalhon, V., Anderson, A.C., Karman, J., Apetoh, L. *et al.* (2010) Tim-3/galectin-9 pathway: regulation of Th1 immunity through promotion of CD11b + Ly-6G+ myeloid cells. *J. Immunol.*, **185**, 1383–1392.
115. Chiba, S., Baghdadi, M., Akiba, H., Yoshiyama, H. *et al.* (2012) Tumor-infiltrating DCs suppress nucleic acid-mediated innate immune responses through interactions between the receptor TIM-3

and the alarmin HMGB1. *Nat. Immunol.*, **13**, 832–842.

116. Ngiow, S.F., Teng, M.W.L., and Smyth, M.J. (2011) Prospects for TIM3-targeted antitumor immunotherapy. *Cancer Res.*, **71**, 6567–6571.

117. Triebel, F., Jitsukawa, S., Baixeras, E., Roman-Roman, S. *et al.* (1990) LAG-3, a novel lymphocyte activation gene closely related to CD4. *J. Exp. Med.*, **171**, 1393–1405.

118. Goldberg, M.V. and Drake, C.G. (2011) LAG-3 in cancer immunotherapy. *Curr. Top. Microbiol. Immunol.*, **344**, 269–278.

119. Grosso, J.F., Kelleher, C.C., Harris, T.J., Maris, C.H. *et al.* (2007) LAG-3 regulates CD8+ T cell accumulation and effector function in murine self- and tumor-tolerance systems. *J. Clin. Invest.*, **117**, 3383–3392.

120. Huang, C.-T., Workman, C.J., Flies, D., Pan, X. *et al.* (2004) Role of LAG-3 in regulatory T cells. *Immunity*, **21**, 503–513.

121. Blackburn, S.D., Shin, H., Haining, W.N., Zou, T. *et al.* (2009) Coregulation of CD8+ T cell exhaustion by multiple inhibitory receptors during chronic viral infection. *Nat. Immunol.*, **10**, 29–37.

122. Hannier, S., Tournier, M., Bismuth, G., and Triebel, F. (1998) CD3/TCR complex-associated lymphocyte activation gene-3 molecules inhibit CD3/TCR signaling. *J. Immunol.*, **161**, 4058–4065.

123. Woo, S.-R., Turnis, M.E., Goldberg, M.V., Bankoti, J. *et al.* (2012) Immune inhibitory molecules LAG-3 and PD-1 synergistically regulate T-cell function to promote tumoral immune escape. *Cancer Res.*, **72**, 917–927.

124. Watanabe, N., Gavrieli, M., Sedy, J.R., Yang, J. *et al.* (2003) BTLA is a lymphocyte inhibitory receptor with similarities to CTLA-4 and PD-1. *Nat. Immunol.*, **4**, 670–679.

125. Gonzalez, L.C., Loyet, K.M., Calemine-Fenaux, J., Chauhan, V. *et al.* (2005) A coreceptor interaction between the CD28 and TNF receptor family members B and T lymphocyte attenuator and herpesvirus entry mediator. *Proc. Natl Acad. Sci. USA*, **102**, 1116–1121.

126. Han, L., Wang, W., Lu, J., Kong, F. *et al.* (2014) AAV-sBTLA facilitates HSP70 vaccine-triggered prophylactic antitumor immunity against a murine melanoma pulmonary metastasis model in vivo. *Cancer Lett.*, **354**, 398–406.

127. Cheung, T.C., Oborne, L.M., Steinberg, M.W., Macauley, M.G. *et al.* (2009) T cell intrinsic heterodimeric complexes between HVEM and BTLA determine receptivity to the surrounding microenvironment. *J. Immunol.*, **183**, 7286–7296.

128. Steinberg, M.W., Turovskaya, O., Shaikh, R.B., Kim, G. *et al.* (2008) A crucial role for HVEM and BTLA in preventing intestinal inflammation. *J. Exp. Med.*, **205**, 1463–1476.

129. Gertner-Dardenne, J., Fauriat, C., Orlanducci, F., Thibult, M.L. *et al.* (2013) The co-receptor BTLA negatively regulates human $V\gamma9V\delta2$ T-cell proliferation: a potential way of immune escape for lymphoma cells. *Blood*, **122**, 922–931.

130. Kaehler, K.C., Piel, S., Livingstone, E., Schilling, B. *et al.* (2010) Update on immunologic therapy with anti-CTLA-4 antibodies in melanoma: identification of clinical and biological response patterns, immune-related adverse events, and their management. *Semin. Oncol.*, **37**, 485–498.

131. Hodi, F.S., Friedlander, P.A., Atkins, M.B., McDermott, D.F.*et al.* (2011) A phase I trial of ipilimumab plus bevacizumab in patients with unresectable stage III or stage IV melanoma. *J. Clin. Oncol.*, **29** (Suppl.)(abstract 8511).

132. Downey, S.G., Klapper, J.A., Smith, F.O., Yang, J.C. *et al.* (2007) Prognostic factors related to clinical response in patients with metastatic melanoma treated by CTL-associated antigen-4 blockade. *Clin. Cancer Res.*, **13**, 6681–6688.

133. Hamid, O., Schmidt, H., Nissan, A., Ridolfi, L. *et al.* (2011) A prospective phase II trial exploring the association between tumor microenvironment biomarkers and clinical activity of ipilimumab in advanced melanoma. *J. Transl. Med.*, **9**, 204.

134. Blank, C.U. and Enk, A. (2015) Therapeutic use of anti-CTLA-4 antibodies. *Int. Immunol.*, **27**, 3–10.

135. Wang, C.J., Kenefeck, R., Wardzinski, L., Attridge, K. *et al.* (2012) Cutting edge: cell-extrinsic immune regulation by CTLA-4 expressed on conventional T cells. *J. Immunol.*, **189**, 1118–1122.

136. Snyder, A., Makarov, V., Merghoub, T., Yuan, J. *et al.* (2014) Genetic basis

for clinical response to CTLA-4 blockade in melanoma. *N. Engl. J. Med.*, **371**, 2189–2199.
137. Kohrt, H., Kowanetz, M., Gettinger, S., Powderly, J. *et al.* (2013) Intratumoral characteristics of tumor and immune cells at baseline and on-treatment correlated with clinical responses to MPDL3280A, an engineered antibody against PD-L1. *J. Immunother. Cancer*, **1**, O12.
138. Herbst, R.S., Soria, J.-C., Kowanetz, M., Fine, G.D. *et al.* (2014) Predictive correlates of response to the anti-PD-L1 antibody MPDL3280A in cancer patients. *Nature*, **515**, 563–567.
139. Sarnaik, A.A., Yu, B., Yu, D., Morelli, D. *et al.* (2011) Extended dose ipilimumab with a peptide vaccine: immune correlates associated with clinical benefit in patients with resected high-risk stage IIIc/IV melanoma. *Clin. Cancer Res.*, **17**, 896–906.
140. Meyer, C., Cagnon, L., Costa-Nunes, C.M., Baumgaertner, P. *et al.* (2014) Frequencies of circulating MDSC correlate with clinical outcome of melanoma patients treated with ipilimumab. *Cancer Immunol. Immunother.*, **63**, 247–257.
141. Couzin-Frankel, J. (2013) Breakthrough of the year 2013. Cancer immunotherapy. *Science*, **342**, 1432–1433.
142. Fridman, W.H., Pagès, F., Sautès-Fridman, C., and Galon, J. (2012) The immune contexture in human tumours: impact on clinical outcome. *Nat. Rev. Cancer*, **12**, 298–306.
143. Galon, J., Costes, A., Sanchez-Cabo, F., Kirilovsky, A. *et al.* (2006) Type, density, and location of immune cells within human colorectal tumors predict clinical outcome. *Science*, **313**, 1960–1964.
144. Galon, J., Mlecnik, B., Bindea, G., Angell, H.K. *et al.* (2014) Towards the introduction of the "Immunoscore" in the classification of malignant tumours. *J. Pathol.*, **232**, 199–209.
145. Curran, M.A., Montalvo, W., Yagita, H., and Allison, J.P. (2010) PD-1 and CTLA-4 combination blockade expands infiltrating T cells and reduces regulatory T and myeloid cells within B16 melanoma tumors. *Proc. Natl Acad. Sci. USA*, **107**, 4275–4280.
146. Tumeh, P.C., Harview, C.L., Yearley, J.H., Shintaku, I.P. *et al.* (2014) PD-1 blockade induces responses by inhibiting adaptive immune resistance. *Nature*, **515**, 568–571.
147. Wolchok, J.D., Kluger, H., Callahan, M.K., Postow, M.A. *et al.* (2013) Nivolumab plus ipilimumab in advanced melanoma. *N. Engl. J. Med.*, **369**, 122–133.
148. Deng, L., Liang, H., Burnette, B., Beckett, M. *et al.* (2014) Irradiation and anti-PD-L1 treatment synergistically promote antitumor immunity in mice. *J. Clin. Invest.*, **124**, 687–695.
149. Li, B., VanRoey, M., Wang, C., Chen, T.H. *et al.* (2009) Anti-programmed death-1 synergizes with granulocyte macrophage colony-stimulating factor--secreting tumor cell immunotherapy providing therapeutic benefit to mice with established tumors. *Clin. Cancer Res.*, **15**, 1623–1634.
150. Melero, I., Grimaldi, A.M., Perez-Gracia, J.L., and Ascierto, P.A. (2013) Clinical development of immunostimulatory monoclonal antibodies and opportunities for combination. *Clin. Cancer Res.*, **19**, 997–1008.
151. Peng, W., Lizée, G., and Hwu, P. (2013) Blockade of the PD-1 pathway enhances the efficacy of adoptive cell therapy against cancer. *Oncoimmunology*, **2**, e22691.
152. Ribas, A., Hodi, F.S., Callahan, M., Konto, C. *et al.* (2013) Hepatotoxicity with combination of vemurafenib and ipilimumab. *N. Engl. J. Med.*, **368**, 1365–1366.
153. Tarhini, A.A., Cherian, J., Moschos, S.J., Tawbi, H.A. *et al.* (2012) Safety and efficacy of combination immunotherapy with interferon alfa-2b and tremelimumab in patients with stage IV melanoma. *J. Clin. Oncol.*, **30**, 322–328.
154. Demaria, S. and Formenti, S.C. (2013) Radiotherapy effects on anti-tumor immunity: implications for cancer treatment. *Front. Oncol.*, **3**, 128.
155. Hodi, F.S., Lawrence, D., Lezcano, C., Wu, X. *et al.* (2014) Bevacizumab plus ipilimumab in patients with metastatic melanoma. *Cancer Immunol. Res.*, **2**, 632–642.
156. Frederick, D.T., Piris, A., Cogdill, A.P., Cooper, Z.A. *et al.* (2013) BRAF inhibition is associated with enhanced melanoma antigen expression and a more favorable tumor microenvironment in patients with metastatic melanoma. *Clin. Cancer Res.*, **19**, 1225–1231.

157. Jiang, X., Zhou, J., Giobbie-Hurder, A., Wargo, J. *et al.* (2013) The activation of MAPK in melanoma cells resistant to BRAF inhibition promotes PD-L1 expression that is reversible by MEK and PI3K inhibition. *Clin. Cancer Res.*, **19**, 598–609.

158. Wilmott, J.S., Long, G.V., Howle, J.R., Haydu, L.E. *et al.* (2012) Selective BRAF inhibitors induce marked T-cell infiltration into human metastatic melanoma. *Clin. Cancer Res.*, **18**, 1386–1394.

20
Immunotherapy with Autologous Cells

Andrew D. Fesnak and Bruce L. Levine
University of Pennsylvania, Department of Pathology and Laboratory Medicine, The Perelman School of Medicine, M6.40 Maloney 3400 Spruce Street, Philadelphia, PA 19104, USA

Translational Medicine Cancer: Advances in Molecular Biology and Medicine,
First Edition. Edited by Robert A. Meyers.

Keywords

Immunotherapy
The use of immune cells, or the manipulation of cells of the immune system, usually through the administration of antibodies or biologic ligands, to effect and augment or suppress the immune system as a treatment for an underlying disease.

Cell therapy
Therapy through the use of cells removed from the body and readministered following minimal manipulation, or following more than minimal manipulation to impart nonhomologous function.

Gene therapy
The use of genetic modification of cells to impart nonhomologous function, or the use of DNA or RNA within a viral vector delivered to cells or directly injected, or the use of DNA or RNA directly administered for the treatment of disease.

Autologous
Customarily used in the context of cell therapies derived from a patient's own cells, in contrast to allogeneic, where cell therapies are derived from a donor.

Adoptive immunotherapy
Commonly used to refer to the transfer of immune cells for the treatment of disease, that may be either autologous or allogeneic.

Transferred autologous cells have been investigated and developed over the past few decades to reconstitute immune function and augment immunity following stem cell transplantation, for the treatment of viral infections, and more recently for the treatment of a wider array of diseases. Cellular immunotherapy begins with either the patient's own (autologous) or donor (allogeneic) cells that have been isolated, from tumor tissue, bone marrow biopsy or apheresis mononuclear cell collection. These cells may then be modified *ex vivo*, expanded and, ultimately, adoptively transferred to the patient. Cell function may be manipulated *ex vivo* through selection, differentiation, and the addition of biologic or chemical mediators. Over the past 25 years, methods of gene delivery to cells to modulate function have improved significantly in efficacy. More recently, the translation of gene editing techniques have enabled targeted gene knockout or the insertion of genetic elements to modulate cell function. With substantial clinical benefit demonstrated in a number of diseases, regulatory approval for more autologous cell therapies can be anticipated. Wide adoption of autologous cell therapies will occur with the further development of *ex-vivo* engineering, automation and the implementation of a source to manufacturing to delivery logistics systems that is an evolution of existing transfusion medicine and organ and tissue transfer networks.

1 Introduction

The term cellular immunotherapy encompasses a diverse field of therapies designed to treat a number of clinical indications. In general, cellular immunotherapy begins with either the patient's own (autologous) or donor (allogeneic) cells that have been isolated, from tumor tissue, from bone marrow biopsy, or by apheresis mononuclear cell collection (Fig. 1). These cells may then be modified *ex vivo*, expanded and, ultimately, adoptively transferred to the same patient, in the case of autologous transfer. Autologous cellular therapies can provide targeted killing of malignant or virally infected cells, and even restore broad immunologic reconstitution in immunodeficient recipients.

Mitchison first demonstrated the antitumor effects of adoptive transfer in rodent tumor models in 1955 [1]. However, it took several decades to appreciate the full potential of adoptive immunotherapy. It was not until the 1970s that the anti-leukemic effect of the reconstituting allogeneic immune system was noted in the post-transplant setting [2]. This observation prompted the transfer of mature donor lymphocytes (donor lymphocyte infusion; DLI) to treat relapsed leukemia following bone marrow transplant [3, 4]. While these allogeneic therapies demonstrate powerful graft-versus-tumor activity, graft-versus-host disease (GVHD) effects limit their use. Therefore, it was postulated that if the specificity of autologous cells could be enriched or altered, it would be possible to selectively target tumor or other antigens, without causing GVHD.

Transferred autologous cells can also be used to combat uncontrolled viral illness. In fact, the first successful clinical use of adoptively transferred antigen specific T cells targeted cytomegalovirus (CMV)-infected cells [5]. In doing so, Greenberg *et al.* demonstrated two important principles that have guided further cell therapy development: (i) that *ex-vivo* expansion was necessary to achieve sufficient clinical effect; and (II) that the specificity of a T-cell product could be manipulated *ex vivo*, in a way that endures after transfer [5]. Virus-specific T cells have also been used to treat Epstein–Barr virus (EBV) -related post-transplant lymphoproliferative disorders [6], systemic adenovirus infection [7], and even to combat human immunodeficiency virus 1 (HIV-1) infection [8, 9].

Adoptive cellular immunotherapy has also been used to restore normal immune function in some immunodeficient hosts. A subset of patients with severe combined immunodeficiency (SCID) possesses a mutation in the common gamma chain, blocking the lymphoid development of T and NK (natural killer) cells. Hematopoietic CD34+ stem cells from these patients were transduced with a retroviral vector expressing this common gamma chain, and the patients experienced a reconstitution of the previously absent immune cell

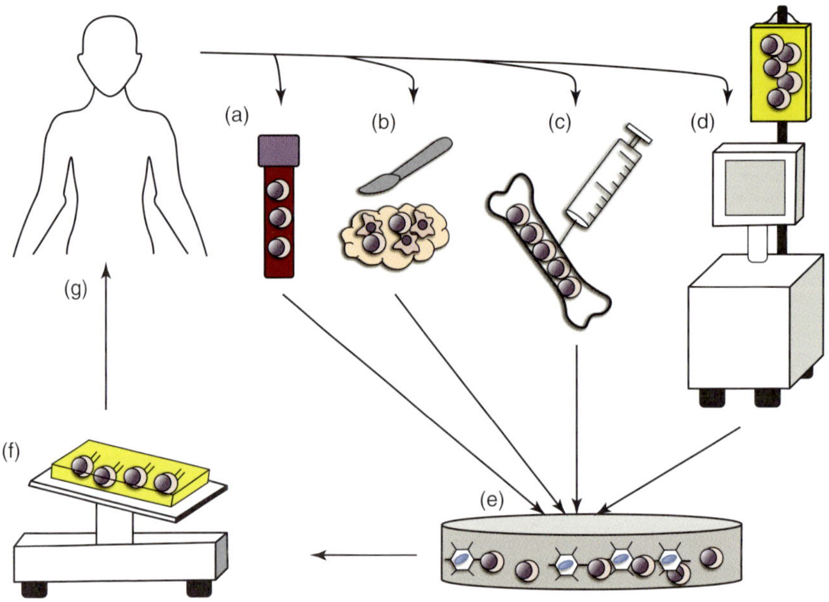

Fig. 1 Overview of approach to adoptive immunotherapy. The patient's own lymphocytes are obtained by: (a) peripheral blood draw; (b) excision of tumor bearing lymphocytes; (c) direct bone marrow biopsy; or (d) apheresis collection of mononuclear cells. Depending on the desired effect, these cells may then undergo (e) *ex-vivo* modifications including gene transfer and (f) expansion prior to (g) reinfusion.

subsets [10, 11]. Though this reconstitution was robust and durable, insertional mutagenesis was seen in some patients, leading to the development of leukemia [12, 13]. While this toxicity has prompted a reconsideration of transduction techniques as well as transduced cell populations, it is important to note that this type of adverse event has not been observed in T-cell therapy trials to date [14, 15]. This is not unexpected given that gene transfer into hematopoietic stem cells has the potential to alter not just lymphoid but rather all hematopoietic lineages. As evidence of this, Ott *et al.* have used gene transfer to correct X-linked chronic granulomatous disease, a defect in myeloid phagocytes [16]. In addition, Braun *et al.* have used gene transfer to correct Wiskott–Aldrich protein expression in lymphoid and myeloid lineages and in platelets [17]. The ability to modify multiple cell lineages is a major advantage of targeting the stem cell compartment for gene transfer.

Adoptive cellular immunotherapy has entailed many different approaches to treat many different diseases over the years. This diversity speaks to the vast potential role of the immune system in combating human disease. As the field moves forward, clinical findings will guide the quest for a deeper understanding of basic biology and laboratory advances translated to improve patient care. In this way, adoptive cellular immunotherapy represents a true bench-to-bedside science.

2 Selecting and Isolating Effector Cell Types

The effector cell type to be isolated, modified, and expanded may vary depending on the desired effect. Two cell lineages have been implicated in transferable immune-mediated cytotoxicity, namely T lymphocytes and NK cells, and both cell types may be targeted to malignant cells or virally infected cells. While it has also been reproducibly observed that the activation and proliferation of both T and NK cells can increase antitumor capacity [18–20], to date the majority of *in-vivo* tumor clearance success has been seen with T lymphocytes. The adoptive transfer of NK cells has also been shown to be safe and feasible [21, 22], and NK cells have been used to treat patients with renal cell carcinoma [23] and malignant glioma [24]. Overall, however, the resulting tumor eradication has been inconsistent [25].

One potential explanation for these findings is that, upon adoptive transfer, NK cells can demonstrate an adjustment of their activation threshold, termed "tuning" [26, 27]. Despite this limitation, active investigations to extend and enhance NK cell *in vivo* antitumor cytotoxicity are being pursued, given the unique ability of NK cells to recognize and kill major histocompatibility complex (MHC) class I-deficient targets [21]. Downregulation of MHC class I is a commonly employed mechanism by which viruses and malignancies have attempted to avoid immunosurveillance. While NK cells may be well suited to clear MHC class I-deficient targets, NK cell activation may need to be better understood and production methods enhanced before this becomes a widely applicable therapy.

Gamma delta T cells – like NK cells but unlike alpha beta T cells – are able to respond to non-MHC-restricted peptide ligands. This rare subset of T cells is able to be activated by MHC-like, stress-induced self-antigens that may be upregulated on the surface of some tumors, such as human glioblastoma multiforme. Consequently, gamma delta T cells represent one potential cellular therapeutic agent to clear these tumors. Currently, efforts are under way to optimize the *ex-vivo* expansion protocols of gamma delta T cells [28, 29], and to date gamma delta T cells have been used in at least 13 human clinical trials [30].

Dendritic cells are professional antigen-presenting cells (APCs) that, while not demonstrating direct cytotoxicity, can present target antigens and direct other target-specific native immune cells to proliferate, activate, and eradicate target cells. Autologous dendritic cells can be loaded with specific antigens or engineered *ex vivo* to present antigens. Once reinfused, these cells can direct tumor-specific immune cells within the patient's existing immune system to target the tumor. Dendritic cell therapy is, therefore, in essence a cellular-based vaccination strategy, and differs in approach to T and NK cell therapy. This cellular vaccine approach forms the basis for Sipuleucel-T (Provenge®), the first Food and Drug Administration (FDA)-approved cell therapy for cancer, and the first FDA-approved dendritic cell therapy. While groundbreaking, this therapy was associated with only a 4.1-month increase in median survival over placebo, leaving room for further progress in the field of anticancer cell therapy [31].

The isolation of T, NK or dendritic cells for autologous immunotherapy can be achieved through a variety of mechanisms. Tumor infiltrating lymphocytes (TILs) can be directly extracted from a patient's tumor tissue; in fact, the tumor may serve as an *in-vivo* enrichment tissue, selecting and preferentially expanding tumor-specific lymphocytes. Immunotherapies targeting diseases with easily accessible and abundant viable tumor tissue, such as melanoma, may derive starting effector cells for culture in this manner. Bone marrow biopsy can provide another source of lymphocytes to use as a starting cell population [32]. Finally, circulating mononuclear cells, including lymphocytes as well as dendritic cells, can be collected peripherally via a phlebotomy or the leukapheresis procedure. Once adequate starting tissue has been collected, the desired cell population can be further enriched (see Sect. 4.1; Preparation of Cells for Culture).

While traditional phlebotomy can provide a small amount of cellular material, the volume required to establish a starting culture is very often in excess of what the patient can tolerate. In those cases, the collection of mononuclear cells by apheresis can provide a large starting cell count. Apheresis collection involves establishing a high blood flow (often through a central catheter) and mixing the blood with citrate anticoagulant; centrifugation is then used to separate the blood components on the basis of their densities (see Fig. 2). The mononuclear cell layer, containing lymphocytes, can be selectively removed during

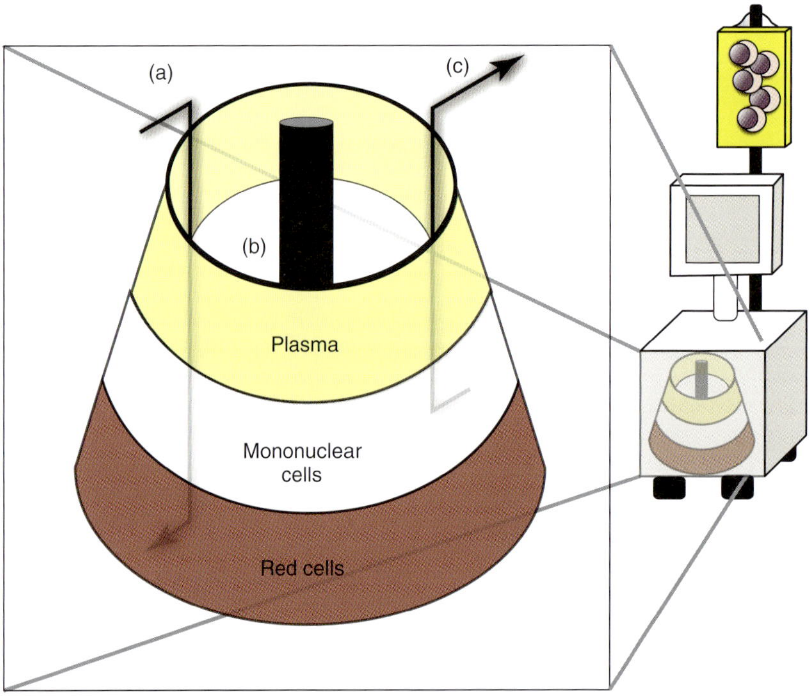

Fig. 2 Mechanism of leukapheresis for mononuclear cell collection. (a) Anticoagulated whole blood enters the machine housing traveling in sterile tubing; (b) Centrifugal force separates the various blood components by their density. In general, separation occurs between plasma (including platelet-rich and platelet-poor plasma), mononuclear cells, and red cell components; (c) By selectively aspirating the desired layer, specific blood components, may be isolated. Mononuclear cell layer removal is depicted here. Adapted from University of Utah School of Medicine WebPath Tutorial (*http://library.med.utah.edu/WebPath/TUTORIAL/BLDBANK/BBAPHER.html*)

leukapheresis. The latter process is advantageous because the uncollected cellular components and plasma can be returned to the patient, allowing the processing of a large blood volume. A typical collection may range from 5 to 15 liters of whole blood, or approximately one to three blood volumes for the average 70-kg patient. Leukapheresis can be performed once daily for several days if necessary, as is the case when collecting circulating hematopoietic stem cells, and the collected products may be pooled to increase the starting cell input. Typically, the procedure is well tolerated, although rare adverse effects associated with the apheresis catheter and/or citrate anticoagulant have been reported.

The efficiency of mononuclear cells removed per unit volume of processed blood ranges from approximately 30% to 60% [33, 34]. Several factors may influence the mononuclear cell yield, including the total volume of processed whole blood, the efficiency of the blood cell separators, the mobilization agents, and the blood flow rate [35–37]. Importantly, contamination with platelets has been shown to limit dendritic cell culture [38]. These data demonstrate that apheresis collection for the purpose of obtaining a cell therapy product is safe and well tolerated. However, further studies are needed to determine the optimal collection parameters for a range of cell therapy products.

At present, none of the techniques used to isolate starting cellular material is able to distinguish amongst the cell subsets, and consequently the final collected cell product will contain a mixture of the patient's mononuclear cells. These will include T, NK and dendritic cells, as well as B cells, other monocytes, neutrophils and, potentially, even malignant cells. Hence, further purification (as described in Sect. 4.1) may be necessary for the optimal culture of a desired subset of cells.

3 Modifying the Effector Cell Population

Collection methods for autologous cell therapy products provide a starting cell population consisting of a mixture of cell types and specificities. The spectrum of available modifications is quite broad. Modern molecular biologic techniques such as viral vector transduction, transposon-mediated transduction and electroporation now allow the alteration of lymphocyte specificity by engineering APCs or by directly re-engineering the lymphocyte cell surface expression. In addition, modifications can be made to a cellular product to generate resistance to viral infection. Finally, these techniques allow the introduction of inducible cell death proteins, providing an "off" switch for the therapy if desired.

3.1 Gene Delivery Techniques

Enhancing the function of primary human cells often requires the introduction of nucleic acid, either DNA or RNA. Several platforms have been shown to be effective in the transfer of genetic material into these cells:

- *Viral vectors* are the most widely used method to transfer large payloads of nucleic acid to primary human cells. This approach is particularly well suited to *ex-vivo* applications, wherein prior immunity to the viral capsid does not limit an efficient transfer.
- *Transposon-mediated transduction* is an alternative method of gene transfer that

utilizes transposase-based integration into the host genome. While transposon biology has been studied for many years, the use of transposons to introduce transgenes into cell therapy products is a relatively new approach.

- *Electroporation*, as opposed to viral vectors or transposon cassettes, employs an electrical pulse to transiently permeabilize the cell membrane, allowing hydrophilic molecules to transfer into the intracellular space [39]. Similar to viral vector transfer, once introduced, the nucleic acids are processed by the cell machinery to express protein.

3.1.1 Viral Vectors

Viral vectors are constructed by the removal of most virally encoded genes and the insertion of a gene of interest. The vector utilizes the viral machinery to introduce the genetic material and, once introduced, the gene of interest is transcribed and translated by the transduced cell machinery. Retrovirus family members are a common choice for the viral vector transduction of human cells. Gamma-retroviral vectors have been investigated in human gene therapy trials [40] and in human T-cell immunotherapy trials [41] since the early 1990s. In this case, the gamma-retrovirus integrates into the host genome, providing long-term gene expression. Gamma-retroviruses must transduce actively dividing cells, however, and insertional mutagenesis has been observed after the transduction of hematopoietic stem cells [13, 42, 43]. To address these issues, lentiviral vectors were developed for human primary cell transduction. Lentiviral vectors transduce nondividing cells [44], appear to have a safer integration site profile [45, 46], and also have the capacity to deliver transgenic material equivalent to gamma-retrovirus vectors. Other viral vectors derived from adenovirus [47, 48] and adeno-associated virus [49] are less commonly employed.

3.1.2 Transposon-Mediated Transduction

Integration-competent, replication-incompetent transposon cassettes allow for the efficient and permanent introduction of transgenes. A DNA transposon system such as Sleeping Beauty (SB) consists of flanking inverted repeats that act as transposase-binding sites [50]. Wild-type transposons encode their own transposes, allowing for replication and integration throughout the genome. For the purpose of cell therapy applications, the SB system has been engineered to replace the coding sequence for the transposase with an expression cassette. During *ex-vivo* processing this cassette, along with *in trans* provision of the transposase protein, allows cut-and-paste transposition of the transgene into the host genome. Of particular note, sites of integration appear to be nearly random. This is a clear advantage over other integrating viral vectors, which are predisposed to integrate at sites of active transcription since integration at actively transcribed sites makes insertional mutagenesis much more likely. The feasibility of transposon-mediated transduction of human peripheral blood lymphocytes has been demonstrated [51], and this technology is currently in use in clinical trials [52].

3.1.3 RNA Electroporation

The newest technology to be explored as a method of modifying human primary T-cell specificity is electroporation. The RNA electroporation of messenger RNA (mRNA) into target cells is a rapid and efficient method for generating transient expression. In humans, this approach is particularly well suited to investigational transgenes if the off-target toxicity is unknown. In addition, genotoxicity is not of concern with RNA electroporation, given that mRNA does not integrate into the genome. For some constructs, RNA electroporation is at least as efficient as retrovirus at gene transfer [53]. Potent anti-tumor activity of RNA-modified cells has been demonstrated against both liquid tumors such as advanced leukemia [54], and solid tumors such as mesothelioma [55] and neuroblastoma [56] in mice. In human trials of chimeric antigen receptor (CAR)-based therapies with potential for off-target toxicity, an RNA-based approach has been shown to be safe [57]. Clinical trials using RNA – CAR-electroporated T cells are currently under way and will provide valuable further information regarding the safety and efficacy of the procedure.

3.2 Enrichment of Antigen-Specific Effector Cells

Within a population of adaptive immune cells, typically a broad repertoire of specificities is observed. During a physiological immune response, those immune cells with sufficient specificity to respond to the presented antigen will preferentially activate and expand. In the context of cellular therapy, advantage can be taken of this situation by selecting a tissue, or recreating an environment, that is rich in the target antigen (Fig. 3). In doing so, target-specific clones will be preferentially expanded and the final product will be enriched for cells with the desired specificity. Enrichment of antigen-specific effector cells enables the production of both virus- and tumor-specific cell products for reinfusion.

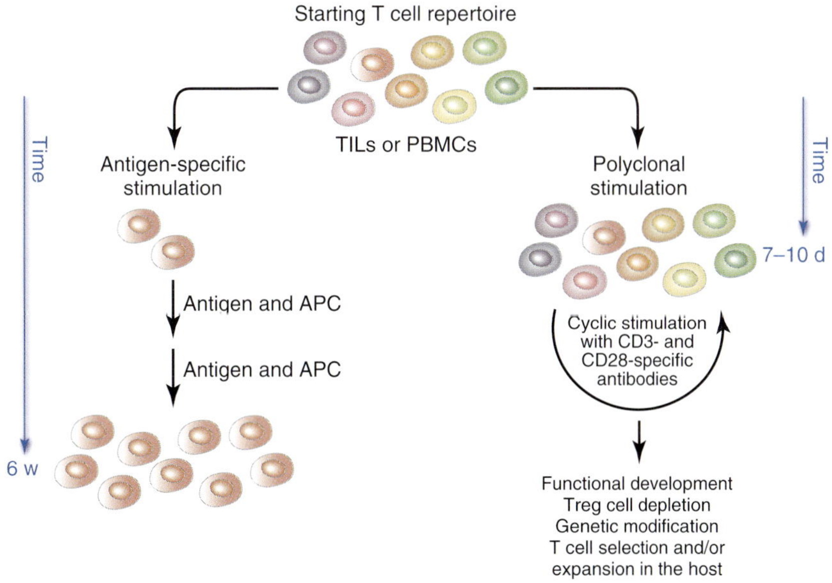

Fig. 3 Selection of antigen-specific clones and expression of engineered T-cell receptors. Reproduced with permission from Ref. [58].

3.2.1 Enrichment of Virus-Specific Lymphocytes

In the setting of bone marrow transplant, viral infection or reactivation represents an enormous cause of morbidity and mortality. Therefore, virus-specific lymphocytes – specifically cytotoxic T cells – are considered one potential avenue for restoring antiviral immunity. However, demonstration of the feasibility of production of these virus-specific T lymphocytes for large clinical trials has been a major challenge. Leen *et al.* have shown that, by modifying an EBV-transformed B-cell line to express viral peptides, a single culture technique could yield CD4+ and CD8+ T lymphocytes with multiple desired specificities [59]. Others have shown the feasibility of virus-specific T-cell products from cord blood products [60]. Importantly, virus-specific T cells produced in this manner are not associated with GVHD [61].

3.2.2 Enrichment of Tumor-Specific Lymphocytes

Within the tumor microenvironment, the presentation of tumor peptides generates an enriched population of tumor-specific TILs. This already enriched starting cellular material may be cultured to increase the number of tumor-specific lymphocytes. Further preferential expansion of TILs *ex vivo* and reactivation of their effector function allows for the reinfusion of a highly specific and effective product. Investigations pioneered by the Rosenberg group at the National Cancer Institute have employed this strategy in the context of melanoma, with striking clinical responses [62, 63]. While early studies demonstrated a limited persistence, the addition of pre-reinfusion recipient lymphodepletion has been shown to increase engraftment and effector function of adoptively transferred cells [64] Patients with melanoma, treated with lymphodepletion, followed by reinfusion of a TIL-based cell therapy, demonstrated clinical response rates between 49% and 72% [64–66], with some patients demonstrating complete remission beyond five years [67].

One potential limitation of this approach is that while enriched for tumor-specific lymphocytes, any tumor tissue will be a mixture of cells. This population could include suppressive T-regulatory cells (T-regs) which may limit the *ex-vivo* expansion of the desired cell effector population or worse, tolerize the tumor. Depleting the tumor tissue of T-regs prior to culture is actively being pursued.

An extended *ex-vivo* clonal expansion may result in cells with limited *in vivo* persistence following transfer and a reduced clinical efficacy [69, 70]. One hypothesis for this decreased efficacy is that the cultured T cells are "exhausted" due to persistent stimulation. In other contexts, such as chronic lymphocytic leukemia (CLL), the attenuation of T-cell function can be reversed via CD28 stimulation [71, 72]. The mechanism, however, of *ex-vivo*-stimulated exhaustion is likely to be very different than that seen in CLL-associated attenuation of function. In contrast, Heslop *et al.* used a co-culture technique to generate functional EBV-specific CTLs capable of persisting up to nine years *in vivo* [6]. Although this discrepancy could be explained by a number of factors, these results demonstrate that the limitations of persistence seen with culture-stimulated cells are not insurmountable.

Many tumors, however, are capable of inhibiting the antigen presentation of tumor peptides, thereby limiting the expansion and activation of anti-tumor lymphocytes. This inhibition can be overcome by forcing APCs, such as dendritic cells, to present specific tumor antigens and drive the expansion of tumor-specific lymphocytes. Upon reinfusion, these modified APCs preferentially stimulate tumor-specific lymphocytes and, in essence, vaccinate the patient against the tumor. Marked clinical and immunologic responses against HER-2/neu (c-erbB2) -positive ductal carcinoma *in situ* breast cancer have been observed after reinfusion of dendritic cells pulsed *ex vivo* with tumor peptides [68].

Several dendritic cell platforms are currently being used to treat human malignancies by augmenting the number and function of antigen-specific T cells. Sipuleucel-T (Provenge®), the first FDA-approved (April 2010) therapeutic cancer vaccine, makes use of such a platform. Sipuleucel-T was approved to treat asymptomatic or minimally symptomatic metastatic castrate-resistant (hormone-refractory) prostate cancer, and is generated from autologous leukapheresis-derived peripheral blood mononuclear cells. The patient's mononuclear cells were co-incubated with a prostate acid phosphatase (PAP) – granulocyte-macrophage colony stimulating factor (GM-CSF) fusion protein. GM-CSF activates the APCs to present PAP, and in response PAP-specific autologous T cells are preferentially expanded; PAP is then expressed on prostate cancer cells. The final product is washed to remove the fusion protein and then reinfused into the patient. In a double-blind, placebo-controlled study in humans, a 50% improvement in three-year mortality was observed over the placebo control (31.7% versus 21.7%, respectively) [31]. This success has driven many groups in both industry [73–77] and academia [78, 79] to expand on the dendritic cell vaccine approach to treat glioblastoma multiforme, prostate cancer, and ovarian cancer.

3.3 Expression of Novel Cell-Surface Receptors

The expression of novel cell-surface receptors can serve to re-engineer lymphocyte specificity. In the case of a modified T cell, this approach can utilize viral vectors or electroporation to express complete T-cell receptors (TCRs) or synthetic CARs. In either case, retargeting cell specificity can limit activation to the desired ligand and even drive the long-term persistence of stably expressed constructs [80].

3.3.1 Non-Endogenous T-Cell Receptors

Specific TCRs have the significant advantage of being able to target a T cell to an intracellular antigen, presented in the context of MHC. On the other hand, this requires that the transduced TCRs be compatible with the patient's human leukocyte antigen (HLA) type. In addition, experiments in mice have raised concerns with regard to transduced-endogenous heterodimer TCR expression. Because the TCR consists of a paired alpha and beta chain, the potential exists for a transduced cells to express a TCR composed of a transduced alpha chain pairs with an endogenous beta chain, or vice versa. The specificity of these heterodimers is unknown, and could generate an autoreactive clone as seen in animal models [81], though this has not been reported in clinical trials to date. Investigators have demonstrated several methods to preferentially express the transgenic TCR and conversely to knock down endogenous TCR expression.

Despite these concerns, the major *in vivo* limitations to T cells redirected with specific TCR has not been autoimmunity, but rather has been on-target, off-tumor activity [82, 83]. The findings of these groups and others [84, 85] demonstrate current limitations in predicting cross-reactive

TCR affinity. However, investigations conducted by Linette *et al.* have provided a strategy to circumvent this limitation by re-engineering induced pluripotent stem cells [84]. Early trials in the adoptive transfer of HLA-restricted, specific TCR-redirected T cells have shown efficacy at tumor clearance [86].

3.3.2 Chimeric Antigen Receptors

Modified T cells, expressing whole, non-endogenous TCRs demonstrate several limitations. These receptors can only recognize peptide presented in the context of MHC, while many tumors may downregulate MHC as an immune evasion mechanism. In addition, a single TCR transmits only the primary activation signal, which is insufficient to drive proliferation. Synthetic receptors termed chimeric antigen receptors are MHC-independent and do not require a second costimulatory ligand for activation. CARs are a highly specific receptors with: (i) an extracellular, single chain variable fragment (scFv); (ii) a TCR-derived transmembrane domain; and (iii) a TCR intracellular signaling domain (often CD3zeta), the primary signal transducing element of the TCR complex. A scFv consists of a fusion of one heavy and one light chain variable domain, linked by a flexible peptide linker. As noted above, unlike native TCRs scFvs can bind to antigen in an MHC-independent manner [87]. First-generation CARs were created by fusing the CD3zeta domain to the transmembrane domain (Fig. 4). Although cells expressing these CARs demonstrated on-target cytotoxicity [88], the *in-vivo* responses were short-lived. It has long been known that primary signal in the absence of costimulation causes the cell to become anergic or unable to activate in response to ligand [89]. CAR-modified T cells also require costimulation to demonstrate a durable response [90]. Second-generation

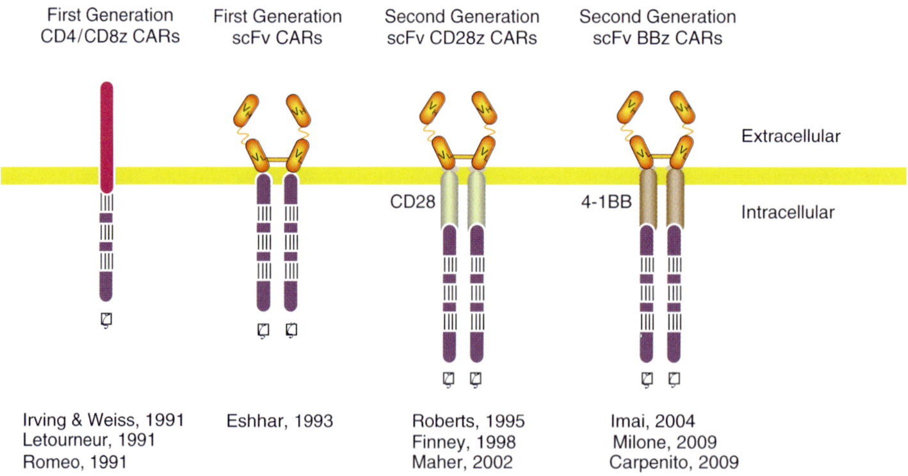

Fig. 4 Multiple generations of CARs. The first-generation CARs were composed of a single extracellular domain with a scFV [88]. Second-generation CARs incorporated intracellular domains of costimulatory molecules such as CD28 [92, 97, 98] or 4-1BB [94, 95, 99] *in cis* with the remainder of the receptor.

CARs address this problem by adding a costimulatory domain, such as the CD28 intracellular domain [91–93] or a tumor necrosis factor (TNF) receptor family domain such as the 4-1BB (CD137) intracellular domain [94] *in cis* with the other CAR domains. CARs with built-in costimulatory domains generate both primary and costimulatory signal in response to a single ligand. Further modifications to CARs, including the addition of multiple costimulatory domains, have demonstrated even greater tumor eradication in animal models [95–97].

CARs have been designed to target a variety of cell-surface proteins. CD19 has been one of the most targeted cell-surface molecules, and one that has seen the most success. CD19 is expressed on B-cell leukemias and lymphomas, and also on healthy B cells. An expansion of CD19-targeted cell therapies, including non-Hodgkin lymphoma (NHL), follicular lymphoma and large cell lymphoma, is being pursued. In addition, receptors with affinity for molecules such as CD20 (acute lymphoblastic leukemia, CLL, NHL), CD30 (NHL and Hodgkin lymphoma), CD33 (acute myeloid leukemia), CD138, and immunoglobulin-kappa light chain (multiple myeloma) are all worthy approaches for future cell therapy trials. Pule *et al.* have reported that EBV-specific T cells, expressing synthetic anti-disialoganglioside GD2 receptors, target neuroblastoma [100], while others have demonstrated the efficacy of anti-mesothelin CAR T cells [57]. CAR-based cell therapy requires an appropriate tumor-specific cell-surface target. In addition, tumor downregulation of target epitopes or the selection of target-negative clones will allow for escape mutants. Future approaches will likely return to a multi-target approach to limit tumor evasion of these engineered products.

3.4 Generation of Resistance to Viral Infection

Human viruses employ several mechanisms to gain entry to a host cell. Binding to a cell-surface receptor (in some cases to multiple receptors) is often required in order for a virus to infect a cell. The HIV-1 virion, for example, binds to CD4 and either the chemokine receptors CXCR4 or CCR5 to infect a T cell. Blocking of viral receptors is a commonly employed approach for the prevention and treatment of systemic viral illness. A permanent depletion of viral entry receptors has the potential to provide an effective therapy for viral infections.

In 1% of Caucasians, there is a naturally occurring mutation termed delta32 that results in the loss of CCR5 expression. An absence of CCR5 is associated with few pathogenic findings, including an increased risk of complications following West Nile virus infection [101]. CCR5 acts as a coreceptor for HIV-1 entry into CD4 T cells, and CCR5 blockade by the drug maraviroc has shown great promise in the treatment of HIV1 [102]. A stem cell transplant from a CCR5-null delta32 donor to an HIV-positive male patient to treat his leukemia resulted in an apparent functional cure [103]. With respect to cell therapy, CCR5-null T cells can be generated from CCR5-positive cells *ex vivo* by zinc finger nuclease gene disruption at the CCR5 locus [104]. These HIV-1 resistant cells can then be reinfused into a patient with HIV-1, providing them with cells that can restore CD4+ function and remain uninfected. To date, Phase I clinical trials have demonstrated the safety and feasibility of adoptively transfer autologous CCR5-null T cells into HIV1-positive subjects [105].

Several novel strategies to combat HIV infection are currently being developed. MazF is an *Escherichia coli*-derived endoribonuclease that recognizes and cleaves the sequence adenine–cytosine–adenine in single-stranded RNA. This sequence is particularly enriched in HIV1 mRNA, and MazF expressing cells could be resistant to infection. *Ex-vivo*-modified MazF-expressing CD4 T cells have been adoptively transferred into Rhesus macaques with success [106], and human trials are currently underway (www:Clinicaltrials.gov, NCT01787994). Recently, several other strategies have been employed to modify autologous cells to become resistant to infection or viral replication, and have shown great promise [107].

4 Manufacturing Cells for Adoptive Transfer

Once a mononuclear cell product has been obtained from a patient via apheresis collection, the product must undergo processing to expand the desired effector cell population. In addition, an appropriate specificity must either be selected or engineered into the cell population to be transferred. Targeted specificity can be obtained by enriching for antigen-specific cells or by using vector-based systems to introduce non-endogenous TCRs or synthetic CARs. The ultimate goal of the manufacturing process is to start with a limited polyclonal population and to generate a large, targeted cell population capable of potent on-target killing and persistence upon transfer.

4.1 Preparing Cells for Culture

Prior to initiating the culture of autologous cells for adoptive transfer obtained by apheresis, the collection product must be washed. If a T- or NK-cell product is to be manufactured, the cells will then undergo monocyte depletion by magnetic beads and/or counterflow centrifugal elutriation. If dendritic cells are being isolated, however, the monocytic fraction will be retained. In order to achieve an isolation of lymphocyte subsets, the cells are incubated with magnetic beads that are covalently linked to monoclonal antibodies specific for cell-surface markers. The cells and beads are passed through a magnetic field and the flow-through, which is now depleted of the magnetically bound cells, is collected and sent for culture. This type of depletion can efficiently remove monocytes as well as any additional undesired cell population prior to culture via anti-CD4, anti-CD8, anti-CD25, and other antibody–bead combinations.

Magnetic bead purification is an ideal approach in the setting of a relatively frequent target cell population, and in which yield and purity are both considerations. Generally, bead purification to provide the starting culture material is sufficient, though interferon-gamma capture can also be considered to enrich infrequent cell populations [108]. Other techniques, such as cell sorting by flow cytometry have been employed, but depending on the number of cells to be prepared this form of isolation can be quite lengthy [109, 110]. Once the desired cell subset has been isolated, this population must be expanded to obtain an effective dose.

4.2 Artificial Antigen-Presenting Cells for *Ex-Vivo* Expansion

Cell expansion is typically achieved by mimicking *in vivo* proliferation signals. In the case of T cells, this includes the provision of both a primary signal through the TCR and a simultaneous costimulatory signal through the CD28 receptor [111]. Simultaneous signaling through immobilized CD3 and CD28 ligands has been shown to induce telomerase activity in CD4+ T cells [112], which suggests that costimulation not only allows proliferation but that the progeny will also retain further proliferative capacity. With other cell types, different protocols are required. For example, the expansion of gamma delta T cells requires CD137L, interleukin (IL)-2 and IL-21 [29], and can be enhanced by the addition of bisphosphonates or CD2 stimulation [113–115]. NK cells, on the other hand, have been stimulated to expand *ex vivo* by whole irradiated tumor cells, among other methods [116].

Ex vivo, the provision of proliferation signals can be achieved by immobilizing ligands on either beads or cell scaffolds. Bead-bound monoclonal antibodies directed against CD3 and CD28 have been used to provide T-cell proliferative signals. These beads are logistically convenient in that they can easily be removed after coincubation and prior to reinfusion. This approach is particularly well suited to the expansion of human polyclonal naïve and central memory T cells, allowing for a more than 10^9-fold expansion of the input cell population [117]. In addition, anti-CD3/anti-CD28 antibody-coated beads do not cross-react with CTLA-4 (a negative costimulatory signal), and this allows for the efficient expansion of T cells. Good manufacturing practice (GMP)-compliant anti-CD3/anti-CD28 beads for use in T-cell expansion have been used in clinical trials since 1996 [118, 119].

Newer cell-based artificial antigen-presenting cell (aAPC) platforms have been developed in the chronic myeloid leukemia (CML) K562 cell line [120–122]. Although K562 cells lack MHC or T costimulatory ligands, they do express intercellular adhesion molecule (ICAM) and lymphocyte function-associated antigen 3 (LFA-3), both of which are enhancers of T cell–APC interaction. In addition, by employing lentiviral vector transduction this cell line can be engineered to express a library of T-cell signaling molecules that includes, but is not limited to, CD32, CD64, CD80, CD83, CD86, CD137L, and CD252 [120–122].

In addition to an ability to coexpress a wide variety of stimulatory molecules, aAPC platforms are more efficient than bead-based techniques at activating and expanding CD8+ T cells and antigen-specific T cells [120–122]. Cell-based aAPCs have been shown to expand the starting cell population approximately 10 000-fold and to maintain effector T-cell function and antigen specificity for more than two months [120, 122]. The efficient expansion of antigen-specific or CD8+ T cells is ideal for many cell therapy applications that aim to target and kill with specificity.

In addition, K562-based aAPCs can be used to expand NK cells for adoptive transfer. Engineered aAPCs expressing CD64/FcγRI, CD86/B7-2, CD137L/4-1BBL, truncated CD19, and membrane-bound IL-21 were shown to expand NK cells from peripheral blood mononuclear cells (PBMCs) from healthy human donors, with a more than 2000-fold expansion over a 14-day culture [123]. Other groups have developed current Good Manufacturing Practices-compliant K562-mediated NK cell expansion systems using IL-15 (mbIL15) and 4-1BB ligand [124]. Whilst the ideal expansion with maintenance of cytotoxicity upon transfer is still being elucidated, *ex-vivo* culture in the presence of membrane-bound IL-21 appears to be important [125, 126].

To enable a wide availability of autologous cell therapies, the next generation of *ex-vivo* expansion is likely to focus on scaling up and automating existing protocols. The transduction efficiency of available retroviral and lentiviral vectors is high even for large products [127–129]. Dynamic wave action bioreactors can accommodate up to 10^{10} or more cells for culture, while closed systems are also available at harvest for the large-scale removal of beads. The clinical-scale magnetic depletion of cell subsets is also available within a closed system. Clinical trials employing protocols for modifying cell products in this fashion are currently under way [130, 131]. These novel protocols may offer further advancements of antitumor cell therapy, or even allow T-cell therapy to branch out for the treatment of nonmalignant conditions such as autoimmune disease.

4.3 Polarizing Effector Cell Phenotypes

As lymphocyte subset analyses have become more refined, it has become clear that effector cell subpopulations demonstrate different capacities. In light of this, it has become desirable at times to skew cell therapy products towards specific cell subsets. Culture conditions represent one method to achieve a skewed cell population at the time of harvest. The *ex-vivo* culture of T cells with rapamycin (a mammalian target of rapamycin; mTOR inhibitor) and IL-4 has been shown to favor T-helper 2 (Th2) subsets, which upon transfer demonstrate favorable graft-versus-tumor outcomes with decreased GVHD in stem cell transplant patients [132]. Others have successfully generated tumor-specific, IL-17-producing primary human T cells by the introduction of a CAR with the inducible costimulatory (ICOS) intracellular domain. These Th17-CAR T cells may demonstrate enhanced persistence [133].

4.4 Expanding Immune Suppressors

CD4+ CD25+ forkhead box protein 3 (FoxP3)-positive T-regs are thymic-derived, *in vivo* suppressors. These cells have the ability to suppress active immune responses and to induce tolerance towards specific antigens [134], thereby playing a crucial role in preventing autoimmunity [135]. In mouse models, *ex-vivo*-expanded donor-derived T-regs have been shown to prevent acute and chronic GVHD while preserving graft-versus-tumor effects [136, 137]. However, in a different clinical setting T-regs may tolerize tumors and inhibit endogenous or adoptively transferred antitumor cytotoxic responses [138, 139]. For these reasons, a great deal of interest has been expressed in both expanding T-reg cell therapies to combat autoimmune disease, and depleting T-regs from antitumor cytotoxic T-cell products prior to reinfusion.

Generating T-regs *ex vivo* poses several obstacles. Typically, peripheral blood T-regs are scarce and there are no definitive specific cell-surface markers. Although most culture conditions favor non-T-reg growth, limiting *ex vivo* expansion [140], Golovina *et al.* showed that, in the presence of CD28 costimulation and mTOR inhibition by rapamycin, T-regs can be expanded more than 1000-fold while retaining their suppressive capacity [141]. Several trials have demonstrated the safety and feasibility of infusing *ex-vivo*-expanded T-regs in humans. However, although a clinical-scale expansion of human T-regs using rapamycin and artificial APCs has been developed, many of these protocols did not employ autologous cells.

Of note, given the capacity for T-regs to suppress, a primary concern is the potential for an increased risk of infection. In a clinical trial involving double umbilical cord blood transplants, 23 patients received enriched, *ex-vivo*-expanded T-regs. Notably, there was no reported increased risk of infection compared to historical controls, and the patients also showed a reduced incidence of grades II–IV acute GVHD [142]. Promising results such as these warrant further investigation into adoptively transferred T-regs as a potential therapy for autoimmune diseases.

5 Clinical Course of Adoptive Transfer

5.1 Preconditioning

The transfer of modified cells into a host with a moderately intact immune system is generally associated with a limited persistence of the transferred cells – an observation that seriously hampered the early clinical trials. In response, many groups began to lymphodeplete recipients immediately prior to reinfusion [65, 66, 143–145], with standard lymphodepletion regimens including cyclophosphamide with or without fludarabine or, at other sites, total body irradiation. Lymphodepletion prior to reinfusion is associated with improvements in the persistence of transferred T cells and of clinical outcome [66, 143, 146]. Several hypotheses exist as to the mechanism of improved outcome. Host lymphodepletion is likely to deplete the malignant cell population, increasing the effector:target ratio. In addition, host T-regs may be responsible for inhibiting the antitumor response by the transferred T cells. Finally, conditioning regimens are associated with an increase in IL-7 and IL-15, both of which cytokines have been associated with improved T-cell expansion post transfer [143]. Pre-transfer lymphodepletion may relieve this host-T-reg-derived inhibition, as well as contributing to long-lived persistence.

5.2 Managing Reinfusion of Autologous Cells

The reinfusion of autologous cells can have a number of on- and off-target effects requiring acute clinical management. Concurrently administered cytokines, such as high-dose IL-2, has resulted in flu-like symptoms and capillary leak with fluid retention [147]. Because modified cells may bear foreign epitopes, these cells may elicit a severe allergic response [148]. One report raised the possibility of adoptively transferred cells exacerbating an existing infection in a patient who demonstrated a sepsis-like syndrome [149]. In addition, screening for both off-target and on-target, off-tumor effects with bioinformatics or pre-clinical studies is imperfect. Unpredicted adverse effects have been observed wherein adoptively transferred, modified cells targeted to tumor antigens, demonstrated an *in vivo* affinity with cardiac muscle [84, 85] or pulmonary endothelium [150].

Once reinfused, adoptively transferred engineered CAR cells appear to be highly effective at lysing leukemic cells. In a clinical trial of two patients with acute lymphoblastic leukemia who were treated with anti-CD19 CAR-T cells, both patients demonstrated an *in vivo* expansion accounting for between 34% and 72% of peripheral CD3+ cells at peak [151]. The patients developed a cytokine release syndrome characterized by fever, with elevations of several cytokines including IL-1β, IL-6, interferon-γ, and TNF-α. The patients experienced a simultaneous reduction in tumor burden and an increase in lactate dehydrogenase consistent with tumor lysis syndrome. Both patients entered clinical remission [146]. This course, including the onset of *in-vivo*-modified cell expansion, cytokine release syndrome and tumor lysis, has been reproducibly observed at a number of centers [151–153]. Among a cohort of 30 relapsed and refractory ALL patients, a 90% complete response rate was observed with anti-CD19 CAR-T cells as a stand-alone therapy [153].

The management of patients as they endure severe tumor lysis and cytokine release syndromes continues to be a challenge. Anti-cytokine therapy has been used with some success [151], though further investigation of the role of these drugs is needed. Advances in cellular engineering may provide to key to safely tolerating the post-reinfusion period of tumor clearance. Di Stasi *et al.* used retroviral transduction to introduce an inducible caspase which, upon exposure to a synthetic dimerization agent, caused transduced cells to rapidly die [154]. One potential caveat to this approach may be that downstream inflammatory events may no longer be dependent on the transferred cells. For example, robust cytokine production initially in response to activated transferred cells may continue even after the elimination of those cells from the patient. Nonetheless, as more targeted cell therapies move into human trials, transduced "suicide genes" are likely to be employed in conjunction with already established anti-cytokine therapies.

While proof of principle has been established with regards to many adoptive cell therapies, further studies are needed to expand and refine targets. Engineered receptors confer a great deal of specificity, but *in-vitro* assays of target specificity have known limitations. At present, the true scope of off-target reactivity *in vivo* cannot be known until transfer to the patient [84, 85]. Even with high specificity, on-target/off-tumor adverse effects have also been observed [150]. Finally, even in the case of the transfer of cells with highly tumor-specific on-target activity, escape mutants have developed [151].

5.3
Promoting Long-Lived Persistence

Adoptively transferred cells have been shown to persist for more than two decades for modified hematopoietic stem cells [155], and for 10 years in the case of CAR-bearing T cells [15]. Not surprisingly, it has also been shown that a longer persistence of transferred cells correlates with a better clinical response [156]. For tumor-targeted modified cells, the *in-vivo* magnitude and expansion kinetics of recipient nonmalignant immune cells, the tumor burden and likely microenvironment, and the subtype of transferred cells influence the degree of persistence. As noted above, preconditioning has been associated with a greatly increased persistence of transferred cells [157, 158]. Several hypotheses exist as to the mechanism of this improved persistence. One possibility is that preconditioning lymphodepletion decreases the number of recipient nonmalignant immune cells, perhaps giving transferred cells a proliferative advantage over the recipient nonmalignant cells. Long-term persistence may also be influenced by the cell subset that is transferred. For instance, in animal models the long-term persistence of adoptively transferred T cells is associated with central memory T cells differentiation. Upon antigen activation, naïve T cells can generate effector memory T cells (TEM; CD8+ CD62L−) and/or long-lived central memory T cells (TCM; CD8+ CD62L−). Most clinical trials with T-cell products have transferred CD8+ TEM cells due to cell culture techniques favoring TEM expansion [159–161]. *In vivo*, however, TCM cells demonstrate superior tumor cytotoxicity [131, 159] and persistence compared to TEM cells [162]. Future studies in humans will likely optimize culture techniques to selectively expand the most effective subset cells for transfer.

6 NextGen Cell Therapy

During recent years, autologous cellular therapy has achieved enormous success, and this has generated an extensive development of clinical trials in humans. Indeed, as of October 2014, over 200 adoptive cellular therapy trials are listed (www.clinicaltrials.gov). Many trials have witnessed dramatic response rates in patients with refractory disease, and cellular therapy is now poised to make a monumental leap forward to the next generation of therapies capable of treating a myriad of diseases and large numbers of patients. Future studies will likely focus on: (i) further improving existing protocols; (ii) applying early success to new diseases; and (iii) making all cell therapies available to as many patients as possible.

An understanding of the management of cytokine release syndrome is critical in order for adoptive cellular therapies to become more widely available. It has yet to be determined which cytokines need to be blocked, and what effect this will have on the long-term persistence of the modified cells. Safety concerns may require a reconsideration of transient expression, as with RNA electroporated cells and/or the introduction of suicide genes. At present, a close monitoring of modified cell persistence and relapse is recommended, as the long term consequences of these therapies are not yet well understood.

One of the most exciting areas to observe over the next decade will be that of adoptive cellular therapies targeting solid tumors. In order to clear solid tumors, the adoptively transferred cells must first traffic and infiltrate the tissue, and then generate cytotoxicity against tumor cells within a solid mass. Each of these requirements for successful eradication present unique challenges to solid tumor-targeting cells, as opposed to their liquid tumor-targeting counterparts. It is very likely that these new demands will require the reconsideration of each step in the process of adoptive cell therapy development, from cell selection to reinfusion and the incorporation of additional immune modulators such as newly approved checkpoint blockade drugs. The inhibition of CTLA4 and/or PD1 has already been shown to be successful in combination with other therapies, and this would serve as a natural starting point to augment adoptive cell therapy [163].

Finally, with the goal of providing hope to those patients whose disease is refractory to current therapies, and regardless of the disease target, automation will emerge as a major challenge during the coming years. As noted above, various equipment already exists for large-scale production in closed systems, though all components have yet to be developed. Some autologous cell therapies may transition to allogeneic or banked products as available methods of gene and cell engineering advance. Likewise, clinical protocols will need to be developed and staff trained to perform techniques that, to date, have only been attempted at the laboratory bench. In addition, regulations may need to be revised or created as new technologies for autologous cell engineering are developed in the laboratory and translated to the clinic. Essentially, the infrastructure of producing and administering a therapeutic will need to be created to form a "next generation" of transfusion medicine. Clearly, this shifting paradigm will require a new generation of physicians and scientists that is committed to translating biology into medicine, and vice versa.

Disclosure Statement

Bruce Levine declares financial interest due to intellectual property and patents in the field of cell and gene therapy. Conflict of interest is managed in accordance with University of Pennsylvania policy and oversight. Dr Levine has sponsored research grant support from Novartis.

References

1. Mitchison, N.A. (1955) Studies on the immunological response to foreign tumor transplants in the mouse. *J. Exp. Med.*, **102** (2), 157–177.
2. Weiden, P.L., Flournoy, N., Thomas, E.D., Prentice, R. *et al.* (1979) Antileukemic effect of graft-versus-host disease in human recipients of allogeneic-marrow grafts. *N. Engl. J. Med.*, **300** (19), 1068–1073.
3. Porter, D.L., Roth, M.S., McGarigle, C., Ferrara, J. *et al.* (1994) Induction of graft-versus-host disease as immunotherapy for relapsed chronic myeloid leukemia. *N. Engl. J. Med.*, **330** (2), 100–106.
4. Kolb, H.J., Mittermuller, J., Clemm, C., Holler, E. *et al.* (1990) Donor leukocyte transfusions for treatment of recurrent chronic myelogenous leukemia in marrow transplant patients. *Blood*, **76** (12), 2462–2465.
5. Greenberg, P.D., Reusser, P., Goodrich, J., and Riddell, S. (1991) Development of a treatment regimen for human cytomegalovirus (CMV) infection in bone marrow transplantation recipients by adoptive transfer of donor-derived CMV-specific T cell clones expanded in vitro. *Ann. N. Y. Acad. Sci.*, **636**, 184–195.
6. Heslop, H.E., Slobod, K.S., Pule, M.A., Hale, G.A. *et al.* (2010) Long-term outcome of EBV-specific T-cell infusions to prevent or treat EBV-related lymphoproliferative disease in transplant recipients. *Blood*, **115** (5), 925–935.
7. Feuchtinger, T., Matthes Martin, S., Richard, C., Lion, T. *et al.* (2006) Safe adoptive transfer of virus-specific T-cell immunity for the treatment of systemic adenovirus infection after allogeneic stem cell transplantation. *Br. J. Haematol.*, **134** (1), 64–76.
8. Borrow, P., Lewicki, H., Hahn, B.H., Shaw, G.M. *et al.* (1994) Virus-specific CD8+ cytotoxic T-lymphocyte activity associated with control of viremia in primary human immunodeficiency virus type 1 infection. *J. Virol.*, **68** (9), 6103–6110.
9. Riddell, S.R., Elliott, M., Lewinsohn, D.A., and Gilbert, M.J. (1996) T-cell-mediated rejection of gene-modified HIV-specific cytotoxic T lymphocytes in HIV-infected patients. *Nat. Med.*, **2** (2), 216–223.
10. Cavazzana-Calvo, M., Hacein-Bey, S., de Saint Basile, G., Gross, F. *et al.* (2000) Gene therapy of human severe combined immunodeficiency (SCID)-X1 disease. *Science*, **288** (5466), 669–672.
11. Hacein-Bey-Abina, S., Le Deist, F., Carlier, F., Bouneaud, C. *et al.* (2002) Sustained correction of X-linked severe combined immunodeficiency by ex vivo gene therapy. *N. Engl. J. Med.*, **346** (16), 1185–1193.
12. Hacein-Bey-Abina, S., von Kalle, C., Schmidt, M., Le Deist, F. *et al.* (2003) A serious adverse event after successful gene therapy for X-linked severe combined immunodeficiency. *N. Engl. J. Med.*, **348** (3), 255–256.
13. Hacein-Bey-Abina, S., Garrigue, A., Wang, G.P., Soulier, J. *et al.* (2008) Insertional oncogenesis in 4 patients after retrovirus-mediated gene therapy of SCID-X1. *J. Clin. Invest.*, **118** (9), 3132–3142.
14. McGarrity, G.J., Hoyah, G., Winemiller, A., Andre, K. *et al.* (2013) Patient monitoring and follow-up in lentiviral clinical trials. *J. Gene Med.*, **15** (2), 78–82.
15. Scholler, J., Brady, T.L., Binder-Scholl, G., Hwang, W.T. *et al.* (2012) Decade-long safety and function of retroviral-modified chimeric antigen receptor T cells. *Sci. Transl. Med.*, **4** (132), 132ra53.
16. Ott, M.G., Schmidt, M., Schwarzwaelder, K., Stein, S. *et al.* (2006) Correction of X-linked chronic granulomatous disease by

gene therapy, augmented by insertional activation of MDS1-EVI1, PRDM16 or SETBP1. *Nat. Med.*, **12** (4), 401–409.
17. Braun, C.J., Boztug, K., Paruzynski, A., Witzel, M. *et al.* (2014) Gene therapy for Wiskott–Aldrich syndrome – long-term efficacy and genotoxicity. *Sci. Transl. Med.*, **6** (227), 227ra33.
18. Korngold, R. and Sprent, J. (1978) Lethal graft-versus-host disease after bone marrow transplantation across minor histocompatibility barriers in mice. *J. Exp. Med.*, **148** (6), 1687–1698.
19. Ruggeri, L., Mancusi, A., Capanni, M., Urbani, E. *et al.* (2007) Donor natural killer cell allorecognition of missing self in haploidentical hematopoietic transplantation for acute myeloid leukemia: challenging its predictive value. *Blood*, **110** (1), 433–440.
20. Venstrom, J.M., Pittari, G., Gooley, T.A., Chewning, J.H. *et al.* (2012) HLA-C – dependent prevention of leukemia relapse by donor activating KIR2DS1. *N. Engl. J. Med.*, **367** (9), 805–816.
21. Miller, J.S., Soignier, Y., Panoskaltsis-Mortari, A., McNearney, S.A. *et al.* (2005) Successful adoptive transfer and in vivo expansion of human haploidentical NK cells in patients with cancer. *Blood*, **105** (8), 3051–3057.
22. Ljunggren, H.-G. and Malmberg, K.-J. (2007) Prospects for the use of NK cells in immunotherapy of human cancer. *Nat. Rev. Immunol.*, 7 (5), 329–339.
23. Arai, S., Meagher, R., Swearingen, M., Myint, H. *et al.* (2008) Infusion of the allogeneic cell line NK-92 in patients with advanced renal cell cancer or melanoma: a phase I trial. *Cytotherapy*, **10** (6), 625–632.
24. Ishikawa, E., Tsuboi, K., Saijo, K., and Harada, H. (2004) Autologous natural killer cell therapy for human recurrent malignant glioma. *Anticancer Res.*, **24** (3b), 1861–1871.
25. Parkhurst, M.R., Riley, J.P., Dudley, M.E., and Rosenberg, S.A. (2011) Adoptive transfer of autologous natural killer cells leads to high levels of circulating natural killer cells but does not mediate tumor regression. *Clin. Cancer Res.*, **17** (19), 6287–6297.
26. Joncker, N.T., Fernandez, N.C., Treiner, E., Vivier, E. *et al.* (2009) NK cell responsiveness is tuned commensurate with the number of inhibitory receptors for self-MHC class I: the rheostat model. *J. Immunol.*, **182** (8), 4572–4580.
27. Vivier, E., Raulet, D.H., Moretta, A., Caligiuri, M.A. *et al.* (2011) Innate or adaptive immunity? The example of natural killer cells. *Science*, **331** (6013), 44–49.
28. Lamb, L.S. Jr., (2009) $\gamma\delta$ T cells as immune effectors against high-grade gliomas. *Immunol. Res.*, **45** (1), 85–95.
29. Deniger, D.C., Maiti, S., Mi, T., Switzer, K. *et al.* (2014) Activating and propagating polyclonal gamma delta T cells with broad specificity for malignancies. *Clin. Cancer Res.*, **20** (22), 5708–5719.
30. Buccheri, S., Guggino, G., and Caccamo, N. (2014) Efficacy and safety of gamma-delta-T cell-based tumor immunotherapy: a meta-analysis. *J. Biol. Regul. Homeost. Agents*, **28** (1), 81–90.
31. Kantoff, P.W., Higano, C.S., Shore, N.D., Berger, E.R. *et al.* (2010) Sipuleucel-T immunotherapy for castration-resistant prostate cancer. *N. Engl. J. Med.*, **363** (5), 411–422.
32. Noonan, K., Matsui, W., Serafini, P., Carbley, R. *et al.* (2005) Activated marrow-infiltrating lymphocytes effectively target plasma cells and their clonogenic precursors. *Cancer Res.*, **65** (5), 2026–2034.
33. Strasser, E.F., Dittrich, S., Weisbach, V., Zimmermann, R. *et al.* (2004) Comparison of two mononuclear cell program settings on two apheresis devices intended to collect high yields of CD14+ and CD3+ cells. *Transfusion*, **44** (7), 1104–1111.
34. Strasser, E.F., Berger, T.G., Weisbach, V., Zimmermann, R. *et al.* (2003) Comparison of two apheresis systems for the collection of CD14+ cells intended to be used in dendritic cell culture. *Transfusion*, **43** (9), 1309–1316.
35. Brown, R.A., Adkins, D., Goodnough, L.T., Haug, J.S. *et al.* (1997) Factors that influence the collection and engraftment of allogeneic peripheral-blood stem cells in patients with hematologic malignancies. *J. Clin. Oncol.*, **15** (9), 3067–3074.
36. Moog, R. (2004) Apheresis techniques for collection of peripheral blood progenitor cells. *Transfus. Apheresis Sci.*, **31** (3), 207–220.
37. Gašová, Z., Marinov, I., Vodvářková, Š., Böhmová, M. *et al.* (2005) PBPC collection techniques: standard versus large

volume leukapheresis (LVL) in donors and in patients. *Transfus. Apheresis Sci.*, **32** (2), 167–176.
38. Glaser, A., Schuler Thurner, B., Feuerstein, B., Zingsem, J. *et al.* (2001) Collection of MNCs with two cell separators for adoptive immunotherapy in patients with stage IV melanoma. *Transfusion*, **41** (1), 117–122.
39. Rols, M.-P. (2008) Mechanism by which electroporation mediates DNA migration and entry into cells and targeted tissues, in *Electroporation Protocols*, Methods in Molecular Biology (ed. S. Li), Humana Press, Totowa, NJ, pp. 19–33.
40. Bordignon, C., Notarangelo, L.D., Nobili, N., Ferrari, G. *et al.* (1995) Gene therapy in peripheral blood lymphocytes and bone marrow for ADA-immunodeficient patients. *Science*, **270** (5235), 470–475.
41. Rosenberg, S.A., Aebersold, P., Cornetta, K., Kasid, A. *et al.* (1990) Gene transfer into humans – immunotherapy of patients with advanced melanoma, using tumor-infiltrating lymphocytes modified by retroviral gene transduction. *N. Engl. J. Med.*, **323** (9), 570–578.
42. Stein, S., Ott, M.G., Schultze-Strasser, S., Jauch, A. *et al.* (2010) Genomic instability and myelodysplasia with monosomy 7 consequent to EVI1 activation after gene therapy for chronic granulomatous disease. *Nat. Med.*, **16** (2), 198–204.
43. Howe, S.J., Mansour, M.R., Schwarzwaelder, K., Bartholomae, C. *et al.* (2008) Insertional mutagenesis combined with acquired somatic mutations causes leukemogenesis following gene therapy of SCID-X1 patients. *J. Clin. Invest.*, **118** (9), 3143–3150.
44. Naldini, L., Blömer, U., Gallay, P., Ory, D. *et al.* (1996) In vivo gene delivery and stable transduction of nondividing cells by a lentiviral vector. *Science*, **272** (5259), 263–267.
45. Schröder, A.R.W., Shinn, P., Chen, H., Berry, C. *et al.* (2002) HIV-1 integration in the human genome favors active genes and local hotspots. *Cell*, **110** (4), 521–529.
46. Biffi, A., Bartolomae, C.C., Cesana, D., Cartier, N. *et al.* (2011) Lentiviral vector common integration sites in preclinical models and a clinical trial reflect a benign integration bias and not oncogenic selection. *Blood*, **117** (20), 5332–5339.
47. Song, W., Kong, H.-L., Carpenter, H., Torii, H. *et al.* (1997) Dendritic cells genetically modified with an adenovirus vector encoding the cDNA for a model antigen induce protective and therapeutic antitumor immunity. *J. Exp. Med.*, **186** (8), 1247–1256.
48. Shayakhmetov, D.M., Papayannopoulou, T., Stamatoyannopoulos, G., and Lieber, A. (2000) Efficient gene transfer into human CD34+ cells by a retargeted adenovirus vector. *J. Virol.*, **74** (6), 2567–2583.
49. Sun, J.-Y., Krouse, R.S., Forman, S.J., Senitzer, D. *et al.* (2002) Immunogenicity of a p210BCR-ABL fusion domain candidate DNA vaccine targeted to dendritic cells by a recombinant adeno-associated virus vector in vitro. *Cancer Res.*, **62** (11), 3175–3183.
50. Ivics, Z., Hackett, P.B., Plasterk, R.H., and Izsvák, Z. (1997) Molecular reconstruction of sleeping beauty, a Tc1-like transposon from fish, and its transposition in human cells. *Cell*, **91** (4), 501–510.
51. Peng, P.D., Cohen, C.J., Yang, S., Hsu, C. *et al.* (2009) Efficient nonviral sleeping beauty transposon-based TCR gene transfer to peripheral blood lymphocytes confers antigen-specific antitumor reactivity. *Gene Ther.*, **16** (8), 1042–1049.
52. Singh, H., Moyes, J.S.E., Huls, M.H., and Cooper, L.J.N. (2015) Manufacture of T cells using the sleeping beauty system to enforce expression of a CD19-specific chimeric antigen receptor. *Cancer Gene Ther.*, **22** (2), 95–100.
53. Birkholz, K., Hombach, A., Krug, C., Reuter, S. *et al.* (2009) Transfer of mRNA encoding recombinant immunoreceptors reprograms CD4+ and CD8+ T cells for use in the adoptive immunotherapy of cancer. *Gene Ther.*, **16** (5), 596–604.
54. Barrett, D.M., Zhao, Y., Liu, X., Jiang, S. *et al.* (2011) Treatment of advanced leukemia in mice with mRNA engineered T cells. *Hum. Gene Ther.*, **22** (12), 1575–1586.
55. Zhao, Y., Moon, E., Carpenito, C., Paulos, C.M. *et al.* (2010) Multiple injections of electroporated autologous T cells expressing a chimeric antigen receptor mediate regression of human disseminated tumor. *Cancer Res.*, **70** (22), 9053–9061.
56. Singh, N., Liu, X., Hulitt, J., Jiang, S. *et al.* (2014) Nature of tumor control by permanently and transiently-modified GD2 chimeric antigen receptor T cells in

xenograft models of neuroblastoma. *Cancer Immunol. Res.*, **2** (11), 1059–1070.

57. Beatty, G.L., Haas, A.R., Maus, M.V., Torigian, D.A. *et al.* (2013) Mesothelin-specific chimeric antigen receptor mRNA-engineered T cells induce antitumor activity in solid malignancies. *Cancer Immunol. Res.*, **2** (2), 112–120.
58. June, C.H. (2007) Principles of adoptive T cell cancer therapy. *J. Clin. Invest.*, **117** (5), 1204–1212.
59. Leen, A.M., Myers, G.D., Sili, U., Huls, M.H. *et al.* (2006) Monoculture-derived T lymphocytes specific for multiple viruses expand and produce clinically relevant effects in immunocompromised individuals. *Nat. Med.*, **12** (10), 1160–1166.
60. Hanley, P.J., Cruz, C., Savoldo, B., Leen, A.M. *et al.* (2009) Functionally active virus-specific T cells that target CMV, adenovirus, and EBV can be expanded from naive T-cell populations in cord blood and will target a range of viral epitopes. *Blood*, **114** (19), 4283–4292.
61. Melenhorst, J.J., Leen, A.M., Bollard, C.M., Quigley, M.F. *et al.* (2010) Allogeneic virus-specific T cells with HLA alloreactivity do not produce GVHD in human subjects. *Blood*, **116** (22), 4700–4702.
62. Rosenberg, S.A., Yannelli, J.R., Yang, J.C., Topalian, S.L. *et al.* (1994) Treatment of patients with metastatic melanoma with autologous tumor-infiltrating lymphocytes and interleukin 2. *J. Natl Cancer Inst.*, **86** (15), 1159–1166.
63. Rosenberg, S.A., Packard, B.S., Aebersold, P.M., Solomon, D. *et al.* (1988) Use of tumor-infiltrating lymphocytes and interleukin-2 in the immunotherapy of patients with metastatic melanoma. A preliminary report. *N. Engl. J. Med.*, **319** (25), 1676–1680.
64. Dudley, M.E., Wunderlich, J.R., Robbins, P.F., Yang, J.C. *et al.* (2002) Cancer regression and autoimmunity in patients after clonal repopulation with antitumor lymphocytes. *Science*, **298** (5594), 850–854.
65. Dudley, M.E., Yang, J.C., Sherry, R., Hughes, M.S. *et al.* (2008) Adoptive cell therapy for patients with metastatic melanoma: evaluation of intensive myeloablative chemoradiation preparative regimens. *J. Clin. Oncol.*, **26** (32), 5233–5239.
66. Wrzesinski, C., Paulos, C.M., Kaiser, A., Muranski, P. *et al.* (2010) Increased intensity lymphodepletion enhances tumor treatment efficacy of adoptively transferred tumor-specific T cells. *J. Immunother.*, **33** (1), 1–7.
67. Rosenberg, S.A., Yang, J.C., Sherry, R.M., Kammula, U.S. *et al.* (2011) Durable complete responses in heavily pretreated patients with metastatic melanoma using T-cell transfer immunotherapy. *Clin. Cancer Res.*, **17** (13), 4550–4557.
68. Czerniecki, B.J., Koski, G.K., Koldovsky, U., Xu, S. *et al.* (2007) Targeting HER-2/neu in early breast cancer development using dendritic cells with staged interleukin-12 burst secretion. *Cancer Res.*, **67** (4), 1842–1852.
69. Ramos, C.A. and Dotti, G. (2011) Chimeric antigen receptor (CAR)-engineered lymphocytes for cancer therapy. *Expert Opin. Biol. Ther.*, **11** (7), 855–873.
70. Berger, C., Turtle, C.J., Jensen, M.C., and Riddell, S.R. (2009) Adoptive transfer of virus-specific and tumor-specific T cell immunity. *Curr. Opin. Immunol.*, **21** (2), 224–232.
71. Riches, J.C., Davies, J.K., McClanahan, F., Fatah, R. *et al.* (2013) T cells from CLL patients exhibit features of T-cell exhaustion but retain capacity for cytokine production. *Blood*, **121** (9), 1612–1621.
72. Bonyhadi, M., Frohlich, M., Rasmussen, A., Ferrand, C. *et al.* (2005) In vitro engagement of CD3 and CD28 corrects T cell defects in chronic lymphocytic leukemia. *J. Immunol.*, **174** (4), 2366–2375.
73. Fishman, M. (2009) A changing world for DCvax: a PSMA loaded autologous dendritic cell vaccine for prostate cancer. *Expert Opin. Biol. Ther.*, **9** (12), 1565–1575.
74. Wheeler, C.J. and Black, K.L. (2009) DCVax®-Brain and DC vaccines in the treatment of GBM. *Expert Opin. Biol. Ther.*, **18** (4), 509–519.
75. Phuphanich, S. and Wheeler, C.J. (2013) Glioma-associated antigens associated with prolonged survival in a phase I study of ICT-107 for patients with newly diagnosed glioblastoma. *J. Immunother.*, **36** (2), 152–157.
76. Phuphanich, S., Wheeler, C.J., Rudnick, J.D., Mazer, M. *et al.* (2013) Phase I trial of a multi-epitope-pulsed dendritic cell vaccine for patients with

newly diagnosed glioblastoma. *Cancer Immunol. Immunother.*, **62** (1), 125–135.
77. Hawkins, E.S., Alken, R., Chandler, J., Fink, KL, *et al.* (2012) A randomized, double-blind, controlled phase IIb study of the safety and efficacy of ICT-107 in newly diagnosed patients with glioblastoma multiforme following resection and chemoradiation. *J. Clin. Oncol.*, **30** (15), Suppl.; abstract TPS2107). Available at: *http://meetinglibrary.asco.org/content/100393-114.*
78. Kandalaft, L.E., Chiang, C.L., Tanyi, J., Motz, G. *et al.* (2013) A phase I vaccine trial using dendritic cells pulsed with autologous oxidized lysate for recurrent ovarian cancer. *J. Transl. Med.*, **11**, 149.
79. Kandalaft, L.E., Powell, D.J. Jr., Chiang, C.L., Tanyi, J. *et al.* (2013) Autologous lysate-pulsed dendritic cell vaccination followed by adoptive transfer of vaccine-primed ex vivo co-stimulated T cells in recurrent ovarian cancer. *Oncoimmunology*, **2** (1), e22664.
80. Cruz, C.R.Y., Micklethwaite, K.P., Savoldo, B., Ramos, C.A. *et al.* (2013) Infusion of donor-derived CD19-redirected virus-specific T cells for B-cell malignancies relapsed after allogeneic stem cell transplant: a phase 1 study. *Blood*, **122** (17), 2965–2973.
81. Bendle, G.M., Linnemann, C., Hooijkaas, A.I., Bies, L. *et al.* (2010) Lethal graft-versus-host disease in mouse models of T cell receptor gene therapy. *Nat. Med.*, **16** (5), 565–570.
82. Johnson, L.A., Morgan, R.A., Dudley, M.E., Cassard, L. *et al.* (2009) Gene therapy with human and mouse T-cell receptors mediates cancer regression and targets normal tissues expressing cognate antigen. *Blood*, **114** (3), 535–546.
83. Morgan, R.A., Chinnasamy, N., Abate-Daga, D.D., Gros, A. *et al.* (2013) Cancer regression and neurologic toxicity following anti-MAGE-A3 TCR gene therapy. *J. Immunother.*, **36** (2), 133–151.
84. Linette, G.P., Stadtmauer, E.A., Maus, M.V., Rapoport, A.P. *et al.* (2013) Cardiovascular toxicity and titin cross-reactivity of affinity-enhanced T cells in myeloma and melanoma. *Blood*, **122** (6), 863–871.
85. Cameron, B.J., Gerry, A.B., Dukes, J., Harper, J.V. *et al.* (2013) Identification of a titin-derived HLA-A1-presented peptide as a cross-reactive target for engineered MAGE A3-directed T cells. *Sci. Transl. Med.*, **5** (197), 197ra103.
86. Chapuis, A.G., Ragnarsson, G.B., Nguyen, H.N., Chaney, C.N. *et al.* (2013) Transferred WT1-reactive CD8+ T cells can mediate antileukemic activity and persist in post-transplant patients. *Sci. Transl. Med.*, **5** (174), 174ra27.
87. Chmielewski, M., Hombach, A.A., and Abken, H. (2013) Antigen-specific T-cell activation independently of the MHC: chimeric antigen receptor-redirected T cells. *Front. Immunol.*, **4**, 154–160.
88. Eshhar, Z., Waks, T., Gross, G., and Schindler, D.G. (1993) Specific activation and targeting of cytotoxic lymphocytes through chimeric single chains consisting of antibody-binding domains and the gamma or zeta subunits of the immunoglobulin and T-cell receptors. *Proc. Natl Acad. Sci. USA*, **90** (2), 720–724.
89. Bretscher, P. and Cohn, M. (1970) A theory of self-nonself discrimination paralysis and induction involve the recognition of one and two determinants on an antigen, respectively. *Science*, **169** (3950), 1042–1049.
90. Brocker, T. (2000) Chimeric Fv-ζ or Fv-ε receptors are not sufficient to induce activation or cytokine production in peripheral T cells. *Blood*, **96** (5), 1999–2001.
91. Finney, H.M., Akbar, A.N., and Lawson, A.D.G. (2004) Activation of resting human primary T cells with chimeric receptors: costimulation from CD28, inducible costimulator, CD134, and CD137 in series with signals from the TCRζ chain. *J. Immunol.*, **172** (1), 104–113.
92. Finney, H.M., Lawson, A.D.G., Bebbington, C.R., and Weir, A.N.C. (1998) Chimeric receptors providing both primary and costimulatory signaling in T cells from a single gene product. *J. Immunol.*, **161** (6), 2791–2797.
93. Krause, A., Guo, H.-F., Latouche, J.-B., Tan, C. *et al.* (1998) Antigen-dependent CD28 signaling selectively enhances survival and proliferation in genetically modified activated human primary T lymphocytes. *J. Exp. Med.*, **188** (4), 619–626.
94. Imai, C., Mihara, K., Andreansky, M., Nicholson, I.C. *et al.* (2004) Chimeric

receptors with 4-1BB signaling capacity provoke potent cytotoxicity against acute lymphoblastic leukemia. *Leukemia*, **18** (4), 676–684.
95. Carpenito, C., Milone, M.C., Hassan, R., Simonet, J.C. *et al.* (2009) Control of large, established tumor xenografts with genetically retargeted human T cells containing CD28 and CD137 domains. *Proc. Natl Acad. Sci. USA*, **106** (9), 3360–3365.
96. Zhong, X.-S., Matsushita, M., Plotkin, J., Rivière, I. *et al.* (2010) Chimeric antigen receptors combining 4-1BB and CD28 signaling domains augment PI3kinase/AKT/Bcl-XL activation and CD8+ T cell-mediated tumor eradication. *Mol. Ther.*, **18** (2), 413–420.
97. Roberts, M.R. (1998) Chimeric receptor molecules for delivery of co-stimulatory signals. US Patent Office 5,712,149; Cell Genesys, Inc., original assignee.
98. Maher, J., Brentjens, R.J., Gunset, G., Rivière, I. *et al.* (2002) Human T-lymphocyte cytotoxicity and proliferation directed by a single chimeric TCRζ /CD28 receptor. *Nat. Biotechnol.*, **20** (1), 70–75.
99. Milone, M.C., Fish, J.D., Carpenito, C., Carroll, R.G. *et al.* (2009) Chimeric receptors containing CD137 signal transduction domains mediate enhanced survival of T cells and increased antileukemic efficacy in vivo. *Mol. Ther.*, **17** (8), 1453–1464.
100. Pule, M.A., Savoldo, B., Myers, G.D., Rossig, C. *et al.* (2008) Virus-specific T cells engineered to coexpress tumor-specific receptors: persistence and antitumor activity in individuals with neuroblastoma. *Nat. Med.*, **14** (11), 1264–1270.
101. Lim, J.K., McDermott, D.H., Lisco, A., Foster, G.A. *et al.* (2010) CCR5 deficiency is a risk factor for early clinical manifestations of West Nile virus infection but not for viral transmission. *J. Infect. Dis.*, **201** (2), 178–185.
102. Fätkenheuer, G., Pozniak, A.L., Johnson, M.A., Plettenberg, A. *et al.* (2005) Efficacy of short-term monotherapy with maraviroc, a new CCR5 antagonist, in patients infected with HIV-1. *Nat. Med.*, **11** (11), 1170–1172.
103. Hütter, G., Nowak, D., Mossner, M., Ganepola, S. *et al.* (2009) Long-term control of HIV by CCR5Delta32/Delta32 stem-cell transplantation. *N. Engl. J. Med.*, **360** (7), 692–698.
104. Perez, E.E., Wang, J., Miller, J.C., Jouvenot, Y. *et al.* (2008) Establishment of HIV-1 resistance in CD4+ T cells by genome editing using zinc-finger nucleases. *Nat. Biotechnol.*, **26** (7), 808–816.
105. Tebas, P., Stein, D., Tang, W.W., Frank, I. *et al.* (2014) Gene editing of CCR5in autologous CD4 T cells of persons infected with HIV. *N. Engl. J. Med.*, **370** (10), 901–910.
106. Saito, N., Chono, H., Shibata, H., Ageyama, N. *et al.* (2014) CD4+ T cells modified by the endoribonuclease MazF are safe and can persist in SHIV-infected Rhesus macaques. *Mol. Ther. Nucleic Acids*, **3** (6), e168.
107. DiGiusto, D.L., Krishnan, A., Li, L., Li, H. *et al.* (2010) RNA-based gene therapy for HIV with lentiviral vector-modified CD34+ cells in patients undergoing transplantation for AIDS-related lymphoma. *Sci. Transl. Med.*, **2** (36), 36ra43.
108. Becker, C., Pohla, H., Frankenberger, B., Schüler, T. *et al.* (2001) Adoptive tumor therapy with T lymphocytes enriched through an IFN-gamma capture assay. *Nat. Med.*, **7** (10), 1159–1162.
109. Edinger, M. and Hoffmann, P. (2011) Regulatory T cells in stem cell transplantation: strategies and first clinical experiences. *Curr. Opin. Immunol.*, **23** (5), 679–684.
110. Putnam, A.L., Safinia, N., Medvec, A., Laszkowska, M. *et al.* (2013) Clinical grade manufacturing of human alloantigen-reactive regulatory T cells for use in transplantation. *Am. J. Transplant.*, **13** (11), 3010–3020.
111. Riley, J.L. and June, C.H. (2005) The CD28 family: a T-cell rheostat for therapeutic control of T-cell activation. *Blood*, **105** (1), 13–21.
112. Weng, N.P., Levine, B.L., June, C.H., and Hodes, R.J. (1995) Human naive and memory T lymphocytes differ in telomeric length and replicative potential. *Proc. Natl Acad. Sci. USA*, **92** (24), 11091–11094.
113. Lopez, R.D., Xu, S., Guo, B., Negrin, R.S. *et al.* (2000) CD2-mediated IL-12-dependent signals render human γδ-T cells resistant to mitogen-induced apoptosis, permitting the large-scale ex vivo expansion of functionally distinct lymphocytes: implications for the development of adoptive immunotherapy strategies. *Blood*, **96** (12), 3827–3837.

114. Sato, K., Kimura, S., Segawa, H., Yokota, A. *et al.* (2005) Cytotoxic effects of γδ T cells expanded ex vivo by a third generation bisphosphonate for cancer immunotherapy. *Int. J. Cancer*, **116** (1), 94–99.
115. Kondo, M., Sakuta, K., Noguchi, A., Ariyoshi, N. *et al.* (2008) Zoledronate facilitates large-scale ex vivo expansion of functional γδ T cells from cancer patients for use in adoptive immunotherapy. *Cytotherapy*, **10** (8), 842–856.
116. Lim, S.A., Kim, T.-J., Lee, J.E., Sonn, C.H. *et al.* (2013) Ex vivo expansion of highly cytotoxic human NK cells by cocultivation with irradiated tumor cells for adoptive immunotherapy. *Cancer Res.*, **73** (8), 2598–2607.
117. Levine, B.L., Bernstein, W.B., Connors, M., Craighead, N. *et al.* (1997) Effects of CD28 costimulation on long-term proliferation of CD4+ T cells in the absence of exogenous feeder cells. *J. Immunol.*, **159** (12), 5921–5930.
118. Carroll, R.G., Riley, J.L., Levine, B.L., Blair, P.J. *et al.* (1998) The role of co-stimulation in regulation of chemokine receptor expression and HIV-1 infection in primary T lymphocytes. *Semin. Immunol.*, **10** (3), 195–202.
119. Levine, B.L., Bernstein, W.B., Aronson, N.E., Schlienger, K. *et al.* (2002) Adoptive transfer of costimulated CD4+ T cells induces expansion of peripheral T cells and decreased CCR5 expression in HIV infection. *Nat. Med.*, **8** (1), 47–53.
120. Maus, M.V., Thomas, A.K., Leonard, D.G.B., Allman, D. *et al.* (2002) Ex vivo expansion of polyclonal and antigen-specific cytotoxic T lymphocytes by artificial APCs expressing ligands for the T-cell receptor, CD28 and 4-1BB. *Nat. Biotechnol.*, **20** (2), 143–148.
121. Thomas, A.K., Maus, M.V., Shalaby, W.S., June, C.H. *et al.* (2002) A cell-based artificial antigen-presenting cell coated with anti-CD3 and CD28 antibodies enables rapid expansion and long-term growth of CD4 T lymphocytes. *Clin. Immunol.*, **105** (3), 259–272.
122. Suhoski, M.M., Golovina, T.N., Aqui, N.A., and Tai, V.C. (2007) Engineering artificial antigen-presenting cells to express a diverse array of co-stimulatory molecules. *Mol. Ther.*, **15** (5), 981–988.
123. Liu, Y., Wu, H.-W., Sheard, M.A., Sposto, R. *et al.* (2013) Growth and activation of natural killer cells ex vivo from children with neuroblastoma for adoptive cell therapy. *Clin. Cancer Res.*, **19** (8), 2132–2143.
124. Fujisaki, H., Kakuda, H., Shimasaki, N., Imai, C. *et al.* (2009) Expansion of highly cytotoxic human natural killer cells for cancer cell therapy. *Cancer Res.*, **69** (9), 4010–4017.
125. Denman, C.J., Senyukov, V.V., Somanchi, S.S., Phatarpekar, P.V. *et al.* (2012) Membrane-bound IL-21 promotes sustained ex vivo proliferation of human natural killer cells. *PLoS One*, **7** (1), e30264.
126. Wang, X., Lee, D.A., Wang, Y., Wang, L. *et al.* (2013) Membrane-bound interleukin-21 and CD137 ligand induce functional human natural killer cells from peripheral blood mononuclear cells through STAT-3 activation. *Clin. Exp. Immunol.*, **172** (1), 104–112.
127. Mitsuyasu, R.T., Anton, P.A., Deeks, S.G., Scadden, D.T. *et al.* (2000) Prolonged survival and tissue trafficking following adoptive transfer of CD4ζ gene-modified autologous CD4+ and CD8+ T cells in human immunodeficiency virus- infected subjects. *Blood*, **96** (3), 785–793.
128. Levine, B.L., Humeau, L.M., and Boyer, J. (2006) Gene transfer in humans using a conditionally replicating lentiviral vector. *Proc. Natl Acad. Sci. USA*, **103** (46), 17372–17377.
129. Walker, R.E., Bechtel, C.M., Natarajan, V., Baseler, M. *et al.* (2000) Long-term in vivo survival of receptor-modified syngeneic T cells in patients with human immunodeficiency virus infection. *Blood*, **96** (2), 467–474.
130. Wang, X., Naranjo, A., Brown, C.E., Bautista, C. *et al.* (2012) Phenotypic and functional attributes of lentivirus-modified CD19-specific human CD8+ central memory T cells manufactured at clinical scale. *J. Immunother.*, **5** (9), 689–701.
131. Berger, C., Jensen, M.C., Lansdorp, P.M., Gough, M. *et al.* (2008) Adoptive transfer of effector CD8+ T cells derived from central memory cells establishes persistent T cell memory in primates. *J. Clin. Invest.*, **118** (1), 294–305.
132. Fowler, D.H., Mossoba, M.E., Steinberg, S.M., Halverson, D.C. *et al.* (2013) Phase

2 clinical trial of rapamycin-resistant donor CD4+ Th2/Th1 (T-Rapa) cells after low-intensity allogeneic hematopoietic cell transplantation. *Blood*, **121** (15), 2864–2874.

133. Guedan, S., Chen, X., Madar, A., Carpenito, C. *et al.* (2014) ICOS-based chimeric antigen receptors program bipolar TH17/TH1 cells. *Blood*, **124** (7), 1070–1080.
134. Sakaguchi, S., Yamaguchi, T., Nomura, T., and Ono, M. (2008) Regulatory T cells and immune tolerance. *Cell*, **133** (5), 775–787.
135. Sakaguchi, S., Takahashi, T., and Nishizuka, Y. (1982) Study on cellular events in post-thymectomy autoimmune oophoritis in mice. II. Requirement of Lyt-1 cells in normal female mice for the prevention of oophoritis. *J. Exp. Med.*, **156** (6), 1577–1586.
136. Taylor, P.A., Lees, C.J., and Blazar, B.R. (2002) The infusion of ex vivo activated and expanded CD4+CD25+ immune regulatory cells inhibits graft-versus-host disease lethality. *Blood*, **99** (10), 3493–3499.
137. Edinger, M., Hoffmann, P., Ermann, J., Drago, K. *et al.* (2003) CD4+CD25+ regulatory T cells preserve graft-versus-tumor activity while inhibiting graft-versus-host disease after bone marrow transplantation. *Nat. Med.*, **9** (9), 1144–1150.
138. Lee, J.C., Hayman, E., Pegram, H.J., Santos, E. *et al.* (2011) In vivo inhibition of human CD19-targeted effector T cells by natural T regulatory cells in a xenotransplant murine model of B cell malignancy. *Cancer Res.*, **71** (8), 2871–2881.
139. Kofler, D.M., Chmielewski, M., Rappl, G., Hombach, A. *et al.* (2011) CD28 costimulation impairs the efficacy of a redirected T-cell antitumor attack in the presence of regulatory T cells which can be overcome by preventing lck activation. *Mol. Ther.*, **19** (4), 760–767.
140. Bluestone, J.A. and Tang, Q. (2005) How do CD4+CD25+ regulatory T cells control autoimmunity? *Curr. Opin. Immunol.*, **17** (6), 638–642.
141. Golovina, T.N., Mikheeva, T., Suhoski, M.M., Aqui, N.A. *et al.* (2008) CD28 costimulation is essential for human T regulatory expansion and function. *J. Immunol.*, **181** (4), 2855–2868.
142. Brunstein, C.G., Miller, J.S., Cao, Q., McKenna, D.H. *et al.* (2011) Infusion of ex vivo expanded T regulatory cells in adults transplanted with umbilical cord blood: safety profile and detection kinetics. *Blood*, **117** (3), 1061–1070.
143. Gattinoni, L., Finkelstein, S.E., Klebanoff, C.A., Antony, P.A. *et al.* (2005) Removal of homeostatic cytokine sinks by lymphodepletion enhances the efficacy of adoptively transferred tumor-specific CD8+ T cells. *J. Exp. Med.*, **202** (7), 907–912.
144. Powell, D.J., Dudley, M.E., Hogan, K.A., Wunderlich, J.R., and Rosenberg, S.A. (2006) Adoptive transfer of vaccine-induced peripheral blood mononuclear cells to patients with metastatic melanoma following lymphodepletion. *J. Immunol.*, **177** (9), 6527–6539.
145. Muranski, P., Boni, A., Wrzesinski, C., Citrin, D.E. *et al.* (2006) Increased intensity lymphodepletion and adoptive immunotherapy – how far can we go? *Nat. Rev. Clin. Oncol.*, **3** (12), 668–681.
146. Porter, D.L., Levine, B.L., Kalos, M., Bagg, A. *et al.* (2011) Chimeric antigen receptor-modified T cells in chronic lymphoid leukemia. *N. Engl. J. Med.*, **365** (8), 725–733.
147. Rosenberg, S.A., Restifo, N.P., Yang, J.C., Morgan, R.A. *et al.* (2008) Adoptive cell transfer: a clinical path to effective cancer immunotherapy. *Nat. Rev. Cancer*, **8** (4), 299–308.
148. Maus, M.V., Haas, A.R., Beatty, G.L., Albelda, S.M. *et al.* (2013) T cells expressing chimeric antigen receptors can cause anaphylaxis in humans. *Cancer Immunol. Res.*, **1** (1), 26–31.
149. Brentjens, R., Yeh, R., Bernal, Y., Rivière, I. *et al.* (2010) Treatment of chronic lymphocytic leukemia with genetically targeted autologous T cells: case report of an unforeseen adverse event in a phase I clinical trial. *Mol. Ther.*, **18** (4), 666–668.
150. Morgan, R.A., Yang, J.C., Kitano, M., Dudley, M.E. *et al.* (2010) Case report of a serious adverse event following the administration of T cells transduced with a chimeric antigen receptor recognizing ERBB2. *Mol. Ther.*, **18** (4), 843–851.
151. Grupp, S.A., Kalos, M., Barrett, D., Aplenc, R. *et al.* (2013) Chimeric antigen receptor-modified T cells for acute

lymphoid leukemia. *N. Engl. J. Med.*, **368** (16), 1509–1518.
152. Lee, D.W., Kochenderfer, J.N., Stetler-Stevenson, M., Cui, Y.K. *et al.* (2014) T cells expressing CD19 chimeric antigen receptors for acute lymphoblastic leukaemia in children and young adults: a phase 1 dose-escalation trial. *Lancet*, **385**, 517–528.
153. Maude, S.L., Frey, N., Shaw, P.A., Aplenc, R. *et al.* (2014) Chimeric antigen receptor T cells for sustained remissions in leukemia. *N. Engl. J. Med.*, **371** (16), 1507–1517.
154. Di Stasi, A., Tey, S.-K., Dotti, G., Fujita, Y. *et al.* (2011) Inducible apoptosis as a safety switch for adoptive cell therapy. *N. Engl. J. Med.*, **365** (18), 1673–1683.
155. Candotti, F., Shaw, K.L., Muul, L., Carbonaro, D. *et al.* (2012) Gene therapy for adenosine deaminase-deficient severe combined immune deficiency: clinical comparison of retroviral vectors and treatment plans. *Blood*, **120** (18), 3635–3646.
156. Robbins, P.F., Dudley, M.E., Wunderlich, J., El-Gamil, M. *et al.* (2004) Cutting edge: persistence of transferred lymphocyte clonotypes correlates with cancer regression in patients receiving cell transfer therapy. *J. Immunol.*, **173** (12), 7125–7130.
157. Rosenberg, S.A.,. and Dudley, M.E. (2009) Adoptive cell therapy for the treatment of patients with metastatic melanoma. *Curr. Opin. Immunol.*, **21** (2), 233–40.
158. Brentjens, R.J., Rivière, I., Park, J.H., Davila, M.L. *et al.* (2011) Safety and persistence of adoptively transferred autologous CD19-targeted T cells in patients with relapsed or chemotherapy refractory B-cell leukemias. *Blood*, **118** (18), 4817–4828.
159. Klebanoff, C.A., Gattinoni, L., Torabi-Parizi, P., Kerstann, K. *et al.* (2005) Central memory self/tumor-reactive CD8+ T cells confer superior antitumor immunity compared with effector memory T cells. *Proc. Natl Acad. Sci. USA*, **102** (27), 9571–9576.
160. Monteiro, J., Batliwalla, F., Ostrer, H., and Gregersen, P.K. (1996) Shortened telomeres in clonally expanded CD28-CD8+ T cells imply a replicative history that is distinct from their CD28+CD8+ counterparts. *J. Immunol.*, **156** (10), 3587–3590.
161. Van den Hove, L.E., Van Gool, S.W., Vandenberghe, P., Boogaerts, M.A. *et al.* (1998) CD57+/CD28- T cells in untreated hemato-oncological patients are expanded and display a Th1-type cytokine secretion profile, ex vivo cytolytic activity and enhanced tendency to apoptosis. *Leukemia*, **12** (10), 1573–1582.
162. Louis, C.U., Savoldo, B., Dotti, G., Pule, M. *et al.* (2011) Antitumor activity and long-term fate of chimeric antigen receptor-positive T cells in patients with neuroblastoma. *Blood*, **118** (23), 6050–6056.
163. Pardoll, D.M. (2012) The blockade of immune checkpoints in cancer immunotherapy. *Nat. Rev. Cancer*, **12** (4), 252–264.

21
Pharmaco-Epigenomics to Improve Cancer Therapies*

Bart Claes[1,2], *Bernard Siebens*[1,2], *and Diether Lambrechts*[1,2]
[1] *VIB, Vesalius Research Center, Herestraat 49, Box 912, Leuven 3000, Belgium*
[2] *KU Leuven Campus Gasthuisberg, Vesalius Research Center, Herestraat 49, Box 912, Leuven 3000, Belgium*

Keywords

Epigenomics
The study of epigenetic changes on a genome-wide level, where epigenetic refers to heritable changes in gene expression and activity that are not due to changes in the DNA sequence, but are caused by covalent modifications of DNA and histones.

*This chapter has previously been published in: Meyers, R.A. (Ed.) Epigenetic Regulation and Epigenomics, 2012, 978-3-527-32682-2.

Translational Medicine Cancer: Advances in Molecular Biology and Medicine,
First Edition. Edited by Robert A. Meyers.

Pharmaco-epigenomics
A recently emerged field of research that uses epigenetic insights to improve pharmacological therapies. The goals are to develop therapies targeted at the epigenome, to identify epigenomic biomarkers that can be used for disease diagnosis, and to predict disease progression as well as the response to a certain therapy.

DNA methylation
The addition of a methyl group to the 5-carbon of the base cytosine in the DNA.

Histone modifications
Post-translational modifications to histones, which are the main protein components of chromatin. Histone modifications include methylation and acetylation.

Cancer therapy
Chemical, physical, or biological treatment of cancer patients in order to inhibit the growth of cancer cells. The most widely used cancer therapies are chemotherapy, radiotherapy, molecularly targeted therapy (which targets a certain protein that is important for cancer growth) and epigenetic therapy.

Angiogenesis
The growth of new blood vessels from pre-existing vessels. Tumors exploit this process to recruit new blood vessels that fuel their growth. Anti-angiogenic therapies represent an effective therapy to slow down tumor growth.

Epigenetic modifications such as aberrant DNA methylation or altered histone modifications play an important role in tumor development and progression. Consequently, therapeutic strategies aimed at reversing epigenetic changes in cancer cells have been developed, and their great promise confirmed in recent studies. Epigenetic modifications can also be used as biomarkers to improve cancer diagnosis or to predict disease prognosis. Importantly, the first examples of how epigenetic alterations can be used as predictive markers for the outcome of conventional chemotherapies or targeted therapies have recently emerged. In this chapter, the recent advances in the field of pharmaco-epigenomics will be highlighted, and a review included of the most recently discovered and promising epigenetic therapies and biomarkers aimed at improving cancer diagnosis and treatment. In particular, attention will be focused on how these epigenetic therapies or biomarkers might change daily clinical practice.

1 Introduction

Despite major advances in its treatment, mortality resulting from cancer remains high, with numbers of cancer-associated deaths worldwide expected to rise to an estimated 12 million in 2030. Much of the recent progress in cancer therapy can be attributed to an improved understanding of cancer as a genetic disease, with several somatic alterations in oncogenes and tumor suppressor genes having been described, functionally validated, and successfully targeted by therapies in the clinic. However, the activity of genes in cancer cells can also be altered by epigenetic mechanisms, which are generally defined as heritable changes in gene expression that are not caused by changes in the DNA sequence itself. Epigenetics encompasses a complex range of reversible changes, which can be categorized into either modifications of the DNA itself or into modifications of the histones. Whilst methylation is the most extensively studied and best understood epigenetic modification of DNA, in the case of histones many different types of post-translational modification have been described, including acetylation, phosphorylation, ubiquitylation, and sumoylation. In general, histone modifications are less well understood than DNA modifications, and have been reported to occur in different histone proteins, to affect various histone amino acid residues, and to exhibit different degrees of modification (e.g., mono-, di-, and trimethylation).

Besides modifications of the histones and DNA, the microRNAs (miRNAs) represent an additional layer of epigenetic regulation. These small, noncoding RNAs are present endogenously in cells, and function as post-transcriptional silencers of certain sets of target genes [1]. Since the miRNAs induce changes in gene expression that can be heritable, they are also considered to be part of the epigenetic landscape. Moreover, crosstalk between miRNAs and other epigenetic pathways has recently been demonstrated: for example, miRNA expression can be subjected to epigenetic control [2], while in contrast miRNAs can regulate the expression of key components of the epigenetic machinery [3].

An improved understanding of which epigenetic changes occur in tumors, and how they drive tumor growth, have led to numerous attempts at improving existing cancer therapies. In this chapter, attention is focused first on the fundamental role of epigenetic changes in cancer, in terms of the two most well-established mechanisms, namely DNA methylation and histone modifications. The contribution of miRNAs to the deregulation of the epigenetic machinery in cancer has been less clearly established to date, and so will be mentioned only briefly at this point. A description will be provided of how an improved understanding of epigenetic processes has led to the development of epigenetic therapies aimed at reversing the changes that accumulate in cancer cells. On this basis, the first clinical results obtained with DNA methyltransferase (DNMT) and histone deacetylase (HDAC) inhibitors in the treatment of cancer are described, and suggestions made as to how epigenetic modifications could be used not only to assist in cancer diagnosis but also to predict the prognosis and/or response to a specific therapy. The means by which the methylation status of DNA repair genes might represent a promising biomarker to predict the response to various chemotherapeutic treatments is also discussed. Finally, details are provided of how the tumor stroma can accumulate epigenetic changes and, in contrast, how epigenetic therapies may also target the stroma. In particular, the role of epigenetic changes in blood vessels will be discussed with regards to the anti angiogenic therapies currently applied on a routine basis in the clinic.

2
The Epigenetic Origin of Cancer

Epigenetic changes occur both early and ubiquitously during the process of carcinogenesis, and continue to accumulate during tumor progression. The two main epigenetic hallmarks of cancer are an altered DNA methylation and histone modifications. Although the exact origin of these alterations is unclear, environmental influences are clearly implicated in their manifestation. For example, whilst monozygotic twins are epigenetically indistinguishable during their early years of life, at an older age they may show remarkable differences in their overall content and distribution of DNA methylation and histone modifications, accounting for the different environmental exposures encountered during their lives [4].

Whereas the role of most epigenetic modifications is poorly understood – especially in the context of cancer – the role of DNA methylation has been investigated more extensively. DNA methylation changes involve both global DNA hypomethylation and local DNA hypermethylation [5]:

- Global DNA *hypomethylation* is frequently observed in tumors, and mostly affects the intergenic regions of the DNA, in particular the repetitive DNA sequences and transposable DNA elements. Hypomethylation is generally believed to result in chromosomal instability and increased mutation events, thereby contributing to tumorigenesis [6], although it has also been associated with the activation of many growth-promoting genes, such as *HRAS*, c-*MYC*, or c-*JUN* [7, 8].
- In contrast, the local DNA *hypermethylation* of CpG islands is found in many promoters of tumor suppressor genes, and is well accepted to repress transcription by inhibiting the binding of specific transcription factors (TFs) and, more indirectly, also by recruiting methyl-CpG-binding proteins and their associated chromatin-remodeling complexes (see Fig. 1) [9].

In cancer, promotor hypermethylation is mostly associated with tumor-suppressor gene silencing, such as the retinoblastoma (*RB*), cyclin-dependent kinase inhibitor 2A (*CDKN2A*), Von Hippel–Lindau (*VHL*), breast cancer 1 (*BRCA1*), or human MutL homolog 1 (*hMLH1*) genes.

CpG island hypermethylation occurs at different stages of cancer development, and affects those genes that are involved in pathways which regulate the cell cycle, apoptosis, DNA repair, cell–cell communication, and angiogenesis, all of which become deregulated during tumor progression [10, 11]. Furthermore, CpG island hypermethylation profiles are specific for each cancer type and, similar to the cancer genome, each type can be assigned a cancer methylome [12, 13]. Recently, miRNA genes have also been reported to be inactivated by altered DNA methylation in cancer. For instance, *miR-127* expression is strongly reduced in cancer cells, but can be restored upon treatment with the demethylating agent 5-aza-2′-deoxycytidine. The restoration of a normal expression pattern leads to translational downregulation of the *miR-127* target BCL6, which is a proto-oncogene [14]. In another study, it was shown that the epigenetic inactivation of *miR-124a* in different human tumor types is correlated with the activation of cyclin D kinase 6, a crucial component of complexes that control progression through the cell cycle [15]. These recent findings illustrate how cancer cells can epigenetically silence miRNAs that act as tumor suppressors in order to promote their uncontrolled growth.

Similar to DNA methylation, histone modifications are commonly disrupted in cancer cells. In general, histone modifications determine how tightly or loosely the DNA is wrapped around the histones and, as such, play an important regulatory

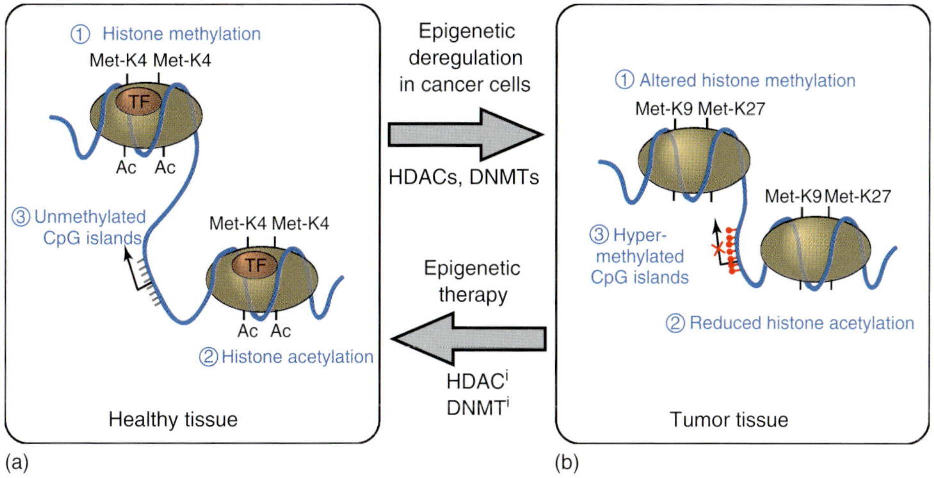

Fig. 1 Restoring epigenetic changes in cancer cells with epigenetic therapy. (a) Chromatin is composed of double-stranded DNA (blue) wrapped around histones (brown). Gene expression is regulated by histone modifications and DNA methylation of CpG islands in the promotor. In normal cells, chromatin surrounding transcriptionally active genes is characterized by specific modifications, such as the methylation of Lys4 of histone H3 ①, histone hyperacetylation ②, and unmethylated CpG islands ③. This leads to an open chromatin state that allows the binding of transcription factors (TFs) and RNA polymerase complexes; (b) In tumor cells, the epigenetic status of chromatin regions becomes altered by the disruption of normal DNA methylation and histone modification patterns. Epigenetic inactivation of tumor suppressor genes is associated with altered histone methylation ①, such as methylation of Lys9 and Lys27 of histone H3, histone hypoacetylation ②, and hypermethylation of CpG islands in promotors ③, leading to a more compact chromatin structure that represses transcription. By interfering with DNA and histone-modifying enzymes, epigenetic drugs can remove inactivation markers, such as DNA methylation, and can induce active markers, such as histone acetylation, leading to the restoration of disrupted epigenetic patterns. AC, acetylation; DNMTs, DNA methyltransferases; HDACs, histone deacetylases; DNMTi, DNMT inhibitor; HDACi, HDAC inhibitor; Met-K4, methylation of Lys4; Met-K9, methylation of Lys9; Met-K27, methylation of Lys27.

role in determining gene expression [16]. For instance, acetyl groups neutralize positive charges on the histone tails, leading to weakened electrostatic interactions between the histones and the negatively charged phosphate backbone of DNA [17]. This correlates with an active or open chromatin state, which in turn allows TFs to access the promoters of target genes. By contrast, HDAC-induced deacetylation results in a compaction of the chromatin and an inactivation of genes (see Fig. 1). A global loss of the monoacetylation and trimethylation of histone H4 can be considered a common hallmark of human tumor cells, while altered histone modifications constitute a mechanism for the inactivation of tumor suppressor genes, as illustrated by the hypermethylation of Lys9 in histone H3 of the *CDKN2A* gene [18–20].

Another example of altered HDAC activity is found in acute promyelocytic leukemia (PML). The retinoic acid receptor (RAR) is an important regulator of myeloid cell differentiation, and induces transcriptional repression in the absence of retinoid ligands. Upon the binding of RAR to the retinoic acid response elements (RAREs), the HDACs are recruited to silence the RAR target genes. In PML, translocations cause the production of fusion genes consisting of the *RARα* and *PML* gene (translocation t(15;17)) or *RARα* and *PLZF* (t(11;17)) [21, 22]. The resulting fusion proteins bind to the RAREs and recruit HDACs, but are no longer responsive to retinoids, which results in a permanent and stable silencing of the RAR target genes. These examples demonstrate the powerful oncogenic potential of aberrant HDAC activity. Furthermore, the disruption of histone modifications occurs early during tumorigenesis and accumulates during tumor progression, thus underlining its importance as an essential epigenetic driver of cancer development [5].

Currently, there is emerging evidence that the different epigenetic processes – including DNA methylation and histone modification – are interdependent [23]. Although they involve different chemical reactions and require different sets of enzymes, the biological relationship between these two systems is essential for the modulation of gene expression. It appears that histone methylation can determine specific DNA methylation patterns, while DNA methylation can serve as a template for histone modifications after DNA replication [23]. This mechanistic connection can be accomplished through direct interactions between DNA and the histone-modifying complexes. For instance, methylated DNA has been shown to attract methylated-DNA-binding proteins, such as the transcriptional repressor MeCP2, which recruits chromatin-remodeling complexes to methylated CpG islands. These in turn modify histones and shape the higher-order chromatin structure [16, 24, 25]. Taken together, this interdependency of epigenetic pathways and their deregulation in cancer offers great potential for clinical applications, as will be discussed below.

3
Pharmaco-Epigenomics: Translating Epigenetics into Clinical Practice

As the fundamental role of epigenetic changes in cancer etiology become more clearly understood, the obvious next step would be to translate such knowledge into improved cancer therapies. This novel and emerging research field, which aims at translating epigenetic insights into clinical practice, is referred to as "pharmaco-epigenomics," and is focused on two main areas of interest:

- To develop cancer therapies that reverse the epigenetic changes which accumulate in cancer cells. In contrast to the genetic alterations that accumulate in tumors (such as somatic mutations and chromosomal rearrangements), altered histone modifications and DNA methylation profiles are reversible in nature, and this makes them particularly attractive targets for therapeutic intervention.
- To identify epigenetic biomarkers that could be used to diagnose cancer, as well as to predict disease progression and the response of a specific tumor to a therapy.

Both areas of research, as well as their clinical implications, are discussed in detail in the following sections.

4
Epigenetic Therapies for Cancer

Insights into the fundamental role that epigenetic alterations play during tumorigenesis have led to the development of novel treatment strategies for cancer, which in general are aimed at blocking or reversing the epigenetic alterations that promote malignancy and allow cancer cells to adapt to changes in the microenvironment (see Fig. 1). Whilst the two gene families to be most frequently targeted are the HDACs and the DNMTs, the exact mechanisms underlying the antitumor activity of drugs that target the HDACs or DNMTs have not been elucidated; however, given the vast influences that DNA methylation and histone modifications exert on gene expression, many cellular pathways are likely to be involved. For example, the HDAC inhibitors valproic acid and sodium butyrate have been reported to induce the expression of *CDKN1A*, a cyclin-dependent kinase (CDK) inhibitor that controls cell-cycle arrest and cell differentiation [26]. Valproic acid and sodium butyrate may also activate multiple apoptotic pathways, involving factors such as nuclear factor-κB (NF-κB), c-Jun N-terminal kinase (c-JNK), and B-cell lymphoma 2 (BCL2) [27–29]. Furthermore, HDAC inhibitors have been shown to alter expression of angiogenesis and metastasis-associated genes [30, 31]. Similarly, the DNMT inhibitor decitabine (5-aza-2′-deoxycytidine) reactivates epigenetically silenced tumor suppressor genes, such as *CDKN2A*, or genes implicated in metastasis, such as tissue inhibitor of metalloproteinases 3 (*TIMP3*) [32, 33]. The results of recent studies have also indicated that a DNA-demethylating treatment with decitabine can rescue the growth-inhibitory effects of certain miRNAs [14].

4.1 Epigenetic Therapies Used in Clinical Practice

In the meantime, it has been established that certain tumor types respond well to DNMT and HDAC inhibitor treatments, with the best clinical efficacy seen in hematologic malignancies. The DNMT inhibitor decitabine, for instance, has been approved for the treatment of patients with myelodysplastic syndrome or acute myeloid leukemia [34, 35]. The structure of decitabine mimics that of cytosine, which allows it to replace single cytosine molecules during DNA replication; it is also believed to form covalent adducts with DNMTs, thereby reducing the overall cellular DNMT activity. This interferes with the normal role of several DNMTs, which can no longer reproduce the methylation pattern of the cell in the DNA of daughter cells, and this leads to a reactivation of epigenetically silenced regions [36, 37].

At present, several clinical trials of epigenetic therapy for solid tumors are under way, and although decitabine has not yet been proven effective, it has been reported to stabilize the disease in lung cancer patients [38, 39]. Skepticism regarding the use of decitabine persists, however, since the clinical benefits attributed to it have not been shown to be caused directly by an interference with the epigenetic machinery. The obvious next step would be to examine epigenetic markers and gene expression patterns in patients before and after decitabine treatment, and to evaluate any correlations with benefit from epigenetic therapy. Other DNMT inhibitors, such as the orally active zebularine, are currently also under development [40], while the wide-spectrum HDAC inhibitor trichostatin A (TSA) also appears to be effective against leukemia. A potential synergy between HDAC inhibitors and DNMT inhibitors (e.g., TSA and decitabine) is also currently under investigation [41, 42]. The HDAC inhibitor vorinostat (suberoylanilide hydroxamic acid) has also recently been approved for the treatment of cutaneous T-cell lymphoma in patients with progressive, persistent, or recurrent disease [43].

Another clinically used HDAC inhibitor, phenylbutyrate, has achieved partial responses in hematologic malignancies [44].

An overview of the different DNMT and HDAC inhibitors currently used in clinical practice, or which are still under development, is provided in Table 1.

4.2 Synergism between Epigenetic and Conventional Therapies

Since epigenetic therapy can induce cancer cell reprogramming, it is possible that HDAC and DNMT inhibitors might act synergistically with conventional chemotherapy. This would not only allow the chemotherapy to be applied at lower dosages, but would also result in a reduced toxicity, while the efficacy of the combined therapy would still be increased compared to that of monotherapy. Chemotherapy often induces both genetic and epigenetic alterations that result in the selection of resistant cell clones. Interestingly, however, tumors that have become resistant to the initial treatment with chemotherapy due to epigenetic changes might become sensitive again to the drug when exposed to epigenetic therapies [61]. This principle appears to apply to interferon, which is the standard treatment for melanoma. Although interferon induces tumor cell apoptosis, it has been shown recently that the selection of apoptosis-resistant clones may occur during treatment via epigenetic mechanisms [62]. The epigenetic silencing of genes involved in signaling downstream of interferon, such as interferon regulatory factor 8 (*IRF8*) and XIAP-associated factor 1 (*XAF1*), was found to trigger resistance against interferon therapy [62, 63]. Furthermore, the injection of decitabine into nude mice carrying melanoma xenografts led to a re-sensitization to interferon treatment [62]. A Phase I clinical trial conducted in melanoma patients, in which decitabine was administered together with interleukin-2, further demonstrated an objective response in 31% of the patients [64]. Likewise, preclinical data have suggested that epigenetic therapy can induce radiosensitization and enhance the efficacy of current conventional radiotherapy treatment regimens [65]. Whilst the molecular mechanisms that underlie this radiosensitizing potential are not yet fully understood, they can be partially explained by the silencing of DNA repair genes, such as *Ku70* and *Ku86* [66]. Similarly, treatment with the HDAC inhibitor Vorinostat can prolong the appearance of repair foci identified by phosphorylated histone 2AX (γ-H2AX), which is indicative of a reduced repair efficiency and an increased radiosensitivity [66]. Epigenetic cancer therapy can thus be considered a very promising approach to cancer treatment, based on its synergy with chemotherapy, its resensitization of chemoresistant tumors, and its ability to increase the efficacy of radiotherapy.

4.3 Potential Side Effects of Epigenetic Therapies

Despite its great potential, the nonspecific effects of epigenetic drugs represent an area of concern for clinical application in patients. For instance, given the effect of global DNA hypomethylation on genomic stability, therapy-induced hypomethylation might promote tumor formation in the long run, although this hypothesis requires verification [67, 68]. Epigenetic therapy might also cause an activation of imprinted or silenced genes, and has indeed been shown to be mutagenic [69] and possibly even carcinogenic [70]. These concerns should not be exaggerated, however, as

Tab. 1 Clinical trials of epigenetic cancer therapies.

Agent	*Study phase*	*Disease*	*Details*	*Reference(s)*
DNMT inhibitor				
5-Azacytidine	3/FDA-approved	AML	Complete remission in 10–17% Hematological improvement in 23–36%	[45]
	3/FDA-approved	MDS	63% improvement in OS over standard of care	[46]
5-Aza-2′-deoxycytidine	3/FDA- approved	MDS, CML	Safe toxicity profile 34% of patients achieved complete response 73% had objective response	[47]
Arabinosyl-5-azacytidine	1/2	CML, cervical carcinoma		[48, 49]
MG98 (DNMT1 antisense)	2	Metastatic solid tumors		[50]
HDAC inhibitor				
Phenylbutyrate	1	MDS, AML	Well tolerated No partial or complete remissions 15% showed hematological improvement	[51]
Vorinostat (SAHA)	2/FDA- approved	CTCL	30% objective response rate	[52]
	1	AML,CLL, MDS, ALL, and CML	23% showed hematological improvement 6% complete responses	
	1	Advanced solid and hematologic cancers	4% partial responses in CTCL	[53]
Valproic acid (VPA)	1	AML, leukemia		[54, 55]
Belinostat	1	Multiple myeloma		[56]
Romidepsin	1	Lung cancer		[56]
	2	CTCL		

(*continued overleaf*)

Tab. 1 *(Continued)*

Agent	*Study phase*	*Disease*	*Details*	*Reference(s)*
CI-994	1/2	Lung and pancreatic cancer		[56]
MS-275	1/2	Lung, lymphoma, and leukemia		[56]
MGCD0103	1/2	Solid cancers, lymphoma, and leukemia		[56]
DNMT and HDAC inhibitor combined				
5-Aza-cytidine and VPA	1	Advanced solid cancers	Combination has safe toxicity profile 25% show stable disease (median 6 months)	[57]
5-Aza-cytidine and phenylbutyrate	1	Refractory solid cancers	Combination has safe toxicity profile No clinical benefits observed	[58]
HDAC inhibitor and chemotherapy combined				
Vorinostat, carboplatin, and paclitaxel	1	Advanced NSCLC	2.7-fold improved response, improved PFS (6 vs. 4.1 months) improved OS (13 vs. 9.7 months) compared to chemotherapy alone	[59]
Vorinostat and doxorubicin	1	Solid cancers	8% showed partial response 8% showed stable disease for over eight months	[60]

a) ALL, acute lymphoblastic leukemia; AML, acute myelogenous leukemia; CLL, chronic lymphocytic leukemia; CTCL, cutaneous T-cell lymphoma; MDS, myelodysplastic syndrome; NSCLC, non-small-cell lung carcinoma; OS, overall survival; PFS, progression-free survival; SAHA, suberoylanilide hydroxamic acid.

DNMT inhibitors act only on dividing cells while leaving other cells mostly unaffected. Furthermore, current evidence suggests that epigenetic drugs have a tendency to activate genes that have become abnormally silenced [71, 72]. Although no mechanism has yet been demonstrated to explain this, it is possible that the chromatin structure of aberrantly silenced genes is more susceptible to reactivation when compared to genes silenced under normal physiological conditions. However, it should be noted that patients receiving HDAC or DNMT inhibitors in the clinic have not yet suffered from any major toxicities, nor any unexplained long-term adverse effects [73, 74]. Although caution is warranted, the currently available clinical evidence suggests that epigenetic therapy is reasonably safe.

With regards to any potential side effects, there is great promise for specific epigenetic therapies targeted at particular genes via the use of promotor-specific TFs [75]. This strategy has, for instance, been shown specifically to reactivate *Maspin*, a tumor suppressor gene that is silenced by promoter methylation in aggressive epithelial tumors [76]. Consequently, Beltran *et al.* constructed an artificial transcription factor (ATF) that consisted of six zinc-finger domains targeting unique 18 bp sequences in the *Maspin* promoter, linked to an activator domain [76]. This ATF reactivated the epigenetically silenced *Maspin*, induced the apoptosis of cancer cells *in vitro*, and also suppressed tumor growth in a xenograft breast cancer model in nude mice. Hence, despite some concern being expressed regarding long-term safety, epigenetic cancer therapies clearly hold great potential. It is hoped that the next generation of targeted therapies will overcome any possible pitfalls and improve both the clinical efficacy and safety of epigenetic drugs.

5
Epigenetic Biomarkers for Cancer Therapy

Since epigenetic alterations play an important role in determining the behavior of various tumors, it is likely that epigenetic alterations – such as DNA methylation and histone modifications – can also be used for the diagnosis and molecular classification of cancer, and to predict cancer progression or response to therapy. Indeed, although the epigenetic mapping of genes within a clinical research setting is challenging due to the poor preservation of chromatin structure in clinical samples, there exists a tight correlation between methylation patterns, chromatin structure, and gene expression [77]. DNA methylation reflects the chromatin structure of a gene, and can be considered as a stable covalent DNA mark for gene activity [78]. As DNA methylation is better preserved compared to histone modification and chromatin structure (even in poor-quality samples), clinical epigenetic cancer research currently relies on DNA methylation for biomarker identification. However, this could change in the near future with improved sample collection, better storage methods, and novel analytical techniques. Currently, three potential applications may be distinguished for epigenetic markers in cancer management: as complementary diagnostic tools; as prognostic markers of disease progression; and as predictive markers of treatment response.

5.1
Methylated DNA Sequences for Improved Cancer Diagnosis

Epigenetic alterations can be used to complement existing diagnostic tools for cancer detection. Sensitive polymerase chain reaction (PCR)-based methods have been developed to detect hypermethylated CpG islands in DNA from various sources, such as blood, urine, sputum, or tumor biopsies [11]. These approaches have stimulated the discovery of abnormally methylated DNA sequences as tumor markers across multiple cancer types. For instance, the hypermethylation of glutathione *S*-transferase 1 (*GSTP1*) is seen in 80–90% of prostate cancer patients, while benign hyperplastic prostate tissue is not hypermethylated [79, 80]. Consequently, the presence of *GSTP1* methylation in prostate biopsies or urine samples could assist in the diagnosis of malignant prostate cancer [81]. In another study, a panel detecting several hypermethylated genes in breast ductal fluids correctly identified twice as many breast cancers than did classic cytological techniques [82]. More importantly, emerging evidence has indicated that epigenetic alterations occur early in carcinogenesis, before other biomarkers are detectable. For example, a substantial hypermethylation of the tumor suppressor *CDKN2A* can be detected in the bronchial pre-neoplastic epithelium of smokers [83].

Likewise, Melotte *et al.* reported a new biomarker for colorectal cancer (CRC) in stool samples [84]. These authors reported that N-Myc downstream-regulated gene 4 (*NDRG4*), a tumor suppressor candidate, is frequently silenced by promotor hypermethylation in CRC. Thus, by using a methylation-specific PCR assay for *NDRG4*, they could successfully identify 53% of CRC cases, and also correctly predicted which of the individuals were free from cancer [84]. In another recent study, a panel of four genes – namely bone morphogenetic protein 3 (*BMP3*), eyes absent homolog 2 (*EYA2*), aristaless-like homeobox-4 (*ALX4*), and vimentin – was identified that could be methylated with high specificity in colorectal carcinomas and adenomas, but rarely in normal epithelium [85]. These markers would clearly be very attractive for use in diagnostic tests to detect early-stage cancers, where the odds of survival were highest. Another potential plasma biomarker for CRC could lie in the methylation status of the septin 9 (*SEPT9*) gene which, according to the results of a recent study, can identify 72% of CRCs at a specificity of 92% [86].

As diagnostic tests within a routine clinical setting should preferably be noninvasive, the detection of epigenetic biomarkers in blood, stool, or urine would require that methods with improved sensitivity be developed, so as to allow the detection of degraded and diluted epigenetic biomarkers in these body fluids. In this context, Li and colleagues have developed a new technique based on next-generation sequencing, that is capable of detecting one methylated molecule among up to 5000 unmethylated molecules [87]. By using vimentin gene methylation as a plasma biomarker, it was possible to detect 52% of early-stage CRCs, compared to only 14% detection with the classical carcinoembryonic antigen (CEA) test. The myriad of ongoing studies nevertheless clearly demonstrates the enormous potential of epigenetic biomarkers for cancer detection.

5.2 Epigenetic Changes as Prognostic Markers

Epigenetic modifications could also serve as biomarkers to predict the disease prognosis of cancer patients. As patients with histologically similar cancers display a remarkable heterogeneity in their disease prognosis, biomarkers that could identify patients with an increased risk of an aggressive tumor and an associated poor survival would allow the selection of a treatment to achieve the best possible clinical risk : benefit ratio. Although current clinical practice is based mainly on immunohistological analyses, good progress has been made recently to improve risk stratification by using either gene-expression [88] or somatic mutation signatures [89, 90]. Yet, epigenetic biomarkers could possibly complement these existing tools. For instance, hypermethylation of the tumor suppressor genes *APC* and *CDKN2A* were shown to be associated with a poor prognosis in breast cancer and CRC, respectively [91, 92]; similarly, patients with lung cancer also have a poor prognosis if hypermethylation of *CDKN2A* is observed in the tumor [93]. In another study, the prognostic value of APC in breast cancer was confirmed, while the Ras association domain-containing protein 1 (*RASSF1A*) gene was also identified as a potential prognostic marker [94]. Furthermore, the gene for secreted frizzled-related protein 1 (*SFRP1*) has also been associated with a poor overall survival in breast cancer patients. This gene encodes a soluble Wnt antagonist, and defects in the Wnt signaling pathway have been implicated in breast cancer pathogenesis [95]. These examples indicate that future prognostic models will most likely incorporate epigenetic markers for a number of key pathways that are activated in a specific tumor type.

It has also been shown that global histone modification profiles, such as histone lysine methylation and acetylation marks, are correlated with clinical and pathological parameters of prostate cancer, and may serve as a significant predictor of prostate cancer recurrence [96]. Although histone modifications have been less well studied than DNA methylation in a clinical context (due mainly to protein degradation issues in clinical samples), the results of this study showed that histone modification profiles also hold great promise to serve as prognostic markers. Further research and clinical validation is needed, however, before DNA and histone modification profiles can be used to complement current pathological analyses.

5.3 Epigenetic Changes as Predictive Markers for Cancer Therapies

Epigenetic alterations may also function as predictive markers to assess the response to a particular cancer therapy. With the increasing recognition that each tumor has its own genetic profile that requires its own specific therapy, it is to be expected that future cancer therapies will become tailored to individual patients. The paradigm of personalized medicine, as illustrated by the recent approval for a test that determines *KRAS* mutations to predict the response to the epidermal growth factor receptor (EGFR) inhibitor cetuximab in CRC patients, is an example that could also apply to epigenetic markers [97]. It is essential, therefore, to identify any epigenetic differences that might explain any inter-individual variations in response to a certain therapy (see Fig. 2).

A well-established example is the DNA repair gene O^6-methylguanine–DNA methyltransferase (*MGMT*), which reverses guanine alkyl adduct formation induced by alkylating drugs, and thus prohibits the formation of lethal DNA crosslinks [101]; this mechanism explains why *MGMT*-expressing tumors are often resistant to alkylating chemotherapy. There is, however, a substantial variation in *MGMT* expression according to tumor type. For instance, almost half of the gliomas lack *MGMT* expression, and these would be expected to have an increased sensitivity to alkylating drugs. Interestingly, *MGMT* inactivation is rarely the result of a mutation or deletion, but rather results from

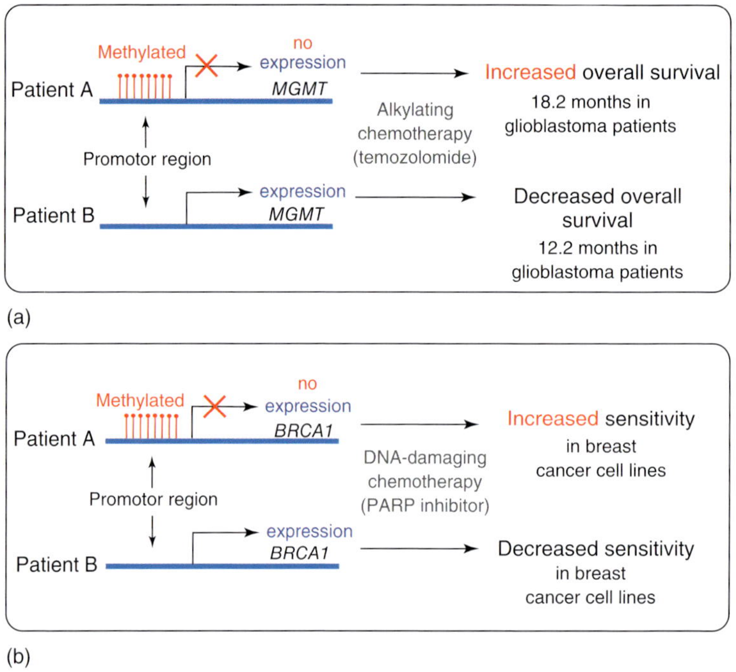

Fig. 2 Prediction of chemotherapeutic response by epigenetic biomarkers. (a) The *MGMT* methylation status can vary remarkably between different cancer patients. About half of glioblastoma patients have CpG island hypermethylation in the *MGMT* promotor (represented by patient A), while the other half does not have *MGMT* promotor methylation and therefore express MGMT (represented by patient B) [98]. The methylation status of *MGMT* in tumors is a predictive biomarker for response to alkylating chemotherapy, since MGMT specifically repairs DNA damage induced by this type of drug. Patients carrying methylated *MGMT* are unable to repair chemotherapeutic damage, which leads to an increased benefit from alkylating drugs and an increased survival compared to unmethylated *MGMT* carriers; (b) Similar heterogeneity exists in *BRCA1* methylation status between breast cancer patients. About 40% of breast cancer patients carry tumors with CpG island hypermethylation in the *BRCA1* promotor (represented by patient A), while the remaining 60% does not and therefore expresses *BRCA1* (represented by patient B) [99]. The methylation status of *BRCA1* in tumors is a potential predictive biomarker for response to drugs that inhibit the DNA repair enzyme poly(ADP)-ribose polymerase (PARP). As both BRCA1 and PARP are involved in DNA repair, *BRCA1*-deficient cells treated with PARP inhibitors no longer have DNA repair activity, accumulate DNA damage and die. Conversely, *BRCA1*-expressing cancer cells treated with PARP inhibitors can still repair DNA damage, and are therefore less sensitive to PARP inhibition. Although this concept has only very recently been proven in breast cancer cell lines [99], it will most likely also be confirmed in cancer patients, as *BRCA1* inactivation via mutation has already been shown to be a predictive biomarker for response to PARP inhibition [100].

hypermethylation of the CpG island in its promoter. Two independent studies have shown that the *MGMT* promoter methylation status can predict the response to the alkylating chemotherapies carmustine or temozolomide, and is associated with tumor regression and a prolonged disease-free and overall survival [98, 102]. The predictive value of epigenetic *MGMT* silencing is currently under investigation in several prospective randomized Phase III clinical studies in glioblastoma patients, in which standard chemoradiotherapy is combined with therapies targeted at important angiogenic molecules, such as bevacizumab [an anti-VEGF (vascular endothelial growth factor) antibody] or cilengitide (an integrin inhibitor). It is expected that the *MGMT* status will predict which tumors are sensitive to the standard chemotherapy, and that the addition of anti-angiogenic therapies will enhance the effects of chemotherapy in these tumors. If the results from these trials prove to be positive, then *MGMT* status will become the first predictive epigenetic biomarker to be applied in a clinical setting, hopefully paving the way for many others.

In fact, monitoring the methylation status of the DNA repair genes as biomarkers for chemotherapeutic response might in time become a common theme, with other studies having described the effects of *MGMT* on cyclophosphamide [101], of *hMLH1* on cisplatin [103], of reduced folate carrier (*RFC*) on methotrexate [104, 105], and of Werner syndrome RecQ helicase-like (*WRN*) on irinotecan [105]. Two other DNA repair genes that have been well studied in this respect are the so-called "breast cancer genes," *BRCA1* and *BRCA2*, which are frequently inactivated by genetic and epigenetic mechanisms in several cancer types [106–108]. These genes are required for the DNA double-strand break repair processes through homologous recombination. As a consequence, cancer cells carrying inactivated *BRCA1* or *BRCA2* genes are no longer capable of repairing DNA damage induced by, for instance platinum-based compounds. Intriguingly, these cancers have recently been shown to respond well to poly(ADP-ribose) polymerase (PARP) inhibitors [100]. The PARP enzyme is essential for the repair of DNA single-strand breaks; thus, PARP inhibitors may enhance the cytotoxic effects of DNA-damaging agents by selectively targeting cells defective in the *BRCA1/2*-dependent DNA repair pathway and inhibiting their PARP-dependent repair mechanisms [100]. While this concept had already been proven for *BRCA1/2* mutation carriers, the results of a very recent study conducted by Veeck *et al.* showed that the inactivation of *BRCA1* by CpG island hypermethylation is also a powerful predictor of sensitivity towards PARP inhibitors [99]. The data reported by Veeck *et al.* showed clearly the value of epigenetic biomarkers, and further illustrated that future studies should focus on the integration of genomic and epigenomic information, so as to establish a complete picture of *BRCA1/2* loss-of-function, in order that these results might be successfully translated into clinical practice (see Fig. 2).

6
Epigenetic Therapy beyond Cancer Cells

While cancer research initially has focused on the growth autonomy of cancer cells, it is becoming apparent that the stroma that surrounds the cancer cells, such as fibroblasts, endothelial cells, and inflammatory cells, also plays a key role in driving tumor cell proliferation [109]. The anti-tumoral properties of novel epigenetic therapies have thus far largely been attributed to the reactivation of silenced tumor suppressor genes in tumor cells. However, given their universal gene regulatory effects, it is likely that epigenetic therapy will also affect cancer stromal cells. There is a variety of stromal cell types that surround and interact with cancer cells, such as endothelial cells, fibroblasts, bone-marrow-derived precursor cells, and infiltrating inflammatory cells, of which several are known to be epigenetically altered under the influence of tumor cells [110–112]. For instance, Hu *et al.* have shown that distinct epigenetic alterations occur in a tumor stage- and cell type-specific manner during breast carcinogenesis, which suggests that epigenetic changes are important for maintenance of the abnormal cellular microenvironment [110]. The results of another study, performed by Chang *et al.*, showed that tumor-associated macrophages (TAMs) display a deacetylation of histones associated with the promotor of *CIITA*, a master regulator of major histocompatability class (MHC)-II expression [113]. This results in a downregulation of *CIITA* expression, a reduced MHC-II-dependent antigen presentation, and thus shows that tumors can induce an immunosuppressive phenotype in TAMs via epigenetic mechanisms.

Within the tumor microenvironment, probably the best-studied cells are the tumor endothelial cells, which not only play a key role during blood vessel formation (angiogenesis) but also respond rapidly to environmental changes, such as tumor hypoxia. Hypoxia is known to induce epigenetic changes, such as a global decrease in H3K9 acetylation, most likely because of HDAC upregulation [114, 115]. Recently, Johnson *et al.* showed that hypoxia-activated genes, such as early growth response 1 (*EGR1*) and *VEGF*, undergo a specific remodeling of the chromatin structure at their core promotors in response to hypoxia, in order to regulate their transcriptional activation [115]. This confirms that epigenetic changes could be important regulators of the response to chronic hypoxia, as often occurs in tumors. Given the critical role of angiogenesis in tumor progression and the widespread clinical application of anti-angiogenic therapies, the epigenetic alterations that occur in tumor vessels, and the means by which epigenetic therapies could target tumor angiogenesis, will now be discussed.

6.1 Epigenetic Changes in the Tumor Vasculature

Angiogenesis is a remarkable dynamic process that is tightly controlled by a balance of stimulatory and inhibitory angiogenic signals. Consequently, an imbalance in these signals will result either in a shortage or an excess of blood vessels, which will contribute to ischemic or malignant disorders, respectively [116]. The first angiogenesis inhibitors have been widely used since 2004 in the first-line treatment of various solid tumors, in combination with chemotherapy [117]. Whilst tumor angiogenesis alone does not initiate malignancy, it does play a critical role in cancer by promoting tumor progression and metastasis [116]. Activation of the so-called "angiogenic switch" is considered as one of the hallmarks of cancer that promotes tumor growth and metastasis [109]. Growing evidence indicates that epigenetic changes of the genes involved in angiogenesis are involved in this switch, and may cause tumors to recruit new blood vessels and sustain their growth [118]. In a recent study reported by Lu *et al.*, it was shown elegantly that cancer cells would induce epigenetic changes in endothelial cells, which in turn promoted tumor angiogenesis. Lu and coworkers identified the methyltransferase EZH2 as a key regulator of tumor angiogenesis, and showed that VEGF produced by cancer cells increased the expression of endothelial EZH2, which methylates and silences vasohibin1, thereby promoting a pro-angiogenic phenotype [119]. It was also shown that the delivery of EZH2 siRNA into tumor and tumor-associated endothelial cells decreased angiogenesis and resulted in a growth inhibition of an orthotopic ovarian cancer model. Taken together, these data confirmed that targeting epigenetic changes in tumor endothelial cells would represent a promising new anti-angiogenic strategy. Several clinical studies in humans have confirmed that epigenetic changes in tumors affect genes involved in the regulation of tumor angiogenesis. Glioblastomas, for instance, are typically characterized by an excessive blood vessel development, and frequently display an epigenetic inactivation of the anti-angiogenic thrombospondin-1 (*THBS-1*) gene [120]. *THBS-1* is also suppressed early in breast carcinogenesis by histone modifications [121], while *THBS-1* silencing via methylation is observed in a significant proportion of primary colorectal adenomas [122]. Interestingly, oxygen–glucose deprivation, which frequently occurs in tumors, was shown to increase *THBS-1* promoter methylation and subsequent silencing; this transcriptional inactivation could be reversed by reoxygenation [123].

6.2 Anti-Angiogenic Effects of Epigenetic Therapies

The fact that epigenetic alterations also occur in tumor endothelial cells, and specifically alter the balance of stimulatory and inhibitory angiogenic signals, suggests that interfering with the epigenetic machinery could be used to reactivate silenced anti-angiogenic factors and inhibit new blood vessel growth. Indeed, several epigenetic therapies have been assigned with anti-angiogenic activities, and some of the mechanisms underlying these have already been elucidated. For instance, the HDAC inhibitor TSA impairs blood vessel formation both *in vitro* and *in vivo* by the downregulation of pro-angiogenic signaling factors, such as VEGF, and the upregulation of anti-angiogenic factors, such as ADAMTS-1 [124–126]. Furthermore, TSA induces the expression of tumor suppressors p53 and VHL, and downregulates HIF-1α, a TF that activates hypoxia-induced angiogenic signaling pathways [127]. However, since p53, VHL, and HIF-1α also have important roles in tumor cells, a major challenge lies in determining whether the effects of HDAC inhibitors act directly on the blood vessels, or indirectly via the tumor cells. Hellebrekers and coworkers recently identified several downregulated genes in tumor-conditioned endothelial cells compared to normal endothelial cells, which included the anti-angiogenic genes *clusterin*, *fibrillin 1*, and *quiescin Q6* [128]. In this case, it was shown that the expression of these genes could be reactivated by treatment with the HDAC inhibitor, TSA. These findings confirm that the anti-angiogenic effects of HDAC inhibitors can, at least in part, be explained by their direct influence on endothelial gene expression. Another recently identified mechanism of action of HDAC inhibitors appears to be an impairment of endothelial progenitor cell function [129]. Adult progenitor cells possess stem cell-like properties, and are able to differentiate into endothelial cells that assist in the growth of new blood vessels [129, 130]. HDAC inhibitors can block their differentiation into endothelial cells via a repression of the TF HoxA9, a master regulator of expression for endothelial-committed genes, such as *eNOS*, *VEGFR-2*, and *VE-cadherin* [129].

Similar to HDAC inhibitors, the DNMT inhibitors can also reactivate epigenetically silenced genes in tumors, and decrease tumor growth both *in vitro* and *in vivo* [32, 131]. Again, these results cannot be interpreted without considering the effect of DNMT inhibitors on blood vessels. Indeed, the specific inhibitors decitabine and zebularine can decrease vessel formation and inhibit the proliferation of tumor-conditioned endothelial cells by reactivation of growth-inhibiting genes, such as *THBS-1, JUNB,* and *IGFBP3,* which are known to be silenced in tumor-conditioned endothelial cells [33]. Furthermore, these compounds can restore expression of the epigenetically silenced intercellular adhesion molecule-1 (ICAM-1) on tumor-conditioned endothelial cells *in vitro* and *in vivo* by a reversal of histone modifications in the ICAM-1 promoter. This results in a restored leukocyte-endothelial cell adhesion and an enhanced leukocyte infiltration, and also expands the therapeutic effects of DNMTs to the immune system [132]. Together, the HDAC and DNMT inhibitors appear to act on multiple cell types, most notably on tumor and endothelial cells, thereby affecting tumor angiogenesis as well as cancer cell survival.

6.3
Epigenetic Biomarkers for Anti-Angiogenic Therapies

Since epigenetic alterations frequently occur in tumor endothelial cells, it is possible that these epigenetic changes might be a potential predictor of response to the anti-angiogenic treatment. As such, a study by Rini *et al.* investigated the predictive effect of genetic and epigenetic inactivation of the *VHL* gene, a negative regulator of VEGF, in renal cell carcinoma patients treated with the monoclonal anti-VEGF antibody bevacizumab [133]. Patients with *VHL* inactivation had an improved response to the anti-angiogenic therapy, and displayed a strong trend towards a prolonged time to disease progression. Although still speculative, it is possible that *VHL* inactivation in tumors renders these patients more dependent on VEGF for initiating and sustaining angiogenesis, and therefore makes them more susceptible to VEGF inhibition. Although these results require further validation, they clearly show that epigenetic biomarkers could become a valuable predictive tool for anti-VEGF therapy. In a similar study, the predictive value of genetic and epigenetic inactivation of the *VHL* gene was investigated in metastatic renal cell carcinoma patients treated with sunitinib, a tyrosine kinase inhibitor that targets amongst others the VEGF receptors [134]. Patients with *VHL*-inactivating mutations exhibited an increased response rate of 52%, compared to 31% for those with wild-type *VHL.* Likewise, the epigenetic inactivation of *VHL* led to a response rate of 41%, versus 31% for those with wild-type *VHL.* Although the effect of epigenetic inactivation was less pronounced than for *VHL* mutations in this study, the observed effects were very similar to those seen with bevacizumab.

A further anti-angiogenic strategy that is currently being evaluated in clinical trials is that of anti-integrin therapy. The integrins, as cell-surface receptors, are involved in angiogenesis and therefore represent an attractive target for the inhibition of blood vessel growth in tumors. A potential biomarker for anti-integrin therapy might be the *ADAM23* gene, which acts as a negative regulator of the integrin$\alpha_V\beta_{III}$ receptor. *ADAM23* is frequently silenced by promoter hypermethylation and, since its silencing correlates with tumor progression, it might be associated with the acquisition of an angiogenic and metastatic phenotype [135]. Tumors characterized by *ADAM23* silencing might therefore depend more on activated α_V integrin signaling for their blood vessel growth, which would make them more eligible for anti-integrin therapy. These interesting findings warrant further study to evaluate the epigenetic profiles of multiple pro- and anti-angiogenic factors, in order to identify genes that commonly are altered in tumor angiogenesis.

7
Future Perspectives

The field of cancer pharmaco-epigenomics has emerged from an increased understanding of how epigenetic alterations can "drive" cancer growth. Whilst several epigenetic therapies have already been approved for hematological malignancies, it is expected that newer and improved therapies will lead to even better results. Indeed, current epigenetic drugs remain largely unspecific and might be complemented or replaced by more targeted epigenetic therapies with increased specificities for particular key genes. Currently, studies to systematically evaluate epigenetic profiles on a genome-wide level are ongoing, and should lead to an unbiased identification of the most common and relevant epigenetically altered genes for a certain cancer type. These epigenetic drivers of the tumor can then be specifically targeted. A combination of current reference chemotherapy or radiotherapy treatments with epigenetic therapy is also expected to lead to an increased efficacy, while reducing the delivered dose of the chemotherapy and, thereby, reducing also any toxicity. A combination of epigenetic therapies with chemotherapy or radiotherapy might also delay resistance to the chemotherapy, which is frequently observed in high-dose chemotherapeutic treatments. Large-scale efforts to map all of the epigenetic alterations that occur in a given cancer type are ongoing, and it is expected that these will also yield tumor type-specific biomarkers for disease diagnosis, prognosis, or prediction of response to a therapy. For instance, The Cancer Genome Atlas (TCGA) project has identified methylation signatures with a strong prognostic value in glioblastoma, while many other cancer types – such as serous ovarian carcinoma – are currently being investigated. Most importantly, as detection methods to identify the epigenetic changes in tumor DNA are improved, epigenetic biomarkers capable of predicting the efficacy of conventional chemotherapies or targeted therapies will also become applicable in a clinical context. Clearly, pharmaco-epigenomics has the potential to become an essential part of clinical practice in the future. Ultimately, the epigenomic profile of a patient might translate into personalized epigenomic medicine.

References

1. Bartel, D.P. (2004) MicroRNAs: genomics, biogenesis, mechanism, and function. *Cell*, **116**, 281–297.
2. Nomura, T., Kimura, M., Horii, T., Morita, S., Soejima, H., Kudo, S., and Hatada, I. (2008) MeCP2-dependent repression of an imprinted miR-184 released by depolarization. *Hum. Mol. Genet.*, **17**, 1192–1199.
3. Szulwach, K.E., Li, X., Smrt, R.D., Li, Y., Luo, Y., Lin, L., Santistevan, N.J., Li, W., Zhao, X., and Jin, P. (2010) Crosstalk between microRNA and epigenetic regulation in adult neurogenesis. *J. Cell Biol.*, **189**, 127–141.
4. Fraga, M.F., Ballestar, E., Paz, M.F., Ropero, S., Setien, F., Ballestar, M.L., Heine-Suner, D., Cigudosa, J.C., Urioste, M., Benitez, J., Boix-Chornet, M., Sanchez-Aguilera, A., Ling, C., Carlsson, E., Poulsen, P., Vaag, A., Stephan, Z., Spector, T.D., Wu, Y.Z., Plass, C., and Esteller, M. (2005) Epigenetic differences arise during the lifetime of monozygotic twins. *Proc. Natl Acad. Sci. USA*, **102**, 10604–10609.
5. Esteller, M. (2008) Epigenetics in cancer. *N. Engl. J. Med.*, **358**, 1148–1159.
6. Wilson, A.S., Power, B.E., and Molloy, P.L. (2007) DNA hypomethylation and human

diseases. *Biochim. Biophys. Acta*, **1775**, 138–162.
7. Feinberg, A.P. and Tycko, B. (2004) The history of cancer epigenetics. *Nat. Rev. Cancer*, **4**, 143–153.
8. Feinberg, A.P. (2007) Phenotypic plasticity and the epigenetics of human disease. *Nature*, **447**, 433–440.
9. Sasaki, H. and Matsui, Y. (2008) Epigenetic events in mammalian germ-cell development: reprogramming and beyond. *Nat. Rev. Genet.*, **9**, 129–140.
10. Esteller, M. (2007) Cancer epigenomics: DNA methylomes and histone-modification maps. *Nat. Rev. Genet.*, **8**, 286–298.
11. Herman, J.G. and Baylin, S.B. (2003) Gene silencing in cancer in association with promoter hypermethylation. *N. Engl. J. Med.*, **349**, 2042–2054.
12. Costello, J.F., Fruhwald, M.C., Smiraglia, D.J., Rush, L.J., Robertson, G.P., Gao, X., Wright, F.A., Feramisco, J.D., Peltomaki, P., Lang, J.C., Schuller, D.E., Yu, L., Bloomfield, C.D., Caligiuri, M.A., Yates, A., Nishikawa, R., Su Huang, H., Petrelli, N.J., Zhang, X., O'Dorisio, M.S., Held, W.A., Cavenee, W.K., and Plass, C. (2000) Aberrant CpG-island methylation has non-random and tumour-type-specific patterns. *Nat. Genet.*, **24**, 132–138.
13. Esteller, M., Corn, P.G., Baylin, S.B., and Herman, J.G. (2001) A gene hypermethylation profile of human cancer. *Cancer Res.*, **61**, 3225–3229.
14. Saito, Y., Liang, G., Egger, G., Friedman, J.M., Chuang, J.C., Coetzee, G.A., and Jones, P.A. (2006) Specific activation of microRNA-127 with downregulation of the proto-oncogene BCL6 by chromatin-modifying drugs in human cancer cells. *Cancer Cell*, **9**, 435–443.
15. Lujambio, A., Ropero, S., Ballestar, E., Fraga, M.F., Cerrato, C., Setien, F., Casado, S., Suarez-Gauthier, A., Sanchez-Cespedes, M., Git, A., Spiteri, I., Das, P.P., Caldas, C., Miska, E., and Esteller, M. (2007) Genetic unmasking of an epigenetically silenced microRNA in human cancer cells. *Cancer Res.*, **67**, 1424–1429.
16. Jenuwein, T. and Allis, C.D. (2001) Translating the histone code. *Science*, **293**, 1074–1080.
17. Margueron, R., Trojer, P., and Reinberg, D. (2005) The key to development: interpreting the histone code? *Curr. Opin. Genet. Dev.*, **15**, 163–176.
18. Fraga, M.F., Ballestar, E., Villar-Garea, A., Boix-Chornet, M., Espada, J., Schotta, G., Bonaldi, T., Haydon, C., Ropero, S., Petrie, K., Iyer, N.G., Perez-Rosado, A., Calvo, E., Lopez, J.A., Cano, A., Calasanz, M.J., Colomer, D., Piris, M.A., Ahn, N., Imhof, A., Caldas, C., Jenuwein, T., and Esteller, M. (2005) Loss of acetylation at Lys16 and trimethylation at Lys20 of histone H4 is a common hallmark of human cancer. *Nat. Genet.*, **37**, 391–400.
19. Nguyen, C.T., Weisenberger, D.J., Velicescu, M., Gonzales, F.A., Lin, J.C., Liang, G., and Jones, P.A. (2002) Histone H3-lysine 9 methylation is associated with aberrant gene silencing in cancer cells and is rapidly reversed by 5-aza-2′-deoxycytidine. *Cancer Res.*, **62**, 6456–6461.
20. Seligson, D.B., Horvath, S., Shi, T., Yu, H., Tze, S., Grunstein, M., and Kurdistani, S.K. (2005) Global histone modification patterns predict risk of prostate cancer recurrence. *Nature*, **435**, 1262–1266.
21. Marks, P., Rifkind, R.A., Richon, V.M., Breslow, R., Miller, T., and Kelly, W.K. (2001) Histone deacetylases and cancer: causes and therapies. *Nat. Rev. Cancer*, **1**, 194–202.
22. Johnstone, R.W. (2002) Histone- deacetylase inhibitors: novel drugs for the treatment of cancer. *Nat. Rev. Drug Discov.*, **1**, 287–299.
23. Cedar, H. and Bergman, Y. (2009) Linking DNA methylation and histone modification: patterns and paradigms. *Nat. Rev. Genet.*, **10**, 295–304.
24. Turner, B.M. (2007) Defining an epigenetic code. *Nat. Cell Biol.*, **9**, 2–6.
25. Jones, P.L., Veenstra, G.J., Wade, P.A., Vermaak, D., Kass, S.U., Landsberger, N., Strouboulis, J., and Wolffe, A.P. (1998) Methylated DNA and MeCP2 recruit histone deacetylase to repress transcription. *Nat. Genet.*, **19**, 187–191.
26. Rocchi, P., Tonelli, R., Camerin, C., Purgato, S., Fronza, R., Bianucci, F., Guerra, F., Pession, A., and Ferreri, A.M. (2005) p21Waf1/Cip1 is a common target induced by short-chain fatty acid HDAC inhibitors (valproic acid, tributyrin and sodium butyrate) in neuroblastoma cells. *Oncol. Rep.*, **13**, 1139–1144.

27. Shetty, S., Graham, B.A., Brown, J.G., Hu, X., Vegh-Yarema, N., Harding, G., Paul, J.T., and Gibson, S.B. (2005) Transcription factor NF-kappaB differentially regulates death receptor 5 expression involving histone deacetylase 1. *Mol. Cell. Biol.*, **25**, 5404–5416.
28. Dai, Y., Rahmani, M., Dent, P., and Grant, S. (2005) Blockade of histone deacetylase inhibitor-induced RelA/p65 acetylation and NF-kappaB activation potentiates apoptosis in leukemia cells through a process mediated by oxidative damage, XIAP downregulation, and c-Jun N-terminal kinase 1 activation. *Mol. Cell. Biol.*, **25**, 5429–5444.
29. Duan, H., Heckman, C.A., and Boxer, L.M. (2005) Histone deacetylase inhibitors down-regulate bcl-2 expression and induce apoptosis in t(14;18) lymphomas. *Mol. Cell. Biol.*, **25**, 1608–1619.
30. Michaelis, M., Michaelis, U.R., Fleming, I., Suhan, T., Cinatl, J., Blaheta, R.A., Hoffmann, K., Kotchetkov, R., Busse, R., Nau, H., Cinatl, J., Jr., (2004) Valproic acid inhibits angiogenesis in vitro and in vivo. *Mol. Pharmacol.*, **65**, 520–527.
31. Joseph, J., Mudduluru, G., Antony, S., Vashistha, S., Ajitkumar, P., and Somasundaram, K. (2004) Expression profiling of sodium butyrate (NaB)-treated cells: identification of regulation of genes related to cytokine signaling and cancer metastasis by NaB. *Oncogene*, **23**, 6304–6315.
32. Suzuki, H., Gabrielson, E., Chen, W., Anbazhagan, R., van Engeland, M., Weijenberg, M.P., Herman, J.G., and Baylin, S.B. (2002) A genomic screen for genes upregulated by demethylation and histone deacetylase inhibition in human colorectal cancer. *Nat. Genet.*, **31**, 141–149.
33. Hellebrekers, D.M., Jair, K.W., Vire, E., Eguchi, S., Hoebers, N.T., Fraga, M.F., Esteller, M., Fuks, F., Baylin, S.B., van Engeland, M., and Griffioen, A.W. (2006) Angiostatic activity of DNA methyltransferase inhibitors. *Mol. Cancer Ther.*, **5**, 467–475.
34. Hackanson, B., Robbel, C., Wijermans, P., and Lubbert, M. (2005) In vivo effects of decitabine in myelodysplasia and acute myeloid leukemia: review of cytogenetic and molecular studies. *Ann. Hematol.*, **84** (Suppl. 1), 32–38.
35. Silverman, L.R. and Mufti, G.J. (2005) Methylation inhibitor therapy in the treatment of myelodysplastic syndrome. *Nat. Clin. Pract. Oncol.*, **2** (Suppl. 1), S12–S23.
36. Klose, R.J. and Bird, A.P. (2006) Genomic DNA methylation: the mark and its mediators. *Trends Biochem. Sci.*, **31**, 89–97.
37. Goll, M.G. and Bestor, T.H. (2005) Eukaryotic cytosine methyltransferases. *Annu. Rev. Biochem.*, **74**, 481–514.
38. Momparler, R.L., Bouffard, D.Y., Momparler, L.F., Dionne, J., Belanger, K., and Ayoub, J. (1997) Pilot phase I-II study on 5-aza-2′-deoxycytidine (Decitabine) in patients with metastatic lung cancer. *Anticancer Drugs*, **8**, 358–368.
39. Schrump, D.S., Fischette, M.R., Nguyen, D.M., Zhao, M., Li, X., Kunst, T.F., Hancox, A., Hong, J.A., Chen, G.A., Pishchik, V., Figg, W.D., Murgo, A.J., and Steinberg, S.M. (2006) Phase I study of decitabine-mediated gene expression in patients with cancers involving the lungs, esophagus, or pleura. *Clin. Cancer Res.*, **12**, 5777–5785.
40. Cheng, J.C., Matsen, C.B., Gonzales, F.A., Ye, W., Greer, S., Marquez, V.E., Jones, P.A., and Selker, E.U. (2003) Inhibition of DNA methylation and reactivation of silenced genes by zebularine. *J. Natl Cancer Inst.*, **95**, 399–409.
41. Shaker, S., Bernstein, M., Momparler, L.F., and Momparler, R.L. (2003) Preclinical evaluation of antineoplastic activity of inhibitors of DNA methylation (5-aza-2′-deoxycytidine) and histone deacetylation (trichostatin A, depsipeptide) in combination against myeloid leukemic cells. *Leuk. Res.*, **27**, 437–444.
42. Cameron, E.E., Bachman, K.E., Myohanen, S., Herman, J.G., and Baylin, S.B. (1999) Synergy of demethylation and histone deacetylase inhibition in the re-expression of genes silenced in cancer. *Nat. Genet.*, **21**, 103–107.
43. Khan, O., La Thangue, N.B. (2008) Drug Insight: histone deacetylase inhibitor-based therapies for cutaneous T-cell lymphomas. *Nat. Clin. Pract. Oncol.*, **5**, 714–726.
44. Gilbert, J., Baker, S.D., Bowling, M.K., Grochow, L., Figg, W.D., Zabelina, Y., Donehower, R.C., and Carducci, M.A. (2001) A phase I dose escalation and bioavailability study of oral sodium phenylbutyrate in patients with refractory solid

tumor malignancies. *Clin. Cancer Res.*, 7, 2292–2300.

45. Silverman, L.R., McKenzie, D.R., Peterson, B.L., Holland, J.F., Backstrom, J.T., Beach, C.L., and Larson, R.A. (2006) Further analysis of trials with azacitidine in patients with myelodysplastic syndrome: studies 8421, 8921, and 9221 by the Cancer and Leukemia Group B. *J. Clin. Oncol.*, **24**, 3895–3903.
46. Fenaux, P., Mufti, G.J., Hellstrom-Lindberg, E., Santini, V., Finelli, C., Giagounidis, A., Schoch, R., Gattermann, N., Sanz, G., List, A., Gore, S.D., Seymour, J.F., Bennett, J.M., Byrd, J., Backstrom, J., Zimmerman, L., McKenzie, D., Beach, C., and Silverman, L.R. (2009) Efficacy of azacitidine compared with that of conventional care regimens in the treatment of higher-risk myelodysplastic syndromes: a randomised, open-label, phase III study. *Lancet Oncol.*, **10**, 223–232.
47. Kantarjian, H.M., O'Brien, S., Shan, J., Aribi, A., Garcia-Manero, G., Jabbour, E., Ravandi, F., Cortes, J., Davisson, J., and Issa, J.P. (2007) Update of the decitabine experience in higher risk myelodysplastic syndrome and analysis of prognostic factors associated with outcome. *Cancer*, **109**, 265–273.
48. Wilhelm, M., O'Brien, S., Rios, M.B., Estey, E., Keating, M.J., Plunkett, W., Sorenson, M., and Kantarjian, H.M. (1999) Phase I study of arabinosyl-5-azacytidine (fazarabine) in adult acute leukemia and chronic myelogenous leukemia in blastic phase. *Leuk. Lymphoma*, **34**, 511–518.
49. Manetta, A., Blessing, J.A., Mann, W.J., and Smith, D.M. (1995) A phase II study of fazarabine (NSC 281272) in patients with advanced squamous cell carcinoma of the cervix. A Gynecologic Oncology Group study. *Am. J. Clin. Oncol.*, **18**, 439–440.
50. Winquist, E., Knox, J., Ayoub, J.P., Wood, L., Wainman, N., Reid, G.K., Pearce, L., Shah, A., and Eisenhauer, E. (2006) Phase II trial of DNA methyltransferase 1 inhibition with the antisense oligonucleotide MG98 in patients with metastatic renal carcinoma: a National Cancer Institute of Canada Clinical Trials Group investigational new drug study. *Invest. New Drugs*, **24**, 159–167.
51. Gore, S.D., Weng, L.J., Zhai, S., Figg, W.D., Donehower, R.C., Dover, G.J., Grever, M., Griffin, C.A., Grochow, L.B., Rowinsky, E.K., Zabalena, Y., Hawkins, A.L., Burks, K., and Miller, C.B. (2001) Impact of the putative differentiating agent sodium phenylbutyrate on myelodysplastic syndromes and acute myeloid leukemia. *Clin. Cancer Res.*, **7**, 2330–2339.
52. Garcia-Manero, G., Yang, H., Bueso-Ramos, C., Ferrajoli, A., Cortes, J., Wierda, W.G., Faderl, S., Koller, C., Morris, G., Rosner, G., Loboda, A., Fantin, V.R., Randolph, S.S., Hardwick, J.S., Reilly, J.F., Chen, C., Ricker, J.L., Secrist, J.P., Richon, V.M., Frankel, S.R., and Kantarjian, H.M. (2008) Phase 1 study of the histone deacetylase inhibitor vorinostat (suberoylanilide hydroxamic acid [SAHA]) in patients with advanced leukemias and myelodysplastic syndromes. *Blood*, **111**, 1060–1066.
53. Kelly, W.K., O'Connor, O.A., Krug, L.M., Chiao, J.H., Heaney, M., Curley, T., MacGregore-Cortelli, B., Tong, W., Secrist, J.P., Schwartz, L., Richardson, S., Chu, E., Olgac, S., Marks, P.A., Scher, H., and Richon, V.M. (2005) Phase I study of an oral histone deacetylase inhibitor, suberoylanilide hydroxamic acid, in patients with advanced cancer. *J. Clin. Oncol.*, **23**, 3923–3931.
54. Chavez-Blanco, A., Segura-Pacheco, B., Perez-Cardenas, E., Taja-Chayeb, L., Cetina, L., Candelaria, M., Cantu, D., Gonzalez-Fierro, A., Garcia-Lopez, P., Zambrano, P., Perez-Plasencia, C., Cabrera, G., Trejo-Becerril, C., Angeles, E., Duenas-Gonzalez, A. (2005) Histone acetylation and histone deacetylase activity of magnesium valproate in tumor and peripheral blood of patients with cervical cancer. A phase I study. *Mol. Cancer*, **4**, 22.
55. Raffoux, E., Chaibi, P., Dombret, H., and Degos, L. (2005) Valproic acid and all-trans retinoic acid for the treatment of elderly patients with acute myeloid leukemia. *Haematologica*, **90**, 986–988.
56. Glaser, K.B. (2007) HDAC inhibitors: clinical update and mechanism-based potential. *Biochem. Pharmacol.*, **74**, 659–671.
57. Braiteh, F., Soriano, A.O., Garcia-Manero, G., Hong, D., Johnson, M.M., Silva Lde, P., Yang, H., Alexander, S., Wolff, J., and Kurzrock, R. (2008) Phase I study of epigenetic modulation with 5-azacytidine and valproic acid in patients with advanced cancers. *Clin. Cancer Res.*, **14**, 6296–6301.
58. Lin, J., Gilbert, J., Rudek, M.A., Zwiebel, J.A., Gore, S., Jiemjit, A., Zhao, M.,

Baker, S.D., Ambinder, R.F., Herman, J.G., Donehower, R.C., and Carducci, M.A. (2009) A phase I dose-finding study of 5-azacytidine in combination with sodium phenylbutyrate in patients with refractory solid tumors. *Clin. Cancer Res.*, **15**, 6241–6249.

59. Ramalingam, S.S., Maitland, M.L., Frankel, P., Argiris, A.E., Koczywas, M., Gitlitz, B., Thomas, S., Espinoza-Delgado, I., Vokes, E.E., Gandara, D.R., and Belani, C.P. (2010) Carboplatin and paclitaxel in combination with either vorinostat or placebo for first-line therapy of advanced non-small-cell lung cancer. *J. Clin. Oncol.*, **28**, 56–62.

60. Munster, P.N., Marchion, D., Thomas, S., Egorin, M., Minton, S., Springett, G., Lee, J.H., Simon, G., Chiappori, A., Sullivan, D., and Daud, A. (2009) Phase I trial of vorinostat and doxorubicin in solid tumours: histone deacetylase 2 expression as a predictive marker. *Br. J. Cancer*, **101**, 1044–1050.

61. Smith, L.T., Otterson, G.A., and Plass, C. (2007) Unraveling the epigenetic code of cancer for therapy. *Trends Genet.*, **23**, 449–456.

62. Reu, F.J., Bae, S.I., Cherkassky, L., Leaman, D.W., Lindner, D., Beaulieu, N., MacLeod, A.R., and Borden, E.C. (2006) Overcoming resistance to interferon-induced apoptosis of renal carcinoma and melanoma cells by DNA demethylation. *J. Clin. Oncol.*, **24**, 3771–3779.

63. Yang, D., Thangaraju, M., Greeneltch, K., Browning, D.D., Schoenlein, P.V., Tamura, T., Ozato, K., Ganapathy, V., Abrams, S.I., and Liu, K. (2007) Repression of IFN regulatory factor 8 by DNA methylation is a molecular determinant of apoptotic resistance and metastatic phenotype in metastatic tumor cells. *Cancer Res.*, **67**, 3301–3309.

64. Gollob, J.A., Sciambi, C.J., Peterson, B.L., Richmond, T., Thoreson, M., Moran, K., Dressman, H.K., Jelinek, J., and Issa, J.P. (2006) Phase I trial of sequential low-dose 5-aza-2′-deoxycytidine plus high-dose intravenous bolus interleukin-2 in patients with melanoma or renal cell carcinoma. *Clin. Cancer Res.*, **12**, 4619–4627.

65. Munshi, A., Kurland, J.F., Nishikawa, T., Tanaka, T., Hobbs, M.L., Tucker, S.L., Ismail, S., Stevens, C., and Meyn, R.E. (2005) Histone deacetylase inhibitors radiosensitize human melanoma cells by suppressing DNA repair activity. *Clin. Cancer Res.*, **11**, 4912–4922.

66. Munshi, A., Tanaka, T., Hobbs, M.L., Tucker, S.L., Richon, V.M., and Meyn, R.E. (2006) Vorinostat, a histone deacetylase inhibitor, enhances the response of human tumor cells to ionizing radiation through prolongation of gamma-H2AX foci. *Mol. Cancer Ther.*, **5**, 1967–1974.

67. Eden, A., Gaudet, F., Waghmare, A., and Jaenisch, R. (2003) Chromosomal instability and tumors promoted by DNA hypomethylation. *Science*, **300**, 455.

68. Yang, A.S., Estecio, M.R., Garcia-Manero, G., Kantarjian, H.M., and Issa, J.P. (2003) Comment on "Chromosomal instability and tumors promoted by DNA hypomethylation" and "Induction of tumors in nice by genomic hypomethylation". *Science*, **302**, 1153; author reply 1153.

69. Jackson-Grusby, L., Laird, P.W., Magge, S.N., Moeller, B.J., and Jaenisch, R. (1997) Mutagenicity of 5-aza-2′-deoxycytidine is mediated by the mammalian DNA methyltransferase. *Proc. Natl Acad. Sci. USA*, **94**, 4681–4685.

70. Carr, B.I., Rahbar, S., Asmeron, Y., Riggs, A., and Winberg, C.D. (1988) Carcinogenicity and haemoglobin synthesis induction by cytidine analogues. *Br. J. Cancer*, **57**, 395–402.

71. Karpf, A.R., Peterson, P.W., Rawlins, J.T., Dalley, B.K., Yang, Q., Albertsen, H., and Jones, D.A. (1999) Inhibition of DNA methyltransferase stimulates the expression of signal transducer and activator of transcription 1, 2, and 3 genes in colon tumor cells. *Proc. Natl Acad. Sci. USA*, **96**, 14007–14012.

72. Liang, G., Gonzales, F.A., Jones, P.A., Orntoft, T.F., and Thykjaer, T. (2002) Analysis of gene induction in human fibroblasts and bladder cancer cells exposed to the methylation inhibitor 5-aza-2′-deoxycytidine. *Cancer Res.*, **62**, 961–966.

73. Yoo, C.B. and Jones, P.A. (2006) Epigenetic therapy of cancer: past, present and future. *Nat. Rev. Drug Discov.*, **5**, 37–50.

74. Jones, P.A. and Baylin, S.B. (2002) The fundamental role of epigenetic events in cancer. *Nat. Rev. Genet.*, **3**, 415–428.

75. Moore, M. and Ullman, C. (2003) Recent developments in the engineering of zinc finger proteins. *Briefings Funct. Genomic. Proteomic.*, **1**, 342–355.
76. Beltran, A., Parikh, S., Liu, Y., Cuevas, B.D., Johnson, G.L., Futscher, B.W., and Blancafort, P. (2007) Re-activation of a dormant tumor suppressor gene maspin by designed transcription factors. *Oncogene*, **26**, 2791–2798.
77. Szyf, M. (2004) Toward a discipline of pharmacoepigenomics. *Curr. Pharmacogenomics*, **2**, 357–377.
78. Geiman, T.M. and Robertson, K.D. (2002) Chromatin remodeling, histone modifications, and DNA methylation – how does it all fit together? *J. Cell. Biochem.*, **87**, 117–125.
79. Esteller, M., Corn, P.G., Urena, J.M., Gabrielson, E., Baylin, S.B., and Herman, J.G. (1998) Inactivation of glutathione *S*-transferase P1 gene by promoter hypermethylation in human neoplasia. *Cancer Res.*, **58**, 4515–4518.
80. Jeronimo, C., Usadel, H., Henrique, R., Oliveira, J., Lopes, C., Nelson, W.G., and Sidransky, D. (2001) Quantitation of GSTP1 methylation in non-neoplastic prostatic tissue and organ-confined prostate adenocarcinoma. *J. Natl Cancer Inst.*, **93**, 1747–1752.
81. Cairns, P., Esteller, M., Herman, J.G., Schoenberg, M., Jeronimo, C., Sanchez-Cespedes, M., Chow, N.H., Grasso, M., Wu, L., Westra, W.B., and Sidransky, D. (2001) Molecular detection of prostate cancer in urine by GSTP1 hypermethylation. *Clin. Cancer Res.*, **7**, 2727–2730.
82. Fackler, M.J., Malone, K., Zhang, Z., Schilling, E., Garrett-Mayer, E., Swift-Scanlan, T., Lange, J., Nayar, R., Davidson, N.E., Khan, S.A., and Sukumar, S. (2006) Quantitative multiplex methylation-specific PCR analysis doubles detection of tumor cells in breast ductal fluid. *Clin. Cancer Res.*, **12**, 3306–3310.
83. Belinsky, S.A., Nikula, K.J., Palmisano, W.A., Michels, R., Saccomanno, G., Gabrielson, E., Baylin, S.B., and Herman, J.G. (1998) Aberrant methylation of p16(INK4a) is an early event in lung cancer and a potential biomarker for early diagnosis. *Proc. Natl Acad. Sci. USA*, **95**, 11891–11896.
84. Melotte, V., Lentjes, M.H., van den Bosch, S.M., Hellebrekers, D.M., de Hoon, J.P., Wouters, K.A., Daenen, K.L., Partouns-Hendriks, I.E., Stessels, F., Louwagie, J., Smits, K.M., Weijenberg, M.P., Sanduleanu, S., Khalid-de Bakker, C.A., Oort, F.A., Meijer, G.A., Jonkers, D.M., Herman, J.G., de Bruine, A.P., and van Engeland, M. (2009) N-Myc downstream-regulated gene 4 (NDRG4): a candidate tumor suppressor gene and potential biomarker for colorectal cancer. *J. Natl Cancer Inst.*, **101**, 916–927.
85. Zou, H., Harrington, J.J., Shire, A.M., Rego, R.L., Wang, L., Campbell, M.E., Oberg, A.L., and Ahlquist, D.A. (2007) Highly methylated genes in colorectal neoplasia: implications for screening. *Cancer Epidemiol. Biomarkers Prev.*, **16**, 2686–2696.
86. de Vos, T., Tetzner, R., Model, F., Weiss, G., Schuster, M., Distler, J., Steiger, K.V., Grutzmann, R., Pilarsky, C., Habermann, J.K., Fleshner, P.R., Oubre, B.M., Day, R., Sledziewski, A.Z., Lofton-Day, C. (2009) Circulating methylated SEPT9 DNA in plasma is a biomarker for colorectal cancer. *Clin. Chem.*, **55**, 1337–1346.
87. Li, M., Chen, W.D., Papadopoulos, N., Goodman, S.N., Bjerregaard, N.C., Laurberg, S., Levin, B., Juhl, H., Arber, N., Moinova, H., Durkee, K., Schmidt, K., He, Y., Diehl, F., Velculescu, V.E., Zhou, S., Diaz, L.A., Jr., Kinzler, K.W., Markowitz, S.D., and Vogelstein, B. (2009) Sensitive digital quantification of DNA methylation in clinical samples. *Nat. Biotechnol.*, **27**, 858–863.
88. Sotiriou, C. and Pusztai, L. (2009) Gene-expression signatures in breast cancer. *N. Engl. J. Med.*, **360**, 790–800.
89. Pharoah, P.D., Day, N.E., and Caldas, C. (1999) Somatic mutations in the p53 gene and prognosis in breast cancer: a meta-analysis. *Br. J. Cancer*, **80**, 1968–1973.
90. Schmidt, M.K., Tollenaar, R.A., de Kemp, S.R., Broeks, A., Cornelisse, C.J., Smit, V.T., Peterse, J.L., van Leeuwen, F.E., Van't, and Veer, L.J. (2007) Breast cancer survival and tumor characteristics in premenopausal women carrying the CHEK2*1100delC germline mutation. *J. Clin. Oncol.*, **25**, 64–69.
91. Muller, H.M., Widschwendter, A., Fiegl, H., Ivarsson, L., Goebel, G., Perkmann, E.,

Marth, C., and Widschwendter, M. (2003) DNA methylation in serum of breast cancer patients: an independent prognostic marker. *Cancer Res.*, **63**, 7641–7645.

92. Wettergren, Y., Odin, E., Nilsson, S., Carlsson, G., and Gustavsson, B. (2008) p16INK4a gene promoter hypermethylation in mucosa as a prognostic factor for patients with colorectal cancer. *Mol. Med.*, **14**, 412–421.
93. Esteller, M., Gonzalez, S., Risques, R.A., Marcuello, E., Mangues, R., Germa, J.R., Herman, J.G., Capella, G., and Peinado, M.A. (2001) K-ras and p16 aberrations confer poor prognosis in human colorectal cancer. *J. Clin. Oncol.*, **19**, 299–304.
94. Muller, H.M., Fiegl, H., Widschwendter, A., and Widschwendter, M. (2004) Prognostic DNA methylation marker in serum of cancer patients. *Ann. N. Y. Acad. Sci.*, **1022**, 44–49.
95. Veeck, J., Niederacher, D., An, H., Klopocki, E., Wiesmann, F., Betz, B., Galm, O., Camara, O., Durst, M., Kristiansen, G., Huszka, C., Knuchel, R., and Dahl, E. (2006) Aberrant methylation of the Wnt antagonist SFRP1 in breast cancer is associated with unfavourable prognosis. *Oncogene*, **25**, 3479–3488.
96. Ellinger, J., Kahl, P., von der, Gathen, J., Rogenhofer, S., Heukamp, L.C., Gutgemann, I., Walter, B., Hofstadter, F., Buttner, R., Muller, S.C., Bastian, P.J., von Ruecker, A. (2009) Global levels of histone modifications predict prostate cancer recurrence. *Prostate*, **70**, 61–69.
97. Normanno, N., Tejpar, S., Morgillo, F., De Luca, A., Van Cutsem, E., and Ciardiello, F. (2009) Implications for KRAS status and EGFR-targeted therapies in metastatic CRC. *Nat. Rev. Clin. Oncol.*, **6**, 519–527.
98. Esteller, M., Garcia-Foncillas, J., Andion, E., Goodman, S.N., Hidalgo, O.F., Vanaclocha, V., Baylin, S.B., and Herman, J.G. (2000) Inactivation of the DNA-repair gene MGMT and the clinical response of gliomas to alkylating agents. *N. Engl. J. Med.*, **343**, 1350–1354.
99. Veeck, J., Ropero, S., Setien, F., Gonzalez-Suarez, E., Osorio, A., Benitez, J., Herman, J.G., and Esteller, M. (2010) BRCA1 CpG island hypermethylation predicts sensitivity to poly(adenosine diphosphate)-ribose polymerase inhibitors. *J. Clin. Oncol.* **28**, e563–e564; author reply e565–e566.
100. Fong, P.C., Boss, D.S., Yap, T.A., Tutt, A., Wu, P., Mergui-Roelvink, M., Mortimer, P., Swaisland, H., Lau, A., O'Connor, M.J., Ashworth, A., Carmichael, J., Kaye, S.B., Schellens, J.H., de Bono, J.S. (2009) Inhibition of poly(ADP-ribose) polymerase in tumors from BRCA mutation carriers. *N. Engl. J. Med.*, **361**, 123–134.
101. Esteller, M., Gaidano, G., Goodman, S.N., Zagonel, V., Capello, D., Botto, B., Rossi, D., Gloghini, A., Vitolo, U., Carbone, A., Baylin, S.B., and Herman, J.G. (2002) Hypermethylation of the DNA repair gene O(6)-methylguanine DNA methyltransferase and survival of patients with diffuse large B-cell lymphoma. *J. Natl Cancer Inst.*, **94**, 26–32.
102. Hegi, M.E., Diserens, A.C., Gorlia, T., Hamou, M.F., de Tribolet, N., Weller, M., Kros, J.M., Hainfellner, J.A., Mason, W., Mariani, L., Bromberg, J.E., Hau, P., Mirimanoff, R.O., Cairncross, J.G., Janzer, R.C., and Stupp, R. (2005) MGMT gene silencing and benefit from temozolomide in glioblastoma. *N. Engl. J. Med.*, **352**, 997–1003.
103. Strathdee, G., MacKean, M.J., Illand, M., and Brown, R. (1999) A role for methylation of the hMLH1 promoter in loss of hMLH1 expression and drug resistance in ovarian cancer. *Oncogene*, **18**, 2335–2341.
104. Ferreri, A.J., Dell'Oro, S., Capello, D., Ponzoni, M., Iuzzolino, P., Rossi, D., Pasini, F., Ambrosetti, A., Orvieto, E., Ferrarese, F., Arrigoni, G., Foppoli, M., Reni, M., and Gaidano, G. (2004) Aberrant methylation in the promoter region of the reduced folate carrier gene is a potential mechanism of resistance to methotrexate in primary central nervous system lymphomas. *Br. J. Haematol.*, **126**, 657–664.
105. Agrelo, R., Cheng, W.H., Setien, F., Ropero, S., Espada, J., Fraga, M.F., Herranz, M., Paz, M.F., Sanchez-Cespedes, M., Artiga, M.J., Guerrero, D., Castells, A., von Kobbe, C., Bohr, V.A., and Esteller, M. (2006) Epigenetic inactivation of the premature aging Werner syndrome gene in human cancer. *Proc. Natl Acad. Sci. USA*, **103**, 8822–8827.
106. Tapia, T., Smalley, S.V., Kohen, P., Munoz, A., Solis, L.M., Corvalan, A., Faundez, P., Devoto, L., Camus, M., Alvarez, M., and

Carvallo, P. (2008) Promoter hypermethylation of BRCA1 correlates with absence of expression in hereditary breast cancer tumors. *Epigenetics*, **3**, 157–163.
107. Lee, M.N., Tseng, R.C., Hsu, H.S., Chen, J.Y., Tzao, C., Ho, W.L., and Wang, Y.C. (2007) Epigenetic inactivation of the chromosomal stability control genes BRCA1, BRCA2, and XRCC5 in non-small cell lung cancer. *Clin. Cancer Res.*, **13**, 832–838.
108. Birgisdottir, V., Stefansson, O.A., Bodvarsdottir, S.K., Hilmarsdottir, H., Jonasson, J.G., and Eyfjord, J.E. (2006) Epigenetic silencing and deletion of the BRCA1 gene in sporadic breast cancer. *Breast Cancer Res.*, **8**, R38.
109. Hanahan, D. and Weinberg, R.A. (2000) The hallmarks of cancer. *Cell*, **100**, 57–70.
110. Hu, M., Yao, J., Cai, L., Bachman, K.E., van den Brule, F., Velculescu, V., and Polyak, K. (2005) Distinct epigenetic changes in the stromal cells of breast cancers. *Nat. Genet.*, **37**, 899–905.
111. Hanson, J.A., Gillespie, J.W., Grover, A., Tangrea, M.A., Chuaqui, R.F., Emmert-Buck, M.R., Tangrea, J.A., Libutti, S.K., Linehan, W.M., and Woodson, K.G. (2006) Gene promoter methylation in prostate tumor-associated stromal cells. *J. Natl Cancer Inst.*, **98**, 255–261.
112. Chung, I., Karpf, A.R., Muindi, J.R., Conroy, J.M., Nowak, N.J., Johnson, C.S., and Trump, D.L. (2007) Epigenetic silencing of CYP24 in tumor-derived endothelial cells contributes to selective growth inhibition by calcitriol. *J. Biol. Chem.*, **282**, 8704–8714.
113. Chang, Y.C., Chen, T.C., Lee, C.T., Yang, C.Y., Wang, H.W., Wang, C.C., and Hsieh, S.L. (2008) Epigenetic control of MHC class II expression in tumor-associated macrophages by decoy receptor 3. *Blood*, **111**, 5054–5063.
114. Johnson, A.B. and Barton, M.C. (2007) Hypoxia-induced and stress-specific changes in chromatin structure and function. *Mutat. Res.*, **618**, 149–162.
115. Johnson, A.B., Denko, N., and Barton, M.C. (2008) Hypoxia induces a novel signature of chromatin modifications and global repression of transcription. *Mutat. Res.*, **640**, 174–179.
116. Carmeliet, P. (2005) Angiogenesis in life, disease and medicine. *Nature*, **438**, 932–936.
117. Kerbel, R.S. (2008) Tumor angiogenesis. *N. Engl. J. Med.*, **358**, 2039–2049.
118. Buysschaert, I., Schmidt, T., Roncal, C., Carmeliet, P., and Lambrechts, D. (2008) Genetics, epigenetics and pharmaco-(epi)genomics in angiogenesis. *J. Cell. Mol. Med.*, **12**, 2533–2551.
119. Lu, C., Han, H.D., Mangala, L.S., Ali-Fehmi, R., Newton, C.S., Ozbun, L., Armaiz-Pena, G.N., Hu, W., Stone, R.L., Munkarah, A., Ravoori, M.K., Shahzad, M.M., Lee, J.W., Mora, E., Langley, R.R., Carroll, A.R., Matsuo, K., Spannuth, W.A., Schmandt, R., Jennings, N.B., Goodman, B.W., Jaffe, R.B., Nick, A.M., Kim, H.S., Guven, E.O., Chen, Y.H., Li, L.Y., Hsu, M.C., Coleman, R.L., Calin, G.A., Denkbas, E.B., Lim, J.Y., Lee, J.S., Kundra, V., Birrer, M.J., Hung, M.C., Lopez-Berestein, G., and Sood, A.K. (2010) Regulation of tumor angiogenesis by EZH2. *Cancer Cell* **18**, 185–197.
120. Li, Q., Ahuja, N., Burger, P.C., and Issa, J.P. (1999) Methylation and silencing of the Thrombospondin-1 promoter in human cancer. *Oncogene*, **18**, 3284–3289.
121. Hinshelwood, R.A., Huschtscha, L.I., Melki, J., Stirzaker, C., Abdipranoto, A., Vissel, B., Ravasi, T., Wells, C.A., Hume, D.A., Reddel, R.R., and Clark, S.J. (2007) Concordant epigenetic silencing of transforming growth factor-beta signaling pathway genes occurs early in breast carcinogenesis. *Cancer Res.*, **67**, 11517–11527.
122. Rojas, A., Meherem, S., Kim, Y.H., Washington, M.K., Willis, J.E., Markowitz, S.D., and Grady, W.M. (2008) The aberrant methylation of TSP1 suppresses TGF-beta1 activation in colorectal cancer. *Int. J. Cancer*, **123**, 14–21.
123. Hu, C.J., Chen, S.D., Yang, D.I., Lin, T.N., Chen, C.M., Huang, T.H., and Hsu, C.Y. (2006) Promoter region methylation and reduced expression of thrombospondin-1 after oxygen-glucose deprivation in murine cerebral endothelial cells. *J. Cereb. Blood Flow Metab.*, **26**, 1519–1526.
124. Deroanne, C.F., Bonjean, K., Servotte, S., Devy, L., Colige, A., Clausse, N., Blacher, S., Verdin, E., Foidart, J.M., Nusgens, B.V., and Castronovo, V. (2002) Histone deacetylases inhibitors as anti-angiogenic agents altering vascular endothelial growth factor signaling. *Oncogene*, **21**, 427–436.

125. Chou, C.W. and Chen, C.C. (2008) HDAC inhibition upregulates the expression of angiostatic ADAMTS1. *FEBS Lett.*, **582**, 4059–4065.
126. Rossig, L., Li, H., Fisslthaler, B., Urbich, C., Fleming, I., Forstermann, U., Zeiher, A.M., and Dimmeler, S. (2002) Inhibitors of histone deacetylation downregulate the expression of endothelial nitric oxide synthase and compromise endothelial cell function in vasorelaxation and angiogenesis. *Circ. Res.*, **91**, 837–844.
127. Kim, M.S., Kwon, H.J., Lee, Y.M., Baek, J.H., Jang, J.E., Lee, S.W., Moon, E.J., Kim, H.S., Lee, S.K., Chung, H.Y., Kim, C.W., and Kim, K.W. (2001) Histone deacetylases induce angiogenesis by negative regulation of tumor suppressor genes. *Nat. Med.*, **7**, 437–443.
128. Hellebrekers, D.M., Melotte, V., Vire, E., Langenkamp, E., Molema, G., Fuks, F., Herman, J.G., Van Criekinge, W., Griffioen, A.W., van Engeland, M. (2007) Identification of epigenetically silenced genes in tumor endothelial cells. *Cancer Res.*, **67**, 4138–4148.
129. Rossig, L., Urbich, C., Bruhl, T., Dernbach, E., Heeschen, C., Chavakis, E., Sasaki, K., Aicher, D., Diehl, F., Seeger, F., Potente, M., Aicher, A., Zanetta, L., Dejana, E., Zeiher, A.M., and Dimmeler, S. (2005) Histone deacetylase activity is essential for the expression of HoxA9 and for endothelial commitment of progenitor cells. *J. Exp. Med.*, **201**, 1825–1835.
130. Young, P.P., Vaughan, D.E., and Hatzopoulos, A.K. (2007) Biologic properties of endothelial progenitor cells and their potential for cell therapy. *Prog. Cardiovasc. Dis.*, **49**, 421–429.
131. Baylin, S.B. (2004) Reversal of gene silencing as a therapeutic target for cancer-roles for DNA methylation and its interdigitation with chromatin. *Novartis Found. Symp.* **259**, 226–233; discussion 234–237, 285–288.
132. Hellebrekers, D.M., Castermans, K., Vire, E., Dings, R.P., Hoebers, N.T., Mayo, K.H., Oude Egbrink, M.G., Molema, G., Fuks, F., van Engeland, M., Griffioen, A.W. (2006) Epigenetic regulation of tumor endothelial cell anergy: silencing of intercellular adhesion molecule-1 by histone modifications. *Cancer Res.*, **66**, 10770–10777.
133. Rini, B.I., Jaeger, E., Weinberg, V., Sein, N., Chew, K., Fong, K., Simko, J., Small, E.J., and Waldman, F.M. (2006) Clinical response to therapy targeted at vascular endothelial growth factor in metastatic renal cell carcinoma: impact of patient characteristics and Von Hippel–Lindau gene status. *Br. J. Urol. Int.*, **98**, 756–762.
134. Choueiri, T.K., Vaziri, S.A., Jaeger, E., Elson, P., Wood, L., Bhalla, I.P., Small, E.J., Weinberg, V., Sein, N., Simko, J., Golshayan, A.R., Sercia, L., Zhou, M., Waldman, F.M., Rini, B.I., Bukowski, R.M., and Ganapathi, R. (2008) von Hippel-Lindau gene status and response to vascular endothelial growth factor targeted therapy for metastatic clear cell renal cell carcinoma. *J. Urol.* **180**, 860–865; discussion 865–866.
135. Verbisck, N.V., Costa, E.T., Costa, F.F., Cavalher, F.P., Costa, M.D., Muras, A., Paixao, V.A., Moura, R., Granato, M.F., Ierardi, D.F., Machado, T., Melo, F., Ribeiro, K.B., Cunha, I.W., Lima, V.C., Maciel Mdo, S., Carvalho, A.L., Soares, F.F., Zanata, S., Sogayar, M.C., Chammas, R., and Camargo, A.A. (2009) ADAM23 negatively modulates alpha(v)beta(3) integrin activation during metastasis. *Cancer Res.*, **69**, 5546–5552.

22
Radioactive Microbes: Cancer Therapy Approaches

Ekaterina Dadachova
Albert Einstein College of Medicine, 1695A Eastchester Road, Bronx, New York 10461, USA

Keywords

Radiovirotherapy
The use of viruses to deliver radioisotopic treatment into infected cells.

Sodium iodide symporter (NIS)
A plasma membrane glycoprotein that mediates the active uptake of I^- in the thyroid, which is the crucial first step in thyroid hormone biosynthesis.

Oncolytic virus/bacterium
A type of virus or bacterium capable of destroying the tumor cells post-infection of the tumor.

^{131}Iodine
A β-particle-emitting radionuclide used extensively in cancer therapy.

Translational Medicine Cancer: Advances in Molecular Biology and Medicine,
First Edition. Edited by Robert A. Meyers.

With the majority of cancer patients succumbing to the metastatic form of the disease, there is an ongoing search for ways to block metastatic spread and render patients free from cancer. Radiovirotherapy involves the use of viruses to deliver radioisotopic treatment into infected cells, and several types of virus (e.g., measles virus, herpes simplex type 1 virus, adenovirus, and vaccinia virus), all of which possess oncolytic properties, have been utilized to deliver transgenes encoding for the human sodium iodide symporter (NIS). As the virus does not carry any radioactivity, radiovirotherapy is always a two-step strategy, with viral gene delivery followed by systemic administration of the radionuclide. The NIS transports ^{131}I, which is used to treat thyroid diseases, and radiovirotherapy has been investigated using this radionuclide in experimental cancers, with encouraging results. In spite of some problems with viral delivery, radiovirotherapy has been consistently more efficient in treating experimental cancers than virotherapy alone. Yet another approach to radioactive microbe cancer therapy, radioactive bacteria (radiolisteria), employs a combinatorial agent capable of selectively infecting tumor cells and delivering a therapeutic radionuclide inside the tumor cells. The tumoricidal effect of such therapy results from the Listeria organisms killing tumor cells via the production of reactive oxygen species and a direct impact of radiation on cancer cells. Whilst the field of radioactive microbes for cancer therapy is still in its infancy, it is hoped that radioactive microbe therapy will be extended to clinical trials in the near future, hopefully with encouraging results.

1 Introduction

With the majority of cancer patients succumbing to the metastatic form of the disease, intense investigations continue to be conducted aimed at blocking metastatic spread and rendering patients free from cancer. Currently, two main approaches are employed to treat metastatic cancers: (i) chemotherapy; and (ii) external beam radiation therapy (EBRT). Unfortunately, the success of both modalities – whether used separately or in combination – has been limited, with neither approach having a lasting impact on the survival of patients with metastatic disease. In addition, chemotherapy is associated with profound short- and long-term toxic effects. Clearly, alternative methods of combating metastatic cancer are required, and an overview of the emerging field of radioactive microbes for cancer therapy is presented in the following sections.

2 Radiovirotherapy

In a recent review by Touchefeu *et al.*, radiovirotherapy was defined as the use of viruses to deliver radioisotopic treatment into infected cells [1]. It has long been known that oncolytic viruses can selectively target and kill cancer cells, and that a combination of these viruses and EBRT can have synergistic antitumor effects. In such applications the viruses most likely act as radiosensitizers, while the ionizing radiation can also enhance any virus-mediated oncolytic effects on the tumor. A more recently developed approach to combine oncolytic viruses with the cytocidal power of ionizing radiation has been a virally-directed radioisotope therapy, since it is known that certain viruses can selectively target tumor cells and deliver transgenes, the expression of which would lead to the production of proteins capable of targeting therapeutic radionuclides into the tumor cells. It should be emphasized here that the virus does not itself carry any radioactivity, and consequently radiovirotherapy will always be a two-step strategy, with viral gene delivery followed by systemic administration of the radionuclide.

The sodium iodide symporter (NIS) is the main transgene that has been investigated in this approach. NIS is a plasma membrane glycoprotein that mediates the active uptake of I^- in the thyroid, which is the crucial first step in thyroid hormone biosynthesis [2]. NIS also mediates I^- uptake in other tissues, such as salivary glands, gastric mucosa, and lactating mammary gland. On the other hand, radiovirotherapy capitalizes on the fact that exogenous NIS gene transfer has been carried out successfully into a variety of other cell lines and tumors [3]. Thus, ^{131}I, which is a low-energy beta-emitter with a half-life of 8 days and has been used in the therapy of thyroid diseases for more than 60 years, can be utilized for the "radio" part of radiovirotherapy. As an additional benefit, ^{131}I emits imageable gamma rays, so that the viral spread can be followed non-invasively by using conventional nuclear medicine imaging techniques. In order to avoid any unnecessary radiation exposure to the subject during the imaging phase of the study, alternatives to ^{131}I radioisotopes, such as ^{123}I or ^{124}I, can be used instead; alternatively, the iodine radioisotope can be replaced with pertechnetate ($^{99m}TcO_4^-$), which is another anion transported by the NIS [3]. For the "viro" part of radiovirotherapy, four viruses have been of most interest to date: measles virus; herpes simplex type 1 virus; adenovirus; and vaccinia virus.

2.1 Measles Virus

Miest *et al.* [4] have reported the details of a vaccine-identical measles virus (MV) as an oncolytic agent against mantle cell lymphoma (MCL), an aggressive B-cell non-Hodgkin's lymphoma, which they armed with NIS. In this case, a homogeneous virus seeding and spreading was observed that was restricted to the perfused tissue. High-resolution single photon emission computed tomography (SPECT) *in-vivo* imaging, assisted by *ex-vivo* imaging, allowed for visualization of the infection spread in both the primary and metastatic tumors for over two weeks after therapy. Interestingly, seeding of the metastases with the virus was faster and more profound than that of primary tumors. By using the thyroid uptake as a baseline, a threefold radioiodine uptake in the tumor was observed but, most importantly, radiovirotherapy resulted in a more impressive therapy than either virotherapy or radioiodine alone (Fig. 1). The same research group also conducted some mechanistic investigations of MV entry into the cells. It is known that, in addition to ubiquitous CD46, vaccine MV retains cell entry through its immune cell-specific receptor signaling lymphocytic activation molecule (SLAM). Thus, it was shown that SLAM, but not CD46-dependent entry, sustained an efficient viral spread and tumor regression, and thereby prolonged cell survival. These mechanistic observations highlighted the importance of tissue-specific receptors for both cell entry and activation of responses assisting viral replication during radiovirotherapy.

Touchefeu *et al.* applied MV-NIS radioviral therapy to head and neck and colorectal

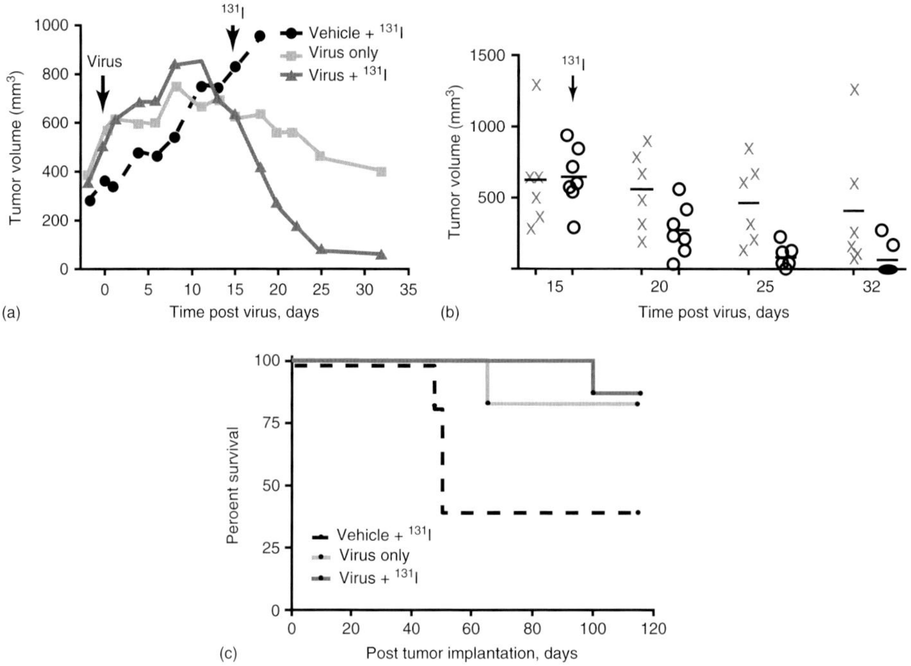

Fig. 1 Radiovirotherapy can be curative. (a) Tumor-response curves for animals treated intravenously (IV) with MV^{vac2}NIS or vehicle, and then treated or not with 1 mCi ^{131}I administered by intraperitoneal (IP) injection, as indicated. Data points represent means; error bars were excluded for clarity; (b) Individual data points are shown for days 15, 20, 25, and 32 after virus administration for animals treated with virus alone (gray X) or combination virus plus ^{131}I (black circles); (c) Kaplan–Meier survival curves for animals treated with MV^{vac2}NIS ($n = 6$), MV^{vac2}NIS + ^{131}I ($n = 7$), or vehicle + ^{131}I ($n = 5$). MV, measles virus.

cancer models [5], and also combined MV-NIS with EBRT and specific checkpoint kinase 1 (Chk1) inhibition. These agents functioned most efficiently in combination, with a synergy being observed between EBRT and MV-NIS *in vitro*, and with EBRT increasing NIS expression in infected cells. The Chk1 inhibitor SAR-020106 displayed synergy with both EBRT and MV-NIS in HN5 head and neck cancer cells, while MV-NIS mediated a ^{131}I-induced cytotoxicity in HN5 head and neck cells and HCT116 colon cancer cells which, in turn, was enhanced by SAR-020106. These *in-vitro* observations were confirmed in studies with mice, when MV-NIS, EBRT and Chk1 inhibition were each effective in reducing the size of HCT116 xenografts.

2.2
Herpes Simplex Virus Type 1

Oncolytic herpes simplex virus type 1 (oHSV-1) has been safely administered to patients in several clinical trials. Li *et al.* aimed at monitoring viral replication in tumor tissue with the goal of improving the oncolytic potency of oHSV-1 in future clinical trials [6]. In

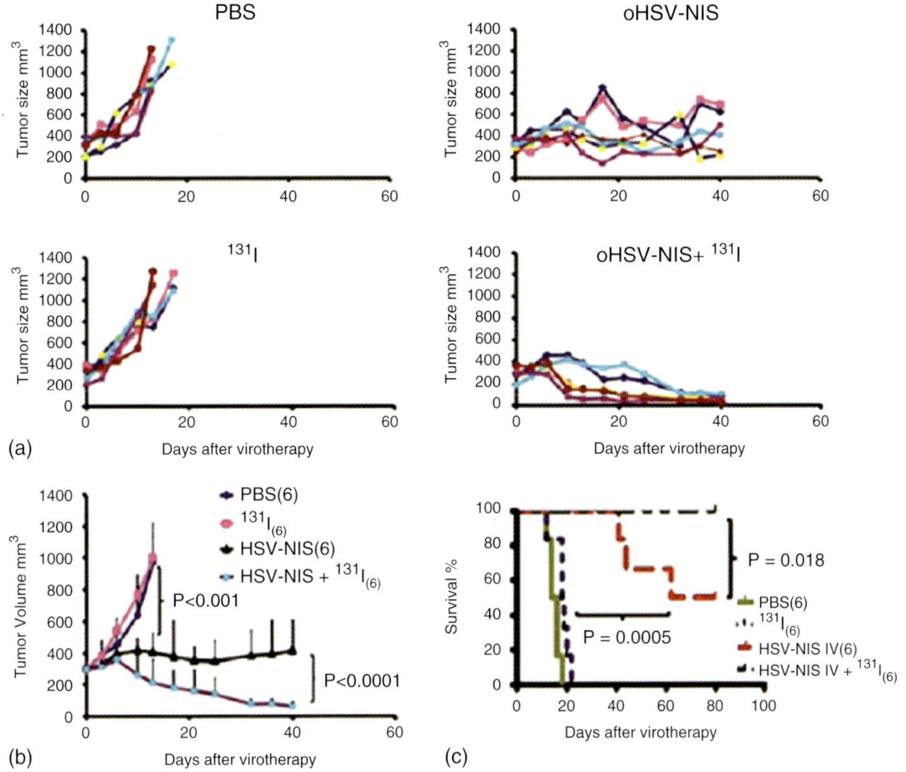

Fig. 2 Radiovirotherapy with systemically administered oncolytic herpes simplex virus-sodium iodide symporter (oHSV-NIS) and radioiodine (^{131}I) in the LNCap prostate cancer xenograft model. LNCap tumor-bearing mice were treated with phosphate-buffered saline (PBS) or oHSV-NIS (12 mice per group). Seven days later, half of the animals in each group received 1 mCi ^{131}I by intraperitoneal injection. (a) Tumor growth for individual mice treated with PBS, oHSV-NIS, PBS + ^{131}I and oHSV-NIS + ^{131}I in each group; (b) Average tumor growth curves plotted for all groups; (c) Corresponding Kaplan–Meier survival curves.

order to achieve a noninvasive *in-vivo* monitoring of the viral spread, as well as to increase its anti-tumor potency, Li and coworkers incorporated the *NIS* gene into the recombinant oHSV genome. The incorporation of *NIS* allowed the *in-vivo* replication of oncolytic herpes simplex virus-sodium iodide symporter (oHSV-NIS) in tumors to be followed, by using SPECT/computed tomography (CT) and pertechnetate ($^{99m}TcO_4^-$). In an initial phase of the therapeutic experiments with oHSV-NIS, xenografts of a human prostate cancer cell line LNCap in nude mice were eradicated by the intratumoral administration of oHSV-NIS. When given systemically, oHSV-NIS, followed by ^{131}I administration after the intratumoral spread of the virus had peaked, prolonged the survival of tumor-bearing mice (Fig. 2). These results confirmed the potential of oHSV-NIS to enhance the outcomes of standard therapy for patients with prostate cancer.

2.3 Adenovirus

Oneal *et al.* generated two adenovirus (Ad) vectors for radiovirotherapy, namely Ad5/3PB-ADP-hNIS and Ad5/3PB-hNIS, which includes a hybrid Ad5/3 fiber knob to improve transduction efficiency and also expresses NIS from the endogenous major late promoter to restrict NIS expression to target cells [7]. Ad5/3PB-ADP-hNIS also encodes for the adenovirus death protein (ADP), which speeds up release of the mature virus after assembly. These two vectors were tested in androgen receptor-expressing prostate cancer cell lines, and found specifically to induce radioisotope uptake, cytopathic effect, and viral replication. Interestingly, Ad5/3PB-ADP-hNIS showed an earlier ^{131}I uptake and cytolysis at a low multiplicity of infection (MOI). Comparative SPECT/CT imaging of xenografted tumors infected with Ad5/3PB-hNIS on days 1, 4, 11, and 14 demonstrated a steady uptake of radioiodine, while infection with Ad5/3PB-ADP-hNIS led to an increasing uptake that was indicative of the viral spread (Fig. 3). Although the

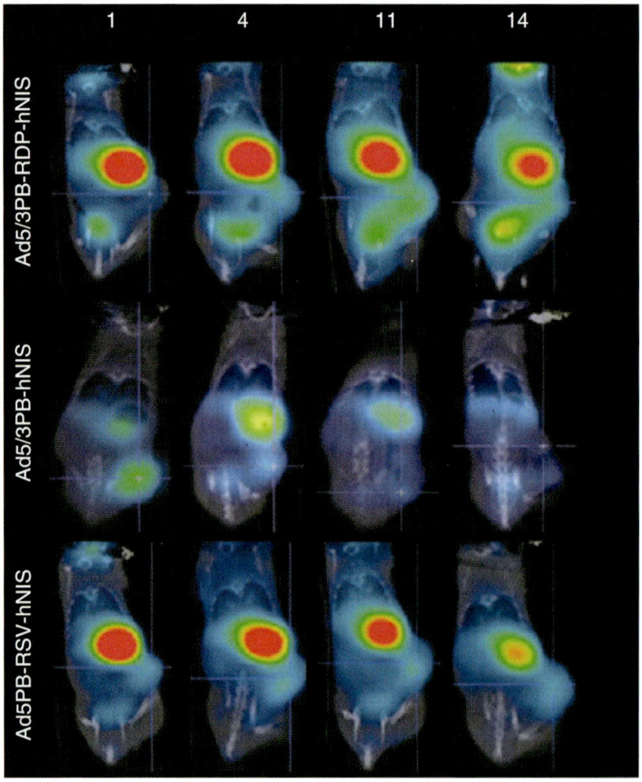

Fig. 3 *In-vivo* radioisotope uptake of LNCaP xenografts. Tumor-bearing mice ($n = 4$) were imaged at the time points shown (1, 4, 11, and 14 days) after injection of 10^9 pfu of virus. At each time point, mice were injected intraperitoneally with 0.5 mCi of $^{99m}TcO_4^-$ and imaged by SPECT/CT 1 h later. Intratumoral uptake was recorded by volume-of-interest analysis. The cross-sectional views of intratumoral isotope concentration are shown. The red areas indicate high intensity radioisotope uptake.

radiovirotherapy of LNCaP xenografts with Ad5/3PB-ADP-hNIS resulted in a significant prolongation of animal survival compared to controls ($p = 0.001$), there was no significant difference between radiovirotherapy and virotherapy alone in this model. The same research group continued to investigate the potential of an Ad-NIS combination in radiovirotherapy in terms of viral dose, increased viral replication, infectivity enhancement, radioiodide uptake, and delayed efflux of radioiodide from the xenografts. A direct correlation was also found with the efficacy of treatment, and that the effect of replication speed required both tumor- and model-specific testing [8, 9]. Finally, it was noted that the radioiodine doses needed to achieve significant and lasting effects of radiovirotherapy in prostate cancer could be scaled up to levels comparable to those used to treat thyroid cancer in patients [10].

Grünwald *et al.* used a different modification of a conditionally replicating oncolytic Ad by using a replication-selective adenovirus in which the *E1a* gene is driven by the mouse alpha-fetoprotein (AFP) promoter and the human *NIS* gene is inserted into the E3 region (Ad5-E1/AFP-E3/NIS) [11]. Grünwald *et al.* then investigated the efficacy of the virus in mice carrying human hepatocellular carcinoma cell (HuH7) xenografts. Although the study was somewhat limited by the intratumoral injection of the virus, at three days after the injection of Ad5-E1/AFP-E3/NIS, followed by ^{123}I, nuclear imaging showed that the HuH7 xenografts has taken up 25% of the injected dose per gram of tumor tissue. Subsequently, it was shown that radiovirotherapy of mice bearing HuH7 xenografts had resulted in a markedly improved animal survival compared to virotherapy alone.

2.4 Improvements in Systemic Delivery of the Theranostic NIS Gene

Certain factors limit the potential applications of Ad-mediated gene therapy, such as a high prevalence of neutralizing antibodies, a widespread expression of the coxsackie-adenovirus receptor (CAR), and adenovirus sequestration by the liver [12]. In a series of experiments, Grünwald *et al.* investigated whether the coating of adenovirus with synthetic dendrimers could be useful in overcoming these difficulties, so as to enable oncolytic virotherapy and radiovirotherapy [12, 13]. In a first series of experiments, the replication-selective Ad5-E1/AFP-E3/NIS virus was coated with poly(amidoamine) dendrimers generation 5 (PAMAM-G5) to study the transduction efficacy and altered tropism of these coated virus particles (using nuclear medicine imaging), and also to evaluate their therapeutic potential for systemic radiovirotherapy in a liver cancer xenograft mouse model [12]. In a second series of experiments to further improve the shielding and targeting, the Ad-NIS constructs were coated with a conjugate that consisted of the same poly(amidoamine) (PAMAM) dendrimer but was linked to the peptidic epidermal growth factor receptor (EGFR)-specific ligand GE11 [13]. The results of these experiments demonstrated an efficient liver detargeting and tumor retargeting of adenoviral vectors after coating with synthetic dendrimers, thus confirming the promise of this strategy to improve viral delivery during radiovirotherapy.

2.5 Vaccinia

Belin *et al.* investigated the ability of an oncolytic vaccinia virus expressing NIS to provide real-time monitoring of viral therapy of malignant pleural mesothelioma (MPM) [14]. The vaccinia-NIS virus used was GLV-1h153 which is a replication-competent, recombinant vaccinia virus (VACV) derived from the parent VACV LIVP strain and contains NIS and enhanced green fluorescent protein (eGFP) transgenes. To achieve real-time, noninvasive monitoring of virotherapy, the research group studied the infectivity and cytotoxic effects of GLV-1h153 on mesothelioma cell lines of all histologic subtypes. It was observed that GLV-1h153 was able to infect and kill all cell lines in both time- and concentration-dependent manner, with the viral replication increasing more than 2.5-log in titer over four days. Although the potential for radiovirotherapy in the treatment of MPM was not investigated in this study, the results showed that the infection of orthotopic MPM xenografts with GLV-1h153, followed by the administration of ^{131}I, allowed for the monitoring of viral growth, using SPECT/CT.

In a follow-up study, the same group investigated the actual radiovirotherapy of GLV-1h153 in an orthotopic triple-negative breast cancer (TNBC) murine model [15]. Mice carrying orthotopic xenografts (MDA-MB-231 cells) were injected intratumorally with either GLV-1h153 or phosphate-buffered saline (PBS). Treatment with ^{131}I (ca. 5 mCi dose) was given systemically at one week after the viral injection. The administration of ^{131}I in combination with GLV-1h153 resulted in a sixfold tumor regression when compared to virotherapy alone (Fig. 4). It was concluded that radiovirotherapy with vaccinia-NIS virus could be a promising therapeutic modality for TNBC.

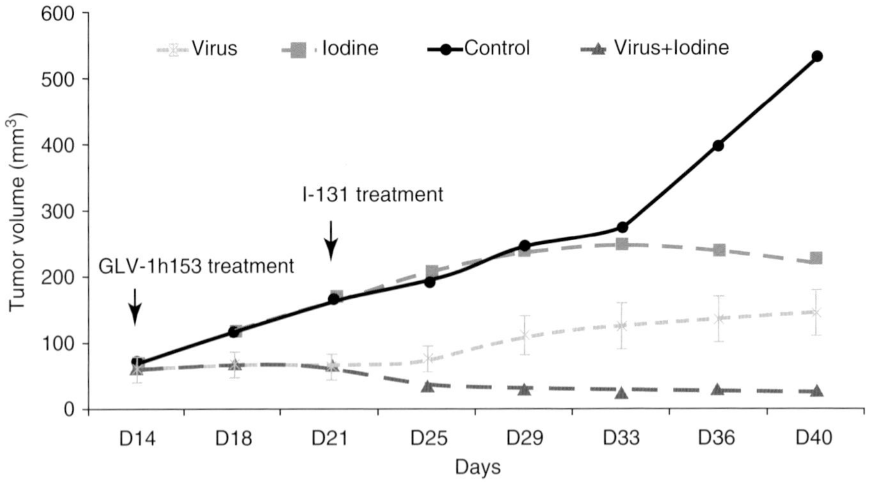

Fig. 4 Combination treatment with GLV1h-153 and ^{131}I *in vivo* achieves greater tumor growth suppression than single therapy with GLV-1h153 or ^{131}I. At 14 days after xenograft implantation, mice were treated with intratumoral injections of GLV-1h153 or PBS. After one week (D21), virus-treated xenografts were randomized into virus alone and virus iodine groups; PBS-treated xenografts were randomized into PBS alone (control) and iodine alone groups (four animals per treatment group). Tumor volume growth rates and differences in the combination treatment were analyzed using a Kruskal–Wallis one-way ANOVA analysis and a Wilcoxon test.

3 Radiolisteria

Listeria monocytogenes is a Gram-positive, facultative intracellular bacterium which infects antigen-presenting cells (APCs), and is currently being used as a cancer vaccine to deliver tumor antigens directly to the APC [16]. Recently, Kim *et al.* discovered that a highly attenuated *Listeria* (Listeria^{at}), which originally was used to deliver tumor-associated antigens into the APCs, also effectively infected tumor cells *in vitro* and *in vivo* [17]. Although Listeria^{at} was efficiently cleared by the immune system in the normal tissues within three to five days, immune suppression in the tumor microenvironment allowed these bacteria to accumulate in metastases and primary tumors, and to kill tumor cells through high levels of reactive oxygen species (ROS) [17].

The hypothesis for the follow-up study was that Listeria^{at} could be used to deliver therapeutic radionuclides emitting cytocidal beta-particles in a specific manner to the microenvironment of metastases and primary tumors, and into tumor cells. To achieve this, Quispe-Tintaya *et al.* coupled rhenium-188 (^{188}Re), a high-energy, beta-emitting radionuclide with a short physical half-life of 16.9 h, to Listeria^{at} by utilizing a *Listeria*-specific antibody [18]. Thus, the first truly radioactive microbe for the treatment of cancer, radiolisteria (RL), was created. The same research team then demonstrated, using a highly metastatic mouse pancreatic tumor model (Panc-02), that RL was capable of delivering radioactivity to the metastases and less abundantly to primary tumors *in vivo*, without harming normal cells. Such specific delivery without toxic side effects was possible because RL was efficiently cleared by the immune system in normal tissues, but not in the heavily immune-suppressed microenvironment of metastases and primary tumor. Eleven treatments with low doses (10^4 colony-forming units) of RL resulted in a dramatic decrease in the number of metastases (>90%) compared to control groups in the Panc-02 model (Fig. 5). In commenting on these studies, Stritzker and Szalay stated that "... taking these data together, the study by Quispe-Tintaya *et al.* clearly demonstrates the therapeutic benefit of a strategy combining the tumor targeting and biotherapeutic effects of *L. monocytogenes* with the cytotoxic effects of ^{188}Re-labeled antibodies administered as a single agent. Still to be elucidated, however, is the exact mechanism by which this exciting unique combinatorial therapeutic approach works and how it can be potentially improved. The work by Quispe-Tintaya *et al.* certainly is a major step forward in the search for Paul Ehrlich's "magic bullet" in the war against cancer" [19].

4 Summary

The field of radioactive microbes for cancer therapy is still in its infancy, but continues to emerge. In the case of radiovirotherapy, a better understanding of the regulation of the NIS will open up pathways to potentiate its functional expression [1]. Today, viral delivery protocols are constantly being improved, such that therapeutic radionuclides which are more powerful than ^{131}I and are also transported by NIS, such as astatine-211 (^{211}At) and ^{188}Re [3], might replace ^{131}I in some applications of radiovirotherapy. In the case of radiobacteria, the ways forward include a simplification of the generation of radioactive bacterial species and the possible use of other, Gram-negative bacterial strains also known

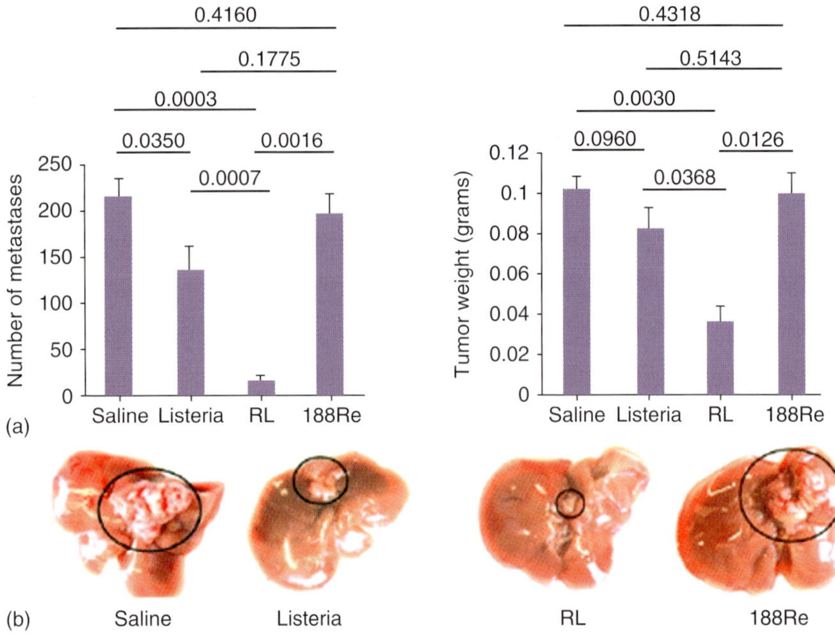

Fig. 5 Treatments with radiolisteria (RL) markedly reduced the number of metastases in the Panc-02 model. (a) Mice were therapeutically immunized 11 times with the low dose (10^4 cfu) of RL (Listeria^{at}-^{188}Re), Listeria^{at}, ^{188}Re, or saline and then euthanized 21 days later. Numbers of metastases and tumor weights were determined. Sample size was five mice per group. This experiment was repeated three times, and the results were averaged. $P < 0.05$ was statistically significant (Mann–Whitney). The error bars represent SEM; (b) Representatives from each group of metastases in the portal liver.

to infect tumor cells. The radioactive microbe approach could also be combined with other anticancer therapies such as chemotherapy and EBRT to achieve synergistic effects. It is hoped that radioactive microbe therapy will translate into clinical trials in the near future, and eventually will be added to the armamentarium of cancer-fighting drugs.

References

1. Touchefeu, Y., Franken, P., and Harrington, K.J. (2012) Radiovirotherapy: principles and prospects in oncology. *Curr. Pharm. Des.*, **18** (22), 3313–3320.
2. Dai, G., Levy, O., and Carrasco, N. (1996) Cloning and characterization of the thyroid iodide transporter. *Nature*, **379** (6564), 458–460.
3. Dadachova, E. and Carrasco, N. (2004) The Na/I symporter (NIS): imaging and therapeutic applications. *Semin. Nucl. Med.*, **34** (1), 23–31.
4. Miest, T.S., Frenzke, M., and Cattaneo, R. (2013) Measles virus entry through the signaling lymphocyte activation molecule governs efficacy of mantle cell lymphoma radiovirotherapy. *Mol. Ther.*, **21** (11), 2019–2031.
5. Touchefeu, Y., Khan, A.A., Borst, G., Zaidi, S.H., McLaughlin, M., Roulstone, V., Mansfield, D., Kyula, J., Pencavel, T., Karapanagiotou, E.M., Clayton, J., Federspiel, M.J., Russell, S.J., Garrett, M., Collins, I., and Harrington, K.J. (2013) Optimising measles virus-guided radiovirotherapy with external beam radiotherapy and specific checkpoint

kinase 1 inhibition. *Radiother. Oncol.*, **108** (1), 24–31.
6. Li, H., Nakashima, H., Decklever, T.D., Nace, R.A., and Russell, S.J. (2013) HSV-NIS, an oncolytic herpes simplex virus type 1 encoding human sodium iodide symporter for preclinical prostate cancer radiovirotherapy. *Cancer Gene Ther.*, **20** (8), 478–485.
7. Oneal, M.J., Trujillo, M.A., Davydova, J., McDonough, S., Yamamoto, M., and Morris, J.C. III, (2012) Characterization of infectivity-enhanced conditionally replicating adenovectors for prostate cancer radiovirotherapy. *Hum. Gene Ther.*, **23** (9), 951–959.
8. Trujillo, M.A., Oneal, M.J., McDonough, S.J., and Morris, J.C. (2013) Viral dose, radioiodide uptake, and delayed efflux in adenovirus-mediated NIS radiovirotherapy correlates with treatment efficacy. *Gene Ther.*, **20** (5), 567–574.
9. Oneal, M.J., Trujillo, M.A., Davydova, J., McDonough, S., Yamamoto, M., and Morris, J.C. III, (2013) Effect of increased viral replication and infectivity enhancement on radioiodide uptake and oncolytic activity of adenovirus vectors expressing the sodium iodide symporter. *Cancer Gene Ther.*, **20** (3), 195–200.
10. Trujillo, M.A., Oneal, M.J., McDonough, S., Qin, R., and Morris, J.C. (2012) A steep radioiodine dose response scalable to humans in sodium-iodide symporter (NIS)-mediated radiovirotherapy for prostate cancer. *Cancer Gene Ther.*, **19** (12), 839–844.
11. Grünwald, G.K., Klutz, K., Willhauck, M.J., Schwenk, N., Senekowitsch-Schmidtke, R., Schwaiger, M., Zach, C., Göke, B., Holm, P.S., and Spitzweg, C. (2013) Sodium iodide symporter (NIS)-mediated radiovirotherapy of hepatocellular cancer using a conditionally replicating adenovirus. *Gene Ther.*, **20** (6), 625–633.
12. Grünwald, G.K., Vetter, A., Klutz, K., Willhauck, M.J., Schwenk, N., Senekowitsch-Schmidtke, R., Schwaiger, M., Zach, C., Wagner, E., Göke, B., Holm, P.S., Ogris, M., and Spitzweg, C. (2013) Systemic image-guided liver cancer radiovirotherapy using dendrimer-coated adenovirus encoding the sodium iodide symporter as theranostic gene. *J. Nucl. Med.*, **54** (8), 1450–1457.
13. Grünwald, G.K., Vetter, A., Klutz, K., Willhauck, M.J., Schwenk, N., Senekowitsch-Schmidtke, R., Schwaiger, M., Zach, C., Wagner, E., Göke, B., Holm, P.S., Ogris, M., and Spitzweg, C. (2013) EGFr-targeted adenovirus dendrimer coating for improved systemic delivery of the theranostic NIS gene. *Mol. Ther. Nucleic Acids*, **2**, e131.
14. Belin, L.J., Ady, J.W., Lewis, C., Marano, D., Gholami, S., Mojica, K., Eveno, C., Longo, V., Zanzonico, P.B., Chen, N.G., Szalay, A.A., and Fong, Y. (2013) An oncolytic vaccinia virus expressing the human sodium iodine symporter prolongs survival and facilitates SPECT/CT imaging in an orthotopic model of malignant pleural mesothelioma. *Surgery*, **154** (3), 486–495.
15. Gholami, S., Chen, C.H., Lou, E., Belin, L.J., Fujisawa, S., Longo, V.A., Chen, N.G., Gönen, M., Zanzonico, P.B., Szalay, A.A., and Fong, Y. (2014) Vaccinia virus GLV-1h153 in combination with 131I shows increased efficiency in treating triple-negative breast cancer. *FASEB J.*, **28** (2), 676–682.
16. Singh, R. and Wallecha, A. (2011) Cancer immunotherapy using recombinant *Listeria monocytogenes.* Transition from bench to clinic. *Hum. Vaccin.*, **7**, 497–505.
17. Kim, S.H., Castro, F., Paterson, Y., and Gravekamp, C. (2009) High efficacy of a Listeria-based vaccine against metastatic breast cancer reveals a dual mode of action. *Cancer Res.*, **69** (14), 5860–5866.
18. Quispe-Tintaya, W., Chandra, D., Jahangir, A., Harris, M., Casadevall, A., Dadachova, E., and Gravekamp, C. (2013) A non-toxic radioactive Listeria^{at} is a highly effective therapy against metastatic pancreatic cancer. *Proc. Natl Acad. Sci. USA*, **110** (21), 8668–8673.
19. Stritzker, J. and Szalay, A.A. (2013) Single-agent combinatorial cancer therapy. *Proc. Natl Acad. Sci. USA*, **110** (21), 8325–8326.

23
The Molecular Genetics of Retinoblastoma

Timothy W. Corson[1] *and Helen Dimaras*[2,3,4]
[1] *Indiana University School of Medicine, Department of Ophthalmology, 1160 West Michigan Street, Indianapolis, IN 46202, USA*
[2] *University of Toronto, Department of Ophthalmology and Vision Sciences, 340 College Street, Toronto ON M5T 3A9, Canada*
[3] *The Hospital for Sick Children, 555 University Avenue, Toronto ON M5G 1X8, Canada*
[4] *The University of Nairobi, Department of Human Pathology, Nairobi, Kenya*

Translational Medicine Cancer: Advances in Molecular Biology and Medicine,
First Edition. Edited by Robert A. Meyers.

Keywords

Aneuploidy
The condition in which a cell contains more or fewer chromosomes than a multiple of its haploid number of chromosomes.

Apoptosis
The process of programmed cell death.

Chromosome instability (CIN)
The unequal distribution of DNA to daughter cells during mitosis.

E2F
A transcription factor family that regulates S-phase genes.

LOH
Loss of heterozygosity.

MYCN
An oncogene, encoding transcription factor N-Myc, that drives a fraction of retinoblastoma.

Proband
The very first individual in a family to be diagnosed with a condition.

RB1
Retinoblastoma tumor susceptibility gene, encoding the cell-cycle regulatory protein pRB.

Retinoblast
A precursor cell that becomes a retinal cell.

Retinoma
A benign precursor to retinoblastoma.

SV40
Simian virus 40 DNA tumor virus.

TAg
Simian virus 40 large T antigen oncoprotein.

Tumor suppressor
A gene that protects a cell from becoming a tumor.

Retinoblastoma is a rare, malignant, childhood tumor that is primarily initiated by the inactivation of both alleles of the retinoblastoma tumor susceptibility gene, *RB1*, in a developing human retinal cell. A rare subset of retinoblastoma is initiated by somatic amplification of the *MYCN* oncogene in a predisposing retinal cell. Surprisingly the retinoblastoma protein (pRB), encoded by *RB1*, is an important transcription factor. Cell-cycle control by pRB is mainly accomplished by transcriptional repression of the genes required for cell-cycle progression. Control of differentiation by pRB is achieved by the activation of transcription. Through extensive post-translational modifications and interactions with other proteins, pRB and family members also influence senescence, chromosomal stability, and apoptosis. Almost every type of tumor has disruption in the retinoblastoma pathway associated with tumor progression, but germline mutation of the *RB1* gene predisposes children to a 95% specific risk of developing retinoblastoma and a significantly increased risk of second primary tumors, such as osteosarcoma and melanoma. However, retinoblastoma is also characterized by other genomic changes subsequent to *RB1* mutation.

1 Introduction

The molecular genetics of retinoblastoma is a complex topic, incorporating clinical ophthalmology and oncology, fundamentals of cancer genetics, protein biochemistry, and cell-cycle biology, plus countless cellular activities including differentiation, development and genome stability. The aim of this chapter is to present an overview of these topics sufficiently detailed to allow an appreciation of the nature of retinoblastoma and the important functions of retinoblastoma protein (pRB), the protein product of the retinoblastoma gene, and with appropriate references for deeper investigation [1–5]. The chapter begins with a summary of the clinical features of this cancer, followed by a discussion of the genetic basis of the disease, though a detailed discussion of the diagnosis, prognosis and management is beyond the scope of this chapter [6–8]. The structure and functions of pRB are presented next in the context of the different cellular processes in which it participates, although a comprehensive presentation of pRB biochemistry is not the intent [9–13]. A discussion of pRB in mouse development then provides necessary background for understanding the modeling of retinoblastoma in mice, which is an active area of current research. Returning to human cancer genetics, additional genetic events in retinoblastoma progression are discussed before closing with some broader considerations about the cell of origin and tissue specificity of retinoblastoma, and why this cancer has an importance to biology and medicine far in excess of its frequency in humans [14–16].

2
Retinoblastoma: A Childhood Ocular Cancer

Retinoblastoma (Fig. 1) most commonly is diagnosed only when tumor fills one or both eyes, and the parents notice a white gleam from the pupil of the baby's eye [15]. Often, the diagnosis is definitive based on clinical examination and imaging that reveals calcification within the tumor (Fig. 1a). When only one eye is affected, the preferred treatment is almost always removal of the eye, which is usually curative. When both eyes are involved, the choice of treatment for each eye depends on the size and extent of tumor, the risk of tumor spreading outside the eye, and the risk to vision of the treatment. The International Intraocular Retinoblastoma Classification (IIRC) stages each eye with retinoblastoma and predicts the likelihood of treatment curing the eye [22].

Eyes with advanced tumor at risk to spread outside the eye, and with virtually no possibility of useful vision even if the tumor is controlled (IIRC Group E), are removed [15, 23]. Eyes with medium or large tumors (IIRC Groups B–D), no risk of extraocular tumor and some possibility of vision, are now commonly treated by systemic or local chemotherapy, combined with laser and cryotherapy to destroy physically any surviving or chemotherapy-resistant tumor cells [15, 23]. Intra-arterial chemotherapy, involving high-dose melphalan introduced directly to the ophthalmic artery, has been successfully used in recent years for larger tumors [24–26]. However, in the absence of rigorous clinical trial data for this technique, evidence-based indications for its use are lacking. Intravitreal chemotherapy has shown promise to treat residual tumor seeds in the vitreous after control of the main tumors has been achieved by other means [27]. Eyes with small tumors not immediately adjacent to the center of vision or the optic nerve (IIRC Group A) may be treated with laser and cryotherapy alone [15]. Late recurrences are rare. Where resources and expertise are available, the fraction of children who lose both eyes to retinoblastoma is now small, and most children will ultimately have vision in at least one eye. This is not the case globally, however, as most children with retinoblastoma live in resource-limited countries where the chance of survival, let alone vision salvage, is low [15].

Retinoblastoma tumor cells may escape from the eye and spread through the optic nerve to the brain or via the blood to grow preferentially in the bone marrow [28, 29]. An accurate histology of enucleated (removed) eyes (Fig. 1b) is essential to determine if the tumor has invaded the structures of the eye, and to what extent [15]. The American Joint Committee on Cancer (AJCC) pTNM staging is used to help clinicians decide if adjuvant chemotherapy and/or surveillance is required [30]. Metastatic retinoblastoma is difficult to cure. Some reports indicate that long-term survival can be achieved by intensive systemic chemotherapy with or

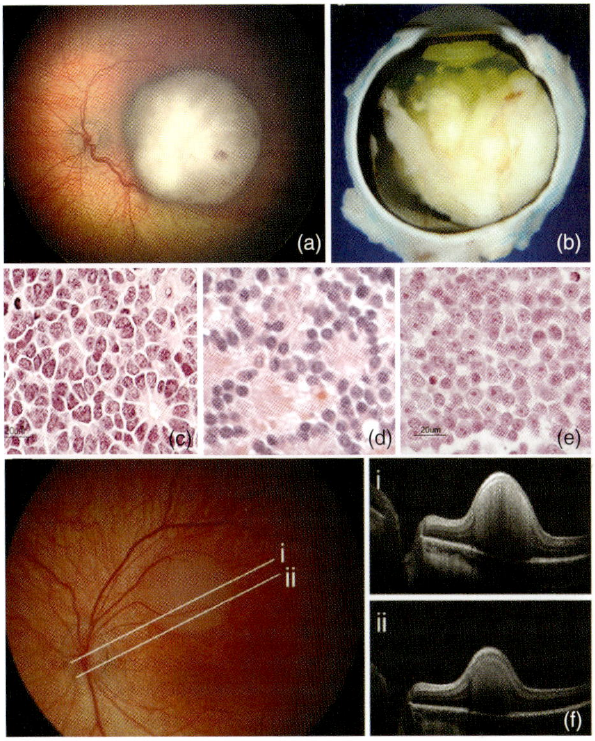

Fig. 1 Clinical features of retinoblastoma. (a) Photograph of the retina showing a medium-sized retinoblastoma (white mass) arising adjacent to the optic nerve; (b) Gross pathology of an enucleated (removed) eye, cut open to reveal a large retinoblastoma filling the vitreous space; (c) Histology of retinoblastoma, showing nuclear molding and a Flexner–Wintersteiner rosette at lower right; (d) Histology of retinoma, showing photoreceptor-like fleurette differentiation; (e) Distinctive histology of *MYCN*-amplified, $RB1^{+/+}$ retinoblastoma showing neuroblastoma-like round nuclei with prominent nucleoli; (f) Optical coherence tomography (OCT) of a small retinoblastoma. Panels (i) and (ii) indicate OCT planes on the funduscopic (retina) image; to the right are OCT images through the retina showing a round tumor situated in the middle layers of the retina. (a) Adapted with permission from Ref. [17]; © 2014, John Wiley & Sons, Ltd; (b) Adapted with permission from Ref. [18]; © 2013, Macmillan Publishers Ltd; (c, e) Adapted with permission from Ref. [19]; © 2013, Elsevier; (d) Adapted with permission from Ref. [20]; © 2008, Oxford University Press; (f) Adapted with permission from Ref. [21]; © 2013, BMJ Publishing Group Limited.

without radiation, by chemotherapy given into the meningeal space, and by megadoses of chemotherapy to destroy any surviving tumor cells, followed by autologous stem cell transplant to reconstitute the bone marrow [31–33].

Radiation therapy is very effective in controlling retinoblastoma tumor growth [34]. However, radiation of children with germline *RB1* mutant alleles greatly increases the risk of a second primary cancer later in life, from the already high risk that the mutant *RB1* allele imposes [35].

Rare cases known as trilateral retinoblastoma are marked by the presence of a brain tumor (pineal or supra-sellar) in addition to eye tumor(s) [36], and require more extensive treatment and surveillance than intraocular retinoblastoma [37]. The brain tumor in these cases is not the result of metastasis, but rather is an independent event. The brain tumor has been shown to display similar clinical and histological features to ocular retinoblastoma [36, 38]. Long-term survival from trilateral retinoblastoma can be achieved with early detection, followed by treatment with high-dose chemotherapy and autologous stem cell rescue [37].

3 Retinoblastoma: The Prototypic Genetic Cancer

The tendency for retinoblastoma to be transmitted in families as an autosomal dominant trait was recognized in the mid-twentieth century [39–42]. The location of the predisposing gene on chromosome 13q was suggested by observations made during the early 1960s that large deletions involving chromosome 13q14 were associated with retinoblastoma and other abnormalities [43–46]. However, bilateral retinoblastoma was thought to be a unique special case of cancer in which genetics played some role, while non-genetic mechanisms were presumed – erroneously – to underlie common forms of cancer.

In 1971, Alfred Knudson proposed – from a simple analysis of age of diagnosis of bilateral versus unilateral retinoblastoma – that both hereditary (usually bilateral) and nonhereditary (always unilateral) retinoblastoma are initiated by two mutational events, M1 and M2 [47]. In 1973, Comings proposed the possibility that these two mutations could be in the two alleles of one gene, the normal function of which would be the suppression of tumors [48].

3.1 Genetic Locus of the Retinoblastoma Gene

Proof that retinoblastoma tumors had mutation of both alleles of the 13q14 predisposing gene was obtained when genetic markers were developed to distinguish the two alleles. Although the *RB1* gene was not yet cloned, markers near the retinoblastoma chromosomal locus showed that the heterozygous 13q14 allele associated with the presence of retinoblastoma in members of a family was often the only retained allele in retinoblastoma tumors (loss of heterozygosity, LOH) [49]. LOH for 13q14 was observed in both hereditary and unilateral (likely not hereditary) retinoblastoma tumors, thus proving the predictions of Comings correct. Both, hereditary and nonhereditary retinoblastomas are nearly always initiated by mutation of the same tumor suppressor gene, the *RB1* gene (see Sect. 3.3 below for an exception to this rule) [50–52]. The retinoblastoma studies opened the way to find other tumor suppressor genes, mutation of which initiates a wide variety of other cancers [53]. Throughout the 1980s, numerous reports were made of the chromosomal localization of disease-specific tumor suppressor genes by searching for LOH in tumors. Cancer was clearly demonstrated to be the most common somatic genetic disease.

In individuals with sporadic bilateral retinoblastoma, M1 occurs either *de novo* in a parent and is transmitted via the germ cells, or during embryonal development [15]. Retinoblastoma is initiated when the remaining normal allele is inactivated (M2) in a developing retinal cell carrying M1. Sporadic unilateral retinoblastoma occurs in children with normal *RB1* constitutional alleles, who develop rare somatic mutations of both alleles (M1/M2) in a developing retinal cell [14].

Bilaterally affected individuals all have one germline *RB1* mutant allele, which results in the autosomal dominant transmission of the predisposition to retinoblastoma. As unilaterally affected patients have only a 15% chance that they carry a germline *RB1* mutant allele [54], the standard approach for unilateral patients is to test the retinoblastoma tumor DNA to identify the two mutant *RB1* alleles and then screen the blood DNA for each identified mutant allele [55]. In this way, a negative result generally rules out heritable retinoblastoma in the unilateral proband, although mosaicism is never excluded. About 10% of new *RB1* germline mutations can be shown to be mosaic, present in less than 50% of the *RB1* DNA and in only a fraction of the cells of the proband [56]. Clinically, mosaicism may become obvious when a parent has more than one child with retinoblastoma but the parent does not show in blood the mutant *RB1* allele present in the children. Mosaicism cannot be inherited as the predisposing mutation must occur during development of the embryo. By DNA analysis, antecedent relatives are excluded from risk, and future offspring of the unilateral patient can be tested for the mutant alleles found in the tumor of their parent, just in case the parent had undetected germline mosaicism [56].

The majority of children with heritable retinoblastoma have newly acquired *RB1* mutations, and thus no family history of the disease. Most frequently, it is the paternal *RB1* allele that is mutated, presumably because of the increased risk of mutation of the sperm relative to the ova [57].

3.2 Retinoma

Nonprogressing retinal lesions were originally considered to represent the "spontaneous regression" of retinoblastoma [58]. Now known as retinoma (sometimes called retinocytoma), this lesion clinically presents as a translucent gray retinal mass, frequently associated with calcification and hyperplasia of the retinal pigment epithelium, unlike the actively proliferating retinoblastoma that shows calcification but appears opaque white [58, 59]. On histopathology (Fig. 1c,d), retinoma shows a varied arrangement of nonproliferative neuronal cells with photoreceptor differentiation in the form of fleurettes [38, 60]. Unlike regressed tumors, retinoma does not show any evidence of necrosis [38]. In contrast, retinoblastoma is highly proliferative, its cells crowding together in a cobblestone pattern, and sometimes showing differentiation in the form of Flexner–Wintersteiner or Homer Wright rosettes [18, 38, 61, 62]. Retinoblastoma also often shows pseudorosettes, which are radial arrangements of cells around the blood vessels; cells too far away from the blood supply tend to necrose [18].

Although retinomas are benign, they can undergo malignant progression to retinoblastoma [63]. Particularly in a young child, such tumors must be inspected regularly for any indication of transition to retinoblastoma [59]. *RB1* germline mutations predispose to both retinoma and retinoblastoma in the same patient, and also in other members of a family who have the mutant *RB1* allele. Careful molecular analyses have shown that retinomas do indeed have mutation of two *RB1* alleles, but lack some of the subsequent genetic changes seen in retinoblastoma (see Sect. 9) [20], which strongly suggests that they are a "premalignant" lesion rather than spontaneous regression. Indeed, retinomas express senescence markers p130 and $p16^{INK4A}$, which suggests that this process must be bypassed *en route* to malignancy [20].

3.3
MYCN-Amplified Retinoblastoma

Approximately 1% of unilateral tumors are not initiated by mutation in *RB1*, but by amplification of the *MYCN* oncogene [19]. N-Myc, encoded by the *MYCN* gene, is a transcription factor and chromatin regulator homologous to the commonly activated oncogene, c-Myc. *MYCN* amplification was first detected in the childhood nervous system tumor neuroblastoma, in which it is associated with a poor prognosis [64]. *MYCN* genomic gain or amplification in retinoblastoma can occur as a secondary event after *RB1* mutation [14], but it is overrepresented in unilateral retinoblastoma lacking detectable *RB1* mutations, strongly suggesting that it “drives” this subset of tumors. *MYCN*-amplified retinoblastomas form a distinct clinical group, with very early onset and distinct histology (Fig. 1e), but they are unifocal and not hereditary [19]. There is a mechanistic link between *MYCN*-amplified retinoblastoma and retinal development: survival and proliferation of retinal progenitors in mice requires Mycn in the absence of the pRB-binding proteins E2f1-3 (see Sect. 6.2 below), although Mycn is not required for wild-type retinal development [65].

4
Spectrum of *RB1* Mutant Alleles

RB1 mutations are spread throughout the gene including the promoter, most exons, and splicing regions of introns (Fig. 2) [66]. More than 50% of *RB1* mutations have been found only once, and the majority of germline mutations originate in the proband [67]. Molecular identification of *RB1* mutations supports accurate determination of the risk of retinoblastoma in relatives [66].

RB1 mutations are most frequently “null” mutations that are out of frame and result in premature STOP codons (Fig. 2) [66]. The premature truncation of translation in turn

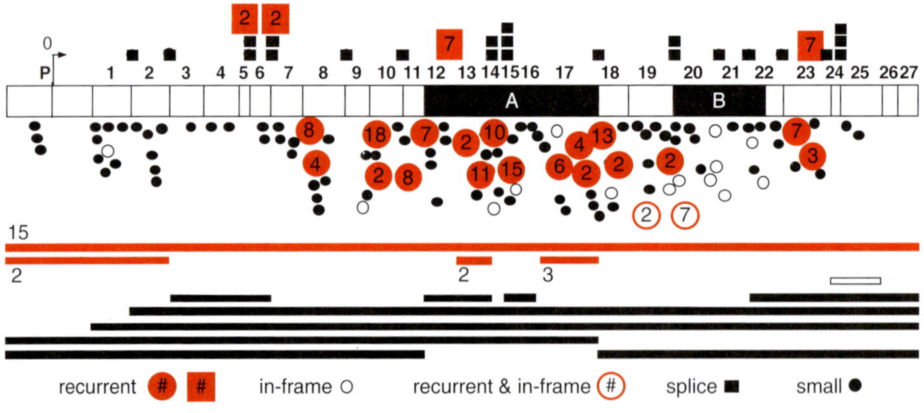

Fig. 2 Retinoblastoma-initiating mutations are distributed throughout the *RB1* gene. A total of 230 example germline mutations is shown, out of more than 1000 that have been identified, many only in a single proband. An up-to-date catalog of *RB1* mutations may be found at: *http://rb1-lovd.d-lohmann.de/home.php*. Promoter (P) methylation is not shown. Recurrent mutations are shown in red, with the number of occurrences indicated. Squares above gene (exons numbered) indicate splice mutations. Circles below gene indicate small point or few nucleotide mutations, with in-frame mutations shown as white circles. Lines indicate lengths of large deletions.

triggers nonsense-mediated decay, unstable mutant mRNA, and no detectable pRB protein. The only significant hot spots for *RB1* mutations are the "recurrent" mutations, most commonly CGA (arginine) mutating to TGA (STOP) because CpG-containing codons are prone to mutate to TpG. Missense mutations almost invariably involve the A/B "pocket" domain of pRB, which is critical in many protein interactions of pRB (see Sect. 6). In one recent study, rare tumors were identified that displayed focal chromothripsis (several thousand clustered chromosomal rearrangements in a distinct genomic region) at the *RB1* locus, suggesting yet another route to *RB1* inactivation [68].

In bilateral probands, 93% of the germline mutations are null, 6% are in-frame, and 1% are promoter. In the uncommon unilateral germline probands (both familial and sporadic), the ratios of null (57%), in-frame (40%), and promoter (3%) mutations are very different [66]. A unilateral tumor seen in a patient with germline *RB1* mutation indicates a reduced expressivity of the germline *RB1* mutation, manifesting as fewer than expected tumors; this is less likely with a null allele [66]. Reduced penetrance/expressivity *RB1* mutant alleles retain enough activity to prevent some, but not all, retinoblastoma tumors. The types of mutant allele identified in retinoblastoma tumors, presumably of somatic origin (89% null, 3% in frame, and 8% promoter methylation), are different again [66].

The second *RB1* allele (M2) is lost by LOH in 52% of sporadic retinoblastoma tumors, with the remainder including gene conversion, mitotic recombination, small deletions, and point mutations [66]. The likelihood that the M2 event is LOH also varies with the type of M1 event. While 50% of the tumors with M1 whole gene and exonic deletions show LOH, 88% of tumors with methylation of the promoter show LOH [66]. The biological reasons for these differences are unknown.

5
The Retinoblastoma Gene and Protein

The *RB1* gene was cloned by the identification of a chromosome 13 DNA clone that was totally deleted from a retinoblastoma tumor. This was correctly interpreted to be LOH for a deleted allele, and the clone turned out to be located within an exon of the *RB1* gene [50]. The 190 kbp *RB1* gene transcribes a 4.7 kb RNA in 27 exons.

Comparison of the human and mouse *RB1* genes shows 80% homology in the first 235 bp upstream of the initiating methionine, 89% homology in the coding sequence, and 77% homology in the 3′ untranslated region. The human and mouse promoters contain binding sites for the transcriptional elements ATF, SP1, and E2F, but are devoid of TATA and CAAT boxes. The human and mouse proteins show 91% identity, suggesting that the function of this protein is highly conserved. pRB is a nuclear phosphoprotein of 928 amino acids.

RB1 function is abrogated either by deletion or mutation as an initiating event in most retinoblastomas (see Sect. 4). In most other tumor types, *RB1* is mutated or functionally inactivated, confirming that *RB1* is a tumor suppressor gene [10]. The RB pathway can be inactivated by hyperactivity of pRB-inactivating factors such as cyclin-dependent kinase (CDK) 4 or cyclin D (see Sect. 6.1). Alternatively, proteins that inhibit CDK activity, such as $p16^{INK4a}$, may be inactivated in cancer, with the same net effect [69]. The reintroduction of *RB1* into certain cells that have nonfunctional *RB1* alleles can suppress proliferation [70, 71]. The ability of pRB to reverse the malignant phenotype of tumor cells may depend on what other genetic alterations have occurred in the cell.

5.1 pRB Structure

pRB is divided into four regions, the N terminus, the A and B domains separated by a spacer region, and the C terminus (Fig. 3) [9]. The disruption of growth suppression by pRB requires the A and B domains as well as portions of the C-terminal domain [72, 73]. The nuclear c-Abl protein binds to the C terminus of pRB, promoting proliferation [74]. The C terminus is required for growth suppression along with the A and B domains [75].

The A/B region of pRB binds to the LXCXE (Leucine-X-Cysteine-X-Glutamic acid, where X is any amino acid) epitope of the viral oncoproteins, simian virus 40 large T antigen (TAg), adenovirus E1A, and human papilloma virus-16 E7 [76–78]. The importance of the A/B domain is evident by the large number of endogenous proteins binding to it, and the number of tumorigenic missense mutations affecting it (Fig. 2). The A domain is required for stable folding of the B region, the site of binding of LXCXE proteins [79]. Even when the A and B regions of pRB are expressed on separate proteins, they are able to form a repressor motif that is regulated by phosphorylation [80]. Another highly conserved region is the intervening sequence between the A and B regions, suggesting an additional protein binding site, or a critical role in folding [79]. Indeed, the E2F transcription factor (see Sect. 6.2) binds to the highly conserved interface between the A and B regions of pRB. The ability of pRB to

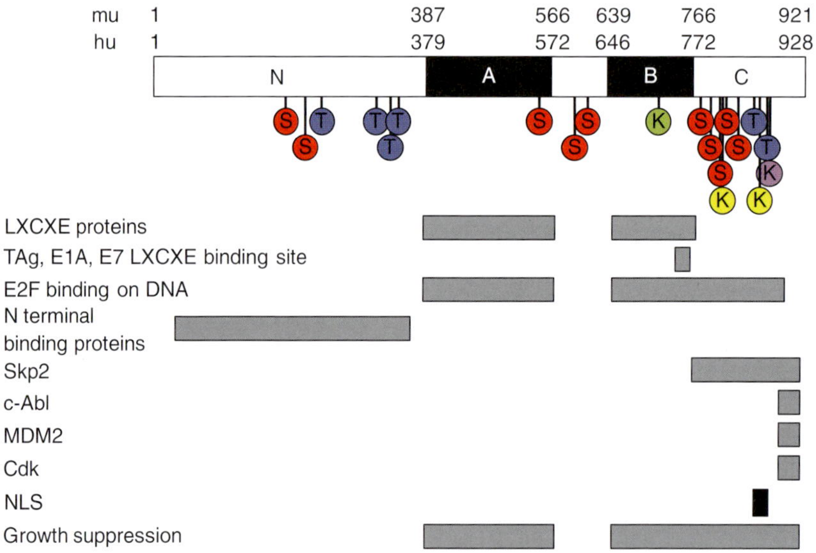

Fig. 3 Structure of pRB. Numbers above the schematic designate the amino acids for murine (mu) and human (hu) pRB. The N and C termini and the A and B regions are required for binding the SV40 large T antigen. Positions of the 16 phosphorylation sites in murine pRB are shown as S (serine; red) and T (threonine; blue). Selected lysines (K) are modified by sumoylation (green), methylation (yellow), and/or acetylation (purple). Areas of pRB required for different functions, including the nuclear localization signal (NLS), are indicated below the schematic.

enter the nucleus is regulated by a bipartite nuclear localization signal located in the C terminus (Fig. 3). The N-terminal domain is also highly conserved, with tandem cyclin folds similar to the A and B regions, and can harbor disease-causing mutations [81]. It can interact with the A/B domain, and can be regulated by occupancy of the LXCXE binding site, and by phosphorylation of the C-terminal domain.

5.2
RB Family Members

Two genes/proteins homologous to pRB, p107 (encoded by *RBL1*), and p130 (encoded by *RBL2*), which had previously been observed to bind E1A, were identified by sequence comparison [82–84]. All three RB family proteins, sometimes known as "pocket proteins," are found in both hypophosphorylated and hyperphosphorylated forms. However, while p107 and p130 bind to cyclin A and cyclin E, pRB does not [85]. Cyclin A binds to the A/B spacer region of p107, which is highly conserved between p107 and p130 [86].

The RB family members are expressed differently during the cell cycle. p130 is expressed during quiescence but disappears as cells are stimulated to re-enter the cell cycle. Quiescent cells do not express p107, but as cells re-enter the cell cycle, p107 accumulates. In contrast, pRB levels are generally constant throughout the cell cycle, although there is a small increase in the level of pRB following growth stimulation [11].

While tumor suppression was defined first by the powerful function of pRB, there is not much evidence that the other RB family members can independently act as tumor suppressors. However, to some extent the RB family compensate for each other. The exogenous overexpression of p107 can suppress the growth of C33A cells, which have a mutant pRB, even though wild-type pRB cannot [83]. Growth suppression in C33A cells by p107 is mediated through its cyclin/CDK-binding domain in addition to the A/B and C domains. p130 has been shown to suppress the growth of only one cell line, the human glioblastoma cell line T98G, which is refractory to suppression by pRB or p107 [87]. Few mutations of *RBL1* or *RBL2* have been observed in human tumors.

6
pRB Functions

6.1
pRB Post-Translational Modifications

One of the first clues to the function of pRB came following the discovery that the phosphorylation status of pRB depends on the cell cycle [88, 89]. pRB is underphosphorylated in G_0/G_1 phase and hyperphosphorylated in the S- and G_2/M phases in all cell types [90]. pRB phosphorylation occurs on serine (S) and threonine (T) residues [91], but no phosphotyrosine is evident. Following mitogenic stimulation, pRB is phosphorylated in three stages during cell-cycle entry and progression: mid-G_1-phase, S-phase, and G_2/M (Figs 3 and 4) [92]. The dephosphorylation of pRB occurs after the beginning of mitosis, and may be regulated by differentiation signals [93].

Because of its significance in growth regulation, great attention has been paid to the regulation of pRB by phosphorylation. TAg specifically binds unphosphorylated (active) pRB to transform cells, possibly by releasing cellular-binding proteins [94]. The inactivation of pRB by phosphorylation occurs through the action of CDKs. The

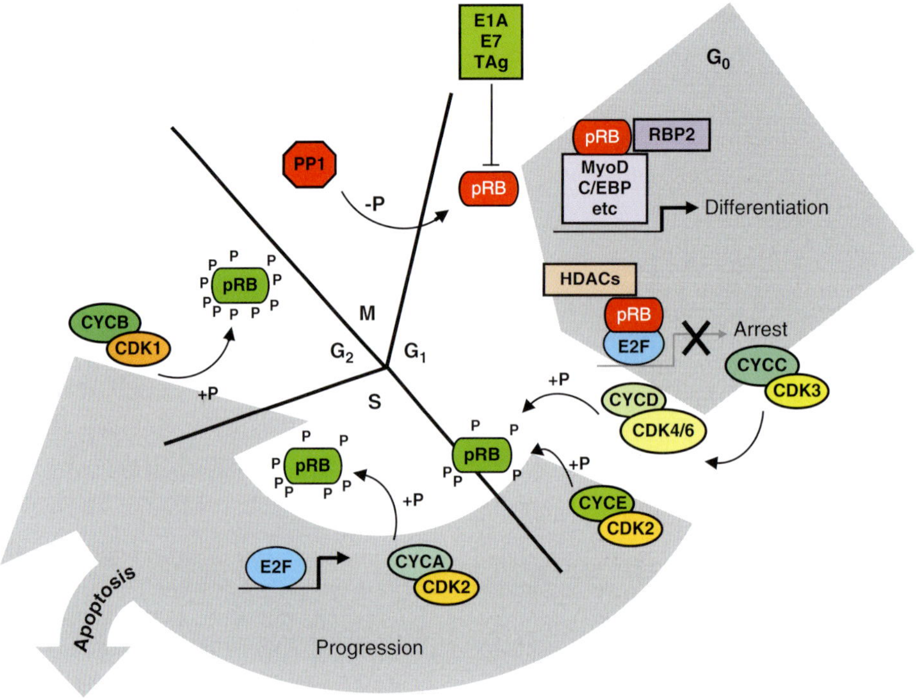

Fig. 4 pRB regulates cell-cycle progression, differentiation, and apoptosis. In G_0/G_1, hypophosphorylated pRB represses genes necessary for S-phase entrance by repressing transcriptional activation by E2F and recruiting HDACs, causing cell-cycle arrest. pRB also activates genes necessary for differentiation by binding cell type-specific transcription factors and the histone demethylase RBP2. Phosphorylation of pRB by cyclin (CYC) C-CDK3 kinase stimulates exit from G_0, and in late G_1 phosphorylation by cyclin D-CDK4/6 and cyclin E-CDK2 kinases releases E2F from pRB, allowing cell-cycle progression. Continual phosphorylation of pRB occurs by cyclin A-CDK2 and cyclin B-CDK1 complexes in S and G_2 phases. pRB is activated in late mitosis by dephosphorylation by PP1c; viral oncoproteins sequester otherwise active, hypophosphorylated pRB. Inappropriate entry into S-phase in the absence of pRB induces apoptosis.

CDK family member, $p34^{cdc2}$ (CDK1), was first found to bind to and phosphorylate pRB [95]. CDK1 is the active subunit of the M-phase-promoting factor, which phosphorylates specific S or T residues in its consensus site Z-S/T-P-X-Z (where X is polar and Z is generally basic). pRB contains 16 potential CDK1 phosphorylation sites, of which at least nine have been shown to be phosphorylated *in vivo* (Fig. 3) [91, 96]. However, CDK1 is activated in late G_2- and M-phases, after which a homologous kinase, CDK2, first phosphorylates pRB, which is consistent with the phosphorylation of pRB to promote exit from G_1 [97]. Despite multiple phosphorylation sites potentially offering a subtle regulation of pRB transcriptional inhibition and protein–protein interactions, crystal structures have indicated that T373 and S608 are particularly crucial residues. The phosphorylation of T373 stimulates a global conformational change, while S608 phosphorylation orders a flexible loop to block binding to the E2F transactivation domain [98].

CDKs are activated following binding to cyclins and phosphorylation of a conserved threonine by the CDK-activating kinase [99]. The expression of cyclins varies during the cell cycle, with cyclin D and E expressed during G_1, cyclin A during S-phase, and cyclin B during mitosis. Cyclins E and A form a functional unit by binding to CDK2 in G_1- or S-phase respectively, whereas cyclin D forms a complex with CDK4 or CDK6 in G_1 [100]. Cyclin B binds to CDK1 in late G_2 (Fig. 4).

While complexes of each of these cyclins can phosphorylate pRB, only cyclins A, D and E are able to reverse the ability of pRB to block cells in G_1. Different cyclin–CDK complexes regulate different phosphorylation sites in a collaborative effort to completely inactivate pRB [101]. In U2OS cells, pRB is partially phosphorylated by cyclin D kinase activity [102], with complete phosphorylation requiring cyclin E activity [103, 104]. The phosphorylation of pRB only by cyclin D kinase did not release pRB from its interaction with E2F1 [105]. Continued phosphorylation of pRB on specific sites in S-phase by cyclin A-CDK2 activity may also be important for the completion of DNA synthesis.

pRB is also regulated by another kinase, cyclin C-CDK3, at the G_0/G_1 transition. The phosphorylation of pRB at specific substrate sites is required for cells to exit G_0 efficiently (Fig. 4). Cyclin C directs phosphorylation of pRB in a temporal pattern that precedes pRB phosphorylation by cyclin D or cyclin E complexes. Therefore, cyclin C plays an important role in regulating the G_0-to-G_1 transition in a pRB-dependent manner [106].

The dephosphorylation (reactivation) of pRB is dependent on the activity of phosphoprotein phosphatase type 1 (PP1c), which associates with pRB *in vivo* during mitosis (Fig. 4) [107]. PP1c competes for CDK binding to the same binding site at the C terminus of pRB [108]. pRB can also be acetylated at its C terminus by the transcriptional coactivator p300; SIRT1 can remove these acetyl groups. Acetylation inhibits phosphorylation [109]. Likewise, the methyltransferases Set7/9 and SMYD2 can methylate the pRB C terminus (Fig. 3) [110]. This methylation enhances pRB's transcriptional repressive activity by recruiting the p53-binding protein 53BP1, and may form a link between pRB activity and the DNA damage response [111]. Adding further complexity, pRB can be ubiquitinated by MDM2 in its C-terminal region, leading to its proteasomal degradation [112].

6.2
RB and E2F Protein Families in Cell Cycle Progression

Viral oncoproteins induce the proliferation of mammalian cells by interfering with the binding of pRB to cellular proteins [113]. The first such cellular protein to be identified was the transcription factor, E2F [114], which regulates the activity of promoters containing an E2F-binding site, originally identified as the sequence element in the adenovirus E2 promoter that allowed transcription in mammalian cells [115]. The E2F-binding site (TTTSSCGC, where S is C or G) has been identified in promoters of genes encoding S-phase-regulatory proteins such as DNA polymerase α, thymidylate synthase (TS), proliferating cell nuclear antigen (PCNA), and ribonucleotide reductase (RR), and cell-cycle progression genes such as those encoding cyclin A, cyclin E, CDK1 and B-Myb [116]. The *E2F1* and *RB1* genes are also regulated by E2F activity, suggesting the existence of an autoregulatory loop in the RB/E2F pathway [115, 117]. pRB mutants that bind free E2F but do not repress transcription are unable to block S-phase entry [118, 119]. The coexpression of E2F1 with pRB releases the G_1 block and pushes cells into S-phase [120]. Thus, the primary way in which pRB blocks the cell cycle in the G_1 phase is through a repression of E2F-responsive genes [121].

The E2F family of proteins consists of eight members (E2F1, -2, -3, -4, -5, -6, -7, and -8) [122]. The E2F1–5 proteins each contain a DNA-binding domain, a transactivation domain, and a C-terminal region that binds pRB or other pRB family members. The ability of E2F1–6 proteins to regulate transcription is also dependent on binding to the dimerization partner (DP) family of proteins [123]. Three mammalian DP proteins (DP1, -2, and -4) bind E2F to enhance interaction with the E2F-binding sites and activate E2F-responsive promoters [124]. However, not all genes with an E2F site are activated by E2F. *In vivo*, E2F is released from the E2F-binding site of the B-Myb promoter at the same time that B-Myb transcription is activated, which suggests that the E2F complex suppresses this promoter [125]. E2F6 contains neither the transactivation domain nor the C-terminal pRB-binding region, but represses transcription by complexing with DP proteins to either actively repress or replace other E2F/DP complexes [122]. E2F7 and E2F8 have two DNA-binding domains, but bind the E2F-binding consensus site independently of DP proteins. They too lack a transactivation domain and the pRB-binding region, but function as repressors of certain E2F-regulated genes [126]. A single gene may be regulated by both activating and repressive E2Fs.

Different RB family proteins bind E2F family proteins at different phases of the cell cycle [127]. pRB binds E2F1–4, whereas p107 and p130 bind only E2F4 and -5. E2F4 or -5 are predominantly bound to p130 in quiescent G_0 cells, and to p107 in proliferating cells [128]. pRB binds and regulates E2F1, -2, and -3 in G_1, but binds to E2F4 when cells are induced to re-enter the cell cycle. p130 (or sometimes p107) forms part of a so-called "DREAM" complex [129], consisting of a DP, RB family member, E2F, and multi-vulval class B (MuvB) protein core [130]. The DREAM complex acts during quiescence to silence transcription at E2F responsive sites (recognized by E2F4 or 5), and a site called the cell cycle genes homology region (CHR), recognized by the multiprotein MuvB core of the DREAM complex. The complex is activated by the phosphorylation of MuvB components by dual-specificity tyrosine-phosphorylation-regulated kinase 1A (DYRK1A), which in turn is activated through the Hippo pathway [131]. This process is required for senescence. After the DREAM complex dissociates, and when p130 is phosphorylated on exit from G_0, the MuvB core also interacts with B-Myb and FOXM1 to become a transcriptional activator at G_2/M [132].

The activity of E2F is also regulated in a cell-cycle-dependent manner by the degradation of free E2F protein. E2F1 and -4 have C-terminal sequences that target them for degradation via the ubiquitin proteasome pathway. The binding of pRB to E2F1, or p107 to E2F4, increases the half-life of the E2F protein [133, 134]. Endogenous E2F1 is more stable in G_0/G_1 when bound to pRB than in S-phase, which suggests that the RB family may maintain cells in G_1 by assuring that E2F is present to repress promoters. E2F activity is also stabilized by acetylation [135], and target specificity is modulated by neddylation [136].

6.3 Other pRB Interacting Proteins

A diversity of action is evident from the large number of proteins that bind to pRB with different functions (Table 1) [10, 137, 138]. In addition to E2F, pRB binds a number of general transcription factors (ATF-2 and SP1) and cell type-specific transcription factors (Elf-1, SPI1, MyoD, and C/EBP). Generally, the binding of pRB represses transcriptional activity; however, binding to factors such as c-Jun, MyoD, and C/EBP has a positive effect on differentiation of specific cell types (see Sect. 6.4).

pRB may repress the activity of factors such as E2F1 by the recruitment of histone deacetylase (HDAC1) to the promoter [139]. HDAC1 facilitates the formation of nucleosome structures that block the access of positive transcription factors to the promoter. In addition, pRB binds a number of corepressors, including RBBP4, RBP1 and RBBP8, which function by directly repressing transcription or by recruiting HDAC1 [2]. However, pRB also represses the transcription of promoters that are not sensitive to HDAC activity, by unknown mechanisms [140]. Thus, pRB can actively repress transcription factors, block access of transcription factors to

Tab. 1 Some of the proteins that bind pRB, selected to show the diversity of functional classes that interact with pRB.

Function	*pRB-binding proteins*
CDK inhibitor	p21
Chromatin modulator	BRG1
Corepressor	RBBP8, RBBP4, RBP1, HBP1
Cyclin	CCND1, 2, 3, CCNA1, CCNB1, CCNC
Deacetylase	SIRT1
E3 ligase	SKP2, APC/C
Histone deacetylase	HDAC1, 2, 3
Histone demethylase	RBP2
Methyltransferase	DNMT1
Molecular chaperone	HSP75
Nuclear matrix component	Lamin A, p84
Phosphatase	PP-1a2
Pol I transcription factor	UBF
Pol II transcription factor	TAFII250
Pol III transcription factor	TFIIIB
Regulator of p53 stability	MDM2, MDM4
Replication licensing factor	MCM7
Ser/Thr kinase	CDK1, CDK2
Signaling molecule	Raf-1
Transcription factor	ATF-2, c-Jun, c-Myc, N-Myc, C/EBP, E2F1-3, Elf-1, hBRM, ID2, MyoD, Myogenin, SPI1, SP1, Pax-3, PHox, Chx10, RIZ
Tumor suppressor	Prohibitin, BRCA1
Tyrosine kinase	c-Abl
Viral oncoprotein	E1A, E7, TAg
Viral transcription factor	EBNA-5, HCMV IE2

basal transcriptional machinery, or directly inhibit the basal transcriptional machinery.

In addition to the regulation of polymerase II transcription, pRB regulates polymerase I genes by binding and inhibiting the ability of an upstream binding factor to bind the ribosomal DNA promoter. pRB binds the general polymerase III transcription factor TFIIIB to regulate genes such as tRNA, 5S RNA, and TATA-box containing genes such as U6 [141]. Polymerase III transcription is also regulated by the binding of p107 and p130 to the BRF subunit of TFIIIB.

The signaling molecule Raf-1 binds pRB, directly linking mitogenic signaling and cell-cycle regulation, which may be one mechanism of transducing Ras cell-surface signals to the cell cycle [142]. The inactivation of Ras in $RB1^{+/+}$ cells efficiently inhibits DNA synthesis [143]. In $RB1^{-/-}$ cells, the inhibition of DNA synthesis is defective; thus, pRB is required for cell-cycle progression downstream of Ras.

The pRB C terminus also interacts with the F-box protein SKP2 [144]. This interaction blocks SKP2 from promoting degradation of the CDK inhibitor p27, providing a route to negative regulation of the cell cycle, independent of the pRB–E2F interaction. pRB also recruits the anaphase-promoting complex/cyclosome (APC/C), enabling the ubiquitylation and degradation of SKP2 itself, and again leading to p27 stabilization [145]. In *Rb1* heterozygous mice, pituitary tumor formation is suppressed by *Skp2* knockdown; knockdown also suppresses human retinoblastoma cell proliferation via a p27-dependent pathway [146]. Surprisingly, this suppression is partly mediated through an increased p53-mediated apoptosis, but Skp2 remains tumor-suppressive even in the absence of p53 [147].

6.4 Differentiation by pRB

pRB plays a role in determining the fate of the cell. pRB positively regulates terminal adipocyte differentiation of murine embryonic fibroblasts by binding to and activating the CCAAT/enhancer-binding proteins (C/EBPs), a family of transcription factors necessary for adipocyte differentiation [148]. When the cell is induced to differentiate, pRB binds to C/EBP. The production and maintenance of a terminally differentiated muscle cell phenotype requires the binding of pRB to MyoD, both *in vitro* and *in vivo* [149]. Differentiating human keratinocytes contain pRB/c-Jun complexes that facilitate the binding of c-Jun to an AP-1 consensus site [150]. In bone development, pRB interacts with an osteoblast transcription factor, CBFA1, resulting in a synergistic transactivation of an osteoblast-specific reporter [151]. This finding is significant since osteosarcoma is the second most common tumor after retinoblastoma in $RB1^{+/-}$ individuals, and loss of pRB occurs in 60% of sporadic osteosarcomas, suggesting a bone-specific tumor suppressor function of pRB [152]. These culture experiments are consistent with the idea that differentiation involves both activation and repression by pRB of genes that are needed for cell type-specific differentiation. Indeed, this hypothesis has been supported *in vivo*. In mice, pRB interacts with the key osteoblast transcription factor Runx2 to promote an osteoblast lineage while repressing the adipocyte lineage in an E2F-dependent manner by occupying peroxisome proliferator-activated receptor (PPAR) γ-regulated promoters [153].

The exact mechanisms of pRB's differentiation effects are not yet clear. However, its interaction with lineage-specific transcription factors such as ATF2, Pax8 and

Tbx2 suggest that it can act as a transcriptional coactivator [2]. Its chromatin modifier activities are likely also at play. For instance, the pRB interactor RBP2 is a histone demethylase that promotes repressive chromatin states. pRB's interaction with RBP2 relieves this activity, thus activating the promoters of genes involved in differentiation, such as osteocalcin in osteosarcoma cells [154]. Knockdown of RBP2 also promoted myotube formation and adipocyte differentiation [155].

6.5 Apoptosis and pRB

Among the diverse functions of pRB is the ability to protect cells from apoptosis. The first evidence of this function was observed in the $Rb1^{-/-}$ mouse (see Sect. 7), which showed increased apoptosis in the tissues of the lens, nervous system, and skeletal muscle [156–159].

It is counterintuitive to think that a tumor suppressor would naturally promote survival. However, it has become clear that *RB1* loss-induced apoptosis is an indirect consequence of cell-cycle dysregulation and acquired genotoxic stress (see Sect. 6.6). Apoptosis induced following the loss of pRB is dependent on the activity of E2F1, which regulates the apoptotic peptidase-activating factor APAF1 and caspases 3, 7, 8, and 9. The overexpression of E2F1 pushes cells into untimely S-phase, followed immediately by an induction of apoptosis [160]. Thus, pRB protects cells from apoptosis by complexing with E2F1 [160]. A special E2F binding domain in the C-terminal tail of pRB serves the unique purpose of regulating E2F1-induced apoptosis, without interfering with other functions of E2F1 [161].

Emerging evidence also points to a pro-apoptotic function for pRB, more consistent with its role as a tumor suppressor. *RB1*-deficient prostate cancer cells resisted treatment with ceramide and radiation, and exhibited resistance to apoptosis, which was reversed upon the introduction of normal pRB [162]. Studies in actively proliferating cells have shown that the presence of DNA damage can stimulate a transcriptionally active pRB–E2F1 complex to promote the expression of proapoptotic genes, such as p73 and caspase 7 [163]. pRB is also known to downregulate anti-apoptotic proteins such as survivin [164], which is associated with a poor prognosis for several tumor types. pRB also has a pro-apoptotic role outside the nucleus, stimulating the apoptotic activity of Bax directly at the mitochondrion [165].

It is likely that the switch between pro- and anti-apoptotic functions of pRB is cell-type and developmental-stage specific, and dependent on its phosphorylation status. In addition to the repression of E2F1, hyperphosphorylated pRB has been shown to additionally interfere with apoptosis by inhibiting ANP32A, a pro-apoptotic factor [166]. In contrast, dephosphorylated, active pRB has been observed during the initial stages of cell death. Further delineation of the role of pRB in apoptosis holds promise for the development of novel cancer therapies.

6.6 Maintenance of Genome Integrity by pRB

The inactivation of *RB1* has been linked to an increased chromosome instability (CIN). Indeed, this is a hallmark of cancer, and the resulting aneuploidy is likely necessary for the malignant progression of retinoblastoma (see Sect. 9). CIN in the absence of pRB is likely due to the cumulative effect of the loss of a number of different pRB functions [5].

One major route to CIN occurs through an interference with pRB's regulation of E2Fs, which are known to regulate several genes involved in mitosis. One such E2F target is *MAD2*, with a major role in spindle-assembly checkpoint (SAC) control [167]. MAD2 is thought to regulate the kinetochore–microtubule attachments that form spindle poles during mitosis, the deregulation of which could contribute to CIN. Another role of *MAD2* is to negatively regulate the anaphase-promoting complex/cyclosome–cell division cycle 20 complex (APC/C-CDC20). APC/C-CDC20 is an E3 ubiquitin ligase complex that is the key downstream effector after the SAC, responsible for the ubiquitination of securin [168]. On securin degradation, separase is freed to cleave cohesin and permit anaphase. Overexpression of MAD2 causes a hyperactive checkpoint, altering the mitotic sequence, and paradoxically promoting chromosome mis-segregation [169]. However, pRB depletion might not be dependent on MAD2 to promote genome instability at the level of chromosome segregation: it causes depletion of histone methyl marks that recruit cohesin to pericentromeric heterochromatin; this reduced cohesin binding is necessary for CIN formation on pRB depletion, although condensin II may also play a role [170]. Several other genes downstream of pRB inactivation and subsequent E2F dysregulation could also contribute to CIN, such as *NEK2*, *PRC1*, and *BUB1*, with roles in chromosome segregation, spindle assembly, and mitotic checkpoint control, respectively.

Another consequence of E2F dysregulation by pRB inactivation is replication fork stalling, stress, and subsequent DNA damage, which suggests that pRB may have a more direct role in normal DNA replication [171]. Perhaps most importantly, pRB has been shown to play a major role in the regulation of chromatin structure. pRB recruits complexes that are important for the maintenance of heterochromatin, and thus essentially controls the establishment of gene expression patterns related to cell-cycle control, differentiation, and development [172]. In the absence of functional pRB, centromere structure and function are compromised, with serious consequences to genome stability [173]. pRB has also been shown to bind Type II topoisomerases and to recruit DNA repair complexes to damaged sites [174]. More detailed cell- and tumor-specific studies of pRB's varied functions and roles are required to fully elucidate its role in maintaining genome integrity.

7
RB Family Regulation of Murine Development

7.1
Rb1 in Murine Development

Rb1$^{-/-}$ mice show normal early retinal development up to embryonic days (E)13–15, but then die with abnormalities in neurogenesis and erythropoiesis due to placental failure (Table 2) [156–158]. Wild-type trophoblast rescues *Rb1*$^{-/-}$ embryos until birth, when a muscle defect, previously recognized in the partial rescue of *Rb1*$^{-/-}$ embryos, is lethal [175]. The cause of early embryonic lethality is an excessive proliferation of trophoblast cells and a severe disruption of the normal labyrinth architecture in the placenta, resulting in a decreased vascularization and reduction in placental transportation function. *Rb1*$^{-/-}$ mice that are reconstituted with a functionally normal placenta lack many of the phenotypes of *Rb1*$^{-/-}$ mice, and show histologically normal dorsal root ganglia, brain, retina, spinal cord and liver.

Tab. 2 Key retinoblastoma (RB) family phenotypes in mice.

Mouse	*Phenotype*	*Reference(s)*
No retinoblastoma		
$Rb1^{+/-}$	$Rb1^{-/-}$ pituitary and thyroid tumors	[156–158]
$Rb1^{-/-}$	Lethal E13-E15, abnormalities in neurogenesis and erythropoiesis, apoptosis, and inappropriate S-phase entry in nervous system and lens	[156–158]
$Rb1^{-/-}$ with normal placenta	Die at birth with marked defects in muscle and lens, no abnormalities in neurogenesis and erythropoeisis	[175, 176]
$Rb1^{-/-}$; $Rb1^{+/+}$ chimera	Normal; apoptosis of $Rb1^{-/-}$ cells in retina	[177–179]
Nes-Cre1; $Rb1^{lox/lox}$	Apoptosis in lens and some neurons, inappropriate S-phase entry in nervous system and lens	[180]
IRBP-Cre; $Rb1^{lox/lox}$	Neuroendocrine tumors of the pineal and pituitary gland	[181]
$p107^{-/-}$	Normal on C57BL/6J, 50% smaller on BALB/cJ	[182–184]
$p130^{-/-}$	Normal on C57BL/6J, die at E11.5 on BALB/cJ	[182, 185]
$p107^{-/-}$; $p130^{-/-}$	Deregulated chondrocytes; defective endochondral bone formation	[182]
$Rb1^{+/-}$; $p107^{-/-}$	Retinal dysplasia; $Rb1^{-/-}$; $p107^{-/-}$ pituitary and thyroid tumors	[183]
$Rb1^{-/-}$; $p107^{-/-}$	Lethal E11.5	[183]
$Rb1^{+/-}$; $E2f1^{-/-}$	Reduced frequency of pituitary and thyroid tumors, increased life span compared to $Rb1^{+/-}$ mouse	[186, 187]
$Rb1^{-/-}$; $E2f1^{-/-}$	Lethal E17	[186]
$Rb1^{+/-}$; $p53^{-/-}$	Retinal dysplasia, $Rb1^{-/-}$; $p53^{-/-}$ pituitary and thyroid tumors	[188]
$Rb1^{+/-}$; $Arf^{-/-}$	Accelerated pituitary tumor development compared to $Rb1^{+/-}$ mouse	[189]
Papilloma virus E7	Apoptosis in the retina	[190]
Viral oncoprotein-induced retinoblastoma		
SV40 TAg transgene (retina)	Retinoblastoma, brain tumor	[191]
Papillomavirus E7 and E6 transgene (retina)	Retinoblastoma	[192]
Papillomavirus E7; $p53^{-/-}$	Retinoblastoma	[190]
Adenovirus E1A (viral delivery); $p53^{-/-}$	Retinoblastoma	[193]
Chimeric knockout retinoblastoma		
$Rb1^{-/-}$; $p107^{-/-}$ chimera	Retinoblastoma	[194]
$Rb1^{-/-}$; $p130^{-/-}$ chimera	Retinoblastoma, pheochromocytoma	[195]
Targeted $Rb1^{-/-}$ *knockout retinoblastoma*		
Pax6α-Cre; $Rb1^{lox/lox}$; $p107^{-/-}$	Retinoblastoma	[180, 196]
Pax6α-Cre; $Rb1^{lox/lox}$; $p27^{-/-}$	Retinoblastoma	[197]

(*continued overleaf*)

Tab. 2 (*Continued*)

Mouse	*Phenotype*	*Reference(s)*
Pax6α-Cre; $Rb1^{lox/lox}$; $p130^{-/-}$; $Arf^{+/-}$	Metastatic retinoblastoma	[198]
Nes-Cre1; $Rb1^{lox/lox}$; $p130^{-/-}$	Maternal inheritance results in retinoblastoma	[180]
Chx10-Cre; $Rb1^{lox/-}$; $p53^{lox/-}$; $p107^{-/-}$	Metastatic retinoblastoma	[199]
Chx10-Cre; $Rb1^{lox/lox}$; $p107^{+/-}$; $p130^{-/-}$	Metastatic retinoblastoma	[200]

Note that the official gene names for murine *p107*, *p130*, and *p53* are *Rbl1*, *Rbl2*, and *Trp53* respectively. In keeping with conventions in the field, the protein names are used here for clarity.

However, they still die at birth or shortly thereafter, with marked defects in skeletal muscle and apoptosis of the peripheral nervous system and lens [175, 176].

$Rb1^{-/-}$ cells in the brains of chimeric embryos exhibited extensive S-phase entry; however, unlike the widespread apoptosis in the germline $Rb1^{-/-}$ mice, the majority of $Rb1^{-/-}$ cells in the chimera survived and differentiated into neurons. The developing retina of chimeric embryos showed ectopic mitoses and substantial cell degeneration, while the contribution of $Rb1^{-/-}$ cells to the adult retina was much reduced [177, 178]. Therefore, in the central nervous system (CNS), cell-autonomous cell-cycle regulation is dependent on functional pRB, but neuronal cell survival, and fate may be non-cell-autonomous. However, recently acquired data have suggested that the role of pRB in development and differentiation is more important for some cell types than others.

7.2
p107 and *p130* in Murine Development

The importance of the pRB family in differentiation is also demonstrated by $p107^{-/-}$; $p130^{-/-}$ mice, which are viable until birth but show deregulated chondrocyte growth and defective endochondral bone formation (reduced rib cage size and shortened limbs) [182]. The importance of each pRB family member may be dependent on the level of expression in a particular cell type, and may also depend on the strain of mouse, suggesting the existence of strain-specific modifier genes.

BALB/cJ and C57BL/6J mice lacking *p107* were both viable, but the BALB/cJ mice displayed impaired growth, reaching only 50% of the weight of their counterparts by 21 days of age. There was also myeloid hyperplasia in the spleen and liver, and a twofold acceleration of the cell cycle in $p107^{-/-}$ fibroblasts and myoblasts, which suggested a role for *p107* in negatively regulating the overall length of the cell cycle [182–184].

The expression of *p107* and *p130* mRNA helps explain some of these data [201]. While *p130* is expressed at very low levels throughout embryogenesis, there is a notable increase in its expression in developing bone, which would explain the bone phenotype in the $p107^{-/-}$; $p130^{-/-}$ knockout mice. *p107*, on the other hand, is expressed highly in the heart, lung, kidney and intestines, but there is no detectable phenotype in these tissues in the $p107^{-/-}$ mice. There is also high expression of *p107* in the CNS and liver, which would explain the accelerated apoptosis seen in the liver and CNS in the *Rb1*$^{-/-}$; *p107*$^{-/-}$ mice compared to *Rb1*$^{-/-}$ mice.

8 Retinoblastoma in Mice

The generation of *Rb1*$^{-/-}$ mice was expected to provide information regarding the development of retinoblastoma, although embryonic lethality in *Rb1*$^{-/-}$ mice precluded studies of later development of the retina. Unlike humans, *Rb1*$^{+/-}$ mice do not develop retinoblastoma but instead develop pituitary tumors in which both *Rb1* alleles are inactivated (Table 2) [156–158]. Interestingly, the frequency of these tumors is decreased when *E2f1* is deleted [186, 187], and accelerated when *p53* is deleted [188]. The extreme sensitivity of developing human retina to a loss of *RB1* has stimulated intense efforts to mimic human retinoblastoma in mice, but with limited success.

8.1 Viral Oncoprotein-Induced Retinoblastoma

The first transgenic mouse model of retinoblastoma arose when the simian virus 40 large and small T-antigens, expressed from the human luteinizing hormone β subunit (LHβ) promoter in order to produce pituitary tumors, were serendipitously expressed (under the regulation of a still-unknown genomic control region) in a specific Müller cell subtype of retina with malignant potential [191, 202]. These transgenic mice developed focal bilateral retinal tumors that were histologically very similar to human retinoblastoma, caused by a high expression of TAg in the retina. Given its ease of generation, 100% penetrance and histological similarity to the human disease, this TAg-RB model has been widely used for preclinical therapeutic studies of retinoblastoma [203].

Viral oncoproteins such as human papilloma virus (HPV) E7 and TAg inactivate pRB when they bind the A/B domain of pRB. TAg binds the active, hypophosphorylated – but not the inactive, hyperphosphorylated – form of pRB (Fig. 4), and also inactivates p107, p130, p53, and other proteins [204]. However, the expression of TAg or E7 from the retina-specific promoters for interphotoreceptor retinoid-binding protein (IRBP) or opsins does not result in tumors like retinoblastoma [190, 205, 206]. However, IRBP-driven HPV E7 results in retinoblastoma in *p53*$^{-/-}$ mice [190], as does the expression of both HPV E6 (which inactivates p53) and E7 in the retina [192]. Focal retinoblastoma-like tumors can be generated by intravitreal injections of a replication-incompetent retrovirus encoding the adenoviral E1A oncogene in newborn *p53*$^{-/-}$ mice [193].

8.2 Chimeric Knockout Retinoblastoma

The retina was normal in both embryonic *Rb1*$^{-/-}$ mice and chimeric *Rb1*$^{-/-}$; *Rb1*$^{+/+}$ mice, and *Rb1*$^{-/-}$ cells contributed to most tissues. However, *Rb1*-deficient cells were excluded from the outer nuclear layer (ONL) and inner nuclear layer (INL) of the retina by apoptosis, which suggests that pRB is required during specific stages of postnatal retinal differentiation [177–179]. While both the *p107*$^{-/-}$ and *p130*$^{-/-}$ mice are viable with no retinal dysplasia [182–185], the loss of *p107*, and one copy of pRB in the *Rb1*$^{+/-}$; *p107*$^{-/-}$ animals, resulted in retinal dysplasia [183]. Although the *Rb1*$^{-/-}$; *p107*$^{-/-}$ mice die at E11.5 [183], preventing evaluation of the developing retina, the loss of both p107 and pRB in the retinoblasts of *Rb1*$^{-/-}$; *p107*$^{-/-}$ chimeric mice resulted in retinoblastoma [194]. Using expression markers for terminally differentiated retinal cell types, the authors concluded that the retinoblastoma originated in cells committed to the amacrine cell compartment of the INL, but not in cells committed to the ONL. *Rb1*$^{-/-}$; *p130*$^{-/-}$ chimeras developed retinoblastoma at a mean age of three months in some cases. These tumors formed rosettes, and also expressed amacrine and occasionally Müller cell markers, with a nuclear morphology suggestive of an INL origin [195].

8.3 Targeted *Rb1*$^{-/-}$ Knockout Retinoblastoma

The advent of targeted knockout technologies allowed a circumvention of the embryonic lethality of *Rb1* germline knockouts, and allowed more consistent results to be obtained than the technically challenging chimeric strategies. When expressed, the Cre recombinase specifically excises DNA between two *loxP* sites flanking a region in a modified genome. The control of Cre expression therefore permits a control of gene deletions [207]. The conditional loss of *Rb1* (along with *p53*) in appropriate cell types can induce prostate tumors, squamous cell carcinoma, primitive neuroectodermal tumors, osteosarcoma, and others [208, 209]. However, a focus here on retinal tumors is instructive.

Conditional inactivation of *Rb1* by expression of Cre in IRBP expressing photoreceptor cells (*IRBP-Cre*; *Rb1*$^{lox/lox}$ mice) failed to cause retinoblastoma, even when *p107* and *p53* were also inactivated [181]. The conditional inactivation of *Rb1* by the expression of Cre from the Pax6α enhancer in *Rb1*$^{lox/lox}$; *p107*$^{-/-}$ mice resulted in an ectopic proliferation of all retinal precursors, but only the precursors of horizontal, amacrine, and Müller glia cells survived and terminally differentiated. The Pax6α enhancer is active in peripheral retinal progenitors at E10. Some 68% of these mice developed tumors in the peripheral retina as early as postnatal day (P) 8, with larger tumors filling the vitreous at later stages [196]. Markers for amacrine (NeuroD and Math3) and Müller glia (Hes5) were present in the P8 tumor, with few markers for photoreceptor (Crx), bipolar (Chx10), and ganglion cells (Brn3B). Tumors present at P30 expressed markers for amacrine (NeuroD, Math3, Pax6, syntaxin, and calretinin) and Müller glia (Hes5, GFAP, and CRALBP) only. Tumor formation in this so-called α-Cre *Rb1* model was accelerated in a *p130*$^{-/-}$ background compared to a *p107*$^{-/-}$ background. Although the kinetics differed, both *Rb1*$^{lox/lox}$; *p130*$^{-/-}$ and *Rb1*$^{lox/lox}$; *p107*$^{-/-}$ double knockout mice showed early dysplasia in the extreme periphery of the retina, with Homer Wright rosettes, and both increased proliferation and apoptosis in this area. Over time, these tumors filled the eye, invaded locally, and in some cases metastasized to cervical lymph nodes. These metastases stained for an amacrine marker (syntaxin), and variably for horizontal and Müller cell markers. They also showed an amplification of *Mycn*, seen in some human retinoblastomas (see Sect. 3.3) [180].

Interestingly, knockout of the CDK inhibitor p27 in the α-Cre floxed *Rb1* mouse could substitute for p107 or p130 loss, leading to tumorigenesis with 100% penetrance. Since Cdk2 is activated in the absence of p27, it was observed that small-molecule Cdk2 inhibition could prevent tumorigenesis in this model [197].

The conditional inactivation of *Rb1* by expression of Cre from the *Nestin* promoter (*Nes-Cre1*) results in mosaic expression of Cre through all layers of the retina with maternal *Nes-Cre1* inheritance. Breeding these mice with $p130^{-/-}$ mice resulted in retinoblastoma tumor formation [180]. The tumors expressed amacrine markers (syntaxin and calretinin), and a few Müller marker-expressing cells were also present. However, no tumor stained for the photoreceptor-specific gene IRBP, the bipolar cell marker protein kinase C-α (PKC-α), or the retinal ganglion cell marker Brn3b. These mouse tumors contained Homer Wright rosettes but not the classical Flexner–Wintersteiner rosettes that are pathognomonic of human retinoblastoma, nor any evidence of photoreceptor differentiation.

The conditional inactivation of *Rb1* and *p53* by expression of Cre from the *Chx10* promoter in a $p107^{-/-}$ background also led to retinoblastoma with high penetrance; these tumors also invaded the anterior chamber and expressed a combination of progenitor cell and differentiated retinal cell markers [199]. *Chx10-Cre*; $Rb1^{lox/lox}$; $p130^{-/-}$; $p107^{+/-}$ mice also developed tumors, despite a compensatory upregulation of p107 expression. Remarkably, these tumors arose clonally, from differentiated horizontal cells that re-entered the cell cycle but maintained synapses. They metastasized to the lymph nodes, brain and bone marrow [200].

The timing of inactivation of *Rb1* may play an important role in the development of retinoblastoma. Unlike in adult mice, pRB is expressed in all retinal cells at different times during development [210]. Therefore, a loss of pRB at different developmental times results in different retinal phenotypes, which could explain why mice expressing TAg from two different promoters necessary for photoreceptor development have different phenotypes (no tumors with the opsin promoter versus the entire photoreceptor layer developing tumors with the IRBP promoter) [205, 206].

9 Additional Genomic Changes in Retinoblastoma

9.1 Genomic Gains and Losses

Mutation of both copies of the *RB1* gene is required for the initiation of retinoblastoma, with the exception of *MYCN*-amplified retinoblastoma (as discussed in Sect. 3.3) [19]. However, all $RB1^{-/-}$ retinoblastoma tumors also have additional genetic alterations required for full malignant transformation [14, 17] (*MYCN*-amplified retinoblastoma have fewer of these changes [19]). These mutations have been termed M3-Mn, extending on Knudson's classic M1 and M2 two-hit mutational events in the *RB1* gene [14]. Karyotypic analysis and comparative genomic hybridization (CGH) show genomic gains and losses in retinoblastoma tumors. The minimal regions most frequently gained are 1q31 (52%), 6p22 (44%), 2p24-25 (30%) and 13q32-34 (12%), and the minimal region most frequently lost is 16q22 (14%) [211]. Some evidence suggests that gains at 1q, 2p, and the loss of 16q, are restricted to more advanced tumors in older children, possibly correlating with tumor progression [212]. Almost unique to retinoblastoma is a specific pattern of 6p gain, isochromosome *i*(6p), identified in 60% of retinoblastoma tumors [213]. This marker chromosome results in four copies of genes on chromosome 6p, or in low-level amplification. Gains and losses may point to potential oncogenes or tumor suppressor genes necessary for progression to malignant retinoblastoma. Identification of these cancer genes is important as they may form the basis of therapy to prevent the progression of small retinoblastoma tumors.

9.2 Candidate M3-Mn Genes (Excluding *MYCN*-Amplified Retinoblastoma)

The likely oncogene candidates gained on 1q in retinoblastoma include the mitotic kinesin *KIF14* and the p53 inhibitor *MDM4*. The latter is overexpressed and amplified in some retinoblastomas, and its inhibition blocks xenograft growth [214]. Single-nucleotide polymorphisms in both *MDM4* and the related *MDM2* are associated with retinoblastoma [215]. Meanwhile, *KIF14* is gained in the majority of retinoblastomas and shows dramatic overexpression [216, 217]. *KIF14*, first identified as an oncogene candidate in retinoblastoma, has now been shown to be overexpressed and a prognostic factor in a variety of other, common cancers, including those of the breast, lung, brain, liver, and ovary [218].

While *MYCN* is the presumptive target of gain at 2p (even in tumors with *RB1* loss), the minimal region of 6p gain contains the transcriptional regulators *DEK* and *E2F3*, both potential oncogenes in retinoblastoma [219]. Chromosome 6p gain is also common in bladder cancer and is associated with an elevated risk of progression [220].

Cadherin-11 (*CDH11*) was identified as a potential tumor suppressor gene in the region of 16q22 genomic loss in retinoblastoma [221]. *CDH11* is a cell adhesion molecule, the loss of which may promote progression of some $RB1^{-/-}$ retinal cells to malignancy and perhaps vitreous seeding or metastasis [222]. Unusually low levels of the intact form of CDH11 are present in osteosarcoma [223], while the variant form was highly expressed in invasive breast cancer and was associated with the promotion of invasiveness [224].

Although chromosome 9 loss is not common in retinoblastoma, comparative analysis of genomic copy number changes in human and murine (α-*Cre*; $Rb1^{lox/lox}$; $p107^{-/-}$) retinoblastoma identified deletions of the *Cdkn2A* locus in some cases [198]. *Arf* is one of the two genes at the *Cdkn2A* locus, along with $p16^{Ink4a}$, which is also downregulated in human retinoblastoma but did not substantially influence murine tumor formation when deleted. Heterozygous loss of *Arf* promoted metastatic murine retinoblastoma formation with LOH of *Arf* (Table 2) [198], while complete *Arf* deletion accelerated pituitary tumor formation in *Rb1* heterozygotes [189]. This suggests that *ARF* loss, along with *MDM4* gain, may inactivate the p53 pathway in some retinoblastoma. Human retinoblastoma do not show mutations of *p53* itself [225].

Interestingly, whole-genome sequencing of retinoblastoma has suggested a fairly stable genome. Across four tumors sequenced, somatic coding changes were documented in only 11 genes [226]; these included recurrent mutations in the transcriptional corepressor *BCOR*, also seen in acute myeloid leukemia. The paucity of coding mutations prompted a renewed search for epigenetic changes in retinoblastoma, building on findings of multiple hypermethylated genes in retinoblastoma including *RB1* itself, and also *MGMT* and *RASSF1A* [17]. The kinase *SYK* was found to show an activated histone signature in a ChIP-on-chip analysis, and has been effectively targeted in preclinical models of retinoblastoma [226].

MicroRNAs also have demonstrated roles in retinoblastoma. The let-7 family and miR-34 show decreased expression in a fraction of retinoblastomas [17].

Conversely, the oncomiR cluster of miR-17-92 is overexpressed in retinoblastoma [227]. This was found when the orthologous cluster was seen amplified in murine retinal tumors, and the human cluster or its paralog, miR-106b-25 were similarly found to be gained at the DNA level and overexpressed. Overexpression of miR-17-92 promoted murine retinoblastoma, while knockdown inhibited growth of human xenografts [228]. Other microRNAs, and protein-coding genes, continue to be identified as differentially expressed in retinoblastoma [17].

9.3
Causes of Drug Resistance in Retinoblastoma

The multidrug resistance (MDR) phenotype in human cancer is due to an upregulation of the *MDR1* gene with increased expression of P-glycoprotein [229]. P-glycoprotein expression is increased in response to environmental toxicity in many organisms, and broadly reduces intracellular levels of vinca alkaloids (vincristine, vinblastine), epipodophyllotoxins (etoposide, teniposide), and other natural-product antineoplastic agents by functioning as an ATP-dependent plasma membrane pump that expels drugs from human cancer cells. Platinum compounds are not substrates of the P-glycoprotein drug efflux mechanism.

The expression of *MDR1* in retinoblastoma at initial diagnosis and increased expression at relapse after failed treatment may account for frequent chemotherapy failures [230]. It is not possible to correlate P-glycoprotein levels before therapy with the outcome of chemotherapy in intraocular retinoblastoma, since biopsies incur a risk of systemic spread. In metastatic retinoblastoma, P-glycoprotein expression before therapy correlates with failure of therapy, and initially undetectable P-glycoprotein correlates with long-term remission [231].

One chemotherapeutic protocol showing promise to cure retinoblastoma-containing eyes without use of radiation or removal of the eye includes a direct approach to counter P-glycoprotein MDR [232, 233]. High-dose cyclosporine A (CSA) is delivered simultaneously with the chemotherapy to block the ability of P-glycoprotein to protect tumor cells from chemotherapy. An international Phase II Clinical Trial of CSA-modified chemotherapy is currently underway (NIH Clinical trials identifier: NCT00110110).

Intraocular retinoblastoma that failed chemotherapy without CSA showed increased P-glycoprotein when examined by immunohistochemistry [234]. Retinoblastoma that failed chemotherapy despite the administration of CSA did not express P-glycoprotein, but expressed instead the multidrug resistance protein (MRP) [234]. MRP belongs to the same ATP-dependent membrane transporter superfamily as P-glycoprotein, and transfection of the MRP gene confers a similar broad-spectrum pattern of drug resistance to antineoplastic agents [235]. Although CSA inhibits P-glycoprotein *in vitro*, there is presently no effective inhibitor of MRP used in humans [235].

10
The Retinoblastoma Cell of Origin

The elucidation of a tumor's cell of origin is crucial for understanding multiple facets of tumor biology: the developmental stage at which the tumor can arise; the mutational sensitivities that permit this; and perhaps the Achilles' heel of the cell type that could form the basis for targeted therapy. In the case of retinoblastoma, knowledge of the cell of origin (the cell in which *RB1* loss enables tumorigenesis) could also help to address the question of why loss of pRB, a multifunctional and widely expressed cell cycle regulator, primarily initiates retinoblastoma, and not other cancers [16].

There is now strong evidence as to the cell of origin of at least some retinoblastomas. There are seven major cell types in the mature retina: one glial cell (Müller) and six neurons (amacrine, bipolar, ganglion, horizontal, rod, and cone). Retinoblastomas display morphological, immunohistochemical and functional properties of multiple retinal cell types. Early studies implied a photoreceptor origin for retinoblastoma, since the differentiated cells of Flexner–Wintersteiner rosettes resemble photoreceptor outer segments [61]. In more recent studies, the high expression of pRB in cone precursors prompted a close examination of cone markers CRX, RXRγ, and TRβ2, and L/M opsins, all of which were present in retinoblastoma cells where pRB was definitively absent and at least one copy of *RB1* was lost [236]. In addition, retinoblastoma cell lines were sensitive to a depletion of RXRγ and TRβ2. Very recent studies showed conclusively that of all the cell types in human fetal retina, only cone precursors expressing these markers went on to proliferate after *RB1* was knocked down. Proliferation of these cone precursors was dependent on known pathways of retinoblastoma tumorigenesis, including SKP2 and MDM2 (see above). Moreover, these depleted cone precursors could also form well-differentiated, retinoblastoma-like tumors in xenograft models [237].

However, small human tumors appear to arise in the INL (not the site of cone photoreceptors), as shown histologically and more recently by optical coherence tomography in live patients (Fig. 1f) [21]. Nonetheless, displaced cone precursors in this layer that give rise to retinoblastoma are a possibility. Moreover, retinoblastoma cells are promiscuous in their expression of markers for retinal progenitors, precursors (also known as transition cells), interneurons, and photoreceptors. Indeed, defining unambiguous markers for all these cell types even in normal tissues is not always straightforward.

Studies in mouse models have shown that pRB loss can promote tumorigenesis in multiple retinal cell types, including amacrine cells [194], differentiated horizontal cells [200], and a subset of Müller glia [202]. However, these findings cannot be directly applied to the human disease, since the mechanisms and timing of pRB loss do not directly correspond. Nonetheless, neurochemical analysis has shown that at least some human retinoblastomas show features of monoamine oxidase (MAO)-expressing amacrine cells, and MAO inhibitors can block retinoblastoma cell line growth [238].

Furthermore, the examination of a tumor for markers may define some features of the cell of origin, but may be misleading. For example, progenitor cells destined to follow a pathway that is dependent on pRB may adopt a different fate in the absence of pRB. Retinal progenitors go through a series of changes in intrinsic properties that control their competence to create different cell types. If a pRB-related pathway is disturbed, a cell may follow a default pRB-independent pathway, defined by extrinsic cues from its neighbors. Thus, it is possible that induced *RB1* loss in the above human fetal retina experiments aberrantly turned on expression of cone-specific genes [237]. This could explain why the majority of human retinoblastomas show photoreceptor-like rosettes, even if they originate in the INL and not in the ONL.

Consistent with promiscuous marker expression, it has been argued that the cell of origin is a relatively uncommitted stem or progenitor cell, since these cells have more "neutral" chromatin, in which the loss of pRB could profoundly influence transcription [2]. In more differentiated cells, chromatin is largely "locked" in open or closed conformations by multiple mechanisms. This assertion is also supported by the fact that retinoblastoma is purely a developmental disease; there is apparently a window of opportunity for pRB loss in the eye to initiate tumorigenesis that is closed before adulthood. (This is distinct from the ability of pRB pathway inactivation to promote, rather than initiate, tumor development in other tissues.)

Of course, the cell of origin may not be shared between all *RB1*$^{-/-}$ retinoblastomas. Gene expression profiling has suggested that there may be two groups of retinoblastoma: one with cone-like features, and another likely from a less-committed precursor cell [239]. This latter group also showed more of the distinctive chromosomal changes seen in retinoblastoma. Even individual tumors have shown coexisting expression patterns reflecting progenitor lineages and mature retinal cell types, within a single cell [238]. Moreover, *MYCN*-amplified retinoblastoma lacking *RB1* mutations almost certainly has a different, more primitive cell of origin than *RB1*$^{-/-}$ tumors [19].

Factors that might render a cell of origin susceptible in the retina are [240]:

- The complex differentiation programs that these cells undergo, during which numerous genes are activated and inactivated, potentially allowing defects in transcriptional control to have catastrophic effects.
- The evidence from mouse models that certain amacrine, horizontal, and Müller cells are "intrinsically death-resistant," allowing them to avoid the apoptosis usually caused by pRB loss.
- The as-yet-untested possibility that these rapidly differentiating cells might have decreased genome stability.

11 Importance of Retinoblastoma

Although a rare eye cancer, retinoblastoma continues to contribute much more than its "share" to understanding of cancer, biological processes, and use of scientific knowledge to optimize health outcomes. *RB1* was the first tumor suppressor gene to be discovered, and revealed the process (search for LOH) that was used to discover other genes underlying human cancer [53]. The RB family members are key cell cycle regulators, so that their mutation alters development and differentiation and deregulates the cell cycle to induce cancer in susceptible cells. Meanwhile, the search for the retinoblastoma cell of origin has spurred considerable research into normal retinal development [16]. The many ways that *RB1* can be damaged, and the high mutation rate, have forced creative technology development to meet the challenges of clinical diagnostics [66], much of which has been widely generalized to other genetic disease challenges. At a health service level, it is clearly shown that the molecular determination of individual *RB1* mutant alleles contributes importantly to health quality, at less cost [66].

The analysis of genetic changes in retinoblastoma subsequent to the loss of both *RB1* alleles has revealed several potential novel genes involved in the progression of multiple tumor types, which are becoming the targets for potential prevention and therapeutic strategies [226, 241]. Recognition of the key role of the *MDR* gene in retinoblastoma has provided the rational basis for a therapeutic strategy to counter drug resistance [242]. Although basic studies have shown an important role of *MDR* in many cancers, only in retinoblastoma has a therapeutic strategy to alter *MDR* effect shown a positive clinical outcome [232, 233].

Much more is yet to come. Retinoblastoma is one of the best examples of a human disease in which only rare cells at a specific developmental and proliferative stage may be at risk to initiate tumors. The rarity of retinoblastoma has been turned again to advantage by the world participation of major retinoblastoma treatment centers in a global consortium called One Retinoblastoma World (*www.1rbw.org*), showing that global cooperation can lead to future optimal opportunities for all families afflicted with this rare, yet unique and important, cancer.

References

1. Burkhart, D.L. and Sage, J. (2008) Cellular mechanisms of tumour suppression by the retinoblastoma gene. *Nat. Rev. Cancer*, **8**, 671–682.
2. Chinnam, M. and Goodrich, D.W. (2011) *RB1*, development, and cancer. *Curr. Top. Dev. Biol.*, **94**, 129–169.
3. Gordon, G.M. and Du, W. (2011) Conserved RB functions in development and tumor suppression. *Protein Cell*, **2**, 864–878.
4. Manning, A.L. and Dyson, N.J. (2011) pRB, a tumor suppressor with a stabilizing presence. *Trends Cell Biol.*, **21**, 433–441.
5. Manning, A.L. and Dyson, N.J. (2012) RB: mitotic implications of a tumour suppressor. *Nat. Rev. Cancer*, **12**, 220–226.
6. Ramasubramanian, A. and Shields, C.L. (2012) *Retinoblastoma*. Jaypee Brothers Medical Publisher, New Delhi.
7. Rodriguez-Galindo, C. and Wilson, M.W. (2010) *Retinoblastoma*. Springer, New York.
8. McDaid, C., Hartley, S., Bagnall, A.M., Ritchie, G. *et al.* (2005) Systematic review of effectiveness of different treatments for childhood retinoblastoma. *Health Technol. Assess.*, **9**, iii, ix–x, 1–145.
9. Dick, F.A. and Rubin, S.M. (2013) Molecular mechanisms underlying RB protein function. *Nat. Rev. Mol. Cell Biol.*, **14**, 297–306.
10. Goodrich, D.W. (2006) The retinoblastoma tumor-suppressor gene, the exception that proves the rule. *Oncogene*, **25**, 5233–5243.
11. Henley, S.A. and Dick, F.A. (2012) The retinoblastoma family of proteins and their regulatory functions in the mammalian cell division cycle. *Cell Div.*, **7**, 10.
12. Munro, S., Carr, S.M., La Thangue, N.B. (2012) Diversity within the pRb pathway: is there a code of conduct? *Oncogene*, **31**, 4343–4352.
13. Sachdeva, U.M., O'Brien, J.M. (2012) Understanding pRb: toward the necessary development of targeted treatments for retinoblastoma. *J. Clin. Invest.*, **122**, 425–434.
14. Corson, T.W. and Gallie, B.L. (2007) One hit, two hits, three hits, more? Genomic changes in the development of retinoblastoma. *Genes Chromosomes Cancer*, **46**, 617–634.
15. Dimaras, H., Kimani, K., Dimba, E.A., Gronsdahl, P. *et al.* (2012) Retinoblastoma. *Lancet*, **379**, 1436–1446.
16. Dyer, M.A. and Bremner, R. (2005) The search for the retinoblastoma cell of origin. *Nat. Rev. Cancer*, **5**, 91–101.
17. Thériault, B.L., Dimaras, H., Gallie, B.L., and Corson, T.W. (2014) The genomic landscape of retinoblastoma: a review. *Clin. Exp. Ophthalmol.*, **42**, 33–52.
18. Eagle, R.C., Jr, (2013) The pathology of ocular cancer. *Eye*, **27**, 128–136.
19. Rushlow, D.E., Mol, B.M., Kennett, J.Y., Yee, S. *et al.* (2013) Characterisation of retinoblastomas without *RB1* mutations: genomic, gene expression, and clinical studies. *Lancet Oncol.*, **14**, 327–334.
20. Dimaras, H., Khetan, V., Halliday, W., Orlic, M. *et al.* (2008) Loss of *RB1* induces non-proliferative retinoma: increasing

genomic instability correlates with progression to retinoblastoma. *Hum. Mol. Genet.*, **17**, 1363–1372.
21. Rootman, D.B., Gonzalez, E., Mallipatna, A., Vandenhoven, C. *et al.* (2013) Hand-held high-resolution spectral domain optical coherence tomography in retinoblastoma: clinical and morphologic considerations. *Br. J. Ophthalmol.*, **97**, 59–65.
22. Murphree, A.L. (2005) Intraocular retinoblastoma: the case for a new group classification. *Ophthalmol. Clin. North Am.*, **18**, 41–53, viii.
23. Canadian Retinoblastoma Society (2009) National retinoblastoma strategy Canadian guidelines for care: strategie therapeutique du retinoblastome guide clinique canadien. *Can. J. Ophthalmol.*, **44** Suppl. 2, S1–S88.
24. Abramson, D.H., Dunkel, I.J., Brodie, S.E., Kim, J.W. *et al.* (2008) A phase I/II study of direct intraarterial (ophthalmic artery) chemotherapy with melphalan for intraocular retinoblastoma initial results. *Ophthalmology*, **115**, 1398–1404, 1404.e1.
25. Abramson, D.H., Dunkel, I.J., Brodie, S.E., Marr, B. *et al.* (2010) Superselective ophthalmic artery chemotherapy as primary treatment for retinoblastoma (chemosurgery). *Ophthalmology*, **117**, 1623–1629.
26. Suzuki, S., Yamane, T., Mohri, M., and Kaneko, A. (2011) Selective ophthalmic arterial injection therapy for intraocular retinoblastoma: the long-term prognosis. *Ophthalmology*, **118**, 2081–2087.
27. Munier, F.L., Gaillard, M.C., Balmer, A., Soliman, S. *et al.* (2012) Intravitreal chemotherapy for vitreous disease in retinoblastoma revisited: from prohibition to conditional indications. *Br. J. Ophthalmol.*, **96**, 1078–1083.
28. Reese, A.B. and Ellsworth, R.M. (1963) The evaluation and current concept of retinoblastoma therapy. *Trans. Am. Acad. Ophthalmol. Otolaryngol.*, **67**, 164–172.
29. Salsbury, A.J., Bedford, M.A., and Dobree, J.H. (1968) Bone marrow appearances in children suffering from retinoblastoma. *Br. J. Ophthalmol.*, **52**, 388–395.
30. Finger, P.T. and AJCC-UICC Ophthalmic Oncology Task Force (2009) 7th edition AJCC staging system for eye cancer: an international language for ophthalmic oncology. *Arch. Pathol. Lab. Med.*, **133**, 1197–1198.
31. Dimaras, H., Heon, E., Budning, A., Doyle, J.J. *et al.* (2009) Retinoblastoma CSF metastasis cured by multimodality chemotherapy without radiation. *Ophthalmic Genet.*, **30**, 121–126.
32. Dunkel, I.J., Aledo, A., Kernan, N.A., Kushner, B. *et al.* (2000) Successful treatment of metastatic retinoblastoma. *Cancer*, **89**, 2117–2121.
33. Rodriguez-Galindo, C., Wilson, M.W., Haik, B.G., Lipson, M.J. *et al.* (2003) Treatment of metastatic retinoblastoma. *Ophthalmology*, **110**, 1237–1240.
34. Reese, A.B., Merriam, G.R., Jr, and Martin, H.E. (1949) Treatment of bilateral retinoblastoma by irradiation and surgery; report on 15-year results. *Am. J. Ophthalmol.*, **32**, 175–190.
35. Duffner, P.K. (2004) Long-term effects of radiation therapy on cognitive and endocrine function in children with leukemia and brain tumors. *Neurologist*, **10**, 293–310.
36. Jakobiec, F.A., Ts'o, M.O., Zimmerman, L.E., and Danis, P. (1977) Retinoblastoma and intracranial malignancy. *Cancer*, **39**, 2048–2058.
37. de Jong, M.C., Kors, W.A., de Graaf, P., Castelijns, J.A. *et al.* (2014) Trilateral retinoblastoma: a systematic review and meta-analysis. *Lancet Oncol.*, **15**, 1157–1167.
38. Zimmerman, L.E. (1985) Retinoblastoma and retinocytoma, in *Ophthalmic Pathology: An Atlas and Textbook* (ed. Spencer, W.H.), W.B. Saunders, Philadelphia, pp. 1292–1351.
39. Fairclough, W.A. (1947) Direct inheritance in retinoblastoma. *Trans. Ophthalmol. Soc. N. Z.*, **1**, 35.
40. Falls, H.F. (1947) Inheritance of retinoblastoma; two families supplying evidence. *J. Am. Med. Assoc.*, **133**, 171–174.
41. Griffith, A.D. and Sorsby, A. (1944) The genetics of retinoblastoma. *Br. J. Ophthalmol.*, **28**, 279–293.
42. Reese, A.B. (1949) Heredity and retinoblastoma. *Arch. Ophthalmol.*, **42**, 119–122.
43. Wilson, M.G., Melnyk, J., and Towner, J.W. (1969) Retinoblastoma and deletion D (14) syndrome. *J. Med. Genet.*, **6**, 322–327.
44. Van Kempen, C. (1966) A case of retinoblastoma, combined with severe mental retardation and a few other congenital

anomalies, associated with complex aberrations of the karyotype. *Maandschr. Kindergeneeskd.*, **34**, 92–96.
45. Thompson, H. and Lyons, R.B. (1965) Retinoblastoma and multiple congenital anomalies, associated with complex mosaicism with deletion of D-chromosome and probable D/C translocation. *Hum. Chromosome Newsl.*, **15**, 21.
46. Lele, K.P., Penrose, L.S., and Stallard, H.B. (1963) Chromosome deletion in a case of retinoblastoma. *Ann. Hum. Genet.*, **27**, 171–174.
47. Knudson, A.G., Jr, (1971) Mutation and cancer: statistical study of retinoblastoma. *Proc. Natl Acad. Sci. USA*, **68**, 820–823.
48. Comings, D.E. (1973) A general theory of carcinogenesis. *Proc. Natl Acad. Sci. USA*, **70**, 3324–3328.
49. Cavenee, W.K., Dryja, T.P., Phillips, R.A., Benedict, W.F. *et al.* (1983) Expression of recessive alleles by chromosomal mechanisms in retinoblastoma. *Nature*, **305**, 779–784.
50. Friend, S.H., Bernards, R., Rogelj, S., Weinberg, R.A. *et al.* (1986) A human DNA segment with properties of the gene that predisposes to retinoblastoma and osteosarcoma. *Nature*, **323**, 643–646.
51. Lee, W.H., Bookstein, R., Hong, F., Young, L.J. *et al.* (1987) Human retinoblastoma susceptibility gene: cloning, identification, and sequence. *Science*, **235**, 1394–1399.
52. Fung, Y.K., Murphree, A.L., T'Ang, A., Qian, J. *et al.* (1987) Structural evidence for the authenticity of the human retinoblastoma gene. *Science*, **236**, 1657–1661.
53. Knudson, A.G. (2001) Two genetic hits (more or less) to cancer. *Nat. Rev. Cancer*, **1**, 157–162.
54. Lohmann, D.R., Gerick, M., Brandt, B., Oelschlager, U. *et al.* (1997) Constitutional *RB1*-gene mutations in patients with isolated unilateral retinoblastoma. *Am. J. Hum. Genet.*, **61**, 282–294.
55. Lohmann, D., Gallie, B., Dommering, C., Gauthier-Villars, M. (2011) Clinical utility gene card for: retinoblastoma. *Eur. J. Hum. Genet.*, **19**; doi: 10.1038/ejhg.2010.200.
56. Rushlow, D., Piovesan, B., Zhang, K., Prigoda-Lee, N.L. *et al.* (2009) Detection of mosaic *RB1* mutations in families with retinoblastoma. *Hum. Mutat.*, **30**, 842–851.
57. Ejima, Y., Sasaki, M.S., Kaneko, A., and Tanooka, H. (1988) Types, rates, origin and expressivity of chromosome mutations involving 13q14 in retinoblastoma patients. *Hum. Genet.*, **79**, 118–123.
58. Gallie, B.L., Ellsworth, R.M., Abramson, D.H., and Phillips, R.A. (1982) Retinoma: spontaneous regression of retinoblastoma or benign manifestation of the mutation? *Br. J. Cancer*, **45**, 513–521.
59. Gallie, B.L., Phillips, R.A., Ellsworth, R.M., and Abramson, D.H. (1982) Significance of retinoma and phthisis bulbi for retinoblastoma. *Ophthalmology*, **89**, 1393–1399.
60. Radnot, M. (1977) Elements of photoreceptors in retinoblastomas (author's transl). *Albrecht Von Graefes Arch. Klin. Exp. Ophthalmol.*, **202**, 63–70.
61. Ts'o, M.O., Zimmerman, L.E., and Fine, B.S. (1970) The nature of retinoblastoma. I. Photoreceptor differentiation: a clinical and histopathologic study. *Am. J. Ophthalmol.*, **69**, 339–349.
62. Ts'o, M.O. (1980) Clues to the cells of origin in retinoblastoma. *Int. Ophthalmol. Clin.*, **20**, 191–210.
63. Eagle, R.C., Jr, Shields, J.A., Donoso, L., and Milner, R.S. (1989) Malignant transformation of spontaneously regressed retinoblastoma, retinoma/retinocytoma variant. *Ophthalmology*, **96**, 1389–1395.
64. Seeger, R.C., Brodeur, G.M., Sather, H., Dalton, A. *et al.* (1985) Association of multiple copies of the N-myc oncogene with rapid progression of neuroblastomas. *N. Engl. J. Med.*, **313**, 1111–1116.
65. Chen, D., Pacal, M., Wenzel, P., Knoepfler, P.S. *et al.* (2009) Division and apoptosis of E2f-deficient retinal progenitors. *Nature*, **462**, 925–929.
66. Richter, S., Vandezande, K., Chen, N., Zhang, K. *et al.* (2003) Sensitive and efficient detection of *RB1* gene mutations enhances care for families with retinoblastoma. *Am. J. Hum. Genet.*, **72**, 253–269.
67. Lohmann, D.R. (1999) *RB1* gene mutations in retinoblastoma. *Hum. Mutat.*, **14**, 283–288.
68. McEvoy, J., Nagahawatte, P., Finkelstein, D., Richards-Yutz, J. *et al.* (2014) *RB1* gene inactivation by chromothripsis in human retinoblastoma. *Oncotarget*, **5**, 438–450.

69. Choi, Y.J. and Anders, L. (2014) Signaling through cyclin D-dependent kinases. *Oncogene*, **33**, 1890–1903.
70. Qian, Y., Luckey, C., Horton, L., Esser, M. *et al.* (1992) Biological function of the retinoblastoma protein requires distinct domains for hyperphosphorylation and transcription factor binding. *Mol. Cell. Biol.*, **12**, 5363–5372.
71. Huang, H.J., Yee, J.K., Shew, J.Y., Chen, P.L. *et al.* (1988) Suppression of the neoplastic phenotype by replacement of the RB gene in human cancer cells. *Science*, **242**, 1563–1566.
72. Hiebert, S.W. (1993) Regions of the retinoblastoma gene product required for its interaction with the E2F transcription factor are necessary for E2 promoter repression and pRb-mediated growth suppression. *Mol. Cell. Biol.*, **13**, 3384–3391.
73. Qin, X.Q., Chittenden, T., Livingston, D.M., and Kaelin, W.G., Jr (1992) Identification of a growth suppression domain within the retinoblastoma gene product. *Genes Dev.*, **6**, 953–964.
74. Welch, P.J. and Wang, J.Y. (1993) A C-terminal protein-binding domain in the retinoblastoma protein regulates nuclear c-Abl tyrosine kinase in the cell cycle. *Cell*, **75**, 779–790.
75. Knudsen, E.S. and Wang, J.Y. (1997) Dual mechanisms for the inhibition of E2F binding to RB by cyclin-dependent kinase-mediated RB phosphorylation. *Mol. Cell. Biol.*, **17**, 5771–5783.
76. Dyson, N., Howley, P.M., Munger, K., and Harlow, E. (1989) The human papilloma virus-16 E7 oncoprotein is able to bind to the retinoblastoma gene product. *Science*, **243**,934–937.
77. Hu, Q.J., Dyson, N., and Harlow, E. (1990) The regions of the retinoblastoma protein needed for binding to adenovirus E1A or SV40 large T antigen are common sites for mutations. *EMBO J.*, **9**, 1147–1155.
78. Whyte, P., Buchkovich, K.J., Horowitz, J.M., Friend, S.H. *et al.* (1988) Association between an oncogene and an anti-oncogene: the adenovirus E1A proteins bind to the retinoblastoma gene product. *Nature*, **334**, 124–129.
79. Lee, J.O., Russo, A.A., and Pavletich, N.P. (1998) Structure of the retinoblastoma tumour-suppressor pocket domain bound to a peptide from HPV E7. *Nature*, **391**, 859–865.
80. Chow, K.N. and Dean, D.C. (1996) Domains A and B in the Rb pocket interact to form a transcriptional repressor motif. *Mol. Cell. Biol.*, **16**, 4862–4868.
81. Hassler, M., Singh, S., Yue, W.W., Luczynski, M. *et al.* (2007) Crystal structure of the retinoblastoma protein N domain provides insight into tumor suppression, ligand interaction, and holoprotein architecture. *Mol. Cell*, **28**, 371–385.
82. Ewen, M.E., Xing, Y.G., Lawrence, J.B., and Livingston, D.M. (1991) Molecular cloning, chromosomal mapping, and expression of the cDNA for p107, a retinoblastoma gene product-related protein. *Cell*, **66**, 1155–1164.
83. Zhu, L., van den Heuvel, S., Helin, K., Fattaey, A. *et al.* (1993) Inhibition of cell proliferation by p107, a relative of the retinoblastoma protein. *Genes Dev.*, **7**, 1111–1125.
84. Hannon, G.J., Demetrick, D., and Beach, D. (1993) Isolation of the Rb-related p130 through its interaction with CDK2 and cyclins. *Genes Dev.*, **7**, 2378–2391.
85. Li, Y., Graham, C., Lacy, S., Duncan, A.M. *et al.* (1993) The adenovirus E1A-associated 130-kD protein is encoded by a member of the retinoblastoma gene family and physically interacts with cyclins A and E. *Genes Dev.*, **7**, 2366–2377.
86. Woo, M.S., Sanchez, I., and Dynlacht, B.D. (1997) p130 and p107 use a conserved domain to inhibit cellular cyclin-dependent kinase activity. *Mol. Cell. Biol.*, **17**, 3566–3579.
87. Claudio, P.P., Howard, C.M., Baldi, A., De Luca, A. *et al.* (1994) p130/pRb2 has growth suppressive properties similar to yet distinctive from those of retinoblastoma family members pRb and p107. *Cancer Res.*, **54**, 5556–5560.
88. Buchkovich, K., Duffy, L.A., and Harlow, E. (1989) The retinoblastoma protein is phosphorylated during specific phases of the cell cycle. *Cell*, **58**, 1097–1105.
89. DeCaprio, J.A., Ludlow, J.W., Lynch, D., Furukawa, Y. *et al.* (1989) The product of the retinoblastoma susceptibility gene has properties of a cell cycle regulatory element. *Cell*, **58**, 1085–1095.

90. Ludlow, J.W., Shon, J., Pipas, J.M., Livingston, D.M. *et al.* (1990) The retinoblastoma susceptibility gene product undergoes cell cycle-dependent dephosphorylation and binding to and release from SV40 large T. *Cell*, **60**, 387–396.
91. Lees, J.A., Buchkovich, K.J., Marshak, D.R., Anderson, C.W. *et al.* (1991) The retinoblastoma protein is phosphorylated on multiple sites by human cdc2. *EMBO J.*, **10**, 4279–4290.
92. DeCaprio, J.A., Furukawa, Y., Ajchenbaum, F., Griffin, J.D. *et al.* (1992) The retinoblastoma-susceptibility gene product becomes phosphorylated in multiple stages during cell cycle entry and progression. *Proc. Natl Acad. Sci. USA*, **89**, 1795–1798.
93. Ludlow, J.W., Glendening, C.L., Livingston, D.M., and DeCaprio, J.A. (1993) Specific enzymatic dephosphorylation of the retinoblastoma protein. *Mol. Cell. Biol.*, **13**, 367–372.
94. Ludlow, J.W., DeCaprio, J.A., Huang, C.M., Lee, W.H. *et al.* (1989) SV40 large T antigen binds preferentially to an underphosphorylated member of the retinoblastoma susceptibility gene product family. *Cell*, **56**, 57–65.
95. Hu, Q.J., Lees, J.A., Buchkovich, K.J., and Harlow, E. (1992) The retinoblastoma protein physically associates with the human cdc2 kinase. *Mol. Cell. Biol.*, **12**, 971–980.
96. Lin, B.T., Gruenwald, S., Morla, A.O., Lee, W.H. *et al.* (1991) Retinoblastoma cancer suppressor gene product is a substrate of the cell cycle regulator cdc2 kinase. *EMBO J.*, **10**, 857–864.
97. Akiyama, T., Ohuchi, T., Sumida, S., Matsumoto, K. *et al.* (1992) Phosphorylation of the retinoblastoma protein by cdk2. *Proc. Natl Acad. Sci. USA*, **89**, 7900–7904.
98. Burke, J.R., Hura, G.L., and Rubin, S.M. (2012) Structures of inactive retinoblastoma protein reveal multiple mechanisms for cell cycle control. *Genes Dev.*, **26**, 1156–1166.
99. Malumbres, M. (2014) Cyclin-dependent kinases. *Genome Biol.*, **15**, 122.
100. Kato, J., Matsushime, H., Hiebert, S.W., Ewen, M.E. *et al.* (1993) Direct binding of cyclin D to the retinoblastoma gene product (pRb) and pRb phosphorylation by the cyclin D-dependent kinase CDK4. *Genes Dev.*, **7**, 331–342.
101. Horton, L.E., Qian, Y., and Templeton, D.J. (1995) G_1 cyclins control the retinoblastoma gene product growth regulation activity via upstream mechanisms. *Cell Growth Differ.*, **6**, 395–407.
102. Ewen, M.E., Sluss, H.K., Sherr, C.J., Matsushime, H. *et al.* (1993) Functional interactions of the retinoblastoma protein with mammalian D-type cyclins. *Cell*, **73**, 487–497.
103. Lundberg, A.S. and Weinberg, R.A. (1998) Functional inactivation of the retinoblastoma protein requires sequential modification by at least two distinct cyclin-cdk complexes. *Mol. Cell. Biol.*, **18**, 753–761.
104. Hinds, P.W., Mittnacht, S., Dulic, V., Arnold, A. *et al.* (1992) Regulation of retinoblastoma protein functions by ectopic expression of human cyclins. *Cell*, **70**, 993–1006.
105. Brown, V.D., Phillips, R.A., and Gallie, B.L. (1999) Cumulative effect of phosphorylation of pRB on regulation of E2F activity. *Mol. Cell. Biol.*, **19**, 3246–3256.
106. Ren, S. and Rollins, B.J. (2004) Cyclin C/cdk3 promotes Rb-dependent G_0 exit. *Cell*, **117**, 239–251.
107. Durfee, T., Becherer, K., Chen, P.L., Yeh, S.H. *et al.* (1993) The retinoblastoma protein associates with the protein phosphatase type 1 catalytic subunit. *Genes Dev.*, **7**, 555–569.
108. Hirschi, A., Cecchini, M., Steinhardt, R.C., Schamber, M.R. *et al.* (2010) An overlapping kinase and phosphatase docking site regulates activity of the retinoblastoma protein. *Nat. Struct. Mol. Biol.*, **17**, 1051–1057.
109. Chan, H.M., Krstic-Demonacos, M., Smith, L., Demonacos, C. *et al.* (2001) Acetylation control of the retinoblastoma tumour-suppressor protein. *Nat. Cell Biol.*, **3**, 667–674.
110. Carr, S.M., Munro, S., Kessler, B., Oppermann, U. *et al.* (2011) Interplay between lysine methylation and Cdk phosphorylation in growth control by the retinoblastoma protein. *EMBO J.*, **30**, 317–327.
111. Carr, S.M., Munro, S., Zalmas, L.P., Fedorov, O. *et al.* (2014) Lysine methylation-dependent binding of 53BP1 to the pRb tumor suppressor. *Proc. Natl Acad. Sci. USA*, **31**, 11341–11346.

112. Uchida, C., Miwa, S., Kitagawa, K., Hattori, T., *et al.* (2005) Enhanced Mdm2 activity inhibits pRB function via ubiquitin-dependent degradation. *EMBO J.*, **24**, 160–169.
113. Bandara, L.R., La Thangue, N.B. (1991) Adenovirus E1a prevents the retinoblastoma gene product from complexing with a cellular transcription factor. *Nature*, **351**, 494–497.
114. Chellappan, S.P., Hiebert, S., Mudryj, M., Horowitz, J.M. *et al.* (1991) The E2F transcription factor is a cellular target for the RB protein. *Cell*, **65**, 1053–1061.
115. Hamel, P.A., Gill, R.M., Phillips, R.A., and Gallie, B.L. (1992) Transcriptional repression of the E2-containing promoters EIIaE, c-myc, and *RB1* by the product of the *RB1* gene. *Mol. Cell. Biol.*, **12**, 3431–3438.
116. DeGregori, J., Kowalik, T., and Nevins, J.R. (1995) Cellular targets for activation by the E2F1 transcription factor include DNA synthesis- and G1/S-regulatory genes. *Mol. Cell. Biol.*, **15**, 4215–4224.
117. Johnson, D.G., Ohtani, K., and Nevins, J.R. (1994) Autoregulatory control of E2F1 expression in response to positive and negative regulators of cell cycle progression. *Genes Dev.*, **8**, 1514–1525.
118. Hiebert, S.W., Chellappan, S.P., Horowitz, J.M., and Nevins, J.R. (1992) The interaction of RB with E2F coincides with an inhibition of the transcriptional activity of E2F. *Genes Dev.*, **6**, 177–185.
119. Helin, K., Harlow, E., and Fattaey, A. (1993) Inhibition of E2F-1 transactivation by direct binding of the retinoblastoma protein. *Mol. Cell. Biol.*, **13**, 6501–6508.
120. Johnson, D.G., Schwarz, J.K., Cress, W.D., and Nevins, J.R. (1993) Expression of transcription factor E2F1 induces quiescent cells to enter S phase. *Nature*, **365**, 349–352.
121. Flemington, E.K., Speck, S.H., and Kaelin, W.G., Jr (1993) E2F-1-mediated transactivation is inhibited by complex formation with the retinoblastoma susceptibility gene product. *Proc. Natl Acad. Sci. USA*, **90**, 6914–6918.
122. Chen, H.Z., Tsai, S.Y., and Leone, G. (2009) Emerging roles of E2Fs in cancer: an exit from cell cycle control. *Nat. Rev. Cancer*, **9**, 785–797.
123. Krek, W., Livingston, D.M., and Shirodkar, S. (1993) Binding to DNA and the retinoblastoma gene product promoted by complex formation of different E2F family members. *Science*, **262**, 1557–1560.
124. Milton, A., Luoto, K., Ingram, L., Munro, S. *et al.* (2006) A functionally distinct member of the DP family of E2F subunits. *Oncogene*, **25**, 3212–3218.
125. Lam, E.W. and Watson, R.J. (1993) An E2F-binding site mediates cell-cycle regulated repression of mouse B-myb transcription. *EMBO J.*, **12**, 2705–2713.
126. Logan, N., Graham, A., Zhao, X., Fisher, R. *et al.* (2005) E2F-8: an E2F family member with a similar organization of DNA-binding domains to E2F-7. *Oncogene*, **24**, 5000–5004.
127. van den Heuvel, S., and Dyson, N.J. (2008) Conserved functions of the pRB and E2F families. *Nat. Rev. Mol. Cell Biol.*, **9**, 713–724.
128. Sardet, C., Vidal, M., Cobrinik, D., Geng, Y. *et al.* (1995) E2F-4 and E2F-5, two members of the E2F family, are expressed in the early phases of the cell cycle. *Proc. Natl Acad. Sci. USA*, **92**, 2403–2407.
129. Sadasivam, S. and DeCaprio, J.A. (2013) The DREAM complex: master coordinator of cell cycle-dependent gene expression. *Nat. Rev. Cancer*, **13**, 585–595.
130. Litovchick, L., Sadasivam, S., Florens, L., Zhu, X. *et al.* (2007) Evolutionarily conserved multisubunit RBL2/p130 and E2F4 protein complex represses human cell cycle-dependent genes in quiescence. *Mol. Cell*, **26**, 539–551.
131. Litovchick, L., Florens, L.A., Swanson, S.K., Washburn, M.P. *et al.* (2011) DYRK1A protein kinase promotes quiescence and senescence through DREAM complex assembly. *Genes Dev.*, **25**, 801–813.
132. Sadasivam, S., Duan, S., and DeCaprio, J.A. (2012) The MuvB complex sequentially recruits B-Myb and FoxM1 to promote mitotic gene expression. *Genes Dev.*, **26**, 474–489.
133. Hateboer, G., Kerkhoven, R.M., Shvarts, A., Bernards, R. *et al.* (1996) Degradation of E2F by the ubiquitin-proteasome pathway: regulation by retinoblastoma family proteins and adenovirus transforming proteins. *Genes Dev.*, **10**, 2960–2970.
134. Hofmann, F., Martelli, F., Livingston, D.M., and Wang, Z. (1996) The retinoblastoma

gene product protects E2F-1 from degradation by the ubiquitin-proteasome pathway. *Genes Dev.*, **10**, 2949–2959.
135. Martinez-Balbas, M.A., Bauer, U.M., Nielsen, S.J., Brehm, A. *et al.* (2000) Regulation of E2F1 activity by acetylation. *EMBO J.*, **19**, 662–671.
136. Aoki, I., Higuchi, M., and Gotoh, Y. (2013) NEDDylation controls the target specificity of E2F1 and apoptosis induction. *Oncogene*, **32**, 3954–3964.
137. Kaelin, W.G., Jr, Pallas, D.C., DeCaprio, J.A., Kaye, F.J. *et al.* (1991) Identification of cellular proteins that can interact specifically with the T/E1A-binding region of the retinoblastoma gene product. *Cell*, **64**, 521–532.
138. Morris, E.J. and Dyson, N.J. (2001) Retinoblastoma protein partners. *Adv. Cancer Res.*, **82**, 1–54.
139. Brehm, A., Miska, E.A., McCance, D.J., Reid, J.L. *et al.* (1998) Retinoblastoma protein recruits histone deacetylase to repress transcription. *Nature*, **391**, 597–601.
140. Lai, A., Lee, J.M., Yang, W.M., DeCaprio, J.A. *et al.* (1999) RBP1 recruits both histone deacetylase-dependent and -independent repression activities to retinoblastoma family proteins. *Mol. Cell. Biol.*, **19**, 6632–6641.
141. Chu, W.M., Wang, Z., Roeder, R.G., and Schmid, C.W. (1997) RNA polymerase III transcription repressed by Rb through its interactions with TFIIIB and TFIIIC2. *J. Biol. Chem.*, **272**, 14755–14761.
142. Wang, S., Ghosh, R.N., and Chellappan, S.P. (1998) Raf-1 physically interacts with Rb and regulates its function: a link between mitogenic signaling and cell cycle regulation. *Mol. Cell. Biol.*, **18**, 7487–7498.
143. Peeper, D.S., Upton, T.M., Ladha, M.H., Neuman, E. *et al.* (1997) Ras signalling linked to the cell-cycle machinery by the retinoblastoma protein. *Nature*, **386**, 177–181.
144. Ji, P., Jiang, H., Rekhtman, K., Bloom, J. *et al.* (2004) An Rb-Skp2-p27 pathway mediates acute cell cycle inhibition by Rb and is retained in a partial-penetrance Rb mutant. *Mol. Cell*, **16**, 47–58.
145. Binne, U.K., Classon, M.K., Dick, F.A., Wei, W. *et al.* (2007) Retinoblastoma protein and anaphase-promoting complex physically interact and functionally cooperate during cell-cycle exit. *Nat. Cell Biol.*, **9**, 225–232.
146. Wang, H., Bauzon, F., Ji, P., Xu, X. *et al.* (2010) Skp2 is required for survival of aberrantly proliferating *Rb1*-deficient cells and for tumorigenesis in $Rb1^{+/-}$ mice. *Nat. Genet.*, **42**, 83–88.
147. Zhao, H., Bauzon, F., Fu, H., Lu, Z. *et al.* (2013) Skp2 deletion unmasks a p27 safeguard that blocks tumorigenesis in the absence of pRb and p53 tumor suppressors. *Cancer Cell*, **24**, 645–659.
148. Chen, P.L., Riley, D.J., Chen, Y., and Lee, W.H. (1996) Retinoblastoma protein positively regulates terminal adipocyte differentiation through direct interaction with C/EBPs. *Genes Dev.*, **10**, 2794–2804.
149. Gu, W., Schneider, J.W., Condorelli, G., Kaushal, S. *et al.* (1993) Interaction of myogenic factors and the retinoblastoma protein mediates muscle cell commitment and differentiation. *Cell*, **72**, 309–324.
150. Nead, M.A., Baglia, L.A., Antinore, M.J., Ludlow, J.W. *et al.* (1998) Rb binds c-Jun and activates transcription. *EMBO J.*, **17**, 2342–2352.
151. Thomas, D.M., Carty, S.A., Piscopo, D.M., Lee, J.S. *et al.* (2001) The retinoblastoma protein acts as a transcriptional coactivator required for osteogenic differentiation. *Mol. Cell*, **8**, 303–316.
152. Deshpande, A. and Hinds, P.W. (2006) The retinoblastoma protein in osteoblast differentiation and osteosarcoma. *Curr. Mol. Med.*, **6**, 809–817.
153. Calo, E., Quintero-Estades, J.A., Danielian, P.S., Nedelcu, S. *et al.* (2010) Rb regulates fate choice and lineage commitment in vivo. *Nature*, **466**, 1110–1114.
154. Ge, W., Shi, L., Zhou, Y., Liu, Y. *et al.* (2011) Inhibition of osteogenic differentiation of human adipose-derived stromal cells by retinoblastoma binding protein 2 repression of RUNX2-activated transcription. *Stem Cells*, **29**, 1112–1125.
155. Benevolenskaya, E.V., Murray, H.L., Branton, P., Young, R.A. *et al.* (2005) Binding of pRB to the PHD protein RBP2 promotes cellular differentiation. *Mol. Cell*, **18**, 623–635.
156. Clarke, A.R., Maandag, E.R., van Roon, M., van der Lugt, N.M. *et al.* (1992) Requirement for a functional *Rb-1* gene in murine development. *Nature*, **359**, 328–330.

157. Jacks, T., Fazeli, A., Schmitt, E.M., Bronson, R.T. *et al.* (1992) Effects of an Rb mutation in the mouse. *Nature*, **359**, 295–300.
158. Lee, E.Y., Chang, C.Y., Hu, N., Wang, Y.C. *et al.* (1992) Mice deficient for Rb are non-viable and show defects in neurogenesis and haematopoiesis. *Nature*, **359**, 288–294.
159. Morgenbesser, S.D., Williams, B.O., Jacks, T., and DePinho, R.A. (1994) p53-dependent apoptosis produced by Rb-deficiency in the developing mouse lens. *Nature*, **371**, 72–74.
160. Hsieh, J.K., Fredersdorf, S., Kouzarides, T., Martin, K. *et al.* (1997) E2F1-induced apoptosis requires DNA binding but not transactivation and is inhibited by the retinoblastoma protein through direct interaction. *Genes Dev.*, **11**, 1840–1852.
161. Dick, F.A. and Dyson, N. (2003) pRB contains an E2F1-specific binding domain that allows E2F1-induced apoptosis to be regulated separately from other E2F activities. *Mol. Cell*, **12**, 639–649.
162. Bowen, C., Spiegel, S., and Gelmann, E.P. (1998) Radiation-induced apoptosis mediated by retinoblastoma protein. *Cancer Res.*, **58**, 3275–3281.
163. Ianari, A., Natale, T., Calo, E., Ferretti, E. *et al.* (2009) Proapoptotic function of the retinoblastoma tumor suppressor protein. *Cancer Cell*, **15**, 184–194.
164. Jiang, Y., Saavedra, H.I., Holloway, M.P., Leone, G. *et al.* (2004) Aberrant regulation of survivin by the RB/E2F family of proteins. *J. Biol. Chem.*, **279**, 40511–40520.
165. Hilgendorf, K.I., Leshchiner, E.S., Nedelcu, S., Maynard, M.A. *et al.* (2013) The retinoblastoma protein induces apoptosis directly at the mitochondria. *Genes Dev.*, **27**, 1003–1015.
166. Adegbola, O. and Pasternack, G.R. (2005) Phosphorylated retinoblastoma protein complexes with pp32 and inhibits pp32-mediated apoptosis. *J. Biol. Chem.*, **280**, 15497–15502.
167. Hernando, E., Nahle, Z., Juan, G., Diaz-Rodriguez, E. *et al.* (2004) Rb inactivation promotes genomic instability by uncoupling cell cycle progression from mitotic control. *Nature*, **430**, 797–802.
168. Manchado, E., Eguren, M., and Malumbres, M. (2010) The anaphase-promoting complex/cyclosome (APC/C): cell-cycle-dependent and -independent functions. *Biochem. Soc. Trans.*, **38**, 65–71.
169. Schvartzman, J.M., Duijf, P.H., Sotillo, R., Coker, C. *et al.* (2011) Mad2 is a critical mediator of the chromosome instability observed upon Rb and p53 pathway inhibition. *Cancer Cell*, **19**, 701–714.
170. Manning, A.L., Yazinski, S.A., Nicolay, B., Bryll, A. *et al.* (2014) Suppression of genome instability in pRB-deficient cells by enhancement of chromosome cohesion. *Mol. Cell*, **53**, 993–1004.
171. Bester, A.C., Roniger, M., Oren, Y.S., Im, M.M. *et al.* (2011) Nucleotide deficiency promotes genomic instability in early stages of cancer development. *Cell*, **145**, 435–446.
172. Isaac, C.E., Francis, S.M., Martens, A.L., Julian, L.M. *et al.* (2006) The retinoblastoma protein regulates pericentric heterochromatin. *Mol. Cell. Biol.*, **26**, 3659–3671.
173. Manning, A.L., Longworth, M.S., and Dyson, N.J. (2010) Loss of pRB causes centromere dysfunction and chromosomal instability. *Genes Dev.*, **24**, 1364–1376.
174. Xiao, H. and Goodrich, D.W. (2005) The retinoblastoma tumor suppressor protein is required for efficient processing and repair of trapped topoisomerase II-DNA-cleavable complexes. *Oncogene*, **24**, 8105–8113.
175. Wu, L., de Bruin, A., Saavedra, H.I., Starovic, M. *et al.* (2003) Extra-embryonic function of Rb is essential for embryonic development and viability. *Nature*, **421**, 942–947.
176. de Bruin, A., Wu, L., Saavedra, H.I., Wilson, P. *et al.* (2003) Rb function in extraembryonic lineages suppresses apoptosis in the CNS of Rb-deficient mice. *Proc. Natl Acad. Sci. USA*, **100**, 6546–6551.
177. Maandag, E.C., van der Valk, M., Vlaar, M., Feltkamp, C. *et al.* (1994) Developmental rescue of an embryonic-lethal mutation in the retinoblastoma gene in chimeric mice. *EMBO J.*, **13**, 4260–4268.
178. Williams, B.O., Schmitt, E.M., Remington, L., Bronson, R.T. *et al.* (1994) Extensive contribution of Rb-deficient cells to adult chimeric mice with limited histopathological consequences. *EMBO J.*, **13**, 4251–4259.
179. Lipinski, M.M., Macleod, K.F., Williams, B.O., Mullaney, T.L. *et al.* (2001) Cell-autonomous and non-cell-autonomous functions of the Rb tumor suppressor in

developing central nervous system. *EMBO J.*, **20**, 3402–3413.
180. MacPherson, D., Sage, J., Kim, T., Ho, D. *et al.* (2004) Cell type-specific effects of Rb deletion in the murine retina. *Genes Dev.*, **18**, 1681–1694.
181. Vooijs, M., te Riele, H., van der Valk, M., and Berns, A. (2002) Tumor formation in mice with somatic inactivation of the retinoblastoma gene in interphotoreceptor retinol binding protein-expressing cells. *Oncogene*, **21**, 4635–4645.
182. Cobrinik, D., Lee, M.H., Hannon, G., Mulligan, G. *et al.* (1996) Shared role of the pRB-related p130 and p107 proteins in limb development. *Genes Dev.*, **10**, 1633–1644.
183. Lee, M.H., Williams, B.O., Mulligan, G., Mukai, S. *et al.* (1996) Targeted disruption of p107: functional overlap between p107 and Rb. *Genes Dev.*, **10**, 1621–1632.
184. LeCouter, J.E., Kablar, B., Hardy, W.R., Ying, C. *et al.* (1998) Strain-dependent myeloid hyperplasia, growth deficiency, and accelerated cell cycle in mice lacking the Rb-related p107 gene. *Mol. Cell. Biol.*, **18**, 7455–7465.
185. LeCouter, J.E., Kablar, B., Whyte, P.F., Ying, C. *et al.* (1998) Strain-dependent embryonic lethality in mice lacking the retinoblastoma-related p130 gene. *Development*, **125**, 4669–4679.
186. Tsai, K.Y., Hu, Y., Macleod, K.F., Crowley, D. *et al.* (1998) Mutation of E2f-1 suppresses apoptosis and inappropriate S phase entry and extends survival of Rb-deficient mouse embryos. *Mol. Cell*, **2**, 293–304.
187. Yamasaki, L., Bronson, R., Williams, B.O., Dyson, N.J. *et al.* (1998) Loss of E2F-1 reduces tumorigenesis and extends the lifespan of $Rb1^{+/-}$ mice. *Nat. Genet.*, **18**, 360–364.
188. Williams, B.O., Remington, L., Albert, D.M., Mukai, S. *et al.* (1994) Cooperative tumorigenic effects of germline mutations in Rb and p53. *Nat. Genet.*, **7**, 480–484.
189. Tsai, K.Y., MacPherson, D., Rubinson, D.A., Nikitin, A.Y. *et al.* (2002) ARF mutation accelerates pituitary tumor development in $Rb^{+/-}$ mice. *Proc. Natl Acad. Sci. USA*, **99**, 16865–16870.
190. Howes, K.A., Ransom, N., Papermaster, D.S., Lasudry, J.G. *et al.* (1994) Apoptosis or retinoblastoma: alternative fates of photoreceptors expressing the HPV-16 E7 gene in the presence or absence of p53. *Genes Dev.*, **8**, 1300–1310.
191. Windle, J.J., Albert, D.M., O'Brien, J.M., Marcus, D.M. *et al.* (1990) Retinoblastoma in transgenic mice. *Nature*, **343**, 665–669.
192. Griep, A.E., Krawcek, J., Lee, D., Liem, A. *et al.* (1998) Multiple genetic loci modify risk for retinoblastoma in transgenic mice. *Invest. Ophthalmol. Vis. Sci.*, **39**, 2723–2732.
193. Laurie, N.A., Gray, J.K., Zhang, J., Leggas, M. *et al.* (2005) Topotecan combination chemotherapy in two new rodent models of retinoblastoma. *Clin. Cancer Res.*, **11**, 7569–7578.
194. Robanus-Maandag, E., Dekker, M., van der Valk, M., Carrozza, M.L. *et al.* (1998) p107 is a suppressor of retinoblastoma development in pRb-deficient mice. *Genes Dev.*, **12**, 1599–1609.
195. Dannenberg, J.H., Schuijff, L., Dekker, M., van der Valk, M. *et al.* (2004) Tissue-specific tumor suppressor activity of retinoblastoma gene homologs p107 and p130. *Genes Dev.*, **18**, 2952–2962.
196. Chen, D., Livne-Bar, I., Vanderluit, J.L., Slack, R.S. *et al.* (2004) Cell-specific effects of RB or RB/p107 loss on retinal development implicate an intrinsically death-resistant cell-of-origin in retinoblastoma. *Cancer Cell*, **5**, 539–551.
197. Sangwan, M., McCurdy, S.R., Livne-Bar, I., Ahmad, M. *et al.* (2012) Established and new mouse models reveal E2f1 and Cdk2 dependency of retinoblastoma, and expose effective strategies to block tumor initiation. *Oncogene*, **31**, 5019–5028.
198. Conkrite, K., Sundby, M., Mu, D., Mukai, S. *et al.* (2012) Cooperation between Rb and Arf in suppressing mouse retinoblastoma. *J. Clin. Invest.*, **122**, 1726–1733.
199. Zhang, J., Schweers, B., and Dyer, M.A. (2004) The first knockout mouse model of retinoblastoma. *Cell Cycle*, **3**, 952–959.
200. Ajioka, I., Martins, R.A., Bayazitov, I.T., Donovan, S. *et al.* (2007) Differentiated horizontal interneurons clonally expand to form metastatic retinoblastoma in mice. *Cell*, **131**, 378–390.
201. Jiang, Z., Zacksenhaus, E., Gallie, B.L., and Phillips, R.A. (1997) The retinoblastoma gene family is differentially expressed during embryogenesis. *Oncogene*, **14**, 1789–1797.

202. Pajovic, S., Corson, T.W., Spencer, C., Dimaras, H. *et al.* (2011) The TAg-RB murine retinoblastoma cell of origin has immunohistochemical features of differentiated Müller glia with progenitor properties. *Invest. Ophthalmol. Vis. Sci.*, **52**, 7618–7624.
203. Nair, R.M. and Vemuganti, G.K. (2015) Transgenic models in retinoblastoma research. *Ocul. Oncol. Pathol.*, **1**, 207–213.
204. DeCaprio, J.A., Ludlow, J.W., Figge, J., Shew, J.Y. *et al.* (1988) SV40 large tumor antigen forms a specific complex with the product of the retinoblastoma susceptibility gene. *Cell*, **54**, 275–283.
205. al-Ubaidi, M.R., Font, R.L., Quiambao, A.B., Keener, M.J. *et al.* (1992) Bilateral retinal and brain tumors in transgenic mice expressing simian virus 40 large T antigen under control of the human interphotoreceptor retinoid-binding protein promoter. *J. Cell Biol.*, **119**, 1681–1687.
206. al-Ubaidi, M.R., Hollyfield, J.G., Overbeek, P.A., and Baehr, W. (1992) Photoreceptor degeneration induced by the expression of simian virus 40 large tumor antigen in the retina of transgenic mice. *Proc. Natl Acad. Sci. USA*, **89**, 1194–1198.
207. Bouabe, H. and Okkenhaug, K. (2013) Gene targeting in mice: a review. *Methods Mol. Biol.*, **1064**, 315–336.
208. Zhou, Z., Flesken-Nikitin, A., Corney, D.C., Wang, W. *et al.* (2006) Synergy of p53 and Rb deficiency in a conditional mouse model for metastatic prostate cancer. *Cancer Res.*, **66**, 7889–7898.
209. Martinez-Cruz, A.B., Santos, M., Lara, M.F., Segrelles, C. *et al.* (2008) Spontaneous squamous cell carcinoma induced by the somatic inactivation of retinoblastoma and Trp53 tumor suppressors. *Cancer Res.*, **68**, 683–692.
210. Spencer, C., Pajovic, S., Devlin, H., Dinh, Q.D. *et al.* (2005) Distinct patterns of expression of the RB gene family in mouse and human retina. *Gene Expr. Patterns*, **5**, 687–694.
211. Chen, D., Gallie, B.L., and Squire, J.A. (2001) Minimal regions of chromosomal imbalance in retinoblastoma detected by comparative genomic hybridization. *Cancer Genet. Cytogenet.*, **129**, 57–63.
212. Lillington, D.M., Kingston, J.E., Coen, P.G., Price, E. *et al.* (2003) Comparative genomic hybridization of 49 primary retinoblastoma tumors identifies chromosomal regions associated with histopathology, progression, and patient outcome. *Genes Chromosomes Cancer*, **36**, 121–128.
213. Squire, J., Phillips, R.A., Boyce, S., Godbout, R. *et al.* (1984) Isochromosome 6p, a unique chromosomal abnormality in retinoblastoma: verification by standard staining techniques, new densitometric methods, and somatic cell hybridization. *Hum. Genet.*, **66**, 46–53.
214. Laurie, N.A., Donovan, S.L., Shih, C.S., Zhang, J. *et al.* (2006) Inactivation of the p53 pathway in retinoblastoma. *Nature*, **444**, 61–66.
215. Castera, L., Sabbagh, A., Dehainault, C., Michaux, D. *et al.* (2010) MDM2 as a modifier gene in retinoblastoma. *J. Natl Cancer Inst.*, **102**, 1805–1808.
216. Bowles, E., Corson, T.W., Bayani, J., Squire, J.A. *et al.* (2007) Profiling genomic copy number changes in retinoblastoma beyond loss of *RB1*. *Genes Chromosomes Cancer*, **46**, 118–129.
217. Corson, T.W., Huang, A., Tsao, M.S., and Gallie, B.L. (2005) *KIF14* is a candidate oncogene in the 1q minimal region of genomic gain in multiple cancers. *Oncogene*, **24**, 4741–4753.
218. Thériault, B.L. and Corson, T.W. (2015) KIF14: A clinically relevant kinesin and potential target for cancer therapy, in *Kinesins and Cancer* (ed. Kozielski, F.), Springer, Heidelberg, pp. 149–170.
219. Orlic, M., Spencer, C.E., Wang, L., and Gallie, B.L. (2006) Expression analysis of 6p22 genomic gain in retinoblastoma. *Genes Chromosomes Cancer*, **45**, 72–82.
220. Evans, A.J., Gallie, B.L., Jewett, M.A., Pond, G.R. *et al.* (2004) Defining a 0.5-mb region of genomic gain on chromosome 6p22 in bladder cancer by quantitative-multiplex polymerase chain reaction. *Am. J. Pathol.*, **164**, 285–293.
221. Marchong, M.N., Chen, D., Corson, T.W., Lee, C. *et al.* (2004) Minimal 16q genomic loss implicates cadherin-11 in retinoblastoma. *Mol. Cancer Res.*, **2**, 495–503.
222. Gratias, S., Rieder, H., Ullmann, R., Klein-Hitpass, L. *et al.* (2007) Allelic loss in a minimal region on chromosome 16q24 is associated with vitreous seeding of retinoblastoma. *Cancer Res.*, **67**, 408–416.

223. Kawaguchi, J., Takeshita, S., Kashima, T., Imai, T. *et al.* (1999) Expression and function of the splice variant of the human cadherin-11 gene in subordination to intact cadherin-11. *J. Bone Miner. Res.*, **14**, 764–775.
224. Feltes, C.M., Kudo, A., Blaschuk, O., and Byers, S.W. (2002) An alternatively spliced cadherin-11 enhances human breast cancer cell invasion. *Cancer Res.*, **62**, 6688–6697.
225. Kato, M.V., Shimizu, T., Ishizaki, K., Kaneko, A. *et al.* (1996) Loss of heterozygosity on chromosome 17 and mutation of the p53 gene in retinoblastoma. *Cancer Lett.*, **106**, 75–82.
226. Zhang, J., Benavente, C.A., McEvoy, J., Flores-Otero, J. *et al.* (2012) A novel retinoblastoma therapy from genomic and epigenetic analyses. *Nature*, **481**, 329–334.
227. Conkrite, K., Sundby, M., Mukai, S., Thomson, J.M. *et al.* (2011) miR-17∼92 cooperates with RB pathway mutations to promote retinoblastoma. *Genes Dev.*, **25**, 1734–1745.
228. Nittner, D., Lambertz, I., Clermont, F., Mestdagh, P. *et al.* (2012) Synthetic lethality between Rb, p53 and Dicer or miR-17-92 in retinal progenitors suppresses retinoblastoma formation. *Nat. Cell Biol.*, **14**, 958–965.
229. Sheps, J.A. and Ling, V. (2007) Preface: the concept and consequences of multidrug resistance. *Pflugers Arch.*, **453**, 545–553.
230. Chan, H.S., DeBoer, G., Thorner, P.S., Haddad, G. *et al.* (1994) Multidrug resistance. Clinical opportunities in diagnosis and circumvention. *Hematol. Oncol. Clin. North Am.*, **8**, 383–410.
231. Chan, H.S., Thorner, P.S., Haddad, G., and Gallie, B.L. (1991) Multidrug-resistant phenotype in retinoblastoma correlates with P-glycoprotein expression. *Ophthalmology*, **98**, 1425–1431.
232. Chan, H.S., DeBoer, G., Thiessen, J.J., Budning, A. *et al.* (1996) Combining cyclosporin with chemotherapy controls intraocular retinoblastoma without requiring radiation. *Clin. Cancer Res.*, **2**, 1499–1508.
233. Gallie, B.L., Budning, A., DeBoer, G., Thiessen, J.J. *et al.* (1996) Chemotherapy with focal therapy can cure intraocular retinoblastoma without radiotherapy. *Arch. Ophthalmol.*, **114**, 1321–1328.
234. Chan, H.S., Lu, Y., Grogan, T.M., Haddad, G. *et al.* (1997) Multidrug resistance protein (MRP) expression in retinoblastoma correlates with the rare failure of chemotherapy despite cyclosporine for reversal of P-glycoprotein. *Cancer Res.*, **57**, 2325–2330.
235. Cole, S.P. (2014) Targeting multidrug resistance protein 1 (MRP1, ABCC1): past, present, and future. *Annu. Rev. Pharmacol. Toxicol.*, **54**, 95–117.
236. Xu, X.L., Fang, Y., Lee, T.C., Forrest, D. *et al.* (2009) Retinoblastoma has properties of a cone precursor tumor and depends upon cone-specific MDM2 signaling. *Cell*, **137**, 1018–1031.
237. Xu, X.L., Singh, H.P., Wang, L., Qi, D.L. *et al.* (2014) Rb suppresses human cone-precursor-derived retinoblastoma tumours. *Nature*, **514**, 385–388.
238. McEvoy, J., Flores-Otero, J., Zhang, J., Nemeth, K. *et al.* (2011) Coexpression of normally incompatible developmental pathways in retinoblastoma genesis. *Cancer Cell*, **20**, 260–275.
239. Kapatai, G., Brundler, M.A., Jenkinson, H., Kearns, P. *et al.* (2013) Gene expression profiling identifies different sub-types of retinoblastoma. *Br. J. Cancer*, **109**, 512–525.
240. Pacal, M. and Bremner, R. (2006) Insights from animal models on the origins and progression of retinoblastoma. *Curr. Mol. Med.*, **6**, 759–781.
241. Basavarajappa, H.D. and Corson, T.W. (2012) KIF14 as an oncogene in retinoblastoma: a target for novel therapeutics? *Future Med. Chem.*, **4**, 2149–2152.
242. Chan, H.S., Canton, M.D., and Gallie, B.L. (1989) Chemosensitivity and multidrug resistance to antineoplastic drugs in retinoblastoma cell lines. *Anticancer Res.*, **9**, 469–474.

24
RNA Interference in Cancer Therapy

Barbara Pasculli[1,2] *and George A. Calin*[2,3,4]
[1] *University of Bari "Aldo Moro", Department of Biosciences, Biotechnology and Pharmacological Sciences, Via Orabona 4, Bari 70125, Italy*
[2] *The University of Texas MD Anderson Cancer Center, Department of Experimental Therapeutics, Unit 1950, 1881 East Road, Houston, TX 77054, USA*
[3] *The University of Texas MD Anderson Cancer, Department of Leukemia, 1515 Holcombe Blvd, Houston, TX 77030, USA*
[4] *The University of Texas MD Anderson Cancer Center, Center for RNA Interference and Non-Coding RNAs, Unit 1950, 1881 East Road, Houston, TX 77030, USA*

Translational Medicine Cancer: Advances in Molecular Biology and Medicine,
First Edition. Edited by Robert A. Meyers.

Keywords

Cancer
Pathologic phenotype defined by abnormal and uncontrolled cell division, arising from the accumulation of epigenetic and genetic alterations affecting the expression profile of both protein-coding genes and noncoding RNAs.

RNA interference
A natural, sequence-specific mechanism of post-transcriptional gene silencing in animal and plants, triggered and mediated by small, double-stranded RNAs.

RISC (RNA-induced silencing complex)
A multicomponent ribonucleoprotein complex consisting of a member of the Argonaute (AGO) family proteins, and a mature miRNA/siRNA incorporated into AGO, together with a number of accessory factors, acting as the molecular effectors of translational repression/degradation of miRNA/siRNA -targeted transcripts.

microRNAs (miRNAs; miRs)
Small, endogenous, single-stranded, noncoding RNAs, 19–25 nucleotides in length, which enter the RNAi apparatus to mediate the post-transcriptional regulation of gene expression by specific but imperfect binding to target mRNAs.

siRNAs
Small-interfering, double-stranded RNAs, 21–23 nucleotides in length, which can be endogenously produced or exogenously provided (synthetic oligonucleotides) and, similar

to miRNAs, are able to associate with the RISC complex and mediate post-transcriptional gene silencing.

antimiRs
Antisense oligonucleotides which sequester, by complementary binding, the mature miRNA in competition with cellular target mRNAs, leading to a functional inhibition of the miRNA and derepression of target mRNA translation.

Personalized medicine
Healthcare approach which designs a therapeutic regimen and selects the patients to treat, according to the individual clinical, molecular, and environmental information.

RNAi-based therapy
The repertoire of strategies harnessing the RNAi endogenous apparatus by using small RNA molecules to target any gene and inhibit oncogenic pathways supporting the tumor.

Current research in cancer therapeutics is leading the way in exploiting the growing amounts of genetic and molecular information on human oncobiology to design new anticancer agents to target specific tumor-related signaling pathways, and to treat patients according to their clinical characteristics and molecular profile. In this context, RNA interference (RNAi), a ubiquitous cellular pathway of post-transcriptional gene regulation, provides an intriguing tool for an innovative rational cancer drug design. Among the endogenous mediators of RNAi, microRNAs (miRNAs) represent the most important class of small RNAs whose global dysregulation is a typical feature of human tumors. Harnessing of the RNAi machinery by using small, synthetic RNAs that target or mimic endogenous miRNAs offers the opportunity to reach virtually any gene and pathway relevant to tumor maintenance. However, along with assessments of the safety profile and effectiveness, strategies for assuring the specificity of these drugs, in terms of their effective delivery to cancer cells, must be devised before RNAi formulations can be made into ideal personalized therapeutics for the clinic.

1
Introduction: Cancer and RNAi

Cancer is a leading cause of death worldwide, second only to cardiovascular disorders. It is a genetic disease involving mutational and epigenetic alterations that cumulate over time, and cause activation of oncogenes and/or loss of function of tumor suppressor genes. More importantly, the past decade of research has revealed that the widespread deregulation of physiological mechanisms supporting the transformation of normal cells into malignant cells is due not only to alterations in protein-coding genes but also to global changes in noncodingRNAs (ncRNAs), such as the microRNAs (miRNAs) profile. miRNAs are involved in the regulation of most biological functions, including development, life span and metabolism, and their deregulation contributes to abnormal and uncontrolled cell division and survival, as well as the capability of migration and metastatization that characterizes the cancer phenotype. The mechanism through which miRNAs exert their regulatory control of gene expression in cells has been recognized as endogenous RNA interference (RNAi). Originally, this pathway was first described as a process triggered by double-stranded RNAs (dsRNAs), able to induce sequence-specific gene silencing [1]. Having arisen in the context of genetic manipulation experiments, the discovery of RNAi implied the existence of specific mechanisms within the cells, from worms to mammals, mediating the unwinding of dsRNAs. Hence, the search for complementary base-pairing partners, sequestration from the intracellular pool of nucleic acid sequences, and the activation of downstream processes culminated in translational repression. Since its discovery, RNAi and its related pathways have been exploited to identify gene functions and new putative targets for cancer therapy. Indeed, the endogenous RNAi machinery can be properly harnessed by providing exogenous triggers that enter the pathway and, similar to miRNAs, promote gene-silencing. If translated into a disease context such as cancer, RNAi is likely to provide an appealing tool that can be used to switch off the causative genes and/or restore deregulated expression patterns.

Whole-genome sequencing and integrated "omics" analyses have greatly speeded up the recent advances in dissecting the molecular basis of cancer, and have pointed out the marked heterogeneity of human tumors, as well as the need to design and develop new rational therapeutic strategies according to the patient's molecular profile. In this regard, RNA interference (RNAi) represents an emerging, powerful approach for personalized cancer medicine that can be used to tailor medical intervention to an individual's genetic and physiological background.

2
RNA Interference: Breakthrough of the Year 2002

RNA interference refers to an evolutionarily conserved cellular process that is mediated by dsRNAs. In worms, plants and flies, RNAi likely provides an innate defense to both endogenous parasitic and exogenous pathogenic nucleic acids (DNA or RNA) [1], whereas its primary role in mammals and human involves regulating the expression of protein-coding genes at the post-transcriptional level [1].

The finding that a dsRNA could activate a sequence-specific silencing, and the whole set of intracellular mediators and functional activities orchestrating the process, was such a high-impact discovery that it was hailed as the biggest "Breakthrough of the Year" in 2002 by the journal *Science* [2], and honored with the 2006 Nobel Prize in Physiology or Medicine to Andrew Z. Fire and Craig C. Mello.

Following on from their studies in the nematode *Caenorhabditis elegans* [3], several research groups have moved quickly to large mammals [4, 5], to start investigating the potential for harnessing RNAi to address biological questions and treat human diseases. Indeed, by enabling reversible gene silencing and modulating the expression of any gene, RNAi provides an appealing approach for loss-of-function gene analyses in vertebrate systems and, more importantly, for clinical purposes.

2.1
RNA Interference: a Brief Historical Digression

Retracing the history of RNAi discovery, the first hints were derived from experiments on engineering transgenic petunias, when Richard Jorgensen was attempting to alter pigmentation and obtain a more intensely purple flower strain [6]. Unexpectedly, the introduction of exogenous pigment-producing transgenes resulted in totally white or variegated pigmentation. As the endogenous loci and "foreign" transgenes appeared to be turning each other off, this phenomenon was called "co-suppression." Similar outcomes were subsequently derived from other plants, fungi (*Neurospora crassa*) [7], and metazoans (*Drosophila* and *C. elegans*) [8–10], where transmission of the silencing effect was also observed through several generations.

At that time, 1995, Guo and Kemphus were investigating the function of the *C. elegans par-1* gene by using an "antisense-mediated silencing" approach, in which a small, synthetic RNA oligonucleotide, complementary to the *par-1* sequence, once delivered to the cells, was expected to bind the messenger RNA (mRNA) and block its translation. Guo and Kemphus found that both the exogenous antisense and sense oligonucleotides (the latter being used as negative control) were able to produce the same lethal effect by inducing *par-1* mRNA silencing [11]. The turning point came when Fire, Mello and colleagues, during their experiments of manipulation of gene expression in *C. elegans*, found that the dsRNA mixture was 10-fold more potent in inducing translational repression than sense or antisense RNAs were alone [3]. This effect became known as RNA interference.

Shortly thereafter, this effect – which became variously known as post-transcriptional gene silencing (PTGS) [12], co-suppression [6], quelling [7], and RNAi – was demonstrated to be even more widespread, occurring in flies [8] as well as mammals [4, 5]. The role of RNAi pathways in the normal regulation of endogenous protein-coding genes was suggested by a flurry of many results. Small RNAs were found to be produced in plants undergoing PTGS [13, 14], while studies in *C. elegans* revealed that these small RNAs – which were now known as short interfering RNAs (siRNAs) were produced by the dsRNA-processing enzyme Dicer [15] and promoted gene silencing by triggering the assembly of a nuclease effector known as the RNA-induced silencing complex (RISC) [8]. Subsequently, a class of natural hairpin dsRNAs [16, 17], now called miRNAs, was shown to be processed by Dicer [18, 19]. Finally, the role of RNAi in chromatin regulation in yeast [20], and chromosomal rearrangement during development of the somatic macronucleus in *Tetrahymena* [21], led to a definite identification of the RNAi machinery as a natural developmental gene regulatory mechanism.

2.2
Short Regulatory RNAs: miRNA and siRNA Biogenesis

When attempting to manipulate and take advantage of RNAi for any experimental application, it is critical to have an overview of the molecular basics of this pathway.

Since its discovery, more insights have been attained about the genes and the molecules implicated in the endogenous RNAi machinery.

RNAi, also defined as a sequence-specific, PTGS process [1, 12] is mainly performed by small RNAs including endogenous miRNAs [22], short interfering RNAs (siRNAs) [4], and exogenous chemically synthetized short hairpin RNAs (shRNAs) [23].

Although similar in function, a distinction can be made between miRNAs and siRNAs based on their different biogenesis [24, 25] and mechanisms of recognition of the target RNAs [26].

miRNAs are initially transcribed in the nucleus by an RNA Polymerase

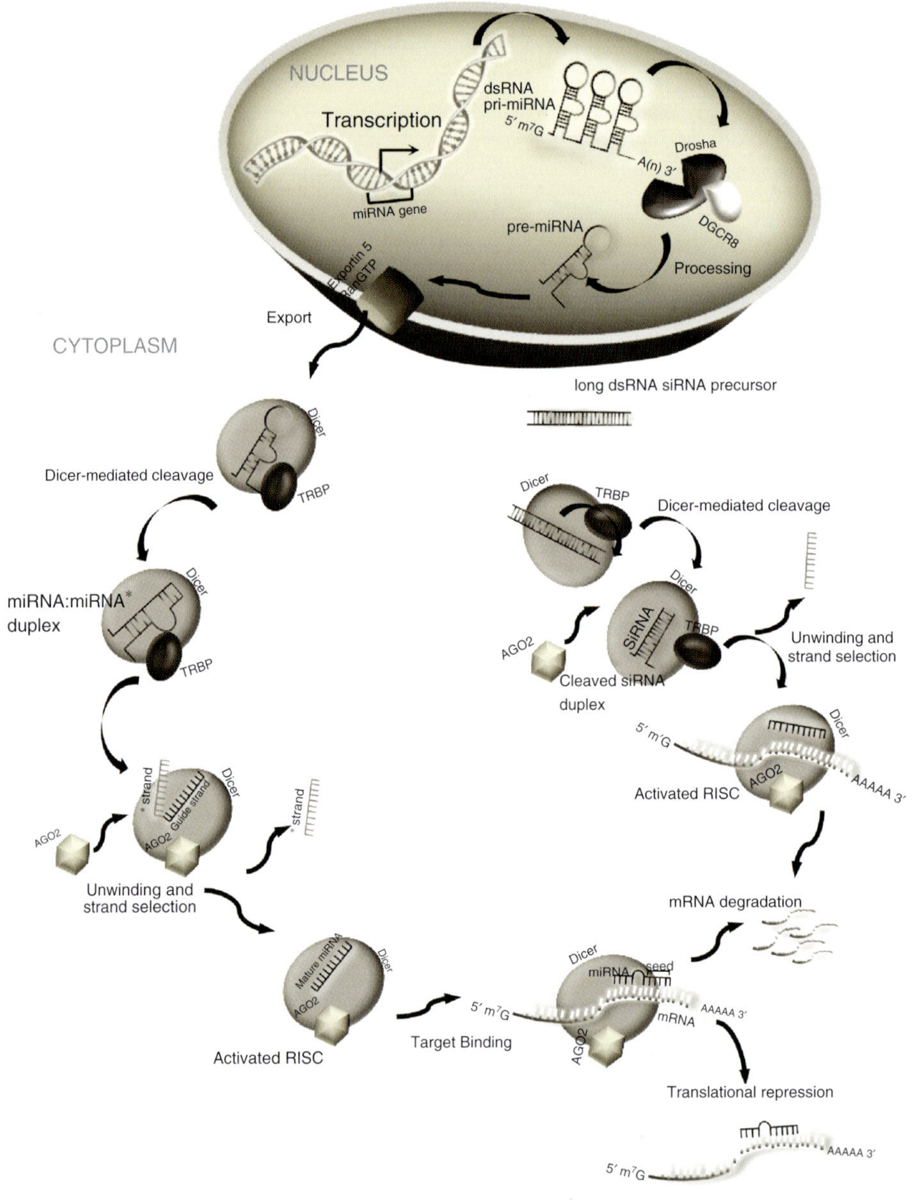

II or III as long primary transcripts (pri-miRNAs), containing single or clustered double-stranded hairpins bearing single-stranded 5′- and 3′-terminal overhangs, and 10 nt distal loops [26]. The pri-miRNA is processed into a 70- to 100-nt, imperfectly base-paired hairpin precursor RNAs (pre-miRNAs) by the Microprocessor Complex, which consists of the RNase III enzyme Drosha and a protein called Pasha in *Drosophila* or DGCR8 (Di George Syndrome critical region gene 8) in

Fig. 1 The miRNA and siRNA pathways of RNAi. Primary miRNAs (pri-miRNAs) are transcribed in the nucleus by RNA Polymerase II as long, capped, and polyadenylated (A-(n)) precursors, and are trimmed by the microprocessor complex (including Drosha and DCGR8) into 70- to 100-nucleotide hairpin precursors, usually containing interspersed mismatches along the duplex, called pre-miRNAs. Pre-miRNAs associate with the Exportin/RanGTP complex and are translocated to the cytoplasm (left side of figure), where they are further processed into a 19- to 25-nucleotide duplex (miRNA:miRNA*) by the Dicer–TRBP complex. Finally, the duplex interacts with an Argonaute (AGO2) protein into the RISC (RNA-induced silencing complex) apparatus: one strand of the duplex (the passenger * strand) is removed, whereas the guide strand, the mature miRNA, remains stably associated within the RISC, and directs the complex to the target mRNA for the post-transcriptional gene silencing. The interactions between miRNA *seed* sequence and the 3′-UTR of the mRNA generally culminate in translation repression. However, cleavage and degradation of the mRNA have also been documented. Long, perfectly base-paired, dsRNA (right side of figure) are also processed by the Dicer–TRBP complex into small siRNAs. Once they enter the RISC complex, an Argonaute protein cleaves the passenger strand, and the guide strand is used to bind the complementary mRNA target. SiRNAs usually anneals to the mRNAs by a perfect complementarity, mainly leading to their degradation.

mammals [27–29]. This initial cleavage is then followed by the Exportin-5/RanGTP-mediated pre-miRNA translocation to the cytoplasm for further processing into a 19- to 25-nt duplex by the RNase III endonuclease Dicer and TRBP (Human immunodeficiency virus-1 transactivating response RNA-binding protein) [30]. Final processing by Dicer is likely to culminate in incorporation of the miRNA into the RISC, which acts as effector of RNAi [31] (Fig. 1).

In contrast, siRNAs are produced from long, perfectly base-paired dsRNA precursors which can be produced endogenously or provided exogenously (Fig. 1).

Indeed, siRNAs are often referred to either genetic material deriving from virus infections or artificial inhibitory products. However, large-scale analyses in plants, fungi and animals have detected small RNAs that lack hairpin-loop precursors, and thus do not belong to miRNA species [32–34]. The larger part of these RNAs is encoded by repetitive elements of the genome, as trasponsons, especially located at heterochromatic regions such as centromeres and telomeres, and confirming that the RNAi machinery contributes to the establishment of a repressed chromatin state at these regions [20]. Other siRNAs are instead transcribed from the antisense strand of chromosome regions that often contain sequences for longer noncoding RNAs [32, 35]. Although their function is far from being completely understood, several organisms (including yeast, flies, plants, and protozoa) have been found to contain many of these siRNAs.

Nevertheless, once in the cytoplasm, the siRNA processing, as well as its assembly into the RISC complex, is mostly Dicer-dependent [36] (Fig. 1).

Thus, both for miRNA and siRNA precursors, the resulting dsRNA is a duplex of 19- to 25-nt strands bearing a 2-nt overhang at its 3′ terminus and a phosphate group at its 5′ terminus.

2.3 The RNAi Machinery

The key component of the RISC complex is an Argonaute (AGO) protein. AGO proteins have been consistently found in bacteria, archaea and eukaryotes [37]. In humans, among the eight AGO-classified proteins [38], AGO2 is the only member which exhibits endonuclease activity associated with its RNaseH-like domain [39] and accomplishes an miRNA/siRNA-induced silencing of target mRNAs. AGO2, together with Dicer and TRBP, represent the minimal RISC-loading complex (RLC) [26]. Additional proteins as Vasa intronic gene (VIG) protein, the Tudor-SN protein, Fragile X-related protein, the putative RNA helicases Dmp68, and Gemin3, as well as GW182 and TTP [40–43], have been found to associate with the RLC, although their functions are still not fully understood.

Once the diced dsRNAs interact with AGO2, one strand of the duplex – the "guide" strand – is selected to be loaded into the RISC complex, while the other strand – the "passenger" strand (often indicated by *) – is discarded. Selection of the guide strand, which generally is based on thermodynamic properties, specifically addresses the silencing activity of the RISC complex to cytoplasmic single-stranded RNAs (ssRNAs), as endogenous mRNAs, through either perfect or imperfect complementarity to the AGO2-bound guide strand. Indeed, the guide strand nucleotides 2–8 (counting from the 5′ end) constitute the "*seed* region," that is able to directly bind the target mRNAs. The degree of complementarity can be either perfect (as occurs mostly for siRNAs) or partial (as is usual for miRNAs), and determines subsequent AGO2-mediated target cleavage or, more frequently (if partial complementarity occurs) a non-endonucleolytic translational repression [26] (Fig. 1).

3
MicroRNAome: The Biological Toolkit for the Regulation of Gene Expression

A large number of miRNA genes (>1000) have been predicted to exist in the human genome, accounting for 1–5% of all predicted human genes. Altogether, miRNAs in human cells are estimated to regulate the expression of between 30% and 90% of human genes [44]. In particular, miRNAs mainly operate the PTGS by repressing translation or accelerating mRNA decay [45]. In both cases, after being loaded into the RISC complex, binding of the target mRNAs to the 3′ untranslated regions (UTRs) does not require perfect complementarity with the "*seed*" sequence. Thus, in contrast to siRNAs, a single miRNA may regulate multiple messenger RNAs and, in turn, each gene can be regulated by different miRNAs [45]. Moreover, recent studies have reported that miRNAs can also bind to the 5′ UTR or the open reading frame [46–49] and, even more surprisingly, they can upregulate translation upon growth arrest conditions [50]. It has also been noted that mature miRNAs may localize in the nucleus, where they may regulate pre-mRNA processing, act as chaperones modifying mRNA structures, or modulate mRNA–protein interactions [51]. The observations that miRNAs can be imported into the nucleus [52] or even secreted from the cells [53] suggest that the mechanistic details and the number of cellular functions of miRNAs are still far from being fully unraveled.

3.1
microRNA Deregulation is a Molecular Feature of Cancer

As data have accumulated, miRNAs have been found to be involved in a number of cellular processes such as proliferation, differentiation, survival, and apoptosis. Hence, it is not surprising that alterations affecting miRNA expression and function may promote the pathogenesis and progression of human tumors. Indeed, like classical genes, miRNAs can be either overexpressed or downregulated and act as oncogenes or tumor-suppressors, depending on their downstream targets [54]. Abnormalities in their expression or functions are associated with the typical hallmarks of cancer, including increased cell proliferation, abrogated apoptosis, enhanced cell motility and invasiveness, and neoangiogenesis [54]. For instance, miR-21 is highly upregulated in the majority of cancer tissues, and the repression of pro-apoptotic genes, such as *PTEN* (*phosphatase and tensin homolog*) or *PDCD4* (*programmed cell death 4*), stimulates proliferation and tumor initiation [55, 56]. Conversely, miR-15a and miR-16-1 were the first tumor-suppressor miRNAs to be described that were expressed from the fragile 13q14 region that is frequently deleted in patients with chronic lymphocytic leukemia (CLL) [57]. Their loss, by releasing the inhibition of tumor-promoting genes, such as *BCL2, BMI1, CCND2,* and *CCND1,* is able to promote cell growth and tumor progression [57–60].

Following the initial discovery of the miR-15a/16-1 cluster, a steadily growing number of reports have established that miRNA expression is globally deregulated in neoplastic cells compared to the corresponding normal tissue [61]. The progressive development of different high-throughput platforms for assessing miRNA expression in normal and diseased tissues (microarrays, bead-based flow cytometry, and deep sequencing) has allowed the miRNA expression profiling of several malignancies, including CLL [62], breast cancer [63], thyroid papillary carcinoma [64], lung cancer [65], glioblastoma [66], pancreatic tumors [67], hepatocellular carcinoma [68], prostate cancer [69], and gastric cancers [70]. This multi-combined approach has revealed the ability of miRNA signatures to not only differentiate between normal and cancerous tissues, but also to identify the tissue of origin. The latter point would be especially relevant when the tumor has already spread to distant metastatic sites, as it would allow the discrimination of different subtypes of a particular cancer even more successfully than previous mRNA panels [71]. Notably, the exciting discovery that miRNAs can be also detected at high levels in body fluids has spurred a rush to investigate miRNAs as potential diagnostic and prognostic biomarkers, as well as a novel class of drug targets, within the clinical setting of human malignancies [54].

3.2
Mechanisms of microRNA Deregulation in Cancer

Overall, large-scale miRNA profiling studies have revealed that the cancer phenotype is characterized by a globally reduced miRNA level relative to normal tissues [72], which suggests a role for miRNAs in the maintenance of a differentiated cell state and homeostasis.

The cause of the widespread differential expression of miRNA genes between neoplastic and normal cells can be linked to different mechanisms, including chromosomal rearrangements of miRNA genes, DNA point mutations, epigenetic mechanisms, or alterations in the machinery responsible for miRNAs biogenesis [54, 73]. Along with the typical CLL breakpoint at the miR-15a-16-1 locus in the 13q14 chromosomal region, members of the let-7 family of tumor suppressor miRNAs map to fragile sites, as 3p2 (let-7g/let-7a-1), 9q22 (let-7f), 11q24 (let-7a-2), and 21q21 (let-7c), are frequently deleted in breast, lung, ovarian and cervical cancers [74]. Conversely, the oncogenic miR17~92 cluster in 13q31, encoding for six different miRNAs (miR-17, miR-18a, miR-19a, miR-20a, miR-19b-1, and miR-92-1), is amplified in several hematopoietic malignancies [54]. Interestingly, miR-17-5p and miR-20 are known to specifically repress the cell-cycle regulator E2F1, which mediates cell-cycle progression through G_1/S checkpoints, and whose expression is also regulated by the transcription factor c-MYC [75]. *In-vivo* studies in Eμ-Myc transgenic mice (a well-established mouse model of B-cell lymphoma) showed that the miR17~92 promoter also contains c-Myc E-box binding sites, and suggests that c-MYC is likely to fine-tune the cell cycle by regulating the expression of both mRNAs and miRNAs.

In addition to structural genetic alterations, miRNA expression in cancer can be also affected by epigenetic changes, such as altered DNA methylation [76]. Indeed, it is likely that half of the genomic sequences of miRNA genes contain CpG islands [77]. In this regard, the hypermethylation of miR-127 has been found to induce its downregulation, and to significantly increase the target proto-oncogene *BCL6* in bladder cancer cells [78]. Similarly, miR-9-1 [79] and the clustered miR-34b and miR-34c [80] were also found to be modulated by DNA methylation. Alternatively, defects in the epigenetic machinery, including DNA (cytosine-5)-methyltransferase 1 (DNMT1) and DNA (cytosine-5)-methyltransferase 3β (DNMT3β), also affect miR-124a expression, as found in colorectal cancer cells [81]. Conversely, the overexpression of putative oncogenic miRNAs in cancer can be due to DNA hypomethylation [82], and alterations affecting other chromatin-remodeling processes [83–85].

Single nucleotide polymorphisms (SNPs) in miRNAs, in either a heterozygous or homozygous configuration, represent another type of genetic variability that can affect gene and protein expression, and may be responsible for the interindividual variability of the cancer phenotype. Indeed, sequence variations in miRNA genes, including pri-miRNAs, pre-miRNAs and mature miRNAs, could also influence the processing and/or target selection of miRNAs.

A C → T germline alteration in the primary transcript of miR-15a/miR-16 reduces their expression levels, and has been found in patients with familial CLL [86]. miRNA precursor-related SNPs, such as pre-miR-196a2, pre-miR-499 and pre-miR-146a, have been associated with the risk of breast cancer [87] and papillary thyroid carcinoma [88]. Moreover, the SNP rs11614913 in pre-miR-196a2 has also shown prognostic significance correlating with survival in patients with non-small-cell lung cancer (NSCLC) [89].

SNPs in the miRNA *seed* region can impair – either by weakening/abolishing or, conversely, by strengthening – miRNA–mRNA interaction, and thus differentially affect the expression of miRNA targets [90, 91]. Despite the low (<1%) probability of SNP occurrence in a miRNA *seed* region, a polymorphism here would significantly impair a miRNA function, as compared to a polymorphism present in a 3′-mismatch-tolerant region (3′-MTR) binding region which, conversely, is very sensitive to mismatches. And last, but not least, miR-polymorphisms can alter the epigenetic regulation of a miRNA (either methylation or acetylation) and promote disease progression [79].

Several reports have also shown that changes in the miRNA biogenesis and processing machinery can be responsible for miRNA dysregulation in cancer [90]. For instance, a generalized downregulation of Dicer and/or Drosha has been significantly correlated to a poor clinical outcome in different tumors such as lung, ovarian, colon, prostate and breast cancer, nasopharyngeal carcinoma, and neuroblastoma [92–100]. In addition, the *DICER1* gene is commonly featured as a haploin-sufficient tumor suppressor because frequently it is single-copy-deleted in cancer [101]. Interestingly, a somatic hot-spot missense mutation in *DICER1* gene restricted to non-epithelial ovarian tumors has also been reported [102]. Inactivating mutations and a consequent loss of the *TARBP2* gene, encoding for TRBP, have been shown to destabilize *DICER1* and impair miRNA processing in cancers bearing microsatellite instability [103]. These tumors are also characterized by mutations of the *XPO5* gene, coding for the Expotin-5 transporter, which generate a truncated protein that is unable to associate with pre-miRNA molecules and exit the nucleus; this decreases the trafficking of pre-miRNAs in the cytoplasm and reduces the pool of diced mature miRNAs [104].

A dysregulation of AGO proteins also occurs in cancer. The loss of AGO1 (*EIF2C1*), AGO3 (*EIF2C3*) and AGO4 (*EIF2C4*) is a recurrent event in Wilms tumor of the kidney and in neuroectodermal tumors [37, 73]. However, the expression of AGO proteins, such as AGO2 (*EIF2C2*), is reported to be reduced in some cases, for example in melanoma samples (primary and metastatic) compared to normal epidermal melanocytes, but elevated in other tumors such as breast and colon cancer; this suggests that expression can be regulated in a cell context-dependent manner [105]. The RNA-binding proteins LIN28/LIN28b represent further potential oncogenic factors; they are normally highly expressed in hematopoietic stem cells (HSC), and have the unique ability to revert human somatic cells to pluripotent cells when coexpressed with the reprogramming factors OCT4, NANOG, and SOX2. The oncogenic roles of LIN28 and LIN28b have been suggested by elevated levels in cancer stem cells from various cancer types, including ovarian, breast and NSCLC [105].

Experimental findings have also documented how miRNA processing can be affected by other miRNAs, either directly, whereby the mouse miR-709 is able to bind a recognition element on pri-miR-15a/16-1 in the nucleus [106], or indirectly, whereby the miR-103-107 family targets Dicer and leads to a downstream downregulation of miRNAs as the miR-200 family, and to epithelial-to-mesenchymal transition (EMT) [100].

Finally, the deregulation of miRNA expression, as either increased or decreased transcription, can also result from an altered transcription factor activity. Indeed, miRNAs can be either positively or negatively regulated by transcription factors such as p53, activating miR-34a [107] and miR-205 [108], MYC, activating miR-17~92 cluster [75] but repressing let-7 [109] and miR-29 family members [110], and HIF-1α, regulating miR-210 [111] or ZEB1, and repressing the transcription of members of the miR-200 family, which in turn are able to directly target ZEB1 and ZEB2 [112].

3.3 Role of microRNAs in Cancer

As noted above, aberrant expressed miRNAs can act either as oncogenes (oncomiRs) or tumor suppressor genes (tumor suppressor miRNAs), and be crucially involved in the initiation and progression of the neoplastic disease in all human tumors.

Of note, their role is strictly dependent on the cellular context and tumor system: miR-221 and miR-222, for example, can function as tumor suppressor miRNAs in leukemia by targeting the *KIT* oncogene. [113]. Furthermore, they are also known for silencing tumor suppressors like *PTEN* or *TIMP3* (tissue inhibitor of metalloproteinases 3), playing the role of oncomiRs in several solid tumors including breast and lung cancer, hepatocellular carcinoma, or glioblastoma [114].

Here, a few examples of miRNAs classified as oncomiRs (Table 1) or tumor suppressor miRs (Table 2) are reported, according to the main role that they play in the majority of cancers.

3.3.1 OncomiRs

miR-155 miR-155, encoded by the B-cell integration cluster (BIC) ncRNA, exerts its role in regulating the response and adaptation of cells to the tumor microenvironment, in particular when hypoxic conditions occur. Indeed, miR-155 mediates the downregulation of FOXO3A upon HIF1α activation [115]. In this context, other targets of miR-155 are SOC1 and RhoA, which collectively act in the signaling pathways promoting EMT [116, 117].

A high expression of miR-155 has been reported in several solid tumors such as breast cancer, colon cancer, ovarian cancer, pancreatic ductal adenocarcinoma, thyroid carcinoma, and lung cancer [71], where it is also considered a marker of poor prognosis. Similarly, miR-155 levels are increased in various B-cell malignancies, including Hodgkin lymphoma, some subtypes of non-Hodgkin lymphoma, and acute myeloid leukemia (AML) [71]. Interestingly, alongside the contribution to cancer genesis and progression, *in-vitro* studies have shown that miR-155 overexpression can confer resistance to radiation therapy, and the administration of antisense oligonucleotides (ASOs) against miR-155 can overturn this effect, which suggests that miR-155 may serve as a predictive biomarker and putative therapeutic target for treatment-resistant tumors [118].

The miR-17~92 Cluster and Paralogs The polycistronic miR-17~92 cluster is located within approximately 1 kb of an intron of the C13orf25 locus on human chromosome 13q31.3, a region that is frequently amplified in several types of B-cell lymphoma and solid tumors [119]. The transcriptional regulation of the miR-17~92 cluster is mostly mediated by the *MYC* oncogene [75], to inhibit the apoptotic activity of E2F1, and induce cell proliferation. Overexpression of the cluster is observed in a variety of human cancers, including small cell lung cancer (SCLC), colon and gastric cancer, retinoblastoma, neuroblastoma, medulloblastoma, and osteosarcoma [71]. The cluster, which is processed to produce mature miR-17, miR-18a, miR-19a, miR-20a, miR-19b-1, and miR-92a-1, can target several factors involved in cancer initiation (HIF1α), cell proliferation (PTEN, E2F1-3, TNFα, RAB14), survival (BIM, TGFBR2), and angiogenesis [TSP1, connective tissue growth factor (CTGF)] [54, 120, 121]. The paralog miR-106b-25 cluster, comprising highly conserved miR-106b, miR-93 and miR-25, has shown to be similarly regulated with its host gene *MCM7* by E2F1, building a negative feedback loop where the cluster controls in turn the E2F1 intracellular level. An overexpression of miR-106b/25 has been reported in cancers of the prostate and pancreas, as well as in neuroblastoma, multiple myeloma [71], and is also likely to be responsible for the development of transforming growth factor-beta (TGF-β) resistance by suppressing p21 and BIM in gastric cancer [122].

Tab. 1 MicroRNAs which act mainly as oncomiRs and are upregulated in cancer.

microRNA	*Locus*	*Cancer type(s)*	*microRNA deregulation effects*	*Key targets*
miR-155	21q21.3	CLL, AML, B-lymphomas; lung, breast, colon, gastric, pancreatic, thyroid cancers	Resisting growth suppressors	FOXO3A, SOCS1, SHIP1
			Supporting inflammation	CEBPB, MEIS1, ETS1, C-MAF, CUTL1
			Mismatch repair impairment	hMSH2, hMSH6, hMLH1, WEE1
			Invasion and metastasis	RhoA
miR-17~92	13q31.3	Leukemias/lymphomas; lung, breast, colon, prostate, pancreatic, cancers; retinoblastoma, glioblastoma	Sustaining proliferation	PTEN, E2F1-3, TNF-α, RAB14, p63
			Resisting growth suppressors	BIM, TGFBR2
			Promoting angiogenesis	TSP1, CTGF, TGF-β, SMAD4
			Hypoxia response	HIF-1α
miR-221/222	Xp11.3	Lung, breast, prostate, papillary thyroid cancers; hepatocellular carcinoma; glioblastoma	Sustaining proliferation	PTEN, TIMP3, DICER
			Resisting growth suppressors	KIT, $p27^{Kip1}$, $p57^{Kip2}$, FOXO3A
			Evading apoptosis	PUMA
			EMT	TRSP1
miR-21	17q23.2	CLL, AML, myeloma; lung, breast, gastric, prostate pancreatic cancers; glioblastoma	Sustaining proliferation and ECM remodeling	PTEN, SPRY1, SPRY2
			Resisting growth suppressor	APAF1
			Evading apoptosis	PDCD4
			Metastasis	TPM-1, TPM-3, RECK, TIMP3

AML, acute myeloid leukemia; CLL, chronic lymphocytic leukemia; EMT, epithelial-to-mesenchymal transition.

The miR-222/221 Cluster Deregulation of the miR-222/221 cluster is mainly involved in promoting cell survival and the metastatic spread of cancer cells. miR-222/221 have been found highly upregulated in hepatocarcinoma, thyroid cancer, melanoma, estrogen receptor-negative breast cancer, and glioblastoma [123–127]. Cell-cycle control and anti-apoptotic activity are related to the

Tab. 2 MicroRNAs which act mainly as tumor suppressor miRNAs and are downregulated in cancer.

microRNA	*Locus*	*Cancer type(s)*	*microRNA deregulation effects*	*Key targets*
miR-15a-16-1	13q14.2	CLL, multiple myeloma, lymphomas; lung, breast, colon prostate, pancreatic, ovarian cancers	Sustaining cell proliferation	CDC2, JUN, FGF-2, FGFR1 CCND-1, CHK1
			Evading apoptosis	BCL2, SIRT1
			Promoting angiogenesis	VEGF, VEGFR2, FGFR1
let-7 family	*LET7A1* 9q22.32	Lymphomas; lung, breast, colon, prostate, gastric, liver cancers	Sustaining cell proliferation	KRAS, NRAS, CDC25A, c-MYC
	LET7A2 11q24.1		Evading apoptosis	BCL-XL
	LET7A3,-7B 22q13.31		Metastasis	HMGA2, TWIST1
	LET7C 21q21.1			
	LET7D,-7F1 9q22.32			
	LET7E 19q13.41			
	LET7F2 Xp11.22			
miR-34a	1p36.22	Lymphoma; lung, breast, colon, renal, bladder pancreatic cancers; neuroblastoma, glioblastoma	Sustaining cell proliferation	CDC25A, CDK4, CDK6, c-MYC
			Evading apoptosis	BCL2, SIRT1
			EMT and metastasis	MET, SNAIL1
miR-200 family	*MIR200A, MIR200B, MIR429* 1p36.33	Lung, breast, endometrial, ovarian, nasopharyngeal, bladder cancers	EMT and metastasis	ZEB1, ZEB2, CTNNB1, BMI-1, PLCγ1, VEGFR1, FN1, LEPR, MALM2, MALM3
	MIR200C, MIR141 12p13.31			

FGF, fibroblast growth factor; LEPR, leptine receptor; NRAS, neuroblastoma rat sarcoma viral oncogene homolog; VEGFR1, vascular endothelial growth factor receptor 1.

repression of kit, p27Kip1, PTEN, and PUMA [128]. It has also been reported that miR-221 and miR-222 transcription is induced by FOSL1, leading to a repression of TRPS1 and an E-cadherin decrease, contributing to the invasive phenotype of basal-like breast cancer [129]. Recently, it was reported that miR-221 and miR-222, controlled by both MET and epidermal growth factor (EGF) receptors, may play a pivotal role in developing tyrosine kinase inhibitor resistance in NSCLC [130].

miR-21 The well-documented miR-21, transcribed from the 17q23.2 genomic locus, is typically highly expressed in virtually all human malignancies, including those derived from breast, colon, liver, brain, pancreas, and prostate.

The wide array of molecular targets, including the tumor suppressors maspin (*SERPINB5*), PDCD4, tropomyosin 1 (TPM1) and PTEN [131, 132], negative regulators of the RAS (rat sarcoma viral oncogene homolog) signaling axis, Spry1, Spry2 and Btg2 [133], or the metalloproteases RECK (reversion-inducing-cysteine-rich protein with kazal motifs) and TIMP3 (tissue inhibitor of metalloproteinase 3) [134] allocates miR-21 within the regulation of several oncogenic mechanisms supporting the aberrant growth and survival of cancer cells, as well as invasion and metastatic spread.

The central role of miR-21 is supported by a number of cell culture and animal model-based studies, which also indicate that the inhibition of miR-21 with ASOs represents a good strategy for reducing cell cancer proliferation and inducing apoptosis. For example, Medina and colleagues used Cre and Tet-off technologies to generate mice which, as a consequence of conditional expression of miR-21, developed a pre-B-cell malignant lymphoid-like phenotype, while subsequent miR-21 inactivation in the same model led to apoptosis and tumor regression [56]. This suggested that human cancers might be treated through the pharmacological inactivation of miRNAs such as miR-21.

The microarray-based profiling of tumor samples from breast cancer patients found that miR-21 was consistently upregulated in several subtypes of breast cancer, and suggested a role for this miRNA in mammary tumor initiation or progression [63]. Similarly, lung cancer profiling detected not only an elevated expression of miR-21 but also a direct correlation between its expression level and the presence of mutations in the epidermal growth factor receptor (*EGFR*) gene in patients [135]. Moreover, the blockade of miR-21 with ASO was able to enhance phosphorylation of epidermal growth factor receptor (p-EGFR) and restore apoptotic mechanisms in glioblastoma cell culture [136].

3.3.2 Tumor Suppressor miRs

The miR-15a/16-1 Cluster The link between miRNAs and cancer was first established with miR-15a and miR-16-1 [57], when it was discovered that they are often deleted and downregulated in CLL and, later, in prostate, lung cancer, and multiple myeloma [137–139], which suggested a main role in tumor suppression. Loss of their expression has been associated with the dysregulation of several pathways controlling cell proliferation, survival, migration and invasion, in which they regulate the levels of BCL-2, CDC2, ETS1, PDCD4, FGF-2, FGFR1 and other key cascade effectors. Moreover, miR-15a/16-1 was shown to regulate the response to cisplatin by targeting WEE1 and CHK1, which are commonly overexpressed in cisplatin-resistant cells, and also the restoration of normal miR-15/16 re-induced sensitivity to the drug [140], supporting the potential of a therapeutic solution to suppress aberrant growth in a variety of cancers.

let-7 Family The let-7 family includes 12 human homologs mapping to fragile sites associated with lung, breast, urothelial and ovarian cancers [71]. The let-7 family plays an important role in controlling cellular growth by regulating the Kirsten rat sarcoma viral oncogene homolog (KRAS) [141], and the specific downregulation of family members is a frequent event associated with a poor prognosis in lung cancer and the presence of lymph node metastases in breast cancer [71]. Moreover, in both types of tumor, polymorphisms in the 3′ UTR of KRAS mRNA are able to impair the let-7 regulating action and predict a worse prognosis [142, 143]. Along with KRAS, let-7 targets comprise LIN28, HMGA2 [54], MYC [144], and cell-cycle-related factors such as cyclinD2, CDK6, and CDC25A [145], allocating let-7 to every process involved in tumor growth and progression.

miR-34 Family The miR-34 family includes three miRNAs – miR-34a, miR-34b and miR-34c – under the direct transcriptional control of p53 [146]. In particular, p53 activation promotes a positive feedback loop in which an increased expression of miR-34 results in the downregulation of E2F, MET and SIRT1 and an indirect activation of p53 [141]. The promoter region of miR-34 family genes also contains CpG islands, and aberrant CpG methylation has shown to reduce miR-34 expression in multiple cancer types [80]. Other studies have progressively identified a number of targets, including BCL2, MYCN, GMNN and HDAC1 of the apoptotic cascade, cell-cycle crucial effectors such as CDC25A, CDK4 and c-MYC, or even WNT, HMGA2, and SNAIL1 that modulate migration and the invasion of cancer cells. For instance, a lower miR-34b expression in NSCLC has been correlated with a higher incidence of lymph node metastasis and a poor prognosis [54].

The miR-200 Family The miR-200 family members (miR-200a, miR-200b, miR-200c, miR-141, and miR-429) are downregulated in human tumors, where they play a critical role in the suppression of EMT, the cellular process through which epithelial cells lose their polarity and cell–cell adhesion, by targeting the key factors ZEB1 and ZEB2 [147] and promoting invasion. In lung cancer cells, an induced overexpression of the miR-200 family promotes mesenchymal-to-epithelial transition by repressing ZEB1 and increasing E-cadherin [148]. Accordingly, a loss of miR-200 family members correlates with a lack of E-cadherin expression in invasive breast cancer cell lines and breast tumor specimens, supporting an *in-vivo* role for the miR-200 family in EMT repression. Interestingly, ZEB1 and SIP1 are likely to recognize an E-box proximal minimal promoter element, generating a potential double-negative feedback loop in which the repression of a primary transcript and mature miR-200 expression allows the maintenance of high ZEB1/SIP1 levels that favor the mesenchymal phenotype of human breast cancer cells [149]. In addition, *in-vitro* studies have described a type of Akt–miR-200–E-cadherin axis, where Akt1 and Akt2 are able to control the abundance of miR-200 and E-cadherin and thus regulate breast cancer metastasis. This has been confirmed by the ratio of Akt1 to Akt2, and the abundance of miR-200 family members and E-cadherin in a set of primary and metastatic human breast cancers [150]. The deregulation of miR-220c/141 has also been linked to aberrant DNA methylation [151]. In conclusion, the miR-200 family prevalently assumes a tumor-suppressor role, and downregulation of its family members is significantly associated with an aggressive cancer cell phenotype.

When taken together, these data and the novel information that is continuously collected from the increasing number of molecular profiling studies in human cancers, highlight "miRNA addiction" as a key feature of cancer cells for developing and maintaining the malignant phenotype. Investigating miRNAs and the molecular circuitries that mediate their regulatory function in cancer can provide new clues about the nature of the cancer itself, and the molecules or mechanisms which may be targeted therapeutically to improve the clinical management of cancer patients.

4
Current Prospects for RNA Interference-Based Therapy for Cancer

To date, it is well documented that cancers can build sophisticated biological networks supporting their ability to progress and, moreover, to evade treatment. The promise of personalized medicine is to streamline clinical decision-making, according to each patient's genetic and physiological profile, distinguishing those who can benefit from a treatment from those who can experience side effects without relief. Indeed, distinct gene expression patterns may affect the body's response to medications, and an improved comprehension of the signaling cascades regulating tumor survival has determined the transition from a more promiscuously chemotherapeutic approach to highly selective targeted pharmaceutics. Initially, the antibody-based trastuzumab (Herceptin®) represented the first targeted therapy specifically for HER2-positive metastatic breast cancer (in 1997). Later, the small molecule imatinib mesylate (Gleevec™) became the first approved kinase inhibitor for targeting BCR/ABL in chronic myeloid leukemia (CML), and then for the treatment of gastrointestinal stromal tumors (GISTs), targeting c-kit.

A number of gain/loss-of-function experiments, in combination with target prediction analyses, have demonstrated that perturbation in miRNAs expression can affect not only every step of the tumorigenic process, but also the sensibility to drugs. Hence, the targeting of miRNAs, and the mechanisms orchestrating the endogenous RNAi, may undoubtedly represent a rich soil for the development of novel anti-cancer targeted therapies. RNAi-induced gene silencing mirrors the inhibitory effects of conventional drugs, such as protein-based compounds (e.g., antibodies and vaccines) and small molecules, whose inhibitory effect involves blocking the function of their targets. However, some disease-related molecules do not have enzymatic function, or they have a conformation that is not accessible to conventional drugs or small molecules, and so they are considered "undruggable." Conversely, RNAi technologies can be designed to virtually target any gene, including such undruggable molecules [152–154], with an exclusively allele-specific gene silencing [155]. Furthermore, antagonizing a target gene expression rather than its function has a more powerful downstream effect, as a single mRNA molecule is translated in multiple copies of a protein. Finally, compared to traditional drugs, the synthesis and production of RNAi modulators is more rapid, does not need the use of cellular expression systems or refolding steps, and can guarantee a higher selectivity and potency.

Nevertheless, to date, there are still some pending issues regarding the efficiency of the delivery systems, the choice of RNAi targets and the safety profile, that need to be addressed before allowing the widespread use of RNAi strategy *in vivo* [156].

4.1 Harnessing the Endogenous RNAi Apparatus for Clinical Purposes

The endogenous RNAi machinery can be engaged by designing exogenous triggers that enter the pathway at different points. Several synthetic modulators and strategies have been implemented to boost and optimize the effects on cancer-related pathways.

Modulating the RNAi machinery consists of either replenishing the levels of downregulated tumor-suppressor miRNAs to block tumor-promoting mRNAs, or sequestering oncogenic, overexpressed miRNAs to allow the transcription of tumor-suppressor proteins. All of the examples described below are simplified in Figures 2 and 3.

4.1.1 Restoring miRNA's Endogenous Regulation of Transcription: siRNAs, shRNAs, and Derivatives

The replacement of tumor suppressor miRNAs (let-7 family, miR-34a, miR-24, miR-26) [51, 74] finds its rationale in the global miRNA downregulation that characterizes the neoplastic phenotype. This strategy would permit the potential inhibition of oncogene transcripts that are aberrantly increased in the cancer state.

Short-interfering RNAs (siRNAs) are synthetic, short (usually 19–23 bp) dsRNAs which enter miRNA biogenesis at the cytoplasmic stage, when the pre-miRNA is chopped by Dicer, to be directly incorporated into the RISC complex. The siRNA guide-strand binds to and cleaves the complementary mRNA by a perfect-complementary match. Importantly, the guide-strand–RISC complex has the ability to be recycled, allowing the translational repression of several mRNA molecules, and propagation of the induced gene-silencing activity.

However, as the half-life of a siRNA is relatively short [157], shRNAs have been designed to enter the nucleus and be transcribed by RNA polymerase II or III from an external expression vector bearing a pre-designed, short, double-stranded DNA sequence. This means that shRNAs are constantly synthesized in the host cells, and achieve a longer gene-silencing effect. The primary transcript generated from the RNA polymerase II promoter contains a hairpin-like stem–loop structure that is regularly processed by Drosha, and translocated into the cytoplasm as pre-shRNA, to be cleaved by Dicer and incorporated into RISC, followed by the same cytoplasmic RNAi process as in siRNA [157].

A new class of bi-functional short hairpin RNAs (bi-shRNAs) has been developed for a more efficient and prolonged target-gene knockdown [158]. These consist of a vector construct for the expression of two (bi) stem–loop shRNAs for each targeted mRNA: one with perfect-matching stem sequences (as siRNAs), and a second with mismatches at the central location (bases 9–12) and additional locations of the stem. Compared to siRNAs, this strategy would promote the loading of mature shRNAs onto both cleavage-dependent (AGO2-mediated) and cleavage-independent (blocked-AGO2 by the mismatch) RISCs, thus achieving a higher and more effective shut down of the target mRNA by the concurrent involvement of both degradation and translation inhibition mechanisms [158].

ssRNAs may also be used as RNAi triggers but, although they have a greater structural flexibility (ssRNA versus dsRNA) and a more amphiphilic nature [159],

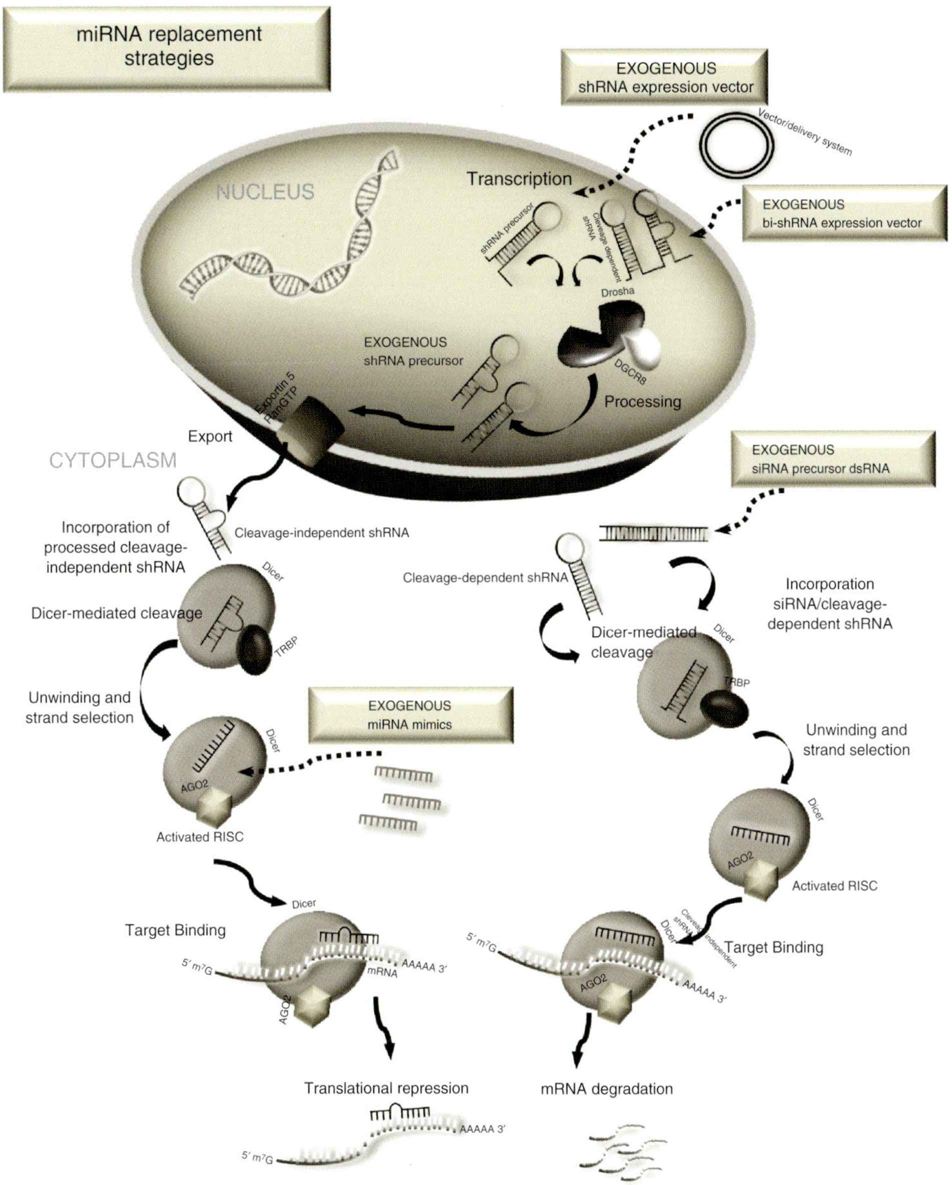

Fig. 2 MicroRNAs as targeted therapeutics. When aiming to repress oncogenic proteins, the expression levels of tumor suppressor microRNAs can be restored by using different strategies including siRNAs, shRNAs (short hairpin RNAs), bi-shRNAs (bifunctional siRNA), targeting by perfect base-complementarity oncogenic mRNAs, or miRNA mimics, synthetic analogs of endogenous miRNAs.

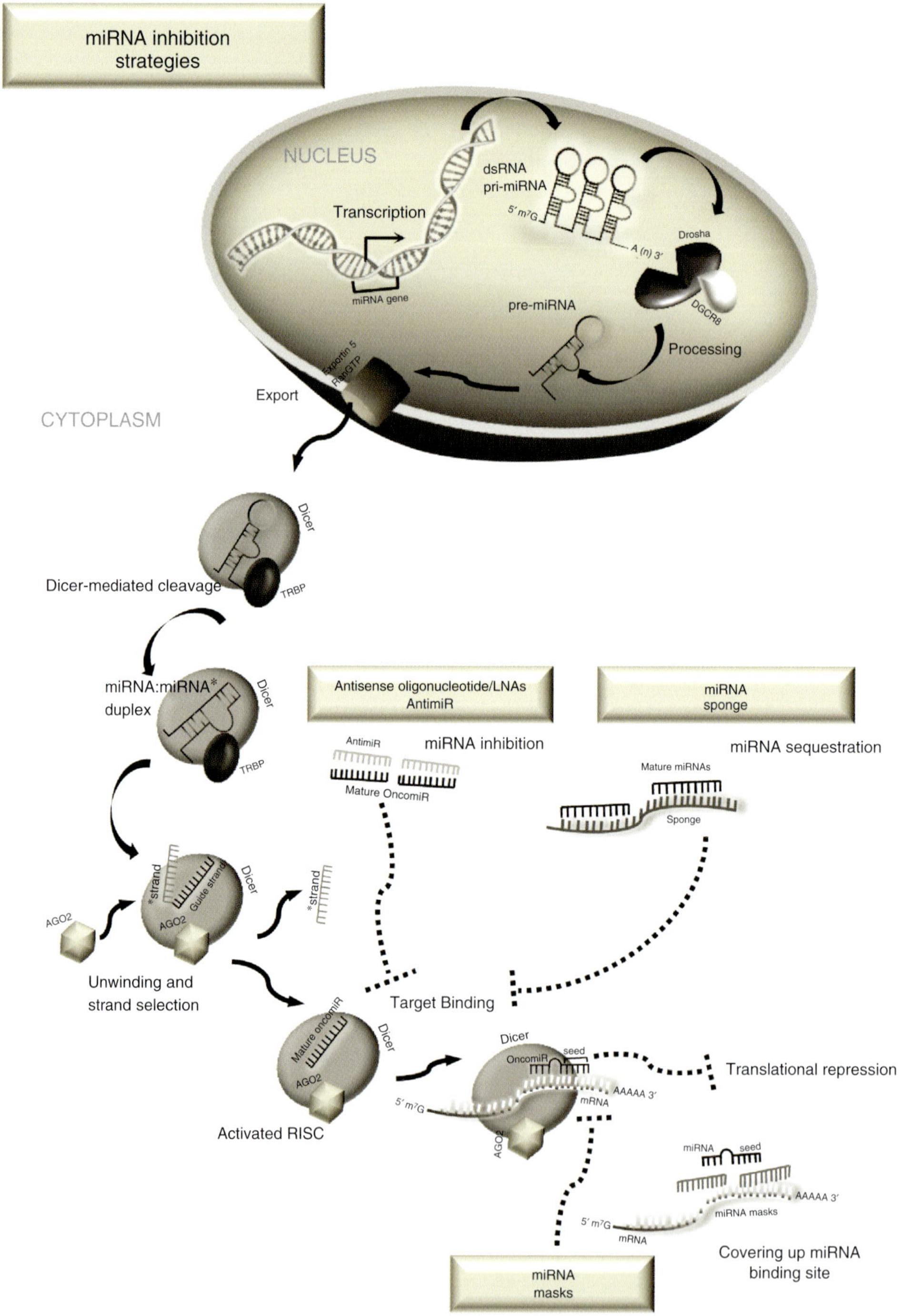
miRNA inhibition strategies
NUCLEUS
Transcription
miRNA gene
dsRNA
pri-miRNA
Drosha
DGCR8
pre-miRNA
Processing
Exportin 5
RanGTP
Export
CYTOPLASM
Dicer
TRBP
Dicer-mediated cleavage
miRNA:miRNA*
duplex
*strand
Guide strand
AGO2
Unwinding and
strand selection
Antisense oligonucleotide/LNAs
AntimiR
miRNA inhibition
Mature OncomiR
miRNA
sponge
miRNA sequestration
Mature miRNAs
Sponge
Target Binding
Mature oncomiR
Activated RISC
OncomiR
seed
AAAAA 3′
mRNA
Translational repression
miRNA masks
Covering up miRNA
binding site
miRNA
masks

Fig. 3 MicroRNAs as therapeutic targets. Aberrantly overexpressed microRNAs (oncomiRs) can conversely be inhibited by using: antimiRs (ASO, antisense oligonucleotides; LNAs, locked nucleic acids), designed to be complementary to the endogenous miRNAs; miRNA sponges, which contain multiple binding sites for microRNAs and prevent their loading onto the RISC complex; miRNA masks, small antisense oligonucleotides, which bind the target mRNA at the 3′-UTR miRNA-target sites, avoiding the interaction between mRNA and endogenous miRNAs, and transcriptional repression.

they are not efficiently recognized by the miRNA biogenesis apparatus (~100-fold less efficient) as siRNAs and shRNAs.

Moreover, a new method to restore the expression levels of endogenous, downregulated miRNAs (single or cluster) employs double-stranded, short artificial miRNAs, termed miRNA *mimics*, which are analogs of endogenous miRNAs and repress mRNA molecules by an imperfect binding to the 3′-UTR following the same mechanism of action [51]. Indeed, the ectopic expression of synthetic miRNAs mimics with tumor suppressor function as miR-15-a and miR-29 in cancer cells has shown to induce apoptosis in prostate and AML cell lines, respectively [160]. The first *in-vivo* evidence derives from Kota and colleagues, who recently cloned tumor suppressor miR-26 into an adeno-associated virus (AAV) vector for intravenous injection in an established MYC-dependent liver cancer mouse model. As expected, the restoration of miR-26 expression, which generally was reduced in liver cancer cells compared to normal cells, resulted in a suppression of tumorigenicity through the downregulation of cyclins D2 and E2 and a consequent suppression of tumor growth, without signs of toxicity [161].

Of note, it is likely that the mimic strategy would prevent the development of resistance to an RNAi approach by cancer cells better than siRNA formulations. Indeed, by mimicking endogenous miRNAs, mimics would impair the downstream function of multiple genes at the same time; thus, it would be more difficult for cancer cells to rearrange the sequence of several genes simultaneously to escape the mimic targeting and inhibition.

4.1.2 Inhibition of oncomiRs: AntimiRs, miRNA Sponges, and miR-Masks

The inverse approach consists of the inhibition of overexpressed miRNAs with oncogenic roles as miR-17~92 cluster, miR-155, by using small, single-stranded ASO drugs or the new generation of locked nucleic acids (LNAs) [51, 74]. These compounds belong to the antimiRs category, the properties of which are complementary to a target miRNA, with an ability to bind with high affinity and specificity and to produce a functional inhibition, thereby derepressing the mRNA targets of that miRNA [162].

A novel technology for managing miRNAs involves the use of expressing vector-carrying tandem miRNA target sites, to saturate an endogenous miRNA, and prevent the binding to its natural targets. This approach, known as "decoy," "sponge," or "eraser," employs different systems of gene delivery such as plasmids and adenovirus-, lentivirus- or retrovirus-based vectors that are normally used for gene knock-down studies. Since the inhibition of one miRNA can be compensated by the expression of other miRNA genes or family members, the tandem target sites built on these vectors can be assembled to catch different miRNAs sharing the same binding site, thus amplifying the effect of miRNA inhibition. Moreover, the sponges have been shown to be stable, they do not require repetitive rounds of administration, and they assure cell-type specificity compared to ASOs. Notably, the problem of a safety profile (i.e., toxicity), the partial interference that can occur, and the unknown targets that still must be identified for each putative miRNA, has led to the use of these constructs being a major challenge that is still in progress [163].

As an alternative to miRNA sponges, MiRNA-Masking Antisense oligonucleotides (miR-Masks) are single-stranded 2′-*O*-methyl-modified ASOs, that are fully complementary to predicted miRNA binding sites in the 3′-UTR of the target mRNA [164]. They compete with an endogenous miRNA by covering up the miRNA binding site and de-repressing its target mRNA. As this approach has the advantage of being gene-specific, it can significantly reduce any off-target effects [164]; however, it does not allow the targeting of multiple pathways.

5 Rational Design of an RNAi-Based Therapeutic Approach in Cancer

5.1 RNAi Target Selection

A successful RNAi clinical pipeline first relies on the rational selection of targets, according to parameters such as tumor specificity, tumor survival relevance, and the predictivity of tumor biological and clinical responses.

Whole-genome RNAi screening has been applied to human research for the first time for the identification of host genes affecting influenza A virus replication [165] and West Nile virus infection [166]. Using this approach, Tiedemann and colleagues identified crucial survival genes for the non-curative myeloma cancer after transfection of siRNA constructs of myeloma human cell [167]. To date, *in-vitro* whole-genome RNAi screening, combined with mutational signatures from different types of human cancer [156], has provided valuable clues on putative molecules to conveniently hit for disturbing oncogenic pathways and inducing tumor regression. For instance, GATA2 and BRCA1 have been identified as potential therapeutic targets in tumors with *KRAS* or *CCNE-1* mutations [168, 169], respectively. Furthermore, RNAi screening in animal models can provide information about the normal physiological functions of the putative RNAi-targets, and predict toxic effects that can derive from their incorrect targeting, compromising the clinical outcome. In the near future, the implementation of RNAi screening with "omics" data from clinical samples will provide a robust framework for prioritizing targets according to their translational potentials (Fig. 4).

5.2 *In-Vivo* RNAi Challenges

Overall, RNAi has important advantages over traditional pharmaceuticals such as protein-based drugs and small molecules for treating cancer, as it can be used to silence any target gene involved in tumor initiation, growth, angiogenesis, metastases formation, and chemoresistance [170]. The design of specific RNAi-effectors can rely on the growing amount of information that is continuously being collected on potential mRNA targets, in terms of corresponding sequences and mechanisms of action. Moreover, their robustness and specificity makes them more efficient and less toxic than traditional drugs. However, despite a number of optimistic preclinical studies having been conducted

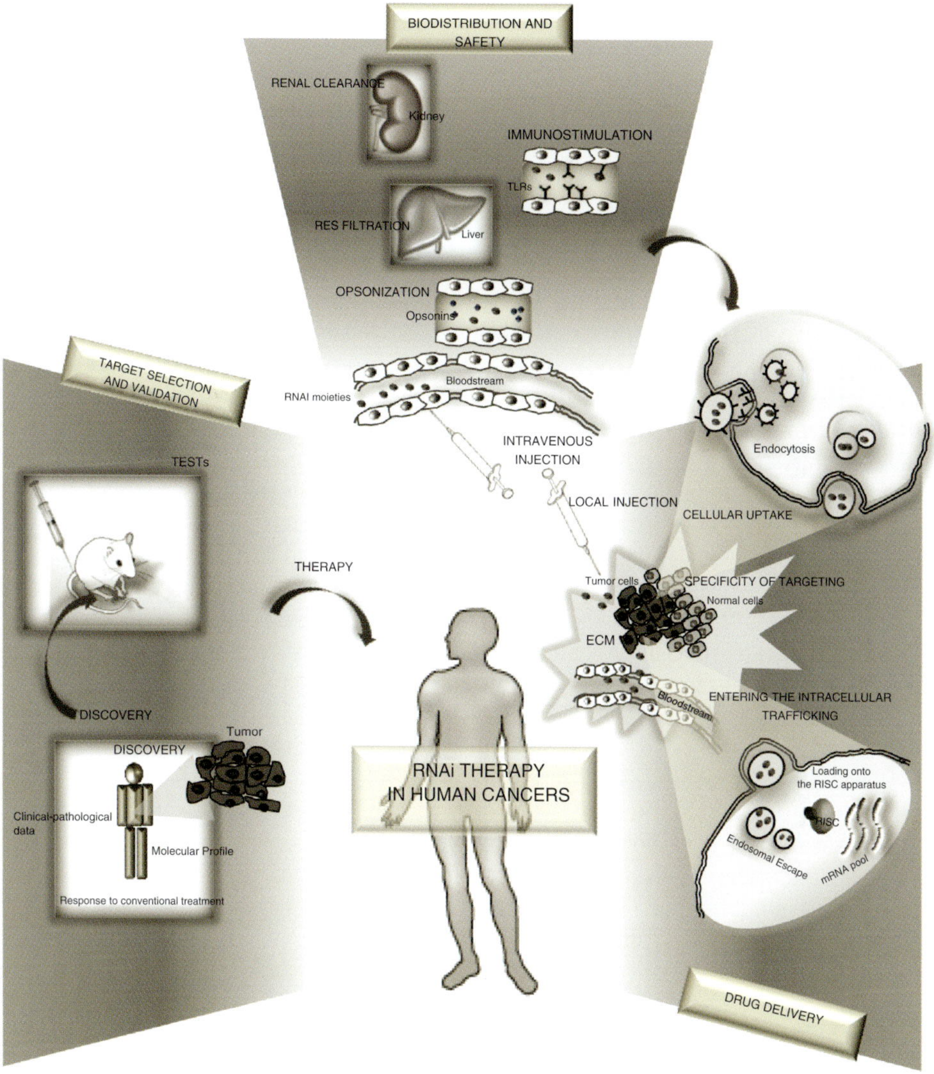

Fig. 4 Rationale design and challenges of RNAi therapeutics for human cancers. Schematic showing the critical issues to be taken into account when designing an RNAi-based strategy for cancer treatment: selection of the targets according to the tumor characteristics and patients' information from the clinic; the biological barriers affecting biostability and safety profile of RNAi therapeutics; and an efficient drug delivery culminating in the correct cellular uptake by tumor cells.

and ongoing clinical trials, several barriers must first be overcome before RNAi molecules can be moved to the clinical setting. For instance, the efficient intra-tumoral delivery and correct cellular uptake of RNAi molecules remain critical issues; in addition, despite their sequence-specificity, the potential to knock-down non-targeted genes is an off-targeting effect that must be estimated before *in-vivo* administration. Finally, RNAi has also been shown to induce an innate immune response and to saturate the endogenous miRNA pathway, leading to cytokine-associated damage and cytotoxicity, respectively [170].

5.2.1 Stability and Bioavailability Issues

When systemic delivery methods are adopted, and before reaching their target tissue and cells, small RNA drugs need to overcome several biological barriers that derive from: (i) their intrinsic nature (superficial negative charge, and high molecular weight for crossing cellular membranes); (ii) the bloodstream, as renal clearance, nucleases digestion, opsonization, and consequent reticuloendothelial system (RES)-mediated elimination; and (iii) the target tissue itself, as the extracellular matrix (ECM), cellular uptake and endosome internalization [171] (Fig. 4).

Double-stranded RNAs are more stable compared to other oligonucleotides (ssRNA or ssDNA) [172] but once they reach the circulation, unmodified (naked) siRNAs have shown to be susceptible to nuclease degradation. Some studies have also shown that when administered, naked siRNAs accumulate preferentially in the kidney to achieve levels 40-fold higher than in other organs, and are excreted into the urine within 1 h, due to their relatively small size and effective glomerular pore size. Hence, the correct target-site accumulation of an effective dosage of RNAi agents may be one of several major concerns for their therapeutic use [173].

To this end, the insertion of chemical modifications has been explored as a strategy to increase siRNA stability without compromising the siRNA silencing potency. Most of the chemical modifications have been carried out on the 2′-OH group of the ribose moiety as it is not essential for siRNA silencing activity [174]. Such modifications are commonly known as 2′-modifications, and mainly referred to as 2′-O,2′-Methyl-nucleoside (2′-OMe) and 2′-deoxyflouridine (2′-F) modifications [175, 176]. Another modification of the 2′ position of the ribose consisting of a methylene linkage between the 2′ and 4′ positions generated LNAs [177]. Indeed, LNA-based oligonucleotides, as miRNA mimics or inhibitors, appear to be highly resistant to nuclease degradation and displays low toxicity in biological systems. An additional type of chemical modification is the phosphoromonothioate (PS) linkage to the 3′ terminus of siRNA's phosphate backbone [178]. AntagomiRs are among the first antimiRs with proven *in vivo* efficacy and, similar to siRNAs, are 2′-O-Me oligonucleotide PS-substitution-modified constructs with the addition of a 3′-cholesteryl moiety [179].

5.2.2 Multiple Off-Target Effects

Although the mRNA cleavage induced by siRNAs and the translational repression by miRNAs are distinct pathways with different requirements [9], the two RNAi mechanisms share common cofactors and entry points, such as the RISC processing machinery. A "miRNA-like off-target gene silencing" can occur when siRNAs or shRNAs are loaded onto the RISC complex in a miRNA-like conformation, due to similarities between the core region at the 5′ end of a siRNA and the *seed* region of miRNAs, both of which guide recognition of the complementary 3′ UTR of the target mRNA and stimulate the expression of often hundreds of off-target genes [180, 181]. Such adverse side effects are a safety concern, especially when RNAi therapeutics are administered over extended periods of time [180]. A substantial reduction in off-targeting may be achieved by combining and implementing different strategies of rational drug design. Bioinformatics, which considers potential interactions between the siRNA *seed* region and the genome-wide repertoire of 3′-UTRs during the *in-silico* stage of RNAi agent selection, can be considered a rational approach to bypass such off-targeting. In addition, chemical and structural modifications can reduce guide strand-mediated off-targeting, by penalizing *seed*-only interactions via more extensive base-pairing (e.g., by including 2′-O-methyl, unlocked nucleic acid-modifications in the *seed*), and eliminate passenger strand-mediated off-targeting by 5′ blocking or by shortening the length of the passenger strand, without compromising the specific gene knockdown [180, 182].

The risk of saturation of the endogenous miRNA machinery by exogenous small RNAs [183] is another type of potential RNAi toxicity. The saturation mechanism involves several components, such as the Exportin-5 [184], and the RISC complex [183], which become less accessible to native miRNAs, especially in the case of a sustained expression of shRNAs [183]. Currently, the best strategy to avoid such complications refers to the use of lower concentrations of siRNA or, even better, to siRNA pools rather than single siRNAs [185, 186].

5.2.3 Immune Response

Besides delivery, acute immune responses from activating innate immune receptors or the complement system are critical "checkpoints" when assessing the safety profile of RNAi compounds. Both, RNAi-related "flu-like symptoms" with inflammatory cytokine elevations and the activation of alternative complement pathways have been observed as adverse effects in different clinical trials [187].

Indeed, synthetic, exogenous, RNAi modulators can induce an innate immune response via interaction with RNA-binding proteins such as Toll-like receptors (TLRs) and protein kinase receptors (PKRs) located on the cell surface or within the cell (in the cytoplasm or endosome compartments) [188] (Fig. 4). The innate immunity activation leads to a release of type I interferon (IFN) [189] and pro-inflammatory cytokines, such as interleukin (IL)-6 and tumor necrosis factor-alpha (TNFα).

It is now known that siRNA immune-stimulation can occur in a sequence-independent manner, as in the induction of TLR3 [190] and IFN pathways [189], as well as a sequence-dependent manner, when shRNAs or siRNAs contain a 5′-GUCCUUCAA-3′ motif or similar GU-rich sequences, and meet TLR7 and TLR8 [191, 192]. Moreover, the length of dsRNAs seems to influence the reactivity of the immune system independently of their nucleotide sequence; that is, a longer dsRNA will markedly increase cytokine production.

Adjustments of the RNAi trigger structure by chemical modifications, or reducing the concentration of siRNA preparation to mitigate immunogenicity, have been applied with positive results [193], although occasionally activation of the immune system can be partly beneficial as an adjuvant effect and improve the efficacy of a treatment against viral infections and some cancers.

5.2.4 Delivery Issues

The appropriate delivery of RNAi therapeutics to the target cells represents the major challenge in the development and *in-vivo* application of RNAi-based approaches. Once they have evaded the main systemic barriers (renal clearance, protein binding, and RES filtration), the RNAi particles must pass through the polysaccharide and fibrous network of the ECM to reach the cell surface of the inner tissues. Although the sinusoidal vasculature supporting organs such as liver, spleen, lymph nodes and bone marrow can facilitate the passive passage of particles up to 100 nm in size from the blood, continuous capillaries are more tightly interwoven (brain, muscle) and require active vascular escape strategies. Some of these strategies take advantage of the natural transvascular transportation that occurs within the endothelial cells, by conjugating specific ligands or bifunctional antibodies for receptors and molecular shuttles to RNAi moieties [187]. Subsequently, in order to allow the localization of RNAi compounds within close proximity of the target cells, some systemic delivery systems for the treatment of solid cancers benefit from an enhanced permeability and retention (EPR) effect, but an incorrect cellular uptake can still affect the success of the intervention. For example, RNAi drugs such as naked moieties do not readily cross the cell membranes due to their negative charge and large size [171]. Compared to direct cell penetration by lytic chemistries that risk inducing cell damage, receptor-mediated endocytosis is likely to be the more convenient and efficient pathway to be exploited. ApoE/LDL (low-density lipoprotein)- or folate/folate–receptor interactions in the case of stable nucleic acid particle lipid-based nanoparticle (SNALP LNP; see below) have demonstrated very productive internalizing pathways because of the wide distribution and elevated turnover of the receptors involved. On the other hand, cationic polyplexes/lipoplexes generally enter the cells by non-receptor-mediated physical association with the outer membrane and macropinocytosis [187]. Finally, internalization through endocytosis is normally followed by trapping in the endosome trafficking, culminating in an enzyme-mediated degradation into the lysosome. The RNAi agents must escape into the cytoplasm to reach and be loaded onto the RISC. To this end, delivery strategies including ionizable liposomes (long cationic polymers) are designed to enable endosomal release before maturation into lysosomes. Most of these strategies exploit the acidification that occurs naturally through endosomal maturation because of proton influx. The neutral charge of some compounds at physiologic pH in the blood becomes positive in the endosomes, facilitating disruption of the endosomal membrane. The low pH can also unmask membrano-lytic groups conjugated to the backbone of the RNAi compounds. Moreover, enzymes in the early endosomes can be harnessed to shed any chemistries linked to the RNAi trigger and that might otherwise interfere with RISC incorporation (e.g., ligands) [187].

In conclusion, to allow the implementation of RNAi-based therapy into the clinic, delivery strategies must overcome several technical and biological challenges. Herein are reported disparate systems that have been developed so far in attempts to satisfy not only efficacy but also safety requirements to support and expand the therapeutic application of RNAi molecules.

5.3
RNAi Implementation with Nanomedicine

To summarize, RNAi regulators of downstream gene expression include miRNA mimics and (mostly employed) siRNAs/shRNAs, whose purpose is to restore downregulated tumor suppressor miRNAs. Alternatively, antimiRs can be used for specific miRNA-silencing when the target miRNAs show oncogenic activity in the tumor.

The rational design of RNAi therapeutics for *in-vivo* applications aims to provide safety and effectiveness, by reducing renal clearance and immunogenicity, and optimizing stability, biocompatibility and site-specific delivery of the synthetic agents. These criteria are imperative not only for naked RNAi compounds but also (as explained later) carrier-formulated systems. In the latter case a low tendency for aggregation, which otherwise could induce toxicity and impair delivery activity, solubility in water and endosomal escape ability, are additional features of adequate drug design.

5.3.1 Chemical Modifications

As noted above, chemical modifications to siRNA strands (ribose, nucleotide bases, phosphorous acid, strand termini or backbone of both sense and antisense strands) can significantly improve the performance of naked RNAi molecules, preventing nuclease-based degradation and prolonging the siRNA half-life in the serum from a few minutes to two to three days [175]. Accordingly, conjugation to the strand ends of an siRNA of small molecules (such as cholesterol, folic acid, and galactose), polymers such as polyethylenimine (PEI) and polyethyleneglycol (PEG), antibodies, peptides [e.g., cell-penetrating peptides (CPPs) and transcriptional activator (TAT) peptide] or aptamers is able to increase cellular uptake by promoting caveolin or clathrin (glycoreceptor-mediated) endocytosis [175], thus enhancing the potency of siRNAs. For instance, the intravenous injection of cholesterol-siRNA conjugates targeting apolipoprotein B (ApoB) in transgenic mice led to a successful downregulation of plasma ApoB protein levels, as well as a total reduction of cholesterol levels in the liver and jejunum [194, 195]. Furthermore, siRNA–PEG conjugates have shown active function, prolonged circulation times, and reduced siRNA urinary excretion following intravenous injection [196]. Collectively, these chemical alterations, along with their combination [197], have been shown to ameliorate the *in-vitro* and *in-vivo* stability, bioavailability and duration of action of RNAi molecules, while remaining associated with a discrete effect of gene silencing.

5.3.2 Nanocarriers

Carrier-mediated transport represents a more robust strategy to overcome biological barriers and optimize uptake by the target cells, correctly locating the RNAi compounds into the expected site of action.

Carrier-mediated delivery strategies can be categorized into viral and nonviral vectors. The viral category includes retroviruses, lentiviruses, baculoviruses, adenoviruses, and AAVs, which are mostly known for their high delivery and transfection efficiency, and are distinguished by size and transgene capacity, target, performance, and duration of required expression (Table 3).

Nevertheless, critical limitations that include the immunogenicity and safety profile of viral carriers have boosted the development of alternative technologies based on nonviral vectors which fulfill the safety, because of a low risk of infection, and large-scale production requirements for their use. Promising nanocarrier systems for the clinical applications of RNAi include metallic, cationic lipid-based and polymer-based systems, collectively referred as nanoparticles (NPs). Nanoparticles are nanoscaled devices (generally in the range of 1 to 1000 nm), the size, shape and chemical surface configuration of which can be modified according to the purpose of their use [171]. To date, several types of NPs have been engineered to protect RNAi

Tab. 3 Viral-delivery systems.

Vectors	*Description*	*Target cells*	*Integration into the host genome*	*Expression status*	
Lentiviral vectors (LVs)	Retroviral-based vector, modified to infect several types of cells. The new generation include replication defective and self-inactivating viruses	Dividing cells	Yes	Transient or stable	***Pros:*** Effective in a wide variety of cell lines and tissues
		Nondividing cells			***Cons***: Immunogenicity and oncogenic potentiality following integration
Retrovirus vectors (RVs)	Enveloped viruses containing a single-stranded RNA molecule converted into a double-stranded DNA once in the infected cell	Dividing cells	Yes	Stable	***Pros:*** Long-lasting gene expression
					Cons: Limited to dividing cells, receptor-mediated internalization, and oncogenic potentiality due to random integration
Adenoviral vectors (AVs)	Non-enveloped virus (d = 60–90 nm) containing a linear double-stranded DNA genome	Dividing cells	No	Transient	***Pros:*** Efficient transfection in several cell types
		Nondividing cells			***Cons:*** High immunogenicity, strong inflammatory stimulation, and low loading capacity (7.5 kb)
Adenovirus-associated vectors (AVVs)	Single-stranded DNA human parvoviruses	Dividing cells	No	Transient or stable	***Pros:*** Nonpathogenic
		Nondividing cells			***Cons:*** Small packaging size (4.5 kb)

agents from serum nuclease degradation and the immune system, and to allow their uptake and cytosolic release by target cells. The most extensively studied systems are detailed in Table 4.

Additional intriguing alternatives for the targeted and intracellular delivery of RNAi compounds are represented by extracellular vesicles (EVs) and minicells. The EVs function as natural carriers of cellular components, including miRNAs, which mediate cell-to-cell communication and systemic trafficking of molecules [198]. They include exosomes, activation- or apoptosis-induced microvesicles (MVs)/microparticles, and apoptotic bodies. Although exosomes and microvesicles differ in size and biogenesis, both are produced by different cell types such as dendritic cells, intestinal epithelial cells, T cells and B cells, as well as cancer cells [199]. The detection of miRNAs in body fluids such as plasma, saliva and breast milk (despite the presence of RNase in the circulation) suggests that they can be (selectively) released by the cells into EVs in order to prevent degradation and reach distant tissues. Thus, compared to other conventional devices, EVs possess natural transporting properties that may be relevant when searching for a biocompatible and efficient system of delivery. Moreover, they are amenable to surface and content manipulation, for improving stability, reducing potential immunogenicity, and favoring the release of the RNAi moieties within the cells. An example of this was reported by Ohno and colleagues, who demonstrated that HEK293-derived EVs, expressing GE11 or EGF on their surface, were able to release let-7a to EGFR-expressing xenografted breast cancer tissue in immune-deficient mice. Similar results were obtained when a siRNA was used as alternative to miRNA [198]. Hence, despite some pending clues – such as stability in the circulation, purification protocols, mechanisms of internalization, composition, and immunogenicity in host organisms – the first experimental data seemed to fully encourage the application of EV formulations to the design of delivery systems for RNAi therapeutics.

Minicells are non-living, anucleate, bacteria-derived cells that are approximately 400 nm in diameter and are produced by mutants in which division is uncoupled from chromosomal replication [200]. They can be emptied of their endogenous RNA and protein contents, and packaged with therapeutic amounts of drugs that include chemotherapeutics, siRNAs and shRNAs. Compared to other carriers (such as liposomes), minicells have the properties of a simple drug packaging with different compounds (independent of their structure, charge, hydrophobicity and solubility) following a one-step coincubation, a higher payload capacity ($1-10 \times 10^6$ molecules versus 10 000 within each liposome), an ability to deliver the drug within a tumor cell without leakage during systemic circulation, an ease of conjugation to affinity ligands for tumor targeting, and no off-target effects [201]. Importantly, as minicells are of bacterial origin, purification procedures to eliminate free endotoxin and free bacterial components must be carried out to minimize the potential for toxic side effects that can derive from the inflammatory responses activated by TLRs.

In their earlier report, MacDiarmid *et al.* successfully demonstrated the *in-vivo* efficacy of minicells as targeting vehicles for an array of chemotherapeutic drugs, including 5-fluoracil, carboplatin, cisplatin, doxorubicin (DOX), irinotecan, paclitaxel (PTX), and vinblastine. Targeting was mediated by bispecific antibody (BsAb) conjugates that recognized both O-polysaccharide and a tumor antigen. As a consequence,

Tab. 4 Non-viral delivery systems: NPs-based technologies.

Delivery system	*Description*	*Common formulations*	*Targeting mechanisms*	*Pros and Cons*
Lipid-based delivery				
Lipid-based nanoparticles (LNPs)/lipoplexes (LPPs)	Lipid-based neutral micelles (d = 100 nm), consisting of lipid bilayer surrounding a hydrophilic aqueous core	• DOPC liposomes • DOPE liposomes • DOTAP liposomes	EPR	***Pros:*** • Protection of nucleic acids from enzymatic degradation • Low immunogenicity • Low toxicity • Biocompatibility • Ease of preparation
Cationic LNPs	"Stealth-coated" cationic lipid-based micelles (d = 100 – 150 nm) able to interact with the negative charge of RNAi moieties. The net surface charge, due to high risk of opsonization and immunogenicity, is generally neutralized by coating with PEG polymer	• SNALP • SLN	MPS	***Pros:*** • Increased half-life • Enhanced cellular uptake • Enhanced endosomal escape ***Cons:*** • Opsonization (SNALP) • Potential immunogenicity (SNALP)
Lipidoids	Non-glycerol cationic lipid-based micelles derived from conjugation with amines to acrylate and acrylamide	• Epoxide-based lipidoids	MPS	***Pros:*** • Suitability for siRNAs pool
Bubble liposomes (BLs)	PEG-modified liposomes containing echo-contrast gas, which can function as a gene and siRNA delivery tool with ultrasound exposure	• DOTAP liposomes • DSDAP liposomes	EPR	***Pros:*** • Ultrasound-enhanced delivery • Stability and long half-life • Biocompatibility

Polymer-based delivery				
Polyethylenimines (PEIs)	High cationic charge density polymers, presenting linear or branched structures with variable molecular weights (MWs)	• PEG-PEI • DA-PEI	Mucoadhesion	***Pros:*** • High transfection efficiency (high MW and branches forms > low MW and linear forms) • Endosomal pH resistance (proton sponge effect). • Physical protection of nucleic acid from degradation. ***Cons:*** • Proven cytotoxicity: cell apoptosis (high MW and branches forms > low MW and linear forms) • Activation of the complement system
Chitosan	High positive-charged mucopolysaccharide. The degree of deacetylation and MW can influence the capability to interact with nucleic acid. Both low degree of acetylation and MW decrease the interaction with RNAi moieties and the stability of nanocomplexes	Chitosan	Mucoadhesion	***Pros:*** • Biodegradability • Low immunogenicity

(*continued overleaf*)

Tab. 4 *(Continued)*

Delivery system	*Description*	*Common formulations*	*Targeting mechanisms*	*Pros and Cons*
Poly($_{D,L}$-lactic-*co*-glycolic acid) (PLGA)	Small lactic and glycolic acids polymer	• PLGA • PLGA-PEG	Mucoadhesion	***Pros:*** • Biodegradability • Rapid tissue penetration (small size) • Physical protection of RNAi molecules from RNase degradation
Dendrimers and dendritic polymers	Highly branched, globular, nanoscaled macromolecules consisting of a core, an interior dendritic backbone (the branches), and an exterior surface with functional groups	• PAMAM • PPI • Polyglycerols • Peptide dendritic polymers	EPR	***Pros:*** • Solubility, and very low intrinsic viscosities • High transfection efficiency, and drug delivery (endosomal escape) • Ease of preparation and commercial availability ***Cons:*** Low degradation rate and toxicity
Magnetic nanoparticles (MNPs)	A class of NP consisting of a magnetic core of a pure metal (Co, Mn, Ni, Fe), or their alloys and oxides usually coated by a shell, isolating the core against the environment	Coated MNPs (silane shell, gold shell, polymeric shell, dendrimeric shell): SPIONs, CPMNs	EPR	***Pros:*** • Tumor targeting by the aid of an external magnetic field • Visualization opportunity by MRI • Enhanced cellular uptake

				Cons: • Tendency to aggregation • Opsonization • Liver, spleen, and brain accumulation
Others				
Carbon nanotubes (CNTs)	Nonspherical nano-cylindrical particles, 50–100 nm in length (d = 1–100 nm)	• SWCNT • MWCNT	MPS	***Pros:*** • Ease of functionalization (PEGylation, Tween-20 coating, Ab-conjugation) • Good biodistribution (ease of shape manipulation to escape renal clearance) ***Cons:*** Non-biodegradable (pending issue)

DOPC, 1,2-oleoyl-*sn*-glycero-3-phosphatidylcholine; DOPE, dioleoylphosphatidylethanolamine; DOTAP, *N*-[1-(2,3-dioleoyloxy)-propyl]-*N*,*N*,*N*-trimethylammoniummethylsulfate; EPR, enhanced permeability and retention; MPS, mononuclear phagocytic system; SNALP, stable lipid-nucleic acid particle; SLNs solid lipid nanoparticles; PEG, polyethyleneglycol; DSDAP, 1,2-distearoyl-3-dimethylammonium-propane; PEI, polyethyleneimine; DA, deoxycholic acid; PAMAM, polyamidoamine; PPI, poly(propyleneimine); SPIONs, superparamagnetic iron oxide nanoparticles; CPMNs, cell-penetrating magnetic nanoparticles; MRI, magnetic resonance imaging; SWCNTs, one-dimensional single-walled carbon nanotubes; MWCNTs, multiwalled carbon nanotubes.

targeted minicell-mediated drug delivery was shown to produce a highly significant inhibition and even a regression of tumor growth in different xenograft models of human breast, ovarian and lung cancer, and also leukemia [201]. More recently, the same authors demonstrated the feasibility of using minicells for the siRNA-mediated treatment of mouse xenograft models. In an attempt to treat drug-resistant tumors, they applied a dual sequential treatment that consisted of an initial administration of si/shRNA-containing minicells linked to BsAb that targeted a known drug resistance mechanism. After allowing a sufficient knockdown of the drug resistance-mediating protein, BsAb-targeted minicells packaged with cytotoxic drug were then administered intravenously. This approach, which required a markedly small amount of drug, siRNA and antibody than was needed for a conventional systemic administration of free therapeutics, led to the complete survival of mice with tumor xenografts, with no signs of toxicity [202]. Overall, similar to EVs (caution must be taken here regarding the immunogenic profile and *in-vivo* efficacy of this system), the data acquired supported the potential of a minicell-based anticancer therapeutic for transfer to the clinical setting.

5.4
Administration

Both, local and systemic modalities of RNAi administration have been assessed in animal models, and are currently under investigation in several clinical trials [202]. Local delivery through intraperitoneal, intranasal, intramuscular, intrahepatic, intraretinal, intratesticular or subcutaneous injection, and through inhalation certainly offers the advantage of site-specificity and high bioavailability with small amounts of even naked RNAi drugs in the target tissue [203]. For instance, effective gene silencing and tumor suppression have been achieved by the direct subcutaneous and intraperitoneal injection of siRNA complexed with a commercial transfection reagent to induce specific gene silencing of human papillomavirus (HPV) 16 E6 oncogene in a model of cervical cancer [172]. Nonetheless, such a methodology is manageable only for certain technically accessible sites, such as mucosal and subcutaneous tissues. Indeed, the RNAi molecules must be distributed throughout a sufficiently large interstitial space, as occurs in the skin. Moreover, a local delivery and passive diffusion of the drugs can be achieved if energy-dependent mechanisms such as pressure, local heat gradients or muscle-related pressure gradients are present. More importantly, local delivery cannot be successfully applied to hematological malignancies or solid tumors that have already metastasized.

Alternatively, a less well-documented modality of administration is represented by electropermeabilization (EP), which involves the application of a calibrated electric field pulse to target cells or tissues, which in turn allows the RNAi molecules to rapidly "catch" their targets. To date, this approach has been used for various *in-vitro* and *in-vivo* evaluations of drug and nucleic acid deliveries. Similar to local administration, it allows the cellular uptake to bypass external barriers such as clearance, and internal barriers such as endosomes/lysosome internalization. EP-mediated siRNA or the introduction of LNA/DNA oligomers has been shown not only to be effective in silencing the gene expression [204] but also to allow a homogeneous distribution of the siRNA in the cytoplasm, facilitating its entry to the RISC complex as it is directly localized in the cytoplasm of the recipient cell

[205, 206]. Successful results have also been obtained by *in-vivo* direct siRNA delivery into the tissues of different animal models [207]. However, a lack of knowledge of the biophysical mechanisms that support reorganization of the cellular membranes after induced permeabilization can limit the wide application of this methodology.

In this regard, the systemic delivery of RNAi molecules (mainly by intravenous infusion) is the best-suited and multifaceted approach, and disparate formulations and targeting systems have been explored in this regard. Along with the above-mentioned chemical modifications, strategies such as the physical encapsulation of RNAi molecules, including nanocarrier functionalization with target moieties for intracellular diffusion, have been assessed as a means of breaking through the biological barriers and maintaining the targeting of diseased tissues. But even in this context, precautionary measures must be taken into account; indeed, as with other small molecules the intravenous route of administration of RNAi therapeutics can be accompanied by infusion reactions. Experience obtained from clinical trials has shown that these effects are dose-dependent and are especially related to the first infusion; consequently, a slowing rate of infusion, coupled with one-monthly optimized formulations (such as SNALP LNPs), have been shown to make adverse reactions significantly more manageable and clinically endurable [187].

5.5 Targeting

5.5.1 Passive Targeting

The unique abnormal architecture of the tumor vasculature can be exploited to enable nanodrugs to specifically accumulate in tumor tissues. Typically, tumor vessels are poorly aligned, with defective endothelial cells and wide fenestrations among them, together with a compromised lymphatic drainage. The "leaky" vascularization, which relates to the EPR effect, allows the extravasation and accumulation of macromolecules up to 400 nm greater in diameter in tumor tissue than in normal tissue [208]. The coating of cationic or hydrophobic nanocarriers with hydrophilic polymers allows them to elude opsonization and serum nuclease degradation, and also to localize in the tumor vessels. PEGylated liposomes (as SNALPs), which represent advanced liposome vehicles for passive siRNA delivery, show a much longer half-life in plasma than do traditional cationic liposomes [208].

As noted above, once the RNAi carriers access in the ECM they need to reach the target cells. The higher expression level of $\alpha_\upsilon\beta_3$ integrin in neoplastic cells compared to normal epithelial cells can be exploited by linking RGD (Arg-Gly-Asp) peptides to the NP surface to facilitate its binding to the tumor ECM, with consequent accumulation [168]. Lastly, because of the high metabolic rate of rapidly growing tumor cells, the microenvironment surrounding the tumor tissue is characterized by a lower pH; hence, some liposome preparations have been designed to be stable at physiological pH (7.4), but will degrade to release drug molecules at an acidic pH [209].

Despite these promising manipulations, the passive delivery process for RNAi therapeutics is difficult to control, and the lack of EPR in certain tumors as well as the heterogeneous permeability of vessels throughout a single tumor make passive delivery difficult to be applied extensively [209].

5.5.2 Active Targeting: Antibodies, Aptamers, Small Molecules

This strategy involves the attachment of affinity ligands (antibodies, peptides, aptamers, or small molecules) to naked RNAi or NPs–RNAi complex molecules [205–213], for molecules that are highly expressed on the cell surface of the diseased tissue. This would improve the bioavailability of the exogenous compounds, minimize nonspecific delivery, and induce cellular uptake with minimal doses of compounds.

Immuno-NP with monoclonal antibodies (mAbs), and engineered fragment antibodies (Fabs) or single chain fragment variable antibodies (scFv) has been widely used, and represents an effective method for targeting specific cells HIV-1 gp160 and HER2 represent peculiar examples. A protamine–antibody fusion protein (F105-P) against the HIV-1 envelope specifically delivered a fluorescein isothiocyanate (FITC)-silencing siRNA to HIV-1 infected and HIV-1 envelope-transfected cells, and induced silencing only in those cells enriched in gp160 (+) [205]. Similarly, a Polo-like Kinase 1 (PLK1)-targeting siRNA was injected intravenously into HER2(+) breast cancer xenografts by using an anti-Her2 ScFv-protamine fusion protein. When using this approach, the complex demonstrated a correct delivery of the siRNA to the HER-positive cells, without releasing to bystander tissues, while the siRNA reached a full PLK1 silencing, culminating in tumor growth retardation, metastasis reduction, and prolonged survival of mice [214].

Aptamers are synthetic, short, single-strand oligonucleotides with several advantages compared to full-size or single-chain antibodies that include a small size that allows for a better tissue internalization, a low immunogenicity and a high stability, as well as an amenability for modification and thus a capability to bind specific and different ligands with secondary or tertiary structures (proteins, receptors) [174, 208]. Aptamers can be easily linked covalently or noncovalently to both naked RNAi molecules (i.e., Aptamer siRNA chimera) and NPs-formulated constructs. For example, in a xenograft model of human prostate cancer, an A10 aptamer–NP complex was shown to recognize and bind the prostate membrane-specific antigen (PMSA) with high specificity, allowing its conjugated siRNAs into the cells to induce BCL2 and PLK1 silencing [215]. In another study, a dual inhibitory anti-gp120 aptamer/siRNA chimera, in which both the aptamer and the siRNA were directed against HIV genes, demonstrated effective silencing.

Cell-penetrating peptides (CPPs) are cationic oligopeptides, derived from natural sequences of viral, insect or mammalian proteins [216], that can be linked to RNAi moieties or NPs by covalent bonds or by electrostatic interactions, in order to accelerate their internalization by increasing cellular plasma membrane permeability. A typical example is the HIV–TAT family, penetratin and the chimeric peptide transportan, and oligoarginines (arginine-repeated unit) [217], whose ability to mediate siRNA silencing has been proven in the central nervous system of mice.

Small molecules can also function as effective ligands for increasing siRNA stability and cellular uptake. Cholesterol and its derivatives increase the binding of RNAi compounds to serum albumin, but reduce the interaction with LDLs and high-density lipoproteins (HDLs), with a consequent improved biodistribution in certain target tissues, such as liver [194, 197, 218]. The folate receptor is a membrane glycoprotein that is highly overexpressed in a number of human tumors, including ovarian, colorectal and breast cancer, but is almost absent in all other normal tissues, except kidney [197, 219]. Similarly, transferrin receptor-mediated delivery has been extensively studied in a variety of targets, including tumors, brain and endothelial cells [220]. Their several advantages include small size, a lack of immunogenicity, convenient availability, and easy chemical conjugation [218, 219].

5.5.3 Alternative Targeting

Alternative approaches include magnetic targeting and ultrasound (US)-enhanced delivery. Magnetic targeting utilizes the magnetic properties of metallic NPs (e.g., silica or gold NPs) to induce their specific accumulation in a target tissue by applying external, local, strong magnetic fields following systemic injection [221]. Once in the circulation, magnetic nanoparticles (MNPs) can be easily monitored using real-time magnetic resonance imaging (MRI) to follow their distribution and cellular uptake [222]. Disparate formulations of siRNA nanovectors exploiting magnetic targeting have been developed for the *in-vivo* imaging of delivered siRNA and gene silencing in tumors, and many others are currently under investigation. For instance, cell-penetrating magnetic nanoparticles (CPMNs) containing CPP-coated MNPs, conjugated with a low-molecular-weight protamine peptide, have consistently been shown to improve transfection efficiency and intracellular siRNA release *in vitro* [223]. In addition, functionalized super-paramagnetic iron oxide nanoparticles (SPIONs), combined with chitosan and siRNAs, have been used to couple the SPION suitability for MRI and/or thermal therapy to the ability of chitosan to increase the stability of associated siRNA [224].

A combination of microbubbles and US has been initially proposed as a less-invasive and tissue-specific method for gene delivery. Following infusion, the application of US produces transient changes in the permeability of the cell membrane, due to pressure oscillations, and allows for the site-specific intracellular delivery of combined molecules. Due to the limits of the microbubbles, in terms of their size, stability and targeting functionality, Suzuki and colleagues have developed different formulations of so-called "bubble liposomes"; these are PEG-modified liposomes in which an US imaging gas is trapped and can function as a novel plasmid DNA (pDNA), siRNA and, more recently, as a miRNA delivery tool when used with US exposure both *in vitro* and *in vivo* [225–228].

5.6 RNAi-Based Drug Combination Strategy (coRNAi)

Initial studies in viral infection-related diseases [hepatitis B virus, hepatitis C virus (HCV), human immunodeficiency virus, and other human viruses] have prompted the evaluation of a coRNAi technology for the treatment initially of metabolic or blood disorders, and then of cancer [229].

Combination therapy has been recommended in cancer treatment because it allows the use of lower doses of each drug, compared to single-drug administration, but reaches a higher efficacy based on additive or synergistic anticancer effects. This would prevent the toxic side effects that are often related to common chemotherapeutics, as well as the development of multidrug resistance (MDR) within the tumor. In fact, as human tumors tend to accumulate mutations and acquire resistance towards single-agent-based therapies, the concurrent targeting of multiple key factors and pathways would impair the development of secondary resistance. The *MDR-1* gene is primarily responsible for activating mechanisms for the expulsion of chemotherapeutics upon expression of the components of the P-gp pump (as MRP) within the cell membrane, which is likely to be induced following the repeated administration of drugs to oppose the cytotoxic action [230]. In addition, cancer cells can activate anti-apoptotic proteins such as Bcl2 in further protective attempts at survival over chemotherapy [230].

To date, one of the more widely investigated coRNAi strategies to combat chemoresistance in cancer has employed siRNAs targeting key components of the MDR pathways, in combination with standard chemotherapeutics. As an example, an *in-vitro* sensitization of cervical cancer cells has been obtained following the administration of silica NPs containing both an siRNA against the *MRP* gene and DOX, to compromise assembly of the P-gp pump and allow DOX to cross the cell membrane and inhibit cellular growth [231]. Similarly, cationic micelles carrying anti-BCL2 siRNA and docetaxel (DTX) have been used to target the anti-apoptotic pathway in an *in-vivo* model of human breast cancer [232]. A synergistic effect of RNAi triggers, when combined with PTX, has also been documented [233]; in this case, the coadministration of PTX and an siRNA against midkine (MKsiRNA) in human prostate cancer xenografts led to a significant augmentation of the antitumor therapeutic effect.

A more promising and recently emerging coRNAi approach relies on the coadministration of siRNAs and miRNAs (siRNA/miRNA-based coRNAi). Several studies have reported the successful delivery of both molecules into the tumor tissue, with surprising synergistic anticancer effects. Nishimura *et al.* showed that the dual inhibition of EphA2 (which is overexpressed in ovarian cancer cells and patients) by using an anti-EphA2 siRNA, combined with a miR-520d-3p mimic, targeting EphA2 and with high expression levels, correlated with a better prognosis and significantly augmented the therapeutic effect in inhibiting cell proliferation, migration and invasion *in vitro*, and of tumor growth *in vivo*, compared to monotherapy with each of the drugs [234].

Similar results with lung cancer have confirmed the effectiveness of this approach, as well as the relevance of nanotechnology in supporting such experimental and therapeutic design. Indeed, a new class of polymer-based NPs, 7C1, has been used to encapsulate an equal molar ratio of anti-KRAS siRNAs and miR-34a mimic in a single NP formulation, suggesting an easier and more flexible approach for coRNAi application. Both, *in-vitro* and *in-vivo* assessments confirmed the synergistic anticancer effect derived from the RNAi compound combination, and this was even more enhanced after coadministration with cisplatin (CP) [235].

Although miRNA-based combination therapy is still in its infancy, initial experiments seem to have demonstrated great potential. Kasinski *et al.* assessed the efficacy and safety of miRNA mimics coadministration in an aggressive *KRAS*; *TP53*NSCLC mouse model for replacing the expression of two lung tumor suppressor miRNAs, let-7 and miR-34. The systemic delivery of miR-34a and let-7 by lung-targeting NPs – either neutral lipid emulsion (NLE) or NOV34 (a miRNA delivery agent already undergoing clinical trials) – resulted in tumor growth suppression and a survival advantage [236].

In conclusion, although delivery issues and the risk of reciprocal competition between codelivered RNAi drugs still needs to be fully resolved [237], the abovementioned results and ongoing clinical trials have provided great encouragement for coRNAi translation in the clinical setting of human cancer. In particular, it is a less-toxic and more direct method for targeting multiple biologically pathways that are relevant to the genesis and progression of tumors.

5.7
RNAi in the Clinical Trials

The first RNAi-based clinical trial began in 2004, with an intravitreal injection of naked siRNA targeting the vascular endothelial growth factor (VEGF) pathway in patients with age-related macular degeneration (AMD) and diabetic macular edema (DME). Most of the naked siRNAs were locally administered to treat topical diseases that included AMD, DME and non-arteritic anterior ischemic optic neuropathy (NAION) or virus infections such as respiratory syncytial virus and HCV. The clinical was trial terminated due to low efficacy, off-target effects, or company issues.

Following the application of nanotechnology to the design of siRNA delivery vehicles, new RNAi agents based on systemic delivery entered clinical trials. To date, the terminated siRNA agents in clinical trials include naked delivered bevasiranib, AGN-745, and PF-655 (NCT0030694; NCT00363714; NCT1445899; www:ClinicalTrials.gov), for the treatment of eye-related disorders, and the SNALP-based delivery agent TKM–ApoB (NCT00927459) for cardiovascular diseases. Bevasiranib administration was terminated because of low efficacy at Phase III in a clinical trial to treat for AMD/DME, while AGN-745 was terminated during the treatment of AMD due to off-target effects at Phase II of clinical testing. TKM–ApoB was terminated because only a transient reduction of cholesterol level was observed at Phase I in the clinical study. One common adverse effect among these agents was the activation of TLRs; however, although most terminated RNAi agents were naked siRNAs, those to be administered locally will still have the chance to enter further testing as long as they demonstrate appropriate safety and efficacy.

CALAA-01 (Calando Pharmaceuticals; NCT006889065) has been the first active targeted delivery agent to enter clinical trials for the treatment of solid tumors. The agent consisted of an unmodified siRNA targeting Ribonucleotide Reductase M2 (RRM2) into cyclodextrin-containing polymer (CDP) NPs, decorated with transferrin, and administered intravenously to patients with metastatic melanoma. Despite the promising data obtained, the CALAA-01 trial was recently terminated, without any clear explanations.

Atu027 (Silence Therapeutic; NCT00938 574), another siRNA-containing PEG-nanoparticle (AtuPLEX) was the first siRNA to be applied for the inhibition of metastases. No drug-related side effects were observed at Phase I of the clinical trial, in which toxicity was evaluated in 27 of 33 patients with advanced solid tumor. A Phase Ib/IIa trial (NCT01808638), in which patients with locally advanced or metastatic pancreatic adenocarcinoma were recruited has since been undertaken, to evaluate the combination of Atu027 with gemcitabine.

ALN-VSP02 (Alnylam/Tekmira; NCT011 58079) is the first dual-targeted RNAi formulation combining two siRNAs into a lipid-NP carrier system (SNALP) targeting both VEGF and kinesin spindle protein (KSP), respectively. A Phase I clinical evaluation (in 28 patients) reported that ALN-VSP02 was well tolerated, with neither hepatotoxicity nor any elevation of complement proteins. Although increases

Tab. 5 siRNA-based RNAi cancer interventions in clinical trials.

Drug	*Formulation*	*siRNA-target*	*Cancer disease*	*Status*	*Identifier at: ClinicalTrials.gov*
CALAA-01	Systemic CDP-NPs	RRM2: cell proliferation	Advanced solid tumors	Phase I: terminated	NCT00689065
ALN-VSP02	Systemic SNALP liposomes (short circulating lipid). coRNAi with two siRNAs	VEGF: angiogenesis; KSP: cell proliferation	Liver cancer and advanced solid tumors with liver metastases	Phase I: completed	NCT01158079
iP siRNA	*Ex vivo* transfection of mature dendritic cells and vaccination transfected mature dendritic cells	LMP2, LMP7, MECL1: proteasome-mediated antigen-presenting processing	Metastatic melanoma	Phase I: completed	NCT00672542
TKM-PLK1	Systemic SNALP liposomes (long-circulating lipid)	PLK1: cell proliferation	Solid tumors, lymphomas, primary and secondary liver metastases	Phase I/II: recruiting	NCT01262235
Atu027	Systemic PEGylated lipoplex (Atuplex)	PKN3: angiogenesis and metastasis	Advanced solid tumors	Phase I: completed	NCT00938574
Atu 027 + Gemcitabine			Advanced pancreatic adenocarcinoma	Phase I: recruiting	NCT01808638
SIG12D LODER	Local LODER polymer (PLGA-based slow-release formulation)	Mutated KRAS	Advanced pancreatic adenocarcinoma	Phase I: ongoing	NCT01188785
				Phase II: not yet recruiting	NCT01676259
EPHARNA *siRNA-EphA2-DOPC*	Non-PEGylated DOPC liposome	EPHA2 cell proliferation	Advanced solid tumors	Phase I: not yet recruiting	NCT01591356
DCR-MYC	Systemic liposome	MYC: cell proliferation	Solid tumors	Phase I: recruiting	NCT02110563
			Multiple myeloma Non-Hodgkins lymphoma		

Tab. 6 miRNA-based RNAi cancer interventions in clinical trials.

Drug	*Formulation*	*Target*	*Cancer disease*	*Status*	*Identifier at: ClinicalTrials.gov*
MRX34	Systemic NLE (neutral lipid emulsion) liposomal miR-34 mimic	BCL2: cell apoptosis	Primary liver cancer or metastatic cancer with liver involvement. lymphoma, leukemia, multiple myeloma	Phase I: recruiting	NCT01829971
SPC3649 (Miravirsen)	Systemic LNA-ASO (antisense oligonucleotide)	miR-122: inhibition of miR-122–HCV complex formation essential to the stability and propagation of HCV RNA	HCV and hepatocellular carcinoma	Phase I/II: completed	NCT01200420
EZN2968 *anti-HIF-1α LNA AS-ODN*	Systemic LNA-ASO (antisense oligonucleotide)	HIF-1α: hypoxia response	Advanced solid tumors or lymphoma	Phase I: completed	NCT00466583
SPC2996	LNA-ASO	BCL2: cell apoptosis	CLL	Phase I/II: completed	NCT00285103

in cytokine levels were observed these were considered due to higher doses, and immediately resolved within 24 h after administration.

On-going and future trials include the TKM-PLK1 compound (NCT01262235) for the treatment of primary liver cancer or liver metastases, and two promising coRNAi trials for siG12D LODER (Silenceed Ltd; NCT01676259), a siRNA targeting KRAS combined with gemcitabine to treat advanced pancreatic cancer, and siRNA-EphA2-DOPC (1,2-oleoyl-*sn*-glycero-3-phosphatidylcholine) (EPHARNA, MD Anderson Cancer Center; NCT01591356). In this case, nonPEGylated liposomes will be administered to patients with advanced and recurrent solid tumors. Details of other, similar, studies are provided in Tables 5 and 6.

6
Summary

In combining the characterization of RNAi in humans with increasing knowledge of the molecular alterations that support the genesis of human malignancies, RNAi therapeutics represent the most promising approach towards innovative and personalized medical interventions on cancer, and a clear alternative to the limited effects of conventional treatments (e.g., chemotherapy, surgery, or radiotherapy).

During the past decades, gene expression profiling studies of human malignancies have revealed that the global dysregulation of small, noncoding RNAs – that is, miRNAs – which are the genuine users of the endogenous RNAi apparatus for fine-tuning gene expression, has a significant effect on every step of tumorigenesis. Thus, the ability to harness the RNAi machinery by restoring the regulatory function of downregulated miRNAs using synthetic miRNA mimics, siRNAs, viral expression constructs, or inhibiting an oncomiR function by chemically modified antimiR oligonucleotides, miRNA sponges, or miR-masks, offers the chance to target the expression of any gene linked to neoplastic phenomena, including those which have proved so far to be "undruggable."

Although a number of RNAi therapeutics have emerged as putative, effective anticancer drugs, the major drawback in their translation into the clinic remains a lack of safe and efficient delivery systems, and the ability to release the RNAi moieties at a therapeutic concentration specifically to the target cells, without affecting healthy tissues. In this context, during the past few years remarkable achievements have been made by designing NPs-based carriers (cationic lipids/polymers) with high transfection abilities, biocompatibilities, stabilities, and a lack of immunogenicity. Today, encouraging results appear to be emerging from ongoing clinical trials regarding the safety and efficacy of RNAi therapeutics, even when repeated routes of administration are used. However, seminal investigations into the real therapeutic potential of these strategies are still required before entering the clinical setting and supporting patient management.

References

1. Hannon, G.J. (2002) RNA interference. *Nature*, **418**, 244–251.
2. Couzin, J. (2002) Breakthrough of the year. Small RNAs make big splash. *Science*, **298**, 2296–2297.
3. Fire, A., Xu, S., Montgomery, M.K., Kostas, S.A. *et al.* (1998) Potent and specific genetic interference by double-stranded RNA in *Caenorhabditis elegans*. *Nature*, **391**, 806–811.

4. Elbashir, S.M., Harborth, J., Lendeckel, W., Yalcin, A. *et al.* (2001) Duplexes of 21-nucleotide RNAs mediate RNA interference in cultured mammalian cells. *Nature*, **411** (6836), 494–498.
5. Elbashir, S.M., Lendeckel, W., and Tuschl, T. (2001) RNA interference is mediated by 21-and 22-nucleotide RNAs. *Genes Dev.*, **15**, 188–200.
6. Jorgensen, R. (1990) Altered gene expression in plants due to trans interactions between homologous genes. *Trends Biotechnol.*, **8**, 340–344.
7. Romano, N. and Macino, G. (1992) Quelling: transient inactivation of gene expression in *Neurospora crassa* by transformation with homologous sequences. *Mol. Microbiol.*, **6**, 3343–3353.
8. Hammond, S.M., Bernstein, E., Beach, D., and Hannon, G.J. (2000) An RNA-directed nuclease mediates posttranscriptional gene silencing in *Drosophila* cells. *Nature*, **404**, 293–296.
9. Fire, A., Albertson, D., Harrison, S.W., and Moerman, D.G. (1991) Production of antisense RNA leads to effective and specific inhibition of gene expression in *C. elegans* muscle. *Development*, **113**, 503–514.
10. Dernburg, A.F., Zalevsky, J., Colaiacovo, M.P., and Villeneuve, A.M. (2000) Transgene-mediated cosuppression in the *C. elegans* germ line. *Genes Dev.*, **14**, 1578–1583.
11. Guo, S. and Kemphues, K.J. (1995) par-1, a gene required for establishing polarity in *C. elegans* embryos, encodes a putative Ser/Thr kinase that is asymmetrically distributed. *Cell*, **81**, 611–620.
12. de Carvalho, F., Gheysen, G., Kushnir, S., Van Montagu, M. *et al.* (1992) Suppression of beta 1,3 glucanase transgene expression in homozygous plants. *EMBO J.*, **11**, 2595–2602.
13. Waterhouse, P.M., Graham, M.W., and Wang, M.B. (1998) Virus resistance and gene silencing in plants can be induced by simultaneous expression of sense and antisense RNA. *Proc. Natl Acad. Sci. USA*, **A95** (23), 13959–13964.
14. Hamilton, A.J. and Baulcombe, D.C. (1999) A species of small antisense RNA in posttranscriptional gene silencing in plants. *Science*, **286**, 950–952.
15. Bernstein, E., Caudy, A.A., Hammond, S.M., and Hannon, G.J. (2001) Role for a bidentate ribonuclease in the initiation step of RNA interference. *Nature*, **409**, 363–366.
16. Reinhart, B.J., Slack, F.J., Basson, M., Pasquinelli, A.E. *et al.* (2000) The 21-nucleotide let-7 RNA regulates developmental timing in *Caenorhabditis elegans*. *Nature*, **403**, 901–906.
17. Lee, R.C., Feinbaum, R.L., and Ambros, V. (1993) The *C. elegans* heterochromic gene lin-4 encodes small RNAs with antisense complementarity to lin-14. *Cell*, **75**, 843–854.
18. Hutvágner, G., McLachlan, J., Pasquinelli, A.E., Bálint, E. *et al.* (2001) A cellular function for the RNA-interference enzyme Dicer in the maturation of the let-7 small temporal RNA. *Science*, **293**, 834–838.
19. Grishok, A., Pasquinelli, A.E., Conte, D., Li, N. *et al.* (2001) Genes and mechanisms related to RNA interference regulate expression of the small temporal RNAs that control *C. elegans* developmental timing. *Cell*, **106**, 23–34.
20. Volpe, T.A., Kidner, C., Hall, I.M., Teng, G. *et al.* (2002) Regulation of heterochromatic silencing and histone H3 lysine-9 methylation by RNAi. *Science*, **297**, 1833–1837.
21. Mochizuki, K., Fine, N.A., Fujisawa, T., and Gorovsky, M.A. (2002) Analysis of a piwi-related gene implicates small RNAs in genome rearrangement in *Tetrahymena*. *Cell*, **110**, 689–699.
22. Farazi, T.A., Spitzer, J.I., Morozov, P., and Tuschl, T. (2011) miRNAs in human cancer. *J. Pathol.*, **223** (2), 102–115.
23. Siolas, D., Lerner, C., Burchard, J., Ge, W. *et al.* (2005) Synthetic shRNAs as potent RNAi triggers. *Nat. Biotechnol.*, **23** (2), 227–231.
24. Carmell, M.A. and Hannon, G.J. (2004) RNase III enzymes and the initiation of gene silencing. *Nat. Struct. Mol. Biol.*, **11**, 214–218.
25. Kim, V.N. (2005) MicroRNA biogenesis: coordinated cropping and dicing. *Nat. Rev. Mol. Cell Biol.*, **6**, 376–385.
26. Wilson, R.C. and Doudna, J.A. (2013) Molecular mechanisms of RNA interference. *Annu. Rev. Biophys.*, **42**, 217–239.
27. Denli, A.M., Tops, B.B, Plasterk, R.H, Ketting, R.F. *et al.* (2004) Processing of

primary microRNAs by the Microprocessor complex. *Nature*, **432**, 231–235.
28. Han, J., Lee, Y., Yeom, K.H., Kim, Y.K. *et al.* (2004). The Drosha–DGCR8 complex in primary microRNA processing. *Genes Dev.*, **18**, 3016–3027.
29. Landthaler, M., Yalcin, A., and Tuschl, T. (2004) The human Di George syndrome critical region gene 8 and its *D. melanogaster* homolog are required for miRNA biogenesis. *Curr. Biol.*, **14**, 2162–2167.
30. Lund, E., Guttinger, S., Calado, A., Dahlberg, J.E. *et al.* (2004) Nuclear export of microRNA precursors. *Science*, **303**, 95–98.
31. Gregory, R.I., Chendrimada, T.P., Cooch, N., and Shiekhattar, R. (2005) Human RISC couples microRNA biogenesis and post-transcriptional gene silencing. *Cell*, **123**, 631–640.
32. Ambros, V., Lee, R.C., Lavanway, A., Williams, P.T. *et al.* (2003) MicroRNAs and other tiny endogenous RNAs in *C. elegans*. *Curr. Biol.*, **13**, 807–818.
33. Aravin, A.A., Lagos-Quintana, M., Yalcin, A., Zavolan, M. *et al.* (2003) The small RNA profile during *Drosophila melanogaster* development. *Dev. Cell*, **5**, 337–350.
34. Xie, Z., Johansen, L.K., Gustafson, A.M., Kasschau, K.D. *et al.* (2004) Genetic and functional diversification of small RNA pathways in plants. *PLoS Biol.*, **2**, e104.
35. Vazquez, F., Vaucheret, H., Rajagopalan, R., Lepers, C. *et al.* (2004) Endogenous trans-acting siRNAs regulate the accumulation of *Arabidopsis* mRNAs. *Mol. Cell*, **16**, 69–79.
36. Tomari, Y., Matranga, C., Haley, B., Martinez, N. *et al.* (2004). A protein sensor for siRNA asymmetry. *Science*, **306**, 1377–1380.
37. Carmell, M.A., Xuan, Z., Zhang, M.Q., and Hannon, G.J. (2002) The Argonaute family: tentacles that reach into RNAi, developmental control, stem cell maintenance, and tumorigenesis. *Genes Dev.*, **16**, 2733–2742.
38. Hutvagner, G. and Simard, M.J. (2008) Argonaute proteins: key players in RNA silencing. *Nat. Rev. Mol. Cell Biol.*, **9**, 22–32.
39. Liu, J., Carmell, M.A., Rivas, F.V., Marsden, C.G. *et al.* (2004) Argonaute2 is the catalytic engine of mammalian RNAi. *Science*, **305**, 1437–1441.
40. Caudy, A.A., Ketting, R.F., Hammond, S.M., Denli, A.M. *et al.* (2003) A micrococcal nuclease homologue in RNAi effector complexes. *Nature*, **425**, 411–414.
41. Ishizuka, A., Siomi, M.C., and Siomi, H.A. (2002) *Drosophila* fragile X protein interacts with components of RNAi and ribosomal proteins. *Genes Dev.*, **16**, 2497–2508.
42. Mourelatos, Z., Dostie, J., Paushkin, S., Sharma, A. *et al.* (2002) miRNPs: a novel class of ribonucleoproteins containing numerous microRNAs. *Genes Dev.*, **16**, 720–728.
43. Valencia-Sanchez, M.A., Liu, J., Hannon, G.J., and Parker, R. (2006) Control of translation and mRNA degradation by miRNAs and siRNAs. *Genes Dev.*, **20**, 515–524.
44. Friedman, R.C., Farh, K.K., Burge, C.B., and Bartel, D.P. (2009) Most mammalian mRNAs are conserved targets of microRNAs. *Genome Res.*, **19**, 92–105.
45. Bartel, D.P. (2004) MicroRNAs: genomics, biogenesis, mechanism, and function. *Cell*, **116**, 281–297.
46. Ørom, U.A., Nielsen, F.C., and Lund, A.H. (2008) MicroRNA-10a binds the 5'UTR of ribosomal protein mRNAs and enhances their translation. *Mol. Cell*, **30**, 460–471.
47. Lytle, J.R., Yario, T.A., and Steitz, J.A. (2007) Target mRNAs are repressed as efficiently by microRNA-binding sites in the 5′ UTR as in the 3′ UTR. *Proc. Natl Acad. Sci. USA*, **104**, 9667–9672.
48. Moretti, F., Thermann, R., and Hentze, M.W. (2010) Mechanism of translational regulation by miR-2 from sites in the 5′ untranslated region or the open reading frame. *RNA*, **16**, 2493–2502.
49. Qin, W. Shi, Y., Zhao, B., Yao, C. *et al.* (2010) miR-24 regulates apoptosis by targeting the open reading frame (ORF) region of FAF1 in cancer cells. *PLoS One*, **5**, e9429.
50. Vasudevan, S., Tong, Y., and Steitz, J.A. (2007) Switching from repression to activation: microRNAs can up-regulate translation. *Science*, **318**, 1931–1934.
51. Filipowicz, W., Bhattacharyya, S.N., and Sonenberg, N. (2008) Mechanisms of post-transcriptional regulation by microRNAs: are the answers in sight? *Nature*, **9**, 102–114.
52. Hwang, H.W., Wentzel, E.A., and Mendell, J.T. (2007) A hexanucleotide element directs

microRNA nuclear import. *Science*, **315**, 97–100.

53. Valadi, H., Ekström, K., Bossios, A., Sjöstrand, M. *et al.* (2007) Exosome-mediated transfer of mRNAs and microRNAs is a novel mechanism of genetic exchange between cells. *Nat. Cell Biol.*, **9**, 654–659.
54. Berindan-Neagoe, I., Moroig, P.d.C., Pasculli, B., and Calin, G.A. (2014) MicroRNAome genome: a treasure for Cancer Diagnosis and therapy. *CA Cancer J. Clin.*, **64** (5), 311–336.
55. Selcuklu, S.D., Donoghue, M.T., and Spillane, C. (2009) miR-21 as a key regulator of oncogenic processes. *Biochem. Soc. Trans.*, **37**, 918–925.
56. Medina, P.P., Nolde, M., and Slack, F.J. (2010) OncomiR addiction in an in vivo model of miRNA-21-induced pre-B-cell lymphoma. *Nature*, **467**, 86–90.
57. Calin, G.A., Dumitru, C.D., Shimizu, M., Bichi, R. *et al.* (2002) Frequent deletions and down-regulation of micro-RNA genes miR15 and miR16 at 13q14 in chronic lymphocytic leukemia. *Proc. Natl Acad. Sci. USA*, **99**, 15524–15529.
58. Cimmino, A., Calin, G.A., Fabbri, M., Iorio, M.V. *et al.* (2005) miR-15 and miR-16 induce apoptosis by targeting BCL2. *Proc. Natl Acad. Sci. USA*, **102**, 13944–13949.
59. Klein, U., Lia, M., Crespo, M., Siegel, R. *et al.* (2010) The DLEU2/miR-15a/16-1 cluster controls B cell proliferation and its deletion leads to chronic lymphocytic leukemia. *Cancer Cell*, **17**, 28–40.
60. Chen, R.W., Bemis, L.T., Amato, C.M., Myint, H. *et al.* (2008) Truncation of CCND1 mRNA alters miR-16-1 regulation in mantle cell lymphoma. *Blood*, **112**, 822–829.
61. Calin, G.A. and Croce, C.M. (2006) MicroRNA signatures in human cancers. *Nat. Rev. Cancer*, **6**, 857–866.
62. Calin, G.A., Ferracin, M., Cimmino, A., Di Leva, G. *et al.* (2005) A MiRNA signature associated with prognosis and progression in chronic lymphocytic leukemia. *N. Engl. J. Med.*, **353**, 1793–1801.
63. Iorio, M.V., Ferracin, M., Liu, C.G., Veronese, A. *et al.* (2005) MiRNA gene expression deregulation in human breast cancer. *Cancer Res.*, **65**, 7065–7070.
64. He, H., Jazdzewski, K., Li, W., Liyanarachchi, S. *et al.* (2005) The role of miRNA genes in papillary thyroid carcinoma. *Proc. Natl Acad. Sci. USA*, **102**, 19075–19080.
65. Yanaihara, N., Caplen, N., Bowman, E., Seike, M. *et al.* (2006) Unique miRNA molecular profiles in lung cancer diagnosis and prognosis. *Cancer Cell*, **9**, 189–198.
66. Ciafré, S.A., Galardi, S., Mangiola, A., Ferracin, M. *et al.* (2005) Extensive modulation of a set of miRNAs in primary glioblastoma. *Biochem. Biophys. Res. Commun.*, **334**, 1351–1358.
67. Roldo, C., Missiaglia, E., Hagan, J.P., Falconi, M. *et al.* (2006) MiRNA expression abnormalities in pancreatic endocrine and acinar tumors are associated with distinctive pathologic features and clinical behavior. *J. Clin. Oncol.*, **24**, 4677–4684.
68. Murakami, Y., Yasuda, T., Saigo, K., Urashima, T. *et al.* (2006) Comprehensive analysis of miRNA expression patterns in hepatocellular carcinoma and non-tumorous tissues. *Oncogene*, **25**, 2537–2545.
69. Porkka, K.P., Pfeiffer, M.J., Waltering, K.K., Vessella, R.L. *et al.* (2007) MiRNA expression profiling in prostate cancer. *Cancer Res.*, **67**, 6130–6135.
70. Chen, W., Tang, Z., Sun, Y., Zhang, Y. *et al.* (2012) miRNA expression profile in primary gastric cancers and paired lymph node metastases indicates that miR-10a plays a role in metastasis from primary gastric cancer to lymph nodes. *Exp. Ther. Med.*, **3**, 351–356.
71. Di Leva, G., Garofalo, M., and Croce, C.M. (2014) MicroRNAs in cancer. *Annu. Rev. Pathol.*, **9**, 287–314.
72. Lee, E.J., Baek, M., Gusev, Y., Brackett, D.J *et al.* (2008) Systematic evaluation of microRNA processing patterns in tissues, cell lines, and tumors. *RNA*, **14**, 35–42.
73. Esquela-Kerscher, A. and Slack, F.J. (2006) Oncomirs: miRNAs with a role in cancer. *Nat. Rev. Cancer*, **6**, 259–269.
74. Petrocca, F. and Lieberman, J. (2011) Promise and challenge of RNA interference-based therapy for cancer. *J. Clin. Oncol.*, **29** (6), 747–754.
75. O'Donnell, K.A., Wentzel, E.A., Zeller, K.I., Dang, C.V. *et al.* (2005) c-Myc-regulated microRNAs modulate E2F1 expression. *Nature*, **435**, 839–843.

76. Lopez-Serra, P. and Esteller, M. (2012) DNA methylation-associated silencing of tumor suppressor microRNAs in cancer. *Oncogene*, **31**, 1609–1622.
77. Weber, B., Stresemann, C., Brueckner, B., and Lyko, F. (2007) Methylation of human microRNA genes in normal and neoplastic cells. *Cell Cycle*, **6**, 1001–1005.
78. Saito, Y., Liang, G., Egger, G., Friedman, J.M. *et al.* (2006) Specific activation of microRNA-127 with downregulation of the proto-oncogene BCL6 by chromatin-modifying drugs in human cancer cells. *Cancer Cell*, **9**, 435–443.
79. Lehmann, U., Hasemeier, B., Christgen, M., Müller, M. *et al.* (2008) Epigenetic inactivation of microRNA gene has-miR- 9-1 in human breast cancer. *J. Pathol.*, **214**, 17–24.
80. Toyota, M. Suzuki, H., Sasaki, Y., Maruyama, R. *et al.* (2008) Epigenetic silencing of microRNA-34b/c and B-cell translocation gene 4 is associated with CpG island methylation in colorectal cancer. *Cancer Res.*, **68**, 4123–4132.
81. Lujambio, A., Ropero, S., Ballestar, E., Fraga, M.F. *et al.* (2007) Genetic unmasking of an epigenetically silenced microRNA in human cancer cells. *Cancer Res.*, **67**, 1424–1429.
82. Iorio, M.V., Visone, R., Di Leva, G., Donati, V. *et al.* (2007) MicroRNA signatures in human ovarian cancer. *Cancer Res.*, **67**, 8699–8707.
83. Scott, G.K., Mattie, M.D., Berger, C.E., Benz, S.C. *et al.* (2006) Rapid alteration of microRNA levels by histone deacetylase inhibition. *Cancer Res.*, **66**, 1277–1281.
84. Saito, Y., Suzuki, H., Tsugawa, H., Nakagawa, I. *et al.* (2009) Chromatin remodeling at Alu repeats by epigenetic treatment activates silenced microRNA-512-5p with downregulation of Mcl-1 in human gastric cancer cells. *Oncogene*, **28**, 2738–2744.
85. Rhodes, L.V., Nitschke, A.M., and Collins-Burow, B.M. (2012) The histone deacetylase inhibitor trichostatin A alters microRNA expression profiles in apoptosis-resistant breast cancer cells. *Oncol. Rep.*, **27**, 10–16.
86. Raveche, E.S., Salerno, E., Scaglione, B.J., Manohar, V. *et al.* (2007) Abnormal microRNA-16 locus with synteny to human 13q14 linked to CLL in NZB mice. *Blood*, **109**, 5079–5086.
87. Hu, Z., Liang, J., Wang, Z., Tian, T. *et al.* (2009) Common genetic variants in pre-microRNAs were associated with increased risk of breast cancer in Chinese women. *Hum. Mutat.*, **30**, 79–84.
88. Jazdzewski, K., Murray, E.L., Franssila, K., Jarzab, B. *et al.* (2008) Common SNP in pre-miR-146a decreases mature miR expression and predisposes to papillary thyroid carcinoma. *Proc. Natl Acad. Sci. USA*, **105**, 7269–7274.
89. Hu, Z., Chen, J., Tian, T., Zhou, X. *et al.* (2008) Genetic variants of miRNA sequences and non-small cell lung cancer survival. *J. Clin. Invest.*, **118**, 2600–2608.
90. Mishra, P.J., Humeniuk, R., Longo-Sorbello, G.S., Banerjee, D. *et al.* (2007) A miR-24 microRNA binding-site polymorphism in dihydrofolate reductase gene leads to methotrexate resistance. *Proc. Natl Acad. Sci. USA*, **104**, 13513–13518.
91. Tan, Z., Randall, G., Fan, J., Camoretti-Mercado, B. *et al.* (2007) Allele-specific targeting of microRNAs to HLA-G and risk of asthma. *Am. J. Hum. Genet.*, **81**, 829–834.
92. Karube, Y., Tanaka, H, Osada, H., Tomida, S. *et al.* (2005) Reduced expression of Dicer associated with poor prognosis in lung cancer patients. *Cancer Sci.*, **96**, 111–115.
93. Merritt, W.M., Lin, Y.G., Han, L.Y., Kamat, A.A. *et al.* (2008) Dicer, Drosha, and outcomes in patients with ovarian cancer. *N. Engl. J. Med.*, **359**, 2641–2650.
94. Guo, X., Liao, Q., Chen, P., Li, X. *et al.* (2012) The microRNA-processing enzymes: Drosha and Dicer can predict prognosis of nasopharyngeal carcinoma. *J. Cancer Res. Clin. Oncol.*, **138**, 49–56.
95. Lin, R.J., Lin, Y.C., Chen, J., Kuo, H.H. *et al.* (2010) microRNA signature and expression of Dicer and Drosha can predict prognosis and delineate risk groups in neuroblastoma. *Cancer Res.*, **70**, 7841–7850.
96. Grelier, G., Voirin, N., Ay, A.S., Cox, D.G. *et al.* (2009) Prognostic value of Dicer expression in human breast cancers and association with the mesenchymal phenotype. *Br. J. Cancer*, **101**, 673–683.
97. Faber, C., Jelezcova, E., Chandran, U., Acquafondata, M. *et al.* (2011) Overexpression of Dicer predicts poor survival

in colorectal cancer. *Eur. J. Cancer*, **47**, 1414–1419.
98. Chiosea, S., Jelezcova, E., Chandran, U., Acquafondata, M. *et al.* (2006) Up-regulation of dicer, a component of the MicroRNA machinery, in prostate adenocarcinoma. *Am. J. Pathol.*, **169**, 1812–1820.
99. Muralidhar, B., Winder, D., Murray, M., Palmer, R. *et al.* (2011) Functional evidence that Drosha overexpression in cervical squamous cell carcinoma affects cell phenotype and microRNA profiles. *J. Pathol.*, **224**, 496–507.
100. Martello, G., Rosato, A., Ferrari, F., Manfrin, A. *et al.* (2010) A MicroRNA targeting dicer for metastasis control. *Cell*, **141**, 1195–1207.
101. Lambertz, I., Nittner, D., Mestdagh, P., Denecker, G. *et al.* (2010) Monoallelic but not biallelic loss of Dicer1 promotes tumorigenesis in vivo. *Cell Death Differ.*, **17**, 633–641.
102. Heravi-Moussavi, A., Anglesio, M.S., Cheng, S.W., Senz, J. *et al.* (2012) Recurrent somatic DICER1 mutations in nonepithelial ovarian cancers. *N. Engl. J. Med.*, **366**, 234–242.
103. Melo, S.A., Ropero, S., Moutinho, C., Aaltonen, L.A. *et al.* (2009) A TARBP2 mutation in human cancer impairs microRNA processing and DICER1 function. *Nat. Genet.*, **41**, 365–370.
104. Melo, S.A., Moutinho, C., Ropero, S., Calin, G.A. *et al.* (2010) A genetic defect in Exportin-5 traps precursor microRNAs in the nucleus of cancer cells. *Cancer Cell*, **18**, 303–315.
105. Adams, B.D., Kasinski, A.L., and Slack, F.J. (2014) Aberrant regulation and function of MicroRNAs in cancer. *Curr. Biol.*, **24**, R762–R776.
106. Tang, R., Li, L., Zhu, D., Hou, D. *et al.* (2012) Mouse miRNA-709 directly regulates miRNA- 15a/16-1 biogenesis at the posttranscriptional level in the nucleus: evidence for a microRNA hierarchy system. *Cell Res.*, **22**, 504–515.
107. Chang, T.C., Wentzel, E.A., Kent, O.A., Ramachandran, K. *et al.* (2007) Transactivation of miR-34a by p53 broadly influences gene expression and promotes apoptosis. *Mol. Cell*, **26**, 745–752.
108. Piovan, C., Palmieri, D., Di Leva, G., Braccioli, L. *et al.* (2012) Oncosuppressive role of p53-induced miR-205 in triple negative breast cancer. *Mol. Oncol.*, **6** (4), 458–472.
109. Chang, T.C., Zeitels, L.R., Hwang, H.W., Chivukula, R.R. *et al.* (2009) Lin-28B transactivation is necessary for Myc mediated let-7 repression and proliferation. *Proc. Natl Acad. Sci. USA*, **106**, 3384–3389.
110. Mott, J.L., Kurita, S., Cazanave, S.C., Bronk, S.F. *et al.* (2010) Transcriptional suppression of mir-29b-1/mir-29a promoter by c-Myc, hedgehog, and NF-kappaB. *J. Cell. Biochem.*, **110**, 1155–1164.
111. Kulshretha, R., Davuluri, R.V., Calin, G.A., and Ivan, M. (2006) A microRNA component of the hypoxic response. *Cell Death Differ.*, **15**, 667–671.
112. Burk, U., Schubert, J., Wellner, U., Schmalhofer, O. *et al.* (2008) A reciprocal repression between ZEB1 and members of the miR-200 family promotes EMT and invasion in cancer cells. *EMBO Rep.*, **9**, 582–589.
113. Felli, N., Fontana, L., Pelosi, E., Botta, R. *et al.* (2005) MicroRNAs 221 and 222 inhibit normal erythropoiesis and erythroleukemic cell growth via kit receptor down-modulation. *Proc. Natl Acad. Sci. USA*, **102**, 18081–18086.
114. Garofalo, M., Quintavalle, C., Romano, G., Croce, C.M. *et al.* (2012) miR221/222 in cancer: their role in tumor progression and response to therapy. *Curr. Mol. Med.*, **12**, 27–33
115. Babar, I.A., Czochor, J., Steinmetz, A., Weidhaas, J.B. *et al.* (2011) Inhibition of hypoxia-induced miR-155 radiosensitizes hypoxic lung cancer cells. *Cancer Biol. Ther.*, **12**, 908–914.
116. Kong, W., He, L., Coppola, M., Guo, J. *et al.* (2010) MicroRNA-155 regulates cell survival, growth, and chemosensitivity by targeting FOXO3a in breast cancer. *J. Biol. Chem.*, **285**, 17869–17879.
117. Kong, W., Yang, H., He, L., Zhao, J.J. *et al.* (2008) MicroRNA-155 is regulated by the transforming growth factor beta/Smad pathway and contributes to epithelial cell plasticity by targeting RhoA. *Mol. Cell Biol.*, **28**, 6773–6784.
118. Mendell, J.T. (2008) miRiad roles for the miR-17-92 cluster in development and disease. *Cell*, **133**, 217–222.

119. Volinia, S., Calin, G.A., Liu, C.G., Ambs, S. *et al.* (2006) A microRNA expression signature of human solid tumors defines cancer gene targets. *Proc. Natl Acad. Sci. USA*, **103**, 2257–2261.
120. Taguchi, A., Yanagisawa, K., Tanaka, M., Cao, K. *et al.* (2008) Identification of hypoxia-inducible factor-1 alpha as a novel target for miR-17-92 microRNA cluster. *Cancer Res.*, **68**, 5540–5545.
121. Dews, M., Homayouni, A., Yu, D., Murphy, D. *et al.* (2006) Augmentation of tumor angiogenesis by a Myc-activated microRNA cluster. *Nat. Genet.*, **38**, 1060–1065.
122. Petrocca, F., Visone, R., Onelli, M.R., Shah, M.H. *et al.* (2008) E2F1-regulated microRNAs impair TGFβ-dependent cell-cycle arrest and apoptosis in gastric cancer. *Cancer Cell*, **13**, 272–286.
123. Fornari, F., Gramantieri, L., Ferracin, M., Veronese, A. *et al.* (2008) miR-221 controls CDKN1C/p57 and CDKN1B/p27 expression in human hepatocellular carcinoma. *Oncogene*, **27**, 5651–5661.
124. Pallante, P., Visone, R., Ferracin, M., Ferraro, A. *et al.* (2006) MicroRNA deregulation in human thyroid papillary carcinomas. *Endocr. Relat. Cancer*, **13**, 497–508.
125. Felicetti, F., Errico, M.C., Bottero, L., Segnalini, P. *et al.* (2008) The promyelocytic leukemia zinc finger–microRNA-221/-222 pathway controls melanoma progression through multiple oncogenic mechanisms. *Cancer Res.*, **68**, 2745–2754.
126. Di Leva, G., Gasparini, P., Piovan, C., Ngankeu, A. *et al.* (2010) MicroRNA cluster 221-222 and estrogen receptor αinteractions in breast cancer. *J. Natl Cancer Inst.*, **102**, 706–721.
127. Conti, A., Aguennouz, M., La Torre, D., Tomasello, C. *et al.* (2009) miR-21 and 221 upregulation and miR-181b downregulation in human grade II-IV astrocytic tumors. *J. Neurooncol.*, **93**, 325–332.
128. Garofalo, M., Di Leva, G., Romano, G., Nuovo, G. *et al.* (2009) miR-221&222 regulate TRAIL resistance and enhance tumorigenicity through PTEN and TIMP3 downregulation. *Cancer Cell*, **16**, 498–509.
129. Stinson, S., Lackner, M.R., Adai, A.T., Yu, N. *et al.* (2011) miR-221/222 targeting of trichorhinophalangeal (TRPS1) promotes epithelial-to-mesenchymal transition in breast cancer. *Sci. Signal.*, **4**, ra41.
130. Garofalo, M., Romano, G., Di Leva, G., Nuovo, G. *et al.* (2011) EGFR and MET receptor tyrosine kinase-altered microRNA expression induces tumorigenesis and gefitinib resistance in lung cancers. *Nat. Med.*, **18**, 74–82.
131. Zhang, J.G., Wang, J.J., Zhao, F., Liu, Q. *et al.* (2010) MicroRNA-21 (miR-21) represses tumor suppressor PTEN and promotes growth and invasion in non-small cell lung cancer (NSCLC). *Clin. Chim. Acta*, **411**, 846–852.
132. Zhu, S., Si, M.L., Wu, H., and Mo, Y.Y. (2007) MicroRNA-21 targets the tumor suppressor gene tropomyosin 1 (TPM1). *J. Biol. Chem.*, **282**, 14328–14336.
133. Hatley, M.E., Patrick, D.M., Garcia, M.R., Richardson, J.A. *et al.* (2010) Modulation of K-Ras-dependent lung tumorigenesis by MicroRNA-21. *Cancer Cell*, **18**, 282–293.
134. Gabriely, G., Wurdinger, T., Kesari, S., Esau, C.C. *et al.* (2008) MicroRNA 21 promotes glioma invasion by targeting matrix metalloproteinase regulators. *Mol. Cell Biol.*, **28**, 5369–5380.
135. Seike, M., Goto, A., Okano, T., Bowman, E.D. *et al.* (2009) MiR-21 is an EGFR-regulated anti-apoptotic factor in lung cancer in neversmokers. *Proc. Natl Acad. Sci. USA*, **106**, 12085–12090.
136. Chan, J.A., Krichevsky, A.M., and Kosik, K.S. (2005) MicroRNA-21 is an antiapoptotic factor in human glioblastoma cells. *Cancer Res.*, **65**, 6029–6033.
137. Ambs, S., Prueitt, R.L., Yi, M., Hudson, R.S. *et al.* (2008) Genomic profiling of microRNA and messenger RNA reveals deregulated microRNA expression in prostate cancer. *Cancer Res.*, **68**, 6162–6170.
138. Bandi, N., Zbinden, S., Gugger, M., Arnold, M. *et al.* (2009) miR-15a and miR-16 are implicated in cell cycle regulation in a Rb-dependent manner and are frequently deleted or down-regulated in non-small cell lung cancer. *Cancer Res.*, **69**, 5553–5559.
139. Roccaro, A.M., Sacco, A., Thompson, B., Leleu, X. *et al.* (2009) MicroRNAs 15a and 16 regulate tumor proliferation in multiple myeloma. *Blood*, **26**, 6669–6680.
140. Pouliot, L.M., Chen, Y.C., Bai, J., Guha, R. *et al.* (2012) Cisplatin sensitivity mediated

by WEE1 and CHK1 is mediated by miR-155 and the miR-15 family. *Cancer Res.*, **72**, 5945–5955.

141. Johnson, S.M., Grosshans, H., Shingara, J., Byrom, M. *et al.* (2005) RAS is regulated by the let-7 microRNA family. *Cell*, **120**, 635–647.

142. Chin, L.J., Ratner, E., Leng, S., Zhai, R. *et al.* (2008) A SNP in a let-7 microRNA complementary site in the KRAS 3′ untranslated region increases non-small cell lung cancer risk. *Cancer Res.*, **68**, 8535–8540.

143. Paranjape, T., Heneghan, H., Lindner, R., Keane, F.K. *et al.* (2011) A 3′-untranslated region KRAS variant and triple-negative breast cancer: a case-control and genetic analysis. *Lancet Oncol.*, **12**, 377–386.

144. Sampson, V.B., Rong, N.H., Han, J., Yang, Q. *et al.* (2007) MicroRNA let-7a down-regulates MYC and reverts MYC-induced growth in Burkitt lymphoma cells. *Cancer Res.*, **67**, 9762–9770.

145. Johnson, C.D., Esquela-Kerscher, A., Stefani, G., Byrom, M. *et al.* (2007) The let-7 microRNA represses cell proliferation pathways in human cells. *Cancer Res.*, **67**, 7713–7722.

146. Corney, D.C., Flesken-Nikitin, A., Godwin, A.K., Wang, W. *et al.* (2007) MicroRNA-34b and microRNA-34c are targets of p53 and cooperate in control of cell proliferation and adhesion-independent growth. *Cancer Res.*, **67**, 8433–8438.

147. Gregory, P.A., Bert, A.G., Paterson, E.L., Barry, S.C. *et al.* (2008) The miR-200 family and miR-205 regulate epithelial to mesenchymal transition by targeting ZEB1 and SIP1. *Nat. Cell Biol.*, **10**, 593–601.

148. Hurteau, G.J., Carlson, J.A., Spivack, S.D., and Brock, G.J. (2007) Overexpression of the microRNA hsa-miR-200c leads to reduced expression of transcription factor 8 and increased expression of E-cadherin. *Cancer Res.*, **67**, 7972–7976.

149. Bracken, C.P., Gregory, P.A., Kolesnikoff, N., Bert *et al.* (2008) A double-negative feedback loop between ZEB1-SIP1 and the microRNA-200 family regulates epithelial-mesenchymal transition. *Cancer Res.*, **68**, 7846–7854.

150. Iliopoulos, D., Polytarchou, C., Hatziapostolou, M., Kottakis, F. *et al.* (2009) MicroRNAs differentially regulated by Akt isoforms control EMT and stem cell renewal in cancer cells. *Sci. Signal.*, **2**, ra62.

151. Vrba, L., Jensen, T.J., Garbe, J.C., Heimark, R.L. *et al.* (2010) Role for DNA methylation in the regulation of miR-200c and miR-141 expression in normal and cancer cells. *PLoS One*, **5**, e8697.

152. Shen, J., Samul, R., Silva, R.L., Akiyama, H. *et al.* (2006) Suppression of ocular neovascularization with siRNA targeting VEGF receptor 1. *Gene Ther.*, **13** (3), 225–234.

153. Filleur, S., Courtin, A., Ait-Si-Ali, S., Guglielmi, J. *et al.* (2003) SiRNA-mediated inhibition of vascular endothelial growth factor severely limits tumor resistance to antiangiogenic thrombospondin-1 and slows tumor vascularization and growth. *Cancer Res.*, **63** (14), 3919–3922.

154. Gaudilliere, B., Shi, Y., and Bonni, A. (2002) RNA interference reveals a requirement for myocyte enhancer factor 2A in activity-dependent neuronal survival. *J. Biol. Chem.*, **277** (48), 46442–46446.

155. Miller, V.M., Xia, H., Marrs, G.L., Gouvion, C.M. *et al.* (2003) Allele-specific silencing of dominant disease genes. *Proc. Natl Acad. Sci. USA*, **100** (12), 7195–7200.

156. Wu, S.Y., Lopez-Berestein, G., Calin, G.A., and Sood, A.K. (2014) RNAi therapies: drugging the undruggable. *Sci. Transl. Med.*, **6** (240), 240ps7.

157. Rao, D.D., Vorhies, J.S., Senzer, N., and Nemunaitis, J. (2009) siRNA vs. shRNA: similarities and differences. *Adv. Drug Deliv. Rev.*, **61**, 746–759.

158. Lima, W.F., Prakash, T.P., Murray, H.M., Kinberger, G.A. *et al.* (2012) Single-stranded siRNAs activate RNAi in animals. *Cell*, **150**, 883–894.

159. Maples, P.B., Senzer, N., Kumar, P., Wang, Z. *et al.* (2010) Enhanced target gene knockdown by a bifunctional shRNA: a novel approach of RNA interference. *Cancer Gene Ther.*, **17** (11), 780–791.

160. Garzon, R., Marcucci, G., and Croce, C.M. (2010) Targeting microRNAs in cancer: rationale, strategies and challenges. *Nat. Rev. Drug Discov.*, **9** (10), 775–789.

161. Kota, J., Chivukula, R.R., O'Donnell, K.A., Wentzel, E.A. *et al.* (2009) Therapeutic microRNA delivery suppresses tumorigenesis in a murine liver cancer model. *Cell*, **137**, 1005–1017.

162. Stenvang, J., Petri, A., Lindow, M., Obad, S. *et al.* (2012) Inhibition of microRNA function by antimiR oligonucleotides. *Silence*, **3** (1), 1.
163. Brown, B.D. and Naldini, L. (2009) Exploiting and antagonizing microRNA regulation for therapeutic and experimental applications. *Nat. Rev.*, **10** (8), 578–585.
164. Choi, W.Y., Giraldez, A.J., and Schier, A.F. (2007) Target protectors reveal dampening and balancing of Nodal agonist and antagonist by miR-430. *Science*, **318**, 271–274.
165. Karlas, A., Machuy, N., Shin, Y., Pleissner, K.P. *et al.* (2010) Genome-wide RNAi screen identifies human host factors crucial for influenza virus replication. *Nature*, **463** (7282), 818–822.
166. Krishnan, M.N., Ng, A., Sukumaran, B., Gilfoy, F.D. *et al.* (2008) RNA interference screen for human genes associated with West Nile virus infection. *Nature*, **455** (7210), 242–245.
167. Tiedemann, R.E., Zhu, Y.X., Schmidt, J., Shi, C.X. *et al.* (2012) Identification of molecular vulnerabilities in human multiple myeloma cells by RNA interference lethality screening of the druggable genome. *Cancer Res.*, **72** (3), 757–768.
168. Steckel, M., Molina-Arcas, M., Weigelt, B., Marani, M. *et al.* (2012) Determination of synthetic lethal interactions in KRAS oncogene-dependent cancer cells reveals novel therapeutic targeting strategies. *Cell Res.*, **22**, 1227–1245.
169. Etemadmoghadam, D., Weir, B.A., Au-Yeung, G., Alsop, K. *et al.*, Australian Ovarian Cancer Study Group (2013) Synthetic lethality between CCNE1 amplification and loss of BRCA1. *Proc. Natl Acad. Sci. USA*, **110**, 19489–19494.
170. Phalon, C., Rao, D.D., and Nemunaitis, J. (2010) Potential use of RNA interference in cancer therapy. *Expert Rev. Mol. Med.*, **12**, e26.
171. Daka, A. and Peer, D. (2012) RNAi-based nanomedicines for targeted personalized therapy. *Adv. Drug Deliv. Rev.*, **64**, 1508–1521.
172. Niu, X.Y., Peng, Z.L., Duan, W.Q., Wang, H. *et al.* (2006) Inhibition of HPV 16 E6 oncogene expression by RNA interference in vitro and in vivo. *Int. J. Gynecol. Cancer*, **16** (2), 743–751.
173. Guo, P., Coban, O., Snead, N.M., Trebley, J. *et al.* (2010) Engineering RNA for targeted siRNA delivery and medical application. *Adv. Drug Deliv. Rev.*, **62**, 650–666.
174. Chiu, Y.L. and Rana, T.M. (2003) siRNA function in RNAi: a chemical modification analysis. *RNA*, **9** (9), 1034–1048.
175. Morrissey, D.V., Lockridge, J.A., Shaw, L., Blanchard, K. *et al.* (2005) Potent and persistent in vivo anti-HBV activity of chemically modified siRNAs. *Nat. Biotechnol.*, **23** (8), 1002–1007.
176. Morrissey, D.V., Blanchard, K., Shaw, L., Jensen, K. *et al.* (2005) Activity of stabilized short interfering RNA in a mouse model of hepatitis B virus replication. *Hepatology*, **41** (6), 1349–1356.
177. Elmen, J., Thonberg, H., Ljungberg, K., Frieden, M. *et al.* (2005) Locked nucleic acid (LNA) mediated improvements in siRNA stability and functionality. *Nucleic Acids Res.*, **33** (1), 439–447.
178. Bumcrot, D., Manoharan, M., Koteliansky, V., and Sah, D.W. (2006) RNAi therapeutics: a potential new class of pharmaceutical drugs. *Nat. Chem. Biol.*, **2** (12), 711–719.
179. Krutzfeldt, J., Rajewsky, N., Braich, R., Rajeev, K.G *et al.* (2005) Silencing of microRNAs in vivo with 'antagomiRs'. *Nature*, **438**, 685–689.
180. Jackson, A.L., Burchard, J., Schelter, J., Chau, B.N. *et al.* (2006) Widespread siRNA "off-target" transcript silencing mediated by seed region sequence complementarity. *RNA*, **12** (7), 1179–1187.
181. Doench, J.G. and Sharp, P.A. (2004) Specificity of microRNA target selection in translational repression. *Genes Dev.*, **18** (5), 504–511.
182. Dykxhoorn, D.M. and Lieberman, J. (2006) Knocking down disease with siRNAs. *Cell*, **126** (2), 231–235.
183. Grimm, D., Streetz, K.L., Jopling, C.L., Storm, T.A. *et al.* (2006) Fatality in mice due to oversaturation of cellular microRNA/short hairpin RNA pathways. *Nature*, **441** (7092), 537–541.
184. Wang, Z., Rao, D.D., Senzer, N., and Nemunaitis, J. (2011) RNA interference and cancer therapy. *Pharm. Res.*, **28**, 2983–2995.
185. Persengiev, S.P., Zhu, X.C., and Green, M.R. (2004) Nonspecific, concentration-dependent stimulation and repression of mammalian gene expression by small

interfering RNAs (siRNAs). *RNA*, **10** (1), 12–18.

186. Jackson, A.L., Bartz, S.R., Schelter, J., Kobayashi, S.V. *et al.* (2003) Expression profiling reveals off-target gene regulation by RNAi. *Nat. Biotechnol.*, **21** (6), 635–637.

187. Haussecker, D. (2014) Current issues of RNAi therapeutics delivery and development. *J. Control. Release*, **195**, 49–54.

188. Agrawal, S. and Kandimalla, E.R. (2004) Antisense and siRNA as agonists of Toll-like receptors. *Nat. Biotechnol.*, **22** (12), 1533–1537.

189. Sledz, C.A., Holko, M., de Veer, M.J., Silverman, R.H. *et al.* (2003) Activation of the interferon system by short-interfering RNAs. *Nat. Cell Biol.*, **5** (9), 834–839.

190. Kariko, K., Bhuyan, P., Capodici, J., and Weissman, D. (2004) Small interfering RNAs mediate sequence-independent gene suppression and induce immune activation by signaling through toll-like receptor. *J. Immunol.*, **172** (11), 6545–6549.

191. Judge, A.D., Sood, V., Shaw, J.R., Fang, D. *et al.* (2005) Sequence-dependent stimulation of the mammalian innate immune response by synthetic siRNA. *Nat. Biotechnol.*, **23** (4), 457–462.

192. Hornung, V., Nemorin, J.G., Montino, C., Müller, C. *et al.* (2005) Sequence-specific potent induction of IFN-alpha by short interfering RNA in plasmacytoid dendritic cells through TLR7. *Nat. Med.*, **11** (3), 263–270.

193. Forsbach, A., Nemorin, J.G., Montino, C., Müller, C. *et al.* (2008) Identification of RNA sequence motifs stimulating sequence-specific TLR8-dependent immune responses. *J. Immunol.*, **180** (6), 3729–3738.

194. Bramsen, J.B. and Kjems, J. (2011) Chemical modification of small interfering RNA. *Methods Mol. Biol.*, **721**, 77–103.

195. Soutschek, J., Akinc, A., Bramlage, B., Charisse, K. *et al.* (2004) Therapeutic silencing of an endogenous gene by systemic administration of modified siRNAs. *Nature*, **432**, 173–178.

196. Di Figlia, M., Sena-Esteves, M., Chase, K., Sapp, E. *et al.* (2007) Therapeutic silencing of mutant huntingtin with siRNA attenuates striatal and cortical neuropathology and behavioral deficits. *Proc. Natl Acad. Sci. USA*, **104**, 17204–17209.

197. Iversen, F., Yang, C.X., Dagnaes-Hansen, F., Schaffert, D.H. *et al.* (2013) Optimized siRNA–PEG conjugates for extended blood circulation and reduced urine excretion in mice. *Theranostics*, **3**, 201–209.

198. Hagiwara, K., Ochiya, T., and Kosaka, N. (2014) A paradigm shift for extracellular vesicles as small RNA carriers: from cellular waste elimination to therapeutic applications. *Drug Deliv. Transl. Res.*, **4** (1), 31–37.

199. György, B., Szabó, T.G., Pásztói, M., Pál, Z. *et al.* (2011) Membrane vesicles, current state-of-the-art: emerging role of extracellular vesicles. *Cell. Mol. Life Sci.*, **68** (16), 2667–2688.

200. Adler, H.I., Fisher, W.D., Cohen, A., and Hardigree, A.A. (1967) Miniature *Escherichia coli* cells deficient in DNA. *Proc. Natl Acad. Sci. USA*, **57**, 321–326.

201. MacDiarmid, J.A., Mugridge, N.B., Weiss, J.C., Phillips, L. *et al.* (2007) Bacterially derived 400 nm particles for encapsulation and cancer cell targeting of chemotherapeutics. *Cancer Cell*, **11**, 431–445.

202. Deng, Y., Wang, C.C., Choy, K.W., Du, Q. *et al.* (2014) Therapeutic potentials of gene silencing by RNA interference: principles, challenges, and new strategies. *Gene*, **538**, 217–227.

203. Wu, S.Y., Yang, X., Gharpure, K.M., Hatakeyama, H. *et al.* (2014) 2′-OMe-phosphorodithioato-modified siRNAs show increased loading into the RISC complex and enhanced anti-tumour activity. *Nat. Commun.*, **5**, 3459.

204. Wilson, J.A., Jayasena, S., Khvorova, A., Sabatinos, S. *et al.* (2003) RNA interference blocks gene expression and RNA synthesis from hepatitis C replicons propagated in human liver cells. *Proc. Natl Acad. Sci. USA*, **100**, 2783–2788.

205. Paganin-Gioanni, A., Bellard, E., Escoffre, J.M., Rols, M.P. *et al.* (2011) Direct visualization at the single-cell level of siRNA electrotransfer into cancer cells. *Proc. Natl Acad. Sci. USA*, **108**, 10443–10447.

206. Chabot, S., Orio, J., Castanier, R., Bellard, E. *et al.* (2012) LNA-based oligonucleotide electrotransfer for miRNA inhibition. *Mol. Ther.*, **20**, 1590–1598.

207. Chabot, S., Teissié, J., and Golzio, M. (2015) Targeted electro-delivery of oligonucleotides

for RNA interference: siRNA and antimiR. *Adv. Drug Deliv. Rev.*, **81**, 161–168.
208. Zhou, Y., Zhang, C., and Liang, W. (2014) Development of RNAi technology for targeted therapy – A track of siRNA based agents to RNAi therapeutics. *J. Control. Release*, **10** (193), 270–281.
209. Bamrungsap, S., Zhao, Z., Chen, T., Wang, L. *et al.* (2012) Nanotechnology in therapeutics: a focus on nanoparticles as a drug delivery system. *Nanomedicine (Lond.)*, 7 (8), 1253–1271.
210. Song, E., Zhu, P., Lee, S.K., Chowdhury, D. *et al.* (2005) Antibody mediated in vivo delivery of small interfering RNAs via cell-surface receptors. *Nat. Biotechnol.*, **23** (6), 709–717.
211. Arap, W., Pasqualini, R., and Ruoslahti, E. (1998) Cancer treatment by targeted drug delivery to tumor vasculature in a mouse model. *Science*, **279** (5349), 377–380.
212. Wu, Y., Sefah, K., Liu, H., Wang, R. *et al.* (2010) DNA aptamer-micelle as an efficient detection/delivery vehicle toward cancer cells. *Proc. Natl Acad. Sci. USA*, **107**, 5–10.
213. Leamon, C.P. and Reddy, J.A. (2004) Folate-targeted chemotherapy. *Adv. Drug Deliv. Rev.*, **56**, 1127–1141.
214. Yao, Y.D., Sun, T.M., Huang, S.Y., Dou, S. *et al.* (2012) Targeted delivery of PLK1–siRNA by ScFv suppresses Her2+ breast cancer growth and metastasis. *Sci. Transl. Med.*, **4**, 130ra148.
215. McNamara, J.O. II, Andrechek, E.R., Wang, Y., Viles, K.D. *et al.* (2006) Cell type-specific delivery of siRNAs with aptamer-siRNA chimeras. *Nat. Biotechnol.*, **24** (8), 1005–1015.
216. Jones, A.T. and Sayers, E.J. (2012) Cell entry of cell penetrating peptides: tales of tails wagging dogs. *J. Control. Release*, **161**, 582–591.
217. Gupta, B. and Torchilin, V.P. (2006) Trans-activating transcriptional activator-mediated drug delivery. *Expert Opin. Drug Deliv.*, **3** 177–190.
218. Lorenz, C., Hadwiger, P., John, M., Vornlocher, H.P. *et al.* (2004) Steroid and lipid conjugates of siRNAs to enhance cellular uptake and gene silencing in liver cells. *Bioorg. Med. Chem. Lett.*, **14**, 4975–4977.
219. Yu, B., Zhao, X., Lee, L.J., and Lee, R.J. (2009) Targeted delivery systems for oligonucleotide therapeutics. *AAPS J.*, **11**, 195–203.
220. Wilner, S.E., Wengerter, B., Maier, K., Borba Magalhães, M.L. *et al.* (2012) An RNA alternative to human transferrin: a new tool for targeting human cells. *Mol. Ther. Nucleic Acids*, **1**, e21.
221. Scherer, F., Anton, M., Schillinger, U., Henke, J. *et al.* (2002) Magnetofection: enhancing and targeting gene delivery by magnetic force in vitro and in vivo. *Gene Ther.*, **9**, 102–109.
222. Medarova, Z., Pham, W., Farrar, C., Petkova, V. *et al.* (2007) In vivo imaging of siRNA delivery and silencing in tumors. *Nat. Med.*, **13** (3), 372–377.
223. Qi, L., Wu, L., Zheng, S., Wang, Y. *et al.* (2012) Cell-penetrating magnetic nanoparticles for highly efficient delivery and intracellular imaging of siRNA. *Biomacromolecules*, **13** (9), 2723–2730.
224. David, S., Marchais, H., Bedin, D., and Chourpa, I. (2014) Modelling the response surface to predict the hydrodynamic diameters of theranostic magnetic siRNA nanovectors. *Int. J. Pharm.*, **478** (1), 409–415.
225. Suzuki, R., Takizawa, T., Negishi, Y., Hagisawa, K. *et al.* (2007) Gene delivery by combination of novel liposomal bubbles with perfluoropropane and ultrasound. *J. Control. Release*, **117**, 130–136.
226. Suzuki, R., Takizawa, T., Negishi, Y., Utoguchi, N. *et al.* (2008) Tumor specific ultrasound enhanced gene transfer in vivo with novel liposomal bubbles. *J. Control. Release*, **125**, 137–144.
227. Suzuki, R., Takizawa, T., Negishi, Y., Utoguchi, N. *et al.* (2008) Effective gene delivery with novel liposomal bubbles and ultrasonic destruction technology. *Int. J. Pharm.*, **354**, 49–55.
228. Endo-Takahashi, Y., Negishi, Y., Nakamura, A., Ukai, S. *et al.* (2014) Systemic delivery of miR-126 by miRNA-loaded Bubble liposomes for the treatment of hindlimb ischemia. *Sci. Rep.*, **24** (4), 3883.
229. Grimm, D. and Kay, M.A. (2007) Combinatorial RNAi: a winning strategy for the race against evolving targets? *Mol. Ther.*, **15** (5), 878–888.
230. Gandhi, N.S., Tekade, R.K., and Chougule, M.B. (2014) Nanocarrier mediated delivery of siRNA/miRNA in combination with

chemotherapeutic agents for cancer therapy: current progress and advances. *J. Control. Release*, **194C**, 238–256.

231. Meng, H., Liong, M., Xia, T., Li, Z. *et al.* (2010) Engineered design of mesoporous silica nanoparticles to deliver doxorubicin and P-glycoprotein siRNA to overcome drug resistance in a cancer cell line. *ACS Nano*, **4**, 4539–4550.
232. Zheng, C., Zheng, M., Gong, P., Deng, J. *et al.* (2013) Polypeptide cationic micelles mediated co-delivery of docetaxel and siRNA for synergistic tumor. *Biomaterials*, **34** (13), 3431–3438.
233. Takei, Y., Kadomatsu, K., Goto, T., and Muramatsu, T. (2006) Combinational antitumor effect of siRNA against midkine and paclitaxel on growth of human prostate cancer xenografts. *Cancer*, **107**, 864–873.
234. Nishimura, M., Jung, E.J., Shah, M.Y., Lu, C. *et al.* (2013) Therapeutic synergy between microRNA and siRNA in ovarian cancer treatment. *Cancer Discov.*, **3** (11), 1302–1315.
235. Xue, W., Dahlman, J.E., Tammela, T., Khan, O.F. *et al.* (2014) Small RNA combination therapy for lung cancer. *Proc. Natl Acad. Sci. USA*, **111** (34), E3553–E3561.
236. Kasinski, A.L., Kelnar, K., Stahlhut, C., Orellana, E. *et al.* (2014) A combinatorial microRNA therapeutics approach to suppressing non-small cell lung cancer. *Oncogene*. doi: 10.1038/onc.2014.282.
237. Castanotto, D., Sakurai, K., Lingeman, R., Li, H. *et al.* (2007) Combinatorial delivery of small interfering RNAs reduces RNAi efficacy by selective incorporation into RISC. *Nucleic Acids Res.*, **35**, 5154–5164.

25
Stem Cells and Colon Cancer*

Simone Di Franco[1], Antonina Benfante[1], Flora Iovino[1], Sebastiano Bonventre[1], Francesco Dieli[2], Giorgio Stassi[1,3], and Matilde Todaro[1]
[1] University of Palermo, Department of Surgical and Oncological Sciences, Cellular and Molecular Pathophysiology Laboratory, Via Liborio Giuffrè 5, Palermo (PA) 90127, Italy
[2] University of Palermo, Division of Immunology and Immunogenetics, Department of Biotechnology and Medical and Forensic Biopathological (DIBIMEF), Corso Tukory 211, Palermo (PA) 90134, Italy
[3] IRCCS Fondazione Salvatore Maugeri, Department of Cellular and Molecular Oncology, Via Salvatore Maugeri, Pavia (PV) 27100, Italy

Keywords

Stem cell
A cell with self-renewal and pluripotent differentiation capacities.

*This chapter has previously been published in: Meyers, R.A. (Ed.) Stem Cells, 2013, 978-3-527-32925-0.

Translational Medicine Cancer: Advances in Molecular Biology and Medicine,
First Edition. Edited by Robert A. Meyers.

Colon cancer stem cell
A cell that gives rise to colon cancer that shares its stem cell features with normal stem cells.

Crypt
An invagination of the colon epithelium, formed by 2000–3000 cells, surrounded by an absorptive and secretory epithelium.

Niche
The environment that ensures the pluripotency of stem cells; it includes fibroblasts, endothelium, and inflammatory cells.

The current concept of tumorigenesis suggests that cancers arise and are "driven" by cells with stem cell-like properties, known as cancer stem cells (CSCs), which share many functional and molecular features with normal stem cells. Self-renewal key pathways (e.g., Wnt, Notch, and Hedgehog) are tightly regulated in normal stem cells, but are impaired in CSCs. For instance, active Wnt pathway plays a crucial role in colon cancer pathophysiology, where deregulation of the adenomatous polyposis coli (*APC*) gene, a negative regulator of Wnt signaling, represents one of the earliest alterations in the multistep process of colon carcinogenesis, causing early adenoma formation. Normal colon stem cells reside at the base of the crypts, in specialized regions termed "niches," that provide a microenvironment structured to ensure their self-renewal. It is becoming increasingly clear that CSCs are involved in tumor recurrence and resistance to conventional therapies, due to high expression levels of ATP-binding cassette (ABC) transporters, slow cycling rates, and the presence of cytoprotective factors. Future therapeutic strategies should selectively target CSCs for a complete clinical remission.

1 Colorectal Cancer

Colorectal cancer (CRC) is a classic example of a multistep pathogenesis, in which each step characterized by the acquisition of an abnormal function of a proto-oncogene or the loss of function of a tumor suppressor gene. Typically, at least four to five of these mutations must be accumulated before tumor formation, and the creation of invasive carcinoma [1]. Whilst many of these mutations affect common gene sequences and are shared by many patients, other mutations are individual and determine the final phenotype of the tumor [2]. The majority of the information available on colon cancer has arisen from the study of hereditary forms; these represent 5–10% of all cases, in which the adenomatous polyposis coli (*APC*) gene mutation is determinant [3–5].

The sites which are affected preferentially by CRC are the rectum and sigmoid colon (see Fig. 1), with disease progression occurring through two major pathogenetic mechanisms: (i) the APC/β-catenin pathway and chromosomal instability; and (ii) microsatellite instability (MSI).

The first of these mechanisms is characterized by alterations in the wingless/wnt signaling pathway, caused by mutations in

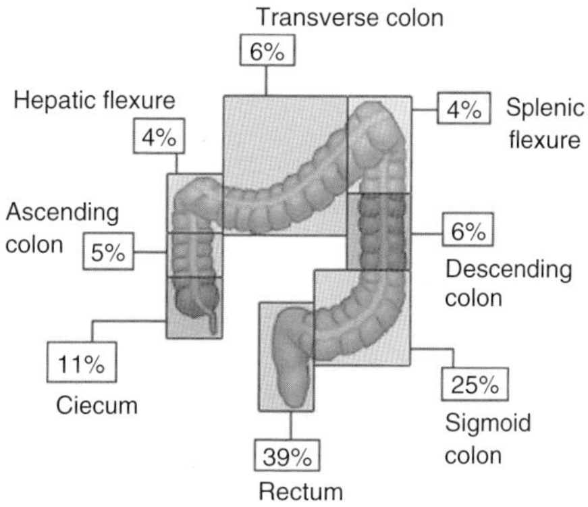

Fig. 1 Graphic representation of the different regions of the colon preferentially affected by colorectal cancer.

APC: the main function here is to regulate the cytoplasmic levels of β-catenin, a protein which is able to migrate into the nucleus and to activate the transcription of genes involved in cell proliferation, differentiation, migration, and apoptosis [6]. The subsequent tumor progression correlates with the mutation in *k-RAS, BRAF, SMAD4-SMAD2,* and cell division control protein 4 (*CDC4*) genes, and the deletion of chromosome 18q; a *TP53* mutation is ultimately the key event for the late adenoma-cancer transition [7] (this progression is shown in Fig. 2).

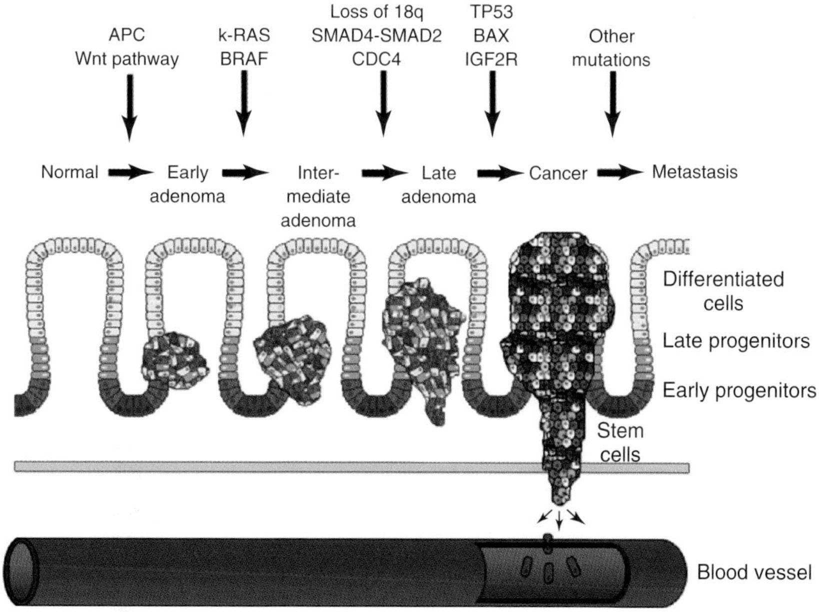

Fig. 2 Schematic representation of colorectal cancer multistep progression.

The mechanism of MSI (which involves 15% of cases) is characterized by chromosomal integrity. In this case, the mutations occur in the mismatch repair machinery, which in turn leads to an accumulation of mutations during mitotic processes. These are particularly related to the microsatellite DNA, which are non-coding repetitive DNA sequences that serve as important sites of modulation of the processes of induction, or the arrest of transcription.

1.1 Stochastic versus Hierarchical Model in Cancer Origin

As with many other tumors, colon cancer is considered to involve a heterogeneous cell population, where the diversity is reflected in the expression of several markers, and also in the different proliferative capacity and tumorigenic potential. Such heterogeneity is due both to genetic and/or epigenetic mutations, and also to the combined effect of the microenvironment that determines the different degrees of differentiation [2, 8].

During the past decade, advances in the biology of cancer have led to radical changes as to how the process of tumorigenesis is viewed. Whereas, previously, cancer was regarded as a genetic disease [9], today two different models of tumorigenesis exist:

- According to the clonal evolution hypothesis (or stochastic model), as described by Nowell in 1976, the tumor is heterogeneous, while every cell within the tumor retains its growth capacity and is capable of reforming a tumor following its implantation into immunocompromised mice [10, 11]. However, this aged concept could not explain why current anti-cancer strategies have demonstrated either limited or no effects during the treatment of cancer.
- In the case of the second model, the results of recent studies have suggested that the ability to initiate and sustain tumor growth is due exclusively to a subpopulation of tumor cells – termed cancer stem cells (CSCs) or "*tumor-initiating cells*" [12, 13]. The CSCs, which characterize the hierarchical model, share stem cell-like features with normal stem cells (i.e., self-renewal and pluripotent differentiation capacity). Moreover, they can generate, through several cycles of division, progenitor cells (PCs) which give rise to non-tumorigenic differentiated population that represent the major part of the tumor. Notably, CSCs can be derived either from normal stem cells or from PCs that have acquired a self-renewal potential [14].

The existence of CSCs in a variety of tumors has helped to explain the phenomenon of tumor chemoresistance, whereby several mutations have been shown to confer to the tumor cells a resistance to anti-cancer drugs, to alter the cell-cycle checkpoints, and to impair the machinery associated with apoptosis. Because of these properties, the CSCs are able to survive current treatments that typically – and preferentially – would kill highly proliferating cells, often giving rise to a minimal residual disease (MRD).

Consequently, a better understanding of the mechanisms that regulate stemness, and also the subsequent characterization of CSCs, may be crucial when selecting the most appropriate targeted strategy.

1.2 Colon Crypt and Cancer Stem Cell Characterization

Although it has long been assumed that neoplastic formation derives from alterations within adult colonic stem cells, the isolation of CSCs and their characterization has, nonetheless, remained elusive.

Anatomically, the wall of the colon is formed by the serosa, the muscularis, the submucosa, and the mucosa (the outer layer). The mucosa, which consists of epithelial and goblet muciparous (mucus-secreting) cells, is surrounded by an absorptive and secretory epithelium that is folded several times to form a set of invaginations (total area ca. 14 000 cm^2 in the adult human colon) that are known as the *crypts of Lieberkuhn*. Each of these crypts incorporates about 2000–3000 cells that comprise different cell populations, including the columnar cells (or *colonocytes*), abundant muciparous cells (or *goblet cells*), and a small fraction (ca. 1%) of enteroendocrine cells [15–18]. Each of these cell types appears to be generated from a colonic stem cell that is located at the base of the crypt, surrounded by mesenchymal cells. This cell, through an asymmetric division, is able to generate an identical daughter cell (self-renewal capacity) as well as a progenitor cell that can proliferate, differentiate, and migrate to the top of the crypt. This so-called "unitarian theory" was first formulated in 1974 [19], and subsequently demonstrated experimentally [20, 21]. These colonic stem cells are responsible for the constant renewal of each crypt which, under normal conditions, occurs about every six days. Notably, the turnover of the cells is greatly increases following tissue damage.

The first evidence relating to the position of the stem cells in the colon was reported by Chang and Leblond [22], and recently confirmed using bromodeoxyuridine (BrdU) DNA-labeling to demonstrate the localization of these slow-cycling stem cells [23]. Typically, the colon stem cells are located in the "stem cell zone" at the bottom of the colonic crypt, where four to six independent stem cells may reside.

Several molecules have been proposed as markers of "stemness": Musashi-1 (Msi-1); Hairy/enhancer-of-split (Hes-1); integrin α_2 and β_1 subunits; Ephrin B (EphB) receptors; Bmi1; Lgr5; and aldehyde dehydrogenase 1 (ALDH1).

Msi-1 is a RNA-binding protein involved in the asymmetric division of *Drosophila melanogaster* [24], and is expressed both in stem cells and transient amplifying PCs. Msi-1 silencing causes mitotic catastrophe and xenograft tumor growth arrest by Notch inhibition and p21 upregulation [25]. Other putative markers include the EphB receptors that regulate migration and proliferation in the intestinal epithelium; however, as EphB2/EphB3 inhibition decreases the number of proliferating cells without affecting stem cell numbers, it is impossible to use EphB for stem cell isolation.

A more promising marker might be Bmi-1, which is involved in hematopoietic, breast, and neural self-renewal. Bmi-1 is expressed predominantly by cells at the base of the crypt [26], where some cells are positive for Lgr5 (Gpr49), a transmembrane protein that is coupled with G-protein. In the murine system, Lgr5 is expressed only in cycling columnar cells that can differentiate into functional colonic epithelium lineages [27]. The Lgr-5$^+$ cells can be propagated *in vitro* by forming crypt–villus organoids that show differentiated cells. Lgr-5 deficiency induces Paneth cell differentiation and the overexpression of Wnt target genes [28].

During recent years, however, the stem cell field has become both complicated and interesting, with new and specific markers having been associated with the CSC phenotype. The first experimental evidence of colorectal cancer stem cells (CR-CSCs) was focused on the role of CD133 as a stemness marker. $CD133^+$ cells are able to grow exponentially *in vitro* as undifferentiated tumor spheres in the absence of serum, and thus initiate tumor growth in mouse models. In this way, it was possible to reproduce the same morphological and antigenic patterns as the original human tumor [29, 30].

Subsequently, additional data have been acquired *in vitro* when, under differentiating conditions on Matrigel, tumorigenic $CD133^+$ cells were shown to give rise to colonies that were similar to crypt-like structures [31].

Unfortunately, CD133 expression in epithelial tissues is not restricted to stem cells or PCs; rather, it is distributed among the differentiated luminal and ductal epithelial cells [32]. In metastatic adenocarcinoma, the expression of CD133 was downregulated, while a $CD133^-$ subset formed a more aggressive tumor than was generated by $CD133^+$ cells. These findings have raised significant controversies regarding the use of CD133 as a CR-CSC marker, since the data suggest that CD133 alone does not exert a critical role in supporting tumor growth, thus confirming the need to identify biomarkers for CR-CSCs. Indeed, many studies have provided proof that, besides CD133, other surface proteins can mark colon cancer stem cells (CCSCs). Within the $CD133^+$ tumor population, there is a subset of $CD44^+$/$EpCAM^{high}$ (epithelial cell adhesion molecule) cells that is capable of generating tumor xenografts upon serial transplantation into non-obese diabetic/severe combined immunodeficient (NOD/SCID) mice. Tumors which originated from $EpCAM^{high}$/$CD44^+$ cells maintained a differentiated phenotype, and reproduced the full morphologic and phenotypic heterogeneity of their parental lesions. Another differentially expressed marker, CD166, was shown capable of enhancing the success of the tumor xenograft [33]. More recently, ALDH1 has been proposed as a promising new marker for normal and malignant human CSC marker, as it is expressed by cells that are positive for $CD44^+$ or $CD133^+$ and located at the base of normal crypts. The sorting of $ALDH1^{high}$ cancer cells and the implantation of as few as 25 cells in NOD/SCID mice led to the generation of tumor xenografts. During tumor progression, the selection of $CD44^+$, $CD133^+$ cells with ALDH activity was increased in number and reached the crypt axis [34]. Details of the putative markers described above are listed in Table 1.

1.3 Stemness Niche: Definition, Role, and Pathways Involved in Its Modulation

The intestinal niche (as represented in Fig. 3), including fibroblasts, the endothelium, and inflammatory cells, is considered to be the environment that ensures the pluripotency of stem cells, as removal from this niche will induce a loss of their properties.

The intestinal subepithelial myofibroblasts (ISEMFs) surround the normal colonic stem cells and regulate organogenesis and tissue repair. The growth of the ISEMFs is regulated by many factors; recently, it has been shown that they secrete growth factors and cytokines, thus controlling the correct balance between stem cell self-renewal and differentiation [35, 36].

The importance of the effect of the ISEMFs and their secreted factors, as well as of the hepatocyte growth factor (HGF), has also been demonstrated in the maintenance of the niche of stem cells in the intestinal crypts. For example, Vermeulen

Tab. 1 Panel representative of putative normal and cancer stem cell markers and their specific roles.

Putative marker	*Alternative name*	*Role(s)*	*Location*
CD34	–	Adhesion molecule, stemness maintenance	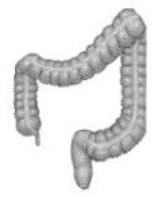
DCAMKL1	–	Kinase, resistance to apoptosis	
EphB receptors	–	Cell migration, stemness maintenance	
Msi-1	–	RNA-binding protein, asymmetric division, proliferation	
Bmi-1	–	Polycomb group repressor, self-renewal, ageing inhibition	
CD24	HSA	Cell adhesion molecule	
CD29	β_1 integrin	Proliferation, matrix–cell interaction	
CD44	–	Cell–cell interaction, hyaluronic acid receptor, cell migration	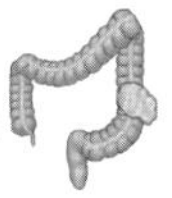
CD133	Prominin 1	Self-renewal, tumorigenesis	
CD166	ALCAM	Cell adhesion molecule	
ESA	EpCAM	Cell adhesion molecule	
Lgr-5	Gpr49	G-protein-coupled receptor, unclear function	
ALDH1	–	Detoxifying enzyme	
Nuclear β-catenin	–	Adherens junctions, cell-cycle regulation, proliferation	

et al. have reported how differentiated cells may be able to restore their stem cell phenotype in response to the presence of myofibroblasts or HGF [37].

Self-renewal pathways, such as Wnt, Notch, and Sonic hedgehog (Shh) are deregulated both in stem cells and CSCs [38, 39].

Defects in the Wnt/β-catenin pathway were identified in leukemic stem cells [40], while activation of the pathway was shown to be mediated by Akt signaling [41]. Defects in Notch pathway were observed in CCSCs, by using the antibody anti DLL4 (which is an important Notch pathway component) to directly inhibit human colon cancer xenograft growth. The Notch pathway is known also to be important in breast [42] and glioblastoma CSCs; in addition, many changes were identified in the Hedgehog (HH) pathway in several cancers, including leukemia [43, 44], pancreas, stomach, prostate, breast [45, 46], glioblastoma [47], and colon [48].

The Wnt activation pathway is one of the most studied processes involved in CSCs progression. The Wnt proteins, which bind the Frizzled receptor, control the stability of β-catenin protein that, in absence of Wnt ligands, binds the APC protein complex. This leads to the phosphorylation, ubiquitination, and finally the degradation of the complex by the proteasome [49]. In contrast, activation of the Wnt pathway promotes an accumulation of β-catenin in the nucleus, which in turn binds the T-cell factor/lymphoid enhancer factor (Tcf/Lef) complex that activates the transcription of several genes involved in cell cycle regulation and proliferation [50]. β-Catenin also regulates expression of the Ephrin receptors EphB1 and EphB2, which in turn regulate the maintenance of stemness, migration, and cell differentiation across the crypt, with the higher level of expression at the top [51, 52].

According to this model, it has been shown that a differential distribution of Wnt members exists along the axis of the crypt. In this case, mRNAs for secreted Fz-related protein (sFRP)-5, Wnt-3, Wnt-6, Wnt-9b, and Fz-5 have each been identified

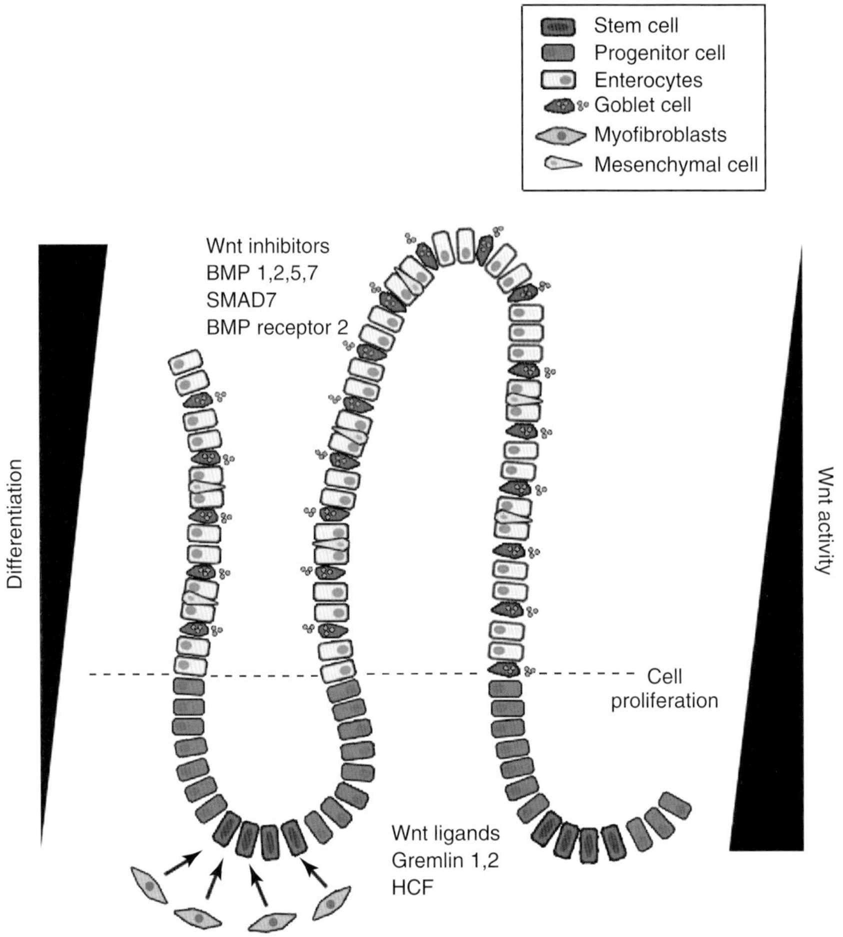

Fig. 3 Graphic representation of the balance between differentiation and stemness along the crypt.

at the base of the crypts, where they contributed to maintaining cell stemness. However, their concentrations were shown to decrease toward the apex of the crypt, where the differentiated cells were shown to reside, and where the Wnt inhibitors were expressed [53].

The bone morphogenetic proteins (BMPs), as members of the transforming growth factor-β (TGF-β) superfamily, are able to trigger several biological processes in CSCs [54] by promoting Smad1/5/8 protein (RSmads) phosphorylation [55]. This permits the latter to translocate, together with Smad4 (co-Smad), into the nucleus where the complex – either alone or in combination with other transcription factors – can promote target gene expression [56, 57].

BMP4 induces the terminal differentiation and apoptosis of colonic epithelial cells, and also increases the conventional therapeutic activity in tumors harboring wild-type SMAD4 and constitutive activation of phosphoinositide 3-kinase (PI3K) [58]. Again, an expression gradient of BMP proteins has been demonstrated, with cells at the top of the crypt expressing high levels of BMP1, BMP2, BMP5, SMAD7, BMP7, and BMP receptor 2, whereas at the base of the crypt (most likely due to the presence of myofibroblasts) the cells produce high levels of BMP antagonists, such as GREM1, GREM2, and chordin-like-1, all of which contribute to stemness maintenance [59, 60].

The contribution of the Notch pathway in the intestinal fate of the stem cells was proven clearly. The Notch pathway includes four different trans-membrane receptors (Notch1 to Notch4) and five ligands (Jagged-1, -2, Delta-like 1, 2, and 4), which are expressed in different combinations in most cell types. The subsequent binding of these ligands induces metalloprotease-dependent (ADAM10 or ADAM17) cleavage of the Notch intracellular domain (NICD). The latter then moves to the nucleus, where it promotes not only the transcription of genes targets but also the creation of transcription factors and coactivators [61].

The Hes-1 proteins represent the best-characterized Notch target genes, as they repress the genes that are involved in epithelial cell fate determination, such as neurogenin (ngn), achaete-Scute, and Math1. The results of recent studies have demonstrated an essential role for both Notch and Math1 in intestinal homeostasis and neoplastic transformation. Moreover, transgenic expression of the NICD blocks cell differentiation, thus inducing an expansion of immature progenitors [62, 63].

Finally, the Shh pathway plays an important role during intestinal organogenesis. Notably, activation of the pathway occurs through the binding of Shh to its receptor, Patched homolog (PTCH), which in turn leads to a release of the G-coupled protein Smoothened (SMO). This, together with the GLI transcription factors, moves to the nucleus in order to induce the activation of target genes [64].

In conclusion, the aberrant activation of the pathways involved in self-renewal of the intestinal epithelium stem cells may promote colon cancer, and highlights the importance of the niche for stem cell proliferation, migration, and invasion [65].

2 Treatment Option Overview: Strategy of Treatment by Stage

The discovery of premalignant lesions (known as *adenomatous polyps*), which are associated with a good survival during the early phase of disease, has led to CRC becoming an ideal candidate for screening analyses. The key aspect in the care of colon cancer patients is that of prevention (in case of symptoms or familiarity), followed by immediate surgery while the tumor is still in its early stages of development; typically, this combined approach is successful in the vast majority of cases. Fecal occult blood (FOB) may be identified symptomatically in about 5% of cases; this should be followed by a fecal occult blood test (FOBT) which, if positive, must be further investigated by colonoscopy. This follow-up approach should be undertaken in both men and women aged over 50 years, at one-to two-year intervals.

Some serological markers may also allow the early detection and diagnosis of CRC. An example is the carcinoembryonic antigen (CEA), which can be used during the preoperative staging and postoperative follow-up of patients with large bowel cancer. Unfortunately, CEA may exhibit a poor predictive value for diagnosis in asymptomatic patients,

Tab. 2 Colon cancer staging accordingly to AJCC and TNM classification.

AJCC stage	*TNM stage*	*TNM stage criteria for colon cancer*
Stage 0	Tis N0 M0	Tis: tumor confined to mucosa; cancer-*in-situ*
Stage 1	T1 N0 M0	T1: invasion of submucosa
Stage 1	T2 N0 M0	T2: invasion of muscularis propria
Stage II-A	T3 N0 M0	T3: invasion of subserosa or beyond (without other organs involvement)
Stage II-B	T4 N0 M0	T4: invasion of adjacent organs or perforation of the visceral peritoneum
Stage III-A	T1–2 N1 M0	N1: metastasis to one to three regional lymph nodes. T1 or T2.
Stage III-B	T3–4 N1 M0	N1: metastasis to one to three regional lymph nodes. T3 or T4.
Stage III-C	Any T, N2 M0	N2: metastasis to four or more regional lymph nodes. Any T.
Stage IV	Any T, any N, M1	M1: distant metastases. Any T, any N.

due to its relatively low sensitivity and specificity [66].

The different stages of colon cancer, according to the American Joint Committee on Cancer (AJCC) and the TNM (tumor, node, metastasis) Classification of Malignant Tumors are listed in Table 2.

Colon cancer staging is based on the level to which the lesion has penetrated the bowel wall; it is useful for both diagnostic and research purposes, and essential to determine the best available treatment. Systems employed for the staging of CRCs depend on the extent of local invasion, the degree of lymph node involvement, and the identification of metastases, and their extent.

Definitive staging can be determined only after surgery has been performed, and a pathology report obtained. A preoperative staging of CRCs can be carried out using endoscopic ultrasound, while for metastasis analysis other successful imaging techniques may include abdominal ultrasound, computed tomography (CT) scanning, and positron emission tomography (PET) scanning.

As the majority of conventional therapies are less effective for the eradication of advanced forms of CRC, surgical resection of the primary and regional lymph nodes remains the standard treatment for patients during the early stages of the disease. The treatment protocols are usually assigned on the basis of the so-called "TNM classification," which is an accurate system employed for treatment selection and prognostic assessment. In combination with surgery, the standard approach to caring for patients at the early stages is that of adjuvant chemotherapy (after primary tumor resection), in order to reduce the risk of relapse.

Adjuvant treatment is recommended for stage III and high-risk stage II CRC patients (even if the definition of "risk" does remain controversial). It is generally assumed that the penetration of cancer through the serosa represents the main means of distinguishing high-risk from low-risk patients, and for this reason stages I and IIA can be considered as low-risk, while stages IIB and III require adjuvant treatments. Patients at stage II may be considered at high-risk if they present with fewer than 12 lymph nodes being involved, a poorly differentiated tumor, vascular or lymphatic or perineural invasion, and tumor with obstruction or perforation. The five-year survival in stage III patients may also depend on the number of lymph nodes involved, as more than four is recognized to reduce the five-year survival to only 26% [66].

2.1
Adjuvant Chemotherapy for Stage II CRC

This stage, which includes heterogeneous tumors, involves 80% of patients with a five-year overall survival (OS) and a lower risk of relapse compared to stage III patients. Both, the IMPACT B2 study and the meta-analyses reported by the Mayo Clinic, have suggested a lack of improvement in OS or disease-free survival (DFS) in patients treated with an adjuvant (5-fluorouracil; F5-FU)-based therapy compared to surgery alone. In contrast, the National Surgical Adjuvant Breast and Bowel (NSABP) Project analysis supported an adjuvant therapy in patients with stage II disease, even in the absence of any OS benefit, because the reduction of recurrence risk was similar in both stage II and III patients [67]. Accordingly, the QUASAR (Quick and Simple and Reliable) study demonstrated a significantly reduced recurrence rate and improved survival in the treatment arm with 5-FU/LV (leucovorin) or levamisole compared to the non-treatment arm [68]. Many studies have included patients with stage II and III CRC; an example was the MOSAIC trial, in which the relative risk of relapse was reduced with FOLFOX 4 (5-FU + LV + oxaliplatin) treatment, while "high-risk" stage II patients characterized by inadequate lymph node sampling, T_4 disease, involvement of the visceral peritoneum, and a poorly differentiated histology were also identified [67]. It is difficult to derive a general conclusion from these studies, however, when taking into consideration the patient heterogeneity among the control and treatment arms of these trials.

2.2
Adjuvant Chemotherapy for Stage III CRC

In stage III CRC, it has been proposed in several studies that the number of lymph nodes present may serve as a prognostic factor. In the case of stage III patients, for many years 5-FU was the only cytotoxic drug to be used in adjuvant protocols, although a combination of oxaliplatin with 5-FU + LV has been shown to improve the OS compared to treatment with only 5-FU + LV. Many years ago, randomized studies with 5-FU in adjuvant therapy failed to show any significant improvements in patient survival [69]. The NSABP study C-01 demonstrated a survival benefit for stage II and III patients treated with MOF (semustine + 5-FU + vincristine) compared to surgical resection alone [70], while a subsequent randomized NCCTG trial, in which surgical resection alone was compared with postoperative levamisole or 5-FU + levamisole, showed a significant improvement in DFS for patients with stage III colon cancer, who had received 5-FU + levamisole [71]. Another, confirmatory intergroup trial, (the INT-0035 trial) showed that 5-FU + levamisole would prolong both DFS and OS, thereby changing the outcome for patients with stage III colon cancer [72]. Subsequently, the combination of 5-FU + LV has been extensively monitored in the adjuvant postsurgical treatment of colon cancer patients; in these cases, the NSABP C-03 trial showed a survival gain with 5-FU + LV versus MOF regimens [73], while the IMPACT study which focused on 5-FU + high-dose LV resulted in a 21% reduction of mortality among stage III patients [74].

Of note, the randomized INT-0089 trial included a total of 3794 patients with either high-risk stage II or stage III CRC, distributed among four treatment arms:

- The Mayo Clinic regimen (5-FU + low-dose LV).
- The Roswell Park regimen (5-FU + high-dose LV).
- The Mayo Clinic regimen + levamisole.
- The levamisole regimen (5-FU + levamisole).

In terms of efficacy, the Roswell Park regimen proved equivalent to that of the Mayo Clinic [75]; moreover, the high toxicity associated with the Mayo Clinic regimen + levamisole led to the Roswell Park regimen becoming elective adjuvant treatment in United States.

A major benefit in the survival rate of patients with advanced CRC was observed with a combination of oxaliplatin and irinotecan at 5-FU + LV treatment, and these positive effects have been demonstrated in two recent trials. In the MOSAIC study, the addition of oxaliplatin to 5-FU + LV (FOLFOX) [76] led to a significant increase in DFS at three years, and reduced the recurrence risk by about 23%. The study follow-up demonstrated a significant improvement in the five-year DFS and a benefit of six years in OS, though only for stage III patients. These results confirmed the benefit of adjuvant treatment with FOLFOX4.

In the NSABP C-07 trial, the efficacy of bolus FU + LV + oxaliplatin (FLOX) treatment was compared to FU + LV alone, and showed the three-year DFS to be significantly improved, with a reduction in recurrence risk that was in accordance with that of the MOSAIC trial [77]. The results of the toxicity spectrum between the MOSAIC and NSABP-C07 trials defined the use of FOLFOX for six months as the standard of care for stage III colon cancer patients. Following validation of the use of capecitabine (an oral fluoropyrimidine), a Phase III trial was conducted to evaluate the safety and efficacy of the latter in combination with oxaliplatin (XELOX), compared to bolus FU + LV (FL) (Mayo Clinic or Roswell Park regimen). The study results highlighted an increase in terms of DFS for the XELOX regimen.

The aborted CALGB-89803 trial, in which irinotecan (CPT-11) was combined with 5-FU (IFL), reported a high mortality and no improvement in terms of either OS or event-free survival for IFL [78]. Along the same lines, the ACCORD trial failed to demonstrate any significant improvement in DFS with irinotecan-based chemotherapy in high-risk stage III patients. Likewise, in the PETACC-3 study, which compared CPT-11/LV5-FU2 (FOLFIRI regimen) with LV5-FU2 in stage III patients, there was no evidence of any benefit for the irinotecan regimen in terms of DFS.

2.3 Adjuvant Chemotherapy for Stage IV and Recurrent Colon Cancer

Stage IV colon cancer is characterized by distant metastases. Indeed, some 50% of such colon cancer patients show hepatic metastases, although ultimately only a small proportion prove to be candidates for surgical resection.

Currently, seven active and approved drugs are available for the treatment of patients with metastatic CRC, namely 5-FU, capecitabine, irinotecan, oxaliplatin, bevacizumab, cetuximab, and panitunumab. At this stage of the disease, biological therapy plays a fundamental role, notably through employing inhibitors of the epidermal growth factor receptors (EGFRs) and of angiogenesis. In fact, preliminary studies with 5-FU, conducted in patients with metastatic disease, showed a prolongation of the time-to-progression (TTP) [79] and an improvement in their quality of life [80].

Prior multi-agent chemotherapy carried out in two randomized studies demonstrated similar efficacies of capecitabine and 5-FU+LV [81, 82]. Two further studies were then conducted to compare the FOLFOX (4/6) and FOLFIRI treatments, and showed identical progression-free survival (PFS) and OS. Subsequently, the use of either FOLFOX or FOLFIRI was accepted as first-line treatment for metastatic CRC patients [83, 84].

When, in the BICC-C trial, the safety and efficacy of three different irinotecan-containing regimens – irinotecan + infusional FU+LV (FOLFIRI), irinotecan +bolus FU+LV (mIFL), and irinotecan +oral capecitabine (CapeIRI) – were evaluated on untreated metastatic CRC patients, those patients treated with FOLFIRI had a better PFS [85]. Following the approval of bevacizumab, the BICC-C trial was modified to test, at random, FOLFIRI/bevacizumab or mIFL/bevacizumab on a further 117 patients. Those patients who received FOLFIRI/bevacizumab enjoyed a significantly better OS, such that FOLFIRI became a first-line treatment for metastatic CRC when using an irinotecan-based regimen [85]. In the Spanish Cooperative Group study, which randomly tested CAPOX against FUOX treatments, it was shown that – in terms of TTP – the CAPOX and FUOX regimens were similar [86].

2.4 Target Therapy: Bevacizumab, Cetuximab/Panitumumab

Currently, anti-angiogenetic therapy is considered to be a promising strategy in the adjuvant treatment of colon cancer. The first such molecule to be approved was bevacizumab, an intravenous recombinant humanized monoclonal antibody against vascular endothelial growth factor (VEGF) [87]. Bevacizumab, when used in association with cytotoxic chemotherapy, modifies the architecture of the tumor blood vessels, thus decreasing the intratumoral hydrostatic pressure and suppressing tumor growth [88]. In the study conducted by Hurwitz, untreated metastatic CRC patients who received IFL + bevacizumab showed a significant increase in PFS and OS when compared to IFL treatment alone [89]. Following the reporting of results from the NCCTG-N9741 study, bevacizumab + FOLFOX was approved as a standard first-line regimen [90]. Accordingly, the results of the ECOG study showed that treatment with FOLFOX4 + bevacizumab led to significant improvements in PFS and OS among patients in whom the tumor had progressed during other treatments [91, 92]. As a consequence of these studies, bevacizumab was shown capable of being combined with either FOLFIRI or FOLFOX for administration to metastatic CRC patients undergoing first-line treatment.

Today, no completed randomized controlled studies have been evaluated to determine whether, following first-line treatment with a bevacizumab regimen, the same regimen might be useful as a second- or third-line treatment. In conclusion, the addition of bevacizumab has been shown to improve the PFS when compared to chemotherapy alone. Moreover, the benefit was greater with irinotecan than with oxaliplatin, and appeared to be more pronounced after first-line therapy compared to second-line, but was not observed in third-line therapy.

Recent efforts have been focused on those agents that target the pathways involved in cell proliferation and metastasis phenomena; an example – cetuximab – is a partially humanized monoclonal antibody that inhibits the human EGFR by preventing its activation [92]. This antibody impairs the dimerization of the receptor, thus promoting its internalization and degradation, and it may also kill target cells via antibody-dependent cell-mediated cytotoxicity (ADCC) and complement fixation [88]. Interestingly, KRAS (an intramembrane serine–threonine kinase activated by EGFR), when in its mutated form, plays a key role as a predictive factor in cetuximab therapy response. In a randomized Phase II study of patients in whom tumor progression had occurred after receiving irinotecan-containing regimens, treatment with irinotecan and cetuximab increased the median TTP in all cases [93]. On the basis of these findings, cetuximab was approved for use in patients with metastatic CRC refractory to 5-FU and irinotecan. In another large randomized study (the Crystal Study), the cetuximab + FOLFIRI regimen was shown to improve PFS, but not OS, in stage IV CRC patients [92]. The results of subsequent retrospective studies suggested that those who responded to cetuximab therapy expressed the wild-type *KRAS* (i.e., tumors that lack activating mutations on the 12th or 13th codons). These results were later confirmed by the Crystal Study, in which tumor response was evaluated in terms of the interaction between the *KRAS* mutation status and treatment. In this case, among patients with *KRAS* wild-type tumors, the hazard ratio (HR) favored the FOLFIRI/cetuximab group [92].

In another analysis, cetuximab-treated patients with a mutated *KRAS* gene showed a significantly decreased PFS compared to wild-type *KRAS* patients (8.1 versus 10.5 months). Moreover, cetuximab-treated patients with mutated *KRAS* tumors had a significantly shorter PFS (8.1 versus 12.5 months) and OS (17.2 versus 24.9 months compared to untreated patients with a mutated *KRAS*) [94].

Some time later, another EGFR inhibitor – panitumumab – was introduced, such that both antibodies could target the EGFR but recognized different isotypes, namely IgG_2 and IgG_1, respectively. Subsequently, an open-label Phase III trial showed that patients with a chemotherapy-refractory disease, when treated with panitumumab, improved their PFS (8 versus 7.3 weeks), but there was no difference in their OS. Following this, panitumumab was approved by the US Food and Drug Administration (FDA) for administration to patients with metastatic CRC that was refractory to chemotherapy [95].

In the PRIME study, FOLFOX4 was tested with or without panitumumab as a first-line therapy for metastatic CRC, in patients with *KRAS* wild-type tumors. The study results confirmed that treatment with panitumumab-FOLFOX4 would significantly improve the PFS, compared to FOLFOX4 alone (9.6 versus 8.0 months), whereas in the case of patients with mutant *KRAS* the addition of panitumumab reduced the PFS (7.3 months for panitumumab-FOLFOX4 versus 8.8 months for FOLFOX4 alone) [96]. Accordingly, the addition of panitumumab to a FOLFOX/bevacizumab regimen resulted in a worse PFS in metastatic patients not selected for *KRAS* mutation (11.4 versus 10.0 months) [97]. In another study on metastatic patients who had already been treated with a fluoropyrimidine regimen, the use of FOLFIRI or FOLFIRI + panitumumab showed patients with a *KRAS* wild-type to significantly improve their PFS, whereas those with mutant *KRAS* did not show any benefits [98]. The validation

Tab. 3 List of therapeutic regimens in colon cancer treatment.

Regimen	*Compound*	*Treatment*
CAPOX	Capecitabine (100 mg m^{-2})	Twice a day on days 1 through 14 plus
	Oxaliplatin (70 mg m^{-2})	On days 1 and 8 every three weeks
FOLFOX4	Oxaliplatin (85 mg m^{-2})	Administered as a 2-h infusion on day 1
	Leucovorin (200 mg m^{-2})	Administered as a 2-h infusion on days 1 and 2
	5-FU (400 mg m^{-2})	Loading dose intravenous bolus
	5-FU (600 mg m^{-2})	Administered via ambulatory pump for a period of 22 h on days 1 and 2 every two weeks
FOLFOX6	Oxaliplatin (85 mg m^{-2})	Administered as a 2-h infusion on day 1
	Leucovorin (400 mg m^{-2})	Administered as a 2-h infusion on days 1 and 2
	5-FU (400 mg m^{-2})	Loading dose intravenous bolus on day 1
	5-FU (2400–3000 mg m^{-2})	Administered via ambulatory pump for a period of 46 h every two weeks
FOLFIRI	Irinotecan (180 mg m^{-2})	Administered as a 2-h infusion on day 1
	Leucovorin (400 mg m^{-2})	Administered as a 2-h infusion on days 1 and 2
	5-FU (400 mg m^{-2})	Loading dose intravenous bolus on day 1
	5-FU (2400–3000 mg m^{-2})	Administered via ambulatory pump for a period of 46 h every two weeks
FUFOX	Oxaliplatin (50 mg m^{-2})	As a 22-h continuous infusion on days 1, 8, 22, and 29 every 36 days
	Leucovorin (500 mg m^{-2})	
	5-FU (2000 mg m^{-2})	
FUOX	5-FU (2250 mg m^{-2})	Continuous infusion during 48 h on days 1, 8, 15, 22, and 29 every 36 days
	Oxaliplatin (85 mg m^{-2})	On days 1, 15, and 29 every six weeks
IFL	Irinotecan (125 mg m^{-2})	—
	5-FU (500 mg m^{-2})	Intravenous bolus
	Leucovorin (20 mg m^{-2})	Intravenous bolus administered weekly for four out of six weeks
XELOX	Oral capecitabine (1000 mg m^{-2})	Twice daily for 14 days
	Oxaliplatin (130 mg m^{-2})	On day 1 every three weeks

of these prognostic/predictive factors are to be expected from a large subset analysis in the context of international trials. Details of all currently available therapeutic regimens for colon cancer are listed in Table 3.

2.5
Alternative Therapies: mTOR, Protein Kinase C, and Src Inhibitors

The mammalian Target of Rapamycin (mTOR) is a serine–threonine kinase, and a member of the phosphatidyl inositol 3′ kinase (PI3K) family. It is activated in response to several growth stimuli, such as nutrients and/or growth factors, while its activation (through phosphorylation) stimulates cell growth and proliferation.

Conversely, mTOR inhibition – through a downregulation of cyclin/CDK (cyclin-dependent kinase) complexes – arrests the cell cycle in late G_1 phase, thus causing an accumulation of the cell-cycle inhibitor, p27. The mTOR inhibitors are analogs of rapamycin, which link the immunophilin FK506/rapamycin-binding protein, block the proliferation of endothelial and vascular smooth muscle cells, inhibit angiogenesis, and induce death via apoptosis [99].

In a Phase I clinical trial, it was shown that among a refractory CRC patient population, the efficacy of a mTOR inhibitor family was greater when administered in combination with standard chemotherapy [100].

The protein kinases C (PKC), a family of serine/threonine kinases, are activated by receptor tyrosine kinases such as the EGFR and vascular epidermal growth factor receptor (VEGFR), and are involved in tumor growth and survival. Moreover, PKC activity is elevated in many types of cancer, including CRC [101].

Enzastaurin, a selective inhibitor that targets PKC, PI3K/AKT, and GSK3, has been tested as either single agent or in combination with 5-FU or bevacizumab, and has demonstrated a disease control rate of about 53%, together with a low toxicity [102].

Src, is a non-receptor tyrosine kinase that has been identified as a proto-oncogene and plays a key role in cancer growth, invasion, angiogenesis, and in apoptosis regulation [103]. Src activity is enhanced in 80% of all CRCs, and its expression level correlates with the metastatic phenotype, a poor prognosis, and resistance to chemotherapy. Currently, the Src inhibitor AZD0530 is undergoing Phase II clinical trials in patients with metastatic CRC.

Unfortunately, none of these above-mentioned therapies is able to completely eradicate tumor cells, an effect which seems due to an inability to eradicate the CSCs subset that possesses a selective survival advantage [104]. Consequently, the discovery of reliable markers to identify CSCs will be necessary in order to better understand the signaling pathways that maintain the balance between the stemness and differentiation of CSCs. In this case, the goal would be to develop combined therapies that would target both the CSCs and their differentiated progeny.

3
Targeting the Differentiation and Survival Signaling in CSCs

As conventional therapies are insufficient to eradicate all tumor cells, the use of molecules that can target a pathway that is responsible for the CSCs resistance or differentiation could overcome this problem.

Todaro *et al.* have shown that CCSCs may survive conventional therapy through interleukin (IL)-4 autocrine production that determines an upregulation of the anti-apoptotic molecules. Moreover, the same group also showed that, by blocking this signaling pathway using specific antibodies, the CoCSCs would become sensitized to chemotherapic agents via a downregulation of anti-apoptotic proteins. The combination of IL-4 antibodies plus standard chemotherapeutic agents led to a reduction in tumor growth also *in vivo* xenograft models [31, 105].

An additional important mechanism by which colon CSCs may be killed would be to induce these cells to develop a more differential status, such that the CoCSCs would become sensitive to conventional therapies.

The treatment of these cells with BMP4, a transforming growth factor-β (TGFβ) superfamily member, induces a differentiation program that sensitizes the CCSCs to 5-FU or oxaliplatin in xenografts, leading to complete and long-term regression. In this case, the BMP4 induces apoptosis, thus reducing β-catenin activation through an inhibition of the PI3K/AKT pathway and an up-modulation of the Wnt-negative regulators [58, 106].

As mentioned above, within the colon crypts the ISEMF cells maintain the stem cells' niche by balancing the opposite signals that promote self-renewal (Wnt and Notch pathways) and differentiation (BMP pathway). On this basis, it is possible that the CSCs niche with the signaling pathway may have a role in maintaining and increasing the CSCs pool [107].

Other signaling pathways, such as hedgehog (HH), are involved in maintaining the CSCs and sustaining the growth of different tumor types, including colon cancer. Here, HH binds the receptor PTCH, leading to the release of SMO and to a downstream activation of the Gli transcription factor family. In advanced/metastatic CCSCs, high levels of Gli1 were found; moreover, the SMO knockdown, by inhibiting the HH signaling, would reduce proliferation *in vitro*, thereby inducing apoptosis while abrogating tumor growth *in vivo* [48].

Notch signaling is also involved in colon CSCs maintenance. Indeed, the inhibition of Notch signaling, by using a human monoclonal antibody, anti-hDLL4 (delta-like ligand), leads to reductions in both tumor growth and CSCs frequency [42].

3.1 Immunotherapy

Immunotherapy represents a new strategy that can be used to exploit the specificity of the immune system while effectively increasing antitumor response. Although cancer cells are less immunogenic than their normal counterparts, the immune system is clearly able to recognize and eliminate them [107]. Recently, the results of various clinical studies have shown that the synergistic use of chemotherapy and immunotherapy is safer than chemotherapy alone [108].

Recently, a number of immunotherapeutic approaches have been used to sensitize the immune system against colon cancer cells, using a T-cell subset ($\gamma\delta$ T cells) with major histocompatibility complex (MHC)-unrestricted lytic activity [109]. Indeed, recent evidence has shown that, within the colon cancer-infiltrating lymphocyte population, there is a subpopulation of T cells, known as $V\gamma 9V\delta 2$. These lymphocytes represent a natural immune response mechanism that can be activated by a variety of agents (e.g., aminobisphosphonates) which show a direct antitumor activity through inhibiting proliferation and inducing apoptosis to target tumor cells. In 2009, Todaro *et al.* showed that the treatment of CCSCs with zoledronate (a bisphosphonate) induced Vγ9 Vδ2 T-cell activation, such that the release of perforin from the T cells caused the CoCSCs to become more susceptible to death. In time, this mechanism may become recognized as an alternative strategy for the eradication of tumorigenic cells.

In contrast to differentiated cells, the CCSCs express a variety of natural killer (NK) cell ligands and undetectable levels of MHC class 1. Recently acquired preliminary data have shown that NK cells can kill CCSCs via a lytic mechanism, thus underlining the role of these cells in immunotherapeutic treatments.

Acknowledgments

Simone Di Franco and Antonina Benfante are both Ph.D. students at the "International Immunopharmacology Course," Palermo, Italy. The authors' special thanks go to Daniele Di Franco (Doctor in Informatics), for his cooperation in the creation of graphic images used in the chapter.

References

1. Vogelstein, B., Fearon, E.R., Hamilton, S.R., Kern, S.E. *et al.* (1988) Genetic alterations during colorectal-tumor development. *N. Engl. J. Med.*, **319** (9), 525–532.
2. Fearon, E.R. and Vogelstein, B. (1990) A genetic model for colorectal tumorigenesis. *Cell*, **61** (5), 759–767.
3. Galiatsatos, P. and Foulkes, W.D. (2006) Familial adenomatous polyposis. *Am. J. Gastroenterol.*, **101** (2), 385–398.
4. Harada, N., Tamai, Y., Ishikawa, T., Sauer, B. *et al.* (1999) Intestinal polyposis in mice with a dominant stable mutation of the beta-catenin gene. *EMBO J.*, **18** (21), 5931–5942.
5. Andreu, P., Colnot, S., Godard, C., Gad, S. *et al.* (2005) Crypt-restricted proliferation and commitment to the Paneth cell lineage following Apc loss in the mouse intestine. *Development*, **132** (6), 1443–1451.
6. Fearnhead, N.S., Britton, M.P., and Bodmer, W.F. (2001) The ABC of APC. *Hum. Mol. Genet.*, **10** (7), 721–733.
7. Kinzler, K.W. and Vogelstein, B. (1996) Lessons from hereditary colorectal cancer. *Cell*, **87** (2), 159–170.
8. Alison, M.R., Lim, S.M., and Nicholson, L.J. (2011) Cancer stem cells: problems for therapy? *J. Pathol.*, **223** (2), 147–161.
9. Vermeulen, L., Sprick, M.R., Kemper, K., Stassi, G. *et al.* (2008) Cancer stem cells – old concepts, new insights. *Cell Death Differ.*, **15** (6), 947–958.
10. Nowell, P.C. (1976) The clonal evolution of tumor cell populations. *Science*, **194** (4260), 23–28.
11. Al-Hajj, M. and Clarke, M.F. (2004) Self-renewal and solid tumor stem cells. *Oncogene*, **23** (43), 7274–7282.
12. Clarke, M.F., Dick, J.E., Dirks, P.B., Eaves, C.J. *et al.* (2006) Cancer stem cells – perspectives on current status and future directions: AACR Workshop on cancer stem cells. *Cancer Res.*, **66** (19), 9339–9344.
13. Dick, J.E. (2003) Stem cells: self-renewal writ in blood. *Nature*, **423** (6937), 231–233.
14. Boman, B.M. and Wicha, M.S. (2008) Cancer stem cells: a step toward the cure. *J. Clin. Oncol.*, **26** (17), 2795–2799.
15. Barker, N., Ridgway, R.A., van Es, J.H., van de Wetering, M. *et al.* (2009) Crypt stem cells as the cells-of-origin of intestinal cancer. *Nature*, **457** (7229), 608–611.
16. Booth, C. and Potten, C.S. (2000) Gut instincts: thoughts on intestinal epithelial stem cells. *J. Clin. Invest.*, **105** (11), 1493–1499.
17. Potten, C.S. and Loeffler, M. (1990) Stem cells: attributes, cycles, spirals, pitfalls and uncertainties. Lessons for and from the crypt. *Development*, **110** (4), 1001–1020.
18. Brittan, M. and Wright, N.A. (2002) Gastrointestinal stem cells. *J. Pathol.*, **197** (4), 492–509.
19. Cheng, H. and Leblond, C.P. (1974) Origin, differentiation and renewal of the four main epithelial cell types in the mouse small intestine. V. Unitarian Theory of the origin of the four epithelial cell types. *Am. J. Anat.*, **141** (4), 537–561.
20. Paulus, U., Potten, C.S., and Loeffler, M. (1992) A model of the control of cellular regeneration in the intestinal crypt after perturbation based solely on local stem cell regulation. *Cell Prolif.*, **25** (6), 559–578.
21. Kirkland, S.C. (1988) Clonal origin of columnar, mucous, and endocrine cell lineages in human colorectal epithelium. *Cancer*, **61** (7), 1359–1363.
22. Chang, W.W. and Leblond, C.P. (1971) Renewal of the epithelium in the descending colon of the mouse. I. Presence of three cell populations: vacuolated-columnar, mucous and argentaffin. *Am. J. Anat.*, **131** (1), 73–99.
23. Kim, S.J., Cheung, S., and Hellerstein, M.K. (2004) Isolation of nuclei from label-retaining cells and measurement of their turnover rates in rat colon. *Am. J. Physiol. Cell Physiol.*, **286** (6), C1464–C1473.
24. Nakamura, M., Okano, H., Blendy, J.A., and Montell, C. (1994) Musashi, a neural RNA-binding protein required for *Drosophila* adult external sensory organ development. *Neuron*, **13** (1), 67–81.
25. Sureban, S.M., May, R., George, R.J., Dieckgraefe, B.K. *et al.* (2008) Knockdown of RNA binding protein musashi-1 leads to

tumor regression in vivo. *Gastroenterology*, **134** (5), 1448–1458.
26. Sangiorgi, E. and Capecchi, M.R. (2008) *Bmi1* is expressed in vivo in intestinal stem cells. *Nat. Genet.*, **40** (7), 915–920.
27. Barker, N., van Es, J.H., Kuipers, J., Kujala, P. *et al.* (2007) Identification of stem cells in small intestine and colon by marker gene *Lgr5*. *Nature*, **449** (7165), 1003–1007.
28. Garcia, M.I., Ghiani, M., Lefort, A., Libert, F. *et al.* (2009) LGR5 deficiency deregulates Wnt signaling and leads to precocious Paneth cell differentiation in the fetal intestine. *Dev. Biol.*, **331** (1), 58–67.
29. O'Brien, C.A., Pollett, A., Gallinger, S., and Dick, J.E. (2007) A human colon cancer cell capable of initiating tumour growth in immunodeficient mice. *Nature*, **445** (7123), 106–110.
30. Ricci-Vitiani, L., Lombardi, D.G., Pilozzi, E., Biffoni, M. *et al.* (2007) Identification and expansion of human colon-cancer-initiating cells. *Nature*, **445** (7123), 111–115.
31. Todaro, M., Alea, M.P., Di Stefano, A.B., Cammareri, P. *et al.* (2007) Colon cancer stem cells dictate tumor growth and resist cell death by production of interleukin-4. *Cell Stem Cell*, **1** (4), 389–402.
32. Shmelkov, S.V., Butler, J.M., Hooper, A.T., Hormigo, A. *et al.* (2008) CD133 expression is not restricted to stem cells, and both CD133+ and CD133- metastatic colon cancer cells initiate tumors. *J. Clin. Invest.*, **118** (6), 2111–2120.
33. Dalerba, P., Dylla, S.J., Park, I.K., Liu, R. *et al.* (2007) Phenotypic characterization of human colorectal cancer stem cells. *Proc. Natl Acad. Sci. USA*, **104** (24), 10158–10163.
34. Huang, E.H., Hynes, M.J., Zhang, T., Ginestier, C. *et al.* (2009) Aldehyde dehydrogenase 1 is a marker for normal and malignant human colonic stem cells (SC) and tracks SC overpopulation during colon tumorigenesis. *Cancer Res.*, **69** (8), 3382–3389.
35. Adegboyega, P.A., Mifflin, R.C., DiMari, J.F., Saada, J.I. *et al.* (2002) Immunohistochemical study of myofibroblasts in normal colonic mucosa, hyperplastic polyps, and adenomatous colorectal polyps. *Arch. Pathol. Lab. Med.*, **126** (7), 829–836.
36. Powell, D.W., Mifflin, R.C., Valentich, J.D., Crowe, S.E. *et al.* (1999) Myofibroblasts. II. Intestinal subepithelial myofibroblasts. *Am. J. Physiol.*, **277** (2, Pt 1), C183–C201.
37. Vermeulen, L., De Sousa e Melo, F., van der Heijden, M., Cameron, K. *et al.* (2010) Wnt activity defines colon cancer stem cells and is regulated by the microenvironment. *Nat. Cell Biol.*, **12** (5), 468–476.
38. Dontu, G., Jackson, K.W., McNicholas, E., Kawamura, M.J. *et al.* (2004) Role of Notch signaling in cell-fate determination of human mammary stem/progenitor cells. *Breast Cancer Res.*, **6** (6), R605–R615.
39. Yilmaz, O.H., Valdez, R., Theisen, B.K., Guo, W. *et al.* (2006) Pten dependence distinguishes haematopoietic stem cells from leukaemia-initiating cells. *Nature*, **441** (7092), 475–482.
40. Jamieson, C.H., Ailles, L.E., Dylla, S.J., Muijtjens, M. *et al.* (2004) Granulocyte-macrophage progenitors as candidate leukemic stem cells in blast-crisis CML. *N. Engl. J. Med.*, **351** (7), 657–667.
41. Korkaya, H., Paulson, A., Charafe-Jauffret, E., Ginestier, C. *et al.* (2009) Regulation of mammary stem/progenitor cells by PTEN/Akt/beta-catenin signaling. *PLoS Biol.*, **7** (6), e1000121.
42. Hoey, T., Yen, W.C., Axelrod, F., Basi, J. *et al.* (2009) DLL4 blockade inhibits tumor growth and reduces tumor-initiating cell frequency. *Cell Stem Cell*, **5** (2), 168–177.
43. Dierks, C., Beigi, R., Guo, G.R., Zirlik, K. *et al.* (2008) Expansion of Bcr-Abl-positive leukemic stem cells is dependent on Hedgehog pathway activation. *Cancer Cell*, **14** (3), 238–249.
44. Zhao, C., Chen, A., Jamieson, C.H., Fereshteh, M. *et al.* (2009) Hedgehog signalling is essential for maintenance of cancer stem cells in myeloid leukaemia. *Nature*, **458** (7239), 776–779.
45. Wicha, M.S., Liu, S., and Dontu, G. (2006) Cancer stem cells: an old idea – a paradigm shift. *Cancer Res.*, **66** (4), 1883–1890; discussion 1895–1896.
46. Liu, S., Dontu, G., Mantle, I.D., Patel, S. *et al.* (2006) Hedgehog signaling and Bmi-1 regulate self-renewal of normal and malignant human mammary stem cells. *Cancer Res.*, **66** (12), 6063–6071.
47. Clement, V., Sanchez, P., de Tribolet, N., Radovanovic, I. *et al.* (2007) HEDGEHOG-GLI1 signaling regulates human glioma

growth, cancer stem cell self-renewal, and tumorigenicity. *Curr. Biol.*, **17** (2), 165–172.
48. Varnat, F., Duquet, A., Malerba, M., Zbinden, M. *et al.* (2009) Human colon cancer epithelial cells harbour active HEDGEHOG-GLI signalling that is essential for tumour growth, recurrence, metastasis and stem cell survival and expansion. *EMBO Mol. Med.*, **1** (6-7), 338–351.
49. Reya, T. and Clevers, H. (2005) Wnt signalling in stem cells and cancer. *Nature*, **434** (7035), 843–850.
50. He, T.C., Sparks, A.B., Rago, C., Hermeking, H. *et al.* (1998) Identification of c-MYC as a target of the APC pathway. *Science*, **281** (5382), 1509–1512.
51. Crosnier, C., Stamataki, D., and Lewis, J. (2006) Organizing cell renewal in the intestine: stem cells, signals and combinatorial control. *Nat. Rev. Genet.*, **7** (5), 349–359.
52. Holmberg, J., Genander, M., Halford, M.M., Annerén, C. *et al.* (2006) EphB receptors coordinate migration and proliferation in the intestinal stem cell niche. *Cell*, **125** (6), 1151–1163.
53. Gregorieff, A., Pinto, D., Begthel, H., Destrée, O. *et al.* (2005) Expression pattern of Wnt signaling components in the adult intestine. *Gastroenterology*, **129** (2), 626–638.
54. Chen, D., Ji, X., Harris, M.A., Feng, J.Q. *et al.* (1998) Differential roles for bone morphogenetic protein (BMP) receptor type IB and IA in differentiation and specification of mesenchymal precursor cells to osteoblast and adipocyte lineages. *J. Cell Biol.*, **142** (1), 295–305.
55. Miyazono, K. (2000) TGF-beta signaling by Smad proteins. *Cytokine Growth Factor Rev.*, **11** (1-2), 15–22.
56. Derynck, R. and Zhang, Y.E. (2003) Smad-dependent and Smad-independent pathways in TGF-beta family signalling. *Nature*, **425** (6958), 577–584.
57. Miyazono, K., Kamiya, Y., and Morikawa, M. (2010) Bone morphogenetic protein receptors and signal transduction. *J. Biochem.*, **147** (1), 35–51.
58. Lombardo, Y., Scopelliti, A., Cammareri, P., Todaro, M. *et al.* (2011) Bone morphogenetic protein 4 induces differentiation of colorectal cancer stem cells and increases their response to chemotherapy in mice. *Gastroenterology*, **140** (1), 297–309.
59. Kosinski, C., Li, V.S., Chan, A.S., Zhang, J. *et al.* (2007) Gene expression patterns of human colon tops and basal crypts and BMP antagonists as intestinal stem cell niche factors. *Proc. Natl Acad. Sci. USA*, **104** (39), 15418–15423.
60. Rider, C.C. and Mulloy, B. (2010) Bone morphogenetic protein and growth differentiation factor cytokine families and their protein antagonists. *Biochem. J.*, **429** (1), 1–12.
61. Artavanis-Tsakonas, S., Rand, M.D., and Lake, R.J. (1999) Notch signaling: cell fate control and signal integration in development. *Science*, **284** (5415), 770–776.
62. Kim, T.H. and Shivdasani, R.A. (2011) Genetic evidence that intestinal Notch functions vary regionally and operate through a common mechanism of Math1 repression. *J. Biol. Chem.*, **286** (13), 11427–11433.
63. van Es, J.H., de Geest, N., van de Born, M., Clevers, H. *et al.* (2010) Intestinal stem cells lacking the Math1 tumour suppressor are refractory to Notch inhibitors. *Nat. Commun.*, **1**, 18.
64. Hegde, G.V., Munger, C.M., Emanuel, K., Joshi, A.D. *et al.* (2008) Targeting of sonic hedgehog-GLI signaling: a potential strategy to improve therapy for mantle cell lymphoma. *Mol. Cancer Ther.*, **7** (6), 1450–1460.
65. Huang, E.H. and Wicha, M.S. (2008) Colon cancer stem cells: implications for prevention and therapy. *Trends Mol. Med.*, **14** (11), 503–509.
66. Labianca, R., Nordlinger, B., Beretta, G.D., Brouquet, A. *et al.* (2010) Primary colon cancer: ESMO Clinical Practice Guidelines for diagnosis, adjuvant treatment and follow-up. *Ann. Oncol.*, **21** (Suppl. 5), v70–v77.
67. Andre, T., Afchain, P., Barrier, A., Blanchard, P. *et al.* (2007) Current status of adjuvant therapy for colon cancer. *Gastrointest Cancer Res.*, **1** (3), 90–97.
68. Gray, R., Barnwell, J., McConkey, C., Hills, R.K. *et al.* Quasar Collaborative Group (2007) Adjuvant chemotherapy versus observation in patients with colorectal cancer: a randomised study. *Lancet*, **370** (9604), 2020–2029.
69. Panettiere, F.J., Goodman, P.J., Costanzi, J.J., Cruz, A.B., Jr. *et al.* (1988) Adjuvant therapy in large bowel adenocarcinoma: long-term

results of a Southwest Oncology Group Study. *J. Clin. Oncol.*, **6** (6), 947–954.
70. Wolmark, N., Fisher, B., Rockette, H., Redmond, C. *et al.* (1988) Postoperative adjuvant chemotherapy or BCG for colon cancer: results from NSABP protocol C-01. *J. Natl. Cancer Inst.*, **80** (1), 30–36.
71. Laurie, J.A., Moertel, C.G., Fleming, T.R., Wieand, H.S. *et al.* (1989) Surgical adjuvant therapy of large-bowel carcinoma: an evaluation of levamisole and the combination of levamisole and fluorouracil. The North Central Cancer Treatment Group and the Mayo Clinic. *J. Clin. Oncol.*, **7** (10), 1447–1456.
72. Moertel, C.G., Fleming, T.R., Macdonald, J.S., Haller, D.G. *et al.* (1990) Levamisole and fluorouracil for adjuvant therapy of resected colon carcinoma. *N. Engl. J. Med.*, **322** (6), 352–358.
73. Wolmark, N., Rockette, H., Fisher, B., Wickerham, D.L. *et al.* (1993) The benefit of leucovorin-modulated fluorouracil as postoperative adjuvant therapy for primary colon cancer: results from National Surgical Adjuvant Breast and Bowel Project protocol C-03. *J. Clin. Oncol.*, **11** (10), 1879–1887.
74. International Multicentre Pooled Analysis of Colon Cancer Trials (IMPACT) investigators (1995) Efficacy of adjuvant fluorouracil and folinic acid in colon cancer. *Lancet*, **345** (8955), 939–944.
75. Haller, D.G., Catalano, P.J., Macdonald, J.S., O'Rourke, M.A. *et al.* (2005) Phase III study of fluorouracil, leucovorin, and levamisole in high-risk stage II and III colon cancer: final report of Intergroup 0089. *J. Clin. Oncol.*, **23** (34), 8671–8678.
76. Andre, T., Boni, C., Mounedji-Boudiaf, L., Navarro, M. *et al.* (2004) Oxaliplatin, fluorouracil, and leucovorin as adjuvant treatment for colon cancer. *N. Engl. J. Med.*, **350** (23), 2343–2351.
77. Kuebler, J.P., Wieand, H.S., O'Connell, M.J., Smith, R.E. *et al.* (2007) Oxaliplatin combined with weekly bolus fluorouracil and leucovorin as surgical adjuvant chemotherapy for stage II and III colon cancer: results from NSABP C-07. *J. Clin. Oncol.*, **25** (16), 2198–2204.
78. Saltz, L.B., Niedzwiecki, D., Hollis, D., Goldberg, R.M. *et al.* (2007) Irinotecan fluorouracil plus leucovorin is not superior to fluorouracil plus leucovorin alone as adjuvant treatment for stage III colon cancer: results of CALGB 89803. *J. Clin. Oncol.*, **25** (23), 3456–3461.
79. Petrelli, N., Douglass, H.O., Jr, Herrera, J., Russell, D. *et al.* (1989) The modulation of fluorouracil with leucovorin in metastatic colorectal carcinoma: a prospective randomized phase III trial. Gastrointestinal Tumor Study Group. *J. Clin. Oncol.*, **7** (10), 1419–1426.
80. Buyse, M., Thirion, P., Carlson, R.W., Burzykowski, T. *et al.* (2000) Relation between tumour response to first-line chemotherapy and survival in advanced colorectal cancer: a meta-analysis. Meta-Analysis Group in Cancer. *Lancet*, **356** (9227), 373–378.
81. Van Cutsem, E., Twelves, C., Cassidy, J., Allman, D. *et al.* (2001) Oral capecitabine compared with intravenous fluorouracil plus leucovorin in patients with metastatic colorectal cancer: results of a large phase III study. *J. Clin. Oncol.*, **19** (21), 4097–4106.
82. Hoff, P.M., Ansari, R., Batist, G., Cox, J. *et al.* (2001) Comparison of oral capecitabine versus intravenous fluorouracil plus leucovorin as first-line treatment in 605 patients with metastatic colorectal cancer: results of a randomized phase III study. *J. Clin. Oncol.*, **19** (8), 2282–2292.
83. Tournigand, C., André, T., Achille, E., Lledo, G. *et al.* (2004) FOLFIRI followed by FOLFOX6 or the reverse sequence in advanced colorectal cancer: a randomized GERCOR study. *J. Clin. Oncol.*, **22** (2), 229–237.
84. Colucci, G., Gebbia, V., Paoletti, G., Giuliani, F. *et al.* (2005) Phase III randomized trial of FOLFIRI versus FOLFOX4 in the treatment of advanced colorectal cancer: a multicenter study of the Gruppo Oncologico Dell'Italia Meridionale. *J. Clin. Oncol.*, **23** (22), 4866–4875.
85. Fuchs, C.S., Marshall, J., Mitchell, E., Wierzbicki, R. *et al.* (2007) Randomized, controlled trial of irinotecan plus infusional, bolus, or oral fluoropyrimidines in first-line treatment of metastatic colorectal cancer: results from the BICC-C Study. *J. Clin. Oncol.*, **25** (30), 4779–4786.
86. Diaz-Rubio, E., Tabernero, J., Gómez-España, A., Massutí, B. *et al.* (2007) Phase III study of capecitabine plus oxaliplatin compared with continuous-infusion fluorouracil plus oxaliplatin as first-line therapy in metastatic colorectal cancer:

final report of the Spanish Cooperative Group for the Treatment of Digestive Tumors Trial. *J. Clin. Oncol.*, **25** (27), 4224–4230.

87. Los, M., Roodhart, J.M., and Voest, E.E. (2007) Target practice: lessons from phase III trials with bevacizumab and vatalanib in the treatment of advanced colorectal cancer. *Oncologist*, **12** (4), 443–450.
88. El Zouhairi, M., Charabaty, A., and Pishvaian, M.J. (2011) Molecularly targeted therapy for metastatic colon cancer: proven treatments and promising new agents. *Gastrointest Cancer Res.*, **4** (1), 15–21.
89. Hurwitz, H., Fehrenbacher, L., Novotny, W., Cartwright, T. *et al.* (2004) Bevacizumab plus irinotecan, fluorouracil, and leucovorin for metastatic colorectal cancer. *N. Engl. J. Med.*, **350** (23), 2335–2342.
90. Sanoff, H.K., Sargent, D.J., Campbell, M.E., Morton, R.F. *et al.* (2008) Five-year data and prognostic factor analysis of oxaliplatin and irinotecan combinations for advanced colorectal cancer: N9741. *J. Clin. Oncol.*, **26** (35), 5721–5727.
91. Giantonio, B.J., Catalano, P.J., Meropol, N.J., O'Dwyer, P.J. *et al.* (2007) Bevacizumab in combination with oxaliplatin, fluorouracil, and leucovorin (FOLFOX4) for previously treated metastatic colorectal cancer: results from the Eastern Cooperative Oncology Group Study E3200. *J. Clin. Oncol.*, **25** (12), 1539–1544.
92. Van Cutsem, E., Köhne, C.H., Hitre, E., Zaluski, J. *et al.* (2009) Cetuximab and chemotherapy as initial treatment for metastatic colorectal cancer. *N. Engl. J. Med.*, **360** (14), 1408–1417.
93. Cunningham, D., Humblet, Y., Siena, S., Khayat, D. *et al.* (2004) Cetuximab monotherapy and cetuximab plus irinotecan in irinotecan-refractory metastatic colorectal cancer. *N. Engl. J. Med.*, **351** (4), 337–345.
94. Tol, J., Koopman, M., Cats, A., Rodenburg, C.J. *et al.* (2009) Chemotherapy, bevacizumab, and cetuximab in metastatic colorectal cancer. *N. Engl. J. Med.*, **360** (6), 563–572.
95. Van Cutsem, E., Peeters, M., Siena, S., Humblet, Y. *et al.* (2007) Open-label phase III trial of panitumumab plus best supportive care compared with best supportive care alone in patients with chemotherapy-refractory metastatic colorectal cancer. *J. Clin. Oncol.*, **25** (13), 1658–1664.
96. Douillard, J.Y., Siena, S., Cassidy, J., Tabernero, J. *et al.* (2010) Randomized, phase III trial of panitumumab with infusional fluorouracil, leucovorin, and oxaliplatin (FOLFOX4) versus FOLFOX4 alone as first-line treatment in patients with previously untreated metastatic colorectal cancer: the PRIME study. *J. Clin. Oncol.*, **28** (31), 4697–4705.
97. Hecht, J.R., Mitchell, E., Chidiac, T., Scroggin, C. *et al.* (2009) A randomized phase IIIB trial of chemotherapy, bevacizumab, and panitumumab compared with chemotherapy and bevacizumab alone for metastatic colorectal cancer. *J. Clin. Oncol.*, **27** (5), 672–680.
98. Peeters, M., Price, T.J., Cervantes, A., Sobrero, A.F. *et al.* (2010) Randomized phase III study of panitumumab with fluorouracil, leucovorin, and irinotecan (FOLFIRI) compared with FOLFIRI alone as second-line treatment in patients with metastatic colorectal cancer. *J. Clin. Oncol.*, **28** (31), 4706–4713.
99. Abraham, R.T. and Gibbons, J.J. (2007) The mammalian target of rapamycin signaling pathway: twists and turns in the road to cancer therapy. *Clin. Cancer Res.*, **13** (11), 3109–3114.
100. Punt, C.J., Boni, J., Bruntsch, U., Peters, M. *et al.* (2003) Phase I and pharmacokinetic study of CCI-779, a novel cytostatic cell-cycle inhibitor, in combination with 5-fluorouracil and leucovorin in patients with advanced solid tumors. *Ann. Oncol.*, **14** (6), 931–937.
101. Mackay, H.J. and Twelves, C.J. (2007) Targeting the protein kinase C family: are we there yet? *Nat. Rev. Cancer*, **7** (7), 554–562.
102. Carducci, M.A., Musib, L., Kies, M.S., Pili, R. *et al.* (2006) Phase I dose escalation and pharmacokinetic study of enzastaurin, an oral protein kinase C beta inhibitor, in patients with advanced cancer. *J. Clin. Oncol.*, **24** (25), 4092–4099.
103. Summy, J.M. and Gallick, G.E. (2006) Treatment for advanced tumors: SRC reclaims center stage. *Clin. Cancer Res.*, **12** (5), 1398–1401.
104. Kemper, K., Grandela, C., and Medema, J.P. (2010) Molecular identification and

targeting of colorectal cancer stem cells. *Oncotarget*, **1** (6), 387–395.

105. Francipane, M.G., Alea, M.P., Lombardo, Y., Todaro, M. *et al.* (2008) Crucial role of interleukin-4 in the survival of colon cancer stem cells. *Cancer Res.*, **68** (11), 4022–4025.

106. Todaro, M., Francipane, M.G., Medema, J.P., and Stassi, G. (2010) Colon cancer stem cells: promise of targeted therapy. *Gastroenterology*, **138** (6), 2151–2162.

107. Sneddon, J.B. and Werb, Z. (2007) Location, location, location: the cancer stem cell niche. *Cell Stem Cell*, **1** (6), 607–611.

108. Blattman, J.N. and Greenberg, P.D. (2004) Cancer immunotherapy: a treatment for the masses. *Science*, **305** (5681), 200–205.

109. Galon, J., Costes, A., Sanchez-Cabo, F., Kirilovsky, A. *et al.* (2006) Type, density, and location of immune cells within human colorectal tumors predict clinical outcome. *Science*, **313** (5795), 1960–1964.

110. Todaro, M., D'Asaro, M., Caccamo, N., Iovino, F. *et al.* (2009) Efficient killing of human colon cancer stem cells by gammadelta T lymphocytes. *J. Immunol.*, **182** (11), 7287–7296.

26
Strategies for Targeting Cancer Stem Cells

R. Allan Mufson
National Cancer Institute, Cancer Immunology/Hematology Branch, 6130 Executive Boulevard, Rockville, MD 20852, USA

Translational Medicine Cancer: Advances in Molecular Biology and Medicine,
First Edition. Edited by Robert A. Meyers.

Keywords

Stem cell
An unspecialized cell characterized by the ability to self-renew by mitosis while in undifferentiated state, and the capacity to give rise to various differentiated cell types by cell differentiation.

Self-renewal
The ability of a cell to divide and give rise to daughter cells, one of which is committed to a differentiated state, while the other daughter cell remains in an undifferentiated state capable of dividing subsequently.

Quiescence
A cell cycle condition in which a cell either withdraws from the cell cycle or remains in a state of extended G_0.

Tumorigenesis
The ability of a cell or population of cells to generate a tumor with all the cellular types found in the parent tumor from which the cells were isolated.

Chemotherapeutic
Any chemical used to treat cancer. This term is normally used to refer to antineoplastic drugs.

Recent tumor biology research has revealed that not all tumor cells have tumor-initiating properties. Indeed, among an increasing number of both hematologic and solid tumor models there is only a subpopulation of cells capable of initiating new tumors. These tumor cells, which appear to have a self-renewal property that allows them to initiate new tumors, have been termed tumor stem cells. Tumor stem cells have many of the properties of normal stem cells, including resistance to radiation and toxic chemicals. These properties make such cells resistant to the normal therapeutic modalities that result in tumor regression. Since it is necessary to eliminate this minority population of cells in order to prevent the re-emergence of tumors after therapy, new modalities for targeting these cells must be developed. Any new therapeutics capable of targeting tumor stem cells must take advantage of an emerging understanding of the unique cell-surface markers, signaling pathways and metabolic characteristics of the these cells.

1 Cancer or Tumor Stem Cells: Identification and Characterization

Most tumors consist of phenotypically and functionally heterogeneous populations of cells, with such heterogeneity having most often been ascribed to a stochastic process associated with clonal evolution. This stochastic concept was based on the idea that all tumor cells possess a complete tumorigenic potential that can be expressed through an accumulation of genetic or epigenetic alterations [1]. Recently, however, a new paradigm has emerged suggesting that a hierarchy exists among tumor cells, with only a minority cell population possessing the potential to recapitulate all of the cell types found in the parent tumor [2, 3]. Such cells exhibit biological similarities to normal tissue stem cell physiology [4, 5]:

- These stem-like cells appear to be more quiescent, with an extended cell cycle.
- They may give rise to shorter-lived, more differentiated cells, as well as being capable of self-renewal.
- The cells may be highly susceptible to signals from the tumor microenvironment.
- They may express surface markers or antigens or specific signaling pathways related to normal stem cell biology.
- They express enhanced levels of ATP binding cassette (ABC) transporters and DNA repair mechanisms, which enhance their resistance to commonly used radiotherapeutic and chemotherapeutic protocols.

Consequently, this population of cells has been termed cancer stem cells, tumor stem cells, or tumor-initiating cells. Such cells have the capacity to self-renew through asymmetric cell division that is characteristic of normal tissue stem cells. In addition, they can generate all of the cell types that comprise the tumor of origin. Tumor stem cells were first identified in acute myeloid leukemia (AML) by employing limiting dilution techniques and transplantation into immune-deficient mice [2, 3]. Similar techniques have been used to identify tumor stem cells in tumors of the brain, colon, bladder, pancreas, liver, lung, ovary and prostate gland.

The initial descriptions of tumor stem cells in AML indicated that they could be identified by cell-surface markers, and that they represented only a small fraction of the tumor cell population. For example, in AML the $CD34^+$, $CD38^-$ tumor stem cells were shown to constitute only 0.02–0.8% of the total leukemic cells [6]. Other tumor stem cells with specific cell-surface markers have been also identified, with varying frequencies. In human breast tumors the CD44+, $CD24^-$ cells constituted about 11–35% of the total tumor cell number, while in human brain the $CD133^+$ stem-like cells constituted about 0.3–25% of the total tumor cells. In colon carcinoma, the $CD133^{+,}$ $EpCAM^{hi}$, $CD44^+$ cells represented about 2.5% of the bulk tumor cells, while in the pancreas $CD44^+$, $CD34^+$, ESA^+ cells represented 0.2–0.8% of the total tumor cells [7–10]. Thus, it is likely that tumor stem cell proportions may vary widely among tumor arising from different tissues types, and cell-surface markers – such as those noted above – may not serve as useful markers for a wide variety of tumors arising from different tissues.

Metabolic markers for tumor stem cells have, however, been identified that may allow the identification of tumor stem cells more broadly. The aldehyde dehydrogenases (ALDHs) are a family of cytosolic enzymes involved in the detoxification of intracellular aldehydes, as well as the conversion of retinol to retinoic acid, which is critical in mammalian development and early stem cell differentiation. Among this family, ALDH1 is involved in the conversion of acetaldehyde to acetic acid, the second step in the metabolism of alcohol [11]. Several studies have correlated an overexpression of ALDH1 with an increased malignant potential of tumor cells of varied histologic origin. ALDH1 was first described as a candidate marker for tumor stem cells in head and neck squamous cell carcinoma

[12, 13], and subsequently was used as a tumor stem cell marker in cancers of the breast, colon, pancreas, prostate and lung [14, 15]. ALDH1 overexpression, in concert with CD44 positivity, correlated with an increased tumorigenesis in comparison with cells marked with CD44 alone in head and neck squamous carcinoma and breast cancers. Subpopulations of both these tumor types, marked by both CD44 expression and ALDH1 overexpression, had a high tumorigenic potential and a poorer prognosis than tumor-initiating cells bearing either marker alone [16, 17]. Pyruvate kinase (PK) M2 is an enzyme that catalyzes the final step of glycolysis to generate ATP. PKM2 is found in all tissues during embryogenesis, and is gradually replaced by other isoforms of PK during development [18]. Tumor cells require a constant supply of glycolytic intermediates for cell proliferation. It has been found that CD44 interacts with PKM2 and enhances the glycolytic phenotype of tumor cells that are either p53-deficient or exposed to hypoxia [19]. CD44 ablation in tumor cells increased the metabolic flux into mitochondrial respiration, and concomitantly inhibited entry into glycolysis and the pentose phosphate shunt. CD44 ablation was shown not only to induce this metabolic shift but also to lead to a decreased glucose uptake in tumor cells and decreased glutathione synthesis [19]. Taken together, these results suggested that CD44-targeted therapies may render such cells more susceptible to reactive oxygen species.

1.1 Tumor Stem Cell Dynamics

In some tumors, it is also possible that tumor stem cells and the non-stem population of tumor cells have a modicum of inconvertibility, depending on the signals in the microenvironment. In the melanoma system, it was found that histone H3 lysine-4 demethylase, JARID1B, was expressed in only a small population of human malignant melanoma cells [20]. These JARID1B^{+} cells cycled only slowly and possessed a high self-renewal potential. The reduction of JARID1B expression by RNA interference resulted in a reduced tumor-initiating ability of these cells, which suggested that they have the characteristics of tumor stem cells. Significantly, it was also reported that JARID1B^{-} melanoma cells could give rise to JARID1B^{+} cells after long-term cell culture, and that these cells possessed tumor-initiating capability after injection into immunopromised host mice. There was, therefore, the possibility of a dynamic interconvertibility between the stem and non-stem phenotype in melanoma. Thus, the stringent hierarchy in which only tumor stem cells can yield both non-stem tumor cells and tumor stem cells may apply only to tumors derived from tissues that have a more hierarchical development pattern, such as those found in the hematopoietic system.

Data from several tumor types, including melanoma in which there is a phenotypic plasticity associated with tumor stem cells, suggest that such plasticity may result from epigenetic modulation by signals from the tumor microenvironment. In fact, melanoma cells exposed to an inflammatory environment rich in CD11B^{+} myeloid cells, secreting TNFα, show the loss of melanocytic antigens on melanoma cells and an increase in nerve growth factor receptor (CD271^{+}) expressing spindle-shaped cells in the inflammatory microenvironment [21]. CD271 has further been shown to be a marker for tumor stem cells in human melanoma. The transplantation of $CD271^+$ human melanoma cells into mice, lacking both adaptive and innate immune cells, resulted in the establishment of melanomas resembling the tumors that were the source of the cells [22]. Cells from these transplanted tumors were capable of metastasis *in vivo*; however, melanomas did not develop when $CD271^-$ cells were transplanted [22]. The $CD271^+$ cells also lacked expression of characteristic melanocytic antigens such as MART1 and MAGE. It has also been shown that TNFα upregulated CD271 expression, and downregulated melanocytic specific antigen expression in both highly and poorly differentiated melanoma cell lines [21]. Interestingly, cutaneous metastases from melanoma patients showed a significant expression of CD271 on subpopulations of cells in close proximity to infiltrating immunocytes [21]. In colon carcinoma, more differentiated tumor cells can acquire a stem cell phenotype after exposure to hepatocytic growth factor (HGF) released from myofibroblasts, while in glioblastoma hypoxia induced a conversion towards a stem-like phenotype in non-stem tumor cells, as evidenced by increases in self-renewal potential and the expression of CD133 [23, 24]. Thus, in particular solid tumors the generation of tumor stem cells may be a dynamic process susceptible to signals from the tumor microenvironment.

1.1.1 Clonal Diversity in the Cancer Stem Cell Model

Recent studies with human B-cell acute lymphocytic leukemia (B-ALL) showed that adult human B-ALL driven by the chromosomal translocations BCR-ABL, and childhood ALL driven by ETV6-RUNX-1 fusion, are genetically diverse [25, 26]. Documentation of the DNA copy number variation in individual leukemic cells, through global single nucleotide polymorphism arrays or multiplex fluorescence *in situ* hybridization, revealed the existence of multiple genetically unique leukemia-initiating clones, as assayed by tumor formation after transfer into immune-compromised non-obese diabetic severe combined immunodeficiency/IL-2 receptor ϒ-deleted (NOD/SCID IL2Rϒ null) mice. These clonal variants are believed to develop through the acquisition of mutations in a multiple branching manner, rather than through linear succession. It also appeared that the pattern of clonal variation evident at the early diagnosis of disease was different from that of overt disease, or in disease relapse. Presumably, only those clones present at diagnosis that were most resistant to chemotherapy and with a high tumor-initiating capacity were able to survive, evolve and cause relapse.

Among the solid tumors, colon carcinoma stem cells also demonstrate functionally distinct clones. By using lentiviral marking of colon carcinoma tumor stem cells, three major subclones were identified [27]. The first subclone maintained self-renewal capacity over a long period of time in serial transplantation assays into NOD/SCID/IL-2Rϒ null mice, while the second clone, with a limited self-renewal capacity, formed tumors only in primary transplants and had little or no serial transplant tumor-generating capacity. A third series of clones was found only in secondary or tertiary transplants. Interestingly, the clone with the long-term self-renewal activity metastasized to bone, and was the only clone capable of metastasis. These results suggest a clonal diversity among tumor stem cells that may result from a differential acquisition of mutations. Although it is unclear whether such clonal diversity among tumor stem cells exists in other tumor types, differential sensitivity to antitumor stem cell therapies merits further investigation.

2
Tumor Stem Cell-Targeted Therapy

The presence of tumor stem cells, with unique stem cell-like properties that differ from those of the bulk of more rapidly dividing tumor cells, can lead to a resistance to commonly used chemotherapeutic agents and radiologic-based therapies. Currently, these therapies are directed against rapidly dividing cells, rather than cells that divide infrequently and may be more quiescent, such as tumor stem cells. Thus, it is clear that if tumor stem cells have a different sensitivity to therapeutic modalities than do the bulk of the tumor cells, therapy will not succeed in completely eradicating tumors, because tumor shrinkage does not reflect the elimination of the tumor stem cells. Therefore, therapies must be developed that target both the more quiescent stem cells, as well as the rapidly dividing, more-differentiated tumor cells. Possible strategies to specifically target tumor stem cells will be considered below (Fig. 1).

2.1
Immunotherapy: Targeting Cell-Surface Markers

2.1.1 Anti CD123 Antibodies
In AML, IL-3 receptor α chain (CD123) expression distinguishes leukemic stem cells from normal hematopoietic stem cells and the bulk of leukemic cells [28]. Based on this observation, a series of antibodies was developed to selectively target the AML stem cells [29]. The anti CD123 neutralizing antibody (7G3) appeared to impair the homing of these cells to bone marrow, and to activate

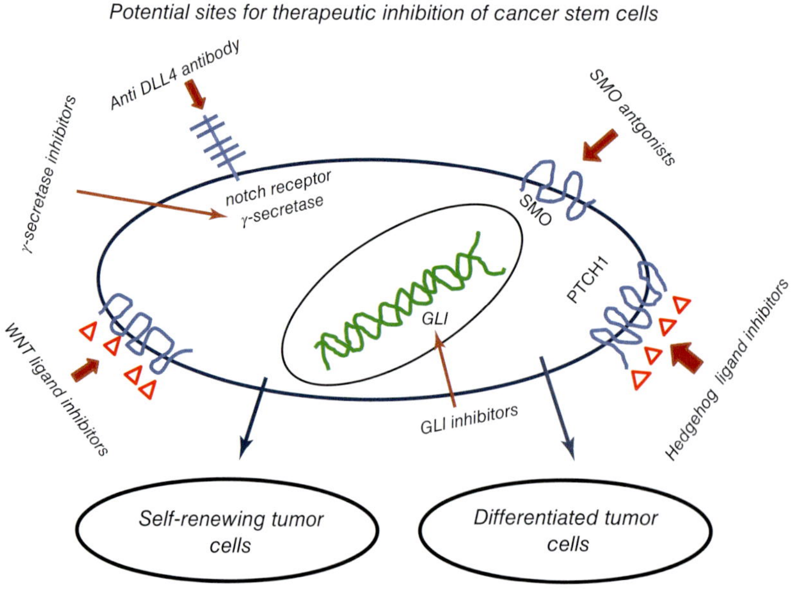

Fig. 1 Potential sites for therapeutic inhibition of cancer stem cells.

innate immunity of tumor-transplanted NOD/SCID mice. The 7G3 antibody inhibited IL-3 mediated intracellular signaling in isolated AML $CD34^+$, $CD38^-$ cells *in vitro*, and reduced cell survival.

2.1.2 Anti CD44 Antibodies

CD44 is a cell-surface glycoprotein with multiple isoforms that functions in cell adhesion, cloning and migration [30]. This antigen has also been shown to have some role in tumor metastasis [31]. CD44 has been identified as a marker for tumor stem cells in a number of tumors, including colon, pancreatic and breast (as noted above). The *in-vivo* targeting of CD44 by small interfering RNA (siRNA) or specific antibodies results in antitumor activity in xenografts of human colon carcinoma and acute leukemia, respectively [32, 33]. It has also been reported that targeting CD44 with monoclonal antibody 245, in an experimental model of human breast cancer xenografts, inhibits tumor growth and post-chemotherapy tumor relapse in both estrogen receptor (ER) -positive and basal-like breast tumors [34].

2.1.3 Anti CD133 Antibodies

Antibodies linked to toxins have also been used to specifically target tumor stem cells. A monoclonal antibody against CD133 has been coupled to a potent antitumor enediyne from the class of calicheamicin antibiotics, a drug used commonly to treat AML [35]. CD133 is expressed on 90% of AMLs, but not on $CD34^+$ normal bone marrow hematopoietic stem cells. The toxin–antibody conjugate binds cell-surface CD133 and is internalized. The chemotherapeutic agent is a potent intercalating molecule, that is released by the low pH of the endosome [35].

2.1.4 Anti EpCAM T-Cell Therapy

EpCAM or CD326 is a 30–40 kDa glycosylated adhesion molecule that is expressed on a number of tumor stem cells. EpCAM has also been designated as tumor-associated signal transducer (TACSTD) 1, and functions as a homotypic cell adhesion molecule, being linked to the WNT signaling pathway through the cadherin catenin interaction [36, 37]. Although EpCAM is also found in normal epithelial cells, in this case it is predominantly located in intercellular spaces in association with epithelial cell tight junctions [38, 39]. The overexpression of EpCAM is associated with a decreased survival for a number of histologic tumor types, including breast, gallbladder, bile duct, ovarian and ampullary pancreatic tumors. Based on its unique cellular distribution and its presence on colon carcinoma tumor stem cells, EpCAM has been seen as a possible therapeutic target for immunotherapy [40]. Engineered bispecific antibodies have been produced that are designed to direct cytotoxic $CD8^+$ T cells against tumor cells expressing this antigen. The antibody contains two variable domains that recognize EpCAM, and two other variable domains to form the binding site for the CD3 complex on T cells [41]. These antibodies induce an immunological synapse between the T cells and tumor cells, similar to those observed in a physiological T-cell attack during which lytic perforins and granzyme are directed through the synapse into the target cell [41]. Preliminary studies showed that these antibodies can lyse colorectal tumor stem cells *in vitro* [42]. *In-vivo* studies using this same bispecific antibody against EpCAM in an orthotopic mouse model of breast cancer in syngeneic mice prevented tumor growth against a 5000-fold excess of a tumor-forming cell dose [42].

2.2
Small Molecules: Targeting Self-Renewal and Embryonic Development Pathways

As mentioned previously, tumor stem cells often exhibit the self-renewal capacity and asymmetric cell division characteristic of normal tissue stem cells. Often, in tumor stem cells these self-renewal pathways are dysregulated. However, it is important to note that such self-renewal pathways are not active in the bulk of more differentiated cells in the tumor. Thus, targeting these dysregulated pathways should allow opportunities to develop small-molecule therapeutics specific for tumor stem cells.

2.2.1 Hedgehog Signaling Pathway

The hedgehog (Hh) signaling pathway is one of the key regulators of metazoan development controlling cell polarity, maintenance of patterning, and maintenance of stem cells during embryogenesis [43]. The signaling pathway takes its name from an intercellular signaling ligand called Hedgehog, originally identified in Drosophila fruit flies. The major molecules in the Hh pathway are the Hh ligands, the receptor Patched (Ptch), and the intracellular transducing molecules Smoothened (Smo) and Glisactin (Gli family of zinc finger transcription factors) [44]. In mammals, one of three Hh pathway ligands (Desert, Indian, and Sonic) binds to the *trans*-membrane receptor Ptch to initiate pathway signaling. In the inactive state, Ptch inhibits the Smoothened signal transducer (Smo) by preventing its localization to the primary cilium, a nonmotile structure present on most vertebrate cells, and no intracellular signaling is initiated. When the Hh ligands bind to the receptor, the complex is internalized and the inhibition of Smo is relieved. Intracellular signaling then occurs, regulating expression of the Gli transcription factor [44]. Patients with mutations in the human homolog of Ptch are highly prone to develop basal cell carcinoma and medulloblastoma [45, 46]. Accumulating evidence indicates involvement of the Hh pathway in tumor stem cell biology. The control of self-renewal pathways in chronic myeloid leukemia (CML) tumor stem cells is clearly regulated by the Hh pathway [47]. In multiple myeloma, there is distinct asymmetry in the distribution of Hh signaling molecules, and sensitivity to Hh pathway inhibitors between B-cell like progenitor cells and mature multiple myeloma plasmacytoid cells [48]. In addition, the Hh pathway inhibitors significantly inhibit the glioblastoma multiforme stem cell cloning efficiency [49], and preferentially deplete pancreatic tumor stem cells [50].

In $CD44^+$, $CD24^{low}$ breast tumor stem cells, Ptch1, GLi1 and Gli2 mRNA levels are increased about one- to seven-, thirty-, and sixfold, respectively, compared to the levels of these molecules in more differentiated mammary tumor cells [51, 52]. In addition, Bmi-1, a polycomb gene repressor, which has a role in hematopoietic and neural stem cell self-renewal, has also been shown to be a downstream target of the Hh pathway and plays a role in normal mammary stem cell self-renewal [52]. Overexpression of Bmi-1 increased the self-renewal capacity of normal mammary stem cells almost twofold. Significantly, the $CD44^+$, $CD24^{low}$ mammary tumor stem cells have fivefold more expression of Bmi-1 mRNA than the bulk of the tumor cells. Inhibition of the pathway with cyclopamine or siRNA against a Gli1 mRNA sequence reduced the proliferation of mammary tumor stem cells, but did not affect the bulk of CD44+, $CD24^+$ cells. The Hh pathway may also have a role in maintaining CML stem cells.

Cyclopamine (11-deoxyjervine), a steroidal alkaloid that binds to and inhibits Smo, is the classic specific inhibitor of the Hh pathway. The molecule showed antitumor activity *in vitro* and *in vivo*, but is no longer in clinical development due to poor oral bioavailability and acid sensitivity. Semi-synthetic and synthetic derivatives have been developed that are better clinical Smo inhibitors [53]; these agents are currently being developed, have an increased potency, and can be delivered orally. Saridegib (also known as IPI-926) is a smoothened receptor inhibitor derived from cyclopamine that is currently undergoing clinical trials for pancreatic cancer and chondrosarcoma [53]. The limited mechanism of action of smoothened inhibitors has led to concern about the emergence of resistance to them. In fact, a case has recently been reported of the emergence of resistance to smoothened inhibitor GDC-0449 in a patient with metastatic medulloblastoma [54]. Of particular interest in light of these resistance findings is the emergence of the antifungal molecule itraconazole as an inhibitor of the Hh pathway [55]. Itraconazole appears to act on smoothened inhibitors, but its mechanism of inhibition is distinct from that of the cyclopamine derivatives, as it appears to decrease the accumulation of smoothened at the primary ciliary structure. Arsenic trioxide is another small-molecule agent that inhibits the Hh pathway, but at the level of the Gli proteins [56]. This compound can inhibit the pathway additively with itraconazole; indeed, a therapeutic regimen of a combination of itraconazole and arsenic trioxide was shown to be active against functional smoothened inhibitor mutants [56]. The additive effects of the combination could allow for lower doses of each drug, while the simultaneous targeting of two different loci within the Hh pathway may more effectively prevent the emergence of drug-resistant tumor stem cell clones [56].

2.2.2 Notch Signaling

The Notch signaling pathway has a central role in regulating intercellular communication during embryonic development, and cellular differentiation and proliferation [57]. Mammalian cell membrane-bound Notch ligands consist of two structurally distinct families: Delta-like ligands (DLLs) 1, 3, and 4; and Jagged ligands 1 and 2 [58]. These ligands interact with four transmembrane Notch receptors (Notch 1–4). Notch receptors are heterodimeric molecules that consist of noncovalently-bound interacting extracellular and transmembrane domains. Once Notch ligand binds to its receptor, after which the receptor undergoes a conformational change that exposes a previously protected site to proteolytic cleavage by metalloproteinase and γ-secretase; this in turn releases an extracellular and intracellular fragment of the receptor. The Notch intracellular domain (NICD) is active in the cytoplasm, and translocates from the cytoplasm to the nucleus where it binds to the transcription initiation complex and core binding factor-1 (CBF-1). This binding regulates Notch target-specific gene expression, such as Myc and p21 [58].

Breast cancer presents a prime model for studying the importance of the Notch pathway to tumor stem cell biology. Ductal carcinoma tissue *in situ* contains a small subpopulation of tumor stem cells capable of self-renewal behavior *in vitro* [59]. Using this model, it was found that the *in-vitro* self-renewal capacity of this subpopulation was reduced after exposure to a γ-secretase inhibitor or a Notch 4-neutralizing antibody [59]. Notch inhibitors have also been found to selectively deplete medulloblastoma tumor stem cells characterized by high CD133 expression, and a similar finding has been made with glioblastoma tumor stem cells [60, 61]. A role for Notch, in concert with Stat 3 and TGFβ, in the maintenance of hepatocellular carcinoma tumor stem cells has also been reported [62]. Tumor stem cells from glioblastoma multiforme exposed to the differentiation-inducing agent retinoic acid underwent both arrest of proliferation and an enhanced expression of lineage-specific differentiation markers [63]. A microarray analysis of such retinoic acid-treated cells showed a molecular profile characterized by the downregulation of expression of Notch pathway molecules. Constitutive activation of the NICD rescued the glioblastoma cells from retinoic acid-induced differentiation and cell death [63].

The inhibition of γ-secretase-mediated Notch cleavage is a primary focus for the development of therapeutics targeting Notch. A γ-secretase inhibitor depleted tumor-initiating cells and the initiation of xenografts by these cells in a preclinical model of glioblastoma [61]. Flow cytometry studies of human pancreatic tumor xenografts treated with gemcitabine and a γ-secretase inhibitor, PF0308014, resulted in a three- to fourfold reduction in pancreatic tumor stem cells [64]. In a highly aggressive orthotopic pancreatic tumor transplant model, PF0308014 administered in combination with gemcitabine was effective in reducing cell proliferation and controlling metastatic dissemination, compared to gemcitabine alone [64].

Another approach to blockade of the Notch pathway involves interfering with Notch ligands. Recently, a monoclonal antibody against Notch ligand Delta-4 was developed [65] which was effective in reducing colon cancer tumor stem cell frequency in colon cancer xenografts, either alone or in combination with the chemotherapeutic agent irinotecan. The anti Delta-4 antibody appeared to block the expression of Notch target genes *HES-1* and *ATOH-1*, and thus appeared to have a direct effect on Notch signaling in these colon cancer stem cells [65].

2.2.3 Wnt Signaling

During embryogenesis, Wnt pathway signaling proteins direct cell fate determinations, and are crucial to the development of multiple organ systems including cardiovascular, central nervous, renal and pulmonary [66]. In the adult organism, Wnt signaling has a crucial role in tissue self-renewal in intestinal crypts, hair follicles and bone growth plates, while the Wnt proteins are presumed to be stem cell growth factors promoting the maintenance and proliferation of normal tissue stem cells. Wnt proteins consist of 19 conserved glycoproteins that serve as ligands for the *trans*-membrane receptor, Frizzled. Wnt binding to Frizzled receptors initiates two unique signal transduction pathways termed canonical and noncanonical. The canonical pathway results in an accumulation of β-catenin in the nucleus, the interaction of β-catenin with the T-cell lymphocyte enhancer factor of transcription factors, and the transcriptional activation of Wnt target genes [67]. These events do not occur in the noncanonical pathway, which is activated by different ligands (Wnt4, 5a, and 11) from the canonical pathway, and is dependent on the Disheveled protein or the regulation of intracellular calcium levels [68]. A microarray analysis of leukemic cells from patients with AML showed that the Wnt pathway is deregulated in the leukemic stem cell fraction, when compared to normal hematopoietic stem cells [69]. In AML leukemic stem cells, the genes coding for the Wnt pathway proteins Axin and adenomatous polyposis coli (APC) are frequently upregulated, and defective Wnt signaling in AML stem cells may contribute to leukemogenesis [70]. Direct evidence linking the Wnt pathway with tumor stem cells derives from studies in colon cancer, which have shown that a high Wnt activity is characteristic of colon cancer tumor stem cells [23].

Inhibitors of the Wnt/β-catenin signal transduction cascade can again be divided into two groups, namely small molecules and biologics. Existing small-molecule inhibitors of the pathway include non-steroidal anti-inflammatory drugs (NSAIDs) and β-catenin antagonists such as ICG-001. The protective effects of NSAIDs against the development and progression of colon carcinoma is well documented. Some investigators have suggested that part of the protective mechanism could be related to an inhibition of Wnt signaling due to increased phosphorylation of protein phosphatase 2A, resulting in the inhibition of its phosphatase activity [71]. The chemopreventive effectiveness of these anti-inflammatory compounds against the development of colon carcinoma has resulted in a specific effort to target colon cancer stem cells. Indirect evidence in support of this possibility derives from studies of mesenchymal stem cells (MSCs), the proliferation of which was inhibited by NSAIDs [72]. Further, the treatment of $CD133^+$ medulloblastoma tumor stem cells *in vitro* with the cyclooxygenase-2 inhibitor celecoxib inhibited the colony-forming ability of these cells and increased the ionizing radiation-induced apoptosis [73]. *In-vivo* studies revealed that celecoxib significantly enhanced the radiosensitivity of the $CD133^+$-initiated tumor xenografts. In fact, celecoxib appeared to inhibit the expression of stem cell-related genes in $CD133^+$ medulloblastoma tumor stem cell grafts in SCID mice [73]. In glioblastoma, another neural tumor in which tumor stem cells appear to play an important role in disease maintenance and progression, celecoxib enhanced the radiosensitivity of the tumor stem cells, as confirmed by a reduced clonogenic survival of the irradiated cells [74].

A specific small molecule targeting the Wnt pathway, designated ICG-001, has been developed which inhibits β-catenin signaling by specifically interfering with the activation of Wnt target genes [75]. IGC-001 appears to selectively induce apoptosis in colon carcinoma cells, but not in normal colon cells. It also appears to be effective in the Min mouse model of colon carcinoma and in xenograft models [75].

Within the realm of biological inhibitors of the Wnt pathway, monoclonal antibodies against Wnt proteins are under development preclinically [76, 77]. Monoclonal antibodies against Wnt induced apoptosis and downstream changes in target protein expression in cells overexpressing Wnt-1. Apoptosis in response to antibody treatment was not detected in cells with little or low Wnt expression. The antibody also suppressed tumor growth *in vivo* [77].

3
Sensitizing Tumor Stem Cells to Chemotherapeutic Agents

The ability of stem cells to protect themselves from damage or destruction due to exposure to toxic molecules for the entire life of the organism is a critical physiological function for these cells. This resistance to toxins or toxic drugs also characterizes tumor stem cells. The ability to overcome this stem cell drug resistance is an important aspect of developing small-molecule therapies to target tumor stem cells.

3.1 Micro-RNAs and Drug Resistance

Micro-RNAs (miRNAs) are a class of 22- to 25-nucleotide RNAs endogenously expressed as noncoding sequences. They are initially transcribed by RNA polymerase II as long primary transcripts that are processed in a series of endonuclease cleavages into the mature miRNA species. The miRNAs are loaded into the RNA-induced silencing complex (RISC) and then either repress the translation of target mRNAs or induce mRNA degradation. In animal cells, miRNAs usually hybridize with target messenger RNAs (mRNAs) that exhibit an imperfect complementarity and generally induce translational repression. However, some degree of RNA destabilization may also occur [78]. In humans, miRNAs are located on all chromosomes except the Y chromosome [79], and play a role in regulating development and differentiation. Hence, it is to be expected that they might also play a role in the development or progression of tumors. Significantly, recent studies have demonstrated that miRNAs regulate the formation of tumor stem cells and the acquisition of the epithelial to mesenchymal transition (EMT) phenotype in a variety of tissues [80].

Experimental evidence has shown directly that miRNAs regulate antitumor drug resistance, which in turn suggests that the targeting of miRNAs might represent a possible therapeutic approach to increasing the sensitivity of otherwise drug-resistant tumor stem cells towards chemotherapeutic agents. Oncogenic miRNAs such as miRNA-21 mediate chemoresistance in glioblastoma cells [81]. MiRNA-215 was shown to regulate EMT, and also be responsible for sensitivity to EGFR therapy in bladder tumor cells [82]. In a breast cancer model, it was shown that members of the miRNA-200 family could prime tumor stem cells for drug sensitivity. In breast tumors containing tumor stem cells with the $CD44^+$, $CD24^-$ phenotype, the introduction of miRNA-128 made the cells more sensitive to doxorubicin compared to parental control cells [83]. The introduction of miRNA-145 into $CD133^+$ glioblastoma tumor stem cells reduced levels of the stem cell-associated genes *SOX2* and *OCT4*, inhibited tumor formation, and differentiated these cells to a $CD133^-$ phenotype [83]. The resulting cells showed an increased apoptotic potential and sensitivity to the chemotherapeutic agent, temozolomide. Together, the results of these studies confirmed the potential value of modulating miRNAs to confer drug sensitivity to otherwise drug-resistant tumor stem cells.

3.1.1 Drug Transporter Inhibitors

One of the unique markers of "stemness" related to both normal tissue and cancer stem cells is the elevated expression of the ABC multidrug efflux pumps or transport proteins [84]. These pumps provide a protective mechanism for expelling a wide spectrum of amphiphilic compounds from tumor cells, including taxanes, anthracyclines and vinca alkaloids. The ability of tumor stem cells to efficiently pump out different compounds is currently exploited in the HOECHST dye efflux assay, a technique used to isolate tumor stem cells by identifying them as a side population during flow cytometry [85, 86]. However, some tumor stem cells, such as those from glioblastoma multiforme, may be an exception [87]. Therapeutics that target transmembrane drug efflux pumps should effectively counteract this fundamental drug resistance of stem cells. One example of such a drug is dofequidar fumarate, an orally active quinolone that inhibits ABCB1/P glycoprotein, ABC1/MDR-associated protein1, or both proteins, to overcome drug efflux. Recent studies have shown that dofequidar may inhibit the efflux of chemotherapeutic drugs, and increase the sensitivity to antitumor compounds in cancer stem cell side populations isolated from various tumor cell lines [88]. Dofequidar treatment also significantly reduced the number of tumor stem cells. Salinomycin, a polyether antibiotic that acts as a highly selective potassium ionophore, was recently shown, using a high-throughput screening assay, to have selective toxicity against breast tumor stem cells [89]. Salinomycin reduced the proportion of breast tumor stem cells by 100-fold compared to paclitaxel, a commonly used breast cancer chemotherapeutic. Although it is not clear by what mechanism salinomycin acts, recent reports have shown it to be a potent inhibitor of multidrug resistance transporter gp170, as confirmed by drug efflux assays in multidrug-resistant cancer cell lines overexpressing P-glycoprotein [90]. It should also be noted that salinomycin induced the expression of plasma membrane E-cadherin, which suggests that the compound also might eliminate tumor stem cells by inducing differentiation. This effect seems to related to the inhibition of the EMT [89].

4 Metformin Targeting Tumor Stem Cells

Metformin is a member of a class of compounds termed biguanides, which were first isolated from the plant Galega officinalis [91]. Globally, metformin is the most widely prescribed drug to treat type 2 diabetes, its effectiveness in diabetes being due to an ability to suppress hepatic gluconeogenesis, which in turn reduces circulating glucose levels [92]. The mechanism of action is presumed to be through an inhibition of mitochondrial oxidative phosphorylation, which results in activation of the AMP kinase (AMPK)/LKB1 pathway [92]. Although the activation of AMPK has been credited with the glucose-lowering effect of metformin, recent studies have indicated that the AMPK/LKB1 pathway may not be required for the metabolic effects of metformin on glucose uptake [93]. Further, metformin has been shown to inhibit the mTORC1 complex in the absence of AMPK [94, 95]. Most recently, metformin has attracted much attention in oncology due to its anticancer properties, with epidemiological studies documenting that metformin reduces the incidence of cancer and prolongs patient survival [96, 97]. Metformin has also increased the number of breast cancer patients obtaining complete remission in response to neoadjuvant therapy [98]. *In-vitro* studies of tumor cell lines have shown that metformin inhibits the proliferation of tumor cell lines derived from a variety of tissues, including lung, breast, pancreas, colon and prostate [99–102]. The mechanism of tumor growth inhibition has been ascribed to an activation of AMPK signaling to downstream pathways that cooperate in inhibiting tumor growth [102]. However, the exact mechanism by which metformin mediates its antitumor activity is unclear, and is only now being elucidated with studies focused specifically on tumor stem cells.

The first evidence that metformin had specific inhibitory activity towards tumor stem cells derived from studies of $CD44^{hi}$ $CD24^{lo}$ breast tumor stem cells [103]. At low concentrations, metformin preferentially eliminated breast cancer stem cells within a large population of transformed breast cancer cells. Further, the treatment of breast cancer cell line stem cells with metformin, for as little as 1 h, eliminated the ability of these cells to form tumors in nude mice, although the cells were not subsequently drug-exposed. The ability of metformin to selectively kill tumor stem cells contrasts with the activity of that of the standard chemotherapeutic agent doxorubicin, which is toxic to the bulk of the tumor cells but not the stem cells [103]. However, metformin cooperates with doxorubicin to reduce both the stem and non-stem cell population in the mixed transformed cell population. Metformin has been also shown to act synergistically with trastuzumab, to suppress the survival and proliferation of tumor stem cells in $Her2^+$ breast tumor cell lines [104].

Recent studies of the effects of metformin on mammary epithelial cell transformation have been performed with an ER-inducible *Src* gene in MCF10A cells, and have implicated the suppression of inflammation in the effectiveness of metformin against tumor stem cells [105]. Within 15 min of induction of the *Src* gene, there was a rapid inflammatory response in the cells, including an activation of NFκB. In this model, a low dose of metformin, that does not inhibit the proliferation of nontransformed cells, blocked morphological transformation and also colony formation in soft agar. The inflammatory transcription profile of cells treated with metformin and the Src inducer is the same as that of untreated cells, which suggests that metformin can reverse most of the Src-induced gene expression changes. In this inducible model, approximately 10% of the transformed cells behave as tumor stem cells, as defined by the $CD44^{hi}$, $CD24^{low}$ cell-surface phenotype, and the ability both to form mammospheres *in vitro* and establish tumor xenografts *in vivo*. Such cells have higher levels of NFκB activation and higher levels of IL-6 and other inflammatory molecules compared to non-stem tumor cells. In these cells, metformin significantly inhibits the expression of the inflammatory gene profile. Metformin reduces NFκB levels in the nucleus and inhibits STAT3 phosphorylation in these cells, but does not have these inhibitory effects on non-stem tumor cells. The ability of metformin to block tumor growth and promote remission, in combination with doxorubicin, is associated with a decreased function of the inflammatory loop [105]. It has been suggested that, as metformin alters energy metabolism, it may also block a metabolic stress response that stimulates the inflammatory pathways associated with a number of tumor types [105].

Metformin has also been reported to reduce the proliferation rate of tumor stem cells isolated from glioblastoma [106]. The drug appeared to impair self-renewal pathways, as it inhibited neurosphere formation *in vitro*. A clonal heterogeneity exists among the glioblastoma stem cells in this model, including a variation in the expression of CD133. Metformin's antiproliferative effects were greatest on the $CD133^+$ tumor stem cells; in fact, glioblastoma cell differentiation greatly reduced the sensitivity to metformin. Interestingly, in these studies metformin had specific effects on tumor stem cells but no effect on normal human stem cells (umbilical-derived MSCs). Most importantly, the effects on the tumor stem cells were associated with a significant inhibition of AKT-dependent survival pathways, but the latter were not affected by metformin in normal stem cells [106]. Interestingly, metformin also inhibited AKT pathways in mammary tumor xenografts containing a population of mammary tumor stem cells [105].

Finally, Metformin has been shown to alter the growth of ovarian tumor stem cells both *in vivo* and *in vitro* [107]. Metformin reduced the percentage of proliferating aldehyde dehydrogenase-positive ovarian tumor stem cells, and inhibited the formation of ovarian tumor spheres from cells of ovarian tumor cell lines and primary patient tumors. *In vivo*, metformin significantly increased the ability of the chemotherapeutic agent cisplatin to restrict the tumor cell line formation of xenografts, and also significantly restricted the growth of aldehyde dehydrogenase-positive tumor stem cell-initiated xenografts.

5
Dopamine Receptor Antagonists Targeting Tumor Stem Cells

By using a novel high-throughput discovery platform that displays differences between neoplastic and normal human pluripotent stem cells, small molecules that induce differentiation and inhibit self-renewal have been identified from libraries of known compounds [108]. One such compound was thioridazine, a known antipsychotic molecule which was found to selectively target and inhibit AML tumor stem cells but to have no effect on normal hematopoietic stem cells. In *in-vivo* human mouse xenograft assays, thioridazine was shown to reduce the engraftment potential of human AML stem cells. Thioridazine is a known antagonist of the family of dopamine receptors. A number of AML tumor stem cells from different sources were examined, and all were found to express dopamine receptors, including $CD44^+$ $CD24^-$ breast tumor stem cells. Dopamine receptors were not detected on normal human $CD34^+$ $CD38^-$ hematopoietic stem cells or $CD34^+$ $CD38^+$ progenitor cells. It is possible that dopamine receptor expression may serve as a cell-surface biomarker for other tumor stem cell populations from other malignancies, and may serve as a unique target to eliminate tumor stem cells.

6
Oncolytic Viruses Targeting Tumor Stem Cells

Oncolytic viruses have been shown to be effective against a wide variety of human tumors in preclinical models [109].

Novel means of virus delivery, including cell-based systems, appear to increase the capacity of these viruses to target distant metastases. However, the question of whether or not these viruses can target and eliminate tumor stem cells is only just beginning to be addressed. Oncolytic viruses would appear to be ideally suited to the elimination of tumor stem cells because their intrinsic cytotoxicity is not limited by drug-resistance molecular pumps, radiation resistance, or altered signaling mechanisms that are present in these stem-like cells.

Several intrinsic characteristics of the adenoviruses make them particularly useful as therapeutic oncolytic viruses. In humans, members of the Adenoviridae generally cause only mild infection of the upper respiratory tract or eyes, or perhaps gastroenteritis. Additionally, adenoviruses do not integrate into the host genome and are unlikely to cause insertional mutations, while the structure of the virus is stable and it can be produced at high titers. The human adenovirus contains a linear DNA genome that is encapsulated by a non-enveloped icosahedral capsid [110]. The primary receptor for most adenovirus serotypes is the Coxsackie-adenovirus receptor [111, 112]. After entering the cytoplasm, the viral genome is transported to the nucleus, where the expression of viral replication is initiated. Viral gene expression leads to the formation of new virus particles which will eventually lyse and kill the cell. An oncolytic adenovirus can be modified to replicate preferentially in tumor cells, and different strategies to achieve this can be employed. For example, a loss of function strategy causes the adenovirus genome to mutate such that the virus can only replicate in cells that compensate for the lost function of the virus. Generally, part of the immediate early (E1A) or early (E1B) adenoviral genes is deleted, and as a consequence of this the adenovirus is unable to express the viral proteins needed for the binding and inhibition of host cellular proteins that inhibit viral DNA replication. The mutated adenovirus will be able to replicate in tumor cells in which cell-cycle machinery defects obviate the binding of these required viral proteins. Most human tumors are believed to be deficient in the p16/retinoblastoma (Rb) pathway, which controls cell-cycle progression [113]. Adenoviruses designated Delta-24 (or D24) have been designed to specifically take advantage of this Rb defect [114, 115]. The D24 virus has a 24 bp deletion in the constant region of E1A and is unable to bind to the Rb protein. However, the D24 adenovirus will replicate in tumor cells with an aberrant Rb pathway [114, 115].

Recently, the D24 adenovirus has been shown to be effective against breast tumor stem cells [116]. For example, the infection of $CD44^{hi}/CD24^{low}$ breast tumor stem cells with the D24 adenovirus prior to implantation into the mammary fat pads of NOD/SCID mice prevented tumor formation. In addition, when established tumors derived from these tumor stem cells were treated with this D24 adenovirus, the tumor growth was arrested [117]. The D24 virus used in these studies employed the serotype 3 receptor, which is well expressed on many tumor types for viral entry. The D24 adenovirus has also been shown to be oncolytic against $CD133^+$ glioma tumor stem cells [117]. The oncolytic effect of the virus appears to be due to replication-induced cell lysis and autophagy [117].

Reoviruses (respiratory enteric orphan viruses) are double-stranded RNA viruses that are also oncolytic and have been considered for activity against tumor stem cells. Although most humans have been exposed to reoviruses, they are relatively benign and do not produce symptoms [118]. The oncolytic potential of reoviruses was suggested by studies which showed that they replicate well in tumor cell lines [118], and have subsequently been shown also to replicate specifically in cells that harbor an activated Ras pathway [119, 120]. Various experiments have been conducted with reoviruses in human breast tumor samples implanted into murine mammary fat pads, when reovirus-mediated infections of the implants induced a regression of patient-derived tumors. In these studies, breast tumor-initiating cells were identified both by the cell-surface phenotype $CD24^-$ $CD44^+$ and a positivity for aldehyde dehydrogenase expression. Notably, the reovirus infected and eradicated the breast tumor stem cells and the non-stem tumor cells to similar degrees [121].

Most recently, studies have progressed on targeting oncolytic viruses specifically at tumor stem cells. CD133, also known as prominin-1, is mainly expressed in the hematopoietic system and on endothelial cells, but a physiological function has not yet been ascribed to the protein. As mentioned above, in glioma a minority subpopulations of cells with enhanced tumorigenic capacity, compared to the bulk of the tumor cells, has been isolated and found to be $CD133^+$. As few as 100 $CD133^+$ cells could induce malignant tumors after transplantation into immunodeficient mice, while 10^5 of the bulk tumor non-stem cells were required for tumorigenesis.

CD133 has also been proposed as a marker for hepatocellular carcinoma cells with a greater colony-forming capacity, a higher proliferative potential, and a greater ability to form tumors *in vivo* [122]. In order to target CD133 cells, a measles virus envelope protein complex was engineered to restrict infection and target the CD133 antigen [123]. This was accomplished by mutating residues that recognized the measles virus receptor and by genetically fusing a single-chain antibody specific for the target receptor of interest to the virus hemagglutinin envelope glycoprotein [124]. The engineered targeted viruses infected and selectively killed $CD133^+$ tumor cells of various histologic origin. The CD133-targeted viruses exerted dramatic antitumor effects on a human hepatocellular carcinoma cell line growing either focally or in a disseminated state in the peritoneum of NOD/SCID mice. To test the oncolytic activity of the CD133-targeted viruses directly on primary tumor cells, cells from patient colon carcinomas and glioblastomas (two other cell types shown to express cell-surface CD133) were cultured in *in-vitro* sphere-forming assays. Spheres formed from the glioblastoma and also from colon carcinoma primary tumors were readily infected by the engineered measles virus, and the $CD133^+$ cells eradicated. The $CD133^-$ tumor cells were not infected by the viruses. Significantly, normal hematopoietic stem cells expressing the CD133 marker were also neither infected nor killed by the virus. This may have been due to an innate immunity to infection of these normal cells, as they may be protected from attack by the measles virus vaccine strains, which carry mutations in the V protein, a viral inhibitor of the β interferon response. A low level of hematopoietic stem cell infection has been reported with clinical measles virus isolates, encoding a fully active V protein capable of counteracting innate cellular immunity. Many tumor cells have defects in these intrinsic protective immune mechanisms, and this may allow the virus to replicate in an unchecked manner. A similar approach, using targeted engineered viruses against specific tumor stem cell-surface markers, could be used to target tumor stem cells expressing other surface markers, such as CD44, EpCAM or CD123. Other recent studies on retargeting adenoviruses by utilizing multiple ligands within the same virion have been reviewed [125]. The induction of cell lysis, combined with the induction of apoptosis, accomplished by incorporating a secreted cytokine into a type 5 adenovirus platform, may serve as another useful approach to killing tumor stem cells [126].

Currently, several problems remain to be overcome to increase the treatment efficacy of these oncolytic virus strategies, since any antitumor approach will encounter obstacles *in vivo* that are not found *in vitro* [127, 128]. The presence of stromal cells, the extracellular matrix, hypoxia and necrotic tissue may each limit the efficient spread of viral particles within the tumor. In the case of adenoviruses, a further *in-vivo* problem may be bioavailability following systemic injection. For example, intravenously introduced adenovirus tends to accumulate in the liver, while macrophages remove adenovirus particles from the circulation very efficiently. Cell-based delivery systems may offer the opportunity to overcome some of the obstacles encountered in oncolytic viral delivery [128]. An ideal cell-based carrier system must: (i) possess an ability to localize to tumors; (ii) be amenable to viral infection, replication and progeny production; and (iii) effectively avoid immune system inactivation. It has been shown recently that normal tissue stem cells exhibit a tropism for neoplasms, appear to seek out tumor foci, and also migrate into tumors; hence, they might help to overcome the problem of ensuring that the virus is spread to distant sites. Stem cells are also immunosuppressive and could help to keep therapeutic viruses hidden from the host immune system. These observations bode well for the use of normal stem cells as a vehicle to circumvent the antiviral response to oncolytic viruses.

7
Conclusions

Recent studies of tumor cell biology have shown that, in both hematologic malignancies and solid tumors, only a subpopulation of tumor cells is tumorigenic and capable of generating new tumors, yet contains all of the cell types present in the original tumor. These so-called "tumor stem cells" appear to possess many of the properties of normal stem cells, including relative quiescence and self-renewal capacity. The distinct properties of tumor stem cells cause this subpopulation to be resistant to the common therapies used to eliminate tumors; indeed, it is common for "standard" therapies to eliminate the bulk of tumor cells without eradicating the tumor stem cells. The complete elimination of malignant disease will require further understanding of the characteristics and dynamics of these "stem-like" malignant cells, while details of their unique signaling pathways, associated mutations, cell-surface markers and/or metabolic requirements will be required to develop new therapeutics for their targeting and elimination. The results of current investigations suggest that combined therapeutic modalities will be essential to eradicate tumor stem cells and the bulk of tumor cells, with a combination of immunotherapy and small-molecule approaches most likely being very effective. High-throughput screening may be useful for identifying new small molecules that are uniquely effective against tumor stem cells. Multimodality therapeutic approaches to tumors, in addition to being most effective for eliminating not only the bulk of tumor cells but also the tumors' stem cells, may have the additional advantage of limiting the emergence of resistance.

References

1. Nowell, P.C. (1976) The clonal evolution of tumor cell populations. *Science*, **194**, (4260), 429–484.
2. Dick, J.E. (2008) Stem cell concepts renew cancer research. *Blood*, **112** (13), 4793–4807.
3. Dick, J.E. (2009) Looking ahead in cancer stem cell research. *Nat. Biotechnol.*, **27** (12), 44–46.
4. Trump, A. and Wrestler, O.D. (2008) Mechanisms of disease: cancer stems cells-targeting the evil twin. *Nat. Clin. Pract. Oncol.*, **5** (12), 337–347.
5. Reya, T., Morrison, J.J., Clarke, M.F., and Weissman, I.L. (2001) Stem cells, cancer and cancer stem cells. *Nature*, **414** (6859), 105–111.
6. Bonnet, D. and Dick, J.E. (1997) Human Acute Myeloid leukemia is organized in a hierarchy that originates from a primitive hematopoietic cell. *Nat. Med.*, **3** (7), 730–737.
7. Al-Hajj, M., Micha, M., Bonita Hernandez, A., Morrison, S. *et al.* (2003) Prospective identification of tumorigenic breast cell. *Proc. Natl Acad. Sci. USA*, **100** (7), 3983–3988.
8. Singh, S., Clarke, I., Terasaki, M., Bonn, V. *et al.* (2003) Identification of cancer stem cell in human brain tumors. *Cancer Res.*, **63** (18), 5821–5828.
9. Ricci-Vitani, L., Lombardi, D., Pilozzi, E., Buffoni, M. *et al.* (2007) Identification and expansion of human colon cancer-imitating cells. *Nature*, **445** (7123), 111–115.
10. Li, C., Heidt, D., Dalerba, P., Burant, C.F. *et al.* (2007) Identification of pancreatic

cancer stem cells. *Cancer Res.*, **67** (13), 1030–1037.

11. Marchitti, S., Brocker, C., and Stagos, D. (2008) Non-P450 aldehyde oxidizing enzymes: the aldehyde dehydrogenase super family. *Expert Opin. Drug Metab. Toxicol.*, **4** (6), 697–720.
12. Visus, C., Ito, D., Amoscato, A., Maciejewska-Franczak, M. *et al.* (2007) Identification of human aldehyde dehydrogenase 1 family member as a novel $CD8^+$ T-cell defined tumor antigen in squamous cell carcinoma of the head and neck. *Cancer Res.*, **67** (21), 10538–10545.
13. Chen, Y.C., Chang, C.J., Hsu, H.S., Tseng, L. *et al.* (2010) Aldehyde dehydrogenase 1 is a putative marker for cancer stem cells in head and neck squamous cancer. *Biochem. Biophys. Res. Commun.*, **385** (3), 307–313.
14. Douville, J., Beaulieu, R., and Balicki, D. (2009) Aldh1 as a functional marker of cancer stem and progenitor cells. *Stem Cells Dev.*, **18** (1), 17–25.
15. Ma, I. and Allan, A. (2011) The role of human aldehyde dehydrogenase in normal and cancer stem cells. *Stem Cell Rev.*, **7** (2), 292–306.
16. Bhaijee, F., Pepper, D., Pitman, K., and Bell, D. (2012) Cancer stem cells in head and neck squamous carcinoma: a review of current knowledge and future applications. *Head Neck*, **34** (6), 894–899.
17. Alison, M.R., Lim, S., and Nicholson, L. (2011) Cancer stem cells: problems for therapy. *J. Pathol.*, **223** (2), 147–161.
18. Gupta, V., Bamezai, R. (2010) Human pyruvate kinase M2 – a multifunctional protein. *Protein Sci.*, **19** (11), 2031–2044.
19. Tamada, M., Nagano, O., Tateyama, S., Ohmura, M. *et al.* (2010) Modulation of glucose metabolism by CD44 contributes to anti-oxidant status and drug resistance in cancer cells. *Cancer Res.*, **72** (12), 1438–1447.
20. Roesch, A., Fukunaga-Kalabis, M., Schmidt, E.C., Zabievowski, S. *et al.* (2010) A temporarily distinct subpopulation of slow cycling melanoma cells is required for continuous tumor growth. *Cell*, **141** (4), 583–594.
21. Landsbert, J., Kohlmeyer, J., Renn, M., Bald, T. *et al.* (2012) Melanomas resist T-cell therapy through inflammation-induced reversible dedifferentiation. *Nature*, **490** (7420), 412–416.
22. Boiko, A., Razorenova, O., van de Rijn, M., Swetter, S. *et al.* (2010) Human melanoma-initiating cells express neural growth factor receptor CD 271. *Nature*, **466** (7302), 133–137.
23. Vermeulen, L., Melo, F., van de Heijden, M., Cameron, K. *et al.* (2010) Wnt activity defines colon cancer stem cells and is regulated by the microenvironment. *Nat. Cell Biol.*, **12** (5), 468–476.
24. Heddlestom, J., Li, Z., McLendon, R., Hjelmland, A. *et al.* (2009) The hypoxic microenvironment maintains glioblastoma stem cells and promotes reprogramming towards a cancer stem cell phenotype. *Cell Cycle*, **8** (20), 3274–3284.
25. Anderson, K., Luz, C., van Delft, F., Bateman, C. *et al.* (2011) Genetic variegation of clonal architecture and propagating cells in leukemia. *Nature*, **469** (7330), 356–361.
26. Notta, F., Mullighan, C., Wang, J., Poeppl, A. *et al.* (2011) Evolution of human BCR^- ABL^{-1} lymphoblastic leukemia-initiating cells. *Nature*, **46** (7330), 362–367.
27. Dieter, S., Ball, C., Hoffman, C., Nowrouzi, A. *et al.* (2011) Distinct types of tumor-initiating cells form human colon cancer tumors and metastases. *Cell Stem Cell*, **9** (4), 357–365.
28. Jordan, C., Upduerch, D., Srivlass, S., Guzman, M. *et al.* (2000) The interleukin-3 receptor alpha chain is a unique marker for human acute myelogenous leukemia stem cells. *Leukemia*, **14** (10), 1177–1184.
29. Jin, L., Lee, E., Ramshaw, H., Busfield, S. *et al.* (2009) Monoclonal antibody-mediated targeting of CD123, IL-3 receptor α chain eliminates human acute myeloid leukemia cells. *Cell Stem Cell*, **5** (1), 31–42.
30. Goodison, S., Urquidi, B., and Tarin, D. (1999) CD44 adhesion molecules. *Mol. Pathol.*, **52** (4), 189–196.
31. Li, E., Tiede, B., Massague, J., and Kang, Y. (2007) Beyond tumorigenesis: cancer stem cells and metastasis. *Cell Res.*, **17** (1), 3–14.
32. Jin, L., Hope, K., Zhai, Q., Smadja-Joffe, F. *et al.* (2006) Targeting CD44 eradicates human acute myeloid leukemic stem cells. *Nat. Med.*, **12** (10), 1167–1174.
33. Subramanian, V., Vincent, J., Gilakian, M., and Jothy, S. (2007) Suppression of

human colon cancer tumors in nude mice by siRNA CD44 gene therapy. *Exp. Mol. Pathol.*, **83** (3), 332–340.

34. Marangoni, E., LeComte, N., Duraind, L., Dinieux, G. *et al.* (2009) CD44 targeting reduces tumor growth and prevents post chemotherapy relapse of human breast cancer xenografts. *Br. J. Cancer*, **100** (6), 918–922.
35. Bross, P., Beitz, J., Chen, G., Chen, X. *et al.* (2001). Approval summary: gemtuzumab ozoagmicin in relapsed acute myeloid leukemia. *Clin. Cancer Res.*, 7 (6), 1490–1496.
36. Munz, M., Baeuerle, P., and Gires, O. (2009). The emerging role of EpCAM in cancer and stem cell signaling. *Nat. Rev. Cancer*, **9** (14), 5627–5629.
37. Gires, O., Klein, C., and Baeuerle, P. (2009). On the abundance of EpCAM on cancer stem cells. *Nat. Rev. Cancer*, **69** (2), 143.
38. Lei, Z., Maeda, T., Nakamura, T., Yamazaki, Y. *et al.* (2012) EpCAM contributes to formation of functional tight junctions in the intestinal epithelium by recruiting claudin proteins. *Dev. Biol.*, **371** (2), 136–145.
39. LaNaour, F. and Zoller, M. (2008) The tumor antigen EpCAM: tetramspanins and the tight junction protein claudin 7, new partners new functions. *Front. Biosci.*, **13**, 5847–5865.
40. Chaudry, M., Sales, K., Ruf, P., Lindhofer, H. *et al.* (2007) EpCAM – an immunotherapeutic target for gastrointestinal malignancy: current experience and future challenges. *Br. J. Cancer*, **96** (7), 1013–1019.
41. Brischwein, K., Schlereth, B., Guller, B., Steiger, C. *et al.* (2006) MT110: a novel bispecific single chain antibody construct with high efficacy in eradicating established tumors. *Mol. Immunol.*, **431** (8), 1129–1143.
42. Amann, M., D'Argoues, S., Lorenczewski, G., Brischwein, K. *et al.* (2009) Antitumor activity of an EpCAM bispecific BiTE antibody during long-term treatment of mice in absence of T-cell anergy and sustained cytokine release. *J. Immunother.*, **32** (5), 452–464.
43. Ingham, P. and MacMahon, A. (2001) Hedgehog signalling in animal development; paradigms and principles. *Genes Dev.*, **15** (23), 3059–3087.
44. Ingham, P., Nakano, Y., and Seger, C. (2011) Mechanisms and functions of Hedgehog signaling across the metazoa. *Nat. Rev. Genet.*, **12** (6), 393–406.
45. Daya-Grosjean, L. and Couve-Privat, S. (2005) Sonic hedgehog signaling in basal cell carcinomas. *Cancer Lett.*, **225** (2), 181–192.
46. Yang, Z., Ellis, T., Markant, S., Read, T. *et al.* (2008) Medulloblastoma can be initiated by deletion of patches in lineage restricted progenitor or stem cells. *Cancer Cell*, **14** (2), 135–145.
47. Zhao, C., Chen, A., Jamiseson, C., Fesreshteh, M. *et al.* (2009) Hedgehog signaling is essential for maintenance of cancer stem cells in myeloid leukemia. *Nature*, **458** (7455), 776–779.
48. Peacock, C., Wang, Q., Gesell, G., Corcoran-Schwartz, I. *et al.* (2007) Hedgehog signaling maintains a tumor stem cell compartment in multiple myeloma. *Proc. Natl Acad. Sci. USA*, **104** (10), 4048–4053.
49. Bar, E., Chaudry, A., Lin, A., Fan, X. *et al.* (2007) Cyclopamine-mediated Hedgehog pathway inhibition depletes stem like cancer cells in glioblastoma. *Stem Cells*, **25** (10), 2524–2533.
50. Feldmann, G., Dhara, S., Fedrich, V., Bedjia, D. *et al.* (2007) Blockade of hedgehog signaling inhibits pancreatic cancer invasion and metastases: a new paradigm for combination therapy in solid cancers. *Cancer Res.*, **67** (5), 2187–2196.
51. Tanaka, H., Nakamura, M., Kameda, C., Kubo, M. *et al.* (2009) The hedgehog signaling pathway plays an essential role in maintaining the $CD44^+$ $CD24^-$ side population of breast cancer cells. *Anticancer Res.*, **29** (6), 2147–2158.
52. Liu, S., Dontu, G., Muntha, I., Patel, S. *et al.* (2006) Hedgehog signaling and Bmi-1 regulates self-renewal of normal and malignant human mammary stem cell. *Cancer Res.*, **66** (12), 6063–6071.
53. Lin, T. and Matsui, W. (2012) Hedgehog pathway as a drug target: smoothened inhibitors in development. *OncoTargets Ther.*, **5** (3), 47–58.
54. Rudin, C.M., Hann, C., Laterra, J., Yauch, R. *et al.* (2009) Treatment of medulloblastoma with hedgehog pathway inhibitor GDC-0449. *N. Engl. J. Med.*, **361** (12), 1173–1178.

55. Kim, J., Tang, T., Gong, R., Kim, J. *et al.* (2010) Itraconazole – a commonly used anti-fungal that inhibits Hedgehog pathway activity and cancer growth. *Cancer Cell*, **17** (4), 388–399.
56. Kim, J., Aftab, B., Blake, A., Tang, J. *et al.* (2013) Itraconazole and arsenic trioxide inhibit hedgehog pathway activation and tumor growth associated with acquired resistance to smoothened antagonists. *Cell*, **23** (1), 23–34.
57. Aravantis-Tsakonis, S., Rand, M., and Lake, R. (1999) Notch signaling: cell fate control and signal integration in development. *Science*, **284** (5415), 770–776.
58. Kopan, R. and Ilagan, M. (2009) The canonical Notch signaling pathway. *Cell*, **137** (2), 216–233.
59. Farnie, G., Clarke, R., Spence, K., Pinnock, N. *et al.* (2007) Novel cell culture technique for primary ductal carcinoma in situ: role of Notch and epidermal growth factor signaling pathways. *J. Natl Cancer Inst.*, **99** (8), 616–627.
60. Fan, X., Matsui, S., Khaki, L., Stearns, D. *et al.* (2006) Notch pathway inhibition depletes stem-like cells and blocks engraftments in embryonal brain tumor. *Cancer Res.*, **66** (15), 7445–7452.
61. Fan, X., Khaki, L., Zhu, T., Soules, M. *et al.* (2010) Notch pathway blockade depletes CD133-positive glioblastoma cells and inhibits growth of tumor neurospheres and xenografts. *Stem Cells*, **28** (1), 5–16.
62. Yao, Z. and Mishra, L. (2009) Cancer stem cells and hepatocellular carcinoma. *Cancer Biol. Ther.*, **8** (18), 1691–1698.
63. Ying, M., Wang, S., Sun, P., Lai, B. *et al.* (2011) Regulation of glioblastoma stem cells by retinoic acid: role for Notch pathway inhibition. *Oncogene*, **30** (3), 3454–3467.
64. Yabuuchi, S., Pai, S., Campbell, N., de Wildes, R. *et al.* (2013) Notch signaling pathway targeted therapy suppresses tumor progression and metastatic spread in pancreatic cancer. *Cancer Lett.*, **335** (1), 41–51.
65. Hoey, T., Yen, W.-C., Axelrod, F., Basi, J. *et al.* (2009) DLL4 blockade inhibits tumor growth and reduces tumor initiating cell frequency. *Cell Stem Cell*, **5** (2), 168–177.
66. Logan, C. and Nusse, R. (2004) The Wnt signaling pathway in development and disease. *Annu. Rev. Cell Dev. Biol.*, **20**, 781.
67. Rao, T. and Kuhl, M. (2010) An updated overview of Wnt signaling pathways: a prelude for more. *Circ. Res.*, **106**, 1798–1806.
68. Sugimura, R. and Li, L. (2010) Non canonical Wnt signaling in vertebrate development, stem cells, and diseases. *Birth Defects Res. C Embryo Today*, **90** (4) 243–256.
69. Majeti, R., Becker, M., Tian, Q., Lee, T.-L. *et al.* (2009) Dysregulated gene expression networks in human acute myelogenous leukemia cells. *Proc. Natl Acad. Sci. USA*, **106** (9), 3396–3401.
70. Muller-Tidow, C., Steffen, B., Cauet, I., Tickenbrock, L. *et al.* (2004) Translocation products in acute myeloid leukemia activates the Wnt Signaling pathway in hematopoietic cells. *Mol. Cell. Biol.*, **24** (7), 2890–2904.
71. Bos, C., Kodach, L., van den Brink, G., Dik, S. *et al.* (2006) Effect of aspirin on the β-catenin pathway is mediated via protein phosphatase 2A. *Oncogene*, **25** (49), 6447–6456.
72. Deng, L., Hu, S., Baydoun, A., Chen, X. *et al.* (2009) Aspirin induces apoptosis in mesenchymal stem cells requiring Wnt/β catenin pathway. *Cell Prolif.*, **42**, (6) 721–730.
73. Chen, K.-H., Hsu, C.-C., Song, W.-S., Huang, C.-S. *et al.* (2010) Celecoxib enhances radiosensitivity in medulloblastoma-derived CD133-positive cells. *Childs Nerv. Syst.*, **26** (1), 1605–1612.
74. Kang, K., Wang, T., Woon, C., Cheah, E. *et al.* (2007) Enhancement of glioblastoma radioresponse by a selective COX-2 inhibitor celecoxib: Inhibition of tumor angiogenesis with extensive tumor necrosis. *Int. J. Radiat. Oncol. Biol. Phys.*, **67** (3), 888–896.
75. Emami, K., Nguyen, C., Ma, H., Kim, D. *et al.* (2004) A small molecule inhibitor of β-catenin/cyclic AMP response element binding protein transcription. *Proc. Natl Acad. Sci. USA*, **101** (34), 12682–12687.
76. Takahashi-Yanaga, F. and Kahn, M. (2010) Targeting Wnt signaling: can we safely eradicate cancer stem cells. *Clin. Cancer Res.*, **16** (12), 3153–3162.
77. He, B., You, L., Uematsu, K., Xu, Z. *et al.* (2004) A monoclonal antibody against

Wnt-1 induces apoptosis in human cancer cells. *Neoplasia*, **6**, 7–14.

78. Rana, T. (2007) Illuminating the silence. Understanding the structure and function of small RNAs. *Nat. Rev. Mol. Cell Biol.*, **8** (1), 23–36.
79. Negrini, M., Ferracin, M., Sabbioni, S., and Croce, C. (2007) MicroRNAs in human cancer from research to therapy. *J. Cell Sci.*, **120** (11), 1833–1840.
80. Xia, H. and Hui, K. (2012) MicroRNAs involved in regulating epithelial-mesenchymal transition and cancer stem cells as molecular targets for cancer therapeutics. *Cancer Gene Ther.*, **19** (11), 723–730.
81. Li, Y., Li, W., Yang, Y., Lu, Y. *et al.* (2009) MicroRNA21 targets LRRFIP1 and contribute to VM-26 resistance in glioblastoma multiforme. *Brain Res.*, **1286**, 13–18.
82. Adan, L., Zhong, M., Choi, W., Qi, W. *et al.* (2009) mi R-200 expression regulates epithelial to mesenchymal transition in bladder cancer cells and reverses resistance to epidermal growth factor receptor therapy. *Clin. Cancer Res.*, **15** (16), 5060–5072.
83. Sakar, F., Li, Y., Wang, Z., and Kong, D. (2010) Implication of microRNAs in drug resistance for designing novel cancer therapy. *Drug Resist. Update*, **13** (3), 57–66.
84. Fletcher, J., Haber, M., Henderson, M., and Norris, M. (2010) ABC transporters in cancer: more than just drug efflux pumps. *Nat. Rev. Cancer*, **10** (2), 147–156.
85. Wulf, G., Wang, R., Kuehle, I., Weidner, D. *et al.* (2001) A leukemia stem cell with intrinsic drug efflux capacity in AML. *Blood*, **98**, 1166–1173.
86. Hirshmann-Jax, C., Foster, A., Wulf, G., Nuchtern, J. *et al.* (2004) A distinct side population of cells with high drug efflux capacity in human tumor cells. *Proc. Natl Acad. Sci. USA*, **101** (39), 14228–14233.
87. Broadley, K., Hunn, M., Farrand, K., Price, K. *et al.* (2011) Side population is not necessary or sufficient for a cancer stem cell phenotype in glioblastoma multiforme. *Stem Cells*, **29** (3), 452–461.
88. Katayama, R., Tsuoro, T., and Fujita, N. (2009) Dofequidar fumarate sensitizes cancer stem-like side population cells to chemotherapeutic drugs by inhibiting ABCG2/BCRP-mediated drug export. *Cancer Sci.*, **100** (11) 2060–2068.
89. Gupta, P., Onder, T., Jiang, G., Kai, T. *et al.* (2009) Identification of selective inhibitors of cancer stem cells by high-throughput screening. *Cell*, **138** (4), 645–659.
90. Riccioni, R., Dupuis, M., Bernabie, M., Petrucci, E. *et al.* (2010) The cancer stem cell selective inhibitor salinomycin is a p-glycoprotein inhibitor. *Blood Cells Mol. Dis.*, **45** (1), 86–92.
91. Witters, L. (2001) The blooming of the French lilac. *J. Clin. Invest.*, **108** (8), 1105–1107.
92. Shaw, R., Lamia, D., Vasquez, D., Kuo, S. *et al.* (2005) The kinase LKB1 mediates glucose homeostasis in liver and therapeutic effects of metformin. *Science*, **310** (5754), 1042–1646.
93. Foretz, M., Hebard, S., LeClerc, J., Zarrinpashneh, E. *et al.* (2010) Metformin inhibits hepatic gluconeogenesis in mice independently of the LKB1/AMPK pathway via a decrease in hepatic energy state. *J. Clin. Invest.*, **120** (7), 2355–2369.
94. Kalender, A., Selavaj, A., Kim, S., Gulati, P. *et al.* (2010) Metformin, independent of AMPK, inhibits mTORc1 in a Ras GTPase-dependent manner. *Cell Metab.*, **11** (5) 390–401.
95. Ben Sahra, I., Regazzati, C., Guillaume, R., Laurent, K. *et al.* (2011) Metformin, independent of AMPK, induces mTOR inhibition and cell cycle arrest through Redd1. *Cancer Res.*, **71** (13), 4366–4372.
96. Evans, J., Donnelly, L., Emslie-Smith, A., Alessi, D. *et al.* (2005) Metformin and reduced risk of cancer in diabetic patients. *Br. Med. J.*, **330** (7503), 1304–1305.
97. Bowker, S., Majumdar, S., Veugelers, P., and Johnson, J. (2006) Increased cancer-related mortality for patients with type 2 diabetes who use sulfonyl ureas or insulin. *Diabetes Care*, **29** (2), 254–258.
98. Jiralerspong, S., Palla, L., Giordano, S., Meric-Bernstam, F. *et al.* (2009) Metformin and pathologic complete responses to neoadjuvant chemotherapy in diabetic patients with breast cancer. *J. Clin. Oncol.*, **27** (20), 3297–3302.
99. Algire, C., Zakihani, M., Blouin, M., Shuai, J. *et al.* (2008) Metformin attenuates the stimulatory effects of a high-energy diet on *in vivo* LLC1 carcinoma growth. *Endocr. Relat. Cancer*, **15** (3), 833–839.

100. Ben Sahara, I., Lauret, K., Loubat, A., Georgetti-Peraldi, S. *et al.* (2008) The anti-diabetic drug metformin exerts an anti-tumoral effect in *in vitro* and *in vivo* through a decrease of cyclin D1 level. *Oncogene*, **27** (25), 3576–3586.
101. Kisfalvi, K., Eibl, G., Sincett-Smith, J., and Rozengurt, E. (2009) Metformin disrupts cross talk between G-protein coupled receptor and insulin receptor and insulin receptor signaling systems and inhibits pancreatic cancer growth. *Cancer Res.*, **69** (16) 6539–6545.
102. Zakikhan, M., Dowling, R., Sonenberg, N., and Pollak, M. (2008) The effects of adiponectin and metformin on prostate and colon neoplasia involve activation of AMP activated protein kinase. *Cancer Prev. Res.*, **1** (5), 369–375.
103. Hirsch, H., Illiopoulos, D., Tsichlis, P., and Struhl, K. (2009) Metformin selectively targets cancer stem cells and acts together with chemotherapy block growth and prolong remission. *Cancer Res.*, **69** (19), 7507–7511.
104. Vasquez-Martin, C., Oliveras-Ferranos, S., Del Barco, J., Martin-Castillo, P. *et al.* (2011) The anti-diabetic drug metformin suppresses self-renewal and proliferation of trastuzamab-resistant tumor initiating breast cancer stem cells. *Breast Cancer Res. Treat.*, **126** (2), 355–364.
105. Hirsch, H., Illiopoulos, D., and Struhl, K. (2013) Metformin inhibits the inflammatory response associated with cellular transformation and cancer stem cell growth. *Proc. Natl Acad. Sci. USA*, **110** (3), 972–977.
106. Wurth, R., Pattarozzi, A., Gatti, M., Bajette, A. *et al.* (2013) Metformin selectively affects human tumor initiating cell viability. *Cell Cycle*, **12** (1), 145–156.
107. Shank, J., Young, K., Ghannam, J., Cabrera, L. *et al.* (2012) Metformin targets ovarian cancer stem cells *in vitro* and *in vivo*. *Gynecol. Oncol.*, **127** (2), 390–397.
108. Elefethrerios, S., Risueno, R., Latonde, S., Shapovalova, Z. *et al.* (2012) Identification of drugs involving dopamine receptor antagonist that selectively target cancer stem cells. *Cell*, **149** (6), 1284–1297.
109. Liu, T.-C. and Kirn, D. (2008) Gene therapy progress and prospects cancer: oncolytic viruses. *Gene Ther.*, **15** (12), 877–884.
110. Russell, W.C. (2008) Update on adenovirus and its vectors. *J. Gen. Virol.*, **81** (Pt. 11), 2573–2604.
111. Tomko, R., Xu, R., and Philipson, L. (1997) HCAR and MCAR: the human and mouse cellular receptors for sub group C adenoviruses and group B Coxsackie viruses. *Proc. Natl Acad. Sci. USA*, **94** (7), 3352–3356.
112. Bergelson, J., Cunningham, J., Drouguett, G., Kurtz-Jones, E. *et al.* (1997) Isolation of a common receptor for Coxsackie B virus and adenoviruses 2 and 5. *Science*, **275** (5304), 1320–1323.
113. Sherr, C. (1996) Cancer cell cycles. *Science*, **274** (5293), 1672–1677.
114. Fueyo, J., Gomer-Manzano, C., Alemany, R., Lee, P. *et al.* (2000) A mutant oncolytic virus adenovirus targeting the Rb pathway produces anti glioma effect *in vivo*. *Oncogene*, **19** (1), 2–12.
115. Heise, C., Hermiston, T., Johnson, L., Brooks, G. *et al.* (2000) An adenovirus E1 mutant that demonstrates potent and selective systemic anti tumoral efficacy. *Nat. Med.*, **6** (12), 1134–1139.
116. Eriksson, M., Guse, K., Bauerschmitz, G., Virkkunen, P. *et al.* (2007) Oncolytic adenoviruses kill breast cancer initiating $CD44^{+}$ $CD24^{low}$ cells. *Mol. Ther.*, **15** (2), 2088–2093.
117. Jiang, H., Gomez-Manzano, C., Aoki, T., Konso, M. *et al.* (2007) Examination of the therapeutic potential of delta-24 RGD in brain tumor stem cells. *J. Natl Cancer Inst.*, **99**(18), 1410–1414.
118. Cripe, T., Wang, P.-Y., Marcato, P., Mahller, Y. *et al.* (2009) Targeting cancer-initiating cells with oncolytic viruses. *Mol. Ther.*, **17** (10), 1677–1682.
119. Coffey, M., Strong, J., Forsyth, P., and Lee, P. (1998) Reovirus therapy of tumors with activated Ras pathway. *Science*, **282** (5392), 1332–1334.
120. Smakman, N., van den Wollenberg, R., Borel-Rinkes, D., Hoeben, R. *et al.* (2005) Sensitization to apoptosis underlies Kras D12-dependent oncolysis of murine C26 colorectal carcinoma cells by reovirus T3 D. *J. Virol.*, **79** (23), 14981–14985.
121. Marcato, P., Dean, C., Gaicomantonio, C., and Lee, P. (2009) Oncolytic reovirus

effectively targets breast cancer stem cells. *Mol. Ther.*, **17** (6), 972–978.

122. Singh, S., Clarke, I., Hide, T., and Dirks, P. (2004) Cancer stem cells in nervous system tumors. *Oncogene*, **2** (43) 7267–7273.

123. Zhu, Z., Hao, X., Yan, M., Yao, M. *et al.* (2010) Cancer stem/progenitor cells are highly enriched in CD133+CD44+ population hepatocellular carcinoma. *Int. J. Cancer*, **126** (9), 2067–2078.

124. Bach, P., Abel, T., Hoffmann, C., and Gal, Z. (2013) Specific elimination of CD133+ tumor cells with targeted oncolytic measles virus. *Cancer Res.*, **73** (2), 865–874.

125. Beatty, M. and Curiel, D. (2012) Adenovirus strategies for tissue specific targeting. *Adv. Cancer Res.*, **115**, 39–67.

126. Das, S., Sarkar, S., Das, R., Dent, P. *et al.* (2012) Cancer Terminator viruses and approaches enhancing therapeutic outcomes. *Adv. Cancer Res.*, **15**, 1–38.

127. Sherise, F., Ahmed, A., Thaci, B., Mercer, R. *et al.* (2010) Crossing the boundaries: stem cells and gene therapy. *Discov. Med.*, **9** (46), 192–196.

128. Nakashima, H., Kaur, B., and Chiocca, E. (2010) Directing systemic oncolytic viral delivery to tumors via carrier cells. *Cytokine Growth Factor Rev.*, **21** (2-3), 119–126.

27
Targeted Therapy: Genomic Approaches

Tim N. Beck[1,2], *Linara Gabitova*[1,3], *and Ilya G. Serebriiskii*[1]
[1] *Fox Chase Cancer Center, Developmental Therapeutics Program, 333 Cottman Avenue, Philadelphia, PA 19111, USA*
[2] *Drexel University College of Medicine, Program in Molecular and Cell Biology and Genetics, 2900 Queen Lane, Philadelphia, PA 19129, USA*
[3] *Kazan Federal University, 18 Kremlevskaya Street, Kazan, Republic of Tatarstan, 420008, Russia*

Translational Medicine Cancer: Advances in Molecular Biology and Medicine,
First Edition. Edited by Robert A. Meyers.

Keywords

Targeted therapy
A biomedical approach to identify and therapeutically exploit entities specific to pathological processes (e.g., a gene, a protein, or RNA), in an effort to minimize adverse effects and maximize therapeutic impact

Precision medicine
A personalized treatment approach based on data collected on an individual basis and thus specific to the individual and a given malady or condition of concern

Genomics
The study of canonical and aberrant nucleotide sequences, gene expression, and chromosome integrity within an organism, and the roles they play in health and disease

Epigenomics
The study of how DNA modifiers and modifications influence gene expression in the absence of DNA sequence alterations

Transcriptomics
The study of RNA, coding and noncoding, within a living organism

Metabolomics
The study of chemical processes involved in the breakdown and utilization of metabolites within an organism and within cells

Proteomics
The study of all proteins within a living organism

Metagenomics
The study of genetic material from the environment, including microorganisms

Bioinformatics
A scientific discipline focused on the research, development, and application of computational tools to acquire, analyze, store, and visualize data, particularly "-omics" data

The crux of targeted therapy is the inhibition, enhancement, correction, reversal, or negation of events to overcome pathological processes, by targeting features or entities specific to the pathology. Genomics has been a critical component of target identification, and has greatly amplified the impact and potential of targeted therapy. Incredibly, genomics has only scratched the surface of this endeavor. Today, progress is driven by pan-omics, integrating genomics, epigenomics, transcriptomics, proteomics, metabolomics, and metagenomics data using bioinformatics and system biology approaches. The advent of advanced "-omics" has provided the molecular foundation for the development of new-targeted therapeutics, including small molecules, biologicals, tools for genomic editing, and more. Targeted therapy is used to treat an array of diverse diseases, ranging from cancer and HIV to atherosclerosis and malaria. In this chapter, the essential aspects of targeted therapy are discussed, including many of the remarkable successes as well as some of the shortcomings. How "-omics" approaches have contributed to the evolution of targeted therapy and what can be expected in the future is also described.

1 Principles of Genomics-Driven Targeted Therapy

1.1 The Promise of Targeted Therapy

The promise of targeted therapy has never been greater, and innovation on this front continues to be driven by the desire to treat and prevent disease with precision and efficacy. Targeted therapy is based on the idea of identifying a unique, pathology-specific, and targetable entity, be it cancer-associated aberrations (Sect. 1.3), foreign organisms (Sect. 1.5), or a normal protein that modulates a disease-associated process (Sect. 1.6). Frequently, a dominant feature necessary for the establishment and maintenance of a disease is targeted. At other times, it is neither possible nor practical to directly target the focal points of a pathological process, and instead modulating entities are targeted (Fig. 1b). A unique target, in theory, provides the opportunity to specifically inhibit – or potentially to enhance – a given process within the pathological milieu, ideally while avoiding the collateral damage of healthy tissue.

Targeted therapy is frequently described in a simplistic linear cause-and-effect schema. A pathological process arises due to a specific target activity; the therapeutic agent inhibits this activity and downstream events are returned to the *status quo ante* (Fig. 1b). The more likely scenario acknowledges the somewhat ambiguous specificity of most targeted agents and the complexity of intracellular processes. Figure 1c suggests that the target marked by the crosshair is indeed solely required for therapeutic benefit; however, it may be the molecular ripple effects of targeted intervention that are critical. In order to achieve a therapeutic effect, factors must add-up to perturb the greater signaling in the correct way. The red circle in Fig. 1d represents the point of optimal therapeutic effectiveness, where drug combination, treatment schedule and dosing are ideal to maximize the impact on the signaling network. It is important to realize that many of the drugs used as targeted therapeutics in actuality have multiple targets (Fig. 2a) that can be divided into four general categories (Fig. 2b). Interestingly, for the vast majority of drugs only a single target is known (Fig. 2c), presumably due to a lack of initial investigation of additional targets, sometimes referred to as "off

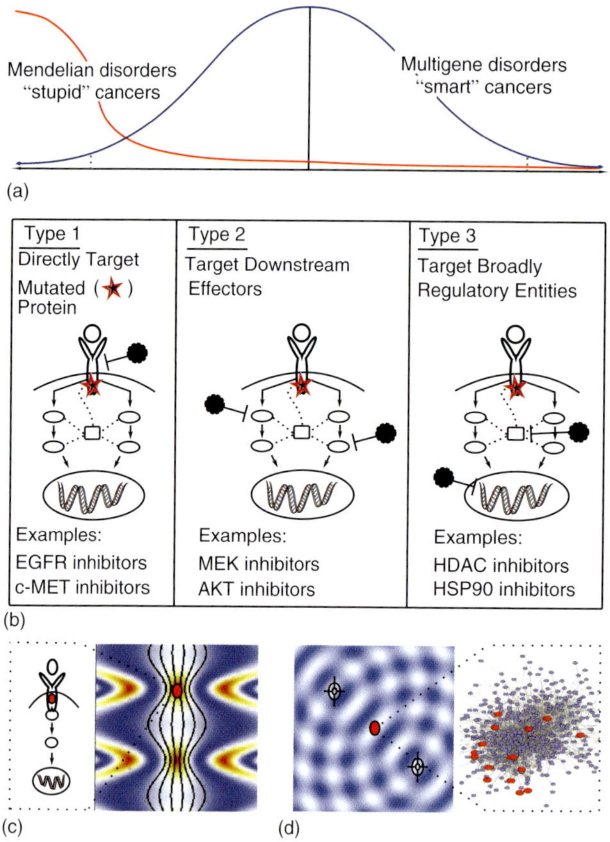

Fig. 1 Targeted therapy approaches. (a) The hypothetical distribution of disorders by the number of underlying interaction is represented by the Poisson curve (blue), with the chances of meaningfully using targeted therapy depicted as a red line. The distribution goes from asymptotically close to zero for complex disorders, and asymptotically closer to one for Mendelian disorders (see text for details); (b) Three different types of targeted therapy approaches are commonly applied; (c,d) As physicists about a century ago moved from considering light as rays to its interpretation as waves (sometimes, at their convenience, using either description), we are moving from the paradigm of thinking about molecular life in terms of linear biochemical and signaling pathways (c, which can be disrupted by a drug with a predetermined consequences), to the more complex picture of biochemical "fabric" (d), where the drugs produce waves or ripples, interacting with those from other drugs or environmental inputs; (c) Common theoretical approach to targeted therapy, wherein signaling events are considered as linear with proposed precision targeting (red circle) and narrowly focused downstream effects; (d) Ripple effects of targeted therapy and the complexity of perturbing the signaling milieu: precisely hitting targets (crosshairs) in a few cases may have the intended therapeutic effect. However, commonly, the ripples effects, perturbation of alternative targets (white spots), under the correct conditions (e.g., correct dosage and drug combination, optimal treatment schedule) induce signaling-network-wide changes for the therapeutic effect (red circles). Wave image (c) modified from the Wolfram Demonstrations Project *http://demonstrations.wolfram.com/TheQuantumHarmonicOscillatorWithTimeDependentBoundaryConditi/*; wave image (d) generated using Wolfram (*http://demonstrations.wolfram.com/WaveInterference/*); network (d) generated using Cytoscape [1].

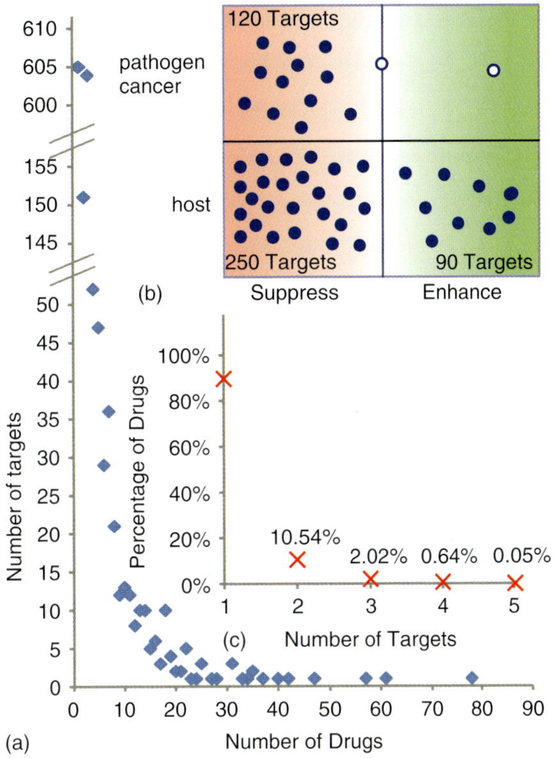

Fig. 2 Target specificity and type based on Therapeutic Targets Database (TTD) data. (a,c) Relationship between the number of drugs and targets reveals that most targets correspond to only a few drugs (a), and, reciprocally, the overwhelming majority of drugs have at most two known targets (c); (b) Targets can be divided into four categories: suppressive effect on cancer/pathogen; suppressive effect on host; enhancing effect on pathogen/cancer; enhancing effect on host. The blue circles represent 10 targets based on TTD data; open blue circles represent manually curated data from the literature, indicating cases where compounds are used either to just visualize the cancer cells to enable the subsequent surgical removal ("neutral" position on the panel; see Sect. 2.5 for details), or to restore/enhance their antigen-presenting function, enabling their subsequent destruction by the immune system (see Sect. 2.3). All information was retrieved from the TTD [2] and sorted as human versus cancer/pathogen based on the target UniProt ID/disease type, and by functionality based on "agonist/activator" versus "inhibitor/antagonist" drug classifiers.

targets". Commonly, knowledge about the specificity of a given drug in terms of target selectivity changes over time as the drug continues to be studied (Fig. 3), adding to the challenge of optimizing treatment conditions. The advent of technologies that make it possible to reliably and rapidly interrogate genomes, transcriptomes, proteomes, metabolomes, and microbiomes, and the necessary bioinformatics resources to analyze the gargantuan amount of data [4–11], have started to outline how pan-omics approaches guide targeted therapy and its paradigmatic impact on precision medicine (Sect. 3) [12–16].

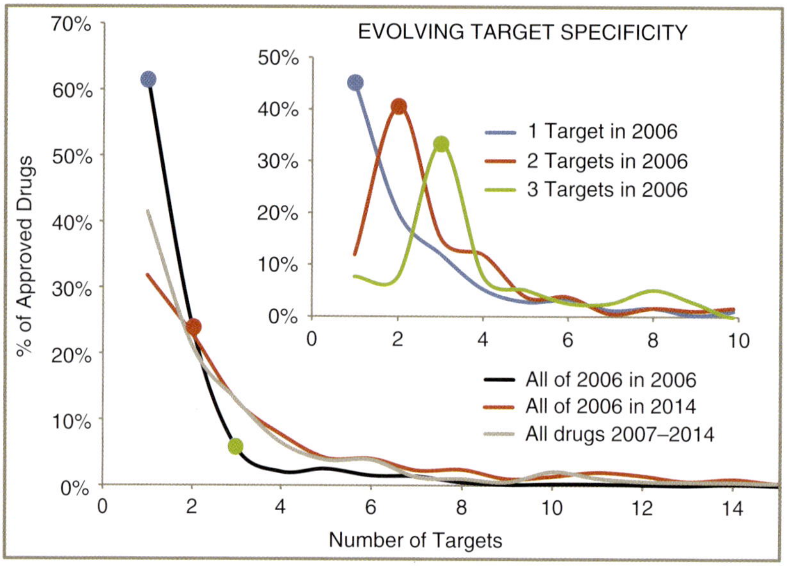

Fig. 3 Evolving target specificity. The number of targets for drugs approved by 2006 and analyzed in 2006 (black line) compared to the same set of drugs reanalyzed in 2014 (red line) highlights the changes in terms of target specificity as drugs are continue to be studied. The gray line represents the number of targets for drugs approved from 2006–2014. The inset shows the reassessment of number of drug targets for drugs originally described to have one target in 2006 (blue line), two targets in 2006 (red line), and three targets in 2006 (green line) when reanalyzed in 2014 (indicated by the lines emanating from each solid sphere). Drug target data were obtained from DrugBank [3].

Although "targeted therapy" is the buzzword, and many prominent examples of targeted drugs exist (Table 1), the specificity of approved drugs is not easy to confirm [81], nor is it easy to identify a trend suggestive of an increase in the development of drugs with greater specificity (Fig. 3). Comparing the distribution of the number of targets per drug for drugs approved by 2006 to drugs approved from 2007 to 2014 demonstrates a higher fraction of drugs with fewer targets. However, a confounding variable is the ever-increasing understanding of a drug's true specificity after it has been approved (Fig. 3). Back in 2006, the distribution of targets per drug showed a higher prevalence of drugs with only one described target compared to data from 2014. In fact, many of these drugs were later demonstrated as having multiple targets (up to 28; see Fig. 3).

Charles Heidelberger and colleagues first introduced targeted therapy to the world with the synthesis of 5-fluorouracil (5-FU), a compound targeting a specific biochemical pathway, namely increased uracil uptake by malignant cells [82, 83]. The number of targeted therapeutics developed since then is extensive, and targeted therapy has seeped into the full spectrum of medical fields and is not only relevant in oncology (Table 1; Sect. 2). Nevertheless, oncology, with the implementation of precision oncology [84, 85], remains a poster child of targeted therapy and is covered extensively in this chapter. The following sections will describe the fundamentals of

Tab. 1 Targeted therapeutics.

Target type	*Target*	*SMDs*	*Biologicals*
Signal transduction cascades	EGFR	Erlotinib [17]	Pertuzumab [18]
		Gefitinib [19, 20]	
		Afatinib [21]	Bevacizumab [22]*
		Sorafenib [23]	
		CO-1686 [24]	Cetuximab [20, 25]
		Lapatinib [26]	
	ERBB2		Trastuzumab [27–29]
	MET	Tivantinib	Onartuzumab
		Cabozantinib	Rilotumumab
		Foretinib [30]	Ficlatuzumab [30]
	MAP2K7	Selumetinib [31]	
		Trametinib [31]	
		PD-0325901 [31]	
		MEK162 [31]	
		PDA-66 [32]*	
	GSK3B	Tideglusib [33]	
	KDR		Bevacizumab [34]
	ESR1		Tamoxifen [35]
			Raloxifen [36]
	ERBB2, FCGR1A		MDX-210 [37]
Cluster of differentiation molecules	CD3		Muromonab-CD3 [38]
	TNFRSF8		Brentuximab vedotin [39]
	CD40		CP-870,893 [40]
			Dacetuzumab [40]
			Chi Lob 7/4 [40]
	TNFRSF8, FCGR3A		HRS-3/A9 bsAb [41]
	CTLA4		Ipilimumab [42, 43]
Fusion proteins	BCR-ABL1	Imatinib [44]	
		Bosutinib [45]	
		Dasatinib [46, 47]	
		Nilotinib [48, 49]	
	EML4-ALK	Crizotinib [50]	
		Ceritinib [51–53]	
Transcription factors	SREBF1, SREBF2	Betulin [54]	
		Fatostatin [55]*	
Metabolic pathways	26S proteasome	Bortezomib [56]	
	20S proteasome	Carfilzomib [57]	
	PCSK9		Evolocumab [58]
			Alirocumab [58]
			SPC5001 [59, 60]
Cell cycle regulators	RNA component of human telomerase RNA (hTR)		Imetelstat [61]
	CDK4, CDK6		PD-0332991 [62]

(*continued overleaf*)

Tab. 1 (*Continued*)

Target type	*Target*	*SMDs*	*Biologicals*
Chaperones	HSP90	Ganetespib [63]	
		17-DMAG [64]	
Post-translational protein modification	FNTA	Tipifarnib	
		Lonafarnib [65]	
Epigenetic therapy	DNMT1	Azacitidine [66]	
	HDAC9	Vorinostat	
		Romidepsin [67, 68]	
	BET	JQ1 [69–72]	
	DOT1L	EPZ-5676 [67, 73, 74]	
Membrane channels and transporters	CFTR	Ataluren [75, 76]	
Structural proteins	TUBA1A		Paclitaxel
			Docetaxel [77]
Extracellular molecules	*Clostridium difficile* endotoxins		Actoxumab/ bezlotoxumab [78, 79]
	Bacillus anthracis anthrax toxin		Raxibacumab [80]

The listed therapeutics are all FDA-approved or are currently being evaluated in clinical trials (as of July 2014). FDA, U.S. Food and Drug Administration; EGFR, epidermal growth factor receptor; ERBB2, erb-b2 receptor tyrosine kinase 2; MET, MET proto-oncogene, receptor tyrosine kinase; MAP2K7, mitogen-activated protein kinase kinase 7; GSK3B, glycogen synthase kinase 3 beta; KDR, kinase insert domain receptor; ESR1, estrogen receptor 1; FCGR1A, Fc fragment of IgG, high affinity Ia, receptor (CD64); CD3, CD3 molecule; TNFRSF8, tumor necrosis factor receptor superfamily, member 8; CD40, CD40 molecule, TNF receptor superfamily member 5; FCGR3A, Fc fragment of IgG, low affinity IIIa, receptor (CD16a); CTLA4, cytotoxic T-lymphocyte-associated protein 4; BCR, breakpoint cluster region; ABL1, ABL proto-oncogene 1, non-receptor tyrosine kinase; EML4, echinoderm microtubule associated protein like 4; ALK, anaplastic lymphoma receptor tyrosine kinase; SREBF1 and SREBF2, sterol regulatory element binding transcription factor 1 and 2, respectively; PCSK9, proprotein convertase subtilisin/kexin type 9; CDK4 and CDK6, cyclin-dependent kinase 4 and 6, respectively; HSP90, heat shock protein 90kDa; FNTA, farnesyltransferase, CAAX box, alpha; DNMT1, DNA (cytosine-5-)-methyltransferase 1; HDAC9, histone deacetylase 9; BET, bromodomain and extra-terminal domain protein family; DOT1L, DOT1-like histone H3K79 methyltransferase; CFTR, cystic fibrosis transmembrane conductance regulator (ATP-binding cassette sub-family C, member 7); TUBA1A, tubulin, alpha 1a; SMDs, small-molecule drugs; *, promising, but not yet in clinical trials.

targeted therapy, its molecular basis, the role of genomics, and pitfalls of current approaches (Sects. 1.2–1.6). Next, the most relevant targeted agents are described: from a mechanistic standpoint as well as in terms of the therapeutic niche each class of agents occupies (Sect. 2). Lastly, the specifics of precision medicine are described (Sect. 4).

1.2 Molecular Basis of Targeted Therapy

The genomic and proteomic landscape of cancer is particularly well suited to describe the molecular basis of targeted therapy. There are approximately 140 individual mutations that are known to initiate tumorigenesis – that is, mutations

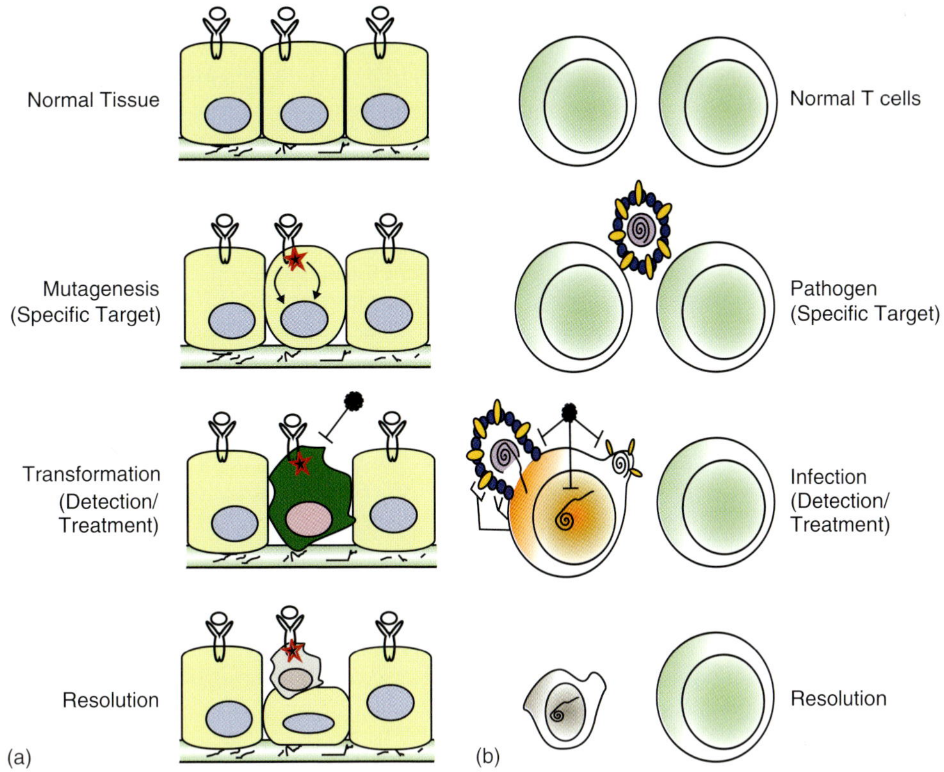

Fig. 4 General biological aspects of targeted therapy. (a) Mutation leads to malignant cell transformation, and at the same time can provide a unique target for therapeutic intervention; (b) Pathogens can present unique targets for therapeutic intervention.

that have the potential to drive normal cells towards the malignant state [86]. Of critical relevance in terms of targeted therapy is the fact that these mutations are unique to the cancer cell; thus, in some cases, providing a target not present in normal cells (Fig. 4a). As an example, the epidermal growth factor receptor (EGFR) can be considered. This receptor is commonly mutated in lung cancer [87, 88] in a way that induces the receptor to constitutively send pro-proliferative signals exclusively within the cancer cells [89]. Human immunodeficiency virus (HIV) is a second example that demonstrates the emerging molecular basis of targeted therapy. In the case of an HIV infection, therapeutics have been designed to specifically target the viral life cycle without harming normal host cells (Fig. 4b) [90]. On a molecular level, both, cancer cells and the viral life cycle, present features that have been successfully targeted (see Table 1).

With regards to targeted therapy, several conceptual components critical to understanding this category of therapeutic agents should be realized (Sect. 2). First, the area of practical applicability and sustained success of targeted therapy is more tilted towards disorders caused by alterations in one or a few genes (Fig. 1a). That being said, even knowing a unique and pharmacologically amenable target does not always make targeted therapy optimal. Cholera, the scourge of humankind for centuries, is a case in point. Its causative agent, *Vibrio cholera*, presents several potentially targetable entities, including the key virulence factors cholera toxin and the toxin coregulated pilus (TCP, required for colonization; [91]). However, simple persistent rehydration lowers the mortality rate almost to zero [92], an achievement unlikely to be improved by targeted agents.

On the other end of the spectrum shown in Fig. 1a are the cases of multiple simultaneous gene alterations, as is commonly observed in cancer [93]. However, this certainly does not preclude the use of targeted therapy. Signaling hubs, points of pro-survival signaling convergence, can potentially be targeted, providing an opportunity to achieve therapeutic goals without individually targeting each upstream mutation (Fig. 1b) [94, 95]. Today, multidrug combinations (Sect. 3.4) are being studied with the same intended goals.

Finally, even knowing a target at the submolecular level is not the same as knowing the underlying phenotypic mechanism(s). The anticoagulant ichorcumab epitomizes this notion. Ichorcumab binds to exosite 1 of thrombin and thus prevents thrombin from cleaving fibrinogen and producing clots; surprisingly, however, other functions of thrombin seem to be retained via mechanisms not understood, significantly reducing the excessive bleeding commonly seen with other anticoagulants [96].

1.3 Purpose of Genomic Approaches and Available Platforms

The advent of next-generation sequencing instruments truly revolutionized the ability to investigate the human genome [7], a fact particularly relevant for the efforts to identify novel therapeutic targets. The first human genome was partially sequenced in 2001 [97, 98], a monumental undertaking that took over a decade to complete and cost around US$ 3 billion. During the second decade of the twenty-first century, it is possible to sequence the entire human genome for close to US$ 1000 and to obtain the data in days rather than years [99], a truly remarkable achievement. Yet, in spite of these incredible achievements, several hurdles must still be overcome (Sect. 1.7).

The crux of next-generation sequencing is the construction of a genome DNA fragment library with synthetic DNA adapters attached to each fragment. The library fragments may then be polymerase-amplified, if necessary, and sequenced [100]. Multiple different next-generation sequencing approaches exist, including reliance on reversible dye terminators (using fluorescent tags) and pH change monitoring (this detects the release of hydrogen ions as nucleotides are being incorporated) [100]. Incredibly, sequencing can now even be performed on the single-cell level, allowing an analysis of the functional state of single cells and determination of cell lineage relationships [101]. From a therapeutic standpoint, the purpose of genomic approaches is to define differences between normal tissue and pathologically relevant entities (Fig. 4), and to anticipate how a pathogen or cancer cell will respond to treatment, which is particularly relevant in the case of biologicals (Sect. 2.2), such as oncolytic viruses [102, 103] that are capable of adjustments after having been administered. Not surprisingly, sequencing is emerging as a cornerstone of clinical care (see Sect. 3; [11, 16]) and drug development [13].

1.3.1 Identification of Genomic Alterations

Cancer can be defined as a disease of the genome, arguably also of the epigenome (Sect. 1.4). Thus, biological studies to elucidate the mechanisms that drive such genomic and epigenomic alterations in cancer, and also an identification of the exact set of alterations in each cancer case, are of immense importance [84]. Next-generation sequencing has emerged as the dominant technique used to identify DNA sequence alterations. The ability to identify point mutations is at an advanced level, and at this point is much more reliable than the identification of large genomic alterations (*https://www.synapse.org/#!Synapse:syn312572*; [7]). The genomic instability of many cancer cells adds to this challenge, by diversifying the genomic landscape of the disease not just from patient to patient (interpatient heterogeneity), but also within a patient (intrapatient heterogeneity), and even within a given tumor (intratumor heterogeneity). Furthermore, heterogeneity is not static, it evolves over time; especially, under the selective pressure of therapeutic treatment [104–107]. The identification of somatic mutations in protein-coding regions (exons) has been a main focus of cancer genomics, and much time and effort has been spent to reliably and accurately identify these mutations. A dominant challenge now, in terms of somatic mutations, is to distinguish between driver events that contribute to oncogenesis and stochastic passenger mutations [108]. The identification of driver genes can be particularly problematic in cases of larger chromosomal alterations, not only because of the large number of genes that are affected simultaneously but also because of technical limitations in terms of detecting large continuous changes in chromosomes [86].

Commonly, individual base-pair alterations cause somatic mutations (Sect. 1.3.1.1), some of which can lead to significant survival and proliferative advantages. These alterations are categorized based on the resulting changes: missense mutations (amino acid substitution resulting in no functional changes, functional hyperactivity, or reduced function); nonsense mutation (producing a premature stop-codon in the mRNA, resulting in a truncated protein); and splice-site mutations (splicing variants). The inferred functional impact of a given mutation type is at times somewhat complicated. Missense mutations, for example, may or may not result in protein expression level changes, depending on the specifics of a given alteration. Even synonymous mutations (the amino acid sequence is not altered, thus, no discernable change in protein sequence occurs) are no longer considered to always be "silent," as they may indeed alter transcription, splicing, mRNA transport, or translation, ultimately leading to phenotypically relevant outcomes [11, 86, 109, 110].

Frame-shift mutations – mutations that are caused by insertions or deletions (indels) of nucleotides that disrupt the normal three-nucleotide codon arrangements – are also observed in cancer [108, 109, 111]. Chromosomal aberrations (Sect. 1.3.1.2) as well as gains and losses in gene copy number (Sect. 1.3.1.3) are additional highly relevant events that contribute to malignant phenotypes [112, 113]. Genomic alterations are further considered to part take in oncogene addiction (e.g., EGFR mutations) or non-oncogene addiction (e.g., CHEK1). Oncogenic addiction describes gene alterations that are essential for the establishment and maintenance of the oncogenic state, while non-oncogenic addiction refers to the broader network of mutations that collectively support the malignant phenotype [114].

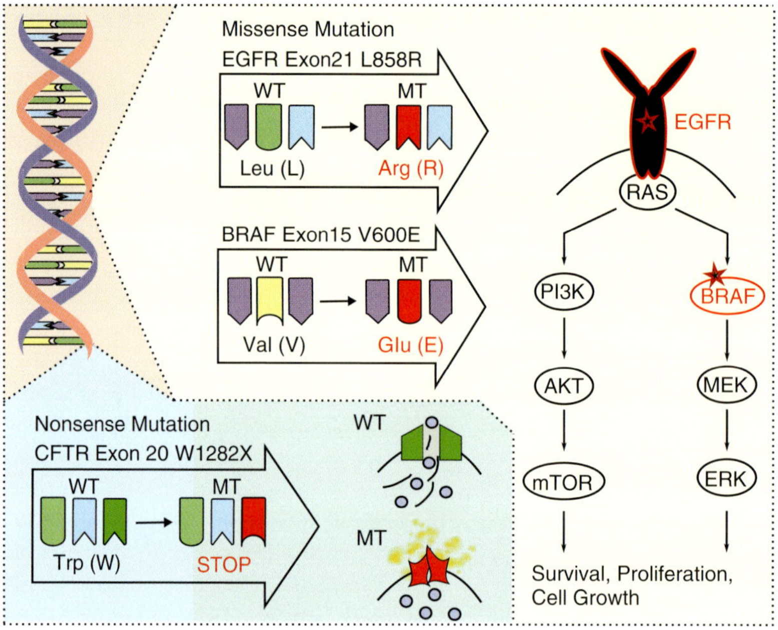

Fig. 5 Pathological point mutations. The missense mutation in EGFR exon 21 (L858R) produces constitutively active receptors; a similar effect is seen with the V600E missense mutation in exon 15 of BRAF. Both mutations result in proteins targeted therapeutically. One of the causes of cystic fibrosis is a nonsense mutation in exon 20 (W1282X) of the cystic fibrosis transmembrane conductance regulator (CFTR).

In summary, the purpose of identifying genomic alterations for targeted therapy is threefold: (i) they can be exploited directly as targets (Fig. 1b – Type 1; mutated proteins or fusion proteins, as discussed below); (ii) they can suggest signaling pathways therapeutically relevant (Fig. 1b – Type 2); or (iii) they can be regarded as base points of vulnerability to explore synthetically lethal combination therapy (Sect. 3.3).

Somatic Mutations Informing Targeted Therapy Among a given tumor type, only very few cancer genes are mutated at high proportions (>20%), with most mutations occurring at intermediate frequencies (2–20%) and only a fraction of these providing druggable targets [84, 86, 93]. Somatic mutations most suited for targeted therapy are activating mutations leading to oncogenic addiction. BRAF (v-raf murine sarcoma viral oncogene homolog B) mutations are an example of oncogenic addiction (Fig. 5). According to the catalog of somatic mutations (COSMIC; [115]), the vast majority (>99%) of BRAF mutations are missense mutations, with codon 600 being one of the most commonly affected sites. BRAF^{V600} mutations are activating mutations and constitutively activate the mitogen-activated protein kinase (MAPK) survival pathway [116, 117] (Fig. 5). The BRAF mutation drives a dominant survival and proliferative signal and presents an attractive therapeutic target, with two targeted agents, vemurafenib [118] and dabrafenib [119], clinically available. Not to oversimplify the targeting of mutant drivers, melanoma with BRAF mutations is quite responsive to BRAF inhibitors, whereas colorectal cancer with BRAF mutations is not [120]. As is the case with most targeted agents, resistance is almost guaranteed, thus necessitating multi-inhibitor treatment approaches. In the case of BRAF mutations, the targeting of additional signaling proteins in the MAPK pathway has provided some additional survival benefits; although, even with dual inhibitors, eventual resistance is observed, most commonly through additional mutations that reactivate the MAPK pathway (Fig. 5) [121, 122].

The altered EGFR in some cases takes on a role similar to that described above for BRAF (Fig. 5). EGFR is a transmembrane receptor tyrosine kinase (RTK) and an example of a class of entities that are frequently targeted therapeutically (Table 1) [123]. Substitutions in exon 21 render the receptor constitutively active, establishing a high cellular dependency on its signaling and providing a sensitive target for targeted therapeutics such as erlotinib or gefitinib [19, 20]. Not surprisingly, resistance to treatment rapidly occurs. Unlike the situation with BRAF mutations, EGFR inhibitor resistance does not necessarily occur via a circumventive reactivation of the associated signaling pathway; instead, a secondary single mutation in the "gatekeeper" residue, T790, which is necessary for inhibitor binding, causes drug resistance [124, 125]. In response, a novel compound – CO-1686 – has been developed to specifically target the T790M drug-resistance mutation (Table 1) [24], epitomizing the continuous race against the Darwinian character of cancer (i.e., selective pressures favoring cancer subclones with the most advantageous alterations) [126, 127].

Nonsense mutations are not frequently targeted therapeutically, mostly because the protein product is frequently non-functional. An exception to this generality involves the cystic fibrosis transmembrane conductance regulator (CFTR; (Fig. 5)). Nonsense mutations in CFTR are the cause of 5–10% of cases (85% in Ashkenazi Jews) of cystic fibrosis (CF) [128]. Several targeted approaches have been proposed as treatment strategies, an important undertaking considering that about 10% of all genetic diseases are due to nonsense mutations [129]. One approach focuses on an induced "readthrough" of premature stopcodons, as has been described for aminoglycosides [130]. The potential "readthrough" drug ataluren has been reported to increase the amount of full-length, functional CFTR [75, 76], and is currently under clinical investigation for the treatment of CF, although its exact mechanism has not been entirely elucidated [131–134]. Gene therapy (see Sect. 2.4), an approach still in its infancy, may be an alternative [135, 136].

Chromosomal Aberrations Extensive aneuploidies are striking characteristics of the cancer genome and generally are caused by chromosomal instability. Deletions and amplifications, as well as chromosomal translocations and chromothripsis (hundreds of genomic rearrangements caused by a single event [137, 138]), are all important mechanisms contributing to chromosomal aberrations [139]. Chromosomal translocations (Fig. 6) arise through chromosome structure instability that is believed to be brought about by the incorrect repair of DNA damage [142]. Cancer-associated chromosomal alterations are of particular interest, as these translocations or inversions potentially result in fusion proteins not found in normal cells, presenting cancer-specific targets [143, 144]. The perhaps best-known chromosomal translocation is termed the Philadelphia

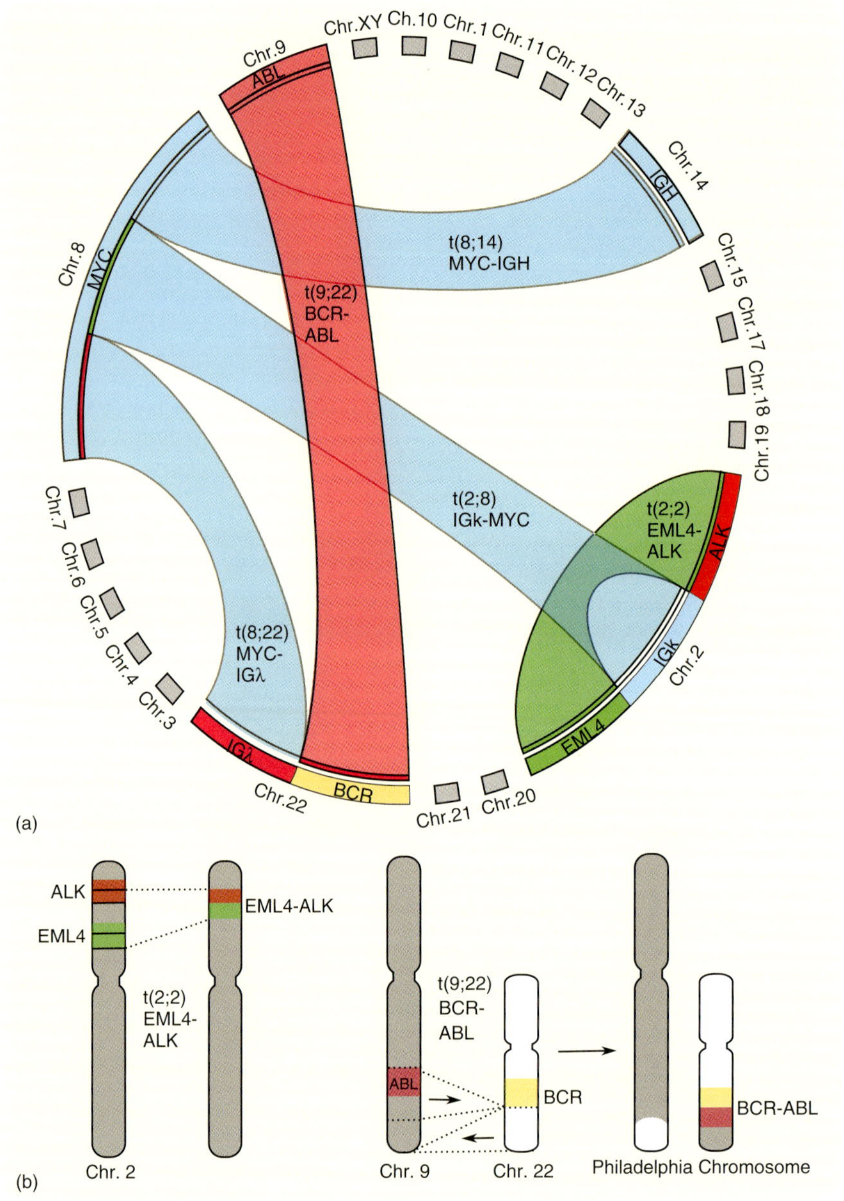

Fig. 6 Chromosomal aberration. (a) Three chromosomal translocations involving MYC are associated with Burkitt's lymphoma (BL): t(8;14), t(2;8), and t(8;22). t(8;14) is associated with 85–90% of BL cases. t(2;8) and t(8;22) are found in 10–15% of cases [140]. (a,b) EML4-ALK is caused by an inversion of Chr. 2 (t(2;2)); the Philadelphia Chromosome comes into existence when a translocation between Chr. 9 and Chr. 22 occurs. BCR-ABL is the oncogenic protein product of the Philadelphia Chromosome. The Circos plot (a) was generated using Circos (*http://circos.ca/*) [141].

Chromosome (Fig. 6), a translocation between the long arms of chromosomes 9 and 22 (t9;22), first described by Norwell and Hungerford [145–147]. The resulting fusion protein of breakpoint cluster region-Abelson tyrosine kinase (BCR-ABL1) is constitutively active and is not found in normal cells. Subsequently, the first selective BCR-ABL1 targeting tyrosine-kinase inhibitor, imatinib, was approved in 2001 for the treatment of cancer [44]. Imatinib has had a remarkable impact on patients with chronic myelogenous leukemia (CML), a cancer that is heavily dependent on BCR-ABL1. Unfortunately, different mutations in the kinase domain of BCR-ABL1 can render cancer cells resistant to imatinib, but this in retrospect not entirely surprising observation led to the development of two additional BCR-ABL1 targeting therapeutics: dasatinib [46, 47] and nilotinib [48, 49], both of which target most imatinib-resistant BCR-ABL1 variants. All three drugs target the ATP-binding pocket, and it seems critical to further diversify available therapeutics to address the threat of broad resistance induced by mutations of the ATP-binding pocket. The first steps in this direction have been taken by investigating the BCR-ABL1 SH2-kinase domain as a potential therapeutic target [148].

Another fusion-gene of high clinical relevance is the echinoderm microtubule-associated protein-like 4-anaplastic lymphoma kinase (EML4-ALK) protein tyrosine kinase produced by an inversion of the short arm of chromosome 2 (Fig. 6) [149, 150]. Importantly, the EML4-ALK fusion protein is indeed only found in malignant cells, thereby providing a unique pathology-specific target. The small-molecule tyrosine kinase inhibitor crizotinib (see Table 1) effectively targets EML4-ALK (as well as several other kinases, including MET and ROS1), and provides significant survival benefits to patients with EML4-ALK-positive lung cancer compared to nontargeted chemotherapy [50]. Regrettably, cancer cells respond rapidly to the change in selective pressure, and secondary mutations that render the cancer cells resistant to crizotinib frequently occur [151]. The next-generation ALK-inhibitor ceritinib has been shown to overcome crizotinib resistance to some degree, although mutations that cause cells to become resistant to ceritinib have already been identified [51–53]. An alternative approach to managing the resistance of ALK-positive lung cancer to ALK-inhibitors is to target the molecular chaperone heat shock protein 90 (HSP90), a protein that is necessary for the correct folding – and thus function – of EML4-ALK (Fig. 6) [152].

Burkitt's lymphoma (BL) [153] is a non-Hodgkin lymphoma derived from germinal center B cells [140]. Remarkably, nearly 100% of BL cases have a translocation of the MYC oncogene (Fig. 6a) [154]. BL is characterized by three translocations, with the most common (85–90% of cases) involving the distal region of the long arm of chromosome 14 and the long arm of chromosome 8 (t(8;14)) [155–157]. All three translocations result in the juxtaposition of MYC to the enhancer element of one of the immunoglobulin genes, increasing the expression of MYC. BL can often be cured with traditional (nontargeted) chemotherapeutics [140, 158], but these are extremely toxic and are all too frequently unavailable in developing countries at the correct doses or combinations. Considering that MYC is one of the most highly amplified oncogenes, targeting MYC directly could potentially have enormous impact, and not only for patients with BL [159]. As a transcription factor, MYC does not present traditionally targetable moieties, which makes specific

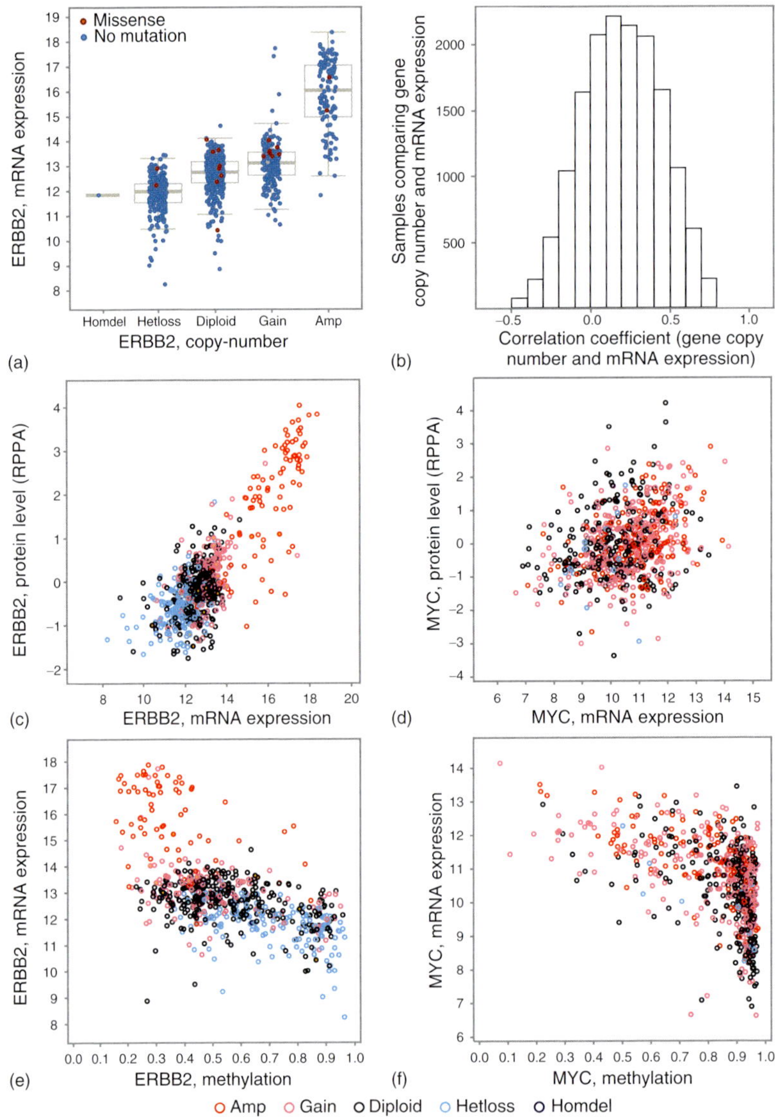

Fig. 7 Variations in gene expression. (a) In the case of ERBB2, gene amplification (Amp) leads to increased mRNA expressions; however, in general, the correlation between gene copy number and mRNA expression is low; (b) Distribution of correlation coefficients between copy number and mRNA levels in lung adenocarcinoma is shown; (c,d) Protein levels for ERBB2 correlate with mRNA expression levels: no such correlation is detectable for MYC; (e) Increased methylation correlates with decreased mRNA expression in the case of ERBB2; in the case of MYC, methylation patterns, and mRNA expression levels do not correlate (f). Data were obtained from the The Cancer Genome Atlas (TCGA; *http://cancergenome.nih.gov/*) to generate panels (a, c–e) using cBioPortal [162, 163] and panel (b) using the Genome Data Analysis Center (*https://confluence.broadinstitute.org/display/GDAC/Home*). Gain, gained; Diploid, normal number of chromosomes; Hetloss, heterozygously deleted; Homdel, homozygously deleted.

drug development particularly challenging [160]. One approach to circumvent the issue of directly targeting MYC would be to focus on associated signaling proteins for which inhibitors are available, such as phosphatidylinositol-4,5-bisphosphate 3-kinase (PI3K), Src family kinases or mTOR [155]. Another interesting approach (see Sect. 1.4) is to target MYC at the epigenetic level [69, 155].

Amplification and Overexpression Amplification on the genomic level is the selective increase in copies of DNA sometimes resulting in overexpression of a cancer gene [161]. (Definitions vary, but amplification is generally considered to be a four- or fivefold increase in copy number relative to an adjacent nonamplified marker on the same chromosome.) A number of databases (e.g., cBioPortal [162, 163], COSMIC [115], Oncomine [164]) that describe the relationship between amplification status and expression levels of a given gene are available to research groups. In the case of cancer genomic amplifications, there generally is not a pathology-specific target; rather, there is a lot more of a given gene product relative to normal cells, and the cancer cells tend to rely on the overexpressed gene product to a greater extent. It is the disproportional presence and dependence that provokes therapeutic interest and creates opportunities for targeted therapy. An important caveat regarding gene copy-number and mRNA expression is that the correlation between the two varies significantly from gene to gene (Fig. 7a and b).

A receptor tyrosine kinase ERBB2 (also known as HER2) is an important example of cancer gene amplification. ERBB2 was the first therapeutic target to be defined as such based on genomic alterations [165], and can be effectively targeted with several therapeutic agents [166, 167] including the monoclonal antibody trastuzumab, which was approved by the FDA in 1998 and initiated the age of targeted cancer therapy [27–29]. Trastuzumab is remarkably efficient when ERBB2 is amplified; although, as is the case for many – if not all – targeted therapeutics (see Sect. 2), resistance inadvertently arises [168]. Mechanisms of resistance include the upregulation of downstream signaling proteins, such as PI3K, increased signaling from related RTKs, or a resistance-associated mutation of ERBB2 itself [168].

ERBB2 is amplified in a high percentage (15–20%) of breast cancer cases [169, 170]. Importantly, there is a significant correlation between the ERBB2 gene copy number and ERBB2 mRNA expression levels (Fig. 7a); however, for the majority of other genes the correlation coefficient relating gene copy number and mRNA expression levels is less than 0.5 (Fig. 7b). The story is further complicated when translation is considered, where ERBB2 is again close to the ideal, most straightforward scenario, with mRNA expression levels correlating with amplification status and protein expression levels subsequently correlating with mRNA levels (Fig. 7c). However, this is frequently not the case (Fig. 7d), and it is important to be aware of potential discordance between gene copy number, mRNA expression levels, and protein expression levels.

Another important example of amplification involves the hepatocyte growth factor receptor (HGFR, or more commonly MET), a receptor of robust therapeutic interest. It has been observed that a significant number of cases of EGFR inhibitor resistance (Sect. 1.3.1.1) acquired MET amplification [171], which specifically increases the sensitivity of cancer cells to MET inhibitors (Table 1) [30] and provides

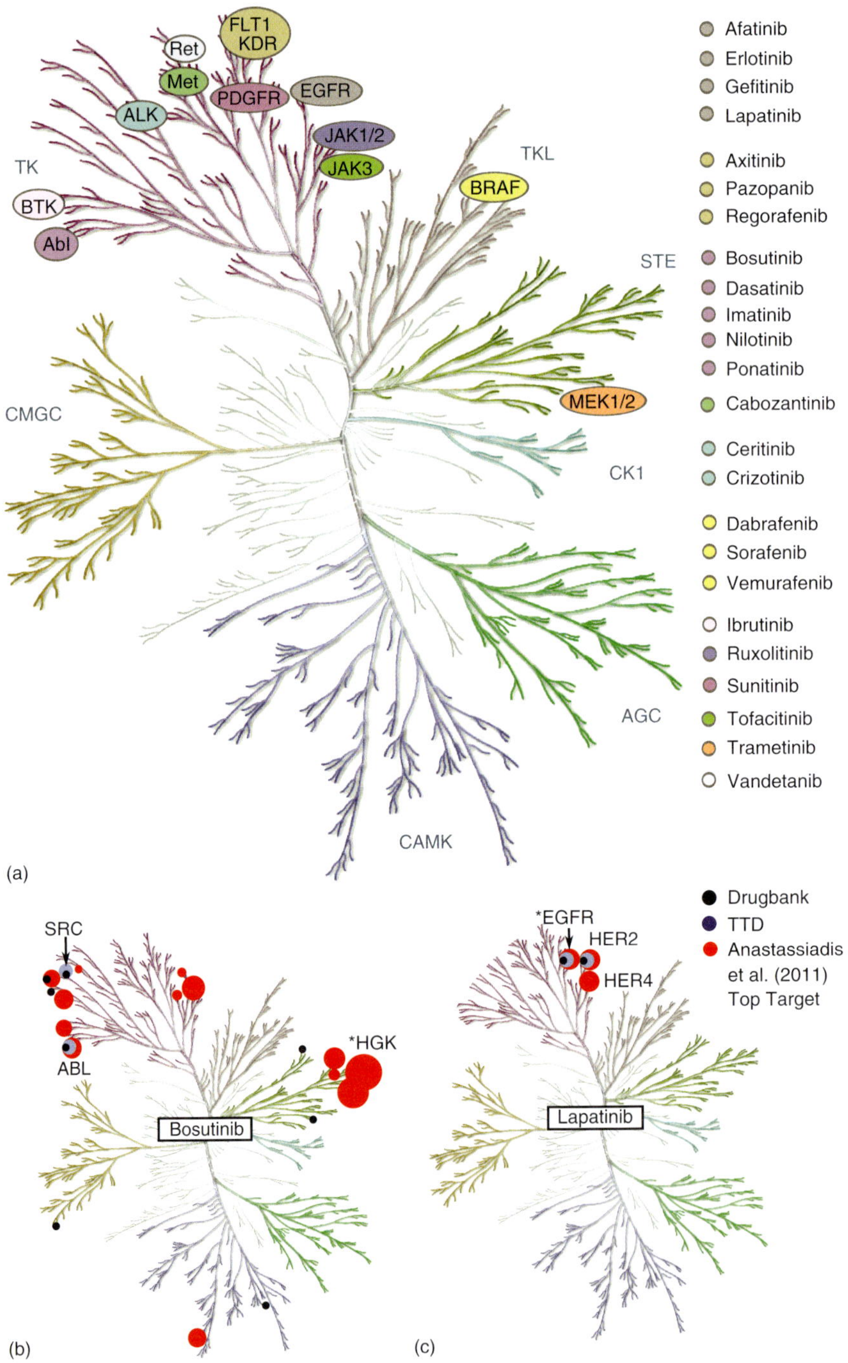
FLT1
KDR
Ret
Met
PDGFR
EGFR
ALK
JAK1/2
JAK3
TK
TKL
BTK
BRAF
Abl
STE
CMGC
MEK1/2
CK1
AGC
CAMK
Afatinib
Erlotinib
Gefitinib
Lapatinib
Axitinib
Pazopanib
Regorafenib
Bosutinib
Dasatinib
Imatinib
Nilotinib
Ponatinib
Cabozantinib
Ceritinib
Crizotinib
Dabrafenib
Sorafenib
Vemurafenib
Ibrutinib
Ruxolitinib
Sunitinib
Tofacitinib
Trametinib
Vandetanib
(a)
SRC
ABL
*HGK
Bosutinib
(b)
*EGFR
HER2
HER4
Lapatinib
(c)
Drugbank
TTD
Anastassiadis et al. (2011) Top Target

Fig. 8 Targeting the human kinome. (a) All FDA-approved small-molecule kinase inhibitors are shown, and the dominant kinase targeted by each inhibitor is indicated on the kinome tree; (b,c) Numbers of drug targets for kinase inhibitors vary widely between different inhibitors; in addition, these numbers are reported differently in different databases and studies. Targets for bosutinib and lapatinib are indicated, as retrieved from the DrugBank (black dots) [3], the Therapeutic Targets Database (TTD; blue circles [2]) and an *in-vitro* study [81] (red circles); (b) Kinases targeted by bosutinib are shown. The size of the red circles indicates the inhibitory activity of the drug against the target (kinase) [81]; (c) Kinase targets for lapatinib are shown (red circle sizes are arbitrary). The human kinome illustration is reproduced courtesy of Cell Signaling Technology, Inc. (*www.cellsignal.com*) [180].

an avenue for targeted therapy, particularly in combination with EGFR-inhibitors if EGFR is expressed [171–173].

Tumor gene expression patterns are important and can be useful in guiding initial therapy [174–176] and understanding biological changes, for example, when cells undergo epithelial-to-mesenchymal transition (a process thought to be necessary for cancer metastasis) [177–179]. Furthermore, and not surprisingly, gene expression patterns and the activation status of proteins will change in response to therapy and are frequently linked to resistance and disease relapse. This is particularly relevant for kinase inhibitors, which represent an important and well-established class of targeted therapeutics (Fig. 8a; Table 1). These compounds generally target the highly conserved ATP binding pocket of kinases [81], and as a result many kinase inhibitors lack the initially envisioned specificity and target many diverse kinases, as is highlighted by the example of bosutinib (Fig. 8b); other kinases inhibitors, such as lapatinib, are more specific (Fig. 8c). The target multiplicity of kinase inhibitors is emblematic of the difficulties of understanding the true relationship between molecular targets and therapeutic effect (see Fig. 1d), and such difficulties are amplified by the still-evolving understanding of many targeted therapeutic agents (see Fig. 3). Interestingly, kinome "reprogramming" has been observed as cells under therapeutically induced selective pressure dramatically change their signaling patterns and gene expression [181, 182]. Indeed, the interpretation and anticipation of these changes in signaling and expression pattern changes represents the biggest challenges to be overcome in order to improve therapeutic outcomes.

1.4 Epigenetics

Epigenetics describes chromatin-based events that regulate DNA-templated processes, that is, the regulation of gene expression changes rather than direct DNA codon alterations [183]. Epigenetics has emerged as an important consideration from a cancer therapeutics standpoint, and impressive advances have recently been made, as exemplified by the creation of the Encyclopedia of DNA Elements (ENCODE) project [184, 185]. Modifications on the epigenetic level modulate DNA events such as transcription, DNA repair, and replication [186]. Four specific epigenomic regulatory processes have been robustly defined, all of which are involved in regulating access to DNA:

- Adding DNA methylation and histone modifiers (performed by so-called enzymatic "writers") [187].
- Removing DNA methylation and histone modifiers (performed by so-called "erasers").

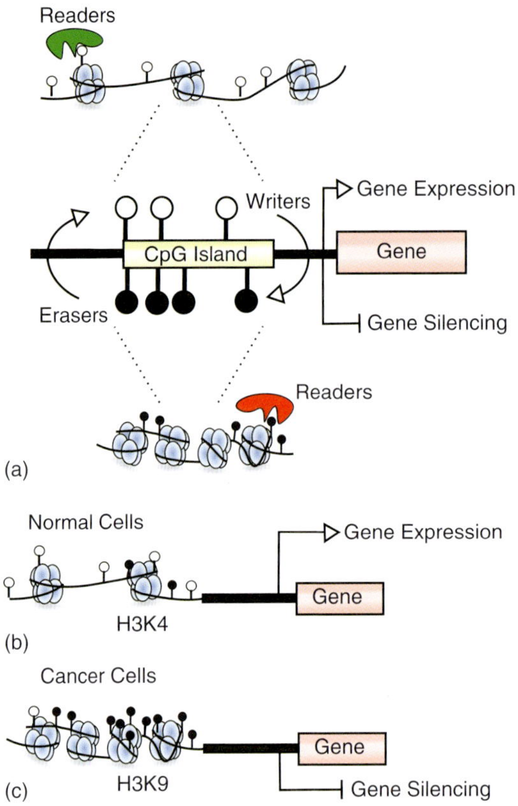

Fig. 9 Epigenetic modulation. (a) Epigenetic modulation of gene expression via methylation status changes are shown; "writers" methylate methylation sites, causing condensation of chromatin and silencing of genes; "erasers" remove methylation, opening up the chromatin and allowing gene expression; (b) Methylation of histone H3 lysine 4 (H3K4) in normal cells is associated with gene expression; (c) Methylation of histone H3 lysine 9 (H3K9) in cancer cells is associated with gene silencing.

- Binding to epigenetic modifiers (performed by so-called "readers").
- Repositioning of nucleosomes across the genome (performed by so-called "movers") (Fig. 9a) [188–190].

DNA promoter cytosine-phosphate-guanine (CpG) island (cytosine in CpG regions can be methylated [191]) hypermethylation, resulting in repressed gene transcription, is particularly frequently observed in cancer (Fig. 9b and c) [192]. Alternatively, the hypomethylation of oncogenes is an additional mechanism of realizing oncogenic potential [188]. ERBB2 expression is an example where a clear correlation exists in terms of methylation status, amplification status, and mRNA expression levels. In cases of ERBB2 amplification, hypomethylation seems to predominate, and this results in increased mRNA expression levels (see Fig. 7e) but, again, this type of robust pattern is infrequent (Fig. 7f). Malignant cells dysregulate epigenetic processes through alterations in DNA/histone modifications, somatic alterations in epigenetic proteins, or altered expression levels of epigenetic proteins [67]. By taking into consideration the fact that the malignant epigenome is indeed very much different from the epigenome of normal tissue, it has been possible to exploit these differences in order to develop targeted therapeutics (Table 1).

Targets of epigenetic therapies in cancer include DNA methyltransferases (DNMTs), histone deacetylases (HDACs), bromodomain and extra-terminal (BET) proteins, DOT1-like histone H3K79 methyltransferase (DOT1-like; DOT1L) and enhancer of zeste homolog 2 (EZH2) [184]. Azacitidine, which originally was designed as a general chemotherapeutic, has been repurposed to function as a targeted agent since the discovery that, at low concentrations, it inhibits all three known DNMTs [66].

The HDACs are another class of attractive epigenetic target, and in general, histone deacetylation is associated with gene repression. Two HDAC inhibitors, vorinostat and romidepsin, are FDA-approved and have achieved some successes in the clinic [67, 68]. EZH2 and DOT1L are both involved in the enzymatic methylation of histone H3, and both have been described as being mutated in cancer [73, 74]; efforts to develop inhibitors to EZH3 and DOT1L are ongoing (Table 1) [67]. Lastly, BET proteins bind to acetylated histones to promote transcription, and interest in this epigenetic regulator was initiated by the discovery of a fusion protein composed of bromodomain containing protein 4 (BRD4) and nuclear protein in testis (NUT) in nuclear protein in testis midline carcinoma (NMC) [184]. The potential to inhibit a novel cancer-specific target led to the development of a specific BET bromodomain inhibitor that currently is undergoing clinical trials [70, 71]. The BET inhibitor JQ1 has shown particular promise by targeting MYC (see Sect. 1.3.1.2) on a transcriptional level rather than directly. Although, from a mechanistic standpoint, the role of JQ1 is not yet completely understood, it has been shown to interfere with RNA polymerase, thus reducing MYC transcription [69, 72].

1.5 Targeting Biological Pathogens

Targeted therapy is not only relevant in the oncological setting but is also a critical tool in the effort to curb bacterial and parasitic infections. Penicillin, discovered by Sir Alexander Fleming [193], revolutionized the treatment of bacterial infections and can be considered a targeted therapeutic because it specifically inhibits transpeptidase proteins that are not found in the human host; thus, penicillin disables the crosslinking of peptidoglycan strands within the bacterial cell wall, resulting in cytolysis, while sparing host cells [194, 195]. As seen with cancer, resistance is quick to arise and multiple means of resistance to penicillin have been described, such as increased efflux of the drug out of bacterial cells, alterations in the transpeptidase structure or, most interesting and most commonly, via the expression of enzymes called β-lactamases that specifically cleave part of the molecular structure of penicillin (the β-lactam backbone) [196]. Again, similar to the efforts described for targeted therapy in cancer, next-generation β-lactam antibiotics are being developed to subvert the impact of β-lactamases [195]. To extend the list of viable targets while limiting the impact of resistance – which is seen as a global threat – genomics approaches have been applied to sequence 74 Gram-negative pathogens and integrate the information with experimental annotations and computational analysis in a platform referred to as antibacTR [197]. genomics approaches were also used to identify PfATP4, a P-Type ATPase candidate on the plasma membrane of *Plasmodium falciparum* (one of the causative agents of malaria [198]) that is needed to enable Na^+ efflux out of the parasite. Spiroindolones specifically disrupt this process and are currently being investigated as novel components of anti-malarial therapeutic regimens [199–201]. Extensive efforts are also ongoing to identify unique targets for other pathogens through comparative genomics (e.g., against *Candida albicans* [202], other fungal pathogens [203], and *Mycobacterium ulcerans* [204]).

Additional targeted approaches to substantiate the currently available antibacterial arsenal include anti-infective monoclonal antibodies (mAbs) (see Sect. 2.2.1) and phage therapy. mAbs against infectious agents are being actively developed, with particular successes achieved by targeting bacterial toxins [205]. *Clostridium difficile* is an example of an infectious agent that is immensely difficult to treat due to its frequent resistance to current treatments. The combination of two antibodies – actoxumab/bezlotoxumab (MK-3415A) – effectively neutralizes the two main pathogenic endotoxins produced by *C. difficile*, and represents a promising new treatment strategy [78, 79]. The first antibody to be approved by the FDA for an antibacterial indication was raxibacumab (ABthrax) [80], in the treatment of inhalation anthrax. Raxibacumab functions by binding to the protective antigen of the anthrax toxin produced by *Bacillus anthracis*; the bound toxin is subsequently incapable of binding to the host cell receptor and therefore is unable to trigger any adverse events [206, 207].

Today, phage therapy is re-emerging as a potential alternative to traditional (and in some cases, rapidly failing) antibiotics [208]. Bacteriophages are viruses that infect bacteria, causing bacterial lysis and the release of virion progeny, and thereby naturally amplifying the therapeutic agent in the presence of bacteria [209]. The exquisite specificity of some bacteriophages to bacteria makes them a class of targeted agent in the truest sense, since bacteriophages neither target host cells nor interfere with the normal gut flora, which is a common problem with traditional antibiotics [209, 210]. Minor immunogenic reactions to bacteriophages are possible, however, and bacteriophages are (at least in theory) capable of modifying bacteria in ways that increase the pathogenicity of the microorganism [209]. However, taking into consideration the recent rise in numbers of multidrug-resistant bacteria, the benefit of further developing bacteriophages as targeted therapeutics should not be underestimated.

1.6 Targeting Low-Density Lipoprotein

Targeted therapy can also be employed to specifically inhibit completely normal proteins in order to prevent or alter pathological processes. Cardiovascular disease caused by atherosclerosis (an imbalanced lipid metabolism and maladaptive immune response that damages the arterial walls) affects millions of people, and elevated levels of low-density lipoprotein (LDL) are a significant contributing factor [211, 212]. Genomic studies have identified an alteration in a gene among a subpopulation of individuals with extraordinarily low blood-cholesterol levels [13, 213–215]. The identified gene is proprotein convertase subtilisin/kexin type 9 (PCSK9), and the realization that loss-of-function mutations in this particular gene lead to low LDL levels has prompted great

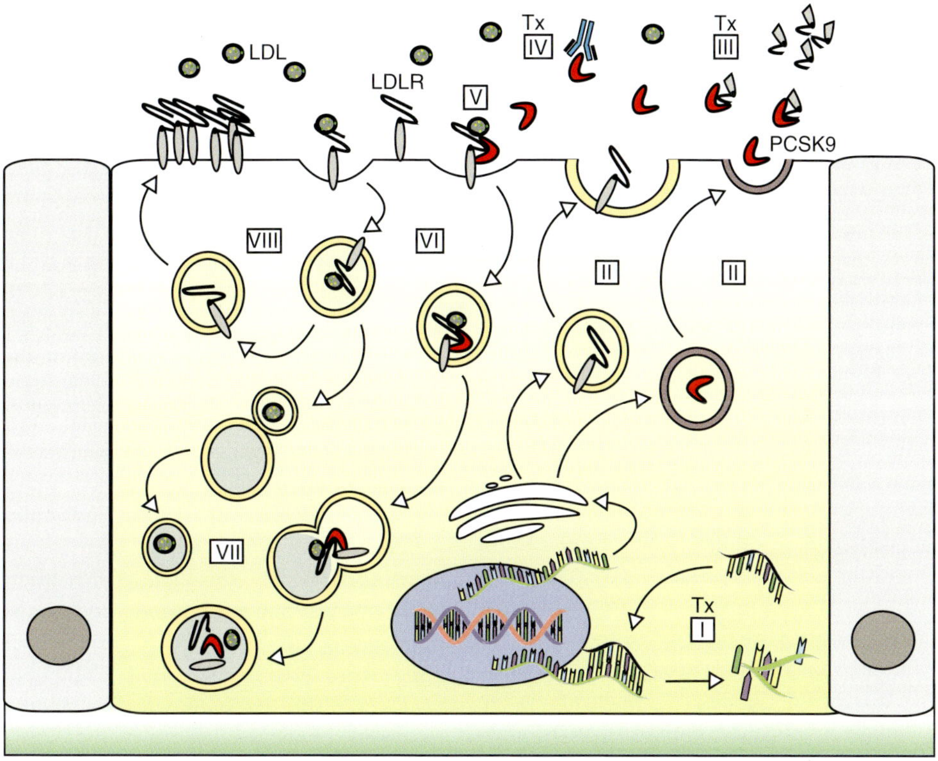

Fig. 10 Targeted therapy against PCSK9. (I) Targeting PCSK9 mRNA with oligonucleotides; (II) exocytosis of LDLR and PCSK9 (red); (III) peptide mimetic binding to PCSK9; (IV) antibody binding to PCSK9; (V) LDL bound to LDLR and PCSK9; (VI) endocytosis of LDL/LDLR/PCSK9 and LDL/LDLR; (VII) lysosomal degradation of LDL and LDL/LDLR/PCSK9; (VIII) recycling of LDLR in the absence of bound PCSK9. PCSK9, proprotein convertase subtilisin/kexin type 9; LDL, low-density lipoprotein; LDLR, low-density lipoprotein receptor; Tx, treatment.

interest in developing PCSK9 antagonists [13, 213, 214]. A reduction in PCSK9 leads to decreased lysosomal destruction of low-density lipoprotein receptors (LDLRs), with a subsequent increase in the absorption of LDL (Fig. 10) [216, 217]. The most promising approach for targeting PCSK9 has been with antibodies (Fig. 10) [218], although additional approaches have included targeting PCSK9 on the transcriptional level with antisense nucleotides or by using mimetic peptides that contain the PCSK9 binding site of LDLR [218]. Currently, SPC5001 is the most advanced of the PCSK9 mRNA targeting antisense therapeutics, and has proven efficacious in lowering LDL levels in nonhuman primates when given subcutaneously (Fig. 10) [59, 60].

1.7
Caveats and Pitfalls

The impact of the age of genomics on targeted therapy development has been rather remarkable (Table 1), though ample challenges remain. The biological effect of a therapeutic agent changes over time (partially due to biological factors, partially due to an initially incomplete understanding of the biology), and the interaction between targeted therapeutics in the context of pathology or when given in combination is often immensely complex (see Figs 1 and 3). Molecularly targeted cancer therapy remains plagued by a high failure rate, a limited therapeutic efficacy, and a relatively small number of patients that significantly benefit from it (10–20%) [29]. Clearly, more targets are needed, better drugs must be developed, and adequate biomarkers are required [219]. Unbiased massive sequencing efforts inherently produce copious amounts of data. One pitfall of such major data acquisition is that the collected data require proper and careful curation and management. It is critical that standards are implemented to ensure that acquired data are handled in the same way by different research groups (e.g., by using standardized analysis pipelines) [178]. Equally important is careful collective data interpretation to minimize the propagation of sequencing and computational errors. Moreover, as noted above, genotypic changes do not always impact phenotypic events as expected, and any conclusions drawn from sequencing studies must consistently be validated via other experimental approaches. Lastly, it is critical that all acquired data are made universally accessible and are comprehensible to the entire research and medical communities [5, 11, 178, 220–223].

2 Therapeutic Agents Used in Targeted Therapy

2.1 Small Molecules

The largest class of targeted agents are the small-molecule drugs (SMDs), also referred to as new chemical entities (NCEs). This class includes molecularly defined nonpeptide organic compounds of low molecular weight (typically <900 Da) that are used to treat a wide range of diseases and conditions (Table 1) [224]. The aforementioned kinase inhibitors are all examples of SMDs (Fig. 8). Currently, the Therapeutic Target Database (TTD) [2] contains information relating to 17 012 SMDs, compared to 14 170 SMDs in the TTD of 2012 [225], thereby highlighting the strong and sustained interest in this class of therapeutics [2]. Examples of the most important small-molecule therapeutics in clinical use are listed in Table 1.

SMDs, by definition, are small molecules – an attribute which allows them to be distributed widely throughout the body, to permeate cell membranes and, most importantly, to reach both intracellular and extracellular targets (Fig. 11) [224]. SMDs are able to: modulate the cellular functions of extracellular growth factors [227], membrane channels and transporters [228], intracellular transcriptional factors [229], proteins lacking enzymatic activities [230], and metabolic pathways [231]; disrupt protein–protein interactions in intracellular signal transduction cascades [31, 32] and cell cycle regulation reactions [232]; deregulate post-translational protein modifications [233]; activate or inactivate autophagy [234]; and cause cell types to adopt features characteristic of other cell types (a process known as transdifferentiation) [235].

Another advantage of SMDs in a general sense is the possibility for chemical modifications and synthesis of core-structure derivatives [236]. Chemical modifications are essential to: (i) improve target specificity once a promising overall molecular structure has been identified [237]; (ii) develop next-generation compounds; (iii) overcome

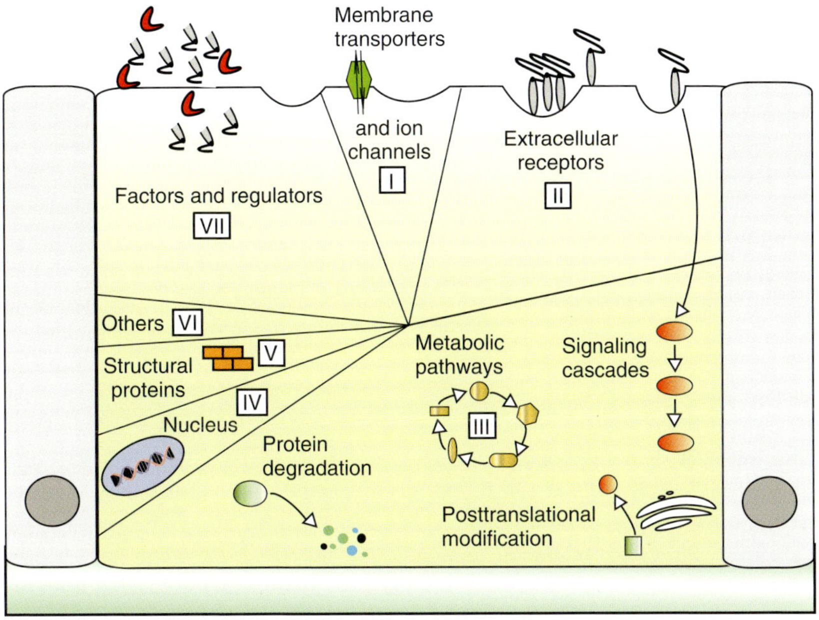

Fig. 11 Distribution of current successful targets. About 6% of targets are membrane transporters and ion channels (I); about 22% are extracellular receptors (II); enzymes, which participate in important cellular processes such as metabolism, protein modification, intracellular signaling, protein degradation, comprise the largest number of targets, approximately 44% (III); only about 4% of targets are contained within the nucleus (IV); and about 4% of targets are structural proteins (V); 19% of targets include factors and regulators (VI); and about 1% of targets are still undetermined (VII). This distribution is based on data from 2006 [226].

acquired resistance [201]; (iv) increase chemical stability [238]; (v) improve means of administration [239]; and (vi) optimize absorption, distribution, metabolism, excretion, and toxicity (ADMET) characteristics [240]. Some SMDs are able not only to act on their direct targets but can also stimulate the immune system to recognize and destroy malignant cells; this process is known as immunogenic cell death [241]. For example, the antitumor effects of the kinase inhibitor dasatinib on c-KIT mutant P815 mastocytoma tumors *in vivo* consist not only of direct tumoricidal activity. Importantly, dasatinib also induces immunomodulatory effects by boosting an antitumor T-cell response [242]. However, unlike other classes of therapeutics (particularly protein-based therapeutics), SMDs are – for the most part – rarely immunogenic [224, 243].

Whilst the advantageous characteristics of SMDs make them potent therapeutic agents, there are several disadvantages. SMDs can be less selective than protein-based therapeutics and can accumulate in specific organs, such as the kidney and liver, resulting in severely toxic side effects [244]. Furthermore, cytochrome P450 enzymes metabolize many of the SMDs, sometimes producing reactive metabolites that can cause unwanted interactions with other medications, such as macrolide antibiotics, azole antifungals, anticonvulsants, protease inhibitors, and others [245]. Another disadvantage is the relatively short half-life (though sometimes this is cited as an advantage as it allows a rapid clearance of the drug) of many SMDs, which commonly necessitates daily administration of the drug [245]. Some of the disadvantages associated with SMDs are avoided by the "biologicals," another extensively investigated group of therapeutic agents (see Sect. 2.2).

2.2 Biologicals

Biomolecular drugs, which are also known as new biological entities (NBEs), biotech drugs, or "biologicals," have emerged as an extremely important class of therapeutic agents, and advances in biotechnology continue to refine the members of this class [224, 246]. Biologicals are substances with fairly large molecular weights (5–200 kDa) that are either prepared from living organisms or derived from substances produced by living organisms [224, 246, 247]. This class of agent is used for the prevention, diagnosis, and treatment of various maladies. Biologicals, as a class, consist of vaccines, cell or gene therapies, hormones, cytokines, tissue growth factors, viruses, and mAbs [103, 224, 248].

The production of biologicals ranges from a reliance on recombinant DNA, to controlled gene expression, to advanced antibody production [249, 250]. For example, modern DNA technology allows the introduction of a functional gene (that encodes the desirable product) into a living organism (such as a bacterium, yeast, or mammalian cells), thus priming the organism to produce copious amounts of the protein [251, 252]. Biologicals are also manufactured using plants and animals. This approach has several important advantages, including correct eukaryotic protein modification (which is difficult to accomplish synthetically), low cost (depending on the specific agent), and high-speed production once a pipeline has been established [253–255].

The development of new biologicals does raise a number of challenges in terms of safety and ethics, however, because it requires the use of cells derived from human tumors or animal models, as well as some unproven technologies to facilitate production [256, 257]. A primary concern here is the limitations of currently used methodologies for the detection of microbial agents and oncogenicity and infectivity of DNA [258], which may pose a potential threat to research workers, patients, and animal models alike.

In general, biologicals are often less well-defined compared to SMDs, predominantly due to post-translational modifications (PTMs) and the presence of impurities inherent to the biological source [259]. These factors make biologicals difficult to analyze both qualitatively and quantitatively [224, 247, 260]. Similar to SMDs, biologicals are useful as either agonists or antagonists of different biological processes, and are generally highly selective and quite potent [259, 261]; however, due to the large molecular size of biologicals, they commonly cannot bypass extracellular barriers and thus can only target cell-surface proteins (Fig. 12) [224, 259]. Biologicals are frequently less stable than SMDs and prone to degradation within the gastrointestinal tract; consequently, they must often be administered by either injection or infusion [262, 263]. In contrast to SMDs, most biologicals are immunogenic. The elicitation of an immune response may not only attenuate the therapeutic effect of a given biological (though in specific cases the immune response may be beneficial [264]), but it can also lead to high toxicity [265].

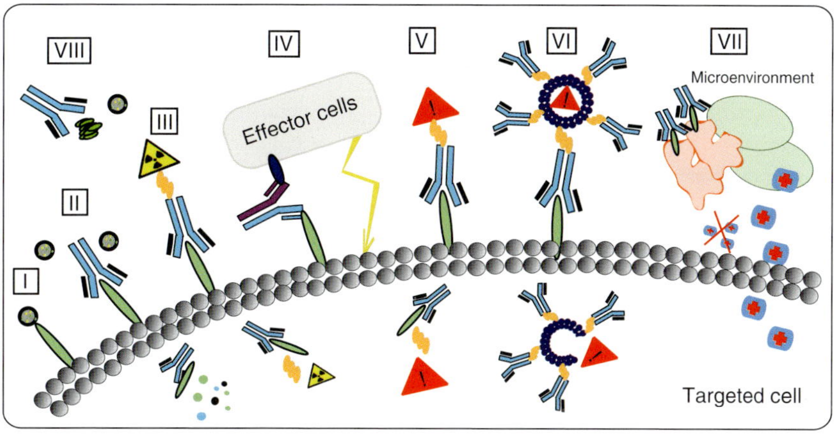

Fig. 12 Monoclonal antibody (mAb)-based therapy. (I) Successful ligand binding to the extracellular receptor; (II) mAb binds to receptor with consequent internalization and degradation of the complex; (III) Immunoconjugate of mAb, linker and radioactive agent binds to receptor, and after internalization the complex is degraded, causing intracellular release of the radioactive agent (radioactive probes can also cause cell death from the membrane surface and do not necessarily need to be internalized); (IV) bispecific monoclonal antibody (bsAb) simultaneously binds to receptors on targeted and immune effector cells; (V) cytotoxic agent is released from the immunoconjugate (similar to (III)); (VI) mAb-covered liposome with cytotoxic cargo is guided to the cell; (VII) targeting and altering the tumor microenvironment using mAbs; (VIII) targeting of extracellular cytokines, toxins and ligands can also be accomplish using mAbs.

2.2.1 Antibody-Based Therapeutics

One important subgroup of biologicals is the large group of antibody-based therapeutics (Table 1). Over the past 30 years, mAb-based therapy has been established as an important therapeutic strategy for the treatment of cancer [266, 267]. One of the first treatment successes, which was reported in 1981, involved a murine mAb against a normal T-cell differentiation antigen to treat T cell leukemia [268]. In 1986, the first mAb, muromonab-CD3 (Orthoclone OKT3), was approved by the FDA not for the treatment of cancer, but for the prevention of acute organ rejection after transplantation, by blocking T-cell function [38]. An initial major obstacle to the increased clinical use of mAbs was the mouse-based production pipeline and the associated high immunogenicity and short half-life of the antibodies [267]. However, the development of new techniques to humanize or chimerize mAbs addressed such obstacles, and this resulted in clinically suitable mAbs that behaved similarly to natural human antibodies [269]. The high molecular weight of mAbs is another obstacle at times, and is the likely explanation for the inefficient crossing of the blood–brain barrier by mAbs; hence, therapeutic mAbs to treat brain cancer are usually delivered intratumorally [270].

mAbs target specific antigens found on the cell surface, such as transmembrane receptors or extracellular growth factors (Fig. 12) [245]. Upon binding, mAbs can exert direct effects [271], with some antibodies having multiple activities. For example, cetuximab blocks the binding of epidermal growth factor (EGF) and of transforming growth factor-α (TGF-α) to EGFR, which in turn leads to an inhibition of receptor activation and cell proliferation, induction of apoptosis, and the sensitization of tumor cells to chemotherapy and radiotherapy [20, 25]. Since mAbs target extracellular antigens, it is presumed that they are more effective against liquid tumors than solid tumors [271].

In some cases, mAbs are conjugated to radioisotopes or toxins (the latter are referred to as "cargo" or "payload"; see Fig. 12) for precision delivery to a specific target [272]. When an antibody binds to an internalizing receptor on the cell surface, receptor-mediated uptake occurs and the antibody is subsequently transferred into the cell, together with its cargo [273]. The cargo is conjugated to the antibody via a linker, which is a critically important part of many immunoconjugates as it should not allow premature release of the cargo yet ensure that, once the target is reached, efficient dissociation occurs [272]. Three mechanisms have been identified by which the cargo can be released upon endocytosis:

- The presence of a reducing environment (intracellular glutathione reduces disulfide bonds).
- A low pH in the lysosomal compartment (this disrupts the hydrazone linkage).
- Lysosomal enzyme cleavage in the cases of peptide linkers [272].

In some cases the linker is noncleavable (e.g., thioether or hindered disulfide bonds), and the whole antibody undergoes degradation within the lysosome in order to release the cytotoxic cargo. Internalization and cleavage are not required if the cargo is a radioisotope, because emitted α- or β-particles can penetrate the cell membrane unaided [39]. An alternative option for utilizing mAbs as drug-delivery vehicles is to cover nanoparticles (e.g., liposomes, which are closed spherical vesicles formed from a lipid bilayer) containing the drug cargo with antibodies or antibody fragments [274].

Antibody-based drug-delivery systems allow the delivery of general cytotoxics with greater accuracy, to maximize efficacy while minimizing side effects [39]. Brentuximab vedotin, which exemplifies the successful clinical application of this type of delivery system, is composed of a chimeric anti-CD30 antibody linked to the microtubule inhibitor monomethyl auristatin E via a protease-cleavable linker [275]. Brentuximab vedotin was approved in 2011 for the treatment of relapsed/refractory Hodgkin lymphoma and systemic anaplastic large cell lymphoma (Table 1) [39].

The direct targeting of cancer cells (see Fig. 12) is only one of the ways in which mAbs are used to treat cancer, as they have also been shown to successfully target and modulate the tumor microenvironment [276]. CD40 agonists (CD40 is expressed by immune cells in the tumor microenvironment [277]) are a prominent example of this treatment strategy. In pancreatic ductal adenocarcinoma (PDA), mAb CD40 agonists induce macrophage activation that subsequently leads to destruction of the PDA tumor stroma [40]. Another successful example is bevacizumab (see Table 1), which binds to vascular endothelial growth factor (VEGF) and prevents tumor-initiated angiogenesis. This limits the tumor's access to vital nutrients and in turn slows down or even stops tumor growth [34].

Another approach to further improve the efficiency of mAbs-based treatments is through the development of bispecific monoclonal antibodies (bsAbs). The bsAbs are mAbs which have been engineered to bind two different targets simultaneously, such that their selectivity is increased by co-targeting two tumor-associated antigens [278]. An example is co-targeting ERBB2/ERBB3 heterodimers with a bispecific single-chain antibody, A5-linker-ML3.9 bs-scFv, which showed high binding selectivity both *in vitro* and *in vivo* [279].

bsAbs can also be engineered to bind both tumor-associated antigen and an antigen on immune effector cells (Fig. 12) [280]. For example, HRS-3/A9 bsAb targets the CD30 and CD16 expressed by lymphoma cells and lymphocytes (natural killer cells and macrophages), respectively [41]. Bispecific T-cell engager (BiTE) antibodies are a second example of bsAbs. In the case of BiTE antibodies, antigens on the surfaces of cancer cells (including antigen CD19, epithelial cell adhesion molecule (EpCAM), ERBB2, EGFR, CD66e, CD33, EphA2, MCSP (melanoma chondroitin sulfate proteoglycan) [281]), and CD3 on T cells are targeted simultaneously [282]. This leads to the recruitment of cytotoxic T cells independent of T-cell receptor selectivity. Upon BiTE-initiated activation, T cells subdue the cancer cells via membrane perforation and subsequent induction of apoptosis [282, 283].

As noted above, one difference between SMDs and mAbs is the ability of mAbs to induce an immune response. Under the correct circumstances, this allows mAbs to exert not only direct inhibitory effects on tumor growth but also to activate indirect accessory antitumor activities, such as antibody-dependent cell-mediated cytotoxicity (ADCC) and complement-dependent cytotoxicity (CDC) [284]; these factors are perhaps responsible for different efficacy profiles of mAbs *in vitro* and *in vivo* [264]. The immunogenicity of mAbs is also responsible for many of the side effects, which include allergic reactions, the induction of autoimmune diseases, organ-specific adverse events such as cardiotoxicity, infections, platelet, and thrombotic disorders, while some mAbs can even contribute to tumor progression [285].

Although the development of therapeutic mAbs involves complex and expensive processes compared to the development of SMDs, interest in mAbs is growing and they remain a promising group of therapeutic agents [286].

2.3 Targeted Immunotherapy

Over the past few years, targeted immunotherapy has emerged as an innovative and successful strategy for the treatment of a multitude of cancers. Immunotherapy generally primes the immune system in different ways: (i) by stimulating a host immune response specifically against cancer cells; or (ii) by broadly boosting the immune system. Cancer cells often have cancer-specific antigens present on the surface (these molecules differ subtly from those expressed by non-cancer cells). These cancer antigens can serve as targets for the immune system, and immunotherapy often focuses on stimulating this process [287].

In addition to antibodies (see Sect. 2.2.1), immunotherapy includes cell-based and cytokine therapies. Cell-based therapies, or cancer vaccines, involve the removal of immune cells from an individual with cancer, followed by *in vitro* activation of the collected immune cells to recognize tumor-specific antigens. Next, *in vitro* expansion of the immune cell population is required, prior to re-injecting the cells into the individual, where the immune cells ideally trigger an immune response against the tumor [288]. Several cell types can be used for this technique, including natural killer cells, lymphokine-activated killer cells, cytotoxic T cells, and dendritic cells [289–291]. Vaccines are often combined with other substances or cells (termed adjuvants) which help to boost the immune response even further [292]. One successful example of a cell-based immunotherapy is the Sipuleucel-T cancer vaccine [293], the successful application of which first requires that dendritic cells are harvested from the patient and then transfected with a viral vector to increase the ability of the dendritic cells to properly present tumor antigens. When the dendritic cells are reintroduced into the patient they interact with lymphocytes to prompt the latter to specifically target cancer cells [294].

Any further advancement of immunotherapy necessitates that several critical points are addressed, specifically the intrinsic heterogeneity and genomic instability of tumors and the immune-suppressive activities of tumors and their microenvironments [295]. To combine therapeutics that affect different aspects of the immune response is a potential solution to some of these challenges [296]. One successful example is "immune checkpoint blockade," which focuses on blocking cytotoxic T-lymphocyte-associated antigen 4 (CTLA-4) and programmed cell death protein 1 (PD1) with mAbs, in combination with autologous granulocyte–macrophage colony-stimulating factor (GM-CSF)-secreting tumor cell vaccines [297]. Both, CTLA-4 and PD1 are co-inhibitory molecules that are expressed on T cells and decrease the antitumor immune response, thus blunting the efficacy of any vaccine-induced effect [297, 298]. "Immune checkpoint blockade" enables an increased activation and effector function of vaccine-induced, tumor-specific T cells [296].

A combination of immunotherapy with gene-based therapy (see Sect. 2.4) offers another means of addressing some of the shortcomings of originally contrived immunotherapy. This strategy, for example, can be directed to restore the correct activity level of transporters associated with antigen processing (TAPs), which are frequently aberrant within tumors [299]. TAPs are necessary for assembly of the major histocompatibility complex (MHC) class I, which is responsible for cell-surface antigen presentation to cytolytic T lymphocytes [300]. Successful reintroduction of the *TAP1* gene into melanoma B16F10 tumors *in vitro* and *in vivo*, using a vector system, had two significant effects: (i) reduction in the number of immunosuppressive CD^{3+}/interleukin (IL)-10-secreting lymphocytes; and (ii) restoration of MHC I on the tumor cell surface. As a result, the treatment of B16F10 tumors in mice with a vaccinia virus vector expressing *TAP1* led to a significant decrease in tumor growth [301].

Currently, plenty of scientific and clinical challenges remain to be overcome before immunotherapy can be considered a clinical mainstay. These include the heterogeneity of antigen distribution of malignant cells, antigen density variation, and the possibility of an unwanted immune response to the treatment [302]. An example of a severe side effect of immunotherapy is the so-called "cytokine storm," which is a massive and systemic release of proinflammatory cytokines that can lead to life-threatening multiorgan failure [303]. Nevertheless, immunotherapy is a rapidly expanding new class of cancer therapeutics, and in 2013 targeted immunotherapy was named "Breakthrough of the Year" in *Science* magazine – an annual and prestigious award reserved for the top scientific achievement worldwide [304].

2.4 Gene Therapy

Gene therapy was defined by the US FDA as products " … that mediate their effects by transcription and/or translation of transferred genetic material and/or by integrating into the host genome and that are administered as nucleic acids, viruses, or genetically engineered microorganisms" [305]. It would not be far-fetched to describe gene therapy as the most precise form of targeted therapy, in cases when applicable. Unlike targeting kinases with SMDs (see Fig. 8) or triggering unforeseen ripple effects with inhibitors (see Fig. 1c), gene therapy replaces the aberrant target with its normal counterpart [306]. According to data made accessible by *The Journal of Gene Medicine* (*http://www.wiley.com/legacy/wileychi/genmed/clinical/*), the majority of gene therapy approaches currently in clinical trials are related to cancer (63.8%), though other important applicable fields include monogenic diseases, infectious diseases, and cardiovascular diseases (Fig. 13). In 2013, a total of 81 clinical trials categorized as involving gene therapy were approved, although, the majority of these are still in Phase I (Fig. 13).

The potential of gene therapy has been demonstrated in a number of elegant studies [306]. For example, in 2009 Aiuti *et al.* used gene therapy to restore the immune function of nine out of ten patients with severe combined immunodeficiency (SCID) due to lack of adenosine deaminase (ADA). This was achieved by transfecting blood stem cells removed from the patients with a retroviral vector containing a functional copy of the *ADA* gene. Vector-treated cells were then returned to the patients [307].

Clustered regularly interspaced short palindromic repeats-associated system (CRISPR-Cas9) genome engineering is a new and very promising approach for targeted gene therapy that has recently been introduced to the world and has already been used to correct defective genes in model organisms and cultured cells [308]. Genome engineering of this nature relies on bacterial type II CRISPR-Cas9 to perform RNA-guided genome editing [309]. Single guided RNA (sgRNA) targets Cas9 to genomic regions that contain a protospacer-adjacent motif and that are complementary to the target region of the sgRNA [308, 310]. Cas9 generates double-stranded DNA breaks at the target loci, and the breaks are subsequently repaired by nonhomologous end-joining or homology-directed repair [308]. Yin *et al.*, in an elegant breakthrough study showed that CRISPR-Cas9 technology can be used to correct a genetic defect in the liver of adult mice [311]. In this study, Yin *et al.* treated mice with hereditary tyrosinemia type I, a disease caused by the lack of a metabolic enzyme required for tyrosine metabolism, using CRIPR-Cas9 technology [311]. Unfortunately, the CRISPR-Cas9 technology is still years away from clinical application, and measures must first be taken to reduce off-target effects and to increase safety by using a system that ensures only a temporary expression of the Cas9 nuclease. Further improving the delivery methods is also quite important, since at present only intravenous administration is possible and only genes found in the liver (where filtration of the blood occurs) can be targeted [309]; this issue is shared with RNA interference (RNAi)-based therapeutics (see Sect. 2.5).

2.5 Emerging Therapeutic Approaches

In addition to gene therapy, nucleic acids are being developed as targeted therapeutic agents in multiply innovative ways. These include the development of vaccines, antisense oligonucleotides, aptamers,

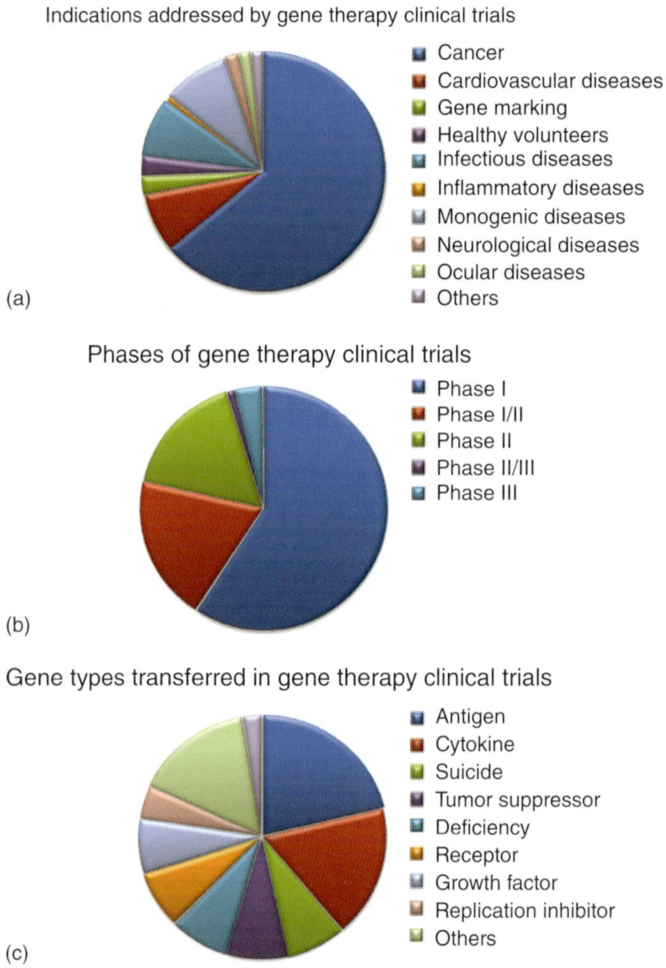

Fig. 13 Gene therapy in clinical trials. (a) Indication treated with gene therapy; (b) phase of gene therapy clinical trials; (c) gene types targeted by gene therapy. Data source: *http://www.wiley.com/legacy/wileychi/genmed/clinical/.*

ribozymes, and DNAzymes [312]. One example of a new therapeutic nucleic acid-based approach with advanced clinical potential is RNAi. This approach is based on the ability of small interfering RNA (siRNA) to identify its complementary mRNAs in the cell's cytoplasm. After binding to the siRNA, the target mRNA undergoes degradation by the RNA-induced silencing complex (RISC), which possesses endonuclease activity [313]. RNAi silences the target genes indirectly, via transcript degradation, with high efficiency and specificity [314]. During the past decade, major advances have been made in the field of RNAi, and its clinical application is ongoing (see Fig. 10). The introduction of new oligonucleotide chemistries, such as 2′-*O*-methyl ribose groups and 2′-fluoro-β-D-arabinonucleotides, significantly improved the serum stability of nucleic acids *in vivo*; new algorithms for siRNA design increased its potency and reduced off-target effects [315]. Encouraging clinical responses in cancer patients after RNAi therapies have been reported, specifically for therapies focused on silencing the translation of mRNAs that encode VEGF and kinesin spindle protein [316], the KRAS G12D mutant [317], and the M2 subunit of nucleotide reductase (RRM2) [318]. However, despite some successes in clinical trials, enthusiasm for RNAi-therapy has been somewhat hampered by the low delivery efficiency of siRNA into tumors, the difficulties in determining appropriate siRNA targets, and the significant risk of adverse side effects [314].

Oncolytic virotherapy is based on using oncolytic viruses as targeted therapy to selectively infect and damage cancer cells [102]. This selectivity is driven by the natural or enhanced cellular tropism of different types of viruses [102]. Recently, a randomized Phase III clinical trial designed to investigate the efficacy of talimogene laherparepvec (T-VEC) therapy to treat metastatic melanoma completed patient accrual. T-VEC is a herpes simplex virus that encodes GM-CSF (stimulates the immune system) and, based on preliminary results, is expected to be the first virotherapeutic agent approved for clinical use in the US [319]. The most relevant remaining obstacles are the optimization and enhancement of systemic virus delivery, intratumoral virus spread, and cross-priming of anticancer immunity [102].

The versatility of SMDs (see Sect. 2.1) has prompted interest in developing compounds akin to the immunoconjugates described in Sect. 2.2.1 (see Fig. 12). These small-molecule drug conjugates (SMDCs) are designed essentially to perform the same function as antibody–drug conjugates, but with different payloads being attached to small molecules that can then deliver the payload to a specific target. Thus far, carbonic anhydrase IX (CAIX) has been identified as a particularly promising target that is robustly expressed by many cancer types. Possible CAIX-focused tumor homing has been demonstrated using an SMDC [320]. Alternatively, an early development of delivering a payload using inhibitors of heat shock protein 90 (HSP90) in the form of an HSP90–inhibitor drug conjugate (HDC) has been reported (*http://www.syntapharma.com/*). HSP90 is selectively targetable in cancer cells [321].

The early detection and adequate visualization of cancer is a critical aspect of precision medicine (see Sect. 3), and targeted agents are increasingly being developed to accommodate this need. Different fluorescent dye probes [322], metallic nanoparticles [323] and semiconductor quantum dots [324] can be conjugated to antibodies or other ligands specific to markers expressed by cancer cells. For example, a rapidly activatable cancer-selective fluorescence imaging probe, γ-glutamyl hydroxymethyl rhodamine green (gGlu-HMRG), was synthesized and tested by Urano *et al.* [325]. gGlu-HMRG is activated by the cleavage of glutamate by the enzyme γ-glutamyltranspeptidase, which is not expressed in normal tissue but is commonly overexpressed on the cell membrane of various cancer cells. This provides targeted specificity of the probe [325] and covers a requirement for successful implementation of the topic addressed in the next section, namely precision medicine.

3
Targeted Therapy and Precision Medicine

3.1
Principles of Precision/Genomic Medicine

A consistent theme throughout the evolution of targeted therapy has been: (i) the need to identify specific targets essential to a pathogen or pathological process; and (ii) that the essentiality of the targeted entity is maintained or that therapy is adjusted to target the next entity which emerges as essential [326]. The over-riding goal of precision medicine is to establish an ability to identify disease susceptibility, to precisely map out the disease course, to personalize treatment strategies, and to optimize the timing of treatment [327]. Currently, several disease-targeted genetic tests are already available, specifically for cancer, cardiac disease, neurological diseases and others, and the number of available tests is likely to increase [16]. It is difficult to predict how precision medicine will impact medical care. Precision medicine may continue to be applied case-by-case as an auxiliary tool, or it may eventually emerge as a lifelong companion, and truly revolutionary theme [12], with the first round of sequencing initiated at birth, followed by periodic updates similar to vaccinations and physical examinations, and being a cardinal tool during times of maladies.

Molecular targeted therapy applications guided by molecular profiling of tumors is currently undergoing clinical investigation and has been determined as safe, feasible, and accomplishable in a timely manner (4–5 weeks) [328]. Current clinical trials have highlighted the upside of precision medicine, in that patients receiving precision medicine-guided targeted therapy tended to fare better than those treated without the guided approach. However, shortcomings such as limited overall responses to therapy, marginally increased survival rates and a significant number of patients who did not respond at all to precision medicine approaches, were more than evident [328–331]. Some of the most important aspects (other than political and financial) concerning precision/genomic medicine are addressed in Sects 3.2–3.4.

3.2
Integrating Genomic Data for Clinical Strategies, Diagnosis, Treatment, and Prognosis

Information integration on a community-wide level (ideally, global) as a basis for understanding information on the individual level is critical. Targeted therapeutics, in many cases, fail inadvertently (see Sect. 3.3), especially, when used as monotherapy [29, 332]. Part of the problem is an inability to appreciate the full complexity of a given disease, be it cancer or an infection caused by microorganisms. For example, as noted above, cancer is an immensely heterogeneous disease with different mutation profiles from case to case, and to some degree from cancer cell subclone population to cancer cell subclone population [93, 333–335]. Encouragingly, the tools available today may suffice to overcome many of these problems, with further refinement and increased precision. The tools currently available accomplish an impressive array of tasks, including genome/exome sequencing, cancer genome sequencing, epigenome analysis, RNA sequencing (transcriptome), proteome analysis (mass spectrometry), metabolome analysis (mass spectrometry), and integrative systems bioinformatics (see Sect. 3.4) [4].

Several solutions addressing disease heterogeneity have recorded some successes, and hold promise to change clinical strategies, diagnosis and, ultimately, treatment design. The detection of circulating nucleic acid [CNA; including cell-free DNA (cfDNA) and RNA], circulating tumor cells (CTCs), and the sequencing of single cells are beginning to emerge as clinically viable options, simplifying the multisample collection process necessary to grasp the heterogeneity of a given disease and to track treatment efficacy [336–340]. It is still being debated how representative CTCs are of a present pathology; nevertheless, the analysis of CTCs from collected blood samples is much more practical than serial tissue biopsies that are separated spatially and temporally, as is necessary to accurately track disease progression [12, 341, 342]. Today, CTCs are actively being evaluated as prognostic tools and as biomarkers for treatment response [343, 344]. Efforts are also ongoing to capture single CTCs for sequencing [345]. Single-cell genome, transcriptome, and even proteome analyses, are already at an advanced stage [338, 339, 346–348], and it has been shown that complex tumor genomics and expression patterns can be inferred from a single cell [349, 350]. The role of cfDNA in the context of precision medicine is similar to that of CTCs. cfDNA is shed by cells during apoptosis and necrosis, and its presence in the bloodstream correlates with tumor stage and can serve as a prognostic factor [337, 351]. A large amount of data can be obtained from cfDNA, including information regarding the presence of single-point mutations, amplifications, rearrangements, and aneuploidy [337, 352]. cfDNA has even been used as an aid in the diagnosis of heart transplant rejection [353]. Once adopted for clinical use, the paradigm guiding precision/genomic medicine is likely to include cfDNA, CTCs, and massive sequential data acquisition.

3.3 Multidrug Therapies

The shortcomings of targeted therapy are evident and monotherapy is seldom applicable if long-term therapeutic effects are desirable [29, 86, 332]. HIV therapy is the prototypical approach for multidrug therapy; it has rendered a disease akin to a death sentence essentially to a mere inconvenience when drugs are available (the collective moral failure of ongoing HIV-related suffering is self-evident; [90]). Based on the mathematical modeling of cancer, it has been proposed that in most cases mono-targeted therapy failing is inevitable and that simultaneous administration of, at a minimum, dual-targeted therapy is required [332]. The same model also proposes that even dual-therapy is bound to fail if a single mutation induces resistance to both agents. Thus, it is necessary to target multiple different signaling pathways simultaneously (Fig. 14), to minimize the probability of resistance; furthermore, it is necessary to track therapy response and disease progression as close to real time as possible (see Sects 3.1 and 3.2). Targeting multiple parallel pathways can be described as "horizontal" targeting; conversely, targeting members of the same pathway is "vertical" targeting (Fig. 14a,b) [354].

Targeted therapeutics are also commonly used in combination with traditional chemotherapeutics in an effort to induce a synergistic effect [355]. For example, combining the EGFR inhibitor lapatinib with the chemotherapeutic capecitabine results in significantly increased progression-free survival for patients with HER2-positive advanced breast cancer [356]. Importantly, the combination has a tolerable toxicity profile, which is one of the benefits commonly attributed to targeted therapeutics. Radiotherapy has also been proposed to

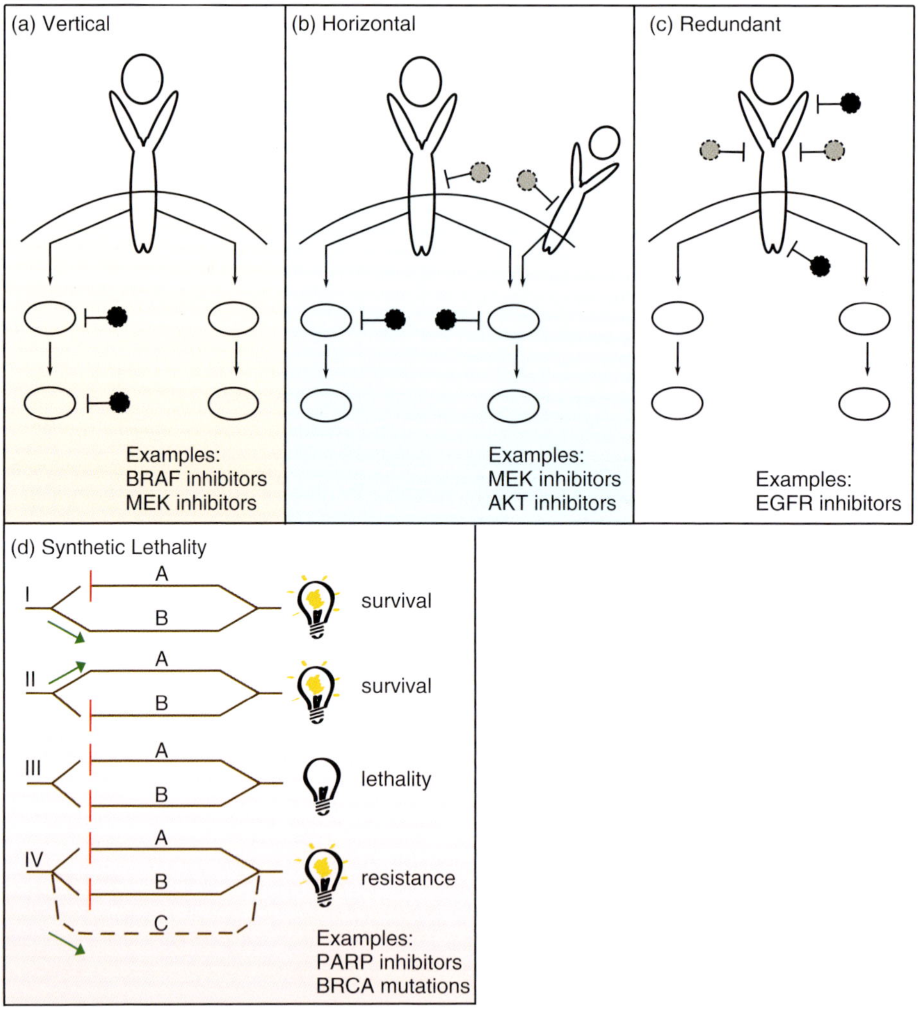

Fig. 14 Multidrug targeted therapy approaches. (a) Vertical treatment approach – inhibition of two or more proteins in the same signaling cascade; (b) Horizontal treatment approach – inhibition of two or more proteins in parallel signaling cascades; (c) Redundant treatment approach – inhibition of the same protein with two or more different therapeutic agents; (d) Synthetic lethality approach: (I and II) single gene disabled (mutation, RNAi, or inhibitor), survival possible; (III) combined inhibition of A and B, synthetic lethality; and (IV) activation of alternative pathways induces resistance (alternatively, A and B can also be reactivated).

benefit from combinatory application in conjunction with targeted therapy. In a murine model, it was suggested that radiation increases EGFR activity, given that EGFR is expressed at high levels [357].

In patients with brain metastases from non-small-cell lung cancer it was observed that, in the presence of EGFR mutations, patients robustly benefited from the combination of erlotinib (Table 1) and whole-brain radiotherapy [358]. The first steps have also been taken to explore the possibility of combining immunotherapy with radiotherapy, with initial interest sparked by observations of the abscopal effect, as first described in 1953 [359]. The abscopal effect is the phenomenon that irradiation of a primary tumor can lead to a size reduction of metastases far outside the irradiated area [360, 361]. Presumably, the immunogenic effect of radiation is the induction of damage-associated molecular patterns (DAMPs) recognizable by pattern recognition receptors expressed by cells of the immune system [362]. The clinical potential of combining radiation with immunotherapy is being explored by combining ipilimumab, a mAb that blocks CTLA-4 (an immune checkpoint molecule that is involved in the downregulation of T-cell activation [42, 43]), with radiotherapy [361].

Multidrug therapies were also anticipated to put into practice the theoretical potential of synthetic lethality. This approach, which was first described by Calvin Bridges, occurs when two genes or gene products are simultaneously perturbed to cause cellular death (Fig. 14d) [363]. The identification of synthetically lethal combinations has predominantly relied on genetic tools (e.g., siRNA, shRNA, and CRISPR [364–368]) and chemical tools [369, 370], generally in high-throughput settings. The most promising domain for synthetic lethality has been in the treatment of BRCA-mutated breast cancer. BRCA1/2 are involved in DNA repair, and cancer cells lacking this critical component of an essential function are close to three orders of magnitude more sensitive to poly (ADP-ribose) polymerase (PARP) inhibitors compared to normal cells (Fig. 14d) [371]. PARP inhibitors, in theory, deliver the second hit, presumably by also interfering with DNA damage repair, which is only lethal in the presence of BRCA1/2 loss-of-function mutations [372, 373]. Disappointingly, thus-far the successes of synthetic lethality observed in model organisms and in high-throughput screens have not been transferrable to the clinic, and even the therapeutic efficacy of PARP inhibitors has been limited [371]. An ability to perfectly align or anticipate the ripple effects of two simultaneous perturbations (see Fig. 1d) to harness the potential of synthetic lethality is likely necessary.

3.4 Integrative Bioinformatics

The apparently insurmountable volume of data produced by the explosion of high-throughput technologies requires adequate bioinformatics systems in place to analyze, normalize, quantify, and help visualize the data and to ensure adequate reproducibility [374]. A vast array of different acquisition and analytical platforms exists to organize and analyze data in different ways [178, 222, 375]. The proposed workflow for molecular profiling involves:

- Obtaining biological samples from the patient (e.g., biopsy, blood sample, CTCs, cfDNA, sputum).
- Data acquisition (proteomics, genomics, epigenomics).
- Bioinformatics (integrate patient data, cross-compare with available data, identify enrichment patterns, identify signaling nodes, and identify appropriate therapeutic approaches).
- Visualizing the data.
- A panel of experts considers the individual report, and the findings are discussed with the patient.
- A treatment regimen is then implemented.

- The whole process is then repeated if needs be.

Bioinformatics resources are plentiful, versatile, and frequently updated. Impressive and near-complete lists of resources are accessible through the Online Bioinformatics Resources Collection (*http://www.hsls.pitt.edu/obrc/*) [376] and through ExPASy (*http://expasy.org/vg/welcome*) [377].

4 Concluding Remarks

At present it is too early to say if targeted therapy will truly revolutionize medicine, or if it will become simply another treatment approach among many. The sobering fact is that the number of drugs approved per billion US dollars spent on research and development has halved approximately every nine years since 1950 [378], partially putting into question the cost–benefit ratio of developing targeted therapeutics. This is of particular concern when considering that only a minority of patients benefit from targeted therapy [29]. There is already a growing understanding that the concept of "one disease, one target, one drug" is not adequate, and the philosophy is turning to the idea of "one drug, multiple targets" – also known as polypharmacology. However, it is also undeniable that individualized treatments based on targeted therapy has already left a significant mark, and that there is no desire to turn back to the "one glove fits all" approach [379]. Genomics specifically, as well as the other "-omics," have started to elucidate the vast complexity of biological systems with remarkable resolution, potentiating the development and applicability of targeted therapy in ways previously unimaginable. The results obtained have led to the first attempts to establish protocols and procedures for true precision medicine, where physicians analyze comprehensive datasets for each patient and then utilize this information to provide personalized, targeted, and more effective care.

Acknowledgments

First and foremost, the authors would like to thank Dr. Erica Golemis for her continuous help and support, and Magda Andrews-Hoke for her assistance. The authors were supported by grant F30 CA180607 from the NIH (to T.N.B.), NIH core grant CA06927 (to the Fox Chase Cancer Center), and Kazan Federal University (to L.G.).

References

1. Saito, R., Smoot, M.E., Ono, K., Ruscheinski, J., *et al.*, (2012) A travel guide to Cytoscape plugins. *Nat. Methods*, **9**(11), 1069–1076.
2. Qin, C., Zhang, C., Zhu, F., Xu, F., *et al.*, (2014) Therapeutic target database update 2014: a resource for targeted therapeutics. *Nucleic Acids Res.*, **42** (Database issue), D1118–D1123.
3. Law, V., Knox, C., Djoumbou, Y., Jewison, T., *et al.*, (2014) DrugBank 4.0: shedding new light on drug metabolism. *Nucleic Acids Res.*, 42 (Database issue), D1091–D1097.
4. Meyer, U.A., Zanger, U.M., and Schwab, M. (2013) Omics and drug response. *Annu. Rev. Pharmacol. Toxicol.*, **53**, 475–502.
5. Berger, B., Peng, J., and Singh, M. (2013) Computational solutions for omics data. *Nat. Rev. Genet.*, **14**(5), 333–346.
6. Patti, G.J., Yanes, O., and Siuzdak, G. (2012) Innovation: metabolomics: the apogee of the omics trilogy. *Nat. Rev. Mol. Cell Biol.*, **13**(4), 263–269.

7. Koboldt, D.C., Steinberg, K.M., Larson, D.E., Wilson, R.K., *et al.*, (2013) The next-generation sequencing revolution and its impact on genomics. *Cell*, **155**(1), 27–38.
8. Schwabe, R.F. and Jobin, C. (2013) The microbiome and cancer. *Nat. Rev. Cancer*, **13**(11), 800–812.
9. Ferte, C., Trister, A.D., Huang, E., Bot, B.M., *et al.*, (2013) Impact of bioinformatic procedures in the development and translation of high-throughput molecular classifiers in oncology. *Clin. Cancer Res.*, **19**(16), 4315–4325.
10. Hu, Y., Yang, X., Qin, J., Lu, N., *et al.*, (2013) Metagenome-wide analysis of antibiotic resistance genes in a large cohort of human gut microbiota. *Nat. Commun.*, **4**, 2151.
11. Chmielecki, J. and Meyerson, M. (2013) DNA sequencing of cancer: what have we learned? *Annu. Rev. Med.*, **65**, 63–79.
12. Garraway, L.A. (2013) Genomics-driven oncology: framework for an emerging paradigm. *J. Clin. Oncol.*, **31**(15), 1806–1814.
13. Kamb, A., Harper, S., and Stefansson, K. (2013) Human genetics as a foundation for innovative drug development. *Nat. Biotechnol.*, **31**(11), 975–978.
14. Mwenifumbo, J.C. and Marra, M.A. (2013) Cancer genome-sequencing study design. *Nat. Rev. Genet.*, **14**(5), 321–332.
15. Van Allen, E.M., Wagle, N., Stojanov, P., Perrin, D.L., *et al.*, (2014) Whole-exome sequencing and clinical interpretation of formalin-fixed, paraffin-embedded tumor samples to guide precision cancer medicine. *Nat. Med.*, **20**(6), 682–688.
16. Rehm, H.L. (2013) Disease-targeted sequencing: a cornerstone in the clinic. *Nat. Rev. Genet.*, **14**(4), 295–300.
17. Wang, P., An, F., Zhuang, X., Liu, J., *et al.*, (2014) Chronopharmacology and mechanism of antitumor effect of erlotinib in Lewis tumor-bearing mice. *PLoS One*, **9**(7), e101720.
18. Reynolds, K. (2014) Sarangi S, Bardia A, Dizon DS, Precision medicine and personalized breast cancer: combination pertuzumab therapy. *Pharmgenomics Pers. Med.*, **7**, 95–105.
19. Pao, W. and Chmielecki, J. (2010) Rational, biologically based treatment of EGFR-mutant non-small-cell lung cancer. *Nat. Rev. Cancer*, **10**(11), 760–774.
20. Liu, H., Cracchiolo, J.R., Beck, T.N., Serebriiskii, I.G, *et al.*, (2014) EGFR Inhibitors as Therapeutic Agents in Head and Neck Cancer, in: Burtness, B., Golemis, E.A. (Eds), Molecular Determinants of Head and Neck Cancer, Springer, *New York, pp.* 55–90.
21. Wislez, M., Malka, D., Bennouna, J., Mortier, L., *et al.*, (2014) A new perspective in the treatment of non-small-cell lung cancer (NSCLC). *Role of afatinib: An oral and irreversible ErbB family blocker. Bull. Cancer*, **101**(6), 647–652.
22. Lauro, S., Onesti, C.E., Righini, R., Marchetti, P. (2014) The use of bevacizumab in non-small cell lung cancer: an update. *Anticancer Res.*, **34**(4), 1537–1545.
23. Kim, S. and G.K. Abou-Alfa (2014) The role of tyrosine kinase inhibitors in hepatocellular carcinoma. *Clin. Adv. Hematol. Oncol.*, **12**(1), 36–41.
24. Walter, A.O., Sjin, R.T., Haringsma, H.J., Ohashi, K., *et al.*, (2013) Discovery of a mutant-selective covalent inhibitor of EGFR that overcomes T790M-mediated resistance in NSCLC. *Cancer Discovery*, **3**(12), 1404–1415.
25. Carter, P.J. (2006) Potent antibody therapeutics by design. *Nat. Rev. Immunol.*, **6**(5), 343–357.
26. de Martino, M., Zhuang, D., Klatte, T., Rieken, M., *et al.*, (2014) Impact of ERBB2 mutations on in vitro sensitivity of bladder cancer to lapatinib. *Cancer Biol. Ther.*, **15**(9), 1239–1247.
27. Hudis, C.A. (2007) Drug therapy: trastuzumab – mechanism of action and use in clinical practice. *N. Engl. J. Med.*, **357**(1), 39–51.
28. Hudziak, R.M., Lewis, G.D., Winget, M., Fendly, B.M., *et al.*, (1989) P185her2 monoclonal antibody has antiproliferative effects in vitro and sensitizes human breast tumor cells to tumor necrosis factor. *Mol. Cell. Biol.*, **9**(3), 1165–1172.
29. Huang, M., Shen, A., Ding, J., and Geng, M. (2014) Molecularly targeted cancer therapy: some lessons from the past decade. *Trends Pharmacol. Sci.*, **35**(1), 41–50.

30. Fasolo, A., Sessa, C., Gianni, L., Broggini, M., *et al.*, (2013) Seminars in clinical pharmacology: an introduction to MET inhibitors for the medical oncologist. *Ann. Oncol.*, **24**(1), 14–20.
31. Shen, C.T., Qiu, Z.L., and Luo, Q.Y. (2014) Efficacy and safety of selumetinib compared with current therapies for advanced cancer: a meta-analysis. *Asian Pac. J. Cancer Prev.*, **15**(5), 2369–2374.
32. Kretzschmar, C., Roolf, C., Langhammer, T.S., Sekora, A., *et al.*, (2014) The novel arylindolylmaleimide PDA-66 displays pronounced antiproliferative effects in acute lymphoblastic leukemia cells. *BMC Cancer*, **14**, 71.
33. Kramer, T., Schmidt, B., and Lo, F. (2012) Monte, Small-molecule inhibitors of GSK-3: structural insights and their application to Alzheimer's disease models. Int. *J. Alzheimers Dis.*, **2012**, 381029.
34. Wu, B., Wu, H., Liu, X., Li, J. *et al.* 2014) Ranibizumab versus bevacizumab for ophthalmic diseases related to neovascularisation: a meta-analysis of randomised controlled trials. *PLoS One*, **9**(7), e101253.
35. Jaitak, V. (2014) Drug target strategies in breast cancer treatment: recent developments. *Anticancer Agents Med. Chem.*, **14**(10), 1414–1427.
36. van Halteren, H., Mulder, D., and Ruijter, E. (2013) Oestrogen receptor-beta as a potential target for treatment in advanced colorectal cancer: a pilot study. *Histopathology*, **64**, 787–790.
37. Chames, P. and Baty, D. (2009) Bispecific antibodies for cancer therapy: the light at the end of the tunnel? *MAbs*, **1**(6), 539–547.
38. Sinnott, J.T., Cullison, J.P., Sweeney, M.S., Weinstein, S.S. (1988) Infections in patients receiving OKT3 monoclonal antibody for cardiac rejection: results of a small clinical trial. *Texas Heart Inst. J.*, **15**(2), 102–106.
39. Palanca-Wessels, M.C. and Press, O.W. (2014) Advances in the treatment of hematologic malignancies using immunoconjugates. *Blood*, **123**(15), 2293–2301.
40. Vonderheide, R.H., Bajor, D.L., Winograd, R., Evans, R.A., *et al.*, (2013) CD40 immunotherapy for pancreatic cancer. *Cancer Immunol. Immunother.*, **62**(5), 949–954.
41. Hartmann, F., Renner, C., Jung, W., da Costa, L., *et al.*, (2001) Anti-CD16/CD30 bispecific antibody treatment for Hodgkin's disease: role of infusion schedule and costimulation with cytokines. *Clin. Cancer Res.*, 7(7), 1873–1881.
42. Hodi, F.S., O'Day, S.J., McDermott, D.F., Weber, R.W., *et al.*, (2010) Improved survival with ipilimumab in patients with metastatic melanoma. *N. Engl. J. Med.*, **363**(8), 711–723.
43. Weber, J. (2007) Review: Anti-CTLA-4 antibody ipilimumab: case studies of clinical response and immune-related adverse events. *Oncologist*, **12**(7), 864–872.
44. Druker, B.J., Sawyers, C.L., Kantarjian, H., Resta, D.J., *et al.*, (2001) Activity of a specific inhibitor of the BCR-ABL tyrosine kinase in the blast crisis of chronic myeloid leukemia and acute lymphoblastic leukemia with the Philadelphia chromosome. *N. Engl. J. Med.*, **344**(14), 1038–1042.
45. Hsyu, P.H., Mould, D.R., Abbas, R., Amantea, M. (2014) Population pharmacokinetic and pharmacodynamic analysis of bosutinib. *Drug Metab. Pharmacokinet.*, advance publication.
46. Talpaz, M., Shah, N.P., Kantarjian, H., Donato, N., *et al.*, (2006) Dasatinib in imatinib-resistant Philadelphia chromosome-positive leukemias. *N. Engl. J. Med.*, **354**(24), 2531–2541.
47. Lombardo, L.J., Lee, F.Y., Chen, P., Norris, D., *et al.*, (2004) Discovery of *N*-(2-chloro-6-methyl-phenyl)-2-(6-(4-(2-hydroxyethyl)-piperazin-1-yl)-2-methylpyrimidin-4-ylamino)thiazole-5-carboxamide (BMS-354825), a dual Src/Abl kinase inhibitor with potent antitumor activity in preclinical assays. *J. Med. Chem.*, **47**(27), 6658–6661.
48. Kantarjian, H., Giles, F., Wunderle, L., Bhalla, K., *et al.*, (2006) Nilotinib in imatinib-resistant CML and Philadelphia chromosome-positive ALL. *N. Engl. J. Med.*, **354**(24), 2542–2551.
49. Weisberg, E., Manley, P.W., Breitenstein, W., Brüggen, J., *et al.*, (2005) Characterization of AMN107, a selective inhibitor of native and mutant Bcr-Abl. *Cancer Cell*, 7(2), 129–141.
50. Shaw, A.T., Kim, D.W., Nakagawa, K., Seto, T., *et al.*, (2013) Crizotinib versus chemotherapy in advanced ALK-positive

lung cancer. *N. Engl. J. Med.*, **368**(25), 2385–2394.

51. Shaw, A.T., Kim, D.W., Mehra, R., Tan, D.S., *et al.*, (2014) Ceritinib in ALK-rearranged non-small-cell lung cancer. *N. Engl. J. Med.*, **370**(13), 1189–1197.
52. Friboulet, L., Li, N., Katayama, R., Lee, C.C., *et al.*, (2014) The ALK inhibitor ceritinib overcomes crizotinib resistance in non-small cell lung cancer. *Cancer Discovery*, **4**(6), 662–673.
53. Ramalingam, S.S. and Khuri, F.R. (2014) Second-generation ALK inhibitors: filling the Non "MET" Gap. *Cancer Discovery*, **4**(6), 634–636.
54. Orchel, A., Kulczycka, A., Chodurek, E., Bebenek, E., *et al.*, (2014) Influence of betulin and 28-*O*-propynoylbetulin on proliferation and apoptosis of human melanoma cells (G-361). *Postepy Hig. Med. Dosw. (Online)*, **68**, 191–197.
55. Li, X., Chen, Y.T., Hu, P., and Huang, W.C. (2014) Fatostatin displays high antitumor activity in prostate cancer by blocking SREBP-regulated metabolic pathways and androgen receptor signaling. *Mol. Cancer Ther.*, **13**(4), 855–866.
56. Weathington, N.M. and Mallampalli, R.K. (2014) Emerging therapies targeting the ubiquitin proteasome system in cancer. *J. Clin. Invest.*, **124**(1), 6–12.
57. Jakubowiak, A.J., Dytfeld, D., Griffith, K.A., Lebovic, D., *et al.*, (2012) A phase 1/2 study of carfilzomib in combination with lenalidomide and low-dose dexamethasone as a frontline treatment for multiple myeloma. *Blood*, **120**(9), 1801–1809.
58. Cicero, A.F., Tartagni, E., and Ertek, S. (2014) Safety and tolerability of injectable lipid-lowering drugs: a review of available clinical data. *Expert Opin. Drug Saf.*, **13** (8), 1023–1030.
59. Lindholm, M.W., Elmén, J., Fisker, N., Hansen, H.F., *et al.*, (2012) PCSK9 LNA antisense oligonucleotides induce sustained reduction of LDL cholesterol in nonhuman primates. *Mol. Ther.*, **20**(2), 376–381.
60. Krieg, A.M. (2012) Targeting LDL cholesterol with LNA. *Mol. Ther. Nucleic Acids*, **1**, e6.
61. Burchett, K.M., Yan, Y., and Ouellette, M.M. (2014) Telomerase inhibitor Imetelstat (GRN163L) limits the lifespan of human pancreatic cancer cells. *PLoS One*, **9**(1), e85155.
62. Li, C., Qi, L., Bellail, A.C., Hao, C., *et al.*, (2014) PD-0332991 induces G1 arrest of colorectal carcinoma cells through inhibition of the cyclin-dependent kinase-6 and retinoblastoma protein axis. *Oncol. Lett.*, **7**(5), 1673–1678.
63. Nagaraju, G.P., Long, T.E., Park, W., Landry, J.C., *et al.*, (2014) *Heat shock protein 90 promotes epithelial to mesenchymal transition, invasion, and migration in colorectal cancer*. Mol. Carcinog., (in press).
64. Tukaj, S., Grüner, D., Zillikens, D., and Kasperkiewicz, M. (2014) Hsp90 blockade modulates bullous pemphigoid IgG-induced IL-8 production by keratinocytes. *Cell Stress Chaperones*, **19**(6), 887–894.
65. Appels, N.M., Beijnen, J.H., and Schellens, J.H. (2005) Development of farnesyl transferase inhibitors: a review. *Oncologist*, **10**(8), 565–578.
66. Issa, J.P.J., Kantarjian, H.M., and Kirkpatrick, P. (2005) Azacitidine. *Nat. Rev. Drug Discovery*, **4**(4), 275–276.
67. Campbell, R.M. and Tummino, P.J. (2014) Cancer epigenetics drug discovery and development: the challenge of hitting the mark. *J. Clin. Invest.*, **124**(3), 1419.
68. West, A.C. and Johnstone, R.W. (2014) New and emerging HDAC inhibitors for cancer treatment. *J. Clin. Invest.*, **124**(1), 30–39.
69. Delmore, J.E., Issa, G.C., Lemieux, M.E., Rahl, P.B., *et al.*, (2011) BET bromodomain inhibition as a therapeutic strategy to target c-Myc. *Cell*, **146**(6), 904–917.
70. Filippakopoulos, P., Qi, J., Picaud, S., Shen, Y., *et al.*, (2010) Selective inhibition of BET bromodomains. *Nature*, **468**(7327), 1067–1073.
71. Parikh, S.A., French, C.A., Costello, B.A., Marks, R.S., *et al.*, (2013) NUT midline carcinoma: an aggressive intrathoracic neoplasm. *J. Thorac. Oncol.*, **8**(10), 1335–1338.
72. Fowler, T., Ghatak, P., Price, D.H., Conaway, R. *et al.*, (2014) Regulation of MYC expression and differential JQ1 sensitivity in cancer cells. *PLoS One*, **9**(1), e87003.
73. Feng, Q., Wang, H., Ng, H.H., Erdjument-Bromage, H., *et al.*, (2002) Methylation of H3-lysine 79 is mediated by a new family of HMTases without a SET domain. *Curr. Biol.*, **12**(12), 1052–1058.

74. Kuzmichev, A., Nishioka, K., Erdjument-Bromage, H., Tempst, P., *et al.*, (2002) Histone methyltransferase activity associated with a human multiprotein complex containing the Enhancer of Zeste protein. *Genes Dev.*, **16**(22), 2893–2905.
75. Du, M., Liu, X., Welch, E.M., Hirawat, S., *et al.*, (2008) PTC124 is an orally bioavailable compound that promotes suppression of the human CFTR-G542X nonsense allele in a CF mouse model. *Proc. Natl Acad. Sci. USA*, **105**(6), 2064–2069.
76. Sermet-Gaudelus, I., Boeck, K.D., Casimir, G.J., Vermeulen, F., *et al.*, (2010) Ataluren (PTC124) induces cystic fibrosis transmembrane conductance regulator protein expression and activity in children with nonsense mutation cystic fibrosis. *Am. J. Respir. Crit. Care Med.*, **182**(10), 1262–1272.
77. Loong, H.H. and Yeo, W. (2014) Microtubule-targeting agents in oncology and therapeutic potential in hepatocellular carcinoma. *Oncol. Targets Ther.*, **7**, 575–585.
78. Reichert, J.M. (2013) Which are the antibodies to watch in 2013? *MAbs*, **5**(1), 1–4.
79. Orth, P., Xiao, L., Hernandez, L.D., Reichert, P., *et al.*, (2014) Mechanism of action and epitopes of *Clostridium difficile* toxin B-neutralizing antibody bezlotoxumab revealed by X-ray crystallography. *J. Biol. Chem.*, **289**(26), 18008–18021.
80. Fox, J.L. (2013) Anthrax drug first antibacterial mAb to win approval. *Nat. Biotechnol.*, **31**(1), 8.
81. Anastassiadis, T., Deacon, S.W., Devarajan, K., Ma, H., *et al.*, (2011) Comprehensive assay of kinase catalytic activity reveals features of kinase inhibitor selectivity. *Nat. Biotechnol.*, **29**(11), 1039–1045.
82. DeVita, V.T., Jr, and Chu, E. (2008) *A history of cancer chemotherapy.* Cancer Res., **68**(21), 8643–8653.
83. Heidelberger, C., Chaudhuri, N.K., Danneberg, P., Mooren, D., *et al.*, (1957) Fluorinated pyrimidines, a new class of tumour-inhibitory compounds. *Nature*, **179**(4561), 663–666.
84. Wang, L. and Wheeler, D.A. (2014) Genomic sequencing for cancer diagnosis and therapy. *Annu. Rev. Med.*, **65**, 33–48.
85. Shrager, J. and Tenenbaum, J.M. (2014) Rapid learning for precision oncology. *Nat. Rev. Clin. Oncol.*, **11**(2), 109–118.
86. Vogelstein, B., Papadopoulos, N., Velculescu, V.E., Zhou, S., *et al.*, (2013) Cancer genome landscapes. *Science*, **339**(6127), 1546–1558.
87. da Cunha Santos, G. (2011) F.A. Shepherd, and M.S. Tsao, EGFR mutations and lung cancer. *Annu. Rev. Pathol.*, **6**, 49–69.
88. Sharma, S.V., Bell, D.W., Settleman, J., and Haber, D.A. (2007) Epidermal growth factor receptor mutations in lung cancer. *Nat. Rev. Cancer*, 7(3), 169–181.
89. Gazdar, A.F. (2009) Activating and resistance mutations of EGFR in non-small-cell lung cancer: role in clinical response to EGFR tyrosine kinase inhibitors. *Oncogene*, **28** (Suppl. 1), S24–S31.
90. Bock, C. and Lengauer, T. (2012) Managing drug resistance in cancer: lessons from HIV therapy. *Nat. Rev. Cancer*, **12**(7), 494–501.
91. Nelson, E.J., Harris, J.B., Morris, J.G., Jr, Calderwood, S.B., *et al.*, (2009) Cholera transmission: the host, pathogen and bacteriophage dynamic. *Nat. Rev. Microbiol.*, 7(10), 693–702.
92. Harris, J.B., LaRocque, R.C., Qadri, F., Ryan, E.T., and Calderwood, S.B. (2012) *Cholera. Lancet*, **379**(9835), 2466–2476.
93. Lawrence, M.S., Stojanov, P., Mermel, C.H., Robinson, J.T., *et al.*, (2014) Discovery and saturation analysis of cancer genes across 21 tumour types. *Nature*, **505**(7484), 495–501
94. Behar, M., Barken, D., Werner, S.L., and Hoffmann, A. (2013) The dynamics of signaling as a pharmacological target. *Cell*, **155**(2), 448–461.
95. Trepel, J., Mollapour, M., Giaccone, G., Neckers, L., *et al.*, (2010) Targeting the dynamic HSP90 complex in cancer. *Nat. Rev. Cancer*, **10**(8), 537–549.
96. Bouchie, A., Allison, M., Webb, S., and DeFrancesco, L. (2014) *Nature Biotechnology*'s academic spinouts of 2013. *Nat. Biotechnol.*, **32**(3), 229–238.
97. Lander, E.S., Linton, L.M., Birren, B., Nusbaum, C., *et al.*, (2001) Initial sequencing and analysis of the human genome. *Nature*, **409**(6822), 860–921.
98. Venter, J.C., Adams, M.D., Myers, E.W., Li, P.W. *et al.*, (2001) The sequence of the human genome. *Science*, **291**(5507), 1304–1351.

99. Hayden, E.C. (2014) Technology: the $1,000 genome. *Nature*, **507**(7492), 294–295.
100. Mardis, E.R. (2013) Next-generation sequencing platforms. *Annu. Rev. Anal. Chem. (Palo Alto)*, **6**, 287–303.
101. Shapiro, E., Biezuner, T., and Linnarsson, S. (2013) Single-cell sequencing-based technologies will revolutionize whole-organism science. *Nat. Rev. Genet.*, **14**(9), 618–630.
102. Russell, S.J., Peng, K.W., and Bell, J.C. (2012) Oncolytic virotherapy. *Nat. Biotechnol.*, **30**(7), 658–670.
103. Bell, J. and McFadden, G. (2014) Viruses for tumor therapy. *Cell Host Microbe*, **15**(3), 260–265.
104. Almendro, V., Marusyk, A., and Polyak, K. (2013) Cellular heterogeneity and molecular evolution in cancer. *Annu. Rev. Pathol.*, **8**, 277–302.
105. Gerlinger, M., Rowan, A.J., Horswell, S., Larkin, J., *et al.*, (2012) Intratumor heterogeneity and branched evolution revealed by multiregion sequencing. *N. Engl. J. Med.*, **366**(10), 883–892.
106. Meacham, C.E. and Morrison, S.J. (2013) Tumour heterogeneity and cancer cell plasticity. *Nature*, **501**(7467), 328–337.
107. Burrell, R.A., McGranahan, N., Bartek, J., Swanton, C. (2013) The causes and consequences of genetic heterogeneity in cancer evolution. *Nature*, **501**(7467), 338–345.
108. Zack, T.I., Schumacher, S.E., Carter, S.L., Cherniack, A.D., *et al.*, (2013) Pan-cancer patterns of somatic copy number alteration. *Nat. Genet.*, **45**(10), 1134–U257.
109. Hartl, D. and Ruvolo, M. (2011) *Genetics: Analysis of Genes and Genomes*. 8th edn , Burlington, MA, Jones & Barlett Learning.
110. Supek, F., Miñana, B., Valcárcel, J., Gabaldón, T., *et al.*, (2014) Synonymous mutations frequently act as driver mutations in human cancers. *Cell*, **156**(6), 1324–1335.
111. Weinstein, J.N., Collisson, E.A., Mills, G.B., Shaw, K.R., *et al.*, (2013) The cancer genome atlas pan-cancer analysis project. *Nat. Genet.*, **45**(10), 1113–1120.
112. Watson, I.R., Takahashi, K., Futreal, P.A., Chin, L. (2013) Emerging patterns of somatic mutations in cancer. *Nat. Rev. Genet.*, **14**(10), 703–718.
113. Garraway, L.A. and Lander, E.S. (2013) Lessons from the cancer genome. *Cell*, **153**(1), 17–37.
114. Luo, J., Solimini, N.L., and Elledge, S.J. (2009) Principles of cancer therapy: oncogene and non-oncogene addiction. *Cell*, **136**(5), 823–837.
115. Forbes, S.A., Bindal, N., Bamford, S., Cole, C., *et al.*, (2011) COSMIC: mining complete cancer genomes in the Catalogue of Somatic Mutations in Cancer. *Nucleic Acids Res.*, **39** (Database issue), D945–D950.
116. Bollag, G., Tsai, J., Zhang, J., Zhang, C., *et al.*, (2012) Vemurafenib: the first drug approved for BRAF-mutant cancer. *Nat. Rev. Drug Discov.*, **11**(11), 873–886.
117. Berger, M.F., Hodis, E., Heffernan, T.P., Deribe, Y.L., *et al.*, (2012) Melanoma genome sequencing reveals frequent PREX2 mutations. *Nature*, **485**(7399), 502–506.
118. Sosman, J.A., Kim, K.B., Schuchter, L., Gonzalez, R., *et al.*, (2012) Survival in BRAF V600-mutant advanced melanoma treated with vemurafenib. *N. Engl. J. Med.*, **366**(8), 707–714.
119. Hauschild, A., Grob, J.J., Demidov, L.V., Jouary, T., *et al.*, (2012) Dabrafenib in BRAF-mutated metastatic melanoma: a multicentre, open-label, phase 3 randomised controlled trial. *Lancet*, **380**(9839), 358–365.
120. Prahallad, A., Sun, C., Huang, S., Di Nicolantonio, F., *et al.*, (2012) Unresponsiveness of colon cancer to BRAF(V600E) inhibition through feedback activation of EGFR. *Nature*, **483**(7387), 100–103.
121. Konieczkowski, D.J., Johannessen, C.M., Abudayyeh, O., Kim, J.W., *et al.*, (2014) A melanoma cell state distinction influences sensitivity to MAPK pathway inhibitors. *Cancer Discovery*, **4**(7), 816–827.
122. Wagle, N., Van Allen, E.M., Treacy, D.J., Frederick, D.T., *et al.*, (2014) MAP kinase pathway alterations in BRAF-mutant melanoma patients with acquired resistance to combined RAF/MEK inhibition. *Cancer Discovery*, **4**(1), 61–68.
123. Lemmon, M.A. and Schlessinger, J. (2010) Cell signaling by receptor tyrosine kinases. *Cell*, **141**(7), 1117–1134.
124. Yun, C.H., Mengwasser, K.E., Toms, A.V., Woo, M.S., *et al.*, (2008) The T790M mutation in EGFR kinase causes drug resistance by increasing the affinity for ATP. *Proc. Natl Acad. Sci. USA*, **105**(6), 2070–2075.
125. Arteaga, C.L. and Engelman, J.A. (2014) ERBB receptors: from oncogene discovery

to basic science to mechanism-based cancer therapeutics. *Cancer Cell*, **25**(3), 282–303.

126. Nowell, P.C. (1976) The clonal evolution of tumor cell populations. *Science*, **194**(4260), 23–28.

127. Greaves, M. and Maley, C.C. (2012) Clonal evolution in cancer. *Nature*, **481**(7381), 306–313.

128. Rowe, S.M., Varga, K., Rab, A., Bebok, Z., *et al.*, (2007) Restoration of W1282X CFTR activity by enhanced expression. *Am. J. Respir. Cell Mol. Biol.*, **37**(3), 347–356.

129. Roberts, R.G. (2013) A read-through drug put through its paces. *PLoS Biol.*, **11**(6), e1001458.

130. Kellermayer, R. (2006) Translational readthrough induction of pathogenic nonsense mutations. *Eur. J. Med. Genet.*, **49**(6), 445–450.

131. Sheridan, C. (2013) Doubts raised over 'read-through' Duchenne drug mechanism. *Nat. Biotechnol.*, **31**(9), 771–773.

132. Peltz, S.W., Morsy, M., Welch, E.M., Jacobson, A. (2013) Ataluren as an agent for therapeutic nonsense suppression. *Annu. Rev. Med.*, **64**, 407–425.

133. Kerem, E., Konstan, M.W., De Boeck, K., Accurso, F.J., *et al.*, (2014) Ataluren for the treatment of nonsense-mutation cystic fibrosis: a randomised, double-blind, placebo-controlled phase 3 trial. *Lancet Respir. Med.*, **2**(7), 539–547.

134. Derichs, N. (2013) Targeting a genetic defect: cystic fibrosis transmembrane conductance regulator modulators in cystic fibrosis. *Eur. Respir. Rev.*, **22**(127), 58–65.

135. Schwank, G., Koo, B.K., Sasselli, V., Dekkers, J.F., *et al.*, (2013) Functional repair of CFTR by CRISPR/Cas9 in intestinal stem cell organoids of cystic fibrosis patients. *Cell Stem Cell*, **13**(6), 653–658.

136. Alton, E.W., Boyd, A.C., Cheng, S.H., Cunningham, S., *et al.*, (2013) A randomised, double-blind, placebo-controlled phase IIB clinical trial of repeated application of gene therapy in patients with cystic fibrosis. *Thorax*, **68**(11), 1075–1077.

137. Stephens, P.J., Greenman, C.D., Fu, B., Yang, F., *et al.*, (2011) Massive genomic rearrangement acquired in a single catastrophic event during cancer development. *Cell*, **144**(1), 27–40.

138. Kloosterman, W.P., Tavakoli-Yaraki, M., van Roosmalen, M.J., van Binsbergen, E., *et al.*, (2012) Constitutional chromothripsis rearrangements involve clustered double-stranded DNA breaks and nonhomologous repair mechanisms. *Cell Rep.*, **1**(6), 648–655.

139. Davoli, T., Xu, A.W., Mengwasser, K.E., Sack, L.M., *et al.*, (2013) Cumulative haploinsufficiency and triplosensitivity drive aneuploidy patterns and shape the cancer genome. *Cell*, **155**(4), 948–962.

140. Molyneux, E.M., Rochford, R., Griffin, B., Newton, R., *et al.*, (2012) Burkitt's lymphoma. *Lancet*, **379**(9822), 1234–1244.

141. Krzywinski, M., Schein, J., Birol, I., Connors, J., *et al.*, (2009) Circos: an information aesthetic for comparative genomics. *Genome Res.*, **19**(9), 1639–1645.

142. Thompson, S.L. and Compton, D.A. (2011) Chromosomes and cancer cells. *Chromosome Res.*, **19**(3), 433–444.

143. Kirkpatrick, M. (2010) How and why chromosome inversions evolve. *PLoS Biol.*, **8**(9), e1000501.

144. Nambiar, M., Kari, V., and Raghavan, S.C. (2008) Chromosomal translocations in cancer. *Biochim. Biophys. Acta*, **1786**(2), 139–152.

145. Nowell, P.C. and Hungerford, D.A. (1960) Chromosome studies on normal and leukemic human leukocytes. *J. Natl Cancer Inst.*, **25**, 85–109.

146. Rowley, J.D. (1973) Letter: A new consistent chromosomal abnormality in chronic myelogenous leukaemia identified by quinacrine fluorescence and Giemsa staining. *Nature*, **243**(5405), 290–293.

147. Nowell, P. and Hungerford, D. (1960) A minute chromosome in human chronic granulocytic leukemia. *Science*, **132**, 1497.

148. Grebien, F., Hantschel, O., Wojcik, J., Kaupe, I., *et al.*, (2011) Targeting the SH2-kinase interface in Bcr-Abl inhibits leukemogenesis. *Cell*, **147**(2), 306–319.

149. Soda, M., Choi, Y.L., Enomoto, M., Takada, S., *et al.*, (2007) Identification of the transforming EML4-ALK fusion gene in non-small-cell lung cancer. *Nature*, **448**(7153), 561–566.

150. Choi, Y.L., Soda, M., Yamashita, Y., Ueno, T., *et al.*, (2010) EML4-ALK mutations in lung cancer that confer resistance to ALK inhibitors. *N. Engl. J. Med.*, **363**(18), 1734–1739.

151. Katayama, R., Shaw, A.T., Khan, T.M., Mino-Kenudson, M., *et al.*, (2012) Mechanisms of acquired crizotinib resistance in ALK-rearranged lung cancers. *Sci. Transl. Med.*, **4**(120), 120ra17.

152. Sang, J., Acquaviva, J., Friedland, J.C., Smith, D.L., *et al.*, (2013) Targeted inhibition of the molecular chaperone Hsp90 overcomes ALK inhibitor resistance in non-small cell lung cancer. *Cancer Discovery*, **3**(4), 430–443.

153. Burkitt, D. (1958) A sarcoma involving the jaws in African children. *Br. J. Surg.*, **46**(197), 218–223.

154. Jaffe, E.S. and Pittaluga, S. (2011) Aggressive B-cell lymphomas: a review of new and old entities in the WHO classification. *Hematology Am. Soc. Hematol. Educ. Program*, **2011**, 506–514.

155. Schmitz, R., Ceribelli, M., Pittaluga, S., Wright, G., *et al.*, (2014) Oncogenic mechanisms in Burkitt lymphoma. *Cold Spring Harbor Perspect. Med.*, **4**(2), a014282.

156. Dalla-Favera, R., Bregni, M., Erikson, J., Patterson, D., *et al.*, (1982) Human c-myc oncogene is located on the region of chromosome 8 that is translocated in Burkitt lymphoma cells. *Proc. Natl Acad. Sci. USA*, **79**(24), 7824–7827.

157. Taub, R., Kirsch, I., Morton, C., Lenoir, G., *et al.*, (1982) Translocation of the C-myc gene into the immunoglobulin heavy-chain locus in human Burkitt lymphoma and murine plasmacytoma cells. *Proc. Natl Acad. Sci. USA*, **79**(24), 7837–7841.

158. Patte, C., Auperin, A., Michon, J., Behrendt, H., Leverger, G., *et al.*, (2001) The Societé Francaise d'Oncologie Pediatrique LMB89 protocol: highly effective multiagent chemotherapy tailored to the tumor burden and initial response in 561 unselected children with B-cell lymphomas and L3 leukemia. *Blood*, **97**(11), 3370–3379.

159. Beroukhim, R., Mermel, C.H., Porter, D., Wei, G., *et al.*, (2010) The landscape of somatic copy-number alteration across human cancers. *Nature*, **463**(7283), 899–905.

160. Dang, C.V. (2012) MYC on the path to cancer. *Cell*, **149**(1), 22–35.

161. Santarius, T., Shipley, J., Brewer, D., Stratton, M.R., *et al.*, (2010) Epigenetics and genetics: a census of amplified and overexpressed human cancer genes. *Nat. Rev. Cancer*, **10**(1), 59–64.

162. Cerami, E., Gao, J., Dogrusoz, U., Gross, B.E., *et al.*, (2012) The cBio cancer genomics portal: an open platform for exploring multidimensional cancer genomics data. *Cancer Discovery*, **2**(5), 401–404.

163. Gao, J.J., Aksoy, B.A., Dogrusoz, U., Dresdner, G., *et al.*, (2013) Integrative analysis of complex cancer genomics and clinical profiles using the cBioPortal. *Sci. Signal.*, **6**(269), pl1.

164. Rhodes, D.R., Kalyana-Sundaram, S., Mahavisno, V., Varambally, R., *et al.*, (2007) Oncomine 3.0: genes, pathways, and networks in a collection of 18,000 cancer gene expression profiles. *Neoplasia*, **9**(2), 166–180.

165. King, C.R., Kraus, M.H., and Aaronson, S.A. (1985) Amplification of a novel V-erbb-related gene in a human mammary carcinoma. *Science*, **229**(4717), 974–976.

166. Dziadziuszko, R. and Jassem, J. (2012) Epidermal growth factor receptor (EGFR) inhibitors and derived treatments. *Ann. Oncol.*, **23** (Suppl. 10), x193–x196.

167. Nielsen, D.L., Kümler, I., Palshof, J.A., Andersson, M. (2013) Efficacy of HER2-targeted therapy in metastatic breast cancer. *Monoclonal antibodies and tyrosine kinase inhibitors. Breast*, **22**(1), 1–12.

168. Gajria, D. and Chandarlapaty, S. (2011) HER2-amplified breast cancer: mechanisms of trastuzumab resistance and novel targeted therapies. *Expert Rev. Anticancer Ther.*, **11**(2), 263–275.

169. Owens, M.A., Horten, B.C., and Da Silva, M.M. (2004) HER2 amplification ratios by fluorescence in situ hybridization and correlation with immunohistochemistry in a cohort of 6556 breast cancer tissues. *Clin. Breast Cancer*, **5**(1), 63–69.

170. Shiu, K.K., Wetterskog, D., Mackay, A., Natrajan, R., *et al.*, (2014) Integrative molecular and functional profiling of ERBB2-amplified breast cancers identifies new genetic dependencies. *Oncogene*, **33**(5), 619–631.

171. Engelman, J.A., Zejnullahu, K., Mitsudomi, T., Song, Y., *et al.*, (2007) MET amplification leads to gefitinib resistance in lung cancer by activating ERBB3 signaling. *Science*, **316**(5827), 1039–1043.

172. Seiwert, T.Y., Jagadeeswaran, R., Faoro, L., Janamanchi, V., *et al.*, (2009) The MET receptor tyrosine kinase is a potential novel therapeutic target for head and neck squamous cell carcinoma. *Cancer Res.*, **69**(7), 3021–3031.
173. Seiwert, T.Y., Beck, T.N., and Salgia, R. (2014) The Role of HGF/c-MET in Head and Neck Squamous Cell Carcinoma, in: Burtness, B., Golemis, E.A. (Eds), *Molecular Determinants of Head and Neck Cancer*, Springer, *New York, pp.* **91–111**.
174. O'Neill, F., Madden, S.F., Clynes, M., Crown, J., *et al.*, (2013) A gene expression profile indicative of early stage HER2 targeted therapy response. *Mol. Cancer*, **12**, 69.
175. Yang, W., Soares, J., Greninger, P., Edelman, E.J., *et al.*, (2013) Genomics of Drug Sensitivity in Cancer (GDSC): a resource for therapeutic biomarker discovery in cancer cells. *Nucleic Acids Res.*, **41** (Database issue), D955–D961.
176. Bateman, A.R., El-Hachem, N., Beck, A.H., Aerts, H.J., *et al.*, (2014) Importance of collection in gene set enrichment analysis of drug response in cancer cell lines. *Sci. Rep.*, **4**, 4092.
177. Tam, W.L. and Weinberg, R.A. (2013) The epigenetics of epithelial-mesenchymal plasticity in cancer. *Nat. Med.*, **19**(11), 1438–1449.
178. Beck, T.N., Chikwem, A.J., Solanki, N.R., Golemis, E.A. (2014) Bioinformatic approaches to augment study of Epithelial-to-Mesenchymal Transition (EMT) in lung cancer. *Physiol. Genomics*, **46**(19), 699–724.
179. Creighton, C.J., Gibbons, D.L., and Kurie, J.M. (2013) The role of epithelial-mesenchymal transition programming in invasion and metastasis: a clinical perspective. *Cancer Manag. Res.*, **5**, 187–195.
180. Manning, G., Whyte, D.B., Martinez, R., Hunter, T., *et al.*, (2002) The protein kinase complement of the human genome. *Science*, **298**(5600), 1912–1934.
181. Graves, L.M., Duncan, J.S., Whittle, M.C., Johnson, G.L. (2013) The dynamic nature of the kinome. *Biochem. J.*, **450**(1), 1–8.
182. Duncan, J.S., Whittle, M.C., Nakamura, K., Abell, A.N., *et al.*, (2012) Dynamic reprogramming of the kinome in response to targeted MEK inhibition in triple-negative breast cancer. *Cell*, **149**(2), 307–321.
183. Dawson, M.A. and Kouzarides, T. (2012) Cancer epigenetics: from mechanism to therapy. *Cell*, **150**(1), 12–27.
184. Jones, P.A. (2014) At the tipping point for epigenetic therapies in cancer. *J. Clin. Invest.*, **124**(1), 14–16.
185. Bernstein, B.E., Birney, E., Dunham, I., Green, E.D., *et al.*, (2012) An integrated encyclopedia of DNA elements in the human genome. *Nature*, **489**(7414), 57–74.
186. Goldberg, A.D., Allis, C.D., and Bernstein, E. (2007) Epigenetics: a landscape takes shape. *Cell*, **128**(4), 635–638.
187. Lister, R., Pelizzola, M., Dowen, R.H., Hawkins, R.D., *et al.*, (2009) Human DNA methylomes at base resolution show widespread epigenomic differences. *Nature*, **462**(7271), 315–322.
188. Ahuja, N., Easwaran, H., and Baylin, S.B. (2014) Harnessing the potential of epigenetic therapy to target solid tumors. *J. Clin. Invest.*, **124**(1), 56–63.
189. Kelly, T.K., Miranda, T.B., Liang, G., Berman, B.P., *et al.*, (2010) H2A.Z maintenance during mitosis reveals nucleosome shifting on mitotically silenced genes. Mol. Cell, **39**(6), 901–911.
190. Maxmen, A. (2012) Cancer research: open ambition. *Nature*, **488**(7410), 148–150.
191. Jones, P.A. (2012) Functions of DNA methylation: islands, start sites, gene bodies and beyond. *Nat. Rev. Genet.*, **13**(7), 484–492.
192. Baylin, S.B. and Jones, P.A. (2011) A decade of exploring the cancer epigenome – biological and translational implications. *Nat. Rev. Cancer*, **11**(10), 726–734.
193. Fleming, A. (1929) On the antibacterial action of cultures of a *Penicillium*, with special reference to their use in the isolation of *B. influenzæ*. *Br. J. Exp. Pathol.*, **10**(3), 226–236.
194. Yocum, R.R., Rasmussen, J.R., and Strominger, J.L. (1980) The mechanism of action of penicillin. Penicillin acylates the active site of *Bacillus stearothermophilus* D-alanine carboxypeptidase. *J. Biol. Chem.*, **255**(9), 3977–3986.
195. Kostova, M.B., Myers, C.J., Beck, T.N., Plotkin, B.J., *et al.*, (2011) C4-Alkylthiols with activity against *Moraxella catarrhalis* and *Mycobacterium tuberculosis*. *Bioorg. Med. Chem.*, **19**(22), 6842–6852.

196. Davies, J. and Davies, D. (2010) Origins and evolution of antibiotic resistance. *Microbiol. Mol. Biol. Rev.*, **74**(3), 417–4133.
197. Panjkovich, A., Gibert, I., and Daura, X. (2014) antibacTR: dynamic antibacterial-drug-target ranking integrating comparative genomics, structural analysis and experimental annotation. *BMC Genomics*, **15**, 36.
198. Crompton, P.D., Moebius, J., Portugal, S., Waisberg, M., *et al.*, (2014) Malaria immunity in man and mosquito: insights into unsolved mysteries of a deadly infectious disease. *Annu. Rev. Immunol.*, **32**, 157–187.
199. Spillman, N.J., Allen, R.J., McNamara, C.W., Yeung, B.K., *et al.*, (2013) Na^+ regulation in the malaria parasite *Plasmodium falciparum* involves the cation ATPase PfATP4 and is a target of the spiroindolone antimalarials. *Cell Host Microbe*, **13**(2), 227–237.
200. Horn, D. and Duraisingh, M.T. (2014) Antiparasitic chemotherapy: from genomes to mechanisms. *Annu. Rev. Pharmacol. Toxicol.*, **54**, 71–94.
201. Rottmann, M., McNamara, C., Yeung, B.K., Lee, M.C., *et al.*, (2010) Spiroindolones, a potent compound class for the treatment of malaria. *Science*, **329**(5996), 1175–1180.
202. Tripathi, H., Luqman, S., Meena, A., Khan, F. (2014) Genomic identification of potential targets unique to *Candida albicans* for the discovery of antifungal agents. *Curr. Drug Targets*, **15**(1), 136–149.
203. Abadio, A.K., Kioshima, E.S., Teixeira, M.M., Martins, N.F., *et al.*, (2011) Comparative genomics allowed the identification of drug targets against human fungal pathogens. *BMC Genomics*, **12**, 75.
204. Butt, A.M. (2012) Nasrullah I, Tahir S, Tong Y, Comparative genomics analysis of *Mycobacterium ulcerans* for the identification of putative essential genes and therapeutic candidates. *PLoS One*, **7**(8), e43080.
205. Fox, J.L. (2013) Anti-infective monoclonals step in where antimicrobials fail. *Nat. Biotechnol.*, **31**(11), 952–954.
206. Kummerfeldt, C.E. (2014) Raxibacumab: potential role in the treatment of inhalational anthrax. *Infect. Drug Resist.*, **7**, 101–109.
207. Migone, T.S., Subramanian, G.M., Zhong, J., Healey, L.M., *et al.*, (2009) Raxibacumab for the treatment of inhalational anthrax. *N. Engl. J. Med.*, **361**(2), 135–144.
208. Reardon, S. (2014) Phage therapy gets revitalized. *Nature*, **510**(7503), 15–16.
209. Abedon, S.T., Kuhl, S.J., Blasdel, B.G., Kutter, E.M. (2011) Phage treatment of human infections. *Bacteriophage*, **1**(2), 66–85.
210. Loc-Carrillo, C. and Abedon, S.T. (2011) Pros and cons of phage therapy. *Bacteriophage*, **1**(2), 111–114.
211. Brautbar, A. and Ballantyne, C.M. (2011) Pharmacological strategies for lowering LDL cholesterol: statins and beyond. *Nat. Rev. Cardiol.*, **8**(5), 253–265.
212. Weber, C. and Noels, H. (2011) Atherosclerosis: current pathogenesis and therapeutic options. *Nat. Med.*, **17**(11), 1410–1422.
213. Cohen, J., Pertsemlidis, A., Kotowski, I.K., Graham, R., *et al.*, (2005) Low LDL cholesterol in individuals of African descent resulting from frequent nonsense mutations in PCSK9. *Nat. Genet.*, **37**(2), 161–165.
214. Zhao, Z., Tuakli-Wosornu, Y., Lagace, T.A., Kinch, L., *et al.*, (2006) Molecular characterization of loss-of-function mutations in PCSK9 and identification of a compound heterozygote. *Am. J. Hum. Genet.*, **79**(3), 514–523.
215. Varret, M., Rabès, J.P., Saint-Jore, B., Cenarro, A., *et al.*, (1999) A third major locus for autosomal dominant hypercholesterolemia maps to 1p34.1-p32. *Am. J. Hum. Genet.*, **64**(5), 1378–1387.
216. Grefhorst, A., McNutt, M.C., Lagace, T.A., Horton, J.D.. *et al.*, (2008) Plasma PCSK9 preferentially reduces liver LDL receptors in mice. *J. Lipid Res.*, **49**(6), 1303–1311.
217. Lagace, T.A., Curtis, D.E., Garuti, R., McNutt, M.C., *et al.*, (2006) Secreted PCSK9 decreases the number of LDL receptors in hepatocytes and in livers of parabiotic mice. *J. Clin. Invest.*, **116**(11), 2995–3005.
218. Do, R.Q., Vogel, R.A., and Schwartz, G.G. (2013) PCSK9 inhibitors: potential in cardiovascular therapeutics. *Curr. Cardiol. Rep.*, **15**(3), 345.
219. Mendelsohn, J. (2013) Personalizing oncology: perspectives and prospects. *J. Clin. Oncol.*, **31**(15), 1904–1911.
220. Garraway, L.A. and Janne, P.A. (2012) Circumventing cancer drug resistance in

the era of personalized medicine. *Cancer Discovery*, **2**(3), 214–226.

221. Lawrence, M.S., Stojanov, P., Polak, P., Kryukov, G.V., *et al.*, (2013) Mutational heterogeneity in cancer and the search for new cancer-associated genes. *Nature*, **499**(7457), 214–218.
222. Liu, H., Beck, T.N., Golemis, E.A., Serebriiskii, I.G. (2014) Integrating in silico resources to map a signaling network. *Methods Mol. Biol.*, **1101**, 197–245.
223. Mudge, J.M., Frankish, A., and Harrow, J. (2013) Functional transcriptomics in the post-ENCODE era. *Genome Res.*, **23**(12), 1961–1973.
224. Cho, M.J. and Juliano, R. (1996) Macromolecular versus small-molecule therapeutics: drug discovery, development and clinical considerations. *Trends Biotechnol.*, **14**(5), 153–158.
225. Zhu, F., Shi, Z., Qin, C., Tao, L., *et al.*, (2012) Therapeutic target database update 2012: a resource for facilitating target-oriented drug discovery. *Nucleic Acids Res.*, **40** (Database issue), D1128–D1136.
226. Zheng, C.J., Han, L.Y., Yap, C.W., Ji, Z.L., *et al.*, (2006) Therapeutic targets: progress of their exploration and investigation of their characteristics. *Pharmacol. Rev.*, **58**(2), 259–279.
227. Incorvati, J.A., Shah, S., Mu, Y., Lu, J. (2013)hor> Targeted therapy for HER2 positive breast cancer. *J. Hematol. Oncol.*, **6**, 38.
228. Subasinghe, N.L., Wall, M.J., Winters, M.P., Qin, N., *et al.*, (2012) A novel series of pyrazolylpiperidine N-type calcium channel blockers. *Bioorg. Med. Chem. Lett.*, **22**(12), 4080–4083.
229. Pencheva, N., Buss, C.G., Posada, J., Merghoub, T., *et al.*, (2014) Broad-spectrum therapeutic suppression of metastatic melanoma through nuclear hormone receptor activation. *Cell*, **156**(5), 986–1001.
230. Peng, T., Wu, J.-R., Tong, L.-J., Li, M.-Y., *et al.*, (2014) Identification of DW532 as a novel anti-tumor agent targeting both kinases and tubulin. *Acta Pharmacol. Sin.*, **35**(7), 916–928.
231. Lim, J.H., Luo, C., Vazquez, F., Puigserver, P. (2014) Targeting mitochondrial oxidative metabolism in melanoma causes metabolic compensation through glucose and glutamine utilization. *Cancer Res.*, **74**, 3535–3545.
232. Caldwell, J.T., Edwards, H., Buck, S.A., Ge, Y.,*et al.*, (2014) Targeting the wee1 kinase for treatment of pediatric Down syndrome acute myeloid leukemia. *Pediatr. Blood Cancer*, **61**, 1767–1773.
233. Reikvam, H., Brenner, A.K., Nepstad, I., Sulen, A., *et al.*, (2014) Heat shock protein 70 – the next chaperone to target in the treatment of human acute myelogenous leukemia? *Expert Opin. Ther. Targets*, **18**, 1–16.
234. Sarkar, S., Perlstein, E.O., Imarisio, S., Pineau, S., *et al.*, (2007) Small molecules enhance autophagy and reduce toxicity in Huntington's disease models. *Nat. Chem. Biol.*, **3**(6), 331–338.
235. Fomina-Yadlin, D., Kubicek, S., Walpita, D., Dancik, V., *et al.*, (2010) Small-molecule inducers of insulin expression in pancreatic alpha-cells. *Proc. Natl Acad. Sci. USA*, **107**(34), 15099–15104.
236. Schreiber, S.L. (2011) Organic synthesis toward small-molecule probes and drugs. *Proc. Natl Acad. Sci. USA*, **108**(17), 6699–6702.
237. Ledford, H. (2010) Complex synthesis yields breast-cancer therapy. *Nature*, **468**(7324), 608–609.
238. Lourido, S., Zhang, C., Lopez, M.S., Tang, K., *et al.*, (2013) Optimizing small molecule inhibitors of calcium-dependent protein kinase 1 to prevent infection by *Toxoplasma gondii*. *J. Med. Chem.*, **56**(7), 3068–3077.
239. Ishikawa, M. and Hashimoto, Y. (2011) Improvement in aqueous solubility in small molecule drug discovery programs by disruption of molecular planarity and symmetry. *J. Med. Chem.*, **54**(6), 1539–1554.
240. Huttunen, K.M., Raunio, H., and Rautio, J. (2011) Prodrugs – from serendipity to rational design. *Pharmacol. Rev.*, **63**(3), 750–771.
241. Vacchelli, E., Aranda, F., Eggermont, A., Galon, J., *et al.*, (2014) Trial watch: chemotherapy with immunogenic cell death inducers. *Oncoimmunology*, **3**(1), e27878.
242. Yang, Y., Liu, C., Peng, W., Lizee, G. *et al.*, (2012) Antitumor T-cell responses contribute to the effects of dasatinib on c-KIT mutant murine mastocytoma and are potentiated by anti-OX40. *Blood*, **120**(23), 4533–4543.

243. Bugelski, P.J. (2005) Genetic aspects of immune-mediated adverse drug effects. *Nat. Rev. Drug Discov.*, **4**(1), 59–69.
244. Atici, S., Cinel, I., Cinel, L., Doruk, N., *et al.*, (2005) Liver and kidney toxicity in chronic use of opioids: an experimental long term treatment model. *J. Biosci.*, **30**(2), 245–252.
245. Gerber, D.E. (2008) Targeted therapies: a new generation of cancer treatments. *Am. Fam. Physician*, **77**(3), 311–319.
246. Sathish, J.G., Sethu, S., Bielsky, M.C., de Haan, L., *et al.*, (2013) Challenges and approaches for the development of safer immunomodulatory biologics. *Nat. Rev. Drug Discov.*, **12**(4), 306–324.
247. Shi, S. (2014) Biologics: an update and challenge of their pharmacokinetics. *Curr. Drug Metab.*, **15**(3), 271–290.
248. Ferreira, I. and Isenberg, D. (2014) Vaccines and biologics. *Ann. Rheum. Dis.*, **73**(8), 1446–1454.
249. Kelley, B. (2009) Industrialization of mAb production technology: the bioprocessing industry at a crossroads. *MAbs*, **1**(5), 443–452.
250. Niederwieser, D. and Schmitz, S. (2011) Biosimilar agents in oncology/haematology: from approval to practice. *Eur. J. Haematol.*, **86**(4), 277–288.
251. Liu, D.T., F.T. Gates, III, and Goldman, N.D. (1985) Quality control of biologicals produced by r-DNA technology. *Dev. Biol. Stand.*, **59**, 161–166.
252. Kamionka, M. (2011) Engineering of therapeutic proteins production in *Escherichia coli*. *Curr. Pharm. Biotechnol.*, **12**(2), 268–274.
253. Chen, Q., Santi, L., and Zhang, C. (2014) Plant-made biologics. *Biomed. Res. Int.*, **2014**, 418064.
254. Huang, W., He, T., Chai, C., Yang, Y., *et al.*, (2012) Triptolide inhibits the proliferation of prostate cancer cells and down-regulates SUMO-specific protease 1 expression. *PLoS One*, **7**(5), e37693.
255. Virulizin. BioDrugs, **16**(5), 374–375. Available at: *http://www.ncbi.nlm.nih.gov/pubmed/12408741.*
256. Grachev, V., Magrath, D., Griffiths, E., Petricciani, J.C., *et al.*, (1998) WHO requirements for the use of animal cells as in vitro substrates for the production of biologicals – (requirements for biological substances no. 50) (Reprinted from WHO Technical Report Series, No. 878, 1998). *Biologicals*, **26**(3), 175–193.
257. Cardoso, C.V. and A.E. de Almeida (2014) Laboratory animal: biological reagent or living being? *Braz. J. Med. Biol. Res.*, **47**(1), 19–23.
258. Knezevic, I., Stacey, G., Petricciani, J., Sheets, R., *et al.*, (April 2009) Evaluation of cell substrates for the production of biologicals: revision of WHO recommendations. Report of the WHO study group on cell substrates for the production of biologicals, 22-23 , Bethesda, USA. *Biologicals, 2010*, **38**(1), 162–169.
259. Morrow, T. and Felcone, L.H. (2004) Defining the difference: what makes biologics unique. *Biotechnol. Healthcare*, **1**(4), 24–29.
260. Gatto, B. (2006) Biologics targeted at TNF: design, production and challenges. *Reumatismo*, **58**(2), 94–103.
261. Yao, S., Zhu, Y., and Chen, L. (2013) Advances in targeting cell surface signalling molecules for immune modulation. *Nat. Rev. Drug Discov.*, **12**(2), 130–146.
262. Chung, S.W., Hil-lal, T.A., and Byun, Y. (2012) Strategies for non-invasive delivery of biologics. *J. Drug Target.*, **20**(6), 481–501.
263. Skalko-Basnet, N. (2014) Biologics: the role of delivery systems in improved therapy. *Biologics*, **8**, 107–114.
264. Kurai, J., Chikumi, H., Hashimoto, K., Takata, M., *et al.*, (2012) Therapeutic antitumor efficacy of anti-epidermal growth factor receptor antibody, cetuximab, against malignant pleural mesothelioma. *Int. J. Oncol.*, **41**(5), 1610–1618.
265. Wolbink, G.J., Aarden, L.A., and Dijkmans, B.A. (2009) Dealing with immunogenicity of biologicals: assessment and clinical relevance. *Curr. Opin. Rheumatol.*, **21**(3), 211–215.
266. Oldham, R.K. (1983) Monoclonal antibodies in cancer therapy. *J. Clin. Oncol.*, **1**(9), 582–590.
267. Oldham, R.K. and Dillman, R.O. (2008) Monoclonal antibodies in cancer therapy: 25 years of progress. *J. Clin. Oncol.*, **26**(11), 1774–1777.
268. Miller, R.A., Maloney, D.G., McKillop, J., Levy, R. (1981) In vivo effects of murine hybridoma monoclonal antibody in a

patient with T-cell leukemia. *Blood*, **58**(1), 78–86.

269. Ahmadzadeh, V., Farajnia, S., Feizi, M.A.H., Nejad, R.A.K. (2014) Antibody humanization methods for development of therapeutic applications. *Monoclon. Antib. Immunodiagn. Immunother.*, **33**(2), 67–73.

270. Butowski, N. and Chang, S.M. (2005) Small molecule and monoclonal antibody therapies in neurooncology. *Cancer Control*, **12**(2), 116–124.

271. Imai, K. and Takaoka, A. (2006) Comparing antibody and small-molecule therapies for cancer. *Nat. Rev. Cancer*, **6**(9), 714–727.

272. Perez, H.L., Cardarelli, P.M., Deshpande, S., Gangwar, S., *et al.*, (2014) Antibody–drug conjugates: current status and future directions. *Drug Discov. Today*, **19**(7), 869–881.

273. Teicher, B.A. and Chari, R.V. (2011) Antibody conjugate therapeutics: challenges and potential. *Clin. Cancer Res.*, **17**(20), 6389–6397.

274. Zhang, J. and Rabbitts, T.H. (2014) Intracellular antibody capture: a molecular biology approach to inhibitors of protein–protein interactions. *Biochim. Biophys. Acta*, **1844**(11), 1970–1976.

275. Senter, P.D. and Sievers, E.L. (2012) The discovery and development of brentuximab vedotin for use in relapsed Hodgkin lymphoma and systemic anaplastic large cell lymphoma. *Nat. Biotechnol.*, **30**(7), 631–637.

276. Scott, A.M., Allison, J.P., and Wolchok, J.D. (2012) Monoclonal antibodies in cancer therapy. *Cancer Immun.*, **12**, 14.

277. Franco, G., Guarnotta, C., Frossi, B., Piccaluga, P.P., *et al.*, (2014) Bone marrow stroma CD40 expression correlates with inflammatory mast cell infiltration and disease progression in splenic marginal zone lymphoma. *Blood*, **123**(12), 1836–1849.

278. Vallera, D.A., Chen, H., Sicheneder, A.R., Panoskaltsis-Mortari, A., *et al.*, (2009) Genetic alteration of a bispecific ligand-directed toxin targeting human CD19 and CD22 receptors resulting in improved efficacy against systemic B cell malignancy. *Leuk. Res.*, **33**(9), 1233–1242.

279. Robinson, M.K., Hodge, K.M., Horak, E., Sundberg, A.L., *et al.*, (2008) Targeting ErbB2 and ErbB3 with a bispecific single-chain Fv enhances targeting selectivity and induces a therapeutic effect in vitro. *Br. J. Cancer*, **99**(9), 1415–1425.

280. Kontermann, R.E. (2012) Dual targeting strategies with bispecific antibodies. *MAbs*, **4**(2), 182–197.

281. Baeuerle, P.A., Kufer, P., and Bargou, R. (2009) BiTE: teaching antibodies to engage T-cells for cancer therapy. *Curr. Opin. Mol. Ther.*, **11**(1), 22–30.

282. Baeuerle, P.A. and Reinhardt, C. (2009) Bispecific T-cell engaging antibodies for cancer therapy. *Cancer Res.*, **69**(12), 4941–4944.

283. Haas, C., Krinner, E., Brischwein, K., Hoffmann, P., *et al.*, (2009) Mode of cytotoxic action of T-cell-engaging BiTE antibody MT110. *Immunobiology*, **214**(6), 441–453.

284. Derer, S., Beurskens, F.J., Rosner, T., Peipp, M., *et al.*, (2014) Complement in antibody-based tumor therapy. *Crit. Rev. Immunol.*, **34**(3), 199–214.

285. Hansel, T.T., Kropshofer, H., Singer, T., Mitchell, J.A., *et al.*, (2010) The safety and side effects of monoclonal antibodies. *Nat. Rev. Drug Discov.*, **9**(4), 325–338.

286. Weiner, L.M., Dhodapkar, M.V., and Ferrone, S. (2009) Monoclonal antibodies for cancer immunotherapy. *Lancet*, **373**(9668), 1033–1040.

287. Ascierto, P.A. and Marincola, F.M. (2014) What have we learned from cancer immunotherapy in the last 3 years? *J. Transl. Med.*, **12**(1), 141.

288. Janikashvili, N., Larmonier, N., and Katsanis, E. (2010) Personalized dendritic cell-based tumor immunotherapy. *Immunotherapy*, **2**(1), 57–68.

289. Badoual, C., Bastier, P.L., Roussel, H., Mandavit, M., *et al.*, (2013) An allogeneic NK cell line engineered to express chimeric antigen receptors: a novel strategy of cellular immunotherapy against cancer. *Oncoimmunology*, **2**(11), e27156.

290. Batchu, R.B. (2014) O.V Gruzdyn, S.T Kung, D.W. Weaver, *et al*, Dendritic cell-based immunotherapy of cancer with cell penetrating domains. Indian *J. Surg. Oncol.*, **5**(1), 3–4.

291. Yano, Y., Ueda, Y., Itoh, T., Fuji, N., *et al.*, (2006) A new strategy using autologous dendritic cells and lymphokine-activated killer cells for cancer immunotherapy: efficient maturation of DCs by co-culture with LAK cells *in vitro*. *Oncol. Rep.*, **16**(1), 147–152.

292. Petrovsky, N. and Aguilar, J.C. (2004) Vaccine adjuvants: current state and future trends. *Immunol. Cell Biol.*, **82**(5), 488–496.
293. Madan, R.A. and Gulley, J.L. (2011) Sipuleucel-T: harbinger of a new age of therapeutics for prostate cancer. *Expert Rev. Vaccines*, **10**(2), 141–150.
294. Di Lorenzo, G., Buonerba, C., and Kantoff, P.W. (2011) Immunotherapy for the treatment of prostate cancer. *Nat. Rev. Clin. Oncol.*, **8**(9), 551–561.
295. Zigler, M., Shir, A., and Levitzki, A. (2013) Targeted cancer immunotherapy. *Curr. Opin. Pharmacol.*, **13**(4), 504–510.
296. Vanneman, M. and Dranoff, G. (2012) Combining immunotherapy and targeted therapies in cancer treatment. *Nat. Rev. Cancer*, **12**(4), 237–251.
297. Hodi, F.S., Butler, M., Oble, D.A., Seiden, M.V., *et al.*, (2008) Immunologic and clinical effects of antibody blockade of cytotoxic T lymphocyte-associated antigen 4 in previously vaccinated cancer patients. *Proc. Natl Acad. Sci. USA*, **105**(8), 3005–3010.
298. Li, B., VanRoey, M., Wang, C., Chen, T.H., *et al.*, (2009) Anti-programmed death-1 synergizes with granulocyte macrophage colony-stimulating factor-secreting tumor cell immunotherapy providing therapeutic benefit to mice with established tumors. *Clin. Cancer Res.*, **15**(5), 1623–1634.
299. Vermeulen, C.F., Jordanova, E.S., ter Haar, N.T., Kolkman-Uljee, S.M., *et al.*, (2007) Expression and genetic analysis of transporter associated with antigen processing in cervical carcinoma. *Gynecol. Oncol.*, **105**(3), 593–599.
300. Blum, J.S., Wearsch, P.A., and Cresswell, P. (2013) Pathways of antigen processing. *Annu. Rev. Immunol.*, **31**, 443–473.
301. Zhang, Q.J., Seipp, R.P., Chen, S.S., Vitalis, T.Z., *et al.*, (2007) TAP expression reduces IL-10 expressing tumor infiltrating lymphocytes and restores immunosurveillance against melanoma. *Int. J. Cancer*, **120**(9), 1935–1941.
302. Zettl, U.K. and Mix, E. (2008) Potentials and limitations of current and emerging immunotherapies. *J. Neurol.*, **255** (Suppl. 6), 1.
303. Frigault, M.J. and June, C.H. (2011) Predicting cytokine storms: it's about density. *Blood*, **118**(26), 6724–6726.
304. Couzin-Frankel, J. (2013) Breakthrough of the year 2013. *Cancer immunotherapy. Science*, **342**(6165), 1432–1433.
305. Wirth, T., Parker, N., and S. Yla-Herttuala (2013) History of gene therapy. *Gene*, **525**(2), 162–169.
306. Kaufmann, K.B., Büning, H., Galy, A., Schambach, A., *et al.*, (2013) Gene therapy on the move. *EMBO Mol. Med.*, **5**(11), 1642–1661.
307. Aiuti, A., Cattaneo, F., Galimberti, S., Benninghoff, U., *et al.*, (2009) Gene therapy for immunodeficiency due to adenosine deaminase deficiency. *N. Engl. J. Med.*, **360**(5), 447–458.
308. Hsu, P.D., Lander, E.S., and Zhang, F. (2014) Development and applications of CRISPR-Cas9 for genome engineering. *Cell*, **157**(6), 1262–1278.
309. High, K., Gregory, P.D., and Gersbach, C. (2014) CRISPR technology for gene therapy. *Nat. Med.*, **20**(5), 476–477.
310. Jinek, M., Chylinski, K., Fonfara, I., Hauer, M., *et al.*, (2012) A programmable dual-RNA-guided DNA endonuclease in adaptive bacterial immunity. *Science*, **337**(6096), 816–821.
311. Yin, H., Xue, W., Chen, S., Bogorad, R.L., *et al.*, (2014) Genome editing with Cas9 in adult mice corrects a disease mutation and phenotype. *Nat. Biotechnol.*, **32**(6), 551–553.
312. Saraswat Pushpendra, P.A., Bhandari, A. (2012) Nucleic Acids as Therapeutics, in: Erdmann, V.A., Barciszewski, J. (Eds), *From Nucleic Acids Sequences to Molecular Medicine, Springer*, Berlin, Heidelberg.
313. Motavaf, M., Safari, S., and Alavian, S.M. (2012) Therapeutic potential of RNA interference: a new molecular approach to antiviral treatment for hepatitis C. *J. Viral Hepat.*, **19**(11), 757–765.
314. Wu, S.Y., Lopez-Berestein, G., Calin, G.A., Sood, A.K. (2014) RNAi therapies: drugging the undruggable. *Sci. Transl. Med.*, **6**(240), 240ps7.
315. Is this really the RNAissance? (2014) *Nat. Biotechnol.*, **32**(3), 201. Available at: *http://www.nature.com/nbt/journal/v32/n3/full/nbt.2853.html.*

316. Tabernero, J., Shapiro, G.I., LoRusso, P.M., Cervantes, A., *et al.*, (2013) First-in-humans trial of an RNA interference therapeutic targeting VEGF and KSP in cancer patients with liver involvement. *Cancer Discovery*, **3**(4), 406–417.
317. Golan, T., Hubert, A., Shemi, A., Segal, A., *et al.*, (2013) A phase I trial of a local delivery of siRNA against k-ras in combination with chemotherapy for locally advanced pancreatic adenocarcinoma. *J. Clin. Oncol.*, **31**(15), 4037.
318. Davis, M.E., Zuckerman, J.E., Choi, C.H., Seligson, D., *et al.*, (2010) Evidence of RNAi in humans from systemically administered siRNA via targeted nanoparticles. *Nature*, **464**(7291), 1067–1070.
319. Senzer, N.N., Kaufman, H.L., Amatruda, T., Nemunaitis, M., *et al.*, (2009) Phase II clinical trial of a granulocyte-macrophage colony-stimulating factor-encoding, second-generation oncolytic herpesvirus in patients with unresectable metastatic melanoma. *J. Clin. Oncol.*, **27**(34), 5763–5771.
320. Krall, N., Pretto, F., Decurtins, W., Bernardes, G.J. L., *et al.*, (2014) A small-molecule drug conjugate for the treatment of carbonic anhydrase IX-expressing tumors. *Angew. Chem. Int. Ed.*, **53**(16), 4231–4235.
321. Proia, D.A. and Bates, R.C. (2014) Ganetespib and HSP90: translating preclinical hypotheses into clinical promise. *Cancer Res.*, **74**(5), 1294–1300.
322. Pu, Y., Wang, W.B., Achilefu, S., Alfano, R.R. (2011) Study of rotational dynamics of receptor-targeted contrast agents in cancerous and normal prostate tissues using time-resolved picosecond emission spectroscopy. *Appl. Opt.*, **50**(10), 1312–1322.
323. Xue, J., Shan, L., Chen, H., Li, Y., *et al.*, (2013) Visual detection of STAT5B gene expression in living cell using the hairpin DNA modified gold nanoparticle beacon. *Biosens. Bioelectron.*, **41**, 71–77.
324. Xing, Y., Smith, A.M., Agrawal, A., Ruan, G., *et al.*, (2006) Molecular profiling of single cancer cells and clinical tissue specimens with semiconductor quantum dots. *Int. J. Nanomedicine*, **1**(4), 473–481.
325. Urano, Y., Sakabe, M., Kosaka, N., Ogawa, M., *et al.*, (2011) Rapid cancer detection by topically spraying a gamma-glutamyltranspeptidase-activated fluorescent probe. *Sci. Transl. Med.*, **3**(110), 110ra119.
326. Chabner, B.A., Ellisen, L.W., and Iafrate, A.J. (2013) Personalized medicine: hype or reality. *Oncologist*, **18**(6), 640–643.
327. Chisholm, R.L. (2013) The opportunities and challenges of implementing genomics-informed personalized medicine. *Clin. Pharmacol. Ther.*, **94**(2), 179–180.
328. Le Tourneau, C., Paoletti, X., Servant, N., Bièche, I., *et al.*, (2014) Randomised proof-of-concept phase II trial comparing targeted therapy based on tumour molecular profiling vs conventional therapy in patients with refractory cancer: results of the feasibility part of the SHIVA trial. *Br. J. Cancer*, **111**(1), 17–24.
329. Kim, E.S., Herbst, R.S., Wistuba, I.I., Lee, J.J., *et al.*, (2011) The BATTLE trial: personalizing therapy for lung cancer. *Cancer Discovery*, **1**(1), 44–53.
330. Tsimberidou, A.M., Iskander, N.G., Hong, D.S., Wheler, J.J., *et al.*, (2012) Personalized medicine in a phase I clinical trials program: the MD Anderson Cancer Center initiative. *Clin. Cancer Res.*, **18**(22), 6373–6383.
331. Tsimberidou, A.M., Wen, S., Hong, D.S., Wheler, J.J., *et al.*, (2014) Personalized medicine for patients with advanced cancer in the Phase I program at MD Anderson: validation and landmark analyses. *Clin. Cancer Res.*, **20**, 4827–4836.
332. Bozic, I., Reiter, J.G., Allen, B., and Antal, T. (2013) Evolutionary dynamics of cancer in response to targeted combination therapy. *Elife*, **2**, e00747.
333. Almendro, V., Marusyk, A., and Polyak, K. (2013) Cellular heterogeneity and molecular evolution in cancer. *Annu. Rev. Pathol.: Mech. Dis.*, **8**, 277–302.
334. Aparicio, S. and Caldas, C. (2013) The implications of clonal genome evolution for cancer medicine. *N. Engl. J. Med.*, **368**(9), 842–851.
335. Kandoth, C., McLellan, M.D., Vandin, F., Ye, K., *et al.*, (2013) Mutational landscape and significance across 12 major cancer types. *Nature*, **502**(7471), 333–339.
336. Balic, M., Williams, A., Lin, H., Datar, R., *et al.*, (2013) Circulating tumor cells: from bench to bedside. *Annu. Rev. Med.*, **64**, 31–44.

337. Diaz, L.A., Jr, and Bardelli, A. (2014) Liquid biopsies: genotyping circulating tumor DNA. *J. Clin. Oncol.*, **32**(6), 579–586.
338. Willison, K.R. and Klug, D.R. (2013) Quantitative single cell and single molecule proteomics for clinical studies. *Curr. Opin. Biotechnol.*, **24**(4), 745–751.
339. Eberwine, J., Sul, J.Y., Bartfai, T., and Kim, J. (2014) The promise of single-cell sequencing. *Nat. Methods*, **11**(1), 25–27.
340. Yu, J., Zhou, J., Sutherland, A., Wei, W., *et al.*, (2014) Microfluidics-based single-cell functional proteomics for fundamental and applied biomedical applications. *Annu. Rev. Anal. Chem. (Palo Alto)*, **7**, 275–295.
341. Krebs, M.G., Metcalf, R.L., Carter, L., Brady, G., *et al.*, (2014) Molecular analysis of circulating tumour cells-biology and biomarkers. *Nat. Rev. Clin. Oncol.*, **11**(3), 129–144.
342. Plaks, V., Koopman, C.D., and Werb, Z. (2013) Cancer. *Circulating tumor cells. Science*, **341**(6151), 1186–1188.
343. Smerage, J.B., Barlow, W.E., Hortobagyi, G.N., Winer, E.P., *et al.*, (2014) Circulating tumor cells and response to chemotherapy in metastatic breast cancer: SWOG S0500. *J. Clin. Oncol.*, **32**(31), 3483–3489.
344. Muller, V., Riethdorf, S., Rack, B., Janni, W., *et al.*, (2012) Prognostic impact of circulating tumor cells assessed with the CellSearch System™ and AdnaTest Breast™ in metastatic breast cancer patients: the DETECT study. *Breast Cancer Res.*, **14**(4), R118.
345. Adalsteinsson, V.A. and Love, J.C. (2014) Towards engineered processes for sequencing-based analysis of single circulating tumor cells. *Curr. Opin. Chem. Eng.*, **4**, 97–104.
346. Macaulay, I.C. and Voet, T. (2014) Single cell genomics: advances and future perspectives. *PLoS Genet.*, **10**(1), e1004126.
347. Streets, A.M., Zhang, X., Cao, C., Pang, Y., *et al.*, (2014) Microfluidic single-cell whole-transcriptome sequencing. *Proc. Natl Acad. Sci. USA*, **111**(19), 7048–7053.
348. Wu, A.R., Neff, N.F., Kalisky, T., Dalerba, P., *et al.*, (2014) Quantitative assessment of single-cell RNA-sequencing methods. *Nat. Methods*, **11**(1), 41–46.
349. Heitzer, E., Auer, M., Gasch, C., Pichler, M., *et al.*, (2013) Complex tumor genomes inferred from single circulating tumor cells by array-CGH and next-generation sequencing. *Cancer Res.*, **73**(10), 2965–2975.
350. Navin, N., Kendall, J., Troge, J., Andrews, P., *et al.*, (2011) Tumour evolution inferred by single-cell sequencing. *Nature*, **472**(7341), 90–94.
351. Murtaza, M., Dawson, S.J., Tsui, D.W., Gale, D., *et al.*, (2013) Non-invasive analysis of acquired resistance to cancer therapy by sequencing of plasma DNA. *Nature*, **497**(7447), 108–112.
352. Dawson, S.J., Tsui, D.W., Murtaza, M., Biggs, H., *et al.*, (2013) Analysis of circulating tumor DNA to monitor metastatic breast cancer. *N. Engl. J. Med.*, **368**(13), 1199–1209.
353. De Vlaminck, I., Valantine, H.A., Snyder, T.M., Strehl, C., *et al.*, (2014) Circulating cell-free DNA enables noninvasive diagnosis of heart transplant rejection. *Sci. Transl. Med.*, **6**(241), 241ra77.
354. Al-Lazikani, B., Banerji, U., and Workman, P. (2012) Combinatorial drug therapy for cancer in the post-genomic era. *Nat. Biotechnol.*, **30**(7), 679–692.
355. Bagnyukova, T., Serebriiskii, I.G., Zhou, Y., Hopper-Borge, E.A., *et al.*, (2010) Chemotherapy and signaling how can targeted therapies supercharge cytotoxic agents? *Cancer Biol. Ther.*, **10**(9), 843–857.
356. Geyer, C.E., Forster, J., Lindquist, D., Chan, S., *et al.*, (2006) Lapatinib plus capecitabine for HER2-positive advanced breast cancer. *N. Engl. J. Med.*, **355**(26), 2733–2743.
357. Akimoto, T., Hunter, N.R., Buchmiller, L., Mason, K., *et al.*, (1999) Inverse relationship between epidermal growth factor receptor expression and radiocurability of murine carcinomas. *Clin. Cancer Res.*, **5**(10), 2884–2890.
358. Welsh, J.W., Komaki, R., Amini, A., Munsell, M.F., *et al.*, (2013) Phase II trial of erlotinib plus concurrent whole-brain radiation therapy for patients with brain metastases from non-small-cell lung cancer. *J. Clin. Oncol.*, **31**(7), 895–902.
359. Mole, R.H. (1953) Whole body irradiation; radiobiology or medicine? *Br. J. Radiol.*, **26**(305), 234–241.
360. Durante, M., Reppingen, N., and Held, K.D. (2013) Immunologically augmented cancer treatment using modern radiotherapy. *Trends Mol. Med.*, **19**(9), 565–582.

361. Kalbasi, A., June, C.H., Haas, N., and Vapiwala, N. (2013) Radiation and immunotherapy: a synergistic combination. *J. Clin. Invest.*, **123**(7), 2756–2763.
362. Krysko, D.V., Garg, A.D., Kaczmarek, A., Krysko, O., *et al.*, (2012) Immunogenic cell death and DAMPs in cancer therapy. *Nat. Rev. Cancer*, **12**(12), 860–875.
363. Nijman, S.M.B. (2011) Synthetic lethality: general principles, utility and detection using genetic screens in human cells. *FEBS Lett.*, **585**(1), 1–6.
364. Astsaturov, I., Ratushny, V., Sukhanova, A., Einarson, M.B., *et al.*, (2010) Synthetic lethal screen of an EGFR-centered network to improve targeted therapies. *Sci. Signal.*, **3**(140), ra67.
365. Hoffman, G.R., Rahal, R., Buxton, F., Xiang, K., *et al.*, (2014) Functional epigenetics approach identifies BRM/SMARCA2 as a critical synthetic lethal target in BRG1-deficient cancers. *Proc. Natl Acad. Sci. USA*, **111**(8), 3128–3133.
366. Bassik, M.C., Kampmann, M., Lebbink, R.J., Wang, S., *et al.*, (2013) A systematic mammalian genetic interaction map reveals pathways underlying ricin susceptibility. *Cell*, **152**(4), 909–922.
367. Koike-Yusa, H., Li, Y., Tan, E-P, Velasco-Herrera, M.D.C., *et al.*, (2014) Genome-wide recessive genetic screening in mammalian cells with a lentiviral CRISPR-guide RNA library. *Nat. Biotechnol.*, **32**(3), 267–273.
368. Luo, J., Emanuele, M.J., Li, D., Creighton, C.J., *et al.*, (2009) A genome-wide RNAi screen identifies multiple synthetic lethal interactions with the ras oncogene. *Cell*, **137**(5), 835–848.
369. Matheny, C.J., Wei, M.C., Bassik, M.C., Donnelly, A.J., *et al.*, (2013) Next-generation NAMPT inhibitors identified by sequential high-throughput phenotypic chemical and functional genomic screens. *Chem. Biol.*, **20**(11), 1352–1363.
370. Chan, D.A. and Giaccia, A.J. (2011) Harnessing synthetic lethal interactions in anticancer drug discovery. *Nat. Rev. Drug Discovery*, **10**(5), 351–364.
371. Nijman, S.M.B. and Friend, S.H. (2013) Potential of the synthetic lethality principle. *Science*, **342**(6160), 809–811.
372. Lord, C.J. and Ashworth, A. (2012) The DNA damage response and cancer therapy. *Nature*, **481**(7381), 287–294.
373. Polyak, K. and Garber, J. (2011) Targeting the missing links for cancer therapy. *Nat. Med.*, **17**(3), 283–284.
374. Servant, N., Roméjon, J., Gestraud, P., La Rosa, P., *et al.*, (2014) Bioinformatics for precision medicine in oncology: principles and application to the SHIVA clinical trial. *Front. Genet.*, **5**, 152.
375. Gehlenborg, N., O'Donoghue, S.I., Baliga, N.S., Goesmann, A., *et al.*, (2010) Visualization of omics data for systems biology. *Nat. Methods*, **7**(3), S56–S68.
376. Chen, Y.B., Chattopadhyay, A., Bergen, P., Gadd, C., *et al.*, (2007) The online bioinformatics resources collection at the University of Pittsburgh Health Sciences Library System – a one-stop gateway to online bioinformatics databases and software tools. *Nucleic Acids Res.*, **35**, D780–D785.
377. Artimo, P., Jonnalagedda, M., Arnold, K., Baratin, D., *et al.*, (2012) ExPASy: SIB bioinformatics resource portal. *Nucleic Acids Res.*, **40**(W1), W597–W603.
378. Scannell, J.W., Blanckley, A., Boldon, H., and Warrington, B. (2012) Diagnosing the decline in pharmaceutical R&D efficiency. *Nat. Rev. Drug Discov.*, **11**(3), 191–200.
379. Anighoro, A., Bajorath, J., and Rastelli, G. (2014) Polypharmacology: challenges and opportunities in drug discovery. *J. Med. Chem.*, **57**(19), 7874–7887.

28
Therapeutics against Cancer Stem Cells: Targeting the Root of Cancer*

Kristen M. Smith[1,2] *and Catriona H. M. Jamieson*[2]
[1] *University of California, San Diego, Department of Medicine and Moores UCSD Cancer Center, La Jolla, California 92037, USA*
[2] *Eclipse Therapeutics, San Diego, California 92121, USA*

Keywords

Cancer stem cell
A cell within a tumor that is capable of self-renewal and that can generate and maintain all cell types of the original tumor from which it was derived.

*This chapter has previously been published in: Meyers, R.A. (Ed.) Stem Cells, 2013, 978-3-527-32925-0.

Translational Medicine Cancer: Advances in Molecular Biology and Medicine,
First Edition. Edited by Robert A. Meyers.

Self-renewal
The property whereby at least one of the daughter cells retains the multipotent capacity of the parent cell upon cell division. In the case of cancer stem cells, this is demonstrated by an *in vivo* serial transplantation ability.

The cancer stem cell hypothesis provides a useful framework with which to investigate the mechanisms of cancer initiation, progression, and treatment. Cells with stem cell-like properties have been implicated in tumor initiation, metastatic spread and disease evolution, as well therapeutic failure. While the roles of cancer stem cells continue to be defined, known stem cell signaling pathways and stem cell niche-regulated mechanisms of survival are emerging as targets to selectively eliminate cancer stem cells. In this chapter, these mechanisms and approaches for targeting cancer stem cells through these pathways will be discussed. Strategies will also be addressed for identifying novel pathways involved in cancer stem cell survival, and requirements for the clinical evaluation of cancer stem cell-targeted therapies.

1 Introduction

Cancer is a heterogeneous disease that was first described in a sixteenth-century BC manuscript attributed to the Egyptian physician Imhotep [1]. Over the centuries, cancer has claimed the lives of countless individuals, and today more than 1.5 million new patients are diagnosed each year [2]. Modern cancer medicine aims for a complete annihilation of the tumor by any means necessary, even when those treatments also have significant adverse side effects for the patient and consist of surgical approaches championed by William Halsted, radiation therapy initiated by Rene Gilbert and extended by Vera Peters and Henry Kaplan and chemotherapy pioneered by Sidney Farber [3]. In many patients, these strategies are initially successful at restraining the aberrant proliferation of tumor cells; however, with a few notable exceptions, all have been inadequate to cure most cancers.

Although the postulated reasons behind the inability to cure most cancers are myriad, a recently described model may hold the key to curative cancer therapies with fewer adverse side effects. The cancer stem cell (CSC) model was suggested by early studies in which a large number of tumor cells were required to initiate tumor growth [4]. During the early 1990s, John Dick and colleagues demonstrated that a subpopulation of leukemia cells from human acute myeloid leukemia (AML) uniquely possessed the capacity to engraft immunodeficient mice and could recapitulate the phenotypic heterogeneity of the original leukemia. The leukemia-initiating subpopulation was defined as $CD34^+/CD38^-$ and shared similar cell-surface marker expression patterns with normal hematopoietic progenitor cells [5, 6]. These results advanced the hypothesis that leukemia is organized in a functional hierarchy analogous to normal hematopoietic development and is initiated or maintained by a subset of cancer cells possessing stem cell properties.

These leukemia-initiating cells, which have also been termed CSCs, are responsible for generating and maintaining the bulk of the cancer, but may also be the key to permanently halting cancer growth and metastasis. Modern cancer therapeutics target the rapidly dividing cells that may be generated by CSCs but do not effectively kill the CSCs themselves. Yet, if the CSCs within a cancer could be effectively depleted, the cancer should ultimately recede due to the limited life span of more differentiated tumor cells [7].

2 Cancer Stem Cells

2.1 Properties of Cancer Stem Cells

A CSC is defined functionally as a cell that is capable of self-renewal, as demonstrated by serial transplantation ability, and which can generate and maintain all cell types of the original tumor from which it was derived [8]. Since the initial definition of a CSC population in AML, CSCs have been identified in other blood cancers, including blast crisis chronic myeloid leukemia (CML) [9], acute lymphoid leukemia (ALL) [10] and multiple myeloma (MM) [11]. Quickly following on the heels of the hematopoietic malignancies came descriptions of CSCs in solid tumors. Studies in brain tumors identified a $CD133^+$ subpopulation from medulloblastoma and glioblastoma multiforme (GBM) tumors that reproduced the heterogeneity of the original tumor in xenograft models [12–15]. Importantly, these studies reintroduced the serum-free sphere culture system as a surrogate assay of tumor-initiating ability, allowing for robust *in vitro* studies of CSCs [16, 17]. Outside of the neural system, investigation conducted by Clarke *et al.* led to the identification of a $CD44^+$ $CD24^{-/low}$ CSC subpopulation in human breast cancers [18]. Subsequently, CSCs have now been described in cancers of the colon [19–22], prostate [23, 24], pancreas [25, 26], head and neck [27, 28], lung [29, 30], liver [31, 32], ovary [33–36] and skin [37–39].

An examination of the biological properties of CSCs requires an ability to prospectively identify and isolate CSCs as well as to perform functional assays to validate putative CSC populations. A significant number of studies have been focused on the problem of CSC identity with two major approaches having been used to isolate CSCs in prospective fashion: (i) the enrichment of CSCs using cell-surface molecules; and (ii) assays that measure the functional characteristics of stem cells. A variety of cell-surface markers have been examined as potential CSC markers for prospective isolation. For example, CD34 and CD38 expression has been used to subdivide leukemia populations [5, 6, 9], while CD133 has been shown to mark glioma and medulloblastoma CSCs [12–14], as well as CSCs in colon, pancreatic and lung cancer [21, 25, 29]. Breast CSCs were described as $CD44^+CD24^{-/low}$, and CD44 has also been shown to mark CSCs in colon, head and neck, pancreatic and prostate cancers [40]. Additional CSC markers include CD15, CD20, CD29, CD49f, CD117, CD166, EpCAM, ESA and LGR5 [41]. The "cocktail" of surface markers used to prospectively identify and isolate CSCs from different tumors continues to be refined as pure CSC populations have not yet been identified. A complementary approach has focused on the functional properties possessed by stem cell populations. One such example is side-population analysis [42], which takes advantage of the elevated expression of ATP binding cassette transporters (e.g.,

ABCB1/MDR1 and ABCG2) in stem cells that results in an enhanced ability to efflux dyes and drugs [43]. Side-population cells are marked by their ability to efflux either Hoechst 33342 or Rhodamine, 23 dyes, and thus appear as an unlabeled side-population by flow cytometric analysis. Side-populations have also been identified in a number of primary solid and liquid tumors within established cancer cell lines and are often enriched for sphere formation ability *in vitro* and tumor formation *in vivo*. An additional example of the functional approach to identifying CSCs utilizes high aldehyde dehydrogenase 1A1 (ALDH1) expression in normal stem cell populations [44, 45]. In addition, ALDH1 enzymatic activity has been used as a marker of CSCs from multiple tumors [46].

2.2 Complexities of the Cancer Stem Cell Model

As investigations have progressed on methods to identify CSCs, the complexities of the CSC model have also been exposed [47]. The use of cell-surface markers to identify and isolate CSC populations has been contentious. The most prominent example is CD133, where early studies identified CSCs exclusively within the $CD133^+$ cell population isolated from central nervous system (CNS) tumors [14, 15]. Additional studies have demonstrated a greater diversity amongst patient samples than was suggested by the original studies [48]. Furthermore, the assays used to functionally validate CSC populations *in vitro* and *in vivo* have been challenged [38, 49, 50]. CSC populations identified by surface marker expression appear to be functionally heterogeneous. For example, AML appears to be organized in a hierarchy similar to normal hematopoiesis, with at least three distinct CSC populations: short-term, long-term and long-term quiescent [51]. The results of recent studies with glioblastoma and colon CSCs have suggested that there may be a similar range of tumor-initiating cells within solid tumors [52, 53]. In addition to the potential for multiple CSC populations, these populations may vary with disease subtype or stage and may evolve during disease progression [54–57] thus confounding efforts to define CSCs by static marker expression rather than cellular function.

One critical question here is the role of CSCs in treatment failure and disease relapse with compelling evidence having suggested that CSCs are resistant to current therapies. Indeed, studies of radiation resistance in glioma patients have demonstrated that the $CD133^+$ CSC population was enriched after radiation [58, 59]. Both radiation and chemotherapy resistance have been observed in breast CSCs [60]. The neoadjuvant therapy of patients resulted in an enrichment of $CD44^+CD24^{-/low}$ breast CSCs at 3 and 12 weeks post-treatment [61]. Likewise, CSCs from other tumors exhibit an increased resistance to chemotherapy: $CD34^+CD38^-$ AML CSCs were resistant to daunorubicin treatment as compared to the $CD34^+$ $CD38^+$ cell population [62] while AML CSCs that remained after chemotherapy retained an ability to serially transplant leukemia [63]. Colon CSCs were shown to be resistant to a number of treatments that included cyclophosphamide, irinotecan [64], oxaliplatin and 5-fluorouracil [65]. Chemotherapeutic treatment resulted in the enrichment of an ESA^+CD44^+ colon CSC subpopulation despite there being a decreased tumor volume in treated animals. Similar results have been reported for CML [66], melanoma [67], liver [31] and pancreas [25]: indeed, chemotherapeutic resistance has even been used as a strategy to enrich for CSCs *in vitro* [61]. Thus, although the mainstays of cancer therapy have a measurable effect on the bulk tumor, they do not seem to affect the root of the tumor nor the inherent tumorigenicity of CSCs.

The corollary question here is whether CSCs are predictive of disease outcome. In AML, leukemic stem cells (LSCs) have been functionally defined by their ability to serially transplant disease. Gene-expression profiles derived from these leukemic CSCs are highly correlated with overall patient survival [68]. In laryngeal cancer, an assessment of CSC density by CD44 expression can be used to predict the outcome of radiotherapy with the odds of recurrence being shown to correlate positively with CD44 expression [69]. The percentage of $CD44^+$ ovarian CSCs was positively associated with a shorter progression-free survival in early-stage ovarian cancers [70]. In breast cancer, there was a negative correlation between CD24 expression and outcome, but there was no correlation between the percentage of $CD44^+CD24^{-/low}$ cells and breast cancer prognosis or disease progression [71]. Likewise, in glioma the presence of $CD133^+$ cell clusters was a significant prognostic factor for a worse progression-free survival and overall survival [72] while other studies found no such correlation [73]. The prognostic role of CSC remains contentious and may reflect our incomplete understanding of CSC identity and behavior.

While the role of CSCs in tumor initiation, maintenance and relapse after therapy continues to be defined, various research groups are developing strategies to selectively target CSC survival and proliferation pathways while sparing normal stem cells. The resistance of CSCs to radiation and chemotherapy is mediated by a number of mechanisms including the overexpression of stem cell-related signaling molecules and niche-regulated mechanisms such as quiescence and the modulation of apoptotic pathways. At this point, the mechanisms and approaches for targeting CSCs through these and other pathways will be discussed as will the various strategies for identifying novel CSC pathways that might be targeted therapeutically, and the mechanisms for analyzing the clinical effects of therapeutics on CSC populations (see Table 1).

3
Stem Cell-Related Signaling Pathways

Developmental pathways, including the Wnt, Hedgehog (HH) and Notch signaling pathways, are evolutionarily conserved and are able to direct essential developmental processes by dictating cell fate decisions and cell migration. While these pathways are tightly controlled and generally inactive in adult tissues, they play an important role in normal adult stem cells. Increasingly, it has become clear that these pathways are aberrantly activated in a wide range of human cancers.

3.1
Wnt/β-Catenin

Since the first description of vertebrate Wnt signaling [74], its central role in a wide spectrum of developmental processes, including embryonic axial patterning and organ formation, cell proliferation, cell migration and stem cell self-renewal, have come to be appreciated. The Wnt family of secreted proteins is composed of at least 19 isoforms that are activating ligands for two distinct Wnt signaling pathways. In the non-canonical Wnt signaling pathway, a subset of Wnt family members activates the Planar Cell Polarity (PCP) and Wnt/Calcium pathways in a β-catenin-independent mechanism to influence cell

Tab. 1 Selected CSC therapeutic agents.

	Therapeutic agent	*Proposed mechanism of action*	*Reference(s)*
Wnt/β-catenin	IWPs	Porcupine *O*-acyltransferase inhibitor	[125]
	Anti-Wnt1 or Wnt 2 monoclonal antibodies	Block receptor-Wnt ligand interaction	[117, 118]
	Anti-LRP6 monoclonal antibodies	Block formation of active receptor signaling complex	[123]
	Anti-Frizzled monoclonal antibodies (OMP-18R5)	Block formation of active receptor signaling complex	[394], *www.clinicaltrials.gov*
	XAV939	Tankyrase 1/2 inhibitor	[124]
	IWR	Axin stabilizer	[125]
	PKF115-584	TCF/β-catenin interaction inhibitor	[101, 126]
	CGP049090	TCF/β-catenin interaction inhibitor	[126]
	ICG-001	CREB/β-catenin interaction inhibitor	[127]
	NSAIDs	COX-dependent and -independent mechanisms	[129–135]
Hedgehog	Cyclopamine and derivatives	SMO inhibitors	[164, 165, 172]
	Vismodegib (GDC-0049)	SMO inhibitor	[169, 170]
	PF-04449913	SMO inhibitor	*www.clinicaltrials.gov*
	BMS-833923 (XL139)	SMO inhibitor	*www.clinicaltrials.gov*
	IPI-926	SMO inhibitor	*www.clinicaltrials.gov*
	LDE-225	SMO inhibitor	*www.clinicaltrials.gov*
	LEQ506	SMO inhibitor	*www.clinicaltrials.gov*
	TAK-441	SMO inhibitor	*www.clinicaltrials.gov*
	Quinazolinone and derivatives	SMO inhibitors	[171]
	Itraconazole and derivatives	SMO inhibitors	[174]
	anti-Hedgehog monoclonal antibodies (5E1)	Block receptor–Hh ligand interaction	[395]
	GANT-61	GLI DNA binding inhibitor	[176]
	Arsenic trioxide	GLI DNA binding inhibitor	[173, 177]
Notch	DFK167	γ-secretase inhibitor	[396]
	L-685458	γ-secretase inhibitor	[397]
	DAPT	γ-secretase inhibitor	[398]
	LY450139	γ-secretase inhibitor	[399]
	LY411575	γ-secretase inhibitor	[255]
	GSI-18	γ-secretase inhibitor	[400]
	MRK-003	γ-secretase inhibitor	[401, 402]

Tab. 1 *(Continued)*

	Therapeutic agent	*Proposed mechanism of action*	*Reference(s)*
	Anti-Notch1 monoclonal antibodies	Block cleavage of Notch receptor	[257, 262]
	Anti-Notch2 monoclonal antibodies	Block cleavage of Notch receptor	[257]
	Anti-Notch3 monoclonal antibodies	Block cleavage of Notch receptor	[263]
	Anti-Notch2/3 monoclonal antibodies (OMP-595R)	Block cleavage of Notch receptor	[403]
	Anti-DLL4 monoclonal antibodies (OMP-21M18, REGN421)	Block binding of DLL4 to Notch receptor	[243, 404], *www.clinicaltrials.gov*
NF-κB	AS602868	IKK inhibition	[279]
	Dexamethasone	Induce expression of IκBα	[404, 405]
	Parthenolide	NF-κB inhibitor	[277]
	DMAPT	NF-κB inhibitor	[278]
Telomerase	GRN-163	Telomerase template antagonist	[436]
	Imetelstat/GRN-163L	Telomerase template antagonist	[333]
	Vaccination (GV1001)	Vaccination with hTERT peptide (611-626)	[330, 437]
mTOR/PI3K/ Akt	Rapamycin	mTORC1 inhibitor	[406, 407]
	Temsirolimus (CCl-779)	mTORC1 inhibitor	[408]
	Everolimus (RAD-001)	mTORC1 inhibitor	[409]
	Ridaforolimus (AP23573, MK-8669)	mTORC1 inhibitor	[410]
	OSI-027	mTORC1/2 dual inhibitor	[411, 412]
	PP242	mTORC1/2 dual inhibitor	[413, 414]
	PP30	mTORC1/2 dual inhibitor	[414]
	WAY-600	mTORC1/2 dual inhibitor	[304]
	WYE-687	mTORC1/2 dual inhibitor	[304]
	WYE-354	mTORC1/2 dual inhibitor	[304]
	AZD8055	mTORC1/2 dual inhibitor	[415]
	INK128	mTORC1/2 dual inhibitor	[416]
	Perifosine (KRX-0401, D-21266)	AKT inhibitor	[417, 418]
	MK-2206	AKT inhibitor	[419]
	GSK690693	AKT inhibitor	[420, 421]
	GSK2141795	AKT inhibitor	[422]
	GDC-0068	AKT inhibitor	[423]
	LY2780301	AKT inhibitor	[424]
	NVP-BKM120	PI3K inhibitor	[425]
	XL147 (SAR245408)	PI3K inhibitor	[426]
	PX-866	PI3K inhibitor	[427]
	GDC-0941	PI3K inhibitor	[428]

(continued overleaf)

Tab. 1 *(Continued)*

	Therapeutic agent	*Proposed mechanism of action*	*Reference(s)*
	CAL-101 (GS-1101)	PI3K inhibitor	[429]
	SF1126	PI3K/mTOR dual inhibitor	[430]
	NVP-BEZ235	PI3K/mTOR dual inhibitor	[431]
	XL765	PI3K/mTOR dual inhibitor	[432]
	GDC-0980	PI3K/mTOR dual inhibitor	[433, 434]
	PI-103	PI3K/mTOR dual inhibitor	[435]
	Metformin	Inhibits mTOR activation via AMPK	[308]
Niche	Anti-CD44 monoclonal antibodies	CD44 activation	[364, 365]
	Pepducins	CXCR4 antagonist	[368]
	Plerixafor (AMD3100)	CXCR4 antagonist	[369, 438]
	Anti-integrin α_6 monoclonal antibodies	Block integrin α_6 binding	[371]
	GLPG0187	Integrin α_v peptide antagonist	[373]
	Cilengitide (EMD 121974)	Integrin $\alpha_v\beta_3$ and $\alpha_v\beta_5$ peptide antagonist	[439, 440]
	Volociximab	Monoclonal antibody targeting integrin $\alpha_5\beta_1$	[441]
	ATN-161	Integrin $\alpha_5\beta_1$ peptide antagonist	[442]
	Bevacizumab (Avastin)	Block VEGF-A/VEGFR binding	[375, 443]
	Cediranib/AZD2171	VEGFR2 tyrosine kinase inhibitor	[376, 444]
	Oblimersen sodium (G3139)	Bcl2 antisense oligonucleotide	[445, 446]
	AT-101 ((−)-gossypol)	Bcl2, Mcl1 inhibitor	[447, 448]
	ABT-263/ABT-737	Pan-Bcl2 inhibitor	[449]
	Obatoclax (GX15-070)	Pan-Bcl2 inhibitor	[450]
	Sabutoclax (BI-97C1)	Pan-Bcl2 inhibitor	[451]

polarity, asymmetric cell division and cell migration during development [75] (Fig. 1). The canonical Wnt signaling pathway functions through β-catenin to affect pattern formation during development and has been extensively studied in tumorigenesis and cancer progression [76] (Fig. 1). In the canonical pathway, the absence of secreted Wnt proteins results in a rapid phosphorylation of β-catenin by the GSK3β/Axin/adenomatous polyposis coli (APC) protein complex, with subsequent ubiquitin-dependent proteosomal degradation. This degradation complex keeps free β-catenin levels low ensuring that the T-cell factor (TCF) and lymphoid enhancer factor (LEF) proteins interact with transcriptional corepressors to block Wnt-dependent gene expression. Canonical Wnt signaling is initiated upon binding of the secreted Wnt proteins to Frizzled receptors in the presence of the coreceptor LRP5/6

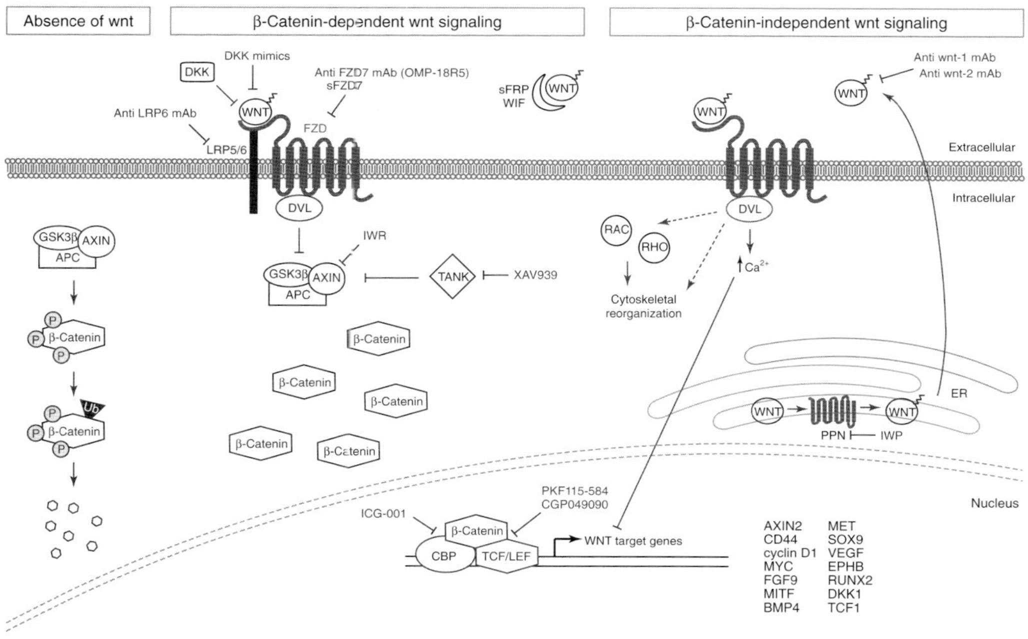

Fig. 1 Schematic of Wnt signaling pathways. In the noncanonical Wnt signaling pathway depicted on the right, Wnt binding to Frizzled (FZD) receptors activates the Planar Cell Polarity (PCP) and Wnt/calcium pathways to influence cell polarity, asymmetric cell division, and cell migration. The canonical Wnt signaling pathway is illustrated on the left. In the absence of secreted Wnt, β-catenin is phosphorylated and targeted for ubiquitin-dependent proteosomal degradation by the GSK3β/Axin/adenomatous polyposis coli (APC) protein complex. Signaling is initiated upon binding of secreted Wnt proteins to FZD in the presence of the coreceptor LRP5/6. Wnt-bound FZD activates Disheveled (DVL) leading to inhibition of the β-catenin degradation complex. Free β-catenin accumulates and translocates to the nucleus, where it acts as a potent transcription factor in complex with TCF/LEF family members. In addition to multiple Wnt family proteins and their identified receptors, several secreted proteins have been identified as negative regulators of canonical Wnt signaling, including the Dickkopf (DKK) family, secreted Frizzled-related proteins (sFRPs), and Wnt inhibitory factor-1 (WIF-1). Inhibitors of various pathways components are indicated in red. (Adapted from Ref. [75].)

(low-density lipoprotein receptor-related protein 5/6). Wnt-bound Frizzled receptors activate Disheveled (DVL) which inhibits formation of the β-catenin degradation complex. Free β-catenin accumulates and translocates to the nucleus where it acts as a potent transcription factor in complex with TCF/LEF family members to initiate the transcription of a variety of Wnt-responsive target genes including *MYC, CCND1* (cyclin D1), *AXIN2, MET* and *VEGF* [77]. In addition to multiple Wnt family proteins and their identified receptors, several secreted proteins have been identified as negative regulators of canonical Wnt signaling, including the Dickkopf (DKK) family, secreted Frizzled-related proteins (sFRPs) and Wnt inhibitory factor-1 (WIF-1) [78].

With its critical function in cell fate and tissue patterning, the Wnt signaling pathway is tightly regulated during early embryogenesis as well as later in development. In adult tissues, Wnt signaling plays a role in the self-renewal of multiple stem cell populations especially those of the gastrointestinal (GI) tract, the skin and hair follicles and the hematopoietic system. The GI tract epithelium is organized into villi and crypts of Lieberkühn, with stem and progenitor cells found in the bottom two-thirds of the crypts, while their differentiated progeny comprise the remainder of the crypt and the villi. These stem/progenitor cells are responsible for the continuous replacement of intestinal epithelial cells throughout the life of the organism and Wnt/β-catenin appears to be the major signaling pathway involved in controlling stem cell self-renewal and cell fate decisions during intestinal homeostasis. Active nuclear β-catenin is observed throughout the crypts [79] and the inhibition of Wnt signaling by Tcf4 conditional knockout in neonatal mice [80] or overexpression of Dkk-1 in adult mice [81, 82] results in crypt loss and a progressive degeneration of villi intestinal stem cells. Like the GI tract, the hair follicle represents a continuously self-renewing organ fueled by multipotent stem cells located in the bulge region [83]. Bulge stem cells are activated during anagen (active hair growth) by the signals from the dermal papilla and generate the hair cell lineages as well as sebocytes and stem cells of the interfollicular dermis [84, 85]. Both, Lef and Tcf transcription factors and β-catenin have been linked to hair follicle stem cell function. The loss of Lef1, mutation of β-catenin or overexpression of the Wnt inhibitor, Dkk1, in mouse skin eliminates hair development [86–88] while an overexpression of Lef-1 increases hair follicle formation [89] and results in the development of ectopic sebaceous glands [90]. The Tcf3 transcription factor is specifically expressed in bulge stem cells, and Tcf-dependent transgenic reporter constructs are induced in bulge stem cells at the start of the hair cycle [91, 92]. Finally, the overexpression of active β-catenin in the adult interfollicular dermis induces *de novo* hair follicle formation [93]. Taken together, these data demonstrate an essential role for Wnt/β-catenin signals in hair follicle differentiation and the fate decisions of bulge stem cells. Lastly, the hematopoietic system is heavily dependent on stem cell self-renewal and this self-renewal appears to be dependent at least in part on Wnt signaling. *In vitro*, β-catenin as well as purified Wnt family members promote both murine and human hematopoietic stem cell (HSC) and progenitor expansion and inhibit differentiation [94–96]. Wnt ligands are produced by the HSCs themselves as well as their supportive microenvironment [94, 97] and Wnt reporter activity has been observed in the marrow niche, suggesting that HSCs respond directly to Wnt signals *in vivo* [95, 98].

Considering the cardinal role that Wnt signaling plays in embryonic development and the maintenance of multiple stem cell populations, it is not surprising that aberrations in Wnt/β-catenin signaling also play a pivotal role in malignant transformation. The Wnt pathway was first implicated in mouse mammary tumor virus-induced murine breast cancer [99]. In human tumor development, β-catenin activation was first described in colon cancer, where the mutation of APC results in an inactivation of the degradation complex and constitutive Wnt/β-catenin pathway activation [100]. In addition to APC mutation, activation of the Wnt signaling pathway in cancer has been observed through an overexpression of Wnt pathway components, including Wnt ligands, Frizzled receptors and DVL family members or elevated levels of nuclear β-catenin [101]. The latter may result from changes in isoform splicing that alter GSK3β expression [102] or epigenetic silencing of Wnt inhibitors such as SFRP [103] and WIF [101]. In addition to colorectal cancer, altered Wnt/β-catenin signaling has also been described in gastric, esophageal, breast, liver, head and neck, endometrial, pancreatic, ovarian, prostate, lung, skin and hematopoietic cancers, as well as in Wilms' tumor and subtypes of medulloblastoma that arise from ventricular zone stem or progenitor cells [101, 104].

Aberrant Wnt signaling can be observed in the CSC population of many of these tumors. In the intestine, the activation of Wnt signaling by *Apc* deletion in *Lgr5*-positive intestinal stem cells but not in transit-amplifying progenitor cells drives rapid adenoma formation [105]. Overexpression of the Wnt family members results in an expansion of the mammary stem cell pool and an increased cancer susceptibility [106, 107]. Increased β-catenin expression has been observed in prostate CSCs [108] and CSC-like cells from established prostate cancer cell lines [109]. Exogenous Wnt exposure increased the self-renewal of prostate CSCs as well as the percentage of cells that express the CSC markers, CD44 and CD133 [110]. β-catenin signaling is also critical in a murine model of squamous cell carcinoma [111]. In this model, tumors are sustained by $CD34^+$ CSCs with characteristics of bulge stem cells and high levels of nuclear active β-catenin. When β-catenin is ablated in these cells, the tumors lose the $CD34^+$ CSC population, regress and are no longer able to generate secondary tumors, which suggests that β-catenin signaling is essential for the maintenance of cutaneous CSCs. Finally, β-catenin has been shown to play an important role in liquid tumors as might be predicted from its function in normal hematopoietic progenitors. β-catenin was shown to be necessary for the self-renewal of normal HSC as well as CSCs in a mouse model of CML [112] where the CSC is enriched in the $CD34^+CD38^+Lin^-$ granulocyte macrophage progenitor (GMP) population [9, 102]. Although β-catenin activation is usually observed in the more rare HSC population, a marked increase in activated β-catenin is observed in the GMP cell population from blast-crisis patients [102]. Likewise, in murine models of AML, the Wnt/β-catenin signaling pathway is required for the self-renewal of LSCs derived from either HSC or GMP cell populations [113].

The Wnt signaling pathway is clearly important for the self-renewal and maintenance of both normal and CSCs as discussed above. However, the clinical targeting of this pathway may pose many challenges in large part due to the complexity of the signaling pathway. Currently, at least 19 Wnt isoforms have been identified, as well as a multitude of receptors, coreceptors, and proteins that modulate Wnt signals. In addition to redundancy in this system, there is also the potential for toxicity in normal stem cells that rely on Wnt signaling for their survival and self-renewal. Despite these challenges, there are a number of steps that could be targeted to develop effective Wnt pathway inhibitors. At the ligand level, inhibitors of Porcupine (PPN) would limit Wnt ligand production. The palmitolyation of nascent Wnt molecules by PPN [114] is required for efficient transport from the endoplasmic reticulum and secretion as well as for the binding of Wnt molecules to their receptors [115, 116]. Likewise, monoclonal antibodies targeting Wnt ligands block the Wnt signaling pathway by preventing ligand–receptor interaction and induce apoptosis in cells with APC as well as β-catenin mutations [117, 118]. Inhibition of the Frizzled/LRP5/6/Wnt receptor complex itself could be achieved using a DKK-like molecule that would bind LRP5/6 and block Wnt signaling by preventing Frizzled/LRP5/6 association [119]. Alternatively, monoclonal antibodies or small peptides targeting Frizzled [120–122] or LRP5/6 [123] could block Wnt signaling at the receptor level. Phase I clinical trials of OMP-18R5, a monoclonal antibody targeting Frizzled 7, have been initiated in solid tumors. Strategies that modulate β-catenin levels within the cytoplasm could block or blunt Wnt signaling. For example, Tankyrase 1 and 2 promote Axin2 degradation, which results in a loss of the GSK3β/Axin/APC degradation complex and an accumulation of β-catenin protein. The inhibition of Tankyrase 1 and 2 with a small molecule stabilized Axin2 and promoted β-catenin degradation in cell lines expressing either mutant or wild-type APC [124]. Small molecules that directly bind and stabilize Axin may have a similar function [125]. Within the nucleus, β-catenin transcriptional activity could be inhibited by preventing formation of the β-catenin/LEF/TCF complex [126] or by the inhibition of β-catenin coactivators such as CBP [127]. Finally, Wnt pathway target genes could provide new therapeutic targets. Recent studies have focused on identifying Wnt target genes that are highly overexpressed in cancer cells with aberrant Wnt pathway activation [128]. As additional CSC markers are identified and CSC populations are more highly purified, it may be possible to identify novel Wnt pathway components or Wnt-induced proteins that are required for CSC self-renewal and survival.

Unfortunately, the majority of reported Wnt pathway inhibitors are still at the preclinical development stage due to an inadequate understanding of their mechanisms of action as well as a lack of specificity. Furthermore, only limited evidence is available that these Wnt pathway inhibitors can eradicate CSCs without also depleting the normal hematopoietic and GI stem cells that are dependent on Wnt signals. The only approved therapeutics with demonstrated activity against the Wnt/β-catenin family are non-steroidal anti-inflammatory drugs (NSAIDs) such as aspirin, indomethacin and sulindrac [129–131] and second-generation cyclooxygenase 2 (COX2) inhibitors such as celecoxib [132]. In colon cancer, NSAID treatment decreases the level of nuclear active β-catenin and can significantly reduce polyp formation in familial adenomatous polyposis (FAP) [133, 134]. Recent studies with an animal model of FAP have demonstrated that NSAID treatment induced apoptosis selectively in CSCs [135], thus supporting the continued development of novel Wnt inhibitors as viable therapeutic options.

3.2 Hedgehog

The HH signaling pathway was originally described in *Drosophila* as a mediator of body patterning during development [136] and was subsequently shown to regulate cellular proliferation, migration, and differentiation in a finely regulated spatial, temporal, and concentration-dependent manner. The mammalian HH family is composed of three members: Sonic Hedgehog (SHH); Indian Hedgehog (IHH); and Desert Hedgehog (DHH). The HH proteins, which can act as either paracrine or autocrine factors, are secreted proteins that initiate signaling by binding to the Patched (PTCH) cell-surface receptor (Fig. 2). Binding of HH proteins to PTCH relieves the PTCH-dependent inhibition of the seven-transmembrane protein, Smoothened (SMO), and triggers nuclear translocation of the GLI family of transcriptional regulators [137]. GLI1 acts as a positive effector of HH signaling by inducing the transcription of target genes including the cyclin D1 (*CCND1*) and *c-MYC* genes which are involved in cell cycle progression as well *GLI1* and *PTCH* which permit autocrine signaling [138]. GLI2 can act as either a positive or negative regulator of transcription, depending on post-transcriptional and post-translational modifications, while GLI3 acts as negative transcriptional regulator. The extent and duration of HH signaling is regulated by ubiquitin-dependent proteolytic processing of GLI proteins [139].

While HH proteins function primarily as a morphogens during development, HH signaling in adult cells predominantly drives proliferation and is critical for the maintenance of several adult stem cell populations, including hematopoietic [140–142], neural [143, 144] and skin and hair follicle stem cells [145, 146]. Aberrant HH signaling has been implicated in the development of several human tumors, and the HH pathway can be activated at several key points. Loss of function germline mutations in *PTCH* result in constitutive HH-independent pathway activation and are found in Gorlin syndrome, a disorder characterized by craniofacial and limb defects and a predisposition to cancers of the CNS, skin and skeletal muscle [147]. GLI1, the intracellular regulator of HH signaling, was originally identified in gliomas where HH signaling is driven by GLI1 overexpression [148]. An acquired mutation of *SMO* or *PTCH* or silencing of *PTCH* through methylation has also been identified in a significant percentage of sporadic basal cell carcinomas and medulloblastomas [149]. Furthermore, aberrant activation of the pathway through an increased expression of the SMO or HH ligand has been described in liquid tumors such as CML and MM as well as in solid tumors including esophageal, breast, lung, prostate, stomach, colon and pancreatic cancers [149].

In multiple tumor types, HH pathway signaling directly affects CSC survival and proliferation. Aberrations in several components of the HH pathway have been observed in glioma CSC. In addition to an overexpression of *GLI1* and *PTCH1* [150], SHH was found to be highly expressed in $CD133^+$ CSC as compared to $CD133^-$ cells [59, 151]. Inhibition of the HH pathway downregulated expression of *GLI1* and *PTCH1* as well as the pluripotency genes, *NANOG*, *OCT4* and *SOX2*, solely in $CD133^+$ cells [151], reduced the percentage of $CD133^-$ and nestin-expressing CSCs *in vitro* and *in vivo* [59], blocked CSC

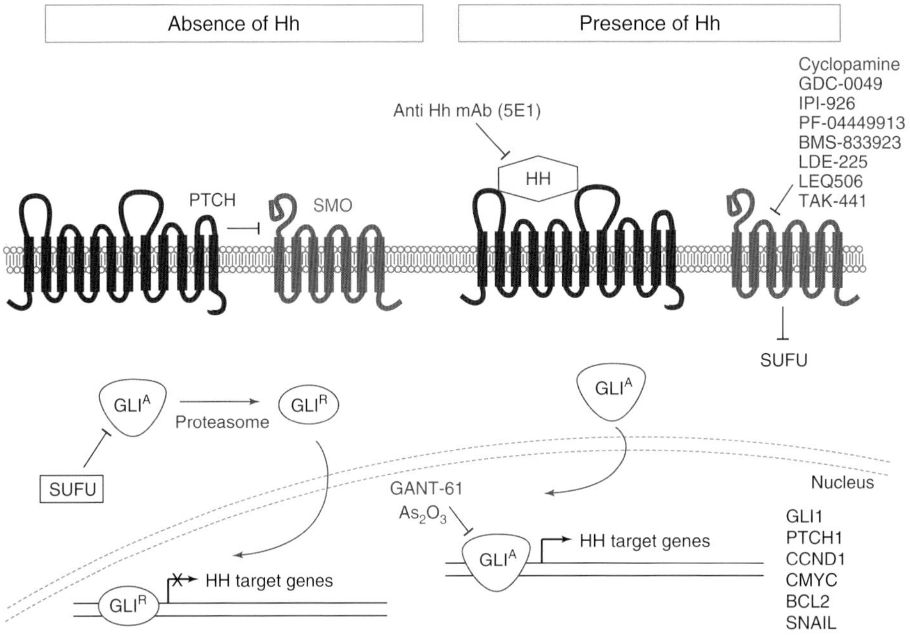

Fig. 2 Schematic of Hedgehog (HH) signaling pathways. In the absence of HH (left), PTCH inhibits signaling from the seven-transmembrane protein, Smoothened (SMO). The GLI family of transcription factors are proteolytically cleaved, allowing them to act as transcriptional repressors. HH binding to PTCH relieves SMO inhibition and triggers nuclear translocation of transcriptionally active forms of GLI. Inhibitors of various pathways components are indicated in red. (Adapted from Ref. [139].)

self-renewal (as assayed by *in vitro* sphere formation) [16, 59, 151] and blocked CSC engraftment and tumor growth in xenograft models [59, 151]. In colorectal cancer, HH signaling is increased within tumors as the disease progresses, with increased GLI1 levels observed in $CD133^+$ CSC from patients with advanced or metastatic disease [152]. The inhibition of SMO activity by genetic or pharmacological means abrogated growth of xenografted primary human colon tumors. Importantly, the continued administration of SMO inhibitors after initial tumor regression resulted in complete tumor remission in mice which suggested that HH inhibition efficiently targets the CSC required for disease relapse in this model [152]. HH activation has been reported in prostate CSC [153]. While there have been some conflicting observations [154], a constitutive activation of the HH signaling pathway has been shown to transform prostate progenitor cells which consequently become metastatic [155]. In pancreatic cancer, genetic alterations of the HH pathway components have been identified in all tested pancreatic cancer samples [156]. Pancreatic CSC, as defined by the $CD24^+/CD44^+/ESA^+$ cell subpopulation [26], overexpress *SHH* and demonstrate increased HH pathway activity [26, 157]. Interestingly, while pharmacological inhibition of the HH pathway in a xenograft model of pancreatic cancer did not reduce the growth of the primary xenograft tumor, it dramatically decreased the number of CSC and resulted in a decreased metastatic activity [158, 159]. Furthermore, a combination of HH pathway inhibition with gemcitabine eliminated the $CD133^+/CXCR4^+$ migrating CSC population that drives metastatic spread in pancreatic cancer [159].

HH activation has also been described in liquid tumor CSCs. MM is characterized by abnormal proliferation of $CD138^+$ plasma cells, and is maintained by $CD19^+CD138^-$ CSC [11]. The expression of both *SMO* and *GLI1* was upregulated in the $CD19^+CD138^-$ cell population isolated from either established cell lines or primary patient samples, and this resulted in HH pathway activation and an expansion of the MM CSC population [160]. A pharmacologic or genetic blockade of HH signaling decreased the number of MM CSCs as well as their clonogenic capacity. Aberrant activation of the HH pathway has also been examined in CML stem cells [161–163]. Microarray studies have revealed that the overexpression of HH pathway components correlates with CD34 expression in CML and suggests that the pathway is preferentially activated in CML CSCs. Additionally, HH pathway activation appears to increase as CML progresses from chronic phase to blast crisis [162]. In mouse models of CML, the HH pathway can be activated in CML CSCs by an overexpression of SMO resulting in an expansion of CML CSCs and accelerated CML progression. A loss of SMO expression by knockout or pharmacological inhibition significantly reduced the percentage of CML CSCs and delayed the development of CML [161, 163].

While the HH pathway can become activated at several points, most drugs targeting the HH pathway are directed against the SMO transmembrane protein. The first identified inhibitor of the HH pathway was cyclopamine [164, 165], a corn lily-derived alkaloid teratogen that causes cyclopenia, phenocopying the embryonic defects observed in *Shh* mutant embryos [166, 167]. Cyclopamine directly binds SMO to efficiently block the HH signaling pathway and decrease CSC numbers in preclinical models of several tumors. However, the clinical use of cyclopamine has been limited by its low oral bioavailability and poor pharmacokinetics [168]. The first targeted HH pathway antagonist to be clinically approved was vismodegib (GDC-0049), a small-molecule inhibitor of SMO. In Phase I trials, vismodegib shrank tumors or healed visible lesions in patients with locally advanced basal cell carcinoma or with metastatic disease [169, 170] and was recently approved by the FDA for the treatment of basal cell carcinoma following positive Phase II trial results. Several alternative SMO-targeted small molecules are currently being clinically evaluated, including PF-04449913, BMS-833923 (XL139), IPI-926, LDE-225, LEQ506 and TAK-441 (*www.clinicaltrials.gov*). Additional SMO-targeted therapies are presently under development, including cyclopamine-related alkaloids [171, 172] and itraconazole and its derivatives, which target SMO through a mechanism distinct from that of cyclopamine and vismodegib [173, 174].

The HH signaling pathway can also be targeted via its upstream ligands or its downstream effectors, the GLI family of transcriptional regulators. Antibodies targeting SHH have been shown to inhibit the clonogenic potential of CSCs [160]. Likewise, targeting the downstream GLI proteins genetically has been shown to inhibit CSC proliferation and tumor growth [152, 175]. Small-molecule inhibitors of GLI DNA binding have been described that include GANT-61 [176] and arsenic trioxide [173, 177]. These molecules may be especially important for inhibiting CSC self-renewal in tumors with mutations in signaling components downstream of SMO, or which have acquired resistance to SMO-targeted inhibitors.

While the preliminary clinical results with SMO-targeted inhibitors have been promising, there have been some notable failures. One of the early SMO inhibitors, Cur-61414, had shown great promise as a topical agent in preclinical mouse models of basal cell carcinoma but this did not translate to success in human patients [178]. Likewise, saridegib (IPI-926) had been tested against metastatic pancreatic cancers but the Phase II trial was recently halted due to a lack of efficacy. These clinical failures highlight the need to target multiple points in the HH signaling pathway and to effectively counter any rapidly developing resistance to SMO antagonists [179].

3.3 Notch

Notch signaling was first described in *Drosophila* where loss of function of the *Notch* gene resulted in notches at the wing margin [180]. Notch encodes a large single-pass transmembrane receptor [181] that serves to transduce signals between adjacent cells [182]. In mammals, the Notch family is composed of four members (Notch 1–4), the signaling of which is regulated through a series of precisely controlled proteolytic steps. Upon binding of ligand (Delta-like-1, -3 and -4 or Jagged-1 and -2) to the Notch receptor on an adjacent cell, the extracellular domain of Notch is cleaved by ADAM/TACE metalloproteases [183] (Fig. 3). This ligand-dependent cleavage product is further processed by the intramembrane γ-secretase/presenilin complex that results in a cytoplasmic release of the Notch intracellular domain (NICD) [184]. The NICD translocates to the nucleus where it associates with the transcription factor, CBF1/CSL, and coactivators of the Mastermind-like family resulting in an increased transcription of Notch target genes including Notch family members, *HES1* and cell-cycle regulators such as *MYC*, Cyclin D1, *CDK2* and *CDKN1A* ($p21^{CIP1/WAF1}$). Notch pathway activity can be modulated at multiple levels including: (i) ligand activation; (ii) receptor activation; (iii) receptor modification by the Fringe family (Lunatic, Manic, Radical); (iv) proteolysis of the Notch receptors; and (v) ubiquitin-mediated degradation of Notch which determines the intensity and duration of the Notch signal [185].

Noncanonical Notch signaling has also been documented but remains less characterized. There are at least three types of noncanonical Notch signaling: (i) NICD-dependent and CBF1/CSL-independent, that involves Notch ligand-dependent cleavage and translocation of activation signals independent of CBF1/CSL; (ii) NICD- and CBF1/CSL-independent, that involves the activation of Notch target genes independent of Notch cleavage; (iii) NICD-independent and CBF1/CSL-dependent gene activation without Notch receptor cleavage and NICD release [186]. Several signaling pathways have been implicated in noncanonical Notch signaling including HH, JAK/STAT, RTK, TGF, WNT, PI3K/AKT, mTOR/AKT, JNK, MEK/ERK and NFκ-B, thus representing an important point of crosstalk between other intracellular signaling pathways [186, 187].

The Notch signaling pathway is conserved from nematodes to humans and is pivotally involved in cell survival, proliferation, and differentiation during development. One

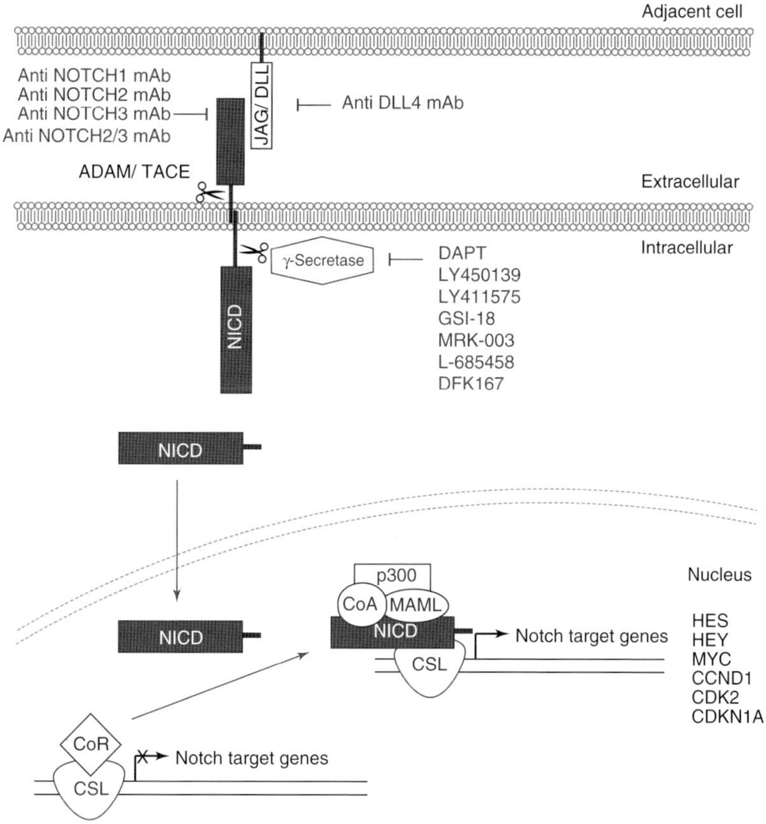

Fig. 3 Schematic of NOTCH signaling pathways. Notch signaling is regulated through a series of precisely controlled proteolytic steps. Upon binding of ligand (JAG/DLL) to the Notch receptor on an adjacent cell, the extracellular domain of Notch is cleaved by ADAM/TACE metalloproteases. This cleavage product is further processed by the intramembrane γ-secretase/presenilin complex, which results in a cytoplasmic release of the NOTCH intracellular domain (NICD). The NICD translocates to the nucleus where it associates with the transcription factor, CBF1/CSL, and coactivators of the Mastermind-like family, resulting in the increased transcription of Notch target genes. Inhibitors of various pathways components are indicated in red. (Adapted from Ref. [452].)

of the most well-studied Notch roles is in the switch from symmetric to asymmetric cell division during neural development [188]. Notch has been shown to inhibit neural stem cell differentiation and to promote their survival and proliferation by maintaining symmetric cell divisions [189]. In adult rodent brains damaged by ischemic injury, transient Notch signaling by intracranial administration of the Notch ligand, Delta-like 4 (Dll4), increased the number of newly generated neural precursor cells and promoted motor skill recovery [190]. Depending on the cell and developmental context, Notch signaling can also drive cell fate decisions, as observed in neural crest stem cells where transient Notch activation causes neuroblasts to adopt a glial fate [191]. In addition to its well-described role in neural development, Notch receptors and ligands are widely expressed in the hematopoietic system. During embryonic development, Notch signaling is required for the development of the definitive HSC [192, 193]. Notch signals are also essential for T-cell fate determination and maturation [194, 195]. The role of Notch signaling in the adult hematopoietic system is less clear. While exogenous Notch signaling can induce the proliferation of hematopoietic progenitors as demonstrated in gain-of-function experiments [195, 196], loss-of-function studies have suggested that Notch signaling is not required for maintenance of the adult HSC under homeostatic conditions [197, 198]. Notch signaling also plays an important role in mammary stem cell self-renewal through the regulation of asymmetric cell division [199]. In addition, Notch signal transduction has been implicated in stem and progenitor cell maintenance in the intestine [200, 201], lung [202], liver [203], prostate [204], skeletal muscle [205] and hair follicle [206].

Given the pivotal function of Notch signaling in multiple tissues, it is not surprising that aberrant Notch signaling has been described in a large number of tumors. The first evidence of a role for Notch signaling in cancer was described in acute T-cell lymphoblastic leukemia (T-ALL), where a translocation event fusing the T-cell receptor β chain to *NOTCH1* resulted in constitutive expression of the NICD [207]. Additionally, more than 50% of T-ALL patients have activating mutations in the NOTCH receptor [208]. Aberrations in Notch signaling have been described in other hematopoietic cancers [209–216] as well as solid tumors including breast [217], neural [218], kidney [219], prostate [220], pancreas [221], liver [222], cervix [223], lung [224], salivary gland [225], ovary [226] and skin [227].

In many of these tumors, there is evidence that Notch signaling is active in the CSC population and promotes CSC self-renewal and therapeutic resistance [60, 228, 229]. Importantly, it appears that the inhibition of Notch signaling can selectively destroy CSCs. While the phenotypic identity of the T-ALL CSCs is still being rigorously defined [230, 231], there is ample evidence that the *in vitro* self-renewal of the T-ALL CSCs is dependent on Notch activation [230–232] and that the inhibition of Notch signaling reduces and in certain cases, eliminates CSC activity in T-ALL models [232, 233]. In breast cancer models, an inhibition of Notch signaling decreased the sphere formation capacity of cells isolated from either primary human ductal carcinoma *in situ* samples or from ERBB2-driven mouse tumors, which suggested that Notch activity is necessary for the self-renewal of breast CSCs [234, 235]. Additional studies have focused on the $ESA^+/CD44^+/CD24^{low}$ breast CSCs and shown that NOTCH4 is overexpressed in this cell population. Both, genetic and/or pharmacologic inhibition of Notch signaling in the $ESA^+/CD44^+/CD24^{low}$ CSC-enriched population reduced both the number of phenotypic CSCs and tumor proliferation *in vivo* [236]. In glioblastoma, the $CD133^+$ CSCs exhibited an increased Notch activity as evidenced by overexpression of the Notch target gene, *HES1* [237]. Knockdown of Notch activity by either genetic or pharmacologic means decreased the expression of stem cell markers, nestin and CD133 and reduced the sphere formation capacity [238]. In addition, Notch inhibition in primary human glioblastoma samples reduced xenograft tumor growth and diminished CSC self-renewal by promoting the cell cycle exit of CSCs and differentiation [237, 239]. Intriguing work suggests that Notch inhibition can also block the capacity of CSCs to differentiate into endothelial cells that comprise the tumor vasculature [240]. In this manner, Notch inhibition may be doubly effective by targeting both the CSCs and the tumor vasculature that it produces [241, 242]. In medulloblastoma models, Notch inhibition induced the apoptosis of nestin-positive cells, which resulted in a depletion of the $CD133^+$ CSC population, reduced CSC self-renewal *in vitro* and reduced tumor growth [243]. Recent studies with colon cancer has shown that the Notch pathway is activated in the colon CSCs in addition to aberrant Wnt activation [244, 245] and that the inhibition of Notch signaling reduced xenograft tumor growth as well as the percentage of phenotypically defined ($ESA^+CD44^+CD166^+$) and functionally defined colon CSCs in those tumors [244]. Finally, emerging evidence has suggested that Notch signaling components are overexpressed in putative lung, pancreatic, prostate and melanoma CSC populations and are required for their self-renewal and tumor-initiation capacity [24, 246–251].

Because the self-renewal of numerous CSC populations requires Notch pathway activation, there is considerable interest in developing Notch-targeted therapies. Notch pathway inhibitors could be targeted at several steps. First it should be possible to disrupt or modulate the interaction between Notch and its ligands. Second, Notch signaling depends on proteolytic cleavage that could be inhibited. Finally, the interruption of protein–protein interactions that take place in the nucleus following NICD translocation could also inhibit Notch signaling. Inhibitors of the first two steps are currently the most clinically advanced strategies.

Activation of the Notch signaling pathway depends on two proteases: TACE/ADAM-10 and γ-secretase. While ADAM-10 inhibitors are still being sought, γ-secretase inhibitors (GSIs) are a clinical reality. Due to the role of γ-secretase complex in amyloid precursor protein (APP) cleavage in the brains of Alzheimer disease patients, GSIs have been developed as potential therapies for Alzheimer disease. The first-generation GSIs fell into several classes including those that mimicked the enzymatic transition state (DFK167, L-685458), helical peptides and dipeptide analogs including DAPT and LY450139 [252]. Preclinical studies in leukemia, glioblastoma, medulloblastoma and breast cancer models with DAPT [234, 236, 238, 253], GSI-18 [237, 243] or MRK-003 [235, 254] have demonstrated a selective targeting of CSCs and reduced tumor sizes and recurrences after treatment. Similar results were not seen in prostate cancer, however, where DAPT treatment had no effect on prostate CSC self-renewal or tumor formation [255]. Unfortunately, clinical trials in Alzheimer patients quickly revealed that chronic γ-secretase inhibition had undesirable on-target effects including severe secretory diarrhea, GI bleeding and immunosuppression [201, 256] and this resulted in a discontinuation of the Phase III trial for LY411575. The γ-secretase intramembrane complex targets the proteolysis of multiple proteins, and it was shown that an inhibition of NOTCH1 and NOTCH2 cleavage in particular was responsible for the systemic toxicity seen with chronic GSI treatment [256–258]. Despite this negative outcome and concerns raised that chronic Notch inhibition could increase cancer risk [259], several clinical trials examining GSI therapy in Notch-dependent tumors are ongoing (*www.clinicaltrials.gov*). The adverse side effects of GSI therapy observed in previous cancer trials could be mitigated by altering the dosing schedules [260] or by coadministration with GI-protective agents [261]. Thus far, Phase I trial results in pediatric refractory CNS tumors [262] have reported manageable adverse event profiles but few positive anti-tumor responses. This lack of clinical success may be due to unforeseen cross-talk with other signaling pathways. Consequently, additional research into the interaction of Hh, Wnt and Notch pathways and especially the role of noncanonical Notch signaling in CSC is required.

An alternative approach to pan-Notch inhibition with GSI has been to effect a selective inhibition of individual Notch pathway components by using antagonist antibodies. These include antibodies targeting NOTCH1 [258, 263], NOTCH2 [258] and NOTCH3 [264] as well as antibodies that target both NOTCH2 and NOTCH3 (OMP-595R; OncoMed Pharmaceuticals). While these antibodies have shown efficacy against cancer cell lines, there is currently no information available regarding their effects on CSCs. In addition to Notch receptor-specific antibodies, antibodies targeting the Notch ligand DLL4 (OMP-21M18 and REGN421) have been developed. Although DLL4 has been implicated in tumor angiogenesis [265, 266], the result of recent studies have suggested that DLL4 might also have direct effects on CSCs, with DLL4 treatment having been shown previously to expand $CD133^+$ and $nestin^+$ medulloblastoma CSCs [267]. The antibody blockade of DLL4 reduced the number of colon CSCs observed in xenograft tumors, as well as the *in vitro* sphere growth and *in vivo* tumor growth of cells isolated from treated tumors [244]. Currently, several Phase I trials are under way to explore the safety of targeting DLL4 in humans.

4 Additional Cancer Stem Cell Targets

Ancillary to the signaling pathways discussed above, signaling pathways and mechanisms that have been implicated in CSC survival or self-renewal include the NF-κB, PI3K/AKT/mTOR, BMI1 [268] and bone morphogenetic protein (BMP [269]) pathways, as well as alterations in telomerase expression, multidrug resistance membrane transporter expression [43], epigenetic modification [270, 271] and response to DNA damage or oxidative stress [58, 272]. Several pathways for which pharmacological intervention has been demonstrated will be discussed in the following subsections.

4.1 NF-κB

NF-κB signaling is associated primarily with regulating the immune response to inflammatory stimuli but also plays an important role in cell differentiation. Although it was originally identified as a B-cell specific transcription factor [273], it is in fact expressed ubiquitously. NF-κB signaling is mediated by dimers of the NF-κB family of transcription factors (RelA/p65, c-Rel, RelB, NF-κB1, NF-κB2) that are sequestered in the cytoplasm by inhibitors of NF-κB (IκB). Canonical NF-κB signaling is triggered by viral or microbial infection, proinflammatory cytokines or T-cell receptor engagement, resulting in an activation of the IκB kinase (IKK) complex. Activated IKK phosphorylates IκB proteins and targets them for ubiquitin-mediated degradation. NF-κB dimers are then able to translocate to the nucleus and activate transcription. Noncanonical NF-κB signaling is initiated through TNF-receptor superfamily members. Ligand binding stabilizes the NF-κB inducing kinase (NIK) which results in phosphorylation and ubiquitin-dependent proteolytic processing of the p100 NF-κB2 precursor protein into the mature p52 subunit. Subsequent nuclear translocation of p52/RelB dimers activates the transcription of a unique set of genes. There is considerable crosstalk between the canonical and noncanonical signaling pathways [274]; for example, activation of the canonical pathway induces transcription of p100 NF-κB2 as well as RelB. In addition to the integration of various stress stimuli responses, NF-κB transcription factors have been implicated in cancer initiation, proliferation and progression [275, 276]. Aberrant NF-κB activation has been described in liquid tumors including MM and various forms of leukemia and lymphomas as well as breast, skin, pancreas, prostate, ovarian, lung, colon and renal cancers [276].

There has been increasing evidence that certain CSC populations are dependent on activated NF-κB signaling for survival and self-renewal. Some of the first indications for this were derived from studies in patients with AML. The inhibition of NF-κB with a proteasome inhibitor [277] or with the naturally occurring sesquiterpene lactone, parthenolide [278], selectively induced apoptosis in $CD34^+$ AML CSCs *in vitro*. A parthenolide derivative, DMAPT, reduced the number of AML CSCs *in vivo* and also demonstrated cytotoxicity towards stem and progenitor cell populations isolated from B-ALL ($CD34^+CD10^-$) and blast crisis CML ($CD34^+CD38^-$) primary patient samples while having little effect on normal HSCs [279]. The self-renewal of AML progenitors cells could also be inhibited by AS602868, an IKK2 inhibitor [280]. Furthermore, NF-κB activity could be inhibited with anti-inflammatory agents such as dexamethasone in human AML patients with no observed adverse effects [281]. In addition to liquid tumors, CSC populations from solid tumors have also been shown to aberrantly activate NF-κB signaling and can be targeted by NF-κB-directed therapies. Parthenolide treatment selectively inhibited the *in vitro* self-renewal of neuroblastoma CSCs at doses that had little effect on normal stem cells [282]. The genetic inhibition of NF-κB signaling reduced breast CSC self-renewal *in vitro* and significantly decreased the percentage of $CD44^+CD24^{-/low}$ breast CSCs in murine breast cancer models [257], thus supporting continued research into the role of NF-κB signaling in CSCs.

While more than 750 modulators of NF-κB signaling have been described [283, 284], the vast majority of these compounds lack specificity. In addition to an anti-CSC effect, they are likely to present undesirable side effects due to their inhibition of normal NF-κB signaling in immune and inflammation processes, as well as the inhibition of other signaling cascades. To date, no specific NF-κB inhibitors have been approved for clinical use. Rather, NF-κB inhibition may be more effective as a short-term adjuvant therapy. As both chemotherapy and radiation treatment induce a NF-κB pro-survival response [285] and enrich for CSCs in several tumors, concomitant NF-κB pathway inhibition and chemotherapy may overcome therapy resistance in CSC populations. For example, the pharmacological inhibition of NF-κB signaling sensitized $CD44^+/CD24^-$ pancreatic CSCs to apoptotic stimuli but had little effect on normal fibroblasts [286, 287]. In an alternative approach, the NF-κB pathway is activated in glioblastoma CSCs upon exposure to differentiating cues [288]. However, rather than proceeding to a mature postmitotic phenotype as occurs during normal differentiation, the differentiating CSCs remain proliferative despite a decreased expression of stem cell markers and an increased expression of lineage-specific markers. Either genetic or small-molecule inhibition of NF-κB caused differentiating glioblastoma CSCs to enter a senescent state rather than to remain proliferative [288] and may provide a novel approach to the treatment of this devastating disease.

4.2 PI3K/AKT/mTOR

The PI3K family of lipid kinases are activated by a wide range of receptor tyrosine kinases, G-protein-coupled receptors and proto-oncogenes, and their phosphorylated lipid products regulate a network of signals controlling gene transcription, mRNA translation and subsequent cell survival, proliferation and differentiation [289]. PI3K activation results in an activation of AKT which promotes cell survival through an inhibition of MDM2 and BAD, enhances cell proliferation and metabolism through inhibition of GSK3 and the FOXO family of transcription factors and regulates protein synthesis and cell growth through activation of the mammalian target of rapamycin (mTOR) signaling cascade [290] (Fig. 4). This pathway is one of the most frequently mutated in cancer and mutations or amplifications of every major component of this signaling network have been identified [291].

The PI3K/AKT/mTOR pathway was initially implicated in CSC generation and survival in a leukemia model. Mice with a conditional deletion in *Pten*, the phosphatase that negatively regulates PI3K signaling, rapidly developed leukemias that could be serially transplanted. Leukemia-initiating cells were enriched within the $Flk2^{-}Sca1^{+}Lin^{-}c\text{-}Kit^{+}CD48^{-}$ stem cell population and could be depleted by mTOR inhibition [292]. Subsequent studies have identified an important role for this signal network in a variety of solid tumor CSCs. In studies of xenograft-maintained primary glioblastoma samples, AKT inhibition decreased the number of viable $CD133^{+}$ CSCs and reduced sphere formation *in vitro*. The treatment of glioblastoma cells prior to orthotopic injection significantly prolonged survival [293]. A dual inhibition of PI3K and mTOR in glioblastoma CSCs had similar effects [294]. $Nestin^{+}$ medulloblastoma CSCs located in the perivascular niche are protected from radiotherapy by activation of the PI3K/AKT pathway and AKT inhibition could sensitize these cells to radiation [295]. Neuroblastoma CSCs had an increased activity of the AKT/mTOR pathway relative to normal stem cells and pharmacologic or genetic inhibition of mTOR signaling decreased proliferation, sphere formation, and *in vivo* CSC survival [282]. Breast CSCs have also been shown to depend on PI3K/mTOR signaling for *in vitro* cell survival and proliferation. mTOR inhibition reduced the percentage of CSCs identified by side population staining, as well as the *in vitro* proliferation of side population cells and *in vivo* tumorigenicity, while a genetic knockdown of PTEN had the opposite effect [296]. $CD133^{+}CD44^{+}$ prostate CSCs identified in PC3 or Du145 prostate cancer cell lines or xenograft tumors could be depleted by dual PI3K/mTOR inhibition [297]. Interestingly, this study also identified PTEN-inactivating mutations in the CSC population of xenograft tumors that relapsed after treatment cessation thus highlighting potential mechanisms of therapy resistance. While inhibition of the PI3K/AKT/mTOR pathway has been shown to inhibit the self-renewal and survival of several CSC populations, mTOR inhibition

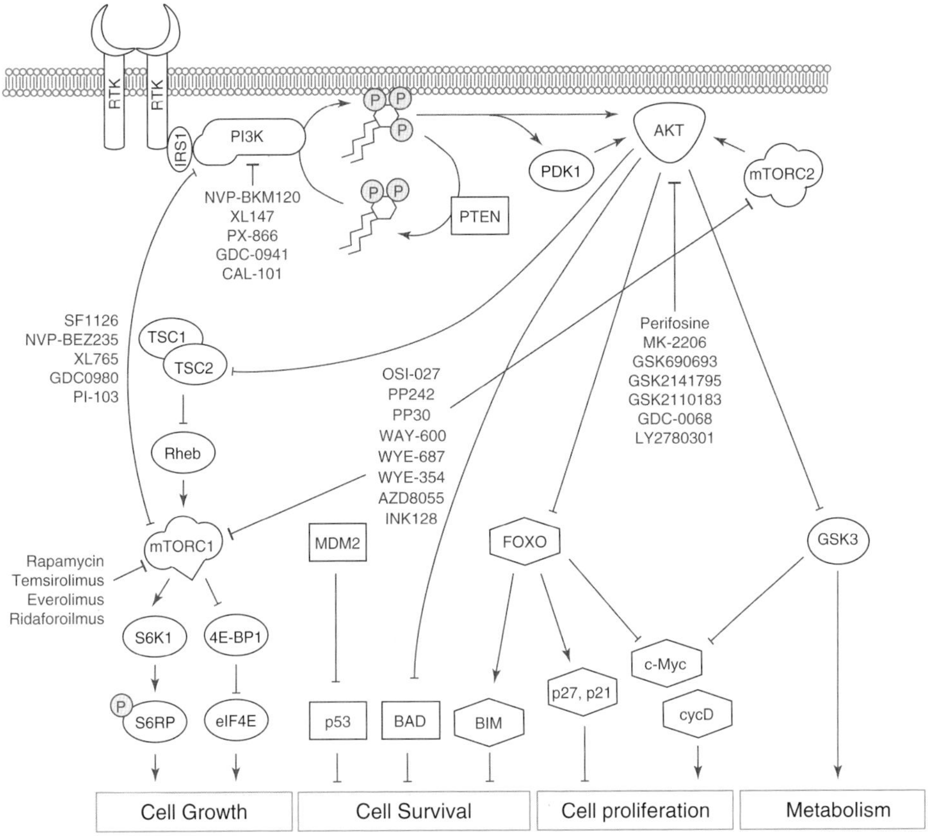

Fig. 4 Schematic of PI3K/AKT/mTOR pathways. Upon activation by a wide range of receptor tyrosine kinases, G-protein-coupled receptors and proto-oncogenes, PI3K phosphorylates the lipid (phosphatidylinositol-4,5-bisphosphate) to generate phosphatidylinositol-3,4,5-triphosphate, resulting in the activation of AKT. AKT regulates protein synthesis and cell growth through the mammalian target of rapamycin (mTOR) signaling cascade cell (left), promotes cell survival through inhibition of MDM2 and BAD (middle), and enhances cell proliferation and metabolism through inhibition of GSK3 and the FOXO family of transcription factors (right). Inhibitors of various pathways components are indicated in red. (Adapted from Ref. [306].)

was shown to increase the percentage of $CD133^+$ liver CSCs both *in vitro* and *in vivo* and to enhance the tumorigenicity of treated tumor cells [298]. This may result from activation of the AKT pathway that is observed with mTOR inhibition [299, 300] and emphasizes the difficulty in effectively targeting this complex signaling network.

There are multiple potential targets within the PI3K/AKT/mTOR pathway for cancer therapy. Unfortunately, there are also significant challenges due to cross-talk and feed-back loops regulating this signaling network. An early focus has been on mTOR-specific inhibitors. Rapamycin and rapamycin-related drugs, such as temsirolimus (CCI-779) and everolimus (RAD001), are allosteric inhibitors of the mTORC1 complex that have been shown to inhibit the self-renewal of several CSC populations [282, 292, 296] but the clinical use of these inhibitors as single agents has achieved limited success [301–303]. In order to target mTOR more effectively, novel ATP-competitive inhibitors have been developed that inhibit both mTORC1 and mTORC2 complexes [304]. Currently, PI3K inhibitors (both pan- and isoform-specific) are being developed and several have entered clinical trials [305, 306] despite little being known of their specific effects on CSCs. AKT inhibitors with efficacy against CSCs [295] are also being tested in clinical trials [305, 306]. Another promising approach has been the development of combination strategies. For example, dual PI3K/mTOR inhibitors have been shown to be effective against CSCs [294, 297] and several are currently undergoing Phase I clinical trials [306]. A combination of PI3K/AKT/mTOR inhibitors with stem cell pathway inhibitors and chemotherapy may also prove to be effective. Rapamycin in combination with a Shh inhibitor and gemcitabine significantly reduced the percentage of pancreatic CSCs from primary patient samples *in vitro* and also induced the regression of established tumors. Moreover, CSCs were undetectable in those tumors that remained after this combination treatment [159].

One final therapeutic candidate under examination is metformin. Metformin is the current front-line therapy for type II diabetes mellitus and extended exposure to metformin has been linked to an overall decrease in cancer incidence [307]. Metformin treatment blocks mTOR activation through an activation of AMP-activated protein kinase [308]. The treatment of established breast or pancreatic cancer cell lines with metformin reduced the percentage of CSCs and decreased *in vitro* self-renewal [309, 310]. Metformin treatment, in combination with chemotherapy, also prolonged survival in breast, prostate and lung cancer xenograft tumors with some models exhibiting durable tumor regression after the cessation of treatment [310, 311].

4.3
Telomerase

Telomeres are regions of repetitive nucleotide sequences [312] that protect the ends of chromosomes from being recognized as double-strand breaks by the DNA repair system. In the absence of telomerase activity, human telomere length is decreased by approximately 50 nucleotide with each cell division due to the inability of DNA polymerases to replicate the ends of linear chromosomes [313]. This gradual telomere shortening limits the proliferative capacity of normal somatic cells which enter a senescent state upon critical chromosome shortening [314, 315]. The vast majority of adult cells lack telomerase activity, with the exception of stem cells in which telomerase activity is tightly regulated [316]. The longest telomeres are found in stem cells of highly proliferative tissues such as the bone marrow, gut, and skin [317]. A careful regulation of telomerase in stem cells is critical for genomic integrity during self-renewal and also appears to be quite important as mutations in either the enzymatic (TERT) or RNA (TERC) component of the telomerase enzyme complex in dyskeratosis congenita patients result in the degenerative failure of multiple organs (including the bone marrow) due to stem cell depletion [318]. In contrast to the absent or low activity of telomerase observed in normal cells, up to 90% of all cancers reactivate telomerase activity. The diverse array of tumors with activated telomerase in combination with a low expression in normal tissue, makes telomerase inhibition an attractive therapeutic target.

The therapeutic value of targeting telomerase activity in CSCs is an area of active research. Increasingly, telomerase activity has been reported within the CSC population. In high-grade gliomas, $CD133^+$ CSCs showed elevated telomerase activity and longer telomeres than the $CD133^-$ cell population [319]. When telomerase activity was blocked in glioma CSCs, a decreased expression of CSC markers, CD133, nestin and CD15 was observed as well as an increased expression of cell differentiation markers and an increased sensitivity to chemotherapeutics [320, 321]. Likewise, $CD15^+$ CSCs from pediatric gliomas and tumor-initiating cells from neuroblastoma tumors exhibited higher telomerase activities than tissue-matched normal stem cells [322]. Telomerase inhibition blocked the self-renewal of these CSCs both *in vitro* and *in vivo* [322] and similar results have also been reported in CML [323], prostate cancer [324, 325], lung cancer [30, 326] and melanoma [327]. However, in other studies similar or even lower levels of telomerase activity were found in CSCs when compared to the non-CSC fraction. Despite this, the inhibition of telomerase activity depleted CSCs in models of MM [328], breast [329], prostate [321] and pancreatic cancers [329].

The major approaches to the therapeutic targeting of telomerase activity are immunotherapy and oligonucleotide-based hTERT inhibitors. Additional telomerase-targeted approache reviewed in Refs [204, 324]).

Immunotherapy approaches are predicated on the observation that telomerase-positive cancer cells display TERT peptide fragments on their cell surface. At present, ongoing trials are being conducted to examine both the direct vaccination and *ex vivo* activation of antigen-presenting cells or dendritic cells. To date, the results of Phase I/II vaccination trials with GV1001 have demonstrated good tolerability but minimal survival benefit [330–332]. Unfortunately none of these trials have examined CSC-specific effects.

The most widely studied oligonucleotide-based telomerase inhibitors are GRN163 and the more potent, lipid-attached derivative, GRN163L (imetelstat) [333]. The treatment of glioma, neuroblastoma, lung or prostate CSCs with imetelstat led to a reduction in the self-renewal capacity of these cells as assessed by *in-vitro* sphere formation. In addition, imetelstat proved to be effective against xenograft tumors and also reduced the number of CSCs remaining in treated tumors [321, 322, 326]. Similar results were observed in CSCs isolated from breast and pancreatic cancer cell lines [329]. Imetelstat is currently undergoing trials for the treatment of chronic lymphoproliferative disease, MM, breast cancer, non-small-cell lung cancer and other treatment-refractory solid tumors. Of note, [by using a telomeric recombination method known as alternative lengthening of telomeres (ALT) [334]], a small number of tumors have been reported to maintain telomere length. To date, reports of cancer cells that exploit this strategy have been limited, however, it remains a potential mechanism of resistance to be considered when applying telomerase-based therapies.

5
Cancer Stem Cell Niche

As discussed above, multiple signaling pathways affect CSC survival and self-renewal. In adult stem cell populations, these pathways are frequently coordinated within the stem cell niche. Niches are specialized microenvironments defined by cell–cell contacts, extracellular matrix adhesion and soluble factors as well as the physical constraints of the cellular milieu. The concept of a stem cell niche has been well explored in the hematopoietic system [335], skin [336], brain [337], GI tract [338] and reproductive tract [339]. The stem cell niche provides a microenvironment that precisely balances self-renewal and differentiation through a variety of regulatory factors, including factors secreted by the stromal microenvironment such as Hh, Wnt, BMP, transforming growth factor (TGF) and vascular endothelial growth factor (VEGF) and local cell–cell signaling through Notch [340]. Furthermore, the niche provides protection to stem cells either by a physical sequestration of stem cells away from differentiation or toxic signals, by enhanced survival signaling [341, 342] or by the maintenance of stem cells in a quiescent state [343]. Niches are dynamic systems that are responsive to both external signals such as inflammation or tissue injury as well as signals from the stem cells themselves.

A substantial number of reports have described the positive effect of tumor stromal environment on tumor growth and its negative impact on treatment success [344]. More recently, this discussion has focused on the concept of a CSC niche that provides similar environmental support as that described for normal adult stem cells [345]. While there has been much speculation about whether CSCs hijack the normal tissue stem cell niche or generate a *de novo* niche, the end result is similar, namely, CSC self-renewal and survival support in a privileged microenvironment. The identification of CSC niches has been complicated by the relative rarity of CSCs and the lack of definitive stem cell markers in many tumors. The most well-described CSC niches are perivascular and endosteal. Brain tumor CSCs, identified by CD133 and nestin expression, are located close to capillaries in a perivascular niche composed of fibroblasts, endothelial cells and inflammatory cells [346]. These niche components enhance the tumor-initiating capacity of brain tumor CSCs, as the cotransplantation of CSCs with endothelial cells accelerates tumor initiation and progression [346]. $CD133^+$ brain tumor CSCs promote angiogenesis via VEGF signaling [347] and appear to contribute directly to the formation of a CSC niche [241, 242]. Likewise, skin tumor CSCs are located in a perivascular niche and are strongly influenced by VEGF signaling. Disruption of VEGF signaling decreased the tumor vasculature but additionally disrupted a VEGF-Nrp1 autocrine loop. This resulted in a reduction of CSC numbers and their *in vivo* self-renewal [348]. In head and neck tumors, approximately 80% of CSCs are found in perivascular niches and endothelial-secreted factors increased their self-renewal while the ablation of endothelial cells decreased CSC numbers [349]. In leukemia, there is increasing evidence that CSCs home to normal osteoblastic and vascular HSC niches and displace normal HSCs as a consequence of their increased proclivity for these niches [350]. CSC niches have also been suggested at the tumor–stroma junction in squamous cell carcinoma [351] and bladder cancer [352]. Similar junctional CSC niches may be present in colorectal cancer where Wnt signaling in CSCs requires costimulation by the HGF (hepatocyte growth factor/scatter factor) produced by stromal fibroblasts [353] or those tumors which depend on paracrine Hh signaling from the tumor stromal microenvironment [161, 354–356]. Finally, tumor metastases may represent *de novo* CSC niches [357].

The resistance of CSCs to therapy may derive from their niche residence. Exposure to soluble factors such as cytokines [granulocyte-macrophage colony-stimulating factor (GM-CSF), interleukin (IL)-4, -6, and -8], angiogenic factors (VEGF, nitric oxide) and TGF-β present in the niche environment can influence the response to chemotherapy [65, 342, 346, 358–360]. Additionally, interactions with the stromal microenvironment can alter the survival pathways activated in CSCs. For example, coculture with stromal cells upregulated expression of the prosurvival proteins, BCL2 and BCL-X_L in $CD34^+$ leukemia CSCs thus diminishing apoptosis after exposure to chemotherapy [341, 361]. Finally, cellular quiescence induced by niche localization may also protect cells against radiation or chemotherapy [66]. In leukemias, CSCs engraft the endosteal niche where they are quiescent [63, 362, 363]. Treatment with GM-CSF induced leukemia CSCs to re-enter the cell cycle and also increased their sensitivity to chemotherapy [363].

Two predominant approaches have been identified for targeting the CSC niche. First, targeting of the adhesion molecules and signals that recruit and retain CSCs in the niche would displace CSCs from the protective microenvironment. Preclinical studies have identified the importance of a putative CSC marker, CD44, in regulating the interaction of CSCs with the bone marrow niche and the subsequent maintenance of their stem cell properties. The inhibition of CD44 with monoclonal antibodies resulted in a reduction of the AML or CML CSC burden in mice [364, 365]. Inhibition of CD44 or its ligands may be useful when targeting other $CD44^+$ CSC populations, such as colon, breast, pancreatic and prostate CSCs. The interaction of CXCR4 with its ligand, SDF-1 (CXCL12), is crucial for HSC homing to the bone marrow and the quiescence of HSCs [366, 367]. An inhibition of this interaction in leukemia models mobilized CSCs from the bone marrow niche and increased their sensitivity to chemotherapeutics [368, 369]. Moreover, integrin-mediated adhesion may be an important target, particularly for metastatic CSC niches [370]. Integrins have been used prospectively to identify CSCs in solid tumors [24, 371, 372] with recent evidence having suggested that the antagonism of integrins may reduce the number of CSCs as well as the tumor burden [371, 373]. Thus far, clinical trials of integrin antagonists have shown few objective responses, though none of these studies has specifically examined the CSC population [374]. A second strategy is aimed at targeting the signaling pathways initiated or supported by niche residence. Methods for targeting the Hh, Wnt and Notch pathways were discussed above but targeting the perivascular niche with anti-angiogenic agents has met with some success. Bevacizumab, which targets VEGF, depletes CSCs from gliomas and medulloblastomas [346] and could prove useful for targeting CSCs in other tumors [348]. A promising efficacy of anti-angiogenesis therapy has been observed in clinical trials for GBM patients with bevacizumab [375] or cediranib (AZD2171, a VEGFR inhibitor) [376]. Additionally, it should be possible to target aberrantly activated survival pathways. A number of BCL2 family inhibitors are currently in preclinical and clinical development [377] in addition to agents that impact both the intrinsic and extrinsic apoptosis pathways [378].

Quiescence regulation represents yet another potential target. While a significant amount of research is still required to fully elucidate the mechanism of stem cell quiescence within the niche and the parallel mechanisms utilized by CSCs, preliminary evidence has suggested that a mobilization of CSCs is possible and that re-awakened CSCs are more sensitive to chemotherapeutics [363]. Therefore, quiescence-targeted anti-CSC strategies may be more effective when combined with chemotherapies that target stem cell pathways or cell division. Alternatively, it may be possible to develop therapeutic agents that are active in nondividing, quiescent cells.

6
CSCs in Drug Discovery

The therapeutic targets discussed above were identified based on the hypothesis that CSCs are dependent on many of the same survival and self-renewal pathways that are utilized by both embryonic and adult stem cells. While this candidate approach has yielded a plentitude of targets, significant concerns remain regarding the potential for toxicity in normal stem cell populations required for tissue maintenance and repair after injury especially GI, hematopoietic, neural and skin stem cell populations. To that end, unbiased cell-based approaches have immense potential to identify both CSC-specific signaling pathways and compounds that target CSCs while sparing normal stem cells.

Conventionally, the aim would be to develop a cell-based, high-throughput screen to identify compounds that specifically affect CSC survival. However, this strategy has been complicated by the relative rarity of CSCs and the difficulty in extensively expanding CSCs *in vitro* while maintaining their stem cell-like properties [379]. In a limited number of tumors, it has been possible to extensively expand CSCs isolated from primary patient samples by propagation under carefully monitored culture conditions. Glioblastoma CSCs can be stably expanded in both adherent monolayer cultures and nonadherent sphere cultures under serum-free conditions [12, 380]. Adherent culture conditions were ideal for the development of a high-content image-based screen to detect cytotoxic or cytostatic small molecules targeting glioblastoma CSCs [380, 381]. Any identified compounds were counterscreened against normal fibroblasts or normal neural stem cells to identify CSC-selective compounds. These studies identified several novel compounds, including modulators of the monoamine-signaling pathway. Perhaps more importantly, they also highlighted interpatient variability and the need to confirm the effects of identified compounds in multiple patient-derived lines.

Cell-based, high-throughput screens utilizing nonadherent sphere cultures have also been developed to identify small molecules that selectively target glioblastoma [382] or neuroblastoma CSCs [282, 383]. To identify compounds that reduced the proliferation of glioblastoma CSCs in sphere conditions, Visnyei and colleagues screened CSCs from multiple patients. Counterscreens were conducted using immortalized fibroblasts, and any CSC-selective hits were furthered triaged by gene expression array. As with the experiments described above, these studies highlighted the intra-patient variability in glioblastoma CSCs and the need for an extensive validation of screening hits [382]. In order to identify novel neuroblastoma CSC-targeted therapeutics, CSCs from high-risk neuroblastoma patients were enriched under serum-free sphere growth conditions and screened against small-molecule libraries in parallel with normal pediatric neural crest-like stem cells. These screens identified several novel CSC-specific inhibitors of neuroblastoma CSC survival, including rapamycin, Polo-like kinase 1 and mitochondrial electron transport modulators that significantly inhibited both *in vitro* and *in vivo* CSC survival and self-renewal but had little effect on normal age- and tissue-matched stem cells [282, 383]. Importantly, the screening of CSCs maintained under sphere-forming culture conditions led to the identification of a distinct spectrum of potential therapeutic compounds from those identified when screening cell lines originally established under adherent conditions in the presence of serum [282]. It has long been known that cells in three-dimensional (3-D) culture can behave very differently from those in two-dimensional (2-D) culture [384]; this suggests that screening assays conducted on 3-D spheres may provide more biologically relevant therapeutic leads [385].

While high-throughput screens have been possible for CSCs isolated from glioblastoma or neuroblastoma patients, it has not been possible to isolate or stably expand sufficient quantities of CSCs from other tumors for *in-vitro* screening assays. To overcome this limitation, several groups have developed representative cell lines with CSC characteristics. In order to model breast CSCs, Gupta and colleagues utilized immortalized human mammary epithelial cells that had been genetically engineered to overexpress the Twist transcription factor or to repress E-cadherin expression through a short hairpin RNA-mediated inhibition of the *CDH1* gene. These modifications reproduced an epithelial-to-mesenchymal transition state that resulted in an enrichment of breast CSC-like cells. High-throughput screening was used to identify one compound, the ionophore salinomycin, which selectively targeted CSC-like cells in both model cell lines and at doses that had little effect on the non-CSC parental cell lines. Salinomycin significantly reduced the percentage of phenotypic $CD44^{+}CD24^{-/low}$ breast CSCs in several established breast cancer cell lines as well as *in-vitro* sphere formation and *in-vivo* tumor formation [386].

Human AML has been modeled using the TEX and M9-ENL cell lines in which normal lineage-negative hematopoietic stem/progenitor cells from human cord blood were transduced with retroviruses encoding the *TLS-ERG* or *MLL-ENL* oncogenic fusion genes. Škrtić and colleagues screened these cell lines against a library of approved drugs focused mainly on antimicrobials and metabolic regulators. Interestingly, two compounds, salinomycin (also identified by Gupta and colleagues) and tigecycline, were identified that targeted both AML CSC-like cell lines but had no significant effect on normal HSCs [387]. The same group subsequently showed that tigecycline had activity against multiple leukemia cell lines and that treatment reduced both the engraftment of primary AML patient samples in immunocompromised mice and the serial transplantation of treated leukemia cells. Tigecycline treatment inhibited mitochondrial translation and its the CSC-selective action may be due to the increased metabolic activity and dependence on mitochondrial function observed in AML CSCs. McDermott and colleagues screened a small-molecule collection with the same cell lines using a counterscreen with normal human cord blood stem and progenitor cells to identify CSC-selective compounds [388, 389]. In this way, two novel agents, ciclopirox olamine, an iron-chelator [388], and kinetin riboside [389], an adenosine derivative, were identified and characterized.

In addition to the high-throughput screens designed to identify novel chemical agents that target CSCs, assays can be developed to identify novel signaling pathways involved in CSC survival, self-renewal or differentiation. Screens conducted with large collections of novel small molecules that may not be suitable for clinical use offer the opportunity to identify pathways that could be more directly interrogated in subsequent assays. Alternatively, genomic screens could be developed to identify modulators of CSC function. For example, Wurdak and colleagues conducted a kinome-wide RNA interference screen to identify kinases that induced the differentiation of glioblastoma CSCs [390]. Upon examining CSCs from multiple glioblastoma patients, the inhibition of *TRRAP* expression provoked the strongest and most reproducible differentiation response. Although the functional role of TRRAP in glioblastoma CSCs remains to fully elucidated, its knockdown reduced the sphere-formation ability of CSCs, induced cell cycle arrest and neural differentiation *in vitro* and also inhibited tumor formation *in vivo* [390].

Cell-based screening strategies provide the means to identify novel pathways or CSC-selective therapeutic agents. However, there are a number of challenges that remain to be addressed. First, a major limiting factor is the ability to isolate sufficient quantities of CSCs from primary tumors or to expand CSC populations *in vitro* while maintaining CSC properties. Significant differences in the immunophenotype, copy number variation and frequency of CSCs between primary tumors and their corresponding xenografted tumors have been observed [391]. Thus, effective protocols for the large-scale isolation and *in vitro* expansion and maintenance of CSCs from primary tumor specimens will need to be identified and validated. Engineered CSC-like cell lines [386–389] may provide an alternative source of cell material for high-throughput screens but the biological and clinical relevance of these cell lines to endogenous CSCs remains to be demonstrated. Second, there is increasing evidence that CSCs are not a homogeneous subpopulation of any given tumor. In several cases, multiple phenotypically and genetically distinct CSCs have been identified that contribute to both short- and long-term tumor maintenance and metastases [25, 52, 53, 55]. The existence of unique pools of CSCs, as well as their inherent instability [57, 392] may require the combination of multiple anti-CSC therapies for effective treatment. Third, the majority of screens discussed above have identified CSC-selective agents based on a lack of cytotoxic activity against one or two normal stem or differentiated cell populations. Because multiple adult stem cell populations share similar survival, self-renewal and differentiation pathways, it will be necessary to determine the effects of identified compounds against a much broader range of adult stem cell populations. Finally, intra-patient variability in CSC phenotype and clinical response will require extensive validation of the screening hits against much larger samples than have been previously tested. The prediction of patient response to identified therapies would be enhanced by a large-scale interrogation of the genome and transcriptome of multiple CSC populations, as well as post-transcriptional and post-translational modifications specific to CSCs.

7 Concluding Remarks

Overall, the experiences with CSC-based high-throughput screens, as well as the candidate approaches discussed above, illustrate the requirements to identify CSC-selective agents. First, the cell population for *in-vitro* testing must reflect CSC biology and be sufficiently expandable and stable to allow for multiple rounds of testing, screening, and validation. Second, countertesting against normal stem cell populations must be conducted to eliminate those compounds with unacceptable cytotoxicity. Preferably, compounds should be tested against hematopoietic, intestinal, skin and neural stem cells. Third, the identified compounds must be tested against primary tumor tissues from multiple patients, using both *in vitro* and *in vivo* CSC functional assays including sphere formation, phenotypic marker expression and tumor-initiating ability. It is important to recognize that longer-term studies may be required to fully capture and characterize the effect of therapeutics on CSCs. Fourth, multiple signaling pathways contribute to CSC survival and proliferation, and many of these pathways are cross-regulated. Combination treatment strategies may be required for the effective targeting of CSC populations. Finally, appropriate clinical trials must be designed to assess the efficacy of CSC-targeted therapies in human patients. A traditional endpoint criterion for clinical trials has been a reduction in overall tumor burden. However, this approach may need to be revised to account for the unique role of CSCs within tumors. It is possible that the selective targeting of CSCs will not appreciably reduce tumor burden in the trial time frame, and therefore additional primary or secondary endpoints should be considered. These might include progression-free survival or time to progression, time to metastasis, or an assessment of CSC content or *in-vitro* activity [393]. It will also be necessary to validate the relationship between CSC content or functional assays and patient survival for use in future clinical trials.

References

1. Breasted, J.H. (2010) *The Edwin Smith Surgical Papyrus: Hieroglyphic Transliteration, Translation and Commentary*, Kessinger Publishing, LLC.
2. Howlader, N.N.A., Krapcho, M., Neyman, N., Aminou, R. *et al.* (Eds) (2011) *SEER Cancer Statistics Review, 1975–2008*, National Cancer Institute, Bethesda.
3. Mukherjee, S. (2010) *The Emperor of All Maladies: A Biography of Cancer*, Schribner, New York.
4. Bomken, S., Fiser, K., Heidenreich, O., and Vormoor, J. (2010) Understanding the cancer stem cell. *Br. J. Cancer*, **103**, 439–445.
5. Bonnet, D. and Dick, J.E. (1997) Human acute myeloid leukemia is organized as a hierarchy that originates from a primitive hematopoietic cell. *Nat. Med.*, **3**, 730–737.
6. Lapidot, T., Sirard, C., Vormoor, J., Murdoch, B. *et al.* (1994) A cell initiating human acute myeloid leukaemia after transplantation into SCID mice. *Nature*, **367**, 645–648.
7. Sehl, M., Zhou, H., Sinsheimer, J.S., and Lange, K.L. (2011) Extinction models for cancer stem cell therapy. *Math. Biosci.*, **234**, 132–146.
8. Clarke, M.F., Dick, J.E., Dirks, P.B., Eaves, C.J. *et al.* (2006) Cancer stem cells – perspectives on current status and future directions: AACR Workshop on cancer stem cells. *Cancer Res.*, **66**, 9339–9344.
9. Jamieson, C.H.M., Ailles, L.E., Dylla, S.J., Muijtjens, M. *et al.* (2004) Granulocyte-macrophage progenitors as candidate leukemic stem cells in blast-crisis CML. *N. Engl. J. Med.*, **351**, 657–667.
10. Castor, A., Nilsson, L., Astrand-Grundström, I., Buitenhuis, M. *et al.* (2005) Distinct patterns of hematopoietic stem cell involvement in acute lymphoblastic leukemia. *Nat. Med.*, **11**, 630–637.
11. Matsui, W., Huff, C.A., Wang, Q., Malehorn, M.T. *et al.* (2004)

Characterization of clonogenic multiple myeloma cells. *Blood*, **103**, 2332–2336.
12. Galli, R., Binda, E., Orfanelli, U., Cipelletti, B. *et al.* (2004) Isolation and characterization of tumorigenic, stem-like neural precursors from human glioblastoma. *Cancer Res.*, **64**, 7011–7021.
13. Hemmati, H.D., Nakano, I., Lazareff, J.A., Masterman-Smith, M. *et al.* (2003) Cancerous stem cells can arise from pediatric brain tumors. *Proc. Natl Acad. Sci. USA*, **100**, 15178–15183.
14. Singh, S.K., Clarke, I.D., Terasaki, M., Bonn, V.E. *et al.* (2003) Identification of a cancer stem cell in human brain tumors. *Cancer Res.*, **63**, 5821–5828.
15. Singh, S.K., Hawkins, C., Clarke, I.D., Squire, J.A. *et al.* (2004) Identification of human brain tumour initiating cells. *Nature*, **432**, 396–401.
16. Reynolds, B.A. and Rietze, R.L. (2005) Neural stem cells and neurospheres – re-evaluating the relationship. *Nat. Methods*, **2**, 333–336.
17. Reynolds, B.A. and Vescovi, A.L. (2009) Brain cancer stem cells: think twice before going flat. *Stem Cell*, **5**, 466–467.
18. Al-Hajj, M., Wicha, M.S., Benito- Hernandez, A., Morrison, S.J. *et al.* (2003) Prospective identification of tumorigenic breast cancer cells. *Proc. Natl Acad. Sci. USA*, **100**, 3983–3988.
19. Chu, P., Clanton, D.J., Snipas, T.S., Lee, J. *et al.* (2009) Characterization of a subpopulation of colon cancer cells with stem cell-like properties. *Int. J. Cancer*, **124**, 1312–1321.
20. Dalerba, P., Dylla, S.J., Park, I.-K., Liu, R. *et al.* (2007) Phenotypic characterization of human colorectal cancer stem cells. *Proc. Natl Acad. Sci. USA*, **104**, 10158–10163.
21. O'Brien, C.A., Pollett, A., Gallinger, S., and Dick, J.E. (2007) A human colon cancer cell capable of initiating tumour growth in immunodeficient mice. *Nature*, **445**, 106–110.
22. Ricci-Vitiani, L., Lombardi, D.G., Pilozzi, E., Biffoni, M. *et al.* (2006) Identification and expansion of human colon-cancer-initiating cells. *Nature*, **445**, 111–115.
23. Collins, A.T., Berry, P.A., Hyde, C., Stower, M.J. *et al.* (2005) Prospective identification of tumorigenic prostate cancer stem cells. *Cancer Res.*, **65**, 10946–10951.
24. Patrawala, L., Calhoun-Davis, T., Schneider-Broussard, R., and Tang, D.G. (2007) Hierarchical organization of prostate cancer cells in xenograft tumors: the $CD44^+$ 2 1^+ cell population is enriched in tumor-initiating cells. *Cancer Res.*, **67**, 6796–6805.
25. Hermann, P.C., Huber, S.L., Herrler, T., Aicher, A. *et al.* (2007) Distinct populations of cancer stem cells determine tumor growth and metastatic activity in human pancreatic cancer. *Cell Stem Cell*, **1**, 313–323.
26. Li, C., Heidt, D.G., Dalerba, P., Burant, C.F. *et al.* (2007) Identification of pancreatic cancer stem cells. *Cancer Res.*, **67**, 1030–1037.
27. Chen, Y.-C., Chen, Y.-W., Hsu, H.-S., Tseng, L.-M. *et al.* (2009) Aldehyde dehydrogenase 1 is a putative marker for cancer stem cells in head and neck squamous cancer. *Biochem. Biophys. Res. Commun.*, **385**, 307–313.
28. Prince, M.E., Sivanandan, R., Kaczorowski, A., Wolf, G.T. *et al.* (2007) Identification of a subpopulation of cells with cancer stem cell properties in head and neck squamous cell carcinoma. *Proc. Natl Acad. Sci. USA*, **104**, 973–978.
29. Eramo, A., Lotti, F., Sette, G., Pilozzi, E. *et al.* (2008) Identification and expansion of the tumorigenic lung cancer stem cell population. *Cell Death Differ.*, **15**, 504–514.
30. Ho, M.M., Ng, A.V., Lam, S., and Hung, J.Y. (2007) Side population in human lung cancer cell lines and tumors is enriched with stem-like cancer cells. *Cancer Res.*, **67**, 4827–4833.
31. Haraguchi, N., Ishii, H., Mimori, K., Tanaka, F. *et al.* (2010) CD13 is a therapeutic target in human liver cancer stem cells. *J. Clin. Invest.*, **120**, 3326–3339.
32. Yang, Z.F., Ho, D.W., Ng, M.N., Lau, C.K. *et al.* (2008) Significance of $CD90^+$ cancer stem cells in human liver cancer. *Cancer Cell*, **13**, 153–166.
33. Alvero, A.B., Chen, R., Fu, H.-H., Montagna, M. *et al.* (2009) Molecular phenotyping of human ovarian cancer stem cells unravels the mechanisms for repair and chemoresistance. *Cell Cycle*, **8**, 158–166.
34. Bapat, S.A., Mali, A.M., Koppikar, C.B., and Kurrey, N.K. (2005) Stem and progenitor-like cells contribute to the aggressive

behavior of human epithelial ovarian cancer. *Cancer Res.*, **65**, 3025–3029.
35. Curley, M.D., Therrien, V.A., Cummings, C.L., Sergent, P.A. *et al.* (2009) CD133 expression defines a tumor initiating cell population in primary human ovarian cancer. *Stem Cells*, **27**, 2875–2883.
36. Zhang, S., Balch, C., Chan, M.W., Lai, H.C. *et al.* (2008) Identification and characterization of ovarian cancer-initiating cells from primary human tumors. *Cancer Res.*, **68**, 4311–4320.
37. Fang, D., Nguyen, T.K., Leishear, K., Finko, R. *et al.* (2005) A tumorigenic subpopulation with stem cell properties in melanomas. *Cancer Res.*, **65**, 9328–9337.
38. Quintana, E., Shackleton, M., Sabel, M.S., Fullen, D.R. *et al.* (2008) Efficient tumour formation by single human melanoma cells. *Nature*, **456**, 593–598.
39. Schatton, T., Murphy, G.F., Frank, N.Y., Yamaura, K. *et al.* (2008) Identification of cells initiating human melanomas. *Nature*, **451**, 345–349.
40. Zöller, M. (2011) CD44: can a cancer-initiating cell profit from an abundantly expressed molecule? *Nat. Rev. Cancer*, **11**, 254–267.
41. Alison, M.R., Lim, S.M.L., and Nicholson, L.J. (2011) Cancer stem cells: problems for therapy. *J. Pathol.*, **223**, 147–161.
42. Goodell, M.A., Brose, K., Paradis, G., Conner, A.S. *et al.* (1996) Isolation and functional properties of murine hematopoietic stem cells that are replicating in vivo. *J. Exp. Med.*, **183**, 1797–1806.
43. Chen, K.G. and Sikic, B.I. (2012) Molecular pathways: regulation and therapeutic implications of multidrug resistance. *Clin. Cancer Res.*, **18**, 1863–1869.
44. Armstrong, L., Stojkovic, M., Dimmick, I., Ahmad, S. *et al.* (2004) Phenotypic characterization of murine primitive hematopoietic progenitor cells isolated on basis of aldehyde dehydrogenase activity. *Stem Cells*, **22**, 1142–1151.
45. Hess, D.A., Wirthlin, L., Craft, T.P., Herrbrich, P.E. *et al.* (2006) Selection based on CD133 and high aldehyde dehydrogenase activity isolates long-term reconstituting human hematopoietic stem cells. *Blood*, **107**, 2162–2169.
46. Boman, B.M. and Wicha, M.S. (2008) Cancer stem cells: a step toward the cure. *J. Clin. Oncol.*, **26**, 2795–2799.
47. Magee, J.A., Piskounova, E., and Morrison, S.J. (2012) Cancer stem cells: Impact, heterogeneity, and uncertainty. *Cancer Cell*, **21**, 283–296.
48. Prestegarden, L. and Enger, P.O. (2010) Cancer stem cells in the central nervous system – a critical review. *Cancer Res.*, **70**, 8255–8258.
49. Kelly, P.N., Dakic, A., Adams, J.M., Nutt, S.L. *et al.* (2007) Tumor growth need not be driven by rare cancer stem cells. *Science*, **317**, 337–337.
50. Singec, I., Knoth, R., Meyer, R.P., Maciaczyk, J. *et al.* (2006) Defining the actual sensitivity and specificity of the neurosphere assay in stem cell biology. *Nat. Methods*, **3**, 801–806.
51. Hope, K.J., Jin, L., and Dick, J.E. (2004) Acute myeloid leukemia originates from a hierarchy of leukemic stem cell classes that differ in self-renewal capacity. *Nat. Immunol.*, **5**, 738–743.
52. Dieter, S.M., Ball, C.R., Hoffmann, C.M., Nowrouzi, A. *et al.* (2011) Distinct types of tumor-initiating cells form human colon cancer tumors and metastases. *Cell Stem Cell*, **9**, 357–365.
53. Chen, R., Nishimura, M.C., Bumbaca, S.M., Kharbanda, S. *et al.* (2010) A hierarchy of self-renewing tumor-initiating cell types in glioblastoma. *Cancer Cell*, **17**, 362–375.
54. Dalerba, P., Cho, R.W., and Clarke, M.F. (2007) Cancer stem cells: models and concepts. *Annu. Rev. Med.*, **58**, 267–284.
55. Notta, F., Mullighan, C.G., Wang, J.C.Y., Poeppl, A. *et al.* (2011) Evolution of human BCR-ABL1 lymphoblastic leukaemia-initiating cells. *Nature*, **469**, 362–367.
56. Roesch, A., Fukunaga-Kalabis, M., Schmidt, E.C., Zabierowski, S.E. *et al.* (2010) A temporarily distinct subpopulation of slow-cycling melanoma cells is required for continuous tumor growth. *Cell*, **141**, 583–594.
57. Quintana, E., Shackleton, M., Foster, H.R., Fullen, D.R. *et al.* (2010) Phenotypic heterogeneity among tumorigenic melanoma cells from patients that is reversible and not hierarchically organized. *Cancer Cell*, **18**, 510–523.

58. Bao, S., Wu, Q., McLendon, R.E., Hao, Y. *et al.* (2006) Glioma stem cells promote radioresistance by preferential activation of the DNA damage response. *Nature*, **444**, 756–760.
59. Bar, E.E., Chaudhry, A., Lin, A., Fan, X. *et al.* (2007) Cyclopamine-mediated hedgehog pathway inhibition depletes stem-like cancer cells in glioblastoma. *Stem Cells*, **25**, 2524–2533.
60. Phillips, T.M., McBride, W.H., and Pajonk, F. (2006) The response of CD24(-/low)/CD44+ breast cancer-initiating cells to radiation. *J. Natl Cancer Inst.*, **98**, 1777–1785.
61. Li, H.-Z., Yi, T.-B., and Wu, Z.-Y. (2008) Suspension culture combined with chemotherapeutic agents for sorting of breast cancer stem cells. *BMC Cancer*, **8**, 135.
62. Costello, R.T., Mallet, F., Gaugler, B., Sainty, D. *et al.* (2000) Human acute myeloid leukemia CD34+/CD38- progenitor cells have decreased sensitivity to chemotherapy and Fas-induced apoptosis, reduced immunogenicity, and impaired dendritic cell transformation capacities. *Cancer Res.*, **60**, 4403–4411.
63. Ishikawa, F., Yoshida, S., Saito, Y., Hijikata, A. *et al.* (2007) Chemotherapy-resistant human AML stem cells home to and engraft within the bone-marrow endosteal region. *Nat. Biotechnol.*, **25**, 1315–1321.
64. Dylla, S.J., Park, I.-K., Lazetic, S., Clarke, M.F. *et al.* (2008) Colorectal cancer stem cells are enriched in xenogeneic tumors following chemotherapy. *PLoS ONE*, **3**, e2428.
65. Todaro, M., Alea, M.P., Di Stefano, A.B., Cammareri, P. *et al.* (2007) Colon cancer stem cells dictate tumor growth and resist cell death by production of interleukin-4. *Cell Stem Cell*, **1**, 389–402.
66. Graham, S.M., Jorgensen, H.G., Allan, E., Pearson, C. *et al.* (2002) Primitive, quiescent, Philadelphia-positive stem cells from patients with chronic myeloid leukemia are insensitive to STI571 in vitro. *Blood*, **99**, 319–325.
67. Held, M.A., Curley, D.P., Dankort, D., McMahon, M. *et al.* (2010) Characterization of melanoma cells capable of propagating tumors from a single cell. *Cancer Res.*, **70**, 388–397.
68. Eppert, K., Takenaka, K., Lechman, E.R., Waldron, L. *et al.* (2011) Stem cell gene expression programs influence clinical outcome in human leukemia. *Nat. Med.*, **17**, 1086–1093.
69. De Jong, M.C., Pramana, J., Van Der Wal, J.E., Lacko, M. *et al.* (2010) CD44 expression predicts local recurrence after radiotherapy in larynx cancer. *Clin. Cancer Res.*, **16**, 5329–5338.
70. Steffensen, K.D., Alvero, A.B., Yang, Y., Waldstrøm, M. *et al.* (2011) Prevalence of epithelial ovarian cancer stem cells correlates with recurrence in early-stage ovarian cancer. *J. Oncol.*, **2011**, 1–12.
71. Al-Ejeh, F., Smart, C.E., Morrison, B.J., Chenevix-Trench, G. *et al.* (2011) Breast cancer stem cells: treatment resistance and therapeutic opportunities. *Carcinogenesis*, **32**, 650–658.
72. Zeppernick, F., Ahmadi, R., Campos, B., Dictus, C. *et al.* (2008) Stem cell marker CD133 affects clinical outcome in glioma patients. *Clin. Cancer Res.*, **14**, 123–129.
73. Christensen, K., Schrøder, H.D., and Kristensen, B.W. (2008) CD133 identifies perivascular niches in grade II–IV astrocytomas. *J. Neurooncol.*, **90**, 157–170.
74. Rijsewijk, F., Schuermann, M., Wagenaar, E., Parren, P. *et al.* (1987) The *Drosophila* homolog of the mouse mammary oncogene int-1 is identical to the segment polarity gene *wingless*. *Cell*, **50**, 649–657.
75. Chien, A.J., Conrad, W.H., and Moon, R.T. (2009) A Wnt survival guide: from flies to human disease. *J. Invest. Dermatol.*, **129**, 1614–1627.
76. Reya, T. and Clevers, H. (2005) Wnt signalling in stem cells and cancer. *Nature*, **434**, 843–850.
77. Angers, S. and Moon, R.T. (2009) Proximal events in Wnt signal transduction. *Nat. Rev. Mol. Cell Biol.*, **10**, 468–477.
78. Kikuchi, A., Yamamoto, H., and Kishida, S. (2007) Multiplicity of the interactions of Wnt proteins and their receptors. *Cell. Signalling*, **19**, 659–671.
79. van de Wetering, M., Sancho, E., Verweij, C., de Lau, W. *et al.* (2002) The beta-catenin/TCF-4 complex imposes a crypt progenitor phenotype on colorectal cancer cells. *Cell*, **111**, 241–250.
80. Korinek, V., Barker, N., Moerer, P., van Donselaar, E. *et al.* (1998) Depletion of

epithelial stem-cell compartments in the small intestine of mice lacking Tcf-4. *Nat. Genet.*, **19**, 379–383.

81. Kuhnert, F., Davis, C.R., Wang, H.T., Chu, P. *et al.* (2004) Essential requirement for Wnt signaling in proliferation of adult small intestine and colon revealed by adenoviral expression of Dickkopf-1. *Proc. Natl Acad. Sci. USA*, **101**, 266–271.
82. Pinto, D., Gregorieff, A., Begthel, H., and Clevers, H. (2003) Canonical Wnt signals are essential for homeostasis of the intestinal epithelium. *Genes Dev.*, **17**, 1709–1713.
83. Blanpain, C., Horsley, V., and Fuchs, E. (2007) Epithelial stem cells: turning over new leaves. *Cell*, **128**, 445–458.
84. Ito, M., Liu, Y., Yang, Z., Nguyen, J. *et al.* (2005) Stem cells in the hair follicle bulge contribute to wound repair but not to homeostasis of the epidermis. *Nat. Med.*, **11**, 1351–1354.
85. Zhang, Y.V., Cheong, J., Ciapurin, N., McDermitt, D.J. *et al.* (2009) Distinct self-renewal and differentiation phases in the niche of infrequently dividing hair follicle stem cells. *Cell Stem Cell*, **5**, 267–278.
86. Andl, T., Reddy, S.T., Gaddapara, T., and Millar, S.E. (2002) WNT signals are required for the initiation of hair follicle development. *Dev. Cell*, **2**, 643–653.
87. Huelsken, J., Vogel, R., Erdmann, B., Cotsarelis, G. *et al.* (2001) Beta-catenin controls hair follicle morphogenesis and stem cell differentiation in the skin. *Cell*, **105**, 533–545.
88. van Genderen, C., Okamura, R.M., Farinas, I., Quo, R.G. *et al.* (1994) Development of several organs that require inductive epithelial-mesenchymal interactions is impaired in LEF-1-deficient mice. *Genes Dev.*, **8**, 2691–2703.
89. Zhou, P., Byrne, C., Jacobs, J., and Fuchs, E. (1995) Lymphoid enhancer factor 1 directs hair follicle patterning and epithelial cell fate. *Genes Dev.*, **9**, 700–713.
90. Petersson, M., Brylka, H., Kraus, A., John, S. *et al.* (2011) TCF/Lef1 activity controls establishment of diverse stem and progenitor cell compartments in mouse epidermis. *EMBO J.*, **30**, 3004–3018.
91. DasGupta, R. and Fuchs, E. (1999) Multiple roles for activated LEF/TCF transcription complexes during hair follicle development and differentiation. *Development*, **126**, 4557–4568.
92. Merrill, B.J., Gat, U., DasGupta, R., and Fuchs, E. (2001) Tcf3 and Lef1 regulate lineage differentiation of multipotent stem cells in skin. *Genes Dev.*, **15**, 1688–1705.
93. Gat, U., DasGupta, R., Degenstein, L., and Fuchs, E. (1998) De novo hair follicle morphogenesis and hair tumors in mice expressing a truncated beta-catenin in skin. *Cell*, **95**, 605–614.
94. Austin, T.W., Solar, G.P., Ziegler, F.C., Liem, L. *et al.* (1997) A role for the Wnt gene family in hematopoiesis: expansion of multilineage progenitor cells. *Blood*, **89**, 3624–3635.
95. Reya, T., Duncan, A.W., Ailles, L., Domen, J. *et al.* (2003) A role for Wnt signalling in self-renewal of haematopoietic stem cells. *Nature*, **423**, 409–414.
96. Willert, K., Brown, J.D., Danenberg, E., Duncan, A.W. *et al.* (2003) Wnt proteins are lipid-modified and can act as stem cell growth factors. *Nature*, **423**, 448–452.
97. Van Den Berg, D.J., Sharma, A.K., Bruno, E., and Hoffman, R. (1998) Role of members of the Wnt gene family in human hematopoiesis. *Blood*, **92**, 3189–3202.
98. Fleming, H.E., Janzen, V., Lo Celso, C., Guo, J. *et al.* (2008) Wnt signaling in the niche enforces hematopoietic stem cell quiescence and is necessary to preserve self-renewal in vivo. *Cell Stem Cell*, **2**, 274–283.
99. Tsukamoto, A.S., Grosschedl, R., Guzman, R.C., Parslow, T. *et al.* (1988) Expression of the int-1 gene in transgenic mice is associated with mammary gland hyperplasia and adenocarcinomas in male and female mice. *Cell*, **55**, 619–625.
100. Kinzler, K.W., Nilbert, M.C., Su, L.K., Vogelstein, B. *et al.* (1991) Identification of FAP locus genes from chromosome 5q21. *Science*, **253**, 661–665.
101. Barker, N. and Clevers, H. (2006) Mining the Wnt pathway for cancer therapeutics. *Nat. Rev. Drug Discovery*, **5**, 997–1014.
102. Abrahamsson, A.E., Geron, I., Gotlib, J., Dao, K.-H.T. *et al.* (2009) Glycogen synthase kinase 3beta missplicing contributes to leukemia stem cell generation. *Proc. Natl Acad. Sci. USA*, **106**, 3925–3929.
103. Suzuki, H., Watkins, D.N., Jair, K.-W., Schuebel, K.E. *et al.* (2004) Epigenetic inactivation of SFRP genes allows constitutive

WNT signaling in colorectal cancer. *Nat. Genet.*, **36**, 417–422.
104. de Bont, J.M., Packer, R.J., Michiels, E.M., den Boer, M.L. *et al.* (2008) Biological background of pediatric medulloblastoma and ependymoma: a review from a translational research perspective. *Neuro-Oncology*, **10**, 1040–1060.
105. Barker, N., Ridgway, R.A., Van Es, J.H., van de Wetering, M. *et al.* (2009) Crypt stem cells as the cells-of-origin of intestinal cancer. *Nature*, **457**, 608–612.
106. Li, Y., Welm, B., Podsypanina, K., Huang, S. *et al.* (2003) Evidence that transgenes encoding components of the Wnt signaling pathway preferentially induce mammary cancers from progenitor cells. *Proc. Natl Acad. Sci. USA*, **100**, 15853–15858.
107. Liu, B.Y., McDermott, S.P., Khwaja, S.S., and Alexander, C.M. (2004) The transforming activity of Wnt effectors correlates with their ability to induce the accumulation of mammary progenitor cells. *Proc. Natl Acad. Sci. USA*, **101**, 4158–4163.
108. Blum, R., Gupta, R., Burger, P.E., Ontiveros, C.S. *et al.* (2009) Molecular signatures of prostate stem cells reveal novel signaling pathways and provide insights into prostate cancer. *PLoS ONE*, **4**, e5722.
109. Hurt, E.M., Kawasaki, B.T., Klarmann, G.J., Thomas, S.B. *et al.* (2008) CD44+ CD24(-) prostate cells are early cancer progenitor/stem cells that provide a model for patients with poor prognosis. *Br. J. Cancer*, **98**, 756–765.
110. Bisson, I. and Prowse, D.M. (2009) WNT signaling regulates self-renewal and differentiation of prostate cancer cells with stem cell characteristics. *Cell Res.*, **19**, 683–697.
111. Malanchi, I., Peinado, H., Kassen, D., Hussenet, T. *et al.* (2008) Cutaneous cancer stem cell maintenance is dependent on β-catenin signalling. *Nature*, **452**, 650–653.
112. Zhao, C., Blum, J., Chen, A., Kwon, H.Y. *et al.* (2007) Loss of β-catenin impairs the renewal of normal and CML stem cells in vivo. *Cancer Cell*, **12**, 528–541.
113. Wang, Y., Krivtsov, A.V., Sinha, A.U., North, T.E. *et al.* (2010) The Wnt/beta-catenin pathway is required for the development of leukemia stem cells in AML. *Science*, **327**, 1650–1653.
114. Takada, R., Satomi, Y., Kurata, T., Ueno, N. *et al.* (2006) Monounsaturated fatty acid modification of Wnt protein: its role in Wnt secretion. *Dev. Cell*, **11**, 791–801.
115. Chen, Z., Li, J., Li, Q.S., Fan, J.Q. *et al.* (2008) Suppression of PPN/MG61 attenuates Wnt/β-catenin signaling pathway and induces apoptosis in human lung cancer. *Oncogene*, **27**, 3483–3488.
116. Zhai, L., Chaturvedi, D., and Cumberledge, S. (2004) *Drosophila* Wnt-1 undergoes a hydrophobic modification and is targeted to lipid rafts, a process that requires porcupine. *J. Biol. Chem.*, **279**, 33220–33227.
117. He, B., You, L., Uematsu, K., Xu, Z. *et al.* (2004) A monoclonal antibody against Wnt-1 induces apoptosis in human cancer cells. *Neoplasia*, **6**, 7–14.
118. You, L., He, B., Xu, Z., Uematsu, K. *et al.* (2004) An anti-Wnt-2 monoclonal antibody induces apoptosis in malignant melanoma cells and inhibits tumor growth. *Cancer Res.*, **64**, 5385–5389.
119. Semënov, M.V., Tamai, K., Brott, B.K., Kühl, M. *et al.* (2001) Head inducer Dickkopf-1 is a ligand for Wnt coreceptor LRP6. *Curr. Biol.*, **11**, 951–961.
120. Nambotin, S.B., Lefrancois, L., Sainsily, X., Berthillon, P. *et al.* (2011) Pharmacological inhibition of Frizzled-7 displays anti-tumor properties in hepatocellular carcinoma. *J. Hepatol.*, **54**, 288–299.
121. Pode-Shakked, N., Harari-Steinberg, O., Haberman-Ziv, Y., Rom-Gross, E. *et al.* (2011) Resistance or sensitivity of Wilms' tumor to anti-FZD7 antibody highlights the Wnt pathway as a possible therapeutic target. *Oncogene*, **30**, 1664–1680.
122. Wei, W., Chua, M.-S., Grepper, S., and So, S.K. (2011) Soluble Frizzled-7 receptor inhibits Wnt signaling and sensitizes hepatocellular carcinoma cells towards doxorubicin. *Mol. Cancer*, **10**, 16.
123. Ettenberg, S.A., Charlat, O., Daley, M.P., Liu, S. *et al.* (2010) Inhibition of tumorigenesis driven by different Wnt proteins requires blockade of distinct ligand-binding regions by LRP6 antibodies. *Proc. Natl Acad. Sci. USA*, **107**, 15473–15478.
124. Huang, S.-M.A., Mishina, Y.M., Liu, S., Cheung, A. *et al.* (2009) Tankyrase inhibition stabilizes axin and antagonizes Wnt signalling. *Nature*, **461**, 614–620.
125. Chen, B., Dodge, M.E., Tang, W., Lu, J. *et al.* (2009) Small molecule–mediated disruption of Wnt-dependent signaling in

tissue regeneration and cancer. *Nat. Chem. Biol.*, **5**, 100–107.
126. Lepourcelet, M., Chen, Y.-N.P., France, D.S., Wang, H. *et al.* (2004) Small-molecule antagonists of the oncogenic Tcf/beta-catenin protein complex. *Cancer Cell*, **5**, 91–102.
127. Emami, K.H., Nguyen, C., Ma, H., Kim, D.H. *et al.* (2004) A small molecule inhibitor of beta-catenin/CREB-binding protein transcription. *Proc. Natl Acad. Sci. USA*, **101**, 12682–12687.
128. Van der Flier, L.G., Sabates-Bellver, J., Oving, I., Haegebarth, A. *et al.* (2007) The intestinal Wnt/TCF signature. *Gastroenterology*, **132**, 628–632.
129. Boon, E.M.J., Keller, J.J., Wormhoudt, T.A.M., Giardiello, F.M. *et al.* (2004) Sulindac targets nuclear β-catenin accumulation and Wnt signalling in adenomas of patients with familial adenomatous polyposis and in human colorectal cancer cell lines. *Br. J. Cancer*, **90**, 224–229.
130. Dihlmann, S., Siermann, A., and von Knebel Doeberitz, M. (2001) The nonsteroidal anti-inflammatory drugs aspirin and indomethacin attenuate beta-catenin/TCF-4 signaling. *Oncogene*, **20**, 645–653.
131. Hawcroft, G., D'Amico, M., Albanese, C., Markham, A.F. *et al.* (2002) Indomethacin induces differential expression of beta-catenin, gamma-catenin and T-cell factor target genes in human colorectal cancer cells. *Carcinogenesis*, **23**, 107–114.
132. Tuynman, J.B., Vermeulen, L., Boon, E.M., Kemper, K. *et al.* (2008) Cyclooxygenase-2 inhibition inhibits c-Met kinase activity and Wnt activity in colon cancer. *Cancer Res.*, **68**, 1213–1220.
133. Steinbach, G., Lynch, P.M., Phillips, R.K., Wallace, M.H. *et al.* (2000) The effect of celecoxib, a cyclooxygenase-2 inhibitor, in familial adenomatous polyposis. *N. Engl. J. Med.*, **342**, 1946–1952.
134. Thun, M.J., Namboodiri, M.M., and Heath, C.W. (1991) Aspirin use and reduced risk of fatal colon cancer. *N. Engl. J. Med.*, **325**, 1593–1596.
135. Qiu, W., Wang, X., Leibowitz, B., Liu, H. *et al.* (2010) Chemoprevention by nonsteroidal anti-inflammatory drugs eliminates oncogenic intestinal stem cells via SMAC-dependent apoptosis. *Proc. Natl Acad. Sci. USA*, **107**, 20027–20032.
136. Nüsslein-Volhard, C. and Wieschaus, E. (1980) Mutations affecting segment number and polarity in *Drosophila*. *Nature*, **287**, 795–801.
137. Kalderon, D. (2000) Transducing the hedgehog signal. *Cell*, **103**, 371–374.
138. Lee, J., Platt, K.A., Censullo, P., and Ruiz i Altaba, A. (1997) Gli1 is a target of Sonic hedgehog that induces ventral neural tube development. *Development*, **124**, 2537–2552.
139. Dodge, M.E. and Lum, L. (2011) Drugging the cancer stem cell compartment: lessons learned from the hedgehog and Wnt signal transduction pathways. *Annu. Rev. Pharmacol. Toxicol.*, **51**, 289–310.
140. Bhardwaj, G., Murdoch, B., Wu, D., Baker, D.P. *et al.* (2001) Sonic hedgehog induces the proliferation of primitive human hematopoietic cells via BMP regulation. *Nat. Immunol.*, **2**, 172–180.
141. Gering, M. and Patient, R. (2005) Hedgehog signaling is required for adult blood stem cell formation in zebrafish embryos. *Dev. Cell*, **8**, 389–400.
142. Trowbridge, J.J., Scott, M.P., and Bhatia, M. (2006) Hedgehog modulates cell cycle regulators in stem cells to control hematopoietic regeneration. *Proc. Natl Acad. Sci. USA*, **103**, 14134–14139.
143. Machold, R., Hayashi, S., Rutlin, M., Muzumdar, M.D. *et al.* (2003) Sonic hedgehog is required for progenitor cell maintenance in telencephalic stem cell niches. *Neuron*, **39**, 937–950.
144. Ruiz i Altaba, A., Mas, C., and Stecca, B. (2007) The Gli code: an information nexus regulating cell fate, stemness and cancer. *Trends Cell Biol.*, **17**, 438–447.
145. Brownell, I., Guevara, E., Bai, C.B., Loomis, C.A. *et al.* (2011) Nerve-derived sonic hedgehog defines a niche for hair follicle stem cells capable of becoming epidermal stem cells. *Stem Cell*, **8**, 552–565.
146. Chiang, C., Swan, R.Z., Grachtchouk, M., Bolinger, M. *et al.* (1999) Essential role for Sonic hedgehog during hair follicle morphogenesis. *Dev. Biol.*, **205**, 1–9.
147. Hahn, H., Wicking, C., Zaphiropoulous, P.G., Gailani, M.R. *et al.* (1996) Mutations of the human homolog of *Drosophila* patched in the nevoid basal cell carcinoma syndrome. *Cell*, **85**, 841–851.

148. Kinzler, K.W., Bigner, S.H., Bigner, D.D., Trent, J.M. *et al.* (1987) Identification of an amplified, highly expressed gene in a human glioma. *Science*, **236**, 70–73.
149. Ng, J.M.Y. and Curran, T. (2011) The Hedgehog's tale: developing strategies for targeting cancer. *Nat. Rev. Cancer*, **11**, 493–501.
150. Sarangi, A., Valadez, J.G., Rush, S., Abel, T.W. *et al.* (2009) Targeted inhibition of the Hedgehog pathway in established malignant glioma xenografts enhances survival. *Oncogene*, **28**, 3468–3476.
151. Clement, V., Sanchez, P., de Tribolet, N., Radovanovic, I. *et al.* (2007) HEDGEHOG-GLI1 signaling regulates human glioma growth, cancer stem cell self-renewal, and tumorigenicity. *Curr. Biol.*, **17**, 165–172.
152. Varnat, F., Duquet, A., Malerba, M., Zbinden, M. *et al.* (2009) Human colon cancer epithelial cells harbour active HEDGEHOG-GLI signalling that is essential for tumour growth, recurrence, metastasis and stem cell survival and expansion. *EMBO Mol. Med.*, **1**, 338–351.
153. Klarmann, G.J., Hurt, E.M., Mathews, L.A., Zhang, X. *et al.* (2009) Invasive prostate cancer cells are tumor initiating cells that have a stem cell-like genomic signature. *Clin. Exp. Metastasis*, **26**, 433–446.
154. Maitland, N.J. and Collins, A.T. (2008) Prostate cancer stem cells: a new target for therapy. *J. Clin. Oncol.*, **26**, 2862–2870.
155. Karhadkar, S.S., Bova, G.S., Abdallah, N., Dhara, S. *et al.* (2004) Hedgehog signalling in prostate regeneration, neoplasia and metastasis. *Nature*, **431**, 707–712.
156. Jones, S., Zhang, X., Parsons, D.W., Lin, J.C.H. *et al.* (2008) Core signaling pathways in human pancreatic cancers revealed by global genomic analyses. *Science*, **321**, 1801–1806.
157. Singh, B.N., Fu, J., Srivastava, R.K., and Shankar, S. (2011) Hedgehog signaling antagonist GDC-0449 (Vismodegib) inhibits pancreatic cancer stem cell characteristics: molecular mechanisms. *PLoS ONE*, **6**, e27306.
158. Feldmann, G., Dhara, S., Fendrich, V., Bedja, D. *et al.* (2007) Blockade of hedgehog signaling inhibits pancreatic cancer invasion and metastases: a new paradigm for combination therapy in solid cancers. *Cancer Res.*, **67**, 2187–2196.
159. Mueller, M.T., Hermann, P.C., Witthauer, J., Viqueira, B.R. *et al.* (2009) Combined targeted treatment to eliminate tumorigenic cancer stem cells in human pancreatic cancer. *Gastroenterology*, **137**, 1102–1113.
160. Peacock, C.D., Wang, Q., Gesell, G.S., Corcoran-Schwartz, I.M. *et al.* (2007) Hedgehog signaling maintains a tumor stem cell compartment in multiple myeloma. *Proc. Natl Acad. Sci. USA*, **104**, 4048–4053.
161. Dierks, C., Beigi, R., Guo, G.-R., Zirlik, K. *et al.* (2008) Expansion of Bcr-Abl-positive leukemic stem cells is dependent on hedgehog pathway activation. *Cancer Cell*, **14**, 238–249.
162. Radich, J.P., Dai, H., Mao, M., Oehler, V. *et al.* (2006) Gene expression changes associated with progression and response in chronic myeloid leukemia. *Proc. Natl Acad. Sci. USA*, **103**, 2794–2799.
163. Zhao, C., Chen, A., Jamieson, C.H., Fereshteh, M. *et al.* (2009) Hedgehog signalling is essential for maintenance of cancer stem cells in myeloid leukaemia. *Nature*, **458**, 776–779.
164. Cooper, M.K., Porter, J.A., Young, K.E., and Beachy, P.A. (1998) Teratogen-mediated inhibition of target tissue response to Shh signaling. *Science*, **280**, 1603–1607.
165. Incardona, J.P., Gaffield, W., Kapur, R.P., and Roelink, H. (1998) The teratogenic Veratrum alkaloid cyclopamine inhibits sonic hedgehog signal transduction. *Development*, **125**, 3553–3562.
166. Chiang, C., Litingtung, Y., Lee, E., Young, K.E. *et al.* (1996) Cyclopia and defective axial patterning in mice lacking Sonic hedgehog gene function. *Nature*, **383**, 407–413.
167. Roessler, E., Belloni, E., Gaudenz, K., Jay, P. *et al.* (1996) Mutations in the human Sonic Hedgehog gene cause holoprosencephaly. *Nat. Genet.*, **14**, 357–360.
168. Scales, S.J. and De Sauvage, F.J. (2009) Mechanisms of Hedgehog pathway activation in cancer and implications for therapy. *Trends Pharmacol. Sci.*, **30**, 303–312.
169. Rudin, C.M., Hann, C.L., Laterra, J., Yauch, R.L. *et al.* (2009) Treatment of medulloblastoma with hedgehog pathway inhibitor GDC-0449. *N. Engl. J. Med.*, **361**, 1173–1178.

170. Von Hoff, D.D., LoRusso, P.M., Rudin, C.M., Reddy, J.C. *et al.* (2009) Inhibition of the hedgehog pathway in advanced basal-cell carcinoma. *N. Engl. J. Med.*, **361**, 1164–1172.
171. Brunton, S.A., Stibbard, J.H.A., Rubin, L.L., Kruse, L.I. *et al.* (2008) Potent inhibitors of the hedgehog signaling pathway. *J. Med. Chem.*, **51**, 1108–1110.
172. Lipinski, R.J., Dengler, E., Kiehn, M., Peterson, R.E. *et al.* (2007) Identification and characterization of several dietary alkaloids as weak inhibitors of hedgehog signaling. *Toxicol. Sci.*, **100**, 456–463.
173. Kim, J., Lee, J.J., Kim, J., Gardner, D. *et al.* (2010) Arsenic antagonizes the Hedgehog pathway by preventing ciliary accumulation and reducing stability of the Gli2 transcriptional effector. *Proc. Natl Acad. Sci. USA*, **107**, 13432–13437.
174. Shi, W., Nacev, B.A., Aftab, B.T., Head, S. *et al.* (2011) Itraconazole side chain analogues: structure–activity relationship studies for inhibition of endothelial cell proliferation, vascular endothelial growth factor receptor 2 (VEGFR2) glycosylation, and hedgehog signaling. *J. Med. Chem.*, **54**, 7363–7374.
175. Mine, T., Matsueda, S., Gao, H., Li, Y. *et al.* (2010) Created Gli-1 duplex short-RNA (i-Gli-RNA) eliminates CD44 Hi progenitors of taxol-resistant ovarian cancer cells. *Oncol. Rep.*, **23**, 1537–1543.
176. Lauth, M., Bergström, A., Shimokawa, T., and Toftgård, R. (2007) Inhibition of GLI-mediated transcription and tumor cell growth by small-molecule antagonists. *Proc. Natl Acad. Sci. USA*, **104**, 8455–8460.
177. Beauchamp, E.M., Ringer, L., Bulut, G., Sajwan, K.P. *et al.* (2011) Arsenic trioxide inhibits human cancer cell growth and tumor development in mice by blocking Hedgehog/GLI pathway. *J. Clin. Invest.*, **121**, 148–160.
178. Tang, T., Tang, J.Y., Li, D., Reich, M. *et al.* (2011) Targeting superficial or nodular basal cell carcinoma with topically formulated small molecule inhibitor of smoothened. *Clin. Cancer Res.*, **17**, 3378–3387.
179. Metcalfe, C. and De Sauvage, F.J. (2011) Hedgehog fights back: mechanisms of acquired resistance against smoothened antagonists. *Cancer Res.*, **71**, 5057–5061.
180. Dexter, J.S. (1914) The analysis of a case of continuous variation in *Drosophila* by a study of its linkage relations. *Am. Nat.*, **48**, 712–758.
181. Wharton, K.A., Johansen, K.M., Xu, T., and Artavanis-Tsakonas, S. (1985) Nucleotide sequence from the neurogenic locus notch implies a gene product that shares homology with proteins containing EGF-like repeats. *Cell*, **43**, 567–581.
182. Fehon, R.G., Kooh, P.J., Rebay, I., Regan, C.L. *et al.* (1990) Molecular interactions between the protein products of the neurogenic loci Notch and Delta, two EGF-homologous genes in *Drosophila*. *Cell*, **61**, 523–534.
183. Brou, C., Logeat, F., Gupta, N., Bessia, C. *et al.* (2000) A novel proteolytic cleavage involved in Notch signaling: the role of the disintegrin-metalloprotease TACE. *Mol. Cell*, **5**, 207–216.
184. Das, I., Craig, C., Funahashi, Y., Jung, K.-M. *et al.* (2004) Notch oncoproteins depend on gamma-secretase/presenilin activity for processing and function. *J. Biol. Chem.*, **279**, 30771–30780.
185. Chillakuri, C.R., Sheppard, D., Lea, S.M., and Handford, P.A. (2012) Notch receptor–ligand binding and activation: Insights from molecular studies. *Semin. Cell Dev. Biol.*, **23**, 421–428.
186. Sanalkumar, R., Dhanesh, S.B., and James, J. (2010) Non-canonical activation of Notch signaling/target genes in vertebrates. *Cell. Mol. Life Sci.*, **67**, 2957–2968.
187. Talora, C., Campese, A.F., Bellavia, D., Felli, M.P. *et al.* (2008) Notch signaling and diseases: an evolutionary journey from a simple beginning to complex outcomes. *Biochim. Biophys. Acta - Mol. Basis Dis.*, **1782**, 489–497.
188. Egger, B., Gold, K.S., and Brand, A.H. (2010) Notch regulates the switch from symmetric to asymmetric neural stem cell division in the *Drosophila* optic lobe. *Development*, **137**, 2981–2987.
189. Gaiano, N. and Fishell, G. (2002) The role of notch in promoting glial and neural stem cell fates. *Annu. Rev. Neurosci.*, **25**, 471–490.
190. Androutsellis-Theotokis, A., Leker, R.R., Soldner, F., Hoeppner, D.J. *et al.* (2006)

Notch signalling regulates stem cell numbers *in vitro* and *in vivo*. *Nature*, **442**, 823–826.
191. Morrison, S.J., Perez, S.E., Qiao, Z., Verdi, J.M. *et al.* (2000) Transient Notch activation initiates an irreversible switch from neurogenesis to gliogenesis by neural crest stem cells. *Cell*, **101**, 499–510.
192. Hadland, B.K., Huppert, S.S., Kanungo, J., Xue, Y. *et al.* (2004) A requirement for Notch1 distinguishes 2 phases of definitive hematopoiesis during development. *Blood*, **104**, 3097–3105.
193. Kumano, K., Chiba, S., Kunisato, A., Sata, M. *et al.* (2003) Notch1 but not Notch2 is essential for generating hematopoietic stem cells from endothelial cells. *Immunity*, **18**, 699–711.
194. Radtke, F., Wilson, A., Mancini, S.J.C., and MacDonald, H.R. (2004) Notch regulation of lymphocyte development and function. *Nat. Immunol.*, **5**, 247–253.
195. Yuan, J.S., Kousis, P.C., Suliman, S., Visan, I. *et al.* (2010) Functions of notch signaling in the immune system: consensus and controversies. *Annu. Rev. Immunol.*, **28**, 343–365.
196. Delaney, C., Heimfeld, S., Brashem-Stein, C., Voorhies, H. *et al.* (2010) Notch-mediated expansion of human cord blood progenitor cells capable of rapid myeloid reconstitution. *Nat. Med.*, **16**, 232–236.
197. Maillard, I., Koch, U., Dumortier, A., Shestova, O. *et al.* (2008) Canonical notch signaling is dispensable for the maintenance of adult hematopoietic stem cells. *Cell Stem Cell*, **2**, 356–366.
198. Mancini, S.J.C., Mantei, N., Dumortier, A., Suter, U. *et al.* (2005) Jagged1-dependent Notch signaling is dispensable for hematopoietic stem cell self-renewal and differentiation. *Blood*, **105**, 2340–2342.
199. Izrailit, J. and Reedijk, M. (2012) Developmental pathways in breast cancer and breast tumor-initiating cells: therapeutic implications. *Cancer Lett.*, **317**, 115–126.
200. Fre, S., Huyghe, M., Mourikis, P., Robine, S. *et al.* (2005) Notch signals control the fate of immature progenitor cells in the intestine. *Nature*, **435**, 964–968.
201. Van Es, J.H., Van Gijn, M.E., Riccio, O., Van Den Born, M. *et al.* (2005) Notch/γ-secretase inhibition turns proliferative cells in intestinal crypts and adenomas into goblet cells. *Nature*, **435**, 959–963.
202. Singh, S. and Chellappan, S.P. (2011) *Stem Cells and Human Diseases*, Springer, Dordrecht, pp. 509–532.
203. Zong, Y., Panikkar, A., Xu, J., Antoniou, A. *et al.* (2009) Notch signaling controls liver development by regulating biliary differentiation. *Development*, **136**, 1727–1739.
204. Crea, F., Mathews, L.A., Farrar, W.L., and Hurt, E.M. (2009) Targeting prostate cancer stem cells. *Anticancer Agents Med. Chem.*, **9**, 1105–1113.
205. Conboy, I.M. and Rando, T.A. (2002) The regulation of Notch signaling controls satellite cell activation and cell fate determination in postnatal myogenesis. *Dev. Cell*, **3**, 397–409.
206. Watt, F.M., Estrach, S., and Ambler, C.A. (2008) Epidermal Notch signalling: differentiation, cancer and adhesion. *Curr. Opin. Cell Biol.*, **20**, 171–179.
207. Ellisen, L.W., Bird, J., West, D.C., Soreng, A.L. *et al.* (1991) TAN-1, the human homolog of the *Drosophila* notch gene, is broken by chromosomal translocations in T lymphoblastic neoplasms. *Cell*, **66**, 649–661.
208. Weng, A.P., Ferrando, A.A., Lee, W., Morris, J.P. *et al.* (2004) Activating mutations of NOTCH1 in human T cell acute lymphoblastic leukemia. *Science*, **306**, 269–271.
209. Houde, C., Li, Y., Song, L., Barton, K. *et al.* (2004) Overexpression of the NOTCH ligand JAG2 in malignant plasma cells from multiple myeloma patients and cell lines. *Blood*, **104**, 3697–3704.
210. Jundt, F., Probsting, K.S., Anagnostopoulos, I., Muehlinghaus, G. *et al.* (2004) Jagged1-induced Notch signaling drives proliferation of multiple myeloma cells. *Blood*, **103**, 3511–3515.
211. Klinakis, A., Lobry, C., Abdel-Wahab, O., Oh, P. *et al.* (2011) A novel tumour-suppressor function for the Notch pathway in myeloid leukaemia. *Nature*, **473**, 230–233.
212. Lee, S.-Y., Kumano, K., Nakazaki, K., Sanada, M. *et al.* (2009) Gain-of-function mutations and copy number increases of Notch2 in diffuse large B-cell lymphoma. *Cancer Sci.*, **100**, 920–926.
213. Nakahara, F., Sakata-Yanagimoto, M., Komeno, Y., Kato, N. *et al.* (2010) Hes1

immortalizes committed progenitors and plays a role in blast crisis transition in chronic myelogenous leukemia. *Blood*, **115**, 2872–2881.

214. Nefedova, Y., Sullivan, D.M., Bolick, S.C., Dalton, W.S. *et al.* (2008) Inhibition of Notch signaling induces apoptosis of myeloma cells and enhances sensitivity to chemotherapy. *Blood*, **111**, 2220–2229.
215. Ramakrishnan, V., Ansell, S., Haug, J., Grote, D. *et al.* (2012) MRK003, a γ-secretase inhibitor exhibits promising in vitro pre-clinical activity in multiple myeloma and non-Hodgkin's lymphoma. *Leukemia*, **26**, 340–348.
216. Xu, D., Hu, J., Xu, S., De Bruyne, E. *et al.* (2011) Dll1/Notch activation accelerates multiple myeloma disease development by promoting CD138+ MM-cell proliferation. *Leukemia*, **26**, 1402–1405.
217. Reedijk, M., Odorcic, S., Chang, L., Zhang, H. *et al.* (2005) High-level coexpression of JAG1 and NOTCH1 is observed in human breast cancer and is associated with poor overall survival. *Cancer Res.*, **65**, 8530–8537.
218. Stockhausen, M.T., Kristoffersen, K., and Poulsen, H.S. (2010) The functional role of Notch signaling in human gliomas. *Neuro-Oncology*, **12**, 199–211.
219. Sjölund, J., Manetopoulos, C., Stockhausen, M.-T., and Axelson, H. (2005) The Notch pathway in cancer: differentiation gone awry. *Eur. J. Cancer*, **41**, 2620–2629.
220. Leong, K.G. and Gao, W.-Q. (2008) The Notch pathway in prostate development and cancer. *Differentiation*, **76**, 699–716.
221. Miyamoto, Y., Maitra, A., Ghosh, B., Zechner, U. *et al.* (2003) Notch mediates TGF alpha-induced changes in epithelial differentiation during pancreatic tumorigenesis. *Cancer Cell*, **3**, 565–576.
222. Gao, J., Song, Z., Chen, Y., Xia, L. *et al.* (2008) Deregulated expression of Notch receptors in human hepatocellular carcinoma. *Dig. Liver Dis.*, **40**, 114–121.
223. Maliekal, T.T., Bajaj, J., Giri, V., Subramanyam, D., and Krishna, S. (2008) The role of Notch signaling in human cervical cancer: implications for solid tumors. *Oncogene*, **27**, 5110–5114.
224. Haruki, N., Kawaguchi, K.S., Eichenberger, S., Massion, P.P. *et al.* (2005) Dominant-negative Notch3 receptor inhibits mitogen-activated protein kinase pathway and the growth of human lung cancers. *Cancer Res.*, **65**, 3555–3561.
225. Tonon, G., Modi, S., Wu, L., Kubo, A. *et al.* (2003) t(11;19)(q21;p13) translocation in mucoepidermoid carcinoma creates a novel fusion product that disrupts a Notch signaling pathway. *Nat. Genet.*, **33**, 208–213.
226. Rose, S.L., Kunnimalaiyaan, M., Drenzek, J., and Seiler, N. (2010) Notch 1 signaling is active in ovarian cancer. *Gynecol. Oncol.*, **117**, 130–133.
227. Pinnix, C.C. and Herlyn, M. (2007) The many faces of Notch signaling in skin-derived cells. *Pigment Cell Res.*, **20**, 458–465.
228. Wang, J., Wakeman, T.P., Lathia, J.D., and Hjelmeland, A.B. *et al.* (2009) Notch promotes radioresistance of glioma stem cells. *Stem Cells*, **28**, 17–28.
229. Wang, Z., Li, Y., Kong, D., Banerjee, S. *et al.* (2009) Acquisition of epithelial-mesenchymal transition phenotype of gemcitabine-resistant pancreatic cancer cells is linked with activation of the notch signaling pathway. *Cancer Res.*, **69**, 2400–2407.
230. Chiu, P.P.L., Jiang, H., and Dick, J.E. (2010) Leukemia-initiating cells in human T-lymphoblastic leukemia exhibit glucocorticoid resistance. *Blood*, **116**, 5268–5279.
231. Gerby, B., Clappier, E., Armstrong, F., Deswarte, C. *et al.* (2011) Expression of CD34 and CD7 on human T-cell acute lymphoblastic leukemia discriminates functionally heterogeneous cell populations. *Leukemia*, **25**, 1249–1258.
232. Ma, W., Gutierrez, A., Goff, D.J., Geron, I. *et al.* (2012) NOTCH1 signaling promotes human T-Cell acute lymphoblastic leukemia initiating cell regeneration in supportive niches. *PLoS ONE*, **7**, e39725.
233. Tatarek, J., Cullion, K., Ashworth, T., Gerstein, R. *et al.* (2011) Notch1 inhibition targets the leukemia-initiating cells in a Tal1/Lmo2 mouse model of T-ALL. *Blood*, **118**, 1579–1590.
234. Farnie, G., Clarke, R.B., Spence, K., Pinnock, N. *et al.* (2007) Novel cell culture technique for primary ductal carcinoma in situ: role of notch and epidermal growth factor receptor

signaling pathways. *J. Natl Cancer Inst.*, **99**, 616–627.

235. Kondratyev, M., Kreso, A., Hallett, R.M., Girgis-Gabardo, A. *et al.* (2012) Gamma-secretase inhibitors target tumor-initiating cells in a mouse model of ERBB2 breast cancer. *Oncogene*, **31**, 93–103.
236. Harrison, H., Farnie, G., Howell, S.J., Rock, R.E. *et al.* (2010) Regulation of breast cancer stem cell activity by signaling through the Notch4 receptor. *Cancer Res.*, **70**, 709–718.
237. Fan, X., Khaki, L., Zhu, T.S., Soules, M.E. *et al.* (2010) Notch pathway blockade depletes CD133-positive glioblastoma cells and inhibits growth of tumor neurospheres and xenografts. *Stem Cells*, **28**, 5–16.
238. Jeon, H.M., Jin, X., Lee, J.S., Oh, S.Y. *et al.* (2008) Inhibitor of differentiation 4 drives brain tumor-initiating cell genesis through cyclin E and notch signaling. *Genes Dev.*, **22**, 2028–2033.
239. Hu, Y.-Y., Zheng, M.-H., Cheng, G., Li, L. *et al.* (2011) Notch signaling contributes to the maintenance of both normal neural stem cells and patient-derived glioma stem cells. *BMC Cancer*, **11**, 82.
240. Hovinga, K.E., Shimizu, F., Wang, R., Panagiotakos, G. *et al.* (2010) Inhibition of notch signaling in glioblastoma targets cancer stem cells via an endothelial cell intermediate. *Stem Cells*, **28**, 1019–1029.
241. Ricci-Vitiani, L., Pallini, R., Biffoni, M., Todaro, M. *et al.* (2010) Tumour vascularization via endothelial differentiation of glioblastoma stem-like cells. *Nature*, **468**, 824–828.
242. Wang, R., Chadalavada, K., Wilshire, J., Kowalik, U. *et al.* (2010) Glioblastoma stem-like cells give rise to tumour endothelium. *Nature*, **468**, 829–833.
243. Fan, X., Matsui, W., Khaki, L., Stearns, D. *et al.* (2006) Notch pathway inhibition depletes stem-like cells and blocks engraftment in embryonal brain tumors. *Cancer Res.*, **66**, 7445–7452.
244. Hoey, T., Yen, W.-C., Axelrod, F., Basi, J. *et al.* (2009) DLL4 blockade inhibits tumor growth and reduces tumor-initiating cell frequency. *Cell Stem Cell*, **5**, 168–177.
245. Sikandar, S.S., Pate, K.T., Anderson, S., Dizon, D. *et al.* (2010) NOTCH signaling is required for formation and self-renewal of tumor-initiating cells and for repression of secretory cell differentiation in colon cancer. *Cancer Res.*, **70**, 1469–1478.
246. Bertolini, G., Roz, L., Perego, P., Tortoreto, M. *et al.* (2009) Highly tumorigenic lung cancer CD133+ cells display stem-like features and are spared by cisplatin treatment. *Proc. Natl Acad. Sci. USA*, **106**, 16281–16286.
247. Huynh, C., Poliseno, L., Segura, M.F., Medicherla, R. *et al.* (2011) The novel gamma secretase inhibitor RO4929097 reduces the tumor initiating potential of melanoma. *PLoS ONE*, **6**, e25264.
248. Ji, Q., Hao, X., Zhang, M., Tang, W. *et al.* (2009) MicroRNA miR-34 inhibits human pancreatic cancer tumor-initiating cells. *PLoS ONE*, **4**, e6816.
249. Kong, D., Banerjee, S., Ahmad, A., Li, Y. *et al.* (2010) Epithelial to mesenchymal transition is mechanistically linked with stem cell signatures in prostate cancer cells. *PLoS ONE*, **5**, e12445.
250. Sullivan, J.P., Spinola, M., Dodge, M., Raso, M.G. *et al.* (2010) Aldehyde dehydrogenase activity selects for lung adenocarcinoma stem cells dependent on notch signaling. *Cancer Res.*, **70**, 9937–9948.
251. Yong, T., Sun, A., Henry, M.D., Meyers, S. *et al.* (2011) Down regulation of CSL activity inhibits cell proliferation in prostate and breast cancer cells. *J. Cell. Biochem.*, **112**, 2340–2351.
252. Groth, C. and Fortini, M.E. (2012) Therapeutic approaches to modulating Notch signaling: current challenges and future prospects. *Semin. Cell Dev. Biol.*, **23**, 465–472.
253. Gal, H., Amariglio, N., Trakhtenbrot, L., Jacob-Hirsh, J. *et al.* (2006) Gene expression profiles of AML derived stem cells; similarity to hematopoietic stem cells. *Leukemia*, **20**, 2147–2154.
254. Pandya, K., Meeke, K., Clementz, A.G., Rogowski, A. *et al.* (2011) Targeting both Notch and ErbB-2 signalling pathways is required for prevention of ErbB-2-positive breast tumour recurrence. *Br. J. Cancer*, **105**, 796–806.
255. Rajasekhar, V.K., Studer, L., Gerald, W., Socci, N.D. *et al.* (2011) Tumour-initiating stem-like cells in human prostate cancer exhibit increased NF-κB signalling. *Nat. Commun.*, **2**, 162.

256. Wong, G.T., Manfra, D., Poulet, F.M., Zhang, Q. *et al.* (2004) Chronic treatment with the gamma-secretase inhibitor LY-411,575 inhibits beta-amyloid peptide production and alters lymphopoiesis and intestinal cell differentiation. *J. Biol. Chem.*, **279**, 12876–12882.
257. Liu, M., Sakamaki, T., Casimiro, M.C., Willmarth, N.E. *et al.* (2010) The canonical NF-κB pathway governs mammary tumorigenesis in transgenic mice and tumor stem cell expansion. *Cancer Res.*, **70**, 10464–10473.
258. Wu, Y., Cain-Hom, C., Choy, L., Hagenbeek, T.J. *et al.* (2010) Therapeutic antibody targeting of individual Notch receptors. *Nature*, **464**, 1052–1057.
259. Liu, S.K., Bham, S.A.S., Fokas, E., Beech, J. *et al.* (2011) Delta-like ligand 4-notch blockade and tumor radiation response. *J. Natl Cancer Inst.*, **103**, 1778–1798.
260. Rizzo, P., Osipo, C., Foreman, K., Golde, T. *et al.* (2008) Rational targeting of Notch signaling in cancer. *Oncogene*, **27**, 5124–5131.
261. Wei, P., Walls, M., Qiu, M., Ding, R. *et al.* (2010) Evaluation of selective gamma-secretase inhibitor PF-03084014 for its antitumor efficacy and gastrointestinal safety to guide optimal clinical trial design. *Mol. Cancer Ther.*, **9**, 1618–1628.
262. Fouladi, M., Stewart, C.F., Olson, J., Wagner, L.M. *et al.* (2011) Phase I trial of MK-0752 in children with refractory CNS malignancies: a pediatric brain tumor consortium study. *J. Clin. Oncol.*, **29**, 3529–3534.
263. Aste-Amézaga, M., Zhang, N., Lineberger, J.E., Arnold, B.A. *et al.* (2010) Characterization of Notch1 antibodies that inhibit signaling of both normal and mutated Notch1 receptors. *PLoS ONE*, **5**, e9094.
264. Li, K., Li, Y., Wu, W., Gordon, W.R. *et al.* (2008) Modulation of Notch signaling by antibodies specific for the extracellular negative regulatory region of NOTCH3. *J. Biol. Chem.*, **283**, 8046–8054.
265. Noguera-Troise, I., Daly, C., Papadopoulos, N.J., Coetzee, S. *et al.* (2006) Blockade of Dll4 inhibits tumour growth by promoting non-productive angiogenesis. *Nature*, **444**, 1032–1037.
266. Ridgway, J., Zhang, G., Wu, Y., Stawicki, S. *et al.* (2006) Inhibition of Dll4 signalling inhibits tumour growth by deregulating angiogenesis. *Nature*, **444**, 1083–1087.
267. Pistollato, F., Rampazzo, E., Persano, L., Abbadi, S. *et al.* (2010) Interaction of hypoxia-inducible factor-1α and Notch signaling regulates medulloblastoma precursor proliferation and fate. *Stem Cells*, **28**, 1918–1929.
268. Lessard, J. and Sauvageau, G. (2003) Bmi-1 determines the proliferative capacity of normal and leukaemic stem cells. *Nature*, **423**, 255–260.
269. Piccirillo, S.G.M., Reynolds, B.A., Zanetti, N., Lamorte, G. *et al.* (2006) Bone morphogenetic proteins inhibit the tumorigenic potential of human brain tumour-initiating cells. *Nature*, **444**, 761–765.
270. Lee, J., Son, M.J., Woolard, K., Donin, N.M. *et al.* (2008) Epigenetic-mediated dysfunction of the bone morphogenetic protein pathway inhibits differentiation of glioblastoma-initiating cells. *Cancer Cell*, **13**, 69–80.
271. Yi, J.M., Tsai, H.-C., Glöckner, S.C., Lin, S. *et al.* (2008) Abnormal DNA methylation of CD133 in colorectal and glioblastoma tumors. *Cancer Res.*, **68**, 8094–8103.
272. Diehn, M., Cho, R.W., Lobo, N.A., Kalisky, T. *et al.* (2009) Association of reactive oxygen species levels and radioresistance in cancer stem cells. *Nature*, **458**, 780–783.
273. Sen, R. and Baltimore, D. (1986) Inducibility of kappa immunoglobulin enhancer-binding protein Nf-kappa B by a posttranslational mechanism. *Cell*, **47**, 921–928.
274. Oeckinghaus, A., Hayden, M.S., and Ghosh, S. (2011) Crosstalk in NF-κB signaling pathways. *Nat. Immunol.*, **12**, 695–708.
275. Naugler, W.E. and Karin, M. (2008) NF-κB and cancer – identifying targets and mechanisms. *Curr. Opin. Genet. Dev.*, **18**, 19 26.
276. Prasad, S., Ravindran, J., and Aggarwal, B.B. (2009) NF-κB and cancer: how intimate is this relationship? *Mol. Cell. Biochem.*, **336**, 25–37.
277. Guzman, M.L., Swiderski, C.F., Howard, D.S., Grimes, B.A. *et al.* (2002) Preferential induction of apoptosis for primary human leukemic stem cells. *Proc. Natl Acad. Sci. USA*, **99**, 16220–16225.
278. Guzman, M.L., Rossi, R.M., Karnischky, L., Li, X. *et al.* (2005) The sesquiterpene lactone parthenolide induces apoptosis

of human acute myelogenous leukemia stem and progenitor cells. *Blood*, **105**, 4163–4169.

279. Guzman, M.L., Rossi, R.M., Neelakantan, S., Li, X. *et al.* (2007) An orally bioavailable parthenolide analog selectively eradicates acute myelogenous leukemia stem and progenitor cells. *Blood*, **110**, 4427–4435.
280. Griessinger, E., Frelin, C., Cuburu, N., Imbert, V. *et al.* (2008) Preclinical targeting of NF-κB and FLT3 pathways in AML cells. *Leukemia*, **22**, 1466–1469.
281. Strair, R.K., Gharibo, M., Schaar, D., Rubin, A. *et al.* (2008) Nuclear factor-kappaB modulation in patients undergoing induction chemotherapy for acute myelogenous leukemia. *Clin. Cancer Res.*, **14**, 7564–7568.
282. Smith, K.M., Datti, A., Fujitani, M., Grinshtein, N. *et al.* (2010) Selective targeting of neuroblastoma tumour-initiating cells by compounds identified in stem cell-based small molecule screens. *EMBO Mol. Med.*, **2**, 371–384.
283. Gilmore, T.D. and Herscovitch, M. (2006) Inhibitors of NF-κB signaling: 785 and counting. *Oncogene*, **25**, 6887–6899.
284. Gupta, S.C., Sundaram, C., Reuter, S., and Aggarwal, B.B. (2010) Inhibiting NF-κB activation by small molecules as a therapeutic strategy. *Biochim. Biophys. Acta, Gene Regul. Mech.*, **1799**, 775–787.
285. Wang, C.Y., Guttridge, D.C., Mayo, M.W., and Baldwin, A.S. (1999) NF-kappaB induces expression of the Bcl-2 homologue A1/Bfl-1 to preferentially suppress chemotherapy-induced apoptosis. *Mol. Cell. Biol.*, **19**, 5923–5929.
286. Kallifatidis, G., Rausch, V., Baumann, B., Apel, A. *et al.* (2009) Sulforaphane targets pancreatic tumour-initiating cells by NF-κB-induced antiapoptotic signalling. *Gut*, **58**, 949–963.
287. Rausch, V., Liu, L., Kallifatidis, G., Baumann, B. *et al.* (2010) Synergistic activity of sorafenib and sulforaphane abolishes pancreatic cancer stem cell characteristics. *Cancer Res.*, **70**, 5004–5013.
288. Nogueira, L., Ruiz-Ontañon, P., Vazquez-Barquero, A., Lafarga, M. *et al.* (2011) Blockade of the NFκB pathway drives differentiating glioblastoma-initiating cells into senescence both in vitro and in vivo. *Oncogene*, **30**, 3537–3548.
289. Vadas, O., Burke, J.E., Zhang, X., Berndt, A. *et al.* (2011) Structural basis for activation and inhibition of class I phosphoinositide 3-kinases. *Sci. Signal.*, **4**, (195) re2.
290. Manning, B.D. and Cantley, L.C. (2007) AKT/PKB signaling: navigating downstream. *Cell*, **129**, 1261–1274.
291. Yuan, T.L. and Cantley, L.C. (2008) PI3K pathway alterations in cancer: variations on a theme. *Oncogene*, **27**, 5497–5510.
292. Yilmaz, Ö.H., Valdez, R., Theisen, B.K., Guo, W. *et al.* (2006) Pten dependence distinguishes haematopoietic stem cells from leukaemia-initiating cells. *Nature*, **441**, 475–482.
293. Eyler, C.E., Foo, W.-C., LaFiura, K.M., McLendon, R.E. *et al.* (2008) Brain cancer stem cells display preferential sensitivity to Akt inhibition. *Stem Cells*, **26**, 3027–3036.
294. Sunayama, J., Sato, A., Matsuda, K., Tachibana, K. *et al.* (2010) Dual blocking of mTor and PI3K elicits a prodifferentiation effect on glioblastoma stem-like cells. *Neuro-Oncology*, **12**, 1205–1219.
295. Hambardzumyan, D., Becher, O.J., Rosenblum, M.K., Pandolfi, P.P. *et al.* (2008) PI3K pathway regulates survival of cancer stem cells residing in the perivascular niche following radiation in medulloblastoma in vivo. *Genes Dev.*, **22**, 436–448.
296. Zhou, J., Wulfkuhle, J., Zhang, H., Gu, P. *et al.* (2007) Activation of the PTEN/mTOR/STAT3 pathway in breast cancer stem-like cells is required for viability and maintenance. *Proc. Natl Acad. Sci. USA*, **104**, 16158–16163.
297. Dubrovska, A., Elliott, J., Salamone, R.J., Kim, S. *et al.* (2010) Combination therapy targeting both tumor-initiating and differentiated cell populations in prostate carcinoma. *Clin. Cancer Res.*, **16**, 5692–5702.
298. Yang, Z., Zhang, L., Ma, A., Liu, L. *et al.* (2011) Transient mTOR inhibition facilitates continuous growth of liver tumors by modulating the maintenance of CD133+ cell populations. *PLoS ONE*, **6**, e28405.
299. Huang, J. and Manning, B.D. (2009) A complex interplay between Akt, TSC2 and the two mTOR complexes. *Biochem. Soc. Trans.*, **37**, 217.
300. O'Reilly, K.E., Rojo, F., She, Q.-B., Solit, D. *et al.* (2006) mTOR inhibition induces

upstream receptor tyrosine kinase signaling and activates Akt. *Cancer Res.*, **66**, 1500–1508.
301. Hess, G., Herbrecht, R., Romaguera, J., Verhoef, G. *et al.* (2009) Phase III study to evaluate temsirolimus compared with investigator's choice therapy for the treatment of relapsed or refractory mantle cell lymphoma. *J. Clin. Oncol.*, **27**, 3822–3829.
302. Motzer, R.J., Escudier, B., Oudard, S., Hutson, T.E. *et al.* (2010) Phase 3 trial of everolimus for metastatic renal cell carcinoma. *Cancer*, **116**, 4256–4265.
303. Yao, J.C., Shah, M.H., Ito, T., Bohas, C.L. *et al.* (2011) Everolimus for advanced pancreatic neuroendocrine tumors. *N. Engl. J. Med.*, **364**, 514–523.
304. Yu, K., Toral-Barza, L., Shi, C., Zhang, W.G. *et al.* (2009) Biochemical, cellular, and in vivo activity of novel ATP-competitive and selective inhibitors of the mammalian target of rapamycin. *Cancer Res.*, **69**, 6232–6240.
305. Engelman, J.A. (2009) Targeting PI3K signalling in cancer: opportunities, challenges and limitations. *Nat. Rev. Cancer*, **9**, 550–562.
306. Markman, B., Dienstmann, R., and Tabernero, J. (2010) Targeting the PI3K/Akt/mTOR pathway – beyond rapalogs. *Oncotarget*, **1**, 530–543.
307. Evans, J.M.M., Donnelly, L.A., Emslie-Smith, A.M., Alessi, D.R. *et al.* (2005) Metformin and reduced risk of cancer in diabetic patients. *Br. Med. J.*, **330**, 1304–1305.
308. Dowling, R.J.O., Zakikhani, M., Fantus, I.G., Pollak, M. *et al.* (2007) Metformin inhibits mammalian target of rapamycin-dependent translation initiation in breast cancer cells. *Cancer Res.*, **67**, 10804–10812.
309. Bao, B., Wang, Z., Ali, S., Ahmad, A. *et al.* (2012) Metformin inhibits cell proliferation, migration and invasion by attenuating CSC function mediated by deregulating miRNAs in pancreatic cancer cells. *Cancer Prev. Res.*, **5**, 355–364.
310. Hirsch, H.A., Iliopoulos, D., Tsichlis, P.N., and Struhl, K. (2009) Metformin selectively targets cancer stem cells, and acts together with chemotherapy to block tumor growth and prolong remission. *Cancer Res.*, **69**, 7507–7511.
311. Iliopoulos, D., Hirsch, H.A., and Struhl, K. (2011) Metformin decreases the dose of chemotherapy for prolonging tumor remission in mouse xenografts involving multiple cancer cell types. *Cancer Res.*, **71**, 3196–3201.
312. Blackburn, E.H. and Gall, J.G. (1978) A tandemly repeated sequence at the termini of the extrachromosomal ribosomal RNA genes in *Tetrahymena. J. Mol. Biol.*, **120**, 33–53.
313. Harley, C.B., Futcher, A.B., and Greider, C.W. (1990) Telomeres shorten during ageing of human fibroblasts. *Nature*, **345**, 458–460.
314. Hayflick, L. and Moorhead, P.S. (1961) The serial cultivation of human diploid cell strains. *Exp. Cell Res.*, **25**, 585–621.
315. Shay, J.W. and Roninson, I.B. (2004) Hallmarks of senescence in carcinogenesis and cancer therapy. *Oncogene*, **23**, 2919–2933.
316. Shay, J.W. and Wright, W.E. (2010) Telomeres and telomerase in normal and cancer stem cells. *FEBS Lett.*, **584**, 3819–3825.
317. Flores, I., Canela, A., Vera, E., Tejera, A. *et al.* (2008) The longest telomeres: a general signature of adult stem cell compartments. *Genes Dev.*, **22**, 654–667.
318. Armanios, M., Chen, J.-L., Chang, Y.-P.C., Brodsky, R.A. *et al.* (2005) Haploinsufficiency of telomerase reverse transcriptase leads to anticipation in autosomal dominant dyskeratosis congenita. *Proc. Natl Acad. Sci. USA*, **102**, 15960–15964.
319. Beier, F., Beier, C.P., Aschenbrenner, I., Hildebrandt, G.C. *et al.* (2011) Identification of CD133(-)/telomerase(low) progenitor cells in glioblastoma-derived cancer stem cell lines. *Cell. Mol. Neurobiol.*, **31**, 337–343.
320. Beck, S., Jin, X., Sohn, Y.-W., Kim, J.-K. *et al.* (2010) Telomerase activity-independent function of TERT allows glioma cells to attain cancer stem cell characteristics by inducing EGFR expression. *Mol. Cells*, **31**, 9–15.
321. Marian, C.O., Cho, S.K., McEllin, B.M., Maher, E.A. *et al.* (2010) The telomerase antagonist, imetelstat, efficiently targets glioblastoma tumor-initiating cells leading to decreased proliferation and tumor growth. *Clin. Cancer Res.*, **16**, 154–163.
322. Castelo-Branco, P., Zhang, C., Lipman, T., Fujitani, M. *et al.* (2011) Neural tumor-initiating cells have distinct telomere maintenance and can be safely targeted

for telomerase inhibition. *Clin. Cancer Res.*, **17**, 111–121.

323. Drummond, M.W., Hoare, S.F., Monaghan, A., Graham, S.M. *et al.* (2005) Dysregulated expression of the major telomerase components in leukaemic stem cells. *Leukemia*, **19**, 381–389.
324. Xu, Y., He, K., and Goldkorn, A. (2011) Telomerase targeted therapy in cancer and cancer stem cells. *Clin. Adv. Hematol. Oncol.*, **9**, 442–455.
325. Xu, T., He, K., Wang, L., and Goldkorn, A. (2011) Prostate tumor cells with cancer progenitor properties have high telomerase activity and are rapidly killed by telomerase interference. *The Prostate*, **71**, 1390–1400.
326. Serrano, D., Bleau, A.-M., Fernandez-Garcia, I., Fernandez-Marcelo, T. *et al.* (2011) Inhibition of telomerase activity preferentially targets aldehyde dehydrogenase-positive cancer stem-like cells in lung cancer. *Mol. Cancer*, **10**, 96.
327. Keshet, G.I., Goldstein, I., Itzhaki, O., Cesarkas, K. *et al.* (2008) MDR1 expression identifies human melanoma stem cells. *Biochem. Biophys. Res. Commun.*, **368**, 930–936.
328. Brennan, S.K., Wang, Q., Tressler, R., Harley, C. *et al.* (2010) Telomerase inhibition targets clonogenic multiple myeloma cells through telomere length-dependent and independent mechanisms. *PLoS ONE*, **5**, e12487.
329. Joseph, I., Tressler, R., Bassett, E., Harley, C. *et al.* (2010) The telomerase inhibitor imetelstat depletes cancer stem cells in breast and pancreatic cancer cell lines. *Cancer Res.*, **70**, 9494–9504.
330. Bernhardt, S.L., Gjertsen, M.K., Trachsel, S., Moller, M. *et al.* (2006) Telomerase peptide vaccination of patients with non-resectable pancreatic cancer: a dose escalating phase I/II study. *Br. J. Cancer*, **95**, 1474–1482.
331. Brunsvig, P.F., Kyte, J.A., Kersten, C., Sundstrom, S. *et al.* (2011) Telomerase peptide vaccination in NSCLC: a phase II trial in stage III patients vaccinated after chemoradiotherapy and an 8-year update on a phase I/II trial. *Clin. Cancer Res.*, **17**, 6847–6857.
332. Greten, T.F., Forner, A., Korangy, F., N'Kontchou, G. *et al.* (2010) A phase II open label trial evaluating safety and efficacy of a telomerase peptide vaccination in patients with advanced hepatocellular carcinoma. *BMC Cancer*, **10**, 209.
333. Herbert, B.-S., Gellert, G.C., Hochreiter, A., Pongracz, K. *et al.* (2005) Lipid modification of GRN163, an N3′ → P5′ thio-phosphoramidate oligonucleotide, enhances the potency of telomerase inhibition. *Oncogene*, **24**, 5262–5268.
334. Cesare, A.J. and Reddel, R.R. (2010) Alternative lengthening of telomeres: models, mechanisms and implications. *Nat. Rev. Genet.*, **11**, 319–330.
335. Scadden, D.T. (2007) The stem cell niche in health and leukemic disease. *Best Pract. Res. Clin. Haematol.*, **20**, 19–27.
336. Blanpain, C., Lowry, W.E., Geoghegan, A., Polak, L. *et al.* (2004) Self-renewal, multipotency, and the existence of two cell populations within an epithelial stem cell niche. *Cell*, **118**, 635–648.
337. Miller, F.D. and Gauthier-Fisher, A. (2009) Home at last: neural stem cell niches defined. *Stem Cell*, **4**, 507–510.
338. Shaker, A. and Rubin, D.C. (2010) Intestinal stem cells and epithelial-mesenchymal interactions in the crypt and stem cell niche. *Transl. Res.*, **156**, 180–187.
339. Losick, V.P., Morris, L.X., Fox, D.T., and Spradling, A. (2011) *Drosophila* stem cell niches: a decade of discovery suggests a unified view of stem cell regulation. *Dev. Cell*, **21**, 159–171.
340. Lander, A.D., Kimble, J., Clevers, H., Fuchs, E. *et al.* (2012) What does the concept of the stem cell niche really mean today? *BMC Biol.*, **10**, 19.
341. Konopleva, M., Konoplev, S., Hu, W., Zaritskey, A.Y. *et al.* (2002) Stromal cells prevent apoptosis of AML cells by up-regulation of anti-apoptotic proteins. *Leukemia*, **16**, 1713–1724.
342. Nishimura, E.K., Suzuki, M., Igras, V., Du, J. *et al.* (2010) Key roles for transforming growth factor beta in melanocyte stem cell maintenance. *Cell Stem Cell*, **6**, 130–140.
343. Orford, K.W. and Scadden, D.T. (2008) Deconstructing stem cell self-renewal: genetic insights into cell-cycle regulation. *Nat. Rev. Genet.*, **9**, 115–128.
344. Hanahan, D. and Coussens, L.M. (2012) Accessories to the crime: functions of cells recruited to the tumor microenvironment. *Cancer Cell*, **21**, 309–322.

345. Sneddon, J.B. and Werb, Z. (2007) Location, location, location: the cancer stem cell niche. *Cell Stem Cell*, **1**, 607–611.
346. Calabrese, C., Poppleton, H., Kocak, M., Hogg, T.L. *et al.* (2007) A perivascular niche for brain tumor stem cells. *Cancer Cell*, **11**, 69–82.
347. Bao, S., Wu, Q., Sathornsumetee, S., Hao, Y. *et al.* (2006) Stem cell-like glioma cells promote tumor angiogenesis through vascular endothelial growth factor. *Cancer Res.*, **66**, 7843–7848.
348. Beck, B., Driessens, G., Goossens, S., Youssef, K.K. *et al.* (2011) A vascular niche and a VEGF-Nrp1 loop regulate the initiation and stemness of skin tumours. *Nature*, **478**, 399–403.
349. Krishnamurthy, S., Dong, Z., Vodopyanov, D., Imai, A. *et al.* (2010) Endothelial cell-initiated signaling promotes the survival and self-renewal of cancer stem cells. *Cancer Res.*, **70**, 9969–9978.
350. Colmone, A., Amorim, M., Pontier, A.L., Wang, S. *et al.* (2008) Leukemic cells create bone marrow niches that disrupt the behavior of normal hematopoietic progenitor cells. *Science*, **322**, 1861–1865.
351. Atsumi, N., Ishii, G., Kojima, M., Sanada, M. *et al.* (2008) Podoplanin, a novel marker of tumor-initiating cells in human squamous cell carcinoma A431. *Biochem. Biophys. Res. Commun.*, **373**, 36–41.
352. He, X., Marchionni, L., Hansel, D.E., Yu, W. *et al.* (2009) Differentiation of a highly tumorigenic basal cell compartment in urothelial carcinoma. *Stem Cells*, **27**, 1487–1495.
353. Vermeulen, L., de Sousa, E.M.F., Van Der Heijden, M., Cameron, K. *et al.* (2010) Wnt activity defines colon cancer stem cells and is regulated by the microenvironment. *Nature*, **12**, 468–476.
354. Theunissen, J.W. and De Sauvage, F.J. (2009) Paracrine hedgehog signaling in cancer. *Cancer Res.*, **69**, 6007–6010.
355. Tian, H., Callahan, C.A., DuPree, K.J., Darbonne, W.C. *et al.* (2009) Hedgehog signaling is restricted to the stromal compartment during pancreatic carcinogenesis. *Proc. Natl Acad. Sci. USA*, **106**, 4254–4259.
356. Yauch, R.L., Gould, S.E., Scales, S.J., Tang, T. *et al.* (2008) A paracrine requirement for hedgehog signalling in cancer. *Nature*, **455**, 406–410.
357. Li, F., Tiede, B., Massagué, J., and Kang, Y. (2007) Beyond tumorigenesis: cancer stem cells in metastasis. *Cell Res.*, **17**, 3–14.
358. Besançon, R., Valsesia-Wittmann, S., Puisieux, A., de Fromentel, C.C. *et al.* (2009) Cancer stem cells: the emerging challenge of drug targeting. *Curr. Med. Chem.*, **16**, 394–416.
359. Charles, N., Ozawa, T., Squatrito, M., Bleau, A.-M. *et al.* (2010) Perivascular nitric oxide activates notch signaling and promotes stem-like character in PDGF-induced glioma cells. *Cell Stem Cell*, **6**, 141–152.
360. Gilbert, L.A. and Hemann, M.T. (2010) DNA damage-mediated induction of a chemoresistant niche. *Cell*, **143**, 355–366.
361. Konopleva, M., Zhao, S., Hu, W., Jiang, S. *et al.* (2002) The anti-apoptotic genes Bcl-X(L) and Bcl-2 are over-expressed and contribute to chemoresistance of non-proliferating leukaemic CD34+ cells. *Br. J. Haematol.*, **118**, 521–534.
362. Guan, Y., Gerhard, B., and Hogge, D.E. (2002) Detection, isolation, and stimulation of quiescent primitive leukemic progenitor cells from patients with acute myeloid leukemia (AML). *Blood*, **101**, 3142–3149.
363. Saito, Y., Uchida, N., Tanaka, S., Suzuki, N. *et al.* (2010) Induction of cell cycle entry eliminates human leukemia stem cells in a mouse model of AML. *Nat. Biotechnol.*, **28**, 275–280.
364. Jin, L., Hope, K.J., Zhai, Q., Smadja-Joffe, F. *et al.* (2006) Targeting of CD44 eradicates human acute myeloid leukemic stem cells. *Nat. Med.*, **12**, 1167–1174.
365. Krause, D.S., Lazarides, K., von Andrian, U.H., and Van Etten, R.A. (2006) Requirement for CD44 in homing and engraftment of BCR-ABL-expressing leukemic stem cells. *Nat. Med.*, **12**, 1175–1180.
366. Nie, Y., Han, Y.C., and Zou, Y.R. (2008) CXCR4 is required for the quiescence of primitive hematopoietic cells. *J. Exp. Med.*, **205**, 777–783.
367. Sugiyama, T., Kohara, H., Noda, M., and Nagasawa, T. (2006) Maintenance of the hematopoietic stem cell pool by CXCL12-CXCR4 chemokine signaling in bone marrow stromal cell niches. *Immunity*, **25**, 977–988.
368. O'Callaghan, K., Lee, L., Nguyen, N., Hsieh, M.Y. *et al.* (2012) Targeting CXCR4 with cell-penetrating pepducins in lymphoma

and lymphocytic leukemia. *Blood*, **119**, 1717–1725.
369. Weisberg, E., Azab, A.K., Manley, P.W., Kung, A.L. *et al.* (2011) Inhibition of CXCR4 in CML cells disrupts their interaction with the bone marrow microenvironment and sensitizes them to nilotinib. *Leukemia*, **26**, 985–990.
370. Deonarain, M.P., Kousparou, C.A., and Epenetos, A.A. (2009) Antibodies targeting cancer stem cells: a new paradigm in immunotherapy? *MAbs*, **1**, 12–25.
371. Lathia, J.D., Gallagher, J., Heddleston, J.M., Wang, J. *et al.* (2010) Integrin alpha 6 regulates glioblastoma stem cells. *Cell Stem Cell*, **6**, 421–432.
372. Vaillant, F., Asselin-Labat, M.L., Shackleton, M., Forrest, N.C. *et al.* (2008) The mammary progenitor marker CD61/3 integrin identifies cancer stem cells in mouse models of mammary tumorigenesis. *Cancer Res.*, **68**, 7711–7717.
373. van der Horst, G., van den Hoogen, C., Buijs, J.T., Cheung, H. *et al.* (2011) Targeting of α(v)-integrins in stem/progenitor cells and supportive microenvironment impairs bone metastasis in human prostate cancer. *Neoplasia*, **13**, 516–525.
374. Carter, A. (2010) Integrins as target: first phase III trial launches, but questions remain. *J. Natl Cancer Inst.*, **102**, 675–677.
375. Friedman, H.S., Prados, M.D., Wen, P.Y., Mikkelsen, T. *et al.* (2009) Bevacizumab alone and in combination with irinotecan in recurrent glioblastoma. *J. Clin. Oncol.*, **27**, 4733–4740.
376. Batchelor, T.T., Sorensen, A.G., di Tomaso, E., Zhang, W.-T. *et al.* (2007) AZD2171, a pan-VEGF receptor tyrosine kinase inhibitor, normalizes tumor vasculature and alleviates edema in glioblastoma patients. *Cancer Cell*, **11**, 83–95.
377. Kang, M.H. and Reynolds, C.P. (2009) Bcl-2 inhibitors: targeting mitochondrial apoptotic pathways in cancer therapy. *Clin. Cancer Res.*, **15**, 1126–1132.
378. Ghobrial, I.M., Witzig, T.E., and Adjei, A.A. (2005) Targeting apoptosis pathways in cancer therapy. *CA: Cancer J. Clin.*, **55**, 178–194.
379. Winquist, R.J., Furey, B.F., and Boucher, D.M. (2010) Cancer stem cells as the relevant biomass for drug discovery. *Curr. Opin. Pharmacol.*, **10**, 385–390.
380. Pollard, S.M., Yoshikawa, K., Clarke, I.D., Danovi, D. *et al.* (2009) Glioma stem cell lines expanded in adherent culture have tumor-specific phenotypes and are suitable for chemical and genetic screens. *Cell Stem Cell*, **4**, 568–580.
381. Danovi, D., Falk, A., Humphreys, P., Vickers, R. *et al.* (2010) Imaging-based chemical screens using normal and glioma-derived neural stem cells. *Biochem. Soc. Trans.*, **38**, 1067.
382. Visnyei, K., Onodera, H., Damoiseaux, R., Saigusa, K. *et al.* (2011) A molecular screening approach to identify and characterize inhibitors of glioblastoma stem cells. *Mol. Cancer Ther.*, **10**, 1818–1828.
383. Grinshtein, N., Datti, A., Fujitani, M., Uehling, D. *et al.* (2011) Small molecule kinase inhibitor screen identifies polo-like kinase 1 as a target for neuroblastoma tumor-initiating cells. *Cancer Res.*, **71**, 1385–1395.
384. Weaver, V.M., Petersen, O.W., Wang, F., Larabell, C.A. *et al.* (1997) Reversion of the malignant phenotype of human breast cells in three-dimensional culture and in vivo by integrin blocking antibodies. *J. Cell Biol.*, **137**, 231–245.
385. Pampaloni, F., Reynaud, E.G., and Stelzer, E.H.K. (2007) The third dimension bridges the gap between cell culture and live tissue. *Nat. Rev. Mol. Cell Biol.*, **8**, 839–845.
386. Gupta, P.B., Onder, T.T., Jiang, G., Tao, K. *et al.* (2009) Identification of selective inhibitors of cancer stem cells by high-throughput screening. *Cell*, **138**, 645–659.
387. Škrtiç, M., Sriskanthadevan, S., Jhas, B., Gebbia, M. *et al.* (2011) Inhibition of mitochondrial translation as a therapeutic strategy for human acute myeloid leukemia. *Cancer Cell*, **20**, 674–688.
388. Eberhard, Y., McDermott, S.P., Wang, X. *et al.* (2009) Chelation of intracellular iron with the antifungal agent ciclopirox olamine induces cell death in leukemia and myeloma cells. *Blood*, **114**, 3064–3073.
389. McDermott, S.P., Eppert, K., Notta, F., Isaac, M. *et al.* (2012) A small molecule screening strategy with validation on human leukemia stem cells uncovers the therapeutic efficacy of kinetin riboside. *Blood*, **119**, 1200–1207.
390. Wurdak, H., Zhu, S., Romero, A., Lorger, M. *et al.* (2010) An RNAi screen identifies

TRRAP as a regulator of brain tumor-initiating cell differentiation. *Cell Stem Cell*, **6**, 37–47.
391. Visvader, J.E., and Lindeman, G.J. (2012) Cancer stem cells: current status and evolving complexities. *Cell Stem Cell*, **10**, 717–728.
392. Schober, M., and Fuchs, E. (2011) Tumor-initiating stem cells of squamous cell carcinomas and their control by TGF-β and integrin/focal adhesion kinase (FAK) signaling. *Proc. Natl Acad. Sci. USA*, **108**, 10544–10549.
393. Vermeulen, L., de Sousa, E.M.F., Richel, D.J., and Medema, J.P. (2012) The developing cancer stem-cell model: clinical challenges and opportunities. *Lancet Oncol.*, **13**, e83–e89.
394. Gurney, A.L. (2011) Frizzled-binding agents and uses thereof. US Patent 20,137,041.
395. Ericson, J., Morton, S., Kawakami, A., Roelink, H. *et al.* (1996) Two critical periods of Sonic Hedgehog signaling required for the specification of motor neuron identity. *Cell*, **87**, 661–673.
396. Wolfe, M.S., Citron, M., Diehl, T.S., Xia, W. *et al.* (1998) A substrate-based difluoro ketone selectively inhibits Alzheimer's gamma-secretase activity. *J. Med. Chem.*, **41**, 6–9.
397. Li, Y.M., Lai, M.T., Xu, M., Huang, Q. *et al.* (2000) Presenilin 1 is linked with gamma-secretase activity in the detergent solubilized state. *Proc. Natl Acad. Sci. USA*, **97**, 6138–6143.
398. Dovey, H.F., John, V., Anderson, J.P., Chen, L.Z. *et al.* (2001) Functional gamma-secretase inhibitors reduce beta-amyloid peptide levels in brain. *J. Neurochem.*, **76**, 173–181.
399. Lanz, T.A., Karmilowicz, M.J., Wood, K.M., Pozdnyakov, N. *et al.* (2006) Concentration-dependent modulation of amyloid-beta in vivo and in vitro using the gamma-secretase inhibitor, LY-450139. *J. Pharmacol. Exp. Ther.*, **319**, 924–933.
400. Lewis, S.J., Smith, A.L., Neduvelil, J.G., Stevenson, G.I. *et al.* (2005) A novel series of potent gamma-secretase inhibitors based on a benzobicyclo[4.2.1]nonane core. *Bioorg. Med. Chem. Lett.*, **15**, 373–378.
401. Lewis, H.D., Leveridge, M., Strack, P.R., Haldon, C.D. *et al.* (2007) Apoptosis in T cell acute lymphoblastic leukemia cells after cell cycle arrest induced by pharmacological inhibition of notch signaling. *Chem. Biol.*, **14**, 209–219.
402. Sparey, T., Beher, D., Best, J., Biba, M. *et al.* (2005) Cyclic sulfamide gamma-secretase inhibitors. *Bioorg. Med. Chem. Lett.*, **15**, 4212–4216.
403. Osherovich, L. (2011) Notch nicked again. *SciBX*, **4**, 1–3.
404. Auphan, N., DiDonato, J.A., Rosette, C., Helmberg, A. *et al.* (1995) Immunosuppression by glucocorticoids: inhibition of NF-kappa B activity through induction of I kappa B synthesis. *Science*, **270**, 286–290.
405. Scheinman, R.I., Cogswell, P.C., Lofquist, A.K., and Baldwin, A.S. (1995) Role of transcriptional activation of I kappa B alpha in mediation of immunosuppression by glucocorticoids. *Science*, **270**, 283–286.
406. Heitman, J., Movva, N.R., and Hall, M.N. (1991) Targets for cell cycle arrest by the immunosuppressant rapamycin in yeast. *Science*, **253**, 905–909.
407. Sehgal, S.N., Baker, H., and Vezina, C. (1975) Rapamycin (AY-22,989), a new antifungal antibiotic II. Fermentation, isolation and characterization. *J. Antibiot.*, **28**, 727–731.
408. Hidalgo, M. and Rowinsky, E.K. (2000) The rapamycin-sensitive signal transduction pathway as a target for cancer therapy. *Oncogene*, **19**, 6680–6686.
409. Sedrani, R., Cottens, S., Kallen, J., and Schuler, W. (1998) Chemical modification of rapamycin: the discovery of SDZ RAD. *Transplant. Proc.*, **30**, 2192–2194.
410. Mita, M.M., Mita, A.C., Chu, Q.S., Rowinsky, E.K. *et al.* (2008) Phase I trial of the novel mammalian target of rapamycin inhibitor deforolimus (AP23573; MK-8669) administered intravenously daily for 5 days every 2 weeks to patients with advanced malignancies. *J. Clin. Oncol.*, **26**, 361–367.
411. Bhagwat, S.V., Gokhale, P.C., Crew, A.P., Cooke, A. *et al.* (2011) Preclinical characterization of OSI-027, a potent and selective inhibitor of mTORC1 and mTORC2: distinct from rapamycin. *Mol. Cancer Ther.*, **10**, 1394–1406.
412. Carayol, N., Vakana, E., Sassano, A., Kaur, S. *et al.* (2010) Critical roles for mTORC2- and rapamycin-insensitive mTORC1-complexes in growth and survival of

BCR-ABL-expressing leukemic cells. *Proc. Natl Acad. Sci. USA*, **107**, 12469–12474.
413. Apsel, B., Blair, J.A., Gonzalez, B., Nazif, T.M. *et al.* (2008) Targeted polypharmacology: discovery of dual inhibitors of tyrosine and phosphoinositide kinases. *Nat. Chem. Biol.*, **4**, 691–699.
414. Feldman, M.E., Apsel, B., Uotila, A., Loewith, R. *et al.* (2009) Active-site inhibitors of mTOR target rapamycin-resistant outputs of mTORC1 and mTORC2. *PLoS Biol.*, **7**, e38.
415. Chresta, C.M., Davies, B.R., Hickson, I., Harding, T. *et al.* (2010) AZD8055 is a potent, selective, and orally bioavailable ATP-competitive mammalian target of rapamycin kinase inhibitor with in vitro and in vivo antitumor activity. *Cancer Res.*, **70**, 288–298.
416. Hsieh, A.C., Liu, Y., Edlind, M.P., Ingolia, N.T. *et al.* (2012) The translational landscape of mTOR signalling steers cancer initiation and metastasis. *Nature*, **485**, 55–61.
417. Hilgard, P., Klenner, T., Stekar, J., Nössner, G. *et al.* (1997) D-21266, a new heterocyclic alkylphospholipid with antitumour activity. *Eur. J. Cancer*, **33**, 442–446.
418. Kondapaka, S.B., Singh, S.S., Dasmahapatra, G.P., Sausville, E.A. *et al.* (2003) Perifosine, a novel alkylphospholipid, inhibits protein kinase B activation. *Mol. Cancer Ther.*, **2**, 1093–1103.
419. Hirai, H., Sootome, H., Nakatsuru, Y., Miyama, K. *et al.* (2010) MK-2206, an allosteric Akt inhibitor, enhances antitumor efficacy by standard chemotherapeutic agents or molecular targeted drugs in vitro and in vivo. *Mol. Cancer Ther.*, **9**, 1956–1967.
420. Heerding, D.A., Rhodes, N., Leber, J.D., Clark, T.J. *et al.* (2008) Identification of 4-(2-(4-amino-1,2,5-oxadiazol-3-yl)-1-ethyl-7-[(3*S*)-3-piperidinylmethyl]oxy-1*H*-imidazo[4,5-*c*]pyridin-4-yl)-2-methyl-3-butyn-2-ol (GSK690693), a novel inhibitor of AKT kinase. *J. Med. Chem.*, **51**, 5663–5679.
421. Rhodes, N., Heerding, D.A., Duckett, D.R., Eberwein, D.J. *et al.* (2008) Characterization of an Akt kinase inhibitor with potent pharmacodynamic and antitumor activity. *Cancer Res.*, **68**, 2366–2374.
422. Pal, S.K., Reckamp, K., Yu, H., and Figlin, R.A. (2010) Akt inhibitors in clinical development for the treatment of cancer. *Expert Opin. Investig. Drugs*, **19**, 1355–1366.
423. Tabernero, J., Saura, C., Roda Perez, D., Dienstmann, R. *et al.* (2011) First-in-human phase I study evaluating the safety, pharmacokinetics (PK), and intratumor pharmacodynamics (PD) of the novel, oral, ATP-competitive Akt inhibitor GDC-0068. *J. Clin. Oncol.*, **29** (Suppl.), abstract 3022.
424. Calvo, E., Benhadji, K.A., Azaro, A., Duran, I. *et al.* (2012) First-in-human phase I study of LY2780301, an oral P70S6K/AKT inhibitor, in patients with refractory solid tumors. *J. Clin. Oncol.*, **30** (Suppl.), abstract 3005.
425. Maira, S.M., Pecchi, S., Huang, A., Burger, M. *et al.* (2012) Identification and characterization of NVP-BKM120, an orally available pan-class I PI3-kinase inhibitor. *Mol. Cancer Ther.*, **11**, 317–328.
426. Chakrabarty, A., Sánchez, V., Kuba, M.G., Rinehart, C. *et al.* (2012) Feedback upregulation of HER3 (ErbB3) expression and activity attenuates antitumor effect of PI3K inhibitors. *Proc. Natl Acad. Sci. USA*, **109**, 2718–2723.
427. Ihle, N.T., Williams, R., Chow, S., Chew, W. *et al.* (2004) Molecular pharmacology and antitumor activity of PX-866, a novel inhibitor of phosphoinositide-3-kinase signaling. *Mol. Cancer Ther.*, **3**, 763–772.
428. Folkes, A.J., Ahmadi, K., Alderton, W.K., Alix, S. *et al.* (2008) The identification of 2-(1*H*-Indazol-4-yl)-6-(4-methanesulfonyl-piperazin-1-ylmethyl)-4-morpholin-4-yl-thieno[3,2-*d*]pyrimidine (GDC-0941) as a potent, selective, orally bioavailable inhibitor of class I PI3 kinase for the treatment of cancer. *J. Med. Chem.*, **51**, 5522–5532.
429. Herman, S.E.M., Gordon, A.L., Wagner, A.J., Heerema, N.A. *et al.* (2010) Phosphatidylinositol 3-kinase-δ inhibitor CAL-101 shows promising preclinical activity in chronic lymphocytic leukemia by antagonizing intrinsic and extrinsic cellular survival signals. *Blood*, **116**, 2078–2088.
430. Garlich, J.R., De, P., Dey, N., Su, J.D. *et al.* (2008) A vascular targeted pan phosphoinositide 3-kinase inhibitor prodrug, SF1126, with antitumor and antiangiogenic activity. *Cancer Res.*, **68**, 206–215.

431. Maira, S.M., Stauffer, F., Brueggen, J., Furet, P. *et al.* (2008) Identification and characterization of NVP-BEZ235, a new orally available dual phosphatidylinositol 3-kinase/mammalian target of rapamycin inhibitor with potent in vivo antitumor activity. *Mol. Cancer Ther.*, **7**, 1851–1863.
432. Prasad, G., Sottero, T., Yang, X., Mueller, S. *et al.* (2011) Inhibition of PI3K/mTOR pathways in glioblastoma and implications for combination therapy with temozolomide. *Neuro-Oncology*, **13**, 384–392.
433. Sutherlin, D.P., Bao, L., Berry, M., Castanedo, G. *et al.* (2011) Discovery of a potent, selective, and orally available class I phosphatidylinositol 3-kinase (PI3K)/mammalian target of rapamycin (mTOR) kinase inhibitor (GDC-0980) for the treatment of cancer. *J. Med. Chem.*, **54**, 7579–7587.
434. Wallin, J.J., Edgar, K.A., Guan, J., Berry, M. *et al.* (2011) GDC-0980 is a novel class I PI3K/mTOR kinase inhibitor with robust activity in cancer models driven by the PI3K pathway. *Mol. Cancer Ther.*, **10**, 2426–2436.
435. Knight, Z.A., Gonzalez, B., Feldman, M.E., Zunder, E.R. *et al.* (2006) A pharmacological map of the PI3-K family defines a role for p110alpha in insulin signaling. *Cell*, **125**, 733–747.
436. Asai, A., Oshima, Y., Yamamoto, Y., Uochi, T. *et al.* (2003) A novel telomerase template antagonist (GRN163) as a potential anticancer agent. *Cancer Res.*, **63**, 3931.
437. Kyte, J.A. (2009) Cancer vaccination with telomerase peptide GV1001. *Expert Opin. Investig. Drugs*, **18**, 687–694.
438. Donzella, G.A., Schols, D., Lin, S.W., Esté, J.A. *et al.* (1998) AMD3100, a small molecule inhibitor of HIV-1 entry via the CXCR4 co-receptor. *Nat. Med.*, **4**, 72–77.
439. Burke, P.M., DeNardo, S.J., Miers, L.A., Lamborn, K.R. *et al.* (2002) Cilengitide targeting of alpha(v)beta(3) integrin receptor synergizes with radioimmunotherapy to increase efficacy and apoptosis in breast cancer xenografts. *Cancer Res.*, **62**, 4263–4272.
440. Eskens, F.A.L.M., Dumez, H., Hoekstra, R., Perschl, A. *et al.* (2003) Phase I and pharmacokinetic study of continuous twice weekly intravenous administration of Cilengitide (EMD 121974), a novel inhibitor of the integrins alphavbeta3 and alphavbeta5 in patients with advanced solid tumours. *Eur. J. Cancer*, **39**, 917–926.
441. Ricart, A.D., Tolcher, A.W., Liu, G., Holen, K. *et al.* (2008) Volociximab, a chimeric monoclonal antibody that specifically binds alpha5beta1 integrin: a phase I, pharmacokinetic, and biological correlative study. *Clin. Cancer Res.*, **14**, 7924–7929.
442. Stoeltzing, O., Liu, W., Reinmuth, N., Fan, F. *et al.* (2003) Inhibition of integrin alpha5beta1 function with a small peptide (ATN-161) plus continuous 5-FU infusion reduces colorectal liver metastases and improves survival in mice. *Int. J. Cancer*, **104**, 496–503.
443. Gordon, M.S., Margolin, K., Talpaz, M., Sledge, G.W. *et al.* (2001) Phase I safety and pharmacokinetic study of recombinant human anti-vascular endothelial growth factor in patients with advanced cancer. *J. Clin. Oncol.*, **19**, 843–850.
444. Wedge, S.R., Kendrew, J., Hennequin, L.F., Valentine, P.J. *et al.* (2005) AZD2171: a highly potent, orally bioavailable, vascular endothelial growth factor receptor-2 tyrosine kinase inhibitor for the treatment of cancer. *Cancer Res.*, **65**, 4389–4400.
445. Jansen, B., Schlagbauer-Wadl, H., Brown, B.D., Bryan, R.N. *et al.* (1998) bcl-2 antisense therapy chemosensitizes human melanoma in SCID mice. *Nat. Med.*, **4**, 232–234.
446. Webb, A., Cunningham, D., Cotter, F., Clarke, P.A. *et al.* (1997) BCL-2 antisense therapy in patients with non-Hodgkin lymphoma. *Lancet*, **349**, 1137–1141.
447. Paoluzzi, L., Gonen, M., Gardner, J.R., Mastrella, J. *et al.* (2008) Targeting Bcl-2 family members with the BH3 mimetic AT-101 markedly enhances the therapeutic effects of chemotherapeutic agents in in vitro and in vivo models of B-cell lymphoma. *Blood*, **111**, 5350–5358.
448. Wang, G., Nikolovska-Coleska, Z., Yang, C.-Y., Wang, R. *et al.* (2006) Structure-based design of potent small-molecule inhibitors of anti-apoptotic Bcl-2 proteins. *J. Med. Chem.*, **49**, 6139–6142.
449. Tse, C., Shoemaker, A.R., Adickes, J., Anderson, M.G. *et al.* (2008) ABT-263: a potent and orally bioavailable

Bcl-2 family inhibitor. *Cancer Res.*, **68**, 3421–3428.

450. Trudel, S., Stewart, A.K., Li, Z., Shu, Y. *et al.* (2007) The Bcl-2 family protein inhibitor, ABT-737, has substantial antimyeloma activity and shows synergistic effect with dexamethasone and melphalan. *Clin. Cancer Res.*, **13**, 621–629.

451. Wei, J., Stebbins, J.L., Kitada, S., Dash, R. *et al.* (2010) BI-97C1, an optically pure Apogossypol derivative as pan-active inhibitor of antiapoptotic B-cell lymphoma/leukemia-2 (Bcl-2) family proteins. *J. Med. Chem.*, **53**, 4166–4176.

452. Pannuti, A., Foreman, K., Rizzo, P., Osipo, C. *et al.* (2010) Targeting notch to target cancer stem cells. *Clin. Cancer Res.*, **16**, 3141–3152.

Index

Translational Medicine Cancer: Advances in Molecular Biology and Medicine,
First Edition. Edited by Robert A. Meyers.

d

n

u